Algorithms and Theory of Computation Handbook

Second Edition

General Concepts and Techniques

Edited by

Mikhail J. Atallah
Marina Blanton

CRC Press
Taylor & Francis Group
Boca Raton London New York

CRC Press is an imprint of the
Taylor & Francis Group an **informa** business

A CHAPMAN & HALL BOOK

Chapman & Hall/CRC
Applied Algorithms and Data Structures Series

Series Editor
Samir Khuller
University of Maryland

Aims and Scopes

The design and analysis of algorithms and data structures form the foundation of computer science. As current algorithms and data structures are improved and new methods are introduced, it becomes increasingly important to present the latest research and applications to professionals in the field.

This series aims to capture new developments and applications in the design and analysis of algorithms and data structures through the publication of a broad range of textbooks, reference works, and handbooks. We are looking for single authored works and edited compilations that will:

- Appeal to students and professionals by providing introductory as well as advanced material on mathematical, statistical, and computational methods and techniques

- Present researchers with the latest theories and experimentation

- Supply information to interdisciplinary researchers and practitioners who use algorithms and data structures but may not have advanced computer science backgrounds

The inclusion of concrete examples and applications is highly encouraged. The scope of the series includes, but is not limited to, titles in the areas of parallel algorithms, approximation algorithms, randomized algorithms, graph algorithms, search algorithms, machine learning algorithms, medical algorithms, data structures, graph structures, tree data structures, and more. We are willing to consider other relevant topics that might be proposed by potential contributors.

Proposals for the series may be submitted to the series editor or directly to:

Randi Cohen
Acquisitions Editor
Chapman & Hall/CRC Press
6000 Broken Sound Parkway NW, Suite 300
Boca Raton, FL 33487

Algorithms and Theory of Computation Handbook, Second Edition

Algorithms and Theory of Computation Handbook, Second Edition: General Concepts and Techniques

Algorithms and Theory of Computation Handbook, Second Edition: Special Topics and Techniques

Chapman & Hall/CRC
Taylor & Francis Group
6000 Broken Sound Parkway NW, Suite 300
Boca Raton, FL 33487-2742

© 2010 by Taylor and Francis Group, LLC
Chapman & Hall/CRC is an imprint of Taylor & Francis Group, an Informa business

No claim to original U.S. Government works

International Standard Book Number: 978-1-58488-822-2 (Hardback)

Library of Congress Cataloging-in-Publication Data

Algorithms and theory of computation handbook. General concepts and techniques / editors, Mikhail J. Atallah and Marina Blanton. -- 2nd ed.
       p. cm. -- (Chapman & Hall/CRC applied algorithms and data structures series)
     Includes bibliographical references and index.
     ISBN 978-1-58488-822-2 (alk. paper)
       1. Computer algorithms. 2. Computer science. 3. Computational complexity. I. Atallah, Mikhail J. II. Blanton, Marina. III. Title. IV. Series.

QA76.9.A43A432 2009
005.1--dc22   2009017979

Visit the Taylor & Francis Web site at
http://www.taylorandfrancis.com

and the CRC Press Web site at
http://www.crcpress.com

Contents

Preface

This handbook aims to provide a comprehensive coverage of algorithms and theoretical computer science for computer scientists, engineers, and other professionals in related scientific and engineering disciplines. Its focus is to provide a compendium of fundamental topics and techniques for professionals, including practicing engineers, students, and researchers. The handbook is organized along the main subject areas of the discipline and also contains chapters from application areas that illustrate how the fundamental concepts and techniques come together to provide efficient solutions to important practical problems.

The contents of each chapter were chosen in such a manner as to help the computer professional and the engineer in finding significant information on a topic of his or her interest. While the reader may not find all the specialized topics in a given chapter, nor will the coverage of each topic be exhaustive, the reader should be able to find sufficient information for initial inquiries and a number of references to the current in-depth literature. In addition to defining terminology and presenting the basic results and techniques for their respective topics, the chapters also provide a glimpse of the major research issues concerning the relevant topics.

Compared to the first edition, this edition contains 21 new chapters and therefore provides a significantly broader coverage of the field and its application areas. This, together with the updating and revision of many of the chapters from the first edition, has made it necessary to move into a two-volume format.

It is a pleasure to extend our thanks to the people and organizations who made this handbook possible: first and foremost the chapter authors, whose dedication and expertise are at the core of this handbook; the universities and research laboratories with which the authors are affiliated for providing the computing and communication facilities and the intellectual environment for this project; Randi Cohen and her colleagues at Taylor & Francis for perfect organization and logistics that spared us the tedious aspects of such a project and enabled us to focus on its scholarly side; and, last but not least, our spouses and families, who provided moral support and encouragement.

Mikhail Atallah
Marina Blanton

Editors

Mikhail Atallah obtained his PhD from The Johns Hopkins University in 1982 and immediately thereafter joined the computer science department at Purdue University, West Lafayette, Indiana, where he currently holds the rank of distinguished professor of computer science. His research interests include information security, distributed computing, algorithms, and computational geometry. A fellow of both the ACM and the IEEE, Dr. Atallah has served on the editorial boards of top journals and on the program committees of top conferences and workshops. He was a keynote and invited speaker at many national and international meetings, and a speaker in the Distinguished Colloquium Series of top computer science departments on nine occasions. In 1999, he was selected as one of the best teachers in the history of Purdue and was included in a permanent wall display of Purdue's best teachers, past and present.

Marina Blanton is an assistant professor in the computer science and engineering department at the University of Notre Dame, Notre Dame, Indiana. She holds a PhD from Purdue University. Her research interests focus on information security, privacy, and applied cryptography, and, in particular, span across areas such as privacy-preserving computation, authentication, anonymity, and key management. Dr. Blanton has numerous publications at top venues and is actively involved in program committee work.

Contributors

Eric Allender
Department of Computer Science
Rutgers University
Piscataway, New Jersey

Alberto Apostolico
College of Computing
Georgia Institute of Technology
Atlanta, Georgia

Lars Arge
Department of Computer Science
University of Aarhus
Aarhus, Denmark

Ricardo Baeza-Yates
Yahoo! Research
Spain

and

Department of Computer Science
University of Chile
Santiago, Chile

Bryan Cantrill
Sun Microsystems, Inc.
Santa Clara, California

Vijay Chandru
National Institute of Advanced Studies
Indian Institute of Science Campus
Bangalore, India

Maxime Crochemore
Department of Computer Science
King's College London
London, United Kingdom

and

Institut Gaspard-Monge
Université Paris-Est
Marne-la-Vallée, France

Camil Demetrescu
Department of Computer and
  Systems Science
University of Rome "La Sapienza"
Rome, Italy

Rodney G. Downey
School of Mathematical and
  Computing Sciences
Victoria University
Wellington, New Zealand

Ioannis Z. Emiris
Department of Informatics and
  Telecommunications
National and Kapodistrian University of Athens
Athens, Greece

David Eppstein
Department of Computer Science
University of California
Irvine, California

Vladimir Estivill-Castro
Institute for Integrated and Intelligent Systems
Griffith University
Meadowbrook, Queensland, Australia

Eli Gafni
Department of Computer Science
University of California
Los Angeles, California

Zvi Galil
Department of Computer Science
Tel Aviv University
Tel Aviv, Israel

Sally A. Goldman
Department of Computer Science
  and Engineering
Washington University
St. Louis, Missouri

Raymond Greenlaw
Department of Information, Computing,
  and Engineering
Armstrong Atlantic State University
Savannah, Georgia

Christophe Hancart
Laboratory of Computer, Information
  Processing and Systems
University of Rouen
Mont-Saint-Aignan, France

H. James Hoover
Department of Computing Science
University of Alberta
Edmonton, Alberta, Canada

Giuseppe F. Italiano
Dipartimento di Informatica, Sistemie
  Produzione
University of Rome "Tor Vergata"
Rome, Italy

Florian Jarre
Mathematical Institute
Universität Düsseldorf
Düsseldorf, Germany

Tao Jiang
Department of Computer Science and
  Engineering
University of California
Riverside, California

Rick Kazman
Information Technology Management
University of Hawaii
Honolulu, Hawaii

and

Software Engineering Institute
Carnegie Mellon University
Pittsburgh, Pennsylvania

Samir Khuller
Department of Computer Science
University of Maryland
College Park, Maryland

Philip N. Klein
Department of Computer Science
Brown University
Providence, Rhode Island

Thierry Lecroq
Computer Science Department and
  LITIS EA 4108 Faculty of Science
University of Rouen
Mont-Saint-Aignan, France

Ming Li
David R. Cheriton School of
  Computer Science
University of Waterloo
Waterloo, Ontario, Canada

Michael C. Loui
Department of Electrical and
  Computer Engineering
University of Illinois at
  Urbana-Champaign
Urbana-Champaign, Illinois

Catherine McCartin
School of Engineering and Advanced
  Technology
Massey University
Palmerston North, New Zealand

Rajeev Motwani (Deceased)
Department of Computer Science
Stanford University
Stanford, California

Victor Y. Pan
Department of Mathematics and
  Computer Science
City University of New York
Bronx, New York

Patricio V. Poblete
Department of Computer Science
University of Chile
Santiago, Chile

Balaji Raghavachari
Department of Computer Science
University of Texas at Dallas
Dallas, Texas

Prabhakar Raghavan
Yahoo! Research
Sunnyvale, California

Rajeev Raman
Department of Computer Science
University of Leicester
Leicester, United Kingdom

M.R. Rao
Indian School of Business
Gachibowli
Hyderabad, India

Bala Ravikumar
Department of Computer and
  Engineering Science
Sonoma State University
Rohnert Park, California

Kenneth W. Regan
Department of Computer Science
  and Engineering
State University of New York at Buffalo
Buffalo, New York

Edward M. Reingold
Department of Computer Science
Illinois Institute of Technology
Chicago, Illinois

Atri Rudra
Department of Computer Science
  and Engineering
State University of New York at Buffalo
Buffalo, New York

Hanan Samet
Department of Computer Science
University of Maryland
College Park, Maryland

Wojciech Szpankowski
Department of Computer Science
Purdue University
West Lafayette, Indiana

Roberto Tamassia
Department of Computer Science
Brown University
Providence, Rhode Island

Elias P. Tsigaridas
INRIA Sophia Antipolis - Méditerranée
Sophia Antipolis
Nice, France

Stephen A. Vavasis
Department of Combinatorics and
  Optimization
University of Waterloo
Waterloo, Ontario, Canada

Samuel S. Wagstaff, Jr.
Department of Computer Science
Purdue University
West Lafayette, Indiana

Neal E. Young
Department of Computer Science
  and Engineering
University of California
Riverside, California

Norbert Zeh
Faculty of Computer Science
Dalhousie University
Halifax, Nova Scotia, Canada

Albert Y. Zomaya
School of Information Technologies
The University of Sydney
Sydney, New South Wales, Australia

Algorithm Design and Analysis Techniques

Edward M. Reingold
Illinois Institute of Technology

We outline the basic methods of algorithm design and analysis that have found application in the manipulation of discrete objects such as lists, arrays, sets, graphs, and geometric objects such as points, lines, and polygons. We begin by discussing **recurrence relations** and their use in the analysis of algorithms. Then we discuss some specific examples in algorithm analysis, **sorting** and **priority queues**. In Sections 1.3 through 1.6, we explore three important techniques of algorithm design—**divide-and-conquer**, **dynamic programming**, and **greedy heuristics**. Finally, we examine establishing lower bounds on the cost of any algorithm for a problem.

1.1 Analyzing Algorithms

It is convenient to classify algorithms based on the relative amount of time they require: how fast does the time required grow as the size of the problem increases? For example, in the case of arrays, the "size of the problem" is ordinarily the number of elements in the array. If the size of the problem is measured by a variable n, we can express the time required as a function of n, $T(n)$. When this function $T(n)$ grows rapidly, the algorithm becomes unusable for large n; conversely, when $T(n)$ grows slowly, the algorithm remains useful even when n becomes large.

We say an algorithm is $\Theta(n^2)$ if the time it takes quadruples (asymptotically) when n doubles; an algorithm is $\Theta(n)$ if the time it takes doubles when n doubles; an algorithm is $\Theta(\log n)$ if the time it takes increases by a constant, independent of n, when n doubles; an algorithm is $\Theta(1)$ if its time does not increase at all when n increases. In general, an algorithm is $\Theta(T(n))$ if the time it requires on

TABLE 1.1   Common Growth Rates of Times of Algorithms

Rate of Growth	Comment	Examples
$\Theta(1)$	Time required is constant, independent of problem size	Expected time for hash searching
$\Theta(\log \log n)$	Very slow growth of time required	Expected time of interpolation search of n elements
$\Theta(\log n)$	Logarithmic growth of time required—doubling the problem size increases the time by only a constant amount	Computing x^n; binary search of an array of n elements
$\Theta(n)$	Time grows linearly with problem size—doubling the problem size doubles the time required	Adding/subtracting n-digit numbers; linear search of an n-element array
$\Theta(n \log n)$	Time grows worse than linearly, but not much worse—doubling the problem size somewhat more than doubles the time required	Merge sort or heapsort of n elements; lower bound on comparison-based sorting of n elements
$\Theta(n^2)$	Time grows quadratically—doubling the problem size quadruples the time required	Simple-minded sorting algorithms
$\Theta(n^3)$	Time grows cubically—doubling the problem size results in an eightfold increase in the time required	Ordinary matrix multiplication
$\Theta(c^n)$	Time grows exponentially—increasing the problem size by 1 results in a c-fold increase in the time required; doubling the problem size *squares* the time required	Some traveling salesman problem algorithms based on exhaustive search

problems of size n grows proportionally to $T(n)$ as n increases. Table 1.1 summarizes the common growth rates encountered in the analysis of algorithms.

The analysis of an algorithm is often accomplished by finding and solving a recurrence relation that describes the time required by the algorithm. The most commonly occurring families of recurrences in the analysis of algorithms are linear recurrences and divide-and-conquer recurrences. In Section 1.1.1 we describe the "method of operators" for solving linear recurrences; in Section 1.1.2 we describe how to obtain an asymptotic solution to divide-and-conquer recurrences by transforming such a recurrence into a linear recurrence.

1.1.1 Linear Recurrences

A linear recurrence with constant coefficients has the form

$$c_0 a_n + c_1 a_{n-1} + c_2 a_{n-2} + \cdots + c_k a_{n-k} = f(n), \tag{1.1}$$

for some constant k, where each c_i is constant. To solve such a recurrence for a broad class of functions f (i.e., to express a_n in closed form as a function of n) by the method of operators, we consider two basic operators on sequences: S, which shifts the sequence left,

$$S \langle a_0, a_1, a_2, \ldots \rangle = \langle a_1, a_2, a_3, \ldots \rangle,$$

and C, which, for any constant C, multiplies each term of the sequence by C:

$$C \langle a_0, a_1, a_2, \ldots \rangle = \langle Ca_0, Ca_1, Ca_2, \ldots \rangle.$$

These basic operators on sequences allow us to construct more complicated operators by sums and products of operators. The sum $(A + B)$ of operators A and B is defined by

$$(A + B) \langle a_0, a_1, a_2, \ldots \rangle = A \langle a_0, a_1, a_2, \ldots \rangle + B \langle a_0, a_1, a_2, \ldots \rangle.$$

The product AB is the composition of the two operators:

$$(AB) \langle a_0, a_1, a_2, \ldots \rangle = A \left(B \langle a_0, a_1, a_2, \ldots \rangle \right).$$

Thus, for example,

$$(\mathcal{S}^2 - 4) \langle a_0, a_1, a_2, \ldots \rangle = \langle a_2 - 4a_0, a_3 - 4a_1, a_4 - 4a_2, \ldots \rangle,$$

which we write more briefly as

$$(\mathcal{S}^2 - 4) \langle a_i \rangle = \langle a_{i+2} - 4a_i \rangle.$$

With the operator notation, we can rewrite Equation 1.1 as

$$P(\mathcal{S}) \langle a_i \rangle = \langle f(i) \rangle,$$

where,

$$P(\mathcal{S}) = c_0 \mathcal{S}^k + c_1 \mathcal{S}^{k-1} + c_2 \mathcal{S}^{k-2} + \cdots + c_k$$

is a polynomial in $\mathcal{S}$.

Given a sequence $\langle a_i \rangle$, we say that the operator A annihilates $\langle a_i \rangle$ if $A\langle a_i \rangle = \langle 0 \rangle$. For example, $\mathcal{S}^2 - 4$ annihilates any sequence of the form $\langle u2^i + v(-2)^i \rangle$, with constants u and v. Here are two important facts about annihilators:

FACT 1.1 *The sum and product of operators are associative, commutative, and product distributes over sum. In other words, for operators A, B, and C,*

$$\begin{aligned}
(A + B) + C &= A + (B + C) & (AB)C &= A(BC), \\
A + B &= B + A & AB &= BA,
\end{aligned}$$

and

$$A(B + C) = AB + AC.$$

As a consequence, if A annihilates $\langle a_i \rangle$, then A annihilates $B\langle a_i \rangle$ for any operator B. This implies that the product *of two annihilators annihilates the* sum *of the sequences annihilated by the two operators—that is, if A annihilates $\langle a_i \rangle$ and B annihilates $\langle b_i \rangle$, then AB annihilates $\langle a_i + b_i \rangle$.*

FACT 1.2 *The operator $(\mathcal{S} - c)$, when applied to $\langle c^i \times p(i) \rangle$ with $p(i)$ a polynomial in i, results in a sequence $\langle c^i \times q(i) \rangle$ with $q(i)$ a polynomial of degree one less than $p(i)$. This implies that the operator $(\mathcal{S} - c)^{k+1}$ annihilates $\langle c^i \times (a\ polynomial\ in\ i\ of\ degree\ k) \rangle$.*

These two facts mean that determining the annihilator of a sequence is tantamount in determining the sequence; moreover, it is straightforward to determine the annihilator from a recurrence relation. For example, consider the Fibonacci recurrence

$$\begin{aligned}
F_0 &= 0, \\
F_1 &= 1, \\
F_{i+2} &= F_{i+1} + F_i.
\end{aligned}$$

The last line of this definition can be rewritten as $F_{i+2} - F_{i+1} - F_i = 0$, which tells us that $\langle F_i \rangle$ is annihilated by the operator

$$\mathcal{S}^2 - \mathcal{S} - 1 = (\mathcal{S} - \phi)(\mathcal{S} + 1/\phi),$$

where $\phi = (1+\sqrt{5})/2$. Thus we conclude from Fact 1.1 that $\langle F_i \rangle = \langle a_i + b_i \rangle$ with $(\mathcal{S} - \phi)\langle a_i \rangle = \langle 0 \rangle$ and $(\mathcal{S} - 1/\phi)\langle b_i \rangle = \langle 0 \rangle$. Fact 1.2 now tells us that

$$F_i = u\phi^i + v(-\phi)^{-i},$$

for some constants u and v. We can now use the initial conditions $F_0 = 0$ and $F_1 = 1$ to determine u and v: These initial conditions mean that

$$u\phi^0 + v(-\phi)^{-0} = 0,$$
$$u\phi^1 + v(-\phi)^{-1} = 1,$$

and these linear equations have the solution

$$u = v = 1/\sqrt{5},$$

and hence

$$F_i = \phi^i/\sqrt{5} + (-\phi)^{-i}/\sqrt{5}.$$

In the case of the similar recurrence,

$$G_0 = 0,$$
$$G_1 = 1,$$
$$G_{i+2} = G_{i+1} + G_i + i,$$

the last equation tells us that

$$\left(\mathcal{S}^2 - \mathcal{S} - 1\right)\langle G_i \rangle = \langle i \rangle,$$

so the annihilator for $\langle G_i \rangle$ is $(\mathcal{S}^2 - \mathcal{S} - 1)(\mathcal{S} - 1)^2$, since $(\mathcal{S} - 1)^2$ annihilates $\langle i \rangle$ (a polynomial of degree 1 in i) and hence the solution is

$$G_i = u\phi^i + v(-\phi)^{-i} + \text{(a polynomial of degree 1 in } i),$$

that is,

$$G_i = u\phi^i + v(-\phi)^{-i} + wi + z.$$

Again, we use the initial conditions to determine the constants u, v, w, and z.

In general, then, to solve the recurrence (Equation 1.1), we factor the annihilator

$$P(\mathcal{S}) = c_0 \mathcal{S}^k + c_1 \mathcal{S}^{k-1} + c_2 \mathcal{S}^{k-2} + \cdots + c_k,$$

multiply it by the annihilator for $\langle f(i) \rangle$, write down the form of the solution from this product (which is the annihilator for the sequence $\langle a_i \rangle$), and then use the initial conditions for the recurrence to determine the coefficients in the solution.

1.1.2 Divide-and-Conquer Recurrences

The divide-and-conquer paradigm of algorithm construction that we discuss in Section 1.3 leads naturally to divide-and-conquer recurrences of the type

$$T(n) = g(n) + uT(n/v),$$

TABLE 1.2 Rate of Growth of the Solution to the Recurrence $T(n) = g(n) + uT(n/v)$, the Divide-and-Conquer Recurrence Relations

$g(n)$	u, v	Growth Rate of $T(n)$
$\Theta(1)$	$u = 1$	$\Theta(\log n)$
	$u \neq 1$	$\Theta(n^{\log_v u})$
$\Theta(\log n)$	$u = 1$	$\Theta[(\log n)^2]$
	$u \neq 1$	$\Theta(n^{\log_v u})$
$\Theta(n)$	$u < v$	$\Theta(n)$
	$u = v$	$\Theta(n \log n)$
	$u > v$	$\Theta(n^{\log_v u})$
$\Theta(n^2)$	$u < v^2$	$\Theta(n^2)$
	$u = v^2$	$\Theta(n^2 \log n)$
	$u > v^2$	$\Theta(n^{\log_v u})$

Note: The variables u and v are positive constants, independent of n, and $v > 1$.

for constants u and v, $v > 1$, and sufficient initial values to define the sequence $\langle T(0), T(1), T(2), \ldots \rangle$. The growth rates of $T(n)$ for various values of u and v are given in Table 1.2. The growth rates in this table are derived by transforming the divide-and-conquer recurrence into a linear recurrence for a subsequence of $\langle T(0), T(1), T(2), \ldots \rangle$.

To illustrate this method, we derive the penultimate line in Table 1.2. We want to solve

$$T(n) = n^2 + v^2 T(n/v),$$

so we want to find a subsequence of $\langle T(0), T(1), T(2), \ldots \rangle$ that will be easy to handle. Let $n_k = v^k$; then,

$$T(n_k) = n_k^2 + v^2 T(n_k/v),$$

or

$$T\left(v^k\right) = v^{2k} + v^2 T\left(v^{k-1}\right).$$

Defining $t_k = T(v^k)$,

$$t_k = v^{2k} + v^2 t_{k-1}.$$

The annihilator for t_k is then $(\mathcal{S} - v^2)^2$, and thus

$$t_k = v^{2k}(ak + b),$$

for constants a and b. Since $n_k = v^k$, $k = \log_v n_k$, so we can express the solution for t_k in terms of $T(n)$,

$$T(n) \approx t_{\log_v n} = v^{2 \log_v n} \left(a \log_v n + b\right) = an^2 \log_v n + bn^2,$$

or

$$T(n) = \Theta\left(n^2 \log n\right).$$

1.2 Some Examples of the Analysis of Algorithms

In this section we introduce the basic ideas of algorithms analysis by looking at some practical problems of maintaining a collection of n objects and retrieving objects based on their relative size. For example, how can we determine the smallest of the elements? Or, more generally, how can we determine the kth largest of the elements? What is the running time of such algorithms in the worst case? Or, on the average, if all $n!$ permutations of the input are equally likely? What if the set of items is dynamic—that is, the set changes through insertions and deletions—how efficiently can we keep track of, say, the largest element?

1.2.1 Sorting

How do we rearrange an array of n values $x[1], x[2], \ldots, x[n]$ so that they are in perfect order—that is, so that $x[1] \leq x[2] \leq \cdots \leq x[n]$? The simplest way to put the values in order is to mimic what we might do by hand: take item after item and insert each one into the proper place among those items already inserted:

```
1       void insert (float x[], int i, float a) {
2       // Insert a into x[1] ... x[i]
3       // x[1] ... x[i-1] are sorted;  x[i] is unoccupied
4       if (i == 1 || x[i-1] <= a)
5           x[i] = a;
6       else {
7           x[i] = x[i-1];
8           insert(x, i-1, a);
9       }
10      }
11
12      void insertionSort (int n, float x[]) {
13      // Sort  x[1] ... x[n]
14      if (n > 1) {
15          insertionSort(n-1, x);
16          insert(x, n, x[n]);
17      }
18      }
```

To determine the time required in the worst case to sort n elements with insertionSort, we let t_n be the time to sort n elements and derive and solve a recurrence relation for t_n. We have

$$t_n = \begin{cases} \Theta(1) & \text{if } n = 1, \\ t_{n-1} + s_{n-1} + \Theta(1) & \text{otherwise,} \end{cases}$$

where s_m is the time required to insert an element in place among m elements using insert. The value of s_m is also given by a recurrence relation:

$$s_m = \begin{cases} \Theta(1) & \text{if } m = 1, \\ s_{m-1} + \Theta(1) & \text{otherwise.} \end{cases}$$

The annihilator for $\langle s_i \rangle$ is $(\mathcal{S} - 1)^2$, so $s_m = \Theta(m)$. Thus the annihilator for $\langle t_i \rangle$ is $(\mathcal{S} - 1)^3$, so $t_n = \Theta(n^2)$. The analysis of the average behavior is nearly identical; only the constants hidden in the Θ-notation change.

We can design better sorting methods using the divide-and-conquer idea of Section 1.3. These algorithms avoid $\Theta(n^2)$ worst-case behavior, working in time $\Theta(n \log n)$. We can also achieve time $\Theta(n \log n)$ by using a clever way of viewing the array of elements to be sorted as a tree: consider x[1] as the root of the tree and, in general, x[2*i] is the root of the left subtree of x[i] and x[2*i+1] is the root of the right subtree of x[i]. If we further insist that parents be greater than or equal to children, we have a **heap**; Figure 1.1 shows a small example.

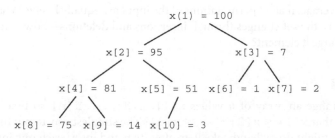

FIGURE 1.1 A heap—that is, an array, interpreted as a binary tree.

A heap can be used for sorting by observing that the largest element is at the root, that is, x[1]; thus to put the largest element in place, we swap x[1] and x[n]. To continue, we must restore the heap property which may now be violated at the root. Such restoration is accomplished by swapping x[1] with its larger child, if that child is larger than x[1], and the continuing to swap it downward until either it reaches the bottom or a spot where it is greater or equal to its children. Since the tree-cum-array has height $\Theta(\log n)$, this restoration process takes time $\Theta(\log n)$. Now, with the heap in x[1] to x[n-1] and x[n] the largest value in the array, we can put the second largest element in place by swapping x[1] and x[n-1]; then we restore the heap property in x[1] to x[n-2] by propagating x[1] downward—this takes time $\Theta(\log(n-1))$. Continuing in this fashion, we find we can sort the entire array in time

$$\Theta\left(\log n + \log(n-1) + \cdots + \log 1\right).$$

To evaluate this sum, we bound it from above and below, as follows. By ignoring the smaller half of the terms, we bound it from below:

$$\log n + \log(n-1) + \cdots + \log 1 \geq \underbrace{\log\frac{n}{2} + \log\frac{n}{2} + \cdots + \log\frac{n}{2}}_{\frac{n}{2}\text{ times}}$$

$$= \frac{n}{2}\log n$$

$$= \Theta(n\log n);$$

and by overestimating all of the terms we bound it from above:

$$\log n + \log(n-1) + \cdots + \log 1 \leq \underbrace{\log n + \log n + \cdots + \log n}_{n\text{ times}}$$

$$= n\log n$$

$$= \Theta(n\log n).$$

The initial creation of the heap from an unordered array is done by applying the above restoration process successively to x[n/2], x[n/2-1], ..., x[1], which takes time $\Theta(n)$.

Hence, we have the following $\Theta(n\log n)$ sorting algorithm:

```
1     void heapify (int n, float x[], int i) {
2        // Repair heap property below x[i] in x[1] ... x[n]
3        int largest = i;   // largest of x[i], x[2*i], x[2*i+1]
4        if (2*i <= n && x[2*i] > x[i])
5           largest = 2*i;
6        if (2*i+1 <= n && x[2*i+1] > x[largest])
7           largest = 2*i+1;
8        if (largest != i) {
9           // swap x[i] with larger child and repair heap below
10          float t = x[largest]; x[largest] = x[i]; x[i] = t;
11          heapify(n, x, largest);
12       }
13    }
14
15    void makeheap (int n, float x[]) {
16       // Make x[1] ... x[n] into a heap
17       for (int i=n/2; i>0; i--)
18          heapify(n, x, i);
```

```
19      }
20
21      void heapsort (int n, float x[]) {
22          // Sort  x[1] ... x[n]
23          float t;
24          makeheap(n, x);
25          for (int i=n; i>1; i--) {
26              // put x[1] in place and repair heap
27              t = x[1]; x[1] = x[i]; x[i] = t;
28              heapify(i-1, x, 1);
29          }
30      }
```

We will see in Section 1.6 that no sorting algorithm can be guaranteed always to use time less than $\Theta(n \log n)$. Thus, in a theoretical sense, heapsort is "asymptotically optimal" (but there are algorithms that perform better in practice).

1.2.2   Priority Queues

Aside from its application to sorting, the heap is an interesting data structure in its own right. In particular, heaps provide a simple way to implement a priority queue—a priority queue is an abstract data structure that keeps track of a dynamically changing set of values allowing the following operations:

create: Create an empty priority queue

insert: Insert a new element into a priority queue

decrease: Decrease the value of an element in a priority queue

minimum: Report the smallest element in a priority queue

deleteMinimum: Delete the smallest element in a priority queue

delete: Delete an element in a priority queue

merge: Merge two priority queues

A heap can implement a priority queue by altering the heap property to insist that parents are less than or equal to their children, so that the smallest value in the heap is at the root, that is, in the first array position. Creation of an empty heap requires just the allocation of an array, an $\Theta(1)$ operation; we assume that once created, the array containing the heap can be extended arbitrarily at the right end. Inserting a new element means putting that element in the $(n + 1)$st location and "bubbling it up" by swapping it with its parent until it reaches either the root or a parent with a smaller value. Since a heap has logarithmic height, insertion to a heap of n elements thus requires worst-case time $O(\log n)$. Decreasing a value in a heap requires only a similar $O(\log n)$ "bubbling up." The smallest element of such a heap is always at the root, so reporting it takes $\Theta(1)$ time. Deleting the minimum is done by swapping the first and last array positions, bubbling the new root value downward until it reaches its proper location, and truncating the array to eliminate the last position. Delete is handled by decreasing the value so that it is the least in the heap and then applying the deleteMinimum operation; this takes a total of $O(\log n)$ time.

The merge operation, unfortunately, is not so economically accomplished—there is little choice but to create a new heap out of the two heaps in a manner similar to the makeheap function in heapsort. If there are a total of n elements in the two heaps to be merged, this re-creation will require time $O(n)$.

There are better data structures than a heap for implementing priority queues, however. In particular, the Fibonacci heap provides an implementation of priority queues in which the delete and `deleteMinimum` operations take $O(\log n)$ time and the remaining operations take $\Theta(1)$ time, provided we consider the time required for a sequence of priority queue operations, rather than the individual times of each operation. That is, we must consider the cost of the individual operations amortized over the sequence of operations: Given a sequence of n priority queue operations, we will compute the total time $T(n)$ for all n operations. In doing this computation, however, we do not simply add the costs of the individual operations; rather, we subdivide the cost of each operation into two parts, the immediate cost of doing the operation and the long-term savings that result from doing the operation—the long-term savings represent costs *not* incurred by later operations as a result of the present operation. The immediate cost minus the long-term savings give the amortized cost of the operation.

It is easy to calculate the immediate cost (time required) of an operation, but how can we measure the long-term savings that result? We imagine that the data structure has associated with it a bank account; at any given moment the bank account must have a nonnegative balance. When we do an operation that will save future effort, we are making a deposit to the savings account and when, later on, we derive the benefits of that earlier operation we are making a withdrawal from the savings account. Let $\mathcal{B}(i)$ denote the balance in the account after the ith operation, $\mathcal{B}(0) = 0$. We define the **amortized cost** of the ith operation to be

$$
\begin{aligned}
\text{Amortized cost of } i\text{th operation} \ &= \ (\text{immediate cost of } i\text{th operation}) \\
&\quad + (\text{change in bank account}) \\
&= \ (\text{immediate cost of } i\text{th operation}) + (\mathcal{B}(i) - \mathcal{B}(i-1)).
\end{aligned}
$$

Since the bank account $\mathcal{B}$ can go up or down as a result of the ith operation, the amortized cost may be less than or more than the immediate cost. By summing the previous equation, we get

$$
\begin{aligned}
\sum_{i=1}^{n} (\text{amortized cost of } i\text{th operation}) \ &= \ \sum_{i=1}^{n} (\text{immediate cost of } i\text{th operation}) \\
&\quad + (\mathcal{B}(n) - \mathcal{B}(0)) \\
&= \ (\text{total cost of all } n \text{ operations}) + \mathcal{B}(n) \\
&\geq \ \text{total cost of all } n \text{ operations} \\
&= \ T(n),
\end{aligned}
$$

because $\mathcal{B}(i)$ is nonnegative. Thus defined, the sum of the amortized costs of the operations gives us an upper bound on the total time $T(n)$ for all n operations.

It is important to note that the function $\mathcal{B}(i)$ is not part of the data structure, but is just our way to measure how much time is used by the sequence of operations. As such, we can choose any rules for $\mathcal{B}$, provided $\mathcal{B}(0) = 0$ and $\mathcal{B}(i) \geq 0$ for $i \geq 1$. Then, the sum of the amortized costs defined by

$$
\text{Amortized cost of } i\text{th operation} = (\text{immediate cost of } i\text{th operation}) + (\mathcal{B}(i) - \mathcal{B}(i-1))
$$

bounds the overall cost of the operation of the data structure.

Now, to apply this method to priority queues. A Fibonacci heap is a list of heap-ordered trees (not necessarily binary); since the trees are heap ordered, the minimum element must be one of the roots and we keep track of which root is the overall minimum. Some of the tree nodes are marked. We define

$$
\begin{aligned}
\mathcal{B}(i) = \ &(\text{number of trees after the } i\text{th operation}) \\
&+ K \times (\text{number of marked nodes after the } i\text{th operation}),
\end{aligned}
$$

where K is a constant that we will define precisely during the discussion below.

The clever rules by which nodes are marked and unmarked, and the intricate algorithms that manipulate the set of trees, are too complex to present here in their complete form, so we just briefly describe the simpler operations and show the calculation of their amortized costs:

`create`: To create an empty Fibonacci heap we create an empty list of heap-ordered trees. The immediate cost is $\Theta(1)$; since the numbers of trees and marked nodes are zero before and after this operation, $\mathcal{B}(i) - \mathcal{B}(i-1)$ is zero and the amortized time is $\Theta(1)$.

`insert`: To insert a new element into a Fibonacci heap we add a new one-element tree to the list of trees constituting the heap and update the record of what root is the overall minimum. The immediate cost is $\Theta(1)$. $\mathcal{B}(i) - \mathcal{B}(i-1)$ is also 1 since the number of trees has increased by 1, while the number of marked nodes is unchanged. The amortized time is thus $\Theta(1)$.

`decrease`: Decreasing an element in a Fibonacci heap is done by cutting the link to its parent, if any, adding the item as a root in the list of trees, and decreasing its value. Furthermore, the marked parent of a cut element is itself cut, and this process of cutting marked parents propagates upward in the tree. Cut nodes become unmarked, and the unmarked parent of a cut element becomes marked. The immediate cost of this operation is no more than kc, where c is the number of cut nodes and $k > 0$ is some constant. Now, letting $K = k + 1$, we see that if there were t trees and m marked elements before this operation, the value of $\mathcal{B}$ before the operation was $t + Km$. After the operation, the value of $\mathcal{B}$ is $(t + c) + K(m - c + 2)$, so $\mathcal{B}(i) - \mathcal{B}(i-1) = (1 - K)c + 2K$. The amortized time is thus no more than $kc + (1 - K)c + 2K = \Theta(1)$ since K is constant.

`minimum`: Reporting the minimum element in a Fibonacci heap takes time $\Theta(1)$ and does not change the numbers of trees and marked nodes; the amortized time is thus $\Theta(1)$.

`deleteMinimum`: Deleting the minimum element in a Fibonacci heap is done by deleting that tree root, making its children roots in the list of trees. Then, the list of tree roots is "consolidated" in a complicated $O(\log n)$ operation that we do not describe. The result takes amortized time $O(\log n)$.

`delete`: Deleting an element in a Fibonacci heap is done by decreasing its value to $-\infty$ and then doing a `deleteMinimum`. The amortized cost is the sum of the amortized cost of the two operations, $O(\log n)$.

`merge`: Merging two Fibonacci heaps is done by concatenating their lists of trees and updating the record of which root is the minimum. The amortized time is thus $\Theta(1)$.

Notice that the amortized cost of each operation is $\Theta(1)$ except `deleteMinimum` and `delete`, both of which are $O(\log n)$.

1.3 Divide-and-Conquer Algorithms

One approach to the design of algorithms is to decompose a problem into subproblems that resemble the original problem, but on a reduced scale. Suppose, for example, that we want to compute x^n. We reason that the value we want can be computed from $x^{\lfloor n/2 \rfloor}$ because

$$
x^n = \begin{cases} 1 & \text{if } n = 0, \\ (x^{\lfloor n/2 \rfloor})^2 & \text{if } n \text{ is even}, \\ x \times (x^{\lfloor n/2 \rfloor})^2 & \text{if } n \text{ is odd}. \end{cases}
$$

This recursive definition can be translated directly into

```
1      int power (int x, int n) {
2          // Compute the n-th power of x
```

```
3        if (n == 0)
4          return 1;
5        else {
6          int t = power(x, floor(n/2));
7          if ((n % 2) == 0)
8            return t*t;
9          else
10           return x*t*t;
11       }
12     }
```

To analyze the time required by this algorithm, we notice that the time will be proportional to the number of multiplication operations performed in lines 8 and 10, so the divide-and-conquer recurrence

$$T(n) = 2 + T(\lfloor n/2 \rfloor),$$

with $T(0) = 0$, describes the rate of growth of the time required by this algorithm. By considering the subsequence $n_k = 2^k$, we find, using the methods of the previous (Section 1.2), that $T(n) = \Theta(\log n)$. Thus above algorithm is considerably more efficient than the more obvious

```
1      int power (int k, int n) {
2        // Compute the n-th power of k
3        int product = 1;
4        for (int i = 1; i <= n; i++)
5          // at this point power is k*k*k*...*k (i times)
6          product = product * k;
7        return product;
8      }
```

which requires time $\Theta(n)$.

An extremely well-known instance of divide-and-conquer algorithm is binary search of an ordered array of n elements for a given element—we "probe" the middle element of the array, continuing in either the lower or upper segment of the array, depending on the outcome of the probe:

```
1    int binarySearch (int x, int w[], int low, int high) {
2        // Search for x among sorted array w[low..high]. The integer
3        // returned is either the location of x in w, or the location
4        // where x belongs.
5        if (low > high) // Not found
6          return low;
7        else {
8          int middle = (low+high)/2;
9          if (w[middle] < x)
10           return binarySearch(x, w, middle+1, high);
11         else if (w[middle] == x)
12           return middle;

13         else
14           return binarySearch(x, w, low, middle-1);
15       }
16   }
```

The analysis of binary search in an array of n elements is based on counting the number of probes used in the search, since all remaining work is proportional to the number of probes. But, the number of probes needed is described by the divide-and-conquer recurrence

$$T(n) = 1 + T(n/2),$$

with $T(0) = 0$, $T(1) = 1$. We find from Table 1.2 (the top line) that $T(n) = \Theta(\log n)$. Hence, binary search is much more efficient than a simple linear scan of the array.

To multiply two very large integers x and y, assume that x has exactly $n \geq 2$ decimal digits and y has at most n decimal digits. Let $x_{n-1}, x_{n-2}, \ldots, x_0$ be the digits of x and $y_{n-1}, y_{n-2}, \ldots, y_0$ be the digits of y (some of the most significant digits at the end of y may be zeros, if y is shorter than x), so that

$$x = 10^{n-1}x_{n-1} + 10^{n-2}x_{n-2} + \cdots + x_0,$$

and

$$y = 10^{n-1}y_{n-1} + 10^{n-2}y_{n-2} + \cdots + y_0.$$

We apply the divide-and-conquer idea to multiplication by chopping x into two pieces, the most significant (leftmost) l digits and the remaining digits:

$$x = 10^l x_{\text{left}} + x_{\text{right}},$$

where $l = \lfloor n/2 \rfloor$. Similarly, chop y into two corresponding pieces:

$$y = 10^l y_{\text{left}} + y_{\text{right}},$$

because y has at most the number of digits that x does, y_{left} might be 0. The product $x \times y$ can be now written

$$
\begin{aligned}
x \times y &= \left(10^l x_{\text{left}} + x_{\text{right}}\right) \times \left(10^l y_{\text{left}} + y_{\text{right}}\right), \\
&= 10^{2l} x_{\text{left}} \times y_{\text{left}} \\
&\quad + 10^l \left(x_{\text{left}} \times y_{\text{right}} + x_{\text{right}} \times y_{\text{left}}\right) \\
&\quad + x_{\text{right}} \times y_{\text{right}}.
\end{aligned}
$$

If $T(n)$ is the time to multiply two n-digit numbers with this method, then

$$T(n) = kn + 4T(n/2);$$

the kn part is the time to chop up x and y and to do the needed additions and shifts; each of these tasks involves n-digit numbers and hence $\Theta(n)$ time. The $4T(n/2)$ part is the time to form the four needed subproducts, each of which is a product of about $n/2$ digits.

The line for $g(n) = \Theta(n)$, $u = 4 > v = 2$ in Table 1.2 tells us that $T(n) = \Theta(n^{\log_2 4}) = \Theta(n^2)$, so the divide-and-conquer algorithm is no more efficient than the elementary-school method of multiplication. However, we can be more economical in our formation of subproducts:

$$
\begin{aligned}
x \times y &= \left(10^n x_{\text{left}} + x_{\text{right}}\right) \times \left(10^n y_{\text{left}} + y_{\text{right}}\right), \\
&= 10^{2n} A + 10^n C + B,
\end{aligned}
$$

where

$$
\begin{aligned}
A &= x_{\text{left}} \times y_{\text{left}} \\
B &= x_{\text{right}} \times y_{\text{right}} \\
C &= \left(x_{\text{left}} + x_{\text{right}}\right) \times \left(y_{\text{left}} + y_{\text{right}}\right) - A - B.
\end{aligned}
$$

The recurrence for the time required changes to

$$T(n) = kn + 3T(n/2).$$

The kn part is the time to do the two additions that form $x \times y$ from A, B, and C and the two additions and the two subtractions in the formula for C; each of these six additions/subtractions involves n-digit numbers. The $3T(n/2)$ part is the time to (recursively) form the three needed products, each of which is a product of about $n/2$ digits. The line for $g(n) = \Theta(n)$, $u = 3 > v = 2$ in Table 1.2 now tells us that

$$T(n) = \Theta\left(n^{\log_2 3}\right).$$

Now

$$\log_2 3 = \frac{\log_{10} 3}{\log_{10} 2} \approx 1.5849625\cdots,$$

which means that this divide-and-conquer multiplication technique will be faster than the straight-forward $\Theta(n^2)$ method for large numbers of digits.

Sorting a sequence of n values efficiently can be done using the divide-and-conquer idea. Split the n values arbitrarily into two piles of $n/2$ values each, sort each of the piles separately, and then merge the two piles into a single sorted pile. This sorting technique, pictured in Figure 1.2, is called merge sort. Let $T(n)$ be the time required by merge sort for sorting n values. The time needed to do the merging is proportional to the number of elements being merged, so that

$$T(n) = cn + 2T(n/2),$$

because we must sort the two halves (time $T(n/2)$ for each half) and then merge (time proportional to n). We see by Table 1.2 that the growth rate of $T(n)$ is $\Theta(n \log n)$, since $u = v = 2$ and $g(n) = \Theta(n)$.

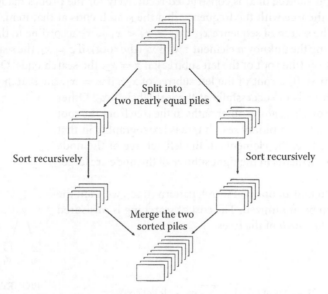

FIGURE 1.2 Schematic description of merge sort.

1.4 Dynamic Programming

In the design of algorithms to solve optimization problems, we need to make the optimal (lowest cost, highest value, shortest distance, and so on) choice among a large number of alternative solutions; dynamic programming is an organized way to find an optimal solution by systematically exploring all possibilities without unnecessary repetition. Often, dynamic programming leads to efficient, polynomial-time algorithms for problems that appear to require searching through exponentially many possibilities.

Like the divide-and-conquer method, dynamic programming is based on the observation that many optimization problems can be solved by solving similar subproblems and then composing the solutions of those subproblems into a solution for the original problem. In addition, the problem is viewed as a sequence of decisions, each decision leading to different subproblems; if a wrong decision is made, a suboptimal solution results, so all possible decisions need to be accounted for.

As an example of dynamic programming, consider the problem of constructing an optimal search pattern for probing an ordered sequence of elements. The problem is similar to searching an array—in Section 1.3 we described binary search in which an interval in an array is repeatedly bisected until the search ends. Now, however, suppose we know the frequencies with which the search will seek various elements (both in the sequence and missing from it). For example, if we know that the last few elements in the sequence are frequently sought—binary search does not make use of this information—it might be more efficient to begin the search at the right end of the array, not in the middle. Specifically, we are given an ordered sequence $x_1 < x_2 < \cdots < x_n$ and associated frequencies of access $\beta_1, \beta_2, \ldots, \beta_n$, respectively; furthermore, we are given $\alpha_0, \alpha_1, \ldots, \alpha_n$ where α_i is the frequency with which the search will fail because the object sought, z, was missing from the sequence, $x_i < z < x_{i+1}$ (with the obvious meaning when $i = 0$ or $i = n$). What is the optimal order to search for an unknown element z? In fact, how should we describe the optimal search order?

We express a search order as a **binary search tree**, a diagram showing the sequence of probes made in every possible search. We place at the root of the tree the sequence element at which the first probe is made, say x_i; the left subtree of x_i is constructed recursively for the probes made when $z < x_i$ and the right subtree of x_i is constructed recursively for the probes made when $z > x_i$. We label each item in the tree with the frequency that the search ends at that item. Figure 1.3 shows a simple example. The search of sequence $x_1 < x_2 < x_3 < x_4 < x_5$ according to the tree of Figure 1.3 is done by comparing the unknown element z with x_4 (the root); if $z = x_4$, the search ends. If $z < x_4$, z is compared with x_2 (the root of the left subtree); if $z = x_2$, the search ends. Otherwise, if $z < x_2$, z is compared with x_1 (the root of the left subtree of x_2); if $z = x_1$, the search ends. Otherwise, if $z < x_1$, the search ends unsuccessfully at the leaf labeled α_0. Other results of comparisons lead along other paths in the tree from the root downward. By its nature, a binary search tree is lexicographic in that for all nodes in the tree, the elements in the left subtree of the node are smaller and the elements in the right subtree of the node are larger than the node.

Because we are to find an optimal search pattern (tree), we want the cost of searching to be minimized. The cost of searching is measured by the *weighted path length* of the tree:

$$\sum_{i=1}^{n} \beta_i \times [1 + \text{level}\,(\beta_i)] + \sum_{i=0}^{n} \alpha_i \times \text{level}\,(\alpha_i),$$

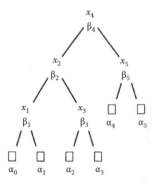

FIGURE 1.3 A binary search tree.

defined formally as

$$W\left(\Box\right) = 0,$$

$$W\left(T = \bigwedge_{T_l \ T_r}\right) = W(T_l) + W(T_r) + \sum \alpha_i + \sum \beta_i,$$

where the summations $\sum \alpha_i$ and $\sum \beta_i$ are over all α_i and β_i in T. Since there are exponentially many possible binary trees, finding the one with minimum weighted path length could, if done naïvely, take exponentially long.

The key observation we make is that a **principle of optimality** holds for the cost of binary search trees: subtrees of an optimal search tree must themselves be optimal. This observation means, for example, if the tree shown in Figure 1.3 is optimal, then its left subtree must be the optimal tree for the problem of searching the sequence $x_1 < x_2 < x_3$ with frequencies $\beta_1, \beta_2, \beta_3$ and $\alpha_0, \alpha_1, \alpha_2, \alpha_3$. (If a subtree in Figure 1.3 were *not* optimal, we could replace it with a better one, reducing the weighted path length of the entire tree because of the recursive definition of weighted path length.) In general terms, the principle of optimality states that subsolutions of an optimal solution must themselves be optimal.

The optimality principle, together with the recursive definition of weighted path length, means that we can express the construction of an optimal tree recursively. Let $C_{i,j}$, $0 \leq i \leq j \leq n$, be the cost of an optimal tree over $x_{i+1} < x_{i+2} < \cdots < x_j$ with the associated frequencies $\beta_{i+1}, \beta_{i+2}, \ldots, \beta_j$ and $\alpha_i, \alpha_{i+1}, \ldots, \alpha_j$. Then,

$$C_{i,i} = 0,$$

$$C_{i,j} = \min_{i < k \leq j} \left(C_{i,k-1} + C_{k,j}\right) + W_{i,j},$$

where

$$W_{i,i} = \alpha_i,$$

$$W_{i,j} = W_{i,j-1} + \beta_j + \alpha_j.$$

These two recurrence relations can be implemented directly as recursive functions to compute $C_{0,n}$, the cost of the optimal tree, leading to the following two functions:

```
1       int W (int i, int j) {
2         if (i == j)
3           return alpha[j];
4         else
5           return W(i,j-1) + beta[j] + alpha[j];
6       }
7
8       int C (int i, int j) {
9         if (i == j)
10          return 0;
11        else {
12            int minCost = MAXINT;
13          int cost;
14          for (int k = i+1; k <= j; k++) {
15              cost = C(i,k-1) + C(k,j) + W(i,j);
16              if (cost < minCost)
```

```
17                minCost = cost;
18            }
19          return minCost;
20        }
21      }
```

These two functions correctly compute the cost of an optimal tree; the tree itself can be obtained by storing the values of k when cost < minCost in line 16.

However, the above functions are unnecessarily time consuming (requiring exponential time) because the same subproblems are solved repeatedly. For example, each call W(i,j) uses time $\Theta(j - i)$ and such calls are made repeatedly for the same values of i and j. We can make the process more efficient by caching the values of W(i,j) in an array as they are computed and using the cached values when possible:

```
1       int w[n][n];
2       for (int i = 0; i < n; i++)
3         for (int j = 0; j < n; j++)
4           w[i][j] = MAXINT;
5
6       int W (int i, int j) {
7         if (w[i][j] == MAXINT)
8           if (i == j)
9             w[i][j] = alpha[j];
10          else
11            w[i][j] =  W(i,j-1) + beta[j] + alpha[j];
12        return w[i][j];
13      }
```

In the same way, we should cache the values of C(i,j) in an array as they are computed:

```
1       int c[n][n];
2       for (int i = 0; i < n; i++)
3         for (int j = 0; j < n; j++)
4           c[i][j] = MAXINT;
5
6       int C (int i, int j) {
7         if (c[i][j] == MAXINT)
8           if (i == j)
9             c[i][j] = 0;
10          else {
11            int minCost = MAXINT;
12            int cost;
13            for (int k = i+1; k <= j; k++) {
14              cost = C(i,k-1) + C(k,j) + W(i,j);
15              if (cost < minCost)
16                minCost = cost;
17            }
18            c[i][j] = minCost;
19          }
20        return c[i][j];
21      }
```

The idea of caching the solutions to subproblems is crucial in making the algorithm efficient. In this case, the resulting computation requires time $\Theta(n^3)$; this is surprisingly efficient, considering that an optimal tree is being found from among exponentially many possible trees.

By studying the pattern in which the arrays C and W are filled in, we see that the main diagonal c[i][i] is filled in first, then the first upper superdiagonal c[i][i+1], then the second upper superdiagonal c[i][i+2], and so on until the upper right corner of the array is reached. Rewriting the code to do this directly, and adding an array R[][] to keep track of the roots of subtrees, we obtain

```
1        int w[n][n];
2        int R[n][n];
3        int c[n][n];
4
5        // Fill in the main diagonal
6        for (int i = 0; i < n; i++) {
7            w[i][i] = alpha[i];
8            R[i][i] = 0;
9            c[i][i] = 0;
10       }
11
12       int minCost, cost;
13       for (int d = 1; d < n; d++)
14           // Fill in d-th upper super-diagonal
15           for (i = 0; i < n-d; i++) {
16               w[i][i+d] = w[i][i+d-1] + beta[i+d] + alpha[i+d];
17               R[i][i+d] = i+1;
18               c[i][i+d] = c[i][i] + c[i+1][i+d] + w[i][i+d];
19               for (int k = i+2; k <= i+d; k++) {
20                   cost = c[i][k-1] + c[k][i+d] + w[i][i+d];
21                   if (cost < c[i][i+d]) {
22                       R[i][i+d] = k;
23                       c[i][i+d] = cost;
24                   }
25               }
26           }
```

which more clearly shows the $\Theta(n^3)$ behavior.

As a second example of dynamic programming, consider the **traveling salesman problem** in which a salesman must visit n cities, returning to his starting point, and is required to minimize the cost of the trip. The cost of going from city i to city j is $C_{i,j}$. To use dynamic programming we must specify an optimal tour in a recursive framework, with subproblems resembling the overall problem. Thus we define

$$T\left(i;j_1,j_2,\ldots,j_k\right) = \begin{cases} \text{cost of an optimal tour from city } i \text{ to city } 1 \\ \text{that goes through each of the cities } j_1, j_2, \ldots, \\ j_k \text{ exactly once, in any order, and through no} \\ \text{other cities.} \end{cases}$$

The principle of optimality tells us that

$$T\left(i;j_1,j_2,\ldots,j_k\right) = \min_{1 \le m \le k} \left\{C_{i,j_m} + T\left(j_m;j_1,j_2,\ldots,j_{m-1},j_{m+1},\ldots,j_k\right)\right\},$$

where, by definition,

$$T(i; j) = C_{i,j} + C_{j,1}.$$

We can write a function T that directly implements the above recursive definition, but as in the optimal search tree problem, many subproblems would be solved repeatedly, leading to an algorithm requiring time $\Theta(n!)$. By caching the values $T(i; j_1, j_2, \ldots, j_k)$, we reduce the time required to $\Theta(n^2 2^n)$, still exponential, but considerably less than without caching.

1.5   Greedy Heuristics

Optimization problems always have an objective function to be minimized or maximized, but it is not often clear what steps to take to reach the optimum value. For example, in the optimum binary search tree problem of Section 1.4, we used dynamic programming to examine systematically all possible trees; but perhaps there is a simple rule that leads directly to the best tree—say by choosing the largest β_i to be the root and then continuing recursively. Such an approach would be less time-consuming than the $\Theta(n^3)$ algorithm we gave, but it does not necessarily give an optimum tree (if we follow the rule of choosing the largest β_i to be the root, we get trees that are no better, on the average, than a randomly chosen trees). The problem with such an approach is that it makes decisions that are locally optimum, though perhaps not globally optimum. But, such a "greedy" sequence of locally optimum choices does lead to a globally optimum solution in some circumstances.

Suppose, for example, $\beta_i = 0$ for $1 \leq i \leq n$, and we remove the lexicographic requirement of the tree; the resulting problem is the determination of an optimal prefix code for $n + 1$ letters with frequencies $\alpha_0, \alpha_1, \ldots, \alpha_n$. Because we have removed the lexicographic restriction, the dynamic programming solution of Section 1.4 no longer works, but the following simple greedy strategy yields an optimum tree: Repeatedly combine the two lowest-frequency items as the left and right subtrees of a newly created item whose frequency is the sum of the two frequencies combined. Here is an example of this construction; we start with five leaves with weights

First, combine leaves $\alpha_0 = 25$ and $\alpha_5 = 21$ into a subtree of frequency $25 + 21 = 46$:

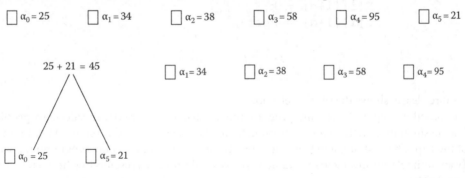

Then combine leaves $\alpha_1 = 34$ and $\alpha_2 = 38$ into a subtree of frequency $34 + 38 = 72$:

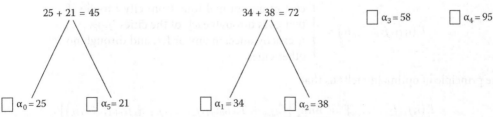

Next, combine the subtree of frequency $\alpha_0 + \alpha_5 = 45$ with $\alpha_3 = 58$:

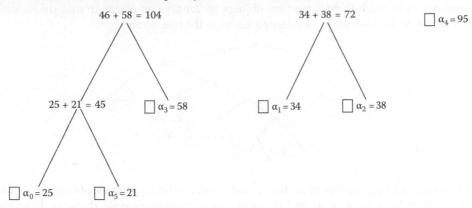

Then, combine the subtree of frequency $\alpha_1 + \alpha_2 = 72$ with $\alpha_4 = 95$:

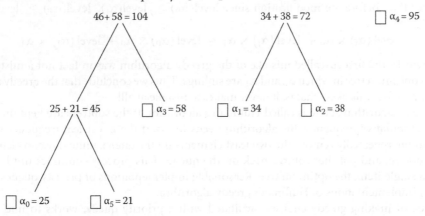

Finally, combine the only two remaining subtrees:

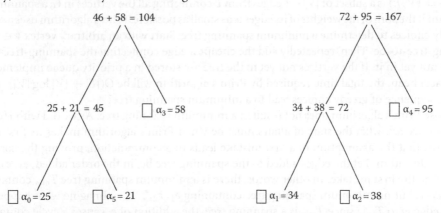

How do we know that the above-outlined process leads to an optimum tree? The key to proving that the tree is optimum is to assume, by way of contradiction, that it is not optimum. In this case, the greedy strategy must have erred in one of its choices, so let us look at the first error this strategy made. Since all previous greedy choices were not errors, and hence lead to an optimum tree, we can assume that we have a sequence of frequencies $\alpha_0, \alpha_1, \ldots, \alpha_n$ such that the first greedy choice is erroneous—without loss of generality assume that α_0 and α_1 are two smallest frequencies, those combined erroneously by the greedy strategy. For this combination to be erroneous, there must be no optimum tree in which these two αs are siblings, so consider an optimum tree, the locations of α_0 and α_1, and the location of the two deepest leaves in the tree, α_i and α_j:

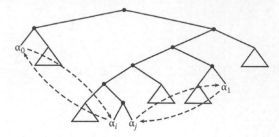

By interchanging the positions of α_0 and α_i and α_1 and α_j (as shown), we obtain a tree in which α_0 and α_1 are siblings. Because α_0 and α_1 are the two lowest frequencies (because they were the greedy algorithm's choice) $\alpha_0 \leq \alpha_i$ and $\alpha_1 \leq \alpha_j$, thus the weighted path length of the modified tree is no larger than before the modification since $\text{level}(\alpha_0) \geq \text{level}(\alpha_i)$, $\text{level}(\alpha_1) \geq \text{level}(\alpha_j)$ and hence

$$\text{Level}(\alpha_i) \times \alpha_0 + \text{level}(\alpha_j) \times \alpha_1 \leq \text{level}(\alpha_0) \times \alpha_0 + \text{level}(\alpha_1) \times \alpha_1.$$

In other words, the first so-called mistake of the greedy algorithm was in fact not a mistake, since there is an optimum tree in which α_0 and α_1 are siblings. Thus we conclude that the greedy algorithm never makes a first mistake—that is, it never makes a mistake at all!

The greedy algorithm above is called Huffman's algorithm. If the subtrees are kept on a priority queue by cumulative frequency, the algorithm needs to insert the $n + 1$ leaf frequencies onto the queue, and the repeatedly remove the two least elements on the queue, unite those to elements into a single subtree, and put that subtree back on the queue. This process continues until the queue contains a single item, the optimum tree. Reasonable implementations of priority queues will yield $O(n \log n)$ implementations of Huffman's greedy algorithm.

The idea of making greedy choices, facilitated with a priority queue, works to find optimum solutions to other problems too. For example, a **spanning tree** of a weighted, connected, undirected graph $G = (V, E)$ is a subset of $|V| - 1$ edges from E connecting all the vertices in G; a spanning tree is minimum if the sum of the weights of its edges is as small as possible. Prim's algorithm uses a sequence of greedy choices to determine a minimum spanning tree: Start with an arbitrary vertex $v \in V$ as the spanning-tree-to-be. Then, repeatedly add the cheapest edge connecting the spanning-tree-to-be to a vertex not yet in it. If the vertices not yet in the tree are stored in a priority queue implemented by a Fibonacci heap, the total time required by Prim's algorithm will be $O(|E| + |V| \log |V|)$. But why does the sequence of greedy choices lead to a minimum spanning tree?

Suppose Prim's algorithm does not result in a minimum spanning tree. As we did with Huffman's algorithm, we ask what the state of affairs must be when Prim's algorithm makes its first mistake; we will see that the assumption of a first mistake leads to a contradiction, proving the correctness of Prim's algorithm. Let the edges added to the spanning tree be, in the order added, $e_1, e_2, e_3, \ldots,$ and let e_i be the first mistake. In other words, there is a minimum spanning tree $T_{\min}$ containing e_1, $e_2, \ldots, e_{i-1}$, but no minimum spanning tree containing $e_1, e_2, \ldots, e_i$. Imagine what happens if we add the edge e_i to $T_{\min}$: since $T_{\min}$ is a spanning tree, the addition of e_i causes a cycle containing e_i. Let $e_{\max}$ be the highest-cost edge on that cycle not among $e_1, e_2, \ldots, e_i$. There must be such an $e_{\max}$ because $e_1, e_2, \ldots, e_i$ are acyclic, since they are in the spanning tree constructed by Prim's algorithm. Moreover, because Prim's algorithm always makes a greedy choice—that is, chooses the lowest-cost available edge—the cost of e_i is no more than the cost of any edge available to Prim's algorithm when e_i is chosen; the cost of $e_{\max}$ is at least that of one of those unchosen edges, so it follows that the cost of e_i is no more than the cost of $e_{\max}$. In other words, the cost of the spanning tree $T_{\min} - \{e_{\max}\} \cup \{e_i\}$

is at most that of $T_{\min}$; that is, $T_{\min} - \{e_{\max}\} \cup \{e_i\}$ is also a minimum spanning tree, contradicting our assumption that the choice of e_i is the first mistake. Therefore, the spanning tree constructed by Prim's algorithm must be a minimum spanning tree.

We can apply the greedy heuristic to many optimization problems, and even if the results are not optimal, they are often quite good. For example, in the n-city traveling salesman problem, we can get near-optimal tours in time $O(n^2)$ when the intercity costs are symmetric ($C_{i,j} = C_{j,i}$ for all i and j) and satisfy the triangle inequality($C_{i,j} \leq C_{i,k} + C_{k,j}$ for all i, j, and k). The closest insertion algorithm starts with a "tour" consisting of a single, arbitrarily chosen city, and successively inserts the remaining cities to the tour, making a greedy choice about which city to insert next and where to insert it: the city chosen for insertion is the city not on the tour but closest to a city on the tour; the chosen city is inserted adjacent to the city on the tour to which it is closest.

Given an $n \times n$ symmetric distance matrix C that satisfies the triangle inequality, let I_n of length $|I_n|$ be the "closest insertion tour" produced by the closest insertion heuristic and let O_n be an optimal tour of length $|O_n|$. Then

$$\frac{|I_n|}{|O_n|} < 2.$$

This bound is proved by an incremental form of the optimality proofs for greedy heuristics we have seen above: we ask not where the first error is, but by how much we are in error at each greedy insertion to the tour—we establish a correspondence between edges of the optimal tour O_n and cities inserted on the closest insertion tour. We show that at each insertion of a new city to the closest insertion tour, the additional length added by that insertion is at most twice the length of corresponding edge of the optimal tour O_n.

To establish the correspondence, imagine the closest insertion algorithm keeping track not only of the current tour, but also of a spider-like configuration including the edges of the current tour (the body of the spider) and pieces of the optimal tour (the legs of the spider). We show the current tour in solid lines and the pieces of optimal tour as dotted lines:

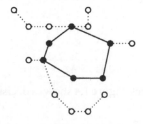

Initially, the spider consists of the arbitrarily chosen city with which the closest insertion tour begins and the legs of the spider consist of all the edges of the optimal tour *except* for one edge eliminated arbitrarily. As each city is inserted into the closest insertion tour, the algorithm will delete from the spider-like configuration one of the dotted edges from the optimal tour. When city k is inserted between cities l and m, the edge deleted is the one attaching the spider to the leg that contains the city inserted (from city x to city y), shown here in bold:

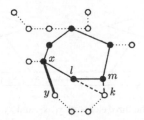

Now,

$$C_{k,m} \leq C_{x,y},$$

because of the greedy choice to add city k to the tour and not city y. By the triangle inequality,

$$C_{l,k} \leq C_{l,m} + C_{m,k},$$

and by symmetry we can combine these two inequalities to get

$$C_{l,k} \leq C_{l,m} + C_{x,y}.$$

Adding this last inequality to the first one above,

$$C_{l,k} + C_{k,m} \leq C_{l,m} + 2C_{x,y},$$

that is,

$$C_{l,k} + C_{k,m} - C_{l,m} \leq 2C_{x,y}.$$

Thus adding city k between cities l and m adds no more to I_n than $2C_{x,y}$. Summing these incremental amounts over the cost of the entire algorithm tells us

$$|I_n| \leq 2\,|O_n|,$$

as we claimed.

1.6   Lower Bounds

In Sections 1.2.1 and 1.3 we saw that we could sort faster than naïve $\Theta(n^2)$ worst-case behavior algorithms: we designed more sophisticated $\Theta(n \log n)$ worst-case algorithms. Can we do still better? No, $\Theta(n \log n)$ is a **lower bound** on sorting algorithms based on comparisons of the items being sorted. More precisely, let us consider only sorting algorithms described by decision boxes of the form

$$x_i : x_j$$

and outcome boxes such as

$$\boxed{x_1 < x_2 < x_3}$$

Such diagrams are called decision trees. Figure 1.4 shows a decision tree for sorting the three elements x_1, x_2, and x_3.

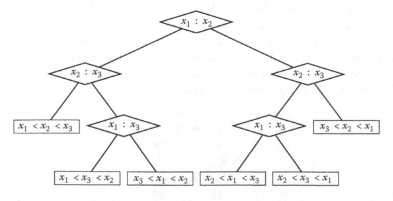

FIGURE 1.4 A decision tree for sorting the three elements x_1, x_2, and x_3.

Restricting ourselves to sorting algorithms represented by decision trees eliminates algorithms not based on comparisons of the elements, but it also appears to eliminate from consideration any of the common sorting algorithms, such as insertion sort, heapsort, and merge sort, all of which use index manipulations in loops, auxiliary variables, recursion, and so on. Furthermore, we have not allowed the algorithms to consider the possibility that some of the elements to be sorted may have equal values. These objections to modeling sorting algorithms on decision trees are serious, but can be countered by arguments that we have not been too restrictive.

For example, disallowing elements that are equal can be defended, because we certainly expect any sorting algorithm to work correctly in the special case that all of the elements are different; we are just examining an algorithm's behavior in this special case—a lower bound in a special case gives a lower bound on the general case. The objection that such normal programming techniques as auxiliary variables, loops, recursion, and so on are disallowed can be countered by the observation that any sorting algorithm based on comparisons of the elements can be stripped of its programming implementation to yield a decision tree. We expand all loops and all recursive calls, ignoring data moves and keeping track only of the comparisons between elements and nothing else. In this way, all common sorting algorithms can be described by decision trees.

We make an important observation about decision trees and the sorting algorithms represented as decision trees: If a sorting algorithm correctly sorts all possible input sequences of n items, then the corresponding decision tree has $n!$ outcome boxes. This observation follows by examining the correspondence between permutations and outcome boxes. Since the decision tree arose by tracing through the algorithm for all possible input sequences (that is, permutations), an outcome box must have occurred as the result of some input permutation or it would not be in the decision tree. Moreover, it is impossible that there are two different permutations corresponding to the same outcome box—such an algorithm cannot sort all input sequences correctly. Since there are $n!$ permutations of n elements, the decision tree has $n!$ leaves (outcome boxes).

To prove the $\Theta(n \log n)$ lower bound, define the *cost* of the ith leaf in the decision tree, $c(i)$, to be the number of element comparisons used by the algorithm when the input permutation causes the algorithm to terminate at the ith leaf. In other words, $c(i)$ is the depth of the ith leaf. This measure of cost ignores much of the work in the sorting process, but the overall work done will be proportional to the depth of the leaf at which the sorting algorithm terminates; because we are concerned only with lower bounds with in the Θ-notation, this analysis suffices.

Kraft's inequality tells us that for any tree with N leaves,

$$\sum_{i=1}^{N} \frac{1}{2^{c(i)}} \leq 1. \tag{1.2}$$

We can prove this inequality by induction on the height of the tree: When the height is zero, there is one leaf of depth zero and the inequality is trivial. When the height is nonzero, the inequality applies inductively to the left and right subtrees; the edges from the root to these subtrees increases the depth of each leaf by one, so the sum over each of the two subtrees is 1/2 and the inequality follows.

We use Kraft's inequality by letting h be the height of a decision tree corresponding to a sorting algorithm applied to n items. Then h is the depth of the deepest leaf, that is, the worst-case number of comparisons of the algorithm: $h \geq c(i)$, for all i. Therefore,

$$\frac{N}{2^h} = \sum_{i=1}^{N} \frac{1}{2^h}$$

$$\leq \sum_{i=1}^{N} \frac{1}{2^{c(i)}}$$

$$\leq 1,$$

and so

$$N \leq 2^h.$$

However, we saw that $N = n!$, so this last inequality can be rewritten as

$$2^h \geq n!.$$

But

$$n! \geq \left(\frac{n}{2}\right)^{n/2},$$

so that

$$h \geq \log_2 n! = \Theta(n \log n),$$

which is what we wanted to prove.

We can make an even stronger statement about sorting algorithms that can be modeled by decision trees: It is impossible to sort in average time better than $\Theta(n \log n)$, if each of the $n!$ input permutations is equally likely to be the input. The average number of decisions in this case is

$$\frac{1}{N} \sum_{i=1}^{N} c(i).$$

Suppose this is less than $\log_2 N$; that is, suppose

$$\sum_{i=1}^{N} c(i) < N \log_2 N.$$

By the arithmetic/geometric mean inequality, we know that

$$\frac{1}{m} \sum_{i=1}^{m} u_i \geq \left(\prod_{i=1}^{m} u_i\right)^{1/m}. \tag{1.3}$$

Applying this inequality, we have

$$\sum_{i=1}^{N} \frac{1}{2^{c(i)}} \geq N \left(\prod_{i=1}^{N} \frac{1}{2^{c(i)}}\right)^{1/N}$$

$$= N \left(2^{-\sum_{i=1}^{N} c(i)}\right)^{1/N}$$

$$> N \left(2^{-N \log_2 N}\right)^{1/N},$$

by assumption,

$$= N \left(N^{-N}\right)^{1/N}$$

$$= 1,$$

contradicting Kraft's inequality.

The lower bounds on sorting are called information theoretic lower bounds, because the rely on the amount of "information" contained in a single decision (comparison); in essence, the best a comparison can do is divide the set of possibilities into two equal parts. Such bounds also apply to many searching problems—for example, such arguments prove that binary search is, in a sense, optimal.

Information theoretic lower bounds do not always give useful results. Consider the **element uniqueness problem**, the problem of determining if there are any duplicate numbers in a set of n numbers, $x_1, x_2, \ldots, x_n$. Since there are only two possible outcomes, yes or no, the information theoretic lower bound says that a single comparison should be sufficient to answer the question. Indeed, that is true: Compare the product

$$\prod_{1 \leq i < j \leq n} (x_i - x_j) \tag{1.4}$$

to zero. If the product is nonzero, there are no duplicate numbers; if it is zero there are duplicates.

Of course, the cost of the one comparison is negligible compared to the cost of computing the product (Equation 1.4). It takes $\Theta(n^2)$ arithmetic operations to determine the product, but we are ignoring this dominant expense. The resulting lower bound is ridiculous.

To obtain a sensible lower bound for the element uniqueness problem, we define an **algebraic computation tree** for inputs $x_1, x_2, \ldots, x_n$ as a tree in which every leaf is either "yes" or "no." Every internal node either is a binary node (i.e., with two children) based on a comparison of values computed in the ancestors of that binary node, or is a unary node (i.e., with one child) that computes a value based on constants and values computed in the ancestors of that unary node, using the operations of addition, subtraction, multiplication, division, and square roots. An algebraic computation tree thus describes functions that take n numbers and compute a yes-or-no answer using intermediate algebraic results. The cost of an algebraic computation tree is its height.

By a complicated argument based on algebraic geometry, one can prove that any algebraic computation tree for the element uniqueness problem has depth at least $\Theta(n \log n)$. This is a much more sensible, satisfying lower bound on the problem. It follows from this lower bound that a simple sort-and-scan algorithm is essentially optimal for the element uniqueness problem.

1.7 Further Information

General discussions of the analysis of algorithms and data structures can be found in [1,4,10], and [9] has a more elementary treatment. Both [3,7] contain detailed treatments of recurrences, especially in regard to the analysis of algorithms. Sorting and searching techniques are explored in depth in [5,6] discusses algorithms for problems such as computing powers, evaluating polynomials, and multiplying large numbers. Reference [12] discusses many important graph algorithms, including several for finding minimum spanning trees. Our discussion of Fibonacci heaps is from [2]; our discussion of the heuristics for the traveling salesman problem is from [11]. A detailed discussion of the lower bound of the element-uniqueness problem is presented in [8,vol. 1, pp. 75–79], along with much other material on algebraic computation trees.

Defining Terms

Algebraic computation tree: A tree combining simple algebraic operations with comparisons of values.

Amortized cost: The cost of an operation considered to be spread over a sequence of many operations.

Average-case cost: The sum of costs over all possible inputs divided by the number of possible inputs.

Binary search tree: A binary tree that is lexicographically arranged so that, for every node in the tree, the nodes to its left are smaller and those to its right are larger.

Binary search: Divide-and-conquer search of a sorted array in which the middle element of the current range is probed so as to split the range in half.

Divide-and-conquer: A paradigm of algorithm design in which a problem is solved by reducing it to subproblems of the same structure.

Dynamic programming: A paradigm of algorithm design in which an optimization problem is solved by a combination of caching subproblem solutions and appealing to the "principle of optimality."

Element uniqueness problem: The problem of determining if there are duplicates in a set of numbers.

Greedy heuristic: A paradigm of algorithm design in which an optimization problem is solved by making locally optimum decisions.

Heap: A tree in which parent–child relationships are consistently "less than" or "greater than."

Information theoretic bounds: Lower bounds based on the rate at which information can be accumulated.

Kraft's inequality: The statement that $\sum_{i=1}^{N} 2^{-c(i)} \leq 1$, where the sum is taken over the N leaves of a binary tree and $c(i)$ is the depth of leaf i.

Lower bound: A function (or growth rate) below which solving a problem is impossible.

Merge sort: A sorting algorithm based on repeated splitting and merging.

Principle of optimality: The observation, in some optimization problems, that components of a globally optimum solution must themselves be globally optimal.

Priority queue: A data structure that supports the operations of creation, insertion, minimum, deletion of the minimum, and (possibly) decreasing the value an element, deletion, or merge.

Recurrence relation: The specification of a sequence of values in terms of earlier values in the sequence.

Sorting: Rearranging a sequence into order.

Spanning tree: A connected, acyclic subgraph containing all of the vertices of a graph.

Traveling salesman problem: The problem of determining the optimal route through a set of cities, given the intercity travel costs.

Worst-case cost: The cost of an algorithm in the most pessimistic input possibility.

Acknowledgments

The comments of Tanya Berger-Wolf, Ken Urban, and an anonymous referee are gratefully acknowledged.

References

1. Cormen, T., Leiserson, C., Rivest, R., and Stein, C., *Introduction to Algorithms*, 2nd edn., MIT Press/ McGraw-Hill, New York, 2001.
2. Fredman, M.L. and Tarjan, R.E., Fibonacci heaps and their use in improved network optimization problems, *J. ACM*, **34**, 596–615, 1987.
3. Greene, D.H. and Knuth, D.E., *Mathematics for the Analysis of Algorithms*, 3rd edn., Birkhäuser, Boston, MA, 1990.
4. Knuth, D.E., *The Art of Computer Programming*, Volume 1: *Fundamental Algorithms*, 3rd edn., Addison-Wesley, Reading, MA, 1997.

5. Knuth, D.E., *The Art of Computer Programming*, Volume 2: *Seminumerical Algorithms*, 3rd edn., Addison-Wesley, Reading, MA, 1997.

6. Knuth, D.E., *The Art of Computer Programming*, Volume 3: *Sorting and Searching*, 2nd edn., Addison-Wesley, Reading, MA, 1997.

7. Lueker, G.S., Some techniques for solving recurrences, *Comput. Surv.*, **12**, 419–436, 1980.

8. Mehlhorn, K., *Data Structures and Algorithms 1: Sorting and Searching*, Springer-Verlag, Berlin, 1984.

9. Reingold, E.M. and Hansen, W.J., *Data Structures in Pascal*, Little, Brown and Company, Boston, MA, 1986.

10. Reingold, E.M., Nievergelt, J., and Deo, N., *Combinatorial Algorithms, Theory and Practice*, Prentice-Hall, Englewood Cliffs, NJ, 1977.

11. Rosencrantz, D.J., Stearns, R.E., and Lewis, P.M., An analysis of several heuristics for the traveling salesman problem, *SIAM J. Comp.*, **6**, 563–581, 1977.

12. Tarjan, R.E., *Data Structures and Network Algorithms*, Society of Industrial and Applied Mathematics, Philadelphia, PA, 1983.

2
Searching

Ricardo Baeza-Yates
Yahoo! Research and University of Chile

Patricio V. Poblete
University of Chile

2.1 Introduction

Searching is one of the main computer applications in all other fields, including daily life. The basic problem consists of finding a given object in a set of objects of the same kind. Databases are perhaps the best examples where searching is the main task involved, and also where its performance is crucial.

We use the dictionary problem as a generic example of searching for a key in a set of keys. Formally, we are given a set S of n distinct keys* $x_1, \ldots, x_n$, and we have to implement the following operations, for a given key x:

$$
\begin{array}{ll}
\text{SEARCH:} & x \in S? \\
\text{INSERT:} & S \leftarrow S \cup \{x\} \\
\text{DELETE:} & S \leftarrow S - \{x\}
\end{array}
$$

Although, for simplicity, we treat the set S as just a set of keys, in practice it would consist of a set of records, one of whose fields would be designated as the key. Extending the algorithms to cover this case is straightforward.

Searches always have two possible outcomes. A search can be successful or unsuccessful, depending on whether the key was found or not in the set. We will use the letter U to denote the cost

---

* We will not consider in detail the case of nondistinct keys. Most of the algorithms work in that case too, or can be extended without much effort, but the performance may not be the same, especially in degenerate cases.

of an unsuccessful search and S to denote the cost of a successful search. In particular, we will use the name U_n (respectively, S_n) to denote the random variable "cost of an unsuccessful (respectively, successful) search for a random element in a table built by random insertions." Unless otherwise noted, we assume that the elements to be accessed are chosen with uniform probability. The notations C'_n and C_n have been used in the literature to denote the expected values of U_n and S_n, respectively [Knu75]. We use the notation $\mathbf{E}X$ to denote the expected value of the random variable X.

In this chapter, we cover the most basic searching algorithms which work on fixed-size arrays or tables and linked lists. They include techniques to search an array (unsorted or sorted), self-organizing strategies for arrays and lists, and hashing. In particular, hashing is a widely used method to implement dictionaries. Here, we cover the basic algorithms and provide pointers to the related literature. With the exception of hashing, we emphasize the SEARCH operation, because updates require $O(n)$ time. We also include a summary of other related searching problems.

2.2   Sequential Search

Consider the simplest problem: search for a given element in a set of n integers. If the numbers are given one by one (this is called an online problem), the obvious solution is to use sequential search. That is, we compare every element, and, in the worst case, we need n comparisons (either it is the last element or it is not present). Under the traditional RAM model, this algorithm is optimal. This is the algorithm used to search in an unsorted array storing n elements, and is advisable when n is small or when we do not have enough time or space to store the elements (for example in a very fast communication line). Clearly, $U_n = n$. If finding an element in any position has the same probability, then $\mathbf{E}S_n = \frac{n+1}{2}$.

2.2.1   Randomized Sequential Search

We can improve the worst case of sequential search in a probabilistic sense, if the element belongs to the set (successful search) and we have all the elements in advance (off-line case). Consider the following randomized algorithm. We flip a coin. If it is a head, we search the set from 1 to n, otherwise, from n to 1. The worst case for each possibility is n comparisons. However, we have two algorithms and not only one. Suppose that the element we are looking for is in position i and that the coin is fair (that is, the probability of head or tail is the same). So, the number of comparisons to find the element is i, if it is a head, or $n - i + 1$ if it is a tail. So, averaging over both algorithms (note that we are not averaging over all possible inputs), the expected worst case is

$$\frac{1}{2} \times i + \frac{1}{2} \times (n - i + 1) = \frac{n+1}{2}$$

which is independent of where the element is! This is better than n. In other words, an adversary would have to place the element in the middle position because he/she does not know which algorithm will be used.

2.2.2   Self-Organizing Heuristics

If the probability of retrieving each element is not the same, we can improve a successful search by ordering the elements in the decreasing order of the probability of access, in either an array or a linked list. Let p_i be the probability of accessing element i, and assume without loss of generality that

$p_i > p_{i+1}$. Then, we have that the optimal static order (OPT) has

$$\text{ES}_n^{\text{OPT}} = \sum_{i=1}^{n} ip_i$$

However, most of the time, we do not know the accessing probabilities and in practice they may change over time. For that reason, there are several heuristics to dynamically reorganize the order of the list. The most common ones are move-to-front (MF) where we promote the accessed element to the first place of the list, and transpose (T) where we advance the accessed element one place in the list (if it is not the first). These two heuristics are memoryless in the sense that they work only with the element currently accessed. MF is best suited for a linked list while "T" can also be applied to arrays. A good heuristic, if access probabilities do not change much with time, is the count (C) heuristic. In this case, every element keeps a counter with the number of times it has been accessed and advances in the list by one or more positions when its count is larger than the previous elements in the list. The main disadvantage of "C" is that we need $O(n)$ extra space to store the counters, if they fit in a word. Other more complex heuristics have been proposed, which are hybrids of the basic ones or/and use limited memory. They can also be extended to double-linked lists or more complex data structures as search trees.

Using these heuristics is advisable for small n, when space is severely limited, or when the performance obtained is good enough.* Evaluating how good a self-organizing strategy is with respect to the optimal order is not easily defined, as the order of the list is dynamic and not static. One possibility is to use the asymptotic expected successful search time, that is, the expected search time achieved by the algorithm after a very large sequence of independent accesses averaged over all possible initial configurations and sequences according to stable access probabilities. In this case, we have that

$$\text{ES}_n^{\text{T}} \le \text{ES}_n^{\text{MF}} \le \frac{\pi}{2}\text{ES}_n^{\text{OPT}} \approx 1.57\,\text{ES}_n^{\text{OPT}}$$

and $\text{ES}_n^{\text{C}} = \text{ES}_n^{\text{OPT}}$.

Another possible analysis is to use the worst-case search cost, but usually this is not fair, because, many times, the worst-case situation does not repeat very often. A more realistic solution is to consider the amortized cost. That is, the average number of comparisons over a worst-case sequence of executions. Then, a costly single access can be amortized with cheaper accesses that follow after. In this case, starting with an empty list, we have

$$S^{\text{MF}} \le 2S^{\text{OPT}}$$

and

$$S^{\text{C}} \le 2S^{\text{OPT}}$$

while S^{T} can be as bad as $O(mS^{\text{OPT}})$ for m operations. If we consider a nonstatic optimal algorithm, that is, an algorithm that knows the sequence of the accesses in advance and can rearrange the list with every access to minimize the search cost, then the results change. Under the assumption that the access cost function is convex, that is, if $f(i)$ is the cost of accessing the ith element, $f(i) - f(i-1) \ge f(i+1) - f(i)$. In this case, we usually have $f(i) = i$ and then only MF satisfies the inequality

$$S^{\text{MF}} \le 2S^{\text{OPT}}$$

---

* Also linked lists are an internal component of other algorithms, like hashing with chaining, which is explained later.

for this new notion of optimal algorithm. In this case, T and C may cost $O(m)$ times the cost of the optimal algorithm for m operations. Another interesting measure is how fast a heuristic converges to the asymptotic behavior. For example, T converges more slowly than MF, but it is more stable. However, MF is more robust as seen in the amortized case.

2.3 Sorted Array Search

In the off-line case, we can search faster, if we allow some time to preprocess the set and the elements can be ordered. Certainly, if we sort the set (using $O(n \log n)$ comparisons in the worst case) and store it in an array, we can use the well-known binary search. Binary search uses divide and conquer to quickly discard half of the elements by comparing the searched key with the element in the middle of the array, and if not equal, following the search recursively either on the first half or the second half (if the searched key was smaller or larger, respectively). Using binary search, we can solve the problem using at most $U_n = \lceil \log_2 (n+1) \rceil$ comparisons. Therefore, if we do many searches we can amortize the cost of the initial sorting.

On an average, a successful search is also $O(\log n)$. In practice, we do not have three-way comparisons; so, it is better to search recursively until we have discarded all but one element and then compare for equality. Binary search is optimal for the RAM comparison model in the worst and the average case. However, by assuming more information about the set or changing the model, we can improve the average or the worst case, as shown in the next sections.

2.3.1 Parallel Binary Search

Suppose now that we change the model by having p processors with a shared memory. That is, we use a parallel RAM (PRAM) model. Can we speed up binary search? First, we have to define how the memory is accessed in a concurrent way. The most used model is concurrent read but exclusive write (CREW) (otherwise it is difficult to know the final value of a memory cell after a writing operation). In a CREW PRAM, we can use the following simple parallel binary search. We divide the sorted set into $p + 1$ segments (then, there are p internal segment boundaries). Processor i compares the key to the element stored in the ith boundary and writes in a variable c_i a 0, if it is greater or a 1 if it is smaller (in case of equality, the search ends). All the processors do this in parallel. After this step, there is an index j such that $c_j = 0$ and $c_{j+1} = 1$ (we assume that $c_0 = 0$ and $c_{p+1} = 1$), which indicates in which segment the key should be. Then, processor i compares c_i and c_{i+1} and if they are different writes the new boundaries where the search continues recursively (see Figure 2.1). This step is also done in parallel (processor 1 and p take care of the extreme cases). When the segment is of size p or less, each processor compares one element and the search ends. Then, the worst-case number of parallel key comparisons is given by

$$U_n = 1 + U_{\frac{n}{p+1}}, \quad U_i = 1 \ (i \le p)$$

which gives $U_n = \log_{p+1} n + O(p)$. That is, $U_n = O(\log n / \log(p+1))$. Note that for $p = 1$, we obtain the binary search result, as expected. It is possible to prove that it is not possible to do it better. In the PRAM model, the optimal speedup is when the work done by p processors is p times the work of the optimal sequential algorithm. In this case, the total work is $p \log n / \log p$, which is larger than $\log n$. In other words, searching in a sorted set cannot be solved with optimal speedup. If we restrict the PRAM model also to exclusive reads (EREW), then $U_n = O(\log n - \log p)$, which is even worse. This is because, at every recursive step, if all the processors cannot read the new segment concurrently, we slow down all the processes.

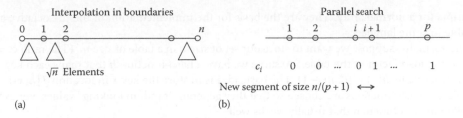

FIGURE 2.1 (a) Binary interpolation and (b) parallel binary search.

2.3.2 Interpolation Search

Assume now that the distribution of the n integers is uniform over a fixed range of $M \gg n$ integers. Then, instead of using binary search to divide the set, we can linearly interpolate the position of the searched element with respect to the smallest and the largest element. In this way, on an average, it is possible to prove that $O(\log \log n)$ comparisons are needed, if the element is in the set. The proof is quite involved and mathematical [PIA78], but there are some variations of interpolation search that have a simpler analysis. The main fact behind the $O(\log \log n)$ complexity is that when we do the first interpolation, with very high probability, the searched element is at a distance $O(\sqrt{n})$. So, the expected number of comparisons is given by the following recurrence

$$ES_n = a + ES_{b\sqrt{n}}, \quad ES_1 = 1$$

for some constants a and b, which give $ES_n = O(\log \log n)$. A simple variation is the following, which is called the binary interpolation search [PR77]. Imagine that we divide the ordered set in $\sqrt{n}$ segments of size approximately $\sqrt{n}$. Then, we use interpolation search on the $\sqrt{n} + 1$ keys that are segment boundaries (including the first and the last key as shown in Figure 2.1a) to find in which segment the key is. After we know the segment, we apply the same algorithm recursively in it. We can think that we have a $\sqrt{n}$-ary search tree, in which in each node we use interpolation search to find the right pointer. By a simple probabilistic analysis, it is possible to show that on an average less than 2.5 comparisons are needed to find in which segment the key is. So, we can use the previous recurrence with $a = 2.5$ and $b = 1$, obtaining less than $2.5 \log_2 \log_2 n$ comparisons on average.

2.4 Hashing

If the keys are drawn from a universe $U = \{0, \ldots, u - 1\}$, where u is a reasonably small natural number, a simple solution is to use a table $T[0..u - 1]$, indexed by the keys. Initially, all the table elements are initialized to a special value empty. When element x is inserted, the corresponding record is stored in the entry $T[x]$.

In the case when all we need to know is whether a given element is present or not, it is enough for $T[x]$ to take only two values: 0 (empty) and 1 (not empty), and the resulting data structure is called a bit vector.

Using this approach, all the three basic operations (INSERT, SEARCH, DELETE) take time $\Theta(1)$ in the worst case.

When the size of the universe is much larger, as is the case for character strings, the same approach could still work in principle, as strings can be interpreted as (possibly very large) natural numbers, but the size of the table would make it impractical. A solution is to map the keys onto a relatively small integer range, using a function called the hash function.

The resulting data structure, called hash tables, makes it possible to use the keys drawn from an arbitrarily large universe as "subscripts," much in the way the small natural numbers are used as subscripts for a normal array. They are the basis for the implementation of the "associative arrays" available in some languages.

More formally, suppose we want to store our set of size n in a table of size m. (The ratio $\alpha = n/m$ is called the load factor of the table.) Assume we have a hash function h that maps each key $x \in U$ to an integer value $h(x) \in [0..m-1]$. The basic idea is to store the key x in location $T[h(x)]$.

Typically, hash functions are chosen so that they generate "random looking" values. For example, the following is a function that usually works well:

$$h(x) = x \bmod m$$

where m is a prime number.

The preceding function assumes that x is an integer. In most practical applications, x is a character string instead. Strings are sequences of characters, each of which has an internal representation as a small natural number (e.g., using the American Standard Code for Information Interchange (ASCII) coding). If a string x can be written as $c_k c_{k-1}, \ldots, c_1 c_0$, where each c_i satisfies $0 \le c_i < C$, then we can compute h as

$$h \leftarrow 0; \quad \text{for } i \text{ in } 0, \ldots, k \text{ do } h \leftarrow (h * C + c_i) \bmod m$$

There is one important problem that needs to be solved. As the keys are inserted in the table, it is possible that we may have collisions between the different keys hashing to the same table slot. If the hash function distributes the elements uniformly over the table, the number of collisions cannot be too large on the average (after all, the expected number of elements per slot is α), but the well-known birthday paradox makes it very likely that there will be at least one collision, even for a lightly loaded table.

There are two basic methods for handling collisions in a hash table: chaining and open addressing.

2.4.1 Chaining

The simplest chaining method stores elements in the table as long as collisions do not occur. When there is a collision, the incoming key is stored in an overflow area, and the corresponding record is appended at the end of a linked list that stores all the elements that hashed to that same location (see Figure 2.2). The original hash table is then called the primary area. Figure 2.2 shows the result of inserting keys $A, B, \ldots, I$ in a hash table using chaining to resolve collisions, with the following hash function:

x	A	B	C	D	E	F	G	H	I
$h(x)$	4	0	4	7	1	8	4	8	1

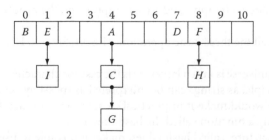

FIGURE 2.2 Hashing with separate chaining.

If the hash function maps elements uniformly, and if the elements are drawn at random from the universe, the expected values for these performance measures U_n and S_n are

$$EU_n = e^{-\alpha} + \alpha + \Theta\left(\frac{1}{m}\right)$$

and

$$ES_n = 1 + \frac{\alpha}{2} + \Theta\left(\frac{1}{m}\right)$$

Note that the search cost is basically independent of the number of elements, and that it depends on the load factor instead. By making the latter low enough, we can have hash tables with very efficient average search times.

The worst case, on the other hand, can be very bad: if all the keys happen to hash to the same location, the search cost is $\Theta(n)$. The probability of this happening is, of course, exceedingly small; so, a more realistic measure of the worst case may be the expected length of the longest chain in a table. This can be shown to be $\Theta\left(\frac{\log m}{\log\log m}\right)$ [Gon81].

Deletions are handled by simply removing the appropriate element from the list. When the element happened to be in the primary area, the first remaining element in the chain must be promoted to the primary area.

The need for an overflow area can be eliminated by storing these records in the table locations that happen to be empty. The resulting method, called coalesced hashing, has slightly larger search times, because of unrelated chains fusing accidentally, but it is still efficient even for a full table ($\alpha = 1$):

$$EU_n = 1 + \frac{1}{4}(e^{2\alpha} - 1 - 2\alpha) + \Theta\left(\frac{1}{m}\right)$$

$$ES_n = 1 + \frac{1}{8\alpha}(e^{2\alpha} - 1 - 2\alpha) + \frac{\alpha}{4} + \Theta\left(\frac{1}{m}\right)$$

Deletions require some care, as simply declaring a given location empty may confuse subsequent searches. If the rest of the chain contains an element that hashes to the now empty location, it must be moved there, and the process must be repeated for the new vacant location, until the chain is exhausted. In practice, this is not as slow as it sounds, as chains are usually short.

The preceding method can be generalized by allocating an overflow area of a given size, and storing the colliding elements there as long as there is space. Once the overflow area (called the cellar in this method) becomes full, the empty slots in the primary area begin to be used. This data structure was studied by Vitter and Chen [Vit80]. By appropriately tuning the relative sizes of the primary and the overflow areas, this method can outperform the other chaining algorithms. Even at a load of 100%, an unsuccessful search requires only 1.79 probes.

Vitter and Chen's analysis of coalesced hashing is very detailed, and also very complex. An alternative approach to this problem has been used by Siegel [Sie95] to obtain a much simpler analysis that leads to more detailed results.

2.4.2 Open Addressing

This is a family of methods that avoids the use of pointers, by computing a new hash value every time there is a collision.

Formally, this can be viewed as using a sequence of hash functions $h_0(x), h_1(x), \ldots$. An insertion probes that sequence of locations until finding an empty slot. Searches follow that same probe sequence, and are considered unsuccessful as soon as they hit an empty location.

The simplest way to generate the probe sequence is by first evaluating the hash function, and then scanning the table sequentially from that location (and wrapping around the end of the table). This is called linear probing, and is reasonably efficient if the load factor is not too high, but, as the table becomes full, it is too slow to be practical:

$$EU_n = \frac{1}{2}\left(1 + \frac{1}{1-\alpha}\right) + \Theta\left(\frac{1}{m}\right)$$

$$ES_n = \frac{1}{2}\left(1 + \frac{1}{(1-\alpha)^2}\right) + \Theta\left(\frac{1}{m}\right)$$

Note that these formulae break down for $\alpha = 1$. For a full table, the unsuccessful and the successful search costs are $\Theta(m)$ and $\Theta(\sqrt{m})$, respectively.

A better method for generating the probe sequences is double hashing. In addition to the original hash function $h(x)$, a second hash function $s(x) \in [1..m-1]$ is used, to provide a "step size." The probe sequence is then generated as

$$h_0(x) = h(x); \quad h_{i+1}(x) = (h_i(x) + s(x)) \bmod m.$$

Figure 2.3 shows the result of inserting keys $A, B, \ldots, I$ using the following hash functions:

x	A	B	C	D	E	F	G	H	I
$h(x)$	4	0	4	7	1	8	4	8	1
$s(x)$	5	1	4	2	5	3	9	2	9

Analyzing double hashing is quite hard [GS78,LM93], and for this reason most mathematical analyses are instead done assuming one of the two simplified models:

- *Uniform probing*: the locations are chosen at random from the set $[0..m-1]$, without replacement, or
- *Random probing*: the locations are chosen at random from the set $[0..m-1]$, with replacement.

For both models, it can be shown that

$$EU_n = \frac{1}{1-\alpha} + \Theta\left(\frac{1}{m}\right)$$

$$ES_n = \frac{1}{\alpha} \ln \frac{1}{1-\alpha} + \Theta\left(\frac{1}{m}\right)$$

Again, for a full table, the above expressions are useless, but we can prove that the search costs are $\Theta(m)$ and $\Theta(\log m)$, respectively.

Deletions cannot be done by simply erasing the given element, because searches would stop there and miss any element located beyond that point in its probe sequence. The solution of marking

0	1	2	3	4	5	6	7	8	9	10
B	E	G	F	A		I	D	C		H

FIGURE 2.3 Open addressing with double hashing.

the location as "dead" (i.e., still occupied for the purposes of searching, but free for the purposes of insertion) works at the expense of deteriorating the search time.

An interesting property of collision resolution in open addressing hash tables is that when two keys collide (one incoming key and one that is already in the table), either of them may validly stay in that location, and the other one has to try its next probe location. The traditional insertion method does not use this degree of freedom, and simply assigns locations to the keys in a "first-come-first-served" (FCFS) fashion.

Several methods have been proposed, that make use of this flexibility to improve the performance of open addressing hash tables. Knuth and Amble [KA74] used it to resolve collisions in favor of the element with the smallest key, with the result that the table obtained is the same as if the keys had been inserted in increasing order. This implies that all keys encountered in a successful search are in increasing order, a fact that can be used to speed up unsuccessful searches.

If we restrict ourselves to methods that arbitrate collisions based only on the past history (i.e., no look ahead), it can be shown that the expected successful search cost does not depend on the rule used (assuming random probing). However, the variance of the successful search cost does depend on the method used, and can be decreased drastically with respect to that of the standard FCFS method.

A smaller variance is important because of at least two reasons. First, a method with a low variance becomes more predictable, and less subject to wide fluctuations in its response time. Second, and more important, the usual method of following the probe sequence sequentially may be improved by replacing it by an optimal search algorithm, that probes the first most likely location, the second most likely, and so on. A reasonable approximation for this is a "mode-centered" search, that probes the most likely location first, and then moves away from it symmetrically.

Perhaps the simplest heuristic in this class is "last-come-first-served" (LCFS) [PM89] that does exactly the opposite from what the standard method does: in the case of a collision, the location is assigned to the incoming key.

For a full table (assuming random probing), the variance of the standard (FCFS) method is $\Theta(m)$. The LCFS heuristic reduces this to $\Theta(\log m)$.

Another heuristic that is much more aggressive in trying to decrease inequalities between the search costs of individual elements is the "Robin Hood" (RH) method [CLM85]. In the case of a collision, it awards the location to the element that has the largest retrieval cost. For a full table (assuming random probing), the variance of the cost of a successful search for RH hashing is ≤ 1.833, and using the optimal search strategy brings the expected retrieval cost down to ≤ 2.57 probes.

These variance reduction techniques can be applied also to linear probing. It can be shown [PVM94] that for a full table, both LCFS and RH decrease the variance from $\Theta(m^{3/2})$ of the standard FCFS method to $\Theta(m)$. In the case of linear probing, it can be shown that for any given set of keys, the RH arrangement minimizes the variance of the search time.

If we wish to decrease the expected search cost itself, and not just the variance, we must look ahead in the respective probe sequences of the keys involved in a collision. The simplest scheme would be to resolve the collision in favor of the key that would have to probe the most locations before finding an empty one. This idea can be applied recursively, and Brent [Bre73] and Gonnet and Munro [GM79] used this to obtain methods that decreased the expected search cost to 2.4921 and 2.13414 probes, respectively, for a full table.

Gonnet and Munro [GM79] considered also the possibility of moving keys backwards in their probe sequences to find the optimal table arrangement for a given set of keys. This problem is mostly of theoretical interest, and there are actually two versions of it, depending on whether the average search cost or the maximum search cost is minimized. Simulation results show that the optimal average search cost for a full table is approximately 1.83 probes.

2.4.3   Choosing a Hash Function

The traditional choice of hash functions suffers from two problems. First, collisions are very likely to occur, and the method has to plan for them. Second, a malicious adversary, knowing the hash function, may generate a set of keys that will make the worst case be $\Theta(n)$.

If the set of keys is known in advance, we may be able to find a perfect hash function, i.e., a hash function that produces no collisions for that set of keys. Many methods have been proposed for constructing perfect hash functions, beginning with the work of Fredman, Komlós, and Szemerédi [FKS84]. Mehlhorn [Meh82] proved a matching upper and lower bound of $\Theta(n)$ bits for the program size of a perfect hash function.

Fox et al. [FQDH91,FHCD92] provide algorithms for finding a minimal perfect hash function (i.e., for a full table) that run in expected linear time on the number of keys involved. Their algorithms have been successfully used on sets of sizes of the order of one million keys. Recent work has improved the practicality of this result to billions of keys in less than 1 hour and just using 1.95 bits per key for the case $m = 1.23n$ [BKZ05]. In the case of minimal functions ($m = n$), this value increases to 2.62 bits per key.

An approach to deal with the worst-case problem was introduced by Carter and Wegman [CW79]. They use a class of hash functions, and choose one function at random from the class for each run of the algorithm. In order for the method to work, the functions must be such that no pair of keys collide very often. Formally, a set $\mathcal{H}$ of hash functions is said to be universal, if for each pair of distinct keys, the number of hash functions $h \in \mathcal{H}$ is exactly $|\mathcal{H}|/m$. This implies that for a randomly chosen h, the probability of a collision between x and y is $1/m$, the same as if h has been assigned truly random hash values for x and y. Cormen, Leiserson, and Rivest [CLR90] show that if keys are composed of $r + 1$ "bytes" $x_0, \ldots, x_r$, each less than m, and $a = \langle a_0, \ldots, a_r \rangle$ is a sequence of elements chosen at random from $[0..m - 1]$, then the set of functions $h_a(x) = \sum_{0 \le i \le r} a_i x_i \bmod m$ is universal.

2.4.4   Hashing in Secondary Storage

All the hashing methods we have covered can be extended to secondary storage. In this setting, keys are usually stored in buckets, each holding a number of keys, and the hash function is used to select a bucket, not a particular key. Instead of the problem of collisions, we need to address the problem of bucket overflow. The analysis of the performance of these methods is notoriously harder than that of the main memory version, and few exact results exist [VP96].

However, for most practical applications, simply adapting the main memory methods is not enough, as they usually assume that the size of the hash table (m) is fixed in advance. Files need to be able to grow on demand, and also to shrink if we want to maintain an adequate memory utilization.

Several methods are known to implement extendible hashing (also called dynamic hash tables). The basic idea is to use an unbounded hash function $h(x) \ge 0$, but to use only its d rightmost bits, where d is chosen to keep overflow low or zero.

Fagin et al. [FNPS79] use the rightmost d bits from the hash function to access a directory of size 2^d, whose entries are pointers to the actual buckets holding the keys. Several directory entries may point to the same bucket.

Litwin [Lit78] and Larson [Lar78,Lar82] studied schemes that do not require a directory. Their methods work by gradually doubling the table size, scanning buckets from left to right. To do this, bucket splitting must be delayed until it is that bucket's turn to be split, and overflow records must be held temporarily using chaining or other similar method.

Perfect hashing functions can also be implemented in external memory. In fact the same results mentioned in the previous section achieve 2.7 keys per bit for $m = 1.23n$, increasing to 3.3 keys per bit in the case of minimal perfect hashing [BZ07].

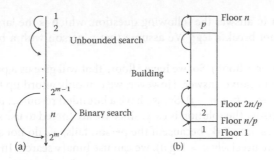

FIGURE 2.4 (a) A_1 unbounded search and (b) the person problem.

2.5 Related Searching Problems

2.5.1 Searching in an Unbounded Set

In most cases, we search in a bounded set. We can also search in an unbounded set. Consider the following game: one person thinks about a positive number and another person has to guess it with questions of the type: the number x is less than, equal to, or larger than the number that you are thinking? This problem was considered in [BY76].

A first obvious solution is to use the sequence $1, 2, \ldots, n$ (i.e., sequential search), using n questions. We can do better by using the "gambler" strategy. That is, we use the sequence $1, 2, 4, \ldots, 2^m$, until we have $2^m \geq n$. In the worst case, we have $m = \lfloor \log n \rfloor + 1$. Next, we can use binary search in the interval $2^{m-1} + 1$ to 2^m to search for n, using in the worst case $m - 1$ questions. Hence, the total number of questions is $2m - 1 = 2\lfloor \log n \rfloor + 1$. This algorithm is depicted in Figure 2.4. That is, only twice a binary search in a finite set of n elements. Can we do better? We can think that what we did is to search the exponent m using sequential search. So, we can use this algorithm, A_1, to search for m using $2\lfloor \log m \rfloor + 1$ questions, and then use binary search, with a total number of questions of $\log n + 2 \log \log n + O(1)$ questions. We could call this algorithm A_2.

In general, we can define algorithm A_k, which uses A_{k-1} to find m and then uses binary search. The complexity of such an algorithm is

$$S_n^k = \log n + \log \log n + \cdots + \log^{(k-1)} n + 2 \log^{(k)} n + O(1)$$

questions, where $\log^{(i)} n$ denotes log applied i times. Of course, if we could know the value of n in advance, there is an optimal value for k of $O(\log^* n)$,* because if k is too large, we go too far. However, we do not known n a priori!

2.5.2 Searching with Bounded Resources

Most of the time, we assume that we can perform an unbounded number of questions when searching. However, in many real situations, we search with bounded resources, For example, gasoline when using a car. As an example, we use a variable cost searching problem, initially proposed in [BB80, Section 3.2], with some changes, but maintaining the same philosophy. Given a building of n floors

---

* $\log^* n$ is the number of times that we have to apply the log function before we reach a value less than or equal to 0.

and k persons, we want to answer the following question: which is the largest floor from where a person can jump and not break a leg? We assume that a person with a broken leg cannot jump again.*

Suppose that the answer is floor j. So, we have j floors that will give us a positive answer and $n - j$ floors that will give us a negative answer. However, we can only afford up to k negative answers to solve the problem (in general $k < n - j$). So, we have a bounded resource: persons.

If we have just one person, the solution is easy, since we are forced to use sequential search to find j. Any other strategy does not work, because if the person fails, we do not solve the problem. If we have many persons (more precisely $k > \log n$), we can use binary search. In both cases, the solution is optimal in the worst case.

If we have two persons, a first solution would be to start using binary search with the first person, and then use the second sequentially in the remaining segment. In the worst case, the first person fails in the first jump, giving a $n/2$ jumps algorithm. The problem is that both persons do not perform the same amount of work. We can balance the work by using the following algorithm: the first person tries sequentially every n/p floors for a chosen p, that is $n/p, 2n/p$, etc. When his/her leg breaks, the second person has a segment of approximately n/p floors to check (see Figure 2.4). In the worst case, the number of floors is p (first person) plus n/p (second person). So, we have

$$U_n^2 = p + n/p + O(1)$$

Balancing the work, we have $p = n/p$, which implies $p = \sqrt{n}$ giving $U_n^2 = 2\sqrt{n} + O(1)$. Note that to succeed, any algorithm has to do sequential search in some segment with the last person.

We can generalize the above algorithm to k persons using the partitioning idea recursively. Every person except the last one partitions the remaining segment into p parts and the last person uses sequential search. In the worst case, every person (except the last one) has to perform p jumps. The last one does sequential search on a segment of size n/p^{k-1}. So, the total cost is approximately

$$U_n^k = (k - 1)p + \frac{n}{p^{k-1}}$$

Balancing the work for every person, we must have $p = n/p^{k-1}$, obtaining $p = n^{1/k}$ (same as using calculus!). Then, the final cost is

$$U_n^k = kn^{1/k}$$

If we consider $k = \log_2 n$, we have

$$U_n^k = kn\, 2^{\log_2(n^{1/k})} = \log_2 n\, 2^{\frac{\log_2 n}{k}} = 2\log_2 n$$

which is almost like a binary search. In fact, taking care of the partition boundaries, and using an optimal partition (related to binomial trees), we can save k jumps, which gives the same as binary search. So, we have a continuum from sequential to binary search as k grows.

We can mix the previous two cases to have an unbounded search with limited resources. The solution mixes the two approaches already given and can be a nice exercise for interested readers.

2.5.3 Searching with Nonuniform Access Cost

In the traditional RAM model, we assume that any memory access has the same cost. However, this is not true, if we consider the memory hierarchy of a computer: registers, cache and main memory, secondary storage, etc. As an example of this case, we use the hierarchical memory model introduced

---

* This is a theoretical example, do not try to solve this problem in practice!

in [AACS87]. That is, the access cost to position x is given by a function $f(x)$. The traditional RAM model is when $f(x)$ is a constant function. Based on the access times of current devices, possible values are $f(x) = \log x$ or $f(x) = x^{\alpha}$ with $0 < \alpha \leq 1$.

Given a set of n integers in a hierarchical memory, two problems are discussed. First, given a fixed order (sorted data), what is the optimal worst-case search algorithm. Second, what is the optimal ordering (implicit structure) of the data to minimize the worst-case search time. This ordering must be described using constant space.

In both cases, we want to have the n elements in n contiguous memory locations starting at some position and only using a constant amount of memory to describe the searching procedure. In our search problem, we consider only successful searches, with the probability of searching for each one of the n elements being the same.

Suppose that the elements are sorted. Let $S(i, j)$ be the optimal worst-case cost to search for an element which is between positions i and j of the memory. We can express the optimal worst-case cost as

$$S(i, j) = \min_{k=i,\ldots,j} \{f(k) + \max(S(i, k - 1), S(k + 1, j))\}$$

for $i \geq j$ or 0 otherwise. We are interested in $S(1, n)$. This recurrence can be solved using dynamic programming in $O(n^2)$ time. This problem was considered in [Kni88], where it is shown that for logarithmic or polynomial $f(x)$, the optimal algorithm needs $O(f(n) \log n)$ comparisons. In particular, if $f(x) = x^{\alpha}$, a lower and upper bound of

$$\frac{n^{\alpha} \log n}{1 + \alpha}$$

for the worst-case cost of searching is given in [Kni88].

In our second problem, we can order the elements to minimize the searching cost. A first approach is to store the data as the implicit complete binary search tree induced by a binary search in the sorted data, such that the last level is compacted to the left (left complete binary tree). That is, we store the root of the tree in position 1 and in general the children of the element in position i in positions $2i$ and $2i + 1$ like in a heap. Nevertheless, there are better addressing schemes that balance as much as possible every path of the search tree.

2.5.4 Searching with Partial Information

In this section, we use a nonuniform cost model plus an unbounded domain. In addition the algorithm does not know all the information of the domain and learns about it while searching. In this case, we are searching for an object in some space under the restriction that for each new "probe" we must pay costs proportional to the distance of the probe position relative to our current probe position and we wish to minimize this cost. This is meant to model the cost in real terms of a robot (or human) searching for an object when the mobile searcher must move about to find the object. It is also the case for many searching problems on secondary memory devices as disk and tapes. This is another example of an online algorithm. An online algorithm is called c-competitive, if the solution to the problem related to the optimal solution when we have all the information at the beginning (off-line case) is bounded by

$$\frac{\text{Solution (online)}}{\text{Optimal (off-line)}} \leq c$$

Suppose that a person wants to find a bridge over a river. We can abstract this problem as finding some distinguished point on a line. Assume that the point is n (unknown) steps away along the line and that the person does not know how far away the point is. What is the minimum number of steps he or she must make to find the point, as a function of n?

The optimal way to find the point (up to lower-order terms) is given by Linear Spiral Search [BYCR93]: execute cycles of steps where the function determining the number of steps to walk before the ith turn starting from the origin is 2^i for all $i \geq 1$. That is, we first walk one step to the left, we return to the origin, then two steps to the right, returning again to the origin, then four steps to the left, etc. The total distance walked is $2 \sum_{i=1}^{\lfloor \log n \rfloor + 1} 2^i + n$ which is no more than 9 times the original distance. That is, this is a 9-competitive algorithm, and this constant cannot be improved.

2.6 Research Issues and Summary

Sequential and binary search are present in many forms in most programs, as they are basic tools for any data structure (either in main memory or in secondary storage). Hashing provides an efficient solution when we need good average search cost. We also covered several variants generalizing the model of computation (randomized, parallel, bounded resources) or the data model (unbounded, nonuniform access cost and partial knowledge).

As for research issues, we list here some of the most important problems that still deserve more work. Regarding hashing, faster and practical algorithms to find perfect hashing functions are still needed. The hierarchical memory model has been extended to cover nonatomic accesses (that is, access by blocks) and other variations. This model still has several open problems. Searching with partial information lead to research on more difficult problems as motion planning.

2.7 Further Information

Additional algorithms and references in the first four sections can be found in [GBY91] and in many algorithm textbooks. More information on self-organizing heuristics can be found on Hester and Hirschberg's survey [HH85]. The amortized-case analysis is presented in [BM85,ST85]. More information on searching with partial information is given in [BYCR93]. More information on searching nonatomic objects is covered in [BY97].

Many chapters of this handbook extend the material presented here. Another example of amortized analysis of algorithms is given in Chapter 1. Analysis of the average case is covered in Chapter 11. Randomized algorithm is covered in Chapters 12 and 25 of *Algorithms and Theory of Computation Handbook, Second Edition: Special Topics and Techniques*, respectively. Searching and updating more complex data structures are explained in Chapters 4 and 5. Searching for strings and subtrees is covered in Chapters 13 and 15, respectively. The use of hashing functions in cryptography is covered in Chapter 9 of *Algorithms and Theory of Computation Handbook, Second Edition: Special Topics and Techniques*.

Defining Terms

Amortized cost: A worst-case cost of a sequence of operations, averaged over the number of operations.

Chaining: A family of hashing algorithms that solves collisions by using pointers to link elements.

CREW PRAM: A computer model that has many processors sharing a memory where many can read at the same time, but only one can write at any given time in a given memory cell.

Hash function: A function that maps keys onto table locations, by performing arithmetic operations on the keys. The keys that hash to the same location are said to collide.

Online algorithm: An algorithm that process the input sequentially.

Open addressing: A family of collision resolution strategies based on computing alternative hash locations for the colliding elements.

Randomized algorithm: An algorithm that makes some random (or pseudorandom) choices.

Self-organizing strategies: A heuristic that reorders a list of elements according to how the elements are accessed.

Universal hashing: A scheme that chooses randomly from a set of hash functions.

References

[AACS87] A. Aggarwal, B. Alpern, K. Chandra, and M. Snir. A model for hierarchical memory. In *Proceedings of the 19th Annual ACM Symposium of the Theory of Computing*, pp. 305–314, New York, 1987.

[BB80] J.L. Bentley and D.J. Brown. A general class of resource trade-offs. In *IEEE Foundations of Computer Science*, Vol. 21, pp. 217–228, Syracuse, NY, Oct. 1980.

[BM85] J.L. Bentley and C.C. McGeoch. Amortized analyses of self-organizing sequential search heuristics. *Communications of the ACM*, 28(4):404–411, Apr. 1985.

[Bre73] R.P. Brent. Reducing the retrieval time of scatter storage techniques. *Communications of the Association for Computing Machinery*, 16(2):105–109, Feb. 1973.

[BY76] J.L. Bentley and A.C-C. Yao. An almost optimal algorithm for unbounded searching. *Information Processing Letters*, 5(3):82–87, Aug. 1976.

[BY97] R. Baeza-Yates. Searching: An algorithmic tour. In A. Kent and J.G. Williams, editors, *Encyclopedia of Computer Science and Technology*, Vol. 37, pp. 331–359. Marcel Dekker, Inc., 1997.

[BYCR93] R.A. Baeza-Yates, J. Culberson, and G. Rawlins. Searching in the plane. *Information and Computation*, 106(2):234–252, Oct. 1993. Preliminary version titled "Searching with Uncertainty" was presented at SWAT'88, Halmstad, Sweden, *LNCS*, 318, pp. 176–189.

[BKZ05] F. Botelho, Y. Kohayakawa, and N. Ziviani. A Practical Minimal Perfect Hashing method. In *4th International Workshop on Efficient and Experimental Algorithms*, pp. 488–500, Santorini, Greece, 2005.

[BZ07] F. Botelho and N. Ziviani. External perfect hashing for very large key sets. In *16th ACM Conference on Information and Knowledge Management*, pp. 653–662, Lisbon, Portugal, 2007.

[CLM85] P. Celis, P.-Å. Larson, and J.I. Munro. Robin Hood hashing. In *26th Annual Symposium on Foundations of Computer Science*, pp. 281–288, Portland, OR, 1985.

[CLR90] T.H. Cormen, C.E. Leiserson, and R.L. Rivest. *The Design and Analysis of Computer Algorithms*. MIT Press, Cambridge, MA, 1990.

[CW79] J.L. Carter and M.N. Wegman. Universal classes of hash functions. *Journal of Computer and System Sciences*, 18(2):143–154, Apr. 1979.

[FHCD92] E.A. Fox, L.S. Heath, Q. Chen, and A.M. Daoud. Minimal perfect hash functions for large databases. *Communications of the Association for Computing Machinery*, 35(1):105–121, Jan. 1992.

[FKS84] M.L. Fredman, J. Komlós, and E. Szemerédi. Storing a sparse table with $O(1)$ worst case access time. *Journal of the Association for Computing Machinery*, 31(3):538–544, July 1984.

[FNPS79] R. Fagin, J. Nievergelt, N. Pippenger, and H.R. Strong. Extendible hashing: A fast access method for dynamic files. *ACM Transactions on Database Systems*, 4(3):315–344, Sep. 1979.

[FQDH91] E.A. Fox, F.C. Qi, A.M. Daoud, and L.S. Heath. Order-preserving minimal perfect hash functions and information retrieval. *ACM Transactions on Information Systems*, 9(3):281–308, 1991.

[GBY91]   G.H. Gonnet and R. Baeza-Yates. *Handbook of Algorithms and Data Structures: In Pascal and C*, 2nd edition. Addison-Wesley, Reading, MA, 1991.

[GM79]   G.H. Gonnet and J.I. Munro. Efficient ordering of hash tables. *SIAM Journal on Computing*, 8(3):463–478, Aug. 1979.

[Gon81]   G.H. Gonnet. Expected length of the longest probe sequence in hash code searching. *Journal of the Association for Computing Machinery*, 28(2):289–304, Apr. 1981.

[GS78]   L.J. Guibas and E. Szemerédi. The analysis of double hashing. *Journal of Computer and System Sciences*, 16(2):226–274, Apr. 1978.

[HH85]   J.H. Hester and D.S. Hirschberg. Self-organizing linear search. *ACM Computing Surveys*, 17(3):295–311, Sep. 1985.

[KA74]   D.E. Knuth and O. Amble. Ordered hash tables. *The Computer Journal*, 17(5):135–142, May 1974.

[Kni88]   W.J. Knight. Search in an ordered array having variable probe cost. *SIAM Journal of Computing*, 17(6):1203–1214, Dec. 1988.

[Knu75]   D.E. Knuth. *The Art of Computer Programming, Sorting and Searching*, 2nd edition. Addison-Wesley, Reading, MA, 1975.

[Lar78]   P. Larson. Dynamic hashing. *BIT*, 18(2):184–201, 1978.

[Lar82]   P. Larson. Performance analysis of linear hashing with partial expansions. *ACM Transactions on Database Systems*, 7(4):566–587, Dec. 1982.

[Lit78]   W. Litwin. Virtual hashing: A dynamically changing hashing. In S.B. Yao, editor, *Fourth International Conference on Very Large Data Bases*, pp. 517–523, ACM Press, West Berlin, Germany, 1978.

[LM93]   G.S. Lueker and Mariko Molodowitch. More analysis of double hashing. *Combinatorica*, 13(1):83–96, 1993.

[Meh82]   K. Mehlhorn. On the program size of perfect and universal hash functions. In *23rd Annual Symposium on Foundations of Computer Science*, pp. 170–175, Chicago, IL, 1982.

[PIA78]   Y. Perl, A. Itai, and H. Avni. Interpolation search—a log log *n* search. *Communications of the ACM*, 21(7):550–553, July 1978.

[PM89]   P.V. Poblete and J.I. Munro. Last-Come-First-Served Hashing. *Journal of Algorithms*, 10(2): 228–248, June 1989.

[PR77]   Y. Perl and E.M. Reingold. Understanding the complexity of interpolation search. *Information Processing Letters*, 6(6):219–221, Dec. 1977.

[PVM94]   P.V. Poblete, A. Viola, and J.I. Munro. The analysis of a hashing scheme by the diagonal poisson transform. *LNCS*, 855:94–105, 1994.

[Sie95]   A. Siegel. On the statistical dependencies of coalesced hashing and their implications for both full and limited independence. In *6th ACM-SIAM Symposium on Discrete Algorithms (SODA)*, pp. 10–19, San Francisco, CA, 1995.

[ST85]   D.D. Sleator and R.E. Tarjan. Amortized efficiency of list update and paging rules. *Communications of the ACM*, 28(2):202–208, Feb. 1985.

[Vit80]   J.S. Vitter. Analysis of coalescing hashing. PhD thesis, Stanford University, Stanford, CA, October 1980.

[VP96]   A. Viola and P.V. Poblete. The analysis of linear probing hashing with buckets. In J. Diaz and M. Serna, editors, *Fourth Annual European Symposium on Algorithms–ESA'96*, pp. 221–233. Springer, Barcelona, Spain, Sep. 1996.

3

Sorting and Order Statistics

Vladimir Estivill-Castro
Griffith University

3.1 Introduction

Sorting is the computational process of rearranging a given sequence of items from some total order into ascending or descending order. Because sorting is a task in the very core of Computer Science, efficient algorithms were developed early. The first practical and industrial application of computers had many uses for sorting. It is still a very frequently occurring problem, often appearing as a preliminary step to some other computational task. A related application to sorting is computing order statistics, for example, finding the median, the smallest or the largest of a set of items. Although finding order statistics is immediate, once the items are sorted, sorting can be avoided and faster algorithms have been designed for the kth largest element, the most practical of which is derived from the structure of a sorting method. Repeated queries for order statistics may be better served by sorting and using an efficient data structure that implements the abstract data type dictionary, with the ranking as the key.

Sorting usually involves data consisting of records in one or several files. One or several fields of the records are used as the criteria for sorting (often a small part of the record) and are called the keys. Usually, the objective of the sorting method is to rearrange the records, so that the keys are arranged in numerical or alphabetical order. However, many times, the actual rearrangement of all the data records is not necessary, but just the logical reorder by manipulating the keys is sufficient.

In many applications of sorting, elementary sorting algorithms are the best alternative. One has to admit that in the era of the Internet and the World Wide Web, network lag is far more noticeable that almost any sorting on small data sets. Moreover, sorting programs are often used once (or only a few times) rather than being repeated many times, one after another. Simple methods are always suitable for small files, say less than 100 elements. The increasing speeds in less expensive computers are enlarging the size for which basic methods are adequate. More advanced algorithms require more careful programming, and their correctness or efficiency is more fragile to the thorough understanding of their mechanisms. Also, sophisticated methods may not take advantage of the existing order in the input, which may already be sorted, while elementary methods usually do.

Finally, elementary sorting algorithms usually have a very desirable property, named stability; that is, they preserve the relative order of the items with equal keys. This is usually expected in applications generating reports from already sorted files, but with a different key. For example, long-distance phone calls are usually recorded in a log in chronological order by date and time of the call. When reporting bills to customers, the carrier sorts by customer name, but the result should preserve the chronological order of the calls made by each particular customer.

Advanced methods are the choice in applications involving a large number of items. Also, they can be used to build robust, general-purpose sorting routines [5–7,11]. Elementary methods should not be used for large, randomly permuted files. An illustrative trade-off between a sophisticated technique that results in general better performance and the simplicity of the programming is the sorting of keys and pointers to records rather than the entire records that we mentioned earlier. Once the keys are sorted, the pointers to the complete records can be used as a pass over to the file to rearrange the data in the desired order. It is usually more laborious to apply this technique because we must construct the auxiliary sequence of keys and pointers (or in the case of long keys, the keys can be kept with the records as well). However, many exchanges and data moves are saved until the final destination of each record is known. In addition, less space is required if, as is common in practice, keys are only a small part of the record. When the keys are a quarter or more of the alphanumeric records, this technique is not worth the effort. However, as records contain multimedia data or other large blobs of information, the sorting of the keys and pointers is becoming the best practice.

In recent years, the study of sorting has been receiving attention [25,26,31–33] because computers now offer a sophisticated memory hierarchy, with CPUs having a large space in a cache faster than random access memory. When the file to be sorted is small enough that all the data fit into the random access memory (main memory) as an array of records or keys, the sorting process is called **internal sorting**. Today, memory sizes are very large, and the situation when the file to be sorted is so large that the data do not entirely fit into the main memory seems to be a matter of the past. However, the fact is that computers have more stratified storage capacity that trades cost for speed and size. Even at the level of microinstruction, there are caches to anticipate code, and as we mentioned, CPUs today manage caches as well. Operating systems are notorious for using virtual memory, and storage devices have increased capacity in many formats for magnetic disks and DVDs. While sequential access in tapes does seem to be a thing of the past, there is now the issue of the data residing on servers over a computer network. Perhaps the most important point is that the algorithms that take advantage of the locality of the reference, both for the instructions as well as for the data, are better positioned to perform well.

Historically, **external sorting** remains the common term for sorting files that are too large to be held in main memory and the rearrangement of records is to happen on disk drives. The availability of large main memories has modified how buffers are allocated to optimize the input/output costs between the records in the main memory and the architectures of disks drives [22,34].

Sorting algorithms can also be classified into two large groups according to what they require about the data to perform the sorting. The first group is called comparison-based. Methods of this class only use the fact that the universe of keys is linearly ordered. Because of this property, the implementation of comparison-based algorithms can be generic with respect to the data type of the keys, and a comparison routine can be supplied as a parameter to the sorting procedure. The second group of algorithms assumes further that keys are restricted to a certain domain or data representation, and uses the knowledge of this information to dissect subparts, bytes, or bits of the keys.

Sorting is also ideal for introducing issues regarding algorithmic complexity. For comparison-based algorithms, it is possible to precisely define an abstract model of computation (namely, decision trees) and show the lower bounds on the number of comparisons any sorting method in this family would require to sort a sequence with n items (in the worst case and in the average case). A comparison-based sorting algorithm that requires $O(n \log n)$ comparisons is said to be optimal, because it matches the $\Omega(n \log n)$ lower bound. Thus, in theory, no other algorithm could be faster. The fact that algorithms that are not comparison-based can result in faster implementations illustrates the relevance of the model of computation with respect to theoretical claims of optimality and the effect that stronger assumptions on the data have for designing a faster algorithm [1,24].

Sorting illustrates randomization (the fact that the algorithm can make a random choice). From the theoretical angle, randomization provides a more powerful machine that has available random bits. In practice, a pseudorandom generator is sufficient and randomization delivers practical value for sorting algorithms. In particular, it provides an easy protection for sophisticated algorithms from special input that may be simple (like almost sorted) but harmful to the efficiency of the method. The most notable example is the use of randomization for the selection of the pivot in quicksort.

In what follows, we will assume that the goal is to sort into ascending order, since sorting in descending order is symmetrical, or can be achieved by sorting in ascending order and reversing the result (in linear time, which is usually affordable). Also, we make no further distinction between the records to be sorted and their keys, assuming that some provision has been made for handling this as suggested before. Through our presentation the keys may not all be different, since one of the main applications of sorting is to bring together records with matching keys. We will present algorithms in pseudocode in the style introduced by Cormen et al. [9], or when more specific detail is convenient, we will use PASCAL code. When appropriate, we will indicate possible trade-offs between clarity and efficiency of the code. We believe that efficiency should not be pursued to the extreme, and certainly not above clarity. The costs of programming and code maintenance are usually larger than the slight efficiency gains of tricky coding. For example, there is a conceptually simple remedy to make every sorting routine **stable**. The idea is to precede it with the construction of new keys and sort according to the lexicographical order of the new keys. The new key for the ith item is the pair (k_i, i), where k_i is the original sorting key. This requires the extra management of the composed keys and adds to the programming effort the risk of a faulty implementation. Today, this could probably be solved by simply choosing a competitive stable sort, perhaps at the expense of slightly more main memory or CPU time.

3.2 Underlying Principles

Divide-and-conquer is a natural, top-down approach for the design of an algorithm for the abstract problem of sorting a sequence $X = \langle x_1, x_2, \ldots, x_n \rangle$ of n items. It consists of dividing the problem into smaller subproblems, hoping that the solution of the subproblems are easier to find, and then composing the partial solutions into the solution of the original problem. A prototype of this idea is mergesort; where the input sequence $X = \langle x_1, x_2, \ldots, x_n \rangle$ is split into two sequences $X_L = \langle x_1, x_2, \ldots, x_{\lfloor n/2 \rfloor} \rangle$ (the left subsequence) and $X_R = \langle x_{\lfloor n/2 \rfloor + 1}, \ldots, x_n \rangle$ (the right subsequence). Finding solutions recursively for sequences with more than one item and terminating the recursion with sequences of only one item (since these are always sorted) provides a solution for the two subproblems. The overall solution is found by describing a method to merge two sorted sequences. In fact, internal sorting algorithms are variations of the two forms of using divide-and-conquer:

Conquer form: Divide is simple (usually requiring constant time), conquer is sophisticated.

Divide form: Divide is sophisticated, conquer is simple (usually requiring constant time).

Again, mergesort is an illustration of the conquer from. The core of the method is the merging of the two sorted sequences, hence its name.

The prototype of the **divide form** is quicksort [19]. Here, one item x_p (called the pivot) is selected from the input sequence X and its key is used to create two subproblems $X_\le$ and $X_\ge$, where $X_\le$ contains items in X with keys less than or equal to the pivot's, while items in $X_\ge$ have items larger than or equal to the pivot's. Now, recursively applying the algorithm for $X_\le$ and $X_\ge$ results in a global solution (with a trivial conquer step that places $X_\le$ before $X_\ge$).

Sometimes, we may want to conceptually simplify the conquer form by not dividing into two subproblems of roughly the same size, but rather divide X into $X' = \langle x_1, \ldots, x_{n-1} \rangle$ and $X'' = \langle x_n \rangle$. This is conceptually simpler in two ways. First, X'' is a trivial subproblem, because it has only one item and thus it is already sorted. Therefore, in a sense we have one subproblem less. Second, the merging of the solutions of X' and X'' is simpler than the merging of the two sequences of almost equal length, because we just need to place x_n in its proper position amongst the sorted items from X'. Because the method is based upon inserting into an already sorted sequence, sorting algorithms with this idea are called **insertion sorts**.

Insertion sorts vary according to how the sorted list for the solution of X' is represented by a data structure that supports insertions. The simplest alternative is to have the sorted list stored in an array, and this method is named insertion sort or straight insertion sort [21]. However, if the sorted list is represented by a level-linked tree with a finger, the method has been named A-sort [24] or local insertion sort [23].

The complementary simplification in the divide form makes one subproblem trivial by selecting a pivot so that $X_\le$ or $X_\ge$ consists of just one item. This is achieved, if we select the pivot as the item with the smallest or largest key in the input. The algorithms under this scheme are called selection sorts, and they vary according to the data structure used to represent X, so that the repeated extraction of the maximum (or minimum) key is efficient. This is typically the requirement of the priority queue abstract data type with the keys as the priorities. When the priority queue is implemented as an array and the smallest key is found by scanning this array, the method is selection sort. However, if the priority queue is an array organized into a heap, the method is called heapsort.

Using divide-and-conquer does not necessarily mean that the division must be into two subproblems. It may divide into several subproblems. For example, if the keys can be manipulated with other operations besides comparisons, bucket sort uses an interpolation formula on the keys to partition the items between m buckets. The buckets are sets of items, which are usually implemented as queues. The queues are usually implemented as linked lists that allow insertion and removal in constant time in first-in-first-out order, as the abstract data type queue requires. The buckets represent subproblems to be sorted recursively. Finally, all the buckets are concatenated together. Shellsort divides the problem of sorting X into several subproblems consisting of interlaced subsequences of X that consist of items d positions apart. Thus, the first subproblem is $\langle x_1, x_{1+d}, x_{1+2d}, \ldots \rangle$ while the second subproblem is $\langle x_2, x_{2+d}, x_{2+2d}, \ldots \rangle$. Shellsort solves the subproblems by applying insertion sort; however, rather than using a **multiway merge** to combine these solutions, it reapplies itself to the whole input with a smaller value d. Careful selection of the sequence of values of d results in a practical sorting method.

We have used divide-and-conquer to conceptually depict the landscape of sorting algorithms (refer to Figure 3.1). Nevertheless, in practice, sorting algorithms are usually not implemented as recursive programs. Instead, a nonrecursive equivalent analog is implemented (although computations may

| | Internal sorting algorithms | |
	Conquer form	Divide form
Comparison based	Mergesort Insertion sort	Quicksort Selection sort Heapsort Shellsort
Restricted universe		Bucket sort Radix sort

FIGURE 3.1 The landscape of internal sorting algorithms.

INSERTION SORT (X, n);

```
1: X[0] ← −∞;
2: for j ← 2 to n do
3:     i ← j − 1;
4:     t ← X[j];
5:     while t < X[i] do
6:         X[i + 1] ← X[i];
7:         i ← i − 1;
8:     end while
9:     X[i + 1] ← t;
10: end for
```

FIGURE 3.2 The sentinel version of insertion sort.

be performed in different order). In applications, the nonrecursive version is more efficient, since the administration of the recursion is avoided. For example, a straightforward nonrecursive version of mergesort proceeds as follows. First, the pairs of lists $\langle x_{2i-1} \rangle$ and $\langle x_{2i} \rangle$ (for $i = 1, \ldots, \lfloor n/2 \rfloor$) are merged to form sorted lists of length two. Next, the pairs of lists $\langle x_{4i-1}, x_{4i-2} \rangle$ and $\langle x_{4i-3}, x_{4i} \rangle$ (for $i = 1, \ldots, \lfloor n/4 \rfloor$) are merged to form lists of length four. The process builds sorted lists twice as long in each round until the input is sorted. Similarly, insertion sort has a practical iterative version illustrated in Figure 3.2.

Placing a sorting method in the landscape provided by divide-and-conquer allows easy computation of its time requirements, at least under the O notation. For example, algorithms of the conquer form have a divide part that takes $O(1)$ time to derive two subproblems of roughly equal size. The solutions of the subproblems may be combined in $O(n)$ time. This results in the following recurrence of the time $T(n)$ to solve a problem of size n:

$$T(n) = \begin{cases} 2T(n)(\lceil n/2 \rceil) + O(n) + O(1) & \text{if } n > 1, \\ O(1) & n = 1. \end{cases}$$

It is not hard to see that each level of recursion takes linear time and that there are at most $O(\log n)$ levels (since n is roughly divided by 2 at each level). This results in $T(n) = O(n \log n)$ time overall. If the divide form splits into one problem of size $n - 1$ and one trivial problem, the recurrence is as follows:

$$T(n) = \begin{cases} T(n-1) + O(1) + O(\text{conquer}(n-1)) & \text{if } n > 1, \\ O(1) & n = 1, \end{cases}$$

where $\text{conquer}(n - 1)$ is the time required for the conquer step. It is not hard to see that there are $O(n)$ levels of recursion (since n is decremented by one at each level). Thus, the solution to the recurrence is $T(n) = O(1) + \sum_{i=1}^{n} O(\text{conquer}(i))$. In the case of insertion sort, the worst case for $\text{conquer}(i)$ is $O(i)$, for $i = 1, \ldots, n$. Thus, we have that insertion sort is $O(n^2)$. However, local insertion sort assures $\text{conquer}(i) = O(\log i)$ time and the result is an algorithm that requires $O(n \log n)$ time. We will not pursue this analysis any further, confident that the reader will be able to find to find enough information here or in the references to identify the time and space complexity of the algorithms presented, at least up to the O notation.

Naturally, one may ask why there are so many sorting algorithms, if they all solve the same problem and fit a general framework. It turns out that, when implemented, each has different properties that makes them more suitable for different objectives.

First, comparison-based sorting algorithms, are ranked by their theoretical performance in the comparison-based model of computation. Thus, an $O(n \log n)$ algorithm should always be preferred over an $O(n^2)$ algorithm, if the files are large. However, theoretical bounds may be for the worst case or the expected case (where the analysis assumes that the keys are pairwise different and all possible permutations of items are equally likely). Thus, an $O(n^2)$ algorithm should be preferred over an $O(n \log n)$ algorithm, if the file is small or if we know that the file is already almost sorted. Particularly, in such a case, the $O(n^2)$ algorithm turns into an $O(n)$ algorithm. For example, insertion sort requires exactly $\text{Inv}(x) + n - 1$ comparisons and $\text{Inv}(X) + 2n - 1$ data moves, where $\text{Inv}(X)$ is the number of inversions in a sequence $X = \langle x_1, x_2, \ldots, x_n \rangle$; that is, the number of pairs (i, j) where $i < j$ and $x_i > x_j$. On the other hand, if quicksort is not carefully implemented, it may degenerate to $\Omega(n^2)$ performance on nearly sorted inputs.

If the theoretical complexities are equivalent, other aspects come into play. Naturally, the next criteria is the size of the constant hidden under the O notation (as well as the size of the hidden minor terms when the file is small). These constants are affected by implementation aspects. The most significant are now listed.

- The relative costs of swaps, comparisons, and all other operations in the computer at hand (and for the data types of the keys and records). Usually, swaps are more costly than comparisons, which in turn are more costly than arithmetic operations; however, comparisons may be just as costly as swaps or an order of magnitude less costly, depending on the length of the keys, records, and strategies to rearrange the records.
- The length of the machine code, so that the code remains in the memory, under an operating system that administers paging or in the cache of the microprocessor.
- Similarly, the locality of references to data or the capacity to place frequently compared keys in a CPU register.

Finally, there may be restrictions that force the choice of one sorting method over another. These include limitations like data structures, holding the data may be a linked list instead of an array, or the space available may be seriously restricted. There may be a need for stability or the programming tool may lack recursion (now, this is unusual). In practice, a hybrid sort is usually the best answer.

3.3 State-of-the-Art and Best Practices

3.3.1 Comparison-Based Internal Sorting

3.3.1.1 Insertion Sort

Figure 3.2 presented insertion sort. This algorithm uses sequential search to find the location, one item at a time, in a portion of the input array already sorted. It is mainly used to sort small arrays.

Besides being one of the simplest sorting algorithms, which results in simple code, it has many desirable properties. From the programming point of view, its loop is very short (usually taking advantage of memory management in the CPU cache or main memory); the key of the inserted element may be placed in a CPU register and access to data exhibits as much locality as it is perhaps possible. Also, if a minimum possible key value is known, a sentinel can be placed at the beginning of the array to simplify the inner loop, resulting in faster execution. Another alternative is to place the item being inserted, itself as a sentinel each time. From the applicability point of view, it is stable; recall that this means that records with equal keys remain in the same relative order after the sort. Its $\Theta(n^2)$ expected-case complexity and worst-case behavior make it only suitable for small files, and thus, it is usually applied in shellsort to sort the interleaved sequences. However, the fact that it is **adaptive** with respect to the measure of disorder "Inv" makes it suitable for almost all sorted files with respect to this measure. Thus, it is commonly used to sort roughly sorted data produced by implementations of quicksort that do not follow recursion calls once the subarray is small. This idea helps quicksort implementations achieve better performance, since the administration of recursive calls for small files is more time-consuming that one call that uses insertion sort on the entire file to complete the sorting.

Insertion sort also requires only constant space; that is, space for a few local variables (refer to Figure 3.2). From the point of view of quadratic sorting algorithms, it is a clear winner. Investigations of theoretical interest have looked at comparison-based algorithms where space requirements are constant, and data moves are linear. Only in this case, insertion sort is not the answer, since selection sort achieves this with equivalent number of comparisons. Thus, when records are very large and no provision is taken for avoiding expensive data moves (by sorting a set of indices rather than the data directly), selection sort should be used.

3.3.1.2 Shellsort

One idea for improving the performance of insertion sort is to observe that each element, when inserted into the sorted portion of the array, travels a distance equal to the number of elements to its left which are greater than itself (the number of elements inverted with it). However, this traveling

SHELLSORT (X, n);

```
1:  d ← n;
2:  repeat
3:      if d < 5 then
4:          d ← 1
5:      else
6:          d ← (5 * d − 1) div 11;
7:      end if
8:      for j ← d + 1 to n do
9:          i ← j − d;
10:         t ← X[j];
11:         while t < X[i] do
12:             X[i + d] ← X[i];
13:             i ← i − d;
14:             if i < d then
15:                 goto 18 17
16:             end if
17:         end while
18:         X[i + d] ← t;
19:     end for
20: until d ≤ 1;
```

FIGURE 3.3 *Shellsort* with increment sequence $\langle \lfloor n\alpha \rfloor$, $\lfloor \lfloor n\alpha \rfloor \alpha \rfloor, \ldots, \rangle$ with $\alpha = 0.4545 < 5/11$.

is done in steps of just adjacent elements and not by exchanges between elements far apart. The idea behind shellsort [21] is to use insertion sort to sort interleaved sequences formed by items d positions apart, thus allowing for exchanges as far as d positions apart. After this, elements far apart are closer to their final destinations, so d is reduced to allow exchanges of closer positions. To ensure that the output is sorted, the final value of d is one.

There are many proposals for the increment sequence [16]; the sequence of values of d. Some proposals are $\langle 2^k − 1, 2^{k-1} − 1, \ldots, 7, 3, 1 \rangle$, $\langle 2^p 3^q, \ldots, 9, 8, 6, 4, 3, 2, 1 \rangle$, and $S_i = \langle \ldots, 40, 13, 4, 1 \rangle$ where $S_i = 3S_{i-1} + 1$. It is possible that a better sequence exists; however, the improvement that they may produce in practice is almost not visible. Algorithm shellsort is guaranteed to be clearly below the quadratic behavior of insertion sort. The exact theoretical complexity remains elusive, but large experiments conjecture $O(n(\log n)^2)$, $O(n^{1.5})$, and $O(n \log n \log \log n)$ and comparisons are required for various increment sequences. Thus, it will certainly be much faster than quadratic algorithms, and for medium-size files, it would remain competitive with $O(n \log n)$ algorithms.

From the programming point of view, shellsort is simple to program. It is insertion sort inside the loop for the increment sequence. It is important not to use a version of insertion sort with sentinels, since for the rounds larger than d, many sentinels would be required. Not using a sentinel demands two exit points for the most inner loop. This can be handled with a clean use of a goto, but in languages which shortcut connectives (typically C, C++, and so on), this feature can be used to avoid gotos. Figure 3.3 shows pseudocode for shellsort.

Unfortunately, shellsort loses some of the virtues of insertion sort. It is no longer stable, and its behavior in nearly sorted files is adaptive, but not as marked as for insertion sort. However, the space requirements remain constant, and its coding is straightforward and usually results in a short program loop. It does not have a bad case and it is a good candidate for a library sorting routine. The usual recommendation when facing a sorting problem is to first try shellsort because a correct implementation is easy to achieve. Only if it proves to be insufficient for the application at hand should a more sophisticated method be attempted.

3.3.1.3 Heapsort

The priority queue abstract data type is an object that allows the storage of items with a key indicating their priority as well as retrieval of the item with the largest key. Given a data structure for this abstract data type, a sorting method can be constructed as follows. Insert each data item in the priority queue with the sorting key as the priority key. Repeatedly extract the item with the largest key from the priority queue to obtain the items sorted in reverse order.

One immediate implementation of the priority queue is an unsorted list (either as an array or as a linked list). Insertion in the priority queue is trivial; the item is just appended to the list. Extraction of the item with the largest key is achieved by scanning the list to find the largest item. If this implementation of the priority queue is used for sorting, the algorithm is called selection sort. Its time complexity is $\Theta(n^2)$, and as was already mentioned, its main virtue is that data moves are minimal.

A second implementation of a priority queue is to keep a list of the items sorted in descending order by their priorities. Now, extraction of the item with the largest priority requires constant time, since it is known that the largest item is at the front of the list, However, inserting a new item into the priority queue implies scanning the sorted list for the position of the new item. Using this implementation of a priority queue for sorting, we observe that we obtain insertion sort once more.

The above implementations of a queue offer constant time either for insertion or extraction of the item with the maximum key, but in exchange for linear time for the other operation. Thus, priority queues implemented like this may result in efficient methods in applications of priority queues where the balance of operations is uneven. However, for sorting, n items are inserted and also n items are extracted; thus, a balance is required between the insert and extract operations. This is achieved by implementing a priority queue as a heap [1,21] that shares the space with the array of data to be sorted. An array $A[1, \ldots, n]$ satisfies the heap property, if $A[\lfloor k/2 \rfloor] \geq A[k]$, for $2 \leq k \leq n$. In this case, $A[\lfloor k/2 \rfloor]$ is called the parent of $A[k]$, while $A[2k]$ and $A[2k + 1]$ are called the children of $A[k]$. However, an item may have no children or only one child, in which case it is called a leaf. The heap is constructed using all the elements in the array and thereafter is located in the lower part of the array. The sorted array is incrementally constructed from the item with the largest key until the element with the smallest key is in the highest part of the array. The first phase builds the heap using all the elements, and careful programming guarantees this requires $O(n)$ time. The second phase repeatedly extracts the item with largest key from the heap. Since the heap shrinks by one element, the space created is used to place the element just extracted. Each of the n updates in the heap takes $O(\log i)$ comparisons, where i is the number of items currently in the heap. In fact, the second phase of heapsort exchanges the first item of the array (the item with the largest key) with the item in the last position of the heap, and sinks the new item at the top of the heap to reestablish the heap property.

Heapsort can be efficiently implemented around a procedure SINK for repairing the heap property (refer to Figure 3.4). Procedure SINK(k, *limit*) moves down the heap, if necessary, exchanging the item at position k with the largest of its two children, and stopping when the item at position k is no longer smaller than one of its children (or when $k > limit$); refer to Figure 3.5. Observe that the loop in SINK has two distinct exits, when item k has no children and when the heap property is reestablished. For our pseudocode, we have decided to avoid the use of gotos. However, the reader can refer to our use of goto in Figure 3.3 for an idea to construct an implementation that actually saves some data moves. Using the procedure SINK, the code for heapsort is simple. From the applicability point of view, heapsort has the disadvantage that it is not stable. However, it is guaranteed to execute in $O(n \log n)$ time in the worst case and requires no extra space.

It is worth revising the analysis of heapsort. First, let us look at SINK. In each pass around its loop, SINK at least doubles the value of k, and SINK terminates when k reaches the limit (or before, if the heap property is reestablished earlier). Thus, in the worst case, SINK requires $O(h)$ where h is the height of the heap. The loop in line 1 and line 2 for heapsort constitute the core of the first phase (in our code, the heap construction is completed after the first execution of line 4). A call to SINK is made for each node. The first $\lfloor n/2 \rfloor$ calls to SINK are for heaps of height 1, the next $\lfloor n/4 \rfloor$ are for

HEAPSORT (X, n);

```
1: for  i ← (n div 2) down-to 2 do
2:     SINK(i,n)
3: end for
4: for  i ← n down-to 2 do
5:     SINK(1,i)
6:     t ← X[1];
7:     X[1] ← X[i];
8:     X[i] ← t;
9: end for
```

FIGURE 3.4 The pseudocode for *Heapsort*.

SINK (*k, limit*);

```
 1: while 2 * k ≤ limit do
 2:     j ← 2k;
 3:     if j < limit then {two children}
 4:         if X[j] < X[j + 1] then
 5:             j ← j + 1;
 6:         end if
 7:     end if
 8:     if X[k] < X[j] then
 9:         t ← X[j];
10:         X[j] ← X[k];
11:         X[k] ← t;
12:         k ← j
13:     else
14:         k ← limit + 1 {force loop exit}
15:     end if
16: end while
```

FIGURE 3.5 The sinking of one item into a heap.

heaps of height 2, and so on. Summing over the heights, we have that the phase for building the heap requires

$$O\left(\sum_{i=1}^{\log n} i\frac{n}{2^i}\right) = O(n)$$

time. Now, the core of the second phase is the loop from line 3. These are n calls to SINK plus a constant for the three assignments. The ith of the SINK calls is in a heap of height $O(\log i)$. Thus, this is $O(n \log n)$ time. We conclude that the first phase of heapsort (building the priority queue) requires $O(n)$ time. This is useful when building a priority queue. The second phase, and thus, the algorithm, requires $O(n \log n)$ time.

Heapsort does not use any extra storage, or does it require a language supplying recursion (recursion is now common but few programmers are aware of the memory requirements of a stack for function calls). For some, it may be surprising that heapsort destroys the order in an already sorted array to re-sort it. Thus, heapsort does not take advantage of the existing order in the input, but it compensates this with the fact that its running time has very little variance across the universe of permutations. Intuitively, items at the leaves of the heap have small keys, which make the sinking usually travel down to a leaf. Thus, almost all the n updates in the heap take at least $\Omega(i)$ comparisons, making the number of comparisons vary very little from one input permutation to another. Although its average case performance may not be as good as quicksort (a constant larger than two is usually the difference), it is rather simple to obtain an implementation that is robust. It is a very good choice for an internal sorting algorithm. Sorting by selection with an array having a heap property is also used for external sorting.

3.3.1.4 Quicksort

For many applications a more realistic measure of the time complexity of an algorithm is its expected time. In sorting, a classical example is quicksort [19,29], which has an optimal expected time complexity of $O(n \log n)$ under the decision tree model, while there are sequences that force it to perform $\Omega(n^2)$ operations (in other words, its worst-case time complexity is quadratic). If the worst-case sequences are very rare, or the algorithm exhibits small variance around its expected case, then this type of algorithm is suitable in practice.

Several factors have made quicksort a very popular choice for implementing a sorting routine. Algorithm quicksort is a simple divide-and-conquer concept; the partitioning can be done in a very short loop and is also conceptually simple; its memory requirements can be guaranteed to be only logarithmic on the size of the input; the pivot can be placed in a register; and, most importantly, the expected number of comparisons is almost half of the worst-case optimal competitors, most notably heapsort. In their presentation, many introductory courses on algorithms favor quicksort. However, it is very easy to implement quicksort in a way that it seems correct and extremely efficient for many sorting situations. But, it may be hiding $O(n^2)$ behavior for a simple case (for example, sorting n equal keys). Users of such library routines will be satisfied initially, only to find out later that on something that seems a simple sorting task, the implementation is consuming too much time to finish the sort.

Fine tuning of quicksort is a delicate issue [5]. Many of the improvements proposed may be compensated by reduced applicability of the method or more fragile and less clear code. Although partitioning is conceptually simple, much care is required to avoid common pitfalls. Among these, we must assure that the selection and placement of the pivot maintains the assumption about the distribution of input sequences. That is, the partitioning must guarantee that all permutations of smaller sequences are equally likely when the permutation of n items are equally likely. The partitioning must also handle extreme cases, which occur far more frequently in practice than the uniformity assumption for the theoretical analysis. These extreme cases include the case where the file is already sorted (either in ascending or descending order) and its subcase, the case in which the keys are all equal, as well as the case in which many keys are replicated. One common application of sorting is bringing together items with equal keys [5].

Recall that quicksort is a prototype of divide-and-conquer with the core of the work performed during the divide phase. The standard quicksort algorithm selects from a fixed location in the array a splitting element or pivot to partition a sequence into two parts. After partitioning, the items of the subparts are in correct order with respect to each other. Most partition schemes result in quicksort not being stable. Figure 3.6 presents a version for partitioning that is correct and assures $O(n \log n)$ performance even if the keys are all equal; it does not require sentinels and the indices i and j never go out of bounds from the subarray. The drawback of using fixed location pivots for the partitioning is when the input is sorted (in descending or ascending order). In these cases, the choice of the pivot drives quicksort to $\Omega(n^2)$ performance. This is still the case for the routine presented here. However, we have accounted for repeated key values; so, if there are e key values equal to the pivot's, then $\lceil e/2 \rceil$ end up in the right subfile. If all the keys are always different, the partition can be redesigned so that it leaves the pivot in its correct position and out of further consideration. Figure 3.7 presents the global view of quicksort. The second subfile is never empty (i.e., $p < r$), and thus, this quicksort always terminates.

The most popular variants to protect quicksort from worst-case behavior are the following. The splitting item is selected as the median of a small sample, typically three items (the first, middle, and last element of the subarray). Many results show that this can deteriorate the expected average time by about 5%–10% (depending on the cost of comparisons and how many keys are different). This

int PARTITION (X, l, r);

```
 1: pivot ← X[l];
 2: i ← l − 1;
 3: j ← r + 1;
 4: loop
 5:     repeat
 6:         j ← j − 1;
 7:     until X[j] ≤ pivot
 8:     repeat
 9:         i ← i + 1;
10:     until X[i] ≥ pivot
11:     if i < j then
12:         exchange X[i] ↔ X[j];
13:     else
14:         return j;
15:     end if
16: end loop
```

FIGURE 3.6 A simple and robust partitioning.

QUICKSORT (X, l, r);

```
1: if l < r then
2:     split ← PARTITION(X,l,r);
3:     QUICKSORT(X,l,split);
4:     QUICKSORT(X,split+1,r);
5: end if
```

FIGURE 3.7 Pseudocode for quicksort. An array is sorted with the call QUICKSORT $(X,1,n)$.

approach assures that a worst case happens with negligible low probability. For this variant, the partitioning can accommodate the elements used in the sample so that no sentinels are required, but there is still the danger of many equal keys.

Another proposal delays selection of the splitting element; instead, a pair of elements that determines the range for the median is used. As the array is scanned, every time an element falls between the pair, one of the values is updated to maintain the range as close to the median as possible. At the end of the partitioning, two elements are in their final partitions, dividing the interval. This method is fairly robust, but it enlarges the inner loop deteriorating performance, there is a subtle loss of randomness, and it also complicates the code significantly. Correctness for many equal keys remains a delicate issue.

Other methods are not truly comparison-based; for example, they use pivots that are the arithmetic averages of the keys. These methods reduce the applicability of the routine and may loop forever on equal keys.

Randomness can be a useful tool in algorithm design, especially if some bias in the input is suspected. A randomization version of quicksort is practical because there are many ways in which the algorithm can proceed with good performance and only a few worst cases. Some authors find displeasing to use a pseudorandom generator for a problem as well studied as sorting. However, we find that the simple partitioning routine presented in Figure 3.7 is robust in many of the aspects that make partitioning difficult to code and can remain simple and robust while also handling worst-case performance with the use of randomization. The randomized version of quicksort is extremely solid and easy to code; refer to Figure 3.8. Moreover, the inner loop of quicksort remains extremely short; it is inside the partition and consists of modifying the integer by 1 (increment or decrement, a very efficient operation in current hardware) and comparing a key with the key of the pivot (this value, along with the indexes i and j, can be placed in a register of the CPU). Also, access to the array exhibits a lot of locality of reference. By using the randomized version, the space requirements become $O(\log n)$ without the need to sort recursively the smallest of the two subfiles produced by the partitioning. If ever in practice the algorithm is taking too long (something with negligible probability), just halting it and running it again will provide a new seed with extremely high probability of reasonable performance.

Further improvements can now be made to tune up the code (of course, sacrificing some simplicity). One of the recursive calls can be eliminated by tail recursion removal, and thus the time for half of the procedure call is saved. Finally, it is not necessary to use a technique such as quicksort to sort small files of less than 10 items by a final call to insertion sort to complete the sorting. Figure 3.8 illustrates the tuned hybrid version of quicksort that incorporates these improvements. To sort a file, first a call is made to ROUGHLY QUICKSORT($X,1,n$) immediately followed by the call INSERTION SORT(X,n). Obviously both calls should be packed under a call for quicksort to avoid accidentally forgetting to make both calls. However, for testing purposes, it is good practice to call them separately. Otherwise, we may receive the impression that the implementation of the quicksort part is correct, while insertion sort is actually doing the sorting.

ROUGHLY QUICKSORT (X, l, r);

```
1:  while r − l > 10 do
2:      i ← RANDOM(l,r);
3:      exchange X[i] ↔ X[l];
4:      split ← PARTITION(X,l,r);
5:      ROUGHLY QUICKSORT(X,l,split);
6:      l ← split +1;
7:  end while
```

FIGURE 3.8 Randomized and tuned version of *Quicksort*.

It is worth revising the analysis of quicksort (although interesting new analyzes have emerged [10,14]). We will do this for the randomized version. This version has no bad inputs. For the same input, each run has different behavior. The analysis computes the expected number of comparisons performed by randomized quicksort on an input X. Because the algorithm is randomized, $T_{RQ}(X)$ is a random variable, and we will be interested in its expected value. For the analysis, we evaluate the largest expected value $T_{RQ}(X)$ over all inputs with n different key values. Thus, we estimate $E[T_{RQ}(n)] = \max\{E[T_{RQ}(X)] \mid \|X\| = n\}$. The largest subproblem for a recursive call is $n - 1$. Thus, the recursive form of the algorithm allows the following derivation, where c is a constant.

$$E[T_{RQ}(n)] = \sum_{i=1}^{n-1} \text{Prob}[i = k]E\begin{bmatrix}\sharp \text{ of comparisons when subcases} \\ \text{are of sizes } i \text{ and } n - i\end{bmatrix} + cn$$

$$\leq cn + \frac{1}{n}\sum_{i=1}^{n-1}\left(E[T_{RQ}(i)] + E[T_{RQ}(n - i)]\right)$$

$$= cn + \frac{2}{n}\sum_{i=1}^{n-1}E[T_{RQ}(i)]. \tag{3.1}$$

Since $E[T_{RQ}(0)] \leq b$ and $E[T_{RQ}(1)] \leq b$ for some constant b, then it is not hard to verify by induction that there is a constant k such that $E[T_{RQ}(n)] \leq kn\log n = O(n\log n)$, for all $n \geq 2$, which is the required result. Moreover, the recurrence (Equation 3.1) can be solved exactly to obtain an expression that confirms that the constant hidden under the O notation is small.

3.3.1.5 Mergesort

Mergesort is not only a prototype of the conquer from in divide-and-conquer, as we saw earlier. Mergesort has two properties that can make it a better choice over heapsort and quicksort in many applications. The first of these properties is that mergesort is naturally a stable algorithm, while additional efforts are required to obtain stable versions of heapsort and quicksort. The second property is that access to the data is sequential; thus, data do not have to be in an array. This makes mergesort an ideal method to sort linked lists. It is possible to use a divide form for obtaining a quicksort version for lists that are also stable. However, the methods for protection against quadratic worst cases still make mergesort a more fortunate choice. The advantages of mergesort are not without cost. Mergesort requires $O(n)$ extra space (for another array or for the pointers in the linked list implementation). It is possible to implement mergesort with constant space, but the gain hardly justifies the added programming effort.

Mergesort is based upon merging two sorted sequences of roughly the same length. Actually, merging is a very common special case of sorting, and it is interesting in its own right. Merging two ordered list (or roughly the same size) is achieved by repeatedly comparing the head elements and moving the one with the smaller key to the output list. Figure 3.9 shows PASCAL code for merging two linked lists. The PASCAL code for mergesort is shown in Figure 3.10. It uses the function for merging of the previous figure. This implementation of mergesort is more general than a sorting procedure for all the items in a linked list. It is a PASCAL function with two parameters, the head of the lists to be sorted, and an integer n indicating how many elements from the head should be included in the sort. The implementation returns as a result the head of the sorted portion, and the head of the original list is a VAR parameter adjusted to point to the $(n + 1)$th element of the original sequence. If n is larger or equal to the length of the list, the pointer returned includes the whole list, and the VAR parameter is set to nil.

```
            type
                list ↑ item;
                item = record k: keytype;
                             next : list;
                          end;

            function merge (X1,X2 :list) : list;
            var head, tail, t: list;
            begin
                  head := nil;
                  while X2 <> nil
                         do
                                if X1 = nil (* reverse roles of X2 and X1 *)
                                then begin X1:=X2; X2:= nil; end;
                                else  begin
                                          if X2↑.k > X1↑.k
                                          then begin t:= X1; X1:=X1↑.next end;
                                          else  begin t:=X2; X2:=X2↑.next end;
                                       t↑.next :=nil;
                                       if head = nil   then head:=t;
                                                           else tail↑.next:=t;
                                       tail:=t;
                                    end;
                         if head = nil   then head:=X1;
                                         else tail↑.next :=X1;
                         merge:=head
            end
```

FIGURE 3.9 PASCAL code for merging two linked lists.

```
            function mergesort( VAR: head; n:integer): list;
            var t: list;
            begin
                  if head = nil
                  then mergesort = nil
                  else if n > 1
                       then mergesort :=merge ( mergesort(head, n div 2), mergesort(head, (n+1) div 2))
                       else begin
                                 t:= head;
                                 head := head↑.next;
                                 t ↑.next := nil;
                                 mergesort := t;
                            end
            end
```

FIGURE 3.10 PASCAL code for a merge function that sorts n items of a linked list.

3.3.2   Restricted Universe Sorts

In this section, we present algorithms that use other aspects about the keys to carry out the sorting. These algorithms were very popular at some point, and were the standard to sort punched cards. With the emergence of comparison-based sorting algorithms, which provided generality as well as elegant analyzes and matching bounds, these algorithms lost popularity. However, their implementation can be much faster than comparison-based sorting algorithms. The choice between comparison-based methods and these types of algorithms may depend on the particular application. For a general sorting routine, many factors must be considered, and the criteria to determine which approach is best suited should not be limited to just running time. If the keys meet the conditions for using these methods, they are certainly a very good alternative. In fact, today's technology has word lengths and memory sizes that make many of the algorithms presented here competitive. These algorithms were considered useful for only small restricted universes. Large restricted universes can be implemented with the current memory sizes and current word sizes for many practical cases. Recent research has shown theoretical [3] and practical [4] improvements on older versions of these methods.

3.3.2.1 Distribution Counting

A special situation of the sorting problems $X = \langle x_1, \ldots, x_n \rangle$ is the sorting of n distinct integers in the range $[1,m]$. If the value $m = O(n)$, then the fact that x_i is a distinct integer allows a very simple sorting method that runs in $O(n)$ time. Use a temporary array $T:[1,m]$ and place each x_i in $T[x_i]$. Scan T to collect th items in sorted order (where T was initialized to hold only 0). Note there that we are using the power of the random access machine to locate an entry in an array in constant time.

This idea can be extended in many ways; the first is to handle the case when the integers are no longer distinct, and the resulting method is called distribution counting. The fundamental idea is to determine, for each x_i, its rank. The rank is the number of elements less than or equal (but before x_i in X) to x_i. The rank can be used to place x_i directly in its final position in the output array OUT. To compute the rank, we use the fact that the set of possible key values is a subset of the integers in $[1,m]$. We count the number E_k of values in X that equal k, for $k = 1, \ldots, m$. Arithmetic sums $\sum_{k=1}^{i} E_k$ can be used to find how many x_j are less than or equal to x_i. Scanning though X, we can now find the destination of x_i, when x_i is reached. Figure 3.11 presents the pseudocode for the algorithm. Observe that two loops are till n and two till m. From this observation, the $O(n + m) = O(n)$ time complexity follows directly. The method has the disadvantage that extra space is required; however, in practice, we will use this method when m fits our available main memory, and in such cases, this extra space is not a problem. A very appealing property of distribution counting is that it is a stable method. Observe that not a single comparison is required to sort. However, we need an array of size m, and the key values must fit the addressing space.

3.3.2.2 Bucket Sort

Bucket sort is an extension of the idea of finding out where in the output array each x_i should be placed. However, the keys are not necessarily integers. We assume that we can apply an interpolation

DISTRIBUTION COUNTING(X,m,OUT);

```
 1: for k ← 1 to m do
 2:     count[k] ← 0;
 3: end for
 4: for i ← 1 to n do
 5:     count[Xᵢ] ← count[Xᵢ] + 1;
 6: end for
 7: for k ← 2 to m do
 8:     count[k] ← count[k]+count[k − 1];
 9: end for
10: for i ← n down-to 1 do
11:     OUT[count[Xᵢ]] ← Xᵢ;
12:     count[Xᵢ] ← count[Xᵢ]-1;
13: end for
```

FIGURE 3.11 Pseudocode for *distribution counting*.

formula to the keys to obtain a new key in the real interval $[0,1)$ which proportionally indicates where the x_i should be relative to the smallest and largest possible keys. The interval $[0,1)$ is partitioned into m equal-sized consecutive intervals each with an associated queue. The item x_i is placed in the queue q_j when the interpolation address from its key lies in the jth interval. The queues are sorted recursively, and then concatenated starting from the lower interval. The first-in last-out properties of the queues assures that, if the recursive method is stable, the overall sorting is stable. In particular, if bucket sort is called recursively, the method is stable. However, in practice, it is expected that the queues will have very few items after one or two partitions. Thus, it is convenient to switch to an alternative stable sorting method to sort the items in each bucket, most preferable insertion sort. For the insertion into the queues, an implementation that allows insertion in constant time should be used. Usually, linked lists with a pointer to their last item is the best alternative. This also assures that the concatenation of the queues is efficient.

The method has an excellent average-case time complexity, namely, it is linear (when $m = \Theta(n)$). However, the assumption is a uniform distribution of the interpolated keys in $[0,1)$. In the worst scenario, the method may send every item to one bucket only, resulting in quadratic performance. The difficulty lies in finding the interpolation function. These functions work with large integers (like the maximum key) and must be carefully programmed to avoid integer overflow.

However, the method has been specialized so that k rounds of it sort k-tuples of integers in $[1,m]$ in $O(k(n + m))$ time, and also sort strings of characters with excellent results [1]. Namely, strings of characters are sorted in $O(n + L)$ time where L is the total length of the strings. In this cases, the alphabet of characters defines a restricted universe and the interpolation formula is just a displacement from the smallest value. Moreover, the number of buckets can be made equal to the different values of the universe. These specializations are very similar to radix sorting, which we discuss next.

3.3.2.3 Radix Sort

Radix sort refers to a family of methods where the keys are interpreted as a representation in some base (usually a power of 2) or a string over a given small but ordered alphabet. The radix sort examines the digits of this representation in as many rounds as the length of the key to achieve the sorting. Thus, radix sort performs several passes over the input, in each pass, performing decisions by one digit only.

The sorting can be done from the most significant digit toward the least significant digit or the other way around. The radix sort version that goes from the most significant toward least significant digit is called top-down radix sort, MSD radix sort, or radix exchange sort [16,21,30]. It resembles bucket sort, and from the perspective of divide-and-conquer is a method of the divide form (Figure 3.12). The most significant digit is used to split the items into groups. Next, the algorithms are applied recursively to the groups separately, with the first digit out of consideration. The sorted groups are collected by the order of increasing values of the splitting digit. Recursion is terminated by groups of size one. If we consider the level of recursion as rounds over the strings of digits of the keys, the algorithm keeps the invariant that after the ith pass, the input is sorted according to the first i digits of the keys.

The radix sort version that proceeds from the least significant digit toward the most significant digit is usually called bottom-up radix sort, straight radix sort, LSD radix sort, or just radix sort [16,21,30]. It could be considered as doing the activities of each round in different order, splitting the items into groups according to the digit under consideration, and grouping the items in order of increasing values of the splitting digit. Apply the algorithm recursively to all the items, but considering the next more significant digit. At first, it may not seem clear why this method is even correct. It has a dual invariant to the top-down sort; however, after the ith pass, the input is sorted according to the last i digits of the keys. Thus, for this bottom-up version to work, it is crucial that the insertion of items in

BUCKET SORT (X,n,m);

```
 1: for i ← 0 to m − 1 do
 2:     Queue[i] ← empty;
 3: end for
 4: for i ← 1 to n do
 5:     insert xᵢ into Queue[⌊ interpol(xᵢ)m⌋];
 6: end for
 7: for i ← 0 to m − 1 do
 8:     SORT(Queue[i]);
 9: end for
10: for i ← 1 to m − 1 do
11:     Concatenate Queue[i] at the back of
        Queue[0];
12: end for
13: return Queue[0];
```

FIGURE 3.12 The code for *Bucket Sort*.

their groups is made in the first-in first-out order of a queue. For the top-down version, this is only required to ensure stability. Both methods are stable though.

The top-down version has several advantages and disadvantages with respect to the bottom-up version. In the top-down version, the algorithm only examines the distinguishing prefixes, while the entire set of digits of all keys are examined by the bottom-up version. However, the top-down version needs space to keep track of the recursive calls generated, while the bottom-up version does not. If the input digits have a random distribution, then both versions of radix sort are very effective. However, in practice this assumption regarding the distribution is not the case. For example, if the digits are the bits of characters, the first leading bit of all lower case letters is the same in most character encoding schemes. Thus, top-down radix sort deteriorates with files with many equal keys (similar to bucket sort).

The bottom-up version is like distribution counting on the digit that is being used. In fact, this is the easiest way to implement it. Thus, the digits can be processed more naturally as groups of digits (and allowing a large array for the distribution counting). This is an advantage of the bottom-up version over the top-down version.

It should be pointed out that radix sort can be considered linear in the size of the input, since each digit of the keys is examined only once. However, other variants of the analysis are possible; these include modifying the assumptions regarding the distribution of the keys or according to considerations of the word size of the machine. Some authors think that the n keys require $\log n$ bits to be represented and stored in memory. From this perspective, radix sorts require $\log n$ passes with $\Omega(n)$ operations on them, still amounting to $O(n \log n)$ time. In any case, radix sorts are a reasonable method for a sorting routine, or a hybrid one. One hybrid method proposed by Sedgewick [30] consists of using the bottom-up version of radix sort, but for the most significant half of the digits of the keys. This makes the file almost sorted, so that the sort can be finished by insertion sort. The result is a linear sorting method for most current word sizes on randomly distributed keys.

3.3.3 Order Statistics

The kth order statistic of a sequence of n items is the kth larger item. In particular, the smallest element is the first-order statistic while the largest element is the nth-order statistic. Finding the smallest or the largest item in a sequence can easily be achieved in linear time. For the smallest item, we just have to scan the sequence, remembering the smallest item seen so far. Obviously, we can find the first and second statistic in linear time by the same procedure, just remembering the two smallest items seen so far. However, as soon as $\log n$ statistics are required, it is best to sort the sequence and retrieve any order statistic required directly.

A common request is to find jointly the smallest and largest items of a sequence of n items. Scanning through the sequence remembering the smallest and largest items seen so far requires that each new item be compared with what is being remembered; thus, $2n + O(1)$ comparisons are required. A better alternative in this case is to form $\lfloor n/2 \rfloor$ pairs of items, and perform the comparisons in pairs. We find the smallest among the smaller items in the pairs, while the largest is found among the larger items of the pairs (a final comparison may be required for an element left out when n is odd). This results in $\lfloor 3n/2 \rfloor + O(1)$ comparisons, which in some applications is worth the programming effort.

The fact that the smallest and largest items can be retrieved in $O(n)$ time without the need for sorting made the quest for linear algorithms for the kth order statistic a very interesting one for some time. Still, today there are several theoreticians researching the possibility of linear selection of the median (the $\lfloor n/2 \rfloor$th item) with a smaller constant factor. As a matter of fact, selection of the kth largest item is another illustration of the use of average case complexity to reflect a practical situation more accurately than worst-case analysis. The theoretical worst-case linear algorithms are so complex that few authors dare to present pseudocode for them. This is perfectly justified, because nobody should implement worst-case algorithms in the light of very efficient algorithms in the expected case, which are far easier conceptually, as they are simpler in programming effort terms and can be protected from worst-case performance (by making such worst case extremely unlikely).

Let us consider divide-and-conquer approaches to finding the kth largest element. If we take the conquer from as in mergesort, it seems difficult to imagine how the kth largest item of the left subsequence and the kth-largest item of the right subsequence relate to the kth largest item of the overall subsequence. However, if we take the divide form, as in quicksort, we see that partitioning divides the input and conceptually splits by the correct rank of the pivot. If the position of the pivot is $i \geq k$, we only need to search for the $X_\leq$ subsequence. Otherwise, we have found i items that we can remove from further consideration, since they are smaller than the kth largest. We just need to find the $k - i$th largest in the subsequence $X_>$. This approach to divide-and-conquer results in only one subproblem to be pursued recursively. The analysis results in an algorithm that requires $O(n)$ time in the expected case. Such a method requires, again, careful protection against the worst case. Moreover, it is more likely that a file that is being analyzed for its order statistics has been inadvertently sorted before, setting up a potential worst case for selection methods whose choice of the pivot is not adequate. In the algorithm presented in Figure 3.13, we use the same partitioning algorithm as Section 3.3.1.4; refer to Figure 3.7.

```
SELECT (X,l,r,k0);
 1: if r = l then
 2:     return X[l];
 3: end if
 4: i ← RANDOM(l,r);
 5: exchange X[i] ↔ X[l];
 6: split ← PARTITION(X,l,r);
 7: if k ≤ split then
 8:     return SELECT(X,l,split,k);
 9: else
10:     return SELECT(X,split+1,r,k-split);
11: end if
```

FIGURE 3.13 Randomized version for selection of the kth largest.

3.3.4 External Sorting

There are many situations where external sorting is required for the maintenance of a well-organized database. Files are often maintained in sorted order with respect to some attributes to facilitate searching and processing. External sorting is not only used to produce organized output, but also to efficiently implement complex operations such as a relational join. Although main memory sizes have consistently been enlarged, they have not been able to keep up with the ability to collect and store data. Fields like knowledge discovery and data mining have demonstrated the need to cleverly manipulate very large volumes of information, and the challenges for managing external storage. External sorting manages the trade-offs of rapid random access of internal memory with the relatively fast sequential access of secondary disks by sorting in two phases: a run creation phase and a merge phase. During the first phase, the file to be sorted is divided into smaller sorted sequences called initial runs or strings [21]. These runs are created by bringing into main memory a fragment of the file. During the second phase, one or more activations of multiway merge are used to combine the initial runs into a single run [27].

Currently, the sorting of files that are too large to be held in the main memory is performed in the disk drives [27]; see Figure 3.14. Situations where only one disk drive is available are now uncommon, since this usually results into very slow sorting processes and complex algorithms, while the problem can be easily solved with another disk drive (which is affordable today) or a disk array with independent read-writing heads. In each pass (one for run-creation and one or more for merging) the input file is read from the *IN* disk drive. The output of one pass is the input for the next, until a single run is formed; thus the *IN* and *OUT* disks swap roles after each pass. While one of the input buffers, say I_i, $i \in \{0, f-1\}$, is being filled, the sorting process reads records from some of the other input buffers $I_0, \ldots, I_{i-1}, I_{i+1}, \ldots, I_{f-1}$. The output file of each pass is written using double buffering. While one of the buffers, say O_i, $i \in \{0, 1\}$, is being filled, the other buffer O_{1-i} is being written to the disk. The roles of O_i and O_{1-i} are interchanged when one buffer is full and the other is empty. In practice, the goal is to produce as much sequential reading and writing as possible, although these days, the operating system may take over the request for reading and writing to disks from other processes in the same machine. Hopefully, several data records are read in an input/output (I/O) operation forming physical blocks, while the capacity of the buffers defines the size of a logical block. For the description of external sorting methods, the use of logical blocks is usually sufficient and we will just name them *blocks*.

During the run-creation phase, the number f of input buffers is two and reading is sequential using double buffering. During the merge pass, the next block to be read is normally from a different run and the disk arm must be repositioned. Thus, reading is normally not sequential. Writing during the merge is, however, faster than reading, since normally it is performed sequentially and no seeks are involved (except for the occasional seek for the next cylinder when the current cylinder is full). In each pass, the output is written sequentially to the disk. Scattered writing during a pass in anticipation of saving seeks because of some sequential reading of the next pass has been shown to be counterproductive [34]. Thus, in the two-disk model (Figure 3.14), the writing during the merge

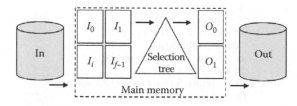

FIGURE 3.14 The model for a pass of external sorting.

will completely overlap with the reading and its time requirements are a minor concern. In contrast to the run-creation phase in which reading is sequential, merging may require a seek each time a data block is read.

Replacement selection usually produces runs that are larger than the available main memory; the larger the initial runs, the faster the overall sorting. Replacement selection allows full overlapping of I/O with sequential reading and writing of the data, and it is standard for the run-creation phase. The classic result on the performance of replacement selection establishes that, when the permutations in the input files are assumed to be equally likely, the asymptotic expected length of the resulting runs is twice the size of the available main memory [21]. Other researchers have modified replacement selection, such that, asymptotically, the expected length of an initial run is more than twice the size of the available main memory. These methods have received limited acceptance because they require more sophisticated I/O operations and prevent full overlapping; hence, the possible benefits hardly justify the added complexity of the methods. Similarly, any attempt to design a new run-creation method that profits from the existing order in the input file almost certainly has inefficient overlapping of I/O operations. More recently, it has been mathematically confirmed that the lengths of the runs created by replacement selection increase as the order of the input files increases [12].

During the run-creation phase, replacement selection consists of a selection tree. This structure is a binary tree where the nodes hold the smaller of their two children. It is called selection tree because the item at the root of the tree holds the smallest key. By tracing the path up of the smallest key from its place at a leaf to the root, we have selected the smallest item among those in the leaves. If we replace the smallest item with another value at the corresponding leaf, we only are required to update the path to the root. Performing the comparisons along this path updates the root as the new smallest item. Selection trees are different from heaps (ordered to extract the item with the smallest keys) in that selection trees have fixed size. During the selection phase, the selection tree is initialized with the first P elements of the input file (where P is the available internal memory). Repeatedly, the smallest item is removed from the selection tree and placed in the output stream, and the next item from the input file is inserted in its place as a leaf in the selection tree. The name replacement selection comes from the fact that the new item from the input file replaces the item just selected to the output stream. To make certain that items enter and leave the selection tree in the proper order, the comparisons are not only with respect to the sorting keys, but also with respect to the current run being output. Thus, the selection tree uses lexicographically the composite keys (r, key), where r is the run-number of the item, and key is the sorting key. The run number of an item entering the selection tree is know by comparing it to the item which it is replacing. If it is smaller than the item just sent to the output stream, the run number is one more than the current run number; otherwise, it is the same run number as the current run.

3.3.4.1 The Merge

In the merging phase of external sorting, blocks from each run are read into the main memory, and the records from each block are extracted and merged into a single run. Replacement selection (implemented with a selection tree) is also used as the process to merge the runs into one [21]. Here, however, each leaf is associated with each of the runs being merged. The order of the merge is the number of leaves in the selection tree. Because main memory is a critical resource here, items in the selection tree are not replicated, but rather a tree of losers is used [21].

There are many factors involved in the performance of disk drives. For example, larger main memories imply larger data blocks and the block-transfer rate is now significant with respect to seek time and rotational latency. Using a larger block size reduces the total number of reads (and seeks) and reduces the overhead of the merging phase. Now, a merge pass requires at least as many buffers as the order ω of the merge. On the other hand, using only one buffer for each run maximizes block size, and if we perform a seek for each block, it reduces the total number of seeks. However, we cannot overlap I/O. On the other hand, using more buffers, say, two buffers for each run, increases the overlap of I/O, but reduces the block size and increases the total number of seeks. Note that because the amount of main memory is fixed during the merge phase, the number of buffers is inversely proportional to their size.

Salzberg [28] found that, for almost all situations, the use of $f = 2w$ buffers assigned as pairs to each merging stream outperforms the use of w buffers. It has been shown that double buffering cannot take advantage of the nearly sorted data [13]. Double buffering does not guarantee full overlap of I/O during merging. When buffers are not fixed to a particular run, but can be reassigned to another run during the merge, they are called floating buffers [20]. Using twice as many floating buffers as the order of the merge provides maximum overlap of I/O [20]. Zheng and Larson [34] combined Knuth's [21] forecasting and floating-buffers techniques and proposed six to ten times as many floating input buffers as the order of the merge, which also require less main memory. Moreover, it was also demonstrated that techniques based on floating buffers profit significantly from nearly sorted files [13].

3.3.4.2 Floating Buffer

The consumption sequence is the particular order in which the merge consumes the blocks from the runs being merged. This sequence can be computed by extracting the highest key (the last key) from each data block (during the previous pass) and sorting them. The time taken to compute the consumption sequence can be overlapped with the output of the last run and the necessary space for the subsidiary internal sort is also available then; thus, the entire consumption sequence can be computed during the previous pass with negligible overhead. The floating-buffers technique exploits the knowledge of the consumption sequence to speed up reading.

We illustrate double buffering and floating buffers with a merging example of four runs that are placed sequentially as shown in Figure 3.15. Let $C = \langle C_1, C_2, \ldots, C_T \rangle$ be the consumption sequence, where C_i identifies a data block with respect to its location on the disk. For example, consider

$$C = \langle 1, 8, 13, 18, 2, 9, 14, 19, 3, 10, 4, 5, 15, 11, 12, 16, 17, 20, 6, 21, 7, 22, 23 \rangle.^*$$

Double buffering uses twice as many buffers as runs, and a seek is required each time a block is needed from a different run. Moreover, even when a block is needed from the same run, this may not be known at exactly the right time; therefore, the disk will continue to rotate and every read has rotational latency. In the example, double buffering reads one block from each run (with a seek in each case) and then it reads a second block from each run (again with a seek in each case). Next, the disk arm travels to block 3 to read a new bock from the first run (one more seek). Afterward, the arm moves to block 10 to get a new block from the second run. Then, it moves to block 4 and reads block 4 and block 5, but a seek is not required for reading block 5 since the run-creation phase places blocks from the same run sequentially on the disk. In total, double buffering performs 19 seeks.

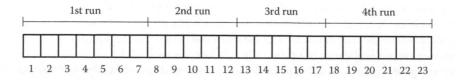

FIGURE 3.15 An example of four runs (written sequentially) on a disk.

---

* Alternatively, the consumption sequence can be specified as $C = \langle c_1, c_2, \ldots, c_T \rangle$ where c_i is the run from which the i-block should be read. For this example, $C = \langle 1, 2, 3, 4, 1, 2, 3, 4, 1, 2, 1, 1, 3, 2, 2, 3, 3, 4, 1, 4, 1, 4, 4 \rangle$.

We now consider the effect of using only seven buffers (of the same size as before) managed as floating buffers [20]. In this case, we use only three more buffers than the number of runs but we use the knowledge of the consumption sequence. The seven buffers are used as follows: four buffers contain a block from each run that is currently being consumed by the merge, two buffer contain look-ahead data, and one buffers is used for reading new data. In the previous example, after the merging of the two data blocks 1 and 7, the buffers are as follows.

block 24	block 2	empty	block 8	block 15	block 23	block 16
buffer 1	buffer 2	buffer 3	buffer 4	buffer 5	buffer 6	buffer 7

Data from buffers 2, 4, 5, and 6 are consumed by the merge and placed in an output buffer. At the same time, one data block is read into buffer 3. As soon as the merge needs a new block from run 3, it is already in buffer 7 and the merge releases buffer 5. Thus, the system enters a new state in which we merge buffers 2, 4, 6, and 7, and we read a new block into buffer 5. Figure 3.16 shows the merge when the blocks are read in the following order:

$$\langle 1, 2, 8, 9, \underline{13}, \underline{18}, \underline{14}, \underline{19}, 3, 4, 5, 10, 11, 12, 15, 16, 17, 6, 7, 20, 21, 22, 23 \rangle. \tag{3.2}$$

The letter e denotes an empty buffer and b_i denotes a buffer that holds data block i. Reading activity is indicated by an arrow into a buffer and merging activity is indicated by an arrow out of a buffer. The ordered sequence in which the blocks are read from the disk into the main memory is called reading sequence. A reading sequence is feasible, if every time the merge needs a block, it is already in the main memory and reading never has to wait because there is no buffer space available in the main memory. Note that the consumption sequence (with two or more floating buffers for each run) is always a feasible sequence [20]. In the example of Figure 3.16, not only is the new reading sequence feasible and provides just–in-time blocks for the merge, but it also requires only 11 seeks and uses even less memory than double buffering!

3.3.4.3 Computing a Feasible Reading Sequence

In Section 3.3.4.2, we have illustrated that floating buffers can save the main memory and the overhead due to seeks. There are, however, two important aspects of using floating buffers. First, floating buffers are effective when knowledge of the consumption sequence is used to compute reading sequences [34]. The consumption sequence is computed by the previous pass (and for the first pass, during the run-creation phase). The consumption sequence is the consumption order of the data blocks in the next pass. Thus, the buffer size for the next pass must be known by the previous pass. Overall, the buffer size and the number f of the input floating buffers for each pass must be chosen before starting the sorting.

Before sorting, we usually know the length $|X_I|$ of the file, and assuming it is in random order, we expect, after run-creation, initial runs of twice the size P of the available main memory. That is, $E[\text{Runs}(X_0)] = |X_I|/(2P)$. Now, we can decide the number of merge passes (most commonly only one) and the order ω of these merge passes. Zheng and Larson [34] follow this approach and recommend the number f of floating buffers to be between 6ω and 10ω. Once the value of f is chosen, the buffer size is determined, and where to partition the input into data blocks is defined. The justification for this strategy is that current memory sizes allow it and an inaccurate estimate of the number of initial runs or their sizes seems not to affect performance [34]. It has been shown that if the input is nearly sorted, the fact that the technique just described may chose f much larger than 10ω does not affect floating buffers. Moreover, for nearly sorted files, reading during the merge becomes almost sequential, and over 80% of seeks can be avoided [13].

The second difficulty consists of computing feasible reading sequences that minimize the number of seeks. Zheng and Larson [34] have related the problem of finding the optimal feasible reading

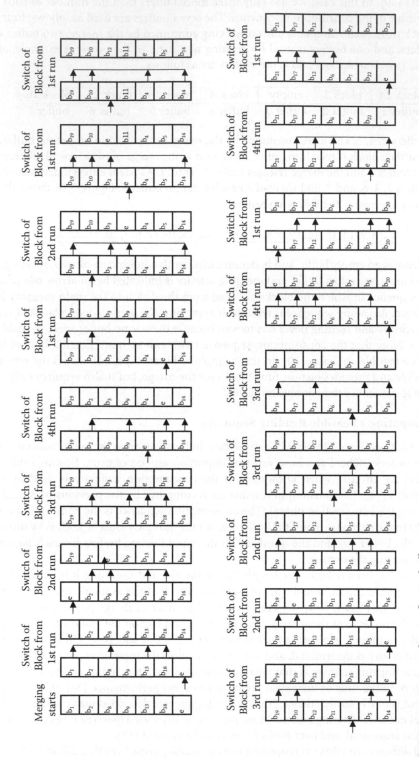

FIGURE 3.16 Merging with seven floating buffers.

sequence to the traveling salesman problem thus; research has concentrated on approximation algorithms.

To describe precisely the problem of finding a feasible reading sequence with fewer seeks, we will partition into what we call groups, the blocks in a read sequence that are adjacent and belong to the same run. The groups are indicated by underlining in the sequence shown as Equation 3.2. A seek is needed at the beginning of each group, because a disk head has to move to a different run. Inside a group, we read blocks from a single run sequentially, as placed by the run-creation phase or a previous merge pass. Note that there is no improvement in reading data blocks from the same run in different order than in the consumption sequence. For long groups, a seek may be required from one cylinder to the next, but such a seek takes minimum time because reading is sequential. Moreover, in this case, the need to read a block from the next cylinder is known early enough to avoid rotational latency. Thus, we want to minimize the number of groups while maintaining feasibility.

We now describe group shifting, an algorithm that computes a feasible reading sequence with fewer seeks. Group shifting starts with a consumption sequence $C = \langle C_1, \ldots, C_T \rangle$ as the initial feasible reading sequence. It scans the groups in the sequence twice. The first scan produces a feasible reading sequence, which is the input for the next scan. A scan builds a new feasible reading sequence incrementally. The first ω groups of the new reading sequence are the first ω groups of the previous reading sequence because, for $i = 1, \ldots, \omega$, the optimal feasible reading sequence for the first i groups consist of the first i groups of the consumption sequence. In each scan, the groups of the previous sequence are analyzed in the order they appear. During the first scan, an attempt is made to move each group in turn forward and catenate it with the previous group from the same run while preserving feasibility. A single group that results from the catenation of groups B_j and B_k is denoted by $(B_j B_k)$. During the second scan an attempt is made to move back the previous group from the same run under analysis, while preserving feasibility. We summarize the algorithm in Figure 3.17.

For an example of a forward move during the first scan, consider $b = 4$, $M^b = \langle \underline{1}, \underline{8}, \underline{13}, \underline{18} \rangle$, and group $b + 1$ is $\underline{2}$. Then, $M^{b+1} = \langle \underline{1, 2}, \underline{8}, \underline{13}, \underline{18} \rangle$. As an example of a backward move during the second scan consider $b = 18$, $M^b = \langle \underline{1, 2}, \underline{8, 9}, \underline{13}, \underline{18}, \underline{14}, \underline{19}, \underline{3, 4, 5}, \underline{10, 11, 12}, \underline{15, 16, 17}, \underline{20, 21}, \underline{6, 7} \rangle$, and group $b + 1$ is $\underline{22, 23}$. Moving $\underline{20, 21}$ over $\underline{6, 7}$ gives the optimal sequence of Figure 3.16.

The algorithm uses the following fact to test that feasibility is preserved. Let $C = \langle C_1, \ldots, C^T \rangle$ be the consumption sequence for ω runs with T data blocks. A reading sequence $R = \langle R_1, \ldots, R_T \rangle$ is feasible for $f > \omega + 1$ floating buffers if and only if, for all k such that $f \leq k \leq T$, we have $\{C_1, C_2, \ldots, C_{k-f+\omega}\} \subset \{R_1, R_2, \ldots, R_{k-1}\}$.

3.4 Research Issues and Summary

We now look at some of the research issues on sorting from the practical point of view. In the area of internal sorting, advances in data structures for the abstract data type dictionary or for the abstract data type priority queue may result in newer or alternative sorting algorithms. The implementation of dictionaries by variants of binary search tress, where the items can easily (in linear time) be recovered in sorted order with an in-order traversal, results in an immediate sorting algorithm. We just insert the items to be sorted into the tree implementing the dictioanry using sorting keys as the dictionary keys. Later, we extract the sorted order from the tree. An insertion sort is obtained for each representation of the dictionary. Some interesting advances, at the theoretical level, but perhaps at the practical level, have been obtained using data structures like fusion trees [2,15]. Although these algorithms are currently somewhat complicated, and they make use of dissecting keys and careful placing of information in memory words, the increase in the word size of computers is making them practically feasible.

Another area of research is the more careful study of alternatives offered by radix sorts. Careful analyzes have emerged for these methods and they take into consideration the effect of nonuniform

GROUP-SHIFTING ($C = \langle C_1, \ldots, C_T \rangle$) original consumptions sequence for a merge pass of order ω)

1: $N \leftarrow \langle, C_1, C_2, \ldots, C_\omega \rangle$;
2: **while** $b < T$ **do**
3:    let B be the $(b+1)$-th group in C and let the sequence N be $N = \langle B_1, B_2, \ldots, B_p \rangle$, where B_l is the last group in N from the same run as B.
4:    **if** group B can be moved forward and joined to B_l preserving feasibility **then**
5:       $N = \langle B_1, \ldots, B_{l-1}, (B_l B), B_{l+1}, \ldots, B_p \rangle$,
6:    **else**
7:       append B to N
8:    **end if**
9: **end while**
10: $M \leftarrow$ first ω groups in N;
11: $b \leftarrow \omega$;
12: **while** $b < |N|$ **do**
13:    Let B be the $(b+1)$-th group in N and let the sequence M be $M = \langle B_1, \ldots, B_p \rangle$, where B_l is the last group in M from the same run as B.
14:    **if** group B_l can be shifted back to the end of N and joined to group B preserving feasibility **then**
15:       $M \leftarrow \langle B_1, \ldots, B_{l-1}, B_{l+1}, \ldots, B_p, (B_l B) \rangle$
16:    **else**
17:       append B to M
18:    **end if**
19: **end while**
20: **return** M

FIGURE 3.17 The algorithm for reducing seeks during the merging phase with floating buffers.

distributions. Moreover, simple combinations of top-down and bottom-up have resulted in hybrid radix sorting algorithms with very good performance [3]. In practical experiments, Andersson and Nilsson observed that their proposed forward radix sort defeats some of the best alternatives offered by the comparison-based approach. When sorting integers in a restricted universe, the theoretical trade-offs of time and space have been pushed extensively lately [17,18].

A third area of research with possible practical implications is the area of adaptive sorting. When the sorting algorithms take advantage of the existing order in the input, the time taken by the algorithm to sort is a smooth growing function of the size of the sequence and the disorder in the sequence. In this case, we say that the algorithm is adaptive [24]. Adaptive sorting algorithms are attractive because nearly sorted sequences are common in practice [21,24,30]; thus, we have the possibility of improving on algorithms that are oblivious to the existing order in the input.

So far we presented insertion sort as an example of this type of algorithm. Adaptive algorithms have received attention for comparison-based sorting. Many theoretical algorithms have been found for many measures of disorder. However, from the practical point of view, these algorithms usually involve more machinery. This additional overhead is unappealing because of its programming effort. Thus, room remains for providing adaptive sorting algorithms that are simple for the practitioner. Although some of these algorithms [8] have been shown to be far more efficient in nearly sorted sequences for just a small overhead on randomly permuted files, they have not received wide acceptance.

Finally, let us summarize the alternatives when facing a sorting problem. First, we must decide if our situation is in the area of external sorting. A model with two disk drives is recommended in this case. Use replacement selection for run-creation and merging, using floating buffers during the second phase. It is possible to tune the sorting for just one pass during the second phase.

If the situation is internal sorting of small files, then insertion sort does the job. If we are sorting integers, or character strings, or some restricted universe, then distribution counting, bucket sort, and radix sort are very good choices. If we are after a stable method, restricted universe sorts are also good options. If we want something more general, the next level up is shellsort, and finally the tuned versions of $O(n \log n)$ comparison-based sorting algorithms. If we have serious grounds to suspect that the inputs are nearly sorted, we should consider adaptive algorithms. Whenever the sorting key is a small portion of the data records, we should try to avoid expensive data moves by sorting a file of keys and indexes. Always preserve a clear and simple code.

3.5 Further Information

For the detailed arguments that provide theoretical lower bounds for comparison-based sorting algorithms, the reader may consult early works on algorithms [1,24]. These books also include a description of sorting strings by bucket sort in time proportional to the total length of the strings.

Sedgewick's book on algorithms [30] provides illustrative descriptions of radix sorts. Other interesting algorithms, for example, linear probing sort, usually have very good performance in practice although they are more complicated to program. They can be reviewed in Gonnet and Baeza-Yates' handbook [16].

We have omitted here algorithms for external sorting with tapes, since they are now very rare. However, the reader may consult classical sources [16,21].

For more information on the advances in fusion trees and radix sort, as well as data structures and algorithms, the reader may wish to review the *ACM Journal of Experimental Algorithms*, the *Proceedings of the IEEE Symposium on Foundations of Computer Science* (FOCS), the *Proceedings of the Annual ACM-SIAM Symposium on Discrete algorithms* (SODA), or the *Proceedings of the ACM Symposium on the Theory of Computing*.

Defining Terms

Adaptive: It is a sorting algorithm that can take advantage of the existing order in the input, reducing its requirements for computational resources as a function of the amount of disorder in the input.

Comparison-based algorithm: It is a sorting method that uses comparisons, and nothing else about the sorting keys, to rearrange the input into ascending or descending order.

Conquer form: It is an instantiation of the divide-and-conquer paradigm for the structure of an algorithm where the bulk of the work is combining the solutions of subproblems into a solution for the original problem.

Divide form: It is an instantiation of the divide-and-conquer paradigm for the structure of an algorithm where the bulk of the work is dividing the problem into subproblems.

External sorting: It is the situation where the file to be sorted is too large to fit into the main memory. The need to consider that random access to data items is limited and sequential access is inexpensive.

Insertion sort: It is the family of sorting algorithms where one item is analyzed at a time and inserted into a data structure holding a representation of a sorted list of previously analyzed items.

Internal sorting: It is the situation when the file to be sorted is small enough to fit into the main memory and using uniform cost for random access is suitable.

Multiway merge: It is the mechanism by which ω sorted runs are merged into a single run. The input runs are usually organized in pairs and merged using the standard method for merging two sorted sequences. The results are paired again, and merged, until just one run is produced. The parameter ω is called the order of the merge.

Restricted universe sorts: These are algorithms that operate on the basis that the keys are members of a restricted set of values. They may not require comparisons of keys to perform the sorting.

Selection sorts: It is a family of sorting algorithms where the data items are retrieved from a data structure, one item at a time, in sorted order.

Sorting arrays: The data to be sorted is placed in an array and access to individual items can be done randomly. The goal of the sorting is that the ascending order matches the order of indices in the array.

Sorting linked lists: The data to be sorted is a sequence represented as a linked list. The goal is to rearrange the pointers of the linked list so that the linked list exhibits the data in a sorted order.

Stable: A sorting algorithm where the relative order of the items with equal keys in the input sequence is always preserved in the sorted output.

References

1. A.V. Aho, J.E. Hopcroft, and J.D. Ullman. *The Design and Analysis of Computer Algorithms*. Addison-Wesley Publishing Co., Reading, MA, 1974.
2. A. Andersson, T. Hagerup, S. Nilsson, and R. Raman. Sorting in linear time? *Journal of Computer and System Sciences*, 57(1):74–93, 1998.
3. A. Andersson and S. Nilsson. A new efficient radix sort. In *Proceedings of the 35th Annual Symposium of Foundations of Computer Science*, pp. 714–721. Santa Fe, New Mexico, 1994, IEEE Computer Society.
4. A. Andersson and S. Nilsson. Implementing radixsort. *ACM Journal of Experimental Algorithms*, 3:7, 1998.
5. J. Bentley and M.D. McIlroy. Engineering a sort function. *Software: Practice and Experience*, 23(11):1249–1265, November 1993.

6. G.S. Brodal, R. Fagerberg, and K. Vinther. Engineering a cache-oblivious sorting algorithms. *ACM Journal of Experimental Algorithmics*, 12 (Article 2.2), 2007.

7. J.-C. Chen. Building a new sort function for a c library. *Software: Practice and Experience*, 34(8):777–795, 2004.

8. C.R. Cook and D.J. Kim. Best sorting algorithms for nearly sorted lists. *Communications of the ACM*, 23:620–624, 1980.

9. T.H. Cormen, C.E. Leiserson, and R.L. Rivest. *Introduction to Algorithms*. MIT Press, Cambridge, MA, 1990.

10. M. Durand. Asymptotic analysis of an optimized quicksort algorithm. *Information Processing Letters*, 85(2):73–77, 2003.

11. S. Edelkamp and P. Stiegeler. Implementing heapsort with $(n \log n - 0.9n)$ and quicksort with $(n \log n + 0.2n)$ comparisons. *ACM Journal of Experimental Algorithms*, 7:5, 2002.

12. V. Estivill-Castro and D. Wood. A survey of adaptive sorting algorithms. *Computing Surveys*, 24:441–476, 1992.

13. V. Estivill-Castro and D. Wood. Foundations for faster external sorting. In *14th Conference on the Foundations of Software Technology and Theoretical Computer Science*, pp. 414–425, Chennai, India, 1994. *Springer Verlag Lecture Notes in Computer Science* 880.

14. J.A. Fill and S. Janson. The number of bit comparisons used by quicksort: An average-case analysis. In *SODA '04: Proceedings of the 15th Annual ACM-SIAM Symposium on Discrete Algorithms*, pp. 300–307, Philadelphia, PA, 2004. Society for Industrial and Applied Mathematics.

15. M.L. Fredman and D.E. Willard. Surpassing the information theoretic bound with fusion trees. *Journal of Computer and System Sciences*, 47(3):424–436, 1993.

16. G.H. Gonnet and R. Baeza-Yates. *Handbook of Algorithms and Data Structures*, 2nd edition. Addison-Wesley Publishing Co., Don Mills, ON, 1991.

17. Y. Han. Improved fast integer sorting in linear space. In *SODA '01: Proceedings of the 12th Annual ACM-SIAM Symposium on Discrete Algorithms*, pp. 793–796, Philadelphia, PA, 2001. Society for Industrial and Applied Mathematics.

18. Y. Han. Deterministic sorting in o(nlog logn) time and linear space. *Journal of Algorithms*, 50(1):96–105, 2004.

19. C.A.R. Hoare. Algorithm 64, Quicksort. *Communications of the ACM*, 4(7):321, July 1961.

20. E. Horowitz and S. Sahni. *Fundamentals of Data Structures*. Computer Science Press, Inc., Woodland Hill, CA, 1976.

21. D.E. Knuth. *The Art of Computer Programming*, Vol. 3: *Sorting and Searching*. Addison-Wesley Publishing Co., Reading, MA, 1973.

22. P.-A. Larson. External sorting: Run formation revisited. *IEEE Transaction on Knowledge and Data Engineering*, 15(4):961–972, July/August 2003.

23. H. Mannila. Measures of presortedness and optimal sorting algorithms. *IEEE Transactions on Computers*, C-34:318–325, 1985.

24. K. Mehlhorn. *Data Structures and Algorithms*, Vol. 1: *Sorting and Searching*. EATCS Monographs on Theoretical Computer Science. Springer-Verlag, Berlin/Heidelberg, 1984.

25. N. Rahman and R. Raman. Analysing cache effects in distribution sorting. *ACM Journal of Experimental Algorithms*, 5:14, 2000.

26. N. Rahman and R. Raman. Adapting radix sort to the memory hierarchy. *ACM Journal of Experimental Algorithms*, 6:7, 2001.

27. B. Salzberg. *File Structures: An Analytic Approach*. Prentice-Hall, Inc., Englewood Cliffs, NJ, 1988.

28. B. Salzberg. Merging sorted runs using large main memory. *Acta Informatica*, 27:195–215, 1989.

29. R. Sedgewick. *Quicksort*. Garland Publishing Inc., New York and London, 1980.

30. R. Sedgewick. *Algorithms*. Addison-Wesley Publishing Co., Reading, MA, 1983.

31. R. Sinha and J. Zobel. Cache-conscious sorting of large sets of strings with dynamic tries. *ACM Journal of Experimental Algorithms*, 9 (Article 1.5), 2004.

32. R. Sinha, J. Zobel, and D. Ring. Cache-efficient string sorting using copying. *ACM Journal of Experimental Algorithms*, 11 (Article 1.2), 2006.

33. R. Wickremesinghe, L. Arge, J. S. Chase, and J. S. Vitter. Efficient sorting using registers and caches. *ACM Journal of Experimental Algorithms*, 7:9, 2002.

34. L.Q. Zheng and P.A. Larson. Speeding up external mergesort. *IEEE Transactions on Knowledge and Data Engineering*, 8(2):322–332, 1996.

4

Basic Data Structures*

Roberto Tamassia
Brown University

Bryan Cantrill
Sun Microsystems, Inc.

4.1 Introduction

The study of data structures, i.e., methods for organizing data that are suitable for computer processing, is one of the classic topics of computer science. At the hardware level, a computer views storage devices such as internal memory and disk as holders of elementary data units (bytes), each accessible through its address (an integer). When writing programs, instead of manipulating the data at the byte level, it is convenient to organize them into higher level entities, called data structures.

4.1.1 Containers, Elements, and Locators

Most data structures can be viewed as **containers** that store a collection of objects of a given type, called the elements of the container. Often a total order is defined among the elements (e.g., alphabetically ordered names, points in the plane ordered by x-coordinate). We assume that the elements of a container can be accessed by means of variables called **locators**. When an object is inserted into the container, a locator is returned, which can be later used to access or delete the object. A locator is typically implemented with a pointer or an index into an array.

---

* The material in this chapter was previously published in *The Computer Science and Engineering Handbook,* Allen B. Tucker, Editor-in-Chief, CRC Press, Boca Raton, FL, 1997.

A data structure has an associated repertory of operations, classified into queries, which retrieve information on the data structure (e.g., return the number of elements, or test the presence of a given element), and updates, which modify the data structure (e.g., insertion and deletion of elements). The performance of a data structure is characterized by the space requirement and the time complexity of the operations in its repertory. The amortized time complexity of an operation is the average time over a suitably defined **sequence** of operations.

However, efficiency is not the only quality measure of a data structure. Simplicity and ease of implementation should be taken into account when choosing a data structure for solving a practical problem.

4.1.2 Abstract Data Types

Data structures are concrete implementations of **abstract data types** (ADTs). A data type is a collection of objects. A data type can be mathematically specified (e.g., real number, directed graph) or concretely specified within a programming language (e.g., int in C, set in Pascal). An ADT is a mathematically specified data type equipped with operations that can be performed on the objects. Object-oriented programming languages, such as C++, provide support for expressing ADTs by means of classes. ADTs specify the data stored and the operations to be performed on them.

4.1.3 Main Issues in the Study of Data Structures

The following issues are of foremost importance in the study of data structures.

Static vs. dynamic: A static data structure supports only queries, while a dynamic data structure supports also updates. A dynamic data structure is often more complicated than its static counterpart supporting the same repertory of queries. A persistent data structure (see, e.g., [9]) is a dynamic data structure that supports operations on past versions. There are many problems for which no efficient dynamic data structures are known. It has been observed that there are strong similarities among the classes of problems that are difficult to parallelize and those that are difficult to dynamize (see, e.g., [32]). Further investigations are needed to study the relationship between parallel and incremental complexity [26].

Implicit vs. explicit: Two fundamental data organization mechanisms are used in data structures. In an explicit data structure, pointers (i.e., memory addresses) are used to link the elements and access them (e.g., a singly linked list, where each element has a pointer to the next one). In an implicit data structure, mathematical relationships support the retrieval of elements (e.g., array representation of a **heap**, see Section 4.3.4.4). Explicit data structures must use additional space to store pointers. However, they are more flexible for complex problems. Most programming languages support pointers and basic implicit data structures, such as arrays.

Internal vs. external memory: In a typical computer, there are two levels of memory: internal memory (RAM) and external memory (disk). The internal memory is much faster than external memory but has much smaller capacity. Data structures designed to work for data that fit into internal memory may not perform well for large amounts of data that need to be stored in external memory. For large-scale problems, data structures need to be designed that take into account the two levels of memory [1]. For example, two-level indices such as B-trees [6] have been designed to efficiently search in large databases.

Space vs. time: Data structures often exhibit a trade-off between space and time complexity. For example, suppose we want to represent a set of integers in the range $[0, N]$ (e.g., for a set of social security numbers $N = 10^{10} - 1$) such that we can efficiently query whether a given element is in the set, insert an element, or delete an element. Two possible data structures for this problem are an N-element bit-array (where the bit in position i indicates the presence of integer i in the set), and a balanced **search tree** (such as a 2–3 tree or a red-black tree). The bit-array has optimal time complexity, since it supports queries, insertions, and deletions in constant time. However, it uses space proportional to the size N of the range, irrespectively of the number of elements actually stored. The balanced search tree supports queries, insertions, and deletions in logarithmic time but uses optimal space proportional to the current number of elements stored.

Theory vs. practice: A large and ever-growing body of theoretical research on data structures is available, where the performance is measured in asymptotic terms ("big-Oh" notation). While asymptotic complexity analysis is an important mathematical subject, it does not completely capture the notion of efficiency of data structures in practical scenarios, where constant factors cannot be disregarded and the difficulty of implementation substantially affects design and maintenance costs. Experimental studies comparing the practical efficiency of data structures for specific classes of problems should be encouraged to bridge the gap between the theory and practice of data structures.

4.1.4 Fundamental Data Structures

The following four data structures are ubiquitously used in the description of discrete algorithms, and serve as basic building blocks for realizing more complex data structures. They are covered in detail in the textbooks listed in Section 4.5 and in the additional references provided.

Sequence: A sequence is a container that stores elements in a certain linear order, which is imposed by the operations performed. The basic operations supported are retrieving, inserting, and removing an element given its position. Special types of sequences include stacks and queues, where insertions and deletions can be done only at the head or tail of the sequence. The basic realization of sequences are by means of arrays and linked lists. Concatenable queues (see, e.g., [18]) support additional operations such as splitting and splicing, and determining the sequence containing a given element. In external memory, a sequence is typically associated with a file.

Priority queue: A priority queue is a container of elements from a totally ordered universe that supports the basic operations of inserting an element and retrieving/removing the largest element. A key application of priority queues is to sorting algorithms. A heap is an efficient realization of a priority queue that embeds the elements into the ancestor/descendant partial order of a binary tree. A heap also admits an implicit realization where the nodes of the tree are mapped into the elements of an array (see Section 4.3.4.4). Sophisticated variations of priority queues include min-max heaps, pagodas, deaps, binomial heaps, and Fibonacci heaps. The buffer tree is efficient external-memory realization of a priority queue.

Dictionary: A dictionary is a container of elements from a totally ordered universe that supports the basic operations of inserting/deleting elements and searching for a given element. **Hash tables** provide an efficient implicit realization of a dictionary. Efficient explicit implementations include skip lists [31], tries, and balanced search trees (e.g., **AVL-trees**, red-black trees, 2–3 trees, 2–3–4 trees, weight-balanced trees, biased search trees, splay trees). The technique of fractional cascading [3] speeds up searching for the same element in a collection of dictionaries. In external memory, dictionaries are typically implemented as B-trees and their variations.

Union-Find: A union-find data structure represents a collection disjoint sets and supports the two fundamental operations of merging two sets and finding the set containing a given element. There is a simple and optimal union-find data structure (rooted tree with path compression) whose time complexity analysis is very difficult to analyze (see, e.g., [15]).

Examples of fundamental data structures used in three major application domains are mentioned below.

Graphs and networks adjacency matrix, adjacency lists, link-cut tree [34], dynamic expression tree [5], topology tree [14], SPQR-tree [8], sparsification tree [11]. See also, e.g., [12,23,35].

Text processing string, suffix tree, Patricia tree. See, e.g., [16].

Geometry and graphics binary space partition tree, chain tree, trapezoid tree, range tree, segment-tree, interval-tree, priority-search tree, hull-tree, quad-tree, R-tree, grid file, metablock tree. See, e.g., [4,10,13,23,27,28,30].

4.1.5 Organization of the Chapter

The rest of this chapter focuses on three fundamental ADTs: sequences, priority queues, and dictionaries. Examples of efficient data structures and algorithms for implementing them are presented in detail in Sections 4.2 through 4.4, respectively. Namely, we cover arrays, singly- and doubly-linked lists, heaps, search trees, (a, b)-trees, AVL-trees, **bucket arrays**, and hash tables.

4.2 Sequence

4.2.1 Introduction

A sequence is a container that stores elements in a certain order, which is imposed by the operations performed. The basic operations supported are:

- INSERTRANK: Insert an element in a given position
- REMOVE: Remove an element

Sequences are a basic form of data organization, and are typically used to realize and implement other data types and data structures.

4.2.2 Operations

Using locators (see Section 4.1.1), we can define a more complete repertory of operations for a sequence S:

SIZE(N) return the number of elements N of S

HEAD(c) assign to c a locator to the first element of S; if S is empty, c is a null locator

TAIL(c) assign to c a locator to the last element of S; if S is empty, a null locator is returned

LOCATERANK(r, c) assign to c a locator to the rth element of S; if $r < 1$ or $r > N$, where N is the size of S, c is a null locator

PREV(c', c'') assign to c'' a locator to the element of S preceding the element with locator c'; if c' is the locator of the first element of S, c'' is a null locator

NEXT(c', c'') assign to c'' a locator to the element of S following the element with locator c'; if c' is the locator of the last element of S, c'' is a null locator

INSERTAFTER(e, c', c'') insert element e into S after the element with locator c', and return a locator c'' to e

INSERTBEFORE(e, c', c'') insert element e into S before the element with locator c', and return a locator c'' to e

INSERTHEAD(e, c) insert element e at the beginning of S, and return a locator c to e

INSERTTAIL(e, c) insert element e at the end of S, and return a locator c to e

INSERTRANK(e, r, c) insert element e in the rth position of S; if $r < 1$ or $r > N + 1$, where N is the current size of S, c is a null locator

REMOVE(c, e) remove from S and return element e with locator c

MODIFY(c, e) replace with e the element with locator c.

Some of the above operations can be easily expressed by means of other operations of the repertory. For example, operations HEAD and TAIL can be easily expressed by means of LOCATERANK and SIZE.

4.2.3   Implementation with an Array

The simplest way to implement a sequence is to use a (one-dimensional) array, where the ith element of the array stores the ith element of the list, and to keep a variable that stores the size N of the sequence. With this implementation, accessing elements takes $O(1)$ time, while insertions and deletions take $O(N)$ time.

Table 4.1 shows the time complexity of the implementation of a sequence by means of an array.

4.2.4   Implementation with a Singly-Linked List

A sequence can also be implemented with a singly-linked list, where each element has a pointer to the next one. We also store the size of the sequence, and pointers to the first and last element of the sequence.

With this implementation, accessing elements takes $O(N)$ time, since we need to traverse the list, while some insertions and deletions take $O(1)$ time.

Table 4.2 shows the time complexity of the implementation of sequence by means of singly-linked list.

4.2.5   Implementation with a Doubly-Linked List

Better performance can be achieved, at the expense of using additional space, by implementing a sequence with a doubly-linked list, where each element has pointers to the next and previous elements. We also store the size of the sequence, and pointers to the first and last element of the sequence.

Table 4.3 shows the time complexity of the implementation of sequence by means of a doubly-linked list.

TABLE 4.1   Performance of a Sequence Implemented with an Array

Operation	Time
SIZE	$O(1)$
HEAD	$O(1)$
TAIL	$O(1)$
LOCATERANK	$O(1)$
PREV	$O(1)$
NEXT	$O(1)$
INSERTAFTER	$O(N)$
INSERTBEFORE	$O(N)$
INSERTHEAD	$O(N)$
INSERTTAIL	$O(1)$
INSERTRANK	$O(N)$
REMOVE	$O(N)$
MODIFY	$O(1)$

Note: We denote with N the number of elements in the sequence at the time the operation is performed. The space complexity is $O(N)$.

TABLE 4.2   Performance of a Sequence Implemented with a Singly-Linked List

Operation	Time
SIZE	$O(1)$
HEAD	$O(1)$
TAIL	$O(1)$
LOCATERANK	$O(N)$
PREV	$O(N)$
NEXT	$O(1)$
INSERTAFTER	$O(1)$
INSERTBEFORE	$O(N)$
INSERTHEAD	$O(1)$
INSERTTAIL	$O(1)$
INSERTRANK	$O(N)$
REMOVE	$O(N)$
MODIFY	$O(1)$

Note: We denote with N the number of elements in the sequence at the time the operation is performed. The space complexity is $O(N)$.

4.3 Priority Queue

4.3.1 Introduction

A priority queue is a container of elements from a totally ordered universe that supports the following two basic operations:

- INSERT: Insert an element into the priority queue
- REMOVEMAX: Remove the largest element from the priority queue

Here are some simple applications of a priority queue:

Scheduling: A scheduling system can store the tasks to be performed into a priority queue, and select the task with highest priority to be executed next.

Sorting: To sort a set of N elements, we can insert them one at a time into a priority queue by means of N INSERT operations, and then retrieve them in decreasing order by means of N REMOVEMAX operations. This two-phase method is the paradigm of several popular sorting algorithms, including Selection-Sort, Insertion-Sort, and Heap-Sort.

TABLE 4.3 Performance of a Sequence Implemented with a Doubly-Linked List

Operation	Time
SIZE	$O(1)$
HEAD	$O(1)$
TAIL	$O(1)$
LOCATERANK	$O(N)$
PREV	$O(1)$
NEXT	$O(1)$
INSERTAFTER	$O(1)$
INSERTAEFORE	$O(1)$
INSERTHEAD	$O(1)$
INSERTTAIL	$O(1)$
INSERTRANK	$O(N)$
REMOVE	$O(1)$
MODIFY	$O(1)$

Note: We denote with N the number of elements in the sequence at the time the operation is performed. The space complexity is $O(N)$.

4.3.2 Operations

Using locators, we can define a more complete repertory of operations for a priority queue Q:

SIZE(N) return the current number of elements N in Q

MAX(c) return a locator c to the maximum element of Q

INSERT(e, c) insert element e into Q and return a locator c to e

REMOVE(c, e) remove from Q and return element e with locator c

REMOVEMAX(e) remove from Q and return the maximum element e from Q

MODIFY(c, e) replace with e the element with locator c.

Note that operation REMOVEMAX(e) is equivalent to MAX(c) followed by REMOVE(c, e).

4.3.3 Realization with a Sequence

We can realize a priority queue by reusing and extending the sequence ADT (see Section 4.2). Operations SIZE, MODIFY, and REMOVE correspond to the homonymous sequence operations.

4.3.3.1 Unsorted Sequence

We can realize INSERT by an INSERTHEAD or an INSERTTAIL, which means that the sequence is not kept sorted. Operation MAX can be performed by scanning the sequence with an iteration of NEXT operations, keeping track of the maximum element encountered. Finally, as observed above, operation REMOVEMAX is a combination of MAX and REMOVE. Table 4.4 shows the time complexity of this realization, assuming that the sequence is implemented with a doubly-linked list.

TABLE 4.4 Performance of a Priority Queue Realized by an Unsorted Sequence, Implemented with a Doubly-Linked List

Operation	Time
SIZE	$O(1)$
MAX	$O(N)$
INSERT	$O(1)$
REMOVE	$O(1)$
REMOVEMAX	$O(N)$
MODIFY	$O(1)$

Note: We denote with N the number of elements in the priority queue at the time the operation is performed. The space complexity is $O(N)$.

4.3.3.2 Sorted Sequence

An alternative implementation uses a sequence that is kept sorted. In this case, operation MAX corresponds to simply accessing the last element of the sequence. However, operation INSERT now requires scanning the sequence to find the appropriate position where to insert the new element. Table 4.5 shows the time complexity of this realization, assuming that the sequence is implemented with a doubly-linked list.

Realizing a priority queue with a sequence, sorted or unsorted, has the drawback that some operations require linear time in the worst case. Hence, this realization is not suitable in many applications where fast running times are sought for all the priority queue operations.

TABLE 4.5 Performance of a Priority Queue Realized by a Sorted Sequence, Implemented with a Doubly-Linked List

Operation	Time
SIZE	$O(1)$
MAX	$O(1)$
INSERT	$O(N)$
REMOVE	$O(1)$
REMOVEMAX	$O(1)$
MODIFY	$O(N)$

Note: We denote with N the number of elements in the priority queue at the time the operation is performed. The space complexity is $O(N)$.

4.3.3.3 Sorting

For example, consider the sorting application (see Section 4.3.1). We have a collection of N elements from a totally ordered universe, and we want to sort them using a priority queue Q. We assume that each element uses $O(1)$ space, and any two elements can be compared in $O(1)$ time. If we realize Q with an unsorted sequence, then the first phase (inserting the N elements into Q) takes $O(N)$ time. However the second phase (removing N times the maximum element) takes time:

$$O\left(\sum_{i=1}^{N} i\right) = O\left(N^2\right).$$

Hence, the overall time complexity is $O(N^2)$. This sorting method is known as Selection-Sort.

However, if we realize the priority queue with a sorted sequence, then the first phase takes time:

$$O\left(\sum_{i=1}^{N} i\right) = O\left(N^2\right),$$

while the second phase takes time $O(N)$. Again, the overall time complexity is $O(N^2)$. This sorting method is known as Insertion-Sort.

4.3.4 Realization with a Heap

A more sophisticated realization of a priority queue uses a data structure called *heap*. A heap is a binary tree T whose internal nodes store each one element from a totally ordered universe, with the following properties (see Figure 4.1):

Level property: All the levels of T are full, except possibly for the bottommost level, which is left-filled;

Partial order property: Let μ be a node of T distinct from the root, and let ν be the parent of μ; then the element stored at μ is less than or equal to the element stored at ν.

The leaves of a heap do not store data and serve only as "placeholders." The level property implies that heap T is a minimum-height binary tree. More precisely, if T stores N elements and has height h, then each level i with $0 \leq i \leq h - 2$ stores exactly 2^i elements, while level $h - 1$ stores between

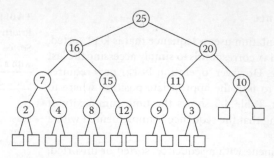

FIGURE 4.1 Example of a heap storing 13 elements.

1 and 2^{h-1} elements. Note that level h contains only leaves. We have

$$2^{h-1} = 1 + \sum_{i=0}^{h-2} 2^i \le N \le \sum_{i=0}^{h-1} 2^i = 2^h - 1,$$

from which we obtain

$$\log_2(N+1) \le h \le 1 + \log_2 N.$$

Now, we show how to perform the various priority queue operations by means of a heap T. We denote with $x(\mu)$ the element stored at an internal node μ of T. We denote with ρ the root of T. We call last node of T the rightmost internal node of the bottommost internal level of T.

By storing a counter that keeps track of the current number of elements, Size consists of simply returning the value of the counter. By the partial order property, the maximum element is stored at the root, and hence, operation Max can be performed by accessing node ρ.

4.3.4.1 Operation Insert

To insert an element e into T, we add a new internal node μ to T such that μ becomes the new last node of T, and set $x(\mu) = e$. This action ensures that the level property is satisfied, but may violate the partial-order property. Hence, if $\mu \neq \rho$, we compare $x(\mu)$ with $x(\nu)$, where ν is the parent of μ. If $x(\mu) > x(\nu)$, then we need to restore the partial order property, which can be locally achieved by exchanging the elements stored at μ and ν. This causes the new element e to move up one level. Again, the partial order property may be violated, and we may have to continue moving up the new element e until no violation occurs. In the worst case, the new element e moves up to the root ρ of T by means of $O(\log N)$ exchanges. The upward movement of element e by means of exchanges is conventionally called upheap.

An example of an insertion into a heap is shown in Figure 4.2.

4.3.4.2 Operation RemoveMax

To remove the maximum element, we cannot simply delete the root of T, because this would disrupt the binary tree structure. Instead, we access the last node λ of T, copy its element e to the root by setting $x(\rho) = x(\lambda)$, and delete λ. We have preserved the level property, but we may have violated the partial order property. Hence, if ρ has at least one nonleaf child, we compare $x(\rho)$ with the maximum element $x(\sigma)$ stored at a child of ρ. If $x(\rho) < x(\sigma)$, then we need to restore the partial order property, which can be locally achieved by exchanging the elements stored at ρ and σ. Again, the partial order property may be violated, and we continue moving down element e until no violation occurs. In the worst case, element e moves down to the bottom internal level of T by means of $O(\log N)$

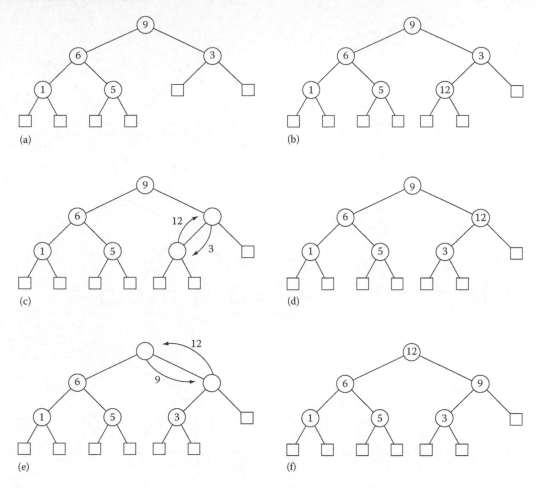

FIGURE 4.2 Example of insertion into a heap. (a) Before insertion. (b) Adding 12. (c–e) Upheap. (f) After insertion.

exchanges. The downward movement of element e by means of exchanges is conventionally called downheap.

An example of operation REMOVEMAX in a heap is shown in Figure 4.3.

4.3.4.3 Operation REMOVE

To remove an arbitrary element of heap T, we cannot simply delete its node μ, because this would disrupt the binary tree structure. Instead, we proceed as before and delete the last node of T after copying to μ its element e. We have preserved the level property, but we may have violated the partial order property, which can be restored by performing either upheap or downheap.

Finally, after modifying an element of heap T, if the partial order property is violated, we just need to perform either upheap or downheap.

4.3.4.4 Time Complexity

Table 4.6 shows the time complexity of the realization of a priority queue by means of a heap. We assume that the heap is itself realized by a data structure for binary trees that supports $O(1)$-time access to the children and parent of a node. For instance, we can implement the heap explicitly with

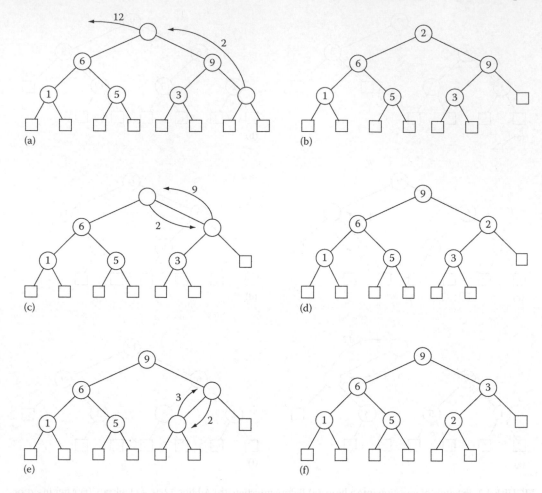

FIGURE 4.3 REMOVEMAX operation in a heap. (a) Removing the maximum element and replacing it with the element of the last node. (b–e) Downheap. (f) After removal.

a linked structure (with pointers from a node to its parents and children), or implicitly with an array (where node i has children $2i$ and $2i + 1$).

Let N the number of elements in a priority queue Q realized with a heap T at the time an operation is performed. The time bounds of Table 4.6 are based on the following facts:

- In the worst case, the time complexity of upheap and downheap is proportional to the height of T
- If we keep a pointer to the last node of T, we can update this pointer in time proportional to the height of T in operations INSERT, REMOVE, and REMOVEMAX, as illustrated in Figure 4.4
- The height of heap T is $O(\log N)$

The $O(N)$ space complexity bound for the heap is based on the following facts:

TABLE 4.6 Performance of a Priority Queue Realized by a Heap, Implemented with a Suitable Binary Tree Data Structure

Operation	Time
SIZE	$O(1)$
MAX	$O(1)$
INSERT	$O(\log N)$
REMOVE	$O(\log N)$
REMOVEMAX	$O(\log N)$
MODIFY	$O(\log N)$

Note: We denote with N the number of elements in the priority queue at the time the operation is performed. The space complexity is $O(N)$.

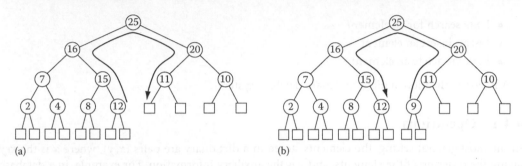

FIGURE 4.4 Update of the pointer to the last node: (a) INSERT; and (b) REMOVE or REMOVEMAX.

- The heap has $2N + 1$ nodes (N internal nodes and $N + 1$ leaves)
- Every node uses $O(1)$ space
- In the array implementation, because of the level property the array elements used to store heap nodes are in the contiguous locations 1 through $2N - 1$

Note that we can reduce the space requirement by a constant factor implementing the leaves of the heap with null objects, such that only the internal nodes have space associated with them.

4.3.4.5 Sorting

Realizing a priority queue with a heap has the advantage that all the operations take $O(\log N)$ time, where N is the number of elements in the priority queue at the time the operation is performed. For example, in the sorting application (see Section 4.3.1), both the first phase (inserting the N elements) and the second phase (removing N times the maximum element) take time:

$$O \left(\sum_{i=1}^{N} \log i \right) = O(N \log N).$$

Hence, sorting with a priority queue realized with a heap takes $O(N \log N)$ time. This sorting method is known as Heap-Sort, and its performance is considerably better than that of Selection-Sort and Insertion-Sort (see Section 4.3.3.3), where the priority queue is realized as a sequence.

4.3.5 Realization with a Dictionary

A priority queue can be easily realized with a dictionary (see Section 4.4). Indeed, all the operations in the priority queue repertory are supported by a dictionary. To achieve $O(1)$ time for operation MAX, we can store the locator of the maximum element in a variable, and recompute it after an update operations. This realization of a priority queue with a dictionary has the same asymptotic complexity bounds as the realization with a heap, provided the dictionary is suitably implemented, e.g., with an (a, b)-tree (see Section 4.4.4) or an AVL-tree (see Section 4.4.5). However, a heap is simpler to program than an (a, b)-tree or an AVL-tree.

4.4 Dictionary

A dictionary is a container of elements from a totally ordered universe that supports the following basic operations:

- FIND: search for an element
- INSERT: insert an element
- REMOVE: delete an element

A major application of dictionaries are database systems.

4.4.1 Operations

In the most general setting, the elements stored in a dictionary are pairs (x, y), where x is the *key* giving the ordering of the elements, and y is the auxiliary information. For example, in a database storing student records, the key could be the student's last name, and the auxiliary information the student's transcript. It is convenient to augment the ordered universe of keys with two special keys: $+\infty$ and $-\infty$, and assume that each dictionary has, in addition to its regular elements, two special elements, with keys $+\infty$ and $-\infty$, respectively. For simplicity, we shall also assume that no two elements of a dictionary have the same key. An insertion of an element with the same key as that of an existing element will be rejected by returning a null locator.

Using locators (see Section 4.1.1), we can define a more complete repertory of operations for a dictionary D:

SIZE(N) return the number of regular elements N of D

FIND(x, c) if D contains an element with key x, assign to c a locator to such an element, otherwise set c equal to a null locator

LOCATEPREV(x, c) assign to c a locator to the element of D with the largest key less than or equal to x; if x is smaller than all the keys of the regular elements, c is a locator the special element with key $-\infty$; if $x = -\infty$, c is a null locator

LOCATENEXT(x, c) assign to c a locator to the element of D with the smallest key greater than or equal to x; if x is larger than all the keys of the regular elements, c is a locator to the special element with key $+\infty$; if $x = +\infty$, c is a null locator

LOCATERANK(r, c) assign to c a locator to the rth element of D; if $r < 1$, c is a locator to the special element with key $-\infty$; if $r > N$, where N is the size of D, c is a locator to the special element with key $+\infty$

PREV(c', c'') assign to c'' a locator to the element of D with the largest key less than that of the element with locator c'; if the key of the element with locator c' is smaller than all the keys of the regular elements, this operation returns a locator to the special element with key $-\infty$

NEXT(c', c'') assign to c'' a locator to the element of D with the smallest key larger than that of the element with locator c'; if the key of the element with locator c' is larger than all the keys of the regular elements, this operation returns a locator to the special element with key $+\infty$

MIN(c) assign to c a locator to the regular element of D with minimum key; if D has no regular elements, c is a null locator

MAX(c) assign to c a locator to the regular element of D with maximum key; if D has no regular elements, c is null a locator

INSERT(e, c) insert element e into D, and return a locator c to e; if there is already an element with the same key as e, this operation returns a null locator

REMOVE(c, e) remove from D and return element e with locator c

MODIFY(c, e) replace with e the element with locator c

Some of the above operations can be easily expressed by means of other operations of the repertory. For example, operation FIND is a simple variation of LOCATEPREV or LOCATENEXT; MIN and MAX are special cases of LOCATERANK, or can be expressed by means of PREV and NEXT.

4.4.2 Realization with a Sequence

We can realize a dictionary by reusing and extending the sequence ADT (see Section 4.2). Operations SIZE, INSERT, and REMOVE correspond to the homonymous sequence operations.

4.4.2.1 Unsorted Sequence

We can realize INSERT by an INSERTHEAD or an INSERTTAIL, which means that the sequence is not kept sorted. Operation FIND(x, c) can be performed by scanning the sequence with an iteration of NEXT operations, until we either find an element with key x, or we reach the end of the sequence. Table 4.7 shows the time complexity of this realization, assuming that the sequence is implemented with a doubly-linked list.

4.4.2.2 Sorted Sequence

We can also use a sorted sequence to realize a dictionary. Operation INSERT now requires scanning the sequence to find the appropriate position where to insert the new element. However, in a FIND operation, we can stop scanning the sequence as soon as we find an element with a key larger than the search key. Table 4.8 shows the time complexity of this realization by a sorted sequence, assuming that the sequence is implemented with a doubly-linked list.

4.4.2.3 Sorted Array

We can obtain a different performance trade-off by implementing the sorted sequence by means of an array, which allows constant-time access to any element of the sequence given its position. Indeed, with this realization we can speed up operation FIND(x, c) using the binary search strategy, as follows. If the dictionary is empty, we are done. Otherwise, let N be the current number of elements in the dictionary. We compare the search key k with the key x_m of the middle element of the sequence, i.e., the element at position $\lfloor N/2 \rfloor$. If $x = x_m$, we have found the element. Else, we recursively search in the subsequence of the elements preceding the middle element if $x < x_m$, or following the middle element if $x > x_m$. At each recursive call, the number of elements of the subsequence being searched halves. Hence, the number of sequence elements accessed and the number of comparisons performed by binary search is $O(\log N)$. While searching takes $O(\log N)$ time, inserting or deleting elements now takes $O(N)$ time.

TABLE 4.7 Performance of a Dictionary Realized by an Unsorted Sequence, Implemented with a Doubly-Linked List

Operation	Time
SIZE	$O(1)$
FIND	$O(N)$
LOCATEPREV	$O(N)$
LOCATENEXT	$O(N)$
LOCATERANK	$O(N)$
NEXT	$O(N)$
PREV	$O(N)$
MIN	$O(N)$
MAX	$O(N)$
INSERT	$O(1)$
REMOVE	$O(1)$
MODIFY	$O(1)$

Note: We denote with N the number of elements in the dictionary at the time the operation is performed.

TABLE 4.8 Performance of a Dictionary Realized by a Sorted Sequence, Implemented with a Doubly-Linked List

Operation	Time
SIZE	$O(1)$
FIND	$O(N)$
LOCATEPREV	$O(N)$
LOCATENEXT	$O(N)$
LOCATERANK	$O(N)$
NEXT	$O(1)$
PREV	$O(1)$
MIN	$O(1)$
MAX	$O(1)$
INSERT	$O(N)$
REMOVE	$O(1)$
MODIFY	$O(N)$

Note: We denote with N the number of elements in the dictionary at the time the operation is performed. The space complexity is $O(N)$.

Table 4.9 shows the performance of a dictionary realized with a sorted sequence, implemented with an array.

4.4.3 Realization with a Search Tree

A search tree for elements of the type (x, y), where x is a key from a totally ordered universe, is a rooted ordered tree T such that

- Each internal node of T has at least two children and stores a nonempty set of elements
- A node μ of T with d children $\mu_1, \ldots, \mu_d$ stores $d - 1$ elements $(x_1, y_1) \cdots (x_{d-1}, y_{d-1})$, where $x_1 \leq \cdots \leq x_{d-1}$
- For each element (x, y) stored at a node in the subtree of T rooted at μ_i, we have $x_{i-1} \leq x \leq x_i$, where $x_0 = -\infty$ and $x_d = +\infty$

TABLE 4.9 Performance of a Dictionary Realized by a Sorted Sequence, Implemented with an Array

Operation	Time
SIZE	$O(1)$
FIND	$O(\log N)$
LOCATEPREV	$O(\log N)$
LOCATENEXT	$O(\log N)$
LOCATERANK	$O(1)$
NEXT	$O(1)$
PREV	$O(1)$
MIN	$O(1)$
MAX	$O(1)$
INSERT	$O(N)$
REMOVE	$O(N)$
MODIFY	$O(N)$

Note: We denote with N the number of elements in the dictionary at the time the operation is performed. The space complexity is $O(N)$.

In a search tree, each internal node stores a nonempty collection of keys, while the leaves do not store any key and serve only as "placeholders." An example of search tree is shown in Figure 4.5a. A special type of search tree is a **binary search tree**, where each internal node stores one key and has two children.

We will recursively describe the realization of a dictionary D by means of a search tree T, since we will use dictionaries to implement the nodes of T. Namely, an internal node μ of T with children $\mu_1, \ldots, \mu_d$ and elements $(x_1, y_1) \cdots (x_{d-1}, y_{d-1})$ is equipped with a dictionary $D(\mu)$ whose regular elements are the pairs $(x_i, (y_i, \mu_i))$, $i = 1, \ldots, d - 1$ and whose special element with key $+\infty$ is $(+\infty, (\cdot, \mu_d))$. A regular element (x, y) stored in D is associated with a regular element $(x, (y, \nu))$ stored in a dictionary $D(\mu)$, for some node μ of T. See the example in Figure 4.5b.

4.4.3.1 Operation FIND

Operation FIND(x, c) on dictionary D is performed by means of the following recursive method for a node μ of T, where μ is initially the root of T (see Figure 4.5b). We execute LOCATENEXT(x, c') on dictionary $D(\mu)$ and let $(x', (y', \nu))$ be the element pointed by the returned locator c'. We have three cases:

- $x = x'$: We have found x and return locator c to (x', y')
- $x \neq x'$ and ν is a leaf: We have determined that x is not in D and return a null locator c
- $x \neq x'$ and ν is an internal node: we set $\mu = \nu$ and recursively execute the method

4.4.3.2 Operation INSERT

Operations LOCATEPREV, LOCATENEXT, and INSERT can be performed with small variations of the above method. For example, to perform operation INSERT(e, c), where $e = (x, y)$, we modify the above cases as follows (see Figure 4.6):

- $x = x'$: An element with key x already exists, and we return a null locator
- $x \neq x'$ and ν is a leaf: We create a new leaf node λ, insert a new element $(x, (y, \lambda))$ into $D(\mu)$, and return a locator c to (x, y)
- $x \neq x'$ and ν is an internal node: We set $\mu = \nu$ and recursively execute the method

Note that new elements are inserted at the "bottom" of the search tree.

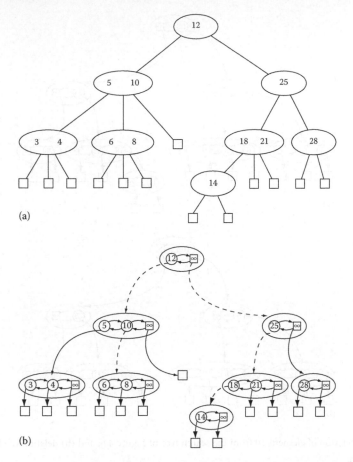

(a)

(b)

FIGURE 4.5 Realization of a dictionary by means of a search tree. (a) A search tree T and (b) realization of the dictionaries at the nodes of T by means of sorted sequences. The search paths for elements 9 (unsuccessful search) and 14 (successful search) are shown with dashed lines.

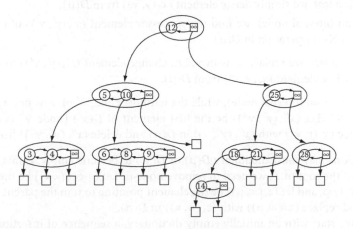

FIGURE 4.6 Insertion of element 9 into the search tree of Figure 4.5.

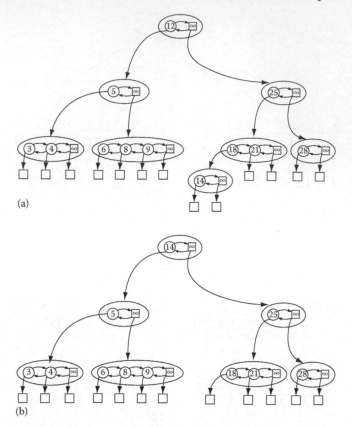

(a)

(b)

FIGURE 4.7 (a) Deletion of element 10 from the search tree of Figure 4.6; and (b) deletion of element 12 from the search tree of part a.

4.4.3.3 Operation REMOVE

Operation REMOVE(e, c) is more complex (see Figure 4.7). Let the associated element of $e = (x, y)$ in T be $(x, (y, \nu))$, stored in dictionary $D(\mu)$ of node μ.

- If node ν is a leaf, we simply delete element $(x, (y, \nu))$ from $D(\mu)$.
- Else (ν is an internal node), we find the successor element $(x', (y', \nu'))$ of $(x, (y, \nu))$ in $D(\mu)$ with a NEXT operation in $D(\mu)$.

  1. If ν' is a leaf, we replace ν' with ν, i.e., change element $(x', (y', \nu'))$ to $(x', (y', \nu))$, and delete element $(x, (y, \nu))$ from $D(\mu)$.

  2. Else (ν' is an internal node), while the leftmost child ν'' of ν' is not a leaf, we set $\nu' = \nu''$. Let $(x'', (y'', \nu''))$ be the first element of $D(\nu')$ (node ν'' is a leaf). We replace $(x, (y, \nu))$ with $(x'', (y'', \nu))$ in $D(\mu)$ and delete $(x'', (y'', \nu''))$ from $D(\nu')$.

The above actions may cause dictionary $D(\mu)$ or $D(\nu')$ to become empty. If this happens, say for $D(\mu)$ and μ is not the root of T, we need to remove node μ. Let $(+\infty, (\cdot, \kappa))$ be the special element of $D(\mu)$ with key $+\infty$, and let $(z, (w, \mu))$ be the element pointing to μ in the parent node π of μ. We delete node μ and replace $(z, (w, \mu))$ with $(z, (w, \kappa))$ in $D(\pi)$.

Note that, if we start with an initially empty dictionary, a sequence of insertions and deletions performed with the above methods yields a search tree with a single node. In Sections 4.4.4 through

4.4.6, we show how to avoid this behavior by imposing additional conditions on the structure of a search tree.

4.4.4 Realization with an (a,b)-Tree

An (a, b)-tree, where a and b are integer constants such that $2 \leq a \leq (b + 1)/2$, is a search tree T with the following additional restrictions:

Level property: All the levels of T are full, i.e., all the leaves are at the same depth

Size property: Let μ be an internal node of T, and d be the number of children of μ; if μ is the root of T, then $d \geq 2$, else $a \leq d \leq b$.

The height of an (a, b) tree storing N elements is $O(\log_a N) = O(\log N)$. Indeed, in the worst case, the root has two children, and all the other internal nodes have a children.

The realization of a dictionary with an (a, b)-tree extends that with a search tree. Namely, the implementation of operations INSERT and REMOVE need to be modified in order to preserve the level and size properties. Also, we maintain the current size of the dictionary, and pointers to the minimum and maximum regular elements of the dictionary.

4.4.4.1 Insertion

The implementation of operation INSERT for search trees given in "Operation INSERT" adds a new element to the dictionary $D(\mu)$ of an existing node μ of T. Since the structure of the tree is not changed, the level property is satisfied. However, if $D(\mu)$ had the maximum allowed size $b - 1$ before insertion (recall that the size of $D(\mu)$ is one less than the number of children of μ), the size property is violated at μ because $D(\mu)$ has now size b. To remedy this overflow situation, we perform the following node-split (see Figure 4.8):

- Let the special element of $D(\mu)$ be $(+\infty, (\cdot, \mu_{b+1}))$. Find the median element of $D(\mu)$, i.e., the element $e_i = (x_i, (y_i, \mu_i))$ such that $i = \lceil (b + 1)/2 \rceil$.
- Split $D(\mu)$ into:

    - dictionary D', containing the $\lceil (b - 1)/2 \rceil$ regular elements $e_j = (x_j, (y_j, \mu_j))$, $j = 1 \cdots i - 1$ and the special element $(+\infty, (\cdot, \mu_i))$;

    - element e; and

    - dictionary D'', containing the $\lfloor (b - 1)/2 \rfloor$ regular elements $e_j = (x_j, (y_j, \mu_j))$, $j = i + 1 \cdots b$ and the special element $(+\infty, (\cdot, \mu_{b+1}))$.

- Create a new tree node κ, and set $D(\kappa) = D'$. Hence, node κ has children $\mu_1 \cdots \mu_i$.
- Set $D(\mu) = D''$. Hence, node μ has children $\mu_{i+1} \cdots \mu_{b+1}$.
- If μ is the root of T, create a new node π with an empty dictionary $D(\pi)$. Else, let π be the parent of μ.
- Insert element $(x_i, (y_i, \kappa))$ into dictionary $D(\pi)$.

After a node-split, the level property is still verified. Also, the size property is verified for all the nodes of T, except possibly for node π. If π has $b + 1$ children, we repeat the node-split for $\mu = \pi$. Each time we perform a node-split, the possible violation of the size property appears at a higher level in the tree. This guarantees the termination of the algorithm for the INSERT operation. We omit the description of the simple method for updating the pointers to the minimum and maximum regular elements.

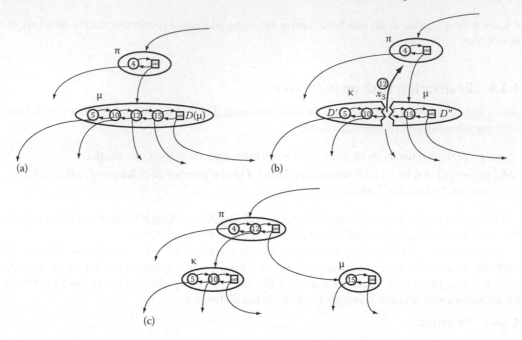

FIGURE 4.8 Example of node-split in a 2–4 tree: (a) initial configuration with an overflow at node μ; (b) split of the node μ into μ' and μ'' and insertion of the median element into the parent node π; and (c) final configuration.

4.4.4.2 Deletion

The implementation of operation REMOVE for search trees given in Section 4.4.3.3 removes an element from the dictionary $D(\mu)$ of an existing node μ of T. Since the structure of the tree is not changed, the level property is satisfied. However, if μ is not the root, and $D(\mu)$ had the minimum allowed size $a - 1$ before deletion (recall that the size of the dictionary is one less than the number of children of the node), the size property is violated at μ because $D(\mu)$ has now size $a - 2$. To remedy this underflow situation, we perform the following node-merge (see Figures 4.9 and 4.10):

- If μ has a right sibling, let μ'' be the right sibling of μ and $\mu' = \mu$; else, let μ' be the left sibling of μ and $\mu'' = \mu$. Let $(+\infty, (\cdot, \nu))$ be the special element of $D(\mu')$.
- Let π be the parent of μ' and μ''. Remove from $D(\pi)$ the regular element $(x, (y, \mu'))$ associated with μ'.
- Create a new dictionary D containing the regular elements of $D(\mu')$ and $D(\mu'')$, regular element $(x, (y, \nu))$, and the special element of $D(\mu'')$.
- Set $D(\mu'') = D$, and destroy node μ'.
- If μ'' has more than b children, perform a node-split at μ''.

After a node-merge, the level property is still verified. Also, the size property is verified for all the nodes of T, except possibly for node π. If π is the root and has one child (and thus, an empty dictionary), we remove node π. If π is not the root and has fewer than $a - 1$ children, we repeat the node-merge for $\mu = \pi$. Each time we perform a node-merge, the possible violation of the size property appears at a higher level in the tree. This guarantees the termination of the algorithm for the REMOVE operation. We omit the description of the simple method for updating the pointers to the minimum and maximum regular elements.

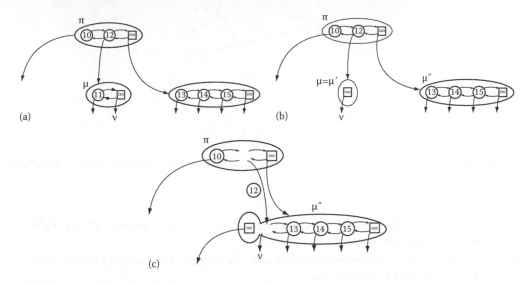

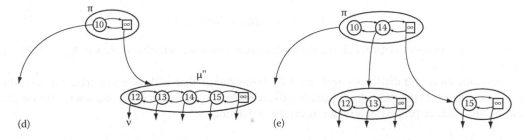

FIGURE 4.9 Example of node merge in a 2–4 tree: (a) initial configuration; (b) the removal of an element from dictionary $D(\mu)$ causes an underflow at node μ; and (c) merging node $\mu = \mu'$ into its sibling μ''.

FIGURE 4.10 Example of node merge in a 2–4 tree: (d) overflow at node μ''; (e) final configuration after splitting node μ''.

4.4.4.3 Complexity

Let T be an (a,b)-tree storing N elements. The height of T is $O(\log_a N) = O(\log N)$. Each dictionary operation affects only the nodes along a root-to-leaf path. We assume that the dictionaries at the nodes of T are realized with sequences. Hence, processing a node takes $O(b) = O(1)$ time. We conclude that each operation takes $O(\log N)$ time.

Table 4.10 shows the performance of a dictionary realized with an (a, b)-tree.

4.4.5 Realization with an AVL-Tree

An AVL-tree is a search tree T with the following additional restrictions:

TABLE 4.10 Performance of a Dictionary Realized by an (a,b)-Tree

Operation	Time
SIZE	$O(1)$
FIND	$O(\log N)$
LOCATEPREV	$O(\log N)$
LOCATENEXT	$O(\log N)$
LOCATERANK	$O(\log N)$
NEXT	$O(\log N)$
PREV	$O(\log N)$
MIN	$O(1)$
MAX	$O(1)$
INSERT	$O(\log N)$
REMOVE	$O(\log N)$
MODIFY	$O(\log N)$

Note: We denote with N the number of elements in the dictionary at the time the operation is performed. The space complexity is $O(N)$.

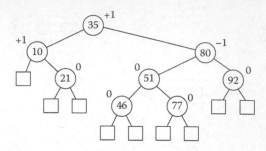

FIGURE 4.11 Example of AVL-tree storing 9 elements. The keys are shown inside the nodes, and the balance factors (see Section 4.4.5.2) are shown next to the nodes.

Binary property: T is a binary tree, i.e., every internal node has two children, (left and right child), and stores one key.

Height-balance property: For every internal node μ, the heights of the subtrees rooted at the children of μ differ at most by one.

An example of AVL-tree is shown in Figure 4.11. The height of an AVL-tree storing N elements is $O(\log N)$. This can be shown as follows. Let N_h be the minimum number of elements stored in an AVL-tree of height h. We have $N_0 = 0, N_1 = 1$, and

$$N_h = 1 + N_{h-1} + N_{h-2}, \text{ for } h \geq 2 \, .$$

The above recurrence relation defines the well-known Fibonacci numbers. Hence, $N_h = \Omega(\phi^N)$, where $1 < \phi < 2$.

The realization of a dictionary with an AVL-tree extends that with a search tree. Namely, the implementation of operations INSERT and REMOVE need to be modified in order to preserve the binary and height-balance properties after an insertion or deletion.

4.4.5.1 Insertion

The implementation of INSERT for search trees given in Section 4.4.3.2 adds the new element to an existing node. This violates the binary property, and hence, cannot be done in an AVL-tree. Hence, we modify the three cases of the INSERT algorithm for search trees as follows:

- $x = x'$: An element with key x already exists, and we return a null locator c
- $x \neq x'$ and ν is a leaf: We replace ν with a new internal node κ with two leaf children, store element (x, y) in κ, and return a locator c to (x, y)
- $x \neq x'$ and ν is an internal node: We set $\mu = \nu$ and recursively execute the method

We have preserved the binary property. However, we may have violated the height-balance property, since the heights of some subtrees of T have increased by one. We say that a node is balanced if the difference between the heights of its subtrees is -1, 0, or 1, and is unbalanced otherwise. The unbalanced nodes form a (possibly empty) subpath of the path from the new internal node κ to the root of T. See the example of Figure 4.12.

4.4.5.2 Rebalancing

To restore the height-balance property, we *rebalance* the lowest node μ that is unbalanced, as follows.

- Let μ' be the child of μ whose subtree has maximum height, and μ'' be the child of μ' whose subtree has maximum height.

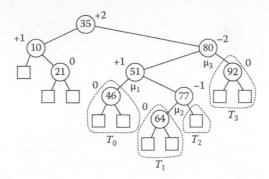

FIGURE 4.12 Insertion of an element with key 64 into the AVL-tree of Figure 4.11. Note that two nodes (with balance factors +2 and −2) have become unbalanced. The dashed lines identify the subtrees that participate in the rebalancing, as illustrated in Figure 4.14.

- Let (μ_1, μ_2, μ_3) be the left-to-right ordering of nodes $\{\mu, \mu', \mu''\}$, and (T_0, T_1, T_2, T_3) be the left-to-right ordering of the four subtrees of $\{\mu, \mu', \mu''\}$ not rooted at a node in $\{\mu, \mu', \mu''\}$.
- Replace the subtree rooted at μ with a new subtree rooted at μ_2, where μ_1 is the left child of μ_2 and has subtrees T_0 and T_1, and μ_3 is the right child of μ_2 and has subtrees T_2 and T_3.

Two examples of rebalancing are schematically shown in Figure 4.14. Other symmetric configurations are possible. In Figure 4.13, we show the rebalancing for the tree of Figure 4.12.

Note that the rebalancing causes all the nodes in the subtree of μ_2 to become balanced. Also, the subtree rooted at μ_2 now has the same height as the subtree rooted at node μ before insertion. This causes all the previously unbalanced nodes to become balanced. To keep track of the nodes that become unbalanced, we can store at each node a balance factor, which is the difference of the heights of the left and right subtrees. A node becomes unbalanced when its balance factor becomes +2 or −2. It is easy to modify the algorithm for operation INSERT such that it maintains the balance factors of the nodes.

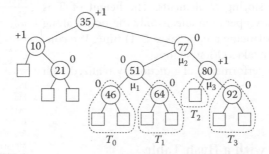

FIGURE 4.13 AVL-tree obtained by rebalancing the lowest unbalanced node in the tree of Figure 4.11. Note that all the nodes are now balanced. The dashed lines identify the subtrees that participate in the rebalancing, as illustrated in Figure 4.14.

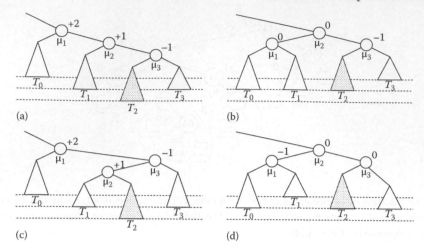

FIGURE 4.14 Schematic illustration of rebalancing a node in the INSERT algorithm for AVL-trees. The shaded subtree is the one where the new element was inserted. (a, b) Rebalancing by means of a "single rotation," and (c, d) Rebalancing by means of a "double rotation."

4.4.5.3 Deletion

The implementation of REMOVE for search trees given in Section 4.4.3 preserves the binary property, but may cause the height-balance property to be violated. After deleting a node, there can be only one unbalanced node, on the path from the deleted node to the root of T.

To restore the height-balance property, we rebalance the unbalanced node using the above algorithm. Notice, however, that the choice of μ'' may not be unique, since the subtrees of μ' may have the same height. In this case, the height of the subtree rooted at μ_2 is the same as the height of the subtree rooted at μ before rebalancing, and we are done. If instead the subtrees of μ' do not have the same height, then the height of the subtree rooted at μ_2 is one less than the height of the subtree rooted at μ before rebalancing. This may cause an ancestor of μ_2 to become unbalanced, and we repeat the rebalancing step. Balance factors are used to keep track of the nodes that become unbalanced, and can be easily maintained by the REMOVE algorithm.

4.4.5.4 Complexity

Let T be an AVL-tree storing N elements. The height of T is $O(\log N)$. Each dictionary operation affects only the nodes along a root-to-leaf path. Rebalancing a node takes $O(1)$ time. We conclude that each operation takes $O(\log N)$ time.

Table 4.11 shows the performance of a dictionary realized with an AVL-tree.

4.4.6 Realization with a Hash Table

The previous realizations of a dictionary make no assumptions on the structure of the keys, and use comparisons between keys to guide the execution of the various operations.

TABLE 4.11 Performance of a Dictionary Realized by an AVL-Tree

Operation	Time
SIZE	$O(1)$
FIND	$O(\log N)$
LOCATEPREV	$O(\log N)$
LOCATENEXT	$O(\log N)$
LOCATERANK	$O(\log N)$
NEXT	$O(\log N)$
PREV	$O(\log N)$
MIN	$O(1)$
MAX	$O(1)$
INSERT	$O(\log N)$
REMOVE	$O(\log N)$
MODIFY	$O(\log N)$

Note: We denote with N the number of elements in the dictionary at the time the operation is performed. The space complexity is $O(N)$.

4.4.6.1 Bucket Array

If the keys of a dictionary D are integers in the range $[1, M]$, we can implement D with a bucket array B. An element (x, y) of D is represented by setting $B[x] = y$. If an integer x is not in D, the location $B[x]$ stores a null value. In this implementation, we allocate a "bucket" for every possible element of D.

Table 4.12 shows the performance of a dictionary realized a bucket array.

The bucket array method can be extended to keys that are easily mapped to integers; e.g., three-letter airport codes can be mapped to the integers in the range $[1, 26^3]$.

4.4.6.2 Hashing

The bucket array method works well when the range of keys is small. However, it is inefficient when the range of keys is large. To overcome this problem, we can use a *hash* function h that maps the keys of the original dictionary D into integers in the range $[1, M]$, where M is a parameter of the hash function. Now, we can apply the bucket array method using the hashed value $h(x)$ of the keys. In general, a collision may happen, where two distinct keys x_1 and x_2 have the same hashed value, i.e., $x_1 \neq x_2$ and $h(x_1) = h(x_2)$. Hence, each bucket must be able to accommodate a collection of elements.

A hash table of size M for a function $h(x)$ is a bucket array B of size M (primary structure) whose entries are dictionaries (secondary structures), such that element (x, y) is stored in the dictionary $B[h(x)]$. For simplicity of programming, the dictionaries used as secondary structures are typically realized with sequences. An example of hash table is shown in Figure 4.15.

If all the elements in the dictionary D collide, they are all stored in the same dictionary of the bucket array, and the performance of the hash table is the same as that of the kind of dictionary used as a secondary structures. At the other end of the spectrum, if no two elements of the dictionary D collide, they are stored in distinct one-element dictionaries of the bucket array, and the performance of the hash table is the same as that of a bucket array.

A typical hash function for integer keys is $h(x) = x \bmod M$. The size M of the hash table is usually chosen as a prime number. An example of hash table is shown in Figure 4.15.

It is interesting to analyze the performance of a hash table from a probabilistic viewpoint. If we assume that the hashed values of the keys are uniformly distributed in the range $[1, M]$, then each bucket holds on average N/M keys, where N is the size of the dictionary. Hence, when $N = O(M)$, the average size of the secondary data structures is $O(1)$.

Table 4.13 shows the performance of a dictionary realized a hash table. Both the worst-case and average time complexity in the above probabilistic model are indicated.

4.5 Further Information

Many textbooks and monographs have been written on data structures, e.g., [2,7,16,17,19–23,27,29,30,33,35,37].

Papers surveying the state of the art in data structures include [4, 15,25,36].

TABLE 4.12 Performance of a Dictionary Realized by Bucket Array

Operation	Time
SIZE	$O(1)$
FIND	$O(1)$
LOCATEPREV	$O(M)$
LOCATENEXT	$O(M)$
LOCATERANK	$O(M)$
NEXT	$O(M)$
PREV	$O(M)$
MIN	$O(M)$
MAX	$O(M)$
INSERT	$O(1)$
REMOVE	$O(1)$
MODIFY	$O(1)$

Note: The keys in the dictionary are integers in the range $[1, M]$. The space complexity is $O(M)$.

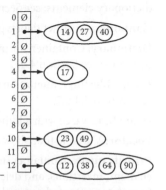

FIGURE 4.15 Example of hash table of size 13 storing 10 elements. The hash function is $h(x) = x \bmod 13$.

TABLE 4.13   Performance of a Dictionary
Realized by a Hash Table of Size M

	Time	
Operation	Worst-Case	Average
Size	$O(1)$	$O(1)$
Find	$O(N)$	$O(N/M)$
LocatePrev	$O(N+M)$	$O(N+M)$
LocateNext	$O(N+M)$	$O(N+M)$
LocateRank	$O(N+M)$	$O(N+M)$
Next	$O(N+M)$	$O(N+M)$
Prev	$O(N+M)$	$O(N+M)$
Min	$O(N+M)$	$O(N+M)$
Max	$O(N+M)$	$O(N+M)$
Insert	$O(1)$	$O(1)$
Remove	$O(1)$	$O(1)$
Modify	$O(1)$	$O(1)$

Note: We denote with N the number of elements in the dictionary at the time the operation is performed. The space complexity is $O(N+M)$. The average time complexity refers to a probabilistic model where the hashed values of the keys are uniformly distributed in the range $[1, M]$.

The LEDA project [24] aims at developing a C++ library of efficient and reliable implementations of sophisticated data structures.

Defining Terms

(a,b)-**tree:** Search tree with additional properties (each node has between a and b children, and all the levels are full); see Section 4.4.4.

Abstract data type: Mathematically specified data type equipped with operations that can be performed on the objects; see Section 4.1.2.

AVL-tree: Binary search tree such that the subtrees of each node have heights that differ by at most one; see Section 4.4.5.

Binary search tree: Search tree such that each internal node has two children; see Section 4.4.3.

Bucket array: Implementation of a dictionary by means of an array indexed by the keys of the dictionary elements; see Section 4.4.6.1.

Container: Abstract data type storing a collection of objects (elements); see Section 4.1.1.

Dictionary: Container storing elements from a sorted universe supporting searches, insertions, and deletions; see Section 4.4.

Hash table: Implementation of a dictionary by means of a bucket array storing secondary dictionaries; see Section 4.4.6.2.

Heap: Binary tree with additional properties storing the elements of a priority queue; see Section 4.3.4.

Locator: Variable that allows to access an object stored in a container; see Section 4.1.1.

Priority queue: Container storing elements from a sorted universe supporting finding the maximum element, insertions, and deletions; see Section 4.3.

Search tree: Rooted ordered tree with additional properties storing the elements of a dictionary; see Section 4.4.3.

Sequence: Container storing object in a certain order, supporting insertions (in a given position) and deletions; see Section 4.2.

References

1. Aggarwal, A. and Vitter, J.S., The input/output complexity of sorting and related problems. *Commun. ACM,* 31, 1116–1127, 1988.
2. Aho, A.V., Hopcroft, J.E., and Ullman J.D., *Data Structures and Algorithms.* Addison-Wesley, Reading, MA, 1983.
3. Chazelle, B. and Guibas, L.J., Fractional cascading: I. A data structuring technique. *Algorithmica,* 1, 133–162, 1986.
4. Chiang, Y.-J. and Tamassia, R., Dynamic algorithms in computational geometry. *Proc. IEEE,* 80(9), 1412–1434, Sep. 1992.
5. Cohen, R.F. and Tamassia, R., Dynamic expression trees. *Algorithmica,* 13, 245–265, 1995.
6. Comer, D., The ubiquitous B-tree. *ACM Comput. Surv.,* 11, 121–137, 1979.
7. Cormen, T.H., Leiserson, C.E., and Rivest, R.L., *Introduction to Algorithms.* The MIT Press, Cambridge, MA, 1990.
8. Di Battista, G. and Tamassia, R., On-line maintenance of triconnected components with SPQR-trees, *Algorithmica,* 15, 302–318, 1996.
9. Driscoll, J.R., Sarnak, N., Sleator, D.D., and Tarjan, R.E., Making data structures persistent. *J. Comput. Syst. Sci.,* 38, 86–124, 1989.
10. Edelsbrunner, H., *Algorithms in Combinatorial Geometry,* volume 10 of *EATCS Monographs on Theoretical Computer Science.* Springer-Verlag, Heidelberg, West Germany, 1987.
11. Eppstein, D., Galil, Z., Italiano, G.F., and Nissenzweig, A., Sparsification: A technique for speeding up dynamic graph algorithms. In *Proceedings of the 33rd Annual IEEE Symposium on Foundations of Computer Science,* Pittsburgh, PA, 60–69, 1992.
12. Even, S., *Graph Algorithms.* Computer Science Press, Potomac, MD, 1979.
13. Foley, J.D., van Dam, A., Feiner, S.K., and Hughes, J.F., *Computer Graphics: Principles and Practice.* Addison-Wesley, Reading, MA, 1990.
14. Frederickson, G.N., A data structure for dynamically maintaining rooted trees. In *Proceedings of 4th ACM-SIAM Symposium on Discrete Algorithms,* Austin, TX, 175–184, 1993.
15. Galil, Z. and Italiano, G.F., Data structures and algorithms for disjoint set union problems. *ACM Comput. Surv.,* 23(3), 319–344, 1991.
16. Gonnet, G.H. and Baeza-Yates, R., *Handbook of Algorithms and Data Structures.* Addison-Wesley, Reading, MA, 1991.
17. Goodrich, M.T. and Tamassia, R., *Data Structures and Algorithms in Java.* John Wiley & Sons, New York, 1998.
18. Hoffmann, K., Mehlhorn, K., Rosenstiehl, P., and Tarjan, R.E., Sorting Jordan sequences in linear time using level-linked search trees. *Inform. Control,* 68, 170–184, 1986.
19. Horowitz, E., Sahni, S., and Metha, D., *Fundamentals of Data Structures in* C_{++}. Computer Science Press, Potomac, MD, 1995.
20. Knuth, D.E., *Fundamental Algorithms,* volume 1 of *The Art of Computer Programming.* Addison-Wesley, Reading, MA, 1968.
21. Knuth, D.E., *Sorting and Searching,* volume 3 of *The Art of Computer Programming.* Addison-Wesley, Reading, MA, 1973.
22. Lewis, H.R. and Denenberg, L., *Data Structures and Their Algorithms.* Harper Collins, New York, 1991.
23. Mehlhorn, K., *Data Structures and Algorithms.* Volumes 1–3. Springer-Verlag, New York, 1984.
24. Mehlhorn, K. and Näher, S., LEDA: A platform for combinatorial and geometric computing. *CACM,* 38, 96–102, 1995.
25. Mehlhorn, K. and Tsakalidis, A., Data structures. In *Algorithms and Complexity,* volume A of *Handbook of Theoretical Computer Science.* J. van Leeuwen (Ed.), Elsevier, Amsterdam, the Netherlands, 1990.

26. Miltersen, P.B., Sairam, S., Vitter, J.S., and Tamassia, R., Complexity models for incremental computation. *Theor. Comput. Sci.,* 130, 203–236, 1994.

27. Nievergelt, J. and Hinrichs, K.H., *Algorithms and Data Structures: With Applications to Graphics and Geometry.* Prentice-Hall, Englewood Cliffs, NJ, 1993.

28. O'Rourke, J., *Computational Geometry in C.* Cambridge University Press, Cambridge, U.K., 1994.

29. Overmars, M.H., *The Design of Dynamic Data Structures,* volume 156 of *LNCS.* Springer-Verlag, Berlin, 1983.

30. Preparata, F.P. and Shamos, M.I., *Computational Geometry: An Introduction.* Springer-Verlag, New York, 1985.

31. Pugh, W., Skip lists: A probabilistic alternative to balanced trees. *Commun. ACM,* 35, 668–676, 1990.

32. Reif, J.H., A topological approach to dynamic graph connectivity, *Inform. Process. Lett.,* 25, 65–70, 1987.

33. Sedgewick, R., *Algorithms in C++.* Addison-Wesley, Reading, MA, 1992.

34. Sleator, D.D. and Tarjan, R.E., A data structure for dynamic tress. *J. Comput. Syst. Sci.,* 26(3), 362–381, 1983.

35. Tarjan, R.E., *Data Structures and Network Algorithms,* volume 44 of *CBMS-NSF Regional Conference Series in Applied Mathematics.* Society for Industrial Applied Mathematics, Philadelphia, PA, 1983.

36. Vitter, J.S. and Flajolet, P., Average-case analysis of algorithms and data structures. In *Algorithms and Complexity,* volume A of *Handbook of Theoretical Computer Science,* J. van Leeuwen (Ed.), Elsevier, Amsterdam, the Netherlands, 1990 pp. 431–524.

37. Wood, D., *Data Structures, Algorithms, and Performance.* Addison-Wesley, Reading, MA, 1993.

5

Topics in Data Structures

Giuseppe F. Italiano
University of Rome "Tor Vergata"

Rajeev Raman
University of Leicester

5.1 Introduction

In this chapter, we describe advanced data structures and algorithmic techniques, mostly focusing our attention on two important problems: set union and persistence. We first describe set union data structures. Their discovery required a new set of techniques and tools that have proved useful in other areas as well. We survey algorithms and data structures for set union problems and attempt to provide a unifying theoretical framework for this growing body of algorithmic tools. **Persistent data structures** maintain information about their past states and find uses in a diverse spectrum of applications. The body of work relating to persistent data structures brings together quite a surprising cocktail of techniques, from real-time computation to techniques from functional programming.

5.1.1 Set Union Data Structures

The set union problem consists of maintaining a collection of disjoint sets under an intermixed sequence of the following two kinds of operations:

{1}   {2}   {3}   {4}   {5}   {6}   {7}   {8}

(a)

{1,3}   {5,2}   {4}   {6}   {7}   {8}

(b)

{4,1,3,7}   {5,2}   {6}   {8}

(c)

{4,1,3,7,5,2}   {6}   {8}

(d)

FIGURE 5.1 Examples of set union operations. (a) The initial collection of disjoint sets; (b) The disjoint sets of (a) after performing union{1,3} and union{5,2}; (c) The disjoint sets of (b) after performing union{1,7} followed by union{4,1}; and (d) The disjoint sets of (c) after performing union{4,5}.

Union(A, B): Combine the two sets A and B into a new set named A

Find(x): Return the name of the set containing element x

The operations are presented on-line, namely, each operation must be processed before the next one is known. Initially, the collection consists of n singleton sets $\{1\}, \{2\}, \ldots, \{n\}$, and the name of set $\{i\}$ is i, $1 \leq i \leq n$. Figure 5.1 illustrates an example of set union operations.

The set union problem has been widely studied and finds applications in a wide range of areas, including Fortran compilers [10,38], property grammars [78,79], computational geometry [49,67,68], finite state machines [4,44], string algorithms [5,48], logic programming and theorem proving [7,8,47,95], and several combinatorial problems such as finding minimum spanning trees [4,53], solving dynamic edge- and vertex-connectivity problems [98], computing least common ancestors in trees [3], solving off-line minimum problems [34,45], finding dominators in graphs [83], and checking flow graph reducibility [82].

Several variants of set union have been introduced, in which the possibility of backtracking over the sequences of unions was taken into account [9,39,59,63,97]. This was motivated by problems arising in logic programming interpreter memory management [40,60,61,96].

5.1.2 Persistent Data Structures

Data structures that one encounters in traditional algorithmic settings are ephemeral; i.e., if the data structure is updated, then the previous state of the data structure is lost. A persistent data structure, on the other hand, preserves old versions of the data structure. Several kinds of persistence can be distinguished based upon what kind of access is allowed to old versions of the data structure. Accesses to a data structure can be of two kinds: updates, which change the information content of the data structure, and queries, which do not. For the sake of ease of presentation, we will assume that queries do not even change the internal representation of the data, i.e., read-only access to a data structure suffices to answer a query.

In the persistent setting we would like to maintain multiple versions of data structures. In addition to the arguments taken by its ephemeral counterparts, a persistent query or update operation takes as an argument the version of the data structure to which the query or update refers. A persistent update also returns a handle to the new version created by the update. We distinguish between three kinds of persistence:

- A partially persistent data structure allows updates only to the latest version of the data structure. All versions of the data structure may be queried, however. Clearly, the versions of a partially persistent data structure exhibit a linear ordering, as shown in Figure 5.2a.

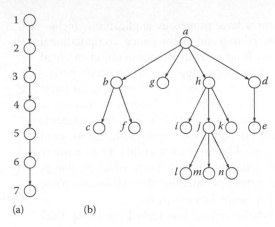

FIGURE 5.2 Structure of versions for (a) partial and (b) full persistence.

- A fully persistent data structure allows all existing versions of the data structure to be queried or updated. However, an update may operate only on a single version at a time—for instance, combining two or more old versions of the data structure to form a new one is not allowed. The versions of a fully persistent data structure form a tree, as shown in Figure 5.2b.

- A **purely functional language** is one that does not allow any destructive operation—one that overwrites data—such as the assignment operation. Purely functional languages are side-effect-free, i.e., invoking a function has no effect other than computing the value returned by the function. In particular, an update operation to a data structure implemented in a purely functional language returns a new data structure containing the updated values, while leaving the original data structure unchanged. Data structures implemented in purely functional languages are therefore persistent in the strongest possible sense, as they allow unrestricted access for both reading and updating all versions of the data structure.

  An example of a purely functional language is pure LISP [64]. Side-effect-free code can also be written in functional languages such as ML [70], most existing variants of LISP (e.g., Common LISP [80]), or Haskell [46], by eschewing the destructive operations supported by these languages.

This section aims to cover a selection of the major results relating to the above forms of persistence. The body of work contains both ad hoc techniques for creating persistent data structures for particular problems as well as general techniques to make ephemeral data structures persistent. Indeed, early work on persistence [17,20,30] focused almost exclusively on the former. Sarnak [75] and Driscoll et al. [28] were the first to offer very efficient general techniques for partial and full persistence. These and related results will form the bulk of the material in this chapter dealing with partial and full persistence. However, the prospect of obtaining still greater efficiency led to the further development of some ad hoc persistent data structures [25,26,41]. The results on functional data structures will largely focus on implementations of individual data structures.

There has also been some research into data structures that support backtrack or rollback operations, whereby the data structure can be reset to some previous state. We do not cover these operations in this section, but we note that fully persistent data structures support backtracking (although sometimes not as efficiently as data structures designed especially for backtracking). Data structures with backtracking for the union–find problem are covered in Section 5.4.

Persistent data structures have numerous applications, including text, program and file editing and maintenance, computational geometry, tree pattern matching, and inheritance in object-oriented programming languages. One elegant application of partially persistent search trees to the classical geometric problem of planar point location was given by Sarnak and Tarjan [76]. Suppose the Euclidean plane is divided into polygons by a collection of n line segments that intersect only at their endpoints (see Figure 5.3), and we want to preprocess the collection of line segments so that, given a query point p, we can efficiently determine the polygon to which p belongs. Sarnak and Tarjan achieve this by combining the well-known plane sweep technique with a persistent data structure.

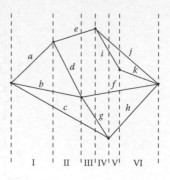

FIGURE 5.3   A planar subdivision.

Imagine moving an infinite vertical line (called the sweep line) from left to right across the plane, beginning at the leftmost endpoint of any line segment. As the sweep line moves, we maintain the line segments currently intersecting the sweep line in a balanced binary search tree, in order of their point of intersection with the sweep line (i.e., of two line segments, the one that intersects the sweep line at a higher location is considered smaller). Figure 5.4 shows the evolution of the search tree as the sweep line continues its progress from left to right. Note that the plane is divided into vertical slabs, within which the search tree does not change.

Given a query point p, we first locate the slab in which the x-coordinate of p lies. If we could remember what our search tree looked like while the sweep line was in this slab, we could query the search tree using the y-coordinate of p to find the two segments immediately above and below p in this slab; these line segments uniquely determine the polygon in which p lies. However, if we maintained the line segments in a partially persistent search tree as the sweep line moves from left to right, all incarnations of the search tree during this process are available for queries.

Sarnak and Tarjan show that it is possible to perform the preprocessing (which merely consists of building up the persistent tree) in $O(n \log n)$ time. The data structure uses $O(n)$ space and can be queried in $O(\log n)$ time, giving a simple optimal solution to the planar point location problem.

5.1.3   Models of Computation

Different models of computation have been developed for analyzing data structures. One model of computation is the **random-access machine**, whose memory consists of an unbounded sequence of

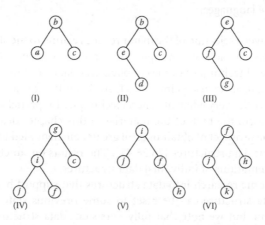

FIGURE 5.4   The evolution of the search tree during the plane sweep. Labels (I) through (VI) correspond to the vertical slabs in the planar subdivision of Figure 5.3.

registers, each of which is capable of holding an integer. In this model, arithmetic operations are allowed to compute the address of a memory register. Usually, it is assumed that the size of a register is bounded by $O(\log n)^*$ bits, where n is the input problem size. A more formal definition of random-access machines can be found in [4]. Another model of computation, known as the **cell probe model** of computation, was introduced by Yao [99]. In the cell probe, the cost of a computation is measured by the total number of memory accesses to a random-access memory with $\lceil \log n \rceil$ bits cell size. All other computations are not accounted for and are considered to be free. Note that the cell probe model is more general than a random-access machine, and thus, is more suitable for proving lower bounds. A third model of computation is the **pointer machine** [13,54,55,77,85]. Its storage consists of an unbounded collection of registers (or records) connected by pointers. Each register can contain an arbitrary amount of additional information but no arithmetic is allowed to compute the address of a register. The only possibility to access a register is by following pointers. This is the main difference between random-access machines and pointer machines. Throughout this chapter, we use the terms random-access algorithms, cell-probe algorithms, and pointer-based algorithms to refer to algorithms respectively for random-access machines, the cell probe model, and pointer machines.

Among pointer-based algorithms, two different classes were defined specifically for set union problems: separable pointer algorithms [85] and nonseparable pointer algorithms [69].

Separable pointer algorithms run on a pointer machine and satisfy the **separability** assumption as defined in [85] (see below). A separable pointer algorithm makes use of a linked data structure, namely, a collection of records and pointers that can be thought of as a directed graph: each record is represented by a node and each pointer is represented by an edge in the graph. The algorithm solves the set union problem according to the following rules [14,85]:

  i. The operations must be performed on line, i.e., each operation must be executed before the next one is known.
  ii. Each element of each set is a node of the data structure. There can be also additional (working) nodes.
  iii. (Separability). After each operation, the data structure can be partitioned into disjoint subgraphs such that each subgraph corresponds to exactly one current set. The name of the set occurs in exactly one node in the subgraph. No edge leads from one subgraph to another.
  iv. To perform find(x), the algorithm obtains the node v corresponding to element x and follows paths starting from v until it reaches the node that contains the name of the corresponding set.
  v. During any operation the algorithm may insert or delete any number of edges. The only restriction is that rule (iii) must hold after each operation.

The class of nonseparable pointer algorithms [69] does not require the separability assumption. The only requirement is that the number of edges leaving each node must be bounded by some constant $c > 0$. More formally, rule (iii) above is replaced by the following rule, while the other four rules are left unchanged:

  iii. There exists a constant $c > 0$ such that there are at most c edges leaving a node.

As we will see later on, often separable and nonseparable pointer-based algorithms admit quite different upper and lower bounds for the same problems.

---

* Throughout this chapter all logarithms are assumed to be to base 2, unless explicitly otherwise specified.

5.2   The Set Union Problem

As defined in Section 5.1, the set union problem consists of performing a sequence of union and find operations, starting from a collection of n singleton sets $\{1\}, \{2\}, \ldots, \{n\}$. The initial name of set $\{i\}$ is i. As there are at most n items to be united, the number of unions in any sequence of operations is bounded above by $(n-1)$. There are two invariants that hold at any time for the set union problem: first, the sets are always disjoint and define a partition of $\{1, 2, \ldots, n\}$; second, the name of each set corresponds to one of the items contained in the set itself. Both invariants are trivial consequences of the definition of union and find operations.

A different version of this problem considers the following operation in place of unions:

> *Unite*(A, B): Combine the two sets A and B into a new set, whose name is either A or B

The only difference between union and unite is that unite allows the name of the new set to be arbitrarily chosen (e.g., at run time by the algorithm). This is not a significant restriction in many applications, where one is mostly concerned with testing whether two elements belong to the same set, no matter what the name of the set can be. However, some extensions of the set union problem have quite different time bounds depending on whether unions or unites are considered. In the following, we will deal with unions unless explicitly specified otherwise.

5.2.1   Amortized Time Complexity

In this section we describe algorithms for the set union problem [84,89] giving the optimal amortized time complexity per operation. We only mention here that the amortized time is the running time per operation averaged over a worst-case sequence of operations, and refer the interested reader to [88] for a more detailed definition of amortized complexity. For the sake of completeness, we first survey some of the basic algorithms that have been proposed in the literature [4,31,38]. These are the quick–find, the weighted quick–find, the quick–union, and the weighted quick–union algorithms. The quick–find algorithm performs find operations quickly, while the quick–union algorithm performs union operations quickly. Their weighted counterparts speed these computations up by introducing some weighting rules during union operations.

Most of these algorithms represent sets as rooted trees, following a technique introduced first by Galler and Fischer [38]. There is a tree for each disjoint set, and nodes of a tree correspond to elements of the corresponding set. The name of the set is stored in the tree root. Each tree node has a pointer to its parent: in the following, we refer to $p(x)$ as the parent of node x.

The quick–find algorithm can be described as follows. Each set is represented by a tree of height 1. Elements of the set are the leaves of the tree. The root of the tree is a special node that contains the name of the set. Initially, singleton set $\{i\}$, $1 \leq i \leq n$, is represented by a tree of height 1 composed of one leaf and one root. To perform a union(A, B), all the leaves of the tree corresponding to B are made children of the root of the tree corresponding to A. The old root of B is deleted. This maintains the invariant that each tree is of height 1 and can be performed in $O(|B|)$ time, where $|B|$ denotes the total number of elements in set B. Since a set can have as many as $O(n)$ elements, this gives an $O(n)$ time complexity in the worst case for each union. To perform a find(x), return the name stored in the parent of x. Since all trees are maintained of height 1, the parent of x is a tree root. Consequently a find requires $O(1)$ time.

A more efficient variant attributed to McIlroy and Morris (see [4]) and known as weighted quick–find uses the freedom implicit in each union operation according to the following weighting rule.

> *Union by size:* Make the children of the root of the smaller tree point to the root of the larger, arbitrarily breaking a tie. This requires that the size of each tree is maintained throughout any sequence of operations.

Although this rule does not improve the worst-case time complexity of each operation, it improves to $O(\log n)$ the amortized bound of a union (see, e.g., [4]).

The quick–union algorithm [38] can be described as follows. Again, each set is represented by a tree. However, there are two main differences with the data structure used by the quick–find algorithm. The first is that now the height of a tree can be greater than 1. The second is that each node of each tree corresponds to an element of a set and therefore there is no need for special nodes. Once again, the root of each tree contains the name of the corresponding set. A union(A, B) is performed by making the tree root of set B a child of the tree root of set A. A find(x) is performed by starting from the node x and by following the pointer to the parent until the tree root is reached. The name of the set stored in the tree root is then returned. As a result, the quick–union algorithm is able to support each union in $O(1)$ time and each find in $O(n)$ time.

This time bound can be improved by using the freedom implicit in each union operation, according to one of the following two union rules. This gives rise to two weighted quick–union algorithms:

> *Union by size:* Make the root of the smaller tree point to the root of the larger, arbitrarily breaking a tie. This requires maintaining the number of descendants for each node, in the following referred to as the size of a node, throughout all the sequence of operations.
>
> *Union by rank:* [89] Make the root of the shallower tree point to the root of the other, arbitrarily breaking a tie. This requires maintaining the height of the subtree rooted at each node, in the following referred to as the rank of a node, throughout all the sequences of operations.

After a union(A, B), the name of the new tree root is set to A. It can be easily proved (see, e.g., [89]) that the height of the trees achieved with either the "union by size" or the "union by rank" rule is never more than $\log n$. Thus, with either rule each union can be performed in $O(1)$ time and each find in $O(\log n)$ time.

A better amortized bound can be obtained if one of the following compaction rules is applied to the path examined during a find operation (see Figure 5.5).

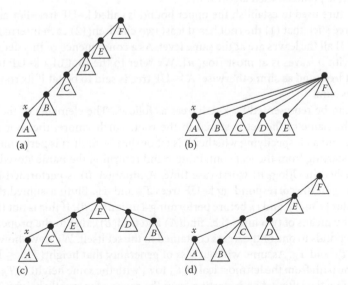

FIGURE 5.5 Illustrating path compaction techniques: (a) the tree before performing a find(x) operation; (b) path compression; (c) path splitting; and (d) path halving.

> *Path compression* [45]: Make every encountered node point to the tree root
>
> *Path splitting* [93,94]: Make every encountered node (except the last and the next to last) point to its grandparent
>
> *Path halving* [93,94]: Make every other encountered node (except the last and the next to last) point to its grandparent

Combining the two choices of a union rule and the three choices of a compaction rule, six possible algorithms are obtained. As shown in [89] they all have an $O(\alpha(m+n, n))$ amortized time complexity, where α is a very slowly growing function, a functional inverse of Ackermann's function [1].

THEOREM 5.1 [89] *The algorithms with either linking by size or linking by rank and either compression, splitting or halving run in $O(n + m\alpha(m+n, n))$ time on a sequence of at most $(n-1)$ unions and m finds.*

No better amortized bound is possible for separable and nonseparable pointer algorithms or in the cell probe model of computation [32,56,89].

THEOREM 5.2 [32,56,89] *Any pointer-based or cell-probe algorithm requires $\Omega(n + m\alpha(m+n, n))$ worst-case time for processing a sequence of $(n-1)$ unions and m finds.*

5.2.2 Single-Operation Worst-Case Time Complexity

The algorithms that use any union and any compaction rule have still single-operation worst-case time complexity $O(\log n)$ [89], since the trees created by any of the union rules can have height as large as $O(\log n)$. Blum [14] proposed a data structure for the set union problem that supports each union and find in $O(\log n / \log \log n)$ time in the worst-case, and showed that this is the actual lower bound for separable pointer-based algorithms.

The data structure used to establish the upper bound is called k–UF tree. For any $k \geq 2$, a k–UF tree is a rooted tree such that (1) the root has at least two children; (2) each internal node has at least k children; and (3) all the leaves are at the same level. As a consequence of this definition, the height of a k–UF tree with n leaves is at most $\lceil \log_k n \rceil$. We refer to the root of a k–UF tree as fat if it has more than k children, and as slim otherwise. A k–UF tree is said to be fat if its root is fat, otherwise it is referred to as slim.

Disjoint sets can be represented by k–UF trees as follows. The elements of the set are stored in the leaves and the name of the set is stored in the root. Furthermore, the root also contains the height of the tree and a bit specifying whether it is fat or slim. A find(x) is performed as described in Section 5.2.1 by starting from the leaf containing x and returning the name stored in the root. This can be accomplished in $O(\log_k n)$ worst-case time. A union(A, B) is performed by first accessing the roots r_A and r_B of the corresponding k–UF trees T_A and T_B. Blum assumed that his algorithm obtained in constant time r_A and r_B before performing a union(A, B). If this is not the case, r_A and r_B can be obtained by means of two finds (i.e., find(A) and find(B)), due to the property that the name of each set corresponds to one of the items contained in the set itself. We now show how to unite the two k–UF trees T_A and T_B. Assume without loss of generality that height(T_B) $\leq$ height(T_A). Let v be the node on the path from the leftmost leaf of T_A to r_A with the same height as T_B. Clearly, v can be located by following the leftmost path starting from the root r_A for exactly height(T_A) − height(T_B) steps. When merging T_A and T_B, only three cases are possible, which give rise to three different types of unions.

Type 1: Root r_B is fat (i.e., has more than k children) and v is not the root of T_A. Then r_B is made a sibling of v.

Type 2: Root r_B is fat and v is fat and equal to r_A (the root of T_A). A new (slim) root r is created and both r_A and r_B are made children of r.

Type 3: This deals with the remaining cases, i.e., either root r_B is slim or $v = r_A$ is slim. If root r_B is slim, then all the children of r_B are made the rightmost children of v, and r_B is deleted. Otherwise, all the children of the slim node $v = r_A$ are made the rightmost children of r_B, and r_A is deleted.

THEOREM 5.3 **[14]** *k–UF trees can support each union and find in $O(\log n / \log \log n)$ time in the worst-case. Their space complexity is $O(n)$.*

PROOF Each find can be performed in $O(\log_k n)$ time. Each union(A, B) can require at most $O(\log_k n)$ time to locate the nodes r_A, r_B, and v as defined above. Both type 1 and type 2 unions can be performed in constant time, while type 3 unions require at most $O(k)$ time, due to the definition of a slim root. Choosing $k = \lceil \log n / \log \log n \rceil$ yields the claimed time bound. The space complexity derives from the fact that a k–UF tree with ℓ leaves has at most $(2\ell - 1)$ nodes. Thus, the forest of k–UF trees requires at most a total of $O(n)$ space to store all the disjoint sets.

Blum showed also that this bound is tight for the class of separable pointer algorithms, while Fredman and Saks [32] showed that the same lower bound holds in the cell probe model of computation.

THEOREM 5.4 **[14,32]** *Every separable pointer or cell-probe algorithm for the disjoint set union problem has single-operation worst-case time complexity at least $\Omega(\log n / \log \log n)$.*

5.2.3 Special Linear Cases

The six algorithms using either union rule and either compaction rule as described in Section 5.2.1 run in $O(n + m\alpha(m, n))$ time on a sequence of at most $(n-1)$ union and m find operations. As stated in Theorem 5.2, no better amortized bound is possible for either pointer-based algorithms or in the cell probe model of computation. This does not exclude, however, that a better bound is possible for a special case of set union. Gabow and Tarjan [34] indeed proposed a random-access algorithm that runs in linear time in the special case where the structure of the union operations is known in advance. Interestingly, Tarjan's lower bound for separable pointer algorithms applies also to this special case, and thus, the power of a random-access machine seems necessary to achieve a linear-time algorithm. This result is of theoretical interest as well as being significant in many applications, such as scheduling problems, the off-line minimum problem, finding maximum matching on graphs, VLSI channel routing, finding nearest common ancestors in trees, and flow graph reducibility [34].

The problem can be formalized as follows. We are given a tree T containing n nodes which correspond to the initial n singleton sets. Denoting by $p(v)$ the parent of the node v in T, we have to perform a sequence of union and find operations such that each union can be only of the form union$(p(v), v)$. For such a reason, T is called the static union tree and the problem will be referred to as the static tree set union. Also the case in which the union tree can dynamically grow by means of new node insertions (referred to as incremental tree set union) can be solved in linear time.

THEOREM 5.5 **[34]** *If the knowledge about the union tree is available in advance, each union and find operation can be supported in $O(1)$ amortized time. The total space required is $O(n)$.*

The same algorithm given for the static tree set union can be extended to the incremental tree set union problem. For this problem, the union tree is not known in advance but is allowed to grow only one node at the time during the sequence of union and find operations. This has application in several algorithms for finding maximum matching in general graphs.

THEOREM 5.6 **[34]** *The algorithm for incremental tree set union runs in a total of $O(m + n)$ time and requires $O(n)$ preprocessing time and space.*

Loebl and Nešetřil [58] presented a linear-time algorithm for another special case of the set union problem. They considered sequences of unions and finds with a constraint on the subsequence of finds. Namely, the finds are listed in a postorder fashion, where a postorder is a linear ordering of the leaves induced by a drawing of the tree in the plane. In this framework, they proved that such sequences of union and find operations can be performed in linear time, thus, getting $O(1)$ amortized time per operation. A preliminary version of these results was reported in [58].

5.3   The Set Union Problem on Intervals

In this section, we describe efficient solutions to the *set* union problem on intervals, which can be defined as follows. Informally, we would like to maintain a partition of a list $\{1, 2, \ldots, n\}$ in adjacent intervals. A union operation joins two adjacent intervals, a find returns the name of the interval containing x and a split divides the interval containing x (at x itself). More formally, at any time we maintain a collection of disjoint sets A_i with the following properties. The A_i's, $1 \le i \le k$, are disjoint sets whose members are ordered by the relation $\le$, and such that $\cup_{i=1}^{k} A_i = \{1, 2, \ldots, n\}$. Furthermore, every item in A_i is less than or equal to all the items in A_{i+1}, for $i = 1, 2, \ldots, n - 1$. In other words, the intervals A_i partition the interval $[1, n]$. Set A_i is said to be adjacent to sets A_{i-1} and A_{i+1}. The set union problem on intervals consists of performing a sequence of the following three operations:

> *Union*(S_1, S_2, S): Given the adjacent sets S_1 and S_2, combine them into a new set $S = S_1 \cup S_2$,
>
> *Find*(x): Given the item x, return the name of the set containing x
>
> *Split*(S, S_1, S_2, x): Partition S into two sets $S_1 = \{a \in S | a < x\}$ and $S_2 = \{a \in S | a \ge x\}$

Adopting the same terminology used in [69], we will refer to the set union problem on intervals as the interval union–split–find problem. After discussing this problem, we consider two special cases: the interval union–find problem and the interval split–find problem, where only union–find and split–find operations are allowed, respectively. The interval union–split–find problem and its subproblems have applications in a wide range of areas, including problems in computational geometry such as dynamic segment intersection [49,67,68], shortest paths problems [6,66], and the longest common subsequence problem [5,48].

5.3.1   Interval Union–Split–Find

In this section we will describe optimal separable and nonseparable pointer algorithms for the interval union–split–find problem. The best separable algorithm for this problem runs in $O(\log n)$ worst-case time for each operation, while nonseparable pointer algorithms require only $O(\log \log n)$ worst-case time for each operation. In both cases, no better bound is possible.

The upper bound for separable pointer algorithms can be easily obtained by means of balanced trees [4,21], while the lower bound was proved by Mehlhorn et al. [69].

THEOREM 5.7 [69] *For any separable pointer algorithm, both the worst-case per operation time complexity of the interval split–find problem and the amortized time complexity of the interval union–split–find problem are $\Omega(\log n)$.*

Turning to nonseparable pointer algorithms, the upper bound can be found in [52,68,91,92]. In particular, van Emde Boas et al. [92] introduced a priority queue which supports among other operations *insert*, *delete*, and *successor* on a set with elements belonging to a fixed universe $S = \{1, 2, \ldots, n\}$. The time required by each of those operation is $O(\log \log n)$. Originally, the space was $O(n \log \log n)$ but later it was improved to $O(n)$. It is easy to show (see also [69]) that the above operations correspond respectively to union, split, and find, and therefore the following theorem holds.

THEOREM 5.8 [91] *Each union, find, and split can be supported in $O(\log \log n)$ worst-case time. The space required is $O(n)$.*

We observe that the algorithm based on van Emde Boas' priority queue is inherently nonseparable. Mehlhorn et al. [69] proved that this is indeed the best possible bound that can be achieved by a nonseparable pointer algorithm.

THEOREM 5.9 [69] *For any nonseparable pointer algorithm, both the worst-case per operation time complexity of the interval split–find problem and the amortized time complexity of the interval union–split–find problem are $\Omega(\log \log n)$.*

Notice that Theorems 5.7 and 5.8 imply that for the interval union–split–find problem the separability assumption causes an exponential loss of efficiency.

5.3.2 Interval Union–Find

The interval union–find problem can be seen from two different perspectives: indeed it is a special case of the union–split–find problem, when no split operations are performed, and it is a restriction of the set union problem described in Section 5.2, where only adjacent intervals are allowed to be joined. Consequently, the $O(\alpha(m + n, n))$ amortized bound given in Theorem 5.1 and the $O(\log n / \log \log n)$ single-operation worst-case bound given in Theorem 5.3 trivially extend to interval union–find. Tarjan's proof of the $\Omega(\alpha(m + n, n))$ amortized lower bound for separable pointer algorithms also holds for the interval union–find problem, while Blum and Rochow [15] have adapted Blum's original lower bound proof for separable pointer algorithms to interval union–find. Thus, the best bounds for separable pointer algorithms are achieved by employing the more general set union algorithms. On the other side, the interval union–find problem can be solved in $O(\log \log n)$ time per operation with the nonseparable algorithm of van Emde Boas [91], while Gabow and Tarjan used the data structure described in Section 5.2.3 to obtain an $O(1)$ amortized time for interval union–find on a random-access machine.

5.3.3 Interval Split–Find

According to Theorems 5.7 through 5.9, the two algorithms given for the more general interval union–split–find problem, are still optimal for the single-operation worst-case time complexity of the interval split–find problem. As a result, each split and find operation can be supported in $\Theta(\log n)$ and in $\Theta(\log \log n)$ time, respectively, in the separable and nonseparable pointer machine model.

As shown by Hopcroft and Ullman [45], the amortized complexity of this problem can be reduced to $O(\log^* n)$, where $\log^* n$ is the iterated logarithm function.* Their algorithm works as follows. The basic data structure is a tree, for which each node at level i, $i \geq 1$, has at most $2^{f(i-1)}$ children, where $f(i) = f(i-1)2^{f(i-1)}$, for $i \geq 1$, and $f(0) = 1$. A node is said to be complete either if it is at level 0 or if it is at level $i \geq 1$ and has $2^{f(i-1)}$ children, all of which are complete. A node that is not complete is called incomplete. The invariant maintained for the data structure is that no node has more than two incomplete children. Moreover, the incomplete children (if any) will be leftmost and rightmost. As in the usual tree data structures for set union, the name of a set is stored in the tree root.

Initially, such a tree with n leaves is created. Its height is $O(\log^* n)$ and therefore a find(x) will require $O(\log^* n)$ time to return the name of the set. To perform a split(x), we start at the leaf corresponding to x and traverse the path to the root to partition the tree into two trees. It is possible to show that using this data structure, the amortized cost of a split is $O(\log^* n)$ [45]. This bound can be further improved to $O(\alpha(m, n))$ as shown by Gabow [33]. The algorithm used to establish this upper bound relies on a sophisticated partition of the items contained in each set.

THEOREM 5.10 [33] *There exists a data structure supporting a sequence of m find and split operations in $O(m\alpha(m, n))$ worst-case time. The space required is $O(n)$.*

La Poutré [56] proved that this bound is tight for (both separable and nonseparable) pointer-based algorithms.

THEOREM 5.11 [56] *Any pointer-based algorithm requires $\Omega(n + m\alpha(m, n))$ time to perform $(n - 1)$ split and m find operations.*

Using the power of a random-access machine, Gabow and Tarjan were able to achieve $\Theta(1)$ amortized time for the interval split–find problem [34]. This bound is obtained by employing a slight variant of the data structure sketched in Section 5.2.3.

5.4   The Set Union Problem with Deunions

Mannila and Ukkonen [59] defined a generalization of the set union problem, which they called set union with deunions. In addition to union and find, the following operation is allowed.

   Deunion: Undo the most recently performed union operation not yet undone

Motivations for studying this problem arise in logic programming, and more precisely in memory management of interpreters without function symbols [40,60,61,96]. In Prolog, for example, variables of clauses correspond to the elements of the sets, unifications correspond to unions and backtracking corresponds to deunions [60].

5.4.1   Algorithms for Set Union with Deunions

The set union problem with deunions can be solved by a modification of Blum's data structure described in Section 5.2.2. To facilitate deunions, we maintain a union stack that stores some bookkeeping information related to unions. Finds are performed as in Section 5.2.2. Unions require some additional work to maintain the union stack. We now sketch which information

---

* $\log^* n = \min\{i \mid \log^{[i]} n \leq 1\}$, where $\log^{[i]} n = \log\log^{[i-1]} n$ for $i > 0$ and $\log^{[0]} n = n$.

is stored in the union stack. For sake of simplicity we do not take into account names of the sets (namely, we show how to handle unite rather than union operations): names can be easily maintained in some extra information stored in the union stack. Initially, the union stack is empty. When a type 1 union is performed, we proceed as in Section 5.2.2 and then push onto the union stack a record containing a pointer to the old root r_B. Similarly, when a type 2 union is performed, we push onto the union stack a record containing a pointer to r_A and a pointer to r_B. Finally, when a type 3 union is performed, we push onto the union stack a pointer to the leftmost child of either r_B or r_A, depending on the two cases.

Deunions basically use the top stack record to invalidate the last union performed. Indeed, we pop the top record from the union stack, and check whether the union to be undone is of type 1, 2, or 3. For type 1 unions, we follow the pointer to r_B and delete the edge leaving this node, thus, restoring it as a root. For type 2 unions, we follow the pointers to r_A and r_B and delete the edges leaving these nodes and their parent. For type 3 unions, we follow the pointer to the node, and move it together with all its right sibling as a child of a new root.

It can be easily showed that this augmented version of Blum's data structure supports each union, find, and deunion in $O(\log n / \log \log n)$ time in the worst-case, with an $O(n)$ space usage. This was proved to be a lower bound for separable pointer algorithms by Westbrook and Tarjan [97]:

THEOREM 5.12 [97] *Every separable pointer algorithm for the set union problem with deunions requires at least* $\Omega(\log n / \log \log n)$ *amortized time per operation.*

All of the union rules and path compaction techniques described in Section 5.2.1 can be extended in order to deal with deunions using the same bookkeeping method (i.e., the union stack) described above. However, path compression with any one of the union rules leads to an $O(\log n)$ amortized algorithm, as it can be seen by first performing $(n - 1)$ unions which build a binomial tree (as defined, for instance, in [89]) of depth $O(\log n)$ and then by repeatedly carrying out a find on the deepest leaf, a deunion, and a redo of that union. Westbrook and Tarjan [97] showed that using either one of the union rules combined with path splitting or path halving yield $O(\log n / \log \log n)$ amortized algorithms for the set union problem with deunions. We now describe their algorithms.

In the following, a union operation not yet undone will be referred to as live, and as dead otherwise. To handle deunions, again a union stack is maintained, which contains the roots made nonroots by live unions. Additionally, we maintain for each node x a node stack $P(x)$, which contains the pointers leaving x created either by unions or by finds. During a path compaction caused by a find, the old pointer leaving x is left in $P(x)$ and each newly created pointer (x, y) is pushed onto $P(x)$. The bottommost pointer on these stacks is created by a union and will be referred to as a union pointer. The other pointers are created by the path compaction performed during the find operations and are called find pointers. Each of these pointers is associated with a unique union operation, the one whose undoing would invalidate the pointer. The pointer is said to be live if the associated union operation is live, and it is said to be dead otherwise.

Unions are performed as in the set union problem, except that for each union a new item is pushed onto the union stack, containing the tree root made nonroot and some bookkeeping information about the set name and either size or rank. To perform a deunion, the top element is popped from the union stack and the pointer leaving that node is deleted. The extra information stored in the union stack is used to maintain set names and either sizes or ranks.

There are actually two versions of these algorithms, depending on when dead pointers are removed from the data structure. Eager algorithms pop pointers from the node stacks as soon as they become dead (i.e., after a deunion operation). On the other hand, lazy algorithms remove dead pointers in a lazy fashion while performing subsequent union and find operations. Combined with the allowed union and compaction rules, this gives a total of eight algorithms. They all have the same time and space complexity, as the following theorem shows.

THEOREM 5.13   [97] *Either union by size or union by rank in combination with either path splitting or path halving gives both eager and lazy algorithms which run in $O(\log n / \log \log n)$ amortized time for operation. The space required by all these algorithms is $O(n)$.*

5.4.2   The Set Union Problem with Unlimited Backtracking

Other variants of the set union problem with deunions have been considered such as set union with arbitrary deunions [36,63], set union with dynamic weighted backtracking [39], and set union with unlimited backtracking [9]. In this chapter, we will discuss only set union with unlimited backtracking and refer the interested readers to the references for the other problems.

As before, we denote a union not yet undone by live, and by dead otherwise. In the set union problem with unlimited backtracking, deunions are replaced by the following more general operation:

> *Backtrack(i):* Undo the last i live unions performed. i is assumed to be an integer, $i \geq 0$.

The name of this problem derives from the fact that the limitation that at most one union could be undone per operation is removed.

Note that this problem is more general than the set union problem with deunions, since a deunion can be simply implemented as backtrack(1). Furthermore, a backtrack(i) can be implemented by performing exactly i deunions. Hence, a sequence of m_1 unions, m_2 finds, and m_3 backtracks can be carried out by simply performing at most m_1 deunions instead of the backtracks. Applying either Westbrook and Tarjan's algorithms or Blum's modified algorithm to the sequence of union, find, and deunion operations, a total of $O((m_1 + m_2) \log n / \log \log n)$ worst-case running time will result. As a consequence, the set union problem with unlimited backtracking can be solved in $O(\log n / \log \log n)$ amortized time per operation. Since deunions are a special case of backtracks, this bound is tight for the class of separable pointer algorithms because of Theorem 5.12.

However, using either Westbrook and Tarjan's algorithms or Blum's augmented data structure, each backtrack(i) can require $\Omega(i \ \log n / \log \log n)$ in the worst-case. Indeed, the worst-case time complexity of backtrack(i) is at least $\Omega(i)$ as long as one insists on deleting pointers as soon as they are invalidated by backtracking (as in the eager methods described in Section 5.4.1, since in this case at least one pointer must be removed for each erased union. This is clearly undesirable, since i can be as large as $(n - 1)$.

The following theorem holds for the set union with unlimited backtracking, when union operations are taken into account.

THEOREM 5.14   [37] *It is possible to perform each union, find and backtrack(i) in $O(\log n)$ time in the worst-case. This bound is tight for nonseparable pointer algorithms.*

Apostolico et al. [9] showed that, when unites instead of unions are performed (i.e., when the name of the new set can be arbitrarily chosen by the algorithm), a better bound for separable pointer algorithms can be achieved:

THEOREM 5.15   [9] *There exists a data structure which supports each unite and find operation in $O(\log n / \log \log n)$ time, each backtrack in $O(1)$ time, and requires $O(n)$ space.*

No better bound is possible for any separable pointer algorithm or in the cell probe model of computation, as it can be shown by a trivial extension of Theorem 5.4.

5.5 Partial and Full Persistence

In this section we cover general techniques for partial and full persistence. The time complexities of these techniques will generally be expressed in terms of slowdowns with respect to the ephemeral query and update operations. The slowdowns will usually be functions of m, the number of versions. A slowdown of $T_q(m)$ for queries means, for example, that a persistent query to a version which is a data structure of size n is accomplished in time $O(T_q(m) \cdot Q(n))$ time, where $Q(n)$ is the running time of an ephemeral query operation on a data structure of size n.

5.5.1 Methods for Arbitrary Data Structures

5.5.1.1 The Fat Node Method

A very simple idea for making any data structure partially persistent is the fat node method, which works as follows. The m versions are numbered by integers from 1 (the first) to m (the last). We will take the convention that if a persistent query specifies version t, for some $1 \le t \le m$, then the query is answered according to the state of the data structure as it was after version t was created but before (if ever) version $t + 1$ was begun.

Each memory location μ in the ephemeral data structure can be associated with a set $C(\mu)$ containing pairs of the form $\langle t, v \rangle$, where v is a value and t is a version number, sometimes referred to as the time stamp of v. A pair $\langle t, v \rangle$ is present in $C(\mu)$ if and only if (a) memory location μ was modified while creating version t and (b) at the completion of version t, the location μ contained the value v. For every memory location μ in the ephemeral data structure, we associate an auxiliary data structure $A(\mu)$, which stores $C(\mu)$ ordered by time stamp.

In order to perform a persistent query in version t we simulate the operation of the ephemeral query algorithm. Whenever the ephemeral query algorithm attempts to read a memory location μ, we query $A(\mu)$ to determine the value of μ in version t. Let t^* be the largest time stamp in $C(\mu)$ which is less than or equal to t. Clearly, the required value is v^* where $\langle t^*, v^* \rangle \in C(\mu)$. Creating version $m + 1$ by modifying version m is also easy: if memory locations $\mu_1, \mu_2, \ldots$ were modified while creating version $m + 1$, and the values of these locations in version $m + 1$ were $v_1, v_2, \ldots$, we simply insert the pair $\langle m + 1, v_i \rangle$ to $A(\mu_i)$ for $i = 1, 2, \ldots$.

If we implement the auxiliary data structures as red-black trees [21] then it is possible to query $A(\mu)$ in $O(\log |C(\mu)|) = O(\log m)$ time and also to add a new pair to $A(\mu)$ in $O(1)$ amortized time (this is possible because the new pair will always have a time stamp greater than or equal to any time stamp in $C(\mu)$). In fact, we can even obtain $O(1)$ worst-case slowdown for updates by using a data structure given in [57]. Note that each ephemeral memory modification performed during a persistent update also incurs a space cost of $O(1)$ (in general this is unavoidable). We thus obtain the following theorem.

THEOREM 5.16 [28] *Any data structure can be made partially persistent with slowdown $O(\log m)$ for queries and $O(1)$ for updates. The space cost is $O(1)$ for each ephemeral memory modification.*

The fat node method can be extended to full persistence with a little work. Again, we will take the convention that a persistent query on version t is answered according to the state of the data structure as it was after version t was created but before (if ever) it was modified to create any descendant version. Again, each memory location μ in the ephemeral data structure will be associated with a set

$C(\mu)$ containing pairs of the form $\langle t, v \rangle$, where v is a value and t is a version (the time stamp). The rules specifying what pairs are stored in $C(\mu)$ are somewhat more complicated. The main difficulty is that the versions in full persistence are only partially ordered. In order to find out the value of a memory location μ in version t, we need to find the deepest ancestor of t in the version tree where μ was modified (this problem is similar to the inheritance problem for object-oriented languages).

One solution is to impose a total order on the versions by converting the version tree into a version list, which is simply a preorder listing of the version tree. Whenever a new version is created, it is added to the version list immediately after its parent, thus inductively maintaining the preordering of the list. We now compare any two versions as follows: the one which is further to the left in the version list is considered smaller.

For example, a version list corresponding to the tree in Figure 5.6 is $[a, b, c, f, g, h, i, j, l, m, n, o, k, d, e]$, and by the linearization, version f is considered to be less than version m, and version j is considered to be less than version l.

Now consider a particular memory location π which was modified in versions b, h, and i of the data structure, with values B, H, and I being written to it in these versions. The following table shows the value of π in each version in the list (a $\perp$ means that no value has yet been written to π and hence its value may be undefined):

Version	a	b	c	f	g	h	i	j	l	m	n	o	k	d	e
Value	$\perp$	B	B	B	$\perp$	H	I	H	H	H	H	H	H	$\perp$	$\perp$

As can be seen in the above example, if π is modified in versions b, h, and i, the version list is divided into intervals containing respectively the sets $\{a\}, \{b, c, f\}, \{g\}, \{h\}, \{i\}, \{j, l, m, n, o, k\}, \{d, e\}$, such that for all versions in that interval, the value of π is the same. In general, the intervals of the version list for which the answer is the same will be different for different memory locations.

Hence, for each memory location μ, we define $C(\mu)$ to contains pairs of the form $\langle t, v \rangle$, where t is the leftmost version in its interval, and v is the value of μ in version t. Again, $C(\mu)$ is stored in an auxiliary data structure $A(\mu)$ ordered by time-stamp (the ordering among versions is as specified by the version list). In the example above, $C(\pi)$ would contain the following pairs:

$$\langle a, \perp \rangle, \langle b, B \rangle, \langle g, \perp \rangle, \langle h, H \rangle, \langle i, I \rangle, \langle j, H \rangle, \langle d, \perp \rangle.$$

In order to determine the value of some memory location μ in version t, we simply search among the pairs stored in $A(\mu)$, comparing versions, until we find the left endpoint of the interval to which t belongs; the associated value is the required answer.

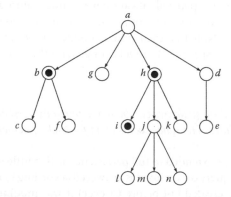

FIGURE 5.6 Navigating in full persistence: an example version tree.

How about updates? Let μ be any memory location, and firstly notice that if a new version is created in which μ is not modified, the value of μ in this new version will be the same as the value of μ in its parent, and the new version will be added to the version list right after its parent. This will simply enlarge the interval to which its parent belongs, and will also not change the left endpoint of the interval. Hence, if μ is not modified in some version, no change need be made to $A(\mu)$. On the other hand, adding a version where μ is modified creates a new interval containing only the new version, and in addition may split an existing interval into two. In general, if μ is modified in k different versions, $C(\mu)$ may contain up to $2k + 1$ pairs, and in each update, up to two new pairs may need to be inserted into $A(\mu)$. In the above example, if we create a new version p as a child of m and modify π to contain P in this version, then the interval $\{j, l, m, n, o, k\}$ splits into two intervals $\{j, l, m\}$ and $\{n, o, k\}$, and the new interval consisting only of $\{p\}$ is created. Hence, we would have to add the pairs $\langle n, H \rangle$ and $\langle p, P \rangle$ to $C(\pi)$.

Provided we can perform the comparison of two versions in constant time, and we store the pairs in say a red-black tree, we can perform a persistent query by simulating the ephemeral query algorithm, with a slowdown of $O(\log |C(\mu)|) = O(\log m)$, where m is the total number of versions. In the case of full persistence, updates also incur a slowdown of $O(\log m)$, and incur a $O(1)$ space cost per memory modification. Maintaining the version list so that two versions can be compared in constant time to determine which of the two is leftward is known as the list order problem, and has been studied in a series of papers [22,90], culminating in an optimal data structure by Dietz and Sleator [24] which allows insertions and comparisons each in $O(1)$ worst-case time. We conclude.

THEOREM 5.17 [28] *Any data structure can be made fully persistent with slowdown $O(\log m)$ for both queries and updates. The space cost is $O(1)$ for each ephemeral memory modification.*

5.5.1.2 Faster Implementations of the Fat Node Method

For arbitrary data structures, the slowdown produced by the fat node method can be reduced by making use of the power of the RAM model. In the case of partial persistence, the versions are numbered with integers from 1 to m, where m is the number of versions, and special data structures for predecessor queries on integer sets may be used. For instance, the van Emde Boas data structure [91,92] processes insertions, deletions, and predecessor queries on a set $S \subseteq \{1, \ldots, m\}$ in $O(\log \log m)$ time each. By using dynamic perfect hashing [27] to minimize space usage, the space required by this data structure can be reduced to linear in the size of the data structure, at the cost of making the updates run in $O(\log \log m)$ expected time. We thus obtain:

THEOREM 5.18 [28,27] *Any data structure can be made partially persistent on a RAM with slowdown $O(\log \log m)$ for queries and expected slowdown $O(\log \log m)$ for updates. The space cost is $O(1)$ per ephemeral memory modification.*

At first sight it does not appear possible to use the same approach for full persistence because the versions are not integers. However, it turns out that algorithms for the list order problem work by assigning integer labels to the elements of the version list such that the labels increase monotonically from the beginning to the end of the list. Furthermore, these labels are guaranteed to be in the range $1, \ldots, m^c$ where m is the number of versions and $c > 1$ is some constant. This means we can once again use the van Emde Boas data structure to search amongst the versions in $O(\log \log m)$ time. Unfortunately, each insertion into the version list may cause many of the integers to be relabeled, and making the changes to the appropriate auxiliary structures may prove expensive. Dietz [23] shows how to combine modifications to the list order algorithms together with standard bucketing techniques to obtain:

THEOREM 5.19   [23] *Any data structure can be made fully persistent on a RAM with slowdown* $O(\log \log m)$ *for queries and expected slowdown* $O(\log \log m)$ *for updates. The space cost is* $O(1)$ *per ephemeral memory modification.*

5.5.2   Methods for Linked Data Structures

The methods discussed above, while efficient, are not optimal and some of them are not simple to code. By placing some restrictions on the class of data structures which we want to make persistent, we can obtain some very simple and efficient algorithms for persistence. One such subclass of data structures is that of linked data structure.

A linked data structure is an abstraction of pointer-based data structures such as linked lists, search trees, etc. Informally, a linked data structure is composed of a collection of nodes, each with a finite number of named fields. Some of these fields are capable of holding an atomic piece of information, while others can hold a pointer to some node (or the value nil). For simplicity we assume the nodes are homogenous (i.e., of the same type) and that all access to the data structure is through a single designated root node. Any version of a linked data structure can be viewed as a directed graph, with vertices corresponding to nodes and edges corresponding to pointers.

Queries are abstracted away as access operations which consist of a series of access steps. The access algorithm has a collection of accessed nodes, which initially contains only the root. At each step, the algorithm either reads information from one of the accessed nodes or follows a non-nil pointer from one of the accessed nodes; the node so reached is then added to the set of accessed nodes. In actual data structures, of course, the information read by the query algorithm would be used to determine the pointers to follow as well as to compute an answer to return. Update operations are assumed to consist of an intermixed sequence of access steps as before and update steps. An update step either creates an explicitly initialized new node or writes a value to a field of some previously accessed node. We now discuss how one might implement persistent access and update operations.

5.5.2.1   Path Copying

A very simple but wasteful method for persistence is to copy the entire data structure after every update. Path copying is an optimization of this for linked data structures, which copies only "essential" nodes. Specifically, if an update modifies a version v by changing values in a set S of nodes, then it suffices to make copies of the nodes in S, together with all nodes that lie on a path from the root of version v to any node in S. The handle to the new version is simply a pointer to the new root. One advantage of this method is that traversing it is trivial: given a pointer to the root in some version, traversing it is done exactly as in the ephemeral case.

This method performs reasonably efficiently in the case of balanced search trees. Assuming that each node in the balanced search tree contains pointers only to its children, updates in balanced search trees such as AVL trees [2] and red-black trees [21] would cause only $O(\log n)$ nodes to be copied (these would be nodes either on the path from the root to the inserted or deleted item, or nodes adjacent to this path). Note that this method does not work as well if the search tree only has an amortized $O(\log n)$ update cost, e.g., in the case of splay trees [87, p. 53 ff]. We therefore get the following theorem, which was independently noted by [74,81].

THEOREM 5.20   *There is a fully persistent balanced search tree with persistent update and query times* $O(\log n)$ *and with space cost* $O(\log n)$ *per update, where n is the number of keys in the version of the data structure which is being updated or queried.*

Of course, for many other data structures, path copying may prove prohibitively expensive, and even in the case of balanced search trees, the space complexity is nonoptimal, as red-black trees with lazy recoloring only modify $O(1)$ locations per update.

5.5.2.2 The Node Copying and Split Node Data Structures

An (ephemeral) bounded-degree linked data structure is one where the maximum in-degree, i.e., the maximum number of nodes that are pointing to any node, is bounded by a constant. Many, if not most, pointer-based data structures have this property, such as linked lists, search trees and so on (some of the data structures covered earlier in this chapter do not have this property). Driscoll et al. [28] showed that bounded-degree linked data structures could be made partially or fully persistent very efficiently, by means of the node copying and split node data structures respectively.

The source of inefficiency in the fat node data structure is searching among all the versions in the auxiliary data structure associated with an ephemeral node, as there is no bound on the number of such versions. The node copying data structure attempts to remedy this by replacing each fat node by a collection of "plump" nodes, each of which is capable of recording a bounded number of changes to an ephemeral node. Again, we assume that the versions are numbered with consecutive integers, starting from 1 (the first) to m (the last). Analogously to the fat node data structure, each ephemeral node x is associated with a set $C(x)$ of pairs $\langle t, r \rangle$, where t is a version number, and r is a record containing values for each of the fields of x. The set $C(x)$ is stored in a collection of plump nodes, each of which is capable of storing $2d + 1$ pairs, where d is the bound on the in-degree of the ephemeral data structure.

The collection of plump nodes storing $C(x)$ is kept in a linked list $L(x)$. Let X be any plump node in $L(x)$ and let X' the next plump node in the list, if any. Let τ denote the smallest time stamp in X' if X' exists, and let $\tau = \infty$ otherwise. The list $L(x)$ is sorted by time stamp in the sense that all pairs in X are sorted by time stamp and all time stamps in X are smaller than τ. Each pair $\langle t, r \rangle$ in X is naturally associated with a valid interval, which is the half-open interval of versions beginning at t, up to, but not including the time stamp of the next pair in X, or τ if no such pair exists. The valid interval of X is simply the union of the valid intervals of the pairs stored in X. The following invariants always hold:

1. For any pair $p = \langle t, r \rangle$ in $C(x)$, if a data field in r contains some value v then the value of the corresponding data field of ephemeral node x during the entire valid interval of p was also v.

    Furthermore, if a pointer field in r contains a pointer to a plump node in $L(y)$ or nil then the corresponding field in ephemeral node x pointed to ephemeral node y or contained nil, respectively, during the entire valid interval of p.

2. For any pair $p = \langle t, r \rangle$ in $C(x)$, if a pointer field in r points to a plump node Y, then the valid interval of p is contained in the valid interval of Y.

3. The handle of version t is a pointer to the (unique) plump node in $L(root)$ whose valid interval contains t.

A persistent access operation on version t is performed by a step-by-step simulation of the ephemeral access algorithm. For any ephemeral node x and version t, let $P(x, t)$ denote the plump node in $L(x)$ whose valid interval contains t. Since the valid intervals of the pairs in $C(x)$ are disjoint and partition the interval $[1, \infty)$, this is well-defined. We ensure that if after some step, the ephemeral access algorithm would have accessed a set S of nodes, then the persistent access algorithm would have accessed the set of plump nodes $\{P(y, t) | y \in S\}$. This invariant holds initially, as the ephemeral algorithm would have accessed only $root$, and by (iv), the handle of version t points to $P(root, t)$.

If the ephemeral algorithm attempts to read a data field of an accessed node x then the persistent algorithm searches among the $O(1)$ pairs in $P(x, t)$ to find the pair whose valid interval contains t, and reads the value of the field from that pair. By (ii), this gives the correct value of the field. If the ephemeral algorithm follows a pointer from an accessed node x and reaches a node y, then the persistent algorithm searches among the $O(1)$ pairs in $P(x, t)$ to find the pair whose valid interval contains t, and follows the pointer specified in that pair. By invariants (1) and (2) this pointer must point to $P(y, t)$. This proves the correctness of the simulation of the access operation.

Suppose during an ephemeral update operation on version m of the data structure, the ephemeral update operation writes some values into the fields of an ephemeral node x. Then the pair $\langle m + 1, r \rangle$ is added to $C(x)$, where r contains the field values of x at the end of the update operation. If the plump node $P(x, m)$ is not full then this pair is simply added to $P(x, m)$. Otherwise, a new plump node that contains only this pair is created and added to the end of $L(x)$. For all nodes y that pointed to x in version m, this could cause a violation of (ii). Hence, for all such y, we add a new pair $\langle m + 1, r' \rangle$ to $C(y)$, where r' is identical to the last record in $C(y)$ except that pointers to $P(x, m)$ are replaced by pointers to the new plump node. If this addition necessitates the creation of a new plump node in $L(y)$ then pointers to $P(m, y)$ are updated as above. A simple potential argument in [28] shows that not only does this process terminate, but the amortized space cost for each memory modification is $O(1)$. At the end of the process, a pointer to the last node in $L(root)$ is returned as the handle to version $m + 1$. Hence, we have that:

THEOREM 5.21   [28] *Any bounded-degree linked data structure can be made partially persistent with worst-case slowdown $O(1)$ for queries, amortized slowdown $O(1)$ for updates, and amortized space cost $O(1)$ per memory modification.*

Although we will not describe them in detail here, similar ideas were applied by Driscoll et al. in the split node data structure which can be used to make bounded-degree linked data structures fully persistent in the following time bounds:

THEOREM 5.22   [28] *Any bounded-degree linked data structure can be made fully persistent with worst-case slowdown $O(1)$ for queries, amortized slowdown $O(1)$ for updates, and amortized space cost $O(1)$ per memory modification.*

Driscoll et al. left open the issue of whether the time and space bounds for Theorems 5.21 and 5.22 could be made worst-case rather than amortized. Toward this end, they used a method called displaced storage of changes to give a fully persistent search tree with $O(\log n)$ worst-case query and update times and $O(1)$ amortized space per update, improving upon the time bounds of Theorem 5.20. This method relies heavily on the property of balanced search trees that there is a unique path from the root to any internal node, and it is not clear how to extract a general method for full persistence from it. A more direct assault on their open problem was made by [25], which showed that all bounds in Theorem 5.21 could be made worst-case on the RAM model. In the same paper it was also shown that the space cost could be made $O(1)$ worst-case on the pointer machine model, but the slowdown for updates remained $O(1)$ amortized. Subsequently, Brodal [11] fully resolved the open problem of Driscoll et al. for partial persistence by showing that all bounds in Theorem 5.21 could be made worst-case on the pointer machine model. For the case of full persistence it was shown in [26] how to achieve $O(\log \log m)$ worst-case slowdown for updates and queries and a worst-case space cost of $O(1)$ per memory modification, but the open problem of Driscoll et al. remains only partially resolved in this case. It should be noted that the data structures of [11,26] are not much more complicated than the original data structures of Driscoll et al.

5.6 Functional Data Structures

In this section we will consider the implementation of data structures in functional languages. Although implementation in a functional language automatically guarantees persistence, the central issue is maintaining the same level of efficiency as in the imperative setting.

The state of the art regarding general methods is quickly summarized. The path-copying method described at the beginning of Section 5.5 can easily be implemented in a functional setting. This means that balanced binary trees (without parent pointers) can be implemented in a functional language, with queries and updates taking $O(\log n)$ worst-case time, and with a suboptimal worst-case space bound of $\Theta(\log n)$. Using the functional implementation of search trees to implement a dictionary which will simulate the memory of any imperative program, it is possible to implement any data structure which uses a maximum of M memory locations in a functional language with a slowdown of $O(\log M)$ in the query and update times, and a space cost of $O(\log M)$ per memory modification.

Naturally, better bounds are obtained by considering specific data structuring problems, and we summarize the known results at the end of this section. First, though, we will focus on perhaps the most fundamental data structuring problem in this context, that of implementing catenable lists. A catenable list supports the following set of operations:

> *Makelist(a)*: Creates a new list containing only the element a
>
> *Head(X)*: Returns the first element of list X. Gives an error if X is empty
>
> *Tail(X)*: Returns the list obtained by deleting the first element of list X without modifying X. Gives an error if X is empty
>
> *Catenate(X, Y)*: Returns the list obtained by appending list Y to list X, without modifying X or Y

Driscoll et al. [29] were the first to study this problem, and efficient but nonoptimal solutions were proposed in [16,29]. We will sketch two proofs of the following theorem, due to Kaplan and Tarjan [50] and Okasaki [71]:

THEOREM 5.23 *The above set of operations can be implemented in $O(1)$ time each.*

The result due to Kaplan and Tarjan is stronger in two respects. Firstly, the solution of [50] gives $O(1)$ worst-case time bounds for all operations, while Okasaki's only gives amortized time bounds. Also, Okasaki's result uses "memoization" which, technically speaking, is a side effect, and hence, his solution is not purely functional. On the other hand, Okasaki's solution is extremely simple to code in most functional programming languages, and offers insight into how to make amortized data structures fully persistent efficiently. In general, this is difficult because in an amortized data structure, some operations in a sequence of operations may be expensive, even though the average cost is low. In the fully persistent setting, an adversary can repeatedly perform an expensive operation as often as desired, pushing the average cost of an operation close to the maximum cost of any operation.

We will briefly cover both these solutions, beginning with Okasaki's. In each case we will first consider a variant of the problem where the catenate operation is replaced by the operation *inject(a, X)* which adds a to the end of list X. Note that *inject(a, X)* is equivalent to *catenate(makelist(a), X)*. Although this change simplifies the problem substantially (this variant was solved quite long ago [43]) we use it to elaborate upon the principles in a simple setting.

5.6.1 Implementation of Catenable Lists in Functional Languages

We begin by noting that adding an element a to the front of a list X, without changing X, can be done in $O(1)$ time. We will denote this operation by $a :: X$. However, adding an element to the end of X involves a destructive update. The standard solution is to store the list X as a pair of lists $\langle F, R \rangle$, with F representing an initial segment of X, and R representing the remainder of X, stored in reversed order. Furthermore, we maintain the invariant that $|F| \geq |R|$.

To implement an inject or tail operation, we first obtain the pair $\langle F', R' \rangle$, which equals $\langle F, a :: R \rangle$ or $\langle tail(F), R \rangle$, as the case may be. If $|F'| \geq |R'|$, we return $\langle F', R' \rangle$. Otherwise we return $\langle F' \mathbin{+\!\!+} reverse(R'), [\ ] \rangle$, where $X \mathbin{+\!\!+} Y$ appends Y to X and $reverse(X)$ returns the reverse of list X. The functions $\mathbin{+\!\!+}$ and reverse are defined as follows:

$$X \mathbin{+\!\!+} Y = Y \text{ if } X = [\ ],$$

$$= head(X) :: (tail(X) \mathbin{+\!\!+} Y) \text{ otherwise.}$$

$$reverse(X) = rev(X, [\ ]), \text{ where} :$$

$$rev(X, Y) = Y \text{ if } X = [\ ],$$

$$= rev(tail(X), head(X) :: Y) \text{ otherwise.}$$

The running time of $X \mathbin{+\!\!+} Y$ is clearly $O(|X|)$, as is the running time of $reverse(X)$. Although the amortized cost of inject can be easily seen to be $O(1)$ in an ephemeral setting, the efficiency of this data structure may be much worse in a fully persistent setting, as discussed above.

If, however, the functional language supports lazy evaluation and memoization then this solution can be used as is. Lazy evaluation refers to delaying calculating the value of expressions as much as possible. If lazy evaluation is used, the expression $F' \mathbin{+\!\!+} reverse(R')$ is not evaluated until we try to determine its head or tail. Even then, the expression is not fully evaluated unless F' is empty, and the list $tail(F' \mathbin{+\!\!+} reverse(R'))$ remains represented internally as $tail(F') \mathbin{+\!\!+} reverse(R')$. Note that reverse cannot be computed incrementally like $\mathbin{+\!\!+}$: once started, a call to reverse must run to completion before the first element in the reversed list is available. Memoization involves caching the result of a delayed computation the first time it is executed, so that the next time the same computation needs to be performed, it can be looked up rather than recomputed.

The amortized analysis uses a "debit" argument. Each element of a list is associated with a number of debits, which will be proportional to the amount of delayed work which must be done before this element can be accessed. Each operation can "discharge" $O(1)$ debits, i.e., when the delayed work is eventually done, a cost proportional to the number of debits discharged by an operation will be charged to this operation. The goal will be to prove that all debits on an element will have been discharged before it is accessed. However, once the work has been done, the result is memoized and any other thread of execution which require this result will simply use the memoized version at no extra cost. The debits satisfy the following invariant. For $i = 0, 1, \ldots$, let $d_i \geq 0$ denote the number of debits on the ith element of any list $\langle F, R \rangle$. Then:

$$\sum_{j=0}^{i} d_i \leq \min\{2i, |F| - |R|\}, \quad \text{for } i = 0, 1, \ldots.$$

Note that the first (zeroth) element on the list always has zero debits on it, and so head only accesses elements whose debits have been paid. If no list reversal takes place during a tail operation, the value of $|F|$ goes down by one, as does the index of each remaining element in the list (i.e., the old $(i + 1)$st element will now be the new ith element). It suffices to pay of $O(1)$ debits at each of the first two locations in the list where the invariant is violated. A new element injected into list R has no delayed computation associated with it, and is give zero debits. The violations of the invariant caused by an inject where no list reversal occurs are handled as above. As a list reversal occurs only if $m = |F| = |R|$ before the operation which caused the reversal, the invariant implies that all debits on the front list have been paid off before the reversal. Note that there are no debits on the rear list. After the reversal, one debit is placed on each element of the old front list (to pay for the delayed incremental $\mathbin{+\!\!+}$ operation) and $m + 1$ debits are placed on the first element of the reversed list (to pay for the reversal), and zero on the remaining elements of the remaining elements of the reversed list, as there is no further delayed computation associated with them. It is easy to verify that the invariant is still satisfied after discharging $O(1)$ debits.

To add catenation to Okasaki's algorithm, a list is represented as a tree whose left-to-right preorder traversal gives the list being represented. The children of a node are stored in a functional queue as described above. In order to perform *catenate(X, Y)* the operation *link(X, Y)* is performed, which adds root of the tree Y is added to the end of the child queue for the root of the tree X. The operation *tail(X)* removes the root of the tree for X. If its children of the root are $X_1, \ldots, X_m$ then the new list is given by $link(X_1, link(X_2, \ldots, link(X_{m-1}, X_m)))$. By executing the *link* operations in a lazy fashion and using memoization, all operations can be made to run in $O(1)$ time.

5.6.2 Purely Functional Catenable Lists

In this section we will describe the techniques used by Kaplan and Tarjan to obtain a purely functional queue. The critical difference is that we cannot assume memoization in a purely functional setting. This appears to mean that the data structures once again have to support each operation in worst-case constant time. The main ideas used by Kaplan and Tarjan are those of data-structural bootstrapping and recursive slowdown. Data-structural bootstrapping was introduced by [29] and refers to allowing a data structure to use the same data structure as a recursive substructure.

Recursive slowdown can be viewed as running the recursive data structures "at a slower speed." We will now give a very simple illustration of recursive slowdown. Let a 2-queue be a data structure which allows the tail and inject operations, but holds a maximum of 2 elements. Note that the bound on the size means that all operations on a 2-queue can be can be trivially implemented in constant time, by copying the entire queue each time. A queue Q consists of three components: a front queue $f(Q)$, which is a 2-queue, a rear queue $r(Q)$, which is also a 2-queue, and a center queue $c(Q)$, which is a recursive queue, each element of which is a pair of elements of the top-level queue. We will ensure that at least one of $f(Q)$ is nonempty unless Q itself is empty.

The operations are handled as follows. An inject adds an element to the end of $r(Q)$. If $r(Q)$ is full, then the two elements currently in $r(Q)$ are inserted as a pair into $c(Q)$ and the new element is inserted into $r(Q)$. Similarly, a tail operation attempts to remove the first element from $f(Q)$. If $f(Q)$ is empty then we extract the first pair from $c(Q)$, if $c(Q)$ is nonempty and place the second element from the pair into $f(Q)$, discarding the first element. If $c(Q)$ is also empty then we discard the first element from $r(Q)$.

The key to the complexity bound is that only every alternate inject or tail operation accesses $c(Q)$. Therefore, the recurrence giving the amortized running time $T(n)$ of operations on this data structure behaves roughly like $2T(n) = T(n/2) + k$ for some constant k. The term $T(n/2)$ represents the cost of performing an operation on $c(Q)$, since $c(Q)$ can contain at most $n/2$ pairs of elements, if n is the number of elements in Q as a whole. Rewriting this recurrence as $T(n) = \frac{1}{2}T(n/2) + k'$ and expanding gives that $T(n) = O(1)$ (even replacing $n/2$ by $n - 1$ in the RHS gives $T(n) = O(1)$).

This data structure is not suitable for use in a persistent setting as a single operation may still take $\Theta(\log n)$ time. For example, if $r(Q), r(c(Q)), r(c(c(Q))) \ldots$ each contain two elements, then a single inject at the top level would cause changes at all $\Theta(\log n)$ levels of recursion. This is analogous to carry propagation in binary numbers—if we define a binary number where for $i = 0, 1, \ldots$, the ith digit is 0 if $c^i(Q)$ contains one element and 1 if it contains two (the 0th digit is considered to be the least significant) then each inject can be viewed as adding 1 to this binary number. In the worst-case, adding 1 to a binary number can take time proportional to the number of digits.

A different number system can alleviate this problem. Consider a number system where the ith digit still has weight 2^i, as in the binary system, but where digits can take the value 0, 1, or 2 [19]. Further, we require that any pair of 2's be separated by at least one 0 and that the rightmost digit, which is not a 1 is a 0. This number system is redundant, i.e., a number can be represented in more than one way (the decimal number 4, for example, can be represented as either 100 or 020). Using this number system, we can increment a value by one in constant time by the following rules: (a) add one by changing the rightmost 0 to a 1, or by changing x 1 to $(x + 1)$ 0; then (b) fixing the rightmost

2 by changing $x\,2$ to $(x+1)\,0$. Now we increase the capacity of $r(Q)$ to 3 elements, and let a queue containing i elements represent the digit $i-1$. We then perform an inject in $O(1)$ worst-case time by simulating the algorithm above for incrementing a counter. Using a similar idea to make tail run in $O(1)$ time, we can make all operations run in $O(1)$ time.

In the version of their data structure which supports catenation, Kaplan and Tarjan again let a queue be represented by three queues $f(Q)$, $c(Q)$, and $r(Q)$, where $f(Q)$ and $r(Q)$ are of constant size as before. The center queue $c(Q)$ in this case holds either (i) a queue of constant size containing at least two elements or (ii) a pair whose first element is a queue as in (i) and whose second element is a catenable queue. To execute $catenate(X, Y)$, the general aim is to first try and combine $r(X)$ and $f(Y)$ into a single queue. When this is possible, a pair consisting of the resulting queue and $c(Y)$ is injected into $c(X)$. Otherwise, $r(X)$ is injected into $c(X)$ and the pair $\langle f(X), c(X) \rangle$ is also injected into $c(X)$. Details can be found in [50].

5.6.3 Other Data Structures

A deque is a list which allows single elements to be added or removed from the front or the rear of the list. Efficient persistent deques implemented in functional languages were studied in [18,35,42], with some of these supporting additional operations. A catenable deque allows all the operations above defined for a catenable list, but also allows deletion of a single element from the end of the list. Kaplan and Tarjan [50] have stated that their technique extends to give purely functional catenable deques with constant worst-case time per operation. Other data structures which can be implemented in functional languages include finger search trees [51] and worst-case optimal priority queues [12]. (See [72,73] for yet more examples.)

5.7 Research Issues and Summary

In this chapter we have described the most efficient known algorithms for set union and persistency.

Most of the set union algorithms we have described are optimal with respect to a certain model of computation (e.g., pointer machines with or without the separability assumption, random-access machines). There are still several open problems in all the models of computation we have considered. First, there are no lower bounds for some of the set union problems on intervals: for instance, for nonseparable pointer algorithms we are only aware of the trivial lower bound for interval union–find. This problem requires $\Theta(1)$ amortized time on a random-access machine as shown by Gabow and Tarjan [34]. Second, it is still open whether in the amortized and the single operation worst-case complexity of the set union problems with deunions or backtracking can be improved for nonseparable pointer algorithms or in the cell probe model of computation.

5.8 Further Information

Research on advanced algorithms and data structures is published in many computer science journals, including *Algorithmica, Journal of ACM, Journal of Algorithms*, and *SIAM Journal on Computing*. Work on data structures is published also in the proceedings of general theoretical computer science conferences, such as the "ACM Symposium on Theory of Computing (STOC)," and the "IEEE Symposium on Foundations of Computer Science (FOCS)." More specialized conferences devoted exclusively to algorithms are the "ACM–SIAM Symposium on Discrete Algorithms (SODA)" and the "European Symposium on Algorithms (ESA)." Online bibliographies for many of these conferences and journals can be found on the World Wide Web.

Galil and Italiano [37] provide useful summaries on the state of the art in set union data structures. A in-depth study of implementing data structures in functional languages is given in [72].

Defining Terms

Cell probe model: Model of computation where the cost of a computation is measured by the total number of memory accesses to a random-access memory with $\lceil \log n \rceil$ bits cell size. All other computations are not accounted for and are considered to be free.

Persistent data structure: A data structure that preserves its old versions. Partially persistent data structures allow updates to their latest version only, while all versions of the data structure may be queried. Fully persistent data structures allow all their existing versions to be queried or updated.

Pointer machine: Model of computation whose storage consists of an unbounded collection of registers (or records) connected by pointers. Each register can contain an arbitrary amount of additional information, but no arithmetic is allowed to compute the address of a register. The only possibility to access a register is by following pointers.

Purely functional language: A language that does not allow any destructive operation—one which overwrites data—such as the assignment operation. Purely functional languages are side-effect-free, i.e., invoking a function has no effect other than computing the value returned by the function.

Random access machine: Model of computation whose memory consists of an unbounded sequence of registers, each of which is capable of holding an integer. In this model, arithmetic operations are allowed to compute the address of a memory register.

Separability: Assumption that defines two different classes of pointer-based algorithms for set union problems. An algorithm is separable if after each operation, its data structures can be partitioned into disjoint subgraphs so that each subgraph corresponds to exactly one current set, and no edge leads from one subgraph to another.

Acknowledgments

The work of the first author was supported in part by the Commission of the European Communities under project no. 20244 (ALCOM-IT) and by a research grant from University of Venice "Ca' Foscari." The work of the second author was supported in part by a Nuffield Foundation Award for Newly-Appointed Lecturers in the Sciences.

References

1. Ackermann, W., Zum Hilbertshen Aufbau der reelen Zahlen, *Math. Ann.,* 99, 118–133, 1928.
2. Adel'son-Vel'skii, G.M. and Landis, E.M., An algorithm for the organization of information, *Dokl. Akad. Nauk SSSR,* 146, 263–266, (in Russian), 1962.
3. Aho, A.V., Hopcroft, J.E., and Ullman, J.D., On computing least common ancestors in trees, *Proceedings of the 5th Annual ACM Symposium on Theory of Computing,* 253–265, Austin, TX, 1973.
4. Aho, A.V., Hopcroft, J.E., and Ullman, J.D., *The Design and Analysis of Computer Algorithms,* Addison-Wesley, Reading, MA, 1974.
5. Aho, A.V., Hopcroft, J.E., and Ullman, J.D., *Data Structures and Algorithms,* Addison-Wesley, Reading, MA, 1983.
6. Ahuja, R.K., Mehlhorn, K., Orlin, J.B., and Tarjan, R.E., Faster algorithms for the shortest path problem, *J. Assoc. Comput. Mach.,* 37, 213–223, 1990.
7. At-Kaci, H., An algebraic semantics approach to the effective resolution of type equations, *Theor. Comput. Sci.,* 45, 1986.
8. At-Kaci, H. and Nasr, R., LOGIN: A logic programming language with built-in inheritance, *J. Logic Program.,* 3, 1986.

9. Apostolico, A., Italiano, G.F., Gambosi, G., and Talamo, M., The set union problem with unlimited backtracking, *SIAM J. Comput.*, 23, 50–70, 1994.

10. Arden, B.W., Galler, B.A., and Graham, R.M., An algorithm for equivalence declarations, *Commun. ACM*, 4, 310–314, 1961.

11. Brodal, G.S., Partially persistent data structures of bounded degree with constant update time, Technical Report BRICS RS-94-35, BRICS, Department of Computer Science, University of Aarhus, Denmark, 1994.

12. Brodal, G.S. and Okasaki, C., Optimal purely functional priority queues, *J. Funct. Program.*, 6(6): 839–857, 1996.

13. Ben-Amram, A.M. and Galil, Z., On pointers versus addresses, *J. Assoc. Comput. Mach.*, 39, 617–648, 1992.

14. Blum, N., On the single operation worst-case time complexity of the disjoint set union problem, *SIAM J. Comput.*, 15, 1021–1024, 1986.

15. Blum, N. and Rochow, H., A lower bound on the single-operation worst-case time complexity of the union–find problem on intervals, *Inform. Process. Lett.*, 51, 57–60, 1994.

16. Buchsbaum, A.L. and Tarjan, R.E., Confluently persistent deques via data-structural bootstrapping, *J. Algorithms*, 18, 513–547, 1995.

17. Chazelle, B., How to search in history, *Inform. Control*, 64, 77–99, 1985.

18. Chuang, T-R. and Goldberg, B., Real-time deques, multihead Turing machines, and purely functional programming, *Proceedings of the Conference of Functional Programming and Computer Architecture*, 289–298, Copenhagen, Denmark, 1992.

19. Clancy, M.J. and Knuth, D.E., A programming and problem-solving seminar, Technical Report STAN-CS-77-606, Stanford University, Stanford, CA, 1977.

20. Cole, R., Searching and storing similar lists, *J. Algorithms*, 7, 202–220, 1986.

21. Cormen, T.H., Leiserson, C.E., and Rivest, R.L., *Introduction to Algorithms*, MIT Press, Cambridge, MA, 1990.

22. Dietz, P.F., Maintaining order in a linked list, *Proceedings of the 14th Annual ACM Symposium on Theory of Computing*, 122–127, San Francisco, CA, 1982.

23. Dietz, P.F., Fully persistent arrays, *Proceedings of the Workshop on Algorithms and Data Structures (WADS '89)*, LNCS, 382, Springer-Verlag, Berlin, 67–74, 1989.

24. Dietz, P.F. and Sleator, D.D., Two algorithms for maintaining order in a list, *Proceedings of the 19th Annual ACM Symposium on Theory of Computing*, 365–372, New York, New York, 1987.

25. Dietz, P.F. and Raman, R., Persistence, amortization and randomization, *Proceedings of the 2nd Annual ACM-SIAM Symposium on Discrete Algorithms*, 77–87, San Francisco, CA, 1991.

26. Dietz, P.F. and Raman, R., Persistence, amortization and parallelization: On some combinatorial games and their applications, *Proceedings of the Workshop on Algorithms and Data Structures (WADS '93)*, LNCS, 709, Springer-Verlag, Berlin, 289–301, 1993.

27. Dietzfelbinger, M., Karlin, A., Mehlhorn, K., Meyer auf der Heide, F., Rohnhert, H., and Tarjan, R.E., Dynamic perfect hashing: Upper and lower bounds, *Proceedings of the 29th Annual IEEE Conference on the Foundations of Computer Science*, 524–531, 1998. The application to partial persistence was mentioned in the talk. A revised version of the paper appeared in *SIAM J. Comput.*, 23, 738–761, 1994.

28. Driscoll, J.R., Sarnak, N., Sleator, D.D., and Tarjan, R.E., Making data structures persistent, *J. Comput. Syst. Sci.*, 38, 86–124, 1989.

29. Driscoll, J.R., Sleator, D.D.K., and Tarjan, R.E., Fully persistent lists with catenation, *J. ACM*, 41, 943–959, 1994.

30. Dobkin, D.P. and Munro, J.I., Efficient uses of the past, *J. Algorithms*, 6, 455–465, 1985.

31. Fischer, M.J., Efficiency of equivalence algorithms, in *Complexity of Computer Computations*, R.E. Miller and J.W. Thatcher (Eds.), Plenum Press, New York, 1972, pp. 153–168.

32. Fredman, M.L. and Saks, M.E., The cell probe complexity of dynamic data structures, *Proceedings of the 21st Annual ACM Symposium on Theory of Computing*, 345–354, Seattle, WA, 1989.

33. Gabow, H.N., A scaling algorithm for weighted matching on general graphs, *Proceedings of the 26th Annual Symposium on Foundations of Computer Science,* 90–100, Portland, OR, 1985.

34. Gabow, H.N. and Tarjan, R.E., A linear time algorithm for a special case of disjoint set union, *J. Comput. Syst. Sci.,* 30, 209–221, 1985.

35. Gajewska, H. and Tarjan, R.E., Deques with heap order, *Inform. Process. Lett.,* 22, 197–200, 1986.

36. Galil, Z. and Italiano, G.F., A note on set union with arbitrary deunions, *Inform. Process. Lett.,* 37, 331–335, 1991.

37. Galil, Z. and Italiano, G.F., Data structures and algorithms for disjoint set union problems, *ACM Comput. Surv.,* 23, 319–344, 1991.

38. Galler, B.A. and Fischer, M., An improved equivalence algorithm, *Commun. ACM,* 7, 301–303, 1964.

39. Gambosi, G., Italiano, G.F., and Talamo, M., Worst-case analysis of the set union problem with extended backtracking, *Theor. Comput. Sci.,* 68, 57–70, 1989.

40. Hogger, G.J., *Introduction to Logic Programming,* Academic Press, London, U.K., 1984.

41. Italiano, G.F. and Sarnak, N., Fully persistent data structures for disjoint set union problems, *Proceedings of the Workshop on Algorithms and Data Structures (WADS '91), LNCS,* 519, Springer-Verlag, Berlin, 449–460, 1991.

42. Hood, R., The Efficient Implementation of Very-High-Level Programming Language Constructs, PhD thesis, Cornell University, Ithaca, NY, 1982.

43. Hood, R. and Melville, R., Real-time operations in pure Lisp, *Inform. Process. Lett.,* 13, 50–53, 1981.

44. Hopcroft, J.E. and Karp, R.M., An algorithm for testing the equivalence of finite automata, TR-71-114, Department of Computer Science, Cornell University, Ithaca, NY, 1971.

45. Hopcroft, J.E. and Ullman, J.D., Set merging algorithms, *SIAM J. Comput.,* 2, 294–303, 1973.

46. Hudak, P., Jones, S.P., Wadler, P., Boutel, B., Fairbairn, J., Fasel, J., Guzman, M.M., Hammond, K., Hughes, J., Johnsson, T., Kieburtz, D., Nikhil, R., Partain, W., and Peterson, J., Report on the functional programming language Haskell, version 1.2, *SIGPLAN Notices,* 27, 1992.

47. Huet, G., Resolutions d'equations dans les langages d'ordre $1, 2, \ldots \omega$, PhD dissertation, Univ. de Paris VII, France, 1976.

48. Hunt, J.W. and Szymanski, T.G., A fast algorithm for computing longest common subsequences, *Commun. Assoc. Comput. Mach.,* 20, 350–353, 1977.

49. Imai, T. and Asano, T., Dynamic segment intersection with applications, *J. Algorithms,* 8, 1–18, 1987.

50. Kaplan, H. and Tarjan, R.E., Persistent lists with catenation via recursive slow-down, *Proceedings of the 27th Annual ACM Symposium on the Theory of Computing,* 93–102, Las Vegas, NV, 1995.

51. Kaplan, H. Tarjan, R.E., Purely functional representations of catenable sorted lists, *Proceedings of the 28th Annual ACM Symposium on the Theory of Computing,* 202–211, Philadelphia, PA, 1996.

52. Karlsson, R.G., Algorithms in a restricted universe, Technical Report CS-84-50, Department of Computer Science, University of Waterloo, Waterloo, ON, 1984.

53. Kerschenbaum, A. and van Slyke, R., Computing minimum spanning trees efficiently, *Proceedings of the 25th Annual Conference of the ACM,* 518–527, San Antonio, TX, 1972.

54. Knuth, D.E., *The Art of Computer Programming,* Vol. 1: *Fundamental Algorithms.* Addison-Wesley, Reading, MA, 1968.

55. Kolmogorv, A.N., On the notion of algorithm, *Uspehi Mat. Nauk.,* 8, 175–176, 1953.

56. La Poutré, J.A., Lower bounds for the union–find and the split–find problem on pointer machines, *Proceedings of the 22nd Annual ACM Symposium on Theory of Computing,* 34–44, Baltimore, MD, 1990.

57. Levcopolous, C. and Overmars, M.H., A balanced search tree with $O(1)$ worst-case update time, *Acta Inform.,* 26, 269–278, 1988.

58. Loebl, M. and Nešetřil, J., Linearity and unprovability of set union problem strategies, *Proceedings of the 20th Annual ACM Symposium on Theory of Computing,* 360–366, Chicago, IL, 1988.

59. Mannila, H. and Ukkonen, E., The set union problem with backtracking, *Proceedings of the 13th International Colloquium on Automata, Languages and Programming (ICALP 86), LNCS,* 226, Springer-Verlag, Berlin, 236–243, 1986.

60. Mannila, M. and Ukkonen, E., On the complexity of unification sequences, *Proceedings of the 3rd International Conference on Logic Programming, LNCS,* 225, Springer-Verlag, Berlin, 122–133, 1986.
61. Mannila, H. and Ukkonen, E., Timestamped term representation for implementing Prolog, *Proceedings of the 3rd IEEE Conference on Logic Programming,* 159–167, Salt Lake City, Utah, 1986.
62. Mannila, H. and Ukkonen, E., Space-time optimal algorithms for the set union problem with backtracking. Technical Report C-1987-80, Department of Computer Science, University of Helsinki, Finland, 1987.
63. Mannila, H. and Ukkonen, E., Time parameter and arbitrary deunions in the set union problem, *Proceedings of the First Scandinavian Workshop on Algorithm Theory* (SWAT 88), *LNCS,* 318, Springer-Verlag, Berlin, 34–42, 1988.
64. McCarthy, J., Recursive functions of symbolic expressions and their computation by machine, *Commun. ACM,* 7, 184–195, 1960.
65. Mehlhorn, K., *Data Structures and Algorithms,* Vol. 1: *Sorting and Searching,* Springer-Verlag, Berlin, 1984.
66. Mehlhorn, K., *Data Structures and Algorithms,* Vol. 2: *Graph Algorithms and NP-Completeness,* Springer-Verlag, Berlin, 1984.
67. Mehlhorn, K., *Data Structures and Algorithms,* Vol. 3: *Multidimensional Searching and Computational Geometry,* Springer-Verlag, Berlin, 1984.
68. Mehlhorn, K. and Näher, S., Dynamic fractional cascading, *Algorithmica,* 5, 215–241, 1990.
69. Mehlhorn, K., Näher, S., and Alt, H., A lower bound for the complexity of the union–split–find problem, *SIAM J. Comput.,* 17, 1093–1102, 1990.
70. Milner, R., Tofte, M., and Harper, R., *The Definition of Standard ML,* MIT Press, Cambridge, MA, 1990.
71. Okasaki, C., Amortization, lazy evaluation, and persistence: Lists with catenation via lazy linking, *Proceedings of the 36th Annual Symposium on Foundations of Computer Science,* 646–654, Milwaukee, WI, 1995.
72. Okasaki, C., Functional data structures, in *Advanced Functional Programming, LNCS,* 1129, Springer-Verlag, Berlin, 67–74, 1996.
73. Okasaki, C., The role of lazy evaluation in amortized data structures, *Proceedings of the 1996 ACM SIGPLAN International Conference on Functional Programming,* 62–72, Philadelphia, PA, 1996.
74. Reps, T., Titelbaum, T., and Demers, A., Incremental context-dependent analysis for language-based editors, *ACM Trans. Program. Lang. Syst.,* 5, 449–477, 1983.
75. Sarnak, N., Persistent Data Structures, PhD thesis, Department of Computer Science, New York University, New York, 1986.
76. Sarnak, N. and Tarjan, R.E., Planar point location using persistent search trees, *Commun. ACM,* 28, 669–679, 1986.
77. Schnage, A., Storage modification machines, *SIAM J. Comput.,* 9, 490–508, 1980.
78. Stearns, R.E. and Lewis, P.M., Property grammars and table machines, *Inform. Control,* 14, 524–549, 1969.
79. Stearns, R.E. and Rosenkrantz, P.M., Table machine simulation, *Conference Record IEEE 10th Annual Symposium on Switching and Automata Theory,* 118–128, Waterloo, Ontario, Canada, 1969.
80. Steele Jr., G.L., *Common Lisp: The Language,* Digital Press, Bedford, MA, 1984.
81. Swart, G.F., Efficient algorithms for computing geometric intersections, Technical Report 85-01-02, Department of Computer Science, University of Washington, Seattle, WA, 1985.
82. Tarjan, R.E., Testing flow graph reducibility, *Proceedings of the 5th Annual ACM Symposium on Theory of Computing,* 96–107, Austin, TX, 1973.
83. Tarjan, R.E., Finding dominators in directed graphs, *SIAM J. Comput.,* 3, 62–89, 1974.
84. Tarjan, R.E., Efficiency of a good but not linear set union algorithm, *J. Assoc. Comput. Mach.,* 22, 215–225, 1975.

85. Tarjan, R.E., A class of algorithms which require nonlinear time to maintain disjoint sets, *J. Comput. Syst. Sci.,* 18, 110–127, 1979.

86. Tarjan, R.E., Application of path compression on balanced trees, *J. Assoc. Comput. Mach.,* 26, 690–715, 1979.

87. Tarjan, R.E., *Data Structures and Network Algorithms,* SIAM, Philadelphia, PA, 1983.

88. Tarjan, R.E., Amortized computational complexity, *SIAM J. Alg. Discr. Meth.,* 6, 306–318, 1985.

89. Tarjan, R.E. and van Leeuwen, J., Worst-case analysis of set union algorithms, *J. Assoc. Comput. Mach.,* 31, 245–281, 1984.

90. Tsakalidis, A.K., Maintaining order in a generalized linked list, *Acta Inform.,* 21, 101–112, 1984.

91. van Emde Boas, P., Preserving order in a forest in less than logarithmic time and linear space, *Inform. Process. Lett.,* 6, 80–82, 1977.

92. van Emde Boas, P., Kaas, R., and Zijlstra, E., Design and implementation of an efficient priority queue, *Math. Syst. Theory,* 10, 99–127, 1977.

93. van Leeuwen, J. and van der Weide, T., Alternative path compression techniques, Technical Report RUU-CS-77-3, Department of Computer Science, University of Utrecht, Utrecht, the Netherlands, 1977.

94. van der Weide, T., *Data Structures: An Axiomatic Approach and the Use of Binomial Trees in Developing and Analyzing Algorithms,* Mathematisch Centrum, Amsterdam, the Netherlands, 1980.

95. Vitter, J.S. and Simons, R.A., New classes for parallel complexity: A study of unification and other complete problems for P, *IEEE Trans. Comput.,* C-35, 403–418, 1989.

96. Warren, D.H.D. and Pereira, L.M., Prolog—the language and its implementation compared with LISP, *ACM SIGPLAN Notices,* 12, 109–115, 1977.

97. Westbrook, J. and Tarjan, R.E., Amortized analysis of algorithms for set union with backtracking, *SIAM J. Comput.,* 18, 1–11, 1989.

98. Westbrook, J. and Tarjan, R.E., Maintaining bridge-connected and biconnected components on-line, *Algorithmica,* 7, 433–464, 1992.

99. Yao, A.C., Should tables be sorted? *J. Assoc. Comput. Mach.,* 28, 615–628, 1981.

6

Multidimensional Data Structures for Spatial Applications*

Hanan Samet
University of Maryland

An overview is presented of a number of representations of multidimensional data that arise in spatial applications. Multidimensional spatial data consists of points as well as objects that have extent such as line segments, rectangles, regions, and volumes. The points may have locational as well as nonlocational attributes. The focus is on spatial data which is a subset of multidimensional data consisting of points with locational attributes and objects with extent. The emphasis is on hierarchical representations based on the "divide-and-conquer" problem-solving paradigm. They are of interest because they enable focusing computational resources on the interesting subsets of data. Thus, there is no need to expend work where the payoff is small. These representations are of use in operations such as range searching and finding nearest neighbors.

6.1 Introduction

The representation of multidimensional spatial data is an important issue in applications in diverse fields that include database management systems, computer graphics, computer vision, computational geometry, image processing, geographic information systems (GIS), pattern recognition, very large scale integrated (VLSI) design, and others. The most common definition of multidimensional data is a collection of points in a higher-dimensional space. These points can represent locations and objects in space as well as more general records. As an example of a record, consider an employee record which has attributes corresponding to the employee's name, address, sex, age, height, weight, and social security number. Such records arise in database management systems and can be treated

---

as points in, for this example, a seven-dimensional space (i.e., there is one dimension for each attribute) albeit the different dimensions have different type units (i.e., name and address are strings of characters, sex is binary; while age, height, weight, and social security number are numbers).

When multidimensional data corresponds to locational data, we have the additional property that all of the attributes have the same unit, which is distance in space. In this case, we can combine the attributes and pose queries that involve proximity. For example, we may wish to find the closest city to Chicago within the two-dimensional space from which the locations of the cities are drawn. Another query seeks to find all cities within 50 mi of Chicago. In contrast, such queries are not very meaningful when the attributes do not have the same type. For example, it is not customary to seek the person with age–weight combination closest to John Jones as we do not have a commonly accepted unit of year–pounds (year–kilograms) or definition thereof. It should be clear that we are not speaking of queries involving boolean combinations of the different attributes (e.g., range queries), which are quite common.

When multidimensional data spans a continuous physical space (i.e., an infinite collection of locations), the issues become more interesting. In particular, we are no longer just interested in the locations of objects, but, in addition, we are also interested in the space that they occupy (i.e., their extent). Some example objects include line segments (e.g., roads, rivers), regions (e.g., lakes, counties, buildings, crop maps, polygons, polyhedra), rectangles, and surfaces. The objects may be disjoint or could even overlap. One way to deal with such data is to store it explicitly by parametrizing it and thereby reducing it to a point in a higher-dimensional space. For example, a line segment in two-dimensional space can be represented by the coordinate values of its endpoints (i.e., a pair of x- and a pair of y-coordinate values), and then stored as a point in a four-dimensional space. Thus, in effect, we have constructed a transformation (i.e., mapping) from a two-dimensional space (i.e., the space from which the line segments are drawn) to a four-dimensional space (i.e., the space containing the representative point corresponding to the line segment).

The transformation approach is fine, if we are just interested in retrieving the data. In particular, it is appropriate for queries about the objects (e.g., determining all line segments that pass through a given point, or that share an endpoint, etc.) and the immediate space that they occupy. However, the drawback of the transformation approach is that it ignores the geometry inherent in the data (e.g., the fact that a line segment passes through a particular region) and its relationship to the space in which it is embedded.

For example, suppose that we want to detect if two line segments are near each other, or, alternatively, to find the nearest line segment to a given line segment.* This is difficult to do in the four-dimensional space, regardless of how the data in it are organized, since proximity in the two-dimensional space from which the line segments are drawn is not necessarily preserved in the four-dimensional space. In other words, although the two line segments may be very close to each other, the Euclidean distance between their representative points may be quite large. This is especially true if there is a great difference in the relative size of the two objects (e.g., a short line segment in

FIGURE 6.1 Example of two objects that are close to each other in the original space but are not clustered in the same region of the transformed space when using a transformation such as the corner transformation.

---

* See [59] for a discussion of different types of queries on databases consisting of collections of line segments.

proximity to a long line segment as in Figure 6.1). On the other hand, when the objects are small (e.g., their extent is small), the method works reasonably well, as the objects are basically point objects (e.g., [126]). For example, suppose that the objects are one-dimensional intervals of the form $[x_1, x_2)$. In this case, the transformation yields a point (x_1, x_2) and it is easy to see that small intervals where $[x_1, x_2) = [x_i, x_j)$ such that $x_i \approx x_j$ implies that the points $[x_i, x_j)$ straddle the line segment $x_1 = x_2$. Unfortunately, we cannot expect that the underlying data will always satisfy this property.

Of course, we could overcome these problems by projecting the line segments back to the original space from which they were drawn, but in such a case, we may ask what was the point of using the transformation in the first place? In other words, at the least, the representation that we choose for the data should allow us to perform operations on the data. Thus, we need special representations for spatial multidimensional data other than point representations. One solution is to use data structures that are based on spatial occupancy.

Spatial occupancy methods decompose the space from which the spatial data is drawn (e.g., the two-dimensional space containing the line segments) into regions called buckets (i.e., bins). They are also commonly known as *bucketing methods*. Traditional bucketing methods such as the grid file [82], BANG file [44], LSD trees [54], buddy trees [117], etc. have been designed for multidimensional point data that need not be locational. In the case of spatial data, these methods have been usually applied to the transformed data (i.e., the representative points). In contrast, we discuss their application to the actual objects in the space from which the objects are drawn (i.e., two dimensions in the case of a collection of line segments).

The bucketing work is rooted in Warnock's hidden-line [136] and hidden-surface [137] algorithms in computer graphics applications that repeatedly subdivide the picture area into successively smaller blocks while simultaneously searching it for areas that are sufficiently simple to be displayed. In this case, the objects in the picture area are also often subdivided as well. It should be clear that the determination of what part of the picture area is hidden or not is equivalent to sorting the picture area with respect to the position of the viewer. Note that the subdivision of the picture area is also equivalent to sorting the spatial data by spatial occupancy. The important point to observe here is that sorting is not explicit in the sense that we do not obtain an absolute ordering. Instead, sorting is implicit in that we only know the ordering of the objects in terms of similarity. The result is very much akin to sorting screws in a hardware store into buckets (i.e., bins) according to size. In other words, sorting is a means of differentiating the objects rather than requiring that they be ordered.

In this chapter, we explore a number of different representations of multidimensional data bearing the above issues in mind. In the case of point data, we examine representations of both locational and nonlocational data, as well as combinations of the two. While we cannot give exhaustive details of all of the data structures, we try to explain the intuition behind their development as well as to give literature pointers to where more information can be found. Many of these representations are described in greater detail in [103,104,106] including an extensive bibliography. Our approach is primarily a descriptive one. Most of our examples are of two-dimensional spatial data although we do touch briefly on three-dimensional data.

At times, we discuss bounds on execution time and space requirements. However, this information is presented in an inconsistent manner. The problem is that such analyses are very difficult to perform for many of the data structures that we present. This is especially true for the data structures that are based on spatial occupancy (e.g., quadtree and R-tree variants). In particular, such methods have good observable average-case behavior but may have very bad worst cases which rarely arise in practice. Their analysis is beyond the scope of this chapter, and usually we do not say anything about it. Nevertheless, these representations find frequent use in applications where their behavior is deemed acceptable, and are often found to be better than that of solutions whose theoretical behavior would appear to be superior. The problem is primarily attributed to the presence of large constant factors which are usually ignored in the big O and Ω analyses [70].

The rest of this chapter is organized as follows. Section 6.2 reviews a number of representations of point data of arbitrary dimensionality. Section 6.3 describes bucketing methods that organize collections of spatial objects (as well as multidimensional point data) by aggregating their bounding boxes. Sections 6.2 and 6.3 are applicable to both spatial and nonspatial data, although all the examples that we present are of spatial data. Section 6.4 focuses on representations of region data, while Section 6.5 discusses a subcase of region data which consists of collections of rectangles. Section 6.6 deals with curvilinear data which also includes polygonal subdivisions and collections of line segments. Section 6.7 contains a summary and a brief indication of some research issues. The "Defining Terms" section reviews some of the definitions of the terms used in this chapter. Note that although our examples are primarily drawn from a two-dimensional space, the representations are applicable to higher-dimensional spaces as well.

6.2   Point Data

Our discussion assumes that there is one record per data point, and that each record contains several attributes or keys (also frequently called fields, dimensions, coordinates, and axes). In order to facilitate retrieval of a record based on some of its attribute values, we also assume the existence of an ordering for the range of values of each of these attributes. In the case of locational attributes, such an ordering is quite obvious as the values of these attributes are numbers. In the case of alphanumeric attributes, the ordering is usually based on the alphabetic sequence of the characters making up the attribute value. Other data such as color could be ordered by the characters making up the name of the color or possibly the color's wavelength. It should be clear that finding an ordering for the range of values of an attribute is generally not an issue; the real issue is what ordering to use!

The representation that is ultimately chosen for the data depends, in part, on answers to the following questions:

1. What operations are to be performed on the data?
2. Should we organize the data or the embedding space from which the data is drawn?
3. Is the database static or dynamic (i.e., can the number of data points grow and shrink at will)?
4. Can we assume that the volume of data is sufficiently small, so that it can all fit in core, or should we make provisions for accessing disk-resident data?

Disk-resident data implies grouping the data (either the underlying space based on the volume—that is, the amount—of the data it contains or the points, hopefully, by the proximity of their values) into sets (termed buckets) corresponding to physical storage units (i.e., pages). This leads to questions about their size, and how they are to be accessed:

1. Do we require a constant time to retrieve a record from a file or is a logarithmic function of the number of records in the file adequate? This is equivalent to asking if the access is via a directory in the form of an array (i.e., direct access) or a tree?
2. How large can the directories be allowed to grow before it is better to rebuild them?
3. How should the buckets be laid out on the disk?

Clearly, these questions are complex, and we cannot address them all here. Some are answered in other sections. In this section, we focus primarily on dynamic data with an emphasis on two dimensions (i.e., attributes) and concentrate on the following queries:

1. Point queries—that is, if a particular point is present.
2. Range queries.
3. Boolean combinations of 1 and 2.

Most of the representations that we describe can be extended easily to higher dimensions, although some, like the priority search tree, are primarily designed for two-dimensional point data. Our discussion and examples are based on the fact that all of the attributes are locational or numeric and that they have the same range, although all of the representations can also be used to handle nonlocational and nonnumeric attributes. When discussing behavior in the general case, we assume a data set of N points and d attributes.

The simplest way to store point data is in a sequential list. Accesses to the list can be sped up by forming sorted lists for the various attributes which are known as inverted lists (e.g., [71]). There is one list for each attribute. This enables pruning the search with respect to the value of one of the attributes. In order to facilitate random access, the lists can be implemented using range trees [15,17].

It should be clear that the inverted list is not particularly useful for range searches. The problem is that it can only speed up the search for one of the attributes (termed the primary attribute). A number of solutions have been proposed. These solutions can be decomposed into two classes. One class of solutions enhances the range tree corresponding to the inverted list to include information about the remaining attributes in its internal nodes. This is the basis of the multidimensional range tree and the variants of the priority search tree [34,76] which are discussed at the end of this section.

The second class of solutions is more widely used and is exemplified by the fixed-grid method [16,71], which is also the standard indexing method for maps. It partitions the space from which the data are drawn into rectangular cells by overlaying it with a grid. Each grid cell c contains a pointer to another structure (e.g., a list) which contains the set of points that lie in c. Associated with the grid is an access structure to enable the determination of the grid cell associated with a particular point p. This access structure acts like a directory and is usually in the form of a d-dimensional array with one entry per grid cell or a tree with one leaf node per grid cell.

There are two ways to build a fixed grid. We can either subdivide the space into equal-sized intervals along each of the attributes (resulting in congruent grid cells) or place the subdivision lines at arbitrary positions that are dependent on the underlying data. In essence, the distinction is between organizing the data to be stored and organizing the embedding space from which the data are drawn [82]. In particular, when the grid cells are congruent (i.e., equal-sized when all of the attributes are locational with the same range and termed a uniform grid), use of an array access structure is quite simple and has the desirable property that the grid cell associated with point p can be determined in constant time. Moreover, in this case, if the width of each grid cell is twice the search radius for a rectangular range query, then the average search time is $O(F \cdot 2^d)$ where F is the number of points that have been found that satisfy the query [18]. Figure 6.2 is an example of a uniform-grid representation for a search radius equal to 10 (i.e., a square of size 20 × 20).*

Use of an array access structure when the grid cells are not congruent requires us to have a way of keeping track of their size, so that we can determine the entry of the array access structure corresponding to the grid cell associated with point p. One way to do this is to make use of what are termed linear scales which indicate the positions of the grid lines (or partitioning hyperplanes in $d > 2$ dimensions). Given a point p, we determine the grid cell in which p lies by finding the "coordinate values" of the appropriate grid cell. The linear scales are usually implemented as one-dimensional trees containing ranges of values.

The use of an array access structure is fine as long as the data are static. When the data are dynamic, it is likely that some of the grid cells become too full while other grid cells are empty. This means that we need to rebuild the grid (i.e., further partition the grid or reposition the grid partition lines or

---

* Note that although the data has three attributes, one of which is nonlocational (i.e., name) and two of which are locational (i.e., the coordinate values), retrieval is only on the basis of the locational attribute values. Thus, there is no ordering on the name, and, therefore, we treat this example as two-dimensional locational point data.

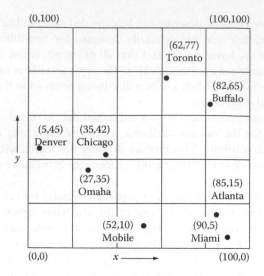

FIGURE 6.2 Uniform-grid representation corresponding to a set of points with a search radius of 20.

hyperplanes), so that the various grid cells are not too full. However, this creates many more empty grid cells as a result of repartitioning the grid (i.e., empty grid cells are split into more empty grid cells). In this case, we have two alternatives. The first is to assign an ordering to all the grid cells and impose a tree access structure on the elements of the ordering that correspond to nonempty grid cells. The effect of this alternative is analogous to using a mapping from d dimensions to one dimension and then applying one of the one-dimensional access structures such as a B-tree, balanced binary tree, etc. to the result of the mapping. There are a number of possible mappings including row, Morton (i.e., bit interleaving or bit interlacing), and Peano-Hilbert* (for more details, see [106] for example, as well as Figure 6.14 and the accompanying discussion in Section 6.4). This mapping alternative is applicable regardless of whether or not the grid cells are congruent. Of course, if the grid cells are not congruent, then we must also record their size in the element of the access structure.

The second alternative is to merge spatially adjacent empty grid cells into larger empty grid cells, while splitting grid cells that are too full, thereby making the grid adaptive. Again, the result is that we can no longer make use of an array access structure to retrieve the grid cell that contains query point p. Instead, we make use of a tree access structure in the form of a k-ary tree where k is usually 2^d. Thus, what we have done is marry a k-ary tree with the fixed-grid method. This is the basis of the point quadtree [37] and the PR quadtree [84,104,106] which are multidimensional generalizations of binary trees.

The difference between the point quadtree and the PR quadtree is the same as the difference between "trees" and "tries" [42], respectively. The binary search tree [71] is an example of the former, since the boundaries of different regions in the search space are determined by the data being stored. Address computation methods such as radix searching [71] (also known as digital searching) are examples of the latter, since region boundaries are chosen from among locations that are fixed regardless of the content of the data set. The process is usually a recursive halving process in one dimension, recursive quartering in two dimensions, etc., and is known as *regular decomposition*.

In two dimensions, a point quadtree is just a two-dimensional binary search tree. The first point that is inserted serves as the root, while the second point is inserted into the relevant quadrant of the

---

* These mappings have been investigated primarily for purely multidimensional locational point data. They cannot be applied directly to the key values for nonlocational point data.

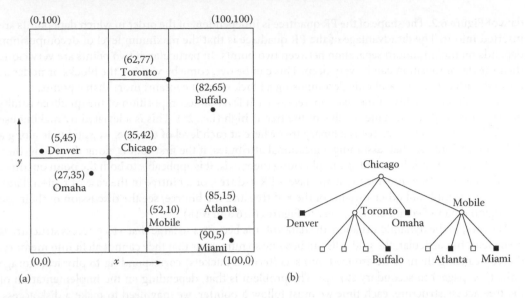

FIGURE 6.3 A point quadtree and the records it represents corresponding to Figure 6.2: (a) the resulting partition of space and (b) the tree representation.

tree rooted at the first point. Clearly, the shape of the tree depends on the order in which the points were inserted. For example, Figure 6.3 is the point quadtree corresponding to the data of Figure 6.2 inserted in the order Chicago, Mobile, Toronto, Buffalo, Denver, Omaha, Atlanta, and Miami.

In two dimensions, the PR quadtree is based on a recursive decomposition of the underlying space into four congruent (usually square in the case of locational attributes) cells until each cell contains no more than one point. For example, Figure 6.4 is the PR quadtree corresponding to the

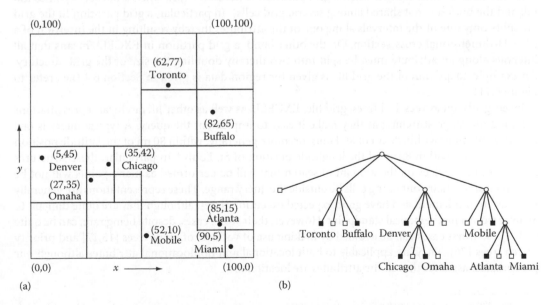

FIGURE 6.4 A PR quadtree and the records it represents corresponding to Figure 6.2: (a) the resulting partition of space, and (b) the tree representation.

data of Figure 6.2. The shape of the PR quadtree is independent of the order in which data points are inserted into it. The disadvantage of the PR quadtree is that the maximum level of decomposition depends on the minimum separation between two points. In particular, if two points are very close, then the decomposition can be very deep. This can be overcome by viewing the blocks or nodes as buckets with capacity c, and only decomposing a block when it contains more than c points.

As the dimensionality of the space increases, each level of decomposition of the quadtree results in many new cells as the fanout value of the tree is high (i.e., 2^d). This is alleviated by making use of a *k-d tree* [13]. The k-d tree is a binary tree where at each level of the tree, we subdivide along a different attribute so that, assuming d locational attributes, if the first split is along the x-axis, then after d levels, we cycle back and again split along the x-axis. It is applicable to both the point quadtree and the PR quadtree (in which case we have a PR k-d tree, or a bintree in the case of region data). For some examples of other variants of the k-d tree and the bintree, see the discussion of their use for region data in Section 6.4 (i.e., refer to Figures 6.16 and 6.18).

At times, in the dynamic situation, the data volume becomes so large that a tree access structure is inefficient. In particular, the grid cells can become so numerous that they cannot all fit into memory, thereby causing them to be grouped into sets (termed buckets) corresponding to physical storage units (i.e., pages) in secondary storage. The problem is that, depending on the implementation of the tree access structure, each time we must follow a pointer, we may need to make a disk access. This has led to a return to the use of an array access structure. The difference from the array used with the static fixed-grid method described earlier is that now the array access structure (termed grid directory) may be so large (e.g., when d gets large) that it resides on the disk as well, and the fact that the structure of the grid directory can be changed as the data volume grows or contracts. Each grid cell (i.e., an element of the grid directory) contains the address of a bucket (i.e., page) that contains the points associated with the grid cell. Notice that a bucket can correspond to more than one grid cell. Thus, any page can be accessed by two disk operations: one to access the grid cell and one more to access the actual bucket.

This results in EXCELL [128] when the grid cells are congruent (i.e., equal-sized for locational point data), and grid file [82] when the grid cells need not be congruent. The difference between these methods is most evident when a grid partition is necessary (i.e., when a bucket becomes too full and the bucket is not shared among several grid cells). In particular, a grid partition in the grid file splits only one of the intervals along one of the attributes thereby resulting in the insertion of a $(d - 1)$-dimensional cross-section. On the other hand, a grid partition in EXCELL means that all intervals along an attribute must be split into two thereby doubling the size of the grid directory. An example adaptation of the grid file is given for region data is given in Section 6.4 (i.e., refer to Figure 6.11).

Fixed-grids, quadtrees, k-d trees, grid file, EXCELL, as well as other hierarchical representations are good for range searching as they make it easy to implement the query. A typical query is one that seeks all cities within 80 mi of St. Louis, or, more generally, within 80 mi of the latitude position of St. Louis and within 80 mi of the longitude position of St. Louis.* In particular, these structures act as pruning devices on the amount of search that will be performed as many points will not be examined since their containing cells lie outside the query range. These representations are generally very easy to implement and have good expected execution times, although they are quite difficult to analyze from a mathematical standpoint. However, their worst cases, despite being rare, can be quite bad. These worst cases can be avoided by making use of variants of range trees [15,17] and priority search trees [76]. They are applicable to both locational and nonlocational attributes although our presentation assumes that all the attributes are locational.

---

* The difference between these two formulations of the query is that the former admits a circular search region, while the latter admits a rectangular search region. In particular, the latter formulation is applicable to both locational and nonlocational attributes, while the former is only applicable to locational attributes.

A one-dimensional range tree is a balanced binary search tree where the data points are stored in the leaf nodes and the leaf nodes are linked in sorted order by use of a doubly linked list. A range search for $[L : R]$ is performed by searching the tree for the node with the smallest value that is $\geq L$, and then following the links until reaching a leaf node with a value greater than R. For N points, this process takes $O(\log_2 N + F)$ time and uses $O(N)$ storage. F is the number of points found that satisfy the query.

A two-dimensional range tree is a binary tree of binary trees. It is formed in the following manner. First, sort all of the points along one of the attributes, say x, and store them in the leaf nodes of a balanced binary search tree, say T. With each nonleaf node of T, say I, associate a one-dimensional range tree, say T_I, of the points in the subtree rooted at I where now these points are sorted along the other attribute, say y. The range tree can be adapted easily to handle d-dimensional point data. In such a case, for N points, a d-dimensional range search takes $O(\log_2^d N + F)$ time, where F is the number of points found that satisfy the query. The d-dimensional range tree uses $O(N \cdot \log_2^{d-1} N)$ storage.

The priority search tree is a related data structure that is designed for solving queries involving semi-infinite ranges in two-dimensional space. A typical query has a range of the form $([L_x : R_x], [L_y : \infty])$. For example, Figure 6.5 is the priority search tree for the data of Figure 6.2. It is built as follows. Assume that no two data points have the same x-coordinate value. Sort all the points along the x-coordinate value and store them in the leaf nodes of a balanced binary search tree (a range tree in our formulation), say T. We proceed from the root node toward the leaf nodes. With each node I of T, associate the point in the subtree rooted at I with the maximum value for its y-coordinate that has not already been stored at a shallower depth in the tree. If such a point does not exist, then leave the node empty. For N points, this structure uses $O(N)$ storage.

It is not easy to perform a two-dimensional range query of the form $([L_x : R_x], [L_y : R_y])$ with a priority search tree. The problem is that only the values of the x-coordinates are sorted. In other words, given a leaf node C that stores the point (x_C, y_C), we know that the values of the x-coordinates

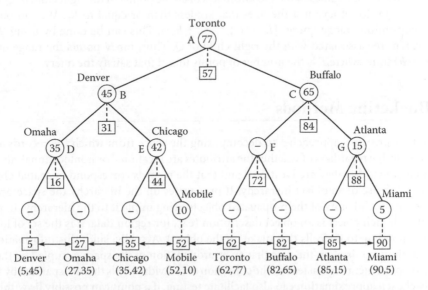

FIGURE 6.5 Priority search tree for the data of Figure 6.2. Each leaf node contains the value of its x-coordinate in a square box. Each nonleaf node contains the appropriate x-coordinate midrange value in a box using a link drawn with a broken line. Circular boxes indicate the value of the y-coordinate of the point in the corresponding subtree with the maximum value for its y-coordinate that has not already been associated with a node at a shallower depth in the tree.

of all nodes to the left of C are smaller than x_C and the values of all those to the right of C are greater than x_C. On the other hand, with respect to the values of the y-coordinates, we only know that all nodes below the nonleaf node D with value y_D have values less than or equal to y_D; the y-coordinate values associated with the remaining nodes in the tree that are not ancestors of D may be larger or smaller than y_D. This is not surprising because a priority search tree is really a variant of a range tree in x and a heap (i.e., priority queue) [71] in y.

A heap enables finding the maximum (minimum) value in $O(1)$ time. Thus, it is easy to perform a semi-infinite range query of the form $([L_x : R_x], [L_y : \infty])$ as all we need to do is descend the priority search tree and stop as soon as we encounter a y-coordinate value that is less than L_y. For N points, performing a semi-infinite range query in this way takes $O(\log_2 N + F)$ time, where F is the number of points found that satisfy the query.

The priority search tree is used as the basis of the range priority tree [34] to reduce the order of execution time of a two-dimensional range query to $O(\log_2 N + F)$ time (but still using $O(N \cdot \log_2 N)$ storage). Define an inverse priority search tree to be a priority search tree S such that with each node of S, say I, we associate the point in the subtree rooted at I with the minimum (instead of the maximum!) value for its y-coordinate that has not already been stored at a shallower depth in the tree. The range priority tree is a balanced binary search tree (i.e., a range tree), say T, where all the data points are stored in the leaf nodes and are sorted by their y-coordinate values. With each nonleaf node of T, say I, which is a left child of its parent, we store a priority search tree of the points in the subtree rooted at I. With each nonleaf node of T, say I, which is a right child of its parent, we store an inverse priority search tree of the points in the subtree rooted at I. For N points, the range priority tree uses $O(N \cdot \log_2 N)$ storage.

Performing a range query for $([L_x : R_x], [L_y : R_y])$ using a range priority tree is done in the following manner. We descend the tree looking for the nearest common ancestor of L_y and R_y, say Q. The values of the y-coordinates of all the points in the left child of Q are less than R_y. We want to retrieve just the ones that are greater than or equal to L_y. We can obtain them with the semi-infinite range query $([L_x : R_x], [L_y : \infty])$. This can be done by using the priority tree associated with the left child of Q. Similarly, the values of the y-coordinates of all the points in the right child of Q are greater than L_y. We want to retrieve just the ones that are less than or equal to R_y. We can obtain them with the semi-infinite range query $([L_x : R_x], [-\infty : R_y])$. This can be done by using the inverse priority search tree associated with the right child of Q. Thus, for N points the range query takes $O(\log_2 N + F)$ time, where F is the number of points found that satisfy the query.

6.3   Bucketing Methods

There are four principal approaches to decomposing the space from which the records are drawn. They are applicable regardless of whether the attributes are locational or nonlocational, although our discussion assumes that they are locational and that the records correspond to spatial objects. One approach makes use of an object hierarchy. It propagates up the hierarchy the space occupied by the objects with the identity of the propagated objects being implicit to the hierarchy. In particular, associated with each object is an object description (e.g., for region data, it is the set of locations in space corresponding to the cells that make up the object). Actually, since this information may be rather voluminous, it is often the case that an approximation of the space occupied by the object is propagated up the hierarchy instead of the collection of individual cells that are spanned by the object. A suitably chosen approximation can also facilitate testing, if a point can possibly lie within the area spanned by the object or group of objects. A negative answer means that no further processing is required for the object or group, while a positive answer means that further tests must be performed. Thus, the approximation serves to avoid wasting work. Equivalently, it serves to differentiate (i.e., "sort") between occupied and unoccupied space.

There are many types of approximations. The simplest is usually the minimum axis-aligned bounding box (AABB) such as the R-tree [12,52], as well as the more general oriented bounding box (OBB) where the sides are orthogonal, while no longer having to be parallel to the coordinate axes (e.g., [48,87]). In both of these cases, the boxes are hyperrectangles. In addition, some data structures use other shapes for the bounding boxes such as hyperspheres (e.g., SS-tree [83,138]), combinations of hyperrectangles and hyperspheres (e.g., SR-tree [66]), truncated tetrahedra (e.g., prism tree [85]), as well as triangular pyramids which are five-sided objects with two parallel triangular faces and three rectangular faces forming a three-dimensional pie slice (e.g., BOXTREE [10]). These data structures differ primarily in the properties of the bounding boxes, and their interrelationships, that they use to determine how to aggregate the bounding boxes, and, of course, the objects. It should be clear that for nonleaf elements of the hierarchy, the associated bounding box b corresponds to the union of the bounding boxes b_i associated with the elements immediately below it, subject to the constraint that b has the same shape as each of b_i.

The R-tree finds much use in database applications. The number of objects or bounding hyperrectangles that are aggregated in each node is permitted to range between $m \leq \lceil M/2 \rceil$ and M. The root node in an R-tree has at least two entries unless it is a leaf node in which case it has just one entry corresponding to the bounding hyperrectangle of an object. The R-tree is usually built as the objects are encountered rather than waiting until all the objects have been input. The hierarchy is implemented as a tree structure with grouping being based, in part, on proximity of the objects or bounding hyperrectangles.

For example, consider the collection of line segment objects given in Figure 6.6 shown embedded in a 4×4 grid. Figure 6.7a is an example R-tree for this collection with $m = 2$ and $M = 3$. Figure 6.7b

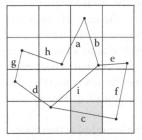

shows the spatial extent of the bounding rectangles of the nodes in Figure 6.7a, with heavy lines denoting the bounding rectangles corresponding to the leaf nodes, and broken lines denoting the bounding rectangles corresponding to the subtrees rooted at the nonleaf nodes. Note that the R-tree is not unique. Its structure depends heavily on the order in which the individual objects were inserted into (and possibly deleted from) the tree.

Given that each R-tree node can contain a varying number of objects or bounding hyperrectangles, it is not surprising that the R-tree was inspired by the B-tree [26]. Therefore, nodes are viewed as analogous to disk pages. Thus, the parameters defining the tree (i.e., m and M) are chosen so that a small number of nodes is visited during a spatial

FIGURE 6.6 Example collection of line segments embedded in a 4×4 grid.

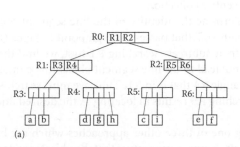

(a)

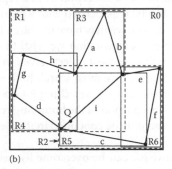

(b)

FIGURE 6.7 (a) R-tree for the collection of line segments with $m = 2$ and $M = 3$, in Figure 6.6 and (b) the spatial extents of the bounding rectangles. Notice that the leaf nodes in the index also store bounding rectangles although this is only shown for the nonleaf nodes.

query (i.e., point and range queries), which means that m and M are usually quite large. The actual implementation of the R-tree is really a B^+-tree [26] as the objects are restricted to the leaf nodes.

As long as the number of objects in each R-tree leaf node is between m and M, no action needs to be taken on the R-tree structure other than adjusting the bounding hyperrectangles when inserting or deleting an object. If the number of objects in a leaf node decreases below m, then the node is said to underflow. In this case, the objects in the underflowing nodes must be reinserted, and bounding hyperrectangles in nonleaf nodes must be adjusted. If these nonleaf nodes also underflow, then the objects in their leaf nodes must also be reinserted. If the number of objects in a leaf node increases above M, then the node is said to overflow. In this case, it must be split and the $M + 1$ objects that it contains must be distributed in the two resulting nodes. Splits are propagated up the tree.

Underflows in an R-tree are handled in an analogous manner to the way with which they are dealt in a B-tree. In contrast, the overflow situation points out a significant difference between an R-tree and a B-tree. Recall that overflow is a result of attempting to insert an item t in node p and determining that node p is too full. In a B-tree, we usually do not have a choice as to the node p that is to contain t since the tree is ordered. Thus, once we determine that p is full, we must either split p or apply a rotation (also termed deferred splitting) process. On the other hand, in an R-tree, we can insert t in any node p, as long as p is not full. However, once t is inserted in p, we must expand the bounding hyperrectangle associated with p to include the space spanned by the bounding hyperrectangle b of t. Of course, we can also insert t in a full node p, in which case we must also split p.

The need to expand the bounding hyperrectangle of p has an effect on the future performance of the R-tree, and thus we must make a wise choice with respect to p. The efficiency of the R-tree for search operations depends on its ability to distinguish between occupied space and unoccupied space (i.e., coverage), and to prevent a node from being examined needlessly due to a false overlap with other nodes. In other words, we want to minimize both coverage and overlap. These goals guide the initial R-tree creation process as well, subject to the previously mentioned constraint that the R-tree is usually built as the objects are encountered rather than waiting until all objects have been input.

The drawback of the R-tree (and any representation based on an object hierarchy) is that it does not result in a disjoint decomposition of space. The problem is that an object is only associated with one bounding hyperrectangle (e.g., line segment c in Figure 6.7 is associated with bounding rectangle R5, yet it passes through R1, R2, R4, R5, and R6, as well as through R0 as do all the line segments). In the worst case, this means that when we wish to determine which object (e.g., an intersecting line in a collection of line segment objects, or a containing rectangle in a collection of rectangle objects) is associated with a particular point in the two-dimensional space from which the objects are drawn, we may have to search the entire collection.

For example, suppose that we wish to determine the identity of the line segment object in the collection of line segment objects given in Figure 6.7 that passes through point Q. Since Q can be in either of R1 or R2, we must search both of their subtrees. Searching R1 first, we find that Q could only be contained in R4. Searching R4 does not lead to the line segment object that contains Q even though Q is in a portion of the bounding rectangle R4 that is in R1. Thus, we must search R2 and we find that Q can only be contained in R5. Searching R5 results in locating i, the desired line segment object.

This drawback can be overcome by using one of three other approaches which are based on a decomposition of space into disjoint cells. Their common property is that the objects are decomposed into disjoint subobjects such that each of the subobjects is associated with a different cell. They differ in the degree of regularity imposed by their underlying decomposition rules, and by the way in which the cells are aggregated into buckets.

The price paid for the disjointness is that in order to determine the area covered by a particular object, we have to retrieve all the cells that it occupies. This price is also paid when we want to delete an object. Fortunately, deletion is not so common in such applications. A related costly consequence of disjointness is that when we wish to determine all the objects that occur in a particular region, we often need to retrieve some of the objects more than once. This is particularly troublesome when the result of the operation serves as input to another operation via composition of functions. For example, suppose we wish to compute the perimeter of all the objects in a given region. Clearly, each object's perimeter should only be computed once. Eliminating the duplicates is a serious issue (see [4] for a discussion of how to deal with this problem for a collection of line segment objects, and [5,30] for a collection of hyperrectangle objects).

The first method based on disjointness partitions the embedding space into disjoint subspaces, and hence the individual objects into subobjects, so that each subspace consists of disjoint subobjects. The subspaces are then aggregated and grouped into another structure, such as a B-tree, so that all subsequent groupings are disjoint at each level of the structure. The result is termed a k-d-B-tree [89]. The R^+-tree [118,127] is a modification of the k-d-B-tree where at each level we replace the subspace by the minimum bounding hyperrectangle of the subobjects or subtrees that it contains. The cell tree [50] is based on the same principle as the R^+-tree except that the collections of objects are bounded by minimum convex polyhedra instead of minimum bounding hyperrectangles.

The R^+-tree (as well as the other related representations) is motivated by a desire to avoid overlap among the bounding hyperrectangles. Each object is associated with all the bounding hyperrectangles that it intersects. All bounding hyperrectangles in the tree (with the exception of the bounding hyperrectangles for the objects at the leaf nodes) are nonoverlapping.* The result is that there may be several paths starting at the root to the same object. This may lead to an increase in the height of the tree. However, retrieval time is sped up.

Figure 6.8 is an example of one possible R^+-tree for the collection of line segments in Figure 6.6. This particular tree is of order (2,3), although, in general, it is not possible to guarantee that all nodes will always have a minimum of two entries. In particular, the expected B-tree performance guarantees are not valid (i.e., pages are not guaranteed to be m/M full) unless we are willing to perform very complicated record insertion and deletion procedures. Notice that line segment objects c, h, and i appear in two different nodes. Of course, other variants are possible since the R^+-tree is not unique.

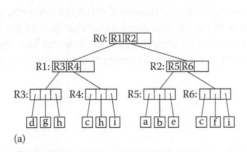
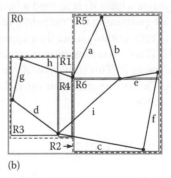

(a)

(b)

FIGURE 6.8 (a) R^+-tree for the collection of line segments in Figure 6.6 with $m = 2$ and $M = 3$, and (b) the spatial extents of the bounding rectangles. Notice that the leaf nodes in the index also store bounding rectangles although this is only shown for the nonleaf nodes.

---

* From a theoretical viewpoint, the bounding hyperrectangles for the objects at the leaf nodes should also be disjoint. However, this may be impossible (e.g., when the objects are planar line segments and if many of the line segments have the same endpoint).

Methods such as the R^+-tree (as well as the R-tree) have the drawback that the decomposition is data-dependent. This means that it is difficult to perform tasks that require composition of different operations and data sets (e.g., set-theoretic operations such as overlay). The problem is that although these methods are good at distinguishing between occupied and unoccupied space in a particular image, they are unable to correlate occupied space in two distinct images, and likewise for unoccupied space in the two images.

In contrast, the remaining two approaches to the decomposition of space into disjoint cells have a greater degree of data-independence. They are based on a regular decomposition. The space can be decomposed either into blocks of uniform size (e.g., the uniform grid [41]) or adapt the decomposition to the distribution of the data (e.g., a quadtree-based approach such as [113]). In the former case, all the blocks are congruent (e.g., the 4×4 grid in Figure 6.6). In the latter case, the widths of the blocks are restricted to be powers of two* and their positions are also restricted. Since the positions of the subdivision lines are restricted, and essentially the same for all images of the same size, it is easy to correlate occupied and unoccupied space in different images.

The uniform grid is ideal for uniformly distributed data, while quadtree-based approaches are suited for arbitrarily distributed data. In the case of uniformly distributed data, quadtree-based approaches degenerate to a uniform grid, albeit they have a higher overhead. Both the uniform grid and the quadtree-based approaches lend themselves to set-theoretic operations and thus they are ideal for tasks which require the composition of different operations and data sets (see also the discussion in Section 6.7). In general, since spatial data is not usually uniformly distributed, the quadtree-based regular decomposition approach is more flexible. The drawback of quadtree-like methods is their sensitivity to positioning in the sense that the placement of the objects relative to the decomposition lines of the space in which they are embedded effects their storage costs and the amount of decomposition that takes place. This is overcome to a large extent by using a bucketing adaptation that decomposes a block only if it contains more than b objects.

In the case of spatial data, all of the spatial occupancy methods discussed above are characterized as employing spatial indexing because with each block the only information that is stored is whether or not the block is occupied by the object or part of the object. This information is usually in the form of a pointer to a descriptor of the object. For example, in the case of a collection of line segment objects in the uniform grid of Figure 6.6, the shaded block only records the fact that a line segment (i.e., c) crosses it or passes through it. The part of the line segment that passes through the block (or terminates within it) is termed a "q-edge." Each q-edge in the block is represented by a pointer to a record containing the endpoints of the line segment of which the q-edge is a part [80]. This pointer is really nothing more than a spatial index and hence the use of this term to characterize this approach. Thus, no information is associated with the shaded block as to what part of the line (i.e., q-edge) crosses it. This information can be obtained by clipping [38] the original line segment to the block. This is important, for often the precision necessary to compute these intersection points is not available.

6.4   Region Data

There are many ways of representing region data. We can represent a region either by its boundary (termed a *boundary-based representation*) or by its interior (termed an *interior-based representation*). In some applications, regions are really objects that are composed of smaller primitive objects by use of geometric transformations and Boolean set operations. Constructive solid geometry (CSG) [88] is a term usually used to describe such representations. They are beyond the scope

---

* More precisely, for arbitrary attributes which can be locational and nonlocational, there exist $j \geq 0$ such that the product of w_i, the width of the block along attribute i, and 2^j is equal to the length of the range of values of attribute i.

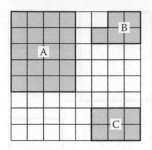

FIGURE 6.9 Example collection of three regions and the cells that they occupy.

of this chapter. Instead, unless noted otherwise, our discussion is restricted to regions consisting of congruent cells of unit area (volume) with sides (faces) of unit size that are orthogonal to the coordinate axes. As an example, consider Figure 6.9 which contains three two-dimensional regions A, B, and C, and their corresponding cells.

Regions with arbitrary boundaries are usually represented by either using approximating bounding hyperrectangles or more general boundary-based representations that are applicable to collections of line segments that do not necessarily form regions. In that case, we do not restrict the line segments to be parallel to the coordinate axes. Such representations are discussed in Section 6.6. It should be clear that although our presentation and examples in this section deal primarily with two-dimensional data, they are valid for regions of any dimensionality.

The region data is assumed to be uniform in the sense that all the cells that comprise each region are of the same type. In other words, each region is homogeneous. Of course, an image may consist of several distinct regions. Perhaps the best definition of a region is as a set of four-connected cells (i.e., in two dimensions, the cells are adjacent along an edge rather than a vertex) each of which is of the same type. For example, we may have a crop map where the regions correspond to the four-connected cells on which the same crop is grown. Each region is represented by the collection of cells that comprise it. The set of collections of cells that make up all of the regions is often termed an image array because of the nature in which they are accessed when performing operations on them. In particular, the array serves as an access structure in determining the region associated with a location of a cell as well as of all remaining cells that comprise the region.

When the region is represented by its interior, then often we can reduce the storage requirements by aggregating identically valued cells into blocks. In the rest of this section, we discuss different methods of aggregating the cells that comprise each region into blocks, as well as the methods used to represent the collections of blocks that comprise each region in the image.

The collection of blocks is usually a result of a space decomposition process with a set of rules that guide it. There are many possible decompositions. When the decomposition is recursive, we have the situation that the decomposition occurs in stages and often, although not always, the results of the stages form a containment hierarchy. This means that a block b obtained in stage i is decomposed into a set of blocks b_j that span the same space. Blocks b_j are, in turn, decomposed in stage $i + 1$ using the same decomposition rule. Some decomposition rules restrict the possible sizes and shapes of the blocks as well as their placement in space. Some examples include:

- Congruent blocks at each stage
- Similar blocks at all stages
- All sides of a block at a given stage are of equal size
- All sides of each block are powers of two, etc.

Other decomposition rules dispense with the requirement that the blocks be rectangular (i.e., there exist decompositions using other shapes such as triangles, etc.), while still others do not require that they be orthogonal, although, as stated before, we do make these assumptions here. In addition, the blocks may be disjoint or be allowed to overlap. Clearly, the choice is large. In the following, we briefly explore some of these decomposition rules. We restrict ourselves to disjoint decompositions, although this need not be the case (e.g., the fieldtree [40]).

The most general decomposition rule permits aggregation along all dimensions. In other words, the decomposition is arbitrary. The blocks need not be uniform or similar. The only requirement is that the blocks span the space of the environment. For example, Figure 6.10 is an arbitrary block decomposition for the collection of regions and cells given in Figure 6.9. We have labeled the blocks corresponding to object O as Oi and the blocks that are not in any of the regions as Wi, using the suffix i to distinguish between them in both cases. The same labeling convention is used throughout this section.

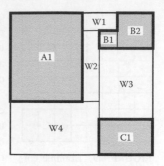

FIGURE 6.10 Arbitrary block decomposition for the collection of regions and cells in Figure 6.9.

The drawback of arbitrary decompositions is that there is little structure associated with them. This means that it is difficult to answer queries such as determining the region associated with a given point, besides performing an exhaustive search through the blocks. Thus, we need an additional data structure known as an index or an access structure. A very simple decomposition rule that lends itself to such an index in the form of an array is one that partitions a d-dimensional space having coordinate axes x_i into d-dimensional blocks by use of h_i hyperplanes that are parallel to the hyperplane formed by $x_i = 0$ $(1 \leq i \leq d)$. The result is a collection of $\prod_{i=1}^{d}(h_i + 1)$ blocks. These blocks form a grid of irregular-sized blocks rather than congruent blocks. There is no recursion involved in the decomposition process. For example, Figure 6.11a is such a block decomposition using hyperplanes parallel to the x- and y-axes for the collection of regions and cells given in Figure 6.9. We term the resulting decomposition as an irregular grid, as the partition lines occur at arbitrary positions in contrast to a uniform grid [41] where the partition lines are positioned so that all of the resulting grid cells are congruent.

Although the blocks in the irregular grid are not congruent, we can still impose an array access structure on them by adding d access structures termed linear scales. The linear scales indicate the position of the partitioning hyperplanes that are parallel to the hyperplane formed by $x_i = 0$ $(1 \leq i \leq d)$. Thus, given a location l in space, say (a,b), in two-dimensional space, the linear scales for the x- and y-coordinate values indicate the column and row, respectively, of the array access structure entry which corresponds to the block that contains l.

For example, Figure 6.11b is the array access structure corresponding to the block decomposition in Figure 6.11a, while Figures 6.11c and 6.11d are the linear scales for the x- and y-axes, respectively. In this example, the linear scales are shown as tables (i.e., array access structures). In fact, they can

Array column	x-Range
1	(0,4)
2	(4,5)
3	(5,6)
4	(6,8)

(c)

Array row	y-Range
1	(0,1)
2	(1,2)
3	(2,5)
4	(5,6)
5	(6,8)

(d)

(a)

A1	W3	W8	B2
A2	W4	B1	B3
A3	W5	W9	W11
W1	W6	W10	W12
W2	W7	C1	C2

(b)

FIGURE 6.11 (a) Block decomposition resulting from the imposition of a grid with partition lines at arbitrary positions on the collection of regions and cells in Figure 6.9 yielding an irregular grid, (b) the array access structure, (c) the linear scale for the x-coordinate values, and (d) the linear scale for the y-coordinate values.

be implemented using tree access structures such as binary search trees, range trees, segment trees, etc. The representation described here is an adaptation for regions of the grid file [82] data structure for points (see Section 6.2).

Perhaps the most widely known decomposition rules for blocks are those referred to by the general terms *quadtree* and *octree* [103,104,106]. They are usually used to describe a class of representations for two- and three-dimensional data (and higher as well), respectively, that are the result of a recursive decomposition of the environment (i.e., space) containing the regions into blocks (not necessarily rectangular) until the data in each block satisfies some condition (e.g., with respect to its size, the nature of the regions that comprise it, the number of regions in it, etc.). The positions and/or sizes of the blocks may be restricted or arbitrary. It is interesting to note that quadtrees and octrees may be used with both interior-based and boundary-based representations, although only the former are discussed in this section.

There are many variants of quadtrees and octrees (see also Sections 6.2, 6.5, and 6.6), and they are used in numerous application areas including high energy physics, VLSI, finite element analysis, and many others. Below, we focus on region quadtrees [63,68] and to a lesser extent on region octrees [62,78]. They are specific examples of interior-based representations for two- and three-dimensional region data (variants for data of higher dimension also exist), respectively, that permit further aggregation of identically valued cells.

Region quadtrees and region octrees are instances of a restricted decomposition rule where the environment containing the regions is recursively decomposed into four or eight, respectively, rectangular congruent blocks until each block is either completely occupied by a region or is empty (recall that such a decomposition process is termed regular). For example, Figure 6.12a is the block decomposition for the region quadtree corresponding to Figure 6.9. Notice that in this case, all the blocks are square, have sides whose size is a power of 2, and are located at specific positions. In particular, assuming an origin at the upper-left corner of the image containing the regions, the coordinate values of the upper-left corner of each block (e.g., (a, b) in two dimensions) of size $2^i \times 2^i$ satisfy the property that $a \bmod 2^i = 0$ and $b \bmod 2^i = 0$. For three-dimensional data, Figure 6.13a is an example of a simple three-dimensional object whose region octree block decomposition is given in Figure 6.13b, and whose tree representation is given in Figure 6.13c.

The traditional, and most natural, access structure for a d-dimensional quadtree corresponding to a d-dimensional image is a tree with a fanout of 2^d (e.g., Figure 6.12b). Each leaf node in the tree corresponds to a different block b and contains the identity of the region associated with b.

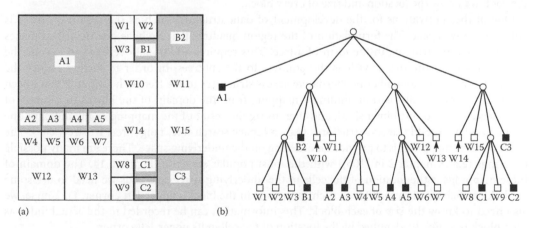

(a)　　　　　　　　　　　　　(b)

FIGURE 6.12 (a) Block decomposition and (b) its tree representation for the region quadtree corresponding to the collection of regions and cells in Figure 6.9.

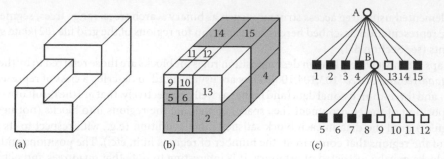

FIGURE 6.13 (a) Example of a three-dimensional object; (b) its region octree block decomposition; and (c) its tree representation.

Each nonleaf node f corresponds to a block whose volume is the union of the blocks corresponding to the 2^d children of f. In this case, the tree is a containment hierarchy and closely parallels the decomposition rule in the sense that they are both recursive processes and the blocks corresponding to nodes at different depths of the tree are similar in shape. The tree access structure captures the characterization of the region quadtree as being of variable resolution in that the underlying space is subdivided recursively until the underlying region satisfies some type of a homogeneity criterion. This is in contrast to a pyramid structure (e.g., [3,133]), which is a family of representations that make use of multiple resolution. This means that the underlying space is subdivided up to the smallest unit (i.e., a pixel) thereby resulting in a complete quadtree, where each leaf node is at the maximum depth and the leaf nodes contain information summarizing the contents of their subtrees.

Determining the region associated with a given point p is achieved by a process that starts at the root of the tree and traverses the links to the children whose corresponding blocks contain p. This process has an $O(m)$ cost, where the image has a maximum of m levels of subdivision (e.g., an image all of whose sides are of length 2^m).

Observe that using a tree with fanout 2^d as an access structure for a regular decomposition means that there is no need to record the size and the location of the blocks as this information can be inferred from the knowledge of the size of the underlying space. This is because the 2^d blocks that result at each subdivision step are congruent. For example, in two dimensions, each level of the tree corresponds to a quartering process that yields four congruent blocks. Thus, as long as we start from the root, we know the location and size of every block.

One of the motivations for the development of data structures such as the region quadtree is a desire to save space. The formulation of the region quadtree that we have just described makes use of an access structure in the form of a tree. This requires additional overhead to encode the internal nodes of the tree as well as the pointers to the subtrees. In order to further reduce the space requirements, a number of alternative access structures to the tree with fanout 2^d have been proposed. They are all based on finding a mapping from the domain of the blocks to a subset of the integers (i.e., to one dimension), and then using the result of the mapping as the index in one of the familiar tree-like access structures (e.g., a binary search tree, range tree, B^+-tree, etc.). The effect of these mappings is to provide an ordering on the underlying space. There are many possible orderings (e.g., Chapter 2 in [106]) with the most popular shown in Figure 6.14. The domain of these mappings is the location of the cells in the underlying space, and thus we need to use some easily identifiable cell in each block such as the one in the block's upper-left corner. Of course, we also need to know the size of each block. This information can be recorded in the actual index as each block is uniquely identified by the location of the cell in its upper-left corner.

Since the size of each block b in the region quadtree can be specified with a single number indicating the depth in the tree at which b is found, we can simplify the representation by incorporating the

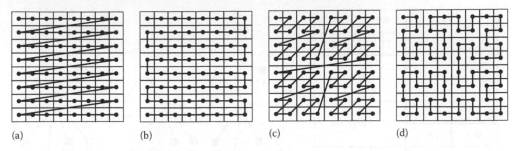

FIGURE 6.14 The result of applying four common different space-ordering methods to an 8×8 collection of cells whose first element is in the upper-left corner: (a) row order, (b) row-prime order, (c) Morton order, and (d) Peano-Hilbert order.

size into the mapping. One mapping simply concatenates the result of interleaving the binary representations of the coordinate values of the upper-left corner (e.g., (a, b) in two dimensions) and i of each block of size 2^i so that i is at the right. The resulting number is termed a locational code and is a variant of the Morton order (Figure 6.14c). Assuming such a mapping, sorting the locational codes in increasing order yields an ordering equivalent to that which would be obtained by traversing the leaf nodes (i.e., blocks) of the tree representation (e.g., Figure 6.12b) in the order NW, NE, SW, SE. The Morton ordering (as well as the Peano-Hilbert ordering shown in Figure 6.14d) is particularly attractive for quadtree-like block decompositions because all cells within a quadtree block appear in consecutive positions in the ordering. Alternatively, these two orders exhaust a quadtree block before exiting it. Therefore, once again, determining the region associated with point p consists of simply finding the block containing p.

Quadtrees and octrees make it easy to implement algorithms for a number of basic operations in computer graphics, image processing, as well as numerous other applications (e.g., [100]). In particular, algorithms have been devised for converting between region quadtrees and numerous representations such as binary arrays [90], boundary codes [32,91], rasters [92,98,120], medial axis transforms [97,99], and terrain models [124], as well as for many standard operations such as connected component labeling [94], perimeters [93], distance [95], and computing Euler numbers [31]. Many of these algorithms are based on the ease with which neighboring blocks can be visited in the quadtree [96] and octree [102]. They are particularly useful in ray tracing applications [47,101] where they enable the acceleration of ray tracing (e.g., [7]) by speeding up the process of finding ray–object intersections. Algorithms have also been devised for converting between region octrees and boundary models [130] and CSG [111].

In some applications, we may require finer (i.e., more) partitions along a subset of the dimensions due to factors such as sampling frequency (e.g., when the blocks correspond to aggregates of point data), while needing coarser (i.e., fewer) partitions along the remaining subset of dimensions. This is achieved by loosening the stipulation that the region quadtree results in 2^d congruent blocks at each subdivision stage, and replacing it by a stipulation that all blocks at the same subdivision stage (i.e., depth) i are partitioned into 2^{c_i} ($1 \leq c_i \leq d$) congruent blocks. We use the term ATree [20] to describe the resulting structure. For example, Figure 6.15a is the block decomposition for the ATree for Figure 6.9, while Figure 6.15b is the corresponding tree access structure.

As the dimensionality of the space (i.e., d) increases, each level of decomposition in the region quadtree results in many new blocks as the fanout value 2^d is high. In particular, it is too large for a practical implementation of the tree access structure. In this case, an access structure termed a *bintree* [69,112,129] with a fanout value of 2 is used. The bintree is defined in a manner analogous to the region quadtree except that at each subdivision stage, the space is decomposed into two equal-sized parts. In two dimensions, at odd stages we partition along the y-axis, and at even

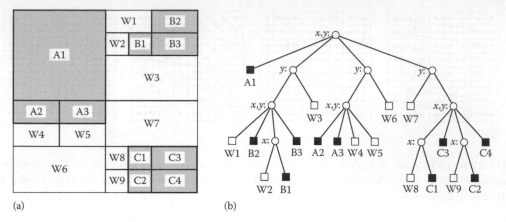

FIGURE 6.15 (a) Block decomposition for the ATree corresponding to the collection of regions and cells in Figure 6.9 and (b) the corresponding tree access structure. The nonleaf nodes are labeled with the partition axis or axes which must be the same for all nodes at the same level.

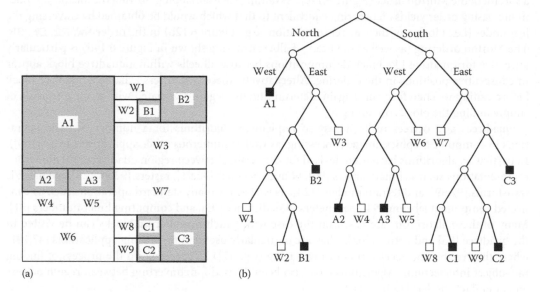

FIGURE 6.16 (a) Block decomposition for the bintree corresponding to the collection of regions and cells in Figure 6.9 and (b) the corresponding tree access structure. The splits alternate between the y- and x-coordinate values with the first split being based on the y-coordinate value.

stages we partition along the x axis. In general, in the case of d dimensions, we cycle through the different axes at every d level in the bintree. The bintree can also be viewed as a special case of the ATree where all blocks at subdivision stage i are partitioned into a predetermined subset of the dimensions into 2^{c_i} blocks where $c_i = 1$. For example, Figure 6.16a is the block decomposition for the bintree for Figure 6.9, while Figure 6.16b is the corresponding tree access structure.

The region quadtree, as well as the bintree, is a regular decomposition. This means that the blocks are congruent—that is, at each level of decomposition, all of the resulting blocks are of the same shape and size. We can also use decompositions where the sizes of the blocks are not restricted in the

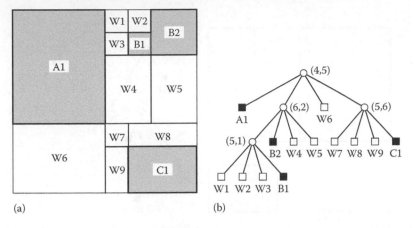

FIGURE 6.17 (a) Block decomposition for the point quadtree corresponding to the collection of regions and cells in Figure 6.9 and (b) the corresponding tree access structure. The (x, y) coordinate values of the locations of the partition points are indicated next to the relevant nonleaf nodes.

sense that the only restriction is that they be rectangular and be a result of a recursive decomposition process. In this case, the representations that we described must be modified so that the sizes of the individual blocks can be obtained. An example of such a structure is an adaptation of the point quadtree [37] to regions. Although the point quadtree was designed to represent points in a higher-dimensional space, the blocks resulting from its use to decompose space do correspond to regions. The difference from the region quadtree is that in the point quadtree, the positions of the partitions are arbitrary, whereas they are a result of a partitioning process into 2^d congruent blocks (e.g., quartering in two dimensions) in the case of the region quadtree. For example, Figure 6.17a is the block decomposition for the point quadtree for Figure 6.9, while Figure 6.17b is the corresponding tree access structure.

As in the case of the region quadtree, as the dimensionality d of the space increases, each level of decomposition in the point quadtree results in many new blocks since the fanout value 2^d is high. In particular, it is too large for a practical implementation of the tree access structure. In this case, we can adapt the k-d tree [13], which has a fanout value of 2, to regions. As in the point quadtree, although the k-d tree was designed to represent points in a higher-dimensional space, the blocks resulting from its use to decompose space do correspond to regions. Thus, the relationship of the k-d tree to the point quadtree is the same as the relationship of the bintree to the region quadtree. In fact, the k-d tree is the precursor of the bintree and its adaptation to regions is defined in a similar manner in the sense that for d-dimensional data we cycle through the d axes at every d level in the k-d tree. The difference is that in the k-d tree, the positions of the partitions are arbitrary, whereas they are a result of a halving process in the case of the bintree. For example, Figure 6.18a is the block decomposition for the k-d tree for Figure 6.9, while Figure 6.18b is the corresponding tree access structure.

The k-d tree can be further generalized so that the partitions take place on the various axes at an arbitrary order, and, in fact, the partitions need not be made on every coordinate axis. In this case, at each nonleaf node of the k-d tree, we must also record the identity of the axis that is being split. We use the term generalized k-d tree to describe this structure. For example, Figure 6.19a is the block decomposition for the generalized k-d tree for Figure 6.9, while Figure 6.19b is the corresponding tree access structure.

The generalized k-d tree is really an adaptation to regions of the adaptive k-d tree [45] and the LSD tree [54] which were originally developed for points. It can also be regarded as a special case of

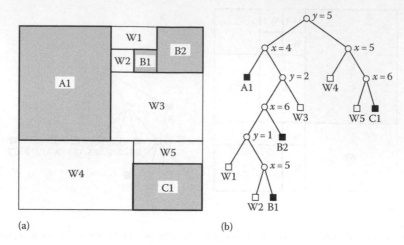

(a)   (b)

FIGURE 6.18 (a) Block decomposition for the k-d tree corresponding to the collection of regions and cells in Figure 6.9 and (b) the corresponding tree access structure. The splits alternate between the *y*- and *x*-coordinate values with the first split being based on the *y*-coordinate value. The locations of the splits are indicated next to the relevant nonleaf nodes.

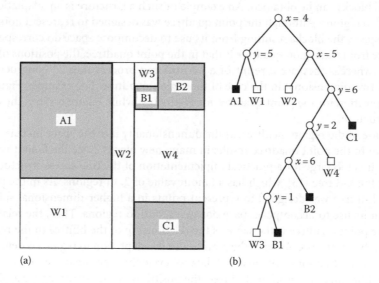

(a)   (b)

FIGURE 6.19 (a) Block decomposition for the generalized k-d tree corresponding to the collection of regions and cells in Figure 6.9 and (b) the corresponding tree access structure. The nonleaf nodes are labeled with the partition axes and the partition values.

the BSP tree (denoting binary space partitioning) [46]. In particular, in the generalized k-d tree, the partitioning hyperplanes are restricted to be parallel to the axes, whereas in the BSP tree they have an arbitrary orientation.

The BSP tree is a binary tree. In order to be able to assign regions to the left and right subtrees, we need to associate a direction with each subdivision line. In particular, the subdivision lines are treated

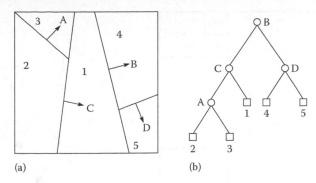

FIGURE 6.20 (a) An arbitrary space decomposition and (b) its BSP tree. The arrows indicate the direction of the positive halfspaces.

as separators between two halfspaces.* Let the subdivision line have the equation $a \cdot x + b \cdot y + c = 0$. We say that the right subtree is the "positive" side and contains all subdivision lines formed by separators that satisfy $a \cdot x + b \cdot y + c \geq 0$. Similarly, we say that the left subtree is "negative" and contains all subdivision lines formed by separators that satisfy $a \cdot x + b \cdot y + c < 0$. As an example, consider Figure 6.20a which is an arbitrary space decomposition whose BSP tree is given in Figure 6.20b. Notice the use of arrows to indicate the direction of the positive halfspaces. The BSP tree is used in computer graphics to facilitate visibility calculations of scenes with respect to a viewer as an alternative to the z-buffer algorithm which makes use of a frame buffer and a z buffer to keep track of the objects that it has already processed. The advantage of using visibility ordering over the z-buffer algorithm is that there is no need to compute or compare the z values.

One of the shortcomings of the generalized k-d tree is the fact that we can only decompose the space into two parts along a particular dimension at each step. If we wish to partition a space into p parts along a dimension i, then we must perform $p - 1$ successive partitions on dimension i. Once these $p - 1$ partitions are complete, we partition along another dimension. The puzzletree [28,29] (equivalent to the X-Y tree [79] and treemap [65,122]) is a further generalization of the k-d tree that decomposes the space into two or more parts along a particular dimension at each step so that no two successive partitions use the same dimension. In other words, the puzzletree compresses all successive partitions on the same dimension in the generalized k-d tree. For example, Figure 6.21a is the block decomposition for the puzzletree for Figure 6.9, while Figure 6.21b is the corresponding tree access structure. Notice that the puzzletree was created by compressing the successive initial partitions on $x = 4$ and $x = 5$ at depth 0 and 1, respectively, and likewise for the successive partitions on $y = 6$ and $y = 2$ at depth 2 and 3, respectively, in Figure 6.19.

At this point, we have seen a progressive development of a number of related methods of aggregating cells into blocks as well as representations of the collections of blocks that comprise each region in the image. As stated earlier, this is motivated, in part, by a desire to save space. As we saw, some of the decompositions have quite a bit of structure, thereby leading to inflexibility in choosing partition lines, etc. In fact, at times, maintaining the original image with an array access structure may be more effective from the standpoint of storage requirements. In the following, we point out some important implications of the use of these aggregations. In particular, we focus on

---

* A (linear) halfspace in d-dimensional space is defined by the inequality $\sum_{i=0}^{d} a_i \cdot x_i \geq 0$ on the $d + 1$ homogeneous coordinates ($x_0 = 1$). The halfspace is represented by a column vector a. In vector notation, the inequality is written as $a \cdot x \geq 0$. In the case of equality, it defines a hyperplane with a as its normal. It is important to note that halfspaces are volume elements; they are not boundary elements.

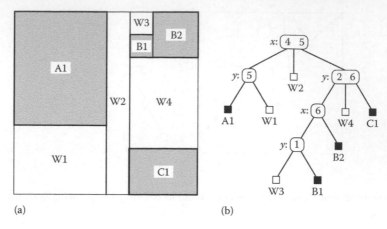

FIGURE 6.21 (a) Block decomposition for the puzzletree corresponding to the collection of regions and cells in Figure 6.9 and (b) the corresponding tree access structure. The nonleaf nodes are labeled with the partition axes and the partition values.

the region quadtree and region octree. Similar results could also be obtained for the remaining block decompositions.

The aggregation of similarly valued cells into blocks has an important effect on the execution time of the algorithms that make use of the region quadtree. In particular, most algorithms that operate on images represented by a region quadtree are implemented by a preorder traversal of the quadtree and, thus, their execution time is generally a linear function of the number of nodes in the quadtree. A key to the analysis of the execution time of quadtree algorithms is the *quadtree complexity theorem* [62] which states that the number of nodes in a region quadtree representation for a simple polygon (i.e., with non-intersecting edges and without holes) is $O(p + q)$ for a $2^q \times 2^q$ image with perimeter p measured in terms of the width of unit-sized cells (i.e., pixels). In all but the most pathological cases (e.g., a small square of unit width centered in a large image), the factor q is negligible, and thus the number of nodes is $O(p)$. The quadtree complexity theorem also holds for three-dimensional data [77] (i.e., represented by a region octree) where perimeter is replaced by surface area, as well as for objects of higher dimensions d for which it is proportional to the size of the $(d − 1)$-dimensional interfaces between these objects.

The most important consequence of the quadtree complexity theorem is that since most algorithms that execute on a region quadtree representation of an image will visit all of the nodes in the quadtree, the fact that the number of nodes in the quadtree is proportional to the perimeter of the image means that the execution time of the algorithms is proportional to the perimeter of the image. In contrast, when the blocks are decomposed into their constituent unit-sized cells, the algorithms still visit all of the unit-sized cells whose number is proportional to the area of the image. Therefore, considering that perimeter is a one-dimensional measure leads us to conclude that region quadtrees and region octrees act like dimension-reducing devices. In its most general case, the quadtree complexity theorem means that the use of a d-dimensional region quadtree, with an appropriate access structure, in solving a problem in a d-dimensional space leads to a solution whose execution time is proportional to a $(d − 1)$-dimensional space measure of the original d-dimensional image (i.e., its boundary). On the other hand, use of the array access structure on the original collection of unit-sized cells results in a solution whose execution time is proportional to the number of unit-sized cells that make up the image (i.e., its volume).

It is also interesting to observe that if we double the resolution of the underlying image represented by a quadtree, then the number of blocks will also double as only the quadtree blocks through which the boundary of the image passes are affected, whereas in the array representation, the resolution doubling affects all of the unit-sized cells whose number is quadrupled, and these unit-sized cells do not get merged into larges-sized blocks.

6.5 Rectangle Data

The rectangle data type lies somewhere between the point and region data types. It can also be viewed as a special case of the region data type in the sense that it is a region with only four sides. Rectangles are often used to approximate other objects in an image for which they serve as the minimum rectilinear enclosing object. For example, bounding rectangles are used in cartographic applications to approximate objects such as lakes, forests, hills, etc. In such a case, the approximation gives an indication of the existence of an object. Of course, the exact boundaries of the object are also stored; but they are only accessed if greater precision is needed. For such applications, the number of elements in the collection is usually small, and most often the sizes of the rectangles are of the same order of magnitude as the space from which they are drawn.

Rectangles are also used in VLSI design rule checking as a model of chip components for the analysis of their proper placement. Again, the rectangles serve as minimum enclosing objects. In this application, the size of the collection is quite large (e.g., millions of components) and the sizes of the rectangles are several orders of magnitude smaller than the space from which they are drawn.

It should be clear that the actual representation that is used depends heavily on the problem environment. At times, the rectangle is treated as the Cartesian product of two one-dimensional intervals with the horizontal intervals being treated in a different manner than the vertical intervals. In fact, the representation issue is often reduced to one of representing intervals. For example, this is the case in the use of the plane-sweep paradigm [86] in the solution of rectangle problems such as determining all pairs of intersecting rectangles. In this case, each interval is represented by its left and right endpoints. The solution makes use of two passes.

The first pass sorts the rectangles in ascending order on the basis of their left and right sides (i.e., x-coordinate values) and forms a list. The second pass sweeps a vertical scan line through the sorted list from left to right halting at each one of these points, say p. At any instant, all rectangles that intersect the scan line are considered active and are the only ones whose intersection needs to be checked with the rectangle associated with p. This means that each time the sweep line halts, a rectangle either becomes active (causing it to be inserted in the set of active rectangles) or ceases to be active (causing it to be deleted from the set of active rectangles). Thus, the key to the algorithm is its ability to keep track of the active rectangles (actually just their vertical sides) as well as to perform the actual one-dimensional intersection test.

Data structures such as the segment tree [14], interval tree [33], and the priority search tree [76] can be used to organize the vertical sides of the active rectangles so that, for N rectangles and F intersecting pairs of rectangles, the problem can be solved in $O(N \cdot \log_2 N + F)$ time. All three data structures enable intersection detection, insertion, and deletion to be executed in $O(\log_2 N)$ time. The difference between them is that the segment tree requires $O(N \cdot \log_2 N)$ space while the interval tree and the priority search tree only need $O(N)$ space.

The key to the use of the priority search tree to solve the rectangle intersection problem is that it treats each vertical side (y_B, y_T) as a point (x, y) in a two-dimensional space (i.e., it transforms the corresponding interval into a point as discussed in Section 6.1). The advantage of the priority search tree is that the storage requirements for the second pass only depend on the maximum number M of the vertical sides that can be actived at any one time. This is achieved by implementing the priority search tree as a red-black balanced binary tree [49], thereby guaranteeing updates in $O(\log_2 M)$ time. This also has an effect on the execution time of the second pass which is $O(N \cdot \log_2 M + F)$ instead of $O(N \cdot \log_2 N + F)$. Of course, the first pass which must sort the endpoints of the horizontal sides still takes $O(N \cdot \log_2 N)$ time for all three representations.

Most importantly, the priority search tree enables a more dynamic solution than either the segment or interval trees as only the endpoints of the horizontal sides need to be known in advance. On the other hand, for the segment and interval trees, the endpoints of both the horizontal and vertical sides must be known in advance. Of course, in all cases, all solutions based on the plane-sweep paradigm are inherently not dynamic as the paradigm requires that we examine all of the data one-by-one. Thus, the addition of even one new rectangle to the database forces the re-execution of the algorithm on the entire database.

In this chapter, we are primarily interested in dynamic problems. The data structures that are chosen for the collection of the rectangles are differentiated by the way in which each rectangle is represented. One solution [55] makes use of the representation discussed in Section 6.1 that reduces each rectangle to a point in a higher-dimensional space, and then treats the problem as if we have a collection of points. Again, each rectangle is a Cartesian product of two one-dimensional intervals where the difference from its use with the plane-sweep paradigm is that each interval is represented by its centroid and extent. In this solution [55], each set of intervals in a particular dimension is, in turn, represented by a grid file [82] which is described in Sections 6.2 and 6.4.

The second representation is region-based in the sense that the subdivision of the space from which the rectangles are drawn depends on the physical extent of the rectangle—not just one point. Representing the collection of rectangles, in turn, with a tree-like data structure has the advantage that there is a relation between the depth of the node in the tree and the size of the rectangle(s) that is (are) associated with it. Interestingly, some of the region-based solutions make use of the same data structures that are used in the solutions based on the plane-sweep paradigm.

There are three types of region-based solutions currently in use. The first two solutions use the R-tree and the R$^+$-tree (discussed in Section 6.3) to store rectangle data (in this case the objects are rectangles instead of arbitrary objects). The third is a quadtree-based approach and uses the MX-CIF quadtree [67].

In the MX-CIF quadtree, each rectangle is associated with the quadtree node corresponding to the smallest block which contains it in its entirety. Subdivision ceases whenever a node's block contains no rectangles. Alternatively, subdivision can also cease once a quadtree block is smaller than a predetermined threshold size. This threshold is often chosen to be equal to the expected size of the rectangle [67]. For example, Figure 6.22b is the MX-CIF quadtree for a collection of rectangles given in Figure 6.22a. Rectangles can be associated with both terminal and nonterminal nodes.

It should be clear that more than one rectangle can be associated with a given enclosing block and, thus, often we find it useful to be able to differentiate between them (this is analogous to a collision in the parlance of hashing). This is done in the following manner [67]. Let P be a quadtree node with centroid (CX, CY), and let S be the set of rectangles that are associated with P. Members of S are

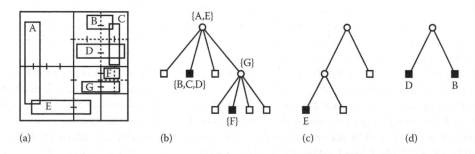

FIGURE 6.22 (a) Collection of rectangles and the block decomposition induced by the MX-CIF quadtree; (b) the tree representation of (a); (c) the binary trees for the y-axes passing through the root of the tree in (b); and (d) the NE child of the root of the tree in (b).

organized into two sets according to their intersection (or collinearity of their sides) with the lines passing through the centroid of P's block—that is, all members of S that intersect the line $x = CX$ form one set and all members of S that intersect the line $y = CY$ form the other set.

If a rectangle intersects both lines (i.e., it contains the centroid of P's block), then we adopt the convention that it is stored with the set associated with the line through $x = CX$. These subsets are implemented as binary trees (really tries), which in actuality are one-dimensional analogs of the MX-CIF quadtree. For example, Figure 6.22c and d illustrate the binary trees associated with the y-axes passing through the root and the NE child of the root, respectively, of the MX-CIF quadtree of Figure 6.22b. Interestingly, the MX-CIF quadtree is a two-dimensional analog of the interval tree described above. More precisely, the MX-CIF is a a two-dimensional analog of the tile tree [75] which is a regular decomposition version of the interval tree. In fact, the tile tree and the one-dimensional MX-CIF quadtree are identical when rectangles are not allowed to overlap.

It is interesting to note that the MX-CIF quadtree can be interpreted as an object hierarchy where the objects appear at different levels of the hierarchy and the congruent blocks play the same role as the minimum bounding boxes. The difference is that the set of possible minimum bounding boxes is constrained to the set of possible congruent blocks. Thus, we can view the MX-CIF quadtree as a variable resolution R-tree. An alternative interpretation is that the MX-CIF quadtree provides a variable number of grids, each one being at half the resolution of its immediate successor, where an object is associated with the grid whose cells have the tightest fit. In fact, this interpretation forms the basis of the filter tree [119] and the multilayer grid file [125], where the only difference from the MX-CIF quadtree is the nature of the access structure for the blocks. In particular, the filter tree uses a hierarchy of grids based on a regular decomposition, while for the multilayer grid file the hierarchy is based on a grid file, in contrast to using a tree structure for the MX-CIF quadtree.

The main drawback of the MX-CIF quadtree is that the size (i.e., width w) of the block c corresponding to the minimum enclosing quadtree block of object o's minimum enclosing bounding box b is not a function of the size of b or o. Instead, it is dependent on the position of o. In fact, c is often considerably larger than b thereby causing inefficiency in search operations due to a reduction in the ability to prune objects from further consideration. This situation arises whenever b overlaps the axes lines that pass through the center of c, and thus w can be as large as the width of the entire underlying space.

A number of ways have been proposed to overcome this drawback. One technique that has been applied in conjunction with the MX-CIF quadtree is to determine the width 2^s of the smallest possible minimum enclosing quadtree block for o and then to associate o with each of the four blocks (assuming that the underlying data is two-dimensional) of width 2^s that could possibly span it (e.g., the expanded MX-CIF quadtree [1] as well as [30]), thereby replicating the references to o. An alternative is to use one of two variants of the fieldtree [39,40]. In particular, the partition fieldtree overcomes the above drawback by shifting the positions of the centroids of quadtree blocks at successive levels of subdivision by one-half the width of the block that is being subdivided, while the cover fieldtree (also known as the loose quadtree and loose octree depending on the dimension of the underlying space [134]) overcomes this drawback by expanding the size of the space that is spanned by each quadtree block c of width w by a block expansion factor p $(p > 0)$, so that the expanded block is of width $(1 + p) \cdot w$. For more details about the ramifications of this expansion, see [108].

6.6 Line Data and Boundaries of Regions

Section 6.4 was devoted to variations on hierarchical decompositions of regions into blocks, an approach to region representation that is based on a description of the region's interior. In this section, we focus on representations that enable the specification of the boundaries of regions, as well as curvilinear data and collections of line segments. The representations are usually based on a series of approximations which provide successively closer fits to the data, often with the aid of bounding rectangles. When the boundaries or line segments have a constant slope (i.e., linear and termed line segments in the rest of this discussion), then an exact representation is possible.

There are several ways of approximating a curvilinear line segment. The first is by digitizing it and then marking the unit-sized cells (i.e., pixels) through which it passes. The second is to approximate it by a set of straight line segments termed a polyline. Assuming a boundary consisting of straight lines (or polylines after the first stage of approximation), the simplest representation of the boundary of a region is the polygon. It consists of vectors which are usually specified in the form of lists of pairs of x- and y-coordinate values corresponding to their starting and ending endpoints. The vectors are usually ordered according to their connectivity. One of the most common representations is the chain code [43] which is an approximation of a polygon's boundary by use of a sequence of unit vectors in the four (and sometimes eight) principal directions.

Chain codes, and other polygon representations, break down for data in three dimensions and higher. This is primarily due to the difficulty in ordering their boundaries by connectivity. The problem is that in two dimensions connectivity is determined by ordering the boundary elements $e_{i,j}$ of boundary b_i of object o, so that the end vertex of the vector v_j corresponding to $e_{i,j}$ is the start vertex of the vector v_{j+1} corresponding to $e_{i,j+1}$. Unfortunately, such an implicit ordering does not exist in higher dimensions as the relationship between the boundary elements associated with a particular object are more complex.

Instead, we must make use of data structures which capture the topology of the object in terms of its faces, edges, and vertices. The winged-edge data structure is one such representation which serves as the basis of the boundary model (also known as BRep [11]). Such representations are not discussed further here.

Polygon representations are very local. In particular, if we are at one position on the boundary, we do not know anything about the rest of the boundary without traversing it element-by-element. Thus, using such representations, given a random point in space, it is very difficult to find the nearest line to it as the lines are not sorted. This is in contrast to hierarchical representations which are global in nature. They are primarily based on rectangular approximations to the data as well as on a regular decomposition in two dimensions. In the rest of this section, we discuss a number of such representations.

In Section 6.3 we already examined two hierarchical representations (i.e., the R-tree and the R^+-tree) that propagate object approximations in the form of bounding hyperrectangles. In this case, the sides of the bounding hyperrectangles had to be parallel to the coordinate axes of the space from which the objects are drawn. In contrast, the strip tree [9] is a hierarchical representation of a single curve (embedded in a two-dimensional space in this discussion) that successively approximates segments of it with bounding rectangles that do not require that the sides be parallel to the coordinate axes. The only requirement is that the curve be continuous; it need not be differentiable.

The strip tree data structure consists of a binary tree whose root represents the bounding rectangle of the entire curve. For example, consider Figure 6.23 where the curve between points P and Q,

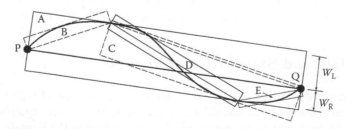

FIGURE 6.23 A curve and its decomposition into strips.

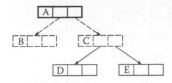

FIGURE 6.24 Strip tree corresponding to Figure 6.23.

at locations (x_P, y_P) and (x_Q, y_Q), respectively, is modeled by a strip tree. The rectangle associated with the root, A in this example, corresponds to a rectangular strip, that encloses the curve, whose sides are parallel to the line joining the endpoints of the curve (i.e., P and Q). The curve is then partitioned into two at one of the locations where it touches the bounding rectangle (these are not tangent points as the curve only needs to be continuous; it need not be differentiable).

Each subcurve is then surrounded by a bounding rectangle and the partitioning process is applied recursively. This process stops when the width of each strip is less than a predetermined value. Figure 6.24 shows the binary tree corresponding to the decomposition into strips in Figure 6.23a.

Figure 6.23 is a relatively simple example. In order to be able to cope with more complex curves such as those that arise in the case of object boundaries, the notion of a strip tree must be extended. In particular, closed curves and curves that extend past their endpoints require some special treatment. The general idea is that these curves are enclosed by rectangles which are split into two rectangular strips, and from now on the strip tree is used as before.

For a related approach, see the arc tree [51]. Its subdivision rule consists of a regular decomposition of a curve based on its length. The latter means that closed curves need no special treatment. In addition, the arc tree makes use of bounding ellipses around each subarc instead of bounding rectangles. The foci of the ellipses are placed at the endpoints of each subarc and the principal axis is as long as the subarc. This means that all subarcs lie completely within each ellipse thereby obviating the need for special treatment for subarcs that extend past their endpoints. The drawback of the arc tree is that we need to be able to compute the length of an arc, which may be quite complex (e.g., if we have a closed form for the curve, then we need an elliptical integral).

Like point and region quadtrees, strip trees are useful in applications that involve search and set operations. For example, suppose that we wish to determine whether a road crosses a river. Using a strip tree representation for these features, answering this query requires that we perform an intersection of the corresponding strip trees. Three cases are possible as is shown in Figure 6.25. Figure 6.25a and 6.25b correspond to the answers NO and YES, respectively, while Figure 6.25c requires us to descend further down the strip tree. Notice the distinction between the task of detecting the possibility of an intersection and the task of computing the actual intersection, if one exists. The strip tree is well suited to the former task. Other operations that can be performed efficiently by using the strip tree data structure include the computation of the length of a curve, areas of closed curves, intersection of curves with areas, point membership, etc.

The strip tree is similar to the point quadtree in the sense that the points at which the curve is decomposed depend on the data. In contrast, a representation based on the region quadtree has fixed decomposition points. Similarly, strip tree methods approximate curvilinear data with rectangles of arbitrary orientation, while methods based on the region quadtree achieve analogous results by use

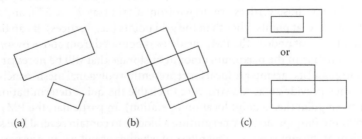

(a) (b) (c)

FIGURE 6.25 Three possible results of intersecting two strip trees: (a) null, (b) clear, and (c) possible.

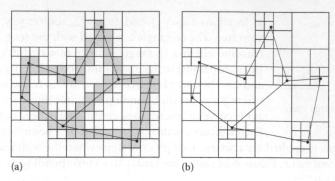

(a)   (b)

FIGURE 6.26 (a) MX quadtree and (b) edge quadtree for the collection of line segments of Figure 6.6.

of a collection of disjoint squares having sides of length of the power of two. In the following, we discuss a number of adaptations of the region quadtree for representing curvilinear data.

The simplest adaptation of the region quadtree is the MX quadtree [62,63]. It is built by digitizing the line segments and labeling each unit-sized cell (i.e., pixel) through which a line segment passes as of type `boundary`. The remaining pixels are marked `WHITE` and are merged, if possible, into larger and larger quadtree blocks. Figure 6.26a is the MX quadtree for the collection of line segment objects in Figure 6.6, where the cells of the type `boundary` are shown shaded. A drawback of the MX quadtree is that it associates a thickness with a line segment. Also, it is difficult to detect the presence of a vertex whenever five or more line segments meet.

The edge quadtree [123,137] is a refinement of the MX quadtree based on the observation that the number of squares in the decomposition can be reduced by terminating the subdivision whenever the square contains a single curve that can be approximated by a single straight line segment. For example, Figure 6.26b is the edge quadtree for the collection of line segment objects in Figure 6.6. Applying this process leads to quadtrees in which long edges are represented by large blocks or a sequence of large blocks. However, small blocks are required in the vicinity of the corners or the intersecting line segments. Of course, many blocks will contain no edge information at all.

The PM quadtree family [80,113] (see also edge-EXCELL [128]) represents an attempt to overcome some of the problems associated with the edge quadtree in the representation of collections of polygons (termed polygonal maps). In particular, the edge quadtree is an approximation because vertices are represented by pixels. There are a number of variants of the PM quadtree. These variants are either vertex-based or edge-based. They are all built by applying the principle of repeatedly breaking up the collection of vertices and edges (forming the polygonal map) until obtaining a subset that is sufficiently simple, so that it can be organized by some other data structure.

The PM_1 quadtree [113] is one example of a vertex-based PM quadtree. Its decomposition rule stipulates that partitioning occurs as long as a block contains more than one line segment unless the line segments are all incident at the same vertex which is also in the same block (e.g., Figure 6.27a). Given a polygonal map whose vertices are drawn from a grid (say $2^m \times 2^m$), and where edges are not permitted to intersect at points other than the grid points (i.e., vertices), it can be shown that the maximum depth of any leaf node in the PM_1 quadtree is bounded from above by $4m + 1$ [110]. This enables the determination of the maximum amount of storage that will be necessary for each node.

The PM_1 quadtree and its variants are ideal for representing polygonal meshes such as triangulations as, for example, they provide an access structure to enable the quick determination of the triangle that contains a given point (i.e., a point location operation). In particular, the PM_2 quadtree [113], which differs from the PM_1 quadtree by permitting a block c to contain several line segments as long as they are incident at the same vertex v regardless of whether or not v is in c, is particularly suitable for representing triangular meshes [27]. For example, to form the PM_2 quadtree for the collection

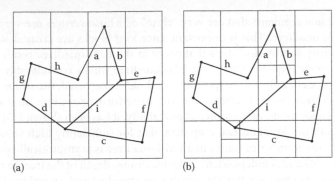

FIGURE 6.27 (a) PM$_1$ quadtree and (b) PMR quadtree for the collection of line segments of Figure 6.6.

of line segments of Figure 6.6, we simply modify its PM$_1$ quadtree in Figure 6.27a so that the NE subquadrant of the SW quadrant of the root is not split as the line segments d and i that pass through it are incident at the same vertex, although this vertex is in another block (i.e., in the SE subquadrant of the SW quadrant of the root).

A similar representation to the PM$_1$ quadtree has been devised for three-dimensional images (e.g., [8] and the references cited in [106]). The decomposition criteria are such that no node contains more than one face, edge, or vertex unless the faces all meet at the same vertex or are adjacent to the same edge. This representation is quite useful since its space requirements for polyhedral objects are significantly smaller than those of a region octree.

The bucket PM quadtree [106] (also termed a bucket PMR quadtree in [73]) and the PMR quadtree [80,81] are examples of edge-based representations where the decomposition criteria only involve the maximum number of line segments b (faces in three dimensions although the discussion below is in terms of two dimensions). In particular, in the bucket PM quadtree, the decomposition halts whenever a block contains b or less line segments, while in the PMR quadtree, a block is decomposed once and only once when it contains more than b line segments. Thus, in the bucket PM quadtree, b acts as a bucket capacity, while in the PMR quadtree, b acts as a splitting threshold. The advantage of the PMR quadtree is that there is no need to split forever when b or more line segments meet at a vertex. An alternative, as exemplified by the PK-tree [105,135], makes use of a lower bound on the number of objects (line segments in this example) that can be associated with each block (termed an instantiation or aggregation threshold).

For example, Figure 6.27b is the PMR quadtree for the collection of line segment objects in Figure 6.6 with a splitting threshold value of 2. The line segments are inserted in alphabetic order (i.e., a-i). It should be clear that the shape of the PMR quadtree depends on the order in which the line segments are inserted. Note the difference from the PM$_1$ quadtree in Figure 6.27a—that is, the NE block of the SW quadrant is decomposed in the PM$_1$ quadtree while the SE block of the SW quadrant is not decomposed in the PM$_1$ quadtree.

On the other hand, a line segment is deleted from a PMR quadtree by removing it from the nodes corresponding to all the blocks that it intersects. During this process, the occupancy of the node and its siblings is checked to see if the deletion causes the total number of line segments in them to be less than the predetermined splitting threshold. If the splitting threshold exceeds the occupancy of the node and its siblings, then they are merged and the merging process is reapplied to the resulting node and its siblings. Notice the asymmetry between the splitting and merging rules.

The PMR quadtree is very good for answering queries such as finding the nearest line to a given point [59] (see [60] for an empirical comparison with hierarchical object representations such as the R-tree and R$^+$-tree). It is preferred over the PM$_1$ quadtree (as well as the MX and edge quadtrees) as it results in far fewer subdivisions. In particular, in the PMR quadtree, there is no need to subdivide in order to separate line segments that are very "close" or whose vertices are very "close," which is the case for the PM_1 quadtree. This is important since four blocks are created at each subdivision step. Thus,, when many subdivision steps that occur in the PM_1 quadtree result in creating many empty blocks, the storage requirements of the PM_1 quadtree will be considerably higher than those of the PMR quadtree. Generally, as the splitting threshold is increased, the storage requirements of the PMR quadtree decrease while the time necessary to perform operations on it will increase.

Using a random image model and geometric probability, it has been shown [73], theoretically and empirically using both random and real map data, that for sufficiently high values of the splitting threshold (i.e., ≥ 4), the number of nodes in a PMR quadtree is asymptotically proportional to the number of line segments and is independent of the maximum depth of the tree. In contrast, using the same model, the number of nodes in the PM_1 quadtree is a product of the number of line segments and the maximal depth of the tree (i.e., n for a $2^n \times 2^n$ image). The same experiments and analysis for the MX quadtree confirmed the results predicted by the quadtree complexity theorem (see Section 6.4) which is that the number of nodes is proportional to the total length of the line segments.

Observe that although a bucket in the PMR quadtree can contain more line segments than the splitting threshold, this is not a problem. In fact, it can be shown [104] that the maximum number of line segments in a bucket is bounded by the sum of the splitting threshold and the depth of the block (i.e., the number of times the original space has been decomposed to yield this block).

6.7 Research Issues and Summary

A review has been presented of a number of representations of multidimensional data. Our focus has been on multidimensional spatial data with extent rather than just multidimensional point data. Moreover, the multidimensional data was not restricted to locational attributes in that the handling of nonlocational attributes for point data was also described. There has been a particular emphasis on hierarchical representations. Such representations are based on the "divide-and-conquer" problem-solving paradigm. They are of interest because they enable focusing computational resources on the interesting subsets of data. Thus, there is no need to expend work where the payoff is small. Although many of the operations for which they are used can often be performed equally as efficiently, or more so, with other data structures, hierarchical data structures are attractive because of their conceptual clarity and ease of implementation.

When the hierarchical data structures are based on the principle of regular decomposition, we have the added benefit that different data sets (often of differing types) are in registration. This means that they are partitioned in known positions which are often the same or subsets of one another for the different data sets. This is true for all the features including regions, points, rectangles, line segments, volumes, etc. In other words, it is easy to correlate occupied and unoccupied space in the two data sets, which is not easy when the positions of the partitions are not constrained as is the case with methods rooted in representations based on object hierarchy even though the resulting decomposition of the underlying space is disjoint. This means that a spatial join query (e.g., [61,64]) such as "finding all cities with more than 20,000 inhabitants in wheat growing regions within 30 miles of the Mississippi River" can be executed by simply overlaying the region (crops), point (i.e., cities), and river maps even though they represent data of different types. Alternatively, we may extract regions such as those within 30 mi of the Mississippi River (e.g., [2]). These operations find use in applications involving spatial data such as GIS.

Current research in multidimensional representations is highly application dependent in the sense that the work is driven by the application. For example, many of the recent developments have been motivated by the interaction with databases. In particular, this has led to a great interest in similarity searching which has in turn fueled much research into techniques for finding nearest neighbors where proximity can also be measured in terms of distance along a graph such as a road network (e.g., [109,114,116]) instead of being restricted to as "the crow flies." Moreover, techniques have been developed for finding neighbors in an incremental manner so that the number of neighbors that are sought need not be known in advance (e.g., [53,56–58,115]). This means that once we have found the k nearest neighbors, we do not have to reinvoke the k nearest neighbor algorithm to find the $k + 1$th nearest object; instead, we continue the process right where we left off after obtaining the kth nearest object.

The choice of a proper representation plays a key role in the speed with which responses are provided to queries. Knowledge of the underlying data distribution is also a factor and research is ongoing to make use of this information in the process of making a choice. Most of the initial applications in which the representation of multidimensional data has been important have involved spatial data of the kind described in this chapter. Such data is intrinsically of low dimensionality (i.e., two and three). Future applications involve higher-dimensional data for applications such as image databases where the data are often points in feature space. The incorporation of the time dimension is also an important issue that confronts many database researchers. In addition, new computing settings such as peer-to-peer (P2P) (e.g., [131,132]) and GPUs (e.g., [72]) are also great interest.

6.8 Further Information

Hands-on experience with some of the representations described in this chapter can be obtained by trying VASCO [21–23,25], a system for Visualizing and Animating Spatial Constructs and Operations. VASCO consists of a set of spatial index JAVA^{TM} (e.g., [6]) applets that enable users on the worldwide web to experiment with a number of hierarchical representations (e.g., [103, 104,106]) for different spatial data types, and see animations of how they support a number of search queries (e.g., nearest neighbor and range queries). The VASCO system can be found at http://cs.umd.edu/~hjs/quadtree/. For an example of their use in a spatial database/GIS, see the SAND Spatial Browser [24,36,107] and the QUILT system [121]. Such systems find use in a number of alternative application domains (e.g., digital government [74]).

It is impossible to give a complete enumeration of where research on multidimensional data structures is published since it is often mixed with the application. Multidimensional spatial data is covered in the texts by Samet [103,104,106]. Their perspective is one from computer graphics, image processing, GIS, databases, solid modeling, as well as VLSI design and computational geometry. A more direct computational geometry perspective can be found in the books by Edelsbrunner [35], Preparata and Shamos [86], and de Berg et al. [19].

New developments in the field of multidimensional data structures for spatial applications are reported in many different conferences, again, since it is so application-driven. Some good starting pointers from the GIS perspective are the annual ACM International Conference on Geographic Information Systems (ACMGIS) sponsored by SIGSPATIAL, the ACM special interest group on spatial information held annually. In addition, there is the Symposium on Spatial and Temporal Databases (SSTD) formerly known as the Symposium on Spatial Databases, and the International Workshop on Spatial Data Handling (SDH), both of which have been held in alternating years. From the standpoint of computational geometry, the annual ACM Symposium on Computational Geometry (SOCG) and the annual ACM-SIAM Symposium on Discrete Algorithms (SODA) are good sources. From the perspective of databases, the annual ACM Conference on the Management of Data (SIGMOD), the Very Large Database Conference (VLDB), the IEEE International Conference on Data Engineering (ICDE), the Symposium on Principles of Database Systems (PODS), and the International Conference on Extending Database Technology (EDBT) often contain a number of papers dealing with the application of such data structures. Other useful sources are the proceedings of the annual ACM SIGGRAPH Conference, the EUROGRAPHICS Conference, and the ACM Solid and Physical Modeling Symposium. In addition, from a pattern recognition and computer vision perspective, the proceedings of the Computer Vision and Pattern Recognition (CVPR) and the International Conference on Pattern Recognition (ICPR) are also of interest.

Journals where results of such research are reported are as varied as the applications. Theoretical results can be found in the *SIAM Journal of Computing* while those from the GIS perspective may be found in the *GeoInformatica, International Journal of Geographical Information Science* (previously known as the *International Journal of Geographical Information Systems*), and *Transactions in GIS*. Many related articles are also found in the computer graphics and computer vision journals such as *ACM Transactions on Graphics*, and the original *Computer Graphics and Image Processing* which over the years has been split and renamed to include *Computer Vision, Graphics and Image Processing, Graphical Models and Image Processing, Graphical Models*, and *Image Understanding*. Other relevant journals include *IEEE Transactions on Pattern Analysis and Machine Intelligence, Visual Computer, Pattern Recognition, Pattern Recognition Letters, Computers & Graphics, Computer Graphics Forum* etc. In addition, numerous articles are found in database journals such as *ACM Transactions on Database Systems, VLDB Journal*, and *IEEE Transaction on Knowledge and Data Engineering*.

Defining Terms

Bintree: A regular decomposition k-d tree for region data.

Boundary-based representation: A representation of a region that is based on its boundary.

Bucketing methods: Data organization methods that decompose the space from which spatial data is drawn into regions called buckets. Some conditions for the choice of region boundaries include the number of objects that they contain or on their spatial layout (e.g., minimizing overlap or coverage).

Fixed-grid method: Space decomposition into rectangular cells by overlaying a grid on it. If the cells are congruent (i.e., of the same width, height, etc.), then the grid is said to be uniform.

Interior-based representation: A representation of a region that is based on its interior (i.e., the cells that comprise it).

k-d tree: General term used to describe space decomposition methods that proceed by recursive decomposition across a single dimension at a time of the space containing the data until some condition is met such as that the resulting blocks contain no more than b objects (e.g., points, line segments, etc.) or that the blocks are homogeneous. The k-d tree is usually a data structure for points which cycle through the dimensions as it decomposes the underlying space.

Multidimensional data: Data that have several attributes. It includes records in a database management system, locations in space, and also spatial entities that have extent such as line segments, regions, volumes, etc.

Octree: A quadtree-like decomposition for three-dimensional data.

Quadtree: General term used to describe space decomposition methods that proceed by recursive decomposition across all the dimensions (technically two dimensions) of the space containing the data until some condition is met such as that the resulting blocks contain no more than b objects (e.g., points, line segments, etc.) or that the blocks are homogeneous (e.g., region data). The underlying space is not restricted to two-dimensions although this is the technical definition of the term. The result is usually a disjoint decomposition of the underlying space.

Quadtree complexity theorem: The number of nodes in a quadtree region representation for a simple polygon (i.e., with nonintersecting edges and without holes) is $O(p + q)$ for a $2^q \times 2^q$ image with perimeter p measured in pixel widths. In most cases, q is negligible and thus the number of nodes is proportional to the perimeter. It also holds for three-dimensional data where the perimeter is replaced by surface area, and in general for d-dimensions where instead of perimeter we have the size of the $(d - 1)$-dimensional interfaces between the d-dimensional objects.

R-Tree: An object hierarchy where associated with each element of the hierarchy is the minimum bounding hyperrectangle of the union of the minimum bounding hyperrectangles of the elements immediately below it. The elements at the deepest level of the hierarchy are groups of spatial objects. The result is usually a nondisjoint decomposition of the underlying space. The objects are aggregated on the basis of proximity and with the goal of minimizing coverage and overlap.

Regular decomposition: A space decomposition method that partitions the underlying space by recursively halving it across the various dimensions instead of permitting the positions of the partitioning lines to vary.

Acknowledgments

This work was supported in part by the National Science Foundation under Grants IRI-9712715, EIA-00-91474, CCF-08-30618, IIS-07-13501, and IIS-0812377, Microsoft Research, NVIDIA, and the University of Maryland General Research Board. The assistance of Gisli Hjaltason and Jagan Sankaranarayanan in the preparation of the figures is greatly appreciated.

References

1. D. J. Abel and J. L. Smith. A data structure and query algorithm for a database of areal entities. *Australian Computer Journal*, 16(4):147–154, November 1984.
2. C.-H. Ang, H. Samet, and C. A. Shaffer. A new region expansion for quadtrees. *IEEE Transactions on Pattern Analysis and Machine Intelligence*, 12(7):682–686, July 1990. Also see *Proceedings of the 3rd International Symposium on Spatial Data Handling*, pp. 19–37, Sydney, Australia, August 1988.
3. W. G. Aref and H. Samet. Efficient processing of window queries in the pyramid data structure. In *Proceedings of the 9th ACM SIGACT-SIGMOD-SIGART Symposium on Principles of Database Systems (PODS)*, pp. 265–272, Nashville, TN, April 1990. Also in *Proceedings of the 5th Brazilian Symposium on Databases*, pp. 15–26, Rio de Janeiro, Brazil, April 1990.
4. W. G. Aref and H. Samet. Uniquely reporting spatial objects: Yet another operation for comparing spatial data structures. In *Proceedings of the 5th International Symposium on Spatial Data Handling*, pp. 178–189, Charleston, SC, August 1992.
5. W. G. Aref and H. Samet. Hashing by proximity to process duplicates in spatial databases. In *Proceedings of the 3rd International Conference on Information and Knowledge Management (CIKM)*, pp. 347–354, Gaithersburg, MD, December 1994.
6. K. Arnold and J. Gosling. *The JAVA™ Programming Language*. Addison-Wesley, Reading, MA, 1996.
7. J. Arvo and D. Kirk. A survey of ray tracing acceleration techniques. In *An Introduction to Ray Tracing*, A. S. Glassner, ed., Chapter 6, pp. 201–262. Academic Press, New York, 1989.
8. D. Ayala, P. Brunet, R. Juan, and I. Navazo. Object representation by means of nonminimal division quadtrees and octrees. *ACM Transactions on Graphics*, 4(1):41–59, January 1985.
9. D. H. Ballard. Strip trees: A hierarchical representation for curves. *Communications of the ACM*, 24(5):310–321, May 1981. Also see corrigendum, *Communications of the ACM*, 25(3):213, March 1982.
10. G. Barequet, B. Chazelle, L. J. Guibas, J. S. B. Mitchell, and A. Tal. BOXTREE: A hierarchical representation for surfaces in 3D. In *Proceedings of the EUROGRAPHICS'96 Conference*, J. Rossignac and F. X. Sillion, eds., pp. 387–396, 484, Poitiers, France, August 1996. Also in *Computer Graphics Forum*, 15(3):387–396, 484, August 1996.
11. B. G. Baumgart. A polyhedron representation for computer vision. In *Proceedings of the 1975 National Computer Conference*, Vol. 44, pp. 589–596, Anaheim, CA, May 1975.

12. N. Beckmann, H.-P. Kriegel, R. Schneider, and B. Seeger. The R*-tree: An efficient and robust access method for points and rectangles. In *Proceedings of the ACM SIGMOD Conference*, pp. 322–331, Atlantic City, NJ, June 1990.

13. J. L. Bentley. Multidimensional binary search trees used for associative searching. *Communications of the ACM*, 18(9):509–517, September 1975.

14. J. L. Bentley. Algorithms for Klee's rectangle problems (unpublished). Computer Science Department, Carnegie-Mellon University, Pittsburgh, PA, 1977.

15. J. L. Bentley. Decomposable searching problems. *Information Processing Letters*, 8(5):244–251, June 1979.

16. J. L. Bentley and J. H. Friedman. Data structures for range searching. *ACM Computing Surveys*, 11(4):397–409, December 1979.

17. J. L. Bentley and H. A. Maurer. Efficient worst-case data structures for range searching. *Acta Informatica*, 13:155–168, 1980.

18. J. L. Bentley, D. F. Stanat, and E. H. Williams Jr. The complexity of finding fixed-radius near neighbors. *Information Processing Letters*, 6(6):209–212, December 1977.

19. M. de Berg, M. van Kreveld, M. Overmars, and O. Schwarzkopf. *Computational Geometry: Algorithms and Applications*. Springer-Verlag, Berlin, Germany, 2nd revised edition, 2000.

20. P. Bogdanovich and H. Samet. The ATree: A data structure to support very large scientific databases. In *Integrated Spatial Databases: Digital Images and GIS*, P. Agouris and A. Stefanidis, eds., Vol. 1737 of *Springer-Verlag Lecture Notes in Computer Science*, pp. 235–248, Portland, ME, June 1999. Also University of Maryland Computer Science Technical Report TR-3435, March 1995.

21. F. Brabec and H. Samet. The VASCO R-tree JAVA^TM applet. In *Visual Database Systems (VDB4). Proceedings of the IFIP TC2//WG2.6 4th Working Conference on Visual Database Systems*, Y. Ioannidis and W. Klas, eds., pp. 147–153, Chapman and Hall, L'Aquila, Italy, May 1998.

22. F. Brabec and H. Samet. Visualizing and animating R-trees and spatial operations in spatial databases on the worldwide web. In *Visual Database Systems (VDB4). Proceedings of the IFIP TC2//WG2.6 4th Working Conference on Visual Database Systems*, Y. Ioannidis and W. Klas, eds., pp. 123–140, Chapman and Hall, L'Aquila, Italy, May 1998.

23. F. Brabec and H. Samet. Visualizing and animating search operations on quadtrees on the worldwide web. In *Proceedings of the 16th European Workshop on Computational Geometry*, K. Kedem and M. Katz, eds., pp. 70–76, Eilat, Israel, March 2000.

24. F. Brabec and H. Samet. Client-based spatial browsing on the world wide web. *IEEE Internet Computing*, 11(1):52–59, January/February 2007.

25. F. Brabec, H. Samet, and C. Yilmaz. VASCO: Visualizing and animating spatial constructs and operations. In *Proceedings of the 19th Annual Symposium on Computational Geometry*, pp. 374–375, San Diego, CA, June 2003.

26. D. Comer. The ubiquitous B-tree. *ACM Computing Surveys*, 11(2):121–137, June 1979.

27. L. De Floriani, M. Facinoli, P. Magillo, and D. Dimitri. A hierarchical spatial index for triangulated surfaces. In *Proceedings of the 3rd International Conference on Computer Graphics Theory and Applications (GRAPP 2008)*, J. Braz, N. Jardim Nunes, and J. Madeiras Pereira, eds., pp. 86–91, Funchal, Madeira, Portugal, January 2008.

28. A. Dengel. Self-adapting structuring and representation of space. Technical Report RR–91–22, Deutsches Forschungszentrum für Künstliche Intelligenz, Kaiserslautern, Germany, September 1991.

29. A. Dengel. Syntactic analysis and representation of spatial structures by puzzletrees. *International Journal of Pattern Recognition and Artificial Intelligence*, 9(3):517–533, June 1995.

30. J.-P. Dittrich and B. Seeger. Data redundancy and duplicate detection in spatial join processing. In *Proceedings of the 16th IEEE International Conference on Data Engineering*, pp. 535–546, San Diego, CA, February 2000.

31. C. R. Dyer. Computing the Euler number of an image from its quadtree. *Computer Graphics and Image Processing*, 13(3):270–276, July 1980. Also University of Maryland Computer Science Technical Report TR-769, May 1979.

32. C. R. Dyer, A. Rosenfeld, and H. Samet. Region representation: Boundary codes from quadtrees. *Communications of the ACM*, 23(3):171–179, March 1980. Also University of Maryland Computer Science Technical Report TR-732, February 1979.

33. H. Edelsbrunner. Dynamic rectangle intersection searching. Institute for Information Processing Technical Report 47, Technical University of Graz, Graz, Austria, February 1980.

34. H. Edelsbrunner. A note on dynamic range searching. *Bulletin of the EATCS*, (15):34–40, October 1981.

35. H. Edelsbrunner. *Algorithms in Combinatorial Geometry*. Springer-Verlag, Berlin, West Germany, 1987.

36. C. Esperança and H. Samet. Experience with SAND/Tcl: A scripting tool for spatial databases. *Journal of Visual Languages and Computing*, 13(2):229–255, April 2002.

37. R. A. Finkel and J. L. Bentley. Quad trees: A data structure for retrieval on composite keys. *Acta Informatica*, 4(1):1–9, 1974.

38. J. D. Foley, A. van Dam, S. K. Feiner, and J. F. Hughes. *Computer Graphics: Principles and Practice*. Addison-Wesley, Reading, MA, 2nd edition, 1990.

39. A. Frank. Problems of realizing LIS: Storage methods for space related data: the fieldtree. Technical Report 71, Institute for Geodesy and Photogrammetry, ETH, Zurich, Switzerland, June 1983.

40. A. U. Frank and R. Barrera. The Fieldtree: A data structure for geographic information systems. In *Design and Implementation of Large Spatial Databases—1st Symposium, SSD'89*, A. Buchmann, O. Günther, T. R. Smith, and Y.-F. Wang, eds., Vol. 409 of *Springer-Verlag Lecture Notes in Computer Science*, pp. 29–44, Santa Barbara, CA, July 1989.

41. W. R. Franklin. Adaptive grids for geometric operations. *Cartographica*, 21(2&3):160–167, Summer and Autumn 1984.

42. E. Fredkin. Trie memory. *Communications of the ACM*, 3(9):490–499, September 1960.

43. H. Freeman. Computer processing of line-drawing images. *ACM Computing Surveys*, 6(1):57–97, March 1974.

44. M. Freeston. The BANG file: A new kind of grid file. In *Proceedings of the ACM SIGMOD Conference*, pp. 260–269, San Francisco, CA, May 1987.

45. J. H. Friedman, J. L. Bentley, and R. A. Finkel. An algorithm for finding best matches in logarithmic expected time. *ACM Transactions on Mathematical Software*, 3(3):209–226, September 1977.

46. H. Fuchs, Z. M. Kedem, and B. F. Naylor. On visible surface generation by a priori tree structures. *Computer Graphics*, 14(3):124–133, July 1980. Also in *Proceedings of the SIGGRAPH'80 Conference*, Seattle, WA, July 1980.

47. A. S. Glassner. Space subdivision for fast ray tracing. *IEEE Computer Graphics and Applications*, 4(10):15–22, October 1984.

48. S. Gottschalk, M. C. Lin, and D. Manocha. OBBTree: A hierarchical structure for rapid interference detection. In *Proceedings of the SIGGRAPH'96 Conference*, pp. 171–180, New Orleans, LA, August 1996.

49. L. J. Guibas and R. Sedgewick. A dichromatic framework for balanced trees. In *Proceedings of the 19th IEEE Annual Symposium on Foundations of Computer Science*, pp. 8–21, Ann Arbor, MI, October 1978.

50. O. Günther. Efficient structures for geometric data management. PhD thesis, Computer Science Division, University of California at Berkeley, Berkeley, CA, 1987. Also Vol. 37 of *Lecture Notes in Computer Science*, Springer-Verlag, Berlin, West Germany, 1988 and Electronics Research Laboratory Memorandum UCB/ERL M87/77.

51. O. Günther and E. Wong. The arc tree: An approximation scheme to represent arbitrary curved shapes. *Computer Vision, Graphics, and Image Processing*, 51(3):313–337, September 1990.

52. A. Guttman. R-trees: A dynamic index structure for spatial searching. In *Proceedings of the ACM SIGMOD Conference*, pp. 47–57, Boston, MA, June 1984.

53. A. Henrich. A distance-scan algorithm for spatial access structures. In *Proceedings of the 2nd ACM Workshop on Geographic Information Systems*, N. Pissinou and K. Makki, eds., pp. 136–143, Gaithersburg, MD, December 1994.

54. A. Henrich, H.-W. Six, and P. Widmayer. The LSD tree: Spatial access to multidimensional point and non-point data. In *Proceedings of the 15th International Conference on Very Large Databases (VLDB)*, P. M. G. Apers and G. Wiederhold, eds., pp. 45–53, Amsterdam, the Netherlands, August 1989.

55. K. Hinrichs and J. Nievergelt. The grid file: A data structure designed to support proximity queries on spatial objects. In *Proceedings of WG'83, International Workshop on Graphtheoretic Concepts in Computer Science*, M. Nagl and J. Perl, eds., pp. 100–113, Trauner Verlag, Haus Ohrbeck (near Osnabrück), West Germany, 1983.

56. G. R. Hjaltason and H. Samet. Ranking in spatial databases. In *Advances in Spatial Databases—4th International Symposium, SSD'95*, M. J. Egenhofer and J. R. Herring, eds., Vol. 951 of *Springer-Verlag Lecture Notes in Computer Science*, pp. 83–95, Portland, ME, August 1995.

57. G. R. Hjaltason and H. Samet. Distance browsing in spatial databases. *ACM Transactions on Database Systems*, 24(2):265–318, June 1999. Also University of Maryland Computer Science Technical Report TR-3919, July 1998.

58. G. R. Hjaltason and H. Samet. Index-driven similarity search in metric spaces. *ACM Transactions on Database Systems*, 28(4):517–580, December 2003.

59. E. G. Hoel and H. Samet. Efficient processing of spatial queries in line segment databases. In *Advances in Spatial Databases—2nd Symposium, SSD'91*, O. Günther and H.-J. Schek, eds., Vol. 525 of *Springer-Verlag Lecture Notes in Computer Science*, pp. 237–256, Zurich, Switzerland, August 1991.

60. E. G. Hoel and H. Samet. A qualitative comparison study of data structures for large line segment databases. In *Proceedings of the ACM SIGMOD Conference*, M. Stonebraker, ed., pp. 205–214, San Diego, CA, June 1992.

61. E. G. Hoel and H. Samet. Benchmarking spatial join operations with spatial output. In *Proceedings of the 21st International Conference on Very Large Data Bases (VLDB)*, U. Dayal, P. M. D. Gray, and S. Nishio, eds., pp. 606–618, Zurich, Switzerland, September 1995.

62. G. M. Hunter. Efficient computation and data structures for graphics. PhD thesis, Department of Electrical Engineering and Computer Science, Princeton University, Princeton, NJ, 1978.

63. G. M. Hunter and K. Steiglitz. Operations on images using quad trees. *IEEE Transactions on Pattern Analysis and Machine Intelligence*, 1(2):145–153, April 1979.

64. E. Jacox and H. Samet. Spatial join techniques. *ACM Transactions on Database Systems*, 32(1):7, March 2007. Also an expanded version in University of Maryland Computer Science Technical Report TR-4730, June 2005.

65. B. Johnson and B. Shneiderman. Tree-maps: A space filling approach to the visualization of hierarchical information structures. In *Proceedings IEEE Visualization'91*, G. M. Nielson and L. Rosenbloom, eds., pp. 284–291, San Diego, CA, October 1991. Also University of Maryland Computer Science Technical Report TR-2657, April 1991.

66. N. Katayama and S. Satoh. The SR-tree: An index structure for high-dimensional nearest neighbor queries. In *Proceedings of the ACM SIGMOD Conference*, J. Peckham, ed., pp. 369–380, Tucson, AZ, May 1997.

67. G. Kedem. The quad-CIF tree: A data structure for hierarchical on-line algorithms. In *Proceedings of the 19th Design Automation Conference*, pp. 352–357, Las Vegas, NV, June 1982. Also University of Rochester Computer Science Technical Report TR-91, September 1981.

68. A. Klinger. Patterns and search statistics. In *Optimizing Methods in Statistics*, J. S. Rustagi, ed., pp. 303–337. Academic Press, New York, 1971.

69. K. Knowlton. Progressive transmission of grey-scale and binary pictures by simple efficient, and lossless encoding schemes. *Proceedings of the IEEE*, 68(7):885–896, July 1980.

70. D. E. Knuth. Big omicron and big omega and big theta. *SIGACT News*, 8(2):18–24, April–June 1976.

71. D. E. Knuth. *The Art of Computer Programming: Sorting and Searching*, Vol. 3. Addison-Wesley, Reading, MA, 2nd edition, 1998.

72. M. D. Lieberman, J. Sankaranarayanan, and H. Samet. A fast similarity join algorithm using graphics processing units. In *Proceedings of the 24th IEEE International Conference on Data Engineering*, pp. 1111–1120, Cancun, Mexico, April 2008.

73. M. Lindenbaum, H. Samet, and G. R. Hjaltason. A probabilistic analysis of trie-based sorting of large collections of line segments in spatial databases. *SIAM Journal on Computing*, 35(1):22–58, September 2005. Also see *Proceedings of the 10th International Conference on Pattern Recognition*, Vol. II, pp. 91–96, Atlantic City, NJ, June 1990 and University of Maryland Computer Science Technical Report TR-3455.1, February 2000.

74. G. Marchionini, H. Samet, and L. Brandt. Introduction to the digital government special issue. *Communications of the ACM*, 46(1):24–27, January 2003.

75. E. M. McCreight. Efficient algorithms for enumerating intersecting intervals and rectangles. Technical Report CSL-80-09, Xerox Palo Alto Research Center, Palo Alto, CA, June 1980.

76. E. M. McCreight. Priority search trees. *SIAM Journal on Computing*, 14(2):257–276, May 1985. Also Xerox Palo Alto Research Center Technical Report CSL-81-5, January 1982.

77. D. Meagher. Octree encoding: A new technique for the representation, manipulation, and display of arbitrary 3-D objects by computer. Electrical and Systems Engineering Technical Report IPL-TR-80-111, Rensselaer Polytechnic Institute, Troy, NY, October 1980.

78. D. Meagher. Geometric modeling using octree encoding. *Computer Graphics and Image Processing*, 19(2):129–147, June 1982.

79. G. Nagy and S. Seth. Hierarchical representation of optically scanned documents. In *Proceedings of the 7th International Conference on Pattern Recognition*, pp. 347–349, Montréal, Canada, July 1984.

80. R. C. Nelson and H. Samet. A consistent hierarchical representation for vector data. *Computer Graphics*, 20(4):197–206, August 1986. Also in *Proceedings of the SIGGRAPH'86 Conference*, Dallas, TX, August 1986.

81. R. C. Nelson and H. Samet. A population analysis for hierarchical data structures. In *Proceedings of the ACM SIGMOD Conference*, pp. 270–277, San Francisco, CA, May 1987.

82. J. Nievergelt, H. Hinterberger, and K. C. Sevcik. The grid file: An adaptable, symmetric multikey file structure. *ACM Transactions on Database Systems*, 9(1):38–71, March 1984.

83. S. M. Omohundro. Five balltree construction algorithms. Technical Report TR-89-063, International Computer Science Institute, Berkeley, CA, December 1989.

84. J. A. Orenstein. Multidimensional tries used for associative searching. *Information Processing Letters*, 14(4):150–157, June 1982.

85. J. Ponce and O. Faugeras. An object centered hierarchical representation for 3d objects: The prism tree. *Computer Vision, Graphics, and Image Processing*, 38(1):1–28, April 1987.

86. F. P. Preparata and M. I. Shamos. *Computational Geometry: An Introduction*. Springer-Verlag, New York, 1985.

87. D. R. Reddy and S. Rubin. Representation of three-dimensional objects. Computer Science Technical Report CMU-CS-78-113, Carnegie-Mellon University, Pittsburgh, PA, April 1978.

88. A. A. G. Requicha. Representations of rigid solids: theory, methods, and systems. *ACM Computing Surveys*, 12(4):437–464, December 1980.

89. J. T. Robinson. The K-D-B-tree: A search structure for large multidimensional dynamic indexes. In *Proceedings of the ACM SIGMOD Conference*, pp. 10–18, Ann Arbor, MI, April 1981.

90. H. Samet. Region representation: Quadtrees from binary arrays. *Computer Graphics and Image Processing*, 13(1):88–93, May 1980. Also University of Maryland Computer Science Technical Report TR-767, May 1979.

91. H. Samet. Region representation: Quadtrees from boundary codes. *Communications of the ACM*, 23(3):163–170, March 1980. Also University of Maryland Computer Science Technical Report TR-741, March 1979.

92. H. Samet. An algorithm for converting rasters to quadtrees. *IEEE Transactions on Pattern Analysis and Machine Intelligence*, 3(1):93–95, January 1981. Also University of Maryland Computer Science Technical Report TR-766, May 1979.

93. H. Samet. Computing perimeters of images represented by quadtrees. *IEEE Transactions on Pattern Analysis and Machine Intelligence*, 3(6):683–687, November 1981. Also University of Maryland Computer Science Technical Report TR-755, April 1979.

94. H. Samet. Connected component labeling using quadtrees. *Journal of the ACM*, 28(3):487–501, July 1981. Also University of Maryland Computer Science Technical Report TR-756, April 1979.

95. H. Samet. Distance transform for images represented by quadtrees. *IEEE Transactions on Pattern Analysis and Machine Intelligence*, 4(3):298–303, May 1982. Also University of Maryland Computer Science Technical Report TR-780, July 1979.

96. H. Samet. Neighbor finding techniques for images represented by quadtrees. *Computer Graphics and Image Processing*, 18(1):37–57, January 1982. Also in *Digital Image Processing and Analysis: Vol. 2: Digital Image Analysis*, R. Chellappa and A. Sawchuck, eds., pp. 399–419, IEEE Computer Society Press, Washington, DC, 1986; and University of Maryland Computer Science Technical Report TR-857, January 1980.

97. H. Samet. A quadtree medial axis transform. *Communications of the ACM*, 26(9):680–693, September 1983. Also see CORRIGENDUM, *Communications of the ACM*, 27(2):151, February 1984 and University of Maryland Computer Science Technical Report TR-803, August 1979.

98. H. Samet. Algorithms for the conversion of quadtrees to rasters. *Computer Vision, Graphics, and Image Processing*, 26(1):1–16, April 1984. Also University of Maryland Computer Science Technical Report TR-979, November 1980.

99. H. Samet. Reconstruction of quadtrees from quadtree medial axis transforms. *Computer Vision, Graphics, and Image Processing*, 29(3):311–328, March 1985. Also University of Maryland Computer Science Technical Report TR-1224, October 1982.

100. H. Samet. An overview of quadtrees, octrees, and related hierarchical data structures. In *Theoretical Foundations of Computer Graphics and CAD*, R. A. Earnshaw, ed., Vol. 40 of *NATO ASI Series F: Computer and System Sciences*, pp. 51–68. Springer-Verlag, Berlin, West Germany, 1988.

101. H. Samet. Implementing ray tracing with octrees and neighbor finding. *Computers & Graphics*, 13(4):445–460, 1989. Also University of Maryland Computer Science Technical Report TR-2204, February 1989.

102. H. Samet. Neighbor finding in images represented by octrees. *Computer Vision, Graphics, and Image Processing*, 46(3):367–386, June 1989. Also University of Maryland Computer Science Technical Report TR-1968, January 1988.

103. H. Samet. *Applications of Spatial Data Structures: Computer Graphics, Image Processing, and GIS*. Addison-Wesley, Reading, MA, 1990.

104. H. Samet. *The Design and Analysis of Spatial Data Structures*. Addison-Wesley, Reading, MA, 1990.

105. H. Samet. Decoupling partitioning and grouping: overcoming shortcomings of spatial indexing with bucketing. *ACM Transactions on Database Systems*, 29(4):789–830, December 2004. Also University of Maryland Computer Science Technical Report TR-4523, August 2003.

106. H. Samet. *Foundations of Multidimensional and Metric Data Structures*. Morgan-Kaufmann, San Francisco, CA, 2006.

107. H. Samet, H. Alborzi, F. Brabec, C. Esperança, G. R. Hjaltason, F. Morgan, and E. Tanin. Use of the SAND spatial browser for digital government applications. *Communications of the ACM*, 46(1):63–66, January 2003.

108. H. Samet and J. Sankaranarayanan. Maximum containing cell sizes in cover fieldtrees and loose quadtrees and octrees. Computer Science Technical Report TR-4900, University of Maryland, College Park, MD, October 2007.

109. H. Samet, J. Sankaranarayanan, and H. Alborzi. Scalable network distance browsing in spatial databases. In *Proceedings of the ACM SIGMOD Conference*, pp. 43–54, Vancouver, Canada, June 2008. Also see University of Maryland Computer Science Technical Report TR-4865, April 2007 (2008 ACM SIGMOD Best Paper Award).

110. H. Samet, C. A. Shaffer, and R. E. Webber. Digitizing the plane with cells of non-uniform size. *Information Processing Letters*, 24(6):369–375, April 1987. Also an expanded version in University of Maryland Computer Science Technical Report TR-1619, January 1986.

111. H. Samet and M. Tamminen. Bintrees, CSG trees, and time. *Computer Graphics*, 19(3):121–130, July 1985. Also in *Proceedings of the SIGGRAPH'85 Conference*, San Francisco, CA, July 1985.

112. H. Samet and M. Tamminen. Efficient component labeling of images of arbitrary dimension represented by linear bintrees. *IEEE Transactions on Pattern Analysis and Machine Intelligence*, 10(4):579–586, July 1988.

113. H. Samet and R. E. Webber. Storing a collection of polygons using quadtrees. *ACM Transactions on Graphics*, 4(3):182–222, July 1985. Also see *Proceedings of Computer Vision and Pattern Recognition'83*, pp. 127–132, Washington, DC, June 1983 and University of Maryland Computer Science Technical Report TR-1372, February 1984.

114. J. Sankaranarayanan and H. Samet. Distance oracles for spatial networks. In *Proceedings of the 25th IEEE International Confenrence on Data Engineering*, pp. 652–663, Shanghai, China, April 2009.

115. J. Sankaranarayanan, H. Alborzi, and H. Samet. Efficient query processing on spatial networks. In *Proceedings of the 13th ACM International Symposium on Advances in Geographic Information Systems*, pp. 200–209, Bremen, Germany, November 2005.

116. J. Sankaranarayanan, H. Samet, and H. Alborzi. Path oracles for spatial networks. In *Proceedings of the 35th International Conference on Very Large Data Bases (VLDB)*, Lyon, France, August 2009.

117. B. Seeger and H.-P. Kriegel. The buddy-tree: An efficient and robust access method for spatial data base systems. In *Proceedings of the 16th International Conference on Very Large Databases (VLDB)*, D. McLeod, R. Sacks-Davis, and H.-J. Schek, eds., pp. 590–601, Brisbane, Australia, August 1990.

118. T. Sellis, N. Roussopoulos, and C. Faloutsos. The R^+-tree: A dynamic index for multi-dimensional objects. In *Proceedings of the 13th International Conference on Very Large Databases (VLDB)*, P. M. Stocker and W. Kent, eds., pp. 71–79, Brighton, United Kingdom, September 1987. Also University of Maryland Computer Science Technical Report TR-1795, 1987.

119. K. Sevcik and N. Koudas. Filter trees for managing spatial data over a range of size granularities. In *Proceedings of the 22nd International Conference on Very Large Data Bases (VLDB)*, T. M. Vijayaraman, A. P. Buchmann, C. Mohan, and N. L. Sarda, eds., pp. 16–27, Mumbai (Bombay), India, September 1996.

120. C. A. Shaffer and H. Samet. Optimal quadtree construction algorithms. *Computer Vision, Graphics, and Image Processing*, 37(3):402–419, March 1987.

121. C. A. Shaffer, H. Samet, and R. C. Nelson. QUILT: A geographic information system based on quadtrees. *International Journal of Geographical Information Systems*, 4(2):103–131, April–June 1990. Also University of Maryland Computer Science Technical Report TR-1885.1, July 1987.

122. B. Shneiderman. Tree visualization with tree-maps: 2-d space-filling approach. *ACM Transactions on Graphics*, 11(1):92–99, January 1992. Also University of Maryland Computer Science Technical Report TR-2645, April 1991.

123. M. Shneier. Two hierarchical linear feature representations: Edge pyramids and edge quadtrees. *Computer Graphics and Image Processing*, 17(3):211–224, November 1981. Also University of Maryland Computer Science Technical Report TR-961, October 1980.

124. R. Sivan and H. Samet. Algorithms for constructing quadtree surface maps. In *Proceedings of the 5th International Symposium on Spatial Data Handling*, Vol. 1, pp. 361–370, Charleston, SC, August 1992.

125. H.-W. Six and P. Widmayer. Spatial searching in geometric databases. In *Proceedings of the 4th IEEE International Conference on Data Engineering*, pp. 496–503, Los Angeles, CA, February 1988.

126. J.-W. Song, K.-Y. Whang, Y.-K. Lee, and S.-W. Kim. The clustering property of corner transformation for spatial database applications. In *Proceedings of the IEEE COMPSAC'99 Conference*, pp. 28–35, Phoenix, AZ, October 1999.

127. M. Stonebraker, T. Sellis, and E. Hanson. An analysis of rule indexing implementations in data base systems. In *Proceedings of the 1st International Conference on Expert Database Systems*, pp. 353–364, Charleston, SC, April 1986.

128. M. Tamminen. The EXCELL method for efficient geometric access to data. *Acta Polytechnica Scandinavica*, 1981. Also *Mathematics and Computer Science Series* No. 34.

129. M. Tamminen. Comment on quad- and octtrees. *Communications of the ACM*, 27(3):248–249, March 1984.

130. M. Tamminen and H. Samet. Efficient octree conversion by connectivity labeling. *Computer Graphics*, 18(3):43–51, July 1984. Also in *Proceedings of the SIGGRAPH'84 Conference*, Minneapolis, MN, July 1984.

131. E. Tanin, A. Harwood, and H. Samet. A distributed quadtree index for peer-to-peer settings. In *Proceedings of the 21st IEEE International Conference on Data Engineering*, pp. 254–255, Tokyo, Japan, April 2005.

132. E. Tanin, A. Harwood, and H. Samet. Using a distributed quadtree index in P2P networks. *VLDB Journal*, 16(2):165–178, April 2007.

133. S. L. Tanimoto and T. Pavlidis. A hierarchical data structure for picture processing. *Computer Graphics and Image Processing*, 4(2):104–119, June 1975.

134. T. Ulrich. Loose octrees. In *Game Programming Gems*, M. A. DeLoura, ed., pp. 444–453. Charles River Media, Rockland, MA, 2000.

135. W. Wang, J. Yang, and R. Muntz. PK-tree: A spatial index structure for high dimensional point data. In *Proceedings of the 5th International Conference on Foundations of Data Organization and Algorithms (FODO)*, K. Tanaka and S. Ghandeharizadeh, eds., pp. 27–36, Kobe, Japan, November 1998. Also University of California at Los Angeles Computer Science Technical Report 980032, September 1998.

136. J. E. Warnock. A hidden line algorithm for halftone picture representation. Computer Science Technical Report TR 4-5, University of Utah, Salt Lake City, UT, May 1968.

137. J. E. Warnock. A hidden surface algorithm for computer generated half tone pictures. Computer Science Technical Report TR 4-15, University of Utah, Salt Lake City, UT, June 1969.

138. D. A. White and R. Jain. Similarity indexing with the SS-tree. In *Proceedings of the 12th IEEE International Conference on Data Engineering*, S. Y. W. Su, ed., pp. 516–523, New Orleans, LA, February 1996.

7

Basic Graph Algorithms

Samir Khuller
University of Maryland

Balaji Raghavachari
University of Texas at Dallas

7.1 Introduction

Graphs provide a powerful tool to model objects and relationships between objects. The study of graphs dates back to the eighteenth century, when Euler defined the Königsberg bridge problem, and since then has been pursued by many researchers. Graphs can be used to model problems in many areas such as transportation, scheduling, networks, robotics, VLSI design, compilers, mathematical biology, and software engineering. Many optimization problems from these and other diverse areas can be phrased in graph-theoretic terms, leading to algorithmic questions about graphs.

Graphs are defined by a set of vertices and a set of edges, where each edge connects two vertices. Graphs are further classified into directed and undirected graphs, depending on whether their edges are directed or not. An important subclass of directed graphs that arises in many applications, such as precedence constrained scheduling problems, are **directed acyclic graphs** (DAG). Interesting subclasses of undirected graphs include **trees**, **bipartite graphs**, and **planar graphs**.

In this chapter, we focus on a few basic problems and algorithms dealing with graphs. Other chapters in this handbook provide details on specific algorithmic techniques and problem areas dealing with graphs, e.g., randomized algorithms (Chapter 12), combinatorial algorithms (Chapter 8), dynamic graph algorithms (Chapter 9), graph drawing (Chapter 6 of *Algorithms and Theory of Computation Handbook, Second Edition: Special Topics and Techniques*), and approximation algorithms (Chapter 34). Pointers into the literature are provided for various algorithmic results about graphs that are not covered in depth in this chapter.

7.2 Preliminaries

An undirected graph $G = (V, E)$ is defined as a set V of vertices and a set E of edges. An edge $e = (u, v)$ is an unordered pair of vertices. A directed graph is defined similarly, except that its edges are ordered pairs of vertices, i.e., for a directed graph, $E \subseteq V \times V$. The terms nodes and vertices are used interchangeably. In this chapter, it is assumed that the graph has neither self loops—edges of the form (v, v)—nor multiple edges connecting two given vertices. The number of vertices of a graph, $|V|$, is often denoted by n. A graph is a **sparse graph** if $|E| \ll |V|^2$.

Bipartite graphs form a subclass of graphs and are defined as follows. A graph $G = (V, E)$ is bipartite if the vertex set V can be partitioned into two sets X and Y such that $E \subseteq X \times Y$. In other words, each edge of G connects a vertex in X with a vertex in Y. Such a graph is denoted by $G = (X, Y, E)$. Since bipartite graphs occur commonly in practice, often algorithms are designed specially for them. Planar graphs are graphs that can be drawn in the plane without any two edges crossing each other. Let K_n be the complete graph on n vertices, and $K_{x,y}$ be the complete bipartite graph with x and y vertices in either side of the bipartite graph, respectively. A homeomorph of a graph is obtained by subdividing an edge by adding new vertices.

A vertex w is adjacent to another vertex v if $(v, w) \in E$. An edge (v, w) is said to be incident to vertices v and w. The neighbors of a vertex v are all vertices $w \in V$ such that $(v, w) \in E$. The number of edges incident to a vertex is called its **degree**. For a directed graph, if (v, w) is an edge, then we say that the edge goes from v to w. The out-degree of a vertex v is the number of edges from v to other vertices. The in-degree of v is the number of edges from other vertices to v.

A **path** $p = [v_0, v_1, \ldots, v_k]$ from v_0 to v_k is a sequence of vertices such that (v_i, v_{i+1}) is an edge in the graph for $0 \le i < k$. Any edge may be used only once in a path. An intermediate vertex (or internal vertex) on a path $P[u, v]$, a path from u to v, is a vertex incident to the path, other than u and v. A path is simple if all of its internal vertices are distinct. A **cycle** is a path whose end vertices are the same, i.e., $v_0 = v_k$. A **walk** $w = [v_0, v_1, \ldots, v_k]$ from v_0 to v_k is a sequence of vertices such that (v_i, v_{i+1}) is an edge in the graph for $0 \le i < k$. A closed walk is one in which $v_0 = v_k$. A graph is said to be connected if there is a path between every pair of vertices. A directed graph is said to be **strongly connected** if there is a path between every pair of vertices in each direction. An acyclic, undirected graph is a **forest**, and a tree is a connected forest. A maximal forest F of a graph G is a forest of G such that the addition of any other edge of G to F introduces a cycle. A directed graph that does not have any cycles is known as a DAG. Consider a binary relation C between the vertices of an undirected graph G such that for any two vertices u and v, uCv if and only if there is a path in G between u and v. C is an equivalence relation, and it partitions the vertices of G into equivalence classes, known as the connected components of G.

Graphs may have weights associated with edges or vertices. In the case of edge-weighted graphs (edge weights denoting lengths), the distance between two vertices is the length of a shortest path between them, where the length of a path is defined as the sum of the weights of its edges. The diameter of a graph is the maximum of the distance between all pairs of vertices.

There are two convenient ways of representing graphs on computers. In the adjacency list representation, each vertex has a linked list; there is one entry in the list for each of its adjacent vertices.

The graph is thus, represented as an array of linked lists, one list for each vertex. This representation uses $O(|V| + |E|)$ storage, which is good for sparse graphs. Such a storage scheme allows one to scan all vertices adjacent to a given vertex in time proportional to the degree of the vertex. In the adjacency matrix representation, an $n \times n$ array is used to represent the graph. The $[i, j]$ entry of this array is 1 if the graph has an edge between vertices i and j, and 0 otherwise. This representation permits one to test if there is an edge between any pair of vertices in constant time. Both these representation schemes extend naturally to represent directed graphs. For all algorithms in this chapter except the all-pairs shortest paths problem, it is assumed that the given graph is represented by an adjacency list.

Section 7.3 discusses various tree traversal algorithms. Sections 7.4 and 7.5 discuss depth-first and breadth-first search techniques, respectively. Section 7.6 discusses the single-source shortest-path problem. Section 7.7 discusses **minimum spanning trees**. Section 7.8 discusses some traversal problems in graphs. Section 7.9 discusses various topics such as **planar graphs**, graph coloring, light approximate shortest path trees, and network decomposition, and Section 7.10 concludes with some pointers to current research on graph algorithms.

7.3 Tree Traversals

A tree is rooted if one of its vertices is designated as the root vertex and all edges of the tree are oriented (directed) to point away from the root. In a rooted tree, there is a directed path from the root to any vertex in the tree. For any directed edge (u, v) in a rooted tree, u is v's parent and v is u's child. The *descendants* of a vertex w are all vertices in the tree (including w) that are reachable by directed paths starting at w. The ancestors of a vertex w are those vertices for which w is a descendant. Vertices that have no children are called **leaves**. A binary tree is a special case of a rooted tree in which each node has at most two children, namely the left child and the right child. The trees rooted at the two children of a node are called the left subtree and right subtree.

In this section we study techniques for processing the vertices of a given binary tree in various orders. It is assumed that each vertex of the binary tree is represented by a record that contains fields to hold attributes of that vertex and two special fields left and right that point to its left and right subtree respectively. Given a pointer to a record, the notation used for accessing its fields is similar to that used in the *C* programming language.

The three major tree traversal techniques are preorder, inorder, and postorder. These techniques are used as procedures in many tree algorithms where the vertices of the tree have to be processed in a specific order. In a preorder traversal, the root of any subtree has to be processed before any of its descendants. In a postorder traversal, the root of any subtree has to be processed after all of its descendants. In an inorder traversal, the root of a subtree is processed after all vertices in its left subtree have been processed, but before any of the vertices in its right subtree are processed. Preorder and postorder traversals generalize to arbitrary rooted trees. The algorithm below shows how postorder traversal of a binary tree can be used to count the number of descendants of each node and store the value in that node. The algorithm runs in linear time in the size of the tree.

```
POSTORDER (T)
1    if T ≠ nil then
2        lc ← POSTORDER(T → left).
3        rc ← POSTORDER(T → right).
4        T → desc ← lc + rc + 1.
5        return (T → desc).
6    else
7        return 0.
8    end-if
end-proc
```

7.4 Depth-First Search

Depth-first search (DFS) is a fundamental graph searching technique developed by Hopcroft and Tarjan [16] and Tarjan [27]. Similar graph searching techniques were given earlier by Even [8]. The structure of DFS enables efficient algorithms for many other graph problems such as biconnectivity, triconnectivity, and planarity [8].

The algorithm first initializes all vertices of the graph as being unvisited. Processing of the graph starts from an arbitrary vertex, known as the root vertex. Each vertex is processed when it is first discovered (also referred to as visiting a vertex). It is first marked as visited, and its adjacency list is then scanned for unvisited vertices. Each time an unvisited vertex is discovered, it is processed recursively by DFS. After a node's entire adjacency list has been explored, that instance of the DFS procedure returns. This procedure eventually visits all vertices that are in the same connected component of the root vertex. Once DFS terminates, if there are still any unvisited vertices left in the graph, one of them is chosen as the root and the same procedure is repeated.

The set of edges that led to the discovery of new vertices forms a maximal forest of the graph, known as the **DFS forest**. The algorithm keeps track of this forest using parent-pointers; an array element $p[v]$ stores the parent of vertex v in the tree. In each connected component, only the root vertex has a nil parent in the DFS tree.

7.4.1 The DFS Algorithm

DFS is illustrated using an algorithm that assigns labels to vertices such that vertices in the same component receive the same label, a useful preprocessing step in many problems. Each time the algorithm processes a new component, it numbers its vertices with a new label.

DFS-Connected-Component (G)
1   $c \leftarrow 0$.
2   **for** all vertices v in G **do**
3       $visited[v] \leftarrow$ **false**.
4       $finished[v] \leftarrow$ **false**.
5       $p[v] \leftarrow$ **nil**.
6   **end-for**
7   **for** all vertices v in G **do**
8   **if not** $visited[v]$ **then**
9           $c \leftarrow c + 1$.
10   DFS (v, c).
11   **end-if**
12 **end-for**
end-proc

DFS (v, c)
1   $visited[v] \leftarrow$ **true**.
2   $component[v] \leftarrow c$.
3   **for** all vertices w in $adj[v]$ **do**
4   **if not** $visited[w]$ **then**
5           $p[w] \leftarrow v$.
6   DFS (w, c).
7   **end-if**
8   **end-for**
9   $finished[v] \leftarrow$ **true**.
end-proc

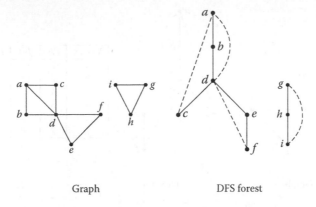

Graph   DFS forest

FIGURE 7.1 Sample execution of DFS on a graph having two connected components.

7.4.2 Sample Execution

Figure 7.1 shows a graph having two connected components. DFS started execution at vertex a, and the DFS forest is shown on the right. DFS visited the vertices b, d, c, e, and f, in that order. It then continued with vertices g, h, and i. In each case, the recursive call returned when the vertex has no more unvisited neighbors. Edges (d, a), (c, a), (f, d), and (i, g) are called back edges, and these edges do not belong to the DFS forest.

7.4.3 Analysis

A vertex v is processed as soon as it is encountered, and therefore at the start of DFS (v), $visited[v]$ is false. Since $visited[v]$ is set to true as soon as DFS starts execution, each vertex is visited exactly once. DFS processes each edge of the graph exactly twice, once from each of its incident vertices. Since the algorithm spends constant time processing each edge of G, it runs in $O(|V| + |E|)$ time.

7.4.4 Classification of Edges

In the following discussion, there is no loss of generality in assuming that the input graph is connected. For a rooted DFS tree, vertices u and v are said to be related, if either u is an ancestor of v, or vice versa.

DFS is useful due to the special nature by which the edges of the graph may be classified with respect to a DFS tree. Note that the DFS tree is not unique, and which edges are added to the tree depends on the order in which edges are explored while executing DFS. Edges of the DFS tree are known as tree edges. All other edges of the graph are known as back edges, and it can be shown that for any edge (u, v), u and v must be related. The graph does not have any cross edges—edges that connect two vertices that are unrelated.

7.4.5 Articulation Vertices and Biconnected Components

One of the many applications of DFS is to decompose a graph into its biconnected components. In this section, it is assumed that the graph is connected. An **articulation vertex** (also known as cut vertex) is a vertex whose deletion along with its incident edges breaks up the remaining graph into two or more disconnected pieces. A graph is called biconnected if it has no articulation vertices. A biconnected component of a **connected graph** is a maximal subset of edges such that

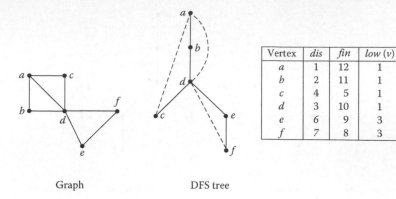

Vertex	dis	fin	low (v)
a	1	12	1
b	2	11	1
c	4	5	1
d	3	10	1
e	6	9	3
f	7	8	3

Graph   DFS tree

FIGURE 7.2 Identifying cut vertices.

the corresponding induced subgraph is biconnected. Each edge of the graph belongs to exactly one biconnected component. Biconnected components can have cut vertices in common.

The graph in Figure 7.2 has two biconnected components, formed by the edge sets $\{(a, b), (a, c),$ $(a, d), (b, d), (c, d)\}$ and $\{(d, e), (d, f), (e, f)\}$. There is a single cut vertex d and it is shared by both biconnected components.

We now discuss a linear-time algorithm, developed by Hopcroft and Tarjan [16] and Tarjan [27], to identify the cut vertices and biconnected components of a connected graph. The algorithm uses the global variable time that is incremented every time a new vertex is visited or when DFS finishes visiting a vertex. Time is initially 0, and is $2|V|$ when the algorithm finally terminates. The algorithm records the value of time when a variable v is first visited in the array location $dis[v]$ and the value of time when $DFS(v)$ completes execution in $fin[v]$. We refer to $dis[v]$ and $fin[v]$ as the discovery time and finish time of vertex v, respectively.

Let T be a DFS tree of the given graph G. The notion of $low(v)$ of a vertex v with respect to T is defined as follows.

$$low(v) = \min(dis[v], dis[w] : (u, w) \text{ is a back edge for some descendant } u \text{ of } v)$$

$low(v)$ of a vertex is the discovery number of the vertex closest to the root that can be reached from v by following zero or more tree edges downward, and at most one back edge upward. It captures how far high the subtree of T rooted at v can reach by using at most one back edge. Figure 7.2 shows an example of a graph, a DFS tree of the graph and a table listing the values of dis, fin, and low of each vertex corresponding to that DFS tree.

Let T be the DFS tree generated by the algorithm, and let r be its root vertex. First, r is a cut vertex if and only if it has two or more children. This follows from the fact that there are no cross edges with respect to a DFS tree. Therefore the removal of r from G disconnects the remaining graph into as many components as the number of children of r. The low values of vertices can be used to find cut vertices that are nonroot vertices in the DFS tree. Let $v \neq r$ be a vertex in G. The following theorem characterizes precisely when v is a cut vertex in G.

THEOREM 7.1   *Let T be a DFS tree of a connected graph G, and let v be a nonroot vertex of T. Vertex v is a cut vertex of G if and only if there is a child w of v in T with $low(w) \geq dis[v]$.*

Computing low values of a vertex and identifying all the biconnected components of a graph can be done efficiently with a single DFS scan. The algorithm uses a stack of edges. When an edge is encountered for the first time it is pushed into the stack irrespective of whether it is a tree edge or a back edge. Each time a cut vertex v is identified because $low(w) \geq dis[v]$ (as in Theorem 7.1), the stack contains the edges of the biconnected component as a contiguous block, with the edge (v, w) at the bottom of this block. The algorithm pops the edges of this biconnected component from the stack, and sets $cut[v]$ to true to indicate that v is a cut vertex.

BICONNECTED COMPONENTS (G)
1   $time \leftarrow 0$.
2   MAKEEMPTYSTACK (S).
3   **for** each $u \in V$ **do**
4       $visited[u] \leftarrow$ **false.**
5       $cut[u] \leftarrow$ **false.**
6       $p[u] \leftarrow$ **nil.**
7   **end-for**
8   Let v be an arbitrary vertex, DFS(v).
end-proc

DFS (v)
1   $visited[v] \leftarrow$ **true.**
2   $time \leftarrow time + 1$.
3   $dis[v] \leftarrow time$.
4   $low[v] \leftarrow dis[v]$.
5   **for** all vertices w in $adj[v]$ **do**
6   **if not** $visited[w]$ **then**
7   PUSH $(S, (v, w))$.
8           $p[w] \leftarrow v$.
9   DFS(w).
10   **if** $(low[w] \geq dis[v])$ **then**
11   **if** $(dis[v] \neq 1)$ **then** $cut[v] \leftarrow$ **true.**   (* v is not the root *)
12   **else if** $(dis[w] > 2)$ **then** $cut[v] \leftarrow$ **true.** (* v is root, and has at least 2 children *)
13   **end-if**
14   OUTPUTCOMP(v, w).
15   **end-if**
16          $low[v] \leftarrow \min(low[v], low[w])$.
17   **else if** $(p[v] \neq w$ **and** $dis[w] < dis[v])$ **then**
18   PUSH $(S, (v, w))$.
19          $low[v] \leftarrow \min(low[v], dis[w])$.
20   **end-if**
21 **end-for**
22 $time \leftarrow time + 1$.
23 $fin[v] \leftarrow time$.
end-proc

OUTPUTCOMP(v, w)
1   PRINT ("New Biconnected Component Found").
2   **repeat**
3       $e \leftarrow$ POP (S).
4   PRINT (e).
5   **until** $(e = (v, w))$.
end-proc

In the example shown in Figure 7.2 when DFS(e) finishes execution and returns control to DFS(d), the algorithm discovers that d is a cut vertex because $low(e) \geq dis[d]$. At this time, the stack contains the edges (d,f), (e,f), and (d,e) at the top of the stack, which are output as one biconnected component.

Remarks: The notion of biconnectivity can be generalized to higher connectivities. A graph is said to be k-connected, if there is no subset of $(k - 1)$ vertices whose removal will disconnect the graph. For example, a graph is triconnected if it does not have any separating pairs of vertices—pairs of vertices whose removal disconnects the graph. A linear-time algorithm for testing whether a given graph is triconnected was given by Hopcroft and Tarjan [15]. An $O(|V|^2)$ algorithm for testing if a graph is k-connected for any constant k was given by Nagamochi and Ibaraki [25]. One can also define a corresponding notion of edge-connectivity, where edges are deleted from a graph rather than vertices. Galil and Italiano [11] showed how to reduce edge connectivity to vertex connectivity.

7.4.6 Directed Depth-First Search

The DFS algorithm extends naturally to directed graphs. Each vertex stores an adjacency list of its outgoing edges. During the processing of a vertex, the algorithm first marks the vertex as visited, and then scans its adjacency list for unvisited neighbors. Each time an unvisited vertex is discovered, it is processed recursively. Apart from tree edges and back edges (from vertices to their ancestors in the tree), directed graphs may also have forward edges (from vertices to their descendants) and cross edges (between unrelated vertices). There may be a cross edge (u, v) in the graph only if u is visited after the procedure call "DFS (v)" has completed execution. The following algorithm implements DFS in a directed graph. For each vertex v, the algorithm computes the discovery time of v ($dis[v]$) and the time at which DFS(v) finishes execution ($fin[v]$). In addition, each edge of the graph is classified as (1) tree edge or (2) back edge or (3) forward edge or (4) cross edge, with respect to the depth-first forest generated.

DIRECTED DFS (G)
1   **for** all vertices v in G **do**
2       $visited[v] \leftarrow$ **false**.
3       $finished[v] \leftarrow$ **false**.
4       $p[v] \leftarrow$ **nil**.
5   **end-for**
6   $time \leftarrow 0$.
7   **for** all vertices v in G **do**
8   **if not** $visited[v]$ **then**
9   DFS (v).
10   **end-if**
11   **end-for**
end-proc

DFS (v)
1   $visited[v] \leftarrow$ **true**.
2   $time \leftarrow time + 1$.
3   $dis[v] \leftarrow time$.
4   **for** all vertices w in $adj[j]$ **do**
5   **if not** $visited[w]$ **then**
6           $p[w] \leftarrow v$.

```
7          PRINT ("Edge from" v "to" w "is a Tree edge").
8          DFS (w).
9     else if not finished[w] then
10         PRINT ("Edge from" v "to" w "is a Back edge").
11    else if dis[v] < dis[w] then
12         PRINT ("Edge from" v "to" w "is a Forward edge").
13    else
14         PRINT ("Edge from" v "to" w "is a Cross edge").
15    end-if
16 end-for
17 finished[v] ← true.
18 time ← time + 1.
19 fin[v] ← time.
end-proc
```

7.4.7 Sample Execution

A sample execution of the directed DFS algorithm is shown in Figure 7.3. DFS was started at vertex
a, and the DFS forest is shown on the right. DFS visits vertices b, d, f, and c, in that order. DFS then
returns and continues with e, and then g. From g, vertices h and i are visited in that order. Observe
that (d, a) and (i, g) are back edges. Edges (c, d), (e, d), and (e, f) are cross edges. There is a single
forward edge (g, i).

7.4.8 Applications of DFS

7.4.8.1 Strong Connectivity

Directed DFS is used to design a linear-time algorithm that classifies the edges of a given directed
graph into its **strongly connected** components—maximal subgraphs that have directed paths connecting any pair of vertices in them. The algorithm itself involves running DFS twice, once on the
original graph, and then a second time on G^R, which is the graph obtained by reversing the direction of all edges in G. During the second DFS, the algorithm identifies all the strongly connected
components. The proof is somewhat subtle, and the reader is referred to [7] for details. Cormen et
al. [7] credit Kosaraju and Sharir for this algorithm. The original algorithm due to Tarjan [27] is
more complicated.

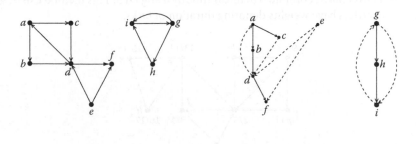

Graph   DFS forest

FIGURE 7.3 Sample execution of DFS on a directed graph.

7.4.8.2 Directed Acyclic Graphs

Checking if a graph is acyclic can be done in linear time using DFS. A graph has a cycle if and only if there exists a back edge relative to its DFS forest. A directed graph that does not have any cycles is known as a **directed acyclic graph** (DAG). DAGs are useful in modeling precedence constraints in scheduling problems, where nodes denote jobs/tasks, and a directed edge from u to v denotes the constraint that job u must be completed before job v can begin execution. Many problems on DAGs can be solved efficiently using dynamic programming (see Chapter 1).

7.4.8.3 Topological Order

A useful concept in DAGs is that of a topological order: a linear ordering of the vertices that is consistent with the partial order defined by its edges. In other words, the vertices can be labeled with distinct integers in the range $[1 \cdots |V|]$ such that if there is a directed edge from a vertex labeled i to a vertex labeled j, then $i < j$. Topological sort has applications in diverse areas such as project management, scheduling and circuit evaluation.

The vertices of a given DAG can be ordered topologically in linear time by a suitable modification of the DFS algorithm. It can be shown that ordering vertices by decreasing finish times (as computed by DFS) is a valid topological order. The DFS algorithm is modified as follows. A counter is initialized to $|V|$. As each vertex is marked finished, the counter value is assigned as its topological number, and the counter is decremented. Since there are no back edges in a DAG, for all edges (u, v), v will be marked finished before u. Thus, the topological number of v will be higher than that of u.

The execution of the algorithm is illustrated with an example in Figure 7.4. Along with each vertex, we show the discovery and finish times, respectively. Vertices are given decreasing topological numbers as they are marked finished. Vertex f finishes first and gets a topological number of 9 ($|V|$); d finishes next and gets numbered 8, and so on. The topological order found by the DFS is $g, h, i, a, b, e, c, d, f$, which is the reverse of the finishing order. Note that a given graph may have many valid topological ordering of the vertices.

Other topological ordering algorithms work by identifying and deleting vertices of in-degree zero (i.e., vertices with no incoming edges) recursively. With some care, this algorithm can be implemented in linear time as well.

7.4.8.4 Longest Path

In project scheduling, a DAG is used to model precedence constraints between tasks. A longest path in this graph is known as a critical path and its length is the least time that it takes to complete the project. The problem of computing the longest path in an arbitrary graph is NP-hard. However, longest paths in a DAG can be computed in linear time by using DFS. This method can be generalized to the case when vertices have weights denoting duration of tasks.

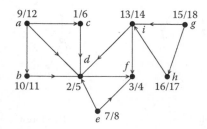

FIGURE 7.4 Example for topological sort. Order in which vertices finish: $f, d, c, e, b, a, i, h, g$.

The algorithm processes the vertices in reverse topological order. Let $P(v)$ denote the length of a longest path coming out of vertex v. When vertex v is processed, the algorithm computes the length of a longest path in the graph that starts at v.

$$P(v) = 1 + \max_{(v,w) \in E} P(w).$$

Since we are processing vertices in reverse topological order, w is processed before v, if (v, w) is an edge, and thus, $P(w)$ is computed before $P(v)$.

7.5 Breadth-First Search

Breadth-first search is another natural way of searching a graph. The search starts at a root vertex r. Vertices are added to a queue as they are discovered, and processed in first-in first-out (FIFO) order.

Initially, all vertices are marked as unvisited, and the queue consists of only the root vertex. The algorithm repeatedly removes the vertex at the front of the queue, and scans its neighbors in the graph. Any neighbor that is unvisited is added to the end of the queue. This process is repeated until the queue is empty. All vertices in the same connected component as the root vertex are scanned and the algorithm outputs a spanning tree of this component. This tree, known as a breadth-first tree, is made up of the edges that led to the discovery of new vertices. The algorithm labels each vertex v by $d[v]$, the distance (length of a shortest path) from the root vertex to v, and stores the BFS tree in the array p, using parent-pointers. Vertices can be partitioned into levels based on their distance from the root. Observe that edges not in the BFS tree always go either between vertices in the same level, or between vertices in adjacent levels. This property is often useful.

7.5.1 The BFS Algorithm

BFS-DISTANCE (G, r)
1   MAKEEMPTYQUEUE (Q).
2   **for** all vertices v in G **do**
3       $visited[v] \leftarrow$ **false**.
4       $d[v] \leftarrow \infty$.
5       $p[v] \leftarrow$ **nil**.
6   **end-for**
7   $visited[r] \leftarrow$ **true**.
8   $d[r] \leftarrow 0$.
9   ENQUEUE (Q, r).
10 **while not** EMPTY (Q) **do**
11      $v \leftarrow$ DEQUEUE (Q).
12   **for** all vertices w in $adj[v]$ **do**
13   **if not** $visited[w]$ **then**
14              $visited[w] \leftarrow$ **true**.
15              $p[w] \leftarrow v$.
16              $d[w] \leftarrow d[v] + 1$.
17   ENQUEUE (w, Q).
18   **end-if**
19   **end-for**
20 **end-while**
end-proc

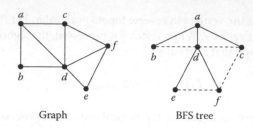

Graph   BFS tree

FIGURE 7.5 Sample execution of BFS on a graph.

7.5.2 Sample Execution

Figure 7.5 shows a connected graph on which BFS was run with vertex a as the root. When a is processed, vertices b, d, and c are added to the queue. When b is processed nothing is done since all its neighbors have been visited. When d is processed, e and f are added to the queue. Finally c, e, and f are processed.

7.5.3 Analysis

There is no loss of generality in assuming that the graph G is connected, since the algorithm can be repeated in each connected component, similar to the DFS algorithm. The algorithm processes each vertex exactly once, and each edge exactly twice. It spends a constant amount of time in processing each edge. Hence, the algorithm runs in $O(|V| + |E|)$ time.

7.5.4 Bipartite Graphs

A simple algorithm based on BFS can be designed to check if a given graph is bipartite: run BFS on each connected component of the graph, starting from an arbitrary vertex in each component as the root. The algorithm partitions the vertex set into the sets X and Y as follows. For a vertex v, if $d[v]$ is odd, then it inserts v into X. Otherwise $d[v]$ is even and it inserts v into Y. Now check to see if there is an edge in the graph that connects two vertices in the same set (X or Y). If the graph contains an edge between two vertices of the same set, say X, then we conclude that the graph is not bipartite, since the graph contains an odd-length cycle; otherwise the algorithm has partitioned the vertex set into X and Y and all edges of the graph connect a vertex in X with a vertex in Y, and therefore by definition, the graph is bipartite. (Note that it is known that a graph is bipartite if and only if it does not have a cycle of odd length.)

7.6 Single-Source Shortest Paths

A natural problem that often arises in practice is to compute the shortest paths from a specified node r to all other nodes in a graph. BFS solves this problem if all edges in the graph have the same length. Consider the more general case when each edge is given an arbitrary, nonnegative length. In this case, the length of a path is defined to be the sum of the lengths of its edges. The distance between two nodes is the length of a shortest path between them. The objective of the shortest path problem is to compute the distance from r to each vertex v in the graph, and a path of that length from r to v. The output is a tree, known as the shortest path tree, rooted at r. For any vertex v in the graph, the unique path from r to v in this tree is a shortest path from r to v in the input graph.

7.6.1 Dijkstra's Algorithm

Dijkstra's algorithm provides an efficient solution to the shortest path problem. For each vertex v, the algorithm maintains an upper bound of the distance from the root to vertex v in $d[v]$; initially $d[v]$ is set to infinity for all vertices except the root, which has d-value equal to zero. The algorithm maintains a set S of vertices with the property that for each vertex $v \in S$, $d[v]$ is the length of a shortest path from the root to v. For each vertex u in $V - S$, the algorithm maintains $d[u]$ to be the length of a shortest path from the root to u that goes entirely within S, except for the last edge. It selects a vertex u in $V - S$ with minimum $d[u]$ and adds it to S, and updates the distance estimates to the other vertices in $V - S$. In this update step it checks to see if there is a shorter path to any vertex in $V - S$ from the root that goes through u. Only the distance estimates of vertices that are adjacent to u need to be updated in this step. Since the primary operation is the selection of a vertex with minimum distance estimate, a priority queue is used to maintain the d-values of vertices (for more information about priority queues, see Chapter 4). The priority queue should be able to handle the DECREASEKEY operation to update the d-value in each iteration. The following algorithm implements Dijkstra's algorithm.

DIJKSTRA-SHORTEST PATHS (G, r)
1   **for** all vertices v in G **do**
2        $visited[v] \leftarrow$ **false**.
3        $d[v] \leftarrow \infty$.
4        $p[v] \leftarrow$ **nil**.
5   **end-for**
6    $d[r] \leftarrow 0$.
7   BUILDPQ (H, d).
8   **while not** EMPTY (H) **do**
9        $u \leftarrow$ DELETEMIN (H).
10       $visited[u] \leftarrow$ **true**.
11   **for** all vertices v in $adj[u]$ **do**
12   RELAX (u, v).
13   **end-for**
14   **end-while**
end-proc

RELAX (u, v)
1   **if not** $visited[v]$ **and** $d[v] > d[u] + w(u, v)$ **then**
2        $d[v] \leftarrow d[u] + w(u, v)$.
3        $p[v] \leftarrow u$.
4   DECREASEKEY $(H, v, d[v])$.
5   **end-if**
end-proc

7.6.2 Sample Execution

Figure 7.6 shows a sample execution of the algorithm. The column titled "Iter" specifies the number of iterations that the algorithm has executed through the while loop in Step 8. In iteration 0 the initial values of the distance estimates are ∞. In each subsequent line of the table, the column marked u shows the vertex that was chosen in Step 9 of the algorithm, and the other columns show the change to the distance estimates at the end of that iteration of the while loop. In the first iteration, vertex r

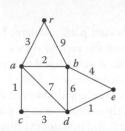

Iter	u	d(a)	d(b)	d(c)	d(d)	d(e)
0	–	∞	∞	∞	∞	∞
1	r	3	9	∞	∞	∞
2	a	3	5	4	10	∞
3	c	3	5	4	7	∞
4	b	3	5	4	7	9
5	d	3	5	4	7	8
6	e	3	5	4	7	8

FIGURE 7.6 Dijkstra's shortest path algorithm.

was chosen, after that a was chosen since it had the minimum distance label among the unvisited vertices, and so on. The distance labels of the unvisited neighbors of the visited vertex are updated in each iteration.

7.6.3 Analysis

The running time of the algorithm depends on the data structure that is used to implement the priority queue H. The algorithm performs $|V|$ DELETEMIN operations and at most $|E|$ DECREASEKEY operations. If a binary heap is used to find the records of any given vertex, each of these operations run in $O(\log |V|)$ time. There is no loss of generality in assuming that the graph is connected. Hence, the algorithm runs in $O(|E| \log |V|)$. If a Fibonacci heap [10] is used to implement the priority queue, the running time of the algorithm is $O(|E| + |V| \log |V|)$. Even though the Fibonacci heap gives the best asymptotic running time, the binary heap implementation is likely to give better running times for most practical instances.

7.6.4 Extensions

Dijkstra's algorithm can be generalized to solve several problems that are related to the shortest path problem. For example, in the bottleneck shortest path problem, the objective is to find, for each vertex v, a path from the root to v in which the length of the longest edge in that path is minimized. A small change to Dijkstra's algorithm (replacing the operation $+$ in RELAX by max) solves this problem. Other problems that can be solved by suitably modifying Dijkstra's algorithm include the following:

- Finding most reliable paths from the root to every vertex in a graph where each edge is given a probability of failure (independent of the other edges)
- Finding the fastest way to get from a given point in a city to a specified location using public transportation, given the train/bus schedules

7.6.5 Bellman–Ford Algorithm

The shortest path algorithm described above directly generalizes to directed graphs, but it does not work if the graph has edges of negative length. For graphs that have edges of negative length, but no cycles of negative length, there is a different algorithm solves due to Bellman and Ford that solves the single-source shortest paths problem in $O(|V||E|)$ time.

In a single scan of the edges, the RELAX operation is executed on each edge. The scan is then repeated $|V| - 1$ times. No special data structures are required to implement this algorithm, and the proof relies on the fact that a shortest path is simple and contains at most $|V| - 1$ edges.

This problem also finds applications in finding a feasible solution to a system of linear equations of a special form that arises in real-time applications: each equation specifies a bound on the difference between two variables. Each constraint is modeled by an edge in a suitably defined directed graph. Shortest paths from the root of this graph capture feasible solutions to the system of equations (for more information, see [7, Chapter 24.5]).

7.6.6 The All-Pairs Shortest Paths Problem

Consider the problem of computing a shortest path between every pair of vertices in a directed graph with edge lengths. The problem can be solved in $O(|V|^3)$ time, even when some edges have negative lengths, as long as the graph has no negative length cycles. Let the lengths of the edges be stored in a matrix A; the array entry $A[i,j]$ stores the length of the edge from i to j. If there is no edge from i to j, then $A[i,j] = \infty$; also $A[i,i]$ is set to 0 for all i. A dynamic programming algorithm to solve the problem is discussed in this section. The algorithm is due to Floyd and builds on the work of Warshall.

Define $P_k[u,v]$ to be a shortest path from u to v that is restricted to using intermediate vertices only from the set $\{1, \ldots, k\}$. Let $D_k[u,v]$ be the length of $P_k[u,v]$. Note that $P_0[u,v] = (u,v)$ since the path is not allowed to use *any* intermediate vertices, and therefore $D_0[u,v] = A[u,v]$. Since there are no negative length cycles, there is no loss of generality in assuming that shortest paths are simple.

The structure of shortest paths leads to the following recursive formulation of P_k. Consider $P_k[i,j]$ for $k > 0$. Either vertex k is on this path or not. If $P_k[i,j]$ does not pass through k, then the path uses only vertices from the set $\{1, \ldots, k-1\}$ as intermediate vertices, and is therefore the same as $P_{k-1}[i,j]$. If k is a vertex on the path $P_k[i,j]$, then it passes through k exactly once because the path is simple. Moreover, the subpath from i to k in $P_k[i,j]$ is a shortest path from i to k that uses intermediate vertices from the set $\{1, \ldots, k-1\}$, as does the subpath from k to j in $P_k[i,j]$. Thus, the path $P_k[i,j]$ is the union of $P_{k-1}[i,k]$ and $P_{k-1}[k,j]$. The above discussion leads to the following recursive formulation of D_k:

$$D_k[i,j] = \begin{cases} \min\left(D_{k-1}[i,j], D_{k-1}[i,k] + D_{k-1}[k,j]\right) & \text{if } k > 0 \\ A[i,j] & \text{if } k = 0 \end{cases}$$

Finally, since $P_n[i,j]$ is allowed to go through any vertex in the graph, $D_n[i,j]$ is the length of a shortest path from i to j in the graph.

In the algorithm described below, a matrix D is used to store distances. It might appear at first glance that to compute the distance matrix D_k from D_{k-1}, different arrays must be used for them. However, it can be shown that in the kth iteration, the entries in the kth row and column do not change, and thus, the same space can be reused.

FLOYD-SHORTEST-PATH (G)

```
1    for i = 1 to |V| do
2        for j = 1 to |V| do
3            D[i,j] ← A[i,j]
4        end-for
5    end-for
6    for k = 1 to |V| do
7        for i = 1 to |V| do
8            for j = 1 to |V| do
9                D[i,j] ← min(D[i,j], D[i,k] + D[k,j]).
10           end-for
11       end-for
12   end for
end-proc
```

7.7   Minimum Spanning Trees

The following fundamental problem arises in network design. A set of sites need to be connected by a network. This problem has a natural formulation in graph-theoretic terms. Each site is represented by a vertex. Edges between vertices represent a potential link connecting the corresponding nodes. Each edge is given a nonnegative cost corresponding to the cost of constructing that link. A tree is a minimal network that connects a set of nodes. The cost of a tree is the sum of the costs of its edges. A minimum-cost tree connecting the nodes of a given graph is called a minimum-cost spanning tree, or simply a **minimum spanning tree** (MST).

The problem of computing a MST arises in many areas, and as a subproblem in combinatorial and geometric problems. MSTs can be computed efficiently using algorithms that are greedy in nature, and there are several different algorithms for finding an MST. One of the first algorithms was due to Boruvka. Two algorithms, popularly known as Prim's algorithm and Kruskal's algorithm, are described here.

We first describe some rules that characterize edges belonging to a MST. The various algorithms are based on applying these rules in different orders. Tarjan [28] uses colors to describe these rules. Initially, all edges are uncolored. When an edge is colored blue it is marked for inclusion in the MST. When an edge is colored red it is marked to be excluded from the MST. The algorithms maintain the property that there is an MST containing all the blue edges but none of the red edges.

A cut is a partitioning of the vertex set into two subsets S and $V - S$. An edge crosses the cut if it connects a vertex $x \in S$ to a vertex $y \in V - S$.

(**Blue rule**) Find a cut that is not crossed by any blue edge and color a minimum weight edge that crosses the cut to be blue.

(**Red rule**) Find a simple cycle containing no red edges and color a maximum weight edge on that cycle to be red.

The proofs that these rules work can be found in [28].

7.7.1   Prim's Algorithm

Prim's algorithm for finding an MST of a given graph is one of the oldest algorithms to solve the problem. The basic idea is to start from a single vertex and gradually "grow" a tree, which eventually spans the entire graph. At each step, the algorithm has a tree of blue edges that covers a set S of vertices. The blue rule is applied by picking the cut $S, V - S$. This may be used to extend the tree to include a vertex that is currently not in the tree. The algorithm selects a minimum-cost edge from the edges crossing the cut and adds it to the current tree (implicitly coloring the edge blue), thereby adding another vertex to S.

As in the case of Dijkstra's algorithm, each vertex $u \in V - S$ can attach itself to only one vertex in the tree so that the current solution maintained by the algorithm is always a tree. Since the algorithm always chooses a minimum-cost edge, it needs to maintain a minimum-cost edge that connects u to some vertex in S as the candidate edge for including u in the tree. A priority queue of vertices is used to select a vertex in $V - S$ that is incident to a minimum-cost candidate edge.

PRIM-MST (G, r)
1   **for** all vertices v in G **do**
2        $visited[v] \leftarrow$ **false**.
3        $d[v] \leftarrow \infty$.
4        $p[v] \leftarrow$ **nil**.
5   **end-for**
6    $d[r] \leftarrow 0$.
7   BUILDPQ (H, d).

```
8   while not Empty(H) do
9       u ← DeleteMin (H).
10      visited[u] ← true.
11      for all vertices v in adj[u] do
12          if not visited[v] and d[v] > w(u, v) then
13              d[v] ← w(u, v).
14              p[v] ← u.
15              DecreaseKey (H, v, d[v]).
16          end-if
17      end-for
18  end-while
end-proc
```

7.7.2 Analysis

First observe the similarity between Prim's and Dijkstra's algorithms. Both algorithms start building the tree from a single vertex and grow it by adding one vertex at a time. The only difference is the rule for deciding when the current label is updated for vertices outside the tree. Both algorithms have the same structure and therefore have similar running times. Prim's algorithm runs in $O(|E| \log |V|)$ time if the priority queue is implemented using binary heaps, and it runs in $O(|E| + |V| \log |V|)$ if the priority queue is implemented using Fibonacci heaps.

7.7.3 Kruskal's Algorithm

Kruskal's algorithm for finding an MST of a given graph is another classical algorithm for the problem, and is also greedy in nature. Unlike Prim's algorithm which grows a single tree, Kruskal's algorithm grows a forest. First the edges of the graph are sorted in nondecreasing order of their costs. The algorithm starts with an empty forest. The edges of the graph are scanned in sorted order, and if the addition of the current edge does not generate a cycle in the current forest, it is added to the forest. The main test at each step is: does the current edge connect two vertices in the same connected component of the current forest? Eventually the algorithm adds $n - 1$ edges to generate a spanning tree in the graph.

The following discussion explains the correctness of the algorithm based on the two rules described earlier. Suppose that as the algorithm progresses, the edges chosen by the algorithm are colored blue and the ones that it rejects are colored red. When an edge is considered and it forms a cycle with previously chosen edges, this is a cycle with no red edges. Since the algorithm considers the edges in nondecreasing order of weight, the last edge is the heaviest edge in the cycle and therefore it can be colored red by the red rule. If an edge connects two blue trees T_1 and T_2, then it is a lightest edge crossing the cut T_1 and $V - T_1$, because any other edge crossing the cut has not been considered yet and is therefore no lighter. Therefore it can be colored blue by the blue rule.

The main data structure needed to implement the algorithm is to maintain connected components. An abstract version of this problem is known as the union–find problem for collection of disjoint sets (Chapters 8, 9, and 34). Efficient algorithms are known for this problem, where an arbitrary sequence of Union and Find operations can be implemented to run in almost linear time (for more information, see [7,28]).

Kruskal-MST(G)
```
1   T ← φ.
2   for all vertices v in G do
```

```
3        p[v] ← v.
4    end-for
5    Sort the edges of G by nondecreasing order of costs.
6    for all edges e = (u, v) in G in sorted order do
7        if FIND (u) ≠ FIND (v) then
8            T ← T ∪ (u, v).
9            UNION (u, v).
10       end-if
11   end-for
end-proc
```

7.7.4 Analysis

The running time of the algorithm is dominated by Step 5 of the algorithm in which the edges of the graph are sorted by nondecreasing order of their costs. This takes $O(|E| \log |E|)$ (which is also $O(|E| \log |V|)$) time using an efficient sorting algorithm such as heap sort. Kruskal's algorithm runs faster in the following special cases: if the edges are presorted, if the edge costs are within a small range, or if the number of different edge costs is bounded. In all these cases, the edges can be sorted in linear time, and Kruskal's algorithm runs in the near-linear time of $O(|E| \alpha(|E|, |V|))$, where $\alpha(m, n)$ is the inverse Ackermann function [28].

7.7.5 Boruvka's Algorithm

Boruvka's algorithm also grows many trees simultaneously. Initially there are $|V|$ trees, where each vertex forms its own tree. At each stage the algorithm keeps a collection of blue trees (i.e., trees built using only blue edges). For convenience, assume that all edge weights are distinct. If two edges have the same weight, they may be ordered arbitrarily. Each tree selects a minimum cost edge that connects it to some other tree and colors it blue. At the end of this parallel coloring step, each tree merges with a collection of other trees. The number of trees decreases by at least a factor of 2 in each step, and therefore after $\log |V|$ iterations there is exactly one tree. In practice, many trees merge in a single step and the algorithm converges much faster. Each step can be implemented in $O(|E|)$ time, and hence, the algorithm runs in $O(|E| \log |V|)$. For the special case of planar graphs, the above algorithm actually runs in $O(|V|)$ time.

Almost linear-time deterministic algorithms for the MST problem in undirected graphs are known [5,10]. Recently, Karger et al. [18] showed that they can combine the approach of Boruvka's algorithm with a random sampling approach to obtain a randomized algorithm with an expected running time of $O(|E|)$. Their algorithm also needs to use as a subroutine a procedure to verify that a proposed tree is indeed an MST [20,21]. The equivalent of MSTs in directed graphs are known as minimum branchings and are discussed in Chapter 8.

7.8 Tour and Traversal Problems

There are many applications for finding certain kinds of paths and tours in graphs. We briefly discuss some of the basic problems.

The **traveling salesman problem** (TSP) is that of finding a shortest tour that visits all the vertices of a given graph with weights on the edges. It has received considerable attention in the literature [22]. The problem is known to be computationally intractable (NP-hard). Several heuristics are known to solve practical instances. Considerable progress has also been made in finding optimal solutions for graphs with a few thousand vertices.

One of the first graph-theoretic problems to be studied, the **Euler tour problem** asks for the existence of a closed walk in a given connected graph that traverses each edge exactly once. Euler proved that such a closed walk exists if and only if each vertex has even degree [12]. Such a graph is known as an **Eulerian graph**. Given an Eulerian graph, an Euler tour in it can be computed using an algorithm similar to DFS in linear time.

Given an edge-weighted graph, the **Chinese postman problem** is that of finding a shortest closed walk that traverses each edge at least once. Although the problem sounds very similar to the TSP problem, it can be solved optimally in polynomial time [1].

7.9 Assorted Topics

7.9.1 Planar Graphs

A graph is called planar if it can be drawn on the plane without any of its edges crossing each other. A planar embedding is a drawing of a planar graph on the plane with no crossing edges. An embedded planar graph is known as a plane graph. A *face* of a plane graph is a connected region of the plane surrounded by edges of the planar graph. The unbounded face is referred to as the exterior face. Euler's formula captures a fundamental property of planar graphs by relating the number of edges, the number of vertices and the number of faces of a plane graph: $|F| - |E| + |V| = 2$. One of the consequences of this formula is that a simple planar graph has at most $O(|V|)$ edges.

Extensive work has been done on the study of planar graphs and a recent book has been devoted to the subject [26]. A fundamental problem in this area is deciding whether a given graph is planar, and if so, finding a planar embedding for it. Kuratowski gave necessary and sufficient conditions for when a graph is planar, by showing that a graph is planar if and only if it has no subgraph that is a homeomorph of K_5 or $K_{3,3}$. Hopcroft and Tarjan [17] gave a linear-time algorithm to test if a graph is planar, and if it is, to find a planar embedding for the graph.

A balanced separator is a subset of vertices that disconnects the graph in such a way, that the resulting components each have at most a constant fraction of the number of vertices of the original graph. Balanced separators are useful in designing "divide-and-conquer" algorithms for graph problems, such as graph layout problems (Chapter 8 of *Algorithms and Theory of Computation Handbook, Second Edition: Special Topics and Techniques*). Such algorithms are possible when one is guaranteed to find separators that have very few vertices relative to the graph. Lipton and Tarjan [24] proved that every planar graph on $|V|$ vertices has a separator of size at most $\sqrt{8|V|}$, whose deletion breaks the graph into two or more disconnected graphs, each of which has at most $2/3|V|$ vertices. Using the property that planar graphs have small separators, Frederickson [9] has given faster shortest path algorithms for planar graphs. Recently, this was improved to a linear-time algorithm by Henzinger et al. [13].

7.9.2 Graph Coloring

A coloring of a graph is an assignment of colors to the vertices, so that any two adjacent vertices have distinct colors. Traditionally, the colors are not given names, but represented by positive integers. The vertex coloring problem is the following: given a graph, to color its vertices using the fewest number of colors (known as the chromatic number of the graph). This was one of the first problems that were shown to be intractable (NP-hard). Recently it has been shown that even the problem of approximating the chromatic number of the graph within any reasonable factor is intractable. But, the coloring problem needs to be solved in practice (such as in the channel assignment problem in cellular networks), and heuristics are used to generate solutions. We discuss a commonly used greedy heuristic below: the vertices of the graph are colored sequentially in an arbitrary order. When a vertex is being processed, the color assigned to it is the smallest positive number that is not used by any of its neighbors that have been processed earlier. This scheme guarantees that if the degree of a vertex is Δ, then its color is at most $\Delta + 1$. There are special classes of graphs, such as planar graphs, in which the vertices can be carefully ordered in such a way that the number of colors used is small. For example, the vertices of a planar graph can be ordered such that every vertex has at most five neighbors that appear earlier in the list. By coloring its vertices in that order yields a six-coloring. There is a different algorithm that colors any planar graph using only four colors.

7.9.3 Light Approximate Shortest Path Trees

To broadcast information from a specified vertex r to all vertices of G, one may wish to send the information along a shortest path tree in order to reduce the time taken by the message to reach the nodes (i.e., minimizing delay). Though the shortest path tree may minimize delays, it may be a much costlier network to construct and considerably heavier than a MST, which leads to the question of whether there are trees that are light (like an MST) and yet capture distances like a shortest path tree. In this section, we consider the problem of computing a light subgraph that approximates a shortest path tree rooted at r.

Let T_{min} be a MST of G. For any vertex v, let $d(r, v)$ be the length of a shortest path from r to v in G. Let $\alpha > 1$ and $\beta > 1$ be arbitrary constants. An (α, β)-light approximate shortest path tree $((\alpha, \beta)$-LAST) of G is a spanning tree T of G with the property that the distance from the root to any vertex v in T is at most $\alpha \cdot d(r, v)$ and the weight of T is at most β times the weight of T_{min}.

Awerbuch et al. [3], motivated by applications in broadcast-network design, made a fundamental contribution by showing that every graph has a shallow-light tree—a tree whose diameter is at most a constant times the diameter of G and whose total weight is at most a constant times the weight of a MST. Cong et al. [6] studied the same problem and showed that the problem has applications in VLSI-circuit design; they improved the approximation ratios obtained in [3] and also studied variations of the problem such as bounding the radius of the tree instead of the diameter.

Khuller et al. [19] modified the shallow-light tree algorithm and showed that the distance from the root to each vertex can be approximated within a constant factor. Their algorithm also runs in linear time if a MST and a shortest path tree are provided. The algorithm computes an $(\alpha, 1 + \frac{2}{\alpha-1})$-LAST.

The basic idea is as follows: initialize a subgraph H to be a MST T_{min}. The vertices are processed in a preorder traversal of T_{min}. When a vertex v is processed, its distance from r in H is compared to $\alpha \cdot d(r, v)$. If the distance exceeds the required threshold, then the algorithm adds to H a shortest path in G from r to v. When all the vertices have been processed, the distance in H from r to any vertex v meets its distance requirement. A shortest path tree in H is returned by the algorithm as the required LAST.

7.9.4 Network Decomposition

The problem of decomposing a graph into clusters, each of which has low diameter, has applications in distributed computing. Awerbuch [2] introduced an elegant algorithm for computing low diameter clusters, with the property that there are few inter-cluster edges (assuming that edges going between clusters are not counted multiply). This construction was further refined by Awerbuch and Peleg [4], and they showed that a graph can be decomposed into clusters of diameter $O(r \log |V|)$ with the property that each r neighborhood of a vertex belongs to some cluster. (An r neighborhood of a vertex is the set of nodes whose distance from the vertex is at most r.) In addition, each vertex belongs to at most $2 \log |V|$ clusters. Using a similar approach Linial and Saks [23] showed that a graph can be decomposed into $O(\log |V|)$ clusters, with the property that each connected component in a cluster has $O(\log |V|)$ diameter. These techniques have found several applications in the computation of approximate shortest paths, and in other distributed computing problems.

The basic idea behind these methods is to perform an "expanding BFS." The algorithm selects an arbitrary vertex, and executes BFS with that vertex as the root. The algorithm continues the search layer by layer, ensuring that the number of vertices in a layer is at least as large as the number of vertices currently in that BFS tree. Since the tree expands rapidly, this procedure generates a low diameter BFS tree (cluster). If the algorithm comes across a layer in which the number of nodes is not big enough, it rejects that layer and stops growing that tree. The set of nodes in the layer that was not added to the BFS tree that was being grown is guaranteed to be small. The algorithm continues by selecting a new vertex that was not chosen in any cluster and repeats the above procedure.

7.10   Research Issues and Summary

We have illustrated some of the fundamental techniques that are useful for manipulating graphs. These basic algorithms are used as tools in the design of algorithms for graphs. The problems studied in this chapter included representation of graphs, tree traversal techniques, search techniques for graphs, shortest path problems, MSTs, and tour problems on graphs.

Current research on graph algorithms focuses on dynamic algorithms, graph layout and drawing, and approximation algorithms. More information about these areas can be found in Chapters 8, 9, and 34 of this book. The methods illustrated in our chapter find use in the solution of almost any graph problem.

The graph isomorphism problem is an old problem in this area. The input to this problem is two graphs and the problem is to decide whether the two graphs are isomorphic, i.e., whether the rows and columns of the adjacency matrix of one of the graphs can be permuted so that it is identical to the adjacency matrix of the other graph. This problem is neither known to be polynomial-time solvable nor known to be NP-hard. This is in contrast to the subgraph isomorphism problem in which the problem is to decide whether there is a subgraph of the first graph that is isomorphic to the second graph. The subgraph isomorphism is known to be NP-complete. Special instances of the **graph isomorphism problem** are known to be polynomially solvable, such as when the graphs are planar, or more generally of bounded genus. For more information on the isomorphism problem, see Hoffman [14].

Another open problem is whether there exists a deterministic linear-time algorithm for computing a MST. Near-linear-time deterministic algorithms using Fibonacci heaps have been known for finding an MST. The newly discovered probabilistic algorithm uses random sampling to find an MST in expected linear time. Much of the recent research in this area is focusing on the design of approximation algorithms for NP-hard problems.

7.11   Further Information

The area of graph algorithms continues to be a very active field of research. There are several journals and conferences that discuss advances in the field. Here we name a partial list of some of the important meetings: ACM Symposium on Theory of Computing (STOC), IEEE Conference on Foundations of Computer Science (FOCS), ACM-SIAM Symposium on Discrete Algorithms (SODA), International Colloquium on Automata, Languages and Programming (ICALP), and European Symposium on Algorithms (ESA). There are many other regional algorithms/theory conferences that carry research papers on graph algorithms. The journals that carry articles on current research in graph algorithms are *Journal of the ACM, SIAM Journal on Computing, SIAM Journal on Discrete Mathematics, Journal of Algorithms, Algorithmica, Journal of Computer and System Sciences, Information and Computation, Information Processing Letters,* and *Theoretical Computer Science.* To find more details about some of the graph algorithms described in this chapter we refer the reader to the books by Cormen et al. [7], Even [8], Gibbons [12], and Tarjan [28].

Defining Terms

Articulation vertex/cut vertex: A vertex whose deletion disconnects a graph into two or more connected components.

Biconnected graph: A graph that has no articulation/cut vertices.

Bipartite graph: A graph in which the vertex set can be partitioned into two sets X and Y, such that each edge connects a node in X with a node in Y.

Branching: A rooted spanning tree in a directed graph, such that the root has a path in the tree to each vertex.

Chinese postman problem: Find a minimum length tour that traverses each edge at least once.

Connected graph: A graph in which there is a path between each pair of vertices.

Cycle: A path in which the start and end vertices of the path are identical.

Degree: The number of edges incident to a vertex in a graph.

DFS forest: A rooted forest formed by depth-first search.

Directed acyclic graph: A directed graph with no cycles.

Euler tour problem: Asks for a traversal of the edges that visits each edge exactly once.

Eulerian graph: A graph that has an Euler tour.

Forest: An acyclic graph.

Graph isomorphism problem: Deciding if two given graphs are isomorphic to each other.

Leaves: Vertices of degree one in a tree.

Minimum spanning tree: A spanning tree of minimum total weight.

Path: An ordered list of distinct edges, $\{e_i = (u_i, v_i) | i = 1, \ldots, k\}$, such that for any two consecutive edges e_i and e_{i+1}, $v_i = u_{i+1}$.

Planar graph: A graph that can be drawn on the plane without any of its edges crossing each other.

Sparse graph: A graph in which $|E| \ll |V|^2$.

Strongly connected graph: A directed graph in which there is a directed path between each ordered pair of vertices.

Topological order: A numbering of the vertices of a DAG such that every edge in the graph that goes from a vertex numbered i to a vertex numbered j satisfies $i < j$.

Traveling salesman problem: Asks for a minimum length tour of a graph that visits all the vertices exactly once.

Tree: A connected forest.

Walk: A path in which edges may be repeated.

Acknowledgments

Samir Khuller's research was supported by NSF Research Initiation Award CCR-9307462 and NSF CAREER Award CCR-9501355. Balaji Raghavachari's research was supported by NSF Research Initiation Award CCR-9409625.

References

1. Ahuja, R.K., Magnanti, T.L., and Orlin, J.B., *Network Flows*. Prentice Hall, Englewood Cliffs, NJ, 1993.
2. Awerbuch, B., Complexity of network synchronization. *J. Assoc. Comput. Mach.*, 32(4), 804–823, 1985.
3. Awerbuch, B., Baratz, A., and Peleg, D., Cost-sensitive analysis of communication protocols. In *Proceedings of the 9th Annual ACM Symposium on Principles of Distributed Computing*, pp. 177–187, Quebec City, Quebec, Canada, August 22–24, 1990.
4. Awerbuch, B. and Peleg, D., Sparse partitions. In *Proceedings of the 31st Annual Symposium on Foundations of Computer Science*, pp. 503–513, St. Louis, MO, October 22–24, 1990.
5. Chazelle, B., A faster deterministic algorithm for minimum spanning trees. In *Proceedings of the 38th Annual Symposium on Foundations of Computer Science*, pp. 22–31, Miami, FL, October 20–22, 1997.
6. Cong, J., Kahng, A.B., Robins, G., Sarrafzadeh, M., and Wong, C.K., Provably good performance-driven global routing. *IEEE Trans. CAD*, 739–752, 1992.
7. Cormen, T.H., Leiserson, C.E., and Rivest, R.L., *Introduction to Algorithms*. The MIT Press, Cambridge, MA, 1989.
8. Even, S., *Graph Algorithms*. Computer Science Press, Rockville, MD, 1979.
9. Frederickson, G.N., Fast algorithms for shortest paths in planar graphs with applications. *SIAM J. Comput.*, 16(6), 1004–1022, 1987.
10. Fredman, M.L. and Tarjan, R.E., Fibonacci heaps and their uses in improved network optimization algorithms. *J. Assoc. Comput. Mach.*, 34(3), 596–615, 1987.
11. Galil, Z. and Italiano, G., Reducing edge connectivity to vertex connectivity. *SIGACT News*, 22(1), 57–61, 1991.
12. Gibbons, A.M., *Algorithmic Graph Theory*. Cambridge University Press, New York, 1985.
13. Henzinger, M.R., Klein, P.N., Rao, S., and Subramanian, S., Faster shortest-path algorithms for planar graphs. *J. Comput. Syst. Sci.*, 55(1), 3–23, 1997.
14. Hoffman, C.M., *Group-Theoretic Algorithms and Graph Isomorphism*. LNCS #136, Springer-Verlag, Berlin, 1982.
15. Hopcroft, J.E. and Tarjan, R.E., Dividing a graph into triconnected components. *SIAM J. Comput.*, 2(3), 135–158, 1973.
16. Hopcroft, J.E. and Tarjan, R.E., Efficient algorithms for graph manipulation. *Commun. ACM*, 16, 372–378, 1973.
17. Hopcroft, J.E. and Tarjan, R.E., Efficient planarity testing. *J. Assoc. Comput. Mach.*, 21(4), 549–568, 1974.
18. Karger, D.R., Klein, P.N., and Tarjan, R.E., A randomized linear-time algorithm to find minimum spanning trees. *J. Assoc. Comput. Mach.*, 42(2), 321–328, 1995.
19. Khuller, S., Raghavachari, B., and Young, N., Balancing minimum spanning trees and shortest-path trees. *Algorithmica*, 14(4), 305–321, 1995.
20. King, V., A simpler minimum spanning tree verification algorithm. *Algorithmica*, 18(2), 263–270, 1997.
21. Komls, J., Linear verification for spanning trees. *Combinatorica*, 5, 57–65, 1985.
22. Lawler, E.l., Lenstra, J.K., Rinnooy Kan, A.H.G., and Shmoys, D.B., *The Traveling Salesman Problem: A Guided Tour of Combinatorial Optimization*. Wiley, New York, 1985.
23. Linial, M. and Saks, M., Low diameter graph decompositions. *Combinatorica*, 13(4), 441–454, 1993.
24. Lipton, R. and Tarjan, R.E., A separator theorem for planar graphs. *SIAM J. Appl. Math.*, 36, 177–189, 1979.
25. Nagamochi, H. and Ibaraki, T., Linear time algorithms for finding sparse k-connected spanning subgraph of a k-connected graph. *Algorithmica*, 7(5/6), 583–596, 1992.

26. Nishizeki, T. and Chiba, N., *Planar Graphs: Theory and Algorithms.* North-Holland, Amsterdam, the Netherlands, 1989.

27. Tarjan, R.E., Depth-first search and linear graph algorithms. *SIAM J. Comput.*, 1(2), 146–160, June 1972.

28. Tarjan, R.E., *Data Structures and Network Algorithms.* Society for Industrial and Applied Mathematics, Philadelphia, PA, 1983.

8

Advanced Combinatorial Algorithms

Samir Khuller
University of Maryland

Balaji Raghavachari
University of Texas at Dallas

8.1 Introduction

The optimization of a given objective, while working with limited resources is a fundamental problem that occurs in all walks of life, and is especially important in computer science and operations research. Problems in discrete optimization vary widely in their complexity, and efficient solutions are derived for many of these problems by studying their combinatorial structure and understanding their fundamental properties. In this chapter, we study several problems and advanced algorithmic techniques for solving them. One of the basic topics in this field is the study of **network flow** and related optimization problems; these problems occur in various disciplines, and provide a fundamental framework for solving problems. For example, the problem of efficiently moving entities, such as bits, people, or products, from one place to another in an underlying network, can be modeled as a network flow problem. Network flow finds applications in many other areas such as **matching**, scheduling, and connectivity problems in networks. The problem plays a central role in the fields of operations research and computer science, and considerable emphasis has been placed on the design of efficient algorithms for solving it.

The network flow problem is usually formulated on directed graphs (which are also known as networks). A fundamental problem is the **maximum flow** problem, usually referred to as the max-flow problem. The input to this problem is a directed graph $G = (V, E)$, a nonnegative **capacity** function $u : E \mapsto \Re^+$ that specifies the capacity of each arc, a source vertex $s \in V$, and a sink vertex $t \in V$. The problem captures the situation when a commodity is being produced at node s, and needs to be shipped to node t, through the network. The objective is to send as many units of flow as possible from s to t, while satisfying flow conservation constraints at all intermediate nodes and capacity constraints on the edges. The problem will be defined formally later.

Many practical combinatorial problems such as the **assignment problem**, and the problem of finding the susceptibility of networks to failures due to faulty links or nodes, are special instances of the max-flow problem. There are several variations and generalizations of the max-flow problem including the vertex capacitated max-flow problem, the minimum-cost max-flow problem, the minimum-cost circulation problem, and the multicommodity flow problem.

Section 8.2 discusses the matching problem. The single commodity maximum flow problem is introduced in Section 8.3. The minimum-cut problem is discussed in Section 8.4. Section 8.5 discusses the min-cost flow problem. The multicommodity flow problem is discussed in Section 8.6. Section 8.7 introduces the problem of computing optimal **branchings**. Section 8.8 discusses the problem of coloring the edges and vertices of a graph. Section 8.9 discusses approximation algorithms for NP-hard problems. At the end, references to current research in graph algorithms are provided.

8.2   The Matching Problem

An entire book [23] has been devoted to the study of various aspects of the matching problem, ranging from necessary and sufficient conditions for the existence of perfect matchings to algorithms for solving the matching problem. Many of the basic algorithms studied in Chapter 7 play an important role in developing various implementations for network flow and matching algorithms.

First the matching problem, which is a special case of the max-flow problem is introduced. Then the assignment problem, a generalization of the matching problem, is studied.

The maximum matching problem is discussed in detail only for bipartite graphs. The same principles are used to design efficient algorithms to solve the matching problem in arbitrary graphs. The algorithms for general graphs are complex due to the presence of odd-length cycles called blossoms, and the reader is referred to [26,Chapter 10 of first edition], or [29,Chapter 9] for a detailed treatment of how blossoms are handled.

8.2.1   Matching Problem Definitions

Given a graph $G = (V, E)$, a matching M is a subset of the edges such that no two edges in M share a common vertex. In other words, the problem is that of finding a set of independent edges, that have no incident vertices in common. The cardinality of M is usually referred to as its size.

The following terms are defined with respect to a matching M. The edges in M are called matched edges and edges not in M are called free edges. Likewise, a vertex is a matched vertex if it is incident to a matched edge. A free vertex is one that is not matched. The mate of a matched vertex v is its neighbor w that is at the other end of the matched edge incident to v. A matching is called perfect if all vertices of the graph are matched in it. (When the number of vertices is odd, we permit one vertex to remain unmatched.) The objective of the maximum matching problem is to maximize $|M|$, the

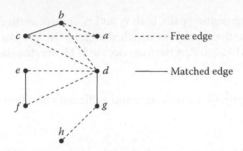

FIGURE 8.1 An augmenting path p with respect to a matching.

size of the matching. If the edges of the graph have weights, then the weight of a matching is defined to be the sum of the weights of the edges in the matching. A path $p = [v_1, v_2, \ldots, v_k]$ is called an alternating path if the edges $(v_{2j-1}, v_{2j}), j = 1, 2, \ldots$ are free, and the edges $(v_{2j}, v_{2j+1}), j = 1, 2, \ldots$ are matched. An **augmenting path** $p = [v_1, v_2, \ldots, v_k]$ is an alternating path in which both v_1 and v_k are free vertices. Observe that an augmenting path is defined with respect to a specific matching. The symmetric difference of a matching M and an augmenting path P, $M \oplus P$, is defined to be $(M - P) \cup (P - M)$. It can be shown that $M \oplus P$ is also a matching. Figure 8.1 shows an augmenting path $p = [a, b, c, d, g, h]$ with respect to the given matching. The symmetric difference operation can be also be used as above between two matchings. In this case, the resulting graph consists of a collection of paths and cycles with alternate edges from each matching.

8.2.2 Applications of Matching

Matchings lie at the heart of many optimization problems and the problem has many applications: assigning workers to jobs, assigning a collection of jobs with precedence constraints to two processors such that the total execution time is minimized, determining the structure of chemical bonds in chemistry, matching moving objects based on a sequence of snapshots, and localization of objects in space after obtaining information from multiple sensors (see [1]).

8.2.3 Matchings and Augmenting Paths

The following theorem gives necessary and sufficient conditions for the existence of a perfect matching in a bipartite graph.

THEOREM 8.1 (Hall's Theorem) *A bipartite graph $G = (X, Y, E)$ with $|X| = |Y|$ has a perfect matching if and only if $\forall S \subseteq X, |N(S)| \geq |S|$, where $N(S) \subseteq Y$ is the set of vertices that are neighbors of some vertex in S.*

Although the above theorem captures exactly the conditions under which a given bipartite graph has a perfect matching, it does not lead to an algorithm for finding perfect matchings directly. The following lemma shows how an augmenting path with respect to a given matching can be used to increase the size of a matching. An efficient algorithm will be described later that uses augmenting paths to construct a maximum matching incrementally.

LEMMA 8.1 Let P be the edges on an augmenting path $p = [v_1, \ldots, v_k]$ with respect to a matching M. Then $M' = M \oplus P$ is a matching of cardinality $|M| + 1$.

PROOF   Since P is an augmenting path, both v_1 and v_k are free vertices in M. The number of free edges in P is one more than the number of matched edges in it. The symmetric difference operator replaces the matched edges of M in P by the free edges in P. Hence, the size of the resulting matching, $|M'|$, is one more than $|M|$.

The following theorem provides a necessary and sufficient condition for a given matching M to be a maximum matching.

THEOREM 8.2   *A matching M in a graph G is a maximum matching if and only if there is no augmenting path in G with respect to M.*

PROOF   If there is an augmenting path with respect to M, then M cannot be a maximum matching, since by Lemma 8.1 there is a matching whose size is larger than that of M. To prove the converse we show that if there is no augmenting path with respect to M then M is a maximum matching. Suppose that there is a matching M' such that $|M'| > |M|$. Consider the subgraph of G induced by the edges $M \oplus M'$. Each vertex in this subgraph has degree at most two, since each node has at most one edge from each matching incident to it. Hence, each connected component of this subgraph is either a path or a simple cycle. For each cycle, the number of edges of M is the same as the number of edges of M'. Since $|M'| > |M|$, one of the paths must have more edges from M' than from M. This path is an augmenting path in G with respect to the matching M, contradicting the assumption that there were no augmenting paths with respect to M.

8.2.4   Bipartite Matching Algorithm

8.2.4.1   High-Level Description

The algorithm starts with the empty matching $M = \emptyset$, and augments the matching in phases. In each phase, an augmenting path with respect to the current matching M is found, and it is used to increase the size of the matching. An augmenting path, if one exists, can be found in $O(|E|)$ time, using a procedure similar to breadth-first search.

The search for an augmenting path proceeds from the free vertices. At each step when a vertex in X is processed, all its unvisited neighbors are also searched. When a matched vertex in Y is considered, only its matched neighbor is searched. This search proceeds along a subgraph referred to as the Hungarian tree.

The algorithm uses a queue Q to hold vertices that are yet to be processed. Initially, all free vertices in X are placed in the queue. The vertices are removed one by one from the queue and processed as follows. In turn, when vertex v is removed from the queue, the edges incident to it are scanned. If it has a neighbor in the vertex set Y that is free, then the search for an augmenting path is successful; procedure AUGMENT is called to update the matching, and the algorithm proceeds to its next phase. Otherwise, add the mates of all the matched neighbors of v to the queue if they have never been added to the queue, and continue the search for an augmenting path. If the algorithm empties the queue without finding an augmenting path, its current matching is a maximum matching and it terminates.

The main data structure that the algorithm uses are the arrays mate and free. The array mate is used to represent the current matching. For a matched vertex $v \in G$, $mate[v]$ denotes the matched neighbor of vertex v. For $v \in X$, $free[v]$ is a vertex in Y that is adjacent to v and is free. If no such vertex exists then $free[v] = 0$. The set A stores a set of directed edges (v, v') such that there is an alternating path of two edges from v to v'. This will be used in the search for augmenting paths from free vertices, while extending the alternating paths. When we add a vertex v' to the queue, we set $label[v']$ to v if we came to v' from v, since we need this information to augment on the alternating path we eventually find.

BIPARTITE MATCHING $(G = (X, Y, E))$
1   **for** all vertices v in G **do**
2       $mate[v] \leftarrow 0.$
3   **end-for**
4   done $\leftarrow$ **false.**
5   **while not** done **do**
6   INITIALIZE.
7   MAKEEMPTYQUEUE $(Q).$
8   **for** all vertices $x \in X$ **do**   (* add unmatched vertices to Q *)
9   **if** $mate[x] = 0$ **then**
10   PUSH $(Q, x).$
11              $label[x] \leftarrow 0.$
12   **end-if**
13   **end-for**
14   found $\leftarrow$ **false.**
15   **while not** found **and not** EMPTY (Q) **do**
16          $x \leftarrow$ POP $(Q).$
17   **if** $free[x] \neq 0$ **then**   (* found augmenting path *)
18   AUGMENT $(x).$
19   found $\leftarrow$ **true.**
20   **else**   (* extend alternating paths from x *)
21   **for** all edges $(x, x') \in A$ **do**
22   **if** $label[x'] = 0$ **then**   (* x' not already in Q *)
23                      $label[x'] \leftarrow x.$
24   PUSH $(Q, x').$
25   **end-if**
26   **end-for**
27   **end-if**
28   **if** EMPTY (Q) **then**
29   done $\leftarrow$ **true.**
30   **end-if**
31   **end-while**
32 **end-while**
end-proc

INITIALIZE
1   **for** all vertices $x \in X$ **do**
2       $free[x] \leftarrow 0.$
3   **end-for**
4   **for** all edges $(x, y) \in E$ **do**
5   **if** $mate[y] = 0$ **then** $free[x] \leftarrow y$
6   **else if** $mate[y] \neq x$ **then** $A \leftarrow A \cup (x, mate[y]).$
7   **end-if**
8   **end-for**
end-proc

AUGMENT (x)
1   **if** $label[x] = 0$ **then**
2        $mate[x] \leftarrow free[x]$.
3        $mate[free[x]] \leftarrow x$
4   **else**
5        $free[label[x]] \leftarrow mate[x]$
6        $mate[x] \leftarrow free[x]$
7        $mate[free[x]] \leftarrow x$
8   AUGMENT $(label[x])$
9   **end-if**
end-proc

8.2.5 Sample Execution

Figure 8.2 shows a sample execution of the matching algorithm. We start with a partial matching and show the structure of the resulting Hungarian tree. In this example, the search starts from the free vertex b. We add c and e to Q. After we explore c, we add d to Q, and then f and a. Since $free[a] = u$, we stop since an augmenting path from vertex b to vertex u is found by the algorithm.

8.2.6 Analysis

If there are augmenting paths with respect to the current matching, the algorithm will find at least one of them. Hence, when the algorithm terminates, the graph has no augmenting paths with respect to the current matching and the current matching is optimal. Each iteration of the main while loop of the algorithm runs in $O(|E|)$ time. The construction of the auxiliary graph A and computation of the array free also take $O(|E|)$ time. In each iteration, the size of the matching increases by one and thus, there are at most $\min(|X|, |Y|)$ iterations of the while loop. Therefore the algorithm solves the matching problem for bipartite graphs in time $O(\min(|X|, |Y|)|E|)$. Hopcroft and Karp (see [26]) showed how to improve the running time by finding a maximal set of disjoint augmenting paths in a single phase in $O(|E|)$ time. They also proved that the algorithm runs in only $O(\sqrt{|V|})$ phases, yielding a worst-case running time of $O(\sqrt{|V|}|E|)$.

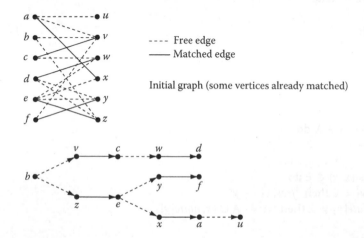

FIGURE 8.2 Sample execution of matching algorithm.

8.2.7 The Matching Problem in General Graphs

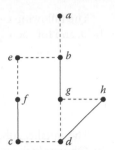

The techniques used to solve the matching problem in bipartite graphs do not extend directly to nonbipartite graphs. The notion of augmenting paths and their relation to maximum matchings (Theorem 8.2) remain the same. Therefore the natural algorithm of starting with an empty matching and increasing its size repeatedly with an augmenting path until no augmenting paths exist in the graph still works. But the problem of finding augmenting paths in nonbipartite graphs is harder. The main trouble is due to odd length cycles known as blossoms that appear along alternating paths explored by the algorithm as it is looking for augmenting paths. We illustrate this difficulty with an example in Figure 8.3. The search for an augmenting path from an unmatched vertex such as e, could go through the following sequence of vertices $[e, b, g, d, h, g, b, a]$.

FIGURE 8.3 Difficulty in dealing with blossoms.

Even though the augmenting path satisfies the "local" conditions for being an augmenting path, it is not a valid augmenting path since it is not simple. The reason for this is that the odd length cycle (g, d, h, g) causes the path to "fold" on itself—a problem that does not arise in the bipartite case. In fact, the matching does contain a valid augmenting path $[e, f, c, d, h, g, b, a]$. In fact, not all odd cycles cause this problem, but odd cycles that are as dense in matched edges as possible, i.e., it depends on the current matching. By "shrinking" blossoms to single nodes, we can get rid of them [26]. Subsequent work focused on efficient implementation of this method.

Edmonds (see [26]) gave the first polynomial-time algorithm for solving the maximum matching problem in general graphs. The current fastest algorithm for this problem is due to Micali and Vazirani [24] and their algorithm runs in $O(|E|\sqrt{|V|})$ steps, which is the same bound obtained by the Hopcroft–Karp algorithm for finding a maximum matching in bipartite graphs.

8.2.8 Assignment Problem

We now introduce the assignment problem—that of finding a maximum-weight matching in a given bipartite graph in which edges are given nonnegative weights. There is no loss of generality in assuming that the graph is a complete bipartite graph, since zero-weight edges may be added between pairs of vertices that are nonadjacent in the original graph without affecting the weight of a maximum-weight matching. The minimization version of the weighted version is the problem of finding a minimum-weight perfect matching in a complete bipartite graph. Both versions of the weighted matching problem are equivalent and we sketch below how to reduce the minimum-weight perfect matching to maximum-weight matching. Choose a constant W that is larger than the weight of any edge, and assign each edge a new weight of $w'(e) = W - w(e)$. Observe that maximum-weight matchings with the new weight function are minimum-weight perfect matchings with the original weights.

In this section, we restrict our attention to the study of the maximum-weight matching problem for bipartite graphs. Similar techniques have been used to solve the maximum-weight matching problem in arbitrary graphs (see [22,26]).

The input is a complete bipartite graph $G = (X, Y, X \times Y)$ and each edge e has a nonnegative weight of $w(e)$. The following algorithm is known as the Hungarian method (see [1,23,26]). The method can be viewed as a primal-dual algorithm in the framework of linear programming [26]. No knowledge of linear programming is assumed here.

A feasible vertex-labeling ℓ is defined to be a mapping from the set of vertices in G to the real numbers such that for each edge (x_i, y_j) the following condition holds:

$$\ell(x_i) + \ell(y_j) \geq w(x_i, y_j).$$

The following can be verified to be a feasible vertex labeling. For each vertex $y_j \in Y$, set $\ell(y_j)$ to be 0, and for each vertex $x_i \in X$, set $\ell(x_i)$ to be the maximum weight of an edge incident to x_i:

$$\ell(y_j) = 0,$$
$$\ell(x_i) = \max_j w(x_i, y_j).$$

The equality subgraph, G_ℓ, is defined to be the spanning subgraph of G which includes all vertices of G but only those edges (x_i, y_j) which have weights such that

$$\ell(x_i) + \ell(y_j) = w(x_i, y_j).$$

The connection between equality subgraphs and maximum-weighted matchings is established by the following theorem.

THEOREM 8.3 *If the equality subgraph, G_ℓ, has a perfect matching, M^*, then M^* is a maximum-weight matching in G.*

PROOF Let M^* be a perfect matching in G_ℓ. By definition,

$$w(M^*) = \sum_{e \in M^*} w(e) = \sum_{v \in X \cup Y} \ell(v).$$

Let M be any perfect matching in G. Then

$$w(M) = \sum_{e \in M} w(e) \leq \sum_{v \in X \cup Y} \ell(v) = w(M^*).$$

Hence, M^* is a maximum-weight perfect matching.

8.2.8.1 High-Level Description

The above theorem is the basis of the following algorithm for finding a maximum-weight matching in a complete bipartite graph. The algorithm starts with a feasible labeling, then computes the equality subgraph and a maximum cardinality matching in this subgraph. If the matching found is perfect, by Theorem 8.3, the matching must be a maximum-weight matching and the algorithm returns it as its output. Otherwise the matching is not perfect, and more edges need to be added to the equality subgraph by revising the vertex labels. The revision should ensure that edges from the current matching do not leave the equality subgraph. After more edges are added to the equality subgraph, the algorithm grows the Hungarian trees further. Either the size of the matching increases because an augmenting path is found, or a new vertex is added to the Hungarian tree. In the former case, the current phase terminates and the algorithm starts a new phase since the matching size has increased. In the latter case, new nodes are added to the Hungarian tree. In $|X|$ phases, the tree includes all the nodes, and therefore there are at most $|X|$ phases before the size of the matching increases.

We now describe in more detail how the labels are updated and which edges are added to the equality subgraph. Suppose M is a maximum matching found by the algorithm. Hungarian trees are

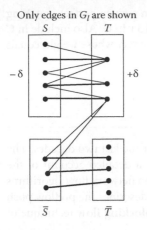

Only edges in G_l are shown

FIGURE 8.4 Sets S and T as maintained by the algorithm.

grown from all the free vertices in X. Vertices of X (including the free vertices) that are encountered in the search are added to a set S and vertices of Y that are encountered in the search are added to a set T. Let $\overline{S} = X - S$ and $\overline{T} = Y - T$. Figure 8.4 illustrates the structure of the sets S and T. Matched edges are shown in bold; the other edges are the edges in G_ℓ. Observe that there are no edges in the equality subgraph from S to $\overline{T}$, even though there may be edges from T to $\overline{S}$. The algorithm now revises the labels as follows. Decrease all the labels of vertices in S by a quantity δ (to be determined later) and increase the labels of the vertices in T by δ. This ensures that edges in the matching continue to stay in the equality subgraph. Edges in G (not in G_ℓ) that go from vertices in S to vertices in $\overline{T}$ are candidate edges to enter the equality subgraph, since one label is decreasing and the other is unchanged. The algorithm chooses δ to be the smallest value such that some edge of $G - G_\ell$ enters the equality subgraph. Suppose this edge goes from $x \in S$ to $y \in \overline{T}$. If y is free then an augmenting path has been found. On the other hand if y is matched, the Hungarian tree is grown by moving y to T and its matched neighbor to S and the process of revising labels is continued.

8.2.9 Weighted Edge-Cover Problem

There are many applications of weighted matchings. One useful application is the problem of finding a minimum-weight edge cover of a graph. Given a graph $G = (V, E)$ with weights on the edges, a set of edges $C \subseteq E$ forms an edge cover of G if every vertex in V is incident to at least one edge in C. The weight of the cover C is the sum of the weights of its edges. The objective is to find an edge cover C of minimum weight. The problem can be reduced to the minimum-weight perfect matching problem as follows: create an identical copy $G' = (V', E')$ of graph G, except the weights of edges in E' are all 0. Add an edge from each $v \in V$ to $v' \in V'$ with weight $w_{\min}(v) = \min_{x \in N(v)} w(v, x)$. The final graph H is the union of G and G', together with the edges connecting the two copies of each vertex; H contains $2|V|$ vertices and $2|E| + |V|$ edges. There exist other reductions from minimum-weight edge-cover to minimum-weight perfect matching; the reduction outlined above has the advantage that it creates a graph with $O(|V|)$ vertices and $O(|E|)$ edges. The minimum weight perfect matching problem may be solved using the techniques described earlier for the bipartite case, but the algorithm is more complex [22,26].

THEOREM 8.4 *The weight of a minimum-weight perfect matching in H is equal to the weight of a minimum-weight edge cover in G.*

PROOF Consider a minimum-weight edge cover C in G. There is no loss of generality in assuming that a minimum-weight edge cover has no path of three edges, since the middle edge can be removed from the cover, thus, reducing its weight further. Hence, the edge cover C is a union of "stars" (trees of height 1). For each star that has a vertex of degree more than one, we can match one of the leaf nodes to its copy in G', with weight at most the weight of the edge incident on the leaf vertex, thus, reducing the degree of the star. We repeat this until each vertex has degree at most one in the edge cover, i.e., it is a matching. In H, select this matching, once in each copy of G. Observe that the cost of the matching within the second copy G' is 0. Thus, given C, we can find a perfect matching in H whose weight is no larger.

To argue the converse, we now show how to construct a cover C from a given perfect matching M in H. For each edge $(v, v') \in M$, add to C a least weight edge incident to v in G. Also include in C any edge of G that was matched in M. It can be verified that C is an edge cover whose weight equals the weight of M.

8.3   The Network Flow Problem

A number of polynomial time flow algorithms have been developed over the last two decades. The reader is referred to the books by Ahuja et al. [1] and Tarjan [29] for a detailed account of the historical development of the various flow methods. An excellent survey on network flow algorithms has been written by Goldberg et al. [13]. The book by Cormen et al. [5] describes the preflow push method, and to complement their coverage an implementation of the **blocking flow** technique of Malhotra et al. (see [26]) is discussed here.

8.3.1   Network Flow Problem Definitions

8.3.1.1   Flow Network

A flow network $G = (V, E)$ is a directed graph with two specially marked nodes, namely, the source s, and the sink t. A capacity function $u : E \mapsto \Re^{+}$ maps edges to positive real numbers.

8.3.1.2   Max-Flow Problem

A flow function $f : E \mapsto \Re$ maps edges to real numbers. For an edge $e = (v, w)$, $f(v, w)$ refers to the flow on edge e, which is also called the net flow from vertex v to vertex w. This notation is extended to sets of vertices as follows: If X and Y are sets of vertices then $f(X, Y)$ is defined to be $\Sigma_{x \in X} \Sigma_{y \in Y} f(x, y)$. A flow function is required to satisfy the following constraints:

- (Capacity constraint) For all edges e, $f(e) \leq u(e)$
- (Skew symmetry constraint) For an edge $e = (v, w)$, $f(v, w) = -f(w, v)$
- (Flow conservation) For all vertices $v \in V - \{s, t\}$, $\Sigma_{w \in V} f(v, w) = 0$

The capacity constraint states that the total flow on an edge does not exceed its capacity. The skew symmetry condition states that the flow on an edge is the negative of the flow in the reverse direction. The flow conservation constraint states that the total net flow out of any vertex other than the source and sink is zero.

The value of the flow is defined to be the net flow out of the source vertex:

$$|f| = \sum_{v \in V} f(s, v).$$

In the maximum flow problem the objective is to find a flow function that satisfies the above three constraints, and also maximizes the total flow value $|f|$.

8.3.1.3   Remarks

This formulation of the network flow problem is powerful enough to capture generalizations where there are many sources and sinks (single-commodity flow), and where both vertices and edges have capacity constraints. To reduce multiple sources to a single source, we add a new source vertex with edges connecting it to the original source vertices. To reduce multiple sinks to a single sink, we add a new sink vertex with edges from the original sinks to the new sink. It is easy to reduce the vertex capacity problem to edge capacities by "splitting" a vertex into two vertices, and making all the incoming edges come into the first vertex and the outgoing edges come out of the second vertex. We then add an edge between them with the capacity of the corresponding vertex, so that the entire flow through the vertex is forced through this edge. The problem for undirected graphs can be solved by treating each undirected edge as two directed edges with the same capacity as the original undirected edge.

First the notion of cuts is defined, and then the max-flow min-cut theorem is introduced. We then introduce residual networks, layered networks and the concept of blocking flows. We then show how to reduce the max-flow problem to the computation of a sequence of blocking flows. Finally, an efficient algorithm for finding a blocking flow is described.

A cut is a partitioning of the vertex set into two subsets S and $V - S$. An edge crosses the cut if it connects a vertex $x \in S$ to a vertex $y \in V - S$. An s-t cut of the graph is a partitioning of the vertex set V into two sets S and $T = V - S$ such that $s \in S$ and $t \in T$. If f is a flow then the net flow across the cut is defined as $f(S, T)$. The capacity of the cut is defined as $u(S, T) = \Sigma_{x \in X} \Sigma_{y \in Y} u(x, y)$. The net flow across a cut may include negative net flows between vertices, but the capacity of the cut includes only nonnegative values, i.e., only the capacities of edges from S to T. The net flow across a cut includes negative flows between vertices (flow from T to S), but the capacity of the cut includes only the capacities of edges from S to T.

Using the flow conservation principle, it can be shown that the net flow across an s-t cut is exactly the flow value $|f|$. By the capacity constraint, the flow across the cut cannot exceed the capacity of the cut. Thus, the value of the maximum flow is no greater than the capacity of a minimum s-t cut. The max-flow min-cut theorem states that the two quantities are actually equal. In other words, if f^* is a maximum flow, then there is some cut $(X, \overline{X})$ with $s \in X$ and $t \in \overline{X}$, such that $|f^*| = u(X, \overline{X})$. The reader is referred to [5,29] for further details.

The residual capacity of an edge (v, w) with respect to a flow f is defined as follows:

$$u'(v, w) = u(v, w) - f(v, w).$$

The quantity $u'(v, w)$ is the number of additional units of flow that can be pushed from v to w without violating the capacity constraints. An edge e is saturated if $u(e) = f(e)$, i.e., if its residual capacity $u'(e) = 0$. The residual graph, $G_R(f)$, for a flow f, is the graph with vertex set V, source and sink s and t, respectively, and those edges (v, w) for which $u'(v, w) > 0$. Figure 8.5 shows a network on the left with a given flow function. Each edge is labeled with two values: the flow through that edge and its capacity. The figure on the right depicts the residual network corresponding to this flow.

An augmenting path for f is a path P from s to t in $G_R(f)$. The residual capacity of P, denoted by $u'(P)$, is the minimum value of $u'(v, w)$ over all edges (v, w) in the path P. The flow can be increased

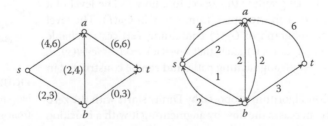

FIGURE 8.5 A network with a given flow function, and its residual network. A label of (f, c) for an edge indicates a flow of f and capacity of c.

by $u'(P)$, by increasing the flow on each edge of P by this amount. Whenever $f(v, w)$ is changed, $f(w, v)$ is also correspondingly changed to maintain skew symmetry.

Most flow algorithms are based on the concept of augmenting paths pioneered by Ford and Fulkerson [8]. They start with an initial zero flow and augment the flow in stages. In each stage, a residual graph $G_R(f)$ with respect to the current flow function f is constructed, and an augmenting path in $G_R(f)$ is found, thus, increasing the value of the flow. Flow is increased along this path until an edge in this path is saturated. The algorithms iteratively keep increasing the flow until there are no more augmenting paths in $G_R(f)$, and return the final flow f as their output. Edmonds and Karp [6] suggested two possible improvements to this algorithm to make it run in polynomial time. The first was to choose shortest possible augmenting paths, where the length of the path is simply the number of edges on the path. This method can be improved using the approach due to Dinitz (see [29]) in which a sequence of blocking flows is found. The second strategy was to select a path which can be used to push the maximum amount of flow. The number of iterations of this method is bounded by $O(|E| \log C)$ where C is the largest edge capacity. In each iteration, we need to find a path on which we can push the maximum amount of flow. This can be done by suitably modifying Dijkstra's algorithm (Chapter 7).

The following lemma is fundamental in understanding the basic strategy behind these algorithms.

LEMMA 8.2   Let f be a flow and f^* be a maximum flow in G. The value of a maximum flow in the residual graph $G_R(f)$ is $|f^*| - |f|$.

PROOF   Let f' be any flow in $G_R(f)$. Define $f + f'$ to be the flow $f(v, w) + f'(v, w)$ for each edge (v, w). Observe that $f + f'$ is a feasible flow in G of value $|f| + |f'|$. Since f^* is the maximum flow possible in G, $|f'| \leq |f^*| - |f|$. Similarly define $f^* - f$ to be a flow in $G_R(f)$ defined by $f^*(v, w) - f(v, w)$ in each edge (v, w), and this is a feasible flow in $G_R(f)$ of value $|f^*| - |f|$, and it is a maximum flow in $G_R(f)$.

8.3.1.4   Blocking Flow

A flow f is a blocking flow if every path from s to t in G contains a saturated edge. Blocking flows are also known as maximal flows. Figure 8.6 depicts a blocking flow. Any path from s to t contains at least one saturated edge. For example, in the path $[s, a, t]$, the edge (a, t) is saturated. It is important to note that a blocking flow is not necessarily a maximum flow. There may be augmenting paths that increase the flow on some edges and decrease the flow on other edges (by increasing the flow in the reverse direction). The example in Figure 8.6 contains an augmenting path $[s, a, b, t]$, which can be used to increase the flow by 3 units.

8.3.1.5   Layered Networks

Let $G_R(f)$ be the residual graph with respect to a flow f. The level of a vertex v is the length of a shortest path from s to v in $G_R(f)$. The level graph L for f is the subgraph of $G_R(f)$ containing vertices reachable from s and only the edges (v, w) such that $level(w) = 1 + level(v)$. L contains all shortest length augmenting paths and can be constructed in $O(|E|)$ time.

The maximum flow algorithm proposed by Dinitz starts with the zero flow, and iteratively increases the flow by augmenting it with a blocking flow in $G_R(f)$ until t is not reachable from s in $G_R(f)$. At each step the current flow is replaced by the sum of the current flow and the blocking flow. This algorithm terminates in $|V| - 1$ iterations, since in

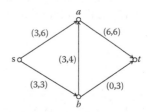

FIGURE 8.6   Example of blocking flow—all paths from s to t are saturated. A label of (f, c) for an edge indicates a flow of f and capacity of c.

each iteration the shortest distance from s to t in the residual graph increases. The shortest path from s to t is at most $|V| - 1$, and this gives an upper bound on the number of iterations of the algorithm.

An algorithm to find a blocking flow that runs in $O(|V|^2)$ time is described here (see [26] for details), and this yields an $O(|V|^3)$ max-flow algorithm. There are a number of $O(|V|^2)$ blocking flow algorithms available, some of which are described in [29].

8.3.2 Blocking Flows

Dinitz's algorithm finds a blocking flow as follows. The main step is to find paths from the source to the sink and push as much flow as possible on each path, and thus, saturate an edge in each path. It takes $O(|V|)$ time to compute the amount of flow that can be pushed on a path. Since there are $|E|$ edges, this yields an upper bound of $O(|V||E|)$ steps on the running time of the algorithm. The following algorithm shows how to find a blocking flow more efficiently.

8.3.2.1 Malhotra–Kumar–Maheshwari Blocking Flow Algorithm

The algorithm has a current flow function f and its corresponding residual graph $G_R(f)$. Define for each node $v \in G_R(f)$, a quantity $tp[v]$ that specifies its maximum throughput, i.e., either the sum of the capacities of the incoming arcs or the sum of the capacities of the outgoing arcs, whichever is smaller. The quantity $tp[v]$ represents the maximum flow that could pass through v in any feasible blocking flow in the residual graph. Vertices with zero throughput are deleted from $G_R(f)$.

The algorithm selects a vertex x with least throughput. It then greedily pushes a flow of $tp[x]$ from x toward t, level by level in the layered residual graph. This can be done by creating a queue which initially contains x, which is assigned the task of pushing $tp[x]$ out of it. In each step, the vertex v at the front of the queue is removed, and the arcs going out of v are scanned one at a time, and as much flow as possible is pushed out of them until v's allocated flow has been pushed out. For each arc (v, w) that the algorithm pushed flow through, it updates the residual capacity of the arc (v, w) and places w on a queue (if it is not already there) and increments the net incoming flow into w. Also $tp[v]$ is reduced by the amount of flow that was sent through it now. The flow finally reaches t, and the algorithm never comes across a vertex that has incoming flow that exceeds its outgoing capacity since x was chosen as a vertex with least throughput. The above idea is repeated to pull a flow of $tp[x]$ from the source s to x. Combining the two steps yields a flow of $tp[x]$ from s to t in the residual network that goes through x. The flow f is augmented by this amount. Vertex x is deleted from the residual graph, along with any other vertices that have zero throughput.

The above procedure is repeated until all vertices are deleted from the residual graph. The algorithm has a blocking flow at this stage since at least one vertex is saturated in every path from s to t. In the above algorithm, whenever an edge is saturated it may be deleted from the residual graph. Since the algorithm uses a greedy strategy to pump flows, at most $O(|E|)$ time is spent when an edge is saturated. When finding flow paths to push $tp[x]$, there are at most $|V|$ times (once for each vertex) when the algorithm pushes a flow that does not saturate the corresponding edge. After this step, x is deleted from the residual graph. Hence, the above algorithm to compute blocking flows terminates in $O(|E| + |V|^2) = O(|V|^2)$ steps.

Karzanov (see [29]) used the concept of preflows to develop an $O(|V|^2)$ time algorithm for developing blocking flows as well. This method was then adapted by Goldberg and Tarjan (see [5]), who proposed a preflow-push method for computing a maximum flow in a network. A preflow is a flow that satisfies the capacity constraints, but not flow conservation. Any node in the network is allowed to have more flow coming into it than there is flowing out. The algorithm tries to move the excess flows towards the sinks without violating capacity constraints. The algorithm finally terminates with a feasible flow that is a max-flow. It finds a max-flow in $O(|V||E| \log(|V|^2)/(|E|))$ time.

8.4   The Min-Cut Problem

Two problems are studied in this section. The first is the $s-t$ min-cut problem: given a directed graph with capacities on the edges and two special vertices s and t, the problem is to find a cut of minimum capacity that separates s from t. The value of a min-cut's capacity can be computed by finding the value of the maximum flow from s to t in the network. Recall that the max-flow min-cut theorem shows that the two quantities are equal. We now show how to find the sets S and T that provides an $s-t$ min-cut from a given $s-t$ max-flow.

The second problem is to find a smallest cut in a given graph G. In this problem on undirected graphs, the vertex set must be partitioned into two sets such that the total weight of the edges crossing the cut is minimized. This problem can be solved by enumerating over all $\{s, t\}$ pairs of vertices, and finding a $s-t$ min-cut for each pair. This procedure is rather inefficient. However, Hao and Orlin [15] showed that this problem can be solved in the same order of running time as taken by a single max-flow computation. Stoer and Wagner [27] have given a simple and elegant algorithm for computing minimum cuts based on a technique due to Nagamochi and Ibaraki [25]. Recently, Karger [19] has given a randomized algorithm that runs in almost linear time for computing minimum cuts without using network-flow methods.

8.4.1   Finding an $s-t$ Min-Cut

Let f^* be a maximum flow from s to t in the graph. Recall that $G_R(f^*)$ is the residual graph corresponding to f^*. Let S be the set of all vertices that are reachable from s in $G_R(f^*)$, i.e., vertices to which s has a directed path. Let T be all the remaining vertices of G. By definition $s \in S$. Since f^* is a max-flow from s to t, $t \in T$. Otherwise the flow can be increased by pushing flow along this path from s to t in $G_R(f^*)$. All the edges from vertices in S to vertices in T are saturated and form a minimum cut.

8.4.2   Finding All-Pair Min-Cuts

For an undirected graph G with $|V|$ vertices, Gomory and Hu (see [1]) showed that the flow values between each of the $|V|(|V| - 1)/2$ pairs of vertices of G can be computed by solving only $|V| - 1$ max-flow computations. Furthermore, they showed that the flow values can be represented by a weighted tree T on $|V|$ nodes. Each node in the tree represents one of the vertices of the given graph. For any pair of nodes (x, y), the maximum flow value from x to y (and hence, the $x-y$ min-cut) in G is equal to the weight of the minimum-weight edge on the unique path in T between x and y.

8.4.3   Applications of Network Flows (and Min-Cuts)

There are numerous applications of the maximum flow algorithm in scheduling problems of various kinds. It is used in open-pit mining, vehicle routing, etc. See [1] for further details.

8.4.3.1   Finding a Minimum-Weight Vertex Cover in Bipartite Graphs

A vertex cover of a graph is a set of vertices that is incident to all its edges. The weight of a vertex cover is the sum of the weights of its vertices. Finding a vertex cover of minimum weight is NP-hard for arbitrary graphs (see Section 8.9 for more details). For bipartite graphs, the problem is solvable efficiently using the maximum flow algorithm.

Given a bipartite graph $G = (X, Y, E)$ with weights on the vertices, we need to find a subset C of vertices of minimum total weight, so that for each edge, at least one of its end vertices is in C.

The weight of a vertex v is denoted by $w(v)$. Construct a flow network N from G as follows: add a new source vertex s, and for each $x \in X$, add a directed edge (s, x), with capacity $w(x)$. Add a new sink vertex t, and for each $y \in Y$, add a directed edge (y, t), with capacity $w(y)$. The edges in E are given infinite capacity and they are oriented from X to Y. Let C be a subset of vertices in G. We use the notation X_C to denote $X \cap C$ and Y_C to denote $Y \cap C$. Let $\overline{X_C} = X - X_C$ and $\overline{Y_C} = Y - Y_C$.

If C is a vertex cover of G, then there are no edges in G connecting $\overline{X_C}$ and $\overline{Y_C}$, since such edges would not be incident to C. A cut in the flow network N, whose capacity is the same as the weight of the vertex cover can be constructed as follows: $S = \{s\} \cup \overline{X_C} \cup Y_C$ and $T = \{t\} \cup X_C \cup \overline{Y_C}$. Each vertex cover of G gives rise to a cut of same weight. Similarly, any cut of finite capacity in N corresponds to a vertex cover of G of same weight. Hence, the minimum-weight s–t cut of N corresponds to a minimum-weight vertex cover of G.

8.4.3.2 Maximum-Weight Closed Subset in a Partial Order

Consider a directed acyclic graph (DAG) $G = (V, E)$. Each vertex v of G is given a weight $w(v)$, that may be positive or negative. A subset of vertices $S \subseteq V$ is said to be closed in G if for every $s \in S$, the predecessors of s are also in S. The predecessors of a vertex s are vertices that have directed paths to s. We are interested in computing a closed subset S of maximum weight. This problem occurs for example in open-pit mining, and it can be solved efficiently by reducing it to computing the minimum cut in the following network N:

- The vertex set of $N = V \cup \{s, t\}$, where s and t are two new vertices
- For each vertex v of negative weight, add the edge (s, v) with capacity $|w(v)|$ to N
- For each vertex v of positive weight, add the edge (v, t) with capacity $w(v)$ to N
- All edges of G are added to N, and each of these edges has infinite capacity

Consider a minimum s–t cut in N. Let S be the positive weight vertices whose edges to the sink t are not in the min-cut. It can be shown that the union of the set S and its predecessors is a maximum-weight closed subset.

8.5 Minimum-Cost Flows

We now study one of the most important problems in combinatorial optimization, namely the minimum cost network flow problem.

We will outline some of the basic ideas behind the problem, and the reader can find a wealth of information in [1] about efficient algorithms for this problem.

The min-cost flow problem can be viewed as a generalization of the max-flow problem, the shortest path problem and the minimum-weight perfect matching problem. To reduce the shortest path problem to the min-cost flow problem, notice that if we have to ship one unit of flow, we will ship it on the shortest path. We reduce the minimum-weight perfect matching problem on a bipartite graph $G = (X, Y, X \times Y)$, to min-cost flow as follows. Introduce a special source s and sink t, and add unit capacity edges from s to vertices in X, and from vertices in Y to t. Compute a flow of value $|X|$ to obtain a minimum-weight perfect matching.

8.5.1 Min-Cost Flow Problem Definitions

The flow network $G = (V, E)$ is a directed graph. There is a capacity function $u : E \mapsto \Re^{+}$ that maps edges to positive real numbers and a cost function $c : E \to R$. The cost function specifies the cost of shipping one unit of flow through the edge. Associated with each vertex v, there is a supply

$b(v)$. If $b(v) > 0$ then v is a supply node, and if $b(v) < 0$ then it is a demand node. It is assumed that $\sum_{v \in V} b(v) = 0$; if the supply exceeds the demand we can add an artificial sink to absorb the excess flow.

A flow function is required to satisfy the following constraints:

- (Capacity Constraint) For all edges e, $f(e) \leq u(e)$
- (Skew Symmetry Constraint) For an edge $e = (v, w)$, $f(v, w) = -f(w, v)$
- (Flow Conservation) For all vertices $v \in V$, $\sum_{w \in V} f(v, w) = b(v)$

We wish to find a flow function that minimizes the cost of the flow.

$$\min z(f) = \sum_{e \in E} c(e) \cdot f(e).$$

Our algorithm again uses the concept of residual networks. As before, the residual network $G_R(f)$ is defined with respect to a specific flow function. The only difference is the following: if there is a flow $f(v, w)$ on edge $e = (v, w)$ then as before, its capacity in the residual network is $u(e) - f(e)$ and its residual cost is $c(e)$. The reverse edge (w, v) has residual capacity $f(e)$, but its residual cost is $-c(e)$. Note that sending a unit of flow on the reverse edge actually reduces the original amount of flow that was sent, and hence, the residual cost is negative. As before, only edges with strictly positive residual capacity are retained in the residual network.

Before studying any specific algorithm to solve the problem, we note that we can always find a feasible solution (not necessarily optimal), if one exists, by finding a max-flow, and ignoring costs. To see this, add two new vertices, a source vertex s and a sink vertex t. Add edges from s to all vertices v with $b(v) > 0$ of capacity $b(v)$, and edges from all vertices w with $b(w) < 0$ to t of capacity $|b(w)|$. Find a max-flow of value $\sum_{b(v)>0} b(v)$. If such flow does not exist then the flow problem is not feasible.

The following theorem characterizes optimal solutions.

THEOREM 8.5   *A feasible solution f is an optimal solution if and only if the residual network $G_R(f)$ has no directed cycles of negative cost.*

We use the above theorem to develop an algorithm for finding a min-cost flow. As indicated earlier, a feasible flow can be found by the techniques developed in the previous section. To improve the cost of the solution, we identify negative-cost cycles in the residual graph and push as much flow around them as possible and reduce the cost of the flow. We repeat this until there are no negative cycles left, and we have found a min-cost flow. A negative-cost cycle may be found by using an algorithm like the Bellman–Ford algorithm (Chapter 7), which takes $O(|E||V|)$ time.

An important issue that affects the running time of the algorithm is the choice of a negative cost cycle. For fast convergence, one should select a cycle that decreases the cost as much as possible, but finding such a cycle in NP-hard. Goldberg and Tarjan [12] showed that selecting a cycle with minimum mean-cost (the ratio of the cost of cycle to the number of arcs on the cycle) yields a polynomial-time algorithm. There is a parametric search algorithm [1] to find a min-mean cycle that works as follows. Let μ^* be the average edge-weight of a min-mean cycle. Then μ^* is the largest value such that reducing the weight of every edge by μ^* does not introduce a negative-weight cycle into the graph. One can search for the value of μ^* using binary search: given a guess μ, decrease the weight of each edge by μ and test if the graph has no negative-weight cycles. If the smallest cycle has zero weight, it is a min-mean cycle. If all cycles are positive, we have to increase our guess μ. If the graph has a negative weight cycle, we need to decrease our guess μ. Karp (see [1]) has given an algorithm based on dynamic programming for this problem that runs in $O(|E||V|)$ time.

The first polynomial-time algorithm for the min-cost flow problem was given by Edmonds and Karp [6]. Their idea is based on scaling capacities and the running time of their algorithm is $O(|E| \log U(|E| + |V| \log |V|))$, where U is the largest capacity. The first strongly polynomial algorithm (whose running time is only a function of $|E|$ and $|V|$) for the problem was given by Tardos [28].

8.6 The Multicommodity Flow Problem

A natural generalization of the max-flow problem is the **multicommodity flow** problem. In this problem, there are a number of commodities $1, 2, \ldots, k$ that have to be shipped through a flow network from source vertices $s_1, s_2, \ldots, s_k$ to sink vertices $t_1, t_2, \ldots, t_k$, respectively. The amount of demand for commodity i is specified by d_i. Each edge and each vertex has a nonnegative capacity that specifies the total maximum flow of all commodities flowing through that edge or vertex. A flow is feasible if all the demands are routed from their respective sources to their respective sinks while satisfying flow conservation constraints at the nodes and capacity constraints on the edges and vertices. A flow is an ϵ-feasible flow for a positive real number ϵ if it satisfies $(d_i)/(1 + \epsilon)$ of demand i. There are several variations of this problem as noted below:

In the simplest version of the multicommodity flow problem, the goal is to decide if there exists a feasible flow that routes all the demands of the k commodities through the network. The flow corresponding to each commodity is allowed to be split arbitrarily (i.e., even fractionally) and routed through multiple paths. The **concurrent flow** problem is identical to the multicommodity flow problem when the input instance has a feasible flow that satisfies all the demands. If the input instance is not feasible, then the objective is to find the smallest fraction ϵ for which there is an ϵ-feasible flow. In the minimum-cost multicommodity flow problem, each edge has an associated cost for each unit of flow through it. The objective is to find a minimum-cost solution for routing all the demands through the network. All of the above problems can be formulated as linear programs, and therefore have polynomial-time algorithms using either the ellipsoid algorithm or the interior point method [20].

A lot of research has been done on finding approximately optimal multicommodity flows using more efficient algorithms. There are a number of papers that provide approximation algorithms for the multicommodity flow problem. For a detailed survey of these results see [16,Chapter 5].

In this section, we will discuss a recent paper that introduced a new algorithm for solving multicommodity flow problems. Awerbuch and Leighton [2] gave a simple and elegant algorithm for the concurrent flow problem. Their algorithm is based on the "liquid-flow paradigm." Flow is pushed from the sources on edges based on the "pressure difference" between the end vertices. The algorithm does not use any global properties of the graph (such as shortest paths) and uses only local changes based on the current flow. Due to its nature, the algorithm can be implemented to run in a distributed network since all decisions made by the algorithm are local in nature. They extended their algorithms to dynamic networks, where edge capacities are allowed to vary [3]. The best implementation of this algorithm runs in $O(k|V|^2|E|\epsilon^{-3} \ln^3(|E|/\epsilon))$ time.

In the integer multicommodity flow problem, the capacities and flows are restricted to be integers. Unlike the single-commodity flow problem, for problems with integral capacities and demands, the existence of a feasible fractional solution to the multicommodity flow problem does not guarantee a feasible integral solution. An extra constraint that may be imposed on this problem is to restrict each commodity to be sent along a single path. In other words, this constraint does not allow one to split the demand of a commodity into smaller parts and route them along independent paths (see the recent paper by Kleinberg [21] for approximation algorithms for this problem). Such constraints are common in problems that arise in telecommunication systems. All variations of the **integer multicommodity flow** problem are NP-hard.

8.6.1   Local Control Algorithm

In this section, we describe Awerbuch and Leighton's local control algorithm that finds an approximate solution for the concurrent flow problem.

First, the problem is converted to the continuous flow problem. In this version of the problem, d_i units of commodity i is added to the source vertex s_i in each phase. Each vertex has queues to store the commodities that arrive at that node. The node tries to push the commodities stored in the queues towards the sink nodes. As the commodities arrive at the sinks, they are removed from the network. A continuous flow problem is stable if the total amount of all the commodities in the network at any point in time is bounded (i.e., independent of the number of phases completed). The following theorem establishes a tight relationship between the multicommodity flow problem (also known as the static problem) and its continuous version.

THEOREM 8.6   *A stable algorithm for the continuous flow problem can be used to generate an ϵ-feasible solution for the static concurrent flow problem.*

PROOF   Let R be the number of phases of the stable algorithm for the continuous flow problem at which $1/(1 + \epsilon)$ fraction of each of the commodities that have been injected into the network have been routed to their sinks. We know that R exists since the algorithm is stable. The average behavior of the continuous flow problem in these R phases generates an ϵ-feasible flow for the static problem. In other words, the flow of a commodity through a given edge is the average flow of that commodity through that edge over R iterations, and this flow is ϵ-feasible.

In order to get an approximation algorithm using the fluid-flow paradigm, there are two important challenges. First, it must be shown that the algorithm is stable, i.e., that the queue sizes are bounded. Second, the number of phases that it takes the algorithm to reach "steady state" (specified by R in the above theorem) must be minimized. Note that the running time of the approximation algorithm is R times the time it takes to run one phase of the continuous flow algorithm.

Each vertex v maintains one queue per commodity for each edge to which it is incident. Initially all the queues are empty. In each phase of the algorithm, commodities are added at the source nodes. The algorithm works in four steps:

1. Add $(1 + \epsilon)d_i$ units of commodity i at s_i. The commodity is spread equally to all the queues at that vertex.
2. Push the flow across each edge so as to balance the queues as much as possible. If Δ_i is the discrepancy of commodity i in the queues at either end of edge e, then the flow f_i crossing the edge in that step is chosen so as to maximize $\Sigma_i f_i(\Delta_i - f_i)/d_i^2$ without exceeding the capacity constraints.
3. Remove the commodities that have reached their sinks.
4. Balance the amount of each commodity in each of the queues at each vertex.

The above four steps are repeated until $1/(1 + \epsilon)$ fraction of the commodities that have been injected into the system have reached their destination.

It can be shown that the algorithm is stable and the number of phases is $O(|E|^2 k^{1.5} L \epsilon^{-3})$, where $|E|$ is the number of edges, k is the number of commodities, and $L \leq |V|$ is the maximum path length of any flow. The proof of stability and the bound on the number of rounds are not included here. The actual algorithm has a few other technical details that are not mentioned here, and the reader is referred to the original paper for further details [2].

8.7 Minimum-Weight Branchings

A natural analog of a spanning tree in a directed graph is a branching (also called an arborescence). For a directed graph G and a vertex r, a branching rooted at r is an acyclic subgraph of G in which each vertex but r has exactly one outgoing edge, and there is a directed path from any vertex to r. This is also sometimes called an in-branching. By replacing "outgoing" in the above definition of a branching to "incoming," we get an out-branching. An optimal branching of an edge-weighted graph is a branching of minimum total weight. Unlike the minimum spanning tree problem, an optimal branching cannot be computed using a greedy algorithm. Edmonds gave the first polynomial-time algorithm to find optimal branchings (see [11]).

Let $G = (V, E)$ be an arbitrary graph, let r be the root of G, and let $w(e)$ be the weight of edge e. Consider the problem of computing an optimal branching rooted at r. In the following discussion, assume that all edges of the graph have a nonnegative weight.

Two key ideas are discussed below that can be converted into a polynomial-time algorithm for computing optimal branchings. First, for any vertex $v \neq r$ in G, suppose that all outgoing edges of v are of positive weight. Let $\epsilon > 0$ be a number that is less than or equal to the weight of any outgoing edge from v. Suppose the weight of every outgoing edge from v is decreased by ϵ. Observe that since any branching has exactly one edge out of v, its weight decreases by exactly ϵ. Therefore an optimal branching of G with the original weights is also an optimal branching of G with the new weights. In other words, decreasing the weight of the outgoing edges of a vertex uniformly, leaves optimal branchings invariant. Second, suppose there is a cycle C in G consisting only of edges of zero weight. Suppose the vertices of C are combined into a single vertex, and the resulting multigraph is made into a simple graph by replacing each multiple edge by a single edge whose weight is the smallest among them, and discarding self loops. Let G_C be the resulting graph. It can be shown that the weights of optimal branchings of G and G_C are the same. An optimal branching of G_C can also be converted into an optimal branching of G by adding sufficiently many edges from C without introducing cycles. The above two ideas can be used to design a recursive algorithm for finding an optimal branching. The fastest implementation of a branching algorithm is due to Gabow et al. [9].

8.8 Coloring Problems

8.8.1 Vertex Coloring

A **vertex coloring** of a graph G is a coloring of the vertices of G such that no two adjacent vertices of G receive the same color. The objective of the problem is to find a coloring that uses as few colors as possible. The minimum number of colors needed to color the vertices of a graph is known as its **chromatic number**. The register allocation problem in compiler design and the map coloring problem are instances of the vertex coloring problem.

The problem of deciding if the chromatic number of a given graph is at most a given integer k is NP-complete. The problem is NP-complete even for fixed $k \geq 3$. For $k = 2$, the problem is to decide if the given graph is bipartite, and this can be solved in linear time using depth-first or breadth-first search. The vertex coloring problem in general graphs is a notoriously hard problem and it has been shown to be intractable even for approximating within a factor of n^ϵ for some constant $\epsilon > 0$ [16, Chapter 10 of first edition]. A greedy coloring of a graph yields a coloring that uses at most $\Delta + 1$ colors, where Δ is the maximal degree of the graph. Unless the graph is an odd cycle or the complete graph, it can be colored with Δ colors (known as Brooks theorem) [11]. A celebrated result on graph coloring is that every planar graph is four-colorable. However, checking if a planar graph is three-colorable is NP-complete [10]. For more information on recent results on approximating the chromatic number, see [18].

8.8.2 Edge Coloring

The **edge coloring** problem is similar to the vertex coloring problem. In this problem, the goal is to color the edges of a given graph using the fewest colors such that no two edges incident to a common vertex are assigned the same color. The problem finds applications in assigning classroom to courses and in scheduling problems. The minimum number of colors needed to color a graph is known as its **chromatic index**. Since any two incident edges must receive distinct colors, the chromatic index of a graph with maximal degree Δ is at least Δ. In a remarkable theorem, Vizing (see [11]) has shown that every graph can be edge-colored using at most $\Delta + 1$ colors. Deciding whether the edges of a given graph can be colored using Δ colors is NP-complete (see [16]). Special classes of graphs such as bipartite graphs and planar graphs of large maximal degree are known to be Δ-edge-colorable.

8.9 Approximation Algorithms for Hard Problems

Many graph problems are known to be NP-complete (see Sections A.1 and A.2 of [10]). The area of approximation algorithms explores intractable (NP-hard) optimization problems and tries to obtain polynomial-time algorithms that generate feasible solutions that are close to optimal. In this section, a few fundamental NP-hard optimization problems in graphs that can be approximated well are discussed. For more information about approximating these and other problems, see the chapter on approximation algorithms (Chapter 35) and a book on approximation algorithms edited by Hochbaum [16].

In the following discussion $G = (V, E)$ is an arbitrary undirected graph and k is a positive integer.

Vertex cover: a set of vertices $S \subset V$ such that every edge in E is incident to at least one vertex of S. The minimum vertex cover problem is that of computing a vertex cover of minimum cardinality. If the vertices of G have weights associated with them, a minimum-weight vertex cover is a vertex cover of minimum total weight. Using the primal-dual method of linear programming, one can obtain a two-approximation.

Dominating set: a set of vertices $S \subset V$ such that every vertex in the graph is either in S or adjacent to some vertex in S. There are several versions of this problem such as the total dominating set (every vertex in G must have a neighbor in S, irrespective of whether it is in S or not), the connected dominating set (induced graph of S must be connected) and the independent dominating set (induced graph of S must be empty). The minimum dominating set problem is to compute a minimum cardinality dominating set. The problems can be generalized to the weighted case suitably. All but the independent dominating set problem can be approximated within a factor of $O(\log n)$.

Steiner tree problem: a tree of minimum total weight that connects a set of terminal vertices S in a graph $G = (V, E)$. There is a two-approximation algorithm for the general problem. There are better algorithms when the graph is defined in Euclidean space. For more information, see [16,17].

Minimum k-connected graph: a graph G is k-vertex-connected, or simply k-connected, if the removal of up to $k - 1$ vertices, along with their incident edges leaves the remaining graph connected. It is k-edge-connected if the removal of up to $k - 1$ edges does not disconnect the graph. In the minimum k-connected graph problem, we need to compute a k-connected spanning subgraph of G of minimum cardinality. The problem can be posed in the context of vertex or edge connectivities, and with edge weights. The edge-connectivity problems can be approximated within a factor of 2 from optimal, and a factor smaller than 2 when the edges do not have weights. The unweighted k-vertex-connected subgraph problem can be approximated within a factor of $1 + 2/k$, and the corresponding weighted problem can be approximated within a factor of $O(\log k)$. When the edges satisfy the triangle inequality, the vertex connectivity problem can be approximated within a factor of about $2 + 2/k$. These problems find applications in fault-tolerant network-design. For more information on approximation algorithms for connectivity problems, see [16,Chapter 6].

Degree constrained spanning tree: a spanning tree of maximal degree k. This problem is a generalization of the traveling salesman path problem. There is a polynomial time approximation algorithm that returns a spanning tree of maximal degree at most $k + 1$ if G has a spanning tree of degree k.

Max-cut: a partition of the vertex set into (V_1, V_2) such that the number of edges in $E \cap (V_1 \times V_2)$ (edges between a vertex in V_1 and a vertex in V_2) is maximized. It is easy to compute a partition in which at least half the edges of the graph cross the cut. This is a two-approximation. Semidefinite programming techniques can be used to derive an approximation algorithm with a performance guarantee of about 1.15.

8.10 Research Issues and Summary

Some of the recent research efforts in graph algorithms have been in the areas of dynamic algorithms, graph layout and drawing, and approximation algorithms. Detailed references about these areas may be found in Chapters 9 and 34 of this book. Many of the methods illustrated in our chapter find use in the solution of almost any optimization problem. For approximation algorithms see the edited book by Hochbaum [16], and more information in the area of graph layout and drawing can be found in an annotated bibliography by Di Battista et al. [4].

A recent exciting result of Karger's [19] is a randomized near-linear time algorithm for computing minimum cuts. Finding a deterministic algorithm with similar bounds is a major open problem.

Computing a maximum flow in $O(|E||V|)$ time in general graphs is still open. Several recent algorithms achieve these bounds for certain graph densities. A recent result by Goldberg and Rao [14] breaks this barrier, by paying an extra $\log U$ factor in the running time, when the edge capacities are in the range $[1 \ldots U]$. Finding a maximum matching in time better than $O(|E|\sqrt{|V|})$ or obtaining nontrivial lower bounds are open problems.

8.11 Further Information

The area of graph algorithms continues to be a very active field of research. There are several journals and conferences that discuss advances in the field. Here we name a partial list of some of the important meetings: ACM Symposium on Theory of Computing (STOC), IEEE Conference on Foundations of Computer Science (FOCS), ACM-SIAM Symposium on Discrete Algorithms (SODA), International Colloquium on Automata, Languages and Programming (ICALP), and the European Symposium on Algorithms (ESA). There are many other regional algorithms/theory conferences that carry research papers on graph algorithms. The journals that carry articles on current research in graph algorithms are *Journal of the ACM, SIAM Journal on Computing, SIAM Journal on Discrete Mathematics, Journal of Algorithms, Algorithmica, Journal of Computer and System Sciences, Information and Computation, Information Processing Letters,* and *Theoretical Computer Science.*

To find more details about some of the graph algorithms described in this chapter we refer the reader to the books by Cormen et al. [5], Even [7], Gibbons [11], and Tarjan [29]. Ahuja et al. [1] have provided a comprehensive book on the network flow problem. The survey chapter by Goldberg et al. [13] provides an excellent survey of various min-cost flow and generalized flow algorithms. A detailed survey describing various approaches for solving the matching and flow problems can be found in Tarjan's book [29]. Papadimitriou and Steiglitz [26] discuss the solution of many combinatorial optimization problems using a primal-dual framework.

Defining Terms

Assignment problem: The problem of finding a matching of maximum (or minimum) weight in an edge-weighted graph.

Augmenting path: A path used to augment (increase) the size of a matching or a flow.

Blocking flow: A flow function in which any directed path from s to t contains a saturated edge.

Blossoms: Odd length cycles that appear during the course of the matching algorithm on general graphs.

Branching: A spanning tree in a rooted graph, such that each vertex has a path to the root (also known as in-branching). An out-branching is a rooted spanning tree in which the root has a path to every vertex in the graph.

Capacity: The maximum amount of flow that is allowed to be sent through an edge or a vertex.

Chromatic index: The minimum number of colors with which the edges of a graph can be colored.

Chromatic number: The minimum number of colors needed to color the vertices of a graph.

Concurrent flow: A multicommodity flow in which the same fraction of the demand of each commodity is satisfied.

Edge coloring: An assignment of colors to the edges of a graph such that no two edges incident to a common vertex receive the same color.

Integer multicommodity flow: A multicommodity flow in which the flow through each edge of each commodity is an integral value. The term is also used to capture the multicommodity flow problem in which each demand is routed along a single path.

Matching: A subgraph in which every vertex has degree at most one.

Maximum flow: The maximum amount of feasible flow that can be sent from a source vertex to a sink vertex in a given network.

Multicommodity flow: A network flow problem involving multiple commodities, in which each commodity has an associated demand and source-sink pairs.

Network flow: An assignment of flow values to the edges of a graph that satisfies flow conservation, skew symmetry and capacity constraints.

s–t cut: A partitioning of the vertex set into S and T such that $s \in S$ and $t \in T$.

Vertex coloring: An assignment of colors to the vertices of a graph such that no two adjacent vertices receive the same color.

Acknowledgments

Samir Khuller's research was supported by NSF Research Initiation Award CCR-9307462 and NSF CAREER Award CCR-9501355. Balaji Raghavachari's research was supported by NSF Research Initiation Award CCR-9409625.

References

1. Ahuja, R.K., Magnanti, T.L., and Orlin, J.B., *Network Flows.* Prentice Hall, Englewood Cliffs, NJ, 1993.
2. Awerbuch, B. and Leighton, T., A simple local-control approximation algorithm for multicommodity flow. In *Proceedings of the 34th Annual Symposium on Foundations of Computer Science,* 459–468, Palo Alto, CA, November 3–5, 1993.

3. Awerbuch, B. and Leighton, T., Improved approximation algorithms for the multi-commodity flow problem and local competitive routing in dynamic networks. In *Proceedings of the 26th Annual ACM Symposium on Theory of Computing*, 487–496, Montreal, QC, Canada, May 23–25, 1994.

4. Di Battista, G., Eades, P., Tamassia, R., and Tollis, I.G., Annotated bibliography on graph drawing algorithms. *Comput. Geom.: Theor. Appl.*, 4, 235–282, 1994.

5. Cormen, T.H., Leiserson, C.E., and Rivest, R.L., *Introduction to Algorithms*. MIT Press, Cambridge, MA, 1989.

6. Edmonds, J. and Karp, R.M., Theoretical improvements in algorithmic efficiency for network flow problems. *J. Assoc. Comput. Mach.*, 19, 248–264, 1972.

7. Even, S., *Graph Algorithms*. Computer Science Press, Rockville, MD, 1979.

8. Ford, L.R. and Fulkerson, D.R., *Flows in Networks*. Princeton University Press, Princeton, NJ, 1962.

9. Gabow, H.N., Galil, Z., Spencer, T., and Tarjan, R.E., Efficient algorithms for finding minimum spanning trees in undirected and directed graphs. *Combinatorica*, 6, 109–122, 1986.

10. Garey, M.R. and Johnson, D.S., *Computers and Intractability: A Guide to the Theory of NP-Completeness*. Freeman, San Francisco, CA, 1979.

11. Gibbons, A.M., *Algorithmic Graph Theory*. Cambridge University Press, New York, 1985.

12. Goldberg, A.V. and Tarjan, R.E., Finding minimum-cost circulations by canceling negative cycles. *J. Assoc. Comput. Mach.*, 36(4), 873–886, October 1989.

13. Goldberg, A.V., Tardos, É., and Tarjan, R.E., Network flow algorithms. In *Algorithms and Combinatorics*, Volume 9: *Flows, Paths and VLSI Layout*, B. Korte, L. Lovsz, H.J. Prmel, and A. Schrijver (Eds.), Springer-Verlag, Berlin, 1990.

14. Goldberg, A. and Rao, S., Beyond the flow decomposition barrier. In *Proceedings of the 38th Annual Symposium on Foundations of Computer Science*, 2–11, Miami Beach, FL, October 20–22, 1997.

15. Hao, J. and Orlin, J.B., A faster algorithm for finding the minimum cut in a directed graph. *J. Algorithm*, 17(3), 424–446, November 1994.

16. Hochbaum, D.S., Ed., *Approximation Algorithms for NP-Hard Problems*. PWS Publishing, Boston, MA, 1996.

17. Hwang, F., Richards, D.S., and Winter, P., *The Steiner Tree Problem*. North-Holland, Amsterdam, The Netherlands, 1992.

18. Karger, D.R., Motwani, R., and Sudan, M., Approximate graph coloring by semidefinite programming. In *Proceedings of the 35th Annual Symposium on Foundations of Computer Science*, 2–13, Santa Fe, NM, November 20–22, 1994.

19. Karger, D.R., Minimum cuts in near-linear time. In *Proceedings of the 28th Annual ACM Symposium on Theory of Computing*, 56–63, Philadelphia, PA, May 22–24, 1996.

20. Karloff, H., *Linear Programming*. Birkhäuser, Boston, MA, 1991.

21. Kleinberg, J. M., Single-source unsplittable flow. In *Proceedings of the 37th Annual Symposium on Foundations of Computer Science*, 68–77, Burlington, VT, October 14–16, 1996.

22. Lawler, E.L., *Combinatorial Optimization: Networks and Matroids*. Holt, Rinehart and Winston, New York, 1976.

23. Lovász, L. and Plummer, M.D., *Matching Theory*. Elsevier Science Publishers B.V., New York, 1986.

24. Micali, S. and Vazirani, V.V., An $O(\sqrt{|V|} \cdot |E|)$ algorithm for finding maximum matching in general graphs. In *Proceedings of the 21st Annual Symposium on Foundations of Computer Science*, 17–27, Syracuse, NY, October 13–15, 1980.

25. Nagamochi, H. and Ibaraki, T., Computing edge-connectivity in multi-graphs and capacitated graphs. *SIAM J. Disc. Math.*, 5, 54–66, 1992.

26. Papadimitriou, C.H. and Steiglitz, K., *Combinatorial Optimization: Algorithms and Complexity*. Prentice Hall, Englewood Cliffs, NJ, 1982.

27. Stoer, M., and Wagner, F., A simple min-cut algorithm. *J. Assoc. Comput. Mach.*, 44(4), 585–590, 1997.
28. Tardos, É., A strongly polynomial minimum cost circulation algorithm. *Combinatorica*, 5, 247–255, 1985.
29. Tarjan, R.E., *Data Structures and Network Algorithms*. Society for Industrial and Applied Mathematics, Philadelphia, PA, 1983.

9

Dynamic Graph Algorithms

Camil Demetrescu
University of Rome "La Sapienza"

David Eppstein
University of California

Zvi Galil
Tel Aviv University

Giuseppe F. Italiano
University of Rome "Tor Vergata"

9.1 Introduction

In many applications of graph algorithms, including communication networks, graphics, assembly planning, and VLSI design, graphs are subject to discrete changes, such as additions or deletions of edges or vertices. In the last decades there has been a growing interest in such dynamically changing graphs, and a whole body of algorithms and data structures for dynamic graphs has been discovered. This chapter is intended as an overview of this field.

In a typical dynamic graph problem one would like to answer queries on graphs that are undergoing a sequence of updates, for instance, insertions and deletions of edges and vertices. The goal of a dynamic graph algorithm is to update efficiently the solution of a problem after dynamic changes, rather than having to recompute it from scratch each time. Given their powerful versatility, it is not surprising that dynamic algorithms and dynamic data structures are often more difficult to design and analyze than their static counterparts.

We can classify dynamic graph problems according to the types of updates allowed. A problem is said to be fully dynamic if the update operations include unrestricted insertions and deletions of edges. A problem is called partially dynamic if only one type of update, either insertions or deletions, is allowed. If only insertions are allowed, the problem is called incremental; if only deletions are allowed it is called decremental.

In this chapter, we describe fully dynamic algorithms for graph problems. We present dynamic algorithms for undirected graphs in Section 9.2. Dynamic algorithms for directed graphs are next described in Section 9.3. Finally, in Section 9.4 we describe some open problems.

9.2   Fully Dynamic Problems on Undirected Graphs

This section describes fully dynamic algorithms for undirected graphs. These algorithms maintain efficiently some property of a graph that is undergoing structural changes defined by insertion and deletion of edges, and/or edge cost updates. For instance, the fully dynamic minimum spanning tree problem consists of maintaining a minimum spanning forest of a graph during the aforementioned operations. The typical updates for a fully dynamic problem will therefore be inserting a new edge, and deleting an existing edge. To check the graph property throughout a sequence of these updates, the algorithms must be prepared to answer queries on the graph property. Thus, a fully dynamic connectivity algorithm must be able to insert edges, delete edges, and answer a query on whether the graph is connected, or whether two vertices are connected. Similarly, a fully dynamic k-edge connectivity algorithm must be able to insert edges, delete edges, and answer a query on whether the graph is k-edge-connected, or whether two vertices are k-edge-connected. The goal of a dynamic algorithm is to minimize the amount of recomputation required after each update. All of the fully dynamic algorithms that we describe in this section are able to dynamically maintain the graph property at a cost (per update operation) which is significantly smaller than the cost of recomputing the graph property from scratch. Many of the algorithms proposed in the literature use the same general techniques, and so we begin by describing these techniques. All of these techniques use some form of graph decomposition, and partition either the vertices or the edges of the graph to be maintained.

The first technique we describe is **sparsification** by Eppstein et al. [16]. This is a divide-and-conquer technique that can be used to reduce the dependence on the number of edges in a graph, so that the time bounds for maintaining some property of the graph match the times for computing in sparse graphs. Roughly speaking, sparsification works as follows. Let $\mathcal{A}$ be an algorithm that maintains some property of a dynamic graph G with m edges and n vertices in time $T(n, m)$. Sparsification maintains a proper decomposition of G into small subgraphs, with $O(n)$ edges each. In this decomposition, each update involves applying algorithm $\mathcal{A}$ to few small subgraphs of G, resulting into an improved $T(n, n)$ time bound per update. Thus, throughout a sequence of operations, sparsification makes a graph looks sparse (i.e., with only $O(n)$ edges): hence, the reason for its name. Sparsification works on top of a given algorithm, and need not to know the internal details of this algorithm. Consequently, it can be applied orthogonally to other data structuring techniques; we will actually see a number of situations in which both clustering and sparsification can be combined to produce an efficient dynamic graph algorithm.

The second technique we present in this section is due to Henzinger and King [39], and it is a combination of a suitable graph decomposition and randomization. We now sketch how this decomposition is defined. Let G be a graph whose spanning forest has to be maintained dynamically. The edges of G are partitioned into $O(\log n)$ levels: the lower levels contain tightly-connected portions of G (i.e., dense edge cuts), while the higher levels contain loosely-connected portions of G (i.e., sparse cuts). For each level i, a spanning forest for the graph defined by all the edges in levels i or below is maintained. If a spanning forest edge e is deleted at some level i, random sampling is used to quickly find a replacement for e at that level. If random sampling succeeds, the forest is reconnected at level i. If random sampling fails, the edges that can replace e in level i form with high probability a sparse cut. These edges are moved to level $(i + 1)$ and the same procedure is applied recursively on level $(i + 1)$.

The last technique presented in this section is due to Holm et al. [46], which is still based on graph decomposition, but derandomizes the previous approach by Henzinger and King. Besides derandomizing the bounds, this technique is able to achieve faster running times.

One particular dynamic graph problem that has been thoroughly investigated is the maintenance of a minimum spanning forest. This is an important problem on its own, but it also has an impact on other problems. Indeed the data structures and techniques developed for dynamic minimum spanning forests have found applications also in other areas, such as dynamic edge and vertex connectivity [16,25,31,37,38,44]. Thus, we will focus our attention to the fully dynamic maintenance of minimum spanning trees.

Extensive computational studies on dynamic algorithms for undirected graphs appear in [1,2,49].

9.2.1 Sparsification

In this section we describe a general technique for designing dynamic graph algorithms, due to Eppstein et al. [16], which is known as sparsification. This technique can be used to speed up many fully dynamic graph algorithms. Roughly speaking, when the technique is applicable it speeds up a $T(n, m)$ time bound for a graph with n vertices and m edges to $T(n, O(n))$; i.e., to the time needed if the graph were sparse. For instance, if $T(n, m) = O(m^{1/2})$, we get a better bound of $O(n^{1/2})$. Sparsification applies to a wide variety of dynamic graph problems, including minimum spanning forests, edge and vertex connectivity. Moreover, it is a general technique and can be used as a black box (without having to know the internal details) in order to dynamize graph algorithms.

The technique itself is quite simple. Let G be a graph with m edges and n vertices. We partition the edges of G into a collection of $O(m/n)$ sparse subgraphs, i.e., subgraphs with n vertices and $O(n)$ edges. The information relevant for each subgraph can be summarized in an even sparser subgraph, which is known as a sparse **certificate**. We merge certificates in pairs, producing larger subgraphs which are made sparse by again computing their certificate. The result is a balanced binary tree in which each node is represented by a sparse certificate. Each update involves $\log(m/n)^*$ graphs with $O(n)$ edges each, instead of one graph with m edges. With some extra care, the $O(\log(m/n))$ overhead term can be eliminated.

We describe two variants of the sparsification technique. We use the first variant in situations where no previous fully dynamic algorithm was known. We use a static algorithm to recompute a sparse certificate in each tree node affected by an edge update. If the certificates can be found in time $O(m + n)$, this variant gives time bounds of $O(n)$ per update. In the second variant, we maintain certificates using a dynamic data structure. For this to work, we need a stability property of our certificates, to ensure that a small change in the input graph does not lead to a large change in the certificates. This variant transforms time bounds of the form $O(m^p)$ into $O(n^p)$. We start by describing an abstract version of sparsification. The technique is based on the concept of a certificate:

DEFINITION 9.1 For any graph property $\mathcal{P}$, and graph G, a certificate for G is a graph G' such that G has property $\mathcal{P}$ if and only if G' has the property.

Nagamochi and Ibaraki [62] use a similar concept, however they require G' to be a subgraph of G. We do not need this restriction. However, this allows trivial certificates: G' could be chosen from two graphs of constant complexity, one with property $\mathcal{P}$ and one without it.

DEFINITION 9.2 For any graph property $\mathcal{P}$, and graph G, a strong certificate for G is a graph G' on the same vertex set such that, for any H, $G \cup H$ has property $\mathcal{P}$ if and only if $G' \cup H$ has the property.

In all our uses of this definition, G and H will have the same vertex set and disjoint edge sets. A strong certificate need not be a subgraph of G, but it must have a structure closely related to that of G. The following facts follow immediately from Definition 9.2.

FACT 9.1 *Let G' be a strong certificate of property $\mathcal{P}$ for graph G, and let G'' be a strong certificate for G'. Then G'' is a strong certificate for G.*

---

* Throughout $\log x$ stands for $\max(1, \log_2 x)$, so $\log(m/n)$ is never smaller than 1 even if $m < 2n$.

FACT 9.2 *Let G' and H' be strong certificates of $\mathcal{P}$ for G and H. Then $G' \cup H'$ is a strong certificate for $G \cup H$.*

A property is said to have sparse certificates if there is some constant c such that for every graph G on an n-vertex set, we can find a strong certificate for G with at most cn edges.

The other key ingredient is a sparsification tree. We start with a partition of the vertices of the graph, as follows: we split the vertices evenly in two halves, and recursively partition each half. Thus we end up with a complete binary tree in which nodes at distance i from the root have $n/2^i$ vertices. We then use the structure of this tree to partition the edges of the graph. For any two nodes α and β of the vertex partition tree at the same level i, containing vertex sets V_α and V_β, we create a node $E_{\alpha\beta}$ in the edge partition tree, containing all edges in $V_\alpha \times V_\beta$. The parent of $E_{\alpha\beta}$ is $E_{\gamma\delta}$, where γ and δ are the parents of α and β respectively in the vertex partition tree. Each node $E_{\alpha\beta}$ in the edge partition tree has either three or four children (three if $\alpha = \beta$, four otherwise). We use a slightly modified version of this edge partition tree as our sparsification tree. The modification is that we only construct those nodes $E_{\alpha\beta}$ for which there is at least one edge in $V_\alpha \times V_\beta$. If a new edge is inserted new nodes are created as necessary, and if an edge is deleted those nodes for which it was the only edge are deleted.

LEMMA 9.1 In the sparsification tree described earlier, each node $E_{\alpha\beta}$ at level i contains edges inducing a graph with at most $n/2^{i-1}$ vertices.

PROOF There can be at most $n/2^i$ vertices in each of V_α and V_β. $\qquad\square$

We say a time bound $T(n)$ is well behaved if for some $c < 1$, $T(n/2) < cT(n)$. We assume well behavedness to eliminate strange situations in which a time bound fluctuates wildly with n. All polynomials are well-behaved. Polylogarithms and other slowly growing functions are not well behaved, but since sparsification typically causes little improvement for such functions we will in general assume all time bounds to be well behaved.

THEOREM 9.1 *[16] Let $\mathcal{P}$ be a property for which we can find sparse certificates in time $f(n, m)$ for some well-behaved f, and such that we can construct a data structure for testing property $\mathcal{P}$ in time $g(n, m)$ which can answer queries in time $q(n, m)$. Then there is a fully dynamic data structure for testing whether a graph has property $\mathcal{P}$, for which edge insertions and deletions can be performed in time $O(f(n, O(n))) + g(n, O(n))$, and for which the query time is $q(n, O(n))$.*

PROOF We maintain a sparse certificate for the graph corresponding to each node of the sparsification tree. The certificate at a given node is found by forming the union of the certificates at the three or four child nodes, and running the sparse certificate algorithm on this union. As shown in Lemmas 9.1 and 9.2, the certificate of a union of certificates is itself a certificate of the union, so this gives a sparse certificate for the subgraph at the node. Each certificate at level i can be computed in time $f(n/2^{i-1}, O(n/2^i))$. Each update will change the certificates of at most one node at each level of the tree. The time to recompute certificates at each such node adds in a geometric series to $f(n, O(n))$. This process results in a sparse certificate for the whole graph at the root of the tree. We update the data structure for property $\mathcal{P}$, on the graph formed by the sparse certificate at the root of the tree, in time $g(n, O(n))$. The total time per update is thus $O(f(n, O(n))) + g(n, cn)$. $\qquad\square$

This technique is very effective in producing dynamic graph data structures for a multitude of problems, in which the update time is $O(n \log^{O(1)} n)$ instead of the static time bounds of $O(m + n \log^{O(1)} n)$. To achieve sublinear update times, we further refine our sparsification idea.

DEFINITION 9.3 Let A be a function mapping graphs to strong certificates. Then A is stable if it has the following two properties:

1. For any graphs G and H, $A(G \cup H) = A(A(G) \cup H)$.
2. For any graph G and edge e in G, $A(G - e)$ differs from $A(G)$ by $O(1)$ edges.

Informally, we refer to a certificate as stable if it is the certificate produced by a stable mapping. The certificate consisting of the whole graph is stable, but not sparse.

THEOREM 9.2 *[16] Let $\mathcal{P}$ be a property for which stable sparse certificates can be maintained in time $f(n, m)$ per update, where f is well behaved, and for which there is a data structure for property $\mathcal{P}$ with update time $g(n, m)$ and query time $q(n, m)$. Then $\mathcal{P}$ can be maintained in time $O(f(n, O(n))) + g(n, O(n))$ per update, with query time $q(n, O(n))$.*

PROOF As before, we use the sparsification tree described earlier. After each update, we propagate the changes up the sparsification tree, using the data structure for maintaining certificates. We then update the data structure for property $\mathcal{P}$, which is defined on the graph formed by the sparse certificate at the tree root.

At each node of the tree, we maintain a stable certificate on the graph formed as the union of the certificates in the three or four child nodes. The first part of the definition of stability implies that this certificate will also be a stable certificate that could have been selected by the mapping A starting on the subgraph of all edges in groups descending from the node. The second part of the definition of stability then bounds the number of changes in the certificate by some constant s, since the subgraph is changing only by a single edge. Thus at each level of the sparsification tree there is a constant amount of change.

When we perform an update, we find these s changes at each successive level of the sparsification tree, using the data structure for stable certificates. We perform at most s data structure operations, one for each change in the certificate at the next lower level. Each operation produces at most s changes to be made to the certificate at the present level, so we would expect a total of s^2 changes. However, we can cancel many of these changes since as described earlier the net effect of the update will be at most s changes in the certificate.

In order to prevent the number of data structure operations from becoming larger and larger at higher levels of the sparsification tree, we perform this cancellation before passing the changes in the certificate up to the next level of the tree. Cancellation can be detected by leaving a marker on each edge, to keep track of whether it is in or out of the certificate. Only after all s^2 changes have been processed do we pass the at most s uncancelled changes up to the next level.

Each change takes time $f(n, O(n))$, and the times to change each certificate then add in a geometric series to give the stated bound. $\square$

Theorem 9.1 can be used to dynamize static algorithms, while Theorem 9.2 can be used to speed up existing fully dynamic algorithms. In order to apply effectively Theorem 9.1 we only need to compute efficiently sparse certificates, while for Theorem 9.2 we need to maintain efficiently stable sparse certificates. Indeed stability plays an important role in the proof of Theorem 9.2. In each level of the update path in the sparsification tree we compute s^2 changes resulting from the s changes in the previous level, and then by stability obtain only s changes after eliminating repetitions and canceling changes that require no update. Although in most of the applications we consider stability can be used directly in a much simpler way, we describe it in this way here for sake of generality.

We next describe the $O(n^{1/2})$ algorithm for the fully dynamic maintenance of a minimum spanning forest given by Eppstein et al. [16] based on sparsification. A minimum spanning forest is not a graph property, since it is a subgraph rather than a Boolean function. However sparsification still applies to this problem. Alternately, sparsification maintains any property defined on the minimum spanning trees of graphs. The data structure introduced in this section will also be an important subroutine in some results described later. We need the following analogue of strong certificates for minimum spanning trees:

LEMMA 9.2 Let T be a minimum spanning forest of graph G. Then for any H there is some minimum spanning forest of $G \cup H$ which does not use any edges in $G - T$.

PROOF   If we use the cycle property on graph $G \cup H$, we can eliminate first any cycle in G (removing all edges in $G - T$) before dealing with cycles involving edges in H.   □

Thus, we can take the strong certificate of any minimum spanning forest property to be the minimum spanning forest itself. Minimum spanning forests also have a well-known property which, together with Lemma 9.2, proves that they satisfy the definition of stability:

LEMMA 9.3   Let T be a minimum spanning forest of graph G, and let e be an edge of T. Then either $T - e$ is a minimum spanning forest of $G - e$, or there is a minimum spanning forest of the form $T - e + f$ for some edge f.

If we modify the weights of the edges, so that no two are equal, we can guarantee that there will be exactly one minimum spanning forest. For each vertex v in the graph let $i(v)$ be an identifying number chosen as an integer between 0 and $n - 1$. Let ϵ be the minimum difference between any two distinct weights of the graph. Then for any edge $e = (u, v)$ with $i(u) < i(v)$ we replace $w(e)$ by $w(e) + \epsilon i(u)/n + \epsilon i(v)/n^2$. The resulting MSF will also be a minimum spanning forest for the unmodified weights, since for any two edges originally having distinct weights the ordering between those weights will be unchanged. This modification need not be performed explicitly–the only operations our algorithm performs on edge weights are comparisons of pairs of weights, and this can be done by combining the original weights with the numbers of the vertices involved taken in lexicographic order. The mapping from graphs to unique minimum spanning forests is stable, since part (1) of the definition of stability follows from Lemma 9.2, and part (2) follows from Lemma 9.3.

We use Frederickson's algorithm [24,25] that maintains a minimum spanning trees in time $O(m^{1/2})$. We improve this bound by combining Frederickson's algorithm with sparsification: we apply the stable sparsification technique of Theorem 9.2, with $f(n, m) = g(n, m) = O(m^{1/2})$.

THEOREM 9.3   [16] *The minimum spanning forest of an undirected graph can be maintained in time $O(n^{1/2})$ per update.*

The dynamic spanning tree algorithms described so far produce fully dynamic connectivity algorithms with the same time bounds. Indeed, the basic question of connectivity can be quickly determined from a minimum spanning forest. However, higher forms of connectivity are not so easy. For edge connectivity, sparsification can be applied using a dynamic minimum spanning forest algorithm, and provides efficient algorithms: 2-edge connectivity can be solved in $O(n^{1/2})$ time per update, 3-edge connectivity can be solved in $O(n^{2/3})$ time per update, and for any higher k, k-edge connectivity can be solved in $O(n \log n)$ time per update [16]. Vertex connectivity is not so easy: for $2 \le k \le 4$, there are algorithms with times ranging from $O(n^{1/2} \log^2 n)$ to $O(n\alpha(n))$ per update [16,38,44].

9.2.2 Randomized Algorithms

All the previous techniques yield efficient deterministic algorithms for fully dynamic problems. Recently, Henzinger and King [39] proposed a new approach that, exploiting the power of randomization, is able to achieve faster update times for some problems. For the fully dynamic connectivity problem, the randomized technique of Henzinger and King yields an expected amortized update time of $O(\log^3 n)$ for a sequence of at least m_0 updates, where m_0 is the number of edges in the initial graph, and a query time of $O(\log n)$. It needs $\Theta(m + n \log n)$ space. We now sketch the main ideas behind this technique. The interested reader is referred to the chapter on Randomized Algorithms for basic definitions on randomized algorithms.

Let $G = (V, E)$ be a graph to be maintained dynamically, and let F be a spanning forest of G. We call edges in F tree edges, and edges in $E \setminus F$ nontree edges. First, we describe a data structure which stores all trees in the spanning forest F. This data structure is based on Euler tours, and allows one to obtain logarithmic updates and queries within the forest. Next, we show how to keep the forest F spanning throughout a sequence of updates. The Euler tour data structure for a tree T is simple, and consists of storing the tree vertices according to an Euler tour of T. Each time a vertex v is visited in the Euler tour, we call this an occurrence of v and we denote it by $o(v)$. A vertex of degree Δ has exactly Δ occurrences, except for the root which has $(\Delta + 1)$ occurrences. Furthermore, each edge is visited exactly twice. Given an n-nodes tree T, we encode it with the sequence of $2n - 1$ symbols produced by procedure ET. This encoding is referred to as $ET(T)$. We now analyze how to update $ET(T)$ when T is subject to dynamic edge operations.

If an edge $e = (u, v)$ is deleted from T, denote by T_u and T_v the two trees obtained after the deletion, with $u \in T_u$ and $v \in T_v$. Let $o(u_1)$, $o(v_1)$, $o(u_2)$ and $o(v_2)$ be the occurrences of u and v encountered during the visit of (u, v). Without loss of generality assume that $o(u_1) < o(v_1) < o(v_2) < o(u_2)$ so that $ET(T) = \alpha o(u_1) \beta o(v_1) \gamma o(v_2) \delta o(u_2) \epsilon$. $ET(T_u)$ and $ET(T_v)$ can be easily computed from $ET(T)$, as $ET(T_v) = o(v_1) \gamma o(v_2)$, and $ET(T_v) = \alpha o(u_1) \beta \delta o(u_2) \epsilon$. To change the root of T from r to another vertex s, we do the following. Let $ET(T) = o(r) \alpha o(s_1) \beta$, where $o(s_1)$ is any occurrence of s. Then, the new encoding will be $o(s_1) \beta \alpha o(s)$, where $o(s)$ is a newly created occurrence of s that is added at the end of the new sequence. If two trees T_1 and T_2 are joined in a new tree T because of a new edge $e = (u, v)$, with $u \in T_1$ and $v \in T_2$, we first reroot T_2 at v. Now, given $ET(T_1) = \alpha o(u_1) \beta$ and the computed encoding $ET(T_2) = o(v_1) \gamma o(v_2)$, we compute $ET(T) = \alpha o(u_1) o(v_1) \gamma o(v_2) o(u) \beta$, where $o(u)$ is a newly created occurrence of vertex u.

Note that all the aforementioned primitives require the following operations: (1) splicing out an interval from a sequence, (2) inserting an interval into a sequence, (3) inserting a single occurrence into a sequence, or (4) deleting a single occurrence from a sequence. If the sequence $ET(T)$ is stored in a balanced search tree of degree d (i.e., a balanced d-ary search tree), then one may insert or splice an interval, or insert or delete an occurrence in time $O(d \log n / \log d)$, while maintaining the balance of the tree. It can be checked in $O(\log n/d)$ whether two elements are in the same tree, or whether one element precedes the other in the ordering. The balanced d-ary search tree that stores $ET(T)$ is referred to as the ET(T)-tree.

We augment ET-trees to store nontree edges as follows. For each occurrence of vertex $v \in T$, we arbitrarily select one occurrence to be the active occurrence of v. The list of nontree edges incident to v is stored in the active occurrence of v: each node in the $ET(T)$-tree contains the number of nontree edges and active occurrences stored in its subtree; thus the root of the $ET(T)$-tree contains the weight and size of T.

Using these data structures, we can implement the following operations on a collection of trees:

- *tree(x)*: return a pointer to $ET(T)$, where T is the tree containing vertex x.
- *non_tree_edges(T)*: return the list of nontree edges incident to T.
- *sample_n_test(T)*: select one nontree edge incident to T at random, where an edge with both endpoints in T is picked with probability $2/w(T)$, and an edge with only endpoint in T is picked with probability $1/w(T)$. Test whether the edge has exactly one endpoint in T, and if so return this edge.
- *insert_tree(e)*: join by edge e the two trees containing its endpoints. This operation assumes that the two endpoints of e are in two different trees of the forest.
- *delete_tree(e)*: remove e from the tree that contains it. This operation assumes that e is a tree edge.
- *insert_non_tree(e)*: insert a nontree edge e. This operation assumes that the two endpoints of e are in a same tree.
- *delete_non_tree(e)*: remove the edge e. This operation assumes that e is a nontree edge.

Using a balanced binary search tree for representing $ET(T)$, yields the following running times for the aforementioned operations: *sample_n_test(T)*, *insert_tree(e)*, *delete_tree(e)*, *insert_non_tree(e)*, and *delete_non_tree(e)* in $O(\log n)$ time, and *non_tree_edges(T)* in $O(w(T) \log n)$ time.

We now turn to the problem of keeping the forest of G spanning throughout a sequence of updates. Note that the hard operation is a deletion of a tree edge: indeed, a spanning forest is easily maintained throughout edge insertions, and deleting a nontree edge does not change the forest. Let $e = (u, v)$ be a tree edge of the forest F, and let T_e be the tree of F containing edge e. Let T_u and T_v be the two trees obtained from T after the deletion of e, such that T_u contains u and T_v contains v. When e is deleted, T_u and T_v can be reconnected if and only if there is a nontree edge in G with one endpoint in T_u and one endpoint in T_v. We call such an edge a replacement edge for e. In other words, if there is a replacement edge for e, T is reconnected via this replacement edge; otherwise, the deletion of e disconnects T into T_u and T_v. The set of all the replacement edges for e (i.e., all the possible edges reconnecting T_u and T_v), is called the candidate set of e.

One main idea behind the technique of Henzinger and King is the following: when e is deleted, use random sampling among the nontree edges incident to T_e, in order to find quickly whether there exists a replacement edge for e. Using the Euler tour data structure, a single random edge adjacent to T_e can be selected and tested whether it reconnects T_e in logarithmic time. The goal is an update time of $O(\log^3 n)$, so we can afford a number of sampled edges of $O(\log^2 n)$. However, the candidate set of e might only be a small fraction of all nontree edges which are adjacent to T. In this case it is unlikely to find a replacement edge for e among the sampled edges. If we find no candidate among the sampled edges, we check explicitly all the nontree edges adjacent to T. After random sampling has failed to produce a replacement edge, we need to perform this check explicitly, otherwise we would not be guaranteed to provide correct answers to the queries. Since there might be a lot of edges which are adjacent to T, this explicit check could be an expensive operation, so it should be made a low probability event for the randomized algorithm. This is not yet true, however, since deleting all edges in a relatively small candidate set, reinserting them, deleting them again, and so on will almost surely produce many of those unfortunate events.

The second main idea prevents this undesirable behavior: we maintain an edge decomposition of the current graph G into $O(\log n)$ edge disjoint subgraphs $G_i = (V, E_i)$. These subgraphs are hierarchically ordered. Each i corresponds to a level. For each level i, there is a forest F_i such that the union $\cup_{i \leq k} F_i$ is a spanning forest of $\cup_{i \leq k} G_i$; in particular the union F of all F_i is a spanning forest of G. A spanning forest at level i is a tree in $\cup_{j \leq i} F_j$. The weight $w(T)$ of a spanning tree T at level i is the number of pairs (e', v) such that e' is a nontree edge in G_i adjacent to the node v in T.

If T_1 and T_2 are the two trees resulting from the deletion of e, we sample edges adjacent to the tree with the smaller weight. If sampling is unsuccessful due to a candidate set which is nonempty but relatively small, then the two pieces of the tree which was split are reconnected on the next higher level using one candidate, and all other candidate edges are copied to that level. The idea is to have sparse cuts on high levels and dense cuts on low levels. Nontree edges always belong to the lowest level where their endpoints are connected or a higher level, and we always start sampling at the level of the deleted tree edge. After moving the candidates one level up, they are normally no longer a small fraction of all adjacent nontree edges at the new level. If the candidate set on one level is empty, we try to sample on the next higher level. There is one more case to mention: if sampling was unsuccessful although the candidate set was big enough, which means that we had bad luck in the sampling, we do not move the candidates to the next level, since this event has a small probability and does not happen very frequently. We present the pseudocode for *replace(u,v,i)*, which is called after the deletion of the forest edge $e = (u, v)$ on level i:

replace(u,v,i)

1. **Let** T_u and T_v be the spanning trees at level i containing u and v, respectively. **Let** T be the tree with smaller weight among T_u and T_v. Ties are broken arbitrarily.

2. **If** $w(T) > \log^2 n$ **then**

   (a) **Repeat** sample_n_test(T) for at most $16\log^2 n$ times. Stop if a replacement edge e is found.

   (b) **If** a replacement edge e is found **then do** delete_non_tree(e), insert_tree(e), and **return**.

3. (a) **Let** S be the set of edges with exactly one endpoint in T.

   (b) **If** $|S| \geq w(T)/(8\log n)$ **then**
   Select one $e \in S$, delete_non_tree(e), and insert_tree(e).

   (c) **Else if** $0 < |S| < w(T)/(8\log n)$ **then**
   Delete one edge e from S, delete_non_tree(e), and insert_tree(e) in level $(i+1)$.
   Forall $e' \in S$ **do** delete_non_tree(e') and insert_non_tree(e') in level $(i+1)$.

   (d) **Else if** $i < l$ **then** replace($u, v, i+1$).

Note that edges may migrate from level 1 to level l, one level at the time. However, an upper bound of $O(\log n)$ for the number of levels is guaranteed, if there are only deletions of edges. This can be proved as follows. For any i, let m_i be the number of edges ever in level i.

LEMMA 9.4 For any level i, and for all smaller trees T_1 on level i, $\sum w(T_1) \leq 2m_i \log n$.

PROOF Let T be a tree which is split into two trees: if an endpoint of an edge is contained in the smaller split tree, the weight of the tree containing the endpoint is at least halved. Thus, each endpoint of a nontree edge is incident to a small tree at most $\log n$ times in a given level i and the lemma follows. □

LEMMA 9.5 For any level i, $m_i \leq m/4^{i-1}$.

PROOF We proceed by induction on i. The lemma trivially holds for $i = 1$. Assume it holds for $(i-1)$. When summed over all small trees T_1 on level $(i-1)$, at most $\sum w(T_1)/(8\log n)$ edges are added to level i. By Lemma 9.4, $\sum w(T_1) \leq 2m_{i-1} \log n$, where m_{i-1} is the number of edges ever in level $(i-1)$. The lemma now follows from the induction step.   □

The following is an easy corollary of Lemma 9.5:

COROLLARY 9.1   The sum over all levels of the total number of edges is $\sum_i m_i = O(m)$.

Lemma 9.5 gives immediately a bound on the number of levels:

LEMMA 9.6   The number of levels is at most $l = \lceil \log m - \log \log n \rceil + 1$.

PROOF   Since edges are never moved to a higher level from a level with less than $2 \log n$ edges, Lemma 9.5 implies that all edges of G are contained in some E_i, $i \leq \lceil \log m - \log \log n \rceil + 1$.   □

We are now ready to describe an algorithm for maintaining a spanning forest of a graph G subject to edge deletions. Initially, we compute a spanning forest F of G, compute $ET(T)$ for each tree in the forest, and select active occurrences for each vertex. For each i, the spanning forest at level i is initialized to F. Then, we insert all the nontree edges with the proper active occurrences into level 1, and compute the number of nontree edges in the subtree of each node of the binary search tree. This requires $O(m+n)$ times to find the spanning forest and initialize level 1, plus $O(n)$ for each subsequent level to initialize the spanning forest at that level. To check whether two vertices x and y are connected, we test if $tree(x) = tree(y)$ on the last level. This can be done in time $O(\log n)$. To update the data structure after the deletion of an edge $e = (u, v)$, we do the following. If e is a nontree edge, it is enough to perform a $delete\_non\_tree(e)$ in the level where e appears. If e is a tree edge, let i be the level where e first appears. We do a $delete\_tree(e)$ at level j, for $j \geq i$, and then call $replace(u, v, i)$. This yields the following bounds.

THEOREM 9.4   [39] *Let G be a graph with m_0 edges and n vertices subject to edge deletions only. A spanning forest of G can be maintained in $O(\log^3 n)$ expected amortized time per deletion, if there are at least $\Omega(m_0)$ deletions. The time per query is $O(\log n)$.*

PROOF   The bound for queries follows from the aforementioned argument. Let e be an edge to be deleted. If e is a nontree edge, its deletion can be taken care of in $O(\log n)$ time via a $delete\_non\_tree$ primitive.

If e is a tree edge, let T_1 and T_2 the two trees created by the deletion of e, $w(T_1) \leq w(T_2)$. During the sampling phase we spend exactly $O(\log^3 n)$ time, as the cost of $sample\_n\_test$ is $O(\log n)$ and we repeat this for at most $16 \log^2 n$ times.

If the sampling is not successful, collecting and testing all the nontree edges incident to T_1 implies a total cost of $O(w(T_1) \log n)$. We now distinguish two cases. If we are unlucky in the sampling, $|S| \geq w(T_1)/(8 \log n)$: this happens with probability at most $(1 - 1/(8 \log n))^{16 \log^2 n} = O(1/n^2)$ and thus contributes an expected cost of $O(\log n)$ per operation. If the cut S is sparse, $|S| < w(T_1)/(8 \log n)$, and we move the candidate set for e one level higher. Throughout the sequence of deletions, the cost incurred at level i for this case is $\sum w(T_1) \log n$, where the sum is taken over the small trees T_1 on level i. By Lemma 9.6 and Corollary 9.1 this gives a total cost of $O(m_0 \log^2 n)$.

In all cases where a replacement edge is found, $O(\log n)$ tree operations are performed in the different levels, contributing a cost of $O(\log^2 n)$. Hence, each tree edge deletion contributes a total of $O(\log^3 n)$ expected amortized cost toward the sequence of updates.   □

If there are also insertions, however, the analysis in Theorem 9.4 does not carry through, as the upper bound on the number of levels in the graph decomposition is no longer guaranteed. To achieve the same bound, there have to be periodical rebuilds of parts of the data structure. This is done as follows. We let the maximum number of levels to be $l = \lceil 2 \log n \rceil$. When an edge $e = (u, v)$ is inserted into G, we add it to the last level l. If u and v were not previously connected, then we do this via a *tree_insert*, otherwise we perform a *non_tree_insert*. In order to prevent the number of levels to grow behind their upper bound, a rebuild of the data structure is executed periodically. A rebuild at level i, $i \geq 1$, is carried out by moving all the tree and nontree edges in level j, $j > i$, back to level i. Moreover, for each $j > i$ all the tree edges in level i are inserted into level j. Note that after a rebuild at level i, $E_j = \emptyset$ and $F_j = F_i$ for all $j > i$, i.e., there are no nontree edges above level i, and the spanning trees on level $j \geq i$ span G.

The crucial point of this method is deciding when to apply a rebuild at level i. This is done as follows. We keep a counter K that counts the number of edge insertions modulo $2^{\lceil 2 \log n \rceil}$ since the start of the algorithm. Let $K_1, K_2, \ldots, K_l$ be the binary representation of K, with K_1 being the most significant bit: we perform a rebuild at level i each time the bit K_i flips to 1. This implies that the last level is rebuilt every other insertion, level $(l-1)$ every four insertions. In general, a rebuild at level i occurs every 2^{l-i+1} insertions. We now show that the rebuilds contribute a total cost of $O(\log^3 n)$ toward a sequence of insertions and deletions of edges.

For the sake of completeness, assume that at the beginning the initialization of the data structures is considered a rebuild at level 1. Given a level i, we define an epoch for i to be any period of time starting right after a rebuild at level j, $j \leq i$, and ending right after the next rebuild at level j', $j' \leq i$. Namely, an epoch for i is the period between two consecutive rebuilds below or at level i: an epoch for i starts either at the start of the algorithm or right after some bit K_j, $j \leq i$, flips to 1, and ends with the next such flip. It can be easily seen that each epoch for i occurs every 2^{l-i} insertions, for any $1 \leq i \leq l$. There are two types of epochs for i, depending on whether it is bit K_i or a bit K_j, $j > i$, that flips to 1: an empty epoch for i starts right after a rebuild at j, $j < i$, and a full epoch for i starts right after a rebuild at i. At the time of the initialization, a full epoch for 1 starts, while for $i \geq 2$, empty epochs for i starts. The difference between these two types of epochs is the following. When an empty epoch for i starts, all the edges at level i have been moved to some level j, $j < i$, and consequently E_i is empty. On the contrary, when a full epoch for i starts, all the edges at level k, $k > i$, have been moved to level i, and thus $E_i \neq \emptyset$.

An important point in the analysis is that for any $i \geq 2$, any epoch for $(i-1)$ consists of two parts, one corresponding to an empty epoch for i followed by another corresponding to a full epoch for i. This happens because a flip to 1 of a bit K_j, $j \leq 2$, must be followed by a flip to 1 of K_i before another bit $K_{j'}$, $j' \leq i$ flips again to 1. Thus, when each epoch for $(i-1)$ starts, E_i is empty. Define m_i' to be the number of edges ever in level i during an epoch for i. The following lemma is the analogous of Lemma 9.4.

LEMMA 9.7 For any level i, and for all smaller trees T_1 on level i searched during an epoch for i, $\sum w(T_1) \leq 2m_i' \log n$.

LEMMA 9.8 $m_i' < n^2/2^{i-1}$.

PROOF To prove the lemma, it is enough to bound the number of edges that are moved to level i during any one epoch for $(i-1)$, as $E_i = \emptyset$ at the start of each epoch for $(i-1)$ and each epoch for i is contained in one epoch for $(i-1)$. Consider an edge e that is in level i during one epoch for $(i-1)$. There are only two possibilities: either e was passed up from level $(i-1)$ because of an edge deletion, or e was moved down during a rebuild at i. Assume that e was moved down during a rebuild at i: since right after the rebuild at i the second part of the epoch for $(i-1)$ (i.e., the full epoch for i) starts, e was moved back to level i still during the empty epoch for i. Note that E_k, for $k \geq i$, was empty at the beginning of the epoch for $(i-1)$; consequently, either e was passed up from E_{i-1} to E_i or e was inserted into G during the empty epoch for i. In summary, denoting by a_{i-1} the maximum number of edges passed up from E_{i-1} to E_i during one epoch for $(i-1)$, and by b_i the number of edges inserted into G during one epoch for i, we have that $m_i' \leq a_{i-1} + b_i$. By definition of epoch, the number of edges inserted during an epoch for i is $b_i = 2^{l-i}$. It remains for us to bound a_{i-1}. Applying the same argument as in the proof of Lemma 9.5, using this time Lemma 9.7, yields that $a_{i-1} \leq m_{i-1}'/4$. Substituting for a_{i-1} and b_i yields $m_i' \leq m_{i-1}'/4 + 2^{l-i}$, with $m_1' \leq n^2$. Since $l = \lceil 2 \log n \rceil$, $m_i' < n^2/2^{i-1}$. $\qquad\Box$

Lemma 9.8 implies that $m_l' \leq 2$, and thus edges never need to be passed to a level higher than l.

COROLLARY 9.2   All edges of G are contained in some level E_i, $i \leq \lceil 2 \log n \rceil$.

We are now ready to analyze the running time of the entire algorithm.

THEOREM 9.5   *[39] Let G be a graph with m_0 edges and n vertices subject to edge insertions and deletions. A spanning forest of G can be maintained in $O(\log^3 n)$ expected amortized time per update, if there are at least $\Omega(m_0)$ updates. The time per query is $O(\log n)$.*

PROOF   There are two main differences with the algorithm for deletions only described in Theorem 9.4. The first is that now the actual cost of an insertion has to be taken into account (i.e., the cost of operation *move_edges*). The second difference is that the argument that a total of $O(m_i \log n)$ edges are examined throughout the course of the algorithm when sparse cuts are moved one level higher must be modified to take into account epochs.

The cost of executing *move_edges(i)* is the cost of moving each nontree and tree edge from E_j to E_i, for all $j > i$, plus the cost of updating all the forests F_k, $i \leq k < j$. The number of edges moved into level i by a *move_edges(i)* is $\sum_{j>i} m_j'$, which by Lemma 9.8, is never greater than $n^2/2^{i-1}$. Since moving one edge costs $O(\log n)$, the total cost incurred by a *move_edges(i)* operation is $O(n^2 \log n/2^i)$. Note that during an epoch for i, at most one *move_edges(i)* can be performed, since that will end the current epoch and start a new one.

Inserting a tree edge into a given level costs $O(\log n)$. Since a tree edge is never passed up during edge deletions, it can be added only once to a given level. This yields a total of $O(\log^2 n)$ per tree edge.

We now analyze the cost of collecting and testing the edges from all smaller trees T_1 on level i during an epoch for i (when sparse cuts for level i are moved to level $(i+1)$). Fix a level $i \leq l$. If $i = l$, since there are $O(1)$ edges in E_l at any given time, the total cost for collecting and testing on level l will be $O(\log n)$. If $i < l$, the cost of collecting and testing edges on all small trees on level i during an epoch for i is $O(2m_i' \log n \times \log n)$ because of Lemma 9.7. By Lemma 9.8, this is $O(n^2 \log^2 n/2^i)$.

In summary, each update contributes $O(\log^3 n)$ per operation plus $O(n^2 \log^2 n/2^i)$ per each epoch for i, $1 \leq i \leq l$. To amortized the latter bound against insertions, during each insertion we distribute $\Theta(\log^2 n)$ credits per level. This sums up to $\Theta(\log^3 n)$ credits per insertion. An epoch for i occurs every $2^{l-i} = \Theta(n^2/2^i)$ insertions, at which time level i has accumulated $\Theta(n^2 \log^2 n/2^i)$ credits to to pay for the cost of *move_edges(i)* and the cost of collecting and testing edges on all small trees. $\qquad\Box$

9.2.3   Faster Deterministic Algorithms

In this section we give a high level description of the fastest deterministic algorithm for the fully dynamic connectivity problem in undirected graphs [46]: the algorithm, due to Holm, de Lichtenberg and Thorup, answers connectivity queries in $O(\log n / \log \log n)$ worst-case running time while supporting edge insertions and deletions in $O(\log^2 n)$ amortized time.

Similarly to the randomized algorithm in [39], the deterministic algorithm in [46] maintains a spanning forest F of the dynamically changing graph G. As aforementioned, we will refer to the edges in F as *tree edges*. Let e be a tree edge of forest F, and let T be the tree of F containing it. When e is deleted, the two trees T_1 and T_2 obtained from T after the deletion of e can be reconnected if and only if there is a nontree edge in G with one endpoint in T_1 and the other endpoint in T_2. We call such an edge a replacement edge for e. In other words, if there is a replacement edge for e, T is reconnected via this replacement edge; otherwise, the deletion of e creates a new connected component in G.

To accommodate systematic search for replacement edges, the algorithm associates to each edge e a level $\ell(e)$ and, based on edge levels, maintains a set of sub-forests of the spanning forest F: for each level i, forest F_i is the sub-forest induced by tree edges of level $\geq i$. If we denote by L denotes the maximum edge level, we have that:

$$F = F_0 \supseteq F_1 \supseteq F_2 \supseteq \cdots \supseteq F_L,$$

Initially, all edges have level 0; levels are then progressively increased, but never decreased. The changes of edge levels are accomplished so as to maintain the following invariants, which obviously hold at the beginning.

Invariant (1): F is a maximum spanning forest of G if edge levels are interpreted as weights.

Invariant (2): The number of nodes in each tree of F_i is at most $n/2^i$.

Invariant (1) should be interpreted as follows. Let (u, v) be a nontree edge of level $\ell(u, v)$ and let $u \cdots v$ be the unique path between u and v in F (such a path exists since F is a spanning forest of G). Let e be any edge in $u \cdots v$ and let $\ell(e)$ be its level. Due to (1), $\ell(e) \geq \ell(u, v)$. Since this holds for each edge in the path, and by construction $F_{\ell(u,v)}$ contains all the tree edges of level $\geq \ell(u, v)$, the entire path is contained in $F_{\ell(u,v)}$, i.e., u and v are connected in $F_{\ell(u,v)}$.

Invariant (2) implies that the maximum number of levels is $L \leq \lfloor \log_2 n \rfloor$.

Note that when a new edge is inserted, it is given level 0. Its level can be then increased at most $\lfloor \log_2 n \rfloor$ times as a consequence of edge deletions. When a tree edge $e = (v, w)$ of level $\ell(e)$ is deleted, the algorithm looks for a replacement edge at the highest possible level, if any. Due to invariant (1), such a replacement edge has level $\ell \leq \ell(e)$. Hence, a replacement subroutine `Replace((u, w), ℓ(e))` is called with parameters e and $\ell(e)$. We now sketch the operations performed by this subroutine.

> `Replace((u, w), ℓ)` finds a replacement edge of the highest level $\leq \ell$, if any. If such a replacement does not exist in level ℓ, we have two cases: if $\ell > 0$, we recurse on level $\ell - 1$; otherwise, $\ell = 0$, and we can conclude that the deletion of (v, w) disconnects v and w in G.

During the search at level ℓ, suitably chosen tree and nontree edges may be promoted at higher levels as follows. Let T_v and T_w be the trees of forest F_ℓ obtained after deleting (v, w) and let, w.l.o.g., T_v be smaller than T_w. Then T_v contains at most $n/2^{\ell+1}$ vertices, since $T_v \cup T_w \cup \{(v, w)\}$ was a tree at level ℓ and due to invariant (2). Thus, edges in T_v of level ℓ can be promoted at level $\ell + 1$ by maintaining the invariants. Nontree edges incident to T_v are finally visited one by one: if an edge does connect T_v and T_w, a replacement edge has been found and the search stops, otherwise its level is increased by 1.

We maintain an Euler tour tree, as described before, for each tree of each forest. Consequently, all the basic operations needed to implement edge insertions and deletions can be supported in $O(\log n)$ time. In addition to inserting and deleting edges from a forest, ET-trees must also support operations such as finding the tree of a forest that contains a given vertex, computing the size of a tree, and, more importantly, finding tree edges of level ℓ in T_v and nontree edges of level ℓ incident to T_v. This can be done by augmenting the ET-trees with a constant amount of information per node: we refer the interested reader to [46] for details.

Using an amortization argument based on level changes, the claimed $O(\log^2 n)$ bound on the update time can be finally proved. Namely, inserting an edge costs $O(\log n)$, as well as increasing its level. Since this can happen $O(\log n)$ times, the total amortized insertion cost, inclusive of level increases, is $O(\log^2 n)$. With respect to edge deletions, cutting and linking $O(\log n)$ forest has a total cost $O(\log^2 n)$; moreover, there are $O(\log n)$ recursive calls to `Replace`, each of cost $O(\log n)$ plus the cost amortized over level increases. The ET-trees over $F_0 = F$ allows it to answer connectivity queries in $O(\log n)$ worst-case time. As shown in [46], this can be reduced to $O(\log n / \log \log n)$ by using a $\Theta(\log n)$-ary version of ET-trees.

THEOREM 9.6   *[46] A dynamic graph G with n vertices can be maintained upon insertions and deletions of edges using $O(\log^2 n)$ amortized time per update and answering connectivity queries in $O(\log n / \log \log n)$ worst-case running time.*

9.3 Fully Dynamic Algorithms on Directed Graphs

In this section, we survey fully dynamic algorithms for maintaining path problems on general directed graphs. In particular, we consider two fundamental problems: fully dynamic transitive closure and fully dynamic all pairs shortest paths (APSP).

In the fully dynamic transitive closure problem we wish to maintain a directed graph $G = (V, E)$ under an intermixed sequence of the following operations:

Insert(x, y): insert an edge from x to y;

Delete(x, y): delete the edge from x to y;

Query(x, y): return *yes* if y is reachable from x, and return *no* otherwise.

In the fully dynamic APSP problem we wish to maintain a directed graph $G = (V, E)$ with real-valued edge weights under an intermixed sequence of the following operations:

Update(x, y, w): update the weight of edge (x, y) to the real value w; this includes as a special case both edge insertion (if the weight is set from $+\infty$ to $w < +\infty$) and edge deletion (if the weight is set to $w = +\infty$);

Distance(x, y): output the shortest distance from x to y.

Path(x, y): report a shortest path from x to y, if any.

Throughout the section, we denote by m and by n the number of edges and vertices in G, respectively.

Although research on dynamic transitive closure and dynamic shortest paths problems spans over more than three decades, in the last couple of years we have witnessed a surprising resurge of interests in those two problems. The goal of this section is to survey the newest algorithmic techniques that have been recently proposed in the literature. In particular, we will make a special effort to abstract some combinatorial properties, and some common data-structural tools that are at the base of those techniques. This will help us try to present all the newest results in a unifying framework so that they can be better understood and deployed also by nonspecialists.

We first list the bounds obtainable for dynamic transitive closure with simple-minded methods. If we do nothing during each update, then we have to explore the whole graph in order to answer reachability queries: this gives $O(n^2)$ time per query and $O(1)$ time per update in the worst case. On the other extreme, we could recompute the transitive closure from scratch after each update; as this task can be accomplished via matrix multiplication [60], this approach yields $O(1)$ time per query and $O(n^\omega)$ time per update in the worst case, where ω is the best known exponent for matrix multiplication (currently $\omega < 2.736$ [5]).

For the incremental version of transitive closure, the first algorithm was proposed by Ibaraki and Katoh [48] in 1983: its running time was $O(n^3)$ over any sequence of insertions. This bound was later improved to $O(n)$ amortized time per insertion by Italiano [50] and also by La Poutré and van Leeuwen [57]. Yellin [72] gave an $O(m^*\delta_{max})$ algorithm for m edge insertions, where m^* is the number of edges in the final transitive closure and δ_{max} is the maximum out-degree of the final graph. All these algorithms maintain explicitly the transitive closure, and so their query time is $O(1)$.

The first decremental algorithm was again given by Ibaraki and Katoh [48], with a running time of $O(n^2)$ per deletion. This was improved to $O(m)$ per deletion by La Poutré and van Leeuwen [57]. Italiano [51] presented an algorithm which achieves $O(n)$ amortized time per deletion on directed acyclic graphs. Yellin [72] gave an $O(m^*\delta_{max})$ algorithm for m edge deletions, where m^* is the initial number of edges in the transitive closure and δ_{max} is the maximum out-degree of the initial graph. Again, the query time of all these algorithms is $O(1)$. More recently, Henzinger and King [39] gave a randomized decremental transitive closure algorithm for general directed graphs with a query time of $O(n/\log n)$ and an amortized update time of $O(n\log^2 n)$.

Despite fully dynamic algorithms were already known for problems on undirected graphs since the earlier 1980s [24], directed graphs seem to pose much bigger challenges. Indeed, the first fully dynamic transitive closure algorithm was devised by Henzinger and King [39] in 1995: they gave a randomized Monte Carlo algorithm with one-side error supporting a query time of $O(n/\log n)$ and an amortized update time of $O(n\widehat{m}^{0.58}\log^2 n)$, where $\widehat{m}$ is the average number of edges in the graph throughout the whole update sequence. Since $\widehat{m}$ can be as high as $O(n^2)$, their update time is $O(n^{2.16}\log^2 n)$. Khanna, Motwani, and Wilson [52] proved that, when a lookahead of $\Theta(n^{0.18})$ in the updates is permitted, a deterministic update bound of $O(n^{2.18})$ can be achieved.

The situation for dynamic shortest paths has been even more dramatic. Indeed, the first papers on dynamic shortest paths date back to 1967 [58,61,64]. In 1985, Even and Gazit [20] and Rohnert [67] presented algorithms for maintaining shortest paths on directed graphs with arbitrary real weights. Their algorithms required $O(n^2)$ per edge insertion; however, the worst-case bounds for edge deletions were comparable to recomputing APSP from scratch. Moreover, Ramalingam and Reps [63] considered dynamic shortest path algorithms with arbitrary real weights, but in a different model. Namely, the running time of their algorithm is analyzed in terms of the output change rather than the input size (output bounded complexity). Frigioni et al. [27,28] designed fast algorithms for graphs with bounded genus, bounded degree graphs, and bounded treewidth graphs in the same model. Again, in the worst case the running times of output-bounded dynamic algorithms are comparable to recomputing APSP from scratch.

Until recently, there seemed to be few dynamic shortest path algorithms which were provably faster than recomputing APSP from scratch, and they only worked on special cases and with small integer weights. In particular, Ausiello et al. [3] proposed an incremental shortest path algorithm for directed graphs having positive integer weights less than C: the amortized running time of their algorithm is $O(Cn\log n)$ per edge insertion. Henzinger et al. [43] designed a fully dynamic algorithm for APSP on planar graphs with integer weights, with a running time of $O(n^{9/7}\log(nC))$ per operation. Fakcharoemphol and Rao in [22] designed a fully dynamic algorithm for single-source shortest paths in planar directed graphs that supports both queries and edge weight updates in $O(n^{4/5}\log^{13/5} n)$ amortized time per operation.

Quite recently, many new algorithms for dynamic transitive closure and shortest path problems have been proposed.

Dynamic transitive closure. For dynamic transitive closure, King and Sagert [55] in 1999 showed how to support queries in $O(1)$ time and updates in $O(n^{2.26})$ time for general directed graphs and $O(n^2)$ time for directed acyclic graphs; their algorithm is randomized with one-side error. The bounds of King and Sagert were further improved by King [55], who exhibited a deterministic algorithm on general digraphs with $O(1)$ query time and $O(n^2 \log n)$ amortized time per update operations, where updates are insertions of a set of edges incident to the same vertex and deletions of an arbitrary subset of edges. All those algorithms are based on reductions to fast matrix multiplication and tree data structures for encoding information about dynamic paths.

Demetrescu and Italiano [11] proposed a deterministic algorithm for fully dynamic transitive closure on general digraphs that answers each query with one matrix look-up and supports updates in $O(n^2)$ amortized time. This bound can be made worst-case as shown by Sankowski in [68]. We observe that fully dynamic transitive closure algorithms with $O(1)$ query time maintain explicitly the transitive closure of the input graph, in order to answer each query with exactly one lookup (on its adjacency matrix). Since an update may change as many as $\Omega(n^2)$ entries of this matrix, $O(n^2)$ seems to be the best update bound that one could hope for this class of algorithms. This algorithm hinges upon the well-known equivalence between transitive closure and matrix multiplication on a closed semiring [23,30,60].

In [10] Demetrescu and Italiano show how to trade off query times for updates on directed acyclic graphs: each query can be answered in time $O(n^\epsilon)$ and each update can be performed in time $O(n^{\omega(1,\epsilon,1)-\epsilon} + n^{1+\epsilon})$, for any $\epsilon \in [0,1]$, where $\omega(1,\epsilon,1)$ is the exponent of the multiplication of an $n \times n^\epsilon$ matrix by an $n^\epsilon \times n$ matrix. Balancing the two terms in the update bound yields that ϵ must satisfy the equation $\omega(1,\epsilon,1) = 1 + 2\epsilon$. The current best bounds on $\omega(1,\epsilon,1)$ [5,47] imply that $\epsilon < 0.575$. Thus, the smallest update time is $O(n^{1.575})$, which gives a query time of $O(n^{0.575})$. This subquadratic algorithm is randomized, and has one-side error. This result has been generalized to general graphs within the same bounds by Sankowski in [68], who has also shown how to achieve an even faster update time of $O(n^{1.495})$ at the expense of a much higher $O(n^{1.495})$ query time. Roditty and Zwick presented an algorithm [65] with $O(m\sqrt{n})$ update time and $O(\sqrt{n})$ query time and another algorithm [66] with $O(m + n \log n)$ update time and $O(n)$ query time.

Techniques for reducing the space usage of algorithms for dynamic path problems are presented in [56]. An extensive computational study on dynamic transitive closure problems appears in [29].

Dynamic shortest paths. For dynamic shortest paths, King [54] presented a fully dynamic algorithm for maintaining APSP in directed graphs with positive integer weights less than C: the running time of her algorithm is $O(n^{2.5}\sqrt{C \log n})$ per update. As in the case of dynamic transitive closure, this algorithm is based on clever tree data structures. Demetrescu and Italiano [12] proposed a fully dynamic algorithm for maintaining APSP on directed graphs with arbitrary real weights. Given a directed graph G, subject to dynamic operations, and such that each edge weight can assume at most S different *real* values, their algorithm supports each update in $O(S \cdot n^{2.5} \log^3 n)$ amortized time and each query in optimal worst-case time. We remark that the sets of possible weights of two different edges need not be necessarily the same: namely, any edge can be associated with a different set of possible weights. The only constraint is that throughout the sequence of operations, each edge can assume at most S different real values, which seems to be the case in many applications. Differently from [54], this method uses dynamic reevaluation of products of real-valued matrices as the kernel for solving dynamic shortest paths. Finally, the same authors [8] have studied some combinatorial properties of graphs that make it possible to devise a different approach to dynamic APSP problems. This approach yields a fully dynamic algorithm for general directed graphs with nonnegative real-valued edge weights that supports any sequence of operations in $O(n^2 \log^3 n)$ amortized time per update and unit worst-case time per distance query, where n is the number of vertices. Shortest paths can be reported in optimal worst-case time. The algorithm is deterministic, uses simple data structures, and appears to be very fast in practice. Using the same approach, Thorup [69] has shown how to achieve $O(n^2(\log n + \log^2((m + n)/n)))$ amortized time per update and $O(mn)$ space. His algorithm works with negative weights as well. In [70], Thorup has shown how to achieve worst-case bounds at the price of a higher complexity: in particular, the update bounds become $\tilde{O}(n^{2.75})$, where $\tilde{O}(f(n))$ denotes $O(f(n) \cdot \text{polylog } n)$.

An extensive computational study on dynamic APSP problems appears in [9].

9.3.1 Combinatorial Properties of Path Problems

In this section we provide some background for the two algorithmic graph problems considered here: transitive closure and APSP. In particular, we discuss two methods: the first is based on a simple doubling technique which consists of repeatedly concatenating paths to form longer paths via matrix multiplication, and the second is based on a Divide and Conquer strategy. We conclude this section by discussing a useful combinatorial property of long paths.

9.3.1.1 Logarithmic Decomposition

We first describe a simple method for computing X^* in $O(n^\mu \cdot \log n)$ worst-case time, where $O(n^\mu)$ is the time required for computing the product of two matrices over a closed semiring. The algorithm is based on a simple path doubling argument. We focus on the $\{+, \cdot, 0, 1\}$ semiring; the case of the $\{\min, +\}$ semiring is completely analogous.

Let $\mathcal{B}_n$ be the set of $n \times n$ Boolean matrices and let $X \in \mathcal{B}_n$. We define a sequence of $\log n + 1$ polynomials $P_0, \ldots, P_{\log n}$ over Boolean matrices as:

$$P_i = \begin{cases} X & \text{if } i = 0 \\ P_{i-1} + P_{i-1}^2 & \text{if } i > 0 \end{cases}$$

It is not difficult to see that for any $1 \leq u, v \leq n$, $P_i[u, v] = 1$ if and only if there is a path $u \rightsquigarrow v$ of length at most 2^i in X. We combine paths of length ≤ 2 in X to form paths of length ≤ 4, then we concatenate all paths found so far to obtain paths of length ≤ 8 and so on. As the length of the longest detected path increases exponentially and the longest simple path is no longer than n, a logarithmic number of steps suffices to detect if any two nodes are connected by a path in the graph as stated in the following theorem.

THEOREM 9.7 *Let X be an $n \times n$ matrix. Then $X^* = I_n + P_{\log n}$.*

9.3.1.2 Long Paths

In this section we discuss an intuitive combinatorial property of long paths. Namely, if we pick a subset S of vertices at random from a graph G, then a sufficiently long path will intersect S with high probability. This can be very useful in finding a long path by using short searches.

To the best of our knowledge, this property was first given in [35], and later on it has been used many times in designing efficient algorithms for transitive closure and shortest paths (see e.g., [12,54,71,73]). The following theorem is from [71].

THEOREM 9.8 *Let $S \subseteq V$ be a set of vertices chosen uniformly at random. Then the probability that a given simple path has a sequence of more than $\frac{cn \log n}{|S|}$ vertices, none of which are from S, for any $c > 0$, is, for sufficiently large n, bounded by $2^{-\alpha c}$ for some positive α.*

As shown in [73], it is possible to choose set S deterministically by a reduction to a hitting set problem [4,59]. A similar technique has also been used in [54].

9.3.1.3 Locality

Recently, Demetrescu and Italiano [8] proposed a new approach to dynamic path problems based on maintaining classes of paths characterized by local properties, i.e., properties that hold for all proper subpaths, even if they may not hold for the entire paths. They showed that this approach can play a crucial role in the dynamic maintenance of shortest paths. For instance, they considered a class of paths defined as follows:

DEFINITION 9.4   A path π in a graph is *locally shortest* if and only if every proper subpath of π is a shortest path.

This definition is inspired by the optimal-substructure property of shortest paths: all subpaths of a shortest path are shortest. However, a locally shortest path may not be shortest.

The fact that locally shortest paths include shortest paths as a special case makes them an useful tool for computing and maintaining distances in a graph. Indeed, paths defined locally have interesting combinatorial properties in dynamically changing graphs. For example, it is not difficult to prove that the number of locally shortest paths that may change due to an edge weight update is $O(n^2)$ if updates are partially dynamic, i.e., increase-only or decrease-only:

THEOREM 9.9   *Let G be a graph subject to a sequence of increase-only or decrease-only edge weight updates. Then the amortized number of paths that start or stop being locally shortest at each update is $O(n^2)$.*

Unfortunately, Theorem 9.9 may not hold if updates are fully dynamic, i.e., increases and decreases of edge weights are intermixed. To cope with pathological sequences, a possible solution is to retain information about the history of a dynamic graph, considering the following class of paths:

DEFINITION 9.5   A *historical shortest path* (in short, *historical path*) is a path that has been shortest at least once since it was last updated.

Here, we assume that a path is updated when the weight of one of its edges is changed. Applying the locality technique to historical paths, we derive locally historical paths:

DEFINITION 9.6   A path π in a graph is *locally historical* if and only if every proper subpath of π is historical.

Like locally shortest paths, also locally historical paths include shortest paths, and this makes them another useful tool for maintaining distances in a graph:

LEMMA 9.9   If we denote by *SP*, *LSP*, and *LHP* respectively the sets of shortest paths, locally shortest paths, and locally historical paths in a graph, then at any time the following inclusions hold: $SP \subseteq LSP \subseteq LHP$.

Differently from locally shortest paths, locally historical paths exhibit interesting combinatorial properties in graphs subject to fully dynamic updates. In particular, it is possible to prove that the number of paths that become locally historical in a graph at each edge weight update depends on the number of historical paths in the graph.

THEOREM 9.10   [8] *Let G be a graph subject to a sequence of update operations. If at any time throughout the sequence of updates there are at most $O(h)$ historical paths in the graph, then the amortized number of paths that become locally historical at each update is $O(h)$.*

To keep changes in locally historical paths small, it is then desirable to have as few historical paths as possible. Indeed, it is possible to transform every update sequence into a slightly longer equivalent sequence that generates only a few historical paths. In particular, there exists a simple smoothing strategy that, given any update sequence Σ of length k, produces an operationally equivalent sequence $F(\Sigma)$ of length $O(k \log k)$ that yields only $O(\log k)$ historical shortest paths between each pair of vertices in the graph. We refer the interested reader to [8] for a detailed description of this smoothing strategy. According to Theorem 9.10, this technique implies that only $O(n^2 \log k)$ locally historical paths change at each edge weight update in the smoothed sequence $F(\Sigma)$.

As elaborated in [8], locally historical paths can be maintained very efficiently. Since by Lemma 9.9 locally historical paths include shortest paths, this yields the fastest known algorithm for fully dynamic APSP.

9.3.2 Algorithmic Techniques

In this section we describe some algorithmic techniques which are the kernel of the best-known algorithms for maintaining transitive closure and shortest paths.

We start with some observations which are common to all the techniques considered here. First of all, we note that the algebraic structure of path problems allows one to support insertions in a natural fashion. Indeed, insertions correspond to the $\oplus$ operation on closed semirings. This perhaps can explain the wealth of fast incremental algorithms for dynamic path problems on directed graphs [48,50,57,72]. However, there seems to be no natural setting for deletions in this algebraic framework. Thus, in designing fully dynamic algorithms it seems quite natural to focus on special techniques and data structures for supporting deletions.

The second remark is that most fully dynamic algorithms are surprisingly based on the same decompositions used for the best static algorithms (see Section 9.3.1). The main difference is that they maintain dynamic data structures at each level of the decomposition. As we will see, the definition of a suitable interface between data structures at different levels plays an important role in the design of efficient dynamic algorithms.

In the remainder of this section we describe data structures that are able to support dynamic operations at each level. In Sections 9.3.3 and 9.3.4, the data structures surveyed here will be combined with the decompositions shown in Section 9.3.1 to obtain efficient algorithms for dynamic path problems on directed graphs.

9.3.2.1 Tools for Trees

In this section we describe a tree data structure for keeping information about dynamic path problems. The first appearance of this tool dates back to 1981, when Even and Shiloach [21] showed how to maintain a breadth-first tree of an undirected graph under any sequence of edge deletions; they used this as a kernel for decremental connectivity on undirected graphs. Later on, Henzinger and King [39] showed how to adapt this data structure to fully dynamic transitive closure in directed graphs. Recently, King [54] designed an extension of this tree data structure to weighted directed graphs for solving fully dynamic APSP.

The Problem. The goal is to maintain information about breadth-first search (BFS) on an undirected graph G undergoing deletions of edges. In particular, in the context of dynamic path problems, we are interested in maintaining BFS trees of depth up to d, with $d \leq n$. For the sake of simplicity, we describe only the case where deletions do not disconnect the underlying graph. The general case can be easily handled by means of "phony" edges (i.e., when deleting an edge that disconnects the graph, we just replace it by a phony edge).

It is well known that BFS partitions vertices into levels, so that there can be edges only between adjacent levels. More formally, let r be the vertex where we start BFS, and let level ℓ_i contains vertices encountered at distance i from r ($\ell_0 = \{r\}$): edges incident to a vertex at level ℓ_i can have their other endpoints either at level ℓ_{i-1}, ℓ_i, or ℓ_{i+1}, and no edge can connect vertices at levels ℓ_i and ℓ_j for $|j - i| > 1$. Let $\ell(v)$ be the level of vertex v.

Given an undirected graph $G = (V, E)$ and a vertex $r \in V$, we would like to support any intermixed sequence of the following operations:

- Delete(x, y): delete edge (x, y) from G.
- Level(u): return the level $\ell(u)$ of vertex u in the BFS tree rooted at r (return $+\infty$ if u is not reachable from r).

In the remainder, to indicate that an operation Y() is performed on a data structure X, we use the notation X.Y().

Data structure. We maintain information about BFS throughout the sequence of edge deletions by simply keeping explicitly those levels. In particular, for each vertex v at level ℓ_i in T, we maintain the following data structures: UP(v), SAME(v), and DOWN(v) containing the edges connecting v to level ℓ_{i-1}, ℓ_i, and ℓ_{i+1}, respectively. Note that for all $v \neq r$, UP(v) must contain at least one edge (i.e., the edge from v to its parent in the BFS tree). In other words, a nonempty UP(v) witnesses the fact that v is actually entitled to belong to that level. This property will be important during edge deletions: whenever UP(v) gets emptied because of deletions, v looses its right to be at that level and must be demoted at least one level down.

Implementation of operations. When edge (x, y) is being deleted, we proceed as follows. If $\ell(x) = \ell(y)$, simply delete (x, y) from SAME(y) and from SAME(x). The levels encoded in UP, SAME, and DOWN still capture the BFS structure of G. Otherwise, without loss of generality let $\ell(x) = \ell_{i-1}$ and $\ell(y) = \ell_i$. Update the sets UP, SAME, and DOWN by deleting x from UP(y) and y from DOWN(x). If UP(y) $\neq \emptyset$, then there is still at least one edge connecting y to level ℓ_{i-1}, and the levels will still reflect the BFS structure of G after the deletion.

The main difficulty is when UP(y) $= \emptyset$ after the deletion of (x, y). In this case, deleting (x, y) causes y to loose its connection to level ℓ_{i-1}. Thus, y has to drop down at least one level. Furthermore, its drop may cause a deeper landslide in the levels below. This case can be handled as follows.

We use a FIFO queue Q, initialized with vertex y. We will insert a vertex v in the queue Q whenever we discover that UP(v) $= \emptyset$, i.e., vertex v has to be demoted at least one level down. We will repeat the following demotion step until Q is empty:

Demotion step:

1. Remove the first vertex in Q, say v.

2. Delete v from its level $\ell(v) = \ell_i$ and tentatively try to place v one level down, i.e., in ℓ_{i+1}.

3. Update the sets UP, SAME, and DOWN consequently:

   a. For each edge (u, v) in SAME(v), delete (u, v) from SAME(u) and insert (u, v) in DOWN(u) and UP(v) (as UP(v) was empty, this implies that UP(v) will be initialized with the old set SAME(v)).

   b. For each edge (v, z) in DOWN(v), move edge (v, z) from UP(z) to SAME(z) and from DOWN(v) to SAME(v); if the new UP(z) is empty, insert z in the queue Q. Note that this will empty DOWN(v).

   c. If UP(v) is still empty, insert v again into Q.

Analysis. It is not difficult to see that applying the Demotion Step until the queue is empty will maintain correctly the BFS levels. Level queries can be answered in constant time. To bound the total time required to process any sequence of edge deletions, it suffices to observe that each time an edge (u, v) is examined during a demotion step, either u or v will be dropped one level down. Thus, edge (u, v) can be examined at most $2d$ times in any BFS levels up to depth d throughout any sequence of edge deletions. This implies the following theorem.

THEOREM 9.11   *Maintaining BFS levels up to depth d requires $O(md)$ time in the worst case throughout any sequence of edge deletions in an undirected graph with m initial edges.*

This means that maintaining BFS levels requires d times the time needed for constructing them. Since $d \leq n$, we obtain a total bound of $O(mn)$ if there are no limits on the depth of the BFS levels.

As it was shown in [39,54], it is possible to extend the BFS data structure presented in this section to deal with weighted directed graphs. In this case, a shortest path tree is maintained in place of BFS levels: after each edge deletion or edge weight increase, the tree is reconnected by essentially mimicking Dijkstra's algorithm rather than BFS. Details can be found in [54].

9.3.3   Dynamic Transitive Closure

In this section we consider algorithms for fully dynamic transitive closure. In particular, we describe the algorithm of King [54], whose main ingredients are the logarithmic decomposition of Section 9.3.1.1 and the tools for trees described in Section 9.3.2.1. This method yields $O(n^2 \log n)$ amortized time per update and $O(1)$ per query. Faster methods have been designed by Demetrescu and Italiano [11] and by Sankowski [68].

We start with a formal definition of the fully dynamic transitive closure problem.

The Problem. Let $G = (V, E)$ be a directed graph and let $TC(G) = (V, E')$ be its transitive closure. We consider the problem of maintaining a data structure for graph G under an intermixed sequence of update and query operations of the following kinds:

- Insert(v, I): perform the update $E \leftarrow E \cup I$, where $I \subseteq E$ and $v \in V$. This operation assumes that all edges in I are incident to v. We call this kind of update a v-CENTERED insertion in G.

- Delete(D): perform the update $E \leftarrow E - D$, where $D \subseteq E$.

- Query(x, y): perform a query operation on $TC(G)$ and return 1 if $(x, y) \in E'$ and 0 otherwise.

We note that these generalized Insert and Delete updates are able to change, with just one operation, the graph by adding or removing a whole set of edges, rather than a single edge. Differently from other variants of the problem, we do not address the issue of returning actual paths between nodes, and we just consider the problem of answering reachability queries.

We now describe how to maintain the Transitive Closure of a directed graph in $O(n^2 \log n)$ amortized time per update operation. The algorithm that we describe has been designed by King [54] and is based on the tree data structure presented in Section 9.3.2.1 and on the logarithmic decomposition described in Section 9.3.1.1. To support queries efficiently, the algorithm uses also a counting technique that consists of keeping a count for each pair of vertices x, y of the number of insertion operations that yielded new paths between them. Counters are maintained so that there is a path from x to y if and only if the counter for that pair is nonzero. The counting technique has been first introduced in [55]: in that case, counters keep track of the number of different distinct paths between pairs of vertices in an acyclic directed graph. We show the data structure used for maintaining the Transitive Closure and how operations `Insert`, `Delete`, and `Query` are implemented.

Data structure. Given a directed graph $G = (V, E)$, we maintain $\log n + 1$ levels. On each level i, $0 \leq i \leq \log n$ we maintain the following data structures:

- A graph $G_i = (V, E_i)$ such that $(x, y) \in E_i$ if there is a path from x to y in G of length $\leq 2^i$. Note that the converse is not necessarily true: i.e., $(x, y) \in E_i$ may not necessarily imply, however, that there is a path from x to y in G of length $\leq 2^i$. We maintain G_0 and $G_{\log n}$ such that $G_0 = G$ and $G_{\log n} = TC(G)$.

- For each $v \in V$, a BFS tree $Out_{i,v}$ of depth 2 of G_i rooted at v and a BFS tree $In_{i,v}$ of depth 2 of $\widehat{G}_i$ rooted at v, where $\widehat{G}_i$ is equal to G_i, except for the orientation of edges, which is reversed. We maintain the BFS trees with instances of the data structure presented in Section 9.3.2.1.

- A matrix $Count_i[x, y] = |\{ v : x \in In_{i,v} \wedge y \in Out_{i,v} \}|$. Note that $Count_i[x, y] > 0$, if there is a path from x to y in G of length $\leq 2^{i+1}$.

Implementation of operations. Operations can be realized as follows:

- `Insert`(v, I): for each $i = 0$ to $\log n$, do the following: add I to E_i, rebuild $Out_{i,v}$ and $In_{i,v}$, updating $Count_i$ accordingly, and add to I any (x, y) such that $Count_i[x, y]$ flips from 0 to 1.

- `Delete`(D): for each $i = 0$ to $\log n$, do the following: remove D from E_i, for each $(x, y) \in D$ do $Out_{i,v}$.`Delete`(x, y) and $In_{i,v}$.`Delete`(x, y), updating $Count_i$ accordingly, and add to D all (x, y) such that $Count_i[x, y]$ flips from positive to zero.

- `Query`(x, y): return 1 if $(x, y) \in E_{\log n}$ and 0 otherwise.

We note that an `Insert` operation simply rebuilds the BFS trees rooted at v on each level of the decomposition. It is important to observe that the trees rooted at other vertices on any level i might not be valid BFS trees of the current graph G_i, but are valid BFS trees of some older version of G_i that did not contain the newly inserted edges. A `Delete` operation, instead, maintains dynamically the tree data structures on each level, removing the deleted edges as described in Section 9.3.2.1 and propagating changes up to the decomposition.

Analysis. To prove the correctness of the algorithm, we need to prove that, if there is a path from x to y in G of length $\leq 2^i$, then $(x, y) \in E_i$. Conversely, it is easy to see that $(x, y) \in E_i$ only if there is a path from x to y in G of length $\leq 2^i$. We first consider `Insert` operations. It is important to observe that, by the problem's definition, the set I contains only edges incident to v for $i = 0$, but this might not be the case for $i > 0$, since entries $Count_i[x, y]$ with $x \neq v$ and $y \neq v$ might flip from 0 to 1 during a v-centered insertion. However, we follow a lazy approach and we only rebuild the BFS trees on each level rooted at v. The correctness of this follows from the simple observation that any new paths that appear due to a v-centered insertion pass through v, so rebuilding the trees rooted at v is enough to keep track of these new paths. To prove this, we proceed by induction. We assume that for any new paths $x \rightsquigarrow v$ and $v \rightsquigarrow y$ of length $\leq 2^i$, $(x, v), (v, y) \in E_i$ at the beginning of loop iteration i. Since $x \in In_{i,v}$ and $y \in Out_{i,v}$ after rebuilding $In_{i,v}$ and $Out_{i,v}$, $Count_i[x, y]$ is increased by one if no path $x \rightsquigarrow v \rightsquigarrow y$ of length $\leq 2^{i+1}$ existed before the insertion. Thus, $(x, y) \in E_{i+1}$ at the beginning of loop iteration $i + 1$. To complete our discussion of correctness, we note that deletions act as "undo" operations that leave the data structures as if deleted edges were never inserted. The running time is established as follows.

THEOREM 9.12   *[54] The fully dynamic transitive closure problem can be solved with the following bounds. Any `Insert` operation requires $O(n^2 \log n)$ worst-case time, the cost of*

`Delete` *operations can be charged to previous insertions, and* `Query` *operations are answered in constant time.*

PROOF The bound for `Insert` operations derives from the observation that reconstructing BFS trees on each of the $\log n + 1$ levels requires $O(n^2)$ time in the worst case. By Theorem 9.11, any sequence of `Delete` operations can be supported in $O(d)$ times the cost of building a BFS tree of depth up to d. Since $d = 2$ in the data structure, this implies that the cost of deletions can be charged to previous insertion operations. The bound for `Query` operations is straightforward. $\square$

The previous theorem implies that updates are supported in $O(n^2 \log n)$ amortized time per operation.

Demetrescu and Italiano [7,11] have shown how to reduce the running time of updates to $O(n^2)$ amortized, by casting this dynamic graph problem into a dynamic matrix problem. Again based on dynamic matrix computations, Sankowski [68] showed how to make the $O(n^2)$ amortized bound worst-case.

9.3.4 Dynamic Shortest Paths

In this section we survey the best-known algorithms for fully dynamic shortest paths. Those algorithms can be seen as a natural evolution of the techniques described so far for dynamic transitive closure. They are not a trivial extension of transitive closure algorithms, however, as dynamic shortest path problems look more complicated in nature. As an example, consider the deletion of an edge (u, v). In the case of transitive closure, reachability between a pair of vertices x and y can be reestablished by *any* replacement path avoiding edge (u, v). In case of shortest paths, after deleting (u, v), we have to look for the *best* replacement path avoiding edge (u, v).

We start with a formal definition of the fully dynamic APSP problem.

The problem. Let $G = (V, E)$ be a weighted directed graph. We consider the problem of maintaining a data structure for G under an intermixed sequence of update and query operations of the following kinds:

- `Decrease`(v, w): decrease the weight of edges incident to v in G as specified by a new weight function w. We call this kind of update a v-CENTERED decrease in G.
- `Increase`(w): increase the weight of edges in G as specified by a new weight function w.
- `Query`(x, y): return the distance between x and y in G.

As in fully dynamic transitive closure, we consider generalized update operations where we modify a whole set of edges, rather than a single edge. Again, we do not address the issue of returning actual paths between vertices, and we just consider the problem of answering distance queries.

Demetrescu and Italiano [8] devised the first deterministic near-quadratic update algorithm for fully dynamic APSP. This algorithm is also the first solution to the problem in its generality. The algorithm is based on the notions of historical paths and locally historical paths in a graph subject to a sequence of updates, as discussed in Section 9.3.1.3.

The main idea is to maintain dynamically the locally historical paths of the graph in a data structure. Since by Lemma 9.9 shortest paths are locally historical, this guarantees that information about shortest paths is maintained as well.

To support an edge weight update operation, the algorithm implements the smoothing strategy mentioned in Section 9.3.1.3 and works in two phases. It first removes from the data structure all maintained paths that contain the updated edge: this is correct since historical shortest paths, in view of their definition, are immediately invalidated as soon as they are touched by an update. This means that also locally historical paths that contain them are invalidated and have to be removed from the data structure. As a second phase, the algorithm runs an all-pairs modification of Dijkstra's algorithm [13], where at each step a shortest path with minimum weight is extracted from a priority queue and it is combined with existing historical shortest paths to form new locally historical paths. At the end of this phase, paths that become locally hustorical after the update are correctly inserted in the data structure.

The update algorithm spends constant time for each of the $O(zn^2)$ new locally historical path (see Theorem 9.10). Since the smoothing strategy lets $z = O(\log n)$ and increases the length of the sequence of updates by an additional $O(\log n)$ factor, this yields $O(n^2 \log^3 n)$ amortized time per update. The interested reader can find further details about the algorithm in [8].

THEOREM 9.13   *[8] The fully dynamic APSP problem can solved in $O(n^2 \log^3 n)$ amortized time per update.*

Using the same approach, but with a different smoothing strategy, Thorup [69] has shown how to achieve $O(n^2(\log n + \log^2(m/n)))$ amortized time per update and $O(mn)$ space. His algorithm works with negative weights as well.

9.4   Research Issues and Summary

In this chapter we have described the most efficient known algorithms for maintaining dynamic graphs. Despite the bulk of work on this area, several questions remain still open. For dynamic problems on directed graphs, can we reduce the space usage for dynamic shortest paths to $O(n^2)$? Second, and perhaps more importantly, can we solve efficiently fully dynamic *single-source* reachability and shortest paths on general graphs? Finally, are there any general techniques for making increase-only algorithms fully dynamic? Similar techniques have been widely exploited in the case of fully dynamic algorithms on undirected graphs [40–42].

9.5   Further Information

Research on dynamic graph algorithms is published in many computer science journals, including *Algorithmica, Journal of ACM, Journal of Algorithms, Journal of Computer and System Science, SIAM Journal on Computing*, and *Theoretical Computer Science*. Work on this area is published also in the proceedings of general theoretical computer science conferences, such as the *ACM Symposium on Theory of Computing* (STOC), the *IEEE Symposium on Foundations of Computer Science* (FOCS), and the *International Colloquium on Automata, Languages and Programming* (ICALP). More specialized conferences devoted exclusively to algorithms are the *ACM–SIAM Symposium on Discrete Algorithms* (SODA) and the *European Symposium on Algorithms* (ESA).

Defining Terms

Certificate: For any graph property P, and graph G, a certificate for G is a graph G' such that G has property P if and only if G' has the property.

Fully dynamic graph problem: Problem where the update operations include unrestricted insertions and deletions of edges.

Partially dynamic graph problem: Problem where the update operations include either edge insertions (incremental) or edge deletions (decremental).

Sparsification: Technique for designing dynamic graph algorithms, which when applicable transform a time bound of $T(n, m)$ into $O(T(n, n))$, where m is the number of edges, and n is the number of vertices of the given graph.

Acknowledgments

The work of the first author has been partially supported by MIUR, the Italian Ministry of Education, University and Research, under Project ALGO-NEXT ("Algorithms for the Next Generation Internet and Web: Methodologies, Design and Experiments").

The work of the fourth author has been partially supported by the Sixth Framework Programme of the EU under Contract Number 507613 (Network of Excellence "EuroNGI: Designing and Engineering of the Next Generation Internet") and by MIUR, the Italian Ministry of Education, University and Research, under Project ALGO-NEXT ("Algorithms for the Next Generation Internet and Web: Methodologies, Design and Experiments").

References

1. D. Alberts, G. Cattaneo, and G. F. Italiano. An empirical study of dynamic graph algorithms. *ACM Journal on Experimental Algorithmics*, 2, 1997.
2. G. Amato, G. Cattaneo, and G. F. Italiano, Experimental analysis of dynamic minimum spanning tree algorithms. In *Proceedings of the 8th ACM-SIAM Annual Symposium on Discrete Algorithms (SODA 97)*, New Orleans, LA, pp. 314–323, January 5–7, 1997.
3. G. Ausiello, G. F. Italiano, A. Marchetti Spaccamela, and U. Nanni. Incremental algorithms for minimal length paths. *Journal of Algorithms*, 12:615–638, 1991.
4. V. Chvátal. A greedy heuristic for the set-covering problem. *Mathematics of Operations Research*, 4(3):233–235, 1979.
5. D. Coppersmith and S. Winograd. Matrix multiplication via arithmetic progressions. *Journal of Symbolic Computation*, 9:251–280, 1990.
6. T. H. Cormen, C. E. Leiserson, R. L. Rivest, and C. Stein. *Introduction to Algorithms*, 2nd edition. The MIT Press, Cambridge, MA, 2001.
7. C. Demetrescu. Fully dynamic algorithms for path problems on directed graphs. PhD thesis, Department of Computer and Systems Science, University of Rome "La Sapienza," Rowa, Italy, February 2001.
8. C. Demetrescu and G. F. Italiano. A new approach to dynamic all pairs shortest paths. *Journal of the Association for Computing Machinery (JACM)*, 51(6):968–992, 2004.
9. C. Demetrescu and G. F. Italiano. Experimental analysis of dynamic all pairs shortest path algorithms. in *ACM Transactions on Algorithms*, 2(4):578–601, 2006.
10. C. Demetrescu and G. F. Italiano. Trade-offs for fully dynamic reachability on dags: Breaking through the $O(n^2)$ barrier. *Journal of the Association for Computing Machinery (JACM)*, 52(2):147–156, 2005.
11. C. Demetrescu and G. F. Italiano. Fully dynamic transitive closure: Breaking through the $O(n^2)$ barrier. In *Proceedings of the 41st IEEE Annual Symposium on Foundations of Computer Science (FOCS'00)*, Re dondo Beach, CA, pp. 381–389, 2000.
12. C. Demetrescu and G. F. Italiano. Fully dynamic all pairs shortest paths with real edge weights. *Journal of Computer and System Sciences*, 72(5):813–837, August 2006.
13. E. W. Dijkstra. A note on two problems in connexion with graphs. *Numerische Mathematik*, 1:269–271, 1959.
14. D. Eppstein. Dynamic generators of topologically embedded graphs. In *Proceedings of the 14th Annual ACM-SIAM Symposium on Discrete Algorithms*, Baltimore, MD, pp. 599–608, 2003.

15. D. Eppstein. All maximal independent sets and dynamic dominance for sparse graphs. In *Proceedings of the 16th Annual ACM-SIAM Symposium on Discrete Algorithms*, Vancouver, BC, pp. 451–459, 2005.

16. D. Eppstein, Z. Galil, G. F. Italiano, and A. Nissenzweig. Sparsification—A technique for speeding up dynamic graph algorithms. *Journal of the ACM*, 44:669–696, 1997.

17. D. Eppstein, Z. Galil, G. F. Italiano, and T. H. Spencer. Separator based sparsification I: Planarity testing and minimum spanning trees. *Journal of Computer and System Science*, 52(1):3–27, 1996.

18. D. Eppstein, Z. Galil, G. F. Italiano, T. H. Spencer. Separator based sparsification II: Edge and vertex connectivity. *SIAM Journal on Computing*, 28:341–381, 1999.

19. D. Eppstein, G. F. Italiano, R. Tamassia, R. E. Tarjan, J. Westbrook, and M. Yung. Maintenance of a minimum spanning forest in a dynamic plane graph. *Journal of Algorithms*, 13:33–54, 1992.

20. S. Even and H. Gazit. Updating distances in dynamic graphs. *Methods of Operations Research*, 49:371–387, 1985.

21. S. Even and Y. Shiloach. An on-line edge-deletion problem. *Journal of the ACM*, 28:1–4, 1981.

22. J. Fakcharoemphol and S. Rao. Planar graphs, negative weight edges, shortest paths, and near linear time. In *Proceedings of the 42nd IEEE Annual Symposium on Foundations of Computer Science (FOCS'01)*, Las Vegas, NV, pp. 232–241, 2001.

23. M. J. Fischer and A. R. Meyer. Boolean matrix multiplication and transitive closure. In *Conference Record 1971 12th Annual Symposium on Switching and Automata Theory*, East Lansing, MI, pp. 129–131, October 13–15 1971.

24. G. N. Frederickson. Data structures for on-line updating of minimum spanning trees. *SIAM Journal on Computing*, 14:781–798, 1985.

25. G. N. Frederickson. Ambivalent data structures for dynamic 2-edge-connectivity and k smallest spanning trees. *SIAM Journal on Computing*, 26:484–538, 1997.

26. M. L. Fredman and M. R. Henzinger. Lower bounds for fully dynamic connectivity problems in graphs. *Algorithmica*, 22(3):351–362, 1998.

27. D. Frigioni, A. Marchetti-Spaccamela, and U. Nanni. Semi-dynamic algorithms for maintaining single source shortest paths trees. *Algorithmica*, 22(3):250–274, 1998.

28. D. Frigioni, A. Marchetti-Spaccamela, and U. Nanni. Fully dynamic algorithms for maintaining shortest paths trees. *Journal of Algorithms*, 34:351–381, 2000.

29. D. Frigioni, T. Miller, U. Nanni, and C. D. Zaroliagis. An experimental study of dynamic algorithms for transitive closure. *ACM Journal of Experimental Algorithms*, 6(9), 2001.

30. M. E. Furman. Application of a method of fast multiplication of matrices in the problem of finding the transitive closure of a graph. *Soviet Mathematics–Doklady*, 11(5):1252, 1970. English translation.

31. Z. Galil and G. F. Italiano. Fully dynamic algorithms for 2-edge-connectivity. *SIAM Journal on Computing*, 21:1047–1069, 1992.

32. Z. Galil and G. F. Italiano. Maintaining the 3-edge-connected components of a graph on-line. *SIAM Journal on Computing*, 22:11–28, 1993.

33. Z. Galil, G. F. Italiano, and N. Sarnak. Fully dynamic planarity testing. *Journal of the ACM*, 48:28–91, 1999.

34. D. Giammarresi and G. F. Italiano. Decremental 2- and 3-connectivity on planar graphs. *Algorithmica*, 16(3):263–287, 1996.

35. D. H. Greene and D. E. Knuth. *Mathematics for the Analysis of Algorithms*. Birkhäuser, Boston, Cambridge, MA, 1982.

36. F. Harary. *Graph Theory*. Addison-Wesley, Reading, MA, 1969.

37. M. R. Henzinger. Fully dynamic biconnectivity in graphs. *Algorithmica*, 13(6):503–538, 1995.

38. M. R. Henzinger. Improved data structures for fully dynamic biconnectivity. *SIAM Journal on Computing*, 29(6):1761–1815, 2000.

39. M. R. Henzinger and V. King. Randomized fully dynamic graph algorithms with polylogarithmic time per operation. *Journal of the ACM*, 46(4):502–536, 1999.

40. M. R. Henzinger and V. King. Fully dynamic biconnectivity and transitive closure. In *Proceedings of the 36th IEEE Symposium Foundations of Computer Science*, Milulavkee, WI, pp. 664–672, 1995.

41. M. Henzinger and V. King. Maintaining minimum spanning forests in dynamic graphs. *SIAM Journal on Computing*, 31(2):364–374, 2001.

42. M. R. Henzinger and V. King. Randomized fully dynamic graph algorithms with polylogarithmic time per operation. *Journal of the ACM*, 46(4):502–516, 1999.

43. M. R. Henzinger, P. Klein, S. Rao, and S. Subramanian. Faster shortest-path algorithms for planar graphs. *Journal of Computer and System Sciences*, 55(1):3–23, August 1997.

44. M. R. Henzinger and J. A. La Poutré. Certificates and fast algorithms for biconnectivity in fully dynamic graphs. In *Proceedings of the 3rd European Symposium on Algorithms. Lecture Notes in Computer Science* 979, Springer-Verlag, Berlin, pp. 171–184, 1995.

45. M. R. Henzinger and M. Thorup. Sampling to provide or to bound: With applications to fully dynamic graph algorithms. *Random Structures and Algorithms*, 11(4):369–379, 1997.

46. J. Holm, K. de Lichtenberg, and M. Thorup. Poly-logarithmic deterministic fully-dynamic algorithms for connectivity, minimum spanning tree, 2-edge, and biconnectivity. *Journal of the ACM*, 48(4):723–760, 2001.

47. X. Huang and V. Y. Pan. Fast rectangular matrix multiplication and applications. *Journal of Complexity*, 14(2):257–299, June 1998.

48. T. Ibaraki and N. Katoh. On-line computation of transitive closure for graphs. *Information Processing Letters*, 16:95–97, 1983.

49. R. D. Iyer Jr., D. R. Karger, H. S. Rahul, and M. Thorup. An experimental study of poly-logarithmic fully-dynamic connectivity algorithms. *ACM Journal of Experimental Algorithmics*, 6, 2001.

50. G. F. Italiano. Amortized efficiency of a path retrieval data structure. *Theoretical Computer Science*, 48:273–281, 1986.

51. G. F. Italiano. Finding paths and deleting edges in directed acyclic graphs. *Information Processing Letters*, 28:5–11, 1988.

52. S. Khanna, R. Motwani, and R. H. Wilson. On certificates and lookahead in dynamic graph problems. *Algorithmica*, 21(4):377–394, 1998.

53. P. N. Klein and S. Sairam. Fully dynamic approximation schemes for shortest path problems in planar graphs. In *Proceedings of the 3rd Workshop Algorithms and Data Structures. Lecture Notes in Computer Science* 709, Springer-Verlag, Berlin, pp. 442–451, 1993.

54. V. King. Fully dynamic algorithms for maintaining all-pairs shortest paths and transitive closure in digraphs. In *Proceedings of the 40th IEEE Symposium on Foundations of Computer Science (FOCS'99)*, New York, pp. 81–99, 1999.

55. V. King and G. Sagert. A fully dynamic algorithm for maintaining the transitive closure. *Journal of Computer and System Sciences*, 65(1):150–167, 2002.

56. V. King and M. Thorup. A space saving trick for directed dynamic transitive closure and shortest path algorithms. In *Proceedings of the 7th Annual International Computing and Combinatorics Conference (COCOON). Lecture Notes in Computer Science* 2108, Springer-Verlag, Berlin, pp. 268–277, 2001.

57. J. A. La Poutré and J. van Leeuwen. Maintenance of transitive closure and transitive reduction of graphs. In *Proceedings of the Workshop on Graph-Theoretic Concepts in Computer Science. Lecture Notes in Computer Science* 314, Springer-Verlag, Berlin, pp. 106–120, 1988.

58. P. Loubal. A network evaluation procedure. *Highway Research Record 205*, pp. 96–109, 1967.

59. L. Lovász. On the ratio of optimal integral and fractional covers. *Discrete Mathematics*, 13:383–390, 1975.

60. I. Munro. Efficient determination of the transitive closure of a directed graph. *Information Processing Letters*, 1(2):56–58, 1971.

61. J. Murchland. The effect of increasing or decreasing the length of a single arc on all shortest distances in a graph. Technical Report, LBS-TNT-26, London Business School, Transport Network Theory Unit, London, U.K., 1967.

62. H. Nagamochi and T. Ibaraki. Linear time algorithms for finding a sparse *k*-connected spanning subgraph of a *k*-connected graph. *Algorithmmica*, 7:583–596, 1992.

63. G. Ramalingam and T. Reps. On the computational complexity of dynamic graph problems. *Theoretical Computer Science*, 158:233–277, 1996.

64. V. Rodionov. The parametric problem of shortest distances. *U.S.S.R. Computational Mathematics and Mathematical Physics*, 8(5):336–343, 1968.

65. L. Roditty and U. Zwick. Improved dynamic reachability algorithms for directed graphs. In *Proceedings of the 43th Annual IEEE Symposium on Foundations of Computer Science (FOCS)*, Vancouver, BC, pp. 679–688, 2002.

66. L. Roditty and U. Zwick. A fully dynamic reachability algorithm for directed graphs with an almost linear update time. In *Proceedings of the 36th Annual ACM Symposium on Theory of Computing (STOC)*, Chicago, IL, pp. 184–191, 2004.

67. H. Rohnert. A dynamization of the all-pairs least cost problem. In *Proceedings of the 2nd Annual Symposium on Theoretical Aspects of Computer Science, (STACS'85). Lecture Notes in Computer Science 182*, Springer-Verlag, Berlin, pp. 279–286, 1985.

68. P. Sankowski. Dynamic transitive closure via dynamic matrix inverse. In *Proceedings of the 45th Annual IEEE Symposium on Foundations of Computer Science (FOCS'04)*, Rome, Italy, pp. 509–517, 2004.

69. M. Thorup. Fully-dynamic all-pairs shortest paths: Faster and allowing negative cycles. In *Proceedings of the 9th Scandinavian Workshop on Algorithm Theory (SWAT'04)*, Humleback, Denmark, pp. 384–396, 2004.

70. M. Thorup. Worst-case update times for fully-dynamic all-pairs shortest paths. In *Proceedings of the 37th ACM Symposium on Theory of Computing (STOC 2005)*, Baltimore, MD, pp. 112–119, 2005.

71. J. D. Ullman and M. Yannakakis. High-probability parallel transitive-closure algorithms. *SIAM Journal on Computing*, 20(1):100–125, 1991.

72. D. M. Yellin. Speeding up dynamic transitive closure for bounded degree graphs. *Acta Informatica*, 30:369–384, 1993.

73. U. Zwick. All pairs shortest paths using bridging sets and rectangular matrix multiplication. *Journal of the ACM*, 49(3):289–317, 2002.

10

External-Memory Algorithms and Data Structures

Lars Arge
University of Aarhus

Norbert Zeh
Dalhousie University

10.1 Introduction

Many modern scientific and business applications, such as, for example, geographic information systems, data mining, and genomic applications, store and process datasets much larger than the main memory of even state-of-the-art high-end computers. In such cases, the input/output (I/O) communication between internal and external memory (such as disks) can become a major performance bottleneck. Therefore, external-memory (or I/O-efficient) algorithms and data structures, which focus on minimizing the number of disk accesses used to solve a given problem, have received considerable attention in recent years.

In this chapter, we discuss some of the most important techniques for developing I/O-efficient algorithms for sorting and searching (Section 10.3), geometric problems (Section 10.4) and graph problems (Section 10.5). These techniques are illustrated using solutions to a number of fundamental problems. We start the discussion with a description of the model used to develop I/O-efficient algorithms (Section 10.2) and some of the challenges it presents. Readers interested in a complete overview of results and techniques are referred to recent surveys and lecture notes (Vitter, 2001; Arge, 2002, 2005; Zeh, 2002a).

10.2   The Input/Output Model

External-memory algorithms are designed and analyzed in the **I/O-model** (Aggarwal and Vitter, 1988). In this model, the computer is equipped with two levels of memory, an internal or main memory and a disk-based external memory; see Figure 10.1. The internal memory is of limited size and can hold up to M data elements. The external memory is of conceptually unlimited size and is divided into blocks of B consecutive data elements. All computation has to happen on data in internal memory. Data is transferred between internal and external memory by means of I/O operations (I/Os); each such operation transfers one block of data. The cost of an algorithm is the number of I/O operations it performs.

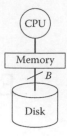

FIGURE 10.1   The I/O-model.

The I/O-model is accurate enough to capture the reality of working with large data sets—limited main memory and blockwise disk access—and simple enough to facilitate the design and analysis of sophisticated algorithms. Note that, exactly to keep the model simple, the computation cost in internal memory is ignored. This is reasonable because a disk access is typically about six orders of magnitude slower than a computation step in memory, making disk accesses, not computation steps, the bottleneck of most large-scale computations.

The first problems that were studied in the I/O-model were searching for a given element in a sorted set (Bayer and McCreight, 1972), sorting and permuting a set of N elements, and scanning a sequence of N elements (Aggarwal and Vitter, 1988). These operations have the following I/O-complexities:

$$\text{Searching bound:} \qquad \Theta\left(\log_B N\right)$$

$$\text{Sorting bound:} \qquad \text{sort}(N) = \Theta\left(\frac{N}{B} \log_{M/B} \frac{N}{B}\right)$$

$$\text{Permutation bound:} \qquad \text{perm}(N) = \Theta(\min(N, \text{sort}(N)))$$

$$\text{Linear or scanning bound:} \qquad \text{scan}(N) = \left\lceil \frac{N}{B} \right\rceil$$

The scanning bound is easy to obtain, as a scan through the sequence can be implemented by performing one I/O for every B consecutive elements that are accessed. In Section 10.3 we discuss how to obtain the sorting and searching bounds. The permutation bound deserves closer scrutiny, as it demonstrates why solving even elementary problems I/O-efficiently is nontrivial.

Assume that we are given N elements in an array A, and they are to be arranged (permuted) in a different order in an array B. In internal memory, this problem is easy to solve in linear time by inspecting each element in A in turn and placing it where it belongs in B.

In external memory, this simple algorithm performs $\Theta(N)$ I/Os in the worst case. Consider the example shown in Figure 10.2. One block of memory is needed to hold the (four-element) block containing the current element in A. If $M = 3B$, this leaves room to hold two more blocks of array B in memory. Placing element 1 into B requires the first block of B to be loaded into memory; placing element 5 into B requires loading the second block into memory. In order to place element 9, the

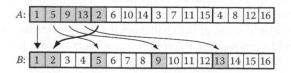

FIGURE 10.2 Illustration of the I/O-complexity of the naive permutation algorithm.

FIGURE 10.3 List ranking. (a) Definition and (b) a list whose nodes are not stored left-to-right.

third block of B has to be brought into memory, which forces one of the two previously loaded blocks to be evicted, say the one containing element 1. Now continue in this fashion until element 2 is about to be placed into B. This element needs to go into the same block as element 1, but this block has been evicted already. Thus, placing element 2 incurs another I/O to place it. Continuing to reason in this fashion, the naive permutation algorithm can be seen to perform $\Theta(N)$ I/Os on this example. It would be better in this case to sort the elements by their target locations, using $\Theta(\text{sort}(N))$ I/Os, as $\text{sort}(N)$ is significantly less than N for all practical values of M and B. Given this choice between permuting naively in $O(N)$ I/Os and sorting the elements by their target locations, a permutation bound of $O(\min(N, \text{sort}(N)))$ follows; Aggarwal and Vitter (1988) showed a matching lower bound.

As another example demonstrating the difficulty in designing I/O-efficient algorithms, consider the list ranking problem; see Figure 10.3a. In this problem, let L be a list with nodes $x_1, x_2, \ldots, x_N$; every node x_i stores the index $\text{succ}(i)$ of its successor $x_{\text{succ}(i)}$ in L; if x_i is the tail of L, then $\text{succ}(i) = \text{nil}$. The *rank* $\text{rank}(x_i)$ of node x_i is its distance from the head x_h of L; that is, $\text{rank}(x_h) = 0$, and the ranks of the other nodes are defined inductively by setting $\text{rank}(x_{\text{succ}(i)}) = \text{rank}(x_i) + 1$. The goal then is to compute the ranks of all nodes in L.

In internal memory, the list ranking problem is easy to solve in linear time: Starting at the head of the list, follow successor pointers and number the nodes in the order they are visited. If the nodes of the list are stored in the order $x_1, x_2, \ldots, x_N$, then the same algorithm takes $O(N/B)$ I/Os in external memory; however, list ranking is often applied to lists whose nodes are stored in an arbitrary order; see Figure 10.3b. In this case, the naive algorithm may cost $\Omega(N)$ I/Os. One could try to arrange the nodes in the order $x_1, x_2, \ldots, x_N$ first—this is essentially the permutation problem—but this would require knowledge of the ranks of the nodes, which is exactly what the algorithm should compute. Therefore, a more sophisticated approach is needed. Section 10.5.1 discusses one such approach.

10.3 Sorting and Searching

In this section, we discuss external-memory solutions for sorting and searching. These solutions highlight some general techniques for designing external-memory algorithms: **multiway merging** and **distribution**, **multiway search trees**, and **buffer trees**.

10.3.1 Sorting

Two different optimal external-memory sorting algorithms can be developed relatively easily based on internal-memory merge and distribution sort (quicksort) and the use of multiway merging/distribution (Aggarwal and Vitter, 1988).

10.3.1.1 Merge Sort

External merge sort works as follows: First, $\Theta(N/M)$ sorted "runs" are formed from the N input elements by repeatedly loading the next $\Theta(M)$ elements into memory, sorting them, and writing them back to external memory. Next, the sorted runs are merged, $M/B - 1$ runs at a time, to obtain $\Theta(N/M \cdot B/M)$ sorted runs; this process continues over $\Theta\left(\log_{M/B} \frac{N}{M}\right) = \Theta\left(\log_{M/B}\left(\frac{N}{B}\frac{B}{M}\right)\right) = \Theta\left(\log_{M/B} \frac{N}{B}\right)$ levels until one sorted run consisting of all N elements is obtained (Figure 10.4).

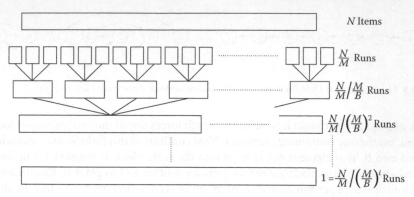

FIGURE 10.4 Merge sort.

Forming the initial runs obviously takes $O(N/B)$ I/Os. Similarly, $M/B - 1$ sorted runs containing N elements in total can be merged in $O(N/B)$ I/Os: Initially, the first block of each run (i.e., $M - B$ elements in total) is loaded into an input buffer; then the B smallest elements in these input buffers are collected in an output buffer, which is then written to disk. This process of collecting and outputting B elements continues until all elements have been output; when an input buffer runs empty while collecting the next B elements from the input buffers, the next block from the corresponding run is read into memory.

In summary, the cost of each level of recursion in Figure 10.4 is $O(N/B)$, and there are $O\left(\log_{M/B} \frac{N}{B}\right)$ levels. Thus, the I/O-complexity of external merge sort is $O\left(\frac{N}{B} \log_{M/B} \frac{N}{B}\right)$.

10.3.1.2 Distribution Sort

External distribution sort is in a sense the reverse of merge sort: First the input elements are distributed into $M/B - 1$ buckets so that all elements in the first bucket are smaller than the elements in the second bucket, which in turn are smaller than all elements in the third bucket, and so on. Each of the buckets is then sorted recursively; a bucket that fits in memory is loaded into main memory, sorted using an internal-memory algorithm, and written back to disk; that is, no further recursion on this bucket is required (Figure 10.5).

Similar to M/B-way merging, M/B-way distribution of N elements takes $O(N/B)$ I/Os, by reading the elements to be distributed into memory in a blockwise fashion, distributing them into one of $M/B - 1$ output buffers, one per bucket, and emptying output buffers whenever they run full. Thus,

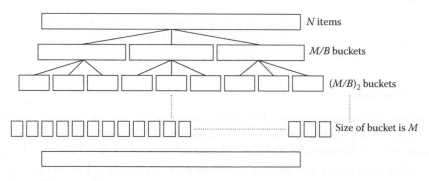

FIGURE 10.5 Distribution sort.

if the distribution is done so that each bucket receives roughly the same number of elements, then there are $O\left(\log_{M/B} \frac{N}{B}\right)$ levels of recursion in the distribution process, and the cost of distribution sort is $O(\text{sort}(N))$ I/Os.

Similar to internal memory quicksort, which uses one pivot to divide the input into two buckets of elements less than and greater than the pivot, respectively, the buckets in distribution sort are defined by $M/B - 2$ pivots; the ith bucket contains all elements between the $(i - 1)$st and the ith pivot. The internal-memory linear-time selection algorithm (Blum et al., 1973), which is used to find a pivot partitioning the input into two buckets of equal size, can easily be modified to use $O(N/B)$ I/Os in external memory. It seems hard, however, to generalize it to find $\Theta(M/B)$ pivot elements in $O(N/B)$ I/Os. Fortunately, since $\log_{\sqrt{M/B}} \frac{N}{B} = O\left(\log_{M/B} \frac{N}{B}\right)$, the cost of distribution sort remains $O(\text{sort}(N))$ if every step partitions the input into only $\Theta(\sqrt{M/B})$ buckets of roughly equal size. This requires only $\Theta(\sqrt{M/B})$ pivots, which can be found as follows: First a sample of the input elements is computed by considering N/M chunks of M elements each; from each such chunk, every $\frac{1}{4}\sqrt{M/B}$th element is chosen to be in the sample. This results in a sample of size $4N/\sqrt{M/B}$. By applying the selection algorithm $\sqrt{M/B}$ times to the sampled elements, every $\frac{4N}{\sqrt{M/B}}/\sqrt{M/B} = \frac{4N}{M/B}$th element of the sample is chosen as a pivot. This requires $\sqrt{M/B} \cdot O(4N/\sqrt{MB}) = O(N/B)$ I/Os. One can show that the resulting $\sqrt{M/B}$ pivot elements define buckets of size at most $\frac{3}{2}N/\sqrt{M/B} = O(N/\sqrt{M/B})$, as required (Aggarwal and Vitter, 1988).

THEOREM 10.1 *(Aggarwal and Vitter, 1988). A set of N elements can be sorted in optimal* $O(\text{sort}(N))$ *I/Os using external merge or distribution sort.*

10.3.2 Searching: B-Trees

The B-tree (Bayer and McCreight, 1972; Comer, 1979; Knuth, 1998) is the external-memory version of a balanced search tree. It uses linear space (i.e., $O(N/B)$ disk blocks) and supports insertions and deletions in $O(\log_B N)$ I/Os. One-dimensional range queries, asking for all elements in the tree in a query interval $[q_1, q_2]$, can be answered in optimal $O(\log_B N + T/B)$ I/Os, where T is the number of reported elements.

B-trees come in several variants, which are all special cases of a more general class of multiway trees called (a, b)-trees (Huddleston and Mehlhorn, 1982):

DEFINITION 10.1 A tree $\mathcal{T}$ is an (a, b)-tree $(a \geq 2, b \geq 2a - 1)$ if the following conditions hold:

- All leaves of $\mathcal{T}$ are on the same level and contain between a and b elements.
- Except for the root, all internal nodes have degree between a and b (contain between $a - 1$ and $b - 1$ elements).
- The root has degree between 2 and b (contains between 1 and $b - 1$ elements).

Normally, the N data elements are stored in sorted order in the leaves of an (a, b)-tree $\mathcal{T}$, and elements in the internal nodes are only used to guide searches (that is, they are routing elements). This way, $\mathcal{T}$ uses linear space and has height $O(\log_a N)$. A range query with interval $[q_1, q_2]$ first searches down $\mathcal{T}$ to locate q_1 and q_2; then the elements in the leaves between the leaves containing q_1 and q_2 are reported.

If parameters a and b are chosen so that $a, b = \Theta(B)$, each node and leaf of an (a, b)-tree can be stored in $O(1)$ disk blocks, the tree occupies $O(N/B)$ disk blocks, and its height is $O(\log_B N)$. A range

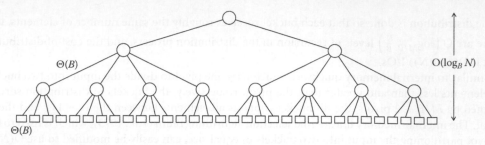

FIGURE 10.6 B-tree. All internal nodes (except possibly the root) have fan-out $\Theta(B)$, and there are $\Theta(N/B)$ leaves. The tree has height $O(\log_B N)$.

query can then be answered in $O(\log_B N + T/B)$ I/Os (Figure 10.6). Multiway search trees (with fan-out $\Theta(B)$, or even $\Theta(B^c)$ for constant $0 < c \le 1$) are therefore often used in external memory.

The insertion of an element x into an (a, b)-tree $\mathcal{T}$ searches down $\mathcal{T}$ for the relevant leaf u and inserts x into u. If u now contains $b + 1$ elements, it is split into two leaves u' and u'' with $\lceil\frac{b+1}{2}\rceil$ and $\lfloor\frac{b+1}{2}\rfloor$ elements, respectively. Both new leaves now have between $\lfloor\frac{b+1}{2}\rfloor \ge a$ and $\lceil\frac{b+1}{2}\rceil \le b$ elements (since $b \ge 2a - 1 \ge 3$). The reference to u in $parent(u)$ is removed and replaced with two references to u' and u'' (that is, a new routing element that separates u' and u'' is inserted into $parent(u)$). If $parent(u)$ now has degree $b + 1$, it is split recursively. This way, the need for a split may propagate up the tree through $O(\log_a N)$ nodes of the tree. When the root splits, a new root of degree-2 is produced, and the height of the tree increases by one.

Similarly, the deletion of an element x finds the leaf u storing x and removes x from u. If u now contains $a - 1$ elements, it is fused with one of its siblings u', that is, u' is deleted and its elements are inserted into u. If this results in u containing more then b (but less than $a - 1 + b < 2b$) elements, u is split into two leaves again. As before, $parent(u)$ is updated appropriately: If u is split, the routing elements in $parent(u)$ are updated, but its degree remains unchanged. Thus, effectively, an appropriate number of elements have been deleted from u' and inserted into u; this is also called a share. If u is not split, the degree of $parent(u)$ decreases by one, and $parent(u)$ may need to be fused with a sibling recursively. As with node splits following an insertion, fuse operations may propagate up the tree through $O(\log_a N)$ nodes. If this results in the root having degree one, it is removed and its child becomes the new root; that is, the height of the tree decreases by one. If the (a, b)-tree satisfies $a, b = \Theta(B)$, this leads to the following result.

THEOREM 10.2   *(Bayer and McCreight, 1972; Comer, 1979). A B-tree uses $O(N/B)$ space and supports updates in $O(\log_B N)$ I/Os and range queries in $O(\log_B N + T/B)$ I/Os.*

10.3.2.1   Weight-Balanced and Persistent B-Trees

A single update in an (a, b)-tree can cause at most $O(\log_a N)$ rebalancing operations (node splits or fuses). If $b = 2a - 1$, it is easy to construct a sequence of updates that causes $\Theta(\log_a N)$ rebalancing operations per update. On the other hand, if $b \ge 2a$, the number of rebalancing operations per update is only $O(1)$ amortized, and if $b > 2a$ only $O\left(\frac{1}{b/2-a}\right)$ (Huddleston and Mehlhorn, 1982).

In more complicated search structures, particularly for geometric search problems, as discussed in Section 10.4, nontrivial secondary structures are often attached to the nodes of the tree. In this case, it is often useful if rebalancing operations are applied only infrequently to nodes with large secondary structures, as such an operation usually entails completely rebuilding the secondary structures of the involved nodes. In particular, since the size of the secondary structure of a node v is often bounded by the number $w(v)$ of elements stored in v's subtree, it is desirable that v be involved in rebalancing operations only once per $\Omega(w(v))$ updates on v's subtree; $w(v)$ is often referred to as the *weight* of v. While (a, b)-trees (and consequently B-trees) do not have this property, **weight-balanced** B-trees do (Arge and Vitter, 2003). Weight-balanced B-trees are similar to B-trees, but balance is enforced by imposing constraints on the weight of each node as a function of its height above the leaf level, rather than by requiring the degree of the node to be within a given range.

THEOREM 10.3 *(Arge and Vitter, 2003). A weight-balanced B-tree uses $O(N/B)$ space and supports updates in $O(\log_B N)$ I/Os and range queries in $O(\log_B N + T/B)$ I/Os. Between two consecutive rebalancing operations on a node v, $\Omega(w(v))$ updates have to be performed on leaves in v's subtree.*

Sometimes it is convenient to have access not only to the current version of a B-tree but also to all earlier versions of the structure. Using general techniques developed in (Driscoll et al., 1989), one can obtain a (partially) **persistent B-tree** (or a multiversion B-tree) (Becker et al., 1996; Varman and Verma, 1997; Arge et al., 2003) that allows for updates in the current B-tree but queries in the current or any previous version.

THEOREM 10.4 *(Becker et al., 1996; Varman and Verma, 1997). After N insertions and deletions on an initially empty persistent B-tree, the structure uses $O(N/B)$ space and supports range queries in the current or any previous version of the structure in $O(\log_B N + T/B)$ I/Os. An update of the current version can be performed in $O(\log_B N)$ I/Os.*

10.3.3 Buffer Trees

In internal memory, N elements can be sorted in optimal $O(N \log N)$ time by performing $\Theta(N)$ operations on a search tree. Using the same algorithm and a B-tree in external memory results in an algorithm that uses $O(N \log_B N)$ I/Os. This is a factor of $\frac{B \log_B N}{\log_{M/B}(N/B)}$ from optimal. In order to obtain an optimal sorting algorithm, a search tree that supports updates in $O\left(\frac{1}{B} \log_{M/B} \frac{N}{B}\right)$ I/Os is needed.

The inefficiency of the B-tree sorting algorithm is a consequence of the B-tree's focus on performing updates and answering queries immediately, as it is designed to be used in an online fashion. This prevents the B-tree from taking full advantage of the large internal memory. In an offline setting, where only the overall I/O use of a series of operations is of interest, the *buffer tree* technique (Arge, 2003) can be used to develop data structures where a series of N operations can be performed in $O(\text{sort}(N))$ I/Os in total. This technique introduces "laziness" in the update and query algorithms and utilizes the large main memory to process a large number of updates and queries simultaneously.

The basic buffer tree is a B-tree with fan-out M/B where each internal node has been augmented with a buffer of size M; every leaf stores $\Theta(B)$ elements (Figure 10.7). The buffers are used to perform operations in a "lazy" manner. For example, an insertion does not search all the way down the tree for the relevant leaf but is simply inserted into the root buffer, which is stored in main memory. When a buffer "runs full," the elements in the buffer are "pushed" one level down to buffers on the next level. Each such buffer emptying process takes $O(M/B)$ I/Os, since the elements in the buffer fit in main memory and need to be distributed over $O(M/B)$ children. If the buffer of any of the children becomes full after emptying the current node's buffer, the buffer emptying process is applied recursively to every child buffer that has run full. Since a buffer emptying process spends $O(M/B)$ I/Os to push $\Theta(M)$ elements one level down the tree, the cost of emptying buffers is $O(1/B)$ amortized per element and level. Since the height of the tree is $O\left(\log_{M/B} \frac{N}{B}\right)$, the cost of N insertions is therefore $O\left(\frac{N}{B} \log_{M/B} \frac{N}{B}\right) = O(\text{sort}(N))$ (disregarding the cost of rebalancing).

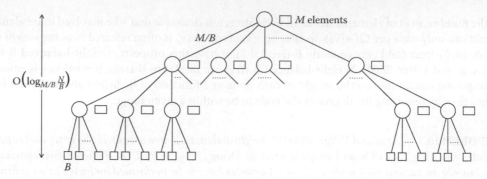

FIGURE 10.7 Buffer tree.

Although somewhat tedious, rebalancing (initiated when emptying buffers of nodes with leaves as children) can also easily be performed using $O(\text{sort}(N))$ I/Os, utilizing that the number of rebalancing operations on an (a, b)-tree with $b > 2a$ is $O\left(\frac{1}{b/2-a}\right)$ amortized per operation (Arge, 2003).

A sequence of N insertions and deletions can also be performed on a buffer tree in $O(\text{sort}(N))$ I/Os. Similar to an insertion, which inserts an "insertion request" into the root buffer, a deletion inserts a deletion request into the root buffer. This request then propagates down the tree until it finds the affected element and removes it. Note that, as a result of the laziness in the structure, there may be several insertions and deletions of the same element pending in buffers of the tree; a time-stamping scheme is used to match corresponding elements (insertions and deletions).

After N operations have been processed on a buffer tree, some of these operations may still be pending in buffers of internal nodes. Thus, in order to use the buffer tree in a simple sorting algorithm, it is necessary to empty all buffers to ensure that all inserted elements are stored at the leaves. At this point, the elements stored at the leaves can be reported in sorted order by traversing the list of leaves left to right. Emptying all buffers involves performing a buffer emptying process on every node of the tree in BFS-order. As emptying one buffer costs $O(M/B)$ I/Os (not counting recursive processes) and as the total number of buffers in the tree is $O((N/B)/(M/B))$, the amortized cost of these final buffer emptying processes is $O(N/B)$.

THEOREM 10.5 *(Arge, 2003). The total cost of a sequence of N update operations on an initially empty buffer tree, as well as the cost of emptying all buffers after all updates have been performed, is $O(\text{sort}(N))$ I/Os.*

COROLLARY 10.1   A set of N elements can be sorted in $O(\text{sort}(N))$ I/Os using a buffer tree.

The buffer tree also supports range queries, in $O\left(\frac{1}{B} \log_{M/B} \frac{N}{B} + T/B\right)$ I/Os amortized. To answer such a query, a query request is inserted into the root buffer; as this element propagates down the tree during buffer emptying processes, the elements in the relevant subtrees are reported. Note that the range queries are batched in the sense that the result of a query is not reported immediately; instead, parts of the result are reported at different times as the query is pushed down the tree. This means that the data structure can only be used in algorithms where future updates and queries do not depend on the result of previous queries.

THEOREM 10.6 *(Arge, 2003). A sequence of N updates and range queries on an initially empty buffer tree can be processed in $O(\text{sort}(N) + T/B)$ I/Os, where T is the total output size of all queries.*

The buffer tree technique can be used to obtain efficient batched versions of a number of other structures. It has, for example, been applied to obtain efficient batched segment trees (Arge, 2003), SB-trees (Arge et al., 1997), and R-trees (van den Bercken et al., 1997; Arge et al., 2002). The technique can also be used to construct a persistent B-tree (obtained by performing N updates) in O(sort(N)) I/Os. Below we discuss how to apply the technique to obtain an efficient external-memory priority queue.

10.3.4 Priority Queues

A "normal" search tree can be used to implement a priority queue supporting INSERT, DELETE, and DELETEMIN operations because the smallest element in a search tree is stored in the leftmost leaf. The same argument does not hold for the buffer tree, as the smallest element may in fact reside in any of the buffers on the path from the root to the leftmost leaf. There is, however, a simple strategy for supporting DELETEMIN operations in O $\left(\frac{1}{B} \log_{M/B} \frac{N}{B} \right)$ I/Os amortized per operation: The first DELETEMIN operation performs a buffer emptying process on each of the nodes on the path from the root to the leftmost leaf; this takes O $\left(\frac{M}{B} \log_{M/B} \frac{N}{B} \right)$ I/Os amortized. The $\Theta(M)$ elements stored in the siblings of the leftmost leaf are now the smallest elements in the tree. These elements are deleted from the tree and loaded into memory. The next $\Theta(M)$ DELETEMIN operations can now be answered without performing any I/Os. Of course, insertions and deletions also have to update the minimal elements in internal memory, but this is easily done without incurring any extra I/Os. Thus, since the buffer emptying cost of O $\left(\frac{M}{B} \log_{M/B} \frac{N}{B} \right)$ I/Os is incurred only every $\Theta(M)$ DELETEMIN operations, the amortized cost per priority queue operation is O $\left(\frac{1}{B} \log_{M/B} \frac{N}{B} \right)$.

THEOREM 10.7 *(Arge, 2003). Using* O(M) *internal memory, an arbitrary sequence of N INSERT, DELETE, and DELETEMIN operations on an initially empty buffer tree can be performed in* O(sort(N)) *I/Os.*

Note that, in the above structure, DELETEMIN operations are processed right away, that is, they are not batched. Another external priority queue with the same bounds as in Theorem 10.7 can be obtained by using the buffer technique on a heap (Fadel et al., 1999).

Internal-memory priority queues usually also support a DECREASEKEY operation, that is, an operation that changes (decreases) the priority of an element already in the queue. If the current priority of the affected element is known, this operation can be performed on the above priority queues by performing a deletion followed by an insertion. Without knowledge of the priority of the affected element, the element cannot be deleted and no DECREASEKEY operation is possible. By applying the buffer technique to a tournament tree, one can obtain a priority queue that supports an UPDATE operation without knowing the key of the element to be updated (Kumar and Schwabe, 1996). This operation is a combination of an INSERT and a DECREASEKEY operation, which behaves like the former if the element is currently not in the priority queue, and like the latter if it is. The cost per operation on this priority queue is O $\left(\frac{1}{B} \log_2 \frac{N}{B} \right)$ I/Os amortized.

10.4 Geometric Problems

In this section, we discuss external-memory solutions to two geometric problems: Planar orthogonal line segment intersection (a batched problem) and planar orthogonal range searching (an online problem). While external-memory algorithms have been developed for a large number of other (and maybe even more interesting) geometric problems, we have chosen these two problems to illustrate some of the most important techniques used to obtain I/O-efficient algorithms for geometric problems; these techniques include **distribution sweeping** (line segment intersection) and dynamic multiway trees with static structure **bootstrapping** (range searching). The reader is referred to recent surveys for a more complete overview of results (e.g. Vitter, 2001; Arge, 2002).

10.4.1   Planar Orthogonal Line Segment Intersection

In this section we illustrate the distribution sweeping technique by utilizing it to solve the planar orthogonal line segment intersection problem, that is, the problem of reporting all intersections between a set of horizontal and a set of vertical line segments. Intuitively, distribution sweeping is a combination of the (internal memory) plane sweep technique and M/B-way distribution. Below we first discuss the optimal internal-memory plane sweep solution to orthogonal line segment intersection; then we discuss the external-memory solution, which uses distribution sweeping.

10.4.1.1   Internal-Memory Plane Sweep Algorithm

In internal memory, a simple optimal solution based on the plane sweep paradigm (Preparata and Shamos, 1985) works as follows (Figure 10.8): Imagine sweeping a horizontal sweep line from $+\infty$ to $-\infty$ while finding the vertical segments intersecting each horizontal segment we meet. The sweep maintains a balanced search tree storing the vertical segments currently crossing the sweep line, ordered by their x-coordinates. For each horizontal segment, its intersections with vertical segments can then be found by performing a range query on the tree with the x-coordinates of its endpoints.

To be more precise, first all the segment endpoints are sorted by their y-coordinates. The sorted sequence of points is then used to perform the sweep, that is, to process the segments in endpoint y-order. When the top endpoint of a vertical segment is reached, the segment is inserted into the search tree; the segment is removed again when its bottom endpoint is reached. When a horizontal segment is reached, a range query is performed on the search tree. As insertions and deletions can be performed in $O(\log_2 N)$ time, and range queries in $O(\log_2 N + T')$ time (where T' is the number of reported intersections), this algorihm takes optimal $O\left(N \log_2 N + T\right)$ time, where N is the total number of segments, and T the total number of reported intersections.

A natural external-memory modification of the plane sweep algorithm would use a B-tree as the tree data structure. However, this would lead to an $O(N \log_B N + T/B)$ I/O solution, while we are looking for an $O\left(\frac{N}{B} \log_{M/B} \frac{N}{B} + T/B\right)$ I/O solution. Using a buffer tree instead of a B-tree, this bound can be achieved, as the buffer tree supports batched range query operations, and the sequence of operations performed on the tree does not depend on the results of the queries (Arge, 2003). Next we describe a solution based on distribution sweeping, which has been used to solve a much large set of geometric problems than the buffer tree.

10.4.1.2   I/O-Efficient Algorithm: Distribution Sweeping

Distribution sweeping (Goodrich et al., 1993) is a powerful technique obtained by combining M/B-way distribution and plane sweeping. To solve a given problem, the plane is divided into M/B vertical slabs, each of which contains $\Theta(N/M)$ input objects (for example points or line segment endpoints). Then a vertical top-down sweep over all the slabs is performed, in order to locate parts of the solution that involve interactions between objects in different slabs, or objects (such as line segments) that completely span one or more slabs. As with M/B-way merging or distribution, the choice of M/B slabs ensures that one block of data from each slab fits in main memory. To find components of

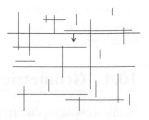

FIGURE 10.8 Solution to the planar orthogonal line segment intersection problem using a plane sweep.

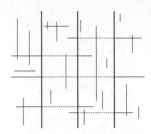

FIGURE 10.9 Solution to the planar orthogonal line segment intersection problem using distribution sweeping.

the solution involving interactions between objects residing in the same slab, the problem is solved recursively in each slab. The recursion stops after $O\left(\log_{M/B} \frac{N}{M}\right) = O\left(\log_{M/B} \frac{N}{B}\right)$ levels, which is when the subproblems are small enough to fit in internal memory. Therefore, in order to obtain an $O(\text{sort}(N))$ algorithm, it must be possible to perform one sweep in $O(N/B)$ I/Os.

To use the general technique outlined above to solve the planar orthogonal line segment intersection problem, the endpoints of all the segments are sorted twice to create two lists: one with the endpoints sorted by their x-coordinates and the other with the endpoints sorted by their y-coordinates. The y-sorted list is used to perform the top-down sweeps; the x-sorted list is used (throughout the algorithm) to locate the pivot elements needed to distribute the input into M/B vertical slabs.

The algorithm now proceeds as follows (Figure 10.9): It divides the plane into M/B slabs and sweeps from top to bottom. When a top endpoint of a vertical segment is reached, the segment is inserted into an active list associated with the slab containing the segment. When a horizontal segment is reached, the active lists associated with all slabs it completely spans are scanned. During this scan, every vertical segment in an active list is either intersected by the horizontal segment or will not be intersected by any of the following horizontal segments and can therefore be removed from the list. This process finds all intersections except those between vertical segments and horizontal segments (or portions of horizontal segments) that do not completely span vertical slabs (the solid parts of the horizontal segments in Figure 10.9). These are found by distributing the segments to the slabs and solving the problem recursively in each slab. A horizontal segment may be distributed to two slabs, namely the slabs containing its endpoints, but will be represented at most twice on each level of the recursion.

By holding a buffer of B elements per active list in memory, the active lists can be read and written in a blockwise fashion, leading to a cost of $O(N/B + T'/B)$ I/Os for the sweep, where T' is the number of intersections reported: $O(1/B)$ I/Os amortized are used every time a vertical segment is processed and every such segment is processed only twice without reporting an intersection, namely when it is distributed to an active list and when it is removed again. Note that it is crucial not to try to remove a vertical segment from an active list when its bottom endpoint is reached, since doing so might require an I/O for each segment. Thus, by the general discussion of distribution sweeping above, all intersections are reported in optimal $O(\text{sort}(N) + T/B)$ I/O operations.

THEOREM 10.8 *(Goodrich et al., 1993). The planar orthogonal line segment intersection problem can be solved in* $O(\text{sort}(N) + T/B)$ *I/Os.*

As mentioned, distribution sweeping has been used to solve a large number of batched problems, including other problems involving axis-parallel objects (see, e.g., Goodrich et al. (1993)) and (in combination with other techniques) more complicated problems involving higher-dimensional or nonaxis-parallel objects (see, e.g., Arge et al., 1998a,b). There are, of course, also a number of batched geometric problems, such as for example computation of Voronoi diagram and Delaunay triangulations (Goodrich et al., 1993), that have been solved using different ideas. Refer to recent surveys for details (Vitter, 2001; Arge, 2002).

10.4.2 Planar Orthogonal Range Searching

As discussed in Section 10.3, efficient external-memory data structures are usually based on multi-way (fan-out $\Theta(B)$) trees. However, when trying to map multilevel internal-memory data structures (that is, data structures where internal nodes contain large secondary data structures) to external memory,

one encounters a number of problems, most of them resulting from
the large fan-out. In this section we illustrate how to overcome some of
these problems through a discussion of the planar orthogonal range
searching problem, that is, the problem of storing a dynamically
changing set S of points in the plane so that all T points contained in an
axis-aligned query rectangle $q = (q_1, q_2, q_3, q_4)$ can be found efficiently
(Figure 10.10).

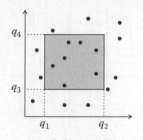

FIGURE 10.10 Orthogonal
range searching.

As discussed in Section 10.3, the one-dimensional range searching
problem can be solved using a B-tree; the structure uses optimal $O(N/B)$
space, answers range queries in $O(\log_B N + T/B)$ I/Os and supports
updates in $O(\log_B N)$ I/Os. We would like to obtain the same bounds in
the two-dimensional case. However, it turns out that, in order to obtain
an $O(\log_B N + T/B)$ query bound, one has to use $\Omega\left(\dfrac{N}{B}\dfrac{\log_B N}{\log_B \log_B N}\right)$
space (Subramanian and Ramaswamy, 1995; Hellerstein et al., 1997;
Arge et al., 1999). In Section 10.4.2.1, we describe the external range
tree, which achieves these bounds. Using only $O(N/B)$ space, the best
query bound that can be achieved is $\Omega\left(\sqrt{N/B} + T/B\right)$ I/Os (Kanth and
Singh, 1999; Agarwal et al., 2001b). In Section 10.4.2.2, we describe the
O-tree, an external kd-tree variant that achieves these bounds.

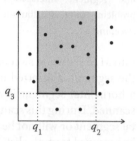

FIGURE 10.11 Three-sided
range searching.

10.4.2.1 External Range Tree

A key part of the external range tree is an $O(N/B)$ space structure for
the 3-sided version of the planar orthogonal range searching problem, where a query $q = (q_1, q_2, q_3)$
asks for all points $(x, y) \in S$ with $q_1 \le x \le q_2$ and $y \ge q_3$ (Figure 10.11). We start with a discussion
of this structure, called an external priority search tree.

10.4.2.1.1 External Priority Search Tree

The static version of the three-sided range searching problem can easily be solved using a sweeping
idea and a persistent B-tree. Consider sweeping the plane with a horizontal line from $+\infty$ to $-\infty$
and inserting the x-coordinates of points from S into a persistent B-tree as they are met. A range
query $q = (q_1, q_2, q_3)$ can then be answered by performing a one-dimensional range query $[q_1, q_2]$
on the B-tree at "time" q_3.

THEOREM 10.9 *A set S of N points in the plane can be stored in a linear-space data structure
such that a three-sided range query can be answered in $O(\log_B N + T/B)$ I/Os. This structure can be
constructed in $O(\text{sort}(N))$ I/Os.*

The above structure is inherently static. A dynamic solution (using the static solution as a
building block) can be obtained using an external version (Arge et al., 1999) of a priority search
tree (McCreight, 1985).

An external priority search tree consists of a base B-tree $\mathcal{T}$ on the x-coordinates of the points in S.
In this tree, each internal node v corresponds naturally to an x-range X_v consisting of the points
below v; this interval is divided into $\Theta(B)$ slabs by the x-ranges of its children. Each node v stores $O(B)$
points from S for each of its $\Theta(B)$ children v_i, namely the B points with the highest y-coordinates in
the x-range X_{v_i} of v_i (if existing) that have not been stored at ancestors of v. The resulting set of $O\left(B^2\right)$
points is stored in the linear-space static structure discussed above—called a B^2-structure—such that
a three-sided query on these points can be answered in $O\left(\log_B B^2 + T_v/B\right) = O\left(1 + T_v/B\right)$ I/Os,
where T_v is the number of points reported. A leaf u of $\mathcal{T}$ stores the points with x-coordinates among
the x-coordinates in u that are not stored higher up in the tree; assuming without loss of generality

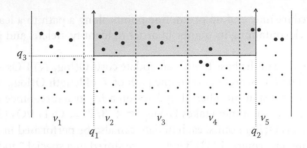

FIGURE 10.12 Internal node v with children $v_1, v_2, \ldots, v_5$. The points in bold are stored in the B^2-structure. To answer a three-sided query $q = (q_1, q_2, q_3)$, we report the relevant of the $O(B^2)$ points and answer the query recursively on v_2, v_3, and v_5. The query is not extended to v_4 because not all of the points from v_4 in the B^2-structure are in q.

that all x-coordinates are distinct, these points fit in a single block. Overall, the external priority search tree uses $O(N/B)$ space, since T uses linear space and every point is stored in precisely one B^2-structure or leaf.

A three-sided query $q = (q_1, q_2, q_3)$ is answered by starting at the root of T and proceeding recursively to the appropriate subtrees: When visiting a node v, its B^2-structure is queried to report all points stored at v that lie in the query range. From v, the search advances to a child v_i if it is either along the search paths to q_1 or q_2, or the entire set of points corresponding to v_i in v's B^2-structure were reported (Figure 10.12). The query procedure reports all points in the query range because it visits every child v_i corresponding to a slab completely spanned by the interval $[q_1, q_2]$ unless at least one of the points in the B^2-structure corresponding to v_i is not in q, which in turn implies that none of the points in the subtree rooted at v_i can be in q.

The cost of a query can be seen to be $O(\log_B N + T/B)$ I/Os as follows. The cost per node v of T visited by the query procedure is $O(1 + T_v/B)$ I/Os. There are $O(\log_B N)$ nodes visited on the search paths in T to the leaves containing q_1 and q_2, and thus the number of I/Os used in these nodes adds up to $O(\log_B N + T/B)$. Each remaining visited node v in T is visited because $\Theta(B)$ points corresponding to v were reported at its parent. Thus, the cost of visiting these nodes adds up to $O(T/B)$, even if a constant number of I/Os are spent at some nodes without finding $\Theta(B)$ points to report.

To insert or delete a point $p = (x, y)$ into or from the external priority search tree, its x-coordinate is first inserted into or deleted from the base tree T. If the update is an insertion, the secondary structures are then updated using the following recursive "bubble-down" procedure starting at the root v: First the (at most) B points in the B^2-structure of v corresponding to the child v_i whose x-range X_{v_i} contains x are found by performing the (degenerate) three-sided query defined by X_{v_i} and $y = -\infty$ on the B^2-structure. If p is below these points (and there are B of them), p is inserted recursively into v_i's subtree. Otherwise, p is inserted into the B^2-structure. If this means that the B^2-structure now contains more than B points corresponding to v_i, the lowest of them (which can again be identified by a simple query) is deleted from the B^2-structure and inserted recursively into v_i's subtree. If v is a leaf, the point is simply stored in the associated block. A deletion first identifies the node v containing p by searching down T for x while querying the B^2-structures of the visited nodes for the (at most) B points corresponding to the relevant child v_i. Point p is then deleted from the B^2-structure. Since this decreases the number of points from X_{v_i} stored at v, a point from v_i needs to be promoted using a recursive "bubble-up" procedure: This procedure first finds the topmost point p' (if existing) stored in v_i using a (degenerate) query on its B^2-structure. This point is then deleted from the B^2-structure of v_i and inserted into the B^2-structure of v. Finally, a point from the child of v_i corresponding to the slab containing p' needs to be promoted recursively to v_i. Thus, the procedure may end up promoting points along a path to a leaf; at a leaf, it suffices to load the single block containing its points, in order to identify, delete, and promote the relevant point.

Disregarding the update of the base tree T, an update costs $O(\log_B N)$ I/Os amortized, which can be seen as follows: An update searches down one path of T of length $O(\log_B N)$ and, at each node, performs a query and a constant number of updates on a B^2-structure. Since each of these queries returns at most B points, each of them takes $O(\log_B B^2 + B/B) = O(1)$ I/Os (Theorem 10.4). Since the B^2-structure contains $O(B^2)$ points, each update can also be performed in $O(1)$ I/Os amortized using global rebuilding (Overmars, 1983): Updates are stored in a special "update block" and, once B updates have been collected, the B^2-structure is rebuilt using $O\left(\frac{B^2}{B}\log_{M/B}\frac{B^2}{B}\right) = O(B)$ I/Os, or $O(1)$ I/Os amortized. The structure continues to be able to answer queries efficiently, as only $O(1)$ extra I/Os are needed to check the update block.

The update of the base tree T also takes $O(\log_B N)$ I/Os (Theorem 10.2), except that we have to consider what happens with the B^2-structures when a rebalancing operation (split or fuse) is performed on a base tree node v. When v is involved in a split or fuse, $parent(v)$ will either gain or lose a slab, and therefore will contain either B points too few or B points too many in its B^2-structures. However, these points can easily be either found by performing B bubble-up operations (for a split) or removed by performing B bubble-down operations (for a fuse). It is relatively easy to realize that, in both cases, these bubble operations can be performed using $O(B\log_B w(v)) = O(w(v))$ I/Os (Arge et al., 1999); thus, if T is implemented as a weight-balanced B-tree (Theorem 10.3), the amortized cost of a rebalancing operation is $O(1)$ I/Os, and rebalancing also takes $O(\log_B N)$ I/Os amortized. For a discussion of how to make the rebalancing bound worst-case, see (Arge et al., 1999).

THEOREM 10.10   *(Arge et al., 1999). An external priority search tree on a set S of N points in the plane uses $O(N/B)$ space and answers three-sided range queries in $O(\log_B N + T/B)$ I/Os. Updates can be performed in $O(\log_B N)$ I/Os.*

The external priority search trees illustrate some of the techniques used to overcome the problems commonly encountered when trying to map multilevel (binary) data structures to (multiway) external-memory data structures: The large fan-out results in the need for secondary data structures (storing $O(B^2)$ points); these structures are implemented using a static structure (based on persistence), which can be updated efficiently with global rebuilding, due to its small size. This idea is often referred to as bootstrapping (Vitter, 2001). Other commonly used techniques are the use of weight-balanced B-trees to support efficient updates of the secondary structures during rebalancing operations, as well as the idea of accounting for some of the query cost by bounding it by the cost of reporting the output (often called filtering (Chazelle, 1986)). The same ideas have been used, for example, to develop an efficient external interval tree (Arge and Vitter, 2003).

10.4.2.1.2   *External Range Tree*

Using the external priority search tree for answering 3-sided range queries, it is easy to obtain a structure for general (4-sided) orthogonal range searching that uses $O\left(\frac{N}{B}\log_2 N\right)$ space: One constructs a binary search tree T over the x-coordinates of the points in S. The nodes of T are packed into blocks on disk in a way similar to a B-tree: Starting at the root, nodes are visited in breadth-first order until a subtree consisting of $\log_2 B - 1$ levels has been traversed; this subtree contains less than $2 \cdot 2^{\log_2 B - 1} = B$ nodes and can therefore be stored in one block. The $O(B)$ trees below this subtree are now packed into blocks recursively. This way the nodes on a root-leaf path in T are stored in $O(\log_B N)$ blocks. Each internal node v of T now has two external priority search trees associated with it. Both structures store all points in the leaves of v's subtree; the first one can answer three-sided queries that are open to the left, the second queries that are open to the right. Since the priority search trees on each level of T use $O(N/B)$ space, the structure uses $O\left(\frac{N}{B} \log_2 N\right)$ space in total.

Answering an orthogonal range query q requires searching down T to find the first node v where the left x-coordinate q_1 and the right x-coordinate q_2 of the query range are contained in different children of v. The points in q are then reported by querying the right-open priority search tree of the left child of v and the left-open priority search tree of the right child of v. This procedure correctly reports all points in the query, since the priority search trees in the two children contain (at least) all the points in the x-range $[q_1, q_2]$ of the query. Since the search in T takes $O(\log_B N)$ I/Os and the queries on the priority search trees take $O(\log_B N + T/B)$ I/Os, the query is answered in $O(\log_B N + T/B)$ I/Os in total.

To improve the space use of the structure, basically the same techniques as in internal memory (Chazelle, 1986) are used. Rather than using a binary base tree T, a tree with fan-out $\log_B N$ is used, which has height $O\left(\log_B N / \log_B \log_B N\right)$. As in the binary case, two external priority search trees are associated with each node v of T. However, in order to still be able to answer a query in $O\left(\log_B N + T/B\right)$ I/Os, two additional linear-space structures (a B-tree and an external priority search tree) are needed for each node v. This way, the structure uses $\Theta\left(\frac{N}{B} \frac{\log_B N}{\log_B \log_B N}\right)$ space. By making T a weight-balanced B-tree, one can show that both queries and updates can be performed I/O-efficiently (Subramanian and Ramaswamy, 1995; Arge et al., 1999).

THEOREM 10.11 *(Arge et al., 1999). An external range tree on a set S of N points in the plane uses* $O\left(\frac{N}{B} \frac{\log_B N}{\log_B \log_B N}\right)$ *space and answers orthogonal range queries in* $O(\log_B N + T/B)$ *I/Os. Updates can be performed in* $O\left(\log_B^2 N / \log_B \log_B N\right)$ *I/Os amortized.*

10.4.2.2 O-Tree

A key part of the $O(N/B)$ space O-tree for answering planar orthogonal range queries is an external version of the kd-tree (Bentley, 1975). This structure, which we describe first, supports queries in $O\left(\sqrt{N/B} + T/B\right)$ I/Os, but updates require $O\left(\log_B^2 N\right)$ I/Os. The O-tree improves the update bound to $O\left(\log_B N\right)$.

10.4.2.2.1 External kd-Tree

An internal-memory kd-tree T on a set S of N points in the plane is a binary tree of height $O\left(\log_2 N\right)$ with the points stored in the leaves of the tree. The internal nodes represent a recursive decomposition of the plane by means of axis-parallel lines that partition the set of points below a node into two subsets of approximately equal size. On even levels of T, horizontal dividing lines are used, on odd levels vertical ones. In this way, a rectangular region R_v is naturally associated with each node v, and the nodes on any particular level of T partition the plane into disjoint regions. In particular, the regions associated with the leaves represent a partition of the plane into rectangular regions containing one point each (Figure 10.13).

An external-memory kd-tree (Robinson, 1981; Procopiuc et al., 2003) is a kd-tree T where the recursive partition stops when the number of points in a region falls below B. This way, the structure has $O(N/B)$ leaves containing $\Theta(B)$ points each. The points in each leaf are stored together in a disk block. Since the number of internal nodes is $O(N/B)$, the structure uses $O(N/B)$ space regardless of how these nodes are stored on disk. However, in order to be able to follow a root-leaf path in $O(\log_B N)$ I/Os, the nodes are stored as described for the (binary) range tree in the previous

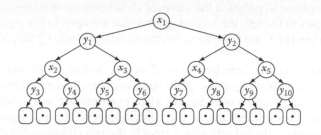

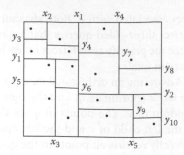

FIGURE 10.13 A kd-tree and the corresponding planar subdivision.

section. One can easily construct the external kd-tree in $O\left(\frac{N}{B}\log_2\frac{N}{B}\right)$ I/Os level-by-level using the $O(N/B)$ I/O median finding algorithm. Using more sophisticated ideas, this can be improved to $O\left(\frac{N}{B}\log_{M/B}\frac{N}{B}\right)$ I/Os (Agarwal et al., 2001a; Procopiuc et al., 2003).

A range query $q = (q_1, q_2, q_3, q_4)$ can be answered using a simple recursive procedure that starts at the root: At a node v the query advances to a child w if q intersects the region R_w associated with w. At a leaf u all points in u that are contained in q are returned. To bound the number of nodes in $\mathcal{T}$ visited when answering a range query q or, equivalently, the number of nodes v where R_v intersects q, it is useful to first bound the number of nodes v where R_v intersects a vertical line l. The region R_r associated with the root r is obviously intersected by l; but, as the regions associated with its two children represent a subdivision of R_r using a vertical line, only the region R_w associated with one of these children w is intersected by l. Because the region R_w is subdivided by a horizontal line, the regions associated with both children of w are intersected. Let $L = O(N/B)$ be the number of leaves in the kd-tree. As the children of w are roots in kd-trees with $L/4$ leaves, the recurrence for the number of regions intersected by l is $Q(L) \le 2 + 2Q(L/4) = O(\sqrt{L}) = O(\sqrt{N/B})$. A similar argument shows that the number of regions intersected by a horizontal line is $O(\sqrt{N/B})$. This means that the number of nodes v with regions R_v intersected by the boundary of a four-sided query q is $O(\sqrt{N/B})$. All the additional nodes visited when answering q correspond to regions completely inside q. Since each leaf contains $\Theta(B)$ points, there are $O(T/B)$ leaves with regions completely inside q. Since the region R_v corresponding to an internal node v is only completely contained in q if the regions corresponding to the leaves below v are contained in q (and since the kd-tree is binary), the total number of regions completely inside q is also $O(T/B)$. Thus, in total, $O(\sqrt{N/B} + T/B)$ nodes are visited and, therefore, a query is answered in $O(\sqrt{N/B} + T/B)$ I/Os.

Insertions into the external kd-tree can be supported using the **logarithmic method** (Bentley, 1979; Overmars, 1983): One maintains a set of $O(\log_2 N)$ static kd-trees $\mathcal{T}_0, \mathcal{T}_1, \dots$ such that $\mathcal{T}_i$ is either empty or has size 2^i. An insertion is performed by finding the first empty structure $\mathcal{T}_i$, discarding all structures $\mathcal{T}_j$ with $j < i$ and building $\mathcal{T}_i$ from the new point and the $\sum_{l=0}^{i-1} 2^l = 2^i - 1$ points in the discarded structures; this takes $O\left(\frac{2^i}{B}\log\frac{2^i}{B}\right)$ I/Os. If this cost is divided between the 2^i points, each of them is charged for $O\left(\frac{1}{B}\log\frac{N}{B}\right)$ I/Os. Because points never move from higher- to lower-indexed structures, each point is charged $O(\log N)$ times. Thus, the amortized cost of an insertion is $O\left(\frac{1}{B}\log\frac{N}{B}\log N\right) = O(\log_B^2 N)$ I/Os.

To answer a query, each of the $O(\log N)$ structures is queried in turn. Querying the ith structure $\mathcal{T}_i$ takes $O\left(1 + \sqrt{2^i/B} + T_i/B\right)$ I/Os, where T_i is the number of reported points. However, if the first $\log B$ structures are kept in memory (using $O(1)$ blocks of main memory), these structures can be queried without performing any I/Os. Thus, the total cost of a query is $\sum_{i=\log B}^{\log N} O\left(\sqrt{2^i/B} + T_i/B\right) = O\left(\sqrt{N/B} + T/B\right)$ I/Os.

Finally, deletions can also be supported in O $\left(\log_B^2 N\right)$ I/Os amortized using partial rebuilding, and both update bounds can be made worst-case using standard deamortization techniques (Overmars, 1983; Arge, 2005).

THEOREM 10.12 *(Procopiuc et al., 2003). An external kd-tree for storing a set S of N points in the plane uses O(N/B) space and supports range queries in* O $\left(\sqrt{N/B} + T/B\right)$ *I/Os. It can be updated in* O $\left(\log_B^2 N\right)$ *I/Os.*

The external kd-tree is an example where a relatively simple modification (early stop of the recursion and blocking of a binary tree) leads to an efficient external structure. The structure is then made dynamic using the logarithmic method (and partial rebuilding). An external version of the logarithmic method (where $\log_B N$ rather than $\log_2 N$ structures are maintained) can be used to improve the update bound slightly; it has also been used in other dynamic structures, for example in dynamic point location (Arge and Vahrenhold, 2004).

10.4.2.2.2 O-Tree Structure

The O-tree (Kanth and Singh, 1999; Arge, 2005), which improves the update bounds of the external kd-tree to $O(\log_B N)$ I/Os, uses ideas similar to ideas utilized in divided kd-trees in internal memory (van Kreveld and Overmars, 1991). The main idea is to divide the plane into $\Theta\left(\frac{N}{B\log_B^2 N}\right)$ regions containing $\Theta\left(B\log_B^2 N\right)$ points each and then construct an external kd-tree on the points in each region. The regions are obtained by dividing the plane into slabs with approximately the same number of points using $\Theta(\sqrt{N/B}/\log_B N)$ vertical lines; each of these slabs is then further divided into regions containing approximately the same number of points using $\Theta\left(\sqrt{N/B}/\log_B N\right)$ horizontal lines. The O-tree then consist of a B-tree on the vertical lines, a B-tree in each slab s_i on the horizontal lines in s_i, and an external kd-tree on the points in each region (Figure 10.14).

To answer a query, external kd-trees corresponding to regions intersected by the query are queried; these regions can easily be found in O $\left(\sqrt{N/B} + T/B\right)$ I/Os using the B-trees. The cost of querying the kd-trees is also O $\left(\sqrt{N/B} + T/B\right)$ I/Os, which basically follows from the fact that only

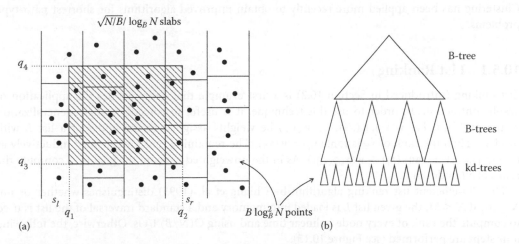

FIGURE 10.14 (a) Division of the plane into regions containing $\Theta(B\log_B^2 N)$ points using $\Theta(\sqrt{N/B}\log_B N)$ vertical lines and $\Theta(\sqrt{N/B}\log_B N)$ horizontal lines in each slab. (b) O-tree: B-tree on vertical lines, B-tree on horizontal lines in each slab, and kd-tree on points in each region. A query $q = (q_1, q_2, q_3, q_4)$ is answered by querying all kd-trees in regions intersected by q.

$O\left(\sqrt{N/B}/\log_B N\right)$ kd-trees corresponding to regions not completely covered by the query need to be inspected and that each of these queries uses $O\left(\sqrt{B\log_B^2 N/B} + T/B\right) = O\left(\log_B N + T/B\right)$ I/Os.

Updates are handled in a straightforward (but tedious) way using a global rebuilding strategy, where the key to obtaining the $O(\log_B N)$ update bound is that updating the external kd-tree corresponding to a region requires only $O\left(\log_B^2(B\log_B^2 N)\right) = O(\log_B N)$ I/Os (Kanth and Singh, 1999; Arge, 2005).

THEOREM 10.13 *(Kanth and Singh, 1999). An O-tree for storing a set of N points in the plane uses linear space and supports range queries in $O(\sqrt{N/B} + T/B)$ I/Os. It can be updated in $O(\log_B N)$ I/Os.*

The O-tree described above is slightly different from the original description (Kanth and Singh, 1999). A very different alternative structure, called the *cross-tree*, achieves the same bounds as the O-tree (Grossi and Italiano, 1999a,b). The external kd-tree, the O-tree, and the cross-tree can all be extended to d-dimensions in a straightforward way to obtain an $O\left((N/B)^{1-1/d} + T/B\right)$ query bound.

10.5 Graph Problems

In this section, we highlight the most important techniques that have been applied successfully to obtain I/O-efficient solutions to a number of fundamental graph problems. The first one, **graph contraction**, is most useful for solving connectivity problems, such as computing the connected components or a minimum spanning tree of a graph. The second one, the **Euler tour** technique, is used to solve fundamental problems on trees and forests. Both techniques have been applied successfully before to obtain efficient parallel algorithms for these types of problems. The other two techniques, **time-forward processing** and **clustering**, do not have equivalents in parallel algorithms. Time-forward processing allows us to solve graph problems that can be expressed as labeling the vertices of a directed acyclic graph based on information passed along its edges. Clustering has been applied more recently to obtain improved algorithms for shortest-path-type problems.

10.5.1 List Ranking

List ranking (introduced in Section 10.2) is a first example that demonstrates the application of graph contraction. In order to use this technique, it is useful to consider a slight generalization of the problem: Let $w(x_1), w(x_2), \ldots, w(x_N)$ be weights assigned to the nodes of a list L with head x_h. Then the rank of x_h is $\text{rank}(x_h) = w(x_h)$. The remaining ranks are defined inductively as $\text{rank}(x_{\text{succ}(x_i)}) = \text{rank}(x_i) + w(x_{\text{succ}(x_i)})$. As in the unweighted version, the goal is to compute the ranks of all nodes in L.

The I/O-efficient list ranking algorithm by Chiang et al. (1995) distinguishes whether or not $N \leq M$. If $N \leq M$, the given list L is loaded into memory and a standard traversal of the list is used to compute the rank of every node in linear time and using $O(N/B)$ I/Os. Otherwise, the following four steps are performed (see Figure 10.15):

1. Compute an independent set I of at least $N/3$ nodes in L.
2. Construct a new list L' by removing all elements in I from L and adding their weights to the weights of their successors.

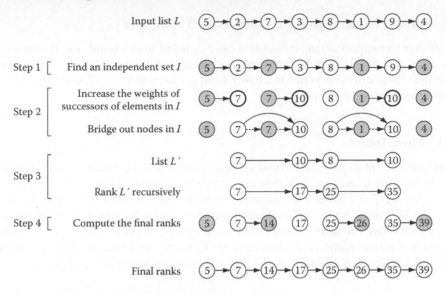

FIGURE 10.15 Illustration of the list-ranking algorithm.

3. Recursively solve the list ranking problem on L'. This assigns ranks to the elements in L'.
4. Compute the ranks of the elements in L: For every $x_i \notin I$, its rank in L is the same as in L'. For every $x_i \in I$, its rank is the rank of its predecessor plus $w(x_i)$.

The correctness of this algorithm is easily verified. The I/O-complexity of the algorithm is $O(\text{sort}(N))$ if Steps 1, 2, and 4 can be implemented using $O(\text{sort}(N))$ I/Os because Step 3 recurses on only two-thirds of the nodes in L.

To implement Step 1, it suffices to consider L as an undirected graph and compute a maximal independent set. Section 10.5.3 discusses how to do this using $O(\text{sort}(N))$ I/Os. Since every vertex in L has degree at most two, every maximal independent set of L has size at least $N/3$.

For every element $x_i \in I$, Step 2 has to add $w(x_i)$ to $w(x_{\text{succ}(i)})$ and, for the node x_j such that $\text{succ}(j) = i$, set $\text{succ}(j) = \text{succ}(i)$. This is easily done by sorting and scanning I and $L \setminus I$ a constant number of times, that is, using $O(\text{sort}(N))$ I/Os. For example, after sorting the elements in I by their successor pointers and the elements in $L \setminus I$ by their own IDs, a scan of the two lists suffices to add $w(x_i)$ to $w(x_{\text{succ}(i)})$, for all $i \in I$.

Step 4 can be implemented using $O(\text{sort}(N))$ I/Os, similar to Step 2, as it requires adding $\text{rank}(x_i)$ to $w(x_{\text{succ}(i)})$, for all $i \in L \setminus I$ such that $x_{\text{succ}(i)} \in I$.

THEOREM 10.14 *(Chiang et al., 1995). The list ranking problem on a list with N nodes can be solved using* $O(\text{sort}(N))$ *I/Os.*

Straightforward generalizations of the list ranking procedure discussed here allow the simultaneous ranking of multiple lists whose nodes are intermixed and the ranking of circular lists. For circular lists, the ranks are well-defined only if the weights together with the "addition" over these weights form an idempotent and commutative semigroup. This is the case, for example, if the addition of two weights returns the minimum of the two weights.

10.5.2   Trees, Forests, and the Euler Tour Technique

The Euler tour technique (Chiang et al., 1995) can be used to solve a number of fundamental problems on trees and forests. This section illustrates the use of this technique by discussing how to compute the distance of every vertex in a tree T from a "root vertex" $r \in T$, and how to identify the connected components of a forest.

10.5.2.1   Euler Tours

An Euler tour $\mathcal{E}(T)$ of a tree T is a circular list that contains two copies of each edge of T, which are directed in opposite directions. To define this tour, one considers the in-edges $y_0 x, y_1 x, \ldots, y_{k-1} x$ of each vertex $x \in T$ and defines the successor of each edge $y_i x$ in the Euler tour to be the edge $x y_{(i+1) \bmod k}$. This construction is illustrated in Figure 10.16a. When applied to a forest instead of a tree, this construction produces a collection of circular lists, one per tree; see Figure 10.16b.

The cost of this construction is easily seen to be $O(\operatorname{sort}(N))$ as it requires the duplication of each edge in T followed by sorting and scanning this edge set a constant number of times.

THEOREM 10.15   *(Chiang et al., 1995). An Euler tour of a tree or forest with N vertices can be computed using* $O(\operatorname{sort}(N))$ *I/Os.*

10.5.2.2   Distances in Trees

The distance of every vertex in T from a root vertex $r \in T$ is computed in two steps: The first step determines, for every edge of T, which of its endpoints is farther away from r. The second step uses this information to compute the distances from r to all vertices of T.

The first step computes an Euler tour $\mathcal{E}(T)$ of T, chooses an arbitrary edge $xr \in \mathcal{E}(T)$ and sets the successor of xr to nil. This makes $\mathcal{E}(T)$ noncircular; next the ranks of the nodes in $\mathcal{E}(T)$ are computed. For an edge xy, $\operatorname{dist}_T(r, y) > \operatorname{dist}_T(r, x)$ if and only if $\operatorname{rank}(xy) < \operatorname{rank}(yx)$; see Figure 10.17a. Thus,

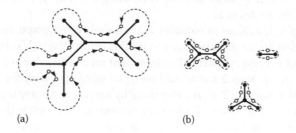

(a)   (b)

FIGURE 10.16   (a) The Euler tour of a tree and (b) the Euler tour of a forest.

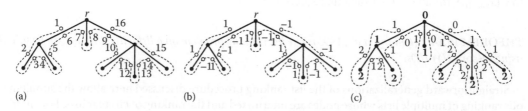

(a)   (b)   (c)

FIGURE 10.17   (a) The ranks of the Euler tour nodes; (b) weights for computing vertex depths; and (c) the weighted ranks represent vertex depths.

after sorting the edges in $\mathcal{E}(T)$ so that edges xy and yx are stored consecutively, for all $xy \in T$, a single scan of $\mathcal{E}(T)$ decides which endpoint of each edge xy is the parent of the other.

The second step now assigns weights to the edges in $\mathcal{E}(T)$ so that edge xy has weight 1 if $\text{dist}_T(r, x) < \text{dist}_T(r, y)$, and weight -1 otherwise; see Figure 10.17b. Then the ranks of the edges in $\mathcal{E}(T)$ w.r.t. these weights are computed. It is easy to see that $\text{dist}_T(r, x) = \text{rank}(yx)$, for every edge yx in $\mathcal{E}(T)$; see Figure 10.17c.

Since each step of this procedure takes $O(\text{sort}(N))$ I/Os, its total I/O-complexity is $O(\text{sort}(N))$.

THEOREM 10.16 *(Chiang et al., 1995). The distances from the root r of an N-vertex tree T to all other vertices in T can be computed using* $O(\text{sort}(N))$ *I/Os.*

10.5.2.3 Connected Components of a Forest

To identify the connected components of a forest F, the first step is again to construct an Euler tour $\mathcal{E}(F)$ of F. For every edge xy in $\mathcal{E}(F)$, its weight is chosen to be equal to $\min(x, y)$, assuming that every vertex of F is represented by a unique integer ID. List ranking applied to $\mathcal{E}(F)$ labels every edge in the circular list representing one tree T of F with the minimum ID of the vertices in T if min is used to add edge weights in $\mathcal{E}(F)$. Given this labeling, sorting and scanning the edges in $\mathcal{E}(F)$ and the vertices of F suffices to label every vertex in F with the ID of the minimum vertex in the tree containing it.

THEOREM 10.17 *(Chiang et al., 1995). The connected components of a forest with N vertices can be computed using* $O(\text{sort}(N))$ *I/Os.*

10.5.3 Time-Forward Processing and Greedy Graph Algorithms

Time-forward processing (Chiang et al., 1995; Arge, 2003) allows us to solve the following *evaluation problem* on directed acyclic graphs (DAGs): Let G be a DAG with vertices $x_1, x_2, \ldots, x_n$, numbered so that $i < j$, for every directed edge $x_i x_j \in G$; that is, the numbering $x_1, x_2, \ldots, x_n$ is a topological ordering of G.[*] Every vertex x_i of G has an associated label $\ell(x_i)$, which is to be replaced with a new label $\ell'(x_i)$ defined as a function of $\ell(x_i)$ and $\ell'(x_{j_1}), \ell'(x_{j_2}), \ldots, \ell'(x_{j_k})$, where $x_{j_1}, x_{j_2}, \ldots, x_{j_k}$ are the in-neighbors of x_i in G. For example, the evaluation of logical circuits can be expressed in this manner, where the initial label of every vertex determines the type of logical gate each vertex represents and the label to be computed for each vertex is the logical value output by this gate as a result of combining the logical values received from its in-neighbors.

In internal memory, this problem is straightforward to solve: Consider the vertices of G in topologically sorted order and compute, for each vertex x_i, its new label $\ell'(x_i)$ from its original label $\ell(x_i)$ and the ℓ'-labels of its in-neighbors. Time-forward processing uses the same strategy; the challenge is to ensure that every vertex x_i knows about the ℓ'-labels of its in-neighbors when computing $\ell'(x_i)$.

Arge's implementation of time-forward processing uses a priority queue Q to pass labels along the edges of G. At any point in time, Q stores an entry $\ell'(x_i)$ with priority j, for every edge $x_i x_j$ in G such that $\ell'(x_i)$ has been computed already, while $\ell'(x_j)$ has not. If there is more than one edge with tail x_i, then $\ell'(x_i)$ is stored several times in Q, once for each such edge, and each of these copies has a different priority, equal to the index of the head of the corresponding edge.

---

[*] In keeping with common practice in graph algorithms, n denotes the number of vertices, and m denotes the number of edges in a graph. The total input size is $N = n + m$ then.

At the time when a vertex x_i is processed, the entries corresponding to its in-edges are the entries with minimum priority in Q and can be retrieved using k DELETEMIN operations, where k is the in-degree of x_i. After using these values to compute $\ell'(x_i)$, x_i makes this label available to its out-neighbors by inserting it into Q, once per out-neighbor.

If Q is implemented as a buffer tree (see Section 10.3.4), every priority queue operation has a cost of $O\left(\frac{1}{B} \log_{M/B} \frac{m}{B}\right)$ I/Os amortized. Since there are two priority queue operations per edge, one INSERT and one DELETEMIN operation, the total cost of all priority queue operations is therefore $O(\text{sort}(m))$. Inspecting the vertices to retrieve their ℓ-labels and adjacency lists costs $O((n + m)/B)$ I/Os if the vertices and their adjacency lists are stored in the order the vertices are inspected. Since this order is known in advance, the latter is easily ensured using $O(\text{sort}(n) + \text{sort}(m))$ I/Os.

THEOREM 10.18   *(Chiang et al., 1995; Arge, 2003). A DAG with n vertices and m edges can be evaluated using* $O(\text{sort}(n) + \text{sort}(m))$ *I/Os, provided a topological ordering of its vertices is known.*

The assumption that the vertices of G are numbered in topologically sorted order is crucial, as no I/O-efficient algorithm for topological sorting is known to date. Nevertheless, time-forward processing is a very useful technique, as many algorithms that use time-forward processing to evaluate DAGs create these DAGs as part of their computation and can use their knowledge of how the DAG was constructed to obtain a topological ordering of the DAG efficiently. The following algorithm for computing a maximal independent set of an undirected graph G (Zeh, 2002b) illustrates this.

In internal memory, this problem is easily solved by inspecting the vertices of G in an arbitrary order, and adding a vertex x to I if none of its neighbors already belongs to I at the time x is inspected. This algorithm can be simulated in external memory using the time-forward processing technique. The first step is to choose an arbitrary order in which to inspect the vertices and number the vertices in this order. Then the edges of G are directed so that every edge has its endpoint with lower number as its tail. This transforms G into a DAG, and the chosen numbering of the vertices is a topological ordering of G. Now the vertices in G are inspected in the chosen order, and a vertex x is added to I if none of its in-neighbors is in I. This correctly computes an independent set because, at the time each vertex is inspected, only its in-neighbors can be in I; there is no need to inspect its out-neighbors. Time-forward processing provides every vertex with the needed information about the membership of its in-neighbors in I.

THEOREM 10.19   *(Zeh, 2002b). A maximal independent set of a graph G can be found using* $O(\text{sort}(n) + \text{sort}(m))$ *I/Os.*

Zeh (2002b) also discusses further applications of this technique to coloring graphs of bounded degree and computing a maximal matching.

10.5.4   Connected Components

The second example illustrating the use of graph contraction is the computation of the connected components of a graph. A number of algorithms for this problem have been proposed (Chiang et al., 1995; Munagala and Ranade, 1999; Abello et al., 2002). The algorithm discussed here is essentially the one by (Chiang et al., 1995).

The classical approach to solving this problem (Tarjan, 1972) uses depth-first search (DFS) or breadth-first search (BFS). Section 10.5.5.2 discusses a simple BFS-algorithm with I/O-complexity $O(n+\text{sort}(m))$. Thus, if $n \leq m/B$, the connected components of G can be computed using $O(\text{sort}(m))$

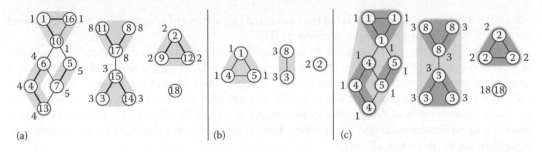

(a)   (b)   (c)

FIGURE 10.18 Computing the connected components of a graph. (a) The graph G. Vertices are labeled with their IDs. The edges chosen in Step 1 are shown in bold. The connected components of the resulting forest G' are highlighted in gray; these components are identified by the component labels assigned to the vertices, shown beside the vertices. (b) The graph G'' constructed in Step 3. Step 4 assigns component labels shown beside the vertices to the vertices of G''. (c) The final labeling of the components computed in Step 5. The label shown inside each vertex is the component label assigned to it in Step 2, the label beside each vertex is its final label identifying the connected component of G containing the vertex.

I/Os. If $n > m/B$, graph contraction is used to reduce this to the case when $n \le m/B$. This approach is again inspired by parallel algorithms for computing connected components.

Assuming again that all vertices have unique IDs, the following five-step algorithm computes the connected components of G:

1. Remove all isolated (i.e., degree-0) vertices from G. For every remaining vertex x, find its neighbor y with minimum ID and add edge xy to an edge list E'. See Figure 10.18a.
2. Compute the connected components of the graph $G' = (V, E')$. See Figure 10.18a.
3. Construct a new graph G'' that contains one vertex v_C per connected component C of G', and an edge $v_{C_1} v_{C_2}$ between two such vertices v_{C_1} and v_{C_2} if there is an edge $xy \in G$ such that $x \in C_1$ and $y \in C_2$. See Figure 10.18b.
4. Recursively compute the connected components of G''. See Figure 10.18b.
5. Label every nonisolated vertex $x \in G$ contained in a component C of G' with the label assigned to v_C in Step 4. For every isolated vertex in G, choose its component label to be equal to its own ID. This is the final labeling representing the connected components of G. See Figure 10.18c.

Each step of this algorithm, excluding the recursive invocation in Step 4, can be implemented using $O(\text{sort}(n) + \text{sort}(m))$ I/Os. Step 1 creates two directed copies xy and yx of every edge $xy \in G$, sorts the resulting list of directed edges by their tails, and scans the sorted list to choose the minimum neighbor of each vertex.

The resulting graph G' can be shown to be a forest (after removing duplicate edges chosen by both their endpoints). Hence, Theorem 10.17 can be used to compute its connected components using $O(\text{sort}(n))$ I/Os.

The construction of graph G'' in Step 3 requires replacing every vertex x with the ID of its connected component in G', which can be done by sorting and scanning the vertex and edge sets of G a constant number of times. Isolated vertices can be removed in the process.

After Step 4, every vertex in G'' stores the ID of the connected component containing it. Every nonisolated vertex x in G stores the ID of the vertex in G'' representing its connected component in G'. Hence, it suffices to sort the vertices of G by the IDs of their representatives in G'', and the vertices in G'' by their own IDs. Then a scan of these two sorted vertex lists suffices to label every nonisolated vertex $x \in G$ with the ID of the connected component of G containing it. Every isolated vertex in G receives its own unique component ID.

As for the cost of the recursive call, it is easy to see that every connected component of G' contains at least two vertices of G. Since G'' contains one vertex for each of these connected components, it contains at most $n/2$ vertices and no more than m edges. Thus, after at most $\log(nB/m)$ recursive calls, the algorithm has produced a graph with at most m/B vertices and at most m edges. The connected components of this graph can be computed using $O(\text{sort}(m))$ I/Os using breadth-first search, and no further recursion is necessary. Hence, the total cost of this connected components algorithm is $O(\text{sort}(m) \log(nB/m))$.

THEOREM 10.20 *(Chiang et al., 1995). The connected components of a graph with n vertices and m edges can be computed using* $O(\text{sort}(n) + \text{sort}(m) \log(nB/m))$ *I/Os.*

The above algorithm can be refined to achieve I/O-complexity $O(\text{sort}(n)+\text{sort}(m) \log \log(nB/m))$ (Munagala and Ranade, 1999), and it can be extended to compute a minimum spanning tree of a weighted graph in the same complexity (Chiang et al., 1995; Arge et al., 2004). Using randomization, both, connected components and minimum spanning trees, can be computed using $O(\text{sort}(n) + \text{sort}(m))$ I/Os with high probability (Chiang et al., 1995; Abello et al., 2002).

10.5.5 Breadth-First Search and Single-Source Shortest Paths in Undirected Graphs

Breadth-first search and single-source shortest paths (SSSP) in undirected graphs with nonnegative edge weights are two problems for which considerable progress toward I/O-efficient solutions has been made recently (Kumar and Schwabe, 1996; Munagala and Ranade, 1999; Meyer, 2001; Mehlhorn and Meyer, 2002; Meyer and Zeh, 2003, 2006). The first algorithm discussed here is a general shortest-path algorithm for undirected graphs due to (Kumar and Schwabe, 1996). This is followed by a discussion of two algorithms for breadth-first search that are faster than the shortest-path algorithm.

10.5.5.1 Single-Source Shortest Paths

In internal memory, the standard algorithm for solving SSSP on graphs with nonnegative edge weights is Dijkstra's algorithm (Dijkstra, 1969). This algorithm maintains tentative distances $d(x)$ of the vertices in G, which are upper bounds on their actual distances $\text{dist}(s, x)$ from s. A vertex is said to be *finished* if its distance from s is known, and active otherwise. An I/O-efficient implementation of Dijkstra's algorithm requires

1. An I/O-efficient priority queue
2. An I/O-efficient means to retrieve the adjacency lists of visited vertices
3. An I/O-efficient means to decide whether a neighbor of a visited vertex is active or finished

The priority queue used by the algorithm discussed here is the tournament tree mentioned in Section 10.3.4. To ensure that the adjacency list of every vertex can be retrieved efficiently, the edges of G are sorted so that the edges in the adjacency list of each vertex x are stored consecutively on disk and that x knows the disk location of the first edge in its adjacency list. The cost of retrieving x's adjacency list is $O(1 + \deg(x)/B)$ then; summed over all vertices, this is $O(n + m/B)$.

The third point is addressed indirectly: When visiting a vertex x, an UPDATE operation is performed on *every* neighbor of x, active or finished. This eliminates the necessity to test the status of each neighbor; however, an UPDATE on a finished neighbor y reinserts y into Q and, thus, causes it to be visited for a second time unless there is a way to remove y from Q before it is visited again.

The method for removing reinserted vertices uses a second priority queue Q' to record UPDATE operations vertices perform on their neighbors. Whenever a vertex x updates the tentative distance of one of its neighbors to some value d, x is inserted into Q' with priority d. Priority queue Q' needs to be implemented as a buffer tree or modified tournament tree so it can store multiple copies of a vertex x with different priorities, one per update x performs.

The information recorded in Q' is used to prevent reinserted vertices to be visited. Every step of Dijkstra's algorithm now retrieves the minimum entries from both Q and Q'. Let these entries be (x, p_x) and (y, p_y), respectively. If $p_x \leq p_y$, (y, p_y) is put back into Q' and vertex x is visited as usual. If, on the other hand, $p_y < p_x$, then (x, p_x) is put back into Q and a DELETE(y) operation is performed on Q. As Kumar and Schwabe (1996) show, this ensures that every vertex is visited only once, as long as no two adjacent vertices have the same distance from the source. Zeh (2002a) sketches a method for dealing with vertices that have the same distance from s.

The cost of this algorithm is equal to the cost of retrieving all adjacency lists, plus the cost of the priority queue operations on Q and Q'. The adjacency lists can be retrieved using $O(n + m/B)$ I/Os. Since $O(m)$ operations are performed on Q and Q', each of which costs at most $O((1/B) \log(n/B))$ I/Os, the cost of all priority queue operations is $O((m/B) \log(n/B))$. This gives the following result.

THEOREM 10.21 *(Kumar and Schwabe, 1996). The single-source shortest-path problem on an undirected graph with n vertices, m edges, and nonnegative edge weights can be solved using $O(n + (m/B) \log_2(n/B))$ I/Os.*

10.5.5.2 Breadth-First Search

While Theorem 10.21 can be used to perform BFS in an undirected graph, there is a simpler and slightly faster algorithm for this special case of shortest-path computations due to Munagala and Ranade (1999). The algorithm builds the BFS-tree level by level; that is, it computes vertex sets $L(0), L(1), \ldots$, where $L(i)$ is the set of vertices that are at distance i from the source vertex s. After setting $L(0) = \{s\}$, each subsequent level $L(i + 1)$ is computed from the previous two levels $L(i - 1)$ and $L(i)$. In particular, $L(i + 1)$ consists of all neighbors of vertices in $L(i)$ that are not in levels $L(i - 1)$ or $L(i)$.

This computation of $L(i + 1)$ from the two previous levels requires retrieving the adjacency lists of all vertices in $L(i)$, sorting the resulting list $X(i)$ of neighbors of vertices in $L(i)$, and scanning $X(i)$, $L(i - 1)$, and $L(i)$ to remove duplicates and vertices that occur already in $L(i - 1)$ or $L(i)$ from $X(i)$. The vertices that remain form $L(i + 1)$, and the algorithm proceeds to computing the next level as long as $L(i + 1) \neq \emptyset$.

Since every adjacency list is retrieved once, the cost of retrieving adjacency lists is $O(n + m/B)$. The total size of all lists $X(0), X(1), \ldots$ is $2m$, so that sorting and scanning these lists takes $O(\text{sort}(m))$ I/Os. Every level $L(i)$ is involved in the construction of only three levels; hence, the total cost of scanning these levels is $O(n/B)$. This proves the following result.

THEOREM 10.22 *(Munagala and Ranade, 1999). Breadth-first search in an undirected graph with n vertices and m edges takes $O(n + \text{sort}(m))$ I/Os.*

10.5.5.3 A Faster BFS-Algorithm

The algorithms from Sections 10.5.5.1 and 10.5.5.2 perform well on dense graphs, but the $O(n)$-term dominates their I/O-complexity on sparse graphs, which means that they are only marginally faster than the standard internal-memory algorithms for BFS and SSSP in this case. This section uses BFS to illustrate a *clustering technique* that has been shown to be useful to speed up shortest-path and BFS computations on undirected graphs.

The BFS-algorithm from Section 10.5.5.2 would have I/O-complexity $O(\text{sort}(m))$ if it didn't spend $O(n+m/B)$ I/Os to retrieve adjacency lists. The improved algorithm by Mehlhorn and Meyer, (2002) is identical to this algorithm, except for its use of a more efficient method to retrieve adjacency lists. This method uses a partition of the given graph G into $O(n/\mu)$ subsets $V_1, V_2, \ldots, V_q$, called *vertex clusters*, such that any two vertices in the same set V_i have distance at most μ from each other. Such a clustering is fairly easy to obtain in $O(\text{MST}(n, m))$ I/Os, where $\text{MST}(n, m)$ is the cost of computing a (minimum) spanning tree of G (see Section 10.5.4): compute a spanning tree T of G and an Euler tour $\mathcal{E}(T)$ of T, partition $\mathcal{E}(T)$ into $2n/\mu$ subtours of length μ, and define the vertices visited by each tour to be in the same cluster. Vertices visited by more than one subtour are assigned arbitrarily to one of the corresponding clusters.

Given such a clustering, the adjacency lists of G are arranged so that the adjacency lists of the vertices in each cluster V_i are stored consecutively. Let E_i denote the concatenation of the adjacency lists of the vertices in V_i. This concatenation is called an *edge cluster*. In addition to arranging adjacency lists in this manner, the algorithm maintains a *hot pool* H of adjacency lists of vertices that will be visited soon by the algorithm. Viewing every edge xy in the adjacency list of a vertex x as directed from x to y, the edges in H are kept sorted by their tails.

Now consider the construction of level $L(i+1)$ from level $L(i)$ in the BFS-algorithm. The first step is the construction of the set $X(i)$. This is done as follows: First $L(i)$ and H are scanned to identify all vertices in $L(i)$ whose adjacency lists are not in H. Let $L'(i)$ be the set of these vertices. The vertices in $L'(i)$ are now sorted by the IDs of the clusters containing them, and the resulting list is scanned to identify all vertex clusters containing vertices in $L'(i)$. For each such cluster V_j, the corresponding edge cluster E_j is retrieved and appended to a list $Y(i)$. The edges in $Y(i)$ are sorted by their tails and then merged into H. Now H contains the adjacency lists of all vertices in $L(i)$ (as well as possibly other adjacency lists belonging to edge clusters loaded into H during the current or any of the previous iterations). Hence, it suffices to scan $L(i)$ and H again to retrieve the adjacency lists of all vertices in $L(i)$ and delete them from H. Adjacency lists in H belonging to vertices not in $L(i)$ remain in H until their corresponding vertices are visited during the construction of a subsequent level. This finishes the construction of $X(i)$, and the construction of $L(i+1)$ continues from this point as described in Section 10.5.5.2.

As already observed, the cost of this algorithm is $O(\text{sort}(m))$, excluding the cost of constructing sets $X(0), X(1), \ldots$. The construction of the set $X(i)$ requires scanning the list $L(i)$ twice. Apart from that, H is scanned a constant number of times, the vertices in $L'(i)$ are sorted and scanned a constant number of times, and a number of edge clusters are loaded into $Y(i)$, which is then sorted and merged into H. Since the total size of all levels $L(i)$ and all sets $L'(i)$ is $O(n)$, sorting and scanning these sets takes $O(\text{sort}(n))$ I/Os. Sorting and scanning $Y(i)$ takes $O(\text{sort}(m))$ I/Os because every edge of G belongs to exactly one such set $Y(i)$.

Here now are the two key observations: Retrieving edge clusters and placing them into $Y(i)$ takes only $O(q + E/B) = O(n/\mu + m/B)$ I/Os, where q is the number of clusters. The cost of scanning H, summed over the construction of all levels, is no more than $O(\mu m/B)$. To see why this is true, consider the vertex $x \in L(i)$ that causes an edge cluster E_j to be loaded into H. Since every vertex in V_j has distance at most μ from x, all vertices in V_j belong to levels $L(i)$ through $L(i + \mu)$; that is, every edge in E_j is scanned at most $O(\mu)$ times before it is removed from H.

In summary, the cost of the BFS-algorithm, excluding the cost of clustering, is $O(n/\mu + \mu m/B + \text{sort}(m))$. The trade-off between the first two terms is balanced if $\mu = \sqrt{nB/m}$, which leads to a complexity of $O(\sqrt{nm/B} + \text{sort}(m))$ I/Os. Clustering adds another $O(\text{MST}(n, m))$ I/Os to the cost of the algorithm, which proves the following result.

THEOREM 10.23 *(Mehlhorn and Meyer, 2002). Breadth-first search in an undirected graph G with n vertices and m edges takes* $O(\sqrt{nm/B} + \text{MST}(n, m))$ *I/Os, where* $\text{MST}(n, m)$ *is the cost of computing a (minimum) spanning tree of G.*

The techniques used to obtain Theorem 10.23 have also been shown to be effective to obtain improved shortest-path algorithms for undirected graphs with nonnegative edge weights (Meyer and Zeh, 2003, 2006).

10.6   Research Issues and Summary

In this chapter we have discussed external-memory algorithms for only a few fundamental problems in order to highlight the most fundamental techniques for developing I/O-efficient algorithms and data structures. However, I/O-efficient algorithms have been developed for a large number of other problems; see e.g., (Vitter, 2001; Arge, 2002). Nevertheless, a number of (even fundamental) problems remain open. For example, space-efficient data structures with an $O(\log_B N)$ query bound still need the be found for many higher-dimensional problems, including the orthogonal range searching problem in more than two dimensions. Obtaining a priority queue that supports a DECREASEKEY operation (without knowing the current key of the element being updated) in $O(\frac{1}{B} \log_{M/B} \frac{N}{B})$ I/Os also remains open. As for graph algorithms, the most important open problem is the development of an efficient DFS-algorithm; as a host of classical graph algorithms crucially rely on DFS as their means to explore the graph, such an algorithm would immediately translate those algorithms into I/O-efficient ones. Another major frontier is the development of algorithms for directed graphs, as even such simple problems as deciding whether a given vertex can reach another given vertex have no I/O-efficient solutions at this point.

10.7   Further Information

In this chapter, we discussed some of the most fundamental techniques for developing I/O-efficient algorithms and data structures through a discussion of a number of fundamental problems. Details of the algorithms, as well as a more complete overview of I/O-efficient results and techniques, can be found in recent surveys and lecture notes (Vitter, 2001; Arge, 2002, 2005; Zeh, 2002a).

Research on I/O-efficient algorithms and data structures is published in many algorithm journals, including *Algorithmica, Journal of the ACM, Journal of Algorithms, SIAM Journal on Computing, Journal on Computer and Systems Sciences,* and *Theoretical Computer Science,* as well as in the proceedings of many algorithms conferences, including *ACM Symposium on Theory of Computation* (STOC), *IEEE Symposium on Foundations of Computer Science* (FOCS), *ACM-SIAM Symposium on Discrete Algorithms* (SODA), *European Symposium on Algorithms* (ESA), *International Colloquium on Automata, Languages, and Programming* (ICALP), *ACM Symposium on Parallel Algorithms and Architectures* (SPAA), and *ACM Symposium on Computational Geometry* (SCG); more practical engineering results often appear in *ACM Journal on Experimental Algorithmics* and in the proceedings of *Workshop on Algorithm Engineering and Experimentation* (ALENEX) or the applied track of ESA. External-memory algorithms and data structures (although often of a heuristic nature) are also published in the database community, including the proceedings of *International Conference on Very Large Databases* (VLDB), *SIGMOD International Conference on Management of Data* (SIGMOD), and *ACM Symposium on Principles of Database Systems* (PODS).

Defining Terms

B-tree (multiway tree): A balanced search tree all of whose leaves have the same depth and whose internal nodes have fan-out $\Theta(B)$.

Bootstrapping: The use of a static structure, made dynamic using global rebuilding, on small point sets, in order to obtain efficient secondary structures to be associated with the nodes of a dynamic structure for the same problem.

Buffer tree: A B-tree of fan-out $\Theta(M/B)$ that performs updates and queries lazily by storing them in buffers associated with its nodes and processing the operations in a buffer whenever a large enough number of operations have been collected. This tree outperforms the B-tree on offline problems.

Distribution sweeping: A combination of multiway distribution and the plane sweep paradigm. At each level of the distribution, a plane sweep is performed to determine all interactions between objects distributed to different lists.

Euler tour: A list representing a depth-first traversal of a tree or forest.

External-memory (I/O-efficient) algorithm: Algorithms designed to minimize the number of disk accesses performed in large-scale computations.

Global rebuilding: A technique for dynamizing static data structures by rebuilding them completely whenever a sufficient number of updates have been collected.

Graph clustering: A technique for partitioning the vertices of the graph into groups so that, in the algorithm at hand, the vertices in each such group can be processed simultaneously or at least in short sequence.

Graph contraction: A technique for reducing the number of vertices in a graph by replacing each of a collection of vertex groups with a single vertex. Recursive application leads to reduced-size graphs that can be processed in memory.

I/O-model: Model used for the design of I/O-efficient algorithms, which enforces blockwise transfer of data between the two memory levels and defines the number of such transfers as the algorithm's complexity.

Logarithmic method: A technique for dynamizing static data structures by maintaining a logarithmic number of exponentially growing structures. These structures are rebuilt periodically, with decreasing frequency as their sizes increase.

Multiway merge/distribution: Recursive merging or distribution of multiple lists at a time.

Persistence: A space-efficient technique for maintaining information about the evolution of a dynamic data structure under a sequence of updates so that queries can be performed on any previous version of the structure.

Time-forward processing: A technique for evaluating directed acyclic graphs.

Weight balancing: An alternate rebalancing strategy for B-trees, which ensures that the frequency of node splits or fusions applied to a given node is inversely proportional to the size of the node's subtree.

Acknowledgments

The work of the first author was supported in part by the US Army Research Office through grant W911NF-04-01-0278, by an Ole Rømer Scholarship from the Danish National Science Research Council, a NABIIT grant from the Danish Strategic Research Council, and by the Danish National Research Foundation. The work of the second author was supported by the Natural Sciences and Engineering Research Council of Canada and the Canadian Foundation for Innovation.

References

J. Abello, A. L. Buchsbaum, and J. Westbrook. A functional approach to external graph algorithms. *Algorithmica*, 32(3):437–458, 2002.

A. Aggarwal and J. S. Vitter. The input/output complexity of sorting and related problems. *Communications of the ACM*, 31(9):1116–1127, 1988.

P. K. Agarwal, L. Arge, O. Procopiuc, and J. S. Vitter. A framework for index bulk loading and dynamization. In *Proceedings of the 28th Annual International Colloquium on Automata, Languages, and Programming, LNCS* 2076, pp. 115–127, Springer-Verlag, Berlin, Germany, 2001a.

P. K. Agarwal, M. de Berg, J. Gudmundsson, M. Hammer, and H. J. Haverkort. Box-trees and R-trees with near-optimal query time. In *Proceedings of the 17th Annual ACM Symposium on Computational Geometry*, pp. 124–133, Medford, MA, 2001b.

L. Arge. External memory data structures. In J. Abello, P. M. Pardalos, and M. G. C. Resende, editors, *Handbook of Massive Data Sets*, pp. 313–358. Kluwer Academic Publishers, Dordrecht, the Netherlands, 2002.

L. Arge. The buffer tree: A technique for designing batched external data structures. *Algorithmica*, 37(1): 1–24, 2003.

L. Arge. External-memory geometric data structures. Lecture notes available at http://www.daimi.au. dk/~large/ioS07/ionotes.pdf, 2005.

L. Arge and J. Vahrenhold. I/O-efficient dynamic planar point location. *Computational Geometry: Theory and Applications*, 29(2):147–162, 2004.

L. Arge and J. S. Vitter. Optimal external memory interval management. *SIAM Journal on Computing*, 32(6):1488–1508, 2003.

L. Arge, P. Ferragina, R. Grossi, and J. Vitter. On sorting strings in external memory. In *Proceedings of the 29th Annual ACM Symposium on Theory of Computation*, pp. 540–548, El Paso, TX, 1997.

L. Arge, O. Procopiuc, S. Ramaswamy, T. Suel, and J. S. Vitter. Theory and practice of I/O-efficient algorithms for multidimensional batched searching problems. In *Proceedings of the 9th Annual ACM-SIAM Symposium on Discrete Algorithms*, pp. 685–694, San Francisco, CA, 1998a.

L. Arge, D. E. Vengroff, and J. S. Vitter. External-memory algorithms for processing line segments in geographic information systems. *Algorithmica*, 47(1):1–25, 1998b.

L. Arge, V. Samoladas, and J. S. Vitter. On two-dimensional indexability and optimal range search indexing. In *Proceedings of the 18th ACM Symposium on Principles of Database Systems*, pp. 346–357, Philadelphia, PA, 1999.

L. Arge, K. H. Hinrichs, J. Vahrenhold, and J. S. Vitter. Efficient bulk operations on dynamic R-trees. *Algorithmica*, 33(1):104–128, 2002.

L. Arge, A. Danner, and S-H. Teh. I/O-efficient point location using persistent B-trees. In *Proceedings of Workshop on Algorithm Engineering and Experimentation*, pp. 82–92, 2003.

L. Arge, G. S. Brodal, and L. Toma. On external-memory MST, SSSP and multi-way planar graph separation. *Journal of Algorithms*, 53(2):186–206, 2004.

R. Bayer and E. McCreight. Organization and maintenance of large ordered indexes. *Acta Informatica*, 1:173–189, 1972.

B. Becker, S. Gschwind, T. Ohler, B. Seeger, and P. Widmayer. An asymptotically optimal multiversion B-tree. *VLDB Journal*, 5(4):264–275, 1996.

J. L. Bentley. Multidimensional binary search trees used for associative searching. *Communications of the ACM*, 18:509–517, 1975.

J. L. Bentley. Decomposable searching problems. *Information Processing Letters*, 8(5):244–251, 1979.

M. Blum, R. W. Floyd, V. Pratt, R. L. Rivest, and R. E. Tarjan. Time bounds for selection. *Journal of Computer and System Sciences*, 7(4):448–461, 1973.

B. Chazelle. Filtering search: A new approach to query-answering. *SIAM Journal on Computing*, 15(3): 703–724, 1986.

Y.-J. Chiang, M. T. Goodrich, E. F. Grove, R. Tamassia, D. E. Vengroff, and J. S. Vitter. External-memory graph algorithms. In *Proceedings of the 6th Annual ACM-SIAM Symposium on Discrete Algorithms*, pp. 139–149, San Francisco, CA, 1995.

D. Comer. The ubiquitous B-tree. *ACM Computing Surveys*, 11(2):121–137, 1979.

E. W. Dijkstra. A note on two problems in connection with graphs. *Numerische Mathematik*, 1(1):269–271, 1959.

J. R. Driscoll, N. Sarnak, D. D. Sleator, and R. Tarjan. Making data structures persistent. *Journal of Computer and System Sciences*, 38:86–124, 1989.

R. Fadel, K. V. Jakobsen, J. Katajainen, and J. Teuhola. Heaps and heapsort on secondary storage. *Theoretical Computer Science*, 220(2):345–362, 1999.

M. T. Goodrich, J.-J. Tsay, D. E. Vengroff, and J. S. Vitter. External-memory computational geometry. In *Proceedings of the 1993 IEEE 34th Annual Foundations of Computer Science*, pp. 714–723, Washington, DC, 1993.

R. Grossi and G. F. Italiano. Efficient cross-tree for external memory. In J. Abello and J. S. Vitter, editors, *External Memory Algorithms and Visualization*, pp. 87–106. American Mathematical Society, Boston, MA, 1999a.

R. Grossi and G. F. Italiano. Efficient splitting and merging algorithms for order decomposable problems. *Information and Computation*, 154(1):1–33, 1999b.

J. M. Hellerstein, E. Koutsoupias, and C. H. Papadimitriou. On the analysis of indexing schemes. In *Proceedings of the 16th ACM Symposium on Principles of Database Systems*, pp. 249–256, Tucson, AZ, 1997.

S. Huddleston and K. Mehlhorn. A new data structure for representing sorted lists. *Acta Informatica*, 17: 157–184, 1982.

K. V. R. Kanth and A. K. Singh. Optimal dynamic range searching in non-replicating index structures. In *Proceedings of the 7th International Conference on Database Theory, LNCS* 1540, pp. 257–276, Springer-Verlag, Berlin, 1999.

D. E. Knuth. *Sorting and Searching*, volume 3 of *The Art of Computer Programming*. Addison-Wesley, Reading, MA, 2nd edition, 1998.

V. Kumar and E. Schwabe. Improved algorithms and data structures for solving graph problems in external memory. In *Proceedings of the 8th IEEE Symposium on Parallel and Distributed Processing*, pp. 169–177, Washington, DC, 1996.

E. M. McCreight. Priority search trees. *SIAM Journal on Computing*, 14(2):257–276, 1985.

K. Mehlhorn and U. Meyer. External-memory breadth-first search with sublinear I/O. In *Proceedings of the 10th Annual European Symposium on Algorithms, LNCS* 2461, pp. 723–735. Springer-Verlag, Berlin, 2002.

U. Meyer. External memory BFS on undirected graphs with bounded degree. In *Proceedings of the 12th Annual ACM-SIAM Symposium on Discrete Algorithms*, pp. 87–88, Washington, DC, 2001.

U. Meyer and N. Zeh. I/O-efficient undirected shortest paths. In *Proceedings of the 11th Annual European Symposium on Algorithms, LNCS* 2832, pp. 434–445. Springer-Verlag, Berlin, 2003.

U. Meyer and N. Zeh. I/O-efficient undirected shortest paths with unbounded edge lengths. In *Proceedings of the 14th Annual European Symposium on Algorithms, LNCS* 4168, pp. 540–551. Springer-Verlag, Berlin 2006.

K. Munagala and A. Ranade. I/O-complexity of graph algorithms. In *Proceedings of the 10th Annual ACM-SIAM Symposium on Discrete Algorithms*, pp. 687–694, Zurich, Switzerland 1999.

M. H. Overmars. *The Design of Dynamic Data Structures, LNCS* 156, Springer-Verlag, New York, 1983.

F. P. Preparata and M. I. Shamos. *Computational Geometry: An Introduction*. Springer-Verlag, New York, 1985.

O. Procopiuc, P. K. Agarwal, L. Arge, and J. S. Vitter. Bkd-tree: A dynamic scalable kd-tree. In *Proceedings of the 8th International Symposium on Spatial and Temporal Databases, LNCS* 2750, pp. 46–65, Springer-Verlag, Berlin 2003.

J.T. Robinson. The K-D-B tree: A search structure for large multidimensional dynamic indexes. In *Proceedings of the 1981 ACM SIGMOD International Conference on Management of Data*, pp. 10–18, Ann Arbor, MI, 1981.

S. Subramanian and S. Ramaswamy. The P-range tree: A new data structure for range searching in secondary memory. In *Proceedings of the 6th Annual ACM-SIAM Symposium on Discrete Algorithms*, pp. 378–387, San Francisco, CA, 1995.

R. E. Tarjan. Depth-first search and linear graph algorithms. *SIAM Journal on Computing*, 1(2):146–160, 1972.

J. van den Bercken, B. Seeger, and P. Widmayer. A generic approach to bulk loading multidimensional index structures. In *Proceedings of the 23rd International Conference on Very Large Databases*, pp. 406–415, San Francisco, CA, 1997.

M. J. van Kreveld and M. H. Overmars. Divided kd-trees. *Algorithmica*, 6:840–858, 1991.

P. J. Varman and R. M. Verma. An efficient multiversion access structure. *IEEE Transactions on Knowledge and Data Engineering*, 9(3):391–409, 1997.

J. S. Vitter. External memory algorithms and data structures: Dealing with MASSIVE data. *ACM Computing Surveys*, 33(2):209–271, 2001.

N. Zeh. I/O-efficient graph algorithms. Lecture notes from EEF Summer School on Massive Datasets available at http://users.cs.dal.ca/~nzeh/Teaching/Summer2007/6104/Notes/zeh02.pdf, 2002a.

N. Zeh. I/O-efficient algorithms for shortest path related problems. PhD thesis, School of Computer Science, Carleton University, Ottawa, ON, 2002b.

11

Average Case Analysis of Algorithms*

Wojciech Szpankowski
Purdue University

11.1 Introduction

An algorithm is a finite set of instructions for a treatment of data to meet some desired objectives. The most obvious reason for analyzing algorithms and data structures is to discover their characteristics in order to evaluate their suitability for various applications, or to compare them with other algorithms for the same application. Needless to say, we are interested in efficient algorithms in order to use efficiently such scarce resources as computer space and time.

Most often algorithm designs aim to optimize the asymptotic worst-case performance, as popularized by Aho et al. [2]. Insightful, elegant, and generally useful constructions have been set up in this endeavor. Along these lines, however, the design of an algorithm is usually targeted at coping efficiently sometimes with unrealistic, even pathological inputs and the possibility is neglected that a

* This research was partially supported by NSF Grants NCR-9206315, 9415491, and CCR-9804760, and NATO Collaborative Grant CRG.950060.

simpler algorithm that works fast "on average" might perform just as well, or even better in practice. This alternative solution, called also a probabilistic approach, became an important issue two decades ago when it became clear that the prospects for showing the existence of polynomial time algorithms for NP-hard problems, were very dim. This fact, and the apparently high success rate of heuristic approaches to solving certain difficult problems, led Richard Karp [34] to undertake a more serious investigation of probabilistic analysis of algorithms. (But, one must realize that there are problems which are also hard "on average" as shown by Levin [42].) In the last decade we have witnessed an increasing interest in the probabilistic (also called average case) analysis of algorithms, possibly due to the high success rate of randomized algorithms for computational geometry, scientific visualization, molecular biology, etc. (e.g., see [46,60]). Finally, we should point out that probabilistic analysis often depends on the input distribution which is usually unknown up front, and this might lead to unrealistic assumptions.

The average case analysis of algorithms can be roughly divided into two categories, namely: analytic in which complex analysis plays a pivotal role, and probabilistic in which probabilistic and combinatorial techniques dominate. The former was popularized by Knuth's monumental three volumes *The Art of Computer Programming* [39–41], whose prime goal was to accurately predict the performance characteristics of an algorithm. Such an analysis often sheds light on properties of computer programs and provides useful insights of combinatorial behaviors of such programs. Probabilistic methods were introduced by Erdös and Rényi and popularized by Alon and Spencer in their book [3]. In general, nicely structured problems are amiable to an analytic approach that usually gives much more precise information about the algorithm under consideration. On the other hand, structurally complex problems are more likely to be first solved by a probabilistic tool that later could be further enhanced by a more precise analytical approach. The average case analysis of algorithms, as a discipline, uses a number of branches of mathematics: combinatorics, probability theory, graph theory, real and complex analysis, and occasionally algebra, geometry, number theory, operations research, and so forth.

In this chapter, we choose one facet of the theory of algorithms, namely that of algorithms and data structures on words (strings) and present a brief exposition on certain analytic and probabilistic methods that have become popular. Our choice of the area stems from the fact that there has been a resurgence of interest in string algorithms due to several novel applications, most notably in computational molecular biology and data compression. Our choice of methods covered here is aimed at closing a gap between analytic and probabilistic methods. There are excellent books on analytic methods (cf. Knuth's three volumes [39–41], Sedgewick and Flajolet [49]) and probabilistic methods (cf. Alon and Spencer [3], Coffman and Lueker [10], and Motwani and Raghavan [46]), however, remarkably very few books have been dedicated to both analytic and probabilistic analysis of algorithms (with possible exceptions of Hofri [28] and Mahmoud [44]). Finally, before we launch our journey through probabilistic and analytic methods, we should add that in recent years several useful surveys on analysis of algorithms have been published. We mentioned here: Frieze and McDiarmid [22], Karp [35], Vitter and Flajolet [59], and Flajolet [16].

This chapter is organized as follows: In the next section we describe some algorithms and data structures on words (e.g., digital trees, suffix trees, edit distance, Lempel–Ziv data compression algorithm, etc.) that we use throughout to illustrate our ideas and methods of analysis. Then, we present probabilistic models for algorithms and data structures on words together with a short review from probability and complex analysis. Section 11.4 is devoted to probabilistic methods and discusses the sieve method, first and second moment methods, subadditive ergodic theorem, techniques of information theory (e.g., entropy and its applications), and large deviations (i.e., Chernoff's bound) and Azuma's type inequality. Finally, in the last section we concentrate on analytic techniques in which complex analysis plays a pivotal role. We shall discuss analytic techniques for recurrences and asymptotics (i.e., Rice's formula, singularity analysis, etc.), Mellin transform and its applications, and poissonization and depoissonization. In the future we plan to expand this chapter to a book.

11.2 Data Structures and Algorithms on Words

As mentioned above, in this survey we choose one facet of the theory of algorithms, namely that of data structures and algorithms on words (strings) to illustrate several probabilistic and analytic techniques of the analysis of algorithms. In this section, we briefly recall certain data structures and algorithms on words that we use throughout this chapter.

Algorithms on words have experienced a new wave of interest due to a number of novel applications in computer science, telecommunications, and biology. Undoubtly, the most popular data structures in algorithms on words are digital trees [41,44] (e.g., tries, PATRICIA, digital search trees (DSTs)), and in particular suffix trees [2,12,54]. We discuss them briefly below, together with general edit distance problem [4,9,12,60], and the shortest common superstring [7,23,58] problem, which recently became quite popular due to possible application to the DNA sequencing problem.

11.2.1 Digital Trees

We start our discussion with a brief review of **digital trees**. The most basic digital tree known as a trie (the name comes from retrieval) is defined first, and then other digital trees are described in terms of the trie.

The primary purpose of a trie is to store a set S of strings (words, keys), say $S = \{X_1, \ldots, X_n\}$. Each word $X = x_1 x_2 x_3 \cdots$ is a finite or infinite string of symbols taken from a finite alphabet $\Sigma = \{\omega_1, \ldots, \omega_V\}$ of size $V = |\Sigma|$. A string will be stored in a leaf of the trie. The trie over S is built recursively as follows: For $|S| = 0$, the trie is, of course, empty. For $|S| = 1$, $trie(S)$ is a single node. If $|S| > 1$, S is split into V subsets $S_1, S_2, \ldots, S_V$ so that a string is in S_j if its first symbol is ω_j. The tries $trie(S_1), trie(S_2), \ldots, trie(S_V)$ are constructed in the same way except that at the kth step, the splitting of sets is based on the kth symbol. They are then connected from their respective roots to a single node to create $trie(S)$. Figure 11.1 illustrates such a construction (cf. [2,41,44]).

There are many possible variations of the trie. The PATRICIA trie eliminates the waste of space caused by nodes having only one branch. This is done by collapsing one-way branches into a single node. In a DST keys (strings) are directly stored in nodes, and hence, external nodes are eliminated. The branching policy is the same as in tries. Figure 11.1 illustrates these definitions (cf. [41,44]).

The suffix tree and the compact suffix tree are similar to the trie and PATRICIA trie, but differ in the structure of the words that are being stored. In suffix trees and compact suffix trees, the words

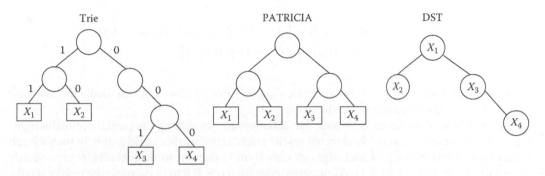

FIGURE 11.1 A trie, PATRICIA trie, and a DST built from the following four strings $X_1 = 11100\ldots$, $X_2 = 10111\ldots$, $X_3 = 00110\ldots$, and $X_4 = 00001\ldots$.

are suffixes of a given string X; that is, the word $X_j = x_j x_{j+1} x_{j+2} \dots$ is the suffix of X which begins at the jth position of X (cf. [2]).

Certain characteristics of tries and suffix trees are of primary importance. Hereafter, we assume that a digital tree is built from n strings or a suffix tree is constructed from a string of length n. The m-depth $D_n(m)$ of the mth leaf in a trie is the number of internal nodes on the path from the root to the leaf. The (typical) depth of the trie D_n then, is the average depth over all its leaves, that is, $\mathrm{Pr}\{D_n \le k\} = \frac{1}{n} \sum_{m=1}^{n} \mathrm{Pr}\{D_n(m) \le k\}$. The path length L_n is the sum of all depths, that is, $L_n = \sum_{m=1}^{n} D_n(m)$. The height H_n of the trie is the maximum depth of a leaf in the trie and can also be defined as the length of the longest path from the root to a leaf, that is, $H_n = \max_{1 \le m \le n}\{D_n(m)\}$. These characteristics are very useful in determining the expected size and shape of the data structures involved in algorithms on words. We study some of them in this chapter.

11.2.2   String Editing Problem

The string editing problem arises in many applications, notably in text editing, speech recognition, machine vision and, last but not least, molecular sequence comparison (cf. [60]). Algorithmic aspects of this problem have been studied rather extensively in the past. In fact, many important problems on words are special cases of string editing, including the longest common subsequence problem (cf. [9,12]) and the problem of approximate pattern matching (cf. [12]). In the following, we review the string editing problem and its relationship to the longest path problem in a special grid graph.

Let Y be a string consisting of ℓ symbols on some alphabet Σ of size V. There are three operations that can be performed on a string, namely deletion of a symbol, insertion of a symbol, and substitution of one symbol for another symbol in Σ. With each operation is associated a weight function. We denote by $W_I(y_i)$, $W_D(y_i)$, and $W_Q(x_i, y_j)$ the weight of insertion and deletion of the symbol $y_i \in \Sigma$, and substitution of x_i by $y_j \in \Sigma$, respectively. An edit script on Y is any sequence of edit operations, and the total weight of it is the sum of weights of the edit operations.

The string editing problem deals with two strings, say Y of length ℓ (for ℓong) and X of length s (for short), and consists of finding an edit script of minimum (maximum) total weight that transforms X into Y. The maximum (minimum) weight is called the edit distance from X to Y, and its is also known as the Levenshtein distance. In molecular biology, the Levenshtein distance is used to measure similarity (homogeneity) of two molecular sequences, say DNA sequences (cf. [60]).

The string edit problem can be solved by the standard dynamic programming method. Let $C_{\max}(i, j)$ denote the maximum weight of transforming the prefix of Y of size i into the prefix of X of size j. Then

$$C_{\max}(i, j) = \max \left\{ C_{\max}(i - 1, j - 1) + W_Q\left(x_i, y_j\right), C_{\max}(i - 1, j) \right.$$
$$\left. + W_D\left(x_i\right), C_{\max}(i, j - 1) + W_I\left(y_j\right) \right\}$$

for all $1 \le i \le \ell$ and $1 \le j \le s$. We compute $C_{\max}(i, j)$ row by row to obtain finally the total cost $C_{\max} = C_{\max}(\ell, s)$ of the maximum edit script.

The key observation for us is to note that interdependency among the partial optimal weights $C_{\max}(i, j)$ induce an $\ell \times s$ grid-like directed acyclic graph, called further a grid graph. In such a graph vertices are points in the grid and edges go only from (i, j) point to neighboring points, namely $(i, j+1)$, $(i+1, j)$ and $(i+1, j+1)$. A horizontal edge from $(i-1, j)$ to (i, j) carries the weight $W_I(y_j)$; a vertical edge from $(i, j-1)$ to (i, j) has weight $W_D(x_i)$; and a diagonal edge from $(i-1, j-1)$ to (i, j) has weight $W_Q(x_i, y_j)$. Figure 11.2 shows an example of such an edit graph. The edit distance is the longest (shortest) path from the point $O = (0, 0)$ to $E = (\ell, s)$.

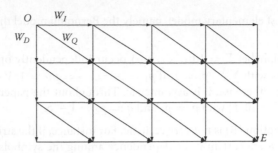

FIGURE 11.2 Example of a grid graph of size $\ell = 4$ and $s = 3$.

11.2.3 Shortest Common Superstring

Various versions of the shortest common superstring (SCS) problem play important roles in data compression and DNA sequencing. In fact, in laboratories DNA sequencing (cf. [60]) is routinely done by sequencing large numbers of relatively short fragments, and then heuristically finding a short common superstring. The problem can be formulated as follows: given a collection of strings, say $X_1, X_2, \ldots, X_n$ over an alphabet Σ, find the shortest string Z such that each of X_i appears as a substring (a consecutive block) of Z.

It is known that computing the SCS is NP-hard. Thus, constructing a good approximation to SCS is of prime interest. It has been shown recently, that a greedy algorithm can compute in $O(n \log n)$ time a superstring that in the worst-case is only β times (where $2 \leq \beta \leq 4$) longer than the SCS [7,58]. Often, one is interested in maximizing total overlap of SCS using a greedy heuristic and to show that such a heuristic produces an overlap O_n^{gr} that approximates well the optimal overlap O_n^{opt} where n is the number of strings.

More precisely, suppose $X = x_1 x_2 \cdots x_r$ and $Y = y_1 y_2 \cdots y_s$ are strings over the same finite alphabet Σ. We also write $|X|$ for the length of X. We define their overlap $o(X, Y)$ by

$$o(X, Y) = \max \left\{ j : \ y_i = x_{r-i+1}, \ 1 \leq i \leq j \right\}.$$

Let $\mathcal{S}$ be a set of all superstrings built over the strings $X_1, \ldots, X_n$. Then

$$O_n^{opt} = \sum_{i=1}^{n} |X_i| - \min_{Z \in \mathcal{S}} |Z|$$

represents the optimal overlap over $\mathcal{S}$.

11.3 Probabilistic Models

In this section, we first discuss a few probabilistic models of randomly generated strings. Then, we briefly review some basic facts from probability theory (e.g., types of stochastic convergence), and finally we provide some elements of complex analysis that we shall use in this chapter.

11.3.1 Probabilistic Models of Strings

As expected, random shape of data structures on words depends on the underlying probabilistic assumptions concerning the strings involved. Below, we discuss a few basic probabilistic models that one often encounters in the analysis of problems on words.

We start with the most elementary model, namely the Bernoulli model that is defined as follows:

(B) BERNOULLI MODEL

Symbols of the alphabet $\Sigma = \{\omega_1, \ldots, \omega_V\}$ occur independently of one another; and $\Pr\{x_j = \omega_i\} = p_i$ with $\sum_{i=1}^{V} p_i = 1$. If $p_1 = p_2 = \cdots = p_V = 1/V$, then the model is called symmetric; otherwise, it is asymmetric. Throughout the paper we only consider binary alphabet $\Sigma = \{0, 1\}$ with $p := p_1$ and $q := p_2 = 1 - p$.

In many cases, assumption (B) is not very realistic. For instance, if the strings are words from the English language, then there certainly is a dependence among the symbols of the alphabet. As an example, h is much more likely to follow an s than a b. When this is the case, assumption (B) can be replaced by

(M) MARKOVIAN MODEL

There is a Markovian dependency between consecutive symbols in a key; that is, the probability $p_{ij} = \Pr\{X_{k+1} = \omega_j | X_k = \omega_i\}$ describes the conditional probability of sampling symbol ω_j immediately after symbol ω_i.

There is another generalization of the Markovian model, namely the mixing model, which is very useful in practice, especially when dealing with problems of data compression or molecular biology when one expects long dependency among symbols of a string.

(MX) MIXING MODEL

Let $\mathcal{F}_m^n$ be a σ-field generated by $\{X_k\}_{k=m}^{n}$ for $m \leq n$. There exists a function $\alpha(\cdot)$ of g such that: (i) $\lim_{g \to \infty} \alpha(g) = 0$, (ii) $\alpha(1) < 1$, and (iii) for any m, and two events $A \in \mathcal{F}_{-\infty}^{m}$ and $B \in \mathcal{F}_{m+g}^{\infty}$ the following holds;

$$(1 - \alpha(g))\Pr\{A\}\Pr\{B\} \leq \Pr\{AB\} \leq (1 + \alpha(g))\Pr\{A\}\Pr\{B\}.$$

In words, model (MX) says that the dependency between $\{X_k\}_{k=1}^{m}$ and $\{X_k\}_{k=m+g}^{\infty}$ is getting weaker and weaker as g becomes larger (note that when the sequence $\{X_k\}$ is i.i.d., then $\Pr\{AB\} = \Pr\{A\}\Pr\{B\}$). The "quantity" of dependency is characterized by $\alpha(g)$ (cf. [8]).

11.3.2 Quick Review from Probability: Types of Stochastic Convergence

We begin with some elementary definitions from probability theory. The reader is referred to [14,15] for more detailed discussions. Let the random variable X_n denote the value of a parameter of interest depending on n (e.g., depth in a suffix tree and/or trie built over n strings). The expected value $E[X_n]$ or mean and the variance $\mathbf{Var}[X_n]$ can be computed as $E[X_n] = \sum_{k=0}^{\infty} k\Pr\{X_n = k\}$ and $\mathbf{Var}[X_n] = \sum_{k=0}^{\infty} (k - E[X_n])^2 \Pr\{X_n = k\}$.

11.3.2.1 Convergence of Random Variables

It is important to note the different ways in which random variables are said to converge. To examine the different methods of convergence, let X_n be a sequence of random variables, and let their distribution functions be $F_n(x)$, respectively.

The first notion of convergence of a sequence of random variables is known as convergence in probability. The sequence X_n converges to a random variable X in probability, denoted $X_n \to X$ (pr.) or $X_n \overset{p}{\to} X$, if for any $\epsilon > 0$,

$$\lim_{n \to \infty} \Pr\{|X_n - X| < \epsilon\} = 1.$$

Note that this does not say that the difference between X_n and X becomes very small. What converges here is the probability that the difference between X_n and X becomes very small. It is, therefore, possible, although unlikely, for X_n and X to differ by a significant amount and for such differences to occur infinitely often.

A stronger kind of convergence that does not allow such behavior is called almost sure convergence or strong convergence. A sequence of random variables X_n converges to a random variable X almost surely, denoted $X_n \to X$ (a.s.) or $X_n \overset{(a.s.)}{\to} X$, if for any $\epsilon > 0$,

$$\lim_{N \to \infty} \Pr\left\{ \sup_{n \geq N} |X_n - X| < \epsilon \right\} = 1.$$

From this formulation of almost sure convergence, it is clear that if $X_n \to X$ (a.s.), the probability of infinitely many large differences between X_n and X is zero. The sequence X_n in this case is said to satisfy the strong law of large numbers. As the term strong implies, almost sure convergence implies convergence in probability.

A simple criterion for almost sure convergence can be inferred from the Borel–Cantelli lemma. We give it in the following corollary.

LEMMA 11.1 **(Borel–Cantelli)** Let $\epsilon > 0$. If $\sum_{n=0}^{\infty} \Pr\{|X_n - X| > \epsilon\} < \infty$, then $X_n \to X$ (a.s.).

PROOF It follows directly from the following chain of inequalities (the reader is referred to Section 11.4.1 for more explanations on these inequalities):

$$\Pr\left\{ \sup_{n \geq N} |X_n - X| \geq \epsilon \right\} = \Pr\left\{ \bigcup_{n \geq N} (|X_n - X| \geq \epsilon) \right\} \leq \sum_{n \geq N} \Pr\{|X_n - X| \geq \epsilon\} \to 0.$$

The last convergence is a consequence of our assumption that $\sum_{n=0}^{\infty} \Pr\{|X_n - X| > \epsilon\} < \infty$.

A third type of convergence is defined on the distribution functions $F_n(x)$. The sequence of random variables X_n converges in distribution or converges in law to the random variable X, denoted $X_n \overset{d}{\to} X$, if $\lim_{n \to \infty} F_n(x) = F(x)$ for each point of continuity of $F(x)$. Almost sure convergence implies convergence in distribution.

Finally, the convergence in mean of order p implies that $E[|X_n - X|^p] \to 0$ as $n \to \infty$, and convergence in moments requires $E[X_n^p] \to E[X^p]$ for any p as $n \to \infty$. It is well known that almost sure convergence and convergence in mean imply the convergence in probability. On the other hand, the convergence in probability leads to the convergence in distribution. If the limiting random variable X is a constant, then the convergence in distribution also implies the convergence in probability (cf. [14]).

11.3.2.2 Generating Functions

The distribution of a random variable can also be described using generating functions. The ordinary generating function $G_n(u)$, and a bivariate exponential generating function $g(z, u)$ are defined as

$$G_n(u) = E\left[u^{X_n} \right] = \sum_{k=0}^{\infty} \Pr\{X_n = k\} u^k$$

and $g(z, u) = \sum_{n=0}^{\infty} G_n(u) \frac{z^n}{n!}$, respectively. These functions are well-defined for any complex numbers z and u such that $|u| < 1$. Observe that

$$E[X_n] = G_n'(1),$$

$$\mathrm{Var}[X_n] = G_n''(1) + G_n'(1) - [G_n'(1)]^2.$$

11.3.2.3 Levy's Continuity Theorem

Our next step is to relate convergence in distribution to convergence of generating functions. The following results, known as Levy's continuity theorem is an archi-fact for most distributional analysis. For our purpose we formulate it in terms of the Laplace transform of X_n, namely $G_n(e^{-t}) = \mathbf{E}[e^{-tX_n}]$ for real t (cf. [14]).

THEOREM 11.1   (Continuity Theorem) *Let X_n and X be random variables with Laplace transforms $G_n(e^{-t})$ and $G(e^{-t})$, respectively. A necessary and sufficient condition for $X_n \xrightarrow{d} X$ is that $G_n(e^{-t}) \to G(e^{-t})$ for all $t \geq 0$.*

The above theorem holds if we set $t = iv$ for $-\infty < v < \infty$ (i.e., we consider characteristic functions). Moreover, if the above holds for t complex number, then we automatically derive convergence in moments due to the fact that an analytical function possesses all its derivatives.

Finally, in order to establish central limit theorem (i.e., convergence to a normal distribution) a theorem by Goncharov (cf. [39], Chap 1.2.10, Ex. 13) is useful (it follows directly from the Continuity Theorem). This theorem states that a sequence of random variables X_n with mean $\mathbf{E}[X_n] = \mu_n$ and standard deviation $\sigma_n = \sqrt{\mathbf{Var}[X_n]}$ approaches a normal distribution if the following holds:

$$\lim_{n \to \infty} e^{-\tau\mu_n/\sigma_n} G_n\left(e^{\tau/\sigma_n}\right) = e^{\tau^2/2}$$

for all $\tau = iv$ and $-\infty < v < \infty$, and X_n converges in moments if τ is a complex number.

11.3.3   Review from Complex Analysis

Much of the necessary complex analysis involves the use of Cauchy's integral formula and Cauchy's residue theorem. We briefly recall a few facts from analytic functions, and then discuss the above two theorems. For precise definitions and formulations the reader is referred to [27]. We shall follow here Flajolet and Sedgewick [21].

A function $f(z)$ of complex variable z is analytic at point $z = a$ if it is differentiable in a neighborhood of $z = a$ or equivalently it has a convergent series representation around $z = a$. Let us concentrate our discussion only on meromorphic functions that are analytical with an exception of a finite number of points called poles. More formally, a meromorphic function $f(z)$ can be represented in a neighborhood of $z = a$ with $z \neq a$ by Laurent series as follows: $f(z) = \sum_{n \geq -M} f_n(z - a)^n$ for some integer M. If the above holds with $f_{-M} \neq 0$, then it is said that $f(z)$ has a *pole* of order M at $z = a$. Cauchy's Integral Theorem states that for any analytical function $f(z)$,

$$f_n := [z^n] f(z) = \frac{1}{2\pi i} \oint f(z) \frac{dz}{z^{n+1}},$$

and the circle is traversed counterclockwise, where throughout the chapter we write $[z^n] f(z)$ for the coefficient of $f(z)$ at z^n.

An important tool frequently used in the analytic analysis of algorithms is residue theory. The residue of $f(z)$ at a point a is the coefficient of $(z - a)^{-1}$ in the expansion of $f(z)$ around a, and it is denoted as $\mathrm{Res}[f(z); z = a] = f_{-1}$. There are many simple rules to evaluate residues and the reader can find them in any standard book on complex analysis (e.g., [27]). Actually, the easiest way to compute a residue of a function is to use the `series` commend in MAPLE that produces a series development of a function. The residue is simply the coefficient at $(z - a)^{-1}$. For example, the following session of MAPLE computes series of $f(z) = \Gamma(z)/(1 - 2^z)$ at $z = 0$ where $\Gamma(z)$ is the Euler gamma function [1]:

```
series(GAMMA(z)/(1-2^z), z=0, 4);
```

$$-\frac{1}{\ln(2)}z^{-2} - \frac{-\gamma - \frac{1}{2}\ln(2)}{\ln(2)}z^{-1}$$

$$-\frac{-\frac{1}{6}\ln(2)^2 + \frac{1}{12}\pi^2 + \frac{1}{2}\gamma^2 + \frac{1}{4}(2\gamma + \ln(2))\ln(2)}{\ln(2)} + O(z)$$

From the above we see that $\mathrm{Res}[f(z); z = 0] = \frac{\gamma}{\log 2} + \frac{1}{2}$.

Residues are very important in evaluating contour integrals. In fact, a well-known theorem in complex analysis, that is, Cauchy's residue theorem states that if $f(z)$ is analytic within and on the boundary of C except at a finite number of poles $a_1, a_2, \ldots, a_N$ inside of C having residues $\mathrm{Res}[f(z); z = a_1], \ldots, \mathrm{Res}[f(z); z = a_N]$, then

$$\oint f(z) dz = 2\pi i \sum_{j=1}^{N} \mathrm{Res}\left[f(z); z = a_j\right],$$

where the circle is traversed counterclockwise.

11.4 Probabilistic Techniques

In this section we discuss several probabilistic techniques that have been successfully applied to the average case analysis of algorithms. We start with some elementary **inclusion–exclusion principle** known also as sieve methods. Then, we present very useful first and second moment methods. We continue with the **subadditive ergodic theorem** that is quite popular for deducing certain properties of problems on words. Next, we turn our attention to some probabilistic methods of information theory, and in particular we discuss entropy and **asymptotic equipartition property**. Finally, we look at some **large deviations** results and Azuma's type inequality. In this section, as well in the next one where analytic techniques are discussed, we adopt the following scheme of presentation: First, we describe the method and give a short intuitive derivation. Then, we illustrate it on some nontrivial examples taken from the problems on words discussed in Section 11.2.

11.4.1 Sieve Method and Its Variations

The inclusion–exclusion principle is one of the oldest tools in combinatorics, number theory (where this principle is known as sieve method), discrete mathematics, and probabilistic analysis. It provides a tool to estimate probability of a union of not disjoint events, say $\bigcup_{i=1}^{n} A_i$ where A_i are events for $i = 1, \ldots, n$. Before we plunge into our discussion, let us first show a few examples of problems on words for which an estimation of the probability of a union of events is required.

Example 11.1

Depth and height in a trie: In Section 11.2.1 we discussed tries built over n binary strings $X_1, \ldots, X_n$. We assume that those strings are generated according to the Bernoulli model with one symbol, say "0," occurring with probability p and the other, say "1," with probability $q = 1 - p$. Let C_{ij}, known as alignment between ith and jth strings, be defined as the length of the longest string that is a prefix of X_i and X_j. Then, it is easy to see that the mth depth $D_n(m)$ (i.e., length of a path in trie from the root to the external node containing X_m), and the height H_n (i.e., the length of the longest path in a trie) can be expressed as follows:

$$D_n(m) = \max_{1 \le i \ne m \le n} \{C_{i,m}\} + 1, \tag{11.1}$$

$$H_n = \max_{1 \le i < j \le n} \{C_{ij}\} + 1. \tag{11.2}$$

Certainly, the alignments C_{ij} are dependent random variables even for the Bernoulli model. The above equations expressed the depth and the height as an order statistic (i.e., maximum of the sequence C_{ij} for $i, j = 1, \ldots, n$). We can estimate some probabilities associated with the depth and the height as a union of properly defined events. Indeed, let $A_{ij} = \{C_{ij} > k\}$ for some k. Then, one finds

$$\Pr\{D_n(m) > k\} = \Pr\left\{ \bigcup_{i=1, \ne m}^{n} A_{i,m} \right\}, \tag{11.3}$$

$$\Pr\{H_n > k\} = \Pr\left\{ \bigcup_{i \ne j=1}^{n} A_{ij} \right\}. \tag{11.4}$$

In passing, we should point out that for the SCS problem (cf. Section 11.2.3) we need to estimate a quantity $M_n(m)$ which is similar to $D_n(m)$ except that C_{im} is defined as the length of the longest string that is a prefix of X_i and suffix of X_m for fixed m. One easily observes that $M_n(m) \overset{d}{=} D_n(m)$, that is, these two quantities are equal *in distribution*.

We have just seen that often we need to estimate a probability of union of events. The following formula is known as inclusion–exclusion formula (cf. [6])

$$\Pr\left\{ \bigcup_{i=1}^{n} A_i \right\} = \sum_{r=1}^{n} (-1)^{r+1} \sum_{|J|=r} \Pr\left\{ \bigcap_{j \in J} A_j \right\}. \tag{11.5}$$

The next example illustrates it on the depth on a trie.

Example 11.2

Generating function of the depth in a trie: Let us compute the generating function of the depth $D_n := D_n(1)$ for the first string X_1. We start with Equation 11.3, and after some easy algebraic manipulation, Equation 11.5 leads to (cf. [31])

$$\Pr\{D_n > k\} = \Pr\left\{ \bigcup_{i=2}^{n} [C_{i,1} \ge k] \right\} = \sum_{r=1}^{n-1} (-1)^{r+1} \binom{n-1}{r} \Pr\{C_{2,1} \ge k, \ldots, C_{r+1,1} \ge k\};$$

since the probability $\Pr\{C_{2,1} \ge k, \ldots, C_{r+1,1} \ge k\}$ does not depend on the choice of strings (i.e., it is the same for any r-tuple of strings selected). Moreover, it can be easily explicitly computed. Indeed, we obtain $\Pr\{C_{2,1} \ge k, \ldots, C_{r+1,1} \ge k\} = (p^{r+1} + q^{r+1})^k$, since r independent binary strings must agree on the first k symbols (we recall that p stands for the probability a symbol, say "0," occurrence and $q = 1 - p$). Thus, the generating function $D_n(u) = \mathbf{E}[u^{D_n}] = \sum_{k \ge 0} \Pr\{D_n = k\} u^k$ becomes

$$\mathbf{E}[u^{D_n}] = 1 + \sum_{r=1}^{n-1} (-1)^{r+1} \binom{n-1}{r} \frac{1-u}{1 - u(p^{r+1} + q^{r+1})}$$

The last formula is a simple consequence of the above, and the following a well-known fact from the theory of generating function $\mathbf{E}[u^X]$ for a random variable X:

$$\mathbf{E}\left[u^X\right] = \frac{1}{1-u} \sum_{k=0}^{\infty} \Pr\{X \leq k\} u^k$$

for $|u| < 1$.

In many real computations, however, one cannot explicitly compute the probability of the events union. Often, one must retreat to inequalities that actually are enough to reach one's goal. The most simple yet still very powerful is the following inequality

$$\Pr\left\{\bigcup_{i=1}^{n} A_i\right\} \leq \sum_{i=1}^{n} \Pr\{A_i\}. \tag{11.6}$$

The latter is an example of a series of inequalities due to Bonferroni which can be formulated as follows: For every even integer $m \geq 0$ we have

$$\sum_{j=1}^{m} (-1)^{j-1} \sum_{1 \leq t_1 < \cdots < t_j \leq n} \Pr\left\{A_{t_1} \cap \cdots \cap A_{t_j}\right\}$$

$$\leq \Pr\left\{\bigcup_{i=1}^{n} A_i\right\}$$

$$\leq \sum_{j=1}^{m+1} (-1)^{j-1} \sum_{1 \leq t_1 < \cdots < t_j \leq n} \Pr\left\{A_{t_1} \cap \cdots \cap A_{t_j}\right\}.$$

In combinatorics (e.g., enumeration problems) and probability the so called inclusion–exclusion principle is very popular and had many successes. We formulate it in a form of a theorem whose proof can be found in Bollobás [6].

THEOREM 11.2 (**Inclusion–Exclusion Principle**) *Let* $A_1, \ldots, A_n$ *be events in a probability space, and let* p_k *be the probability of exactly k of them to occur. Then:*

$$p_k = \sum_{r=k}^{n} (-1)^{r+k} \binom{r}{k} \sum_{|J|=r} \Pr\left\{\bigcap_{j \in J} A_j\right\}.$$

Example 11.3

Computing a distribution through its moments (cf. [6]): Let X be a random variable defined on $\{0, 1, \ldots, n\}$, and let $\mathbf{E}_r[X] = \mathbf{E}X(X-1)\cdots(X-r+1)$ be the rth factorial moment of X. Then

$$\Pr\{X = k\} = \frac{1}{k!} \sum_{r=k}^{n} (-1)^{r+k} \frac{\mathbf{E}_r[X]}{(r-k)!}.$$

Indeed, it suffices to set $A_i = \{X \geq i\}$ for all $i = 1, \ldots, n$, and observe that $\sum_{|J|=r} \Pr\{\bigcap_{j \in J} A_j\} = \mathbf{E}_r[X]/r!$. Since the event $\{X = k\}$ is equivalent to the event that exactly k of A_i occur, a simple application of Theorem 11.2 proves the announced result.

11.4.2 Inequalities: First and Second Moment Methods

In this subsection, we review some inequalities that play a considerable role in probabilistic analysis of algorithms. In particular, we discuss first and second moment methods.

We start with a few standard inequalities (cf. [14]):

Markov Inequality: For a nonnegative random variable X and $\varepsilon > 0$ the following holds:

$$\Pr\{X \geq \varepsilon\} \leq \frac{E[X]}{\varepsilon}.$$

Indeed: let $I(A)$ be the indicator function of A (i.e., $I(A) = 1$ if A occurs, and zero otherwise). Then,

$$E[X] \geq E\left[XI\left(X \geq \varepsilon\right)\right] \geq \varepsilon E\left[I\left(X \geq \varepsilon\right)\right] = \varepsilon \Pr\{X \geq \varepsilon\}.$$

Chebyshev's Inequality: If one replaces X by $|X - E[X]|$ in the Markov inequality, then

$$\Pr\{|X - E[X]| > \varepsilon\} \leq \frac{\text{Var}[X]}{\varepsilon^2}.$$

Schwarz's Inequality (also called Cauchy–Schwarz): Let X and Y be such that $E[X^2] < \infty$ and $E[Y^2] < \infty$. Then

$$E[|XY|]^2 \leq E[X^2]E[Y^2],$$

where $E[X]^2 := (E[X])^2$.

Jensen's Inequality: Let $f(\cdot)$ be a downward convex function, that is, for $\lambda \in (0,1)$ we have $\lambda f(x) + (1 - \lambda)f(y) \geq f(\lambda x + (1 - \lambda)y)$. Then

$$f(E[X]) \leq E[f(X)].$$

The remainder part of this subsection is devoted to the first and the second moment methods that we illustrate on several examples arising in the analysis of digital trees. The first moment method for a nonnegative random variable X states that

$$\Pr\{X > 0\} \leq E[X]. \tag{11.7}$$

This follows directly from Markov's inequality after setting $\varepsilon = 1$. The above inequality implies also the basic Bonferroni inequality Equation 11.6. Indeed, let A_i ($i = 1, \ldots, n$) be events, and set $X = I(A_1) + \cdots + I(A_n)$. Inequality Equation 11.6 follows.

In a typical application of Equation 11.7, we expect to show that $E[X] \to 0$, just $X = 0$ occurs almost always or with high probability (whp). We illustrate it in the next example.

Example 11.4

Upper bound on the height in a trie: In Example 11.1 we showed that the height H_n of a trie is given by Equation 11.2 or 11.4. Thus, using the first moment method we have

$$\Pr\{H_n \geq k + 1\} \leq \Pr\left\{\max_{1 \leq i < j \leq n}\{C_{ij}\} \geq k\right\} \leq n^2 \Pr\{C_{ij} \geq k\}$$

for any integer k. From Example 11.2 we know that $\Pr\{C_{ij} \geq k\} = (p^2 + q^2)^k$. Let $P = p^2 + q^2$, $Q = P^{-1}$, and set $k = 2(1 + \varepsilon)\log_Q n$ for any $\varepsilon > 0$. Then, the above implies

$$\Pr\left\{H_n \geq 2(1 + \varepsilon)\log_Q n + 1\right\} \leq \frac{n^2}{n^{2(1+\varepsilon)}} = \frac{1}{n^{2\varepsilon}} \to 0,$$

thus, $H_n/(2 \log_Q n) \leq 1$ (pr.). Below, in Example 11.5, we will actually prove that $H_n/(2 \log_Q n) = 1$ (pr.) by establishing a matching lower bound.

Let us look now at the second moment method. Setting in the Chebyshev inequality $\varepsilon = E[X]$ we prove that

$$\Pr\{X = 0\} \leq \frac{\text{Var}[X]}{E[X]^2}.$$

But, one can do better (cf. [3,10]). Using Schwarz's inequality for a random variable X we obtain the following chain of inequalities

$$E[X]^2 = E[I(X \neq 0)X]^2 \leq E[I(X \neq 0)]E\left[X^2\right] = \Pr\{I(X \neq 0)\}E\left[X^2\right],$$

which finally implies the second moment inequality

$$\Pr\{X > 0\} \geq \frac{E[X]^2}{E[X^2]}. \tag{11.8}$$

Actually, another formulation of this inequality due to Chung and Erdős is quite popular. To derive it, set in Equation 11.8 $X = I(A_1) + \cdots + I(A_n)$ for a sequence of events $A_1, \ldots, A_n$. Noting that $\{X > 0\} = \bigcup_{i=1}^{n} A_i$, we obtain from Equation 11.8 after some algebra

$$\Pr\left\{\bigcup_{i=1}^{n} A_i\right\} \geq \frac{\left(\sum_{i=1}^{n} \Pr\{A_i\}\right)^2}{\sum_{i=1}^{n} \Pr\{A_i\} + \sum_{i \neq j} \Pr\{A_i \cap A_j\}}. \tag{11.9}$$

In a typical application, if we are able to prove that $\text{Var}[X]/E[X^2] \to 0$, then we can show that $\{X > 0\}$ almost always. The next example—which is a continuation of Example 11.4—illustrates this point.

Example 11.5

Lower bound for the height in a trie: We now prove that $\Pr\{H_n \geq 2(1 - \varepsilon) \log_Q n\} \to 1$ for any $\varepsilon > 0$, just completing the proof that $H_n/(2 \log_Q n) \to 1$ (pr.). We use the Chung–Erdős formulation, and set $A_{ij} = \{C_{ij} \geq k\}$. Throughout this example, we assume $k = 2(1 - \varepsilon) \log_Q n$. Observe that now in Equation 11.9 we must replace the single summation index i by a double summation index (i, j). The following is obvious: $\sum_{1 \leq i < j \leq n} \Pr\{A_{ij}\} = \frac{1}{2}n(n - 1)P^k$, where $P = p^2 + q^2$. The other sum in Equation 11.9 is a little harder to deal with. We must sum over (i, j), (l, m), and we consider two cases: (i) all indices are different, (ii) $i = l$ (i.e., we have (i, j), (i, m)). In the second case we must consider the probability $\Pr\{C_{ij} \geq k, C_{i,m} \geq k\}$. But, as in Example 11.2, we obtain $\Pr\{C_{ij} \geq k, C_{i,m} \geq k\} = (p^3 + q^3)^k$ since once you choose a symbol in the string X_i you must have the same symbol at the same position in X_j, X_m. In summary,

$$\sum_{(ij),(l,m)} \Pr\left\{C_{ij} \geq k, C_{lm} \geq k\right\} \leq \frac{1}{4}n^4 P^{2k} + n^3 \left(p^3 + q^3\right)^k.$$

To complete the derivation, it suffices to observe that

$$\left(p^3 + q^3\right)^{\frac{1}{3}} \leq P^{\frac{1}{2}},$$

which is easy to prove by elementary methods (cf. [60]). Then, Equation 11.9 becomes

$$\Pr\{H_n \geq k+1\} = \Pr\left\{\bigcup_{i=1}^{n} A_i\right\} \geq \frac{1}{n^{-2}p^{-k}+1+4\left(p^3+q^3\right)^k/\left(np^{2k}\right)}$$

$$\geq \frac{1}{1+n^{-2\varepsilon}+4/\left(np^{k/2}\right)} \geq \frac{1}{1+n^{-2\varepsilon}+4n^{-\varepsilon}} \to 1.$$

Thus, we have shown that $H_n/(2\log_Q n) \geq 1$ (pr.), which completes our proof of

$$\lim_{n\to\infty} \Pr\left\{2(1-\varepsilon)\log_Q n \leq H_n \leq 2(1+\varepsilon)\log_Q n\right\} = 1$$

for any $\varepsilon > 0$.

11.4.3 Subadditive Ergodic Theorem

The celebrated ergodic theorem of Feller [15] found many useful applications in computer science. It is used habitually during a computer simulation run or whenever one must perform experiments and collect data. However, for probabilistic analysis of algorithms a generalization of this result due to Kingman [36] is more important. We briefly review it here and illustrate on a few examples.

Let us start with the following well-known fact attributed to Fekete (cf. [50]). Assume a (deterministic) sequence $\{x_n\}_{n=0}^{\infty}$ satisfies the so-called subadditivity property, that is,

$$x_{m+n} \leq x_n + x_m$$

for all integers $m, n \geq 0$. It is easy to see that then (cf. [14])

$$\lim_{n\to\infty} \frac{x_n}{n} = \inf_{m\geq 1} \frac{x_m}{m} = \alpha$$

for some $\alpha \in [-\infty, \infty)$. Indeed, fix $m \geq 0$, write $n = km + l$ for some $0 \leq l \leq m$, and observe that by the above subadditivity property

$$x_n \leq k x_m + x_l.$$

Taking $n \to \infty$ with $n/k \to m$ we finally arrive at $\limsup_{n\to\infty} \frac{x_n}{n} \leq \inf_{m\geq 1} \frac{x_m}{m} \leq \alpha$ where the last inequality follows from arbitrariness of m. This completes the derivation since $\liminf_{n\to\infty} \frac{x_n}{n} \geq \alpha$ is automatic. One can also see that replacing "$\leq$" in the subadditivity property by "$\geq$" (thus, superadditivity property) will not change our conclusion except that $\inf_{m\geq 1} \frac{x_m}{m}$ should be replaced by $\sup_{m\geq 1} \frac{x_m}{m}$.

In the early 1970s people started asking whether the above deterministic subadditivity result could be extended to a sequence of random variables. Such an extension would have an impact on many research problems of those days. For example, Chvatal and Sankoff [9] used ingenious tricks to establish the probabilistic behavior of the Longest Common Subsequence problem (cf. Section 11.2.2 and below) while we show below that it is a trivial consequence of a stochastic extension of the above subadditivity result. In 1976 Kingman [36] presented the first proof of what later will be called Subadditivity Ergodic Theorem. Below, we present an extension of Kingman's result (cf. [50]).

To formulate it properly we must consider a sequence of doubly indexed random variables $X_{m,n}$ with $m \leq n$. One can think of it as $X_{m,n} = (X_m, X_{m+1}, \ldots, X_n)$, that is, as a substring of a single-indexed sequence X_n.

THEOREM 11.3 **(Subadditive Ergodic Theorem [36])** (i) *Let $X_{m,n}$ ($m < n$) be a sequence of nonnegative random variables satisfying the following three properties*

(a) $X_{0,n} \leq X_{0,m} + X_{m,n}$ (*subadditivity*);

(b) $X_{m,n}$ is stationary (i.e., the joint distributions
   of $X_{m,n}$ are the same as $X_{m+1,n+1}$) and ergodic (cf. [14]);

(c) $E[X_{0,1}] < \infty$.

Then,

$$\lim_{n\to\infty} \frac{E[X_{0,n}]}{n} = \gamma \quad \text{and} \quad \lim_{n\to\infty} \frac{X_{0,n}}{n} = \gamma \quad \text{(a.s.)} \tag{11.10}$$

for some constant γ.

(ii) (**Almost Subadditive Ergodic Theorem**) *If the subadditivity inequality is replaced by*

$$X_{0,n} \leq X_{0,m} + X_{m,n} + A_n \tag{11.11}$$

such that $\lim_{n\to\infty} E[A_n/n] = 0$, then Equation 11.10 holds, too.

We must point out, however, that the above result proves only the existence of a constant γ such that Equation 11.10 holds. It says nothing how to compute it, and in fact many ingenious methods have been devised in the past to bound this constant. We discuss it in a more detailed way in the examples below.

Example 11.6

String editing problem: Let us consider the string editing problem of Section 11.2.2. To recall, one is interested in estimating the maximum cost C_{max} of transforming one sequence into another. This problem can be reduced to finding the longest (shortest) path in a special grid graph (cf. Figure 11.2). Let us assume that the weights W_I, W_D, and W_Q are independently distributed, thus, we adopt the Bernoulli model (B) of Section 11.3.1. Then, using the subadditive ergodic theorem it is immediate to see that

$$\lim_{n\to\infty} \frac{C_{max}}{n} = \lim_{n\to\infty} \frac{EC_{max}}{n} = \alpha \quad \text{(a.s.)},$$

for some constant $\alpha > 0$, provided ℓ/s has a limit as $n\to\infty$. Indeed, let us consider the $\ell \times s$ grid with starting point O and ending point E (cf. Figure 11.2). Call it Grid(O,E). We also choose an arbitrary point, say A, inside the grid so that we can consider two grids, namely Grid(O,A) and Grid(A,E). Actually, point A splits the edit distance problem into two subproblems with objective functions $C_{max}(O, A)$ and $C_{max}(A, E)$. Clearly, $C_{max}(O, E) \geq C_{max}(O, A) + C_{max}(A, E)$. Thus, under our assumption regarding weights, the objective function C_{max} is superadditive, and direct application of the Superadditive Ergodic Theorem proves the result.

11.4.4 Entropy and Its Applications

Entropy and mutual information was introduced by Shannon in 1948, and overnight a new field of information theory was born. Over the last 50 years information theory underwent many changes, and remarkable progress was achieved. These days entropy and the Shannon–McMillan–Breiman Theorem are standard tools of the average case analysis of algorithms. In this subsection, we review some elements of information theory and illustrate its usage to the analysis of algorithms.

Let us start with a simple observation: Consider a binary sequence of symbols of length n, say $(X_1, \ldots, X_n)$, with p denoting the probability of one symbol and $q = 1 - p$ the probability of the other symbol. When $p = q = 1/2$, then $\Pr\{X_1, \ldots, X_n\} = 2^{-n}$, and it does not matter what are the actual values of $X_1, \ldots, X_n$. In general, $\Pr\{X_1, \ldots, X_n\}$ is not the same for all possible values of

$X_1, \ldots, X_n$, however, we shall show that a typical sequences $(X_1, \ldots, X_n)$ have "asymptotically" the same probability. Indeed, consider $p \neq q$ in the example above. Then, the probability of a typical sequence is approximately equal to (we use here the central limit theorem for i.i.d. sequences):

$$p^{np+O(\sqrt{n})} q^{nq+O(\sqrt{n})} = e^{-n(-p \log p - q \log q) + O(\sqrt{n})} \sim e^{-nh}$$

where $h = -p \log p - q \log q$ is the entropy of the underlying Bernoulli model. Thus, a typical sequence X_1^n has asymptotically the same probability equal to e^{-nh}.

To be more precise, let us consider a stationary and ergodic sequence $\{X_k\}_{k=1}^{\infty}$, and define $X_m^n = (X_m, X_{m+1}, \ldots, X_n)$ for $m \leq n$ as a substring of $\{X_k\}_{k=1}^{\infty}$. The entropy rate h of $\{X_k\}_{k=1}^{\infty}$ is defined as (cf. [11,14])

$$h := -\lim_{n \to \infty} \frac{\mathbf{E}\left[\log \Pr\{X_1^n\}\right]}{n}, \tag{11.12}$$

where one can prove the limit above exists. We must point out that $\Pr\{X_1^n\}$ is a random variable since X_1^n is a random sequence!

We show now how to derive the Shannon–McMillan–Breiman Theorem in the case of the Bernoulli model and the mixing model, and later we formulate the theorem in its full generality. Consider first the Bernoulli model, and let $\{X_k\}$ be generated by a Bernoulli source. Thus,

$$-\frac{\log \Pr\{X_1^n\}}{n} = -\frac{1}{n} \sum_{i=1}^{n} \log \Pr\{X_i\} \to \mathbf{E}\left[-\log \Pr\{X_1\}\right] = h \qquad \text{(a.s.),}$$

where the last implication follows from the Strong Law of Large Numbers (cf. [14]) applied to the sequence $(-\log \Pr\{X_1\}, \ldots, -\log \Pr\{X_n\})$. One should notice a difference between the definition of the entropy Equation 11.12 and the result above. In Equation 11.12 we take the *average* of $\log \Pr\{X_1^n\}$ while in the above we proved that almost surely for all but finitely sequences the probability $\Pr\{X_1^n\}$ can be closely approximated by e^{-nh}. For the Bernoulli model, we have already seen it above, but we are aiming at showing that the above conclusion is true for much more general probabilistic models.

As the next step, let us consider the mixing model (MX) (that includes as a special case the Markovian model (M)). For the mixing model the following is true:

$$\Pr\{X_1^{n+m}\} \leq c \Pr\{X_1^n\} \Pr\{X_{n+1}^{n+m}\}$$

for some constant $c > 0$ and any integers $n, m \geq 0$. Taking logarithm we obtain

$$\log \Pr\{X_1^{n+m}\} \leq \log \Pr\{X_1^n\} + \log \Pr\{X_{n+1}^{n+m}\} + \log c$$

which satisfies the subadditivity property Equation 11.11 of the *Subadditive Ergodic Theorem* discussed in Section 11.4.3. Thus, by Equation 11.10 we have

$$h = -\lim_{n \to \infty} \frac{\log \Pr\{X_1^n\}}{n} \qquad \text{(a.s.).}$$

Again, the reader should notice the difference between this result and the definition of the entropy.

We are finally ready to state the Shannon–McMillan–Breiman Theorem in its full generality (cf. [14]).

THEOREM 11.4   (Shannon–McMillan–Breiman) *For a stationary and ergodic sequence* $\{X_k\}_{k=-\infty}^{\infty}$ *the following holds:*

$$h = -\lim_{n \to \infty} \frac{\log \Pr\{X_1^n\}}{n} \qquad \text{(a.s.).}$$

where h is the entropy rate of the process $\{X_k\}$.

An important conclusion of this result is the so-called **asymptotic equipartition property** (AEP) which basically asserts that asymptotically typical sequences have the same probability approximately equal to e^{-nh}. More precisely, For a stationary and ergodic sequence X_1^n, the state space Σ^n can be partitioned into two subsets $\mathcal{B}_n^\varepsilon$ ("bad set") and $\mathcal{G}_n^\varepsilon$ ("good set") such that for given $\varepsilon > 0$ there is N_ε so that for $n \geq N_\varepsilon$ we have $\Pr\{\mathcal{B}_n^\varepsilon\} \leq \varepsilon$, and $e^{-nh(1+\varepsilon)} \leq \Pr\{x_1^n\} \leq e^{-nh(1-\varepsilon)}$ for all $x_1^n \in \mathcal{G}_n^\varepsilon$.

Example 11.7

Shortest common superstring or depth in a trie/suffix tree: For concreteness let us consider the SCS discussed in Section 11.2.3, but the same arguments as below can be used to derive the depth in a trie (cf. [48]) or a suffix tree (cf. [54]). Define C_{ij} as the length of the longest suffix of X_i that is equal to the prefix of X_j. Let $M_n(i) = \max_{1 \leq j \leq n, j \neq i}\{C_{ij}\}$. We write M_n for a generic random variable distributed as $M_n(i)$ (observe that $M_n \overset{d}{=} M_n(i)$ for all i, where $\overset{d}{=}$ means "equal in distribution"). We would like to prove that in the mixing model, for any $\varepsilon > 0$,

$$\lim_{n\to\infty} \Pr\left\{(1-\varepsilon)\frac{1}{h}\log n \leq M_n \leq (1+\varepsilon)\frac{1}{h}\log n\right\} = 1$$

provided $\alpha(g) \to 0$ as $g \to \infty$, that is, $M_n/\log n \to h$ (pr.). To prove an upper bound, we take any fixed typical sequence $w_k \in \mathcal{G}_k^\varepsilon$ as defined in AEP above, and observe that $\Pr\{M_n \geq k\} \leq n\Pr\{w_k\} + \Pr\{\mathcal{B}_k\}$. The result follows immediately after substituting $k = (1 + \varepsilon)h^{-1}\log n$ and noting that $\Pr\{w_k\} \leq e^{nh(1-\varepsilon)}$. For a lower bound, let $w_k \in \mathcal{G}_k^\varepsilon$ be any fixed typical sequence with $k = \frac{1}{h}(1 - \varepsilon)\log n$. Define Z_k as the number of strings $j \neq i$ such that a prefix of length k is equal to w_k and a suffix of length k of the ith string is equal to $w_k \in \mathcal{G}_k^\varepsilon$. Since w_k is fixed, the random variables C_{ij} are independent, and hence, by the second moment method (cf. Section 11.4.2)

$$\Pr\{M_n < k\} = \Pr\{Z_k = 0\} \leq \frac{\mathbf{Var}Z_k}{(\mathbf{E}Z_k)^2} \leq \frac{1}{n\Pr\{w_k\}} = O\left(n^{-\varepsilon^2}\right),$$

since $\mathbf{Var}Z_k \leq nP(w_k)$, and this completes the derivation.

In many problems on words another kind of entropy is widely used (cf. [4,5,54]). It is called Rényi entropy and defined as follows: For $-\infty \leq b \leq \infty$, the bth order Rényi entropy is

$$h_b = \lim_{n\to\infty} \frac{-\log\left(\mathbf{E}\left[\Pr\{X_1^n\}^{b-1}\right]\right)}{bn} = \lim_{n\to\infty} \frac{-\log\left(\sum_{w\in\Sigma^n}(\Pr\{w\})^b\right)^{-1/b}}{n}, \tag{11.13}$$

provided the above limit exists. In particular, by inequalities on means we obtain $h_0 = h$ and

$$h_{-\infty} = \lim_{n\to\infty} \frac{\max\{-\log\Pr\{X_1^n\}, \Pr\{X_1^n\} > 0\}}{n},$$

$$h_\infty = \lim_{n\to\infty} \frac{\min\{-\log\Pr\{X_1^n\}, \Pr\{X_1^n\} > 0\}}{n}.$$

For example, the entropy $h_{-\infty}$ appears in the formulation of the shortest path in digital trees (cf. [48,54]), the entropy h_∞ is responsible for the height in PATRICIA tries (cf. [48,54]), while h_2 determines the height in a trie. Indeed, we claim that in the mixing model the height H_n in a trie behaves probabilistically as $H_n/\log n \to 2/h_2$. To prove it, one should follow the footsteps of our discussion in Examples 11.4 and 11.5 (details can be found in [48,54]).

11.4.5 Central Limit and Large Deviations Results

Convergence of a sum of independent, identically distributed (i.i.d.) random variables is central to probability theory. In the analysis of algorithms, we mostly deal with weakly dependent random variables, but often results from the i.i.d. case can be extended to this new situation by some clever tricks. A more systematic treatment of such cases is usually done through generating functions and complex analysis techniques (cf. [30–32,44]), which we briefly discuss in the next section. Hereafter, we concentrate on the i.i.d. case.

Let us consider a sequence $X_1, \ldots, X_n$ of i.i.d. random variables, and let $S_n = X_1 + \cdots + X_n$. Define $\mu := \mathbf{E}[X_1]$ and $\sigma^2 := \mathbf{Var}[X_1]$. We pay particular interest to another random variable, namely

$$s_n := \frac{S_n - n\mu}{\sigma\sqrt{n}}$$

which distribution function we denote as $F_n(x) = \Pr\{s_n \leq x\}$. Let also $\Phi(x)$ be the distribution function of the standard normal distribution, that is,

$$\Phi(x) := \frac{1}{\sqrt{2\pi}} \int_{-\infty}^{x} e^{-\frac{1}{2}t^2} \, dt.$$

The Central Limit Theorem asserts that $F_n(x) \to \Phi(x)$ for continuity points of $F_n(\cdot)$, provided $\sigma < \infty$ (cf. [14,15]). A stronger version is due to Berry-Esséen who proved that

$$|F_n(x) - \Phi(x)| \leq \frac{2\rho}{\sigma^2\sqrt{n}} \tag{11.14}$$

where $\rho = \mathbf{E}[|X - \mu|^3] < \infty$. Finally, Feller [15] has shown that if centralized moments $\mu_2, \ldots, \mu_r$ exist, then

$$F_n(x) = \Phi(x) - \frac{1}{\sqrt{2\pi}} e^{-\frac{1}{2}x^2} \sum_{k=3}^{r} n^{-\frac{1}{2}k+1} R_k(x) + O\left(n^{-\frac{1}{2}r+\frac{1}{2}}\right)$$

uniformly in x, where $R_k(x)$ is a polynomial depending only on $\mu_1, \ldots, \mu_r$ but not on n and r.

One should notice from the above, in particular from Equation 11.14, the weakness of central limit results that are able only to assess the probability of small deviations from the mean. Indeed, the results above are true for $x = O(1)$ (i.e., for $S_n \in (\mu n - O(\sqrt{n}), \mu n + O(\sqrt{n}))$ due to only a polynomial rate of convergence as shown in Equation 11.14. To see it more clearly, we quote a result from Greene and Knuth [25] who estimated

$$\Pr\{S_n = \mu n + r\} = \frac{1}{\sigma\sqrt{2\pi n}} \exp\left(\frac{-r^2}{2\sigma^2 n}\right) \left(1 - \frac{\kappa_3}{2\sigma^4}\left(\frac{r}{n}\right) + \frac{\kappa_3}{6\sigma^6}\left(\frac{r^3}{n^2}\right)\right) + O\left(n^{-\frac{3}{2}}\right) \tag{11.15}$$

where κ_3 is the third cumulant of X_1. Observe now that when $r = O(\sqrt{n})$ (which is equivalent to $x = O(1)$ in our previous formulæ) the error term dominates the leading term of the above asymptotic, thus, the estimate is quite useless.

From the above discussion, one should conclude that the central limit theorem has limited range of application, and one should expect another law for large deviations from the mean, that is, when $x_n \to \infty$ in the above formulæ. The most interesting from the application point of view is the case when $x = O(\sqrt{n})$ (or $r = O(n)$), that is, for $\Pr\{S_n = n(\mu + \delta)\}$ for $\delta \neq 0$. We shall discuss this large deviations behavior next.

Let us first try to "guess" a large deviation behavior of $S_n = X_1 + \cdots + X_n$ for i.i.d. random variables. We estimate $\Pr\{S_n \geq an\}$ for $a > 1$ as $n \to \infty$. Observe that (cf. [14])

$$\Pr\{S_{n+m} \geq (n+m)a\} \geq \Pr\{S_m \geq ma, \, S_{n+m} - S_m \geq na\} = \Pr\{S_n \geq na\} \Pr\{S_m \geq ma\},$$

since S_m and $S_{n+m} - S_m$ are independent. Taking logarithm of the above, and recognizing that $\log \Pr\{S_n \geq an\}$ is a superadditive sequence (cf. Section 11.4.3), we obtain

$$\lim_{n \to \infty} \frac{1}{n} \log \Pr\{S_n \geq na\} = -I(a),$$

where $I(a) \geq 0$. Thus, S_n decays exponentially when far away from its mean, not in a Gaussian way as the central limit theorem would predict! Unfortunately, we obtain the above result from the subadditive property which allowed us to conclude the existence of the above limit, but says nothing about $I(a)$.

In order to take the full advantage of the above derivation, we should say something about $I(a)$ and, more importantly, to show that $I(a) > 0$ under some mild conditions. For the latter, let us first assume that the moment generating function $M(\lambda) = \mathbf{E}[e^{\lambda X_1}] < \infty$ for some $\lambda > 0$. Let also $\kappa(\lambda) = \log M(\lambda)$ be the cumulant function of X_1. Then, by Markov's inequality (cf. Section 11.4.2)

$$e^{\lambda na} \Pr\{S_n \geq na\} = e^{\lambda na} \Pr\left\{e^{\lambda S_n} \geq e^{\lambda na}\right\} \leq \mathbf{E}e^{\lambda S_n}.$$

Actually, due to arbitrariness of $\lambda > 0$, we finally arrive at the so-called Chernoff bound, that is,

$$\Pr\{S_n \geq na\} \leq \min_{\lambda > 0}\left\{e^{-\lambda na}\mathbf{E}\left[e^{\lambda S_n}\right]\right\}. \tag{11.16}$$

We should emphasize that the above bound is true for dependent random variables since we only used Markov's inequality applied to S_n.

Returning to the i.i.d. case, we can rewrite the above as

$$\Pr\{S_n \geq na\} \leq \min_{\lambda > 0}\left\{\exp(-n(a\lambda - \kappa(\lambda)))\right\}.$$

But, under mild conditions the above minimization problem is easy to solve. One finds that the minimum is attended at λ_a which satisfies $a = M'(\lambda_a)/M(\lambda_a)$. Thus, we proved that $I(a) \geq a\lambda_a - \log M(\lambda_a)$. A more careful evaluation of the above leads to the following classical large deviations result (cf. [14])

THEOREM 11.5 *Assume $X_1, \ldots, X_n$ are i.i.d. Let $M(\lambda) = \mathbf{E}[e^{\lambda X_1}] < \infty$ for some $\lambda > 0$, the distribution of X_i is not a point mass at μ, and there exists $\lambda_a > 0$ in the domain of the definition of $M(\lambda)$ such that*

$$a = \frac{M'(\lambda_a)}{M(\lambda_a)}.$$

Then

$$\lim_{n \to \infty} \frac{1}{n} \log \Pr\{S_n \geq na\} = -\left(a\lambda_a - \log M(\lambda_a)\right)$$

for $a > \mu$.

A major strengthening of this theorem is due to Gärtner and Ellis (cf. [13]) who extended it to weakly dependent random variables. Let us consider S_n as a sequence of random variables (e.g., $S_n = X_1 + \cdots + X_n$), and let $M_n(\lambda) = \mathbf{E}[e^{\lambda S_n}]$. The following is known (cf. [13,29]):

THEOREM 11.6 (Gärtner–Ellis) *Let*

$$\lim_{n \to \infty} \frac{\log M_n(\lambda)}{n} = c(\lambda)$$

exist and is finite in a subinterval of the real axis. If there exists λ_a such that $c'(\lambda_a)$ is finite and $c'(\lambda_a) = a$, then

$$\lim_{n \to \infty} \frac{1}{n} \log Pr\{S_n \geq na\} = -(a\lambda_a - c(\lambda_a)).$$

Let us return again to the i.i.d. case and see if we can strengthen Theorem 11.5 which in its present form gives only a logarithmic limit. We explain our approach on a simple example, following Greene and Knuth [25]. Let us assume that $X_1, \ldots, X_n$ are discrete i.i.d. with common generating function $G(z) = E[z^X]$. We recall that $[z^m]G(z)$ denote the coefficient at z^m of $G(z)$. In Equation 11.15 we show how to compute such a coefficient at $m = \mu n + O(\sqrt{n})$ of $G^n(z) = Ez^{S_n}$. We observed also that Equation 11.15 cannot be used for large deviations, since the error term was dominating the leading term in such a case. But, one may shift the mean of S_n to a new value such that Equation 11.15 is valid again. Thus, let us define a new random variable $\widetilde{X}$ whose generating function is $\widetilde{G}(z) = \frac{G(z\alpha)}{G(\alpha)}$ where α is a constant that is to be determined. Observe that $E[\widetilde{X}] = \widetilde{G}'(1) = \alpha G'(\alpha)/G(\alpha)$. Suppose we need large deviations result around $m = n(\mu + \delta)$ where $\delta > 0$. Clearly, Equation 11.15 cannot be applied directly. Now, a proper choice of α can help. Let us select α such that the new $\widetilde{S}_n = \widetilde{X}_1 + \cdots + \widetilde{X}_n$ has mean $m = n(\mu + \delta)$. This results in setting α to be a solution of

$$\frac{\alpha G'(\alpha)}{G(\alpha)} = \frac{m}{n} = \mu + \delta.$$

In addition, we have the following obvious identity

$$[z^m] G^n(z) = \frac{G^n(\alpha)}{\alpha^m} [z^m] \left(\frac{G(\alpha z)}{G(\alpha)}\right)^n. \tag{11.17}$$

But, now we can use Equation 11.15 to the right-hand side of the above, since the new random variable $\widetilde{S}_n$ has mean around m.

To illustrate the above technique that is called shift of mean we present an example.

Example 11.8

Large deviations by "shift of mean" (cf. [25]): Let S_n be binomially distributed with parameter $1/2$, that is, $G^n(z) = ((1+z)/2)^n$. We want to estimate the probability $Pr\{S_n = n/3\}$, which is far away from its mean ($ES_n = n/2$) and central limit result Equation 11.15 cannot be applied. We apply the shift of mean method, and compute α as

$$\frac{\alpha G'(\alpha)}{G(\alpha)} = \frac{\alpha}{1 + \alpha} = \frac{1}{3},$$

thus, $\alpha = 1/2$. Using Equation 11.15 we obtain

$$\left[z^{n/3}\right] \left(\frac{2}{3} + \frac{1}{3}z\right)^n = \frac{3}{2\sqrt{\pi n}} \left(1 - \frac{7}{24n}\right) + O\left(n^{-5/2}\right).$$

To derive the result we want (i.e., coefficient at $z^{n/3}$ of $(z/2 + 1/2)^n$), one must apply Equation 11.17. This finally leads to

$$\left[z^{n/3}\right] (z/2 + 1/2)^n = \left(\frac{3 \cdot 2^{1/3}}{4}\right)^n \frac{3}{2\sqrt{\pi n}} \left(1 - \frac{7}{24n} + O\left(n^{-2}\right)\right),$$

which is a large deviations result (the reader should observe the exponential decay of the above probability).

The last example showed that one may expect a stronger large deviation result than presented in Theorem 11.5. Indeed, under proper mild conditions it can be proved that Theorem 11.5 extends to (cf. [13,29])

$$\Pr\{S_n \geq na\} \sim \frac{1}{\sqrt{2\pi n \sigma_a \lambda_a}} \exp(-nI(a))$$

where
$\sigma_a = M''(\lambda_a)$ with λ_a
$I(a) = a\lambda_a - \log M(\lambda_a)$ as defined in Theorem 11.5

Finally, we deal with an interesting extension of the above large deviations results initiated by Azuma, and recently significantly extended by Talagrand [56]. These results are known in the literature under the name Azuma's type inequality or method of bounded differences (cf. [45]). It can be formulated as follows:

THEOREM 11.7 **(Azuma's Type Inequality)** *Let X_i be i.i.d. random variables such that for some function $f(\cdot, \ldots, \cdot)$ the following is true*

$$\left| f(X_1, \ldots, X_i, \ldots, X_n) - f(X_1, \ldots, X_i', \ldots, X_n) \right| \leq c_i, \tag{11.18}$$

where $c_i < \infty$ are constants, and X_i' has the same distribution as X_i. Then,

$$\Pr\left\{ \left| f(X_1, \ldots, X_n) - \mathbf{E}[f(X_1, \ldots, X_n)] \right| \geq t \right\} \leq 2 \exp\left(-2t^2 \Big/ \sum_{i=1}^{n} c_i^2 \right) \tag{11.19}$$

for some $t > 0$.

We finish this discussion with an application of the Azuma inequality.

Example 11.9

Concentration of mean for the editing problem: Let us consider again the editing problem from Section 11.2.2 the following is true:

$$\Pr\{|C_{max} - EC_{max}| > \varepsilon \mathbf{E}[C_{max}]\} \leq 2 \exp\left(-\varepsilon^2 \alpha n\right),$$

provided all weights are bounded random variables, say $\max\{W_I, W_D, W_Q\} \leq 1$. Indeed, under the Bernoulli model, the X_i are i.i.d. (where X_i, $1 \leq i \leq n = l+s$, represents symbols of the two underlying sequences), and therefore Equation 11.18 holds with $f(\cdot) = C_{max}$. More precisely,

$$\left| C_{max}(X_1, \ldots, X_i, \ldots, X_n) - C_{max}(X_1, \ldots, X_i', \ldots, X_n) \right| \leq \max_{1 \leq i \leq n} \{W_{max}(i)\},$$

where $W_{max}(i) = \max\{W_I(i), W_D(i), W_Q(i)\}$. Setting $c_i = 1$ and $t = \varepsilon E C_{max} = O(n)$ in the Azuma inequality we obtain the desired concentration of mean inequality.

11.5 Analytic Techniques

Analytic analysis of algorithms was initiated by Knuth almost 30 years ago in his magnum opus [39–41], which treated many aspects of fundamental algorithms, seminumerical algorithms, and sorting and searching. A modern introduction to analytic methods can be found in a marvelous book by Sedgewick and Flajolet [49], while advanced analytic techniques are covered in a forthcoming book *Analytical Combinatorics* by Flajolet and Sedgewick [21]. In this section, we only briefly discuss functional equations arising in the analysis of digital trees, complex asymptotics techniques, Mellin transform, and analytic depoissonization.

11.5.1 Recurrences and Functional Equations

Recurrences and functional equations are widely used in computer science. For example, the divide-and-conquer recurrence equations appear in the analysis of searching and sorting algorithms (cf. [41]). Hereafter, we concentrate on recurrences and functional equations that arise in the analysis of digital trees and problems on words.

However, to introduce the reader into the main subject we first consider two well-known functional equations. Let us enumerate the number of unlabeled binary trees built over n vertices. Call this number b_n, and let $B(z) = \sum_{n=0}^{\infty} b_n z^n$ be its ordinary generating function. Since each such a tree is constructed in a recursive manner with left and right subtrees being unlabeled binary trees, we immediately arrive at the following recurrence $b_n = b_0 b_{n-1} + \cdots + b_{n-1} b_0$ for $n \geq 1$ with $b_0 = 1$ by definition. Multiplying by z^n and summing from $n = 1$ to infinity, we obtain $B(z) - 1 = zB^2(z)$ which is a simple functional equation that can be solved to find

$$B(z) = \frac{1 - \sqrt{1 - 4z}}{2z}.$$

To derive the above functional equation, we used a simple fact that the generating function $C(z)$ of the convolution c_n of two sequences, say a_n and b_n (i.e., $c_n = a_0 b_n + a_1 b_{n-1} + \cdots + a_n b_0$), is the product of $A(z)$ and $B(z)$, that is, $C(z) = A(z)B(z)$.

The above functional equation and its solution can be used to obtain an explicit formula on b_n. Indeed, we first recall that $[z^n]B(z)$ denotes the coefficient at z^n of $B(z)$ (i.e., b_n). A standard analysis leads to (cf. [39,44])

$$b_n = \left[z^n\right] B(z) = \frac{1}{n+1} \binom{2n}{n},$$

which is the famous Catalan number.

Let us now consider a more challenging example, namely, enumeration of rooted labeled trees. Let t_n the number of rooted labeled trees, and $t(z) = \sum_{n=0}^{\infty} \frac{t_n}{n!} z^n$ its exponential generating function. It is known that $t(z)$ satisfies the following functional equation (cf. [28,49]) $t(z) = ze^{t(z)}$. To find t_n, we apply Lagrange's Inversion Formula. Let $\Phi(u)$ be a formal power series with $[u^0]\Phi(u) \neq 0$, and let $X(z)$ be a solution of $X = z\Phi(X)$. The coefficients of $X(z)$ or in general $\Psi(X(z))$ where Ψ is an arbitrary series can be found by

$$\left[z^n\right] X(z) = \frac{1}{n} \left[u^{n-1}\right] (\Phi(u))^n,$$

$$\left[z^n\right] \Psi(X(z)) = \frac{1}{n} \left[u^{n-1}\right] (\Phi(u))^n \Psi'(u).$$

In particular, an application of the above to $t(z)$ leads to $t_n = n^{n-1}$, and to an interesting formula (which we encounter again in Example 11.14)

$$t(z) = \sum_{n=1}^{\infty} \frac{n^{n-1}}{n!} z^n. \tag{11.20}$$

After these introductory remarks, we can now concentrate on certain recurrences that arise in problems on words; in particular in digital trees and SCS problems. Let x_n be a generic notation for a quantity of interest (e.g., depth, size or path length in a digital tree built over n strings). Given x_0 and x_1, the following three recurrences originate from problems on tries, PATRICIA tries, and DSTs, respectively (cf. [16,28,37,41,44,49,51–53]):

$$x_n = a_n + \beta \sum_{k=0}^{n} \binom{n}{k} p^k q^{n-k} (x_k + x_{n-k}), \quad n \geq 2 \tag{11.21}$$

$$x_n = a_n + \beta \sum_{k=1}^{n-1} \binom{n}{k} p^k q^{n-k} (x_k + x_{n-k}) - \alpha (p^n + q^n) x_n, \quad n \geq 2 \tag{11.22}$$

$$x_{n+1} = a_n + \beta \sum_{k=0}^{n} \binom{n}{k} p^k q^{n-k} (x_k + x_{n-k}), \quad n \geq 0 \tag{11.23}$$

where
a_n is a known sequence (also called additive term)
α and β are some constants
$p + q = 1$

To solve this recurrences and to obtain explicit or asymptotic expression for x_n we apply exponential generating function approach. We need to know the following two facts: Let a_n and b_n be sequences with $a(z) = \sum_{n=0}^{\infty} \frac{a_n}{n!} z^n$ and $b(z)$ as their exponential generating functions. (Hereafter, we consequently use lower-case letters for exponential generating functions, like $a(z)$, and upper-case letters for ordinary generating functions, like $A(z)$). Then,

- For any integer $h \geq 0$

$$\frac{d^h}{dz^h} a(z) = \sum_{n=0}^{\infty} \frac{a_{n+h}}{n!} z^n.$$

- If $c_n = \sum_{r=0}^{n} \binom{n}{r} a_r b_{n-r}$, then the exponential generating function $c(z)$ of c_n becomes $c(z) = a(z)b(z)$.

Now, we are ready to attack the above recurrences and show how they can be solved. Let us start with the simplest one, namely Equation 11.21. Multiplying it by z^n, summing up, and taking into account the initial conditions, we obtain

$$x(z) = a(z) + \beta x(zp)e^{zq} + \beta x(zq)e^{zp} + d(z), \tag{11.24}$$

where $d(z) = d_0 + d_1 z$ and d_0 and d_1 depend on the initial condition for $n = 0, 1$. The trick is to introduce the so-called Poisson transform $\tilde{X}(z) = x(z)e^{-z}$ which reduces the above functional equation to

$$\tilde{X}(z) = \tilde{A}(z) + \beta \tilde{X}(zp) + \beta \tilde{X}(z) + d(z)e^{-z}. \tag{11.25}$$

Observe that $\tilde{x}_n$ and x_n are related by $x_n = \sum_{k=0}^{n} \binom{n}{k} \tilde{x}_k$. Using this, and comparing coefficients of $\tilde{X}(z)$ at z^n we finally obtain

$$x_n = x_0 + n(x_1 - x_0) + \sum_{k=2}^{n} (-1)^k \binom{n}{k} \frac{\hat{a}_k + kd_1 - d_0}{1 - \beta(p^k + q^k)}, \tag{11.26}$$

where $n![z^n]\tilde{A}(z) = \tilde{a}_n := (-1)^n \hat{a}_n$. In fact, $\hat{a}_n$ and a_n form the so-called binomial inverse relations, i.e.,

$$\hat{a}_n = \sum_{k=0}^{n} \binom{n}{k} (-1)^k a_k, \quad a_n = \sum_{k=0}^{n} \binom{n}{k} (-1)^k \hat{a}_k,$$

and $\hat{\hat{a}}_n = a_n$ (cf. [41]).

Example 11.10

Average path length in a trie: Let us consider a trie in the Bernoulli model, and estimate the average length, ℓ_n, of the external path length, i.e., $\ell_n = \mathbf{E}[L_n]$ (cf. Section 11.2.1). Clearly, $\ell_0 = \ell_1 = 0$ and for $n \geq 2$

$$\ell_n = n + \sum_{k=0}^{n} \binom{n}{k} p^k q^{n-k} \left(\ell_k + \ell_{n-k} \right).$$

Thus, by Equation 11.26

$$\ell_n = \sum_{k=2}^{n} (-1)^k \binom{n}{k} \frac{k}{1 - p^k - q^k}.$$

Below, we shall discuss the asymptotics of ℓ_n (cf. Example 11.15).

Let us now consider recurrence Equation 11.22, which is much more intricate. It has an exact solution only for some special cases (cf. [41,51,53]) that we discuss below. We first consider a simplified version of Equation 11.22, namely

$$x_n \left(2^n - 2 \right) = 2^n a_n + \sum_{k=1}^{n-1} \binom{n}{k} x_k.$$

with $x_0 = x_1 = 0$ (for a more general recurrence of this type see [51]). After multiplying by z^n and summing up we arrive at

$$x(z) = \left(e^{z/2} + 1 \right) x(z/2) + a(z) - a_0 \tag{11.27}$$

where $x(z)$ and $a(z)$ are exponential generating functions of x_n and a_n. To solve this recurrence we must observe that after multiplying both sides by $z/(e^z - 1)$ and defining

$$\check{X}(z) = x(z) \frac{z}{e^z - 1} \tag{11.28}$$

we obtain a new functional equation that is easier to solve, namely

$$\check{X}(z) = \check{X}(z/2) + \check{A}(z),$$

where in the above we assume for simplicity $a_0 = 0$. This functional equation is of a similar type as $\widetilde{X}(x)$ considered above. The coefficient $\check{x}_n$ at z^n of $\check{X}(z)$ can also be easily extracted. One must, however, translate coefficients $\check{x}_n$ into the original sequence x_n. In order to accomplish this, let us introduce the Bernoulli polynomials $B_n(x)$ and Bernoulli numbers $B_n = B_n(0)$, that is, $B_n(x)$ is defined as

$$\frac{z e^{tz}}{e^z - 1} = \sum_{k=0}^{\infty} B_k(t) \frac{z^k}{k!}.$$

Furthermore, we introduce Bernoulli inverse relations for a sequence a_n as

$$\check{a}_n = \sum_{k=0}^{n} \binom{n}{k} B_k a_{n-k} \quad \Longleftrightarrow \quad a_n = \sum_{k=0}^{n} \binom{n}{k} \frac{\check{a}_k}{k+1}.$$

One should know that (cf. [41])

$$a_n = \binom{n}{r} q^n \quad \Longleftrightarrow \quad \check{a}_n = \binom{n}{r} q^r B_{n-r}(q).$$

for $0 < q < 1$. For example, for $a_n = \binom{n}{r}q^n$ the above recurrence has a particular simply solution, namely:

$$x_n = \sum_{k=1}^{n} (-1)^k \binom{n}{k} \frac{B_{k+1}(1-q)}{k+1} \frac{1}{2^{k+1}-1}.$$

A general solution to the above recurrence can be found in [51], and it involves $\breve{a}_n$.

Example 11.11

Unsuccessful search in PATRICIA: Let us consider the number of trials u_n in an unsuccessful search of a string in a PATRICIA trie constructed over the symmetric Bernoulli model (i.e., $p = q = 1/2$). As in Knuth [41]

$$u_n \left(2^n - 2\right) = 2^n \left(1 - 2^{1-n}\right) + \sum_{k=1}^{n-1} \binom{n}{k} u_k$$

and $u_0 = u_1 = 0$. A simple application of the above derivation leads, after some algebra, to (cf. [53])

$$u_n = 2 - \frac{4}{n+1} + 2\delta_{n0} + \frac{2}{n+1} \sum_{k=2}^{n} \binom{n+1}{k} \frac{B_k}{2^{k-1}-1}$$

where $\delta_{n,k}$ is the Kronecker delta, that is, $\delta_{n,k} = 1$ for $n = k$ and zero otherwise.

We were able to solve the functional equations (Equation 11.24) and (Equation 11.27) exactly, since we reduce them to a simple functional equation of the form (Equation 11.25). In particular, Equation 11.27 became Equation 11.25, since luckily $e^z - 1 = (e^{z/2} - 1)(e^{z/2} + 1)$, as already pointed out by Knuth [41], but one cannot expect that much luck with other functional equations. Let us consider a general functional equation

$$F(z) = a(z) + b(z)F(\sigma(z)), \tag{11.29}$$

where $a(z), b(z), \sigma(z)$ are known functions. Formally, iterating this equation we obtain

$$F(z) = \sum_{k=0}^{\infty} a\left(\sigma^{(k)}(z)\right) \prod_{j=0}^{k-1} b\left(\sigma^{(j)}(z)\right),$$

where $\sigma^{(k)}(z)$ is the kth iterate of $\sigma(\cdot)$. When applying the above to solve real problems, one must assure the existence of the infinite series involved. In some cases (cf. [19,38]), we can provide asymptotic solutions to such complicated formulæ by appealing to the Mellin transform which we discuss below in Section 11.5.3.

Finally, we deal with the recurrence Equation 11.23. Multiplying both sides by $z^n/n!$ and using the above discussed properties of exponential generating functions we obtain for $x(z) = \sum_{n \geq 0} x_n \frac{z^n}{n!}$

$$x'(z) = a(z) + x(zp)e^{zq} + x(zq)e^{zp},$$

which becomes after the substitution $\widetilde{X}(z) = x(z)e^{-z}$

$$\widetilde{X}'(z) + \widetilde{X}(z) = \widetilde{A}(z) + \widetilde{X}(zp) + \widetilde{X}(zq). \tag{11.30}$$

The above is a differential–functional equation that we did not discuss so far. It can be solved, since a direct translation of coefficients gives $\widetilde{x}_{n+1} + \widetilde{x}_n = \widetilde{a}_n + \widetilde{x}_n(p^n + q^n)$. Fortunately, this is a simple linear recurrence that has an explicit solution. Taking into account $x_n = \sum_{k=0}^{n} \binom{n}{k} \tilde{x}_k$, we finally obtain

$$x_n = x_0 - \sum_{k=1}^{n}(-1)^k \binom{n}{k} \sum_{i=1}^{k-1} \hat{a}_i \prod_{j=i+1}^{k-1} \left(1 - p^j - q^j\right) = x_0 - \sum_{k=1}^{n}(-1)^k \binom{n}{k} \sum_{i=1}^{k-1} \hat{a}_i \frac{Q_k}{Q_i}, \quad (11.31)$$

where $Q_k = \prod_{j=2}^{k}(1 - p^j - q^j)$, and $\hat{a}_n$ is the binomial inverse of a_n as defined above. In passing, we should observe that solutions of the recurrences (Equations 11.21 through 11.23) have a form of an alternating sum, that is, $x_n = \sum_{k=1}^{n}(-1)^k \binom{n}{k} f_n$ where f_n has an explicit formula. In Section 11.5.3, we discuss how to obtain asymptotics of such an alternating sum.

Example 11.12

Expected path length in a digital search tree: Let ℓ_n be the expected path length in a DSI. Then (cf. [21,41]) for all $n \geq 0$,

$$\ell_{n+1} = n + \sum_{k=1}^{n} \binom{n}{k} p^k q^{n-k} \left(\ell_k + \ell_{n-k}\right)$$

with $\ell_0 = 0$. By Equation 11.31 it has the following solution:

$$\ell_n = \sum_{k=2}^{n}(-1)^k \binom{n}{k} Q_{k-1}$$

where Q_k is defined above.

We were quite lucky when solving the above differential-functional equation, since we could reduce it to a linear recurrence of first order. However, this is not any longer true when we consider the so-called b-DSTs in which one assumes that a node can store up to b strings. Then, the general recurrence (Equation 11.23) becomes

$$x_{n+b} = a_n + \beta \sum_{k=0}^{n} \binom{n}{k} p^k q^{n-k} \left(x_k + x_{n-k}\right) \quad n \geq 0,$$

provided $x_0, \ldots, x_{b-1}$ are given. Our previous approach would lead to a linear recurrence of order b that does not possess a nice explicit solution. The "culprit" lies in the fact that the exponential generating function of a sequence $\{x_{n+b}\}_{n=0}^{\infty}$ is the bth derivative of the exponential generating function $x(z)$ of $\{x_n\}_{n=0}^{\infty}$. On the other hand, if one considers ordinary generating function $X(z) = \sum_{n \geq 0} x_n z^n$, then the sequence $\{x_{n+b}\}_{n=0}^{\infty}$ translates into $z^{-b}(X(z) - x_0 - \cdots - x_{b-1} z^{b-1})$. This observation led Flajolet and Richmond [18] to reconsider the standard approach to the above binomial recurrences, and to introduce ordinary generating function into the play. A careful reader observes, however, that then one must translate into ordinary generating functions sequences such as $s_n = \sum_{k=0}^{n} \binom{n}{k} a_k$ (which were easy under exponential generating functions since they become $a(z)e^z$). But, it is not difficult to see that

$$s_n = \sum_{k=0}^{n} \binom{n}{k} a_k \quad \Longrightarrow \quad S(z) = \frac{1}{1-z} A\left(\frac{z}{1-z}\right).$$

Indeed,

$$\frac{1}{1-z}A\left(\frac{z}{1-z}\right) = \sum_{m=0}^{\infty} a_m z^m \frac{1}{(1-z)^{m+1}} = \sum_{m=0}^{\infty} a_m z^m \sum_{j=0}^{\infty} \binom{m+j}{j} z^j$$

$$= \sum_{n=0}^{\infty} z^n \sum_{k=0}^{\infty} \binom{n}{k} a_k.$$

Thus, the above recurrence for $p = q = 1/2$ and any $b \geq 1$ can be translated into ordinary generating functions as

$$X(z) = \frac{1}{1-z}G\left(\frac{z}{1-z}\right)$$

$$G(z)(1+z)^b = 2z^b G(z/2) + P(z),$$

where $P(z)$ is a function of a_n and initial conditions. But, the latter functional equation falls under Equation 11.29 and its solution was already discussed above.

11.5.2 Complex Asymptotics

When analyzing an algorithm we often aim at predicting its rate of growth of time or space complexity for large inputs, n. Precise analysis of algorithms aims at obtaining precise asymptotics and/or full asymptotic expansions of such performance measures. For example, in the previous subsection we studied some parameters of tries (e.g., path length ℓ_n, unsuccessful search u_n, etc.) that depend on input of size n. We observed that these quantities are expressed by some complicated alternating sums (cf. Examples 11.10 through 11.12). One might be interested in precise rate of growth of these quantities. More precisely, if x_n represents a quantity of interest with input size n, we may look for a simple explicit function a_n (e.g., $a_n = \log n$ or $a_n = \sqrt{n}$) such that $x_n \sim a_n$ (i.e., $\lim_{n\to\infty} x_n/a_n = 1$), or we may be aiming at a very precise asymptotic expansion such as $x_n = a_n^1 + a_n^2 + \cdots + o(a_n^k)$ where for each $1 \leq i \leq k$ we have $a_n^{i+1} = o(a_n^i)$.

The reader is referred to an excellent recent survey by Odlyzko [47] on asymptotic methods. In this subsection, we briefly discuss some elementary facts of asymptotic evaluation, and describe a few useful methods.

Let $A(z) = \sum_{n=0}^{\infty} a_n z^n$ be the generating function of a sequence $\{a_n\}_{n=0}^{\infty}$. In the previous subsection, we look at $A(z)$ as a formal power series. Now, we ask whether $A(z)$ converges, and what is its region of convergence. It turns out that the radius of convergence for $A(z)$ is responsible for the asymptotic behavior of a_n for large n. Indeed, by Hadamard's Theorem [27,57] we know that radius R of convergence of $A(z)$ (where z is a complex variable) is given by

$$R = \frac{1}{\limsup_{n\to\infty} |a_n|^{1/n}}.$$

In other words, for every $\varepsilon > 0$ there exists N such that for $n > N$ we have

$$|a_n| \leq \left(R^{-1} + \varepsilon\right)^n,$$

for infinitely many n we have

$$|a_n| \geq \left(R^{-1} - \varepsilon\right)^n.$$

Informally saying, $\frac{1}{n} \log |a_n| \sim 1/R$, or even less formally the exponential growth of a_n is determined by $(1/R)^n$. In summary, singularities of $A(z)$ determine asymptotic behavior of its coefficients for large n. In fact, from Cauchy's Integral Theorem (cf. Section 11.3.3) we know that

$$|a_n| \le \frac{M(r)}{r^n}$$

where $M(r)$ is the maximum value of $|A(z)|$ over a circle of radius $r < R$.

Our goal now is to make a little more formal our discussion above, and deal with multiple singularities. We restrict ourselves to meromorphic functions $A(z)$, i.e., ones that are analytic with the exception of a finite number of *poles*. To make our discussion even more concrete we study the following function

$$A(z) = \sum_{j=1}^{r} \frac{a_{-j}}{(z-\rho)^j} + \sum_{j=0}^{\infty} a_j (z-\rho)^j.$$

More precisely, we assume that $A(z)$ has the above Laurent expansion around a pole ρ of multiplicity r. Let us further assume that the pole ρ is the closest to the origin, that is, $R = |\rho|$ (and there are no more poles on the circle of convergence). In other words, the sum of $A(z)$ which we denote for simplicity as $A_1(z)$, is analytic in the circle $|z| \le |\rho|$, and its possible radius of convergence $R' > |\rho|$. Thus, coefficients a'_n of $A_1(n)$ are bounded by $|a'_n| = O((1/R' + \varepsilon)^n)$ for any $\varepsilon > 0$. Let us now deal with the first part of $A(z)$. Using the fact that $[z^n](1-z)^{-r} = \binom{n+r-1}{r-1}$ for a positive integer r, we obtain

$$\sum_{j=1}^{r} \frac{a_j}{(z-\rho)^j} = \sum_{j=1}^{r} \frac{a_j(-1)^j}{\rho^j(1-z/\rho)^j}$$

$$= \sum_{j=1}^{r} (-1)^j a_j \rho^{-j} \sum_{n=0}^{\infty} \binom{n+j-1}{n} \left(\frac{z}{\rho}\right)^n$$

$$= \sum_{n=1}^{\infty} z^n \sum_{j=1}^{r} (-1)^j a_j \binom{n}{j-1} \rho^{-(n+j)}.$$

In summary, we prove that

$$[z^n] A(z) = \sum_{j=1}^{r} (-1)^j a_j \binom{n}{j-1} \rho^{-(n+j)} + O\left((1/R' + \varepsilon)^n\right)$$

for $R' > \rho$ and any $\varepsilon > 0$.

Example 11.13

Frequency of a given pattern occurrence: Let H be a given pattern of size m, and consider a random text of length n generated according to the Bernoulli model. An old and well-studied problem of pattern matching (cf. [15]) asks for an estimation of the number O_n of pattern H occurrences in the text. Let $T_r(z) = \sum_{n=0}^{\infty} \Pr\{O_n = r\}z^n$ denote the generating function for $\Pr\{O_n = r\}$ for $|z| \le 1$. It can be proved (cf. [24,26]) that

$$T_r(z) = \frac{z^m P(H)(D(z) + z - 1)^{r-1}}{D^{r+1}(z)},$$

where

$D(z) = P(H)z^m + (1 - z)A_H(z)$

$A(z)$ is the so-called autocorrelation polynomial (a polynomial of degree m)

It is also easy to see that there exists smallest $\rho > 1$ such that $D(\rho) = 0$. Then, an easy application of the above analysis leads to

$$\Pr\{O_n(H) = r\} = \sum_{j=1}^{r+1}(-1)^j a_j \binom{n}{j-1}\rho^{-(n+j)} + O\left(\rho_1^{-n}\right)$$

where $\rho_1 > \rho$ and $a_{r+1} = \rho^m P(H)(\rho - 1)^{r-1}\left(D'(\rho)\right)^{-r-1}$.

The method just described can be called the method of subtracted singularities, and its general description follows: Imagine that we are interested in the asymptotic formula for coefficients a_n of a function $A(z)$, whose circle of convergence is R. Let us also assume that we can find a simpler function, say $\bar{A}(z)$ that has the same singularities as $A(z)$ (e.g., in the example above $\bar{A}(z) = \sum_{j=1}^r \frac{a_{-j}}{(z-\rho)^j}$). Then, $A_1(z) = A(z) - \bar{A}(z)$ is analytical in a larger disk, of radius $R' > R$, say, and its coefficients are not dominant in an asymptotic sense. To apply this method successfully, we need to develop asymptotics of some known functions (e.g., $(1-z)^\alpha$ for any real α) and establish the so-called *transfer theorems* (cf. [17]). This leads us to the so-called singularity analysis of Flajolet and Odlyzko [17], which we discuss next.

We start with the observation that $[z^n]A(z) = \rho^n[z^n]A(z/\rho)$, that is, we need only to study singularities at, say, $z = 1$. The next observation deals with asymptotics of $(1 - z)^{-\alpha}$. Above, we show how to obtain coefficients at z^n of this function when α is a natural number. Then, the function $(1 - z)^{-\alpha}$ has a pole of order α at $z = 1$. However, when $\alpha \neq 1, 2\ldots$, then the function has an algebraic singularity (in fact, it is then a multivalued function). Luckily enough, we can proceed formally as follows:

$$\begin{aligned}
[z^n](1 - z)^{-\alpha} &= \binom{n + \alpha - 1}{n} = \frac{\Gamma(n + \alpha)}{\Gamma(\alpha)\Gamma(n + 1)} \\
&= \frac{n^{\alpha-1}}{\Gamma(\alpha)}\left(1 + \frac{\alpha(\alpha - 1)}{2n} + O\left(\frac{1}{n^2}\right)\right),
\end{aligned}$$

provided $\alpha \notin \{0, -1, -2, \ldots\}$. In the above, $\Gamma(x) = \int_0^\infty e^{-t}x^{t-1}dx$ is the Euler gamma function (cf. [1,27]), and the latter asymptotic expansion follows from the Stirling formula. Even more generally, let

$$A(z) = (1 - z)^{-\alpha}\left(\frac{1}{z}\log\frac{1}{1 - z}\right)^\beta.$$

Then, as shown by Flajolet and Odlyzko [17],

$$a_n = [z^n]A(z) = \frac{n^{\alpha-1}}{\Gamma(\alpha)}\left(1 + C_1\frac{\beta}{\log n} + C_2\frac{\beta(\beta - 1)}{2\log^2 n} + O\left(\frac{1}{\log^3 n}\right)\right), \quad (11.32)$$

provided $\alpha \notin \{0, -1, -2, \ldots\}$, and C_1 and C_2 are constants that can be calculated explicitly.

The most important aspect of the singularity theory comes next: In many instances we do *not* have an explicit expression for the generating function $A(z)$ but only an expansion of $A(z)$ around a singularity. For example, let $A(z) = (1 - z)^{-\alpha} + O(B(z))$. In order to pass to coefficients of a_n we need a "transfer theorem" that will allow us to pass to coefficients of $B(z)$ under the "Big Oh" notation. We shall discuss them below.

We need a definition of Δ-analyticity around the singularity $z = 1$:

$$\Delta = \left\{z : |z| < R, z \neq 1, |\arg(z - 1)| > \phi\right\}$$

for some $R > 1$ and $0 < \phi < \pi/2$ (i.e., the domain Δ is an extended disk around $z = 1$ with a circular part rooted at $z = 1$ deleted). Then (cf. [17]):

THEOREM 11.8　*Let $A(z)$ be Δ-analytic that satisfies in a neighborhood of $z = 1$ either*

$$A(z) = O\left((1-z)^{-\alpha} \log^\beta (1-z)^{-1}\right)$$

or

$$A(z) = o\left((1-z)^{-\alpha} \log^\beta (1-z)^{-1}\right).$$

Then, either

$$[z^n] = O\left(n^{\alpha-1} \log^\beta n\right)$$

or

$$[z^n] = o\left(n^{\alpha-1} \log^\beta n\right),$$

respectively.

A classical example of singularity analysis is the Flajolet and Odlyzko analysis of the height of binary trees (cf. [17]), however, we finish this subsection with a simpler application that quite well illustrates the theory (cf. [55]).

Example 11.14

Certain sums from coding theory: In coding theory the following sum is of some interest:

$$S_n = \sum_{i=0}^{n} \binom{n}{i} \left(\frac{i}{n}\right)^i \left(1 - \frac{i}{n}\right)^{n-i}.$$

Let $s_n = n^n S_n$. If $s(z)$ denotes the exponential generating function of s_n, then by a simple application of the convolution principle of exponential generating functions we obtain $s(z) = (b(z))^2$ where $b(z) = (1 - t(z))^{-1}$ and $t(z)$ is the "tree function" defined in Section 11.5.1 (cf. Equation 11.20). In fact, we already know that this function also satisfies the functional equation $t(z) = z e^{t(z)}$. One observes that $z = e^{-1}$ is the singularity point of $t(z)$, and (cf. [55])

$$t(z) - 1 = -\sqrt{2(1 - ez)} + \frac{2}{3}(1 - ez) - \frac{11\sqrt{2}}{36}(1 - ez)^{3/2}$$
$$+ \frac{43}{135}(1 - ez)^2 + O\left((1 - ez)^{5/2}\right),$$

$$s(z) = \frac{1}{2h(z)\left(1 + \frac{\sqrt{2}}{3}\sqrt{h(z)} + \frac{11}{36}h(z) + O\left(h^{3/2}(z)\right)\right)^2}$$

$$= \frac{1}{2(1 - ez)} + \frac{\sqrt{2}}{3\sqrt{(1 - ez)}} + \frac{1}{36} + \frac{\sqrt{2}}{540}\sqrt{1 - ez} + O(1 - ez).$$

Thus, an application of the singularity analysis leads finally to the following asymptotic expansion

$$S_n = \sqrt{\frac{n\pi}{2}} + \frac{2}{3} + \frac{\sqrt{2\pi}}{24}\frac{1}{\sqrt{n}} - \frac{4}{135}\frac{1}{n} + O\left(1/n^{3/2}\right).$$

For more sophisticated examples the reader is referred to [17,21,55].

11.5.3 Mellin Transform and Asymptotics

In previous sections, we study functional equations such as Equation 11.25 or more generally Equation 11.30. They can be summarized by the following general functional equation:

$$f^{(b)}(z) = a(z) + \alpha f(zp) + \beta f(zq), \tag{11.33}$$

where
$f^{(b)}(z)$ denotes the bth derivative of $f(z)$
α, β are constants
$a(z)$ is a known function

An important point to observe is that in the applications described so far the unknown function $f(z)$ was usually a Poisson transform, that is, $\widetilde{f}(z) = \sum_{n \geq 0} f_n \frac{z^n}{n!} e^{-z}$. We briefly discuss consequences of this point at the end of this subsection where some elementary depoissonization results will be presented. An effective approach to solve asymptotically (either for $z \to 0$ or $z \to \infty$) the above function equation is by the so called Mellin transform which we discuss next. Knuth [41], together with De Bruijn, is responsible for introducing the Mellin transform into the "orbit" of the average case analysis of algorithms; however, it was popularized by Flajolet and his school who applied Mellin transforms to "countably" many problems of analysis of algorithms and analytic combinatorics. We base this subsection mostly on a survey by Flajolet et al. [19].

For a function $f(x)$ on $x \in [0, \infty)$ we define the Mellin transform as

$$\mathcal{M}(f, s) = f^*(s) = \int_0^{\infty} f(x) x^{s-1} dx,$$

where s is a complex number. For example, observe that from the definition of the Euler gamma function, we have $\Gamma(s) = \mathcal{M}(e^x, s)$. The Mellin transform is a special case of the Laplace transform (set $x = e^t$) or the Fourier transform (set $x = e^{i\omega}$). Therefore, using the inverse Fourier transform, one establishes the inverse Mellin transform (cf. [27]), namely,

$$f(x) = \frac{1}{2\pi i} \int_{c-i\infty}^{c+i\infty} f^*(s) x^{-s} ds,$$

provided $f(x)$ is piecewise continuous. In the above, the integration is along a vertical line $\Im(s) = c$, and c must belong to the so-called fundamental strip where the Mellin transform exists (see properly (P1) below).

The usefulness of the Mellin transform to the analysis of algorithms is a consequence of a few properties that we discuss in the sequel.

(P1) FUNDAMENTAL STRIP
Let $f(x)$ be a piecewise continuous function on the interval $[0, \infty)$ such that

$$f(x) = \begin{cases} O(x^{\alpha}) & x \to 0 \\ O(x^{\beta}) & x \to \infty. \end{cases}$$

Then the Mellin transform of $f(x)$ exists for any complex number s in the fundamental strip $-\alpha < \Re(s) < -\beta$, which we will denote $\langle -\alpha; -\beta \rangle$.

(P2) SMALLNESS OF MELLIN TRANSFORMS
Let $s = \sigma + it$. By the Riemann–Lebesgue lemma

$$f^*(\sigma + it) = o\left(|t|^{-r}\right) \quad \text{as } t \to \pm\infty$$

provided $f \in C^r$ where C^r is the set of functions having continuous r derivatives.

(P3) Basic Functional Properties
The following holds in appropriate fundamental strips:

$$f(\mu x) \Leftrightarrow \mu^{-s} f^*(s) \quad (\mu > 0)$$

$$f(x^\rho) \Leftrightarrow \frac{1}{\rho} f^*(s/\rho) \quad (\rho > 0)$$

$$\frac{d}{dx} f(x) \Leftrightarrow -(s-1) f^*(s-1)$$

$$\int_0^x f(t) dt \Leftrightarrow -\frac{1}{s} f^*(s+1)$$

$$f(x) = \sum_{k \geq 0} \lambda_k g(\mu_k x) \Leftrightarrow f^*(s) = g^*(s) \sum_{k \geq 0} \lambda_k \mu_k^{-s} \quad \text{(Harmonic Sum Rule)}$$

(P4) Asymptotics for $x \to 0$ and $x \to \infty$
Let the fundamental strip of $f^*(s)$ be the set of all s such that $-\alpha < \Re(s) < -\beta$ and assume that for $s = \sigma + i\tau$, $f^*(s) = O(|s|^r)$ with $r > 1$ as $|s| \to \infty$. If $f^*(s)$ can be analytically continued to a meromorphic function for $-\beta \leq \Re(s) \leq M$ with finitely many poles λ_k such that $\Re(\lambda_k) < M$, then as $x \to \infty$,

$$F(x) = -\sum_{\lambda_k \in \mathcal{H}} \text{Res} \left\{ F^*(s) x^{-s}, s = \lambda_k \right\} + O\left(x^{-M}\right) \quad x \to \infty$$

where M is as large as we want. (In a similar fashion one can continue the function $f^*(s)$ to the left to get an asymptotic formula for $x \to 0$.)
Sketch of a Proof. Consider the rectangle R with the corners at $c - iA$, $M - iA$, $M + iA$, and $c + iA$. Choose A so that the sides of R do not pass through any singularities of $F^*(s) x^{-s}$. When evaluating

$$\lim_{A \to \infty} \int_R = \lim_{A \to \infty} \left(\int_{c-iA}^{c+iA} + \int_{c+iA}^{M+iA} + \int_{M+iA}^{M-iA} + \int_{M-iA}^{c-iA} \right),$$

the second and fourth integrals contribute very little, since $F^*(s)$ is small for s with a large imaginary part by property (P2). The contribution of the fourth integral is computed as follows:

$$\left| \int_{M+i\infty}^{M-i\infty} F^*(s) x^{-s} ds \right| = \left| \int_\infty^{-\infty} F^*(M+it) x^{-M-it} dt \right| \leq \left| x^{-M} \right| \int_\infty^{-\infty} |F^*(M+it)| \, |x^{-it}| \, dt.$$

But the last integrand decreases exponentially as $|t| \to \infty$, thus, giving a contribution of $O(x^{-M})$. Finally, using Cauchy's residue theorem and taking into account the negative direction of R, we have

$$-\sum_{\lambda_k \in \mathcal{H}} \text{Res} \left\{ F^*(s) x^{-s}, s = \lambda_k \right\} = \frac{1}{2i\pi} \int_{c-i\infty}^{c+i\infty} F^*(s) x^{-s} ds + O\left(x^{-M}\right),$$

which proves the desired result.
  Specifically, the above implies that if the above smallness condition on $f^*(s)$ is satisfied for $-\beta < \Re(s) \leq M$, $(M > 0)$, then

$$f^*(s) = \sum_{k=0}^{K} \frac{d_k}{(s-b)^{k+1}}, \tag{11.34}$$

leads to

$$f(x) = -\sum_{k=0}^{K} \frac{d_k}{k!} x^{-b} (-\log x)^k + O\left(x^{-M}\right) \quad x \to \infty. \tag{11.35}$$

In a similar fashion, if for $-M < \Re(s) < -\alpha$ the smallness condition of $f^*(s)$ holds and

$$f^*(s) = \sum_{k=0}^{K} \frac{d_k}{(s-b)^{k+1}}, \tag{11.36}$$

then

$$f(x) = \sum_{k=0}^{K} \frac{d_k}{k!} x^{-b} (-\log x)^k + O\left(x^{M}\right) \quad x \to 0. \tag{11.37}$$

(P5) MELLIN TRANSFORM IN THE COMPLEX PLANE (cf. [19,33])
If $f(z)$ is analytic in a cone $\theta_1 \leq \arg(z) \leq \theta_2$ with $\theta_1 < 0 < \theta_2$, then the Mellin transform $f^*(s)$ can be defined by replacing the path of the integration $[0, \infty[$ by any curve starting at $z = 0$ and going to ∞ inside the cone, and it is identical with the real transform $f^*(s)$ of $f(z) = F(z)\big|_{z \in \mathcal{R}}$. In particular, if $f^*(s)$ fulfills an asymptotic expansion such as Equation 11.34 or 11.36, then Equation 11.35 or 11.37 for $f(z)$ holds in $z \to \infty$ and $z \to 0$ in the cone, respectively.

Let us now apply Mellin transforms to some problems studies above. For example, consider a trie for which the functional equation (Equation 11.25) becomes

$$\widetilde{X}(z) = \widetilde{A}(z) + \widetilde{X}(zp) + \widetilde{X}(zq),$$

where
$p + q = 1$
$\widetilde{A}(z)$ is the Poisson transform of a known function

Thanks to property (P3) the Mellin transform translates the above functional equation to an algebraic one which can be immediately solved, resulting in

$$X^*(s) = \frac{A^*(s)}{1 - p^{-s} - q^{-s}},$$

provided there exists a fundamental strip for $X^*(s)$ where also $A^*(s)$ is well defined. Now, due to property (P4) we can easily compute asymptotics of $\widetilde{X}(z)$ as $z \to \infty$ in a cone. More formally, we obtain asymptotics for z real, say x, and then either analytically continue our results or apply property (P5) which basically says that there is a cone in which the asymptotic results for real x can be extended to a complex z. Examples of usage of this technique can be found in [21,28,30,32,33,38,41,44].

This is a good plan to attack problems as the above; however, one must translate asymptotics of the Poisson transform $\widetilde{X}(z)$ into the original sequence, say x_n. One would like to have $x_n \sim \widetilde{X}(n)$, but this is not true in general (e.g., take $x_n = (-1)^n$). To assure the above asymptotic equivalence, we briefly discuss the so called **depoissonization** [30,32,33]. We cite below only one result that found many applications in the analysis of algorithms (cf. [32]).

THEOREM 11.9 *Let $\widetilde{X}(z) = \sum_{n \geq 0} x_n \frac{z^n}{n!} e^{-z}$ be the Poisson transform of a sequence x_n that is assumed to be an entire function of z. We postulate that in a cone S_θ ($\theta < \pi/2$) the following two conditions simultaneously hold for some real numbers $A, B, R > 0$, β, and $\alpha < 1$:*

(I) *For $z \in S_\theta$*

$$|z| > R \;\Rightarrow\; \left|\widetilde{X}(z)\right| \le B|z|^\beta \Psi(|z|),$$

where $\Psi(x)$ is a slowly varying function, that is, such that for fixed t $\lim_{x \to \infty} \frac{\Psi(tx)}{\Psi(x)} = 1$ (e.g., $\Psi(x) = \log^d x$ for some $d > 0$);
(O) *For $z \notin S_\theta$*

$$|z| > R \;\Rightarrow\; \left|\widetilde{X}(z)e^z\right| \le A\exp(\alpha|z|).$$

Then,

$$x_n = \widetilde{X}(n) + O\left(n^{\beta-1}\Psi(n)\right) \tag{11.38}$$

or more precisely:

$$x_n = \widetilde{X}(n) - \frac{1}{2}\widetilde{X}''(n) + O\left(n^{\beta-2}\Psi(n)\right).$$

where $\widetilde{X}''(n)$ is the second derivative of $\widetilde{X}(z)$ at $z = n$.

The verification of conditions (I) and (O) is usually not too difficult, and can be accomplished directly on the functional equation at hand through the so called increasing domains method discussed in [32].

Finally, we should say that there is an easier approach to deal with a majority of functional equations of type (Equation 11.25). As we pointed out, such equations possess solutions that can be represented as alternating sums (cf. Equation 11.26 and Examples 11.10 through 11.12). Let us consider a general alternating sum

$$S_n = \sum_{k=m}^n (-1)^k \binom{n}{k} f_k$$

where f_k is a known, but otherwise, general sequence. The following two equivalent approaches (cf. [41,52]) use complex integration (the second one is actually a Mellin-like approach) to simplify the computations of asymptotics of S_n for $n \to \infty$ (cf. [39,52]).

THEOREM 11.10 (Rice's Formula) (i) *Let $f(s)$ be an analytical continuation of $f(k) = f_k$ that contains the half line $[m, \infty)$. Then,*

$$S_n := \sum_{k=m}^n (-1)^k \binom{n}{k} f_k = \frac{(-1)^n}{2\pi i} \int_C f(s) \frac{n!}{s(s-1)\cdots(s-n)}\,ds$$

where C is a positively enclosed curve that encircles $[m, n]$ and does not include any of the integers $0, 1, \ldots, m-1$.

(ii) Let $f(s)$ be analytic left to the vertical line $(\frac{1}{2} - m - i\infty, \frac{1}{2} - m + i\infty)$ with subexponential growth at infinity, then

$$S_n = \frac{1}{2\pi i} \int_{\frac{1}{2}-m-i\infty}^{\frac{1}{2}-m+i\infty} f(-z)B(N+1,z)dz$$

$$= \frac{1}{2\pi i} \int_{\frac{1}{2}-m-i\infty}^{\frac{1}{2}-m+i\infty} f(-z)n^{-z}\Gamma(z)$$

$$\left(1 - \frac{z(z+1)}{2n} + \frac{z(1+z)}{24n^2}(3(1+z)^2 + z - 1) + O\left(n^{-3}\right)\right)dz$$

where $B(x,y) = \Gamma(x)\Gamma(y)/\Gamma(x+y)$ is the Beta function.

The precise growth condition for $f(z)$ of part (ii) can be found in [52].

Example 11.15

Asymptotics of some alternating sums: In Examples 11.10 through 11.12 we deal with alternating sums of the following general type:

$$S_n(r) = \sum_{k=2}^{n}(-1)^k\binom{n}{k}\binom{k}{r}\frac{1}{p^{-k} - q^{-k}},$$

where $p + q = 1$. We now use Theorem 11.10 to obtain asymptotics of S_n as n becomes large and r is fixed. To simplify our computation we use part (ii) of Theorem 11.10, which leads to

$$S_n(r) = \frac{1}{2\pi i}\frac{(-1)^n}{r!} \int_{\frac{1}{2}-[2-r]^+-i\infty}^{\frac{1}{2}-[2-r]^++i\infty} n^{r-z}\Gamma(z)\frac{1}{1 - p^{r-z} - q^{r-z}}dz + e_n.$$

$x^+ = \max\{0, x\}$, where e_n is an error term that we discuss later. The above integral should remind the reader of the integral appearing in the inverse Mellin transform. Thus, we can estimate it using a similar approach. First of all, we observe that the function under the integral has infinitely many poles satisfying

$$1 = p^{r-z} + q^{r-z}.$$

It can be proved (cf. [32]) that these poles, say z_k for $k = 0, \pm 1, \ldots$, lie on a line $\Re(z) = r - 1$ provided $\log p/\log q$ is rational, which we assume to hold. Thus, we can write $z_k = r - 1 + iy_k$ where $y_0 = 0$ and otherwise a real number for $k \neq 0$. Observe also that the line at $\Re(z) = r - 1$ lies right to the line of integration $(\frac{1}{2} - [2 - r]^+ - i\infty, \frac{1}{2} - [2 - r]^+ + i\infty)$. To take advantages of the Cauchy residue theorem, we consider a big rectangle with left side being the line of integration, the right size position at $\Re(z) = M$ (where M is a large number), and bottom and top side position at $\Im(z) = \pm A$, say. We further observe that the right side contributes only $O(n^{r-M})$ due to the factor n^{r-M} in the integral. Both bottom and top sides contribute negligible amounts too, since the gamma function decays exponentially fast with the increase of imaginary part (i.e., when $A \to \infty$). In summary, the integral is equal to a circular integral (around the rectangle) plus a negligible part $O(n^{r-M})$. But, then by Cuachy's residue theorem the latter integral is equal to minus the sum of all residues at z_k, that is,

$$S_n(r) = -\sum_{k-\infty}^{\infty} \text{Res}\left(\frac{n^{r-z}\Gamma(z)}{1 - p^{r-z} - q^{r-z}}, z = z_k\right) + O\left(n^{r-M}\right).$$

We can compute the residues using MAPLE (as shown in Section 11.3.3). Equivalently, for $k = 0$ (the main contribution to the asymptotics comes from $z_0 = r - 1$) we can use the following expansions around $w = z - z_0$:

$$n^{r-z} = n\left(1 - w\ln n + O\left(w^2\right)\right),$$

$$\left(1 - p^{r-z} - q^{r-z}\right)^{-1} = -w^{-1}h^{-1} + \frac{1}{2}h_2 h^{-2} + O(w),$$

$$\Gamma(z) = (-1)^{r+1}\left(w^{-1} - \gamma + \delta_{r,0}\right) + O(w) \quad r = 0, 1$$

where
$h = -p\ln p - q\ln q$
$h_2 = p\ln^2 p + q\ln^2 q$
$\gamma = 0.577215\ldots$ is the Euler constant
$\delta_{r,0}$ is the Kronecker symbol

Considering in addition the residues coming from z_k for $k \neq 0$ we finally arrive at

$$S_n(r) = \begin{cases} \frac{1}{h}n\left(\ln n + \gamma - \delta_{r,0} + \frac{1}{2}h_2\right) + (-1)^r n P_r(n) + e_n & r = 0, 1 \\ n\frac{(-1)^r}{r(r-1)h} + (-1)^r n P_r(n) + e_n & r \geq 2, \end{cases}$$

where the error term can be computed easily to be $e_n = O(1)$ (using the arguments as above and observing that the error term has a similar integral representation but with term n^{-1} in front of it). In the above $P_r(n)$ is a contribution from z_k for $k \neq 0$, and it is a fluctuating function with small amplitude. For example, when $p = q = 1/2$, then

$$P_r(n) = \frac{1}{\ln 2}\sum_{k \neq 0}\Gamma\left(r + 2\pi i k/\log 2\right)\exp\left(-2\pi i k \log_2 n\right)$$

is a periodic function of $\log x$ with period 1, mean 0 and amplitude $\leq 10^{-6}$ for $r = 0, 1$.

11.6 Research Issues and Summary

In this chapter we presented a brief overview of probabilistic and analytic methods of the average-case analysis of algorithms. Among probabilistic methods we discussed the inclusion–exclusion principle, first and second moments methods, subadditive ergodic theorem, entropy and asymptotic equipartition property, central limit theorems, large deviations, and Azuma's type inequality. Analytic tools discussed in this chapter are recurrences, functional equations, complex asymptotics, Mellin transform, and analytic depoissonization. These techniques were illustrated in examples that arose in the design and analysis of algorithms on words.

We can trace back probabilistic techniques to the 1960 famous paper of Erdös and Rènyi "On the Evolution of Random Graphs." The analytic approach "was born" in 1963 when Knuth analyzed successfully the average numbers of probes in the linear hashing. Since then we have witnessed an increasing interest in the average-case analysis and design of algorithms, possibly due to the high success rate of randomized algorithms for computational geometry, scientific visualization, molecular biology, etc. We now see the emergence of combinatorial and asymptotic methods that permit us to classify data structures into broad categories that are susceptible to a unified treatment. Probabilistic methods that have been so successful in the study of random graphs and hard combinatorial optimization problems also play an important role in the field. These developments have two important consequences for the analysis of algorithms: it becomes possible to predict average-case behavior under more general probabilistic models, at the same time it becomes possible to analyze much more structurally complex algorithms.

11.7 Further Information

In this chapter we illustrated probabilistic techniques on examples from "stringology," that is, problems on words. Probabilistic methods found applications in many other facets of computer science, namely, random graphs (cf. [3,22]), computational geometry (cf. [46]), combinatorial algorithms (cf. [10,34,50]), molecular biology (cf. [60]), and so forth. Probabilistic methods are useful in the design of randomized algorithms that make random choices during their executions. The reader interested in these algorithms is referred to [35,46]. Analytic techniques are discussed in Knuth [39–41], recent book of Sedgewick and Flajolet [49], and in a forthcoming new book by the same authors [21].

Finally, a homepage of *Analysis of Algorithms* was recently created. The interested reader is invited to visit http://pauillac.inria.fr/algo/AofA/index.html.

Defining Terms

Asymptotic equipartition property: A set of all strings of a given length can be partitioned into a set of "bad states" of a low probability, and a set of "good states" such that every string in the latter set has approximately the same probability.

Digital trees: Trees that store digital information (e.g., strings, keys, etc.). There are several types of digital trees, namely: tries, PATRICIA tries, DSTs, and suffix trees.

Edit distance: The smallest number of insertions/deletions/substitutions required to change one string onto another.

Inclusion–exclusion principle: A rule that allows to compute the probability of exactly r occurrences of events $A_1, A_2, \ldots, A_n$.

Large deviations: When away by more than the standard deviation from the mean, apply the large deviation principle!

Meromorphic function: A function that is analytic, except for a finite number of poles (at which the function ceases to be defined).

Poissonization and depoissonization: When a deterministic input is replaced by a Poisson process, then the new model is called the Poisson model. After finding a solution to the Poisson model, one must interpret it in terms of the original problem, i.e., depoissonize the Poisson model.

Probabilistic models: Underlying probabilistic models that determine the input distribution (e.g., of generated strings). We discussed the Bernoulli model, the Markovian model, and the mixing model.

Rice's method: A method of complex asymptotics that can handle certain alternating sums arising in the analysis of algorithms.

Shortest common superstring: A shortest possible string that contains as substrings a number of given strings.

Singularity analysis: A complex asymptotic technique for determining the asymptotics of certain algebraic functions.

Subadditive ergodic theorem: If a stationary and ergodic process satisfies the subadditive inequality, then it grows almost surely linearly in time.

Acknowledgments

The author thanks his colleagues P. Jacquet, J. Kieffer, G. Louchard, H. Prodinger, and K. Park for reading earlier versions of this chapter, and for comments that led to improvements in the presentation.

References

1. Abramowitz, M. and Stegun, I., *Handbook of Mathematical Functions*, Dover, New York, 1964.
2. Aho, A., Hopcroft, J., and Ullman, J., *The Design and Analysis of Computer Algorithms*, Addison-Wesley, Reading, MA, 1974.
3. Alon, N. and Spencer, J., *The Probabilistic Method*, John Wiley & Sons, New York, 1992.
4. Arratia, R. and Waterman, M., A phase transition for the score in matching random sequences allowing deletions, *Ann. Appl. Probab.*, 4, 200–225, 1994.
5. Arratia, R., Gordon, L., and Waterman, M., The Erds-Rnyi Law in distribution for coin tossing and sequence matching, *Ann. Stat.*, 18, 539–570, 1990.
6. Bollobās, B., *Random Graphs*, Academic Press, London, U.K., 1985.
7. Blum, A., Jiang, T., Li, M., Tromp, J., and Yannakakis, M., Linear approximation of shortest super-string, *J. ACM*, 41, 630–647, 1994.
8. Bradely, R., Basic properties of strong mixing conditions, In *Dependence in Probability and Statistics*, E. Eberlein and M. Taqqu (Eds.), Birkhäuser, Boston, MA 1986, pp. 165–192.
9. Chvatal V. and Sankoff, D., Longest common subsequence of two random sequences, *J. Appl. Probab.*, 12, 306–315, 1975.
10. Coffman, E. and Lueker, G., *Probabilistic Analysis of Packing and Partitioning Algorithms*, John Wiley & Sons, New York, 1991.
11. Cover, T.M. and Thomas, J.A., *Elements of Information Theory*, John Wiley & Sons, New York, 1991.
12. Crochemore, M. and Rytter, W., *Text Algorithms*, Oxford University Press, New York, 1995.
13. Dembo, A. and Zeitouni, O., *Large Deviations Techniques*, Jones and Bartlett, Boston, MA, 1993.
14. Durrett, R., *Probability: Theory and Examples*, Wadsworth, Belmont, CA, 1991.
15. Feller, W., *An Introduction to Probability Theory and Its Applications*, Vol. II, John Wiley & Sons, New York, 1971.
16. Flajolet, P., Analytic analysis of algorithms, *LNCS*, W. Kuich (Ed.), Vol. 623, Springer-Verlag, Berlin, 1992, pp. 186–210.
17. Flajolet, P. and Odlyzko, A., Singularity analysis of generating functions, *SIAM J. Discr. Methods*, 3, 216–240, 1990.
18. Flajolet, P. and Richmond, B., Generalized digital trees and their difference-differential equations, *Random Struct. Algor.*, 3, 305–320, 1992.
19. Flajolet, P., Gourdon, X., and Dumas, P., Mellin transforms and asymptotics: Harmonic sums, *Theor. Comput. Sci.*, 144, 3–58, 1995.
20. Flajolet, P. and Sedgewick, R., Mellin transforms and asymptotics: Finite differences and Rice's integrals. *Theor. Comput. Sci.*, 144, 101–124, 1995.
21. Flajolet, P. and Sedgewick, R., *Analytical Combinatorics*, in preparation (available also at http://pauillac.inria.fr/algo/flajolet/Publications/publist.html).
22. Frieze, A. and McDiarmid, C., Algorithmic theory of random graphs, *Random Struct. Algor.*, 10, 5–42, 1997.
23. Frieze, A. and Szpankowski, W., Greedy algorithms for the shortest common superstring that are asymptotically optimal, *Algorithmica*, 21, 21–36, 1998.
24. Fudos, I., Pitoura, E., and Szpankowski, W., On Pattern occurrences in a random text, *Inform. Process. Lett.*, 57, 307–312, 1996.

25. Greene, D.H. and Knuth, D.E., *Mathematics for the Analysis of Algorithms*, Birkhauser, Cambridge, MA, 1981.
26. Guibas, L. and Odlyzko, A.M., String overlaps, pattern matching, and nontransitive games, *J. Comb. Theor. Ser. A*, 30, 183–208, 1981.
27. Henrici, P., *Applied and Computational Complex Analysis*, Vols. 1–3, John Wiley & Sons, New York, 1977.
28. Hofri, M., *Analysis of Algorithms. Computational Methods and Mathematical Tools*, Oxford University Press, New York, 1995.
29. Hwang, H-K., Large deviations for combinatorial distributions I: Central limit theorems, *Ann. Appl. Probab.*, 6, 297–319, 1996.
30. Jacquet, P. and Régnier, M., Normal limiting distribution of the size of tries, *Proceedings of Performance'87*, North Holland, Amsterdam, the Netherlands, 1987, pp. 209–223.
31. Jacquet, P. and Szpankowski, W., Autocorrelation on words and its applications. Analysis of suffix trees by string-ruler approach, *J. Comb. Theor. Ser. A*, 66, 237–269, 1994.
32. Jacquet, P. and Szpankowski, W., Asymptotic behavior of the Lempel-Ziv parsing scheme and digital search trees, *Theor. Comput. Sci.*, 144, 161–197, 1995.
33. Jacquet, P. and Szpankowski, W., Analytical depoissonization and its applications, *Theor. Comput. Sci.*, 201, 1–62, 1998.
34. Karp, R., The probabilistic analysis of some combinatorial search algorithms, In *Algorithms and Complexity*, J.F. Traub (Ed.), Academic Press, New York, 1976.
35. Karp, R., An introduction to randomized algorithms, *Discr. Appl. Math.*, 34, 165–201, 1991.
36. Kingman, J.F.C., *Subadditive processes*, In *Ecole d'Eté de Probabilités de Saint-Flour V-1975, Lecture Notes in Mathematics*, Vol. 539, Springer-Verlag, Berlin, 1976.
37. Kirschenhofer, P., Prodinger, H., and Szpankowski, W., On the variance of the external path in a symmetric digital trie, *Discrete Appl. Math.*, 25, 129–143, 1989.
38. Kirschenhofer, P., Prodinger, H., and Szpankowski, W., Analysis of a splitting process arising in probabilistic counting and other related algorithms, *Random Struct. Algor.*, 9, 379–401, 1996.
39. Knuth, D.E., *The Art of Computer Programming. Fundamental Algorithms*, Vol. 1, Addison-Wesley, Reading, MA, 1973.
40. Knuth, D.E., *The Art of Computer Programming. Seminumerical Algorithms*. Vol. 2, Addison Wesley, Reading, MA, 1981.
41. Knuth, D.E., *The Art of Computer Programming. Sorting and Searching*, Vol. 3, Addison-Wesley, Reading, MA, 1973.
42. Levin, L., Average case complete problems, *SIAM J. Comput.*, 15, 285–286, 1986.
43. Louchard, G., Random Walks, Gaussian processes and list structures, *Theor. Comp. Sci.*, 53, 99–124, 1987.
44. Mahmoud, H., *Evolution of Random Search Trees*, John Wiley & Sons, New York 1992.
45. McDiarmid, C., On the method of bounded differences, In *Surveys in Combinatorics*, J. Siemons (Ed.), Vol. 141, *London Mathematical Society Lecture Notes Series*, Cambridge University Press, Cambridge, U.K., 1989, pp. 148–188.
46. Motwani, R. and Raghavan, P., *Randomized Algorithms*, Cambridge University Press, Cambridge, U.K., 1995.
47. Odlyzko, A., Asymptotic enumeration, In *Handbook of Combinatorics*, Vol. II, R. Graham, M. Götschel, and L. Lovász, (Eds.), Elsevier Science, Amsterdam, the Netherlands, 1995, pp. 1063–1229.
48. Pittel, B., Asymptotic growth of a class of random trees, *Ann. Probability*, 18, 414–427, 1985.
49. Sedgewick, R. and Flajolet, P., *An Introduction to the Analysis of Algorithms*, Addison-Wesley, Reading, MA, 1995.
50. Steele, J.M., *Probability Theory and Combinatorial Optimization*, SIAM, Philadelphia, PA, 1997.

51. Szpankowski, W., Solution of a linear recurrence equation arising in the analysis of some algorithms, *SIAM J. Alg. Discr. Methods*, 8, 233–250, 1987.

52. Szpankowski, W., The evaluation of an alternating sum with applications to the analysis of some data structures, *Inform. Process. Lett.*, 28, 13–19, 1988.

53. Szpankowski, W., Patricia tries again revisited, *J. ACM*, 37, 691–711, 1990.

54. Szpankowski, W., A generalized suffix tree and its (un)expected asymptotic behaviors, *SIAM J. Comput.*, 22, 1176–1198, 1993.

55. Szpankowski, W., On asymptotics of certain sums arising in coding theory, *IEEE Trans. Inform. Theor.*, 41, 2087–2090, 1995.

56. Talagrand, M., A new look at independence, *Ann. Appl. Probab.*, 6, 1–34, 1996.

57. Titchmarsh, E.C., *The Theory of Functions*, Oxford University Press, Oxford, U.K., 1944.

58. Ukkonen, E., A linear-time algorithm for finding approximate shortest common superstrings, *Algorithmica*, 5, 313–323, 1990.

59. Vitter, J. and Flajolet, P., Average-case analysis of algorithms and data structures, In *Handbook of Theoretical Computer Science*, J. van Leewen (Ed.), Elsevier Science Publishers, Amsterdam, the Netherlands, 1990, pp. 433–524.

60. Waterman, M., *Introduction to Computational Biology*, Chapman & Hall, London, U.K., 1995.

12

Randomized Algorithms*

Rajeev Motwani
Stanford University

Prabhakar Raghavan
Yahoo! Research

12.1 Introduction

A **randomized algorithm** is one that makes random choices during its execution. The behavior of such an algorithm may thus, be random even on a fixed input. The design and analysis of a randomized algorithm focuses on establishing that it is likely to behave "well" on every input; the likelihood in such a statement depends only on the probabilistic choices made by the algorithm during execution and not on any assumptions about the input. It is especially important to distinguish a randomized algorithm from the average-case analysis of algorithms, where one analyzes an algorithm assuming that its input is drawn from a fixed probability distribution. With a randomized algorithm, in contrast, no assumption is made about the input.

Two benefits of randomized algorithms have made them popular: simplicity and efficiency. For many applications, a randomized algorithm is the simplest algorithm available, or the fastest, or both. In the following text, we make these notions concrete through a number of illustrative examples. We assume that the reader has had undergraduate courses in algorithms and complexity,

---

* Chapter coauthor Rajeev Motwani passed away in a tragic drowning accident on June 5, 2009. He was a Professor of Computer Science at Stanford University and a brilliant researcher and educator who was awarded the prestigious Gödel Prize, and was an early advisor and supporter of many companies including Google and PayPal.

and in probability theory. A comprehensive source for randomized algorithms is the book by the authors [25]. The articles by Karp [18], Maffioli et al. [22], and Welsh [45] are good surveys of randomized algorithms. The book by Mulmuley [26] focuses on randomized geometric algorithms.

Throughout this chapter we assume the RAM model of computation, in which we have a machine that can perform the following operations involving registers and main memory: input–output operations, memory–register transfers, indirect addressing, and branching and arithmetic operations. Each register or memory location may hold an integer that can be accessed as a unit, but an algorithm has no access to the representation of the number. The arithmetic instructions permitted are $+, -, \times$, and $/$. In addition, an algorithm can compare two numbers, and evaluate the square root of a positive number. In this chapter $E[X]$ will denote the expectation of a random variable X, and $Pr[A]$ will denote the probability of an event A.

12.2   Sorting and Selection by Random Sampling

Some of the earliest randomized algorithms included algorithms for sorting a set (S) of numbers, and the related problem of finding the kth smallest element in S. The main idea behind these algorithms is the use of random sampling: a randomly chosen member of S is unlikely to be one of its largest or smallest elements; rather, it is likely to be "near the middle." Extending this intuition suggests that a random sample of elements from S is likely to be spread "roughly uniformly" in S. We now describe randomized algorithms for sorting and selection based on these ideas.

Algorithm RQS:

Input: A set of numbers S.

Output: The elements of S sorted in increasing order.

1. Choose an element y uniformly at random from S: every element in
        S has equal probability of being chosen.

2. By comparing each element of S with y, determine the set S_1 of
        elements smaller than y and the set S_2 of elements larger
        than y.

3. Recursively sort S_1 and S_2. Output the sorted version of S_1,
        followed by y, and then the sorted version of S_2.

Algorithm RQS is an example of a randomized algorithm—an algorithm that makes random choices during execution. It is inspired by the Quicksort algorithm due to Hoare [13], and described in [25]. We assume that the random choice in Step 1 can be made in unit time. What can we prove about the running time of RQS?

We now analyze the expected number of comparisons in an execution of RQS. Comparisons are performed in Step 2, in which we compare a randomly chosen element to the remaining elements. For $1 \leq i \leq n$, let $S_{(i)}$ denote the element of rank i (the ith smallest element) in the set S. Define the random variable X_{ij} to assume the value 1 if $S_{(i)}$ and $S_{(j)}$ are compared in an execution, and the value 0 otherwise. Thus, the total number of comparisons is $\sum_{i=1}^{n}\sum_{j>i}X_{ij}$. By linearity of expectation the expected number of comparisons is

$$E\left[\sum_{i=1}^{n}\sum_{j>i}X_{ij}\right] = \sum_{i=1}^{n}\sum_{j>i}E\left[X_{ij}\right]. \tag{12.1}$$

Let p_{ij} denote the probability that $S_{(i)}$ and $S_{(j)}$ are compared during an execution. Then

$$\mathbf{E}\left[X_{ij}\right] = p_{ij} \times 1 + \left(1 - p_{ij}\right) \times 0 = p_{ij}. \tag{12.2}$$

To compute p_{ij} we view the execution of RQS as a binary tree T each node of which is labeled with a distinct element of S. The root of the tree is labeled with the element y chosen in Step 1, the left subtree of y contains the elements in S_1, and the right subtree of y contains the elements in S_2. The structures of the two subtrees are determined recursively by the executions of RQS on S_1 and S_2. The root y is compared to the elements in the two subtrees, but no comparison is performed between an element of the left subtree and an element of the right subtree. Thus, there is a comparison between $S_{(i)}$ and $S_{(j)}$ if and only if one of these elements is an ancestor of the other.

Consider the permutation π obtained by visiting the nodes of T in increasing order of the level numbers, and in a left-to-right order within each level; recall that the ith level of the tree is the set of all nodes at distance exactly i from the root. The following two observations lead to the determination of p_{ij}.

1. There is a comparison between $S_{(i)}$ and $S_{(j)}$ if and only if $S_{(i)}$ or $S_{(j)}$ occurs earlier in the permutation π than any element $S_{(\ell)}$ such that $i < \ell < j$. To see this, let $S_{(k)}$ be the earliest in π from among all elements of rank between i and j. If $k \notin \{i,j\}$, then $S_{(i)}$ will belong to the left subtree of $S_{(k)}$ while $S_{(j)}$ will belong to the right subtree of $S_{(k)}$, implying that there is no comparison between $S_{(i)}$ and $S_{(j)}$. Conversely, when $k \in \{i,j\}$, there is an ancestor–descendant relationship between $S_{(i)}$ and $S_{(j)}$, implying that the two elements are compared by RQS.

2. Any of the elements $S_{(i)}, S_{(i+1)}, \ldots, S_{(j)}$ is equally likely to be the first of these elements to be chosen as a partitioning element and hence, to appear first in π. Thus, the probability that this first element is either $S_{(i)}$ or $S_{(j)}$ is exactly $2/(j - i + 1)$.

It follows that $p_{ij} = 2/(j-i+1)$. By Equations 12.1 and 12.2, the expected number of comparisons is given by

$$\sum_{i=1}^{n} \sum_{j>i}^{n} p_{ij} = \sum_{i=1}^{n} \sum_{j>i}^{n} \frac{2}{j-i+1}$$

$$\leq \sum_{i=1}^{n-1} \sum_{k=1}^{n-i} \frac{2}{k+1}$$

$$\leq 2 \sum_{i=1}^{n} \sum_{k=1}^{n} \frac{1}{k}.$$

It follows that the expected number of comparisons is bounded above by $2nH_n$, where H_n is the nth harmonic number, defined by $H_n = \sum_{k=1}^{n} 1/k$.

THEOREM 12.1 *The expected number of comparisons in an execution of RQS is at most $2nH_n$.*

Now $H_n = \ln n + \Theta(1)$, so that the expected running time of RQS is $O(n \log n)$. Note that this expected running time holds for every input. It is an expectation that depends only on the random choices made by the algorithm, and not on any assumptions about the distribution of the input.

12.2.1 Randomized Selection

We now consider the use of random sampling for the problem of selecting the kth smallest element in a set S of n elements drawn from a totally ordered universe. We assume that the elements of S are all distinct, although it is not very hard to modify the following analysis to allow for multisets. Let $r_S(t)$ denote the rank of an element t (the kth smallest element has rank k) and recall that $S_{(i)}$ denotes the ith smallest element of S. Thus, we seek to identify $S_{(k)}$. We extend the use of this notation to subsets of S as well. The following algorithm is adapted from one due to Floyd and Rivest [10].

Algorithm LazySelect:

Input: A set S of n elements from a totally ordered universe, and
an integer k in $[1, n]$.

Output: The kth smallest element of S, $S_{(k)}$.

1. Pick $n^{3/4}$ elements from S, chosen independently and uniformly at
   random with replacement; call this multiset of elements R.

2. Sort R in $O\!\left(n^{3/4} \log n\right)$ steps using any optimal sorting algorithm.

3. Let $x = kn^{-1/4}$. For $\ell = \max\{\lfloor x - \sqrt{n}\rfloor, 1\}$ and $h = \min\{\lceil x + \sqrt{n}\rceil, n^{3/4}\}$,
   let $a = R_{(\ell)}$ and $b = R_{(h)}$. By comparing a and b to every
   element of S, determine $r_S(a)$ and $r_S(b)$.

4. if $k < n^{1/4}$, let $P = \{y \in S \mid y \le b\}$ and $r = k$;
   else if $k > n - n^{1/4}$, let $P = \{y \in S \mid y \ge a\}$ and $r = k - r_S(a) + 1$;
   else if $k \in [n^{1/4}, n - n^{1/4}]$, let $P = \{y \in S \mid a \le y \le b\}$ and $r = k - r_S(a) + 1$;
   Check whether $S_{(k)} \in P$ and $|P| \le 4n^{3/4} + 2$. If not, repeat
   Steps 1–3 until such a set P is found.

5. By sorting P in $O\!\left(|P| \log |P|\right)$ steps, identify P_r, which is $S_{(k)}$.

Figure 12.1 illustrates Step 3, where small elements are at the left end of the picture and large ones to the right. Determining (in Step 4) whether $S_{(k)} \in P$ is easy, since we know the ranks $r_S(a)$ and $r_S(b)$ and we compare either or both of these to k, depending on which of the three **if** statements in Step 4 we execute. The sorting in Step 5 can be performed in $O\!\left(n^{3/4} \log n\right)$ steps.

Thus, the idea of the algorithm is to identify two elements a and b in S such that both of the following statements hold with high probability:

1. The element $S_{(k)}$ that we seek is in P, the set of elements between a and b.
2. The set P of elements is not very large, so that we can sort P inexpensively in Step 5.

As in the analysis of RQS we measure the running time of LazySelect in terms of the number of comparisons performed by it. The following theorem is established using the *Chebyshev bound* from elementary probability theory; a full proof may be found in [25].

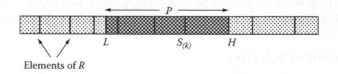

FIGURE 12.1 The LazySelect algorithm.

THEOREM 12.2 *With probability* $1 - o(n^{-1/4})$, *LazySelect finds* $S_{(k)}$ *on the first pass through Steps 1–5, and thus, performs only* $2n + o(n)$ *comparisons.*

This adds to the significance of LazySelect: the best known deterministic selection algorithms use $3n$ comparisons in the worst case, and are quite complicated to implement.

12.3 A Simple Min-Cut Algorithm

Two events $\mathcal{E}_1$ and $\mathcal{E}_2$ are said to be independent if the probability that they both occur is given by

$$\Pr[\mathcal{E}_1 \cap \mathcal{E}_2] = \Pr[\mathcal{E}_1] \times \Pr[\mathcal{E}_2]. \tag{12.3}$$

More generally when $\mathcal{E}_1$ and $\mathcal{E}_2$ are not necessarily independent,

$$\Pr[\mathcal{E}_1 \cap \mathcal{E}_2] = \Pr[\mathcal{E}_1 \mid \mathcal{E}_2] \times \Pr[\mathcal{E}_2] = \Pr[\mathcal{E}_2 \mid \mathcal{E}_1] \times \Pr[\mathcal{E}_1], \tag{12.4}$$

where $\Pr[\mathcal{E}_1 \mid \mathcal{E}_2]$ denotes the conditional probability of $\mathcal{E}_1$ given $\mathcal{E}_2$. When a collection of events is not independent, the probability of their intersection is given by the following generalization of Equation 12.4:

$$\Pr\left[\cap_{i=1}^{k}\mathcal{E}_i\right] = \Pr[\mathcal{E}_1] \times \Pr[\mathcal{E}_2 \mid \mathcal{E}_1] \times \Pr[\mathcal{E}_3 \mid \mathcal{E}_1 \cap \mathcal{E}_2] \cdots \Pr\left[\mathcal{E}_k \mid \cap_{i=1}^{k-1}\mathcal{E}_i\right]. \tag{12.5}$$

Let G be a connected, undirected multigraph with n vertices. A multigraph may contain multiple edges between any pair of vertices. A cut in G is a set of edges whose removal results in G being broken into two or more components. A min-cut is a cut of minimum cardinality. We now study a simple algorithm due to Karger [15] for finding a min-cut of a graph.

We repeat the following step: pick an edge uniformly at random and merge the two vertices at its end points. If as a result there are several edges between some pairs of (newly formed) vertices, retain them all. Remove edges between vertices that are merged, so that there are never any self-loops. This process of merging the two end-points of an edge into a single vertex is called the contraction of that edge. With each contraction, the number of vertices of G decreases by one. Note that as long as at least two vertices remain, an edge contraction does not reduce the min-cut size in G. The algorithm continues the contraction process until only two vertices remain; at this point, the set of edges between these two vertices is a cut in G and is output as a candidate min-cut (Figure 12.2). What is the probability that this algorithm finds a min-cut?

DEFINITION 12.1 For any vertex v in a multigraph G, the *neighborhood* of G, denoted $\Gamma(v)$, is the set of vertices of G that are adjacent to v. The *degree* of v, denoted $d(v)$, is the number of edges incident on v. For a set S of vertices of G, the neighborhood of S, denoted $\Gamma(S)$, is the union of the neighborhoods of the constituent vertices.

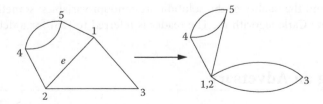

FIGURE 12.2 A step in the min-cut algorithm; the effect of contracting edge $e = (1, 2)$ is shown.

Note that $d(v)$ is the same as the cardinality of $\Gamma(v)$ when there are no self-loops or multiple edges between v and any of its neighbors.

Let k be the min-cut size and let C be a particular min-cut with k edges. Clearly G has at least $kn/2$ edges (otherwise there would be a vertex of degree less than k, and its incident edges would be a min-cut of size less than k). We bound from below the probability that no edge of C is ever contracted during an execution of the algorithm, so that the edges surviving till the end are exactly the edges in C.

For $1 \leq i \leq n - 2$, let $\mathcal{E}_i$ denote the event of not picking an edge of C at the ith step. The probability that the edge randomly chosen in the first step is in C is at most $k/(nk/2) = 2/n$, so that $\mathbf{Pr}[\mathcal{E}_1] \geq 1 - 2/n$. Conditioned on the occurrence of $\mathcal{E}_1$, there are at least $k(n - 1)/2$ edges during the second step so that $\mathbf{Pr}[\mathcal{E}_2 \mid \mathcal{E}_1] \geq 1 - 2/(n - 1)$. Extending this calculation, $\mathbf{Pr}\left[\mathcal{E}_i \mid \cap_{j=1}^{i-1}\mathcal{E}_j\right] \geq 1 - 2/(n - i + 1)$. We now recall Equation 12.5 to obtain

$$\mathbf{Pr}[\cap_{i=1}^{n-2}\mathcal{E}_i] \geq \prod_{i=1}^{n-2}\left(1 - \frac{2}{n - i + 1}\right) = \frac{2}{n(n - 1)}.$$

Our algorithm may err in declaring the cut it outputs to be a min-cut. But the probability of discovering a particular min-cut (which may in fact be the unique min-cut in G) is larger than $2/n^2$, so the probability of error is at most $1 - 2/n^2$. Repeating the aforementioned algorithm $n^2/2$ times making independent random choices each time, the probability that a min-cut is not found in any of the $n^2/2$ attempts is (by Equation 12.3) at most

$$\left(1 - \frac{2}{n^2}\right)^{n^2/2} < 1/e.$$

By this process of repetition, we have managed to reduce the probability of failure from $1 - 2/n^2$ to less than $1/e$. Further executions of the algorithm will make the failure probability arbitrarily small (the only consideration being that repetitions increase the running time). Note the extreme simplicity of this randomized min-cut algorithm. In contrast, most **deterministic algorithms** for this problem are based on network flow and are considerably more complicated.

12.3.1 Classification of Randomized Algorithms

The randomized sorting algorithm and the min-cut algorithm exemplify two different types of randomized algorithms. The sorting algorithm always gives the correct solution. The only variation from one run to another is its running time, whose distribution we study. Such an algorithm is called a **Las Vegas algorithm**.

In contrast, the min-cut algorithm may sometimes produce a solution that is incorrect. However, we prove that the probability of such an error is bounded. Such an algorithm is called a Monte Carlo algorithm. In Section 12.3 we observed a useful property of a **Monte Carlo algorithm:** if the algorithm is run repeatedly with independent random choices each time, the failure probability can be made arbitrarily small, at the expense of running time. In some randomized algorithms both the running time and the quality of the solution are random variables; sometimes these are also referred to as Monte Carlo algorithms. The reader is referred to [25] for a detailed discussion of these issues.

12.4 Foiling an Adversary

A common paradigm in the design of randomized algorithms is that of foiling an adversary. Whereas an adversary might succeed in defeating a deterministic algorithm with a carefully constructed "bad" input, it is difficult for an adversary to defeat a deterministic algorithm in this fashion. Due to the random choices made by the randomized algorithm the adversary cannot, while constructing the input, predict the precise behavior of the algorithm. An alternative view of this process is to think of the randomized algorithm as first picking a series of random numbers which it then uses in the course of execution as needed. In this view, we may think of the random numbers chosen at the start as "selecting" one of a family of deterministic algorithms. In other words a randomized algorithm can be thought of as a probability distribution on deterministic algorithms. We illustrate these ideas in the setting of AND-OR tree evaluation; the following algorithm is due to Snir [39].

For our purposes an AND-OR tree is a rooted complete binary tree in which internal nodes at even distance from the root are labeled AND and internal nodes at odd distance are labeled OR. Associated with each leaf is a Boolean value. The evaluation of the game tree is the following process. Each leaf returns the value associated with it. Each OR node returns the Boolean OR of the values returned by its children, and each AND node returns the Boolean AND of the values returned by its children. At each step an evaluation algorithm chooses a leaf and reads its value. We do not charge the algorithm for any other computation. We study the number of such steps taken by an algorithm for evaluating an AND-OR tree, the worst case being taken over all assignments of Boolean values to the leaves.

Let T_k denote an AND-OR tree in which every leaf is at distance $2k$ from the root. Thus, any root-to-leaf path passes through k AND nodes (including the root itself) and k OR nodes, and there are 2^{2k} leaves. An algorithm begins by specifying a leaf whose value is to be read at the first step. Thereafter, it specifies such a leaf at each step, based on the values it has read on previous steps. In a deterministic algorithm, the choice of the next leaf to be read is a deterministic function of the values at the leaves read so far. For a randomized algorithm, this choice may be randomized. It is not hard to show that for any deterministic evaluation algorithm, there is an instance of T_k that forces the algorithm to read the values on all 2^{2k} leaves.

We now give a simple randomized algorithm and study the expected number of leaves it reads on any instance of T_k. The algorithm is motivated by the following simple observation. Consider a single AND node with two leaves. If the node were to return 0, at least one of the leaves must contain 0. A deterministic algorithm inspects the leaves in a fixed order, and an adversary can therefore always "hide" the 0 at the second of the two leaves inspected by the algorithm. Reading the leaves in a random order foils this strategy. With probability $1/2$, the algorithm chooses the hidden 0 on the first step, so its expected number of steps is $3/2$, which is better than the worst case for any deterministic algorithm. Similarly, in the case of an OR node, if it were to return a 1 then a randomized order of examining the leaves will reduce the expected number of steps to $3/2$. We now extend this intuition and specify the complete algorithm.

To evaluate an AND node v, the algorithm chooses one of its children (a subtree rooted at an OR node) at random and evaluates it by recursively invoking the algorithm. If 1 is returned by the subtree, the algorithm proceeds to evaluate the other child (again by recursive application). If 0 is returned, the algorithm returns 0 for v. To evaluate an OR node, the procedure is the same with the roles of 0 and 1 interchanged. We establish by induction on k that the expected cost of evaluating any instance of T_k is at most 3^k.

The basis ($k = 0$) is trivial. Assume now that the expected cost of evaluating any instance of T_{k-1} is at most 3^{k-1}. Consider first a tree T whose root is an OR node, each of whose children is the root of a copy of T_{k-1}. If the root of T were to evaluate to 1, at least one of its children returns 1. With probability $1/2$ this child is chosen first, incurring (by the inductive hypothesis) an expected cost of at most 3^{k-1} in evaluating T. With probability $1/2$ both subtrees are evaluated, incurring a net cost of at most $2 \times 3^{k-1}$. Thus, the expected cost of determining the value of T is

$$\leq \frac{1}{2} \times 3^{k-1} + \frac{1}{2} \times 2 \times 3^{k-1} = \frac{3}{2} \times 3^{k-1}. \tag{12.6}$$

If on the other hand the OR were to evaluate to 0 both children must be evaluated, incurring a cost of at most $2 \times 3^{k-1}$.

Consider next the root of the tree T_k, an AND node. If it evaluates to 1, then both its subtrees rooted at OR nodes return 1. By the discussion in the previous paragraph and by linearity of expectation, the expected cost of evaluating T_k to 1 is at most $2 \times (3/2) \times 3^{k-1} = 3^k$. On the other hand, if the instance of T_k evaluates to 0, at least one of its subtrees rooted at OR nodes returns 0. With probability $1/2$ it is chosen first, and so the expected cost of evaluating T_k is at most

$$2 \times 3^{k-1} + \frac{1}{2} \times \frac{3}{2} \times 3^{k-1} \le 3^k.$$

THEOREM 12.3   *Given any instance of T_k, the expected number of steps for the aforementioned randomized algorithm is at most 3^k.*

Since $n = 4^k$ the expected running time of our randomized algorithm is $n^{\log_4 3}$, which we bound by $n^{0.793}$. Thus, the expected number of steps is smaller than the worst case for any deterministic algorithm. Note that this is a Las Vegas algorithm and always produces the correct answer.

12.5   The Minimax Principle and Lower Bounds

The randomized algorithm of Section 12.4 has an expected running time of $n^{0.793}$ on any uniform binary AND-OR tree with n leaves. Can we establish that no randomized algorithm can have a lower expected running time? We first introduce a standard technique due to Yao [46] for proving such lower bounds. This technique applies only to algorithms that terminate in finite time on all inputs and sequences of random choices.

The crux of the technique is to relate the running times of randomized algorithms for a problem to the running times of deterministic algorithms for the problem when faced with randomly chosen inputs. Consider a problem where the number of distinct inputs of a fixed size is finite, as is the number of distinct (deterministic, terminating, and always correct) algorithms for solving that problem. Let us define the **distributional complexity** of the problem at hand as the expected running time of the best deterministic algorithm for the worst distribution on the inputs. Thus, we envision an adversary choosing a probability distribution on the set of possible inputs, and study the best deterministic algorithm for this distribution. Let p denote a probability distribution on the set $\mathcal{I}$ of inputs. Let the random variable $C(I_{ismp}, A)$ denote the running time of deterministic algorithm $A \in \mathcal{A}$ on an input chosen according to p. Viewing a randomized algorithm as a probability distribution q on the set $\mathcal{A}$ of deterministic algorithms, we let the random variable $C(I, A_{ismq})$ denote the running time of this randomized algorithm on the worst-case input.

PROPOSITION 12.1   (Yao's minimax principle): For all distributions p over $\mathcal{I}$ and q over $\mathcal{A}$,

$$\min_{A \in \mathcal{A}} \mathbf{E}\left[C(I_{ismp}, A)\right] \le \max_{I \in \mathcal{I}} \mathbf{E}\left[C(I, A_{ismq})\right].$$

In other words, the expected running time of the optimal deterministic algorithm for an arbitrarily chosen input distribution p is a lower bound on the expected running time of the optimal (Las Vegas) randomized algorithm for Π. Thus, to prove a lower bound on the **randomized complexity** it suffices to choose any distribution p on the input and prove a lower bound on the expected running time of deterministic algorithms for that distribution. The power of this technique lies in the flexibility in the choice of p and, more importantly, the reduction to a lower bound on deterministic algorithms. It is important to remember that the deterministic algorithm "knows" the chosen distribution p.

The aforementioned discussion dealt only with lower bounds on the performance of Las Vegas algorithms. We briefly discuss Monte Carlo algorithms with error probability $\epsilon \in [0, 1/2]$. Let us define the distributional complexity with error ϵ, denoted $\min_{A \in \mathcal{A}} \mathbf{E}[C_\epsilon(I_{ismp}, A)]$, to be the minimum expected running time of any deterministic algorithm that errs with probability at most ϵ under the input distribution p. Similarly, we denote by $\max_{I \in \mathcal{I}} \mathbf{E}[C_\epsilon(I, A_{ismq})]$ the expected running time (under the worst input) of any randomized algorithm that errs with probability at most ϵ (again, the randomized algorithm is viewed as a probability distribution q on deterministic algorithms). Analogous to Proposition 12.1, we then have the following:

PROPOSITION 12.2   For all distributions p over $\mathcal{I}$ and q over $\mathcal{A}$ and any $\epsilon \in [0, 1/2]$,

$$\frac{1}{2} \left(\min_{A \in \mathcal{A}} \mathbf{E}\left[C_{2\epsilon}\left(I_{ismp}, A \right) \right] \right) \leq \max_{I \in \mathcal{I}} \mathbf{E}\left[C_\epsilon\left(I, A_{ismq} \right) \right].$$

12.5.1   Lower Bound for Game Tree Evaluation

We now apply Yao's minimax principle to the AND-OR tree evaluation problem. A randomized algorithm for AND-OR tree evaluation can be viewed as a probability distribution over deterministic algorithms, because the length of the computation as well as the number of choices at each step are both finite. We may imagine that all of these coins are tossed before the beginning of the execution.

The tree T_k is equivalent to a balanced binary tree all of whose leaves are at distance $2k$ from the root, and all of whose internal nodes compute the NOR function: a node returns the value 1 if both inputs are 0, and 0 otherwise. We proceed with the analysis of this tree of NORs of depth $2k$.

Let $p = (3 - \sqrt{5})/2$; each leaf of the tree is independently set to 1 with probability p. If each input to a NOR node is independently 1 with probability p, its output is 1 with probability

$$\left(\frac{\sqrt{5} - 1}{2} \right)^2 = \frac{3 - \sqrt{5}}{2} = p.$$

Thus, the value of every node of the NOR tree is 1 with probability p, and the value of a node is independent of the values of all the other nodes on the same level. Consider a deterministic algorithm that is evaluating a tree furnished with such random inputs; let v be a node of the tree whose value the algorithm is trying to determine. Intuitively, the algorithm should determine the value of one child of v before inspecting any leaf of the other subtree. An alternative view of this process is that the deterministic algorithm should inspect leaves visited in a depth-first search of the tree, except of course that it ceases to visit subtrees of a node v when the value of v has been determined. Let us call such an algorithm a depth-first pruning algorithm, referring to the order of traversal and the fact that subtrees that supply no additional information are "pruned" away without being inspected. The following result is due to Tarsi [41].

PROPOSITION 12.3   Let T be a NOR tree each of whose leaves is independently set to 1 with probability q for a fixed value $q \in [0, 1]$. Let $W(T)$ denote the minimum, over all deterministic algorithms, of the expected number of steps to evaluate T. Then, there is a depth-first pruning algorithm whose expected number of steps to evaluate T is $W(T)$.

Proposition 12.3 tells us that for the purposes of our lower bound, we may restrict our attention to depth-first pruning algorithms. Let $W(h)$ be the expected number of leaves inspected by a depth-first pruning algorithm in determining the value of a node at distance h from the leaves, when each leaf is independently set to 1 with probability $(3 - \sqrt{5})/2$. Clearly

$$W(h) = W(h - 1) + (1 - p) \times W(h - 1),$$

where the first term represents the work done in evaluating one of the subtrees of the node, and the second term represents the work done in evaluating the other subtree (which will be necessary if the first subtree returns the value 0, an event occurring with probability $1 - p$). Letting h be $\log_2 n$, and solving, we get $W(h) \geq n^{0.694}$.

THEOREM 12.4   *The expected running time of any randomized algorithm that always evaluates an instance of T_k correctly is at least $n^{0.694}$, where $n = 2^{2^k}$ is the number of leaves.*

Why is our lower bound of $n^{0.694}$ less than the upper bound of $n^{0.793}$ that follows from Theorem 12.3? The reason is that we have not chosen the best possible probability distribution for the values of the leaves. Indeed, in the NOR tree if both inputs to a node are 1, no reasonable algorithm will read leaves of both subtrees of that node. Thus, to prove the best lower bound we have to choose a distribution on the inputs that precludes the event that both inputs to a node will be 1; in other words, the values of the inputs are chosen at random but not independently. This stronger (and considerably harder) analysis can in fact be used to show that the algorithm of Section 12.4 is optimal; the reader is referred to the paper of Saks and Wigderson [34] for details.

12.6   Randomized Data Structures

Recent research into data structures has strongly emphasized the use of randomized techniques to achieve increased efficiency without sacrificing simplicity of implementation. An illustrative example is the randomized data structure for dynamic dictionaries called *skip list* that is due to Pugh [28].

The dynamic dictionary problem is that of maintaining a set of keys X drawn from a totally ordered universe so as to provide efficient support of the following operations: find(q, X)—decide whether the query key q belongs to X and return the information associated with this key if it does indeed belong to X; insert(q, X)—insert the key q into the set X, unless it is already present in X; delete(q, X)—delete the key q from X, unless it is absent from X. The standard approach for solving this problem involves the use of a binary search tree and gives worst-case time per operation that is $O(\log n)$, where n is the size of X at the time the operation is performed. Unfortunately, achieving this time bound requires the use of complex rebalancing strategies to ensure that the search tree remains "balanced," i.e., has depth $O(\log n)$. Not only does rebalancing require more effort in terms of implementation, it also leads to significant overheads in the running time (at least in terms of the constant factors subsumed by the big-oh notation). The skip list data structure is a rather pleasant alternative that overcomes both these shortcomings.

Before getting into the details of randomized skip lists, we will develop some of the key ideas without the use of randomization. Suppose we have a totally ordered data set $X = \{x_1 < x_2 < \cdots < x_n\}$. A gradation of X is a sequence of nested subsets (called *levels*)

$$X_r \subseteq X_{r-1} \subseteq \cdots \subseteq X_2 \subseteq X_1$$

such that $X_r = \emptyset$ and $X_1 = X$. Given an ordered set X and a gradation for it, the level of any element $x \in X$ is defined as

$$L(x) = \max \{i \mid x \in X_i\},$$

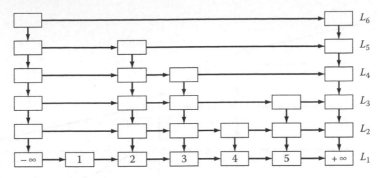

FIGURE 12.3 A skip list.

that is, $L(x)$ is the largest index i such that x belongs to the ith level of the gradation. In what follows, we will assume that two special elements $-\infty$ and $+\infty$ belong to each of the levels, where $-\infty$ is smaller than all elements in X and $+\infty$ is larger than all elements in X.

We now define an ordered list data structure with respect to a gradation of the set X. The first level, X_1 is represented as an ordered linked list, and each node x in this list has a stack of $L(x) - 1$ additional nodes directly above it. Finally, we obtain the skip list with respect to the gradation of X by introducing horizontal and vertical pointers between these nodes as illustrated in Figure 12.3. The skip list in Figure 12.3 corresponds to a gradation of the data set $X = \{1, 3, 4, 7, 9\}$ consisting of the following six levels:

$$X_6 = \emptyset$$
$$X_5 = \{3\}$$
$$X_4 = \{3, 4\}$$
$$X_3 = \{3, 4, 9\}$$
$$X_2 = \{3, 4, 7, 9\}$$
$$X_1 = \{1, 3, 4, 7, 9\}.$$

Observe that starting at the ith node from the bottom in the left-most column of nodes, and traversing the horizontal pointers in order yields a set of nodes corresponding to the elements of the ith level X_i.

Additionally, we will view each level i as defining a set of intervals each of which is defined as the set of elements of X spanned by a horizontal pointer at level i. The sequence of levels X_i can be viewed as successively coarser partitions of X. In Figure 12.3, the levels determine the following partitions of X into a intervals.

$$X_6 = [-\infty, +\infty]$$
$$X_5 = [-\infty, 3] \cup [3, +\infty]$$
$$X_4 = [-\infty, 3] \cup [3, 4] \cup [4, +\infty]$$
$$X_3 = [-\infty, 3] \cup [3, 4] \cup [4, 9] \cup [9, +\infty]$$
$$X_2 = [-\infty, 3] \cup [3, 4] \cup [4, 7] \cup [7, 9] \cup [9, +\infty]$$
$$X_1 = [-\infty, 1] \cup [1, 3] \cup [3, 4] \cup [4, 7] \cup [7, 9] \cup [9, +\infty].$$

An alternate view of the skip list is in terms of a tree defined by the interval partition structure, as illustrated in Figure 12.4 for the aforementioned example. In this tree, each node corresponds to an interval, and the intervals at a given level are represented by nodes at the corresponding level of the

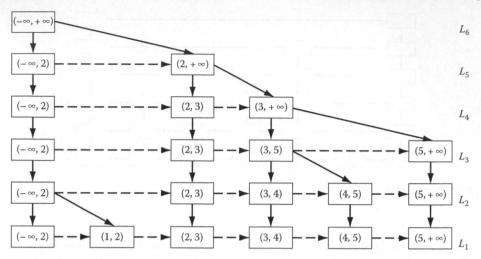

FIGURE 12.4 Tree representation of a skip list.

tree. When an interval J at level $i + 1$ is a superset of an interval I at level i, then the corresponding node J has the node I as a child in this tree. Let $C(I)$ denote the number of children in the tree of a node corresponding to the interval I, i.e., it is the number of intervals from the previous level that are subintervals of I. Note that the tree is not necessarily binary since the value of $C(I)$ is arbitrary. We can view the skip list as a threaded version of this tree, where each thread is a sequence of (horizontal) pointers linking together the nodes at a level into an ordered list. In Figure 12.4, the broken lines indicate the threads, and the full lines are the actual tree pointers.

Finally, we need some notation concerning the membership of an element x in the intervals defined earlier, where x is not necessarily a member of X. For each possible x, let $I_j(x)$ be the interval at level j containing x. In the degenerate case where x lies on the boundary between two intervals, we assign it to the leftmost such interval. Observe that the nested sequence of intervals containing y,

$$I_r(y) \subseteq I_{r-1}(y) \subseteq \cdots \subseteq I_1(y),$$

corresponds to a root-leaf path in the tree corresponding to the skip list.

It remains to specify the choice of the gradation that determines the structure of a skip list. This is precisely where we introduce randomization into the structure of a skip list. The idea is to define a random gradation. Our analysis will show that with high probability, the search tree corresponding to a random skip list is "balanced," and then the dictionary operations can be efficiently implemented.

We define the random gradation for X as follows: given level X_i, the next level X_{i+1} is determined by independently choosing to retain each element $x \in X_i$ with probability $1/2$. The random selection process begins with $X_1 = X$ and terminates when for the first time the resulting level is empty. Alternatively, we may view the choice of the gradation as follows: for each $x \in X$, choose the level $L(x)$ independently from the geometric distribution with parameter $p = 1/2$ and place x in the levels $X_1, \ldots, X_{L(x)}$. We define r to be one more than the maximum of these level numbers. Such a random level is chosen for every element of X upon its insertion and remains fixed until its deletion.

We omit the proof of the following theorem bounding the space complexity of a randomized skip list. The proof is a simple exercise, and it is recommended that the reader verify this to gain some insight into the behavior of this data structure.

THEOREM 12.5 *A random skip list for a set X of size n has expected space requirement* $O(n)$.

We will go into more details about the time complexity of this data structure. The following lemma underlies the running time analysis.

LEMMA 12.1 The number of levels r in a random gradation of a set X of size n has expected value $E[r] = O(\log n)$. Further, $r = O(\log n)$ with high probability.

PROOF We will prove the high probability result; the bound on the expected value follows immediately from this. Recall that the level numbers $L(x)$ for $x \in X$ are i.i.d. (independent and identically distributed) random variables distributed geometrically with parameter $p = 1/2$; notationally, we will denote these random variables by $Z_1, \ldots, Z_n$. Now, the total number of levels in the skip list can be determined as

$$r = 1 + \max_{x \in X} L(x) = 1 + \max_{1 \le i \le n} Z_i,$$

that is, as one more than the maximum of n i.i.d. geometric random variables.

For such geometric random variables with parameter p, it is easy to verify that for any positive real t, $\mathbf{Pr}[Z_i > t] \le (1-p)^t$. It follows that

$$\mathbf{Pr}\left[\max_i Z_i > t\right] \le n(1-p)^t = \frac{n}{2^t},$$

since $p = 1/2$ in this case. For any $\alpha > 1$, setting $t = \alpha \log n$, we obtain that

$$\mathbf{Pr}\left[r > \alpha \log n\right] \le \frac{1}{n^{\alpha - 1}}.$$

We can now infer that the tree representing the skip list has height $O(\log n)$ with high probability. To show that the overall search time in a skip list is similarly bounded, we must first specify an efficient implementation of the find operation. We present the implementation of the dictionary operations in terms of the tree representation; it is fairly easy to translate this back into the skip list representation.

To implement $find(y, X)$, we must walk down the path

$$I_r(y) \subseteq I_{r-1}(y) \subseteq \cdots \subseteq I_1(y).$$

For this, at level j, starting at the node $I_j(y)$, we use the vertical pointer to descend to the leftmost child of the current interval; then, via the horizontal pointers, we move rightward till the node $I_j(y)$ is reached. Note that it is easily determined whether y belongs to a given interval, or to an interval to its right. Further, in the skip list, the vertical pointers allow access only to the leftmost child of an interval, and therefore we must use the horizontal pointers to scan its children.

To determine the expected cost of a $find(y, X)$ operation, we must take into account both the number of levels and the number of intervals/nodes scanned at each level. Clearly, at level j, the number of nodes visited is no more than the number of children of $I_{j+1}(y)$. It follows that the cost of find can be bounded by

$$O\left(\sum_{j=1}^{r} \left(1 + C\left(I_j(y)\right)\right)\right).$$

The following lemma shows that this quantity has expectation bounded by $O(\log n)$.

LEMMA 12.2   For any y, let $I_r(y), \ldots, I_1(y)$ be the search path followed by find(y, X) in a random skip list for a set X of size n. Then,

$$\mathbf{E}\left[\sum_{j=1}^{r}\left(1 + C(I_j(y))\right)\right] = \mathrm{O}(\log n).$$

PROOF   We begin by showing that for any interval I in a random skip list, $\mathbf{E}[C(I)] = \mathrm{O}(1)$. By Lemma 12.1, we are guaranteed that $r = \mathrm{O}(\log n)$ with high probability, and so we will obtain the desired bound. It is important to note that we really do need the high probability bound on Lemma 12.1, since it is incorrect to multiply the expectation of r with that of $1 + C(I)$ (the two random variables need not be independent). However, in the approach we will use, since $r > \alpha \log n$ with probability at most $1/n^{\alpha-1}$ and $\Sigma_j(1 + C(I_j(y))) = \mathrm{O}(n)$, it can be argued that the case $r > \alpha \log n$ does not contribute significantly to the expectation of $\Sigma_j C(I_j(y))$.

To show that the expected number of children of an interval J at level i is bounded by a constant, we will show that the expected number of siblings of J (children of its parent) is bounded by a constant; in fact, we will only bound the number of right siblings since the argument for the number of left siblings is identical. Let the intervals to the right of J be the following:

$$J_1 = [x_1, x_2]\,; J_2 = [x_2, x_3]\,; \ldots ; J_k = [x_k, +\infty]\,.$$

Since these intervals exist at level i, each of the elements $x_1, \ldots, x_k$ belong to X_i. If J has s right siblings, then it must be the case that $x_1, \ldots, x_s \notin X_{i+1}$, and $x_{s+1} \in X_{i+1}$. The latter event occurs with probability $1/2^{s+1}$ since each element of X_i is independently chosen to be in X_{i+1} with probability $1/2$. Clearly, the number of right siblings of J can be viewed as a random variable that is geometrically distributed with parameter $1/2$. It follows that the expected number of right siblings of J is at most 2.

Consider now the implementation of the insert and delete operations. In implementing the operation insert(y, X), we assume that a random level $L(y)$ is chosen for y as described earlier. If $L(y) > r$, then we start by creating new levels from $r + 1$ to $L(y)$ and then redefine r to be $L(y)$. This requires $O(1)$ time per level, since the new levels are all empty prior to the insertion of y. Next we perform find(y, X) and determine the search path $I_r(y), \ldots, I_1(y)$, where r is updated to its new value if necessary. Given this search path, the insertion can be accomplished in time $\mathrm{O}(L(y))$ by splitting around y the intervals $I_1(y), \ldots, I_{L(y)}(y)$ and updating the pointers as appropriate. The delete operation is the converse of the insert operation; it involves performing find(y, X) followed by collapsing the intervals that have y as an end-point. Both operations incur cost that is the cost of a find operation and additional cost proportional to $L(y)$. By Lemmas 12.1 and 12.2, we obtain the following theorem.

THEOREM 12.6   *In a random skip list for a set X of size n, the operations* find, insert, *and* delete *can be performed in expected time* $\mathrm{O}(\log n)$.

12.7   Random Reordering and Linear Programming

The linear programming problem is a particularly notable example of the two main benefits of randomization—simplicity and speed. We now describe a simple algorithm for linear programming based on a paradigm for randomized algorithms known as random reordering. For many problems it is possible to design natural algorithms based on the following idea. Suppose that the input consists of n elements. Given any subset of these n elements, there is a solution to the partial problem defined by these elements. If we start with the empty set and add the n elements of the input one at a time, maintaining a partial solution after each addition, we will obtain a solution to the entire problem when all the elements have been added. The usual difficulty with this approach is that the running time of the algorithm depends heavily on the order in which the input elements are added; for any fixed ordering, it is generally possible to force this algorithm to behave badly. The key idea behind random reordering is to add the elements in a random order. This simple device often avoids the pathological behavior that results from using a fixed order.

The linear programming problem is to find the extremum of a linear objective function of d real variables subject to a set H of n constraints that are linear functions of these variables. The intersection of the n half-spaces defined by the constraints is a polyhedron in d-dimensional space (which may be empty, or possibly unbounded). We refer to this polyhedron as the feasible region. Without loss of generality [35] we assume that the feasible region is nonempty and bounded. (Note that we are not assuming that we can test an arbitrary polyhedron for nonemptiness or boundedness; this is known to be equivalent to solving a linear program.) For a set of constraints S, let $\mathcal{B}(S)$ denote the optimum of the linear program defined by S; we seek $\mathcal{B}(S)$.

Consider the following algorithm due to Seidel [37]: add the n constraints in random order, one at a time. After adding each constraint, determine the optimum subject to the constraints added so far. This algorithm may also be viewed in the following "backward" manner, which will prove useful in the sequel.

Algorithm SLP:

Input: A set of constraints H, and the dimension d.

Output: The optimum $\mathcal{B}(H)$.

0. If there are only d constraints, output $\mathcal{B}(H) = H$.

1. Pick a random constraint $h \in H$;
   Recursively find $\mathcal{B}(H \setminus \{h\})$.

2.1. if $\mathcal{B}(H \setminus \{h\})$ does not violate h, output $\mathcal{B}(H \setminus \{h\})$
   to be the optimum $\mathcal{B}(H)$.

2.2. else project all the constraints of $H \setminus \{h\}$ onto h and recursively
   solve this new linear programming problem of one lower
   dimension.

The idea of the algorithm is simple. Either h (the constraint chosen randomly in Step 1) is redundant (in which case we execute Step 2.1), or it is not. In the latter case, we know that the vertex formed by $\mathcal{B}(H)$ must lie on the hyperplane bounding h. In this case, we project all the constraints of $H \setminus \{h\}$ onto h and solve this new linear programming problem (which has dimension $d - 1$).

The optimum $\mathcal{B}(H)$ is defined by d constraints. At the top level of recursion, the probability that a random constraint h violates $\mathcal{B}(H \setminus \{h\})$ is at most d/n. Let $T(n, d)$ denote an upper bound on the expected running time of the algorithm for any problem with n constraints in d dimensions. Then, we may write

$$T(n, d) \leq T(n - 1, d) + \mathrm{O}(d) + \frac{d}{n}[\mathrm{O}(dn) + T(n - 1, d - 1)]. \tag{12.7}$$

In Equation 12.7, the first term on the right denotes the cost of recursively solving the linear program defined by the constraints in $H \backslash \{h\}$. The second accounts for the cost of checking whether h violates $\mathcal{B}(H \backslash \{h\})$. With probability d/n it does, and this is captured by the bracketed expression, whose first term counts the cost of projecting all the constraints onto h. The second counts the cost of (recursively) solving the projected problem, which has one fewer constraint and dimension. The following theorem may be verified by substitution, and proved by induction.

THEOREM 12.7   *There is a constant b such that the recurrence (Equation 12.7) satisfies the solution* $T(n, d) \leq bnd!$.

In contrast if the choice in Step 1 of SLP were not random, the recurrence (Equation 12.7) would be

$$T(n, d) \leq T(n-1, d) + O(d) + O(dn) + T(n-1, d-1), \qquad (12.8)$$

whose solution contains a term that grows quadratically in n.

12.8   Algebraic Methods and Randomized Fingerprints

Some of the most notable randomized results in theoretical computer science, particularly in complexity theory, have involved a nontrivial combination of randomization and algebraic methods. In this section we describe a fundamental randomization technique based on algebraic ideas. This is the randomized fingerprinting technique, originally due to Freivalds [11], for the verification of identities involving matrices, polynomials, and integers. We also describe how this generalizes to the so-called Schwartz–Zippel technique for identities involving multivariate polynomials (independently due to Schwartz [36] and Zippel [47]; see also DeMillo and Lipton [7]). Finally, following Lovász [21], we apply the technique to the problem of detecting the existence of perfect matchings in graphs.

The fingerprinting technique has the following general form. Suppose we wish to decide the equality of two elements x and y drawn from some "large" universe U. Assuming any reasonable model of computation, this problem has a deterministic complexity $\Omega(\log |U|)$. Allowing randomization, an alternative approach is to choose a random function from U into a smaller space V such that with high probability x and y are identical if and only if their images in V are identical. These images of x and y are said to be their fingerprints, and the equality of fingerprints can be verified in time $O(\log |V|)$. Of course, for any fingerprint function the average number of elements of U mapped to an element of V is $|U|/|V|$; so, it would appear impossible to find good fingerprint functions that work for arbitrary or worst-case choices of x and y. However, as we will show in the following text, when the identity-checking is only required to be correct for x and y chosen from a small subspace S of U, particularly a subspace with some algebraic structure, it is possible to choose good fingerprint functions without any a priori knowledge of the subspace, provided the size of V is chosen to be comparable to the size of S.

Throughout this section we will be working over some unspecified field $\mathcal{F}$. Since the randomization will involve uniform sampling from a finite subset of the field, we do not even need to specify whether the field is finite or not. The reader may find it helpful in the infinite case to assume that $\mathcal{F}$ is the field $\mathcal{Q}$ of rational numbers, and in the finite case to assume that $\mathcal{F}$ is $\mathcal{Z}_p$, the field of integers modulo some prime number p.

12.8.1   Freivalds' Technique and Matrix Product Verification

We begin by describing a fingerprinting technique for verifying matrix product identities. Currently, the fastest algorithm for matrix multiplication (due to Coppersmith and Winograd [6]) has running time $O(n^{2.376})$, improving significantly on the obvious $O(n^3)$ time algorithm; however, the fast matrix multiplication algorithm has the disadvantage of being extremely complicated. Suppose we have an implementation of the fast matrix multiplication algorithm and, given its complex nature, are unsure of its correctness. Since program verification appears to be an intractable problem, we consider the more reasonable goal of verifying the correctness of the output produced by executing the algorithm on specific inputs. (This notion of verifying programs on specific inputs is the basic tenet of the theory of program checking recently formulated by Blum and Kannan [5].) More concretely, suppose we are given three $n \times n$ matrices X, Y, and Z over a field $\mathcal{F}$, and would like to verify that $XY = Z$. Clearly, it does not make sense to use simpler but slower matrix multiplication algorithm for the verification, as that would defeat the whole purpose of using the fast algorithm in the first place. Observe that, in fact, there is no need to recompute Z; rather, we are merely required to verify that the product of X and Y is indeed equal to Z. Freivalds' technique gives an elegant solution that leads to an $O(n^2)$ time randomized algorithm with bounded error probability.

The idea is to first pick a random vector $r \in \{0,1\}^n$, i.e., each component of r is chosen independently and uniformly at random from the set $\{0,1\}$ consisting of the additive and multiplicative identities of the field $\mathcal{F}$. Then, in $O(n^2)$ time, we can compute $y = Yr$, $x = Xy = XYr$, and $z = Zr$. We would like to claim that the identity $XY = Z$ can be verified by merely checking that $x = z$. Quite clearly, if $XY = Z$ then $x = z$; unfortunately, the converse is not true in general. However, given the random choice of r, we can show that for $XY \neq Z$, the probability that $x \neq z$ is at least $1/2$. Observe that the fingerprinting algorithm errs only if $XY \neq Z$ but x and z turn out to be equal, and this has a bounded probability.

THEOREM 12.8 *Let X, Y, and Z be $n \times n$ matrices over some field $\mathcal{F}$ such that $XY \neq Z$; further, let r be chosen uniformly at random from $\{0,1\}^n$ and define $x = XYr$ and $z = Zr$. Then,*

$$\mathbf{Pr}[x = z] \leq 1/2.$$

PROOF Define $W = XY - Z$ and observe that W is not the all-zeroes matrix. Since $Wr = XYr - Zr = x - z$, the event $x = z$ is equivalent to the event that $Wr = 0$. Assume, without loss of generality, that the first row of W has a nonzero entry and that the nonzero entries in that row precede all the zero entries. Define the vector w as the first row of W, and assume that the first $k > 0$ entries in w are nonzero. Since the first component of Wr is $w^T r$, giving an upper bound on the probability that the inner product of w and r is zero will give an upper bound on the probability that $x = z$.

Observe that $w^T r = 0$ if and only if

$$r_1 = \frac{-\sum_{i=2}^{k} w_i r_i}{w_1}. \tag{12.9}$$

Suppose that while choosing the random vector r, we choose $r_2, \ldots, r_n$ before choosing r_1. After the values for $r_2, \ldots, r_n$ have been chosen, the right-hand side of Equation 12.9 is fixed at some value $v \in \mathcal{F}$. If $v \notin \{0,1\}$, then r_1 will never equal v; conversely, if $v \in \{0,1\}$, then the probability that $r_1 = v$ is $1/2$. Thus, the probability that $w^T r = 0$ is at most $1/2$, implying the desired result.

We have reduced the matrix multiplication verification problem to that of verifying the equality of two vectors. The reduction itself can be performed in $O(n^2)$ time and the vector equality can be checked in $O(n)$ time, giving an overall running time of $O(n^2)$ for this Monte Carlo procedure. The error probability can be reduced to $1/2^k$ via k independent iterations of the Monte Carlo algorithm. Note that there was nothing magical about choosing the components of the random vector r from $\{0, 1\}$, since any two distinct elements of $\mathcal{F}$ would have done equally well. This suggests an alternative approach toward reducing the error probability, as follows: each component of r is chosen independently and uniformly at random from some subset S of the field $\mathcal{F}$; then, it is easily verified that the error probability is no more than $1/|S|$.

Finally, note that Freivalds' technique can be applied to the verification of any matrix identity $A = B$. Of course, given A and B, just comparing their entries takes only $\bigcirc(n^2)$ time. But there are many situations where, just as in the case of matrix product verification, computing A explicitly is either too expensive or possibly even impossible, whereas computing Ar is easy. The random fingerprint technique is an elegant solution in such settings.

12.8.2 Extension to Identities of Polynomials

The fingerprinting technique due to Freivalds is fairly general and can be applied to many different versions of the identity verification problem. We now show that it can be easily extended to identity verification for symbolic polynomials, where two polynomials $P_1(x)$ and $P_2(x)$ are deemed identical if they have identical coefficients for corresponding powers of x. Verifying integer or string equality is a special case since we can represent any string of length n as a polynomial of degree n by using the kth element in the string to determine the coefficient of the kth power of a symbolic variable.

Consider first the polynomial product verification problem: given three polynomials $P_1(x)$, $P_2(x)$, $P_3(x) \in \mathcal{F}[x]$, we are required to verify that $P_1(x) \times P_2(x) = P_3(x)$. We will assume that $P_1(x)$ and $P_2(x)$ are of degree at most n, implying that $P_3(x)$ has degree at most $2n$. Note that degree n polynomials can be multiplied in $\bigcirc(n \log n)$ time via fast Fourier transforms, and that the evaluation of a polynomial can be done in $\bigcirc(n)$ time.

The randomized algorithm we present for polynomial product verification is similar to the algorithm for matrix product verification. It first fixes a set $S \subseteq \mathcal{F}$ of size at least $2n + 1$ and chooses $r \in S$ uniformly at random. Then, after evaluating $P_1(r)$, $P_2(r)$, and $P_3(r)$ in $\bigcirc(n)$ time, the algorithm declares the identity $P_1(x)P_2(x) = P_3(x)$ to be correct if and only if $P_1(r)P_2(r) = P_3(r)$. The algorithm makes an error only in the case where the polynomial identity is false but the value of the three polynomials at r indicates otherwise. We will show that the error event has a bounded probability.

Consider the degree $2n$ polynomial $Q(x) = P_1(x)P_2(x) - P_3(x)$. The polynomial $Q(x)$ is said to be identically zero, denoted by $Q(x) \equiv 0$, if each of its coefficients equals zero. Clearly, the polynomial identity $P_1(x)P_2(x) = P_3(x)$ holds if and only if $Q(x) \equiv 0$. We need to establish that if $Q(x) \not\equiv 0$, then with high probability $Q(r) = P_1(r)P_2(r) - P_3(r) \neq 0$. By elementary algebra we know that $Q(x)$ has at most $2n$ distinct roots. It follows that unless $Q(x) \equiv 0$, not more that $2n$ different choices of $r \in S$ will cause $Q(r)$ to evaluate to 0. Therefore, the error probability is at most $2n/|S|$. The probability of error can be reduced either by using independent iterations of this algorithm, or by choosing a larger set S. Of course, when $\mathcal{F}$ is an infinite field (e.g., the reals), the error probability can be made 0 by choosing r uniformly from the entire field $\mathcal{F}$; however, that requires an infinite number of random bits!

Note that we could also use a deterministic version of this algorithm where each choice of $r \in S$ is tried once. But this involves $2n + 1$ different evaluations of each polynomial, and the best known algorithm for multiple evaluations needs $\Theta(n \log^2 n)$ time, which is more than the $\bigcirc(n \log n)$ time requirement for actually performing a multiplication of the polynomials $P_1(x)$ and $P_2(x)$.

This verification technique is easily extended to a generic procedure for testing any polynomial identity of the form $P_1(x) = P_2(x)$ by converting it into the identity $Q(x) = P_1(x) - P_2(x) \equiv 0$. Of course, when P_1 and P_2 are explicitly provided, the identity can be deterministically verified in $\bigcirc(n)$ time by comparing corresponding coefficients. Our randomized technique will take just as long to merely evaluate $P_1(x)$ and $P_2(x)$ at a random value. However, as in the case of verifying matrix identities, the randomized algorithm is quite useful in situations where the polynomials are implicitly specified, e.g., when we only have a "black box" for computing the polynomials with no information about their coefficients, or when they are provided in a form where computing the actual coefficients is expensive. An example of the latter situation is provided by the following problem concerning the determinant of a symbolic matrix. In fact, the determinant problem will require a technique for the verification of polynomial identities of multivariate polynomials that we will discuss shortly.

Consider an $n \times n$ matrix M. Recall that the determinant of the matrix M is defined as follows:

$$\det(M) = \sum_{\pi \in S_n} \text{sgn}(\pi) \prod_{i=1}^{n} M_{i,\pi(i)}, \qquad (12.10)$$

where

S_n is the symmetric group of permutations of order n

$\text{sgn}(\pi)$ is the sign of a permutation π

(The sign function is defined to be $\text{sgn}(\pi) = (-1)^t$, where t is the number of pairwise exchanges required to convert the identity permutation into π.) Although the determinant is defined as a summation with $n!$ terms, it is easily evaluated in polynomial time provided that the matrix entries M_{ij} are explicitly specified. Consider the Vandermonde matrix $M(x_1, \ldots, x_n)$ which is defined in terms of the indeterminates $x_1, \ldots, x_n$ such that $M_{ij} = x_i^{j-1}$, i.e.,

$$M = \begin{pmatrix} 1 & x_1 & x_1^2 & \cdots & x_1^{n-1} \\ 1 & x_2 & x_2^2 & \cdots & x_2^{n-1} \\ & & \vdots & & \\ 1 & x_n & x_n^2 & \cdots & x_n^{n-1} \end{pmatrix}.$$

It is known that for the Vandermonde matrix, $\det(M) = \prod_{i<j}(x_i - x_j)$. Consider the problem of verifying this identity without actually devising a formal proof. Computing the determinant of a symbolic matrix is infeasible as it requires dealing with a summation over $n!$ terms. However, we can formulate the identity verification problem as the problem of verifying that the polynomial $Q(x_1, \ldots, x_n) = \det(M) - \prod_{i<j}(x_i - x_j)$ is identically zero. Based on our discussion of Freivalds' technique, it is natural to consider the substitution of random values for each x_i. Since the determinant can be computed in polynomial time for any specific assignment of values to the symbolic variables $x_1, \ldots, x_n$, it is easy to evaluate the polynomial Q for random values of the variables. The only issue is that of bounding the error probability for this randomized test.

We now extend the analysis of Freivalds' technique for univariate polynomials to the multivariate case. But first, note that in a multivariate polynomial $Q(x_1, \ldots, x_n)$, the degree of a term is the sum of the exponents of the variable powers that define it, and the total degree of Q is the maximum over all terms of the degrees of the terms.

THEOREM 12.9 *Let $Q(x_1, \ldots, x_n) \in \mathcal{F}[x_1, \ldots, x_n]$ be a multivariate polynomial of total degree m. Let S be a finite subset of the field $\mathcal{F}$, and let $r_1, \ldots, r_n$ be chosen uniformly and independently from S. Then,*

$$\mathbf{Pr}\left[Q(r_1, \ldots, r_n) = 0 \mid Q(x_1, \ldots, x_n) \neq 0\right] \leq \frac{m}{|S|}.$$

PROOF We will proceed by induction on the number of variables n. The basis of the induction is the case $n = 1$, which reduces to verifying the theorem for a univariate polynomial $Q(x_1)$ of degree m. But we have already seen for $Q(x_1) \not\equiv 0$, the probability that $Q(r_1) = 0$ is at most $m/|\mathcal{S}|$, taking care of the basis.

We now assume that the induction hypothesis holds for multivariate polynomials with at most $n - 1$ variables, where $n > 1$. In the polynomial $Q(x_1, \ldots, x_n)$ we can factor out the variable x_1 and thereby express Q as

$$Q(x_1, \ldots, x_n) = \sum_{i=0}^{k} x_1^i P_i(x_2, \ldots, x_n),$$

where $k \leq m$ is the largest exponent of x_1 in Q. Given our choice of k, the coefficient $P_k(x_2, \ldots, x_n)$ of x_1^k cannot be identically zero. Note that the total degree of P_k is at most $m - k$. Thus, by the induction hypothesis, we conclude that the probability that $P_k(r_2, \ldots, r_n) = 0$ is at most $(m - k)/|\mathcal{S}|$.

Consider now the case where $P_k(r_2, \ldots, r_n)$ is indeed not equal to 0. We define the following univariate polynomial over x_1 by substituting the random values for the other variables in Q:

$$q(x_1) = Q(x_1, r_2, r_3, \ldots, r_n) = \sum_{i=0}^{k} x_1^i P_i(r_2, \ldots, r_n).$$

Quite clearly, the resulting polynomial $q(x_1)$ has degree k and is not identically zero (since the coefficient of x_1^k is assumed to be nonzero). As in the basis case, we conclude that the probability that $q(r_1) = Q(r_1, r_2, \ldots, r_n)$ evaluates to 0 is bounded by $k/|\mathcal{S}|$.

By the preceding arguments, we have established the following two inequalities:

$$\mathbf{Pr}\,[P_k(r_2, \ldots, r_n) = 0] \leq \frac{m - k}{|\mathcal{S}|}\,;$$

$$\mathbf{Pr}\,[Q(r_1, r_2, \ldots, r_n) = 0 \mid P_k(r_2, \ldots, r_n) \neq 0] \leq \frac{k}{|\mathcal{S}|}.$$

Using the elementary observation that for any two events $\mathcal{E}_1$ and $\mathcal{E}_2$, $\mathbf{Pr}[\mathcal{E}_1] \leq \mathbf{Pr}[\mathcal{E}_1 \mid \overline{\mathcal{E}_2}] + \mathbf{Pr}[\mathcal{E}_2]$, we obtain that the probability that $Q(r_1, r_2, \ldots, r_n) = 0$ is no more than the sum of the two probabilities on the right-hand side of the two obtained inequalities, which is $m/|\mathcal{S}|$. This implies the desired result.

This randomized verification procedure has one serious drawback: when working over large (or possibly infinite) fields, the evaluation of the polynomials could involve large intermediate values, leading to inefficient implementation. One approach to dealing with this problem in the case of integers is to perform all computations modulo some small random prime number; it can be shown that this does not have any adverse effect on the error probability.

12.8.3 Detecting Perfect Matchings in Graphs

We close by giving a surprising application of the techniques from Section 12.8.2. Let $G(U, V, E)$ be a bipartite graph with two independent sets of vertices $U = \{u_1, \ldots, u_n\}$ and $V = \{v_1, \ldots, v_n\}$, and edges E that have one end-point in each of U and V. We define a matching n G as a collection of edges $M \subseteq E$ such that each vertex is an end-point of at most one edge in M; further, a perfect matching is defined to be a matching of size n, i.e., where each vertex occurs as an end-point of exactly one edge in M. Any perfect matching M may be put into a 1-to-1 correspondence with the permutations in $\mathcal{S}_n$, where the matching corresponding to a permutation $\pi \in \mathcal{S}_n$ is given by the collection of edges $\{(u_i, v_{\pi(i)}) \mid 1 \leq i \leq n\}$. We now relate the matchings of the graph to the determinant of a matrix obtained from the graph.

THEOREM 12.10   *For any bipartite graph $G(U, V, E)$, define a corresponding $n \times n$ matrix A as follows:*

$$A_{ij} = \begin{cases} x_{ij} & (u_i, v_j) \in E \\ 0 & (u_i, v_j) \notin E \end{cases}.$$

Let the multivariate polynomial $Q(x_{11}, x_{12}, \ldots, x_{nn})$ denote the determinant $\det(A)$. Then, G has a perfect matching if and only if $Q \not\equiv 0$.

PROOF   We may express the determinant of A as follows:

$$\det(A) = \sum_{\pi \in \mathcal{S}_n} \text{sgn}(\pi) A_{1,\pi(1)} A_{2,\pi(2)} \cdots A_{n,\pi(n)}.$$

Note that there cannot be any cancellation of the terms in the summation since each indeterminate x_{ij} occurs at most once in A. Thus, the determinant is not identically zero if and only if there exists some permutation π for which the corresponding term in the summation is nonzero. Clearly, the term corresponding to a permutation π is nonzero if and only if $A_{i,\pi(i)} \neq 0$ for each i, $1 \leq i \leq n$; this is equivalent to the presence in G of the perfect matching corresponding to π.

The matrix of indeterminates is sometimes referred to as the Edmonds matrix of a bipartite graph. The aforementioned result can be extended to the case of nonbipartite graphs, and the corresponding matrix of indeterminates is called the Tutte matrix. Tutte [42] first pointed out the close connection between matchings in graphs and matrix determinants; the simpler relation between bipartite matchings and matrix determinants was given by Edmonds [8].

We can turn the aforementioned result into a simple randomized procedure for testing the existence of perfect matchings in a bipartite graph (due to Lovász [21]): using the algorithm from Section 12.8.2, determine whether the determinant is identically zero or not. The running time of this procedure is dominated by the cost of computing a determinant, which is essentially the same as the time required to multiply two matrices. Of course, there are algorithms for constructing a maximum matching in a graph with m edges and n vertices in time $O(m\sqrt{n})$ (see Hopcroft and Karp [14], Micali and Vazirani [23,44], and Feder and Motwani [9]). Unfortunately, the time required to compute the determinant exceeds $m\sqrt{n}$ for small m, and so the benefit in using this randomized decision procedure appears marginal at best. However, this technique was extended by Rabin and Vazirani [31,32] to obtain simple algorithms for the actual construction of maximum matchings; although their randomized algorithms for matchings are simple and elegant, they are still slower than the deterministic $O(m\sqrt{n})$ time algorithms known earlier. Perhaps more significantly, this randomized decision procedure proved to be an essential ingredient in devising fast parallel algorithms for computing maximum matchings [19,27].

12.9   Research Issues and Summary

Perhaps the most important research issue in the area of randomized algorithms is to prove or disprove that are problems solvable in polynomial time by either Las Vegas or Monte Carlo algorithms, but cannot be solved in polynomial time by any deterministic algorithm. Another important direction for future work is to devise high quality pseudo-random number generators, which take a small seed of truly random bits and stretch it into a much longer string that can be used as the random string to fuel randomized algorithms.

12.10 Further Information

In this section we give pointers to a plethora of randomized algorithms not covered here. The reader should also note that the examples above are but a (random!) sample of the many randomized algorithms for each of the problems considered. These algorithms have been chosen to illustrate the main ideas behind randomized algorithms, rather than to represent the state of the art for these problems. The reader interested in other algorithms for these problems is referred to the book by Motwani and Raghavan [25].

Randomized algorithms also find application in a number of other areas: in load-balancing [43], approximation algorithms and combinatorial optimization [12,17,24], graph algorithms [2,16], data structures [3], counting and enumeration [38], parallel algorithms [19,20], distributed algorithms [30], geometric algorithms [26], online algorithms [4,33], and number-theoretic algorithms [29,40] (see also [1]). The reader interested in these applications may consult these articles or the book by Motwani and Raghavan [25].

Defining Terms

Deterministic algorithm: An algorithm whose execution is completely determined by its input.

Distributional complexity: The expected running time of the best possible deterministic algorithm over the worst possible probability distribution on the inputs.

Las Vegas algorithm: A randomized algorithm that always produces correct results, with the only variation from one run to another being in its running time.

Monte Carlo algorithm: A randomized algorithm that may produce incorrect results, but with bounded error probability.

Randomized algorithm: An algorithm that makes random choices during the course of its execution.

Randomized complexity: The expected running time of the best possible randomized algorithm over the worst input.

Acknowledgment

Supported in part by NSF Grant ITR-0331640, TRUST (NSF award number CCF-0424422), and grants from Cisco, Google, KAUST, Lightspeed, and Microsoft.

References

1. Agrawal, M., Kayal, N., and Saxena, N., PRIMES is in P. *Annals of Mathematics,* 160(2), 781–793, 2004.
2. Aleliunas, R., Karp, R.M., Lipton, R.J. Lovász, L., and Rackoff, C., Random walks, universal traversal sequences, and the complexity of maze problems. In *Proceedings of the 20th Annual Symposium on Foundations of Computer Science,* pp. 218–223, San Juan, Puerto Rico, Oct. 1979.
3. Aragon, C.R. and Seidel, R.G., Randomized search trees. In *Proceedings of the 30th Annual IEEE Symposium on Foundations of Computer Science,* pp. 540–545, Duke, NC, Oct. 1989.
4. Ben-David, S., Borodin, A., Karp, R.M., Tardos, G., and Wigderson, A., On the power of randomization in on-line algorithms. *Algorithmica,* 11(1), 2–14, 1994.
5. Blum, M. and Kannan, S., Designing programs that check their work. In *Proceedings of the 21st Annual ACM Symposium on Theory of Computing,* pp. 86–97, ACM, New York, May 1989.

6. Coppersmith, D. and Winograd, S., Matrix multiplication via arithmetic progressions. *Journal of Symbolic Computation,* 9, 251–280, 1990.

7. DeMillo, R.A. and Lipton, R.J., A probabilistic remark on algebraic program testing. *Information Processing Letters,* 7, 193–195, 1978.

8. Edmonds, J., Systems of distinct representatives and linear algebra. *Journal of Research of the National Bureau of Standards, 71B,* 4, 241–245, 1967.

9. Feder, T. and Motwani, R., Clique partitions, graph compression and speeding-up algorithms. In *Proceedings of the 25th Annual ACM Symposium on Theory of Computing,* pp. 123–133, ACM, New York, 1991.

10. Floyd, R.W. and Rivest, R.L., Expected time bounds for selection. *Communications of the ACM,* 18, 165–172, 1975.

11. Freivalds, R., Probabilistic machines can use less running time. In *Information Processing 77, Proceedings of IFIP Congress 77,* pp. 839–842, Gilchrist, B., Ed., North-Holland Publishing Company Amsterdam, the Netherlands, Aug. 1977.

12. Goemans, M.X. and Williamson, D.P., 0.878-approximation algorithms for MAX-CUT and MAX-2SAT. In *Proceedings of the 26th Annual ACM Symposium on Theory of Computing,* pp. 422–431, ACM, New York, 1994.

13. Hoare, C.A.R., Quicksort. *Computer Journal,* 5, 10–15, 1962.

14. Hopcroft, J.E. and Karp, R.M., An $n^{5/2}$ algorithm for maximum matching in bipartite graphs. *SIAM Journal on Computing,* 2, 225–231, 1973.

15. Karger, D.R., Global min-cuts in $\mathcal{RNC}$, and other ramifications of a simple min-cut algorithm. In *Proceedings of the 4th Annual ACM-SIAM Symposium on Discrete Algorithms,* pp. 21–30, SIAM, Philadelphia, PA, 1993.

16. Karger, D.R., Klein, P.N., and Tarjan, R.E., A randomized linear-time algorithm for finding minimum spanning trees. *Journal of the ACM,* 42, 321–328, 1995.

17. Karger, D., Motwani, R., and Sudan, M., Approximate graph coloring by semidefinite programming. In *Proceedings of the 35th Annual IEEE Symposium on Foundations of Computer Science,* pp. 2–13, ACM, New York, 1994.

18. Karp, R.M., An introduction to randomized algorithms. *Discrete Applied Mathematics,* 34, 165–201, 1991.

19. Karp, R.M., Upfal, E., and Wigderson, A., Constructing a perfect matching is in random $\mathcal{NC}$. *Combinatorica,* 6, 35–48, 1986.

20. Karp, R.M., Upfal, E., and Wigderson, A., The complexity of parallel search. *Journal of Computer and System Sciences,* 36, 225–253, 1988.

21. Lovász, L., On determinants, matchings and random algorithms. In *Fundamentals of Computing Theory,* Budach, L., Ed., Akademia-Verlag, Berlin, Germany, 1979.

22. Maffioli, F., Speranza, M.G., and Vercellis, C., Randomized algorithms. In *Combinatorial Optimization: Annotated Bibliographies,* pp. 89–105, O'hEigertaigh, M., Lenstra, J.K., and Rinooy Kan, A.H.G., Eds., John Wiley & Sons, New York, 1985.

23. Micali, S. and Vazirani, V.V., An $O(\sqrt{|V|}|e|)$ algorithm for finding maximum matching in general graphs. In *Proceedings of the 21st Annual IEEE Symposium on Foundations of Computer Science,* pp. 1 7–27, Syracuse, NY, 1980.

24. Motwani, R., Naor, J., and Raghavan, P., Randomization in approximation algorithms. In *Approximation Algorithms,* Hochbaum, D., Ed., PWS, Boston, MA, 1996.

25. Motwani, R. and Raghavan, P., *Randomized Algorithms.* Cambridge University Press, New York, 1995.

26. Mulmuley, K., *Computational Geometry: An Introduction through Randomized Algorithms.* Prentice Hall, New York, 1993.

27. Mulmuley, K., Vazirani, U.V., and Vazirani, V.V., Matching is as easy as matrix inversion. *Combinatorica,* 7, 105–113, 1987.

28. Pugh, W., Skip lists: A probabilistic alternative to balanced trees. *Communications of the ACM,* 33(6), 668–676, 1990.

29. Rabin, M.O., Probabilistic algorithm for testing primality. *Journal of Number Theory,* 12, 128–138, 1980.

30. Rabin, M.O., Randomized Byzantine generals. In *Proceedings of the 24th Annual Symposium on Foundations of Computer Science,* pp. 403–409, IEEE, New York, 1983.

31. Rabin, M.O. and Vazirani, V.V., Maximum matchings in general graphs through randomization. Technical Report TR-15-84, Aiken Computation Laboratory, Harvard University, Cambridge, MA, 1984.

32. Rabin, M.O. and Vazirani, V.V., Maximum matchings in general graphs through randomization. *Journal of Algorithms,* 10, 557–567, 1989.

33. Raghavan, P. and Snir, M., Memory versus randomization in on-line algorithms. *IBM Journal of Research and Development,* 38, 683–707, 1994.

34. Saks, M. and Wigderson, A., Probabilistic Boolean decision trees and the complexity of evaluating game trees. In *Proceedings of the 27th Annual IEEE Symposium on Foundations of Computer Science,* pp. 29–38, Toronto, ON, Canada, 1986.

35. Schrijver, A., *Theory of Linear and Integer Programming.* John Wiley & Sons, New York, 1986.

36. Schwartz, J.T., Fast probabilistic algorithms for verification of polynomial identities. *Journal of the ACM,* 27(4), 701–717, 1980.

37. Seidel, R.G., Small-dimensional linear programming and convex hulls made easy. *Discrete and Computational Geometry,* 6, 423–434, 1991.

38. Sinclair, A., *Algorithms for Random Generation and Counting: A Markov Chain Approach. Progress in Theoretical Computer Science.* Birkhäuser, Boston, MA, 1992.

39. Snir, M., Lower bounds on probabilistic linear decision trees. *Theoretical Computer Science,* 38, 69–82, 1985.

40. Solovay, R. and Strassen, V., A fast Monte-Carlo test for primality. *SIAM Journal on Computing,* 6(1), 84–85, 1977. See also *SIAM Journal on Computing,* 7, 118, Feb. 1, 1978.

41. Tarsi, M., Optimal search on some game trees. *Journal of the ACM,* 30, 389–396, 1983.

42. Tutte, W.T., The factorization of linear graphs. *Journal of the London Mathematical Society,* 22, 107–111, 1947.

43. Valiant, L.G., A scheme for fast parallel communication. *SIAM Journal on Computing,* 11, 350–361, 1982.

44. Vazirani, V.V., A theory of alternating paths and blossoms for proving correctness of $O(\sqrt{V}E)$ graph maximum matching algorithms. *Combinatorica,* 14(1), 71–109, 1994.

45. Welsh, D.J.A., Randomised algorithms. *Discrete Applied Mathematics,* 5, 133–145, 1983.

46. Yao, A.C-C., Probabilistic computations: Towards a unified measure of complexity. In *Proceedings of the 17th Annual Symposium on Foundations of Computer Science,* pp. 222–227, Providence, Rhode Island, 1977.

47. Zippel, R.E., Probabilistic algorithms for sparse polynomials. In *Proceedings of EUROSAM 79,* volume 72 of *Lecture Notes in Computer Science,* pp. 216–226, Marseille, France 1979.

13

Pattern Matching in Strings

Maxime Crochemore
King's College London and Université Paris-Est

Christophe Hancart
University of Rouen

13.1 Introduction

The present chapter describes a few standard algorithms used for processing texts. They apply, for example, to the manipulation of texts (text editors), to the storage of textual data (text compression), and to data retrieval systems. The algorithms of the chapter are interesting in different respects. First, they are basic components used in the implementations of practical software. Second, they introduce programming methods that serve as paradigms in other fields of computer science (system or software design). Third, they play an important role in theoretical computer science by providing challenging problems.

Although data are stored variously, text remains the main form of exchanging information. This is particularly evident in literature or linguistics where data are composed of huge corpora and dictionaries. This applies as well to computer science where a large amount of data are stored in linear files. And this is also the case in molecular biology where biological molecules can often be approximated as sequences of nucleotides or amino-acids. Moreover, the quantity of available data in these fields tends to double every 18 months. This is the reason why algorithms should be efficient even if the speed of computers increases regularly.

The manipulation of texts involves several problems among which are pattern matching, approximate pattern matching, comparing strings, and text compression. The first problem is partially treated in the present chapter, in that we consider only one-dimensional objects. Extensions of the methods to higher dimensional objects and solutions to the second problem appear in the Chapter 15. The third problem includes the comparison of molecular sequences, and is developed in the corresponding chapter. Finally, an entire chapter is devoted to text compression.

Pattern matching is the problem of locating a collection of objects (the pattern) inside raw text. This is the opposite of the database approach in which texts are structured in fields which themselves are searched by keywords. In this chapter, texts and elements of patterns are strings, which are finite sequences of symbols over a finite alphabet. Methods for searching patterns described by general regular expressions derive from standard parsing techniques (see Chapter 20). We focus our attention to the case where the pattern represents a finite set of strings. Although the latter case is a specialization of the former case, it can be solved with more efficient algorithms.

Solutions to pattern matching in strings divide in two families. In the first one, the pattern is fixed. This situation occurs for example in text editors for the "search" and "substitute" commands, and in telecommunications for checking tokens. In the second family of solutions, the text is considered as fixed while the pattern is variable. This applies to dictionaries and to databases of molecular sequences, for example, and to full-text databases in general.

The efficiency of algorithms is evaluated by their worst-case running times and the amount of memory space they require. In almost all cases, these are the most objective and consistent criteria to appreciate the efficiency of algorithms. A more realistic measure is to consider the expected running time of programs. But such a computation is machine dependent and, moreover, it is based on an average analysis which is often unreachable. This is partly due to the fact that texts are hard to modelize in a probabilistic framework and that computations are impractible in pertinent models. So, most average-case analyses are on random texts.

The alphabet, the finite set of symbols, is denoted by Σ, and the whole set of strings over Σ by Σ^*. The length of a string u is denoted by $|u|$; it is the length of the underlying finite sequence of symbols. The concatenation of two strings u and v is denoted by uv. A string v is said to be a factor (also called segment, substring, . . .) of a string u if u can be written in the from $u'vu''$ where $u', u'' \in \Sigma^*$; if $i = |u'|$ and $j = |u'v| - 1$, we say that the factor v starts at position i and ends at position j in u; the factor v is also denoted by $u[i . . j]$. The symbol at position i in a string u, that is the $i + 1$th symbol of u, is denoted by $u[i]$; we consider implicitly that $u = u[0 . . |u| - 1]$.

13.2 Matching Fixed Patterns

We consider in this section the two cases where the pattern represents a fixed string or a fixed dictionary (a finite set of strings). Algorithms search for and locate all the occurrences of the pattern in any text.

In the string-matching problem, the first case, it is convenient to consider that the text is examined through a window. The window delimits a factor of the text and has usually the length of the pattern. It slides along the text from left to right. During the search, it is periodically shifted according to rules that are specific to each algorithm. When the window is at a certain position on the text, the algorithm checks whether the pattern occurs there or not, by comparing some symbols in the window with the corresponding aligned symbols of the pattern; if there is a whole match, the position is reported. During this scan operation, the algorithm acquires from the text information which are often used to determine the length of the next shift of the window. Some part of the gathered information can also be memorized in order to save time during the next scan operation.

In the dictionary-matching problem, the second case, methods are based on the use of automata, or related data structures.

13.2.1 The Brute Force Algorithm

The simplest implementation of the sliding window mechanism is the brute force algorithm. The strategy consists here in sliding uniformly the window one position to the right after each scan operation. As far as scans are correctly implemented, this obviously leads to a correct algorithm.

We give below the pseudocode of the corresponding procedure. The inputs are a nonempty string x, its length m (thus $m \geq 1$), a string y, and its length n. The variable p in the procedure corresponds to the current left position of the window on the text. It is understood that the string-to-string comparison in line 2 has to be processed symbol per symbol according to a given order.

BRUTE-FORCE-MATCHER(x, m, y, n)
1   **for** $p \leftarrow 0$ **to** $n - m$ **do**
2   **if** $y[p \mathbin{.\,.} p + m - 1] = x$ **then**
3   **report** p

The time complexity of the brute force algorithm is $O(m \times n)$ in the worst case (for instance when $a^{m-1}b$ is searched in a^n for any two symbol $a, b \in \Sigma$ satisfying $a \neq b$ if we assume that the rightmost symbol in the window is compared last). But its behavior is linear in n when searching in random texts.

13.2.2 The Karp–Rabin Algorithm

Hashing provides a simple method for avoiding a quadratic number of symbol comparisons in most practical situations. Instead of checking at each position p of the window on the text whether the pattern occurs here or not, it seems to be more efficient to check only if the factor of the text delimited by the window, namely, $y[p \mathbin{.\,.} p + m - 1]$, "looks like" x. In order to check the resemblance between the two strings, a hash function is used. But, to be helpful for the string-matching problem, the hash function should be highly discriminating for strings. According to the running times of the algorithms, the function should also have the following properties:

- To be efficiently computable
- To provide an easy computation of the value associated with the next factor from the value associated with the current factor

The last point is met when symbols of alphabet Σ are assimilated with integers and when the hash function, say h, is defined for each string $u \in \Sigma^*$ by

$$h(u) = \left(\sum_{i=0}^{|u|-1} u[i] \times d^{|u|-1-i} \right) \bmod q,$$

where q and d are two constants. Then, for each string $v \in \Sigma^*$, for each symbols $a', a'' \in \Sigma$, $h(va'')$ is computed from $h(a'v)$ by the formula

$$h(va'') = ((h(a'v) - a' \times d^{|v|}) \times d + a'') \bmod q.$$

During the search for pattern x, it is enough to compare the value $h(x)$ with the hash value associated with each factor of length m of text y. If the two values are equal, that is, in case of collision, it is still necessary to check whether the factor is equal to x or not by symbol comparisons.

The underlying string-matching algorithm, which is denoted as the Karp–Rabin algorithm, is implemented below as the procedure KARP–RABIN-MATCHER. In the procedure, the values $d^{m-1} \bmod q$, $h(x)$, and $h(y[0 \mathbin{.\,.} m - 2])$ are first precomputed, and stored respectively in the variables r, s, and t (lines 1–7). The value of t is then recomputed at each step of the search phase (lines 8–12). It is

p	0	1	2	3	4	5	6	7	8	9	10	11	12	13	14	15	16	17	18	19
$y[p]$	n	o	␣	d	e	f	e	n	s	e	␣	f	o	r	␣	s	e	n	s	e
$h(y[p .. p+4])$	8	8	6	28	9	18	28	26	22	12	17	24	16	0	1	9	—	—	—	—

FIGURE 13.1 An illustration of the behavior of the Karp–Rabin algorithm when searching for the pattern $x = $ sense in the text $y = $ no␣defense␣for␣sense. Here, symbols are assimilated with their ASCII codes (hence $c = 256$), and the values of q and d are set, respectively, to 31 and 2. This is valid for example when the maximal integer is $2^{16} - 1$. The value of $h(x)$ is $(115 \times 16 + 101 \times 8 + 110 \times 4 + 115 \times 2 + 101) \bmod 31 = 9$. Since only $h(y[4 .. 8])$ and $h(y[15 .. 19])$ among the defined values of $h(y[p .. p+4])$ are equal to $h(x)$, two string-to-string comparisons against x are performed.

assumed that the value of symbols ranges from 0 to $c - 1$; the quantity $(c - 1) \times q$ is added in line 8 to provide correct computations on positive integers.

KARP–RABIN–MATCHER(x, m, y, n)

```
1   r ← 1
2   s ← x[0] mod q
3   t ← 0
4   for i ← 1 to m − 1 do
5       r ← (r × d) mod q
6       s ← (s × d + x[i]) mod q
7       t ← (t × d + y[i − 1]) mod q
8   for p ← 0 to n − m do
9       t ← (t × d + y[p + m − 1]) mod q
10      if t = s and y[p .. p + m − 1] = x then
11          report p
12      t ← ((c − 1) × q + t − y[p] × r) mod q
```

Convenient values for d are powers of 2; in this case, all the products by d can be computed as shifts on integers. The value of q is generally a large prime (such that the quantities $(q - 1) \times d + c - 1$ and $c \times q - 1$ do not cause overflows), but it can also be the value of the implicit modulus supported by integer operations. An illustration of the behavior of the algorithm is given in Figure 13.1.

The worst-case complexity of the above string-matching algorithm is quadratic, as it is for the brute force algorithm, but its expected running time is $O(m + n)$ if parameters q and d are adequate.

13.2.3 The Knuth–Morris–Pratt Algorithm

This section presents the first discovered linear-time string-matching algorithm. Its design follows a tight analysis of a version of the brute force algorithm in which the string-to-string comparison proceeds from left to right. The brute force algorithm wastes the information gathered during the scan of the text. On the contrary, the Knuth–Morris–Pratt algorithm stores the information with two purposes. First, it is used to improve the length of shifts. Second, there is no backward scan of the text.

Consider a given position p of the window on the text. Assume that a mismatch occurs between symbols $y[p + i]$ and $x[i]$ for some i, $0 \le i < m$ (an illustration is given in Figure 13.2). Thus, we have $y[p .. p + i - 1] = x[0 .. i - 1]$ and $y[p + i] \ne x[i]$. With regard to the information given by $x[0 .. i - 1]$, interesting shifts are necessarily connected with the borders of $x[0 .. i - 1]$. (A border

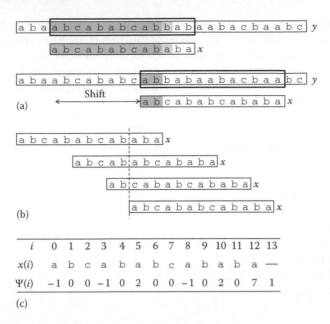

FIGURE 13.2 An illustration of the shift in the Knuth–Morris–Pratt algorithm when searching for the pattern $x = $ abcababcababa. (a) The window on the text y is at position 3. A mismatch occurs at position 10 on x. The matching symbols are shown darkly shaded, and the current analyzed symbols lightly shaded. Avoiding both a backtrack on the text and an immediate mismatch leads to shift the window 8 positions to the right. The string-to-string comparison resumes at position 2 on the pattern. (b) The current shift is the consequence of an analysis of the list of the proper borders of $x[0 \mathinner{..} 9]$ and of the symbol which follow them in x. The prefixes of x that are borders of $x[0 \mathinner{..} 9] = $ abcabacbab are right-aligned along the discontinuous vertical line. String $x[0 \mathinner{..} 4] = $ abcab is a border of $x[0 \mathinner{..} 9]$, but is followed by symbol a which is identical to $x[10]$. String $x[0 \mathinner{..} 1]$ is the expected border, since it is followed by symbol c. (c) The values of the function ψ for pattern x.

of a string u is a factor of u that is both a prefix and a suffix of u). Among the borders of $x[0 \mathinner{..} i-1]$, the longest proper border followed by a symbol different from $x[i]$ is the best possible candidate, subject to the existence of such of a border. (A factor v of a string u is said to be a proper factor of u if u and v are not identical, that is, if $|v| < |u|$.) This introduces the function ψ defined for each $i \in \{0, 1, \ldots, m-1\}$ by

$$\psi[i] = \max\{k \mid (0 \leq k < i, x[i-k \mathinner{..} i-1] = x[0 \mathinner{..} k-1], x[k] \neq x[i]) \text{ or } (k = -1)\}.$$

Then, after a shift of length $i - \psi[i]$, the symbol comparisons can resume with $y[p+i]$ against $x[\psi[i]]$ in the case where $\psi[i] \geq 0$, and $y[p+i+1]$ against $x[0]$ otherwise. Doing so, we miss no occurrence of x in y, and avoid a backtrack on the text. The previous statement is still valid when no mismatch occurs, that is when $i = m$, if we consider for a moment the string $x\$$ instead of x, where $\$$ is a symbol of alphabet Σ occurring nowhere in x. This amounts to completing the definition of function ψ by setting

$$\psi[m] = \max\{k \mid 0 \leq k < m, x[m-k \mathinner{..} m-1] = x[0 \mathinner{..} k-1]\}.$$

The Knuth–Morris–Pratt string-matching algorithm is given in pseudocode below as the procedure Knuth–Morris–Pratt-Matcher. The values of function ψ are first computed by the function Better-Prefix-Function given after. The value of the variable j is equal to $p + i$ in the remainder of the code (the search phase of the algorithm strictly speaking); this simplifies the code, and points out the sequential processing of the text. Observe that the preprocessing phase applies a similar method to the pattern itself, as if $y = x[1 .. m - 1]$.

Knuth–Morris–Pratt-Matcher(x, m, y, n)

```
 1   ψ ← Better-Prefix-Function(x, m)
 2   i ← 0
 3   for j ← 0 to n − 1 do
 4       while i ≥ 0 and y[j] ≠ x[i] do
 5           i ← ψ[i]
 6       i ← i + 1
 7       if i = m then
 8           report j + 1 − m
 9           i ← ψ[m]
```

Better-Prefix-Function(x, m)

```
 1   ψ[0] ← −1
 2   i ← 0
 3   for j ← 1 to m − 1 do
 4       if x[j] = x[i] then
 5           ψ[j] ← ψ[i]
 6       else ψ[j] ← i
 7           do   i ← ψ[i]
 8           while i ≥ 0 and x[j] ≠ x[i]
 9       i ← i + 1
10   ψ[m] ← i
11   return ψ
```

The algorithm has a worst-case running time in $O(m + n)$, and requires $O(m)$ extra-space to store function ψ. The linear running time results from the fact that the number of symbol comparisons performed during the preprocessing phase and the search phase is less than $2m$ and $2n$ respectively. All the previous bounds are independent of the size of the alphabet.

13.2.4 The Boyer–Moore Algorithm

The Boyer–Moore algorithm is considered as the most efficient string-matching algorithm in usual applications. A simplified version of it, or the entire algorithm, is often implemented in text editors for the "search" and "substitute" commands.

The scan operation proceeds from right to left in the window on the text, instead of left to right as in the Knuth–Morris–Pratt algorithm. In case of a mismatch, the algorithm uses two functions to shift the window. These two shift functions are called the better-factor shift function and the bad-symbol shift function. In the next two paragraphs, we explain the goal of the two functions and we give procedures to precompute their values.

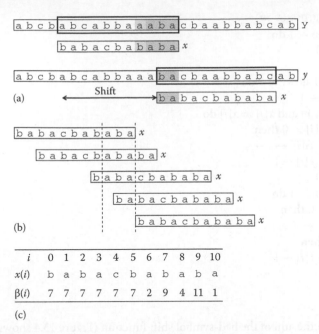

FIGURE 13.3 An illustration of the better-factor shift in the Boyer–Moore algorithm when searching for the pattern $x = \text{babacbababa}$. (a) The window on the text is at position 4. The string-to-string comparison, which proceeds from right to left, stops with a mismatch at position 7 on x. The window is shifted 9 positions to the right to avoid an immediate mismatch. (b) Indeed, the string $x[8 .. 10] = \text{aba}$ is repeated three times in x, but is preceded each time by symbol $x[7] = \text{b}$. The expected matching factor in x is then the prefix ba of x. The factors of x identical with aba and the prefixes of x ending with a suffix of aba are right-aligned along the rightmost discontinuous vertical line. (c) The values of the shift function β for pattern x.

We first explain the aim of the better-factor shift function. Let p be the current (left) position of the window on the text. Assume that a mismatch occurs between symbols $y[p + i]$ and $x[i]$ for some i, $0 \le i < m$ (an illustration is given in Figure 13.3). Then, we have $y[p + i] \ne x[i]$ and $y[p + i + 1 .. p + m - 1] = x[i + 1 .. m - 1]$. The better-factor shift consists in aligning the factor $y[p + i + 1 .. p + m - 1]$ with its rightmost occurrence $x[k + 1 .. m - 1 - i + k]$ in x preceded by a symbol $x[k]$ different from $x[i]$ to avoid an immediate mismatch. If no such factor exists, the shift consists in aligning the longest suffix of $y[p + i + 1 .. p + m - 1]$ with a matching prefix of x. The better-factor shift function β is defined by

$$\beta[i] = \min\{i - k \mid (0 \le k < i, \ x[k + 1 .. m - 1 - i + k] = x[i + 1 .. m - 1], \ x[k] \ne x[i])$$
$$\text{or } (i - m \le k < 0, \ x = x[i - k .. m - 1]x[m - i + k .. m - 1])\}$$

for each $i \in \{0, 1, \ldots, m - 1\}$. The value $\beta[i]$ is then exactly the length of the shift induced by the better-factor shift. The values of function β are computed by the function given below as the function BETTER-FACTOR-FUNCTION. An auxiliary table, namely f, is used; it is an analogue of the function ψ used in the Knuth–Morris–Pratt algorithm, but defined this time for the reverse pattern; it is indexed from 0 to $m - 1$. The running time of the function BETTER-FACTOR-FUNCTION is $O(m)$.

Better-Factor-Function(x, m)

```
 1  for j ← 0 to m − 1 do
 2       β[j] ← 0
 3  i ← m
 4  for j ← m − 1 downto 0 do
 5       f[j] ← i + 1
 6       while i < m and x[j] ≠ x[i] do
 7            if β[i] = 0 then
 8                 β[i] ← i − j
 9                 i ← f[i] − 1
10            i ← i − 1
11  for j ← 0 to m − 1 do
12       if β[j] = 0 then
13            β[j] ← i + 1
14       if j = i then
15            i ← f[i] − 1
16  return β
```

We now come to the aim of the bad-symbol shift function (Figure 13.4 shows an illustration).

Consider again the text symbol $y[p+i]$ that causes a mismatch. Assume first that this symbol occurs in $x[0 .. m − 2]$. Then, let k be the position of the rightmost occurrence of $y[p + i]$ in $x[0 .. m − 2]$. The window can be shifted $i − k$ positions to the right if $k < i$, and only one position otherwise, without missing an occurrence of x in y. Assume now that symbol $y[p + i]$ does not occur in x. Then, no occurrence of x in y can overlap the position $p + i$ on the text, and thus, the window can be shifted $i + 1$ positions to the right. Let δ be the table indexed on alphabet Σ, and defined for each symbol $a \in \Sigma$ by

$$\delta[a] = \min\{m\} \cup \{m − 1 − j \mid 0 \leq j < m − 1,\ x[j] = a\}.$$

According to the above discussion, the bad-symbol shift for the unexpected text symbol a aligned with the symbol at position i on the pattern is the value

$$\gamma[a, i] = \max\{\delta[a] + i − m + 1,\ 1\},$$

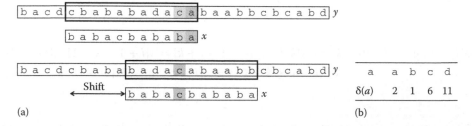

(a)

(b)

FIGURE 13.4 An illustration of the bad-symbol shift in the Boyer–Moore algorithm when searching for the pattern $x = $ babacbababa. (a) The window on the text is at position 4. The string-to-string comparison stops with a mismatch at position 9 on x. Considering only this position and the unexpected symbol occurring at this position, namely symbol $x[9] = $ c, leads to shift the window 5 positions to the right. Notice that if the unexpected symbol were a or d, the applied shift would have been 1 and 10 respectively. (b) The values of the table δ for pattern x when alphabet Σ is reduced to $\{$a, b, c, d$\}$.

which defines the bad-symbol shift function γ on $\Sigma \times \{0, 1, \ldots, m - 1\}$. We give now the code of the function LAST-OCCURRENCE-FUNCTION that computes table δ. Its running time is $O(m + \text{card } \Sigma)$.

LAST-OCCURRENCE-FUNCTION(x, m)

```
1   for each symbol a ∈ Σ do
2       δ[a] ← m
3   for j ← 0 to m − 2 do
4       δ[x[j]] ← m − 1 − j
5   return δ
```

The shift applied in the Boyer–Moore algorithm in case of a mismatch is the maximum between the better-factor shift and the bad-symbol shift. In case of a whole match, the shift applied to the window is m minus the length of the longest proper border of x, that is also the value $\beta[0]$ (this value is indeed what is called "the period" of the pattern). The code of the entire algorithm is given below.

BOYER-MOORE-MATCHER(x, m, y, n)

```
1    β ← BETTER-FACTOR-FUNCTION(x, m)
2    δ ← LAST-OCCURRENCE-FUNCTION(x, m)
3    p ← 0
4    while p ≤ n − m do
5        i ← m − 1
6        while i ≥ 0 and y[p + i] = x[i] do
7            i ← i − 1
8        if i ≥ 0 then
9            p ← p + max{β[i], δ[y[p + i]] + i − m + 1}
10       else report p
11           p ← β[0]
```

The worst-case running time of the algorithm is quadratic. It is surprising however that, when used to search only for the first occurrence of the pattern, the algorithm runs in linear time. Slight modifications of the strategy yield linear-time algorithms. When searching for $a^{m-1}b$ in a^n with $a, b \in \Sigma$ and $a \neq b$, the algorithm considers only $\lfloor n/m \rfloor$ symbols of the text. This bound is the absolute minimum for any string-matching algorithm in the model where only the pattern is preprocessed. Indeed, the algorithm is expected to be extremely fast on large alphabets (relative to the length of the pattern).

13.2.5 Practical String-Matching Algorithms

The bad-symbol shift function introduced in the Boyer–Moore algorithm is not very efficient for small alphabets, but when the alphabet is large compared with the length of the pattern (as it is often the case with the ASCII table and ordinary searches made under a text editor), it becomes very useful. Using only the corresponding table produces some efficient algorithms for practical searches. We describe one of these algorithms below.

Consider a position p of the window on the text, and assume that the symbols $y[p + m − 1]$ and $x[m − 1]$ are identical. If $x[m − 1]$ does not occur in the prefix $x[0 .. m − 2]$ of x, the window can be shifted m positions to the right after the string-to-string comparison between $y[p .. p + m − 2]$ and $x[0 .. m − 2]$ is performed. Otherwise, let k be the position of the rightmost occurrence of $x[m − 1]$ in $x[0 .. m − 2]$; the window can be shifted $m − 1 − k$ positions to the right. This shows that $\delta[y[p + m - 1]]$ is also a valid shift in the case where $y[p + m - 1] = x[m - 1]$. The underlying algorithm is the Horspool algorithm.

The pseudocode of the Horspool algorithm is given below. To prevent two references to the rightmost symbol in the window at each scan and shift operation, table δ is slightly modified: $\delta[x[m - 1]]$ contains the sentinel value 0, after its previous value is saved in variable t. The value of the variable j is the value of the expression $p + m - 1$ in the discussion above.

HORSPOOL-MATCHER(x, m, y, n)
  1   $\delta \leftarrow$ LAST-OCCURRENCE-FUNCTION(x, m)
  2   $t \leftarrow \delta[x[m - 1]]$
  3   $\delta[x[m - 1]] \leftarrow 0$
  4   $j \leftarrow m - 1$
  5   **while** $j < n$ **do**
  6          $s \leftarrow \delta[y[j]]$
  7   **if** $s \neq 0$ **then**
  8                  $j \leftarrow j + s$
  9   **else if** $y[j - m + 1 .. j - 1] = x[0 .. m - 2]$ **then**
 10   **report** $j - m + 1$
 11                  $j \leftarrow j + t$

Just like the brute force algorithm, the Horspool algorithm has a quadratic worst-case time complexity. But its behavior may be at least as good as the behavior of the Boyer–Moore algorithm in practice because the Horspool algorithm is simpler. An example showing the behavior of both algorithms is given in Figure 13.5.

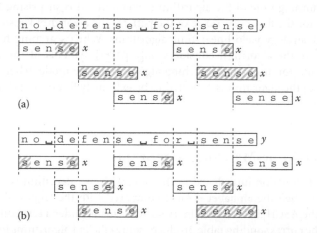

FIGURE 13.5 An illustration of the behavior of two fast string-matching algorithms when searching for the pattern $x =$ sense in the text $y =$ no_defense_for_sense. The successive positions of the window on the text are suggested by the alignments of x with the corresponding factors of y. The symbols of x considered during each scan operation are shown hachured. (a) Behavior of the Boyer–Moore algorithm. The first and second shifts result from the better-shift function, the third and fourth from the bad-symbol function, and the fifth from a shift of the length of x minus the length of its longest proper border (the period of x). (b) Behavior of the Horspool algorithm. We assume here that the four leftmost symbols in the window are compared with the symbols of $x[0 .. 3]$ from left to right.

13.2.6   The Aho–Corasick Algorithm

The UNIX operating system provides standard text-file facilities. Among them is the series of `grep` commands that locate patterns in files. We describe in this section the Aho–Corasick algorithm underlying an implementation of the `fgrep` command of UNIX. It searches files for a finite and fixed set of strings (the dictionary), and can for instance output lines containing at least one of the strings.

If we are interested in searching for all occurrences of all strings of a dictionary, a first solution consists in repeating some string-matching algorithm for each string. Considering a dictionary X containing k strings and a text y, the search runs in that case in time $O(m + n \times k)$, where m is the sum of the length of the strings in X, and n the length of y. But this solution is not efficient, since text y has to be read k times. The solution described in this section provides both a sequential read of the text and a total running time which is $O(m + n)$ on a fixed alphabet. The algorithm can be viewed as a direct extension of a weaker version of the Knuth–Morris–Pratt algorithm.

The search is done with the help of an automaton that stores the situations encountered during the process. At a given position on the text, the current state is identified with the set of pattern prefixes ending here. The state represents all the factors of the pattern that can possibly lead to occurrences. Among the factors, the longest contains all the information necessary to continue the search. So, the search is realized with an automaton, denoted by $\mathcal{D}(X)$, of which states are in one-to-one correspondence with the prefixes of X. Implementing completely the transition function of $\mathcal{D}(X)$ would required a size $O(m \times \text{card } \Sigma)$. Instead of that, the Aho–Corasick algorithm requires only $O(m)$ space. To get this space complexity, a part of the transition function is made explicit in the data, and the other part is computed with the help of a failure function. For the first part, we assume that for any input (p, a), the function denoted by TARGET returns some state q if the triple (p, a, q) is an edge in the data, and the value NIL otherwise. The second part uses the failure function *fail*, which is an analogue of the function ψ used in the Knuth–Morris–Pratt algorithm. But this time, the function is defined on the set of states, and for each state p different from the initial state,

> $fail[p] = $ the state identified with the longest proper suffix of the prefix identified with p
>
> that is also a prefix of a string of X.

The aim of the failure function is to defer the computation of a transition from the current state, say p, to the computation of the transition from the state $fail[p]$ with the same input symbol, say a, when no edge from p labeled by symbol a is in the data; the initial state, which is identified with the empty string, is the default state for the statement. We give below the pseudocode of the function NEXT-STATE that computes the transitions in the representation. The initial state is denoted by i.

NEXT-STATE(p, a, i)
1   **while** $p \neq$ NIL and TARGET(p, a) = NIL **do**
2       $p \leftarrow fail[p]$
3   **if** $p \neq$ NIL **then**
4       $q \leftarrow$ TARGET(p, a)
5   **else** $q \leftarrow i$
6   **return** q

The preprocessing phase of the Aho–Corasick algorithm builds the explicit part of $\mathcal{D}(X)$ including function *fail*. It is divided itself into two phases.

The first phase of the preprocessing phase consists in building a sub-automaton of $\mathcal{D}(X)$. It is the trie of X (the digital tree in which branches spell the strings of X and edges are labeled by symbols)

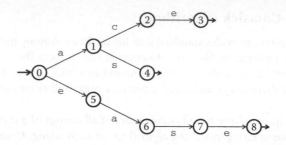

FIGURE 13.6 The trie, tree-like automaton, of the pattern $X = \{\text{ace}, \text{as}, \text{ease}\}$. The initial state is distinguished by a thick ingoing arrow, each terminal state by a thick outgoing arrow. The states are numbered from 0 to 8, according to the order in which they are created by the construction statement described in the present section. State 0 is identified with the empty string, state 1 with a, state 2 with ac, state 3 with ace, and so on. The automaton accepts the language X.

having as initial state the root of the trie and as terminal states the nodes corresponding to strings of X (an example is given in Figure 13.6). It differs from $\mathcal{D}(X)$ in two points:

- It contains only the forward edges
- It accepts only the set X

(An edge (p, a, q) in the automaton is said to be forward if the prefix identified with q is of the form ua where u is the prefix corresponding to p.) The function given below as the function TRIE-LIKE-AUTOMATON computes the automaton corresponding to the trie of X by returning its initial state. The terminal mark of each state r is managed through the attribute $terminal[r]$; the mark is either TRUE or FALSE depending on whether state r is terminal or not. We assume that the function NEW-STATE creates and returns a new state, and that the procedure ADD-EDGE adds a given new edge to the data.

TRIE-LIKE-AUTOMATON(X)
```
 1   i ← NEW-STATE
 2   terminal[i] ← FALSE
 3   for string x from first to last string of X do
 4       p ← i
 5       for symbol a from first to last symbol of x do
 6           q ← TARGET(p, a)
 7           if q = NIL then
 8               q ← NEW-STATE
 9               terminal[q] ← FALSE
10               ADD-EDGE(p, a, q)
11           p ← q
12       terminal[p] ← TRUE
13   return i
```

The second step of the preprocessing phase consists mainly in precomputing the failure function. This is done by a breadth-first traversal of the trie-like automaton. The corresponding pseudocode is given below as the procedure MAKE-FAILURE-FUNCTION.

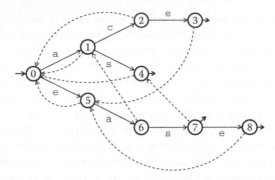

FIGURE 13.7 The explicit part of the automaton $\mathcal{D}(X)$ of the pattern $X = \{\texttt{ace}, \texttt{as}, \texttt{ease}\}$. Compared to the trie-like automaton of X displayed in Figure 13.6, state 7 has been made terminal; this is because the corresponding prefix, namely $\texttt{eas}$, ends with the string $\texttt{as}$ that is in X. The failure function *fail* is depicted with discontinuous nonlabeled directed edges.

MAKE-FAILURE-FUNCTION(i)

```
1   fail[i] ← NIL
2   θ ← EMPTY-QUEUE
3   θ ← ENQUEUE(θ, i)
4   while QUEUE-IS-EMPTY(θ) = FALSE do
5       p ← HEAD(θ)
6       θ ← DEQUEUE(θ)
7       for each symbol a such that TARGET(p, a) ≠ NIL do
8           q ← TARGET(p, a)
9           fail[q] ← NEXT-STATE(fail[p], a, i)
10          if terminal[fail[q]] = TRUE then
11              terminal[q] ← TRUE
12          θ ← ENQUEUE(θ, q)
```

During the computation, some states can be made terminal. This occurs when the state is identified with a prefix that ends with a string of X (an illustration is given in Figure 13.7).

The complete dictionary-matching algorithm, implemented in the pseudocode below as the procedure AHO–CORASICK-MATCHER, starts with the two steps of the preprocessing; the search follows, which simulates automaton $\mathcal{D}(X)$. It is understood that the empty string does not belong to X.

AHO–CORASICK-MATCHER(X, y)

```
1   i ← TRIE-LIKE-AUTOMATON(X)
2   MAKE-FAILURE-FUNCTION(i)
3   p ← i
4   for symbol a from first to last symbol of y do
5       p ← NEXT-STATE(p, a, i)
6       if terminal[p] = TRUE then
7           report an occurrence
```

The total number of tests "TARGET(p, a) = NIL" performed by function **NEXT-STATE** during its calls by procedure MAKE-FAILURE-FUNCTION and during its calls by the search phase of the algorithm are bounded by $2m$ and $2n$ respectively, similarly as the bounds on comparisons in the Knuth–Morris–Pratt algorithm. Using a total ordering on the alphabet, the running time of function TARGET is both $O(\log k)$ and $O(\log \text{card } \Sigma)$, since the maximum number of edges outgoing a state in the data representing automaton $\mathcal{D}(X)$ is bounded both by k and by card Σ. Thus, the entire algorithm runs in time $O(m + n)$ on a fixed alphabet, and in time $O((m + n) \times \log \min\{k, \text{card } \Sigma\})$ in the general case. The algorithm requires $O(m)$ extra-space to store the data and to implement the queue used during the breadth-first traversal executed in procedure MAKE-FAILURE-FUNCTION.

Let us discuss the question of reporting occurrences of pattern X (line 7 of procedure AHO-CORASICK-MATCHER). The simplest way of doing it is to report the ending positions of occurrences. This remains to output the position of the current symbol in the text. A second possibility is to report the whole set of strings in X ending at the current position. To do so, the attribute *terminal* has to be transformed. First, for a state r, *terminal*[r] is the set of the string of X that are suffixes of the string corresponding to r. Second, to avoid a quadratic behavior, sets are manipulated by their identifiers only.

13.2.7 Small Patterns

For most text-searching problems, the length of the pattern is small, and is no more than the word size. Representing the state of the search as an integer, and using binary operations to compute the transitions from state to state give some efficient algorithms easy to implement. We present below one algorithm of this class.

Let X be a dictionary of k strings, and m be the sum of the length of the strings in X. Now, consider the automaton $\mathcal{N}(X)$ obtained from the k straightforward deterministic automata accepting the k strings by

- Merging the k initial states into one initial state, say i
- Adding the edges in the form (i, a, i), for each symbol $a \in \Sigma$

The automaton $\mathcal{N}(X)$ is nondeterministic, and it accepts the language Σ^*X (an example is given in Figure 13.8).

The search for occurrences of strings in X is performed with a simulation of the deterministic automaton recognizing Σ^*X. Indeed, the determinization of $\mathcal{N}(X)$ is not performed, but is just simulated via the subset construction: at a given time, the automaton is not in a given state, but in a set of states. This subset is recomputed whenever necessary in the execution of the search.

Let us number the states of $\mathcal{N}(X)$ from -1 to $m - 1$ using a preorder tree walk (this is the case for the example given in Figure 13.8). Let us code the subsets of the set of states of $\mathcal{N}(X)$ minus the initial state by an integer using the following convention: state number j is in the subset if and only

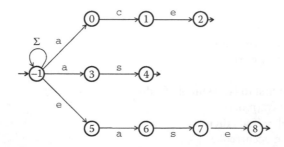

FIGURE 13.8 A straightforward nondeterministic automaton with only one initial state that accepts the language Σ^*X for pattern $X = \{\texttt{ace}, \texttt{as}, \texttt{ease}\}$. The edge labeled by Σ denotes the card Σ edges labeled by the card Σ distinct symbols in Σ.

if the bit at position j of the binary code (starting at position 0) of the integer is 1. Now, let v be the binary value corresponding to the current subset, let a be the current input symbol, and let v' be the binary value corresponding to the next subset. It is then easy to verify that v' is computed from v from following the rule: the jth bit in v' is 1 if and only if

- Either there is an edge labeled by a from the initial state to state number j
- Either there is an edge labeled by a from state number $j - 1$ to state number j and the bit at position $j - 1$ in v is 1

Consider now a binary value f defined for each j, $0 \leq j \leq m - 1$, by:

$$the\,jth\,bit\,of\,f\,is\,1\,if\,and\,only\,if$$
$$there\,exists\,an\,edge\,labeled\,by\,a\,from\,initial\,state\,to\,state\,number\,j,$$

and the binary value table σ indexed on alphabet Σ, and defined for each symbol $a \in \Sigma$ and each bit-position j, $0 \leq j \leq m - 1$, by:

$$the\,jth\,bit\,of\,\sigma[a]\,is\,1\,if\,and\,only\,if$$
$$number\,j\,corresponds\,to\,a\,target\,state\,of\,some\,edge\,labeled\,by\,a.$$

Then, v, v', and a satisfy the relation

$$v' = (2v \vee f) \wedge \sigma[a],$$

denoting respectively by $\vee$ and $\wedge$ the binary operations "or" and "and." It only remains to be able to test whether one of the states represented by v' is a terminal state or not. Let t be a binary value such that, for each j, $0 \leq j \leq m - 1$,

$$the\,jth\,bit\,of\,t\,is\,1\,if\,and\,only\,if\,state\,number\,j\,is\,a\,terminal\,state.$$

The corresponding test is then

$$v' \wedge t \neq 0.$$

An example is given in Figure 13.9.

We give now the code of the function TINY-AUTOMATON that computes binary values f and t, and table σ. Its running time is $O(m + \text{card } \Sigma)$.

TINY-AUTOMATON(X)

```
 1  f ← 0
 2  t ← 0
 3  for each symbol a ∈ Σ do
 4      σ[a] ← 0
 5  r ← 1
 6  for string x from first to last string of X do
 7      f ← f ∨ r
 8      for symbol a from first to last symbol of x do
 9          σ[a] ← σ[a] ∨ r
10          r' ← r
11          r ← 2r
12          t ← t ∨ r'
13  return (f, t, σ)
```

(a)	(b)	a	a c e s x	p	0 1 2 3 4 5
				y[p]	t e a s e s
1	0		1 0 0 0 0		0 0 0 1 0 0 0
0	0		0 1 0 0 0		0 0 0 0 0 0 0
0	1		0 0 1 0 0		0 0 0 0 0 0 0
1	0		1 0 0 0 0		0 0 0 1 0 0 0
0	1	σ[a]	0 0 0 1 0	r	0 0 0 0 1* 0 0
1	0		0 0 1 0 0		0 0 1 0 0 1 0
0	0		1 0 0 0 0		0 0 0 1 0 0 0
0	0		0 0 0 1 0		0 0 0 0 1 0 0
0	1		0 0 1 0 0		0 0 0 0 0 1* 0

(a)   (b)   (c)   (d)

FIGURE 13.9 An illustration of the behavior of the Tiny-Matcher algorithm when searching for the pattern $X = \{\texttt{ace}, \texttt{as}, \texttt{ease}\}$ in the text $y = \texttt{teases}$. (a) The binary value f, which codes potential transitions from the initial state. Bits of a given binary value are written downward, bit at position 0 right at the top. (b) The binary value t, which codes terminal states. (c) The values of the table σ for pattern X. Symbol x means a symbol that does not appear in the strings in X. (d) The successive binary values of the integer r corresponding to the subset of states reached during the search. Asterisked bits indicate that a terminal state is reached, which means that a string in X ends at the current position on the text.

The code of the entire algorithm is given below.

Tiny-Matcher(X, y)
1 $(f, t, \sigma) \leftarrow$ Tiny-Automaton(X)
2 $r \leftarrow f$
3 **for** symbol a from first to last symbol of y **do**
4      $r \leftarrow (2r \vee f) \wedge \sigma[a]$
5   **if** $r \wedge t \neq 0$ **then**
6   **report** an occurrence

The total-running time of the algorithm is $O(m + n + \text{card } \Sigma)$, where n is the length of y. It requires $O(\text{card } \Sigma)$ extra-space to store table σ.

The technique developed in the current section can easily be generalized, for example in allowing don't care symbols.

13.3   Indexing Texts

This section deals with the pattern-matching problem applied to fixed texts. Solutions consist in building an index on the text that speeds up further searches. The indexes that we consider here are data structures that contain all the suffixes and therefore all the factors of the text. Two types of structures are presented: suffix trees and suffix automata. They are both compact representations of suffixes in the sense that their sizes are linear in the length of the text, although the sum of lengths of suffixes of a string is quadratic. Moreover, their constructions take linear time on fixed alphabets. On an arbitrary alphabet Σ on which exists an ordering, a log card Σ factor has to be added to almost all running times given in the following. This corresponds to the branching operation involved in the respective data structures.

Indexes are powerful tools that have many applications. Here is a nonexhaustive list of them, assuming an index on the text y.

- Membership: testing if a string x occurs in y
- Occurrence number: producing the number of occurrences of a string x in y
- List of positions: analogue of the string-matching problem of Section 13.2
- Longest repeated factor: locating the longest factor of y occurring at least twice in y
- Longest common factor: finding a longest string that occurs both in a string x and in y

Solutions to some of these problems are first considered with suffix trees, then with suffix automata. Suffix arrays is another data structure that provides solution running slightly slower. They are shortly described at the end of the section.

13.3.1 Suffix Trees

The suffix tree $T(y)$ of a nonempty string y of length n is a data structure containing all the suffixes of y. In order to simplify the statement, it is assumed that y ends with a special symbol of the alphabet occurring nowhere else in y (this special symbol is denoted by \$ in the examples). The suffix tree of y is a compact trie that satisfies the following properties:

- The branches from the root to the external nodes spell the nonempty suffixes of y, and each external node is marked by the position of the occurrence of the corresponding suffix in y.
- The internal nodes have at least two successors, except if y is a one-length string.
- The edges outgoing an internal node are labeled by factors starting with different symbols.
- Any string that labels an edge is represented by the couple of integers corresponding to its position in y and its length.

(An example of suffix tree is displayed in Figure 13.10.) The special symbol at the end of y avoids marking nodes, and implies that $T(y)$ has exactly n external nodes. The other properties then imply that the total size of $T(y)$ is $O(n)$, which makes it possible to design a linear-time construction of the data structure. The algorithm described in the following and implemented by the procedure SUFFIX-TREE given further has this time complexity.

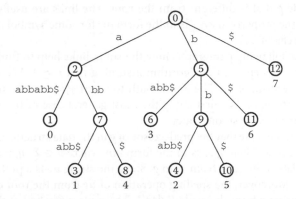

FIGURE 13.10 The suffix tree $T(y)$ of the string $y =$ aabbabb\$. The nodes are numbered from 0 to 12, according to the order in which they are created by the construction algorithm described in the present section. Each of the eight external nodes of the trie is marked by the position of the occurrence of the corresponding suffix in y. Hence, the branch $(0, 5, 9, 4)$, running from the root to an external node, spells the string bbabb\$, which is the suffix of y starting at position 2.

The construction algorithm works as follows. It inserts the nonempty suffixes $y[i \ldots n-1]$, $0 \leq i < n$, of y in the data structure from the longest to the shortest suffix. In order to explain how this is performed, we introduce the two notations

h_i = the longest prefix of $y[i \ldots n-1]$ that is a prefix of some strictly longest suffix of y,

and

t_i = the string w such that $y[i \ldots n-1]$ is identical with $h_i w$,

defined for each $i \in \{1, \ldots, n-1\}$. The strategy to insert the suffixes is precisely based on these definitions. Initially, the data structure contains only the string y. Then, the insertion of the string $y[i \ldots n-1]$, $1 \leq i < n$, proceeds in two steps:

- First, the "head" in the data structure, that is, the node h corresponding to string h_i, is located, possibly breaking an edge.
- Second, a node called the "tail," say t, is created, added as successor of node h, and the edge from h to t is labeled with string t_i.

The second step of the insertion is clearly performed in constant time. Thus, finding the head is critical for the overall performance of the construction algorithm. A brute-force method to find the head consists in spelling the current suffix $y[i \ldots n-1]$ from the root of the trie, giving an $O(|h_i|)$ time complexity for the insertion at step i, and an $O(n^2)$ running time to build the suffix tree $\mathcal{T}(y)$. Adding "short-circuit" links leads to an overall $O(n)$ time complexity, although there is no guarantee that the insertion at any step i is realized in constant time.

Observe that in any suffix tree, if the string corresponding to a given internal node p in the data structure is in the form au with $a \in \Sigma$ and $u \in \Sigma^*$, then there exists an unique internal node corresponding to the string u. From this remark are defined the suffix links by

$link[p]$ = the node q corresponding to the string u

when p corresponds to the string au for some symbol $a \in \Sigma$

for each internal node p that is different from the root. The links are useful when computing h_i from h_{i-1} because of the property: if h_{i-1} is in the form aw for some symbol $a \in \Sigma$ and some string $w \in \Sigma^*$, then w is a prefix of h_i.

We explain in three following paragraphs how the suffix links help to find the successive heads efficiently. We consider a step i in the algorithm assuming that $i \geq 1$. We denote by g the node that corresponds to the string h_{i-1}. The aim is both to insert $y[i \ldots n-1]$ and to find the node h corresponding to the string h_i. We first study the most general case of the insertion of the suffix $y[i \ldots n-1]$. Particular cases are studied later.

We assume in the present case that the predecessor of g in the data structure, say g', is both defined and different from the root. Then h_{i-1} is in the form auv where $a \in \Sigma$, $u, v \in \Sigma^*$, au corresponds to the node g', and v labels the edge from g' to g. Since the string uv is a prefix of h_i, it can be fully spelled from the root. Moreover, the spelling operation of uv from the root can be short-circuited by spelling only the string v from the node $link[g']$. The node q reached at the end of the spelling operation (possibly breaking the last partially taken down edge) is then exactly the node $link[g]$. It remains to spell the string t_{i-1} from q for completely inserting the string $y[i \ldots n-1]$. The spelling stops on the expected node h (possibly breaking again an edge) which becomes the new head in the data structure. The suffix of t_{i-1} that has not been spelled so far, is exactly the string t_i. (An example for the whole previous statement is given in Figure 13.11.)

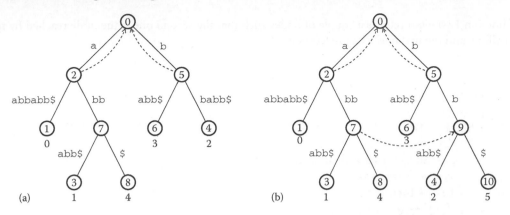

FIGURE 13.11 During the construction of the suffix tree $\mathcal{T}(y)$ of the string $y = $ aabbabb$, the step 5, that is, the insertion of the suffix bb$. The defined suffix link are depicted with discontinuous nonlabeled directed edges. (a) Initially, the head in the data structure is node 7, and its suffix link is not yet defined. The predecessor of node 7, node 2, is different from the root, and the factor of y that is spelled from the root to node 7, namely $h_4 = $ abb, is in the form auv, where $a \in \Sigma$, $u \in \Sigma^*$, and v is the string of Σ^* labeling the edge from node 2 to node 7. Here, $a = $ a, u is the empty string, and $v = $ bb. Then, the string $uv = $ bb is spelled from the node linked with node 2, that is, from node 0; the spelling operation stops on the edge from node 5 to node 4; this edge is broken, which creates node 9. Node 9 is linked to node 7. The string $t_4 = $ \$ is spelled from node 9; the spelling operation stops on node 9, which becomes the new head in the data structure. (b) Node 10 is created, added as successor of node 9, and the edge from node 9 to node 10 is labeled by the string \$, remainder of the last spelling operation.

The second case is when g is a (direct) successor of the root. The string h_{i-1} is then in the form au where $a \in \Sigma$ and $u \in \Sigma^*$. Similar to the above case, the string u can be fully spelled from the root. The spelling of u gives a node q, which is then linked with g. Afterward, the string t_{i-1} is spelled from q.

The last case is when g is the root itself. The string t_{i-1} minus its first symbol has to be spelled from the root: Which ends the study of all the possible cases that can arise.

The important aspect of the algorithm is the use of two different implementations for the two spelling operations pointed out above. The first one, given in the pseudocode below as the function FAST-FIND, deals with the situation where we know in advance that a given factor $y[j . . j + k - 1]$ of y can be fully spelled from a given node p of the trie. It is then sufficient to scan only the first symbols of the labels of the encountered nodes, which justifies the name of the function. The second implementation of the spelling operation spells a factor $y[j . . j + k - 1]$ of y from a given node p too, but, this time, the spelling is performed symbol by symbol. The corresponding function is implemented after as the function SLOW-FIND. Before giving the pseudocode of the functions, we precise the notations used in the following.

- For any input (y, p, j), the function SUCCESSOR-BY-ONE-SYMBOL returns the node q such that q is a successor of the node p and the first symbol of the label of the edge from p to q is $y[j]$; if such a node q does not exist, it returns NIL.
- For any input (p, q), the function LABEL returns the two integers that represent the label of the edge from the node p to the node q.
- The function NEW-NODE creates and returns a new node.
- For any input (p, j, k, q, ℓ), the function NEW-BREAKING-NODE creates and returns the node q' breaking the edge $(p, y[j . . j + k - 1], q)$ at the position ℓ in the label $y[j . . j + k - 1]$. (Which gives the two edges $(p, y[j . . j + \ell - 1], q')$ and $(q', y[j + \ell . . j + k - 1], q)$.)

Function FAST-FIND returns a couple of nodes such that the second one is the node reached by the spelling, and the first one is its predecessor.

FAST-FIND(y, p, j, k)

```
1   p' ← NIL
2   while k > 0 do
3       p' ← p
4       q ← SUCCESSOR-BY-ONE-SYMBOL(y, p, j)
5       (r, s) ← LABEL(p, q)
6       if s ≤ k then
7           p ← q
8           j ← j + s
9           k ← k − s
10      else p ← NEW-BREAKING-NODE(p, r, s, q, k)
11           k ← 0
12  return (p', p)
```

Compared to function FAST-FIND, function SLOW-FIND considers an extra-input that is the predecessor of node p (denoted by p'). It considers in addition two extra-outputs that are the position and the length of the factor that remains to be spelled.

SLOW-FIND(y, p', p, j, k)

```
1   b ← FALSE
2   do  q ← SUCCESSOR-BY-ONE-SYMBOL(y, p, j)
3       if q = NIL then
4           b ← TRUE
5       else (r, s) ← LABEL(p, q)
6           ℓ ← 1
7           while ℓ < s and y[j + ℓ] = y[r + ℓ] do
8               ℓ ← ℓ + 1
9           j ← j + ℓ
10          k ← k − ℓ
11          p' ← p
12          if ℓ = s then
13              p ← q
14          else p ← NEW-BREAKING-NODE(p, r, s, q, ℓ)
15              b ← TRUE
16  while b = FALSE
17  return (p', p, j, k)
```

The complete construction algorithm is implemented as the function SUFFIX-TREE given below. The function returns the root of the constructed suffix-tree. Memorizing systematically the predecessors h' and q' of the nodes h and q avoids considering doubly linked tries. The name of the attribute which marks the positions of the external nodes is made explicit.

SUFFIX-TREE(y, n)

```
 1   p ← NEW-NODE
 2   h′ ← NIL
 3   h ← p
 4   r ← −1
 5   s ← n + 1
 6   for i ← 0 to n − 1 do
 7        if h′ = NIL then
 8             (h′, h, r, s) ← SLOW-FIND(y, NIL, p, r + 1, s − 1)
 9        else (j, k) ← LABEL(h′, h)
10             if h′ = p then
11                  (q′, q) ← FAST-FIND(y, p, j + 1, k − 1)
12             else (q′, q) ← FAST-FIND(y, link[h′], j, k)
13             link[h] ← q
14             (h′, h, r, s) ← SLOW-FIND(y, q′, q, r, s)
15        t ← NEW-NODE
16        ADD-EDGE(h, (r, s), t)
17        position[t] ← i
18   return p
```

The algorithm runs in time $O(n)$ (more precisely $O(n \times \log \operatorname{card} \Sigma)$ if we take into account the branching in the data structure). Indeed, the instruction at line 4 in function FAST-FIND is performed less than $2n$ times, and the number of symbol comparisons done at line 7 in function SLOW-FIND is less than n.

Once the suffix tree of y is built, some operations can be performed rapidly. We describe four applications in the following. Let x be a string of length m.

Testing whether x occurs in y or not can be solved in time $O(m)$ by spelling x from the root of the trie symbol by symbol. If the operation succeeds, x occurs in y. Otherwise, we get the longest prefix of x occurring in y.

Producing the number of occurrences of x in y starts identically by spelling x. Assume that x occurs actually in y. Let p be the node at the extremity of the last taken down edge, or be the root itself if x is empty. The expected number, say k, is then exactly the number of external nodes of the sub-trie of root p. This number can be computed by traversing the sub-trie. Since each internal node of the sub-trie has at least two successors, the total size of the sub-trie is $O(k)$, and the traversal of the sub-trie is performed in time $O(k)$ (independently of Σ). The method can be improved by precomputing in time $O(n)$ (independently of Σ) all the values associated with each internal node; the whole operation is then performed in time $O(m)$, whatever is the number of occurrences of x.

The method for reporting the list of positions of x in y proceeds in the same way. The running time needed by the operation is $O(m)$ to locate x in the trie, plus $O(k)$ to report each of the positions associated with the k external nodes.

Finding the longest repeated factor of y remains to compute the "deepest" internal node of the trie, that is, the internal node corresponding to a longest possible factor in y. This is performed in time $O(n)$.

13.3.2   Suffix Automata

The suffix automaton $S(y)$ of a string y is the minimal deterministic automaton recognizing Suff(y), that is, the set of suffixes of y. This automaton is minimal among all the deterministic automata recognizing the same language, which implies that it is not necessarily complete. An example is given in Figure 13.12.

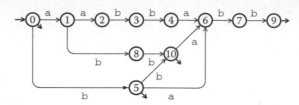

FIGURE 13.12 The suffix automaton $S(y)$ of the string $y =$ aabbabb. The states are numbered from 0 to 10, according to the order in which they are created by the construction algorithm described in the present section. The initial state is state 0, terminal states are states 0, 5, 9, and 10. This automaton is the minimal deterministic automaton accepting the language of the suffixes of y.

The main point about suffix automata is that their size is linear in the length of the string. More precisely, given a string y of length n, the number of states of $S(y)$ is equal to $n + 1$ when $n \leq 2$, and is bounded by $n + 1$ and $2n - 1$ otherwise; as to the number of edges, it is equal to $n + 1$ when $n \leq 1$, it is 2 or 3 when $n = 2$, and it bounded by n and $3n - 4$ otherwise.

The construction of the suffix automaton of a string y of length n can be performed in time $O(n)$, or, more precisely, in time $O(n \times \log \text{card} \, \Sigma)$ on an arbitrary alphabet Σ. It makes use of a failure function *fail* defined on the states of $S(y)$. The set of states of $S(y)$ identifies with the quotient sets

$$u^{-1} \, \text{Suff}(y) = \{v \in \Sigma^* \mid uv \in \text{Suff}(y)\}$$

for the strings u in the whole set of factors of y. One may observe that two sets in the form $u^{-1} \, \text{Suff}(y)$ are either disjoint or comparable. This allows to set

$$\textit{fail}[p] = \text{the smallest quotient set stricly containing the quotient set identified with } p,$$

for each state p of the automaton different from the initial state of the automaton. The function given below as the function SUFFIX-AUTOMATON builds the suffix automaton of y, and returns the initial state, say i, of the automaton. The construction is online, which means that at each step of the construction, just after processing a prefix y' of y, the suffix automaton $S(y')$ is build. Denoting by t the state without outgoing edge in the automaton $S(y')$, terminal states of $S(y')$ are implicitly known by the "suffix path" of t, that is, the list of the states

$$t, \textit{fail}[t], \textit{fail}[\textit{fail}[t]], \ldots, i.$$

The algorithm uses the function *length* defined for each state p of $S(y)$ by

$$\textit{length}[p] = \text{the length of the longest string spelled from } i \text{ to } p.$$

SUFFIX-AUTOMATON(y)

```
 1   i ← NEW-STATE
 2   terminal[i] ← FALSE
 3   length[i] ← 0
 4   fail[i] ← NIL
 5   t ← i
 6   for symbol a from first to last symbol of y do
 7        t ← SUFFIX-AUTOMATON-EXTENSION(i, t, a)
 8   p ← t
 9   do   terminal[p] ← TRUE
10        p ← fail[p]
11   while p ≠ NIL
12   return i
```

The online construction is based on the function SUFFIX-AUTOMATON-EXTENSION that is implemented below. The latter function processes the next symbol, say a, of the string y. If y' is the prefix of y preceding a, it transforms the suffix automaton $\mathcal{S}(y')$ already build into the suffix automaton $\mathcal{S}(y'a)$.

SUFFIX-AUTOMATON-EXTENSION(i, t, a)

```
 1   t' ← t
 2   t ← NEW-STATE
 3   terminal[t] ← FALSE
 4   length[t] ← length[t'] + 1
 5   p ← t'
 6   do   ADD-EDGE(p, a, t)
 7          p ← fail[p]
 8   while p ≠ NIL and TARGET(p, a) = NIL
 9   if p = NIL then
10          fail[t] ← i
11   else q ← TARGET(p, a)
12          if length[q] = length[p] + 1 then
13                 fail[t] ← q
14          else r ← NEW-STATE
15                 terminal[r] ← FALSE
16                 for each symbol b such that TARGET(q, b) ≠ NIL do
17                        ADD-EDGE(r, b, TARGET(q, b))
18                 length[r] ← length[p] + 1
19                 fail[r] ← fail[q]
20                 fail[q] ← r
21                 fail[t] ← r
22                 do   DELETE-EDGE(p, a, TARGET(p, a))
23                        ADD-EDGE(p, a, r)
24                        p ← fail[p]
25                 while p ≠ NIL and TARGET(p, a) = q
26   return t
```

We illustrate the behavior of function SUFFIX-AUTOMATON-EXTENSION in Figure 13.13.

With the suffix automaton $\mathcal{S}(y)$ of y, several operations can be solved efficiently. We describe three of them. Let x be a string of length m.

Membership test solves in time $O(m)$ by spelling x from the initial state of the automaton. If the entire string is spelled, x occurs in y. Otherwise we get the longest prefix of x occurring in y.

Computing the number k of occurrences of x in y (assuming that x is a factor of y) starts similarly. Let p be the state reached after the spelling of x from the initial state. Then k is exactly the number of terminal states accessible from p. The number k associated with each state p can be precomputed in time $O(n)$ (independently of the alphabet) by a depth-first traversal of the graph underlying the automaton. The query for x is then performed in time $O(m)$, whatever is k.

The basis of an algorithm for computing a longest factor common to x and y is implemented in the procedure ENDING-FACTORS-MATCHER given below. This procedure reports at each position in y the length of the longest factor of x ending here. It can obviously be used for string matching. It works as the procedure AHO–CORASICK-MATCHER does in the use of the failure function. The running time of the search phase of the procedure is $O(m)$.

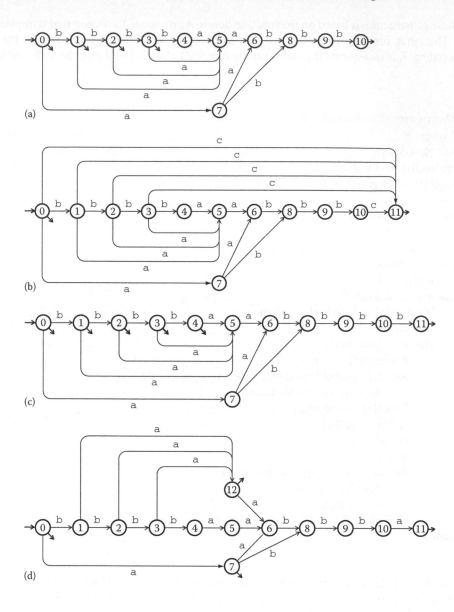

FIGURE 13.13 An illustration of the behavior of function SUFFIX-AUTOMATON-EXTENSION. The function transforms the suffix automaton $\mathcal{S}(y')$ of a string y' in the suffix automaton $\mathcal{S}(y'a)$ for any given symbol a (the terminal states being implicitly known). Let us consider that $y' = $ bbbbaabbb, and let us examine three possible cases according to a, namely $a = $ c, $a = $ b, and $a = $ a. (a) The automaton $\mathcal{S}($bbbbaabbb$)$. The state denoted by t' is state 10, and the suffix path of t' is the list of the states 10, 3, 2, 1, and 0. During the execution of the first loop of the function, state p runs through a part of the suffix path of t'. At the same time, edges labeled by a are created from p the newly created state $t = 11$, unless such an edge already exists, in which case the loop stops. (b) If $a = $ c, the execution stops with an undefined value for p. The edges labeled by c start at terminal states, and the failure of t is the initial state. (c) If $a = $ b, the loop stops on state $p = 3$, because an edge labeled by b is defined on it. The condition at line 12 of function SUFFIX-AUTOMATON-EXTENSION is satisfied, which means that the edge labeled by a from p is not a short-circuit. In this case, the state ending the previous edge is the failure of t. (d) Finally, when $a = $ a, the loop stops on state $p = 3$ for the same reason, but the edge labeled by a from p is a short-circuit. The string bbba is a suffix of the (newly considered) string bbbbaabbba, but bbbba is not. Since these two strings reach state 5, this state is duplicated into a new state $r = 12$ that becomes terminal. Suffixes bba and ba are redirected to this new state. The failure of t is r.

ENDING-FACTORS-MATCHER(y, x)

```
 1   i ← SUFFIX-AUTOMATON(y)
 2   ℓ ← 0
 3   p ← i
 4   for symbol a from first to last symbol of x do
 5       if TARGET(p, a) ≠ NIL then
 6           ℓ ← ℓ + 1
 7           p ← TARGET(p, a)
 8       else do   p ← fail[p]
 9           while p ≠ NIL and TARGET(p, a) ≠ NIL
10           if p = NIL then
11               ℓ ← 0
12               p ← i
13           else ℓ ← length[p] + 1
14               p ← TARGET(p, a)
15       report ℓ
```

Retaining a largest value of the variable ℓ in the procedure (instead of reporting all values) solves the longest common factor problem.

13.3.3 Suffix Arrays

There is a clever and rather simple way to deal with all suffixes of a text of length n: to arrange their list in increasing lexicographic order to be able to perform binary searches on them. The implementation of this idea leads to a data structure called suffix array. It is an efficient representation in the sense that

- It has $O(n)$ size
- It can be constructed in $O(n \log n)$ time
- It allows the computation of the membership test of a string of length m in the text in time $O(m + \log n)$

So, the time required to construct and use the structure is slightly greater than that needed to compute the suffix tree. But suffix arrays have two advantages:

- Their construction is rather simple. It is even commonly admitted that, in practice, it behaves better than the construction of suffix trees.
- It consists of two linear size arrays which, in practice again, take little memory space.

The first array contains the list of positions of lexicographically-sorted suffixes and the second array essentially stores the maximal lengths of prefixes common to consecutive suffixes in the list.

The construction in $O(n \log n)$ time is obviously optimal on a general alphabet because it includes sorting the symbols of the text. To break this bound, a common hypothesis is to consider strings on an integer alphabet for which it is assumed that the number of different letters occurring in any string w is no more than $|w|^c$ for some fixed constant c. This implies that sorting the symbols of w can be achieved in time $O(|w|)$. On such an alphabet, the construction of a suffix array can be done in linear time.

13.4   Research Issues and Summary

String searching by hashing was introduced by Harrison (1971), and later fully analyzed in Karp and Rabin (1987).

The first linear-time string-matching algorithm is due to Knuth et al. (1977). It can be proved that, during the search, the delay, that is, the number of times a symbol of the text is compared to symbols of the pattern, is less than $\lfloor \log_\Phi(m + 1) \rfloor$, where Φ is the golden ratio $(1 + \sqrt{5})/2$. Simon (1993) gives a similar algorithm but with a delay bounded by the size of the alphabet (of the pattern). Hancart (1993) proves that the delay of Simon's algorithm is less than $1 + \lfloor \log_2 m \rfloor$. This paper also proves that this is optimal among algorithms processing the text with a one-symbol buffer. The bound becomes $O(\log \min\{1 + \lfloor \log_2 m \rfloor, \text{card } \Sigma\})$ using an ordering on the alphabet Σ, which is not a restriction in practice.

Galil (1981) gives a general criterion to transform string-matching algorithms that work sequentially on the text into real-time algorithms.

The Boyer–Moore algorithm was designed in Boyer and Moore (1977). The version given in this chapter follows Knuth et al. (1977). This paper contains the first proof on the linearity of the algorithm when restricted to the search of the first occurrence of the pattern. Cole (1994) proves that the maximum number of symbol comparisons is bounded by $3n$ for nonperiodic patterns, and that this bound is tight.

Knuth et al. (1977) considers a variant of the Boyer–Moore algorithm in which all previous matches inside the current window are memorized. Each window configuration becomes the state of what is called the Boyer–Moore automaton. It is still unknown whether the maximum number of states of the automaton is polynomial or not.

Several variants of the Boyer–Moore algorithm avoid the quadratic behavior when searching for all occurrences of the pattern. Among the most efficient in terms of the number of symbol comparisons are the algorithm of Apostolico and Giancarlo (1986), Turbo-BM algorithm by Crochemore et al. (1992) (the two previous algorithms are analyzed in Lecroq (1995)), and the algorithm of Colussi (Colussi, 1994).

The Horspool algorithm is from Horspool (1980). The paper contains practical aspects of string matching that are developed in Hume and Sunday (1991).

The optimal bound on the expected time complexity of string matching is $O(\frac{\log m}{m} n)$ (see Knuth et al. (1977) and the paper of Yao (1980)).

String matching can be solved by linear-time algorithms requiring only a constant amount of memory in addition to the pattern and the (window on the) text. This can be proved by different techniques presented in Crochemore and Rytter (1994). The most recent solution is by Gąsieniec et al. (1995).

Cole et al. (1995) shows that, in the worst case, any string-matching algorithm working with symbol comparisons makes at least $n + \frac{9}{4m}(n - m)$ comparisons during its search phase. Some string-matching algorithms make less than $2n$ comparisons. The presently-known upper bound on the problem is $n + \frac{8}{3(m+1)}(n - m)$, but with a quadratic-time preprocessing phase (see Cole et al. (1995)). With a linear-time preprocessing phase, the current upper bounds are $\frac{4}{3}n - \frac{1}{3}m$ and $n + \frac{4 \log m + 2}{m}(n - m)$ (see, respectively, Galil and Giancarlo (1992) and Breslauer and Galil (1993)). Except in a few cases (patterns of length 3 for example), lower and upper bounds do not meet. So, the problem of the exact complexity of string matching is open.

The Aho–Corasick algorithm is from Aho and Corasick (1975). Commentz-Walter (1979) has designed an extension of the Boyer–Moore algorithm that solves the dictionary-matching problem. It is fully described in Aho (1990).

Ideas of Section 13.2.7 are from Baeza-Yates and Gonnet (1992) and Wu and Manber (1992a). An implementation of the method given in Section 13.2.7 is the `agrep` command of UNIX (Wu and Manber, 1992b).

The suffix-tree construction of Section 13.1 is from McCreight (1976). An online version is by Ukkonen (1992). A previous algorithm by Weiner (1973) relates suffix trees to a data structure close to suffix automata. On an integer alphabet (see end of previous section), Farach-Colton (1997) designed a linear-time construction.

The construction of suffix automata, also described as direct acyclic word graphs and often denoted by the acronym DAWG, is from Blumer et al. (1985) and from Crochemore (1986). An application to data retrieval by the mean of inverted files is described in Blumer et al. (1987).

The alternative data structure of suffix array for indexes given in Section 13.3 is by Manber and Myers (1993). Linear-time suffix sorting algorithms on an integer alphabet have been designed independently by Kärkkäinen and Sanders (2003), by Kim et al. (2003), and by Ko and Aluru (2003). Possibly adding the longest-common-prefix array construction from Kasai et al. (2001) produces the complete suffix array in global linear-time.

13.5   Further Information

Problems and algorithms presented in the chapter are just a sample of questions related to pattern matching. They share the formal methods used to design efficient algorithms. A wider panorama of algorithms on texts may be found in a few books by Stephen (1994), Crochemore and Rytter (1994), Gusfield (1997), Crochemore and Rytter (2002), and Crochemore et al. (2007).

Research papers in pattern matching are disseminated in a few journals, among which are *Communications of the ACM*, *Journal of the ACM*, *Theoretical Computer Science*, *Journal of Algorithms*, *SIAM Journal on Computing*, and *Algorithmica*.

Two main annual conferences present the latest advances of this field of research:

- *Combinatorial Pattern Matching*, which started in 1990 in Paris (France), and was held since in London (England), Tucson (Arizona), Padova (Italy), Asilomar (California), Helsinki (Finland), and Laguna Beach (California).
- *Workshop on String Processing*, which started in 1993 in Belo Horizonte (Brazil), and was held since in Valparaiso (Chile) and Recife (Brazil).

But general conferences in computer science often have sessions devoted to pattern matching.

Information retrieval questions are treated in Frakes and Baeza-Yates (1992), where the reader can find references on the problem.

Several books on the design and analysis of general algorithms contain a chapter devoted to algorithms on texts. Here is a sample of these books: Books by Baase (1988), Cormen et al. (1990), Gonnet and Baeza-Yates (1991), and Sedgewick (1988).

Defining Terms

Border: A string v is a border of a string u if v is both a prefix and a suffix of u. String v is said to be *the* border of u if it is the longest proper border of u.

Factor: A string v is a factor of a string u if $u = u'vu''$ for some strings u' and u''.

Occurrence: A string v occurs in a string u if v is a factor of u.

Pattern: A finite number of strings that are searched for in texts.

Prefix: A string v is a prefix of a string u if $u = vu''$ for some string u''.

Proper: Qualifies a factor of a string that is not equal to the string itself.

Suffix: A string v is a suffix of a string u if $u = u'v$ for some string u'.

Suffix automaton: Smallest automaton accepting the suffixes of a string.

Suffix tree: Trie containing all the suffixes of a string.

Text: A stream of symbols that is searched for occurrences of patterns.

Trie: Digital tree, tree in which edges are labeled by symbols or strings.

Window: Factor of the text that is aligned with the pattern.

References

Aho, A.V. 1990. Algorithms for finding patterns in strings. In *Handbook of Theoretical Computer Science*, ed. J. van Leeuwen, vol. A, chap. 5, pp. 255–300. Elsevier, Amsterdam, the Netherlands.

Aho, A.V. and Corasick, M.J. 1975. Efficient string matching: an aid to bibliographic search. *Comm. ACM* 18:333–340.

Apostolico, A. and Giancarlo, R. 1986. The string searching strategies revisited. *SIAM J. Comput.* 15(1): 98–105.

Baase, S. 1988. *Computer Algorithms – Introduction to Design and Analysis*. Addison-Wesley, Reading, MA.

Baeza-Yates, R. and Gonnet, G.H. 1992. A new approach to text searching. *Comm. ACM* 35:74–82.

Blumer, A., Blumer, J., Ehrenfeucht, A., Haussler, D., Chen, M.T., and Seiferas, J. 1985. The smallest automaton recognizing the subwords of a text. *Theoret. Comput. Sci.* 40:31–55.

Blumer, A., Blumer, J., Ehrenfeucht, A., Haussler, D., and McConnell, R. 1987. Complete inverted files for efficient text retrieval and analysis. *J. ACM* 34:578–595.

Boyer, R.S. and Moore, J.S. 1977. A fast string searching algorithm. *Comm. ACM* 20:762–772.

Breslauer, D. and Galil, Z. 1993. Efficient comparison based string matching. *J. Complexity* 9:339–365.

Cole, R. 1994. Tight bounds on the complexity of the Boyer-Moore pattern matching algorithm. *SIAM J. Comput.* 23:1075–1091.

Cole, R., Hariharan, R., Zwick, U., and Paterson, M.S. 1995. Tighter lower bounds on the exact complexity of string matching. *SIAM J. Comput.* 24:30–45.

Colussi, L. 1994. Fastest pattern matching in strings. *J. Algorithms* 16:163–189.

Commentz-Walter, B. 1979. A string matching algorithm fast on the average. In *Proceedings of the 6th International Colloquium on Automates, Languages and Programming*, ed., H. A. Maurer, number 71 in Lecture Notes in Computer Science, pp. 118–132, Graz, Austria.

Cormen, T.H., Leiserson, C.E., and Rivest, R.L. 1990. *Introduction to Algorithms*. MIT Press, Cambridge, MA.

Crochemore, M. 1986. Transducers and repetitions. *Theoret. Comput. Sci.* 45:63–86.

Crochemore, M. and Rytter, W. 1994. *Text Algorithms*. Oxford University Press, New York.

Crochemore, M. and Rytter, W. 2002. *Jewels of Stringology*. World Scientific Press, Singapore.

Crochemore, M., Hancart, C., and Lecroq T. 2007. *Algorithms on Strings*. Cambridge University Press, New York.

Crochemore, M., Czumaj, A., Sieniec, L.G., Jarominek, S., Lecroq, T., Plandowski, W., and Rytter, W. 1994. Speeding up two string matching algorithms, *Algorithmica*, 12(4/5):247–267.

Farach-Colton, M. 1997. Optimal suffix tree construction with large alphabets. In *Proceedings of the 38th IEEE Annual Symposium on Foundations of Computer Science*, pp. 137–143, Miami Beach, FL.

Frakes, W.B. and Baeza-Yates, R. 1992. *Information Retrieval: Data Structures and Algorithms*. Prentice-Hall, Upper Saddle River, NJ.

Galil, Z. 1981. String matching in real time. *J. ACM* 28:134–149.

Galil, Z. and Giancarlo, R. 1992. On the exact complexity of string matching: upper bounds. *SIAM J. Comput.* 21:407–437.

Gasieniec, L., Plandowski, W., and Rytter, W. 1995. Constant-space string matching with smaller number of comparisons: Sequential sampling. In Z. Galil and E. Ukkonen (eds.), *Proceedings of the Sixth Annual Symposium on Combinatorial Pattern Matching*, Espoo, Finland. *Lecture Notes in Computer Science*, Volume 937, pp. 78–89, Springer-Verlag, Berlin.

Gonnet, G.H. and Baeza-Yates, R.A. 1991. *Handbook of Algorithms and Data Structures*. Addison-Wesley, Reading, MA.

Gusfield D. 1997. *Algorithms on Strings, Trees and Sequences: Computer Science and Computational Biology*. Cambridge University Press, New York.

Hancart, C. 1993. On Simon's string searching algorithm. *Info. Process. Lett.* 47:95–99.

Horspool, R.N. 1980. Practical fast searching in strings. *Software-Pract. Experience* 10:501–506.

Hume, A. and Sunday, D.M. 1991. Fast string searching. *Software-Pract. Experience* 21:1221–1248.

Kärkkäinen, J. and Sanders, P. 2003. Simple linear work suffix array construction. In *Proceedings of the 30th International Colloquium on Automata, Languages, and Programming*, pp. 943–955. *Lecture Notes in Computer Science*, Volume 2719, Springer-Verlag, Berlin.

Karp, R.M. and Rabin, M.O. 1987. Efficient randomized pattern-matching algorithms. *IBM J. Res. Dev.* 31:249–260.

Kasai, T., Lee, G., Arimura, H., Arikawa, S., and Park, K. 2001. Linear-time longest-common-prefix computation in suffix arrays and its applications. In *Proceedings of the 12th Annual Symposium on Combinatorial Pattern Matching*, pp. 181–192. *Lecture Notes in Computer Science*, Volume 2089, Springer-Verlag, Berlin.

Kim, D. K. , Sim, J. S., Park, H., and Park, K. 2003. Linear-time construction of suffix arrays. In *Proceedings of the 14th Annual Symposium on Combinatorial Pattern Matching*, pp. 186–199. *Lecture Notes in Computer Science*, Volume 2676, Springer-Verlag, Berlin.

Knuth, D.E., Morris Jr, J.H., and Pratt, V.R. 1977. Fast pattern matching in strings. *SIAM J. Comput.* 6:323–350.

Ko, P. and Aluru, S. 2003. Space efficient linear time construction of suffix arrays. In *Proceedings of the 14th Annual Symposium on Combinatorial Pattern Matching*, pp. 200–210. *Lecture Notes in Computer Science*, Volume 2676, Springer-Verlag, Berlin.

Lecroq, T. 1995. Experimental results on string-matching algorithms. *Software-Pract. Experience* 25:727–765.

McCreight, E.M. 1976. A space-economical suffix tree construction algorithm. *J. Algorithms* 23:262–272.

Manber, U. and Myers, G. 1993. Suffix arrays: A new method for on-line string searches. *SIAM J. Comput.* 22:935–948.

Sedgewick, R. 1988. *Algorithms*. Addison-Wesley, Reading, MA.

Simon, I. 1993. String matching algorithms and automata. In *1st American Workshop on String Processing*, eds. R. Baeza-Yates and N. Ziviani, pp. 151–157. Universidade Federal de Minas Gerais, Belo Horizonte, Brazil.

Stephen, G.A. 1994. *String Searching Algorithms*. World Scientific Press, Singapore.

Ukkonen, E. 1995. On-line construction of suffix trees. *Algorithmica*, 14(3):249–260.

Weiner, P. 1973. Linear pattern matching algorithm. In *Proceedings of the 14th Annual IEEE Symposium on Switching and Automata Theory*, pp. 1–11, Washington, D.C.

Wu, S. and Manber, U. 1992a. Fast text searching allowing errors. *Comm. ACM* 35:83–91.

Wu, S. and Manber, U. 1992b. Agrep—A fast approximate pattern-matching tool. *Usenix Winter 1992 Technical Conference*, pp. 153–162, San Francisco, CA.

Yao, A.C. 1979. The complexity of pattern matching for a random string. *SIAM J. Comput.*, 8:368–387.

14

Text Data Compression Algorithms

Maxime Crochemore
King's College London and
Université Paris-Est

Thierry Lecroq
University of Rouen

14.1 Text Compression

This chapter describes a few algorithms that compress texts. Compression serves both to save storage space and transmission time. We shall assume that the text is stored in a file. The aim of compression algorithms is to produce a new file, as short as possible, containing the compressed version of the same text. Methods presented here reduce the representation of text without any loss of information, so that decoding the compressed text restores exactly the original data.

The term "text" should be understood in a wide sense. It is clear that texts can be written in natural languages or can be texts usually generated by translators (like various types of compilers). But texts can also be images or other kinds of structures as well provided the data are stored in linear files. Texts considered here are sequence of characters from a finite alphabet Σ of size σ.

The interest in data compression techniques remains important even if mass storage systems improve regularly because the amount of data grows accordingly. Moreover, a consequence of the extension of computer networks is that the quantity of data they exchange grows exponentially. So, it is often necessary to reduce the size of files to reduce proportionally their transmission times. Other advantages in compressing files regard two connected issues: integrity of data and security.

While the first is easily accomplished through redundancy checks during the decompression phase, the second often requires data compression before applying cryptography.

This chapter contains three classical text compression algorithms. Variants of these algorithms are implemented in practical compression software, in which they are often combined together or with other elementary methods. Moreover, we present all-purpose methods, that is, methods in which no sophisticated modeling of the statistics of texts is done. An adequate modeling of a well-defined family of texts may increase significantly the compression when coupled with the coding algorithms of this chapter.

Compression ratios of the methods depend on the input data. However, most often, the size of compressed text vary from 30% to 50% of the size of the input. At the end of this chapter, we present ratios obtained by the methods on several example texts. Results of this type can be used to compare the efficiency of methods in compressing data. But, the efficiency of algorithms is also evaluated by their running times, and sometimes by the amount of memory space they require at run time. These elements are important criteria of choice when a compression algorithm is to be implemented in a telecommunication software.

Two strategies are applied to design the algorithms. The first strategy is a statistical method that takes into account the frequencies of symbols to build a uniquely decipherable code optimal with respect to the compression (Sections 14.2 and 14.3). Section 14.4 presents a refinement of the coding algorithm of Huffman based on the binary representation of numbers. Huffman codes contain new codewords for the symbols occurring in the text. In this method fixed-length blocks of bits are encoded by different codewords. *A contrario* the second strategy encodes variable-length segments of the text (Section 14.5). To put it simply, the algorithm, while scanning the text, replaces some already read segments by just a pointer to their first occurrences. This second strategy often provides better compression ratios.

14.2   Static Huffman Coding

The Huffman method is an optimal statistical coding. It transforms the original code used for characters of the text (ASCII code on 8 bits, for instance). Coding the text is just replacing each symbol (more exactly each occurrence of it) by its new **codeword**. The method works for any length of blocks (not only 8 bits), but the running time grows exponentially with the length. In the following, we assume that symbols are originally encoded on 8 bits to simplify the description.

The Huffman algorithm uses the notion of **prefix code**. A prefix code is a set of words containing no word that is a **prefix** of another word of the set. The advantage of such a code is that decoding is immediate. Moreover, it can be proved that this type of code does not weaken the compression.

A prefix code on the binary alphabet {0, 1} corresponds to a binary tree in which the links from a node to its left and right children are labeled by **0** and **1** respectively. Such a tree is called a (digital) **trie**. Leaves of the trie are labeled by the original characters, and labels of branches are the words of the code (codewords of characters). Working with prefix code implies that codewords are identified with leaves only. Moreover, in the present method codes are complete: they correspond to complete tries, i.e., trees in which all internal nodes have exactly two children.

In the model where characters of the text are given new codewords, the Huffman algorithm builds a code that is optimal in the sense that the compression is the best possible (if the model of the source text is a zero-order Markov process, that is if the probability of symbol occurrence are independent). The length of the encoded text is minimum. The code depends on the input text, and more precisely on the frequencies of characters in the text. The most frequent characters are given shortest codewords while the least frequent symbols correspond to the longest codewords.

14.2.1 Encoding

The complete compression algorithm is composed of three steps: count of character frequencies, construction of the prefix code, and encoding of the text. The last two steps use information computed by their preceding step.

The first step consists in counting the number of occurrences of each character in the original text (see Figure 14.1). We use a special end marker, denoted by END, which virtually appears only once at the end of the text. It is possible to skip this first step if fixed statistics on the alphabet are used. In this case, however, the method is optimal according to the statistics, but not necessarily for the specific text.

The second step of the algorithm builds the tree of a prefix code, called a Huffman tree, using the character frequency *freq(a)* of each character *a* in the following way:

- Create a one-node tree *t* for each character *a*, setting *weight(t)* = *freq(a)* and *label(t)* = *a*,
- Repeat

  ○ Extract the two least weighted trees t_1 and t_2,

  ○ Create a new tree t_3 having left subtree t_1, right subtree t_2, and weight *weight*(t_3) = *weight*(t_1) + *weight*(t_2),

- Until only one tree remains.

The tree is constructed by the algorithm H-BUILD-TREE in Figure 14.2. The implementation uses two linear lists. The first list, *lleaves*, contains the leaves of the future tree associated each with a symbol. The list is sorted in increasing order of weights of leaves (frequencies of symbols). The second list,

---

H-COUNT (*fin*)
1 **for** each character $a \in \Sigma$
2   **do** *freq(a)* ← 0
3 **while** not end of file *fin* **and** *a* is the next symbol
4   **do** *freq(a)* ← *freq(a)* + 1
5 *freq*(END) ← 1

FIGURE 14.1 Counts the character frequencies.

H-BUILD-TREE
1 **for** each $a \in \Sigma \cup \{$END$\}$
2   **do if** *freq(a)* ≠ 0
3   **then** create a new node *t*
4   *weight(t)* ← *freq(a)*
5   *label(t)* ← *a*
6 *lleaves* ← list of all created nodes in increasing order of weight
7 *ltrees* ← empty list
8 **while** LENGTH (*lleaves*) + LENGTH (*ltrees*) > 1
9   **do** (ℓ, *r*) ← extract two nodes of smallest weight (among the two nodes at the
              beginning of *lleaves* and the two nodes at the beginning of *ltrees*)
10   create a new node *t*
11   *weight(t)* ← *weight*(ℓ) + *weight*(*r*)
12   *left(t)* ← ℓ
13   *right(t)* ← *r*
14   insert *t* at the end of *ltrees*
16 **return** *t*

FIGURE 14.2 Builds the Huffman coding tree.

```
H-Build-Code (t, length)
1   if t is not a leaf
2       then temp[length] ← 0
3           H-Build-Code (left(t), length + 1)
4           temp[length] ← 1
5           H-Build-Code (right(t), length + 1)
6       else codeword(label(t)) ← temp[0 . . length − 1]
```

FIGURE 14.3 Builds character codewords from the coding tree.

```
H-Encode-Tree (fout, t)
1   if t is not a leaf
2       then write a 0 in the file fout
3           H-Encode-Tree (fout, left(t))
4           H-Encode-Tree (fout, right(t))
5       else write a 1 in the file fout
6           write the original code of label(t) in the file fout
```

FIGURE 14.4 Stores the coding tree in the compressed file.

```
H-Encode-Text (fin, fout)
1   while not end of file fin and a is the next symbol
2       do write codeword(a) in the file fout
3   write codeword(END) in the file fout
```

FIGURE 14.5 Encodes the characters in the compressed file.

ltrees, contains the newly created trees. The operation of extracting the two least weighted trees is done by checking the two first trees of the list *lleaves* and the two first trees of the list *ltrees*. Each new tree is inserted at the end of the list of the trees. The only tree remaining at the end of the procedure is the coding tree.

After the coding tree is built, it is possible to recover the codewords associated with characters by a simple depth-first-search of the tree (see Figure 14.3); *codeword(a)* denotes the binary codeword associated with the character *a*.

In the third step, the original text is encoded. Since the code depends on the original text, in order to be able to decode the compressed text, the coding tree and the original codewords of symbols must be stored together with the compressed text.

This information is placed in a header of the compressed file, to be read at decoding time just before the decompression starts. The header is written during a depth-first traversal of the tree. Each time an internal node is encountered a **0** is produced. When a leaf is encountered a **1** is produced followed by the original code of the corresponding character on 9 bits (so that the end marker can be equal to 256 if all the 8-bit characters appear in the original text). This part of the encoding algorithm is shown in Figure 14.4.

After the header of the compressed file is made, the encoding of the original text is realized by the algorithm of Figure 14.5.

A complete implementation of the Huffman algorithm, composed of the three steps described above, is given in Figure 14.6.

```
H-Encoding (fin, fout)
1   H-Count (fin)
2   t ← H-Build-Tree
3   H-Build-Code (t, 0)
4   H-Encode-Tree (fout, t)
5   H-Encode-Text (fin, fout)
```

FIGURE 14.6 Complete function for Huffman encoding.

Example 14.1

$y = $ **ACAGAATAGAGA**

Length of $y = 12 \times 8 = 104$ bits (assuming an 8-bit code)

Character frequencies:

A	C	G	T	END
7	1	3	1	1

Different steps during the construction of the coding tree:

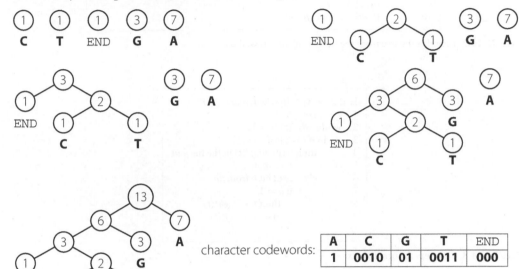

character codewords:

A	C	G	T	END
1	0010	01	0011	000

Encoded tree: **0001**binary(END,9)**01**binary(**C**,9)**1** binary(**T**,9)**1**binary(**G**,9)**1**binary(**A**,9), which produces a header of length 54 bits:

0001 100000000 01 001000011 1 001010100 1 001000111 1 001000001

Encoded text:

1	0010	1	01	1	1	0011	1	01	1	01	1	000
A	C	A	G	A	A	T	A	G	A	G	A	END

of length 24 bits

Total length of the compressed file: 78 bits.

The construction of the tree takes $O(\sigma \log \sigma)$ time if the sorting of the list of the leaves is implemented efficiently. The rest of the encoding process runs in time linear in the sum of the sizes of the original and compressed texts.

14.2.2 Decoding

Decoding a file containing a text compressed by Huffman algorithm is a mere programming exercise. First, the coding tree is rebuilt by the algorithm of Figure 14.7. Then, the original text is recovered by parsing the compressed text with the coding tree. The process begins at the root of the coding tree, and follows a left edge when a **0** is read or a right edge when a **1** is read. When a leaf is encountered, the corresponding character (in fact the original codeword of it) is produced and the parsing resumes

```
H-REBUILD-TREE (fin, t)
1    read bit b from fin
2    if b = 1
3        then make t a leaf
4            label(t) ← symbol corresponding to the 9 next bits read from fin
5        else create a new node ℓ
6            left(t) ← ℓ
7            H-REBUILD-TREE (fin, ℓ)
8            create a new node r
9            right(t) ← r
10           H-REBUILD-TREE (fin, r)
```

FIGURE 14.7 Rebuilds the tree from the header of compressed file.

```
H-DECODE-TEXT (fin, fout, root)
1    t ← root
2    while label(t) ≠ END
3        do if t is a leaf
4            then write label(t) in the file fout
5                t ← root
6            else  read bit b from fin
7                if b = 1
8                    then t ← right(t)
9                    else  t ← left(t)
```

FIGURE 14.8 Recovers the original text.

```
H-DECODING (fin, fout)
1    create a new node root
2    H-REBUILD-TREE (fin, root)
3    H-DECODE-TEXT (fin, fout, root)
```

FIGURE 14.9 Complete function for Huffman decoding.

at the root of the tree. The process ends when the codeword of the end marker is encountered. An implementation of the decoding of the text is presented in Figure 14.8.

The complete decoding program is given in Figure 14.9. It calls the preceding functions. The running time of the decoding program is linear in the sum of the sizes of the texts it manipulates.

14.3   Dynamic Huffman Coding

The two main drawbacks of the static Huffman method are the following: first, if the frequencies of characters the source text are not known a priori, the source text has to be read twice; second, the coding tree must be included in the compressed file. This is avoided by a dynamic method where the coding tree is updated each time a symbol is read from the source text. The current tree is a Huffman tree related to the part of the text that is already treated. The tree evolves exactly in the same way during the decoding process. The efficiency of the method is based on a characterization of Huffman trees, known as the *siblings property*.

Siblings property: Let T be a Huffman tree with n leaves (a complete binary weighted tree built by the procedure H-BUILD-TREE in which all leaves have positive weights). Then the nodes of T can be arranged in a sequence $(x_0, x_1, \ldots, x_{2n-2})$ such that

1. The sequence of weights $(weight(x_0), weight(x_1), \ldots, weight(x_{2n-2}))$ is in decreasing order

2. For any i $(0 \leq i \leq n - 2)$, the consecutive nodes x_{2i+1} and x_{2i+2} are siblings (they have the same parent)

The compression and decompression processes initialize the dynamic Huffman tree by a one-node tree that corresponds to an artificial character, denoted by ART. The weight of this single node is 1.

14.3.1 Encoding

Each time a symbol a is read from the source text, its codeword in the tree is sent. However this happens only if a appeared previously. Otherwise the code of ART is sent followed by the original codeword of a. Afterwards, the tree is modified in the following way: first, if a never occurred before, a new internal node is created and its two children are a new leaf labeled by a and the leaf ART; then, the tree is updated (see below) to get a Huffman tree for the new prefix of text.

14.3.1.1 Implementation

Each node is identified with an integer n, the root is the integer 0. The invariant of compression and decompression algorithms is that, if the tree has m nodes, the sequence of nodes $(m - 1, \ldots, 1, 0)$ satisfies the siblings property. The tree is stored in a table, and we use the next notations, for a node n:

- $parent(n)$ is the parent of n $(parent(root) = \text{UNDEFINED})$
- $child(n)$ is the left child of n (if n is an internal node, otherwise $child(n) = \text{UNDEFINED}$), and $child(n) + 1$ is its right child (this is useful only at decoding time)
- $label(n)$ is the symbol associated with n when n is a leaf
- $weight(n)$ is the weight of n (it is the frequency of $label(n)$ if n is a leaf)

Indeed, the child link is useful only at decoding time so that the implementation may differ between the two phases of the algorithm. But, to simplify the description and give a uniform treatment of the data structure, we assume the same implementation during the encoding and the decoding steps. Weights of nodes are handled by the procedure DH-UPDATE.

The correspondence between symbols and leaves of the tree is done by a table called *leaf*: for each symbol $a \in \Sigma \cup \{\text{END}\}$, *leaf*[$a$] is the corresponding leaf of the tree, if any.

The coding tree initially contains only one node labeled by symbol ART. The initialization is given in Figure 14.10.

```
DH-INIT
1    root ← 0
2    child(root) ← UNDEFINED
3    weight(root) ← 1
4    parent(root) ← UNDEFINED
5    for each letter a ∈ Σ ∪ {END}
6        do   leaf[a] ← UNDEFINED
7    leaf[ART] ← root
```

FIGURE 14.10 Initializes the dynamic Huffman tree.

```
DH-ENCODING (fin, fout)
1    DH-INIT
2    while not end of file fin and a is the next symbol
3        do  DH-ENCODE-SYMBOL (a, fout)
4               DH-UPDATE (a)
5    DH-ENCODE-SYMBOL (END, fout)
```

FIGURE 14.11 Complete function for dynamic Huffman encoding.

```
DH-ENCODE-SYMBOL (a, fout)
1    S ← empty stack
2    n ← leaf[a]
3    if n = UNDEFINED
4        then n ← leaf[ART]
5    while n ≠ root
6        do  if n is odd
7               then PUSH (S, 1)
8               else  PUSH (S, 0)
9            n ← parent(n)
10   SEND (S, fout)
11   if leaf[a] = UNDEFINED
12       then write in fout the original codeword of a on 9 bits
13           DH-ADD-NODE (a)
```

FIGURE 14.12 Encodes one symbol.

```
DH-ADD-NODE (a)
1    transform leaf[ART] into
2        an internal node of weight 1 with
3            a left child of weight 0 for leaf[a],
4            a right child of weight 1 for leaf[ART].
```

FIGURE 14.13 Adds a new symbol in the tree.

Encoding the source text is a succession of three steps: read a symbol a from the source text, encode the symbol a according to the current tree, update the tree. It is described in Figure 14.11.

Encoding a symbol a already encountered consists in accessing its associated leaf $leaf[a]$, then computing its codeword by a bottom-up walk in the tree (following the *parent* links up to the root). Each time a node n ($\neq root$) is encountered if n is odd a **1** is sent (n is the right child of its parent), and if n is even a **0** is sent (n is then the left child of its parent). As the codeword of a is read in reverse direction a stack S is used to temporarily store the bits and send them properly. The procedure SEND $(S, fout)$ send the bits of the stack S in the correct order to the file $fout$.

If a has not been encountered yet, then the code of ART is sent followed by the original codeword of a on 9 bits, and a new leaf is created for a (see Figures 14.12 and 14.13).

14.3.2   Decoding

At decoding time the compressed text is parsed with the coding tree. The current node is initialized with the root like in the encoding algorithm, and then the tree evolves symmetrically. Each time a **0** is read from the compressed file the walk down the tree follows the left link, and it follows the right link if a **1** is read. When the current node is a leaf, its associated symbol is written in the output file and the tree is updated exactly as it is done during the encoding phase.

14.3.2.1 Implementation

As for the encoding process the current Huffman tree is stored in a table, and it is initialized with the artificial symbol ART. The same elements of the data structure are used during the decompression. Note that the next node when parsing a bit b from node n is just $child(n) + b$ with the convention on left–right links and 0–1 bits. The tree is updated by the procedure DH-UPDATE used previously and that is described in the next section. Figures 14.14 and 14.15 display the decoding mechanism.

14.3.3 Updating

During encoding and decoding phases the current tree has to be updated to take into account the correct frequency of symbols. When a new symbol is considered the weight of its associated leaf is incremented by 1, and the weights of ancestors have to be modified correspondingly. The procedure that realizes the operation is shown in Figure 14.16. Its proof of correctness is based on the siblings property.

We explain how the procedure DH-UPDATE works. First, the weight of the leaf n corresponding to a is incremented by 1. Then, if point 1 of the siblings property is no longer satisfied, node n is exchanged with the closest node m ($m < n$) in the list such that $weight(m) < weight(n)$. Doing so, the nodes remain in decreasing order of their weights. It is important here that leaves have positive weights because this guarantees that m is not a parent nor an ancestor of node n. Afterwards, the same operation is repeated on the parent of n until the root of the tree is reached.

The aim of procedure DH-SWAP-NODES used in Figure 14.16 is to exchange the subtrees rooted at its input nodes m and n. In concrete terms, this remains to exchange the records stored at the two nodes in the table. It is meant that nothing is to be done if $m = n$.

DH-DECODING ($fin, fout$)
1 DH-INIT
2 $a \leftarrow$ DH-DECODE-SYMBOL (fin)
3 **while** $a \neq$ END
4   **do** write a in $fout$
5   DH-UPDATE (a)
6       $a \leftarrow$ DH-DECODE-SYMBOL (fin)

FIGURE 14.14 Complete function for dynamic Huffman decoding.

DH-DECODE-SYMBOL (fin)
1 $n \leftarrow root$
2 **while** $child(n) \neq$ UNDEFINED
3   **do** read bit b from fin
4       $n \leftarrow child(n) + b$
5 $a \leftarrow label(n)$
6 **if** $a =$ ART
7   **then** $a \leftarrow$ symbol corresponding to the next 9 bits read from fin
8   DH-ADD-NODE (a)
9 **return** (a)

FIGURE 14.15 Decodes one symbol from the input file.

DH-Update (*a*)
1   *n* ← *leaf*[*a*]
2   **while** *n* ≠ *root*
3   **do**   *weight*(*n*) ← *weight*(*n*) + 1
4   *m* ← *n*
5   **while** *weight*(*m* − 1) < *weight*(*n*)
6   **do**   *m* ← *m* − 1
7   DH-Swap-Nodes (*m*, *n*)
8   *n* ← *parent*(*m*)
9   *weight*(*root*) ← *weight*(*root*) + 1

FIGURE 14.16   Updates the current Huffman tree.

Example 14.2

y = **ACAGAATAGAGA**

Initial tree:

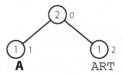

ART

Next symbol is **A**:
The ASCII code of **A** is sent on 9 bits;
Bits sent: **001000001**

Next symbol is **C**:
The code of ART is sent followed by the ASCII code of **C** on 9 bits;
Bits sent: **1 001000011**
Nodes 1 and 2 are swapped.

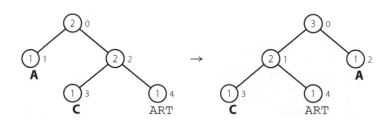

Next symbol is **A**:
The code of **A** is sent;
Bit sent: **1**

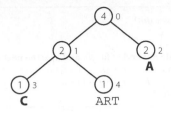

Next symbol is **G**:
The code of ART is sent followed by the ASCII code of **G** on 9 bits;
Bits sent: **01 001000111**
Nodes 3 and 4 are swapped.

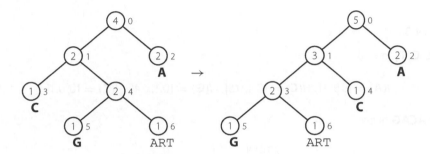

Finally, the entire sequence of bits sent is the following:

$\underbrace{001000001}$	$\underbrace{1}$	$\underbrace{001000011}$	$\underbrace{1}$	$\underbrace{01}$	$\underbrace{001000111}$	$\underbrace{1}$	$\underbrace{1}$	$\underbrace{101}$
A	ART	C	A	ART	G	A	A	ART

$\underbrace{001010100}$	$\underbrace{0}$	$\underbrace{100}$	$\underbrace{0}$	$\underbrace{100}$	$\underbrace{0}$	$\underbrace{111}$	$\underbrace{100000000}$
T	A	G	A	G	A	ART	END

The total length of the compressed text is 66.

14.4 Arithmetic Coding

14.4.1 Encoding

The basic idea of arithmetic coding is to consider symbol as digits of a numeration system, and texts as decimal parts of numbers between 0 and 1. The length of the interval attributed to a digit (it is 0.1 for digits in the usual base 10 system) is made proportional to the frequency of the digit in the text. The encoding is thus assimilated to a change in the base of a numeration system. To cope with precision problems, the number corresponding to a text is handled via a lower bound and an upper bound, which remains to associate with a text a subinterval of $[0, 1[$. The compression comes from the fact that large intervals require less precision to separate their bounds.

More formally, the interval associated with each symbol $a_i \in \Sigma$ $(1 \le i \le \sigma)$ is denoted by $I(a_i) = [\ell_i, h_i[$. The intervals satisfy the conditions: $\ell_1 = 0$, $h_\sigma = 1$, and $\ell_i = h_{i-1}$ for $1 < i \le \sigma$. Note that $I(a_i) \cap I(a_j) = \emptyset$ if $a_i \ne a_j$.

The encoding phase consists in computing the interval corresponding to the input text. The basic step that deals with a symbol a_i of the source text transforms the current interval $[\ell, h[$ into $[\ell', h'[$ where $\ell' = \ell + (h - \ell) * \ell_i$ and $h' = \ell + (h - \ell) * h_i$, starting with the initial

```
AR-ENCODE (fin)
1    ℓ ← 0
2    h ← 1
3    while not end of file fin and aᵢ is the next symbol
4        do  ℓ ← ℓ + (h − ℓ) * ℓᵢ
5              h ← ℓ + (h − ℓ) * hᵢ
6    return(ℓ)
```

FIGURE 14.17 Basic arithmetic encoding.

interval $[0, 1[$ (see Figure 14.17). Indeed, in a theoretical approach, ℓ only is needed to encode the input text.

Example 14.3

$\Sigma = \{A, C, G, T\}, \sigma = 4$

$$I(A) = [0.5, 1[, \quad I(C) = [0.4, 0.5[, \quad I(G) = [0.1, 0.4[, \quad I(T) = [0, 0.1[$$

Encoding **ACAG** gives:

symbol	ℓ	h
	0	1
A	0.5	1
C	0.7	0.75
A	0.725	0.75
G	0.7275	0.735

14.4.2 Decoding

The reverse operation of decoding a text compressed by the previous algorithm theoretically requires only the lower bound ℓ of the interval. Decoding the number ℓ is done as follows: first find the symbol a_i such that $\ell \in I(a_i)$, produced the symbol a_i, and then replace ℓ by ℓ' defined by:

$$\ell' \leftarrow \frac{\ell - \ell_i}{h_i - \ell_i}.$$

The same process is repeated until $\ell = 0$ (see Figure 14.18). Indeed, the implementation of the decoding phase simulates what is done on the current interval considered at encoding time.

```
AR-DECODE (ℓ, fout)
1   while ℓ ≠ 0
2       do  find a_i such that ℓ ∈ I(a_i)
3           write a_i in file fout
4           ℓ ← (ℓ − ℓ_i)/(h_i − ℓ_i)
```

FIGURE 14.18 Basic arithmetic decoding.

Example 14.4

$\Sigma = \{$**A**, **C**, **G**, **T**$\}$, $\sigma = 4$

$$I(\mathbf{A}) = [0.5, 1[, \quad I(\mathbf{C}) = [0.4, 0.5[, \quad I(\mathbf{G}) = [0.1, 0.4[, \quad I(\mathbf{T}) = [0, 0.1[$$

Decoding $\ell = 0.7275$:

ℓ	a_i
0.7275	**A**
0.455	**C**
0.55	**A**
0.1	**G**
0	

14.4.3 Implementation

The main problem when implementing the arithmetic coding compression algorithm is to cope with precision on real number operations. The $[0, 1[$ interval of real numbers is substituted by the interval of integers $[0, 2^N − 1[$, where N is a fixed integer.

So, the algorithms work with integral values of size N. During the process, each time the binary representation of bounds ℓ and h have a common prefix this prefix is sent and ℓ is shifted to the left and filled with **0**'s while h is shifted to the left and filled with **1**'s.

The intervals associated with symbols of the alphabet are computed with the help of symbol frequencies, in a dynamic way: each character frequency is initialized with 1 and is incremented each time the symbol is encountered.

We denote by *freq*[i] the frequency of symbol a_i of the alphabet. We also consider the cumulative frequency of symbols, and set *cum-freq*[i] $= \sum_{j=i+1}^{\sigma}$ *freq*[j]. Then *cum-freq*[0] is the cumulative frequency of all symbols. Note that *cum-freq*[0] $− \sigma − 1$ is the length of the prefix of the input scanned so far. The symbols are maintained in decreasing order of their frequencies. This obviously save on the expected number of operations to update the table *cum-freq*.

Then when a symbol a_i is read from the source text, the current interval $[\ell, h[$ is updated in the following way:

$$\ell \leftarrow \ell + \frac{(h − \ell + 1) * \textit{cum-freq}[i]}{\textit{cum-freq}[0]}$$

$$h \leftarrow \ell + \frac{(h − \ell + 1) * \textit{cum-freq}[i − 1]}{\textit{cum-freq}[0]} − 1$$

The common prefix (if any) of ℓ and h is sent and ℓ and h are shifted and filled (respectively with **0**'s and **1**'s). At this point, if the interval is too short (if $h − \ell < $ *cum-freq*[0]) it is extended to $[2 * (\ell − 2^{N−2}), 2 * (h − 2^{N−2}) + 1[$ and a waiting counter is incremented by 1. And the same

```
AR-ENCODING (fin, fout)
1    Initialize tables freq and cum-freq
2    ℓ ← 0
3    h ← 2^N − 1
4    waiting-counter ← 0
5    while not end of file fin and a_i is the next symbol
6        do   AR-ENCODE-SYMBOL (a_i)
7             Update tables freq and cum-freq
8             Maintain symbols in decreasing order of frequencies
9    AR-ENCODE-SYMBOL (END)
10   waiting-counter ← waiting-counter + 1
11   AR-SEND-BIT (leftmost bit of h)
```

FIGURE 14.19 Complete arithmetic encoding function.

```
AR-ENCODE-SYMBOL (a_i)
1    ℓ ← ℓ + ((h − ℓ + 1) * cum-freq[i])/cum-freq[0]
2    h ← ℓ + ((h − ℓ + 1) * cum-freq[i − 1])/cum-freq[0] − 1
3    repeat
4        if leftmost bit of ℓ = leftmost bit of h
5            then AR-SEND-BIT (leftmost bit of h)
6                 ℓ ← 2 * ℓ
7                 h ← 2 * h + 1
8            else if h − ℓ < cum-freq[0]
9                 then ℓ ← 2 * (ℓ − 2^{N−2})
10                      h ← 2 * (h − 2^{N−2}) + 1
11                      waiting-counter ← waiting-counter + 1
12   until leftmost bit of ℓ ≠ leftmost bit of h and h − ℓ ≥ cum-freq[0]
```

FIGURE 14.20 Encodes one symbol.

operation is repeated as long as the interval is too short. After that, when a bit is sent, the reverse bit is sent the number of times indicated by the waiting counter (see Figures 14.19 through 14.21).

Example 14.5

$\Sigma = \{A, C, G, T\}, N = 8$

		A	C	G	T	END
i	0	1	2	3	4	5
cum-freq	5	4	3	2	1	0
freq	0	1	1	1	1	1

$[0, 255[\xrightarrow{A} [0 + \frac{(255 − 0 + 1)*4}{5}, 0 + \frac{(255 − 0 + 1)*5}{5} − 1[= [204, 255[= [11001100_2, 11111111_2[$
11 is sent and $[00110000_2, 11111111_2[= [48, 255[$ is the next interval;

		A	C	G	T	END
i	0	1	2	3	4	5
cum-freq	6	4	3	2	1	0
freq	0	2	1	1	1	1

$[48, 255[\xrightarrow{C} [96, 231[= [10011000_2, 10111001_2[$
10 is sent and $[01100000_2, 11100111_2[= [96, 231[$ is the next interval;

```
AR-Send-Bit (bit, fout)
1   write bit in fout
2   while waiting-counter > 0
3      do   if bit = 0
4               then write 1 in fout
5                    write 0 in fout
6               waiting-counter ← waiting-counter − 1
```

FIGURE 14.21 Sends one bit followed by the waiting bits, if any.

		A	**C**	**G**	**T**	END
i	0	1	2	3	4	5
cum-freq	7	5	3	2	1	0
freq	0	2	2	1	1	1

$[96, 231[\overset{A}{\longrightarrow} [193, 231[= [11000001_2, 11100111_2[$
11 is sent and $[00000100_2, 10011111_2[= [4, 159[$ is the next interval;

Next symbol is **G**: **001** is sent;
Next symbol is **A**: **1** is sent;
Next symbol is **A**: **1** is sent;
Next symbol is **T**: **0111** is sent;
Next symbol is **A**: **10** is sent;
Next symbol is **G**: nothing is sent and the current interval becomes $[111, 139[$;

$[111, 139[\overset{A}{\longrightarrow} [127, 139[= [01111111_2, 10001011_2[$, nothing is sent.
The interval is too short and replaced by $[2 * (127 - 2^{N-2}), 2 * (139 - 2^{N-2}) + 1[= [126, 151[$ and one bit is waiting, $[01111110, 10010111[= [126, 151[$ is the next interval;

$[126, 151[\overset{G}{\longrightarrow} [134, 138[= [10000110_2, 10001010_2[$
1+0+000 are sent and $[01100000_2, 10101111_2[= [96, 175[$ is the next interval;

$[96, 175[\overset{A}{\longrightarrow} [141, 175[= [10001101_2, 10101111_2[$
10 is sent and $[00110100_2, 10111111_2[= [52, 191[$ is the next interval;

$[52, 191[\overset{END}{\longrightarrow} [52, 59[= [00110100_2, 00111011_2[$
0011 is sent;

1+0 are sent in order to finish the encoding process.

The decoding process is exactly the reverse of the coding process. It uses a window of size N on the compressed file. First, the window is filled with the first N bits of the compressed file and *value* is the corresponding base 2 number. The current interval is initialized with $\ell = 0$ and $h = 2^N - 1$.
Then, the symbol a_i to be produced is the first character such that:

$$cum\text{-}freq[i] > \frac{(value - \ell + 1) * cum\text{-}freq[0] - 1)}{h - \ell + 1},$$

and ℓ and h are then updated exactly in the same way than during the coding process. If the binary representations of ℓ and h have a common prefix of length p they are both shifted p binary position to the left and ℓ is filled by **0**'s, h is filled with **1**'s. The window on the compressed file is shifted p symbols to the right and the variable *value* is updated correspondingly. The tables *freq* and *cum-freq* are updated and the symbols are maintained in decreasing order of the frequencies as in the coding process.
This is repeated until the symbol END is produced (see Figures 14.22 and 14.23).

```
AR-DECODING (fin, fout)
1    Init the tables freq and cum-freq
2    value ← 0
3    for i ← 1 to N
4        do    read bit b from fin
5               value ← 2 * value + b
6    ℓ ← 0
7    h ← 2^N − 1
8    repeat
9        i ← AR-DECODE-SYMBOL (fin)
10       if a_i ≠ END
11           then write a_i in fout
12                Update the tables freq and cum-freq
13                Maintain symbols in decreasing order of frequencies
14   until a_i = END
```

FIGURE 14.22 Complete arithmetic decoding function.

```
AR-DECODE-SYMBOL (fin)
1    cum ← ((value − ℓ + 1) * cum-freq[0] − 1)/(h − ℓ + 1)
2    i ← 1
3    while cum-freq[i] > cum
4        do    i ← i + 1
5    ℓ ← ℓ + ((h − ℓ + 1) * cum-freq[i])/cum-freq[0]
6    h ← ℓ + ((h − ℓ + 1) * cum-freq[i − 1])/cum-freq[0] − 1
7    repeat
8        if leftmost bit of ℓ = leftmost bit of h
9            then ℓ ← 2 * ℓ
10                h ← 2 * h + 1
11                read bit b from fin
12                value ← 2 * value + b
13           else if h − ℓ < cum-freq[0]
14                then ℓ ← 2 * (ℓ − 2^{N−2})
15                     h ← 2 * (h − 2^{N−2}) + 1
16                     read bit b from fin
17                     value ← 2 * (value − 2^{N−2}) + b
18   until leftmost bit of ℓ ≠ leftmost bit of h and h − ℓ ≥ cum-freq[0]
19   return i
```

FIGURE 14.23 Decodes one symbol.

Example 14.6

Decoding the text **11101100111011110100001000111 0**
$\Sigma = \{A, C, G, T\}$

		A	C	G	T	END
i	0	1	2	3	4	5
cum-freq	5	4	3	2	1	0
freq	0	1	1	1	1	1

$value = 236 = 11101100_2$
$cum = 4$
$[0, 255[\overset{A}{\rightarrow} [204, 255[= [11001100_2, 11111111_2[$: shift by 2
The next interval is $[00110000_2, 11111111_2[$ and $value = 10110011_2 = 179$

		A	**C**	**G**	**T**	END
i	0	1	2	3	4	5
cum-freq	6	4	3	2	1	0
freq	0	2	1	1	1	1

$value = 179 = 10110011_2$
$cum = 3$
$[48, 255[\overset{\text{C}}{\rightarrow} [152, 185[= [10011000_2, 10111001_2[$: shift by 2
The next interval is $[01100000_2, 11100111_2[$ and $value = 11001110_2 = 206$

		A	**C**	**G**	**T**	END
i	0	1	2	3	4	5
cum-freq	7	5	3	2	1	0
freq	0	2	2	1	1	1

$value = 206 = 11001110_2$
$cum = 5$
$[96, 231[\overset{\text{A}}{\rightarrow} [193, 231[= [11000001_2, 11100111_2[$: shift by 2
The next interval is $[00000100_2, 10011111_2[$ and $value = 00111011_2 = 59$

Next symbols are **GAATAG** and the current interval becomes $[111, 139[$ and $value = 132 = 10000100_2$
$cum = 10$
$[111, 139[\overset{\text{A}}{\rightarrow} [127, 139[= [01111111_2, 10001011_2[$: no shift
The interval is too short and is replaced by $[01111110_2, 10010111_2[$ and $value = 10001000_2 = 136$

$value = 136 = 10001000_2$
$cum = 6$
$[126, 151[\overset{\text{G}}{\rightarrow} [134, 138[= [10000110_2, 10001010_2[$: shift by 4
The next interval is $[01100000_2, 10101111_2[$ and $value = 10001110_2 = 142$

$value = 142 = 10001110_2$
$cum = 9$
$[96, 175[\overset{\text{A}}{\rightarrow} [141, 175[= [10001101_2, 10101111_2[$: shift by 2
The next interval is $[00110100_2, 10111111_2[$ and $value = 00111000_2 = 56$

$value = 56 = 00111000_2$
$cum = 0$
The symbol is END, the decoding process is over.

Maintaining the symbols in decreasing order of frequencies can be done in $O(\log \sigma)$ using a suitable data structure (see Fenwick, 1994).

14.5   LZW Coding

Ziv and Lempel (1977) designed a compression method using encoding **segments**. These segments of the original text are stored in a dictionary that is built during the compression process. When a segment of the dictionary is encountered later while scanning the text it is substituted by its index in the dictionary. In the model where portions of the text are replaced by pointers on previous occurrences, the Ziv–Lempel compression scheme can be proved to be asymptotically optimal (on large enough texts satisfying good conditions on the probability distribution of symbols).

The dictionary is the central point of the algorithm. It has the property of being prefix-closed (every prefix of a word of the dictionary is in the dictionary), so that it can be implemented efficiently as a trie. Furthermore, a hashing technique makes its implementation efficient. The version described in this section is called the Lempel–Ziv–Welsh method after several improvements introduced by Welsh (Welsh, 1994). The algorithm is implemented by the `compress` command existing under the UNIX operating system.

14.5.1   Encoding

We describe the scheme of the coding method. The dictionary is initialized with all strings of length 1, the characters of the alphabet. The current situation is when we have just read a segment w of the text. Let a be the next symbol (just following the given occurrence w). Then we proceed as follows:

- If wa is not in the dictionary, we write the index of w in the output file, and add wa to the dictionary. We then reset w to a and process the next symbol (following a).
- If wa is in the dictionary we process the next symbol, with segment wa instead of w.

Initially, the segment w is set to the first symbol of the source text, so that it is clear that "w belongs to the dictionary" is an invariant of the operations described above.

Example 14.7

The alphabet is the 8-bit ASCII alphabet, $y = $ **ACAGAATAGAGA**.
The dictionary initially contains the ASCII symbols, their indexes are their ASCII codewords.

A C A G A A T A G A G A	w	written	added
↑	A	65	**AC**, 257
↑	C	67	**CA**, 258
↑	A	65	**AG**, 259
↑	G	71	**GA**, 260
↑	A	65	**AA**, 261
↑	A	65	**AT**, 262
↑	T	84	**TA**, 263
↑	A		
↑	AG	259	**AGA**, 264
↑	A		
↑	AG		
↑	AGA	264	
		256	

14.5.2   Decoding

The decoding method is symmetric to the coding algorithm. The dictionary is recovered while the decompression process runs. It is basically done in this way:

- Read a code c in the compressed file
- Write in the output file the segment w having index c in the dictionary
- Add the word wa to the dictionary where a is the first letter of the next segment.

In this scheme, the dictionary is updated after the next segment is decoded because we need its first symbol a to concatenate at the end of the current segment w. So, a problem occurs if the next index computed at encoding time is precisely the index of the segment wa. Indeed, this happens only in a very special case, in which the symbol a is also the first symbol of w itself. This arises if the rest of the text to encode starts with a segment $azazax$ (a a symbol, z a string) for which az belongs to the dictionary but aza does not. During the compression process the index of az is output, and aza is added to the dictionary. Next, aza is read and its index is output. During the decompression process the index of aza is read while the first occurrence of az has not been completed yet, the segment aza is not already in the dictionary. However, since this is the unique case where the situation occurs, the segment aza is recovered by taking the last segment az added to the dictionary concatenated with its first letter a.

Example 14.8

Decoding the sequence 65, 67, 65, 71, 65, 65, 84, 259, 264, 256
The dictionary initially contains the ASCII symbols, their indexes are their ASCII codewords.

read	written	added
65	**A**	
67	**C**	**AC**, 257
65	**A**	**CA**, 258
71	**G**	**AG**, 259
65	**A**	**GA**, 260
65	**A**	**AA**, 261
84	**T**	**AT**, 262
259	**AG**	**TA**, 263
264	**AGA**	**AGA**, 264
256		

The critical situation occurs when reading the index 264 because, at that moment, no word of the dictionary has this index.

14.5.3 Implementation

We describe how the dictionary, which is the main data structure of the method, can be implemented. It is natural to consider two implementations adapted for the two phases respectively because the dictionary is not manipulated in the same manner during these phases. They have in common a dictionary implemented as a trie stored in a table D. A node p of the trie is just an index on the table D. It has the three following components:

- $parent(p)$, a link to the parent node of p
- $label(p)$, a character
- $code(p)$, the codeword (index in the dictionary) associated with p

In the compression algorithm shown in Figure 14.24, for a node p we need to compute its child according to some letter a. This is done by hashing, with a hashing function defined on pairs in the form (p, a). This provides a fast access to the children of a node.

The function HASH-SEARCH, with input $(D, (p, a))$, returns the node q such that $parent(q) = p$ and $label(q) = a$, if such a node exists and NIL otherwise. The procedure HASH-INSERT, with

```
LZW-CODING (fin, fout)
1    count ← −1
2    for each character a ∈ Σ
3        do count ← count + 1
4            HASH-INSERT (D, (−1, a, count))
5    count ← count + 1
6    HASH-INSERT (D, (−1, END, count))
7    p ← −1
8    while not end of file fin and a is the next symbol
9        q ← HASH-SEARCH (D, (p, a))
10       if q = NIL
11           then write code(p) on 1 + log(count) bits in fout
12                count ← count + 1
13                HASH-INSERT (D, (p, a, count))
14                p ← HASH-SEARCH (D, (−1, a))
15           else p ← q
16   write p on 1 + log(count) bits in fout
17   write code(HASH-SEARCH (D, (−1, END)) in 1 + log(count) bits in fout
```

FIGURE 14.24 LZW encoding algorithm.

input $(D, (p, a, c))$, inserts a new node q in the dictionary D with $parent(q) = p$, $label(q) = a$ and $code(q) = c$.

For the decompression algorithm, no hashing technique is necessary on the table representation of the trie that implements the dictionary. Having the index of the next segment, a bottom-up walk in the trie produces the mirror image of the expected segment. A stack is then used to reverse it. We assume that the function $string(c)$ performs this specific work for a code c. The bottom-up walk follows the parent links of the data structure. The function $first(w)$ gives the first character of the word w. These features are part of the decompression algorithm displayed in Figure 14.25.

```
LZW-DECODING (fin, fout)
1    count ← −1
2    for each character a ∈ Σ
3        do count ← count + 1
4            HASH-INSERT (D, (−1, a, count))
5    count ← count + 1
6    HASH-INSERT (D, (−1, END, count))
7    c ← first code on 1 + log(count) bits from fin
8    write string(c) in fout
9    a ← first(string(c))
10   repeat
11       d ← next code on 1 + log(count) from fin
12       if d > count
13           then count ← count + 1
14                parent(count) ← c
15                label(count) ← a
16                write string(c)a in fout
17                c ← d
18           else a ← first(string(d))
19                if a ≠ END
20                    then count ← count + 1
21                         parent(count) ← c
22                         label(count) ← a
23                         write string(d) in fout
17                         c ← d
24                    else  exit
25   forever
```

FIGURE 14.25 LZW decoding algorithm.

The Ziv–Lempel compression and decompression algorithms run both in time linear in the sizes of the files provided the hashing technique is implemented efficiently. Indeed, it is very fast in practice, except when the table becomes full and should be reset. In this situation, usual implementations also reset the whole dictionary to its initial value.

The main advantage of Ziv–Lempel compression method, compared to Huffman coding, is that it captures long repeated segments in the source file and thus often yields better compression ratios. An even more powerful compression technique is provided by (Ziv and Lempel, 1977) where the implicit dictionary is composed of all segments occurring before the current position in the text to be encoded.

14.6 Mixing Several Methods

We describe simple compression methods and then an example of a combination of several of them, basis of the popular **bzip** software.

14.6.1 Run Length Encoding

The aim of run length encoding (RLE) is to efficiently encode repetitions occurring in the input data. Let us assume that it contains a good quantity of repetitions of the form $aa \ldots a$ for some character a ($a \in \Sigma$). A repetition of k consecutive occurrences of letter a is replaced by $\&ak$, where the symbol $\&$ is a new character ($\& \notin \Sigma$).

The string $\&ak$ that encodes a repetition of k consecutive occurrences of a is itself encoded on the binary alphabet $\{0, 1\}$. In practice, letters are often represented by their ASCII code. Therefore, the codeword of a letter belongs to $\{0, 1\}^k$ with $k = 7$ or 8. Generally there is no problem in choosing or encoding the special character $\&$. The integer k of the string $\&ak$ is also encoded on the binary alphabet, but it is not sufficient to translate it by its binary representation, because we would be unable to recover it at decoding time inside the stream of bits. A simple way to cope with this is to encode k by the string $0^\ell \text{bin}(k)$, where $\text{bin}(k)$ is the binary representation of k, and ℓ is the length of it. This works well because the binary representation of k starts with a 1 so there is no ambiguity to recover ℓ by counting during the decoding phase. The size of the encoding of k is thus roughly $2 \log k$. More sophisticated integer representations are possible, but none is really suitable for the present situation. Simpler solution consists in encoding k on the same number of bits as other symbols, but this bounds values of ℓ and decreases the power of the method.

14.6.2 Move to Front

The move to front (MTF) method may be regarded as an extension of RLE or a simplification of Ziv–Lempel compression. It is efficient when the occurrences of letters in the input text are localized into relatively short segment of it. The technique is able to capture the proximity between occurrences of symbols and to turn it into a short encoded text.

Letters of the alphabet Σ of the input text are initially stored in a list that is managed dynamically. Letters are represented by their rank in the list, starting from 1, rank that is itself encoded as described above for RLE.

Letters of the input text are processed in an online manner. The clue of the method is that each letter is moved to the beginning of the list just after it is translated by the encoding of its rank.

The effect of MTF is to reduce the size of the encoding of a letter that reappears soon after its preceding occurrence.

14.6.3 Integrated Example

Most compression software combine several methods to be able to compress efficiently a large range of input data. We present an example of this strategy, implemented by the UNIX command **bzip**.

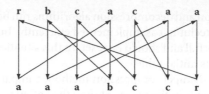

FIGURE 14.26 Example of text $y = \mathbf{baccara}$. Top line is $BW(y)$ and bottom line the sorted list of letters of it. Top-down arrows correspond to succession of occurrences in y. Each bottom-up arrow links the same occurrence of a letter in y. Arrows starting from equal letters do not cross. The circular path is associated with rotations of the string y. If the starting point is known, the only occurrence of letter $\mathbf{b}$ here, following the path produces the initial string y.

Let $y = y[0]y[1]\cdots y[n-1]$ be the input text. The k-th rotation (or conjugate) of y, $0 \le k \le n-1$, is the string $y_k = y[k]y[k+1]\cdots y[n-1]y[0]y[1]\cdots y[k-1]$.

We define the BW transformation as $BW(y) = y[p_0]y[p_1]\cdots y[p_{n-1}]$, where $p_i + 1$ is such that y_{p_i+1} has rank i in the sorted list of all rotations of y.

It is remarkable that y can be recovered from both $BW(y)$ and a position on it, starting position of the inverse transformation (see Figure 14.26). This is possible due to the following property of the transformation. Assume that $i < j$ and $y[p_i] = y[p_j] = a$. Since $i < j$, the definition implies $y_{p_i+1} < y_{p_j+1}$. Since $y[p_i] = y[p_j]$, transferring the last letters of y_{p_i+1} and y_{p_j+1} to the beginning of these words does not change the inequality. This proves that the two occurrences of a in $BW(y)$ are in the same relative order as in the sorted list of letters of y. Figure 14.26 illustrates the inverse transformation.

Transformation BW obviously does not compress the input text y. But $BW(y)$ is compressed more efficiently with simple methods. This is the strategy applied for the command **bzip**. It is a combination of the BW transformation followed by MTF encoding and RLE encoding. Arithmetic coding, a method providing compression ratios slightly better than Huffman coding, may also be used.

14.7 Experimental Results

The table of Figure 14.27 contains a sample of experimental results showing the behavior of compression algorithms on different types of texts. The table is extracted from (Zipstein, 1992).

The source files are French text, C sources, Alphabet, and Random. Alphabet is a file containing a repetition of the line abc...zABC...Z. Random is a file where the symbols have been generated randomly, all with the same probability and independently of each other.

The compression algorithms reported in the table are the Huffman algorithm of Section 14.2, the Ziv–Lempel algorithm of Section 14.5, and a third algorithm called Factor. This latter algorithm encodes segments of the source text as Ziv–Lempel algorithm does. But the segments are taken among all segments already encountered in the text before the current position. The method gives

Source Texts	French	C Sources	Alphabet	Random
Sizes in Bytes	62816	684497	530000	70000
Huffman	53.27%	62.10%	72.65%	**55.58%**
Ziv–Lempel	**41.46%**	34.16%	2.13%	63.60%
Factor	47.43%	**31.86%**	**0.09%**	73.74%

FIGURE 14.27 Sizes of texts compressed with three algorithms.

TABLE 14.1 Compression Results with Three Algorithms: Huffman Coding (**pack**), Ziv–Lempel Coding (**gzip-b**), and Burrows–Wheeler Coding (**bzip2-1**).

| Sizes in Bytes | 111,261 | 768,771 | 377,109 | 513,216 | 39,611 | 93,695 | |
Source Texts	bib	book1	news	pic	progc	trans	Average
pack	5.24	4.56	5.23	1.66	5.26	5.58	4.99
gzip-b	2.51	3.25	3.06	0.82	2.68	1.61	2.69
bzip2-1	2.10	2.81	2.85	0.78	2.53	1.53	2.46

Note: Figures give the number of bits used per character (letter). They show that **pack** is the least efficient method and that **bzip2-1** compresses slightly more than **gzip-b**.

usually better compression ratio but is more difficult to implement. Compression based on arithmetic as presented in this chapter gives compression ratios slightly better than Huffman coding.

The table of Figure 14.27 gives in percentage the sizes of compressed files. Results obtained by Ziv–Lempel and Factor algorithms are similar. Huffman coding gives the best result for the Random file. Finally, experience shows that exact compression methods often reduce the size of data to 30%–50% of their original size.

Table 14.1 contains a sample of experimental results showing the behavior of compression algorithms on different types of texts from the Calgary Corpus: bib (bibliography), book1 (fiction book), news (USENET batch file), pic (black and white fax picture), progc (source code in C), and trans (transcript of terminal session).

The compression algorithms reported in the table are the Huffman coding algorithm implemented by **pack**, the Ziv–Lempel algorithm implemented by **gzip-b** and the compression based on the *BW* transform implemented by **bzip2-1**.

Additional compression results can be found at http://corpus.canterbury.ac.nz.

14.8   Research Issues and Summary

The statistical compression algorithm is from Huffman (1951). The UNIX command pack implements the algorithm.

The dynamic version was discovered independently by Faller (1973) and Gallager (1978). Practical versions were given by Cormack and Horspool (1984) and Knuth (1985). A precise analysis leading to an improvement was presented in (Vitter, 1987). The command compact of UNIX implements the dynamic Huffman coding.

It is unclear to whom precisely should be attributed the idea of data compression using arithmetic coding. It is sometimes refer to Shannon-Fano-Elias coding (Mansuripur, 1987), and has become popular after the publication of the article of Witten et al. (1987). An efficient data structure for the tables of frequencies is due to Fenwick (1994). The main interest in arithmetic coding for text compression is that the two different processes "modeling" the statistics of texts and "coding" can be made independent modules.

Several variants of the Ziv–Lempel algorithm exist. The reader can refer to the books of Bell et al. (1990) or Storer (1988) for a discussion on them.

The *BW* transform is from Burrows and Wheeler (1994).

The books of Held (1991) and Nelson (1992) present practical implementations of various compression algorithms, while the book of Crochemore and Rytter (2002) overflows the strict topic of text compression and describes more algorithms and data structures related to text manipulations.

14.9 Further Information

The set of algorithms presented in this chapter provides the basic methods for data compression. In commercial software they are often combined with other more elementary techniques that are described in textbooks. A wider panorama of data compression algorithms on texts may be found in several books by Bell et al. (1990), Crochemore and Rytter (2002), Held (1991), Nelson (1992), and Storer (1998).

Research papers in text data compression are disseminated in a few journals, among which are *Communications of the ACM, Journal of the ACM, Theoretical Computer Science, Algorithmica, Journal of Algorithms, Journal of Discrete Algorithms, SIAM Journal on Computing,* and *IEEE Trans. Information Theory.*

An annual conference presents the latest advances of this field of research:

- Data Compression Conference, which is regularly held at Snowbird (Utah) in spring.

Two other conferences on pattern matching also present research issues in this domain:

- Combinatorial Pattern Matching (CPM), which started in 1990 and was held in Paris (France), London (England), Tucson (Arizona), Padova (Italy), Asilomar (California), Helsinki (Finland), Laguna Beach (California), Aarhus (Denmark), Piscataway (New Jersey), Warwick University (England), Montreal (Canada), Jerusalem (Israel), Fukuoka (Japan), Morelia, Michocán (Mexico), Istanbul (Turkey), Jeju Island (Korea), Barcelona (Spain), and London, Ontario (Canada).

- Workshop on String Processing (WSP), which started in 1993 and was held in Belo Horizonte (Brasil), Valparaiso (Chile), Recife (Brasil), and Valparaiso (Chile) and became *String Processing and Information Retrieval* (SPIRE) in 1998 and was held in Santa Cruz (Bolivia), Cancun (Mexico), A Coruña (Spain), Laguna de San Rafael (Chile), Lisbon (Portugal), Manaus (Brazil), Padova (Italy), Buenos Aires (Argentina), Glasgow (Scotland), and Santiago (Chile).

And general conferences in computer science often have sessions devoted to data compression algorithms.

Defining Terms

Codeword: Sequence of bits of a code corresponding to a symbol.

Prefix: A word $u \in \Sigma^*$ is a prefix of a word $w \in \Sigma^*$ if $w = uz$ for some $z \in \Sigma^*$.

Prefix code: Set of words such that no word of the set is a prefix of another word contained in the set. A prefix code is represented by a coding tree.

Segment: A word $u \in \Sigma^*$ is a segment of a word $w \in \Sigma^*$ if u occurs in w, i.e., $w = vuz$ for two words $v, z \in \Sigma^*$. (u is also referred to as a factor or a subword of w)

Trie: Tree in which edges are labeled by letters or words.

References

Bell, T.C., Cleary, J.G., and Witten, I.H. 1990. *Text Compression*, Prentice Hall, Englewood Cliffs, NJ.

Burrows, M. and Wheeler, D.J. 1994. A block sorting data compression algorithm. Report 124. Digital System Research Center.

Crochemore, M. and Rytter, W. 2002. *Jewels of Stringology*, World Scientific Publishing, Singapore.

Cormack, G.V. and Horspool, R.N.S. 1984. Algorithms for adaptive Huffman Codes. *Info. Process. Lett.* 18(3):159–165.

Faller, N. 1973. An adaptive system for data compression. In *Record of the 7th Asilomar Conference on Circuits, Systems, and Computers.* pp. 593–597.

Fenwick, P.M. 1994. A new data structure for cumulative frequency tables. *Software Pract. Experience.* 24(7):327–336.

Gallager, R.G. 1978. Variations on a theme by Huffman. *IEEE Trans. Info. Theory.* 24(6):668–674.

Held, G. 1991. *Data Compression*, John Wiley & Sons, New York.

Horspool, R.N. 1980. Practical fast searching in strings. *Softw. Pract. & Exp.* 10(6):501–506.

Huffman, D.A. 1951. A method for the construction of minimum redundancy codes. *Proc. IRE.* 40:1098–1101.

Knuth, D.E. 1985. Dynamic Huffman coding. *J. Algorithms.* 6(2):163–180.

Mansuripur, M. 1987. *Introduction to Information Theory*, Prentice Hall, Englewood Cliffs, New Jersey.

Nelson, M. 1992. *The Data Compression Book*, M&T Books, New York.

Storer, J.A. 1988. *Data Compression: Methods and Theory*, Computer Science Press, Rockville, MD.

Vitter, J.S. 1987. Design and analysis of dynamic Huffman codes. *J. ACM.* 34(4):825–845.

Welsh, T.A. 1994. A technique for high-performance data compression. *IEEE Comput.* 17(6):8–19.

Witten, I.H., Neal, R., and Cleary, J.G. 1987. Arithmetic coding for data compression. *Comm. ACM.* 30(6):520–540.

Zipstein, M. 1992. Data compression with factor automata. *Theor. Comput. Sci.* 92(1):213–221.

Ziv, J. and Lempel, A. 1977. A universal algorithm for sequential data compression. *IEEE Trans. Info. Theory.* 23(3):337–343.

15

General Pattern Matching

Alberto Apostolico
Georgia Institute of Technology

15.1 Introduction

This chapter reviews combinatorial and algorithmic problems related to searching and matching of strings and slightly more complicated structures such as arrays and trees. These problems arise in a vast variety of applications and in connection with various facets of storage, transmission, and retrieval of information. A list would include the design of structures for the efficient management of large repositories of strings, arrays and special types of graphs, fundamental primitives such as the various variants of exact and approximate searching, specific applications such as the identification of periodicities and other regularities, efficient implementations of ancillary functions such as compression and encoding of elementary discrete objects, etc. The main objective of studies in pattern matching is to abstract and identify some primitive manipulations, develop new techniques and efficient algorithms to perform them, both by serial and parallel or distributed computation, and implement the new algorithms.

Some initial pattern matching problems and techniques arose in the early 1970s in connection with emerging technologies and problems of the time, e.g., compiler design. Since then, the range of applications of the tools and methods developed in pattern matching has expanded to include text, image and signal processing, speech analysis and recognition, data compression, computational biology, computational chemistry, computer vision, information retrieval, symbolic computation, computational learning, computer security, graph theory and VLSI, etc. In little more than two decades, an initially sparse set of more or less unrelated results has grown into a considerable body of knowledge. A complete bibliography of string algorithms would contain more than 500 articles. Aho [2] references over 140 papers in his recent survey of string-searching algorithms alone; advanced workshops and schools, books and special issues in major journals have already been dedicated to the subject and more are planned for the future. The interested reader will find a few reference books and conference proceedings in the bibliography of this chapter.

While each application domain presents peculiarities of its own, a number of pattern matching primitives are shared, in nearly identical forms, within wide spectra of distant and diverse areas. For instance, searching for identical or similar substrings in strings is of paramount interest to software development and maintenance, philology or plagiarism detection in the humanities, inference of common ancestries in molecular genetics, comparison of geological evolutions, stereo matching for robot vision, etc. Checking the equivalence (i.e., identity up to a rotation) of circular strings finds use in determining the homology of organisms with circular genomes, comparing closed curves in computer vision, establishing the equivalence of polygons in computer graphics, etc. Finding repeated patterns, symmetries, and cadences in strings is of interest to data compression, detection of recurrent events in symbolic dynamics, genome studies, intrusion detection in distributed computer systems, etc. Similar considerations hold for higher structures. In general, an intermediate objective of studies in pattern matching is to understand and characterize combinatorial structures and properties that are susceptible of exploitation in computational matching and searching on discrete elementary structures.

Most pattern matching issues are still subject to extensive investigation both within serial and parallel models of computation. This survey concentrates on sequential algorithms, but the reader is encouraged to explore for himself the rich repertoire of parallel algorithms developed in recent years. Most of these algorithms bear very little resemblance to their serial counterparts. Similar considerations apply to some algorithms formulated in a probabilistic setting.

Pattern matching problems may be classified according to a number of paradigms. One way is based on the type of structure (strings, arrays, trees, etc.) in terms of which they are posed. Another, is according to the model of computation used, e.g., RAM, PRAM, Turing machine. Yet another one is according to whether the manipulations that one seeks to optimize need be performed online, off-line, in real time, etc. One could distinguish further between matching and searching and, within the latter, between exact and approximate searches, or vice versa. The classification used here is thus somewhat arbitrary. We assume some familiarity of the reader with exact string searching, both on- and off-line, which is covered in a separate chapter of this volume. We start by reviewing some basic variants of string searching where occurrences of the pattern need not be exact. Next, we review algorithms for string comparisons. Then, we consider pattern matching on two-dimensional arrays and finally on rooted trees.

15.2 String Searching with Don't-Care Symbols

As already mentioned, we assume familiarity of the reader with the problem of exact string searching, in which we are interested in finding all the occurrences of a pattern string y into a text string x. One of the natural departures from this formulation consists of assuming that a symbol can (perhaps only at some definite positions) match a small group of other symbols. At one extreme we may have, in addition to the symbols in the input alphabet Σ, a **don't-care** symbol ϕ with the property that ϕ matches any other character in Σ. This gives raise to variants of string searching where, in principle, ϕ appears (1) only in the pattern, (2) only in the text, or (3) both in pattern and text. There seems to be no peculiar result on variant (2), whence we shall consider this as just a special case of (3). The situation is different with variant (1), which warrants the separate treatment which is given next.

15.2.1 Don't Cares in Pattern Only

Fischer and Paterson [20] and Pinter [45] discuss the problem faced if one tried to extend the KMP string searching algorithm [33] in order to accommodate don't cares in the pattern: the obvious transitivity on character equality, that subtends those and other exact string searching, is lost with don't cares. Pinter noted that a partial recovery is possible if the number and positions of don't cares is fixed. In fact, in this case one may resort to ideas used by Aho and Corasick [3] in connection with exact multiple string searching and solve the problem within the same time complexity $O(n + m + r) \log |\Sigma|$ time, where $n = |x|$ is the length of the text string, $m = |y|$ is the length of the pattern, and r is the total number of occurrences of the fragments of the pattern that would be obtained by cleaving the latter at don't cares.

We outline Pinter's approach. Since the don't cares appear in fixed known positions, we may consider the pattern decomposed into **segments** of Σ^+, say, $\hat{y}_1, \hat{y}_2, \ldots, \hat{y}_p$ and ϕ-**blocks** consisting of runs of occurrences of ϕ, respectively. Each $\hat{y}_i$ can be treated as an individual pattern in a multiple pattern matching machine. Through the search, one computes for each $\hat{y}_i$ a list of its occurrences in x. Let d_i be the known distance between the starting positions of $\hat{y}_i$ and $\hat{y}_{i+1}$. We may now merge the occurrence list while keeping track of these distances, using the natural observation that if a match occurred starting at position j, then the $\hat{y}_i$'s will appear in the same order and intersegment distance as they appear in the pattern. Here the merge process takes place after the search. To make his algorithm work in real time applications, Pinter used an array of counters, a data structure originally proposed by R.L. Rivest. Instead of merging lists, counters count from 0 to p while collecting evidence of a pattern occurrence. Specifically, the counting mechanism is as follows. Let the **offset** of a segment be the distance from the beginning of the pattern to the end of that segment. Whenever a segment match is detected ending at position j, then its offset f_j is subtracted from j, thus yielding the starting position $j - f_i$ of a corresponding candidate occurrence of the pattern. Next, the counter assigned to position $1 + (j - f_i) \bmod m$ is incremented by 1. Therefore, a counter initialized to zero reaches p iff the last m characters examined on the text matched the pattern. A check whether a counter has reached p is performed each time that counter is reused.

Manber and Baeza-Yates [39] consider the case where the pattern embeds a string of at most k don't cares, i.e., has the form $y = u\phi^i v$, where $i \leq k$, $u, v \in \Sigma^*$, and $|u| \leq m$ for some given k, m. Their algorithm is off-line in the sense that the text x is preprocessed to build the suffix array [40] associated with it. This operation costs $O(n \log |\Sigma|)$ time in the worst case. Once this is done, the problem reduces to one of efficient implementation of two-dimensional orthogonal range queries (for these latter see, e.g., [16,60]).

One more variant of string searching with don't care in pattern only is discussed in [55]. Also Takeda's algorithm is based on the algorithm in [4]. The problem is stated as follows. Consider a set $A = \{A_1, A_2, \ldots, A_p\}$, where each $A_i \subseteq \Sigma$ is a nonempty set called a picture and pictures are mutually disjoint. While a don't-care symbol matches all symbols, a picture matches a subset of the alphabet. For any pattern y, we have now that $y \in (\Sigma \cup A)^+$. Then, given a set of patterns $Y = \{y^{(1)}, \ldots, y^{(k)}\}$, the problem is to find all occurrences of $y^{(i)}$ for $i = 1, \ldots, k$. Thus, when $A = \Sigma$, the problem reduces to plain string searching with don't cares. A pattern matching machine for such a family can be quite time consuming to build. Takeda managed to improve on time efficiency by saving on the number of explicit "goto" edges created in that machine.

Takeda's variant finds natural predecessors in an even more general class considered by Abrahamson [1]. This latter paradigm applies to an unbounded alphabet Σ, as long as individual symbols have finite encodings. Let $P = \{P_1, P_2, \ldots, P_k\}$ be a set of **pattern elements**, where each pattern element is a subset of Σ. There are positive and negative pattern elements. A positive element, is denoted by $< \sigma_1, \ldots, \sigma_f >$ and has the property of matching each one of the characters $\sigma_1, \sigma_2, \ldots, \sigma_f$. A negative element is denoted by $[\sigma_1, \ldots, \sigma_f]$ and will match any character of Σ

except characters $\sigma_1, \sigma_2, \ldots, \sigma_f$. A pattern $y \in P^+$ identifies now a family of strings, namely, all strings in the form $y_1 y_2 \ldots y_m$ such that $y_i \in \Sigma$ is compatible with the element of P used to identify the ith element of y. Using a time–space tradeoff proof technique due to Borodin, Abrahamson proved that the time-space lower bound on a subproblem with $n = 2m$ is $\Omega(m^2 / \log m)$.

By combining **divide-and-conquer** with an idea of Fischer and Paterson [20] which will be discussed more thoroughly later, Abrahamson designed an algorithm taking time $O(N + M + n\hat{M}^{1/2} \log m \log^{1/2} m)$, where N and M denote the lengths of the encodings (e.g., in bits) of x and y, respectively, and $\hat{M}$ represents the number of distinct elements of Σ which are present in the pattern.

15.2.2 Don't Cares in Pattern and Text

In an elegant, landmark paper, Fischer and Paterson [20] exposed the similarity of string searching to multiplication, thereby obtaining a number of interesting algorithms for exact string searching and some of its variants. It is not difficult to see that string matching problems can be rendered as special cases of a general linear product. Given two vectors X and Y, their **linear product** with respect to two suitable operations $\otimes$ and $\oplus$, is denoted by $X \overset{\otimes}{\underset{\oplus}{}} Y$, and is a vector $Z = Z_0 Z_1 \cdots Z_{m+n}$ where $Z_k = \bigoplus_{i+j=k} X_i \otimes Y_j$ for $k = 0, \ldots, m + n$. If we interpret $\oplus$ as the Boolean $\wedge$ and $\otimes$ as the symbol equivalence $\equiv$, then a match of the reverse Y^R of Y, occurs ending at position k in X, where $m \leq k \leq n$, if $[X_{k-m} \cdots X_k] \equiv [Y_m \cdots Y_0]$, that is, with obvious meaning, if $(X \overset{\equiv}{\underset{\wedge}{}} Y)_k = \text{TRUE}$. This observation brings string searching into the family of Boolean, polynomial, and integer multiplications, and leads quickly to an $O(n \log m \log \log m)$ time solution even in the presence of don't cares, provided that the size of Σ is fixed.

To see this, we show first that string searching can be regarded as a Boolean linear product, i.e., one where $\oplus$ is $\vee$ and $\otimes$ is $\wedge$. Let the text string be specified as $x = x_0 x_1 x_2 \cdots x_n$ and similarly let $y = y_0 y_1 y_2 \cdots y_m$ be the pattern. Recall that we assume a finite alphabet Σ, and that both x and y may contain some don't cares. For each $\rho \in \Sigma$, define $H_\rho(X_i) = 1$ if $x_i = \rho$, and $H_\rho(X_i) = 0$ if $x_i \neq \rho$ or $x_i = \phi$. Assume now that the vector X corresponding to string x contains only symbol σ and ϕ while Y, corresponding to string y, contains only symbol $\tau \neq \sigma$ and ϕ, with both σ and $\tau \in \Sigma$. Then $\bigwedge_{i+j=k} X_i \equiv Y_j$ means that $\bigwedge_{i+j=k} \neg H_\sigma(X_i) \vee \neg H_\tau(Y_j) \Longleftrightarrow \neg \bigvee_{i+j=k} H_\sigma(X_i) \wedge H_\tau(Y_j)$. The last term is a boolean product, whence such a product is not harder than string searching. On the other hand, $Z = X \overset{\equiv}{\underset{\wedge}{}} Y = \neg \bigvee_{\sigma \neq \tau; \sigma, \tau \in \Sigma} H_\sigma(X) \overset{\wedge}{\underset{\vee}{}} H_\tau(Y)$.

As is well known, the Boolean product can be obtained by performing the polynomial product, in which $\oplus$ is $+$ and $\otimes$ is $\times$. For this, just encode TRUE and FALSE as 1 and 0, respectively. One way to compute the polynomial product is to embed the product in a single large integer multiplication. There are well known fast solutions for the latter problem. For the $\{0, 1\}$ string vectors X and Y, the maximum coefficient is $m + 1$, so if we choose r such that $2^r > m + 1$, compute the integers $X(2^r) = \Sigma_{i=0}^n X_i \cdot 2^{ri}$ and $Y(2^r) = \Sigma_{j=0}^m Y_j \cdot 2^{rj}$ and then multiply $X(2^r)$ and $Y(2^r)$, the result will be the product evaluated at 2^r. The consecutive blocks of length r in the binary representation of $Z(2^r)$ will give the coefficients of Z, which can be transferred back to the Boolean product, and from there back to the string matching product. The Schönhage–Strassen [53] algorithm multiplies an N-digit number by an M-digit number in time $O(N \cdot \log M \cdot \log \log M)$, for $N > M$, using a multitape Turing machine. For the string matching product, $N = nr = O(n \log m)$, $M = mr = O(m \log m)$, so that the problem is solved on that machine in time $O(n \log^2 m \log \log m)$. The algorithm as presented assumes that the alphabet finite. For any alphabet Σ of size polynomial in n, however, we can always encode the two input strings in binary at a cost of a multiplicative factor $O(\log(|\Sigma|))$, and then execute just two Boolean products. This results in an extra $O(\log m)$ factor in the time complexity.

Note the adaptation of fast multiplication to string searching provides a basis for counting the mismatches generated by a pattern y at every position of a text x. This results from treating all symbols of Σ separately, and thus in overall time $O(n(|\Sigma|)\log^2 m \log\log m)$. This latter complexity is comparable to the above only for finite Σ. However, we shall see later that better bounds are achievable under this approach.

15.3   String Editing and Longest Common Subsequences

We now introduce three **edit operations** on strings. Namely, given any string w we consider the deletion of a symbol from w, the insertion of a new symbol in w, and the substitution of one of the symbols of w with another symbol from Σ. We assume that each edit operation has an associated nonnegative real number representing the cost of that operation. More precisely, the cost of deleting from w an occurrence of symbol a is denoted by $D(a)$, the cost of inserting some symbol a between any two consecutive positions of w is denoted by $I(a)$, and the cost of substituting some occurrence of a in w with an occurrence of b is denoted by $S(a, b)$. An **edit script** on w is any sequence Γ of viable edit operations on w, and the cost of Γ is the sum of all costs of the edit operations in Γ.

Now, let x and y be two strings of respective lengths $|x| = n$ and $|y| = m \le n$. The **string editing problem** for input strings x and y consists of finding an edit script Γ' of minimum cost that transforms y into x. The cost of Γ' is the **edit distance** from y to x. Edit distances where individual operations are assigned integer or unit costs occupy a special place. Such distances are often called Levenshtein distances, since they were introduced by Levenshtein [38] in connection with error correcting codes. String editing finds applications in a broad variety of contexts, ranging from speech processing to geology, from text processing to molecular biology.

It is not difficult to see that the general (i.e., with unbounded alphabet and unrestricted costs) problem of edit distance computation is solved by a serial algorithm in $\Theta(mn)$ time and space, through dynamic programming. Due to widespread application of the problem, however, such a solution and a few basic variants were discovered and published in literature catering to diverse disciplines (see, e.g., [44,47,49,59]). In computer science, the problem was dubbed "the string-to-string correction problem." The CS literature was possibly the last one to address the problem, but the interest in the CS community increased steadily in subsequent years. By the early 1980s, the problem had proved so pervasive, especially in biology, that a book by Sankoff and Kruskal [46] was devoted almost entirely to it. Special issues of the *Bulletin of Mathematical Biology* and various other books and journals routinely devote significant portions to it.

An $\Omega(mn)$ lower bound was established for the problem by Wong and Chandra [61] for the case where the queries on symbols of the string are restricted to tests of equality. For unrestricted tests, a lower bound $\Omega(n\log n)$ was given by Hirschberg [27]. Algorithms slightly faster than $\Theta(mn)$ were devised by Masek and Paterson [41], through resort to the so-called "Four Russians Trick." The "Four Russians" are Arlazarov et al. [8]. Along these lines, the total execution time becomes $\Theta(n^2/\log n)$ for bounded alphabets and $O(n^2(\log\log n)/\log n)$ for unbounded alphabets. The method applies only to the classical Levenshtein distance metric, and does not extend to general cost matrices. To this date, the problem of finding either tighter lower bounds or faster algorithms is still open.

The criterion that subtends the computation of edit distances by dynamic programming is readily stated. For this, let $C(i, j)$, $(0 \le i \le |y|,\ 0 \le j \le |x|)$ be the minimum cost of transforming the prefix of y of length i into the prefix of x of length j. Let w_k denote the kth symbol of string w. Then $C(0, 0) = 0$, $C(i, 0) = C(i - 1, 0) + D(y_i)$, $(i = 1, 2, \ldots, |y|)$, $C(0, j) = C(0, j - 1) + I(x_j)$ $(j = 1, 2, \ldots, |x|)$, and

$$C(i, j) = \min\left\{C(i - 1, j - 1) + S(y_i, x_j),\ C(i - 1, j) + D(y_i),\ C(i, j - 1) + I(x_j)\right\}$$

for all i, j, $(1 \leq i \leq |y|, 1 \leq j \leq |x|)$. Observe that, of all entries of the C-matrix, only the three entries $C(i-1, j-1)$, $C(i-1, j)$, and $C(i, j-1)$ are involved in the computation of the final value of $C(i, j)$. Hence, $C(i, j)$ can be evaluated row-by-row or column-by-column in $\Theta(|y||x|) = \Theta(mn)$ time. An optimal edit script can be retrieved at the end by backtracking through the local decisions that were made by the algorithm.

A few important problems are special cases of string editing, including the **longest common subsequence** (LCS) problem, **local alignment**, i.e., the detection of local similarities of the kind sought typically in the analysis of molecular sequences such as DNA and proteins, and some important variants of **string searching with errors**, or searching for approximate occurrences of a pattern string in a text string. As highlighted in the following brief discussion, a solution to the general string editing problem implies typically similar bounds for all these special cases.

15.3.1 Longest Common Subsequences

Perhaps the single most widely studied special case of string editing is the so-called LCS problem. The problem is defined as follows. Given a string z over an alphabet $\Sigma = (i_1, i_2, \ldots, i_s)$, a **subsequence** of z is any string w that can be obtained from z by deleting zero or more (not necessarily consecutive) symbols. The LCS problem for input strings $x = x_1 x_2 \cdots x_n$ and $y = y_1 y_2 \cdots y_m$ $(m \leq n)$ consists of finding a third string $w = w_1 w_2 \cdots w_l$ such that w is a subsequence of x and also a subsequence of y, and w is of maximum possible length. In general, string w is not unique.

Like the string editing problem itself, the LCS problem arises in a number of applications spanning from text editing to molecular sequence comparisons, and has been studied extensively over the past. Its relation to string editing can be understood as follows.

Observe that the effect of a given substitution can be always achieved, alternatively, through an appropriate sequence consisting of one deletion and one insertion. When the cost of a nonvacuous substitution (i.e., a substitution of a symbol with a different one) is higher than the global cost of one deletion followed by one insertion, then an optimum edit script will always avoid substitutions and produce instead y from x solely by insertions and deletions of overall minimum cost. Specifically, assume that insertions and deletions have unit costs, and that a cost higher than 2 is assigned to substitutions. Then, the pairs of matching symbols preserved in an optimal edit script constitute a LCS of x and y. It is not difficult to see that the cost e of such an optimal edit script, the length l of an LCS and the lengths of the input strings obey the simple relationship: $e = n + m - 2l$. Similar considerations can be developed for the variant where matching pairs are assigned weights and a heaviest common subsequence is sought (see, e.g., Jacobson and Vo [31]).

Lower bounds for the LCS problem are time $\Omega(n \log n)$ or linear time, according to whether the size s of Σ is unbounded or bounded [27]. Aho et al. [4] showed that, for unbounded alphabets, any algorithm using only "equal–unequal" comparisons must take $\Omega(mn)$ time in the worst case. The asymptotically fastest general solution rests on the corresponding solution by Masek and Paterson [41] to the string editing, hence takes time $O(n^2 \log \log n / \log n)$. Time $\Theta(mn)$ is achieved by the following dynamic programming algorithm from Hirschberg [26].

Let $L[0 \cdots m, 0 \cdots n]$ be an integer matrix initially filled with zeroes.

The following code transforms L in such a way that $L[i, j]$ $(1 \leq i \leq m, 1 \leq j \leq n)$ contains the length of an LCS between $y_1 y_2 \cdots y_i$ and $x_1 x_2 \cdots x_j$.

```
for i = 1 to m do
    for j = 1 to n do if y_i = x_j then L[i, j] = L[i − 1, j − 1] + 1
        else L[i, j] = Max {L[i, j − 1], L[i − 1, j]}
```

The correctness of this strategy follows from the obvious relations:

$$
\begin{array}{ccccc}
L[i-1,j] & \leq & L[i,j] & \leq & L[i-1,j]+1; \\
L[i,j-1] & \leq & L[i,j] & \leq & L[i,j-1]+1; \\
L[i-1,j-1] & \leq & L[i,j] & \leq & L[i-1,j-1]+1.
\end{array}
$$

If only the length l of an LCS is desired, then this code can be adapted to use only linear space. If an LCS is required at the outset, then it is necessary to keep track of the decision made at every step by the algorithm, so that an LCS w can be retrieved at the end by backtracking. The early $\Theta(mn)$ time algorithm by Hirschberg [27] achieved both a linear space bound and the production of an LCS at the outset, through a combination of dynamic programming and divide-and-conquer. Subsequent approaches to the LCS problem achieve time complexities better than $\Theta(mn)$ in favorable cases, though a quadratic performance is always touched and sometimes even exceeded in the worst cases. These approaches exploit in various ways the **sparsity** inherent to the LCS problem. Sparsity allows us to relate algorithmic performances to parameters other than the lengths of the input. Some such parameters are introduced next.

The ordered pair of positions i and j of L, denoted $[i,j]$, is a **match** iff $y_i = x_j$, and we use r to denote the total number of matches between x and y. If $[i,j]$ is a match, and an LCS $w_{i,j}$ of $y_1 y_2 \cdots y_i$ and $x_1 x_2 \cdots x_j$ has length k, then k is the **rank** of $[i,j]$. The match $[i,j]$ is k-dominant if it has rank k and for any other pair $[i',j']$ of rank k either $i' > i$ and $j' \leq j$ or $i' \leq i$ and $j' > j$. A little reflection establishes that computing the **k-dominant** matches ($k = 1, 2, \ldots, l$) is all is needed to solve the LCS problem (see, e.g., [7,26]). Clearly, the LCS of x and y has length l iff the maximum rank attained by a dominant match is l. It is also useful to define, on the set of matches in L, the following partial order relation $\mathcal{R}$: match $[i,j]$ precedes match $[i',j']$ in $\mathcal{R}$ if $i < i'$ and $j < j'$. A set of matches such that in any pair one of the matches always precedes the other in $\mathcal{R}$ constitutes a chain relative to the partial order relation $\mathcal{R}$. A set of matches such that in any pair neither match precedes the other in $\mathcal{R}$ is an **antichain**. Then, the LCS problem translates into the problem of finding a longest chain in the **poset** (partially ordered set) of matches induced by $\mathcal{R}$ (cf. [48]). A decomposition of a poset into antichains is **minimal** if it partitions the poset into the minimum possible number of antichains (refer, e.g., to [12]).

THEOREM 15.1 *[17] A maximal chain in a poset P meets all antichains in a minimal antichain decomposition of P.*

In other words, the number of antichains in a minimal decomposition represents also the length of a longest chain. Even though it is never explicitly stated, most known approaches to the LCS problem in fact compute a minimal antichain decomposition for the poset of matches induced by $\mathcal{R}$. The kth antichain in this decomposition is represented by the set of all matches having rank k. For general posets, a minimal antichain decomposition is computed by flow techniques [12], although not in time linear in the number of elements of the poset. Most LCS algorithms that exploit sparsity have their natural predecessors in either [30] or [26]. In terms of antichain decompositions, the approach of Hirschberg [26] consists of computing the antichains in succession, while that of Hunt and Szymanski [30] consists of extending partial antichains relative to all ranks already discovered, one new symbol of y at a time. The respective time complexities are $O(nl + n \log s)$ and $O(r \log n)$. Thus, the algorithm of Hunt and Szymanski is favorable in very sparse cases, but worse than quadratic when r tends to mn. An important specialization of this algorithm is that to the problem of finding a longest ascending subsequence in a permutation of the integers from 1 to n. Here, the total number of matches is n, which results in a total complexity $O(n \log n)$. Resort to the "fat-tree" structures introduced by van Emde Boas [19] leads to $O(n \log \log n)$ for this problem, a bound which had been shown to be optimal by Fredman [21].

Figure 15.1 illustrates the concepts introduced thus far, displaying the final L-matrix for the strings $x =$ cbadbb and $y =$ abcabccbc. We use circles to represent matches, with bold circles denoting dominant matches. Dotted lines thread antichains relative to $\mathcal{R}$ and also separate regions.

		a	b	c	a	b	c	c	b	c
		1	2	3	4	5	6	7	8	9
c	1	0	0	①	1	1	①	①	1	①
b	2	0	①	1	1	②	2	2	②	2
a	3	①	1	1	②	2	2	2	2	3
d	4	1	1	1	2	2	2	2	2	3
b	5	1	②	2	2	③	3	3	③	3
b	6	1	②	2	2	③	3	3	④	4

FIGURE 15.1　Illustrating antichain decompositions.

15.3.2　Hirschberg's Paradigm: Finding Antichains One at a Time

We outline a $\Theta(mn)$ time LCS algorithm in which antichains of matches relative to the various ranks are discovered one after the other. Consider the dummy pair $[0, 0]$ as a "0-dominant match," and assume that all $(k - 1)$-dominant matches for some k, $0 \le k \le l - 1$ have been discovered at the expense of scanning the part of the L-matrix that would lie above or to the left of the antichain $(k - 1)$, inclusive. To find the kth antichain, scan the unexplored area of the L-matrix from right to left and top-down, until a stream of matches is found occurring in some row i. The leftmost such match is the k-dominant match $[i, j]$ with smallest i-value. The scan continues at next row and to the left of this match, and the process is repeated at successive rows until all of the kth antichain has been identified. The process may be illustrated like in Figure 15.2. The large circles denote "pebbles" used to intercept the matches in an antichain. Initially, the pebbles are positioned on the matches created in the last column by the ad-hoc wildcard symbol \$. Next, pebbles are considered in succession from the top, and each pebble is moved to the leftmost match it can reach without crossing a previously discovered antichain. Once all pebbles have been considered, those contributing to the new antichain are identified easily.

Note that for each k the list of $(k - 1)$-dominant matches is enough to describe the shape of the antichain and also to guide the searches involved at the subsequent stage. Thus, also in this case linear space is sufficient if one wishes to compute only the length of w.

An efficient implementation of this scheme leads to the algorithm by Hirschberg [26], which takes time $O(nl + n \log s)$ and space $O(d + n)$, where d is the number of dominant matches.

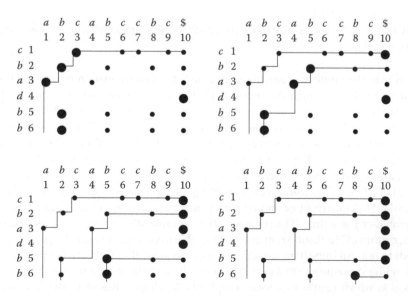

FIGURE 15.2　Hirschberg's paradigm: Discovering one antichain at a time. The positions occupied by the pebbles at the end of consecutive antichain constructions are displayed clockwise from left-top.

15.3.3 Incremental Antichain Decompositions and the Hunt–Szymanski Paradigm

When the number r of matches is small compared to m^2 (or to the expected value of lm), an algorithm with running time bounded in terms of r may be advantageous. Along these lines, Hunt and Szymanski [30] set up an algorithm (HS) with a time bound of $O((n + r) \log n)$. This algorithm works by computing, row after row, the ranks of all matches in each row. The treatment of a new row corresponds thus to extending the antichain decomposition relative to all preceding rows. A same match is never considered more than once. On the other hand, the time required by HS degenerates as r gets close to mn. In these cases this algorithm is outperformed by the algorithm of Hirschberg [26], which exhibits a bound of $O(ln)$ in all situations.

Algorithm HS is reproduced below. Essentially, it scans the list of matching positions $MATCHLIST$ associated with the ith row of L and considers those matches in succession, from right to left. For each match, HS decides whether it is a k-dominant match for some k through a binary search in the array $THRESH$ which maintains the leftmost previously discovered k-dominant match for each k. If the current match forces an update for rank k, then the contents of $THRESH[k]$ is modified accordingly. Observe that considering the matches in reverse order is crucial to the correct operation of HS. The details are found in the code below.

```
Algorithm "HS": element array y[1 : m], x[1 : n];
integer array THRESH[0 : m]; list array MATCHLIST[1 : m];
pointer array LINK[1 : m]; pointer PTR;
begin (PHASE 1: initializations
     for i = 1 to m do
     begin
          MATCHLIST[i] = {j₁, j₂, ..., jₚ}
          such that j₁ > j₂ > · > jₚ
          and yᵢ = xⱼq for 1 ≤ q ≤ p
          THRESH[i] = n + 1 for 1 ≤ i ≤ m;
     end
     THRESH[0] = 0; LINK[0] = null;
PHASE 2: find k-dominant matches)
     for i = 1 to m do
          for j on MATCHLIST[i] do
               begin find k such that
                    THRESH[k − 1] < j ≤ THRESH[k];
               if j < THRESH[k] then
                    begin
                    THRESH[k] = j;
                         LINK[k] = newnode(i, j, LINK[k − 1])
                    end
          end
     (PHASE 3: recover LCS w in reverse order)
     k̂ = largest k such that THRESH[k] ≠ n + 1;
     PTR = LINK[k̂];
     while PTR ≠ null do begin
     print the match [i, j] pointed to by PTR;
     advance PTR end
end.
```

The total time spent by HS is bounded by $O((r + m) \log n + n \log s)$, where the $n \log s$ term is charged by the preprocessing needed to create the lists of matches. The space is bounded by $O(d+n)$. As mentioned, this is good in sparse cases but becomes worse than quadratic for dense r.

15.4 String Searching with Errors

In this section, we assume unit cost for all edit operations. Given a pattern y and a text x, the most general variant of the problem consists of computing, for every position of the text, the best edit distance achievable between y and any substring w of x ending at that position. It is not difficult to express a solution in terms of a suitable adaptation of the recurrence previously introduced in connection with string editing. The first obvious change consists of setting all costs to 1 except that $S(y_i, x_j) = 0$ for $y_i = x_j$. Thus, we have now, for all i, j, $(1 \leq i \leq |y|, 1 \leq j \leq |x|)$,

$$C(i,j) = \min \left\{ C(i - 1, j - 1) + 1, \ C(i - 1, j) + 1, \ C(i, j - 1) + 1 \right\} .$$

A second change consists of setting the initial conditions so that $C(0,0) = 0$, $C(i,0) = i$ ($i = 1, 2, \ldots, m$), $C(0,j) = 0$ ($j = 1, 2, \ldots, n$). This has the effect of setting to zero the cost of prefixing y by any prefix of x. In other words, any prefix of the text can be skipped free of charge in an optimum edit script.

Clearly, the computation of the final value of $C(i, j)$ may proceed as in the general case, and it will still take $\Theta(|y||x|) = \Theta(mn)$ time. Note, however, that we are interested now in the entire last row of matrix C at the outset. Although we assumed unit costs, the validity of the method extends clearly to the case of general positive costs.

In practice, it is often more interesting to locate only those segments of x that present a high similarity with y under the adopted measure. Formally, given a pattern y, a text x and an integer k, this restricted version of the problem consists of locating all terminal positions of substrings w of x such that the edit distance between w and y is at most k. The recurrence given above will clearly produce this information. However, there are more efficient methods to deal with this restricted case. In fact, a time complexity $O(kn)$ and even sublinear expected time are achievable. We refer to Landau and Vishkin [36,37], Sellers [49], Ukkonen [57], Galil and Giancarlo [22], and Chang and Lawler [14] for detailed discussions. In the following, we review some basic principles subtending an $O(kn)$ algorithm for string searching with k differences. Note that when k is a constant the corresponding time complexity is linear.

The crux of the method is to limit computation to $O(k)$ elements in each diagonal of the matrix C. These entries will be called extremal and may be defined as follows: a diagonal entry is d-extremal if it is the deepest entry on that diagonal to be given value d ($d = 1, 2, \ldots, k$). Note that a diagonal might not feature any, say, 1-extremal entry, in which case it would correspond to a perfect match of the pattern. The identification of d-extremal entries proceeds from extension of entries already known to be $(d - 1)$-extremal. Specifically, assume we knew that entry $C(i, j)$ is $(d - 1)$-extremal. Then, any entry reachable from $C(i, j)$ through a unit vertical, horizontal, or diagonal-mismatch step possibly followed by a maximal diagonal stream of matches is d-extremal at worst. In fact, the cost of a diagonal stream of matches is 0, whence the cost of an entry of the type considered cannot exceed d. On the other hand, that cost cannot be smaller than $d - 1$, otherwise this would contradict the assumption $C(i, j) = d - 1$. Let entries reachable from a $(d - 1)$-extremal entry $C(i, j)$ through a unit vertical, horizontal, or diagonal-mismatch step be called d-adjacent. Then the following program encapsulates the basic computations.

Algorithm "KERR" :
element array $x[1 : n], y[1 : m], C[0 : m; 0 : n]$; integer k
    begin

> (*PHASE* 1 : *initializations*)
>> *set first row of C to 0;*
>> *find the boundary set S_0 of 0-extremal entries by exact string searching;*
>
> (*PHASE* 2 : *identify k-extremal entries*)
>> **for** $d = 1$ **to** k **do**
>>> **begin**
>>>> *walk one step horizontally, vertically and (on mismatch) diagonally*
>>>> *from each $(d - 1)$-extremal entry in set $S_{(d-1)}$ to find d-adjacent entries;*
>>>> *from each d-adjacent entry, compute the farthest d-valued*
>>>> *entry reachable diagonally from it;*
>>>
>>> **end**
>>
>> **for** $i = 1$ **to** $n - m + 1$ **do**
>>> **begin**
>>>> *select lowest d-entry on diagonal i*
>>>> *and put it in the set S_d of d-extremal entries*
>>>
>>> **end**

> **end.**

It is easy to check that the algorithm performs k iterations in each one of which it does essentially a constant number of manipulations on each of the n diagonals. In turn, each one of these manipulations takes constant time except at the point where we ask to reach the farthest d-valued entry from some other entry on a same diagonal. We would know how to answer quickly that question if we knew how to handle the following query: given two arbitrary positions i and j in the two strings y and x, respectively, find the longest common prefix between the suffix of y that starts at position i and the suffix of x that starts at position j. In particular, our bound would follow if we knew how to process each query in constant time. It is not known how that could be done without preprocessing becoming somewhat heavy. On the other hand, it is possible to have it such that all queries have a cumulative amortized cost of $O(kn)$. This possibility rests on efficient algorithms for performing **lowest common ancestor** queries in trees. Space limitations do not allow us to belabor this point any further.

Note that the special case where insertions and deletions are forbidden is also solved by an algorithm very similar to the above and within the same time bound. This variant of the problem is often called **string searching with mismatches**. A probabilistic approach to this problem is implicit in [14], one more is described in [9]. When k cannot be considered a constant, an interesting alternative results from Abrahamson's approach to multiple-value string searching.

Specifically, this algorithm of Abrahamson's combines divide-and-conquer with the idea of Fischer and Paterson [20] which was discussed earlier. In divide-and-conquer, the problem is first partitioned into subproblems; these are then solved by ad-hoc techniques, and finally the partial solutions are combined. One possible way to "divide" is to take projections of the pattern into two complementary subsets, another is to split and handle separately the positive and negative portions of the pattern. We have already seen that the adaptation of fast multiplication to string searching leads to a time bound $O(n(|\Sigma|) \log^2 m \log \log m)$.

This performance is good for bounded Σ but quite poor when Σ is unbounded. In this latter case, however, some of the symbols must be very unfrequent. Using this observation, Abrahamson designed a projection into $\Sigma' = \{\sigma \in \Sigma : \sigma \text{ occurs at most } z \text{ times in } y\}$ and the corresponding complement set Σ''. The rare symbols can be handled efficiently by some direct match-counting, since they cannot produce more than zn matches in total. The frequent ones are limited in number to m/z and we can apply multiplication to each one of them separately. The overall result is time $O(nm/z \log^2 m \log \log m)$, which becomes $O(nm^{1/2} \log m \log \log^{1/2} m)$ if we pick $z = m^{1/2} \log m \log \log^{1/2} m$.

15.5 Two-Dimensional Matching

The problem of matching and searching of two-dimensional objects arises in as many applications as there are ways to involve pictures and other planar representations and objects. Just like the full-fledged problem of recognizing the digitized signal of a spoken word in a speech finds a first rough approximation in string searching, the problem of recognizing a particular subject in a scene finds a first, simplistic model in the computational task that we consider in this section: locating occurrences of a small array into a larger one. Even at this level of simplification, this task is enough complicated already that we shall ignore such variants as those allowing for different shapes and rotations, variants that do not appear in the one dimensional searches.

Two-dimensional matching may be exact and approximate just like with strings, but edit operations of insertion and deletion denature the structure of an array and thus may be meaningless in most settings. The literature on two-dimensional searching concentrates on exact matching, and so does the treatment of this section.

15.5.1 Searching with Automata

In exact two-dimensional searching, the input consists of a "text" array $X[n \times n]$ and a "pattern" array $Y[m \times m]$. The output consists of all locations (i, j) in X where there is an occurrence of Y, where the word "occurrence" is to be interpreted in the obvious sense that $X_{i+k,j+l} = Y_{k+1,l+1}$, $0 \leq k, l \leq m - 1$.

The naive attack leads to an $O(n^2 m^2)$ solution for the problem. It is not difficult to reduce this down to $O(n^2 m)$ by resorting to established string searching tools. This may be seen as follows. Imagine to build a linear pattern y where each character consists of one of the consecutive rows of Y. Now, build similarly the family of text strings $x_1^{(i)} x_2^{(i)} \cdots x_{n-m+1}^{(i)}$ ($1 \leq i \leq n$) such that $x_j^{(i)}$ is the character for $X_{i,j} X_{i,j+1} \cdots X_{i,j+m-1}$. Clearly, Y occurs at $X_{i,j}$ iff y occurs at $x_j^{(i)}$. If one could assume a constant cost for comparing a character of y with one of $x^{(i)}$, it would take $O(n)$ time by any of the known fast string searching to find the occurrences on y in each $x^{(i)}$. Hence, it would take optimal time $O(n^2)$ for the n strings in the global problem. Since comparing two strings of m characters each charges in fact m comparisons, then the overall bound becomes $O(n^2 m)$, as stated.

Automata-based techniques were developed along these lines by Bird [11] and Baker [10]. Later efforts exposed also a germane problem which came to be called "dictionary matching" and acquired some independent interest. Some details of such an automata-based two-dimensional searching are given next.

The main idea is to build on the distinct rows of the pattern Y the Aho–Corasick [3] automaton for multiple string searching. Once this connection is made, it becomes possible to solve the problem at a cost of preprocessing time $O(m^2 \log |\Sigma|)$ (to build the automaton for at most m patterns with m characters each), and time $O(n^2 \log |\Sigma| + tocc)$ to scan the text. Here *tocc*, stands for total number of occurrences, i.e., is the size of the output. In multiple string matching, the parameter *tocc* may play havoc with time linearity, since more than one pattern might end and thus have to be outputted at any given position. Here, however, the rows of Y are all of the same size, whence only one such row may occur at any given position.

15.5.2 Periods and Witnesses in Two Dimensions

Automata-based approaches such as those just discussed result in time complexities that carry a dependence to alphabet size. This is caused by the branching of forward transitions that leave the states of the machine in multiple string searching. Single string searching is not affected by this problem. In fact, single string searching found quickly linear solutions without alphabet dependency. In contrast, several years elapse before alphabet dependency was eliminated from two-dimensional searching.

Alphabet dependency was eliminated in steps, first from the search phase only, and finally also from preprocessing. A key factor in the first step of progress was offered by a two-dimensional extension of the notion of a **witness**, a concept first introduced and used by Vishkin [58] in connection with parallel exact string searching. It is certainly rare, and therefore quite remarkable, that a tool devised specifically to speed-up a parallel algorithm would find use in designing a better serial algorithm.

It is convenient to illustrate the idea of a witness on strings. Assume then to be given two copies of a pattern y, reciprocally aligned in such a way that the top copy is displaced, say, d positions ahead of the bottom one. A witness for d, if it exists, is any pair of mismatching characters that would prevent the two superimposed copies of y to coexist. Thus, if we were to be given two d-spaced, overlapping candidate occurrences of y on a text x, and a witness were defined for d, then at least one of the candidate occurrences of y in x will necessarily fail. One alternative way to regard a witness at d is as a counterexample to the claim that d is a period for y. The latter is a necessary, though not sufficient condition for having y occur twice, d positions apart.

The use of witnesses during the search phase presupposes preparation of appropriate tables. These tables essentially provide, for each d where this is true, a mismatch proving the incompatibility of two overlapping matches at a distance of d. The notion of a witness generalizes naturally to higher dimensions. In two dimensions two witnesses tables were introduced by Amir et al. [5] as follows. Witness $Wit_{i,j}$ is any position (p, q) such that $X_{i+p,j+q}$ does not match $X_{i,j}$ or else it is 0. Note that, given an array W, there are essentially only two ways of superpositions one of thet two copies onto the other. These consist, respectively, of shifting one of the copies toward the right and bottom or toward the right and top of the other. These two families correspond to two witness tables that depend on whether $i < 0$ or $i \geq 0$. Amir et al. [5] showed how to build the witness table in time $O(m^2 \log |\Sigma|)$.

Once the table is available, the search phase is performed in two stages that are called respectively **candidate consistency testing** and **candidate verification**. The candidates are the positions of X, interpreted as top-left corners of potential occurrences of the pattern. At the beginning each position is a viable candidate. A pair of candidates is consistent if the pattern could be placed at both places without conflicting with the witness tables. The task of the first phase is to use the witness tables to remove one in each pair of inconsistent candidates. Clearly, one character comparison with the position of the text array that corresponds to the witness suffices to carry out this "duel" between the candidates. Note that a duel might rule out both candidates, however, eliminating one will do.

At the end of the consistency check we can verify the surviving candidates. A same text symbol could belong to several candidates, but all of these candidates must agree on that symbol. Thus, each position in the text can be labeled true or false according to whether or not it complies with what all partecipating candidates surrounding it prescribe for that position. Conversely, whenever a candidate covers a position of the text that is labeled as false, then that candidate can no longer survive. A procedure set up along these lines leads to an $O(n^2)$ search phase, within a model of computation in which character comparisons take constant time and only result in assessing whether the characters are equal or unequal.

The preprocessing in this approach is still dependent on the size of the alphabet. Alphabet independent preprocessing and overall linear time algorithm was achieved by Galil and Park [24]. Like with strings, one may build an index structure based on preprocessing of the text and then run faster queries off-line with varying patterns. Details can be found in, e.g., [25].

15.6   Tree Matching

The discrete structures considered in this section are labeled, rooted trees, with the possible additional constraint that children of each node be ordered. Recall that a **tree** is any undirected, connected and acyclic graph. Choosing one of the vertices as the root makes the tree rooted, and fixing an order among the children of each node makes the tree ordered. Like with other classes of discrete objects, there is exact and approximate searching and matching of trees. We examine both of these issues next.

15.6.1   Exact Tree Searching

In exact tree searching, we are given two ordered trees, namely, a "pattern" tree P with m nodes and a "text" tree T with n nodes, and we are asked to find all occurrences of P in T. An occurrence of P in T is an ordered subtree P' rooted at some node v of T such that P could be rigidly superimposed onto P' without any label mismatch or edge skip. The second condition means that the kth child of a node of P matches precisely the kth child of a node of T.

An $O(nm^{0.75}polylog(m))$ improvement over the trivial $O(mn)$ time algorithm was designed by Kosaraju [34]. A faster, $O(n\sqrt{m})polylog(m))$ algorithm, is due to Dubiner et al. [18]. Their approach is ultimately reminiscent of Abrahamson's pidgeon-hole approach to generalizations of string searching such as those examined earlier in our discussion. It is based on a combination of periodicity properties in strings and some techniques of tree partitioning that achieve succint representations of long paths in the pattern.

Some notable variants of exact tree pattern matching arise in applications such as code generation and unification for logic programming and term-rewriting systems. In this context, a label can be a constant or a variable, where a variable at a leaf may match an entire subtree. In the most general setting, the input consists of a set S of patterns, rather than a single pattern, and of course of a text T. Early analyses and algorithms for the general problem are due to Hoffman and O'Donnel [29]. Two basic families of treatment descend, respectively, from matching the text tree from the root or from the leaves. The bottom-up approach is the more convenient of the two in the context of term-rewriting systems. This approach is heavy on pattern preprocessing, where it may require exponential time and space, although essentially linear in the processing phase. Improvements and special cases are treated by Chase [15], Cai et al. [13], and Thorup [56].

15.6.2   Tree Editing

The editing problem for unordered trees is NP-complete. However, much faster algorithms can be obtained for ordered trees. Early definitions and algorithms may be traced back to Selkow [50] and Tai [54]. In more recent years, the problem and some of its basic variants have been studied extensively by Shasha and Zhang and their coauthors. The outline given below concentrates on some of their work.

Let T be a tree of $|T| = n$ nodes, each node labeled with a symbol from some alphabet Σ. We consider three edit operations on T, consisting, respectively, of the deletion of a node v from T (followed by the reassignment of all children of v to the node of which v was formerly a child), the insertion of a new node along some consecutive arcs departing from a same node of T, and the substitution of the label of one of the nodes of T with another label from Σ. Like with strings, we assume that each edit operation has an associated nonnegative real number representing the cost of that operation. We similarly extend the notion of edit script on T to be any consistent sequence Γ of edit operations on T, and define the cost of Γ as the sum of all costs of the edit operations in Γ. These notions generalize easily to any ordered **forest** of trees.

Now, let F and F' be two forests of respective sizes $|F| = n$ and $|F'| = m$. The **forest editing problem** for input F and F' consists of finding an edit script Γ' of minimum cost that transforms F into F'. The cost of Γ' is the edit distance from F to F'. When F and F' consist each of exactly one tree, then we speak of the **tree editing problem**.

A convenient way to visualize the editing of trees or forests is by means of a mapping of nodes from one of two structures to the other. The map is represented by a set of links between node pairs (ν, ν') such that either these two nodes have precisely the same label—and thus node ν is exactly conserved as ν'—or else the label of ν gets substituted with that of ν'. Each node takes part in at most one link. The unaffected nodes of F (respectively, F') represent deletions (respectively, insertions). A mapping defined along these lines has the property of preserving both ancestor-descendant and sibling orders. In other words, a link from a descendant of ν may only reach a descendant of ν', and, similarly, links from two siblings to two others must not cross each other.

Early dynamic programming solutions for tree editing consume $\Theta(|F|^3 |F'|^3)$ time. Much faster algorithms have been set up subsequently. Some other interesting problems are special cases of forest editing, including "tree alignment," the "largest common subtree" problem, and the problem of "approximate tree matching" between a pattern tree and text tree. While any solution to the general tree editing problem implies similar bounds for all these special cases, some of the latter admit a faster treatment.

We review the criterion that subtends the computation of tree edit distances by dynamic programming after Zhang and Shasha [62]. This leads to an algorithm with time bounded by the product of the squares of the sizes of the trees. A convenient preliminary step is to resort to a linear representation for the trees involved. The discussion on mappings suggests that such a representation consist of assigning to each node its ordinal number in the postorder visit of the tree. Let x and y be the strings representing the postorder visits of two trees T and T'. Then a prefix of, say, x will identify in general some forest of subtrees each rooted at some descendant of the root of T. Note that the leftmost leaf in the leftmost tree is denoted precisely x_1. Let i_1 be the corresponding root, and let $forestdist(i, j)$ represent the cost of transforming the subforest of T corresponding to $x_1, \ldots, x_i$ into the subforest of T' corresponding to $y_1, \ldots, y_j$. Let $treedist(i, j)$ be the cost of transforming the tree rooted at x_i into the tree rooted at y_j. Then, in the most general case, these costs are dictated by the following recurrence:

$$
forestdist\,(i, j) = \min \begin{cases} forestdist\,(i-1, j) + D\,(x_i) \\ forestdist\,(i, j-1) + I\,(y_j) \\ forestdist\,(l(i)-1, l(j)-1) + treedist\,(i, j) \end{cases}
$$

Here $l(i)$ (respectively, $l(j)$) is the index in x (respectively, y) of the leftmost leaf in the subtree rooted at the node x_i (x_j). We leave the initialization conditions for an exercise. Note that treedist is little more than a notational convention, since it is a special case of forestdist, and thus is computed essentially through the same recursion. In fact, a recursion in the above form can be applied to any pair of substrings of x and y, with obvious meaning. In the special case where both forests consist of a single tree, i.e., x_i and y_j have x_1 and y_1 as their respective leftmost leaves, then $treedist(i, j)$ becomes the substitution cost $S(x_i, y_j)$.

Building the algorithm around the above recurrence, and the subtended postorder visits, brings about an important advantage: each time that treedist is invoked, the main ingredients for its computation (namely, the pairwise distances of subtrees thereof) are already in place and thus need not be recomputed from scratch. We illustrate this point using $C(i, j)$, $(0 \le i \le |x|, 0 \le j \le |y|)$ as shorthand for forestdist. Observe that the recurrence above indicates that the value of $C(i, j)$ depends, in addition to the two neighboring values $C(i-1, j)$ and $C(i, j-1)$, on one generally more distant value $C(i', j')$. The pair (i', j') is called the conjugate of pair (i, j). The following facts are easy to check.

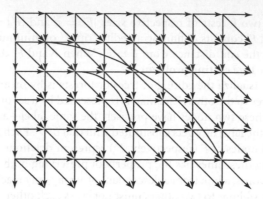

FIGURE 15.3 The upper-left corner of an AGDAG highlights the basic structure of such graphs: a grid with occasional outer-planar edges.

FACT 15.1   *Every pair (i, j) has at most one conjugate.*

FACT 15.2   *If (i, j) has conjugate (i', j'), then, for any pair (k, l) with $i' \leq k \leq i$ and $j' \leq l \leq j$ we also have $i' \leq k' \leq i$ and $j' \leq l' \leq j$.*

Figuratively, Fact 15.1 states that each pair (i, j) of C is associated with exactly one (possibly empty) submatrix of C, with upper-left corner at the conjugate (i', j') of (i, j) (inclusive) and lower right corner at $(i - 1, j - 1)$ (inclusive). Fact 15.2 states that the submatrices defined by two pairs and their corresponding conjugates are either nested or disjoint.

Like in the case of string editing, the "close" interdependencies among the entries of the C-matrix induce an $(|x| + 1) \times (|y| + 1)$ "grid directed acyclic graph" (GDAG for short). String editing can be viewed as a shortest-path problem on a GDAG. To take care also of the interdependencies by conjugacy that appear in tree editing, however, the GDAG must be augented by adding to the grid outerplanar edges connecting pairs of conjugate points.

Formally, an $l_1 \times l_2$ augmented GDAG (or AGDAG) is a directed acyclic graph whose vertices are the $l_1 l_2$ points of an $l_1 \times l_2$ grid, and such that the only edges from point (i, j) are to grid points $(i, j + 1)$, $(i + 1, j)$, $(i + 1, j + 1)$ and $(i' - 1, j' - 1)$, where (i, j) is the conjugate of (i', j'). We refer to Figure 15.3 for an example. We make the convention of drawing the points such that point (i, j) is at the ith row from the top and jth column from the left. The top-left point is $(0, 0)$ and has no edge entering it (i.e., is a "source"), and the bottom-right point is (m, n) and has no edge leaving it (i.e., is a "sink").

We associate an $(|x| + 1) \times (|y| + 1)$ AGDAG G with the tree editing problem in the natural way: the $(|x| + 1)(|y| + 1)$ vertices of G are in one-to-one correspondence with the $(|x| + 1)(|y| + 1)$ entries of the C-matrix. We draw edges connecting a point to its neighbors in the planar grid of the AGDAG, while the edge that is incident on point $(i - 1, j - 1)$ from the unique conjugate of (i, j), if the latter exists, are drawn outerplanar. Clearly, the cost of a grid edge from vertex (k, l) to vertex (i, j) is equal to $I(y_j)$ if $k = i$ and $l = j - 1$, to $D(x_i)$ if $k = i - 1$ and $l = j$, to $S(x_i, y_j)$ if $k = i - 1$ and $l = j - 1$. The cost of an outerplanar edge is the cost of the optimal solution to the submatrix associated with that edge. Thus, edit scripts that transform x into y or vice versa are in one-to-one correspondence to certain weighted paths of G that originate at the source (which corresponds to $C(0, 0)$) and end on the sink (which corresponds to $C(|x|, |y|)$). Specifically, in any such path horizontal or vertical edges can be traversed unconditionally, but the traversal of a diagonal edge from $(i - 1, j - 1)$ to (i, j) is allowed only if it follows the traversal of the outerplanar edge that is incident upon $(i - 1, j - 1)$ (if it exists). The details are left for an exercise.

15.7 Research Issues and Summary

The focus of this chapter is represented by combinatorial and algorithmic issues of searching and matching with strings and other simple structures like arrays and trees. We have reviewed the basic variants of these problems, with the notable exception of exact string searching. The latter is definitely the primeval problem in the set, and has been devoted so much study to warrant a separate chapter in the present Handbook.

We started by reviewing, in Section 15.2, string searching in the presence of don't-care symbols. In Section 15.3, we considered the problem of comparing two strings for similarity, under some basic sets of edit operations. This latter problem subtends the important variants of string searching where the occurrences of the pattern need not be exact; rather, they might be corrupted by a number of mismatches, and possibly by insertions and deletions of symbols as well. We abandoned the realm of one-dimensional pattern matching in Section 15.5, in which we highlighted the comparatively less battered topics of exact searching on two-dimensional arrays. Finally, in Section 15.6, we reviewed exact and approximate searching on rooted trees.

As said at the beginning, most pattern matching issues are still subject to extensive investigation. Meanwhile, new problems and variants continue to arise in application areas that feature, in prominent position, the very information infrastructure under development. In most cases, the goal of current studies is to design better serial algorithms than those previously available. Parallel or distributed versions of the problems are also investigated. Typically, the solutions of such versions may be expected not to resemble in any significant way their serial predecessors. In fact (as exemplified by the previously encountered notion of a witness) they are more likely to expose novel combinatorial properties, some of which of intrinsic interest. Whether a problem be regarded within a serial, parallel, or distributed computational context, algorithms are also sought that display a good expected, rather than worst case, performance. Relatively little work has been performed from this perspective, which requires often a thorough reexamination of the problem and may result in a totally new line of attack, as experienced in such classical instances as the Boyer–Moore string searching algorithm and Quicksort.

An exhausitve list of specific open problems of pattern matching would be impossible. Here we limit mention to a few important ones.

For problems of searching with don't care, string editing, LCS, and variations thereof, there are still wide and little understood gaps between the known, often trivial lower bounds and the efficiency of available algorithms. Likewise, relatively little is known in terms of nontrivial lower bounds for two-dimensional searches with mismatches, and also for both exact and approximate tree matching. Some general problems of fundamental nature remain unexplored across the entire board of pattern structures, problem variations, and computational models. Notable among these is the problem of preprocessing the "text" structure so that "patterns" presented online can be searched quickly thereafter. Such an approach has long been known to be elegantly and efficiently viable for exact searching on strings, but remains largely unexplored for approximate searches of all kind of patterns. The latter represent possibly the most recurrent queries in applications of molecular biology, information retrieval and other fields, so that progress in this direction would be valued enormously.

15.8 Further Information

Most books on design and analysis of algorithms devote one or more chapters to pattern matching. Here, we limit mention to specialized sources.

The collection of essays *Combinatorics on Words,* published in 1982 by Addison-Wesley under a fictitious editorship (M. Lothaire) contains most of the basic properties used in string searching, and more. An early attempt at unified coverage of string algorithmics is found in *Combinatorial Algorithms on Words,* edited by A. Apostolico and Z. Galil in 1985 for Springer-Verlag. *Time*

Warps, String Edits and Macromolecules: The Theory and Practice of Sequence Comparison, edited by D. Sankoff and J. B. Kruskal (Addison-Wesley, 1983), represents still a valuable source for sequence analysis and comparison tools in computational biology and other areas. A few more volumes of recent years are, in order of appearance: *Text Algorithms* by M. Crochemore and W. Rytter (Oxford University Press, 1994), *String Searching Algorithms* by G. A. Stephen (World Scientific, 1994), *Pattern Matching Algorithms,* edited by A. Apostolico and Z. Galil (Oxford University Press, 1997), and *Algorithms on Strings, Trees and Sequences* by D. Gusfield (Cambridge University Press, 1997). This last volume puts particular emphasis on issues arising in computational biology. A broader treatment of this field can be found in *Introduction to Computational Biology* by M. S. Waterman (Chapman & Hall, 1995). *Data Compression, Methods and Theory* by J. A. Storer (Computer Science Press, 1988) describes applications of pattern matching to the important family of compression methods by "textual substitution."

A rich bibliography on "words, automata and algorithms" is maintained by I. Simon of the University of São Paulo (Brazil). One on "sequence analysis and comparison" is maintained by William H. E. Day in Port Maitland, Canada. A collection of "pattern matching pointers" is currently maintained by S. Lonardi at http://www.cs.purdue.edu/homes/stelo/pattern.html.

Papers on the subject of pattern matching appear primarily in archival journals of theoretical computer science, but important contributions are also found in journals of application areas such as computational biology (notably, *CABIOS* and *Journal of Computational Biology*) and various specialties of computer science (cf., e.g., *IEEE Transactions* on Information Theory, *Pattern Recognition, Machine Intelligence, Software, etc.*). Special issues have been dedicated to pattern matching by *Algorithmica* and *Theoretical Computer Science.* Papers on the subject are presented at most major conferences. The International Symposia on Combinatorial Pattern Matching have gathered yearly since 1990. Beginning in 1992, proceedings have been published in the *Lecture Notes in Computer Science Series* of Springer-Verlag (serial numbers of volumes already published: 644, 684, 807, 937, 1075, and 1264). Specifically flavored contributions appear also at conferences such as RECOMB (International Conference on Computational Molecular Biology), the IEEE Annual Data Compression Conference, the South American Workshop on String Processing, and others.

Defining Terms

Antichain: A subset of mutually incomparable elements in a partially ordered set.

Block: A sequence of don't-care symbols.

Candidate consistency testing: The stage of two-dimensional matching where it is checked whether a candidate occurrence of the pattern is checked against the "witness" table.

Candidate verification: The stage of two-dimensional searching where candidate occurrences of the pattern, not ruled out previously as mutually incompatible, are actually tested.

Chain: A linearly ordered subset of a partially ordered set.

d-**adjacent:** An entry reachable from a $(d-1)$-extremal entry through a unit vertical, horizontal, or diagonal-mismatch step.

Divide-and-conquer: One of the basic paradigms of problem solving, in which the problem is decomposed (recursively) into smaller parts; solutions are then sought for the subproblems and finally combined in a solution for the whole.

Don't care: A "wildcard" symbol matching any other symbol of a given alphabet.

Edit operation: On a string, the operation of deletion, or insertion, or substitution, performed on a single symbol. On a tree T, the deletion of a node ν from T followed by the reassignment of all children of ν to the node of which ν was formerly a child, or the insertion of a new node along some consecutive arcs departing from a same node of T, or the substitution of the label of one of the nodes of T with another label from Σ. Each edit operation has an associated nonnegative real number representing its cost.

Edit distance: For two given strings, the cost of a cheapest edit script transforming one of the strings into the other.

Edit script: A sequence of viable edit operations on a string.

Exact string searching: The algorithmic problem of finding all occurrences of a given string usually called "the pattern" in another, larger "text" string.

Extremal: Some of the entries of the auxiliary array used to perform string searching. An entry is d-extremal if it is the deepest entry on its diagonal to be given value d.

Forest: A collection of trees.

Forest editing problem: The problem of transforming one of two given forests into the other by an edit script of minimum cost.

Linear product: For two vectors X and Y, and with respect to two suitable operations $\otimes$ and $\oplus$, is a vector $Z = Z_0 Z_1 \ldots Z_{m+n}$ where $Z_k = \bigoplus_{i+j=k} X_i \otimes Y_j$ ($k = 0, \ldots, m + n$).

Local alignment: The detection of local similarities among two or more strings.

Longest (or heaviest) common subsequence problem: The problem of finding a maximum-length (or maximum weight) subsequence for two or more input strings.

Lowest common ancestor: The deepest node in a tree that is an ancestor of two given leaves.

k-dominant match: A match $[i, j]$ having rank k and such that for any other pair $[i', j']$ of rank k either $i' > i$ and $j' \leq j$ or $i' \leq i$ and $j' > j$.

Match: The result of comparing two instances of a same symbol.

Minimal antichain decomposition: A decomposition of a poset into the minimum possible number of antichains.

Offset: The distance from the beginning of a string to the end of a segment in that string.

Pattern element: A positive (negative) pattern element is a "partial wildcard" presented as a subset of the alphabet Σ, with the symbols in the subset specifying which symbols of Σ are matched (mismatched) by the pattern element.

Picture: A collection of mutually disjoint subsets of an alphabet.

Poset: A set the elements of which are subject to a partial order.

Rank: For a given match, this is the number of matches in a longest chain terminating with that match, inclusive.

Segment: The substring of a pattern delimited by two don't cares or one don't care and one pattern boundary.

Sparsity: Used here to refer to LCS problem instances in which the number of matches is small compared to the product of the lengths of the input strings.

String editing problem: For input strings x and y, is the problem of finding an edit script of minimum cost that transforms y into x.

String searching with errors: Searching for approximate (e.g., up to a predefined number of symbol mismatches, insertions, and deletions) occurrences of a pattern string in a text string.

String searching with mismatches: The special case of string matching with errors where mismatches are the only type of error allowed.

Subsequence: Of a string, is any string that can be obtained by deleting zero or more symbols from that string.

Tree: A graph undirected, connected, and acyclic. In a rooted tree, a special node is selected and called the root: the nodes reachable from a node by crossing arcs in the direction away from the root are the children of that node. In unordered rooted trees, there is no pre-set order among the children of a node. Assuming such an order makes the tree ordered.

Tree editing problem: The problem of transforming one of two given trees into the other by an edit script of minimum cost.

Witness: A mismatch of two symbols of string *y* at a distance of *d* is a "witness" to the fact that in no subject *y* could occur twice at a distance of exactly *d* positions (equivalently, that *d* cannot be a period of *y*).

Acknowledgments

This work was supported in part by NSF Grants CCR-9201078 and CCR-9700276, by NATO Grant CRG 900293, by the National Research Council of Italy, by British Engineering and Physical Sciences Research Council Grant GR/L19362. Xuyan Xu contributed to Section 15.2 through bibliographic searching and drafting. The referees carried out a very careful scrutiny of the manuscript and made many helpful comments.

References

1. Abrahamson, K., Generalized string matching, *SIAM. J. Comput.*, 16(6), 1039–1051, 1987.
2. Aho, A.V., Algorithms for finding patterns in strings, *Handbook of Theoretical Computer Science*, J. van Leeuwen, Ed., Elsevier, Amsterdam, the Netherlands, pp. 255–300, 1990.
3. Aho, A.V. and Corasick, M.J., Efficient string matching: An aid to bibliographic search, *CACM*, 18(6), 333–340, 1975.
4. Aho, A.V., Hirschberg, D.S., and Ullman, J.D., Bounds on the complexity of the longest common subsequence problem, *J. Assoc. Comput. Mach.*, 23(1), 1–12, 1976.
5. Amir, A., Benson, G., and Farach, M., An alphabet independent approach to two dimensional matching, *SIAM J. Comp.*, 23(2), 313–323, 1994.
6. Apostolico, A., Browne, S., and Guerra, C., Fast linear space computations of longest common subsequences, *Theor. Comput. Sci.*, 92(1), 3–17, 1992.
7. Apostolico, A. and Guerra, C., The longest common subsequence problem revisited, *Algorithmica*, 2, 315–336, 1987.
8. Arlazarov, V.L., Dinic, E.A., Kronrod, M.A., and Faradzev, I.A., On economical construction of the transitive closure of a directed graph, *Dokl. Akad. Nauk SSSR*, 194, 487–488 (in Russian). English translation in *Sov. Math. Dokl.*, 11(5), 1209–1210, 1970.
9. Atallah, M.J., Jacquet, P., and Szpankowski, W., A probabilistic approach to pattern matching with mismatches, *Random Struct. Algor.*, 4, 191–213, 1993.
10. Baker, T.P., A technique for extending rapid exact-match string matching to arrays of more than one dimension, *SIAM J. Comp.*, 7(4), 533–541, 1978.
11. Bird, R.S., Two dimensional pattern matching, *Inform. Process. Lett.*, 6(5), 168–170, 1977.
12. Bogart, K.P. *Introductory Combinatorics*, Pitman, New York, 1983.
13. Cai, J., Paige, R., and Tarjan, R., More efficient bottom-up multi-pattern matching in trees, *Theor. Comput. Sci.*, 106(1), 21–60, 1992.
14. Chang, W.I. and Lawler, E.L., Approximate string matching in sublinear expected time, in *Proceedings of the 31st Annual IEEE Symposium on Foundations of Computer Science*, St. Louis, MO, pp. 116–124, 1990.

15. Chase, D., An improvement to bottom-up tree pattern matching, in *Proceedings of the 14th Annual ACM Symposium on Principle of Programming Languages*, Munich, West Germany, pp. 168–177, 1987.

16. Chazelle, B., A functional approach to data structures and its use in multidimensional searching, *SIAM. J. Comput.*, 17(3), pp. 427–462, 1988.

17. Dilworth, R.P., A decomposition theorem for partially ordered sets, *Ann. Math.*, 51, 161–165, 1950.

18. Dubiner, M., Galil, Z., and Magen, E., Faster tree pattern matching, *JACM*, 14(2), 205–213, 1994.

19. van Emde Boas, P., Preserving order in a forest in less than logarithmic time, in *Proceedings of the 16th Symposium on the Foundations of Computer Science*, Berkeley, CA, pp. 75–84, 1975.

20. Fischer, M.J. and Paterson, M., String matching and other products, *Complexity of Computation*, SIAM-AMS Proceedings 7, R., Karp, Ed., New York, 1973.

21. Fredman, M.L., On computing the length of longest increasing subsequences, *Discr. Math.*, 11, 29–35, 1975.

22. Galil Z. and Giancarlo, R., Data structures and algorithms for approximate string matching, *J. Complexity*, 4, 33–72, 1988.

23. Galil, Z. and Park, K., An improved algorithm for approximate string matching, *SIAM J. Comput.*, 19(6), 989–999, 1990.

24. Galil, Z. and Park, K., Truly alphabet-independent two-dimensional pattern matching, in *Proceedings of the 33rd Symposium on the Foundations of Computer Science (FOCS 92)*, Pittsburg, PA, pp. 247–256, 1992.

25. Giancarlo, R. and Grossi, R., On the construction of classes of suffix trees for square matrices: Algorithms and applications, in *Proceedings of 22nd International Colloquium on Automata, Languages, and Programming*, Z. Fulop and F. Gecseg, Eds., *LNCS* 944, Springer-Verlag, London, U.K., pp. 111–122, 1995.

26. Hirschberg, D.S., Algorithms for the longest common subsequence problem, *JACM*, 24(4), 664–675, 1977.

27. Hirschberg, D.S., An information theoretic lower bound for the longest common subsequence problem, *Info. Process. Lett.*, 7(1), 40–41, 1978.

28. Aho, A.V., Hirschberg, D.S., and Ullman, J.D., Bounds on the complexity of the longest common subsequence problem, *JACM*, 23(1), 1–12, 1976.

29. Hoffman, C. and O'Donnel, J., Pattern matching in trees, *JACM*, 29(1), 68–95, 1982.

30. Hunt, J.W. and Szymanski, T.G., A fast algorithm for computing longest common subsequences, *CACM*, 20(5), 350–353, 1977.

31. Jacobson, G. and Vo, K.P., Heaviest increasing/common subsequence problems, *Combinatorial Pattern Matching*, in *Proceedings of the 3rd Annual Symposium*, A. Apostolico, M. Crochemore, Z. Galil and U. Manber, Eds., *LNCS* 644, Springer-Verlag, London, U.K., pp. 52–66, 1992.

32. Jiang, T., Wang, L., and Zhang, K., Alignment of trees—an alternative to tree edit, in *Proceedings of the 5th Symposium on Combinatorial Pattern Matching*, Asilomer, CA, pp. 75–86, 1994.

33. Knuth, D.E., Morris, J.H., and Pratt, V.R., Fast pattern matching in strings, *SIAM. J. Comput.*, 6(2), 323–350, 1977.

34. Kosaraju, S.R., Efficient tree pattern matching, in *Proceedings of the 30th Annual IEEE Symposium on Foundations of Computer Science (FOCS)*, Research Triangle Park, NC, pp. 178–183, 1989.

35. Kumar, S.K. and Rangan, C.P., A linear space algorithm for the LCS problem, *Acta Informatica*, 24, 353–362, 1987.

36. Landau, G.M. and Vishkin, U., Introducing efficient parallelism into approximate string matching and a new serial algorithm, in *Proceedings of the 18th Annual ACM STOC*, New York, pp. 220–230, 1986.

37. Landau, G.M. and Vishkin, U., Fast string matching with k differences, *J. Comput. Syst. Sci.*, 37, 63–78, 1988.

38. Levenshtein, V.I., Binary codes capable of correcting deletions, insertions and reversals, *Sov. Phys. Dokl.,* 10, 707–710, 1966.

39. Manber, U. and Baeza-Yates, R. An algorithm for string matching with a sequence of don't cares, *Info. Process. Lett.,* 37(3), 133–136, 1991.

40. Manber, U. and Myers, E.W., Suffix Array: A new method for on-line string searches, in *Proceedings of the 1st Annual ACM-SIAM Symposium on Discrete Algorithms,* San Francisco, CA, pp. 319–327, 1990.

41. Masek, W.J. and Paterson, M.S., A faster algorithm computing string edit distances, *J. Comput. Syst. Sci.,* 20(1), 18–31, 1980.

42. Muthukrishnan, S. and Hariharan, R., On the equivalence between the string matching with don't cares and the convolution, *Inform. Computation,* 122(1), 140–148, 1995.

43. Myers, E.W., An $O(ND)$ difference algorithm and its variations, *Algorithmica,* 1, 251–266, 1986.

44. Needleman, R.B. and Wunsch, C.D., A general method applicable to the search for similarities in the amino-acid sequence of two proteins, *J. Mol. Biol.,* 48, 443–453, 1983.

45. Pinter, R., Efficient string matching with don't-care patterns, *Combinatorial Algorithms on Words,* Apostolico, A. and Galil, Z., Eds., *NATO ASI Series* 12, Springer-Verlag, Berlin, pp. 11–29, 1985.

46. Sankoff, D. and Kruskal, J., Eds., Time Warps, string edits, and macromolecules, in *The Theory and Practice of Sequence Comparision.* Addison-Wesley, Reading, MA, 1983.

47. Sankoff, D., Matching sequences under deletion-insertion constraints, *Proc. Natl. Acad. Sci. U.S.A.,* 69, 4–6, 1972.

48. Sankoff, D. and Sellers, P.H., Shortcuts, diversions and maximal chains in partially ordered sets, *Discr. Math.,* 4, 287–293, 1973.

49. Sellers, P.H., The theory and computation of evolutionary distance, *SIAM J. Appl. Math.,* 26, 787–793, 1974.

50. Selkow, S.M., The tree-to-tree editing problem, *Info. Process. Lett.,* 6, 184–186, 1977.

51. Shasha, D., Wang, J.T.L., and Zhang, K., Exact and approximate algorithms for unordered tree matching, *IEEE Trans. Syst. Man Cybern.,* 24(4), 668–678, 1994.

52. Shasha, D. and Zhang, K., Fast algorithms for the unit cost editing distance between trees, *J. Algorithm,* 11, 581–621, 1990.

53. Schonhage, A. and Strassen, V., Schnelle Multiplikation grosser Zahlen, *Computing (Arch. Elektron. Rechnen),* 7, 281–292. MR 45 No. 1431, 1971.

54. Tai, K.C., The tree-to-tree correction problem, *J. ACM,* 26, 422–433, 1979.

55. Takeda, M., A fast matching algorithm for patterns with pictures, *Bull. Info. Cyber.,* 25(3–4), 137–153, 1993.

56. Thorup, M., Efficient preprocessing of simple binary pattern forests, in *Proceedings of the 4th Scandinavian Workshop on Algorithm Theory, LNCS* 824, Springer-Verlag, London, U.K., pp. 350–358, 1994.

57. Ukkonen, E., Finding approximate patterns in strings, *J. Algorithm,* 6, 132–137, 1985.

58. Vishkin, U. Optimal parallel pattern matching in strings, *Inform. Control,* 67(1–3), 91–113, 1985.

59. Wagner, R.A. and Fischer, M.J., The string to string correction problem, *J. Assoc. Comput. Mach.,* 21, 168–173, 1974.

60. Willard, D.E., On the application of sheared retrieval to orthogonal range queries, in *Proceedings of the 2nd Annual ACM Symposium on Computational Geometry,* Yorktown Heights, NY, pp. 80–89, 1986.

61. Wong, C.K. and Chandra, A.K., Bounds for the string editing problem, *J. Assoc. Comput. Mach.,* 23(1), 13–16, 1976.

62. Zhang, K. and Shasha, D., Simple fast algorithms for the editing distance between trees and related problems, *SIAM J. Comput.,* 18(6), 1245–1262, 1989.

16

Computational Number Theory

Samuel S. Wagstaff, Jr.
Purdue University

16.1   Introduction

Number theory studies the whole numbers and, to some extent, the rational and algebraic numbers. Computational number theory is the study of computations with these kinds of numbers. Tasks performed by computational number theorists include

- Build tables used to suggest conjectures about integers.
- Write programs to test conjectures about integers.
- Write programs to prove theorems about integers with many cases.
- Invent and analyze algorithms to be used in the aforementioned tasks.

Sometimes algorithmic number theory is a synonym for computational number theory. Programs to test a conjecture may either seek a counterexample to disprove the conjecture or verify the conjecture for many cases, thereby supporting it. Computer algebra systems used by scientists and engineers use algorithms of number theory to perform their basic steps.

Algorithms used in computational number theory often work exclusively with integers, but not always. The analysis of such algorithms often draws on other areas of mathematics. This phenomenon is illustrated in the case study of greatest common divisors in the following section.

16.2   A Case Study: Greatest Common Divisors

The greatest common divisor of two integers is the largest integer that divides both of them. The greatest common divisor of integers a and b is denoted $\gcd(a, b)$. Since every integer divides 0, $\gcd(0, 0)$ is arbitrarily defined to be 0.

The Euclidean algorithm for computing $\gcd(a, b)$ is the oldest nontrivial algorithm. In its definition the notation $a \bmod b$, where a is an integer and b is a positive integer, means the (nonnegative) remainder when a is divided by b. We always have $0 \leq a \bmod b < b$. In terms of the floor function, we have $a \bmod b = a - b \times \lfloor a/b \rfloor$. Here is the Euclidean algorithm:

EUCLID (a, b)
1   **if** $b = 0$
2   **then return** $|a|$
3   **else return** EUCLID$(b, a \bmod |b|)$

The running time of EUCLID is proportional to the number e of mod operations, which is the same as the recursive depth. The worst case occurs when a and b are consecutive Fibonacci numbers. These numbers are defined by $f_0 = 0, f_1 = 1$, and $f_{n+1} = f_n + f_{n-1}$ for $n \geq 1$.

THEOREM 16.1   *(Lamé, 1845). For $e \geq 1$ let a and b be integers with $a > b > 0$ for which* EUCLID(a, b) *requires exactly e mod operations, and a is as small as possible for this to happen. Then* $a = f_{e+2}$ *and* $b = f_{e+1}$.

Lamé's theorem is the oldest one that gives a nontrivial worst case time complexity for an algorithm. Let $\phi = (1 + \sqrt{5})/2$ denote the "Golden Ratio" and let $\psi = (1 - \sqrt{5})/2 = 1 - \phi = -1/\phi$. The next result follows from the fact that $f_n = (\phi^n - \psi^n)/\sqrt{5}$, which is easy to prove by induction.

COROLLARY 16.1   If $0 \leq a, b \leq N$, then the number of mod operations taken by EUCLID(a, b) is $\leq \lceil \log_\phi(\sqrt{5}N) \rceil - 2$.

Since $\lceil \log_\phi(\sqrt{5}N) \rceil - 2 \approx 4.785 \log_{10} N - 0.328$, Corollary 16.1 implies that the number of mod operations performed by EUCLID(a, b) is no more than five times the number of decimal digits in the larger of a and b.

The average running time of EUCLID is more complicated. Let $e(a, b)$ denote the number of mod operations performed by EUCLID(a, b). When b is a fixed positive integer, the value of $e(a, b)$ depends only on $a \bmod b$. The average running time of EUCLID(a, b) is

$$T_b = \frac{1}{b} \sum_{a=0}^{b-1} e(a, b)$$

when b is fixed. Let $T_0 = 0$. To estimate T_b for large b, one might assume that for $0 < a < b$ the value of $b \bmod a$ is "random" in the interval $[0, a-1]$. This assumption would lead to the recurrence formula

$$T_0 = 0, \quad T_b = 1 + \frac{1}{b}\left(T_0 + T_1 + \cdots + T_{b-1}\right), \quad b \geq 1,$$

whose solution is $T_b = 1 + \frac{1}{2} + \cdots + \frac{1}{b} = \ln b + O(1)$, where $\ln b$ is the natural logarithm of b. It turns out that the assumption is pessimistic and that T_b does not grow as fast as $\ln b$ because the average value of $b \bmod a$ during the algorithm is actually less than the naive guess $a/2$.

To obtain a better estimate of T_b, one studies the metric theory of continued fractions. A simple continued fraction is an expression of the form

$$x = q_0 + \cfrac{1}{q_1 + \cfrac{1}{q_2 + \cfrac{1}{q_3 + \cdots + \cfrac{1}{q_k}}}},$$

which we will denote by $[q_0, q_1, q_2, q_3, \ldots, q_k]$. The numbers q_i, except for the last one, q_k, are required to be integers. Every real number x has a simple continued expansion that may be computed by this algorithm:

CONTINUED-FRACTION (x)

```
1    i ← 0
2    q_0 ← ⌊x⌋
3    x ← x − q_0
4    while x > 0 and i < Limit
5        do  i ← i + 1
6            q_i ← ⌊1/x⌋
7            x ← (1/x) − q_i
```

When $x = b/a$ is a rational number and Limit is large enough (say, Limit $= 5\log_{10}(\max(|a|, |b|)))$, the steps of CONTINUED-FRACTION(b/a) correspond in a simple way to those of EUCLID(a, b). Let us assume that $a > b > 0$ for simplicity. Define $a_0 = a$, $b_0 = b$, and $x_0 = b/a = b_0/a_0$. Then CONTINUED-FRACTION(b/a) begins by computing $q_0 = \lfloor b/a \rfloor = 0$. So long as $x_i := b_i/a_i \neq 0$ CONTINUED-FRACTION computes $q_{i+1} = \lfloor a_i/b_i \rfloor$ and

$$x_{i+1} = \frac{a_i}{b_i} - q_{i+1} = \frac{a_i - b_i \lfloor a_i/b_i \rfloor}{b_i} = \frac{a_i \bmod b_i}{b_i}.$$

If we define $a_{i+1} = b_i$ and $b_{i+1} = a_i \bmod b_i$, then we will have $x_i = b_i/a_i$ always. But this is how the recursion in Line 3 of EUCLID(a, b) works.

Thus the average behavior of EUCLID(a, b) for fixed b is determined by that of CONTINUED-FRACTION(x) for $x = 0/b$, $1/b$, $\ldots$, $(b-1)/b$. When b is large, this is essentially the study of CONTINUED-FRACTION(x) for a random x with a uniform distribution in the unit interval $[0, 1)$. For $0 \leq y \leq 1$, define $F_i(y)$ to be the probability that $x_i \leq y$, where x_i is the value of x at Line 4 of CONTINUED-FRACTION, and the initial value of x (at Line 2) in CONTINUED-FRACTION has a uniform distribution in the unit interval $[0, 1)$. Using advanced calculus, one can show that

$$F_i(y) = \lg(1 + y) + O(2^{-i}) \quad \text{as} \quad i \to \infty, \tag{16.1}$$

where $\lg z = \log_2 z$. Then measure theory is used to prove the following result which connects $F_i(y)$ to EUCLID(a, b), via the correspondence with CONTINUED-FRACTION(b/a) described earlier.

THEOREM 16.2 *Let b, i, and q and be positive integers. Let $p_i(q, b)$ denote the probability that the $(i + 1)$st quotient q_{i+1} in CONTINUED-FRACTION(b/a), where a is chosen with a uniform distribution in $\{0, 1, \ldots, b - 1\}$, is equal to q. Then*

$$\lim_{b \to \infty} p_i(q, b) = F_i\left(\frac{1}{q}\right) - F_i\left(\frac{1}{q + 1}\right).$$

It turns out that T_b has an erratic behavior, larger when b is prime and smaller when b has several prime factors. This happens because EUCLID(a, b) behaves just like EUCLID$(a/d, b/d)$ when

$\gcd(a, b) = d$. Smoother average behavior occurs when we average only over the pairs (a, b) for which $\gcd(a, b) = 1$. Define τ_b to be this average, that is,

$$\tau_b = \frac{1}{\varphi(b)} \sum_{\substack{a=1 \\ \gcd(a,b)=1}}^{b-1} e(a, b),$$

where $\varphi(b)$ is the number of a in $1 \le a \le b$ with $\gcd(a, b) = 1$. Then one can show

$$T_b = \frac{1}{b} \sum_{d|b} \varphi(d)\tau_d.$$

To estimate τ_b, note that in the Euclidean algorithm we have for each t

$$\frac{b_0}{a_0} \cdot \frac{b_1}{a_1} \cdots \frac{b_{t-1}}{a_{t-1}} = \frac{b_{t-1}}{a_0}$$

because $a_{i+1} = b_i$. Thus, if $t = e(a, b)$ and if $a = a_0$ and $b = b_0$ are relatively prime (that is, $b_{t-1} = \gcd(a, b) = 1$), then we have

$$x_0 x_1 \cdots x_{t-1} = 1/a.$$

We can compute τ_b from this formula because we know that the distribution function for x_i is $F_i(y)$, and Equation 16.1 estimates this function. The result is that

$$\tau_b \approx \frac{12 \ln 2}{\pi^2} \ln b + 1.47.$$

Using this approximation and the formula for T_b in terms of τ_b, one can prove

$$T_b \approx \frac{12 \ln 2}{\pi^2} \left(\ln b - \sum_{d|b} \frac{\Lambda(d)}{d} \right) + 1.47, \tag{16.2}$$

where $\Lambda(d)$ is von Mangoldt's function given by

$$\Lambda(d) = \begin{cases} \ln p & \text{if } d = p^r \text{ for } p \text{ prime and } r \ge 1, \\ 0 & \text{otherwise.} \end{cases}$$

In summary, we have seen that the number of mod operations performed by EUCLID(a, b) is never more than $4.8 \log_{10} N - 0.32$ when $0 < a, b < N$, and for fixed b is on average

$$\frac{12 \ln 2}{\pi^2} \ln b \approx 0.8428 \ln b \approx 1.941 \log_{10} b,$$

minus a correction term involving the divisors of b shown in Equation 16.2. See Chapter 4 [Knu97] for more details and references.

On a binary computer an integer division or mod operation is often slow. In contrast, addition, subtraction, shifting of binary numbers, and testing whether a binary number is even or odd are fast operations. In 1961, J. Stein [Ste67] created a new gcd algorithm with no division operation. Called the "Binary GCD Algorithm," it uses only subtraction, halving, and parity testing. The pseudocode in the following text presents this algorithm. The first four lines handle the case of a zero input. Lines 5 and 6 remove and remember the highest powers of 2 dividing each input number. After Line 6, a and b are odd integers. The heart of the algorithm is in Lines 7–10, in which the larger of a and b is repeatedly replaced by the difference between the two numbers, with all factors of 2 removed. The procedure ODD-DIVISOR(n) removes and counts all factors of 2 dividing the positive integer n; if 2^k is the highest power of 2 dividing the input n, it returns the pair $(k, n/2^k)$.

ODD-DIVISOR (n)
1     $k \leftarrow 0$
2   **while** n is even
3   **do** $k \leftarrow k + 1$
4                 $n \leftarrow \lfloor n/2 \rfloor$
5   **return** (k, n)
BINARY-GCD (a, b)
1   **if** $a = 0$
2   **then return** $|b|$
3   **if** $b = 0$
4   **then return** $|a|$
5     $(k_a, a) \leftarrow$ ODD-DIVISOR(a)
6     $(k_b, b) \leftarrow$ ODD-DIVISOR(b)
7   **while** $a \neq b$
8   **do** $(k_t, t) \leftarrow$ ODD-DIVISOR$(|a - b|)$
9               $a \leftarrow \min(a, b)$
10              $b \leftarrow t$
11   **return** $2^{\min(k_a, k_b)} a$

A fair measure of the time complexity of BINARY-GCD(a, b) is the number of subtraction operations done in Line 8 of the main loop. One can show that, in the worst case, this number may be as large as $1 + \lfloor \lg \max(a, b) \rfloor$. (One example of the worst case is $a = 2^i - 1$, $b = 1$.)

The average behavior of BINARY-GCD(a, b) is much harder to analyze. Numerical experiments suggest that the average number of subtraction operations done in Line 8 of the main loop is about $0.7 \lfloor \lg \max(a, b) \rfloor$. In 1976, Brent [Bre76] exhibited a dynamical system which simulates BINARY-GCD and predicts that the average number is $\beta \lfloor \lg \max(a, b) \rfloor$, where $\beta \approx 0.705971$. His heuristic analysis of this dynamical system depended on an assumption that a certain sequence of probability distribution functions converges to a limiting distribution function. This hypothesis remained unproved and the average behavior unsettled until 2007, when Maze [Maz07] used advanced mathematical analysis to prove the assumption.

We have seen how natural questions about simple algorithms for integers may lead to thorny problems in other areas of mathematics.

We conclude our discussion of greatest common divisors with one more practical problem. An important theorem in elementary number theory states that if integers a, b are not both 0, then there exist integers x, y so that $ax + by = \gcd(a, b)$. Many algorithms in computational number theory require one or both of x, y. These integers are not unique. Normally, we try to find examples of them with smallest absolute values. The algorithm EXTENDED-EUCLID(a, b) in the following text does this and returns x, y, and $g = \gcd(a, b)$. In this algorithm, q is an integer variable and $\vec{u}$, $\vec{v}$, $\vec{w}$ are triples of integers labeled as $\vec{u} = (u_0, u_1, u_2)$, etc. The subtraction and multiplication in Line 5 are subtraction and scalar multiplication of three-dimensional vectors. In Line 8, the algorithm returns $\vec{u} = (g, x, y)$. The algorithm assumes that the input a and b are nonnegative integers.

EXTENDED-EUCLID (a, b)
1     $\vec{u} \leftarrow (a, 1, 0)$
2     $\vec{v} \leftarrow (b, 0, 1)$
3   **while** $v_0 > 0$
4   **do** $q \leftarrow \lfloor u_0 / v_0 \rfloor$
5              $\vec{w} \leftarrow \vec{u} - q\vec{v}$

6                                  $\vec{u} \leftarrow \vec{v}$
7                                  $\vec{v} \leftarrow \vec{w}$
8   **return** $\vec{u}$

If one ignores the second and third components of the vector variables, then EXTENDED-EUCLID(a, b) is just a nonrecursive version of EUCLID(a, b). Therefore, $u_0 = g = \gcd(a, b)$ when the algorithm returns. The rest of the proof that the algorithm is correct is done by induction, noting the invariants $au_1 + bu_2 = u_0$ and $av_1 + bv_2 = v_0$ at the beginning and end of each pass through the **while** loop. The time complexity of EXTENDED-EUCLID is the same as that of EUCLID.

16.3 Basic Algorithms

Many problems in computational number theory involve integers too large to fit into a standard computer word of 32 or 64 bits. Such integers are stored as arrays of digits in some radix r, usually a power of 2. For example, on a 32 bit computer one might use radix $r = 2^{30}$, placing 30 bits in each word. The "classical" algorithms for arithmetic on such integers are the analogues of the familiar radix 10 algorithms we learn in second or third grade. Thus, we add large integers by adding pairs of their digits from low order to high order while "carrying" the excess above the radix to the next higher order position. Subtraction involves "borrowing" from the next higher order digit when needed. Multiplication is done by multiplying one of the integers by each digit of the other integer and adding the intermediate products, properly shifted. Division is slightly more complicated. To use radix 10 for input and output, one must convert between radix 10 and the internal radix r. By the way, the algorithms for arithmetic with polynomials in one variable are essentially the same as these algorithms for arithmetic with large numbers. The difference is that the digits of large integers must lie between 0 and $r - 1$, while coefficients of polynomials are not restricted. Thus, the polynomial arithmetic algorithms do not carry or borrow.

The classical algorithms perform addition, subtraction, and radix conversion in linear time in the length of the input. Using the classical algorithms, one can multiply two n-digit integers or divide a $2n$-digit integer dividend by an n-digit integer divisor in time $O(n^2)$. (The division produces both an n-digit quotient and an n-digit remainder.)

When the integers are sufficiently large, there are faster algorithms than the classical ones for multiplication and division. For multiplication there is a divide-and-conquer method based on an idea of Karatsuba [Kar62]. Suppose we want to multiply two $2n$-bit integers A and B. We split each number into two halves:

$$A = 2^n a_1 + a_0, \quad B = 2^n b_1 + b_0,$$

where a_0 and a_1 are the low and high order n bits of A, and likewise for B. Note that the formula

$$AB = (2^{2n} + 2^n)a_1 b_1 + 2^n(a_1 - a_0)(b_0 - b_1) + (2^n + 1)a_0 b_0$$

reduces the problem of multiplying one pair of $2n$-bit integers to three multiplications of pairs of n-bit integers, together with some adding and shifting operations. If we let $M(n)$ denote the number of bit operations needed to multiply two n-bit integers, then the formula shows that

$$M(2n) \leq 3M(n) + cn$$

because the adding and shifting can be done in linear time in n. By using the formula recursively, this inequality leads to the estimate $M(n) = O(n^{\lg 3}) = O(n^{1.59})$, much faster than the classical method for large n. More complicated algorithms for large integer multiplication, some of which use fast discrete Fourier transforms, reduce the time complexity $M(n)$ for multiplying two n-bit integers to $O(n \, (\log n) \log \log n)$ bit operations. These algorithms become useful when n is at least a few hundred.

Fast division, say a divided by b, is done by computing $2^k/b$, for some suitable k, by a kind of Newton's method, and then multiplying a times $2^k/b$ by a fast multiplication algorithm. The result is that a $2n$-bit dividend can be divided by an n-bit divisor in time $O(M(n))$ for large n.

The fast exponentiation algorithm is used to evaluate high powers efficiently. This algorithm computes $y = x^n$.

FAST-EXPONENTIATION (x, n)

```
1    e ← n
2    y ← 1
3    z ← x
4    while e > 0
5        do  if e is odd
6                then y ← y · z
7            z ← z · z
8            e ← ⌊e/2⌋
9    return y
```

The fast exponentiation algorithm has been known for at least 1000 years and works when x, y, and z are any mathematical objects that can be multiplied associatively. We give some examples later. In the ith iteration of the **while** loop, after Line 7, z has the value x^{2^i}. Whenever the ith bit of the binary number n (from right to left) is a 1, Line 6 multiplies $x^{2^{i-1}}$ into the partial product y, which produces $y = x^n$ in Line 9.

The number of multiplications performed by FAST-EXPONENTIATION (x, n) is $\lfloor \lg n \rfloor + \nu(n)$, where $\nu(n)$ is the number of 1 bits in the binary representation of n. We always have $1 \le \nu(n) < \lfloor \lg n \rfloor + 1$, so the number of multiplications is between $\lg n$ and $2 \lg n$. Although the FAST-EXPONENTIATION algorithm may appear to be optimally efficient, when the exponent n is large and fixed, Brauer [Bra39] proved that one can construct an addition chain for n to compute x^n in essentially $(1 + \epsilon) \lg n$ multiplications. See Knuth [Knu97] for details.

16.4 Algorithms for Congruences

Congruences are often involved in problems in computational number theory. Sometimes one uses a congruence to solve a problem about whole numbers. Sometimes a congruence, which may be viewed as a statement about whole numbers, is the object of study.

If a and b are integers and m is a positive integer, we say a is congruent to b modulo m and write $a \equiv b \bmod m$ to mean that m divides $a - b$. It is easy to show that "$\equiv$ modulo m" is an equivalence relation and interacts well with addition, subtraction, and multiplication. The latter means that if $a \equiv b \bmod m$ and $c \equiv d \bmod m$, then $a \pm c \equiv b \pm d \bmod m$ and $ac \equiv bd \bmod m$. The congruence "$a \equiv b \bmod m$" is shorthand for the statement, "There exists an integer k so that $a = b + km$."

One type of problem about a congruence is to find the values of an unknown that make it true. For example, we may wish to solve a congruence $f(x) \equiv 0 \bmod m$. This problem is equivalent to asking for all x for which $f(x)$ is divisible by m. In most cases of interest the answer is $x \equiv x_i \bmod m$ for some i in $1 \le i \le j$, where $x_1, \ldots, x_j$ is a finite list of integers.

In case $f(x) = ax - b$, we must solve $ax \equiv b \bmod m$. If we could divide by a, the answer would be $x \equiv b/a \bmod m$. However, since a congruence is a relation about divisibility, it does not always cooperate with division. In the example, if a is relatively prime to m, that is, if $\gcd(a, m) = 1$, we can define an inverse a^{-1} to a modulo m with the property $aa^{-1} \equiv 1 \bmod m$. Then we can multiply both sides of $ax \equiv b \bmod m$ by a^{-1} to get

$$x \equiv 1x \equiv a^{-1}ax \equiv a^{-1}b \equiv ba^{-1} \bmod m.$$

But if a is not relatively prime to m, then the congruence $ax \equiv b \bmod m$ may have no solution or many solutions modulo m.

To solve $ax \equiv b \bmod m$ when a is relatively prime to m, one may use EXTENDED-EUCLID(a, m) to find integers r, s, with $ra + sm = \gcd(a, m) = 1$. This equation implies the congruence $ra \equiv 1 \bmod m$, so r is an inverse for a modulo m. The solution to $ax \equiv b \bmod m$ is $x \equiv br \bmod m$. Recall that EXTENDED-EUCLID(a, m) returns a vector $\vec{u} = (u_0, u_1, u_2)$ with $au_1 + mu_2 = u_0 = \gcd(a, m)$. MOD-INVERSE$(a, m)$ computes $r \equiv a^{-1} \bmod m$.

MOD-INVERSE (a, m)
1    $\vec{u} \leftarrow$ EXTENDED-EUCLID(a, m)
2   **if** $u_0 = 1$
3   **then return** $u_1 \bmod m$
4   **else fail**

The number of integers between 1 and n which are relatively prime to n is denoted $\varphi(n)$, the Euler phi function of n. This function appears in Euler's theorem:

THEOREM 16.3   *(Euler). If a is relatively prime to the positive integer n, then $a^{\varphi(n)} \equiv 1 \bmod n$.*

Before Euler, Fermat had proved the case of Euler's theorem when n is prime. In that case, $\varphi(n) = n - 1$.

THEOREM 16.4   *(Fermat). If p is prime and does not divide a, then $a^{p-1} \equiv 1 \bmod p$.*

One simple application of Euler's theorem is to find an inverse modulo m. If a is relatively prime to m, then $a \cdot a^{\varphi(m)-1} \equiv 1 \bmod m$, so $r = a^{\varphi(m)-1} \bmod m$ is an inverse of a modulo m. This provides an alternate way to solve $ax \equiv b \bmod m$ when a is relatively prime to m. Let MOD-FAST-EXP(x, n, m) denote the algorithm FAST-EXPONENTIATION(x, n) with Lines 6 and 7 changed to

6   **then** $y \leftarrow (y \cdot z) \bmod m$
7              $z \leftarrow (z \cdot z) \bmod m$

Use MOD-FAST-EXP$(a, \varphi(m) - 1, m)$ to compute $r = a^{\varphi(m)-1} \bmod m$. Then the solution to $ax \equiv b \bmod m$ is $x \equiv br \bmod m$.

We have described two methods for computing the inverse of a modulo m, one via MOD-FAST-EXP$(a, \varphi(m) - 1, m)$ and one with EXTENDED-EUCLID(a, m). These two methods have nearly the same time complexity. Which one is faster on a given computer depends on the architecture and the details of the implementation.

Montgomery arithmetic [Mon85] provides an alternate representation of integers to speed a computation that involves many multiplications modulo a fixed positive integer m. The evaluation of a power modulo m is an excellent example of a situation where it is used. When one executes MOD-FAST-EXP(a, n, m), as when finding an inverse to a modulo m, the operations on large integers alternate between multiplying two integers and dividing their product by m to get the remainder. The alternate representation of x is (xR^{-1}) modulo m, where R is a certain power of 2. When two integers in this form are multiplied, their product may be reduced modulo m by a second multiplication followed by a shift operation. The multiplication and shift are considerably faster than a division operation. After the shift, the result is the modular product in the alternate representation, ready for the next multiplication. When all modular multiplications are finished, it is easy to convert the result back to the standard representation.

So far, we have discussed the solution of a single congruence in one variable. We next state the Chinese Remainder Theorem, which treats several congruences in one variable, but with different moduli. It has many uses in computational number theory.

THEOREM 16.5   (*Chinese Remainder Theorem*). *Let* $m_1, \ldots, m_r$ *be positive integers with* $\gcd(m_i, m_j) = 1$ *whenever* $1 \le i < j \le r$. *Let* $a_1, \ldots, a_r$ *be any r integers. Then the r congruences*

$$x \equiv a_i \bmod m_i, \quad 1 \le i \le r,$$

have a common solution. Any two common solutions are congruent modulo $m = m_1 \cdots m_r$.

The standard proof of the Chinese Remainder Theorem gives an efficient algorithm for finding the common solution.

CRT $(m_1, \ldots, m_r, a_1, \ldots, a_r)$
1      $m \leftarrow m_1 \cdots m_r$
2      $x \leftarrow 0$
3   **for** $i \leftarrow 1$ **to** r
4   **do** $n \leftarrow m/m_i$
5                      $x \leftarrow x + n \cdot a_i \cdot$ MOD-INVERSE(n, m_i)
6   **return** $x \bmod m$

The Chinese Remainder Theorem may be used to perform complex calculations with large integers faster than through other means. For example, suppose we want to compute the determinant of a square matrix with integer entries. The determinant is an integer. Hadamard [Had93] showed that the absolute value of the determinant of a $k \times k$ matrix whose entries have absolute value $\le M$ cannot exceed $B = k^{k/2} M^k$. Choose moduli $m_1, \ldots, m_r$ to be distinct primes near 2^{30} whose product exceeds $2B$. For each i, compute the determinant modulo m_i, for example, by Gaussian elimination. Remember that when the algorithm requires dividing by some integer d, one must instead multiply by $d^{-1} \bmod m_i$. Let a_i be the value of the determinant modulo m_i. Then use the Chinese Remainder Theorem to compute the determinant modulo $m = m_1 \cdots m_r$. Call the value x, where $0 \le x < m$. If $x \le m/2$, then the determinant is x, while if $x > m/2$, then the determinant is $x - m < 0$. Note that there are no congruences in either the question or its answer.

An important use of the Chinese Remainder Theorem is the reduction of solving a congruence $f(x) \equiv 0 \bmod m$ for a general integer m to solving $f(x) \equiv 0 \bmod p^e$, where p is prime and e is a positive integer. If $m = p_1^{e_1} \cdots p_r^{e_r}$ is the factorization of m into the product of powers of distinct primes, and if $x \equiv a_i \bmod p_i^{e_i}$ is a solution of $f(x) \equiv 0 \bmod p_i^{e_i}$, then CRT$(p_1^{e_1}, \ldots, p_r^{e_r}, a_1, \ldots, a_r)$ gives a solution of $f(x) \equiv 0 \bmod m$, and all solutions of $f(x) \equiv 0 \bmod m$ arise this way. If $f(x) \equiv 0 \bmod p_i^{e_i}$ has s_i distinct solutions, for $1 \le i \le r$, then $f(x) \equiv 0 \bmod m$ has $s_1 \cdots s_r$ distinct solutions. See the second following paragraph for the case when $f(x)$ is a quadratic polynomial.

We have seen how to solve a linear congruence $ax \equiv b \bmod m$. Sometimes one must solve higher degree congruences. For brevity, we consider only quadratic congruences here. The most general one is

$$ax^2 + bx + c \equiv 0 \bmod m,$$

where a, b, and c are integers. Roughly speaking, one uses the familiar quadratic formula

$$x = \frac{-b \pm \sqrt{b^2 - 4ac}}{2a}.$$

The division by $2a$ is done by multiplying by the inverse of $2a$ modulo m. The new problem here is taking a square root modulo m. We must solve $y^2 \equiv r \bmod m$ where $r = b^2 - 4ac$.

To solve $y^2 \equiv r \bmod m$, first factor $m = p_1^{e_1} \cdots p_k^{e_k}$ into a product of powers of different primes. Solve $z_i^2 \equiv r \bmod p_i$ for each i as described in the following text. Then solve $y_i^2 \equiv r \bmod p_i^{e_i}$ by "lifting" z_i using Hensel's lemma, a form of Newton's method. There may be 0, 1, or 2 solutions z_i and the same number of y_is. If there is no solution for some i, then $y^2 \equiv r \bmod m$ has no solution. If there is at least one solution for each i, then call $y = \text{CRT}(p_1^{e_1}, \ldots, p_k^{e_k}, y_1, \ldots, y_k)$ for every k-tuple of solutions $y_1, \ldots, y_k$ to get all solutions y to $y^2 \equiv r \bmod m$. See Chapter 7 of [Wag03] for Hensel's lemma and more details.

The interesting step in the process just described is solving $x^2 \equiv r \bmod p$, where p is an odd prime. The integer r is called a quadratic residue modulo p if $\gcd(r, p) = 1$ and there is an x such that $x^2 \equiv r \bmod p$. It is called a quadratic nonresidue modulo p if $\gcd(r, p) = 1$ and there is no x with $x^2 \equiv r \bmod p$. Euler's criterion tells whether r is a quadratic residue or nonresidue modulo an odd prime p.

THEOREM 16.6   *(Euler's criterion). If p is an odd prime and r is not a multiple of p, then $r^{(p-1)/2} \equiv \pm 1 \bmod p$. If it is $+1$, then r is a quadratic residue modulo p. If it is -1, then r is a quadratic nonresidue modulo p.*

If p is an odd prime and r is a quadratic residue modulo p, then the congruence $x^2 \equiv r \bmod p$ has exactly two solutions modulo p, namely, $x \equiv \pm t \bmod p$. If p is an odd prime and r is a multiple of p, then the congruence $x^2 \equiv r \bmod p$ has exactly one solution modulo p, namely, $x \equiv 0 \bmod p$. When $p \equiv 3 \bmod 4$ and r is a quadratic residue modulo p, the solutions are $x \equiv \pm \left(r^{(p+1)/4} \right) \bmod p$. For a general odd prime p, when r is a quadratic residue modulo p, one may solve the congruence $x^2 \equiv r \bmod p$ with this randomized algorithm due to Tonelli. The modular exponentiations are done using MOD-FAST-EXP.

TONELLI (r, p)

1   **choose random** n in $1 < n < p$
2   **if** $n^{(p-1)/2} \equiv +1 \bmod p$
3   **then fail**
4    $(e, f) \leftarrow$ ODD-DIVISOR$(p - 1)$
5    $R \leftarrow r^f \bmod p$
6    $N \leftarrow n^f \bmod p$
7    $j \leftarrow 0$
8   **for** $i \leftarrow 1$ **to** $e - 1$
9   **do**   **if** $(RN^j)^{2^{e-i-1}} \equiv -1 \bmod p$
10   **then** $j \leftarrow j + 2^i$
11   **return** $r^{(f+1)/2} N^{j/2} \bmod p$

See Chapter 7 of [BS96] for a proof that the algorithm is correct. The algorithm TONELLI is similar to Algorithm 2.3.8 of Crandall and Pomerance [CP01]. It fails with probability $1/2$ and runs in $O((\lg p)^4)$ bit operations. In Lines 1–3, the algorithm chooses random $n \bmod p$ until it finds a quadratic nonresidue. Each n has probability $1/2$ of working. Therefore, the expected number of n

one must try is 2. No small upper bound has been proved for the maximum number of n that must be tried, but in practice the algorithm is efficient.

The last topic concerning congruences is that of primitive roots. An integer g is a primitive root modulo m if $\varphi(m)$ is the smallest positive integer e for which $g^e \equiv 1 \bmod m$. Not all positive integers m have a primitive root. Only the numbers 2, 4, p^e, and $2p^e$, where p is any odd prime and $e \geq 1$, have primitive roots. Many computational algorithms require a primitive root, usually modulo a prime. Each prime p has $\varphi(p-1)$ primitive roots. We will explain one way to find a primitive root for a prime in the next section. (See the sentence after Lucas' theorem.)

Given a primitive root g modulo m, the discrete logarithm of a is the number x modulo $\varphi(m)$ such that $g^x \equiv a \bmod m$. As the security of many ciphers depends on the difficulty of computing discrete logarithms, the discussion of how to compute them appears in Chapter 11 of *Algorithms and Theory of Computational Handbook, Second Edition: Special Topics and Techniques* rather than here. See [Wag03] for a more complete treatment.

16.5   Primes and Factoring

A prime is an integer larger than 1 with no exact divisor other than 1 and itself. If an integer $n > 1$ is composite (not prime), then n must be divisible by some integer between 2 and $\sqrt{n}$. This fact suggests the well-known trial division method of testing an integer $n > 1$ for being prime. The function Is-Prime(n) uses trial division and returns 1 if n is prime and 0 if n is composite (or < 2).

Is-Prime (n)
1   **if** $n < 2$
2   **then return** 0
3   **for** $i \leftarrow 2$ **to** $\lfloor \sqrt{n} \rfloor$
4   **do   if** $n \bmod i = 0$
5   **then return** 0
6   **return** 1

The time complexity of Is-Prime(n) is $O(\sqrt{n})$ when n is prime, and possibly faster when n is composite. The sieve of Eratosthenes produces a table of primes between 1 and some limit n. When the algorithm finishes, the bit b_i is 1 if i is prime and 0 if i is composite, for $1 \leq i \leq n$.

Eratosthenes (n)
1   **for** $i \leftarrow 2$ **to** n
2   **do** $b_i \leftarrow 1$
3   $b_1 \leftarrow 0$
4   $p \leftarrow 2$
5   **while** $p^2 \leq n$
6   **do   if** $b_p = 1$
7   **then for** $i \leftarrow p$ **to** $\lfloor n/p \rfloor$
8   **do** $b_{ip} \leftarrow 0$
9   **else** $p \leftarrow p + 1$
10 **return** b

Eratosthenes(n) runs in $O(n \lg \lg n)$ arithmetic operations. (It is customary to state the time complexity of sieves in arithmetic operations. However, Eratosthenes(n) takes $O(n(\lg n) \lg \lg n)$ bit operations.) Pritchard [Pri81] devised the first "sublinear sieve." It finds the primes $< n$ in

$O(n/\lg\lg n)$ arithmetic operations, but $O(n(\lg n)/\lg\lg n)$ bit operations. The great efficiency of sieve algorithms is the reason why there are no large published tables of primes—it is just too easy to make them.

Some important computational problems about primes include the following:

1. Determine whether n is prime or composite.
2. Find the next prime larger than x.
3. Compute $\pi(x)$, the number of primes $\leq x$.
4. Find the nth prime.

In many computational problems, such as computing $\varphi(n)$, we need to know the factorization of an integer into a product of primes. Therefore, we add this computational problem to the list aforementioned.

5. Find the prime factorization of n.

All five problems received much attention during the past century. One hundred years ago, IS-PRIME was about the best one could do to solve Problem 1. Early advances introduced the notion of probable prime. We call an integer $n > 1$ a probable prime to base b if $\gcd(n, b) = 1$ and $b^{n-1} \equiv 1 \bmod n$. By Fermat's theorem, every prime is a probable prime. Composite probable primes are rare. The two parts of the definition of probable prime are easy to test with the algorithms EUCLID(n, b) and MOD-FAST-EXP$(b, n-1, n)$. If a large odd integer n is a probable prime to some base b, then it almost certainly is prime. If one wants more assurance that n is prime, one might try several different bases in probable prime tests. However, the increase in assurance is not as great as one might first guess because of the existence of Carmichael numbers, which are composite probable primes n to every base b with $\gcd(n, b) = 1$. It was proved recently that there are infinitely many Carmichael numbers, but they are rare. Stronger probable prime tests were devised that are as fast as probable prime tests and that a composite number n cannot satisfy for more than $n/4$ bases $b < n$.

Strong probable prime tests efficiently separate primes from composites without the complication of Carmichael numbers and with very rare failures. Industrial-grade primes for use in cryptography are often chosen as strong probable primes. When one wants to *prove* that a probable prime really is prime, one might use a theorem like this one.

THEOREM 16.7   *(Lucas, 1876). Suppose $n > 1$ is a probable prime to base b and $b^{(n-1)/q} \not\equiv 1 \bmod n$ holds for every prime factor q of $n - 1$. Then n is prime.*

One nice feature of Lucas' theorem is that, if it is used to prove that n is prime, then b is a primitive root modulo n. Lucas' theorem requires knowledge of the prime factorization of $n - 1$, which is often not available. Later, Pocklington, Brillhart, Lehmer, Selfridge, Williams, and others found similar theorems that require only partial factorization of $n - 1$, $n + 1$, and even of other numbers like $n^2 + 1$.

Pratt [Pra75] used Lucas' theorem to show that every prime has a succinct certificate of its primality. Suppose we know that 2, 3, 5, 7 are all the primes < 10. To form the certificate for a prime $n > 10$, list the prime factors of $n - 1$ and a primitive root b modulo n. For each prime $n_1 > 10$ in the list, list all the prime factors of $n_1 - 1$ and state a primitive root for n_1. Continue this procedure recursively until all primes are < 10. The certificate has the form of a tree rooted at n. Each node contains a prime and a primitive root of that prime. The children of the node with the prime p are nodes for each prime factor of $p - 1$. The leaves are nodes for primes <10. A certificate for the primality of n can be checked in polynomial time because Pratt showed that the number of nodes is $< \lg n$. It should be noted that there is no known fast way to find a succinct certificate for n.

In the twentieth century many authors devised other tests for primality. Some were probabilistic and fast. Others were deterministic but either slow or depended on an unproved, plausible hypothesis.

By 1983, there were deterministic tests that almost ran in polynomial time. Finally, in 2002, Agrawal, Kayal, and Saxena gave a deterministic algorithm that decides whether n is prime or composite in polynomial time in $\log n$. (It took 2 years for their article [AKS04] to appear in print.) The running time of their test, which uses elegant and fairly simple mathematics, is $O((\log n)^{7.5})$. While their result is a great theoretical achievement, their algorithm is not as fast in practice as some older tests. The current champion proof of the primality of a large integer, not of special form, was done by the elliptic curve test we will describe later and which is probabilistic.

For numbers of special form, such as Mersenne numbers $M_p = 2^p - 1$ and Fermat numbers $F_k = 2^{2^k} + 1$, fast deterministic polynomial time primality tests have long been known. For example, Pepin proved that F_k is prime if and only if $3^{(F_k-1)/2} \equiv -1 \bmod F_k$, and this condition is easy to check with MOD-FAST-EXP. Since the exponent $e = (F_k - 1)/2 = 2^{2^k-1}$ is a power of 2, Line 6 of that algorithm is executed only once. As for Mersenne numbers, if M_p could be prime, then p must be prime. Lucas and Lehmer proved that if one defines $S_1 = 4$ and $S_{k+1} = (S_k^2 - 2) \bmod M_p$ for $k \geq 1$, then M_p is prime if and only if $S_{p-1} \equiv 0 \bmod M_p$. Like the computation for Pepin's test, this one is dominated by squaring large integers.

To find the next prime greater than a given number x, try the integers greater than x in turn. Most of them will be obviously not prime because they have small prime factors. Use one of the primality tests mentioned earlier to test the remaining integers for being prime. The average distance between consecutive primes near x is $\ln x$, and this is the expected number of integers, including those obviously composite, that one must try in order to find the first prime $>x$. It is conjectured that the maximum distance between consecutive primes near x is $O((\ln x)^2)$. However, all one can *prove* is that this distance is $O(x^\alpha)$ for some $0 < \alpha < 1$. The smallest value of α for which this statement has been proved is $\alpha = 0.535$. If the Riemann Hypothesis holds, then one could take $\alpha = 0.5$.

The simple form of the Prime Number Theorem tells us that $\pi(x)$, the number of primes $\leq x$, is approximately $x/\ln x$. In fact, it asserts that

$$\lim_{x \to \infty} \frac{\pi(x)}{x/\ln x} = 1.$$

To estimate the difference between $\pi(x)$ and $x/\ln x$, and find a closer approximation to $\pi(x)$, one studies the Riemann zeta function, defined for $s > 1$ by

$$\zeta(s) = \sum_{n=1}^{\infty} \frac{1}{n^s} = \prod_p \left(1 - \frac{1}{p^s}\right)^{-1}.$$

(The product extends over all primes p.) One can extend the definition of $\zeta(s)$ to all complex numbers $s \neq 1$. The key property of the Riemann zeta function for finding a good approximation to $\pi(x)$ is the location of its zeros. It is known that $\zeta(s)$ has infinitely many zeros with real part between 0 and 1, and these are the ones that determine the value of $\pi(x)$. If all of these zeros have real part exactly equal to $1/2$, then we could determine $\pi(x)$ essentially with an error not more than $O(\sqrt{x} \ln x)$, and this is the best approximation for $\pi(x)$ that we could hope to give. The Riemann Hypothesis is the conjecture that all complex zeros of $\zeta(s)$ with real part between 0 and 1 have real part equal to $1/2$. Extensive computations by many people during the past century have verified the hypothesis for billions of these zeros.

Thus, the Riemann Hypothesis is involved in the third problem about primes. It is also connected to the fourth problem. It follows easily from the Prime Number Theorem that the nth prime is approximately $n \ln n$. We could give a better approximation if we knew more about the location of the zeros of $\zeta(s)$ with real part between 0 and 1, and the best possible approximation if the Riemann Hypothesis were known to be true.

We do not know efficient algorithms for computing either $\pi(n)$ or the nth prime exactly when n is large. In each case, the fastest known algorithm has time complexity $\Omega(\sqrt{n})$. Various inequalities and approximate formulas for both numbers depend on the location of (at least some of) the zeros of $\zeta(s)$ with real part between 0 and 1.

For example, Rosser and Schoenfeld [RS62] proved that

$$\frac{n}{\ln n}\left(1 + \frac{1}{2\ln n}\right) < \pi(n) < \frac{n}{\ln n}\left(1 + \frac{3}{2\ln n}\right)$$

for all $n \geq 59$. Rosser [Ros39] proved in 1939 that the nth prime is always $> n \ln n$. For $n \geq 6$ it is always $< n(\ln n + \ln \ln n)$. Let $\mathrm{li}(x) = \int_0^x \frac{dy}{\ln y}$ denote the logarithmic integral. Then $\mathrm{li}(x)$ is a better approximation to $\pi(x)$ than $x/\ln x$. There is an infinite sequence of values of x for which

$$\pi(x) - \mathrm{li}(x) > \frac{\sqrt{x}}{\ln x}$$

and an infinite sequence of values of x for which

$$\pi(x) - \mathrm{li}(x) < -\frac{\sqrt{x}}{\ln x}.$$

Schoenfeld [Sch76] proved that if the Riemann Hypothesis holds, then

$$|\pi(x) - \mathrm{li}(x)| < \frac{\sqrt{x}\ln x}{8\pi}$$

for all $x \geq 2657$. All results in this paragraph depend on the knowledge of some zeros of $\zeta(s)$ with real part $1/2$.

The fifth problem is to factor an integer n. The algorithm IS-PRIME(n) factors a composite n by trial division. If it returns in Line 5, then i is a proper factor of n, that is, one with $1 < i < n$.

In fact, IS-PRIME(n) locates the smallest prime factor i of n. Some integer factoring algorithms locate small prime factors of a composite n before they find larger ones, and their running time increases roughly with the size of the prime factor found. IS-PRIME is an algorithm of this type. Pollard and Williams invented algorithms of this type that are mentioned in Chapter 11 of *Algorithms and Theory of Computational Handbook, Second Edition: Special Topics and Techniques*. Their running times are proportional to the largest prime factors of $p - 1$ and $p + 1$, respectively, where p is the prime factor found. Another important algorithm of this type is the Elliptic Curve Method, ECM, described in Section 16.7.

Other integer factoring algorithms, like the Quadratic and Number Field Sieves, described in Chapter 11 of *Algorithms and Theory of Computational Handbook, Second Edition: Special Topics and Techniques*, have time complexity which depends only on the size of the number n to be factored and not on the size of the factors discovered. The Number Field Sieve is currently the fastest known algorithm for factoring a large integer with no small prime divisor. The RSA cryptosystem depends on the difficulty of factoring integers, and provides one reason for the great interest in this problem.

16.6 Diophantine Equations

One major topic of number theory is the solution of equations in whole numbers, called Diophantine equations. Hundreds of them have been studied. We mention here only a few of them. There is no general method for solving Diophantine equations. In many cases, the equations were first solved numerically, the form of the solutions was noted, conjectures were made about the solutions, and finally theorems were proved about the solutions.

Pell's equation is $x^2 - dy^2 = \pm 1$, where $d > 0$ is an integer parameter and x, y are the integer unknowns. From the point of view of analytic geometry the graph is a hyperbola in the x–y plane.

We want the points (x, y) on the graph whose coordinates are integers. Suppose d is not a square. Expand $\sqrt{d}$ in a simple continued fraction. That is, execute CONTINUED-FRACTION($\sqrt{d}$) to find the first k terms of the continued fraction for $\sqrt{d}$:

$$[q_0, q_1, q_2, q_3, \ldots, q_k] = \frac{x_k}{y_k},$$

say, where x_k and y_k are relatively prime positive integers. It turns out that the sequence $q_1, q_2, \ldots$ is periodic, and if the period length is n, then the solutions of $x^2 - dy^2 = \pm 1$ with positive x and y are $x = x_{jn-1}$ and $y = y_{jn-1}$ for $j = 1, 2, 3, \ldots$. If n is even, then all solutions have $+1$ and there are no solutions with -1. If n is odd, then $x_{n-1}^2 - dy_{n-1}^2 = -1$ and the successive solutions alternate between $+1$ and -1. See Chapter 7 of [BW99] for more about Pell's equation and its colorful history.

The most famous Diophantine equation is $x^n + y^n = z^n$, the subject of Fermat's Last Theorem. Here $n \geq 2$ is a parameter. We seek solutions x, y, z in positive integers having no common factor > 1. When $n = 2$ there is a simple parametric solution: $x = r^2 - s^2$, $y = 2rs$, $z = r^2 + s^2$, where r and s are any integers with $r > s > 0$, $\gcd(r, s) = 1$ and $r + s$ odd. When $n > 2$ the answer is even simpler: there are no solutions. But the proof by A. Wiles [Wil95] is complicated. Before 1995, computational number theorists proved Fermat's Last Theorem exponent-by-exponent using a theorem of the form, "If a condition involving n holds, then Fermat's Last Theorem is true for exponent n." It suffices to prove it for $n = 4$ and for odd primes n. This was done for all n up to a few million. See [BCEM93], for example.

The Diophantine equation in Fermat's Last Theorem is the special case $k = 2$ of the equation $x_1^n + \cdots + x_k^n = y^n$, expressing an nth power as a sum of k nth powers. Wiles proved that there is no solution in positive integers if $k = 2$ and $n > 2$. Euler conjectured that there is also no solution when $2 < k < n$. Lander and Parkin [LP66] disproved Euler's conjecture by using a computer to find the solution

$$27^5 + 84^5 + 110^5 + 133^5 = 144^5$$

to the case $k = 4$, $n = 5$. Later, Elkies [Elk88] exhibited a parametric solution to the case $k = 3$, $n = 4$, whose first example had numbers in the millions. Then, Frye [Elk88] used a supercomputer to find the smallest solution,

$$95800^4 + 217519^4 + 414560^4 = 422481^4.$$

In many cases, tables of numerical solutions to equations like

$$x_1^n + \cdots + x_j^n = y_1^n + \cdots + y_k^n$$

have led to parametric solutions for the equation. See the survey [LPS67].

Consider the Diophantine equation $n = x^2 + y^2$, where n is a positive integer parameter and x, y are unknown integers. This problem is to express n as a sum of two squares. Not all n can be so expressed. One can show that n can be written as a sum of two squares if and only if there is no prime $q \equiv 3 \bmod 4$ and no odd positive integer e so that q^e exactly divides n. The equation

$$(a^2 + b^2)(c^2 + d^2) = (ac + bd)^2 + (ad - bc)^2$$

reduces the problem to that of expressing a prime $p \equiv 1 \bmod 4$ as $x^2 + y^2$ as follows: this equation lets us express yz as a sum of two squares whenever we can express each of y and z as a sum of two squares. By induction, if we can express each of $k > 1$ integers as a sum of two squares, then we can express their product that way. If n can be expressed as a sum of two squares, then $n = 2^r p_1^{e_1} \cdots p_k^{e_k} q_1^{f_1} \cdots q_m^{f_m}$, where the primes $p_i \equiv 1 \bmod 4$, the primes $q_i \equiv 3 \bmod 4$, the exponents are all positive integers (except $r = 0$ is allowed), and the f_j are even. Since $2 = 1^2 + 1^2$, we can express 2^r as a sum of two squares. If we can express p_i as a sum of two squares, then we can express $p_i^{e_i}$ as a sum of two squares. We can write $q_j^{f_j} = \left(q_j^{f_j/2}\right)^2 + 0^2$ because f_j is even. Finally, the equation allows us to express n as a sum of two squares because it is the product of integers we have already represented that way.

To write a prime $p \equiv 1 \bmod 4$ as a sum of two squares we recommend the elegant solution of Brillhart [Bri72]. The number $p-1$ is a quadratic residue modulo p because $p \equiv 1 \bmod 4$. Use TONELLI$(p-1, p)$ to find an integer m with $m^2 \equiv p-1 \bmod p$ and $0 < m < p/2$. Monitor the recursive calls as EUCLID(p, m) runs. As we mentioned in the discussion of CONTINUED-FRACTION(m/p), there will be a decreasing sequence of integers $a_0 = p > a_1 = m > a_2 > a_3 > \cdots$ so that the successive recursive calls to EUCLID will be EUCLID(a_i, a_{i+1}) for $i = 0, 1, 2, \ldots$. Let a_k be the first number in this sequence less than $\sqrt{p}$. Then $p = x^2 + y^2$ with $x = a_k$ and $y = a_{k+1}$ if $a_2 > 1$, and $x = m$ and $y = 1$ if $a_2 = 1$.

16.7 Elliptic Curves

An abelian group G is a set of elements together with a binary operation $+$ such that: (1) for every a, b in G, $a + b$ is a unique element of G; (2) for all a, b, c in G, $a + (b + c) = (a + b) + c$; (3) there is an identity element e such that $a + e = e + a = a$ for every element a of G; (4) every element a of G has a unique inverse $-a$ in G, with the property $a + (-a) = (-a) + a = e$; and (5) for every pair of elements a, b of G, $a + b = b + a$.

An elliptic curve $E_{a,b}$ is the graph of $y^2 = x^3 + ax + b$, where a, b are parameters and x, y are the variables, together with a special point (not on the graph) called the point at infinity and denoted ∞. Assume that $4a^3 + 27b^2 \neq 0$, which implies that $x^3 + ax + b = 0$ does not have a repeated root. One can define a rule for adding two points P, Q on $E_{a,b}$ so that their sum $P + Q$ is a point on $E_{a,b}$ and this addition operation makes $E_{a,b}$ into an abelian group with identity $e = \infty$. The inverse $-P$ of P is the reflection of P in the x-axis. To define $P + Q$ geometrically when $Q \neq P$ and $Q \neq -P$, draw the straight line through P and Q. When $Q = P$, draw the tangent line to the graph at P. In both cases, the line will intersect the graph in exactly one more point. Reflect the third point across the x-axis to get $P + Q$. When $Q = -P$, we have $P + (-P) = \infty$. There are formulas for computing $P + Q$ in terms of the coordinates of P and Q. One of these formulas computes the slope of the line through P and Q. See Washington [Was03] for more details.

An important application considers an elliptic curve modulo a prime p, in which the coordinates of the points are numbers modulo p and the formulas for adding points are used formally, ignoring their geometrical meaning. The slope of the line in these formulas is the ratio of two numbers modulo p. The slope is computed by multiplying its numerator by the multiplicative inverse of its denominator modulo p, and this inverse is computed by MOD-INVERSE. Therefore, the denominator must be relatively prime to p in order for the slope to be defined. When one tries to add points on an elliptic curve modulo a composite number p, MOD-INVERSE may fail because $u_0 > 1$. But then u_0 is a proper factor of p. This pleasant failure is the basis for Lenstra's elliptic curve factoring algorithm in the following text.

If m is a positive integer, let mP denote $P + P + \cdots + P$, where m copies of P are added. A variation of FAST-EXPONENTIATION computes $Q = mP$ efficiently. The additions in Lines 6 and 7 are on the elliptic curve $E_{a,b}$.

FAST-ADD-POINTS (P, m)

1    $e \leftarrow m$
2    $Q \leftarrow \infty$
3    $R \leftarrow P$
4   **while** $e > 0$

```
5         do  if e is odd
6                then Q ← Q + R
7                R ← R + R
8                e ← ⌊e/2⌋
9         return Q
```

When m is a large integer the coordinates of mP may be complicated rational numbers, even when those of P are small integers. The algorithm is used mostly for an elliptic curve modulo p, in which case the coordinates of the points are integers between 0 and $p - 1$.

Lenstra invented a factoring algorithm using elliptic curves, called ECM for Elliptic Curve Method, and shown in the following text. To factor an integer n with ECM, one first (in Line 1) chooses a random elliptic curve $E_{a,b}$ modulo n and a point on it. This may be done as follows: choose random integers a, x_0, y_0 in $[0, n - 1]$. Let $P = (x_0, y_0)$ and $b = y_0^2 - x_0^3 - ax_0$. Most choices will lead to Line 7. The parameter k tells how much effort to exert trying to factor n. Let q_i denote the ith prime power. (So $q_1 = 2$, $q_2 = 3$, $q_3 = 4$, $q_4 = 5$, $q_5 = 7$, $q_6 = 8$, $q_7 = 9$, etc.) Use FAST-ADD-POINTS(P, q_i) to compute q_iP on $E_{a,b}$ modulo n in Line 8. Either it will add the points correctly or else one of its calls to MOD-INVERSE will produce a proper factor of n.

ECM is an analogue of the Pollard $p - 1$ method described in Chapter 11 of *Algorithms and Theory of Computational Handbook, Second Edition: Special Topics and Techniques.*

```
ECM (n, k)
1    choose random elliptic curve E_{a,b} modulo n and point P ≠ ∞ on it
2    g ← gcd(4a³ + 27b², n)
3    if g = n
4        then fail
5    if g > 1
6        then return factor g of n
7    for i ← 1 to k
8        do P ← q_iP on E_{a,b} or factor n
9    fail
```

ECM(n, k) almost never returns in Line 6. It usually fails in Line 9 or factors n in Line 8. If p is a prime factor of n, and m is the smallest positive integer with $mP = \infty$ on $E_{a,b}$ modulo p, then m is a (random) integer usually near p. Roughly speaking, ECM(n, k) will find the factor p of n if the largest prime factor of m is $\leq q_k$. If the algorithm fails in Line 9, try it again with a different elliptic curve. Let $L(x) = \exp(\sqrt{(\ln x) \ln \ln x})$. If we choose $k = L(p)^{\sqrt{2}/2}$, then the expected total number of elliptic curve additions needed for ECM(n, k) to find p (by running it repeatedly) is $L(p)^{\sqrt{2}}$. Of course, we cannot compute this optimal value for k because we do not know p. But one can show that if k begins at a small value and increases slowly each time ECM(n, k) fails, then the expected total number of elliptic curve additions needed to factor n is $O(L(n))$, the same time complexity as the Quadratic Sieve. When the approximate sizes of the prime factors of n are unknown, one should try ECM first and the Quadratic or Number Field Sieve after ECM fails many times. This is so because most composite numbers that are not specially constructed by multiplying primes of prescribed sizes have one large prime factor and one or more much smaller ones. In this situation, one hopes that ECM will discover the small prime factor(s), and the remaining unfactored part will be prime. But when it is known that n has exactly two prime factors and that they are approximately equal, as is the case for RSA public keys, then it is faster to factor n by the Quadratic or Number Field Sieve than by ECM.

There is an analogue of the theorem of Lucas that uses elliptic curves instead of powers to prove that a large prime is prime.

THEOREM 16.8 *[GK99]. Let $n > 1$ and let E be an elliptic curve modulo n. Suppose there exist a prime number $p > (n^{1/4} + 1)^2$ and a point $P \neq \infty$ on E so that $pP = \infty$ on E. Then n is prime.*

The Goldwasser–Killian primality test applies this theorem recursively to p. It is the current champion for proving primality of a large prime without special form.

The elliptic curve discrete logarithm problem is to find an integer m for which $Q = mP$, given two points P, Q of an elliptic curve. This problem is discussed in Chapter 11 of *Algorithms and Theory of Computational Handbook, Second Edition: Special Topics and Techniques*.

For many purposes one must know the total number N of points on an elliptic curve E modulo a prime p. Hasse's theorem gives the bound $p + 1 - 2\sqrt{p} \leq N \leq p + 1 + 2\sqrt{p}$, but sometimes one needs the exact value of N. Schoof [Sch85] gave a fast algorithm for finding N. He computes $N \bmod q$ for many small primes q and applies the Chinese Remainder Theorem to determine N. See Section 4.5 of Washington [Was03].

16.8 Other Topics

D. H. Lehmer used a computer to prove that every set of seven consecutive integers greater than 36 contains either a prime or a multiple of a prime greater than 41. The proof had $2^{13} = 8192$ cases, one for each divisor of the product of all primes $\leq p_{13} = 41$. In each case, the computer solved a Pell equation and checked a condition for the smallest 21 solutions. See [Leh63] for details.

We mentioned Brillhart's algorithm for writing a prime as a sum of two squares of integers. An integer $n > 0$ is a sum of three squares of integers if and only if it is not of the form $4^a(8b + 7)$ for nonnegative integers a, b. Every positive integer is a sum of four squares. See [MW06] for simple, efficient algorithms for finding these representations.

Guy's book [Guy04] lists hundreds of unsolved problems in number theory. In many of them, computation was either the source of the problem or has shed some light on it. These include: Goldbach's conjecture that every even integer ≥ 4 is a sum of two primes (Sinisalo [Sin93] verified it up to 4×10^{11}), the twin prime conjecture (are there infinitely many pairs $(p, p+2)$ of primes?), the distribution of the difference $p_{n+1} - p_n$ between consecutive primes, solutions of $\varphi(n) = \varphi(n+k)$ for fixed k, variable n (can it hold for infinitely many n?), and whether $\varphi(n)$ ever divides $n-1$ for a composite n.

Another interesting basic question is how to tell whether a positive integer is an exact power of another integer. See [Ber98] for a fast algorithm.

Cohen's book [Coh96] discusses many topics in computational algebraic number theory. These include factoring polynomials, lattice reduction and computing class numbers, class groups, fundamental units, and regulators of quadratic fields.

16.9 Some Applications

One hundred years ago it was proved, the Prime Number Theorem was discovered empirically by examining tables of primes. As we mentioned in Section 16.1, computational number theorists construct tables to seek conjectures, which may be proved later.

To give a recent example, let $r_s(n)$ denote the number of solutions to $n = x_1^2 + \cdots + x_s^2$, the number of ways to write n as a sum of s squares of integers. The following table of $r_5(n)$ appears in an exercise in [MW06]. The entry in Row a, Column b shows $r_5(10a + b)$.

	b									
a	0	1	2	3	4	5	6	7	8	9
0	1	10	40	80	90	112	240	320	200	250
1	560	560	400	560	800	960	730	480	1240	1520
2	752	1120	1840	1600	1200	1210	2000	2240	1600	1680
3	2720	3200	1480	1440	3680	3040	2250	2800	3280	4160
4	2800	1920	4320	5040	2800	3472	5920	4480	2960	3370

Most entries in this table are multiples of 10. Ignoring $r_5(0) = 1$, the entries that are not multiples of 10 are $\equiv 2 \bmod 10$ (at $n = 5, 20$, and 45). Can you guess the pattern? A study of these entries led to this theorem.

THEOREM 16.9 *[Wag07]. If p is an odd prime and n is a positive integer, then*

$$r_p(n) \equiv \begin{cases} 2 \bmod 2p & \text{if } n = pt^2 \text{ for some positive integer } t, \\ 0 \bmod 2p & \text{otherwise.} \end{cases}$$

The algorithm for computing the table of $r_5(n)$ simply evaluates the fifth power (by an variation of FAST-EXPONENTIATION) of the expression

$$1 + 2q^1 + 2q^4 + 2q^9 + 2q^{16} + 2q^{25} + 2q^{36} + 2q^{49},$$

which is the beginning of the generating function for the squares. The coefficient of q^n in the fifth power is $r_5(n)$.

The table [BLS+02] of factors of $b^n \pm 1$ for $2 \leq b \leq 12$ has many applications. An integer like 6 or 28 is called perfect because it equals the sum of its divisors less than it: $6 = 1 + 2 + 3$, $28 = 1 + 2 + 4 + 7 + 14$. Euclid knew that $2^{p-1}(2^p - 1)$ is perfect whenever $2^p - 1$ is prime. Note that $6 = 2^{2-1}(2^2 - 1)$ and $28 = 2^{3-1}(2^3 - 1)$. Euler proved that all even perfect numbers have this form. People continue to search for Mersenne primes $2^p - 1$ today. The factors of numbers $b^n \pm 1$ are used to restrict possible odd perfect numbers. For example, it has been proved that any odd perfect number must exceed 10^{300} and have at least 29 prime factors, one of which exceeds 1,000,000. The table of factors of $2^n - 1$ in [BLS+02] is used in constructing linear feedback shift registers with longest possible period. If the first appearance of the prime number p in the table of factors of $10^n - 1$ is as a factor of $10^k - 1$, then k is the length of the period of $1/p$ as a repeating decimal number. For example, 37 divides 999 but not 9 or 99, so $1/37 = 0.027027027\ldots$ has period 3.

A system of congruence classes $a_i \bmod n_i$ $(1 \leq i \leq k)$ with distinct moduli n_i is called a covering congruence if every integer m satisfies $m \equiv a_i \bmod n_i$ for at least one i. One example is: 0 mod 2, 0 mod 3, 1 mod 4, 5 mod 6, and 7 mod 12. It is not known whether covering congruences exist with arbitrarily large least modulus, or with only odd moduli. Choi [Cho71] has computed one with all moduli > 20. Sierpiński [Sie60] used covering congruences to prove that there are infinitely many integers k for which $k \cdot 2^n + 1$ is composite for every positive integer n. The smallest known value of k with this property is $k = 78,557$, but there are several smaller candidates k which are gradually being eliminated by extensive computation to find values of n for which $p = k \cdot 2^n + 1$ is prime. When such an integer p, with $k < 78,557$, passes a probable prime test, it is easy to prove prime using Lucas' theorem because one can quickly factor $p - 1 = k \cdot 2^n$ completely.

The $3x + 1$ problem, also known by the names Syracuse, Collatz, and Kakutani, iterates the mapping:

$$f(n) = \begin{cases} 3n + 1 & \text{if } n \text{ is odd} \\ n/2 & \text{if } n \text{ is even} \end{cases}$$

that is, beginning with n, one computes $f(n), f(f(n)), f(f(f(n)))$, etc. For example, if one begins with $n = 13$, the sequence of iterates is

$$13, \quad 40, \quad 20, \quad 10, \quad 5, \quad 16, \quad 8, \quad 4, \quad 2, \quad 1, \quad 4, \quad \dots$$

The question is whether every starting value reaches the cycle 4, 2, 1. It does for all starting values $n \le 2 \cdot 10^{12}$. Dozens of papers, many of them computational, study this problem and its variations. See [AL95] and Section E16 of [Guy04].

Fermat's theorem states that if p is an odd prime, then $2^{p-1} \equiv 1 \bmod p$. Wieferich primes are those odd primes for which $2^{p-1} \equiv 1 \bmod p^2$. The first two Wieferich primes are 1093 and 3511. No more are known up to 10^{15}. Wieferich primes were once important for Fermat's Last Theorem.

Factoring, primality testing and FAST-EXPONENTIATION have important applications in cryptography. RSA could be broken if one could factor its public key. The RSA, Pohlig-Hellman, and Diffie-Hellman algorithms all use large primes and FAST-EXPONENTIATION.

16.10 Research Issues and Summary

Faster primality testing is a current research topic. Most fast cryptographic algorithms that need primes choose probable primes because probable prime tests are much faster than true prime tests. But real prime tests are becoming faster and may soon be fast enough for use in programs such as the Secure Sockets Layer. Seek papers that refer to [AKS04] to find the latest work.

Another hot topic is algorithms for factoring large integers. The Cunningham Project [BLS$^+$02] is one benchmark for our current ability to factor integers. The fastest general factoring algorithm is the General Number Field Sieve. It begins by selecting a suitable polynomial that determines how long the rest of the algorithm will take. See Kleinjung [Kle06] for a good way to choose the polynomial and Crandall and Pomerance [CP01] for a description of the algorithm.

Research in Diophantine equations continues. There are many more equations to be solved in whole numbers.

Current work in computational algebraic number theory includes calculations in and about number fields of degree higher than 2. See Cohen [Coh00], for example.

Miller and Takloo-Bighash's book [MTB06] exposes some recent computational work in analytic number theory, particularly the use of random matrix theory (from physics) to attack the Riemann Hypothesis.

We have omitted discussion of many important topics in this short article. Even for the topics we mentioned, the citations are far from complete. See Section 16.11 for general references to this vast area of mathematics.

16.11 Further Information

Research on computational number theory is published often in the journals *Mathematics of Computation* and *Experimental Mathematics*. Such research appears occasionally in number theory journals such as *Acta Arithmetica* and *Journal of Number Theory*, and in many other journals.

Several annual or semiannual conferences with published proceedings deal with cryptanalysis. ANTS, the Algorithmic Number Theory Symposium, began in 1994 and is held biannually. ANTS offers the Selfridge Prize for best paper. It is named after the great computational number theorist John Selfridge. ECC, the Workshop on Elliptic Curve Cryptography, began in 1997 and is held annually. Many of its papers deal with computational issues involving elliptic curves. The Western Number Theory Conference, held annually since the 1960s when it was started by Dick and Emma Lehmer at Asilomar, always has a large session on computational number theory, but it has no published proceedings.

The book by Bach and Shallit [BS96] deals with the algorithms of number theory. Bressoud and Wagon [BW99] and Rosen [Ros04] tell much more about the subject of this chapter. Chapter 4 of [Knu97] deals with the basic algorithms for arithmetic with large integers, greatest common divisors, fast exponentiation, and the like. Wagstaff [Wag03] relates many topics of this chapter to cryptanalysis. Washington [Was03] tells much about computing with elliptic curves.

There are many web pages related to computational number theory. You can find them with a search engine. Start with Number Theory Web at http://www.numbertheory.org/ntw/.

Defining Terms

Diophantine equation: An equation with integer coefficients to be solved with integer values for the variables.

Elliptic curve: The graph of an equation like $y^2 = x^3 + ax + b$ with a way of "adding" points on the graph. It forms a group used in cryptography.

Modular inverse: Multiplicative inverse (reciprocal) a^{-1} of a number a modulo m. It satisfies $aa^{-1} \equiv a^{-1}a \equiv 1 \bmod m$.

Probable prime: An integer which is almost certainly prime because it enjoys a property satisfied by all primes and few composites.

Quadratic residue: An integer r relatively prime to m for which there exists a "square root" x with $x^2 \equiv r \bmod m$.

Relatively prime: Of integers, having no common factor greater than 1.

Riemann Hypothesis: The conjecture that all zeros of the Riemann zeta function $\zeta(s)$ with real part between 0 and 1 have real part $1/2$. If it is true, then we could prove much more about the distribution of primes.

Sieve: a number theoretic algorithm in which, for each prime number p in a list, some operation is performed on every pth entry in an array.

References

[AKS04]   M. Agrawal, N. Kayal, and N. Saxena. PRIMES is in P. *Ann. Math.*, 160:781–793, 2004.

[AL95]   D. Applegate and J. C. Lagarias. Density bounds for the $3x+1$ problem, 1 and 2. *Math. Comp.*, 64:411–426 and 427–438, 1995.

[BCEM93] J. P. Buhler, R. E. Crandall, R. Ernvall, and T. Metsänkylä. Irregular primes and the cyclotomic invariants to four million. *Math. Comp.*, 61:151–153, 1993.

[Ber98]   D. J. Bernstein. Detecting perfect powers in essentially linear time. *Math. Comp.*, 67:1253–1283, 1998.

[BLS^{+}02] J. Brillhart, D. H. Lehmer, J. L. Selfridge, B. Tuckerman, and S. S. Wagstaff, Jr. *Factorizations of $b^n \pm 1$, $b = 2, 3, 5, 6, 7, 10, 11, 12$ up to High Powers.* American Mathematical Society, Providence, Rhode Island, Third edition, 2002. Electronic book available at http://www.ams.org/online_bks/conm22.

[Bra39]   A. Brauer. On addition chains. *Bull. Am. Math. Soc.*, 45:736–739, 1939.

[Bre76]   R. P. Brent. Analysis of the binary Euclidean algorithm. In J. F. Traub, editor, *New Directions and Recent Results in Algorithms and Complexity*, Academic Press, New York, pp. 321–355, 1976.

[Bri72]   J. Brillhart. Note on representing a prime as a sum of two squares. *Math. Comp.*, 26:1011–1013, 1972.

[BS96]   E. Bach and J. Shallit. *Algorithmic Number Theory*. MIT Press, Cambridge, MA, 1996.

[BW99]   D. Bressoud and S. Wagon. *A Course in Computational Number Theory*. Key College Publishing, Emeryville, CA, 1999.

[Cho71]   S. L. G. Choi. Covering the set of integers by congruence classes of distinct moduli. *Math. Comp.*, 25:885–895, 1971.

[Coh96]   H. Cohen. *A Course in Computational Algebraic Number Theory*. Springer-Verlag, New York, 1996.

[Coh00]   H. Cohen. *Advanced Topics in Computational Number Theory*. Springer-Verlag, New York, 2000.

[CP01]   R. Crandall and C. Pomerance. *Prime Numbers: A Computational Perspective*. Springer-Verlag, New York, 2001.

[Elk88]   N. Elkies. On $a^4 + b^4 + c^4 = d^4$. *Math. Comp.*, 51:825–835, 1988.

[GK99]   S. Goldwasser and J. Kilian. Primality testing using elliptic curves. *J. ACM*, 46 (4):450–472, 1999.

[Guy04]   R. K. Guy. *Unsolved Problems in Number Theory*, Third edition, Springer-Verlag, New York, 2004.

[Had93]   J. Hadamard. Résolution d'une question relative aux déterminants. *Bull. Sci. Math.*, 17:235–246, 1893.

[Kar62]   A. A. Karatsuba. Multiplication of multidigit numbers by automatic computers. *Doklady Akad. Nauk SSSR*, 145:293–294, 1962.

[Kle06]   T. Kleinjung. On polynomial selection for the general number field sieve. *Math. Comp.*, 75:2037–2047, 2006.

[Knu97]   D. E. Knuth. *The Art of Computer Programming*, volume 2, *Seminumerical Algorithms*, Third edition, Addison-Wesley, Reading, MA, 1997.

[Leh63]   D. H. Lehmer. Some high-speed logic. *Proc. Symp. Appl. Math.*, 15:141–145, 1963.

[LP66]   L. J. Lander and T. R. Parkin. Counterexample to Euler's conjecture on sums of like powers. *Bull. Am. Math. Soc.*, 72:1079, 1966.

[LPS67]   L. J. Lander, T. R. Parkin, and J. L. Selfridge. A survey of equal sums of like powers. *Math. Comp.*, 21:446–459, 1967.

[Maz07]   G. Maze. Existence of a limiting distribution for the binary GCD algorithm. *J. Discrete Algorithm*, 5:176–186, 2007.

[Mon85]   P. L. Montgomery. Modular multiplication without trial division. *Math. Comp.*, 44:519–521, 1985.

[MTB06]   S. J. Miller and R. Takloo-Bighash. *An Invitation to Modern Number Theory*. Princeton University Press, Princeton, NJ, 2006.

[MW06]   C. J. Moreno and S. S. Wagstaff, Jr. *Sums of Squares of Integers*. Chapman & Hall/CRC Press, Boca Raton, FL, 2006.

[Pra75]   V. Pratt. Every prime has a succinct certificate. *SIAM J. Comput.*, 4:214–220, 1975.

[Pri81]   P. Pritchard. A sublinear additive sieve for finding prime numbers. *Comm. ACM*, 24:18–23, 1981.

[Ros39]   B. Rosser. The n-th prime is greater than $n \log n$. *Proc. London Math. Soc. (2)*, 45:21–44, 1939.

[Ros04]   K. H. Rosen. *Elementary Number Theory and Its Applications*, 5th edition, Addison-Wesley, Reading, MA, 2004.

[RS62]   J. B. Rosser and L. Schoenfeld. Approximate formulas for some functions of prime numbers. *Ill. J. Math.*, 6:64–94, 1962.

[Sch76]   L. Schoenfeld. Sharper bounds for the Chebyshev functions $\theta(x)$ and $\psi(x)$. *Math. Comp.*, 30:337–360, 1976.

[Sch85]   R. Schoof. Elliptic curves over finite fields and the computation of square roots mod p. *Math. Comp.*, 44:483–494, 1985.

[Sie60]   W. Sierpiński. Sur une problème concernant les nombres $k \cdot 2^n + 1$. *Elem. Math.*, 15:73–74, 1960.

[Sin93]   M. K. Sinisalo. Checking the Goldbach conjecture up to $4 \cdot 10^{11}$. *Math. Comp.*, 61:931–934, 1993.

[Ste67]   J. Stein. Computational problems associated with Racah algebra. *J. Comput. Phys.*, 1:397–405, 1967.

[Wag03]   S. S. Wagstaff, Jr. *Cryptanalysis of Number Theoretic Ciphers*. Chapman & Hall/CRC Press, Boca Raton, FL, 2003.

[Wag07]   S. S. Wagstaff, Jr. Congruences for $r_s(n)$ modulo $2s$. *J. Number Theory*, 127:326–329, 2007.

[Was03]   L. C. Washington. *Elliptic Curves: Number Theory and Cryptography*. Chapman & Hall/CRC Press, Boca Raton, FL, 2003.

[Wil95]   A. Wiles. Modular elliptic curves and Fermat's last theorem. *Ann. Math.*, 141 (3):443–551, 1995.

17

Algebraic and Numerical Algorithms

Ioannis Z. Emiris
National and Kapodistrian University of Athens

Victor Y. Pan
City University of New York

Elias P. Tsigaridas
INRIA Sophia Antipolis-Méditerranée

17.1 Introduction

Arithmetic manipulation with matrices and polynomials is a common subject for algebraic (or symbolic) and numerical computing. Typical computational problems in these areas include the solution of a polynomial equation and linear and polynomial systems of equations, univariate and multivariate polynomial evaluation, interpolation, factorization and decompositions, rational interpolation, computing matrix factorization and decompositions (which include various triangular and orthogonal factorizations such as LU, PLU, QR, QRP, QLP, CS, LR, Cholesky factorizations and eigenvalue and singular value decompositions [SVD]), computation of the matrix characteristic and minimal polynomials, determinants, Smith and Frobenius normal forms, ranks, (generalized) inverses, univariate and multivariate polynomial resultants, Newton's polytopes, and greatest common divisors and least common multiples as well as manipulation with truncated series and algebraic sets.

Such problems can be solved based on the error-free algebraic (symbolic) computations with infinite precision. This demanding task is achieved in the present day advanced computer library GMP and computer algebra systems such as Maple and Mathematica by employing various nontrivial computational techniques such as the Euclidean algorithm and continuous fraction approximation, Hensel's and Newton's lifting, Chinese remainder algorithm, elimination and resultant methods, and Gröbner bases computation. The price for the achieved accuracy is the increase of the memory space and computer time supporting the computations.

An alternative numerical approach relies on operations with binary numbers truncated or rounded to a fixed precision. Operating with the IEEE standard floating point numbers represented with double precision enables much faster computations that use much less memory space but requires theoretical and/or experimental study of the affect of the rounding errors on the output. The study involves forward and backward error analysis, linear and nonlinear operators, and uses various advanced techniques from approximation and perturbation theories. If necessary, more costly computations with the extended precision are used to yield uncorrupted output. The resulting algorithms are combined in the high performance libraries and packages of subroutines such as MATLAB®, NAG SMP, LAPACK, ScaLAPACK, ARPACK, PARPACK, and MPSolve.

Combining algebraic and numerical methods frequently increases their power and enables more effective computations. In this chapter, we cover some algebraic and numerical algorithms in the large, popular and highly important areas of matrix computations and root-finding for univariate polynomials and systems of multivariate polynomials. We give some pointers to the bibliography on these and adjacent subjects and in Section 17.5 further references on algebraic and numerical algorithms. The bibliography is huge, and whereever possible we try to cite books, surveys, and comprehensive articles with pointers to further references, rather than the original technical articles. Our expositions in Sections 17.2 and 17.3 largely follow the line of the first surveys in this area in [198,199,204,205].

We state the complexity bounds under the random access machine (RAM) model of computation [2]. In most cases, we assume the arithmetic model, that is, we assign a unit cost to addition, subtraction, multiplication, and division of real numbers, as well as to reading or writing them into a memory location. This model is realistic for computations with a fixed (e.g., the IEEE standard double) precision, which fits the size of a computer word. In this case, the arithmetic model turns into the word model [116]. In other cases, we compute with the extended precision and assume the Boolean or bit model, assigning the unit cost to every Boolean or bit operation. This accounts for both arithmetic operations and the length (precision) of the operands. We denote the bounds on this complexity by $\widetilde{\mathcal{O}}_B(\cdot)$. We always specify whether we use the arithmetic, word, or Boolean model unless this is clear from the context. The $\widetilde{\mathcal{O}}_B(\cdot)$ notation, means that we are ignoring the logarithmic factors.

We write **ops** for "arithmetic operations," "section.name" for "Section section.name," and "log" for "$\log_2$" unless specified otherwise.

17.2 Matrix Computations

Matrix computations are the most popular and highly important area of scientific and engineering computing. Most frequently they are performed numerically, with rounding-off or chopping the input to the IEEE standard double precision. This is mostly assumed in the present section unless specified otherwise.

In a chapter of this size, we must omit or just barely touch many important subjects of matrix computations. The reader can find further material and bibliography in the surveys [194,198,199] and the books [8,11,24,26,74,80,83,127,129,136,208,239,250,258,259,266,277]. For more specific subject areas we further refer the reader to [8,74,127,239,259,266,277] on the eigendecompositions and SVDs [11,74,80,127,136,258,266], on other numerical matrix factorizations [28,160], on the overdetermined and underdetermined linear systems, their least-squares solution, and various other numerical computations with singular matrices, [24,127,240] on parallel matrix algorithms, and to [56,63,85,86,116,119,123,128,200,206,211,213,222,227,264,265,276] on "error-free rational matrix computations," including computations in finite fields, rings, and semirings that output the solutions to linear systems of equations, matrix inverses, ranks, determinants, characteristic and minimum polynomials, and Smith and Frobenius normal forms.

17.2.1 Dense, Sparse, and Structured Matrices: Their Storage and Multiplication by Vectors

An $m \times n$ matrix $A = [\, a_{i,j}, \ i = 0, 1, \ldots, m - 1; \ j = 0, 1, \ldots, n - 1 \,]$, also denoted $[a_{i,j}]_{i,j=0}^{m-1,n-1}$ and $[\mathbf{A}_0, \ldots, \mathbf{A}_{m-1}]$, is a 2-dimensional array, with the (i,j)th entry $[A]_{i,j} = a_{i,j}$ and the jth column $\mathbf{A}_j$. A^T is the transpose of A. A is a column vector $\mathbf{A}_0$ of dimension m, if $n = 1$. A is a row vector of dimension n, if $m = 1$. We write $\mathbf{v} = [v_i]_{i=0}^{n-1}$ to denote an nth dimensional column vector and $\mathbf{w} = A\mathbf{v} = [w_i]_{i=0}^{m-1}, w_i = \sum_{j=0}^{n-1} a_{i,j} v_j, i = 0, \ldots, m - 1$, to denote the matrix-by-vector product. The straightforward algorithm computes such a product by using $(2n - 1)m$ ops. This is the sharp bound for a general (that is, dense unstructured) $m \times n$ matrix, represented with its entries. In actual computations, however, matrices are most frequently special and instead of mn entries can be represented with much fewer parameters.

An $m \times n$ matrix is sparse, if it is filled mostly with zeros, that is, if it has only $\phi << mn$ nonzero entries. An important example is banded matrices $[b_{i,j}]_{i,j}$, whose all nonzero entries lie near the diagonal, so that $b_{i,j} = 0$ unless $|i - j| \leq w$ for a small bandwidth $2w + 1$. This class is generalized to sparse matrices associated with graphs that have families of small separators [121,126,166]. A sparse matrix can be stored economically by using appropriate data structures and can be multiplied by a vector fast, theoretically in $2\phi - m$ ops. Sparse matrices arise in many important applications, in particular, to solving ordinary and partial differential equations (ODEs and PDEs) and graph computations.

Typically, dense structured $n \times n$ matrices can be defined by $O(n)$ parameters and multiplied by a vector by using $O(n \log n)$ or $O(n \log^2 n)$ ops. Such matrices are omnipresent in computations in signal and image processing, coding, ODEs, PDEs, integral equations, particle simulation, and Markov chains. Most popular are Toeplitz matrices $T = [t_{i,j}]_{i,j=0}^{m,n}, t_{i,j} = t_{i+1,j+1}$ for all i and j. Such a matrix is defined by $m + n - 1$ entries of its first row and first column. Toeplitz-by-vector product $T \mathbf{v}$ is defined by "vector convolution" (see Chapter 18). It can be computed by using $O((m + n) \log(m + n))$ ops. Close ties between the computations with Toeplitz and other structured matrices and polynomials enable acceleration in both areas [19,21–25,29,92,100, 183–185,199,208,218,220,228,237].

Similar properties of the Hankel, Bézout, Sylvester, Frobenius (companion), Vandermonde, and Cauchy matrices can be extended to more general classes of structured matrices via associating linear displacement operators. (See [24,208] and Chapter 18 for the details and the bibliography.) Finally, dense structured semiseparable (rank structured) matrices generalize banded matrices and are expressed via $O(n)$ parameters and multiplied by vectors in $O(n)$ ops [269].

17.2.2 Matrix Multiplication and Some Extensions

The straightforward algorithm computes the $m \times p$ product AB of $m \times n$ by $n \times p$ matrices by using $2mnp - mp$ ops, which is $2n^3 - n^2$, if $m = n = p$.

The latter upper bound is not sharp. The subroutines for $n \times n$ matrix multiplication on some modern computers, such as CRAY and Connection Machines, rely on algorithms by Strassen 1969 and Winograd 1971 using $O(n^{2.81})$ ops [127,136]. The algorithms of Coppersmith and Winograd in [64] use at most Cn^ω ops for $\omega < 2.376$ and a huge constant C such that $Cn^\omega < 2n^3$ only for extremely large values of n. Coppersmith and Winograd in [64] combine their techniques of arithmetic progression with various previous advanced techniques. Each of these techniques alone contributes a dramatic increase of the overhead constant that makes the resulting algorithms practically noncompetitive. The only exception is the technique of trilinear aggregating that alone supports the exponent 2.7753 (see [161,194]). The recent practical numerical algorithms in [145] rely on this technique. For matrices of reasonable sizes, they use about as many ops as the Strassen's and Winograd's algorithms but need less memory space and are more stable numerically.

One can multiply a pair of $n \times n$ structured matrices in nearly linear arithmetic time, namely, by using $O(n \log n)$ or $O(n \log^2 n)$ ops, where both input and output matrices are represented via their short generator matrices having $O(n)$ entries (see [24,208] or "structured matrices" in Chapter 18).

If the input values are reasonably bounded integers, then matrix multiplication (as well as vector convolution in Chapter 18) can be reduced to a single multiplication of two longer integers, by means of the techniques of binary segmentation (cf. [24, Examples 3.9.1–3.9.3; 195, Section 40; 198]). The Boolean cost of the computations does not decrease, but the techniques can be practically useful where the two longer integers still fit the computer precision.

Many fundamental matrix computations can be reduced to $O(\log n)$ or a constant number of $n \times n$ matrix multiplications [24, Chapter 2]. This includes the evaluation of det A, the **determinant** of an $n \times n$ matrix A; its inverse A^{-1} (where det $A \neq 0$); the coefficients of its **characteristic polynomial** $c_A(x) = \det(xI - A)$ and minimal polynomial $m_A(x)$, for a scalar variable x; the Smith and Frobenius normal forms; the rank, rank A; the solution vector $\mathbf{x} = A^{-1}\mathbf{v}$ to a nonsingular linear system of equations $A\mathbf{x} = \mathbf{v}$; various orthogonal and triangular factorizations of the matrix A, and a submatrix of A having the maximal rank, as well as some fundamental computations with singular matrices.

Furthermore, similar reductions to matrix multiplication are known for some apparently distant combinatorial and graph computations such as computing the transitive closure of a graph [2], computing all pairs with shortest distances in graphs [24, p.222] and pattern recognition. Consequently, all these operations use $O(n^{\omega})$ ops where theoretically $\omega < 2.376$ [2, Chapter 6; 24, Chapter 2].

In practice, however, due to the overhead constants hidden in the "O" notation for $\omega < 2.775$ for matrix multiplication, additional overhead for its extensions, the memory space requirements, and numerical stability problems, all these extensions of matrix multiplication use the order of n^3 ops [127]. Nevertheless, the reduction to matrix multiplication is practically important because it allows to employ block matrix algorithms. Although they use the order of n^3 ops, they are performed on multiprocessors much faster than the straightforward algorithms [127,240]. Most recent development in parallelism and pipelining has strongly accentuated the power of block matrix algorithms (called level-three basic linear algebra subprograms (*BLAS*)).

We conclude this section by demonstrating two basic techniques for the extension of matrix multiplication. Hereafter, we denote by 0 the null matrices (filled with zeros) and by I the identity (square) matrices (which have ones on their diagonals and zeros elsewhere).

One of the basic ideas of block matrix algorithms is to represent the input matrix A as a block matrix and to operate with its blocks (rather than with its entries). For example, compute det A and A^{-1} by first factorizing A as a 2×2 block matrix,

$$A = \begin{bmatrix} I & 0 \\ A_{1,0}A_{0,0}^{-1} & I \end{bmatrix} \begin{bmatrix} A_{0,0} & 0 \\ 0 & S \end{bmatrix} \begin{bmatrix} I & A_{0,0}^{-1}A_{0,1} \\ 0 & I \end{bmatrix}$$

where $S = A_{1,1} - A_{1,0}A_{0,0}^{-1}A_{0,1}$. Note that the 2×2 block triangular factors are readily invertible, det $A = (\det A_{0,0}) \det S$ and $(BCD)^{-1} = D^{-1}C^{-1}B^{-1}$, so that the original problems for the input A are reduced to the same problems for the half-size matrices $A_{0,0}$ and S. It remains to factorize them recursively. The northwestern blocks (such as $A_{0,0}$), called leading principal submatrices, must be nonsingular throughout the recursive process, but this property holds for the large and highly important class of positive definite matrices $A = C^T C$, det $C \neq 0$, and can be always achieved by means of symmetrization, pivoting, or randomization [2, Chapter 6; 24, Chapter 2; 208, Sections. 5.5 and 5.6)]

Another basic technique is the computation of the Krylov sequence or Krylov matrix $[B^i \mathbf{v}]_{i=0}^{k-1}$ for an $n \times n$ matrix B and an n-dimensional vector $\mathbf{v}$ [127,129,250]. The straightforward algorithm uses $(2n - 1)n(k - 1)$ ops, which is about $2n^3$, for $k = n$. An alternative algorithm first computes the matrix powers

$$B^2, B^4, B^8, \ldots, B^{2^s}, \qquad s = \lceil \log k \rceil - 1,$$

and then the products of $n \times n$ matrices B^{2^i} by $n \times 2^i$ matrices, for $i = 0, 1, \ldots, s$:

$$
\begin{aligned}
B &\quad \mathbf{v}, \\
B^2 &\quad [\, \mathbf{v}, B\mathbf{v} \,] = \left[\, B^2\mathbf{v}, B^3\mathbf{v} \,\right], \\
B^4 &\quad \left[\, \mathbf{v}, B\mathbf{v}, B^2\mathbf{v}, B^3\mathbf{v} \,\right] = \left[\, B^4\mathbf{v}, B^5\mathbf{v}, B^6\mathbf{v}, B^7\mathbf{v} \,\right], \\
&\quad \vdots
\end{aligned}
$$

The last step completes the evaluation of the Krylov sequence in $2s + 1$ matrix multiplications, by using $O(n^\omega \log k)$ ops overall.

Special techniques for parallel computation of Krylov sequences for sparse and/or structured matrices A can be found in [20]. According to these techniques, Krylov sequence is recovered from the solution of the associated linear system $(I - A)\, \mathbf{x} = \mathbf{v}$ which is solved fast in the case of a special matrix A.

There are benefits of incorporating block matrix algorithms into Krylov sequences by replacing its basic vector $\mathbf{b}$ with a matrix of an appropriate size.

In the next two subsections, we more closely examine the solution of a linear system of equations, $A\, \mathbf{x} = \mathbf{b}$, which is the most frequent operation in practice of scientific and engineering computing and is highly important theoretically.

17.2.3 Solution of Linear Systems of Equations

General nonsingular linear system of n equations $A\mathbf{x} = \mathbf{b}$ can be solved in $(2/3)n^3 + O(n^2)$ ops by means of Gaussian elimination. One can perform it numerically and (in spite of rounding errors) arrive at an uncorrupted output by applying pivoting, that is, appropriate interchange of the equations (and sometimes also unknowns) to avoid divisions by absolutely smaller numbers. A by-product is factorization of $A = PLU$ (or $A = PLU\,P'$), for lower triangular matrices L and U^T and permutation matrices P (and P').

For sparse and positive definite linear systems, pivoting can be modified to preserve sparseness during the elimination and thus to yield faster solution [80,83,121,124,125,164,200,227]. Gaussian elimination with the (generalized) nested dissection policy of pivoting requires only $O(s(n)^3)$ ops to solve a sparse positive definite linear system of n equations whose associated graph has a family of separators of diameter $s(n)$. $s(n) = O(\sqrt{n})$ for a large and important class of sparse linear systems arising from discretization of ODEs and PDEs. For general sparse linear systems, $s(n)$ can be as large as n, and we also have no formal proof for any better upper bounds than $O(n^3)$ for Gaussian elimination under any other policy of pivoting. Some heuristic policies (such as Markowitz rule), however, substantially accelerate sparse Gaussian elimination according to ample empirical evidence.

Both Gaussian elimination and the (Block) Cyclic Reduction algorithms use $O(nw^2)$ ops for banded linear systems with bandwidth $O(w)$. This is $O(n)$ where the bandwidth is constant, and similarly for the (dense) semiseparable (rank structured) matrices [269].

Likewise, we can dramatically accelerate Gaussian elimination for dense structured input matrices represented with their short generators, defined by the associated displacement operators. This includes Toeplitz, Hankel, Vandermonde, and Cauchy matrices and matrices with similar structures.

By applying the recursive 2×2 block factorization in the previous subsection (with proper care about preserving matrix structure in the recursive process), we arrive at the MBA divide-and-conquer algorithm (due to Morf 1974/1980 and Bitmead and Anderson 1980) that solves nonsingular structured linear systems of n equations in $O(n \log^2 n)$ ops (see [24,208]), although this computation is prone to numerical stability problems unless the input matrix is positive and definite.

For nonsingular Cauchy-like and Vandermonde-like linear systems of n equations, pivoting preserves matrix structure, and Gaussian elimination can be performed by using $O(n^2)$ ops in numerically stable algorithms. The latter property is extended to linear systems with the Toeplitz/Hankel structures. Pivoting destroys their structure, but their solution can be reduced to Cauchy/Vandermonde-like systems by means of "displacement transformation" (see Chapter 18).

A popular alternative to Gaussian elimination is the iterative solution algorithms such as the conjugate gradient and GMRES algorithms [14,55,127,129,250,271]. They compute sufficiently long Krylov sequences (defined in the previous section), approximate the solution with linear combinations $\sum_i c_i B^i \mathbf{b}$ for appropriate coefficients c_i, and stop where the solution is approximated within a desired tolerance to the output errors. Typically, the algorithms perform every iteration step at the cost of multiplying the input matrix and its transpose by two vectors. This cost is small for structured and sparse matrices. (We can even call a matrix sparse and/or structured if and only if it can be multiplied by a vector fast.)

The multilevel methods [109,172,226] are even more effective for some special classes of linear systems arising in discretization of ODEs and PDEs. In the underlying algebraic process (called the algebraic multigrid), one first aggregates an input linear system, then solves the resulting smaller system, and finally disaggregates the solution into the solution of the original system [181]. The power of this technique is accentuated in its recursive multilevel application.

Generally, iterative methods are highly effective for sparse and/or structured linear systems (and become the methods of choice) as long as they converge fast. Special techniques of preconditioning of the input matrices at a low computational cost enable faster convergence of iterative algorithms for many important special classes of sparse and structured linear systems [14,55,129], and more recently, for quite a general as well as structured linear systems [223,224,236].

Even with all known preconditioning techniques, we cannot deduce competitive upper bounds on the worst-case complexity of iterative solution unless we can readily approximate the inverse M^{-1} of the input matrix M. An approximation X_0 serves well as long as the norm ν of the residual matrix $I - MX_0$ is noticeably less than one. Indeed, in this case, we can rapidly refine the initial approximation, e.g., with Newton's iteration, $X_{i+1} = 2X_i - X_i M X_i$, for which we have $I - MX_{i+1} = (I - MX_i)^2 = (I - MX_0)^{2^{i+1}}$ and, therefore, $||I - MX_{i+1}|| \leq \nu^{2^{i+1}}$ for $i = 0, 1, \ldots$. See more on Newton's iteration in [219,229] and the references therein.

A Newton iteration step uses two matrix multiplications. This is relatively costly for general matrices but takes nearly linear time in n for $n \times n$ structured matrices represented with their short displacement generators (see Chapter 18). The multiplications gradually destroy matrix structure, but some advanced techniques in [208, Chapters 4 and 6; 216,225,231,232] counter this problem.

17.2.4 Error-Free Rational Matrix Computations

Rational matrix computations for a rational or integer input (such as the solution of a linear system and computing the determinant) can be performed with no errors. To decrease the computational cost, one should control the growth of the precision of computing. We refer the reader to [10,119] on some special techniques that achieve this in rational Gaussian elimination. A more fundamental tool of symbolic (algebraic) computing is the reduction of the computations modulo one or several fixed primes or prime powers. Based on such a reduction, the rational or integer output values $z = p/q$ (e.g., the solution vector for a linear system) can be computed modulo a sufficiently large integer m.

Then the desired rational values z are recovered from the values $z \bmod m$ by means of the continued fraction approximation algorithm, which is the Euclidean algorithm applied to integers [116,275], in our case to the integers m and $z \bmod m$. If the output z is known to be an integer lying between $-r$ and r and if $m > 2r$, then the integer z is readily recovered from $z \bmod m$ as follows:

$$z = \begin{cases} z \bmod m & \text{if} \quad z \bmod m < r \\ -m + z \bmod m & \text{otherwise} . \end{cases}$$

For example, if we compute integer determinant, we can choose the modulus m based on the Hadamard's bound. The reduction modulo a prime p can turn a nonsingular matrix A and a nonsingular linear system $A\mathbf{x} = \mathbf{v}$ into singular ones, but this can occur only with a low probability for a random choice of the prime p in a fixed sufficiently large interval as well as, say, for a reasonably large power of two and a random integer matrix [222].

The precision of $\log m$ bits for computing the integer $z \bmod m$ can be excessively large for a large m, but one can first compute this integer modulo k with smaller relatively prime integers $m_1, m_2, \ldots, m_k$ (we call them coprimes) such that $m_1 m_2 \ldots m_k = m$ and then can apply the Chinese remainder algorithm. The error-free computations modulo m_i require the smaller precision of $\log m_i$ bits, whereas the computational cost of the subsequent recovery of the value $z \bmod m$ is dominated by the cost of computing the values $z \bmod m_i$ for $i = 1, \ldots, k$.

For matrix and polynomial computations, there are effective alternative techniques of p-adic (Newton–Hensel) lifting, [116]. Moenck and Carter (1979) and Dixon (1982) have elaborated upon them for solving linear systems of equations and matrix inversion, thus creating symbolic counterparts to well-known numerical techniques of Newton's iteration and iterative refinement in linear algebra.

Newton's lifting begins with a prime p, a larger integer k, an integer matrix M, and its inverse $Q = M^{-1} \bmod p$, such that $I - QM \bmod p = 0$. Then one writes $X_0 = Q$, recursively computes the matrices $X_j = 2X_{j-1} - X_{j-1}MX_{j-1} \bmod (p^{2^j})$ observing that $I - X_j M = 0 \bmod (p^{2^j})$ for $j = 1, 2, \ldots, k$, and finally recovers the inverse matrix M^{-1} from $X_k = M^{-1} \bmod p^{2^k}$.

Hensel's lifting begins with the same input complemented with an integer vector $\mathbf{b}$. Then one writes $\mathbf{r}^{(0)} = \mathbf{b}$, recursively computes the vectors

$$\mathbf{u}^{(i)} = Q\mathbf{r}^{(i)} \bmod p, \quad \mathbf{r}^{(i+1)} = (\mathbf{r}^{(i)} - M\mathbf{u}^{(i)})/p, \quad i = 0, 1, \ldots, k-1,$$

and $\mathbf{x}^{(k)} = \sum_{i=0}^{k-1} \mathbf{u}^{(i)} p^i$ such that $M\mathbf{x}^{(k)} = \mathbf{b} \bmod (p^k)$, and finally recovers the solution $\mathbf{x}$ to the linear system $M\mathbf{x} = \mathbf{b}$ from the vector $\mathbf{x}^{(k)} = \mathbf{x} \bmod (p^k)$.

Newton's and Hensel's lifting are particularly powerful where the input matrices M and M^{-1} are sparse and/or structured. Then a lifting step takes $O(n)$ ops up to a constant, log, or polylog factor. This includes, e.g., Toeplitz, Hankel, Vandermonde, Cauchy, banded and semiseparable matrices, and the matrices whose associated graphs have small separator families. Newton's lifting uses fewer steps, but recursively doubles the precision of computing. Hensel's lifting is performed with the precision in $O(\log p)$ and, as proved in [211,212,221,222,236] enables the solution in nearly optimal time under both Boolean and word models. Moreover, the computations can be performed modulo the powers of two, which allows additional practical benefits of applying binary computations.

17.2.5 Computing the Signs and the Values of Determinants

The value and frequently just the sign of $\det A$, the determinant of a square matrix A, are required in some fundamental geometric and algebraic/geometric computations such as the computation of convex hulls, Voronoi diagrams, algebraic curves and surfaces, multivariate and univariate resultants and Newton's polytopes. The faster numerical methods are preferred as long as the correctness of the output can be certified, which is usually the case in actual geometric and algebraic computations. In the customary arithmetic filtering approach, one applies numerical methods as long as they work and, in the rare cases when they fail, shift to the slower algebraic methods.

Numerical computation of $\det A$ can rely on the factorizations $A = PLUP'$ (see Section 17.2.2) or $A = QR$ [58,127]. One can certify the output sign where the matrix A is well conditioned [235]. The advanced preconditioning techniques in [217] can be employed to improve the conditioning of this matrix.

One can bound the precision of the error-free computations by performing the modulo sufficiently on many reasonably bounded coprime moduli m_i and then recovering the value $\det A \bmod m$, $m = \prod_i m_i$, by applying the Chinese remainder algorithm. Some relevant techniques are elaborated upon in [31]. In particular, to minimize the computational cost, one can select random primes or prime powers m_i recursively until the output value modulo their product stabilizes. This signals that the product m is likely to exceed the unknown value $2|\det A|$, so that $(\det A) \bmod m$ is likely to produce the correct value of $\det A$. Typically for many applications, the value $|\det A|$ tends to be much smaller than Hadamard's upper estimate for $|\det A|$. Then fewer moduli m_i and hence less computations are needed.

In an alternative approach in [1,87,196, Appendix; 197] $\det A$ is recovered as a least common denominator of the rational components of the solutions to linear systems $Ay(i) = b(i)$ for random vectors $b(i)$. The power of Hensel's lifting is employed for computing the solution vectors $y(i)$.

Storjohann in [260,261] advances randomized Newton's lifting to yield $\det A$ directly in the optimal asymptotic Boolean time $O(n^{\omega+1})$ for $\omega < 2.376$.

Wiedemann in 1986 and Coppersmith in 1994, followed by a stream of publications by other researchers, extended the Lanczos and block Lanczos classical algorithms to yield $\det A$. The most costly stages are the computation of the Krylov or block Krylov sequence (for the preconditioned matrix A and random vectors or block vectors) and the solution of a Hankel or block Hankel linear system of equations. To the advantage of this approach versus the other ones, including Storjohann's, the Krylov computations are relatively inexpensive for sparse and/or structured matrices. An important application is the computation of multivariate resultants, which are the determinants of sparse and structured matrices associated with the systems of multivariate polynomial equations. Here, the approach becomes particularly attractive because the structure and sparseness enable us to multiply the matrices by vectors fast but hardly allow any other computational benefits. In [101], the extension of the algorithms to the multivariate determinants and resultants has been elaborated upon and analyzed in some detail.

Even for general matrix A, however, the bottleneck was initially at the stage of the solution of the block Hankel linear system [143]. The structured Hensel lifting has finally become both a theoretical and practical way out, which allowed both to decrease the exponent of the randomized Boolean complexity for computing the determinant of a general matrix from $10/3$ in [143] to $16/5$ and to keep all computational blocks practically valid [210,213].

17.3 Univariate Polynomial Root-Finding, Factorization, and Approximate GCDs

17.3.1 Complexity of Univariate Polynomial Root-Finding

Solution of an nth degree polynomial equation,

$$p(x) = \sum_{i=0}^{n} p_i x^i = p_n \prod_{j=1}^{n} (x - z_j) = 0, \quad p_n \neq 0,$$

that is, the computation of the roots $z_1, \ldots, z_n$ for given coefficients $p_0, \ldots, p_n$, is a classical problem that has greatly influenced the development of mathematics throughout four millennia, since the Sumerian times [204]. The problem remains highly important for the theory and practice of the present day algebraic and algebraic/geometric computation, and dozens of new algorithms for its solution appear every year.

Polynomial root-finding requires an input precision exceeding the output precision by the factor of n, so that we need at least $(n+1)nb/2$ bits (and consequently at least $\lceil (n+1)nb/4 \rceil$ bit operations) to represent the input coefficients $p_0, \ldots, p_{n-1}$ to approximate even a single root of a monic polynomial $p(x)$ within error bound 2^{-b}. To see why, consider, for instance, the polynomial $(x - \frac{6}{7})^n$ and perturb its x-free coefficient by 2^{-bn}. Observe the resulting jumps of the root $x = 6/7$ by 2^{-b}, and observe similar jumps where the coefficients p_i are perturbed by $2^{(i-n)b}$ for $i = 1, 2, \ldots, n - 1$. Therefore, to ensure the output precision of b bits, we need an input precision of at least $(n - i)b$ bits for each coefficient p_i, $i = 0, 1, \ldots, n - 1$.

It can be surprising, but we can approximate all n roots within 2^{-b} by using bn^2 Boolean (bit) operations up to a polylogarithmic factor, that is, we can approximate all roots almost as soon as we write down the input. We achieve this by means of the divide-and-conquer algorithms in [202,204,209] (see [149,191,207,251] on the related works). The algorithms first compute a sufficiently wide root-free annulus A on the complex plane, whose exterior and interior contain comparable numbers of the roots (that is, the same numbers up to a fixed constant factor). Then the two factors of $p(x)$ are numerically computed, that is, $F(x)$, having all its roots in the interior of the annulus, and $G(x) = p(x)/F(x)$, having no roots there. The same process is recursively repeated for $F(x)$ and $G(x)$ until factorization of $p(x)$ into the product of linear factors is computed numerically. From this factorization, approximations to all roots of $p(x)$ are obtained.

It is interesting that both lower and upper bounds on the Boolean time decrease to bn (up to polylog factors) [209], if we only seek the factorization, rather than the roots, that is, if instead of all roots z_j, we compute some scalars a_j and b_j such that $||p(x) - \prod_{j=1}^{n}(a_j x - c_j)|| < 2^b$ for the polynomial norm defined by $|| \sum_i q_i x^i || = \sum_i |q_i|$.

Combining these bounds with a simple argument in [251, Section 20] readily supports the record complexity bound of $\tilde{O}_B((\tau+n)n^2)$ on the bit-complexity of the isolation of real roots of a polynomial of degree n with integer coefficients in a range $(-2^\tau, 2^\tau)$.

17.3.2 Root-Finding via Functional Iterations

The record computational complexity estimates for root-finding can be also obtained based on some functional iteration algorithms, if one assumes their convergence rate based on the ample empirical evidence, although never proved formally. The users seem to accept such an evidence instead of the proof and prefer the latter algorithms because they are easy to program, have been carefully implemented, and allow to tune the precision of the computation to the precision required for every output root (which must be chosen higher for clustered and multiple roots than for the single isolated roots).

For approximating a single root z, the current practical champions are modifications of the Newton's iteration, $z(i + 1) = z(i) - a(i)p(z(i))/p'(z(i))$, $a(i)$ being the step-size parameter [169], Laguerre's method [111,133], and the Jenkins–Traub algorithm [138]. One can deflate the input polynomial via its numerical division by $x - z$ to extend these algorithms to approximating the other roots.

To approximate all roots simultaneously, it is even more effective to apply the Durand–Kerner's (actually Weierstrass') algorithm, which is defined by the following recurrence:

$$z_j(l + 1) = z_j(l) - \frac{p\left(z_j(l)\right)}{p_n \prod_{i \neq j} \left(z_j(l) - z_i(l)\right)}, \quad j = 1, \ldots, n, \quad l = 0, 1, \ldots. \quad (17.1)$$

A simple customary choice for the n initial approximations $z_j(0)$ to the n roots of the polynomial $p(x)$ (see [20] for some effective alternatives) is given by $z_j(0) = Z\, t \exp(2\pi\sqrt{-1}/n)$, $j = 1, \ldots, n$. Here, $t > 1$ is a fixed scalar and Z is an upper bound on the root radius, such that all roots z_j lie in the circle on the complex plane having radius Z and centered in the origin. This holds, e.g., for

$$Z = 2\max_{i<n} \left| p_i/p_n \right|^{\frac{1}{n-i}}. \tag{17.2}$$

For a fixed l and for all j, the computation according to Equation 17.1 is simple. We only need the order of n^2 ops for every l or only $O(n \log^2 n)$ ops with deteriorated numerical stability if we use the fast multipoint polynomial evaluation algorithms [2,24,30,208].

We refer the reader to [26,173,175,176,204] on this and other effective functional iteration algorithms and further extensive bibliography and to [20] for one of the most advanced current implementation MPSolve, based on the so-called Aberth's algorithm (published first by Börsch–Supan and then Ehrlich).

17.3.3 Matrix Methods for Polynomial Root-Finding

Some recent highly effective polynomial root-finders rely on matrix methods. The roots are approximated as the eigenvalues of the associated (generalized) companion matrices, that is, the matrices whose characteristic polynomial is exactly the input polynomial. To the advantage of this approach, it employs numerically stable methods and the excellent software of matrix computations. MATLAB's polynomial root-finder applies the QR algorithm to the Frobenius companion matrix. This is effective because the QR algorithm is the present day champion for the eigenvalue computation. Malek and Vaillancourt in 1995 and then Fortune [104] succeeded by applying the same algorithm to other generalized companion matrices. They improve the approximations recursively by alternating the QR algorithm with Durand–Kerner's. Fortune's highly competitive root-finding package EigenSolve relies on this strategy.

The generalized companion matrices can be chosen structured, e.g., one can choose the Frobenius companion matrix or a diagonal plus rank-one (hereafter *DPR1*) matrix. The algorithms in [19,21, 22,214] exploit this structure to accelerate the eigenvalue computations. At first, this was achieved in [19] based on the inverse (power) Rayleigh–Ritz iteration, which turned out to be also closely related to Cardinal's effective polynomial root-finder (cf. [214]). Then in [21,22], the same idea was pursued based on the QR algorithm. All papers [19,21,22] use linear space and linear arithmetic time per iteration step versus quadratic in the general QR algorithm used by MATLAB and Fortune. We refer the reader to [23,218] on various aspects of this approach and on some related directions for its further study.

17.3.4 Extension to Eigen-Solving

According to ample empirical evidence [121], the cited structured eigen-solvers (applied to polynomial root-finding) typically use from 2 to 3 iteration steps to approximate an eigenvalue of a matrix and the associated eigenvectors. For an $n \times n$ DPR1 matrix, this means $O(n^2)$ ops for all eigenvalues and eigenvectors, versus the order of n^3 for general matrix. The paper [215] extends these fast eigen-solvers to generic matrix by defining its relatively inexpensive similarity transform into a DPR1 matrix. (Similarity transforms $A \leftarrow S^{-1}AS$ preserve eigenvalues and allow readily to reconstruct eigenvectors.)

17.3.5 Real Polynomial Root-Finding

In many algebraic and geometric applications, the input polynomial $p(x)$ has real coefficients, and only its real roots must be approximated. When all roots are real, the Laguerre algorithm, its modifications, and some divide-and-conquer matrix methods [25] become highly effective, but all these algorithms do not work that well where the polynomial has also nonreal roots. Frequently, the real roots make up only a small fraction of all roots.

Somewhat surprisingly, the fastest real root-finder in the current practice is still MPSolve, which uses almost the same running time for real roots as for all complex roots.

Some alternative algorithms specialized in approximation of only real roots are based on using the Descartes rule of signs or the Sturm or Sturm–Habicht sequences to isolate all real roots from each other. The record complexity of the isolation is again based on the factorization of a polynomial in the complex domain (see the end of Section 17.3.1).

17.3.6 Extension to Approximate Polynomial GCDs

Polynomial root-finding has been extended in [206] to the computation of approximate univariate polynomial greatest common divisor (GCD) of two polynomials, that is, the GCD of the maximum degree for two polynomials of the same or smaller degrees lying in the ϵ-neighbourhood of the input polynomials for a fixed positive ϵ. When the roots of both polynomials are approximated closely, it remains to compute a maximal matching in the associated bipartite graphs whose two sets of vertices are given by the roots of the two polynomials and whose edges connect the pair of roots that lie near each other [206].

This computational problem is important in control. The Euclidean algorithm is sensitive to input perturbations and can easily fail [95]. Partial remedies rely on other approaches (see the bibliography in [106,206]), notably via computing the singular values of the associated Sylvester (resultant) matrices [65]. These approaches are more sound numerically than the Euclidean algorithm but still treat the GCDs indirectly, via the coefficients, whose perturbation actually affects the GCDs only via its affect on the roots.

17.3.7 Univariate Real Root Isolation

In this section, we cover rational algorithms, that is, exact (error-free) algorithms that operate with rational numbers. Changing the definitions of the previous sections, we consider the polynomial of degree d,

$$f(x) = a_d x^d + \cdots + a_1 x + a_0,$$

with integral coefficients and let $\tau = 1 + \max_{i \leq d}\{\lg |a_i|\}$ denote their maximum bit size.

We isolate a real (possibly multiple) root, if we compute a real line interval with rational endpoints that contain this root and no other real roots. In addition, we may also need to report the multiplicity of this root. Our task is the isolation of all real roots by covering them with disjoint intervals.

It is known, e.g., [177,178,280], that the minimal distance between the roots of f, the separation bound, is at most $d^{-(d+2)/2}(d + 1)^{(1-d)/2}2^{\tau(1-d)}$, or roughly speaking $2^{-\widetilde{\mathcal{O}}(d\tau)}$ provided the coefficients $a_0, a_1, \ldots, a_d$ are integers. Therefore, isolation of all real roots can be achieved via their approximation within less than one half of this bound. This gives us the fastest known solutions, both practically (in terms of actual CPU time) and theoretically (in terms of bit-complexity estimates). In this section, however, we cover isolation by rational algorithms because they are still in use, have long and respected history, and are of independent technical interest.

17.3.7.1 Subdivision Algorithms

The subdivision algorithms mimic binary search and are the most frequently used, among exact algorithms for real root isolation. They repeatedly subdivide an initial interval that contains all real roots until every tested interval contains at most one real root. They differ in the way in which they count the real roots of a polynomial in an interval. Best known are the algorithms *STURM*, *DESCARTES*, and *BERNSTEIN*.

The algorithm *STURM* was introduced by Sturm in 1835 [262] and is the closest to binary search. It starts with a real interval containing all real roots and subdivides it recursively. In each subinterval $[a, b]$, it computes the number of real roots of a polynomial f as $V_a - V_b$ where V_c denotes the number of sign variations in the signed polynomial remainder sequence computed for f and its derivative at a real point c (see also Section 17.4.1). The algorithm subdivides every interval $[a, b]$ that contains more than one real root and reapplies itself on every subinterval. The overall time-complexity is essentially the number of all intervals examined, multiplied by the complexity of evaluating a polynomial remainder sequence at two points. The algorithm is valid even where there exist multiple roots.

One can bound the number of intervals by $\mathcal{O}(d\tau + d\lg d)$ [72,73,82,99]. Then, by using the fast algorithms for the evaluation of the polynomial remainder sequence, [165,242], we obtain an overall complexity of $\widetilde{\mathcal{O}}_B(d^4\tau^2)$ [72,82,99], which also includes the cost of computing the multiplicity of every real root. (cf. the total cost was in $\widetilde{\mathcal{O}}_B(d^7\tau^3)$ in [135].)

DESCARTES algorithm relies on Descartes' rule of sign, which states that the number of sign variations in the coefficient list of a real polynomial f exceeds the number of positive real roots of f by an even number provided there are no multiple roots. In general, this rule overestimates the number of positive real roots. However, when the number of sign variations is zero or one, then we obtain the exact number of positive real roots. In order to count the number of roots of f in a real line interval, we transform the polynomial to another polynomial whose real roots in $(0, \infty)$ correspond to the roots of f in the initial interval. Unless we detect that there is at most one root there, we subdivide the interval and apply the algorithm on each subinterval. It is proved that the subdivision eventually ends up with a polynomial with at most one sign variation over the corresponding interval. The required polynomial transformations can be performed by shifting variable x and rewriting the coefficients in the reverse order.

The complexity of the algorithm is proportional to the number of intervals considered multiplied by the complexity of shifting the polynomial. For fast algorithms implementing polynomial shifts, the reader may refer to [24,115,116]. Correctness and termination of the algorithm rely on Vincent's theorem [270] or on the one and two-circle theorems [3,61,152]. The *DESCARTES* algorithm was presented in its modern form by Collins and Akritas [60], see also [139,146,151].

The *BERNSTEIN* algorithm is also based on Descartes' rule of sign and additionally takes advantage of the good properties of the Bernstein basis polynomial representation. The algorithm was presented by Lane and Riesenfeld [159], but its complexity, as well as its connection to the topological degree computation, was first exhibited in [186]. See also [12,89,140,187] for interesting variants. The complexity of all approaches is $\widetilde{\mathcal{O}}_B(d^6\tau^2)$. More recently, it was proven [90] that the number of steps that both *DESCARTES* and *BERNSTEIN* perform is $\mathcal{O}(d\tau + d\lg d)$, leading to a complexity bound of $\widetilde{\mathcal{O}}_B(d^4\tau^2)$ [90,99]. This holds in the case of multiple roots as well.

17.3.7.2 The Continued Fraction Algorithm

Unlike the subdivision-based algorithms, which bisects a given initial interval, the continued fraction algorithm *CF* computes the continued fraction expansions of the real roots of the polynomial. The first formulation of the algorithm is due to Vincent [270]. His theorem states that repeated transformations of the form $x \mapsto c + \frac{1}{x}$ eventually yields a polynomial with zero (or, respectively, one) sign variation. Then Descartes' rule implies that the transformed polynomial has zero (respectively, one) real root in $(0, \infty)$. If one sign variation is attained, then the inverse transformation computes an isolating interval for the real root. Moreover, the c's that appear in the transformation correspond, hopefully, to the partial quotients of the continued fraction expansion of the real root. However, Vincent's algorithm has exponential complexity [60].

Tsigaridas and Emiris [267] applied some results from the metric theory of the continued fractions to bound the numbers of steps of the *CF* algorithm by $\widetilde{\mathcal{O}}(d\tau)$. This implied an expected complexity in $\widetilde{\mathcal{O}}_B(d^4\tau^2)$. The bound covers the case of multiple roots as well. More recently, Sharma [254] proved that the worst-case complexity of the algorithm is in $\widetilde{\mathcal{O}}_B(d^7\tau^2)$.

To summarize, the arithmetic and bit complexities of *STURM, DESCARTES*, and *BERNSTEIN* are in $\widetilde{\mathcal{O}}(d^2\tau)$ and $\widetilde{\mathcal{O}}_B(d^4\tau^2)$, respectively. The *CF* algorithm has $\widetilde{\mathcal{O}}_B(d^4\tau^2)$ expected and $\widetilde{\mathcal{O}}_B(d^7\tau^2)$ worst-case bit complexity.

There are several open questions concerning the exact algorithms for root isolation. What is the lower bound for the bit complexity of the problem? What is the true expected (arithmetic or bit) complexity of the algorithms? Is the $\widetilde{\mathcal{O}}_B(d^4\tau^2)$ bound optimal for the subdivision algorithms? Does the $\widetilde{\mathcal{O}}_B(d^4\tau^2)$ bound hold for complex root isolation?

17.4 Systems of Nonlinear Equations

Given a system $P = \{p_1(x_1,\ldots,x_n), p_2(x_1,\ldots,x_n),\ldots,p_r(x_1,\ldots,x_n)\}$ of nonlinear polynomials with rational coefficients (each $p_i(x_1,\ldots,x_n)$ is said to be an element of $\mathbf{Q}[x_1,\ldots,x_n]$, the ring of polynomials in $x_1,\ldots,x_n$ over the field of rational numbers), the n-tuple of complex numbers $(a_1,\ldots,a_n)$ is a solution of the system, if $f_i(a_1,\ldots,a_n) = 0$ for each i with $1 \leq i \leq r$. In this section, we explore the problem of solving a well-constrained system of nonlinear equations. We also indicate how an initial phase of exact algebraic computation leads to certain numerical methods that can approximate all solutions; the interaction of symbolic and numeric computation is currently an active domain of research, e.g., [27,98]. We provide an overview and cite references to different symbolic techniques used for solving systems of algebraic (polynomial) equations. In particular, we describe methods involving resultant and Gröbner basis computations.

The Sylvester resultant method is the technique most frequently utilized for determining a common root of two polynomial equations in one variable [150]. However, using the Sylvester method successively to solve a system of multivariate polynomials proves to be inefficient. Successive resultant techniques, in general, lack efficiency as a result of their sensitivity to the ordering of the variables [147]. The question of successive, or iterated resultants, was recently addressed in [43].

It is more efficient to eliminate all the variables together from a set of polynomials, thus, leading to the notion of the multivariate resultant. The three most commonly used multivariate resultant formulations are the Bézout–Dixon [79,96], Sylvester-Macaulay [45,47,167], and the hybrid formulations [77,148]. Concerning the Sylvester-Macaulay type, we shall also emphasize on sparse resultant formulations [46,263].

The theory of Gröbner bases provides powerful tools for performing computations in multivariate polynomial rings. Formulating the problem of solving systems of polynomial equations in terms of polynomial ideals, we will see that a Gröbner basis can be computed from the input polynomial set, thus, allowing for a form of back substitution in order to compute the common roots.

Although not discussed, it should be noted that the characteristic set algorithm can be utilized for polynomial system solving. Ritt [246] introduced the concept of a characteristic set as a tool for studying solutions of algebraic differential equations. In 1984, Wu [279] in search of an effective method for automatic theorem proving, converted Ritt's method to ordinary polynomial rings. Given the aforementioned system P, the characteristic set algorithm transforms P into a triangular form, such that the set of common roots of P is equivalent to the set of roots of the triangular system [147].

Throughout this exposition, we will also see that these techniques used to solve nonlinear equations can be applied to other problems as well, such as computer-aided design, robot kinematics and automatic geometric theorem proving.

17.4.1 The Sylvester Resultant

The question of whether two polynomials $f(x)$, $g(x) \in \mathbf{Q}[x]$,

$$f(x) = f_n x^n + f_{n-1} x^{n-1} + \cdots + f_1 x + f_0,$$
$$g(x) = g_m x^m + g_{m-1} x^{m-1} + \cdots + g_1 x + g_0,$$

have a common root leads to a condition that has to be satisfied by the coefficients of both f and g. Using a derivation of this condition due to Euler, the Sylvester matrix of f and g (which is of order $m + n$) can be formulated. The vanishing of the determinant of the Sylvester matrix, known as the Sylvester resultant, is a necessary and sufficient condition for f and g to have common roots [150].

As a running example let us consider the following system in two variables provided by Lazard [161]:

$$f = x^2 + xy + 2x \quad\quad + y - 1 = 0,$$
$$g = x^2 \quad\quad + 3x - y^2 + 2y - 1 = 0.$$

The Sylvester resultant can be used as a tool for eliminating several variables from a set of equations. Without loss of generality, the roots of the Sylvester resultant of f and g treated as polynomials in y, whose coefficients are polynomials in x, are the x-coordinates of the common roots of f and g. More specifically, the Sylvester resultant of the Lazard system with respect to y is given by the following determinant:

$$\det\left(\begin{bmatrix} x+1 & x^2+2x-1 & 0 \\ 0 & x+1 & x^2+2x-1 \\ -1 & 2 & x^2+3x-1 \end{bmatrix}\right) = -x^3 - 2x^2 + 3x.$$

An alternative matrix formulation named after Bézout yields the same determinant. This formulation is discussed below in the context of multivariate polynomials, in Section 17.4.2.

The roots of the Sylvester resultant of f and g are $\{-3, 0, 1\}$. For each x value, one can substitute the x value back into the original polynomials yielding the solutions $(-3, 1), (0, 1), (1, -1)$.

The method just outlined can be extended recursively, using polynomial GCD computations, to a larger set of multivariate polynomials in $\mathbf{Q}[x_1, \ldots, x_n]$. This technique, however, is impractical for eliminating many variables, due to an explosive growth of the degrees of the polynomials generated in each elimination step.

The Sylvester formulations have led to a subresultant theory, developed simultaneously by G. E. Collins, and W. S. Brown and J. Traub. The subresultant theory produced an efficient algorithm for computing polynomial GCDs and their resultants, while controlling intermediate expression swell [32,59,150].

Polynomial GCD algorithms have been developed that use some kind of implicit representations for symbolic objects and, thus, avoid the computationally costly content and primitive part computations needed in those GCD algorithms for polynomials in explicit representation [76,141,142].

17.4.2 Resultants of Multivariate Systems

The solvability of a set of nonlinear multivariate polynomials can be determined by the vanishing of a generalization of the Sylvester resultant of two polynomials in a single variable. We examine two generalizations, namely, the classical and the sparse resultants. The classical resultant of a system of n homogeneous polynomials in n variables vanishes exactly when there exists a common solution in projective space [67,272]. The sparse (or toric) resultant characterizes solvability over a smaller space which coincides with affine space under certain genericity conditions [120,263]. More general resultants are not analyzed here, see [40,41]. In any case, the main algorithmic question is to construct a matrix whose determinant is the resultant or a nontrivial multiple of it. Macaulay-type formulas give the resultant as the exact quotient of a determinant divided by one of its minors.

Due to the special structure of the Sylvester matrix, Bézout developed a method for computing the resultant as a determinant of order $\max\{m, n\}$ during the eighteenth century. Cayley [54] reformulated Bézout's method leading to Dixon's [79] extension of the bivariate case. This method can be generalized to a set

$$\{p_1(x_1, \ldots, x_n), p_2(x_1, \ldots, x_n), \ldots, p_{n+1}(x_1, \ldots, x_n)\}$$

of $n + 1$ generic polynomials in n variables [96]. The vanishing of the determinant of the Bézout–Dixon matrix is a necessary and sufficient condition for the polynomials to have a nontrivial projective common root, and also a necessary condition for the existence of an affine common root. The Bézout-Dixon formulation gives the resultant up to a multiple, and hence, in the affine case it can happen that the vanishing of the determinant does not necessarily indicate that the equations in question have a common root. A nontrivial multiple, known as the projection operator, can be extracted via a method discussed in [53, theorem 3.3.4]. This article, along with [91], explains the correlation between residue theory and the Bézout–Dixon matrix, which yields an alternative method for studying and approximating all common solutions.

In 1916, Macaulay [167] constructed a matrix whose determinant is a multiple of the classical resultant for n homogeneous polynomials in n variables. The Macaulay matrix simultaneously generalizes the Sylvester matrix and the coefficient matrix of a system of linear equations [67]. As the Dixon formulation, the Macaulay determinant is a multiple of the resultant. Macaulay, however, proved that a certain minor of his matrix divides the matrix determinant so as to yield the exact resultant in the case of generic homogeneous polynomials. Canny [45] has proposed a general method that perturbs any polynomial system and extracts a nontrivial projection operator.

By exploiting the structure of polynomial systems by means of sparse elimination theory [120,263], a matrix formula for computing the sparse resultant of $n + 1$ polynomials in n variables was given by Canny and Emiris [46] and consequently improved in [49,93]. The determinant of the sparse resultant matrix, like the Macaulay and Dixon matrices, only yields a projection operation, not the exact resultant. To address degeneracy issues, two extensions of Canny's perturbation have been proposed in the sparse context [71,244]. D'Andrea [70] extended Macaulay's rational formula for the resultant to the sparse setting, thus defining the sparse resultant as the quotient of two determinants.

Here, sparsity means that only certain monomials in each of the $n + 1$ polynomials have nonzero coefficients. Sparsity is measured in geometric terms, namely, by the Newton polytope of the polynomial, which is the convex hull of the exponent vectors corresponding to nonzero coefficients. The mixed volume of the Newton polytopes of n polynomials in n variables is defined as a certain integer-valued function that bounds the number of affine common roots of these polynomials, according to a theorem of [18]. This remarkable theorem is the cornerstone of sparse elimination. The mixed volume bound is significantly smaller than the classical Bézout bound for polynomials with small Newton polytopes. Since these bounds also determine the degree of the sparse and classical resultants, respectively, the latter has larger degree for sparse polynomials. Last, but not the least, the classical resultant can identically vanish over sparse systems, whereas the sparse resultant does not and, hence, yields information about their common roots. For an example, see [68].

17.4.3 Polynomial System Solving by Using Resultants

Suppose we are asked to find the common roots of a set of n polynomials in n variables $\{p_1(x_1, \ldots, x_n), p_2(x_1, \ldots, x_n), \ldots, p_n(x_1, \ldots, x_n)\}$. By augmenting the polynomial set by a generic linear polynomial [45,48,68], one can construct the u-resultant of a given system of polynomials. The u-resultant is named after the vector of indeterminate u, traditionally used to represent the generic coefficients of the additional linear polynomial. The u-resultant factors into linear factors over the complex numbers, providing the common roots of the given polynomials equations. The u-resultant method relies on the properties of the multivariate resultant, and, hence, can be constructed using either Dixon's, Macaulay's, or sparse formulations. An alternative approach, where we hide a variable in the coefficient field, instead of adding a polynomial, is discussed in [92,96,170].

Consider the previous example augmented by a generic linear form:

$$f_1 = x^2 + xy + 2x \quad + y - 1 = 0,$$
$$f_2 = x^2 \quad + 3x - y^2 + 2y - 1 = 0,$$
$$f_l = \quad ux \quad + vy + w = 0.$$

As described in [47], the following matrix M corresponds to the Macaulay u-resultant of the above system of polynomials, with z being the homogenizing variable:

$$M = \begin{bmatrix} 1 & 0 & 0 & 1 & 0 & 0 & 0 & 0 & 0 & 0 \\ 1 & 1 & 0 & 0 & 1 & 0 & u & 0 & 0 & 0 \\ 2 & 0 & 1 & 3 & 0 & 1 & 0 & u & 0 & 0 \\ 0 & 1 & 0 & -1 & 0 & 0 & v & 0 & 0 & 0 \\ 1 & 2 & 1 & 2 & 3 & 0 & w & v & u & 0 \\ -1 & 0 & 2 & -1 & 0 & 3 & 0 & w & 0 & u \\ 0 & 0 & 0 & 0 & -1 & 0 & 0 & 0 & 0 & 0 \\ 0 & 1 & 0 & 0 & 2 & -1 & 0 & 0 & v & 0 \\ 0 & -1 & 1 & 0 & -1 & 2 & 0 & 0 & w & v \\ 0 & 0 & -1 & 0 & 0 & -1 & 0 & 0 & 0 & w \end{bmatrix}.$$

It should be noted that

$$\det(M) = (u - v + w)(-3u + v + w)(v + w)(u - v)$$

corresponds to the affine solutions $(1, -1)$, $(-3, 1)$, $(0, 1)$, and one solution at infinity.

Resultants can also be applied to reduce polynomial system solving to a regular or generalized eigenproblem (cf. "Matrix Eigenvalues and Singular Values Problems"), thus, transforming the nonlinear question to a problem in linear algebra. This is a classical technique that enables us to approximate all solutions (cf. [6,48,53,92,96]). For demonstration, consider the previous system and its resultant matrix M. The matrix rows are indexed by the following row vector of monomials in the eliminated variables:

$$\mathbf{v} = \left[x^3, x^2 y, x^2, xy^2, xy, x, y^3, y^2, y, 1 \right].$$

Vector $\mathbf{v}M$ expresses the polynomials indexing the columns of M, which are multiples of the three input polynomials by various monomials. Let us specialize variables u and v to random values. Then the matrix M contains a single variable w and is denoted $M(w)$. Solving the linear system

$\mathbf{v}M(w) = \mathbf{0}$ in vector $\mathbf{v}$ and in scalar w is a generalized eigenproblem, since $M(w)$ can be represented as $M_0 + wM_1$, where M_0 and M_1 have numeric entries. If, moreover, M_1 is invertible, we arrive at the following eigenproblem:

$$\mathbf{v}\left(M_0 + wM_1\right) = \mathbf{0} \iff \vec{v}\left(-M_1^{-1}M_0 - wI\right) = \mathbf{0} \iff \mathbf{v}\left(-M_1^{-1}M_0\right) = w\mathbf{v}.$$

For every solution (a, b) of the original system, there is a vector $\mathbf{v}$ among the computed eigenvectors, which we evaluate at $x = a$, $y = b$ and from which the solution can be recovered by means of division (cf. [92]). As for the eigenvalues, they correspond to the values of w at the solutions.

An alternative method for approximating or isolating all real roots of the system is to use the so-called rational univariate representation (RUR) of algebraic numbers [44,248]. This allows us to express each root coordinate as the value of a univariate polynomial, evaluated over a different algebraic number. The latter are all solutions of a single polynomial equation, and can thus be approximated or isolated by the algorithms presented in the preceding sections. The polynomials involved in this approach are derived from the resultant.

The resultant matrices are sparse and have quasi Toeplitz/Hankel structure (also called multilevel Toeplitz/Hankel structure), which enables their fast multiplication by vectors. By combining the latter property with various advanced nontrivial methods of multivariate polynomial root-finding, substantial acceleration of the construction and computation of the resultant matrices and approximation of the system's solutions were achieved in [29,100,101,183–185].

An empirical comparison of the detailed resultant formulations can be found in [97,170]. The multivariate resultant formulations have been used for diverse applications such as algebraic and geometric reasoning [53,170], computer-aided design [154,252], robot kinematics [245,170], computing molecular conformations [94,171] and for implicitization and finding base points [39,57,66,137,170].

17.4.4 Gröbner Bases

Solving systems of nonlinear equations can be formulated in terms of polynomial ideals [13,67,130, 153]. Let us first establish some terminology.

The ideal generated by a system of polynomial equations $p_1, \ldots, p_r$ over $\mathbf{Q}[x_1, \ldots, x_n]$ is the set of all linear combinations

$$(p_1, \ldots, p_r) = \left\{h_1 p_1 + \ldots + h_r p_r \mid h_1, \ldots, h_r \in \mathbf{Q}[x_1, \ldots, x_n]\right\}.$$

The algebraic variety of $p_1, \ldots, p_r \in \mathbf{Q}[x_1, \ldots, x_n]$ is the set of their common roots,

$$V\left(p_1, \ldots, p_r\right) = \left\{(a_1, \ldots, a_n) \in \mathbf{C}^n \mid f_1\left(a_1, \ldots, a_n\right) = \ldots = f_r\left(a_1, \ldots, a_n\right) = 0\right\}.$$

A version of the *Hilbert Nullstellensatz* states that

$$V\left(p_1, \ldots, p_r\right) = \text{the empty set } \emptyset \iff 1 \in \left(p_1, \ldots, p_r\right) \text{ over } \mathbf{Q}[x_1, \ldots, x_n],$$

which relates the solvability of polynomial systems to the ideal membership problem.

A term $t = x_1^{e_1} x_2^{e_2} \ldots x_n^{e_n}$ of a polynomial is a product of powers with $\deg(t) = e_1 + e_2 + \cdots + e_n$. In order to add needed structure to the polynomial ring, we will require that the terms in a polynomial be ordered in an admissible fashion [67,119]. Two of the most common admissible orderings are the **lexicographic order** ($\prec_l$), where terms are ordered as in a dictionary, and the **degree order** ($\prec_d$), where terms are first compared by their degrees with equal degree terms compared lexicographically. A variation to the lexicographic order is the reverse lexicographic order, where the lexicographic order is reversed [73, p. 96].

It is this above-mentioned structure that permits a type of simplification known as polynomial reduction. Much like a polynomial remainder process, the process of polynomial reduction involves subtracting a multiple of one polynomial from another to obtain a smaller degree result [13,67,130,153].

A polynomial g is said to be reducible with respect to a set $P = \{p_1, \ldots, p_r\}$ of polynomials, if it can be reduced by one or more polynomials in P. When g is no longer reducible by the polynomials in P, we say that g is reduced or is a normal form with respect to P.

For an arbitrary set of basis polynomials, it is possible that different reduction sequences applied to a given polynomial g could reduce to different normal forms. A basis $G \subseteq \mathbf{Q}[x_1, \ldots, x_n]$ is a Gröbner basis, if and only if every polynomial in $\mathbf{Q}[x_1, \ldots, x_n]$ has a unique normal form with respect to G. Bruno Buchberger [33–36] showed that every basis for an ideal $(p_1, \ldots, p_r)$ in $\mathbf{Q}[x_1, \ldots, x_n]$ can be converted into a Gröbner basis $\{p_1^*, \ldots, p_s^*\} = GB(p_1, \ldots, p_r)$, concomitantly designing an algorithm that transforms an arbitrary ideal basis into a Gröbner basis. Another characteristic of Gröbner bases is that by using the above mentioned reduction process we have

$$g \in \left(p_1, \ldots, p_r\right) \iff \left(g \bmod p_1^*, \ldots, p_s^*\right) = 0.$$

Further, by using the Nullstellensatz, it can be shown that $p_1, \ldots, p_r$ viewed as a system of algebraic equations is solvable, if and only if $1 \notin GB(p_1, \ldots, p_r)$.

Depending on which admissible term ordering is used in the Gröbner bases construction, an ideal can have different Gröbner bases. However, an ideal cannot have different (reduced) Gröbner bases for the same term ordering.

Any system of polynomial equations can be solved using a lexicographic Gröbner basis for the ideal generated by the given polynomials. It has been observed, however, that Gröbner bases, more specifically lexicographic Gröbner bases, are hard to compute [13,119,158,278]. In the case of zero-dimensional ideals, those whose varieties have only isolated points, Faugère et al. [104] outlined a change of basis algorithm which can be utilized for solving zero-dimensional systems of equations. In the zero-dimensional case, one computes a Gröbner basis for the ideal generated by a system of polynomials under a degree ordering. The change of basis algorithm can then be applied to the degree ordered Gröbner basis to obtain a Gröbner basis under a lexicographic ordering. More recently, significant progress has been achieved in the algorithmic realm of Gröbner basis computations by the work of J.-C. Faugère [102,103].

Another way to find all common real roots is by means of RUR; see the previous section or [44,248]. All polynomials involved in this approach can be derived from the Gröbner basis.

A rather recent development concerns the generalization of Gröbner bases to border bases, which contain all the information required for system solving but can be computed faster and seem to be numerically more stable [153,188,257].

Turning to Lazard's example in the form of a polynomial basis,

$$
\begin{aligned}
f_1 &= x^2 & +xy & +2x & & +y & -1,\\
f_2 &= x^2 & & +3x & -y^2 & +2y & -1,
\end{aligned}
$$

one obtains (under lexicographical ordering with $x \prec_l y$) a Gröbner basis in which the variables are triangularized such that the finitely many solutions can be computed via back substitution:

$$
\begin{aligned}
f_1^* &= x^2 & +3x & +2y & -2,\\
f_2^* &= & xy & -x & -y & +1,\\
f_3^* &= & & y^2 & -1.
\end{aligned}
$$

It should be noted that the final univariate polynomial is of minimal degree and the polynomials used in the back substitution will have degree no larger than the number of roots.

As an example of the process of polynomial reduction with respect to a Gröbner basis, the following demonstrates two possible reduction sequences to the same normal form. The polynomial x^2y^2 is reduced with respect to the previously computed Gröbner basis $\{f_1^*, f_2^*, f_3^*\} = GB(f_1, f_2)$ along the following two distinct reduction paths, both yielding $-3x - 2y + 2$ as the normal form.

$$x^2 y^2$$
$$f_1^*$$

$$-3xy^2 - 2y^3 + 3y^2$$
$$f_2^* \qquad\qquad f_3^*$$

$$-3xy - 2y^3 - y^2 + 3y \qquad -3x - 2y^3 + 2y^2$$
$$f_2^* \qquad\qquad\qquad f_3^*$$

$$-3x - 2y^3 - y^2 + 3 \qquad -3x - 2y^3 + 2y^2$$
$$f_3^* \qquad\qquad\qquad f_3^*$$

$$-3x - y^2 - 2y + 3$$

$$f_3^*$$
$$-3x - 2y + 2$$

There is a strong connection between lexicographic Gröbner bases and the previously mentioned resultant techniques. For some types of input polynomials, the computation of a reduced system via resultants might be much faster than the computation of a lexicographic Gröbner basis.

In a survey article, Buchberger [35] detailed how Gröbner bases can be used as a tool for many polynomial ideal theoretic operations. Other applications of Gröbner basis computations include automatic geometric theorem proving [146,279], multivariate polynomial factorization and GCD computations [122], computer-aided geometric design [137], polynomial interpolation [156,157], coding and cryptography [5,105], and robotics [106].

17.5 Research Issues and Summary

The present day computations in sciences, engineering, and signal and image processing employ both algebraic and numerical approaches. These two approaches rely on distinct techniques and ideas, but in many cases combining the power of both of them enhances the efficiency of the computations. This is frequently the case in matrix computations and root-finding for univariate polynomials and multivariate systems of polynomial equations. We briefly reviewed these three subjects and gave pointers to the extensive bibliography.

Among numerous interesting and important research directions of the topics in Sections 17.2 and 17.3, we wish to cite computations with structured matrices, including semiseparable matrices and their applications to polynomial root-finding, currently of growing interest, new techniques for preconditioning with many promising extensions (which include the computation of determinants), and polynomial root-finding.

Section 17.4 of this chapter has briefly reviewed polynomial system solving based on resultant matrices as well as Gröbner bases. Both approaches are currently active. This includes applications to small- and medium-size systems. Efficient implementations that handle the nongeneric cases, including multiple roots and nonisolated solutions, are probably the most crucial issue today in relation to resultants. Another interesting direction is algorithmic improvement by exploiting matrix structure, for both resultants and Gröbner bases.

17.6   Further Information

The books and journal special issues [2,7,24,26,30,38,73,77,98,119,150,208,257,282] provide a broader introduction to the general subject and further bibliography.

There are well-known libraries and packages of subroutines for the most popular numerical matrix computations, in particular, [81] for solving linear systems of equations, [113,256], ARPACK, and PARPACK for approximating matrix eigenvalues, and [4] for both of the two latter computational problems. Comprehensive treatment of numerical matrix computations can be found in [127,258, 259], with extensive bibliography, and there are several more specialized books on them [8,11,74, 80,121,129,136,238,258,259,266,277] as well as many survey articles [134,193,199] and thousands of research articles. Further applications to the graph and combinatorial computations related to linear algebra are cited in "Some Computations Related to Matrix Multiplication" and [200].

Special (more efficient) parallel algorithms have been devised for special classes of matrices, such as sparse [124,125,200,227], banded [80,230], and dense structured [24,203,269]. We also refer the reader to [223] on simple but effective extension of Brent's principle for improving the processor and work efficiency of parallel matrix algorithms (with applications to path computations in graphs) and to [125,127,134] on practical parallel algorithms for matrix computations.

On symbolic-numeric algorithms, see the books [24,208,274], surveys [198,199,204,205], a special issue [98], and the bibliography therein.

For the general area of exact computation and the theory behind algebraic algorithms and computer algebra, we refer the reader to [12,37,67,68,73,78,116,119,178,179,278,280,282].

There are a lot of generic software packages for exact computation. We simply mention SYNAPS* [186] a C++ open source library devoted to symbolic and numeric computations with polynomials, algebraic numbers and polynomial systems, NTL[†] a high-performance, portable C++ library providing data structures and algorithms for manipulating vectors, matrices, and polynomials over the integers and over finite fields, CORE,[‡] another C++ library that provides an API for computations with different levels of accuracy in order to support the exact geometric computation (EGC) approach for numerically robust algorithms, and EXACUS [15],[§] also a C++ library with algorithm for curves and surfaces that provides exact methods for solving polynomial equations. A highly efficient software tool is FGB/RS,[¶] which contains algorithms for Gröbner basis computations, the RUR, and computing certified real solutions of systems of polynomial equalities and inequalities. Finaly, let us also mention LINBOX[‖] [84], which is a C++ library that provides exact and high-performance implementations of linear algebra algorithms.

This chapter does not cover the area of polynomial factorization. We refer the interested reader to [75, Chapter 15; 116,162], and the bibliography therein.

The *SIAM Journal on Matrix Analysis and Applications* and *Linear Algebra and Its Applications* are specialized on matrix computations. *Mathematics of Computations* and *Numerische Mathematik* are leading among numerous other good journals on numerical computing.

The *Journal of Symbolic Computation* and *Journal of Computational Complexity* specialize on topics in computer algebra, which are also covered in the *Journal of Pure and Applied Algebra* and less regularly in the *Journal of Complexity*. Other two journals are currently dedicated to the subject of the chapter, namely *Mathematics for Computer Science* and *Applicable Algebra in Engineering,*

---

* http://www-sop.inria.fr/galaad/software/synaps/

† http://www.shoup.net/ntl/

‡ http://cs.nyu.edu/exact/

§ http://www.mpi-inf.mpg.de/projects/EXACUS/

¶ http://fgbrs.lip6.fr/salsa/Software/

‖ http://www.linalg.org

Communication and Computing. Theoretical Computer Science has become more open to algebraic–numerical and algebraic–geometric subjects (see particularly [42,98]).

The annual *International Symposium on Symbolic and Algebraic Computation (ISSAC)* is devoted to computer algebra, whose topics are also presented at the annual Conference *MEGA* and the annual *ACM Conference on Computational Geometry*, and also frequently at various Computer Science conferences, including STOC, FOCS, and SODA.

Among many conferences on numerical computing, most comprehensive ones are organized under the auspieces of SIAM and ICIAM. International Conferences on Symbolic-Numeric Algorithms can be traced back to 1997 (SNAP in INRIA, Sophia Antipolis), and resumed in Xi'an, China, in 2005, Timishiora, Romania, in 2006 (supported by IEEE), and London, Ontario, Canada, in 2007 (supported by ACM).

The topics of Symbolic Numerical Computation are also represented at the conferences on the *Foundations of Computational Mathematics (FoCM)* (met every 3 years, next time in 2008 in Hong Kong) and occasionally at the ISSAC.

Defining Terms

Characteristic polynomial: Shift an input matrix A by subtracting a scaled identity matrix xI. The determinant of the resulting matrix is the characteristic polynomial of the matrix A. Its roots coincide with the eigenvalues of the shifted matrix $A - xI$.

Condition number: A scalar κ derived from a matrix that measures its relative nearness to a singular matrix. Very close to singular means a large condition number, in which case numeric inversion becomes an ill-conditioned problem and OUTPUT ERROR NORM $\approx \kappa$ INPUT ERROR NORM.

Degree order: An order on the terms in a multivariate polynomial; for two variables x and y with $x \prec y$ the ascending chain of terms is $1 \prec x \prec y \prec x^2 \prec xy \prec y^2 \cdots$.

Determinant: A polynomial in the entries of a square matrix with the property that its value is nonzero, if and only if the matrix is invertible. (Determinant of a triangular matrix is the product of its diagonal entries, so that the identity matrix has unit determinant. Determinant of a matrix product is the product of the determinants of the factors. $\det A = -\det B$ if the matrix B is obtained by interchanging a pair of adjacent rows or columns of a matrix A.)

Gröbner basis: Given a term ordering, the Gröbner basis of a polynomial ideal is a generating set of this ideal, such that the (multivariate) division of any polynomial by the basis has a unique remainder.

Lexicographic order: An order on the terms in a multivariate polynomial; for two variables x and y with $x \prec y$ the ascending chain of terms is $1 \prec x \prec x^2 \prec \cdots \prec y \prec xy \prec x^2y \cdots \prec y^2 \prec xy^2 \cdots$.

Matrix eigenvector: A column vector v such that, given square matrix A, $A\mathbf{v} = \lambda\mathbf{v}$, where λ is the matrix eigenvalue corresponding to $\mathbf{v}$. A generalized eigenvector v is such that, given square matrices A, B, it satisfies $A\mathbf{v} = \lambda B\mathbf{v}$. Both definitions extend to a row vector which premultiplies the corresponding matrix.

Mixed volume: An integer-valued function of n-convex polytopes in n-dimensional Euclidean space. Under proper scaling, this function bounds the number of toric complex roots of a well-constrained polynomial system, where the convex polytopes are defined to be the Newton polytopes of the given polynomials.

Ops: Arithmetic operations, i.e., additions, subtractions, multiplications, or divisions; as in flops, i.e., floating point operations.

Resultant: A polynomial in the coefficients of a system of n polynomials with $n + 1$ variables, whose vanishing is the minimal necessary and sufficient condition for the existence of a solution of the system.

Separation bound: The minimum distance between two (complex) roots of a univariate polynomial.

Singularity: A square matrix is singular, if there is a nonzero second matrix such that the product of the two is the zero matrix. Singular matrices do not have inverses.

Sparse matrix: A matrix where many of the entries are zero.

Structured matrix: A matrix whose every entry can be derived by a formula depending on a smaller number of parameters, typically in the order of at most $m + n$ parameters for an $m \times n$ matrix, as opposed to its mn entries. For instance, an $m \times n$ Cauchy matrix has $\frac{1}{s_i - t_j}$ as the entry in row i and column j and is defined by $m + n$ parameters s_i and t_j, $1, \ldots, m, j = 1, \ldots, n$.

Acknowledgments

This material is based on work supported in part by IST Programme of the European Union as a Shared-cost RTD (FET Open) Project under Contract No IST-006413-2 (ACS—Algorithms for Complex Shapes) (first and third authors) and by NSF Grant CCR 9732206 and PSC CUNY Awards 67297-0036 and 68291-0037(second author).

References

1. Abbott, J., Bronstein, M., and Mulders, T., Fast deterministic computations of the determinants of dense matrices. In *Proceedings of the 1999 International Symposium on Symbolic and Algebraic Computation (ISSAC'99)*, pp. 197–204. ACM Press, New York, 1999.

2. Aho, A., Hopcroft, J., and Ullman, J., *The Design and Analysis of Algorithms*. Addison-Wesley, Reading, MA, 1974.

3. Alesina, A. and Galuzzi, M., A new proof of Vincent's theorem. *L'Enseignement Mathématique*, 44, 219–256, 1998.

4. Anderson, E., Bai, Z., Bischof, C., Blackford, S., Demmel, J., Dongarra, J., Du Croz, J., Greenbaum, A., Hammarling, S., McKenney, A., and Sorensen, D., *LAPACK Users' Guide*. 3rd edn. SIAM Publications, Philadelphia, PA, 1999.

5. Augot, D., Bardet, M., and Faugére, J.-C. Efficient decoding of (binary) cyclic codes above the correction capacity of the code using Graöbner bases. In *IEEE International Symposium on Information Theory 2003 (ISIT '03)*. IEEE Press, Washington, DC, 2003.

6. Auzinger, W. and Stetter, H.J., An elimination algorithm for the computation of all zeros of a system of multivariate polynomial equations. In *International Series of Numerical Mathematics*, Vol. 86, pp. 12–30. Birkhaäuser, Basel, 1988.

7. Bach, E. and Shallit, J., *Algorithmic Number Theory*, Volume 1: *Efficient Algorithms*. MIT Press, Cambridge, MA, 1996.

8. Bai, Z., Demmel, J., Dongarra, J., Ruhe, A., and van der Vorst, H., eds., *Templates for the Solution of Algebraic Eigenvalue Problems: A Practical Guide*. SIAM, Philadelphia, PA, 2000.

9. Bailey, D., Borwein, P., and Plouffe, S., On the rapid computation of various polylogarithmic constants. *Math. Comp.*, 66, 903–913, 1997. http://mosaic.cecm.sfu.ca/preprints/1995pp.html, 1995.

10. Bareiss, E.H., Sylvester's identity and multistep integers preserving Gaussian elimination. *Math. Comp.*, 22, 565–578, 1968.

11. Barrett, R., Berry, M.W., Chan, T.F., Demmel, J., Donato, J., Dongarra, J., Eijkhout, V., Pozo, R., Romine, C., and Van Der Vorst, H., *Templates for the Solution of Linear Systems: Building Blocks for Iterative Methods*. SIAM, Philadelphia, PA, 1993.

12. Basu, S., Pollack, R., and Roy, M.-F., *Algorithms in Real Algebraic Geometry*, volume 10 of *Algorithms and Computation in Mathematics*. Springer-Verlag, Berlin, Germany, 2003.

13. Becker, T. and Weispfenning, V., *Gröbner Bases: A Computational Approach to Commutative Algebra*. Springer-Verlag, New York, 1993.

14. Benzi, M., Preconditioning techniques for large linear systems: A survey. *J. Comput. Phys.*, 182, 418–477, 2002.

15. Berberich, E., Eigenwillig, A., Hemmer, M., Hert, S., Kettner, L., Mehlhorn, K., Reichel, J., Schmitt, S., Schömer, E., and Wolpert, N., EXACUS: Efficient and exact algorithms for curves and surfaces. In *ESA*, volume 1669 of *LNCS*, pp. 155–166. Springer-Verlag, Berlin, 2005.

16. Berlekamp, E.R., Factoring polynomials over finite fields. *Bell Syst. Tech. J.*, 46, 1853–1859, 1967. Republished in revised form in: Berlekamp, E.R., *Algebraic Coding Theory*, Chapter 6, McGraw-Hill, New York, 1968.

17. Berlekamp, E.R., Factoring polynomials over large finite fields. *Math. Comp.*, 24, 713–735, 1970.

18. Bernstein, D.N., The number of roots of a system of equations. *Funct. Anal. Appl.*, 9(2), 183–185, 1975.

19. Bini, D.A., Gemignani, L., and Pan, V.Y., Inverse power and Durand/Kerner iteration for univariate polynomial root-finding. *Comp. Math. Appli.*, 47(2/3), 447–459, January 2004. (Also Technical Reports TR 2002 003 and 2002 020, CUNY Ph.D. Program in Computer Science, Graduate Center, City University of New York, 2002.)

20. Bini, D.A. and Fiorentino, G., Design, Analysis, and Implementation of a Multiprecision Polynomial Rootfinder. *Numer. Algorithms*, 23, 127–173, 2000.

21. Bini, D.A., Gemignani, L., and Pan, V.Y., Fast and stable QR eigenvalue algorithms for generalized companion matrices and secular equation. *Numerische Math.*, 3, 373–408, 2005. (Also Technical Report 1470, Department of Mathematics, University of Pisa, Pisa, Italy, July 2003.)

22. Bini, D.A., Gemignani, L., and Pan, V.Y., Improved initialization of the accelerated and robust QR-like polynomial root-finding. *Electron. Trans. Numer. Anal.*, 17, 195–205, 2004.

23. Bini, D. and Pan, V.Y., Parallel complexity of tridiagonal symmetric eigenvalue problem. In *Proceedings of the 2nd Annual ACM-SIAM Symposium on Discrete Algorithms*, pp. 384–393. ACM Press, New York, and SIAM Publications, Philadelphia, PA, 1991.

24. Bini, D. and Pan, V.Y., *Polynomial and Matrix Computations*, Volume 1, *Fundamental Algorithms*. Birkhäuser, Boston, MA, 1994.

25. Bini, D. and Pan, V.Y., Computing matrix eigenvalues and polynomial zeros where the output is real. *SIAM J. Comput.*, 27(4), 1099–1115, 1998.

26. Bini, D. and Pan, V.Y., *Polynomial and Matrix Computations*, Volume 2. Birkhäuser, Boston, MA, to appear.

27. Bini, D.A., Pan, V.Y., and Verschelde, J., eds., Special Issue on symbolic–numerical algorithms, *Theor. Comput. Sci.*, 409(2), 255–268, 2008.

28. Björck, Å., *Numerical Methods for Least Squares Problems*. SIAM, Philadelphia, PA, 1996.

29. Bondyfalat, D., Mourrain, B., and Pan, V.Y., Computation of a specified root of a polynomial system of equations using eigenvectors. *Linear Algebra Appl.*, 319, 193–209, 2000. In *Proceedings of the 1998 International Symposium on Symbolic and Algebraic Computation (ISSAC'98)*, pp. 252–259. ACM Press, New York, 1998.

30. Borodin, A. and Munro, I., *Computational Complexity of Algebraic and Numeric Problems*. Elsevier, New York, 1975.

31. Brönnimann, H., Emiris, I.Z., Pan, V.Y., and Pion, S., Sign determination in residue number systems. *Theoretical Computer Science*, 210(1), 173–197, 1999. In *Proceedings of the 13th Annual ACM Symposium on Computational Geometry*, pp. 174–182. ACM Press, New York, 1997.

32. Brown, W.S. and Traub, J.F., On Euclid's algorithm and the theory of subresultants. *J. ACM*, 18, 505–514, 1971.

33. Buchberger, B., A theoretical basis for the reduction of polynomials to canonical form. *ACM SIGSAM Bull.*, 10(3), 19–29, 1976.

34. Buchberger, B., A note on the complexity of constructing Gröbner-bases. In *Proceedings of the EUROCAL '83*, van Hulzen, J.A., ed., *LNCS*, pp. 137–145. Springer-Verlag, Berlin, 1983.

35. Buchberger, B., Gröbner bases: An algorithmic method in polynomial ideal theory. In *Recent Trends in Multidimensional Systems Theory*, Bose, N.K., ed., pp. 184–232. D. Reidel, Dordrecht, the Netherlands, 1985.

36. Buchberger, B., Ein Algorithmus zum Auffinden der Basiselemente des Restklassenringes nach einem nulldimensionalen Polynomideal. Dissertation, University of Innsbruck, Innsbruck, Austria, 1965.

37. Buchberger, B., Collins, G.E., Loos, R., and Albrecht, R., eds., *Computer Algebra: Symbolic and Algebraic Computation*. 2nd edn. Springer-Verlag, Berlin, 1983.

38. Bürgisser, P., Clausen, M., and Shokrollahi, M.A., *Algebraic Complexity Theory*. Springer, Berlin, 1997.

39. Busé, L., and D'Andrea, C. Inversion of parameterized hypersurfaces by means of subresultants, In *Proceedings of the 2004 International Symposium on Symbolic and Algebraic Computation ISSAC '04*, pp. 65–71. ACM, New York, 2004.

40. Busé, L., Elkadi, M., and Mourrain, B. Generalized resultants over unirational algebraic varieties. *J. Symbolic Comp.*, 29, 515–526, 2000.

41. Busé, L., Elkadi, M., and Mourrain, B. Residual resultant of complete intersection. *J. Pure Appl. Algebra*, 164, 35–57, 2001.

42. Busaé, L., Elkadi, M., and Mourrain, B., eds., Special Issue on Algebraic–Geometric Computations, *Theor. Comp. Science*, 392(1–3), 1–178, 2008.

43. Busaé, L. and Mourrain, B. Explicit factors of some iterated resultants and discriminants. Technical Report, INRIA, Sophia-Antipolis, France, 2007. http://hal.inria.fr/inria-00119287/en/

44. Canny, J., Some algebraic and geometric computations in PSPACE. In *Proceedings of the 20th Annual ACM Symposium Theory of Computing*, pp. 460–467. ACM, New York, 1988.

45. Canny, J., Generalized characteristic polynomials. *J. Symbolic Comput.*, 9(3), 241–250, 1990.

46. Canny, J. and Emiris, I., An efficient algorithm for the sparse mixed resultant. In *Proceedings of the AAECC-10*, Cohen, G., Mora, T., and Moreno, O., eds., volume 673 of *LNCS*, pp. 89–104. Springer-Verlag, Berlin, 1993.

47. Canny, J., Kaltofen, E., and Lakshman, Y., Solving systems of non-linear polynomial equations faster. In *Proceedings of the ACM-SIGSAM 1989 International Symposium Symbolic Algebraic Computation ISSAC '89*, pp. 121–128. ACM, New York, 1989.

48. Canny, J. and Manocha, D., Efficient techniques for multipolynomial resultant algorithms. In *Proceedings of the 1991 International Symposium on Symbolic Algebraic Computation ISSAC '91*, Watt, S.M., ed., pp. 85–95. ACM Press, New York, 1991.

49. Canny, J. and Pedersen, P., An algorithm for the Newton resultant. Technical Report 1394, Computer Science Department, Cornell University, Ithaca, NY, 1993.

50. Cantor, D.G., On arithmetical algorithms over finite fields. *J. Comb. Theory A*, 50, 285–300, 1989.

51. Cantor, D., Galyean, P., and Zimmer, H., A continued fraction algorithm for real algebraic numbers. *Math. Comp.*, 26(119), 785–791, July 1972.

52. Cantor, D.G. and Zassenhaus, H., A new algorithm for factoring polynomials over finite fields. *Math. Comp.*, 36, 587–592, 1981.

53. Cardinal, J.-P. and Mourrain, B., Algebraic approach of residues and applications. In *The Mathematics of Numerical Analysis*, Renegar, J., Shub, M., and Smale, S., eds., volume 32 of *Lectures in Applied Mathematics*, pp. 189–210. AMS, Providence, RI, 1996.

54. Cayley, A., On the theory of elimination. *Cambridge and Dublin Math. J.*, 3, 210–270, 1865.

55. Chen, K., *Matrix Preconditioning Techniques and Applications*. Cambridge University Press, Cambridge, U.K., 2005.
56. Chen, Z. and Storjohann, A., A BLAS based C library for exact linear algebra on integer matrices. In *Proceedings of the 2005 International Symposium on Symbolic Algebraic Computation (ISSAC'05)*, Kauers, M., ed., pp. 92–99. ACM Press, New York, 2005.
57. Chionh, E., Base points, resultants and implicit representation of rational surfaces. Ph.D. Thesis, Department Computer Science, University Waterloo, 1990.
58. Clarkson, K.L., Safe and effective determinant evaluation. In *Proceedings of the 33rd Annual IEEE Symposium on Foundations of Computer Science*, pp. 387–395. IEEE Computer Society Press, Los Alamitos, CA, 1992.
59. Collins, G.E., Subresultants and reduced polynomial remainder sequences. *J. ACM*, 14, 128–142, 1967.
60. Collins, G.E. and Akritas, A., Polynomial real root isolation using Descartes' rule of signs. In *SYMSAC '76*, pp. 272–275. New York, ACM Press 1976.
61. Collins, G.E. and Johnson, J., Quantifier elimination and the sign variation method for real root isolation. In *ISSAC*, pp. 264–271. ACM, New York, 1989.
62. Collins, G.E., and Loos, R., Real zeros of polynomials. In *Computer Algebra: Symbolic and Algebraic Computation*, 2nd edn., Buchberger, B., Collins, G.E., and Loos, R., eds., pp. 83–94. Springer-Verlag, Wien, 1982.
63. Coppersmith, D., Solving homogeneous linear equations over GF(2) via block Wiedemann algorithm. *Math. Comp.*, 62(205), 333–350, 1994.
64. Coppersmith, D. and Winograd, S., Matrix multiplication via arithmetic progressions. *J. Symb. Comput.*, 9(3), 251–280, 1990.
65. Corless, R.M., Gianni, P.M., Trager, B.M., and Watt, S.M., The singular value decomposition for polynomial systems. In *Proceedings of International Symposium on Symbolic Algebraic Computation (ISSAC'95)*, pp. 195–207, ACM Press, New York, 1995.
66. Cox, D., Goldman, R., and Zhang, M. On the validity of implicitization by moving quadrics for rational surfaces with no base points. *J. Symb. Comput.*, 29(3), 419–440, 2000.
67. Cox, D., Little, J., and O'Shea, D. *Ideals, Varieties, and Algorithms*, 2nd edn., *Undergraduate Texts in Mathematics*. Springer, New York, 1997.
68. Cox, D., Little, J., and O'Shea, D. *Using Algebraic Geometry*, 2nd edn., *Graduate Texts in Mathematics*. Springer, New York, 2005.
69. Cuppen, J.J.M., A divide and conquer method for the symmetric tridiagonal eigenproblem. *Numer. Math.*, 36, 177–195, 1981.
70. D'Andrea, C. Macaulay-style formulas for the sparse resultant. *Trans. AMS*, 354, 2595–2629, 2002.
71. D'Andrea, C., and Emiris, I.Z. Computing sparse projection operators. In *Symbolic Computation: Solving Equations in Algebra, Geometry, and Engineering*, Green, E.L., Hosten, S., Laubenbacher, R.C., and Powers, V.A., eds., pp. 121–139. AMS, Providence, RI, 2001.
72. Davenport, J.H., Cylindrical algebraic decomposition. Technical Report 88-10, School of Mathematical Sciences, University of Bath, Bath, U.K., available at: http://www.bath.ac.uk/masjhd/, 1988.
73. Davenport, J.H., Tournier, E., and Siret, Y., *Computer Algebra Systems and Algorithms for Algebraic Computation*. Academic Press, London, 1988.
74. Demmel, J.J.W., *Applied Numerical Linear Algebra*. SIAM Publications, Philadelphia, PA, 1997.
75. Daíaz, A., Emiris, I.Z., Kaltofen, E., and Pan, V.Y., Algebraic Algorithms. In *Handbook of Algorithms and Theory of Computation*, Atallah, M.J., ed., Chapter 15. CRC Press, Boca Raton, FL, 1999.
76. Díaz, A. and Kaltofen, E., On computing greatest common divisors with polynomials given by black boxes for their evaluation. In *Proceedings of the 1995 International Symposium on Symbolic Algebraic Computation (ISSAC '95)*, Levelt, A.H.M., ed., pp. 232–239. ACM Press, New York, 1995.
77. Dickenstein, A. and Emiris, I.Z., Multihomogeneous resultant formulae by means of complexes. *J. Symb. Comp.*, 36, 317–342, 2003.

78. Dickenstein, A. and Emiris, I.Z., eds., *Solving Polynomial Equations: Foundations, Algorithms and Applications*. Volume 14 in *Algorithms and Computation in Mathematics*. Springer-Verlag, Berlin, Germany, 2005.

79. Dixon, A.L., The elimination of three quantics in two independent variables. *Proc. London Math. Soc.*, 6, 468–478, 1908.

80. Dongarra, J.J., Duff, I.S., Sorensen, D.C., and Van Der Vorst, H.A., *Numerical Linear Algebra for High-Performance Computers*, SIAM, Philadelphia, PA, 1998.

81. Dongarra, J., Bunch, J., Moler, C., and Stewart, P. *LINPACK Users' Guide*. SIAM Publications, Philadelphia, PA, 1978.

82. Du, Z., Sharma, V., and Yap, C.K., Amortized bound for root isolation via Sturm sequences. In *International Workshop on Symbolic Numeric Computing*, Wang, D. and Zhi, L., eds., pp. 113–129, Birkhauser, Basel, Switzerland, 2005.

83. Duff, I.S., Erisman, A.M., and Reid, J.K., *Direct Methods for Sparse Matrices*. Clarendon Press, Oxford, U.K., 1986.

84. Dumas, J.-G., Gautier, T., Giesbrecht, M., Giorgi, P., Hovinen, B., Kaltofen, E., Saunders, B.D., Turner, W.J., and Villard, G., LinBox: A generic library for exact linear algebra. In *Proceedings of the 1st International Congres Mathematical Software ICMS 2002*, Cohen, A.M., Gao, X.-S., and Takayama, N., eds., pp. 40–50. World Scientific, Singapore, 2002.

85. Dumas, J.-G., Gautier, T., and Pernet, C., Finite field linear algebra subroutines. In *Proceedings of the International Symposium Symbolic Algebraic Computation (ISSAC'02)*, pp. 63–74. ACM Press, New York, 2002.

86. Dumas, J.-G., Giorgi, P., and Pernet, C., Finite field linear algebra package. In *Proceedings of the International Symposium Symbolic Algebraic Computation (ISSAC'04)*, pp. 118–126. ACM Press, New York, 2004.

87. Eberly, W., Giesbrecht, M., and Villard, G., On computing the determinant and Smith form of an integer matrix. In *Proceedings of the 41st Annual Symposium on Foundations of Computer Science (FOCS'2000)*, pp. 675–685. IEEE Computer Society Press, Los Alamitos, CA, 2000.

88. Eberly, W. and Kaltofen, E., On randomized Lanczos algorithms. In *Proceedings of the 1997 International Symposium Symbolic Algebraic Computation (ISSAC'97)*, Kaüchlin, W., ed., pp. 176–183. ACM Press, New York, 1997.

89. Eigenwillig, A., Kettner, L., Krandick, W., Mehlhorn, K., Schmitt, S., and Wolpert, N., A Descartes algorithm for polynomials with bit-stream coefficients. In *CASC*, Ganzha, V., Mayr, E., and Vorozhtsov, E., eds., volume 3718 of *LNCS*, pp. 138–149. Springer-Verlag, Berlin, Germany, 2005.

90. Eigenwillig, A., Sharma, V., and Yap, C.K., Almost tight recursion tree bounds for the Descartes method. In *ISSAC '06: Proceedings of the 2006 International Symposium on Symbolic and Algebraic Computation*, pp. 71–78. ACM Press, New York, 2006.

91. Elkadi, M. and Mourrain, B., Algorithms for residues and Lojasiewicz exponents. *J. Pure Appl. Algebra*, 153, 27–44, 2000.

92. Emiris, I.Z., On the complexity of sparse elimination. *J. Complexity,* 12, 134–166, 1996.

93. Emiris, I.Z. and Canny, J.F., Efficient incremental algorithms for the sparse resultant and the mixed volume. *J. Symb. Comp.,* 20(2), 117–149, 1995.

94. Emiris, I.Z., Fritzilas, E., and Manocha, D., Algebraic algorithms for conformational analysis and docking. *Int. J. Quantum Chem.*, 106, 190–210, 2005.

95. Emiris, I.Z., Galligo, A., and Lombardi, H., Certified approximate univariate GCDs. (Special issue on algorithms for Algebra) *J. Pure Appl. algebra,* 117 & 118, 229–251, 1997.

96. Emiris, I.Z. and Mourrain, B., Matrices in elimination theory. *J. Symb. Comp.,* 28, 3–44, 1999.

97. Emiris, I.Z. and Mourrain, B., Computer algebra methods for studying and computing molecular conformations. *Algorithmica*, 25, 372–402, 1999.

98. Emiris, I.Z., Mourrain, B., and Pan, V.Y., eds., Special issue on algebraic and numerical algorithms, *Theor. Comp. Science*, 315, 307–672, 2004.

99. Emiris, I.Z., Mourrain, B., and Tsigaridas, E.P., Real algebraic numbers: Complexity analysis and experimentation. In *Reliable Implementations of Real Number Algorithms: Theory and Practice*, Hertling, P., Hoffmann, C., Luther, W., and Revol, N., eds., *LNCS*, Springer-Verlag, Berlin, Germany 2007. also available in www.inria.fr/rrrt/rr-5897.html.

100. Emiris, I.Z. and Pan, V.Y., Symbolic and numeric methods for exploiting structure in constructing resultant matrices. *J. Symb. Comp.*, 33, 393–413, 2002.

101. Emiris I.Z. and Pan, V.Y., Improved algorithms for computing determinants and resultants. *J. Complexity*, 21(1), 43–71, 2005. In *Proceedings of the 6th International Workshop on Computer Algebra in Scientific Computing (CASC '03)*, Mayr, E.W., Ganzha, V.G., and Vorozhtzov, E.V., eds., pp. 81–94, Technische University München, München, Germany, 2003.

102. Faugére, J.-C., A new efficient algorithm for computing Gröbner bases (F4). *J. Pure Appl. Algebra*, 139, 61–88, 1999.

103. Faugére, J.-C., A new efficient algorithm for computing Gröbner bases without reduction to zero (F5). In *Proceedings of the 2002 International Symposium on Symbolic Algebraic Computation (ISSAC '02)*, pp. 75–83. ACM Press, New York, 2002.

104. Faugére, J.-C., Gianni, P., Lazard, D., and Mora, T., Efficient computation of zero-dimensional Gröbner bases by change of ordering. *J. Symb. Comput.*, 16(4), 329–344, 1993.

105. Faugère, J.-C. and Joux, A., Algebraic cryptanalysis of hidden field equation (HFE): Cryptosystems using Gröbner bases. In *Proceedings of the CRYPTO 2003, LNCS*, pp. 44–60, Springer-Verlag, Berlin, Germany, 2003.

106. Faugère, J.-C. and Lazard, D., The combinatorial classes of parallel manipulators. *Mech. Mach. Theory*, 30, 765–776, 1995.

107. Ferguson, H.R.P. and Bailey, D.H., Analysis of PSLQ, an integer relation finding algorithm. Technical Report NAS-96-005, NASA Ames Research Center, 1996.

108. Ferguson, H.R.P. and Forcade, R.W., Multidimensional Euclidean algorithms. *J. Reine Angew. Math.*, 334, 171–181, 1982.

109. Fiorentino, G. and Serra, S., Multigrid methods for symmetric positive definite block Toeplitz matrices with nonnegative generating functions. *SIAM J. Sci. Comput.*, 17, 1068–1081, 1996.

110. Fortune, S., An iterated eigenvalue algorithm for approximating roots of univariate polynomials. *J. Symbolic Comput.*, 33 (5), 627–646, 2002. In *Proceedings of the International Symposium on Symbolic and Algebraic Computation (ISSAC'01)*, pp. 121–128. ACM Press, New York, 2001.

111. Foster, L.V., Generalizations of Laguerre's method: Higher order methods. *SIAM J. Numer. Anal.*, 18, 1004–1018, 1981.

112. Gao, S., Kaltofen, E., May, J., Yang, Z., and Zhi, L., Approximate factorization of multivariate polynomial via differential equations. In *Proceedings of the International Symposium on Symbolic and Algebraic Computaion (ISSAC'04)*, pp. 167–174. ACM Press, New York, 2004.

113. Garbow, B.S. et al., *Matrix Eigensystem Routines: EISPACK Guide Extension*. Springer, New York, 1972.

114. von zur Gathen, J. and Gerhard, J., Arithmetic and factorization over F_2. In *ISSAC 96 Proceedings of the 1996 International Symposium on Symbolic Algebraic Computation*, Lakshman, Y.N., ed., pp. 1–9. ACM Press, New York, 1996.

115. von zur Gathen, J. and Gerhard, J., Fast algorithms for Taylor shifts and certain difference equations. In *ISSAC*, pp. 40–47. ACM Press, New York 1997.

116. von zur Gathen, J. and Gerhard, J., *Modern Computer Algebra*. 2nd edn., Cambridge University Press, Cambridge, U.K., 2003.

117. von zur Gathen, J. and Lücking, T., Subresultants revisited. *Theor. Comput. Sci.*, 1–3(297), 199–239, 2003.

118. von zur Gathen, J. and Shoup, V., Computing Frobenius maps and factoring polynomials. *Comput. Complexity*, 2, 187–224, 1992.

119. Geddes, K.O., Czapor, S.R., and Labahn, G., *Algorithms for Computer Algebra*. Kluwer Academic Press, Oxford, U.K., 1992.

120. Gelfand, I.M., Kapranov, M.M., and Zelevinsky, A.V., *Discriminants, Resultants and Multidimensional Determinants*. Birkhäuser Verlag, Boston, MA, 1994.

121. George, A. and Liu, J.W.-H., *Computer Solution of Large Sparse Positive Definite Linear Systems*. Prentice Hall, Englewood Cliffs, NJ, 1981.

122. Gianni, P. and Trager, B., GCD's and factoring polynomials using Gröbner bases. In *Proceedings of EUROCAL '85, Vol. 2, LNCS*, 204, pp. 409–410. Springer-Verlag, Berlin, Germany, 1985.

123. Giesbrecht, M., Nearly optimal algorithms for canonical matrix forms. *SIAM J. Comput.*, 24(5), 948–969, 1995.

124. Gilbert, J.R. and Hafsteinsson, H., Parallel symbolic factorization of sparse linear systems. *Parallel Comput.*, 14, 151–162, 1990.

125. Gilbert, J.R. and Schreiber, R., Highly parallel sparse Cholesky factorization. *SIAM J. Sci. Comp.*, 13, 1151–1172, 1992.

126. Gilbert, J.R. and Tarjan, R.E., The analysis of a nested dissection algorithm. *Numer. Math.*, 50, 377–404, 1987.

127. Golub, G.H. and Van Loan, C.F., *Matrix Computations*, 3rd edn. Johns Hopkins University Press, Baltimore, MD, 1996.

128. Gondran, M. and Minoux, M., *Graphs and Algorithms*. Wiley–Interscience, New York, 1984.

129. Greenbaum, A., *Iterative Methods for Solving Linear Systems*. SIAM Publications, Philadelphia, PA, 1997.

130. Greuel, G.-M. and Pfister, G., A *Singular Introduction to Commutative Algebra*. Springer-Verlag, Berlin, Germany, 2002.

131. Grigoriev, D.Yu. and Lakshman, Y.N., Algorithms for computing sparse shifts for multivariate polynomials. In *Proceedings of the 1995 International Symposium Symbolic Algebraic Computation (ISSAC '95)*, Levelt, A.H.M., ed., pp. 96–103. ACM Press, New York, 1995.

132. Habicht, W., Eine verallgemeinerung des sturmschen wurzelzählverfarens. *Comm. Math. Helvetici*, 21:99–116, 1948.

133. Hansen, E., Patrick, M., and Rusnak, J., Some modifications of Laguerre's method. *BIT*, 17, 409–417, 1977.

134. Heath, M.T., Ng, E., and Peyton, B.W., Parallel algorithms for sparse linear systems. *SIAM Rev.*, 33, 420–460, 1991.

135. Heindel, L.E., Integer arithmetic algorithms for polynomial real zero determination. *J. Assoc. Comput. Mach.*, 18(4), 533–548, October 1971.

136. Higham, N.J., *Accuracy and Stability of Numerical Algorithms*. 2nd edn. SIAM, Philadelphia, PA, 2002.

137. Hoffmann, C.M., Sendra, J.R., and Winkler, F. Special issue on parametric algebraic curves and applications. *J. Symb. Comp.*, 23, 1997, Academic Press.

138. Jenkins, M.A. and Traub, J.F., A three-stage variable-shift iteration for polynomial zeros and its relation to generalized Rayleigh iteration. *Numer. Math.*, 14, 252–263, 1970.

139. Johnson, J. Algorithms for polynomial real root isolation. In *Quantifier Elimination and Cylindrical Algebraic Decomposition*, Cavinsess B. and Johnson, J., eds., pp. 269–299. Springer, New York, 1998.

140. Johnson, J., Krandick, W., Lynch, K., Richardson, D., and Ruslanov, A., High-performance implementations of the Descartes method. In *Proceedings of the 2006 International Symposium on Symbolic and Algebraic Computation (ISSAC '06)*, pp. 154–161, New York, ACM Press, 2006.

141. Kaltofen, E., Greatest common divisors of polynomials given by straight-line programs. *J. ACM*, 35(1), 231–264, 1988.

142. Kaltofen, E. and Trager, B., Computing with polynomials given by black boxes for their evaluations: Greatest common divisors, factorization, separation of numerators and denominators. *J. Symb. Comp.*, 9(3), 301–320, 1990.

143. Kaltofen, E. and Villard, G., On the complexity of computing determinants. In *Proceedings of the 5th Asian Symposium on Computer Mathematics (ASCM 2001)*, Shirayanagi, K. and Yokoyama, K., eds., *Lecture Notes Series on Computing*, 9, pp. 13–27. World Scientific, Singapore, 2001.

144. Kaltofen, E. and Villard, G., Computing the sign or the value of the determinant of an integer matrix, a complexity survey. *J. Comp. Appl. Math.*, 162(1), 133–146, 2004.

145. Kaporin, I., The aggregation and cancellation techniques as a practical tool for faster matrix multiplication. *Theor. Comput. Sci.*, 315(2–3), 469–510, 2004.

146. Kapur, D., Geometry theorem proving using Hilbert's Nullstellensatz. *J. Symb. Comp.*, 2, 399–408, 1986.

147. Kapur, D. and Lakshman, Y.N., Elimination methods an introduction. In *Symbolic and Numerical Computation for Artificial Intelligence*, Donald, B., Kapur, D., and Mundy, J., eds. Academic Press, Orlando, FL, 1992.

148. Khetan, A., The resultant of an unmixed bivariate system. *J. Symb. Comput.* 36, 425–442, 2003.

149. Kirrinnis, P., Polynomial factorization and partial fraction decomposition by simultaneous Newton's iteration. *J. Complexity*, 14, 378–444, 1998.

150. Knuth, D.E., *The Art of Computer Programming*, Vol. 2, *Seminumerical Algorithms*, 2nd edn., Addison-Wesley, Reading, MA, 1981; 3rd edn., 1997.

151. Krandick, W., Isolierung reeller nullstellen von polynomen. In J. Herzberger, ed., *Wissenschaftliches Rechnen*, pp. 105–154. Akademie-Verlag, Berlin, Germany, 1995.

152. Krandick, W., and Mehlhorn, K., New bounds for the Descartes method. *JSC*, 41(1), 49–66, 2006.

153. Kreuzer, M., and Robbiano, L., *Computational Commutative Algebra 1*. Springer-Verlag, Heidelberg, Germany, 2000.

154. Krishnan, S. and Manocha, D., Numeric-symbolic algorithms for evaluating one-dimensional algebraic sets. In *Proceedings of the ACM International Symposium on Symbolic and Algebraic Computation*, pp. 59–67. ACM, New York 1995.

155. Laderman, J., Pan, V.Y. and Sha, H.X., On practical algorithms for accelerated matrix multiplication. *Linear Algebra Appl.*, 162–164, 557–588, 1992.

156. Lakshman, Y.N. and Saunders, B.D., On computing sparse shifts for univariate polynomials. In *Proceedings of the International Symposium on Symbolic Algebraic Computation ISSAC '94*, von zur Gathen, J. and Giesbrecht, M., eds., pp. 108–113. ACM Press, New York, 1994.

157. Lakshman, Y.N. and Saunders, B.D., Sparse polynomial interpolation in non-standard bases. *SIAM J. Comput.*, 24(2), 387–397, 1995.

158. Lakshman, Y.N., On the complexity of computing Gröbner bases for zero-dimensional polynomial ideals. Ph.D. Thesis, Computer Science Department, Rensselaer Polytechnic Institute, Troy, New York, 1990.

159. Lane, J.M., and Riesenfeld, R.F., Bounds on a polynomial. *BIT*, 21:112–117, 1981.

160. Lawson, C.L. and Hanson, R.J., *Solving Least Squares Problems*. Prentice-Hall, Englewood Cliffs, NJ, 1974. Reissued with a survey of recent developments by SIAM, Philadelphia, PA, 1995.

161. Lazard, D., Resolution des systemes d'equation algebriques. *Theor. Comput. Sci.*, 15, 77–110, 1981. In French.

162. Lenstra, A.K., Lenstra, H.W., and Lovász, L., Factoring polynomials with rational coefficients. *Math. Ann.*, 261, 515–534, 1982.

163. Levelt, A.H.M., ed., *Proceedings of the 1995 International Symposium Symbolic Algebraic Comput. (ISSAC'95)*. ACM Press, New York, 1995.

164. Leyland, P., Cunningham project data. Internet document, Oxford University, ftp://sable.ox.ac.uk/pub/math/cunningham/, Nov. 1995.

165. Lickteig, T. and Roy, M.-F., Sylvester-Habicht Sequences and Fast Cauchy Index Computation. *J. Symb. Comput.*, 31(3), 315–341, 2001.

166. Lipton, R.J., Rose, D., and Tarjan, R.E., Generalized nested dissection. *SIAM J. Numer. Anal.*, 16(2), 346–358, 1979.

167. Macaulay, F.S., *Algebraic Theory of Modular Systems*. Cambridge Tracts 19, Cambridge, U.K., 1916.
168. MacWilliams, F.J. and Sloan, N.J.A., *The Theory of Error-Correcting Codes,* North-Holland, New York, 1977.
169. Madsen, K., A root-finding algorithm based on Newton's method. *BIT*, 13, 71–75, 1973.
170. Manocha, D., Algebraic and numeric techniques for modeling and robotics Ph.D. Thesis, Comp. Science Division, Department of Electrical Engineering and Computer Science, University of California, Berkeley, CA, 1992.
171. Manocha, D., Zhu, Y., and Wright, W., Conformational analysis of molecular chains using nano-kinematics. *Comp. Appl. Biol. Sci.*, 11(1), 71–86, 1995.
172. McCormick, S., ed., *Multigrid Methods*. SIAM Publications, Philadelphia, PA, 1987.
173. McNamee, J.M., A bibliography on roots of polynomials. *J. Comput. Appl. Math.*, 47(3), 391–394, 1993.
174. McNamee, J.M., A supplementary bibliography on roots of polynomials. *J. Comp. Appl. Math.*, 78, 1, 1997.
175. McNamee, J.M., An updated supplementary bibliography on roots of polynomials. *J. Comp. Appl. Math.*, 110, 305–306, 1999.
176. McNamee, J.M., A 2000 updated supplementary bibliography on roots of polynomials. *J. Comp. Appl. Math.*, 142, 433–434, 2002.
177. Mignotte, M., Some useful bounds. In *Computer Algebra: Symbolic and Algebraic Computation*, 2nd edn., Buchberger, B., Collins, G.E., and Loos, R., eds., pp. 259–263. Springer-Verlag, Wien, Austria, 1982.
178. M. Mignotte. *Mathematics for Computer Algebra*. Springer-Verlag, Berlin, Germany, 1992.
179. Mignotte, M. and Stefanescu, D., *Polynomials: An Algorithmic Approach*. Springer-Verlag, Singapore, 1999.
180. Miller, V., Factoring polynomials via relation-finding. In *Proceedings of the ISTCS '92*, Dolev, D., Galil, Z., and Rodeh, M., eds., volume 601 of *LNCS*, pp. 115–121. Springer-Verlag, Berlin, Germany, 1992.
181. Miranker, W.L. and Pan, V.Y., Methods of aggregations. *Linear Algebra Appl.*, 29, 231–257, 1980.
182. Monagan, M.B., A heuristic irreducibility test for univariate polynomials. *J. Symb. Comput.*, 13(1), 47–57, 1992.
183. Mourrain, B. and Pan, V.Y., Asymptotic acceleration of solving polynomial systems. In *Proceedings of the 30th Annual ACM Symposium on Theory of Computing*, pp. 488–496. ACM Press, New York, 1998.
184. Mourrain, B. and Pan, V.Y., Multivariate polynomials, duality and structured matrices, *J. Complexity*, 16(1), 110–180, 2000.
185. Mourrain, B., Pan, V.Y., and Ruatta, O., Accelerated solution of multivariate polynomial systems of equations. *SIAM J. Comp.*, 32, 2, 435–454, 2003. In *Proceedings of the Smalefest 2000*, Cucker F. and Rojas M., eds., *Foundations of Computational Mathematics Series*, pp. 267–294. World Scientific, River Edge, New Jersey, 2002.
186. Mourrain, B., Pavone, J.-P., Trébuchet, P., and Tsigaridas, E.P., SYNAPS: A library for symbolic-numeric computing. In *Proceedings of the 8th International Symposium on Effective Methods in Algebraic Geometry (MEGA)*, Italy, May 2005.
187. Mourrain, B., Rouillier, F., and Roy, M.-F., *Bernstein's Basis and Real Root Isolation*, pp. 459–478. Mathematical Sciences Research Institute Publications. Cambridge University Press, Cambridge, U.K., 2005.
188. Mourrain, B. and Traébuchet, P., Solving projective complete intersection faster. *J. Symb. Comput.*, 33(5), 679–699, 2002.
189. Mourrain, B., Vrahatis, M., and Yakoubsohn, J.C., On the complexity of isolating real roots and computing with certainty the topological degree. *J. Complexity*, 18(2), 2002.
190. Musser, D.R., Multivariate polynomial factorization. *J. ACM*, 22, 291–308, 1975.

191. Neff, C.A. and Reif, J.H., An $O(n^{l+\epsilon})$ algorithm for the complex root problem. In *Proceedings of the 34th Annual IEEE Symposium on Foundations of Computer Science (FOCS'94)*, pp. 540–547. IEEE Computer Society Press, Los Alamitos, CA, 1994.

192. Niederreiter, H., New deterministic factorization algorithms for polynomials over finite fields. In *Finite Fields: Theory, Applications and Algorithms*, Mullen, L. and Shiue, P.J.-S., eds., volume 168 of *Contemporary Mathematics*, pp. 251–268. American Mathematical Society, Providence, RI, 1994.

193. Ortega, J.M. and Voight, R.G., Solution of partial differential equations on vector and parallel computers. *SIAM Rev.*, 27(2), 149–240, 1985.

194. Pan, V.Y., How can we speed up matrix multiplication? *SIAM Rev.*, 26(3), 393–415, 1984.

195. Pan, V.Y., *How to Multiply Matrices Faster*, volume 179 of *LNCS*. Springer-Verlag, Berlin, Germany, 1984.

196. Pan, V.Y., Complexity of parallel matrix computations. *Theor. Comp. Sci.*, 54, 65–85, 1987.

197. Pan, V.Y., Computing the determinant and the characteristic polynomials of a matrix via solving linear systems of equations. *Inform. Process. Lett.*, 28, 71–75, 1988.

198. Pan, V.Y., Complexity of algorithms for linear systems of equations. In *Computer Algorithms for Solving Linear Algebraic Equations (State of the Art)*, Spedicato, E., ed., volume 77 of *NATO ASI Series*, Series F: *Computer and Systems Sciences*, pp. 27–56. Springer, Berlin, 1991, and Academic Press, Dordrecht, the Netherlands, 1992.

199. Pan, V.Y., Complexity of computations with matrices and polynomials. *SIAM Rev.*, 34(2), 225–262, 1992.

200. Pan, V.Y., Parallel solution of sparse linear and path systems. In *Synthesis of Parallel Algorithms*, Reif, J.H., ed., Chapter 14, pp. 621–678. Morgan Kaufmann, San Mateo, CA, 1993.

201. Pan, V.Y., Parallel computation of a Krylov matrix for a sparse and structured input. *Math. Comput. Model.*, 21(11), 97–99, 1995.

202. Pan, V.Y., Optimal and nearly optimal algorithms for approximating polynomial zeros. *Comp. Math. Appl.*, 31(12), 97–138, 1996. Proceedings version: *27th Annual ACM STOC*, pp. 741–750. ACM Press, New York, 1995.

203. Pan, V.Y., Parallel computation of polynomial GCD and some related parallel computations over abstract fields. *Theor. Comp. Sci.*, 162(2), 173–223, 1996.

204. Pan, V.Y., Solving a polynomial equation: Some history and recent progress. *SIAM Rev.*, 39(2), 187–220, 1997.

205. Pan, V.Y., Some recent algebraic/numerical algorithms. In *Electronic Proceedings of IMACS/ACAŠ98*, 1998. http:www-troja.fjfi.cvut.cz/aca98/sessions/approximate

206. Pan, V.Y., Numerical computation of a polynomial GCD and extensions. *Information and Computation*, 167(2), 71–85, 2001. In *Proceedings of the 9th Annual ACM-SIAM Symposium on Discrete Algorithms (SODA 98)*, pp. 68–77. ACM Press, New York, and SIAM Publications, Philadelphia, PA, 1998.

207. Pan, V.Y., On approximating complex polynomial zeros: Modified quadtree (Weyl's) construction and improved Newton's iteration. *J. Complexity*, 16(1), 213–264, 2000.

208. Pan, V.Y., *Structured Matrices and Polynomials: Unified Superfast Algorithms*, Birkhäuser/ Springer, Boston, MA/New York, 2001.

209. Pan, V.Y., Univariate polynomials: Nearly optimal algorithms for factorization and rootfinding. *J. Symb. Comp.*, 33(5), 701–733, 2002. In *Proceedings of the International Symposium on Symbolic and Algebraic Computation (ISSAC 01)*, pp. 253–267, ACM Press, New York, 2001.

210. Pan, V.Y., Randomized acceleration of fundamental matrix computations. In *Proceedings of Symposium on Theoretical Aspects of Computer Science (STACS)*, LNCS, 2285, pp. 215–226, Springer, Heidelberg, Germany, 2002.

211. Pan, V.Y., Can we optimize Toeplitz/Hankel computations? In *Proceedings of the 5th International Workshop on Computer Algebra in Scientific Computing (CASC 02)*, Mayr, E.W., Ganzha, V.G., and Vorozhtzov, E.V., eds., pp. 253–264. Technische University, München, Germany, 2002.

212. Pan, V.Y., Nearly optimal Toeplitz/Hankel computations. Technical Reports 2002 001 and 2002 017, Ph.D. Program in Computer Science, the Graduate Center, CUNY, New York, 2002. Pan, V.Y., Murphy, B., and Rosholt, R.E., Nearly Optimal Symbolic-Numerical Algorithms for Structured Integer Matrices and Polynomials. In *Proceedings of International Symposium on Symbolic-Numerical Computations*. ACM Press, New York, 2009.

213. Pan, V.Y., On theoretical and practical acceleration of randomized computation of the determinant of an integer matrix. *Zapiski Nauchnykh Seminarov POMI (in English)*, 316, 163–187, St. Petersburg, Russia, 2004. Also available at http://comet.lehman.cuny.edu/vpan/

214. Pan, V.Y., Amended DSeSC power method for polynomial root-finding. *Comp. Math. Appl.*, 49(9–10), 1515–1524, 2005.

215. Pan, V.Y., Murphy, B., Rosholt, R.E., Tang, Y., Wang, X., and Zheng, A., Eigen-solving via reduction to DPR1 matrices. *Comp. Math. Appl.*, 56, 166–171, 2008.

216. Pan, V.Y., Branham, S., Rosholt, R., and Zheng, A., Newton's iteration for structured matrices and linear systems of equations. In *SIAM Volume on Fast Reliable Algorithms for Matrices with Structure*, Kailath, T. and Sayed, A., eds., Chapter 7, pp. 189–210. SIAM Publications, Philadelphia, PA, 1999.

217. Pan, V. Y., Grady, D., Murphy, B., Qian, G., Rosholt, R.E., and Ruslanov, A., Schur aggregation for linear systems and determinants, *Special Issue on Symbolic–Numerical Algorithms* D.A. Bini, V.Y. Pan, and J. Verschelde, eds., *Theor. Comp. Sci.*, 409, 2, 255–268, 2008.

218. Pan, V.Y., Ivolgin, D., Murphy, B., Rosholt, R.E., Tang, Y., and Wang, X., Root-finding with Eigen-solving. In *Symbolic-Numeric Computation*, Wang, D. and Zhi, L., eds. Birkhäuser, Basel/Boston, MA, 2007.

219. Pan, V.Y., Kunin, M., Rosholt, R.E., and Kodal, H., Homotopic residual correction processes. *Math. Comp.*, 75, 345–368, 2006.

220. Pan, V.Y., Landowne, E., and Sadikou, A., Univariate polynomial division with a remainder by means of evaluation and interpolation. *Inform. Process. Lett.*, 44, 149–153, 1992.

221. Pan, V.Y., Murphy, B., and Rosholt, R.E., Unified nearly optimal algorithms for structured matrices. In *Numerical Methods for Structured Matrices and Applications: Georg Heinigs Memorial Volume*, Birkhauser Verlag, in press.

222. Pan, V.Y., Murphy, B., Rosholt, R.E., and Wang, X., Toeplitz and Hankel meet Hensel and Newton: Nearly optimal algorithms and their practical acceleration with saturated initialization. Technical Report 2004 013, Ph.D. Program in Computer Science, The Graduate Center, City University of New York, New York, 2004.

223. Pan, V.Y. and Preparata, F.P., Work-preserving speed-up of parallel matrix computations. *SIAM J. Comput.*, 24(4), 811–821, 1995.

224. Pan, V.Y. and Qian, G., Solving homogeneous linear systems with randomized preprocessing. *Linear Algebra and Its Applications*, accepted.

225. Pan, V.Y., Rami, Y., and Wang, X., Structured matrices and Newtons iteration: Unified approach. *Linear Algebra Appl.*, 343/344, 233–265, 2002.

226. Pan, V.Y. and Reif, J.H., Compact multigrid. *SIAM J. Sci. Stat. Comp.*, 13(1), 119–127, 1992.

227. Pan, V.Y. and Reif, J.H., Fast and efficient parallel solution of sparse linear systems. *SIAM J. Comp.*, 22(6), 1227–1250, 1993.

228. Pan, V.Y., Sadikou, A., Landowne, E., and Tiga, O., A new approach to fast polynomial interpolation and multipoint evaluation. *Comp. Math. Appl.*, 25(9), 25–30, 1993.

229. Pan, V.Y. and Schreiber, R., An improved Newton iteration for the generalized inverse of a matrix, with applications. *SIAM J. Sci. Stat. Comp.*, 12(5), 1109–1131, 1991.

230. Pan, V.Y., Sobze, I., and Atinkpahoun, A., On parallel computations with band matrices. *Inform. Comput.*, 120(2), 227–250, 1995.

231. Pan, V.Y., Van Barel, M., Wang, X., and Codevico, G., Iterative inversion of structured matrices. (Special Issue on Algebraic and Numerical Algorithms, Emiris, I.Z., Mourrain B., and Pan, V.Y. eds.), *Theor. Comp. Sci.*, 315(2–3), pp. 581–592, 2004.

232. Pan, V.Y. and Wang, X., Inversion of displacement operators. *SIAM J. Matrix Anal. Appl.*, 24(3), 660–677, 2003.

233. Pan, V.Y. and Wang, X., On rational number reconstruction and approximation. *SIAM J. Comput.*, 33(2), 502–503, 2004.

234. Pan, V.Y. and Wang, X., Degeneration of integer matrices modulo an integer. *Linear Algebra Appl.*, 429, 2113–2130, 2008.

235. Pan, V.Y. and Yu, Y., Certification of numerical computation of the sign of the determinant of a matrix. *Algorithmica*, 30, 708–724, 2001. In *Proceedings of the 10th Annual ACM-SIAM Symposium on Discrete Algorithms (SODA 99)*, pp. 715–724. ACM Press, New York, and SIAM Publications, Philadelphia, PA, 1999.

236. V. Y. Pan, X. Yan, Additive preconditioning, eigenspaces, and the inverse iteration. *Linear Algebra Appl.*, 430, 186–203, 2009.

237. Pan, V.Y., Zheng, A.L., Huang, X.H., and Yu, Y.Q., Fast multipoint polynomial evaluation and interpolation via computations with structured matrices. *Ann. Numer. Math.*, 4, 483–510, 1997.

238. Parlett, B., *Symmetric Eigenvalue Problem*. Prentice Hall, Englewood Cliffs, NJ, 1980.

239. Press, W., Flannery, B., Teukolsky, S., and Vetterling, W., *Numerical Recipes in C: The Art of Scientific Computing*. Cambridge University Press, Cambridge, U.K., 1988; 2nd ed., 1992.

240. Quinn, M.J., *Parallel Computing: Theory and Practice*. McGraw-Hill, New York, 1994.

241. Rabin, M.O., Probabilistic algorithms in finite fields. *SIAM J. Comp.*, 9, 273–280, 1980.

242. Reischert, D., Asymptotically fast computation of subresultants. In *Proceedings of the 1997 International Symposium on Symbolic and Algebric Computation*, pp. 233–240, ACM, New York, 1997.

243. Renegar, J., On the worst case arithmetic complexity of approximating zeros of polynomials. *J. Complexity*, 3(2), 90–113, 1987.

244. Rojas, J.M. Solving degenerate sparse polynomial systems faster. (Special Issue on Elimination), *J. Symb. Comput.*, 28, 155–186, 1999.

245. Raghavan M. and Roth., B., Solving polynomial systems for the kinematics analysis and synthesis of mechanisms and robot manipulators. (Special Issue), *Trans. ASME*, 117, 71–79, 1995.

246. Ritt, J.F., *Differential Algebra*. AMS, New York, 1950.

247. Rosen, D., and Shallit, J., A continued fraction algorithm for approximating all real polynomial roots. *Math. Mag.*, 51, 112–116, 1978.

248. Rouillier, F., Solving zero-dimensional systems through the rational univariate representation, *AAECC J.*, 9, 433–461, 1999.

249. Rouillier, F. and Zimmermann, P., Efficient isolation of polynomial's real roots. *J. Comput. Appl. Math.*, 162(1), 33–50, 2004.

250. Saad, Y., *Iterative Methods for Sparse Linear Systems*. PWS Publishing Co., Boston, MA, 1996, 1st edn.; SIAM Publications, Philadelphia, PA, 2003, 2nd edn.

251. Schönhage, A., The fundamental theorem of algebra in terms of computational complexity. Mathematics Department, University of Tübingen, Tübingen, Germany, 1982.

252. Sederberg, T. and Goldman, R., Algebraic geometry for computer-aided design. *IEEE Comput. Graph. Appl.*, 6(6), 52–59, 1986.

253. Sendra, J.R. and Winkler, F., Symbolic parameterization of curves. *J. Symb. Comput.*, 12(6), 607–631, 1991.

254. Sharma, V., Complexity of real root isolation using continued fractions. In *Proceedings of Annual ACM ISSAC*, Brown, C. W., ed., Waterloo, ON, Canada, 2007.

255. Shoup, V., A new polynomial factorization algorithm and its implementation. *J. Symb. Comput.*, 20(4), 363–397, 1995.

256. Smith, B.T., Boyle, J.M., and Dongarra, J.J., *Matrix Eigensystem Routines: EISPACK Guide,* 2nd edn. Springer, New York, 1970.

257. Stetter, H., *Numerical Polynomial Algebra*. SIAM, Philadelphia, PA, 2004.

258. Stewart, G.W., *Matrix Algorithms,* Vol I: *Basic Decompositions.* SIAM, Philadelphia, PA, 1998.
259. Stewart, G.W., *Matrix Algorithms,* Vol II: *Eigensystems.* SIAM, Philadelphia, PA, 1998.
260. Storjohann, A., High order lifting and integrality certificaiton. *J. Symb. Comp.,* 36(3-4), 613-648, 2003.
261. Storjohann, A., The shifted number system for fast linear algebra on integer matrices. *J. Complexity,* 21(4), 609-650, 2005.
262. Sturm, C., Mémoire sur la résolution des equations numériques. *Mém. Savants Étranger,* 6, 271-318, 1835.
263. Sturmfels, B., Sparse elimination theory. In *Proceedings of the Computational Algebraic Geometry and Commutative Algebra,* Eisenbud, D. and Robbiano, L., eds. Cortona, Italy, 1991.
264. Tarjan, R.E., A unified approach to path problems. *J. ACM,* 28(3), 577-593, 1981.
265. Tarjan, R.E., Fast algorithms for solving path problems. *J. ACM,* 28(3), 594-614, 1981.
266. Trefethen, L.N. and Bau III, D., *Numerical Linear Algebra.* SIAM Publications, Philadelphia, PA, 1997.
267. Tsigaridas, E.P. and Emiris, I.Z., Univariate polynomial real root isolation: Continued fractions revisited. In *Proceedings of 14th European Symposium of Algorithms (ESA),* Y. Azar and T. Erlebach, eds., volume 4168 of *LNCS,* pp. 817-828. Springer-Verlag, Zurich. Switzerland, 2006.
268. Uspensky, J.V., *Theory of Equations.* McGraw-Hill, New York, 1948.
269. Vandebril, R., Van Barel, M., Golub, G., and Mastronardi, N., A bibliography on Semiseparable Matrices. *Calcolo,* 42(3-4), 249-270, 2005.
270. Vincent, A.J.H., Sur la résolution des équations numériques. *J. Math. Pures Appl.,* 1, 341-372, 1836.
271. van der Vorst, H.A., *Iterative Krylov Methods for Large Linear Systems.* Cambridge University Press, Cambridge, U.K., 2003.
272. van der Waerden, B.L., *Modern Algebra,* 3rd edn. F. Ungar, New York, 1950.
273. Walsh, P.G., The computation of Puiseux expansions and a quantitative version of Runge's theorem on diophantine equations. Ph.D. Thesis, University Waterloo, Waterloo, ON, Canada, 1993.
274. Wang, D. and Zhi, L. eds., *Symbolic-Numeric Computation* Birkhäuser, Basel/Boston, MA, 2007.
275. Wang, X. and Pan, V.Y., Acceleration of Euclidean algorithm and rational number reconstruction. *SIAM J. Computing,* 32(2), 548-556, 2003.
276. Wiedemann, D., Solving sparse linear equations over finite fields. *IEEE Trans. Info. Theory,* IT-32, 54-62, 1986.
277. Wilkinson, J.H., *The Algebraic Eigenvalue Problem.* Clarendon Press, Oxford, U.K., 1965.
278. Winkler, F., *Polynomial Algorithms in Computer Algebra.* Springer, Wien, Austria, 1996.
279. Wu, W., Basis principles of mechanical theorem proving in elementary geometries. *J. Syst. Sci. Math Sci.,* 4(3), 207-235, 1984.
280. Yap, C.K., *Fundamental Problems of Algorithmic Algebra.* Oxford University Press, New York, 2000.
281. Zassenhaus, H., On Hensel factorization I. *J. Number Theory,* 1, 291-311, 1969.
282. Zippel, R., *Effective Polynomial Computations.* Kluwer Academic, Boston, MA, 1993.

18

Applications of FFT and Structured Matrices

Ioannis Z. Emiris
National and Kapodistrian
University of Athens

Victor Y. Pan
City University of New York

18.1 Introduction

The subject of this chapter lies in the area of theoretical computer science, although it is closely related to algebraic and numerical computations for sciences, engineering, and electrical engineering.

Our central theme is the dramatic acceleration of the most fundamental computations with integers, polynomials, and structured matrices by means of application of the **fast Fourier transform (FFT)**. More precisely, we first apply the FFT to multiply a pair of integers, a pair of polynomials, and a structured matrix by a vector in nearly linear time, versus higher order, typically quadratic time required for the same operation in the classical algorithms. Then we extend this process to other fundamental operations by ultimately reducing them to multiplication.

We state the complexity bounds under the random access machine (RAM) model of computation [1]. In most cases we assume the arithmetic model, that is, we assign a unit cost to addition, subtraction, multiplication, and division of real numbers, as well as to reading or writing them into a memory location. This model is realistic for computations with a fixed (e.g., the IEEE standard double) precision, which fits the size of a computer word. In this case the arithmetic model turns into the word model [50]. In other cases we assume extended precision of computing and the Boolean or bit model, that is, we assign the unit cost to every Boolean or bit operation. This accounts for both arithmetic operations and the length (precision) of the operands. We always specify whether we use the arithmetic, word, or Boolean model unless this is clear from the context.

Section 18.2 examines the fundamental transforms of vectors, in particular, the discrete Fourier transform, its inverse, vector convolution, its extension to wrapped convolutions, and the sine and cosine transforms. We specify the FFT, discuss it in some detail, and apply it as the basic algorithm for computing the aforementioned transforms.

Section 18.3 applies these results to some fundamental operations on univariate polynomials and shows the correlations between the main vector transforms and polynomial arithmetic. The results are carried over to the integers and the polynomials in several variables.

Section 18.4 covers applications to structured matrices, which are omnipresent in scientific and engineering computations and signal and image processing. The mn entries of an $m \times n$ structured matrix are defined via the order of $m + n$ parameters, which means linear memory space versus quadratic for general matrices. Based on the application of the FFT and the correlation between computations with structured matrices and polynomials, one can similarly decrease the arithmetic time complexity of multiplication of such a matrix by a vector and other fundamental operations with structured matrices. This naturally leads us to the unification of structured matrix computations, which in turn helps us to improve the computations further.

We practically omit the lower bound topics, referring the reader to [18,25,73] on some nontrivial results. One can always apply the information lower bounds: since each arithmetic and each Boolean operand has two operations and one output, there must always be at least $\max\{O, I/2\}$ operations, where O and I are the sizes of the output and input, respectively. We also omit certain generalizations of Fourier transforms that are of some interest in theoretical computer science; see, for instance [25]. Many important applications of the FFT to computations in sciences, engineering, and electrical engineering cannot be covered in a chapter of the present size (see the end of Section 18.2.1 and the introduction of Section 18.4).

We prefer to cite books, surveys, or comprehensive articles, rather than the publications that first contained a certain result. There is a tacit understanding that the interested reader will look into the bibliography of the cited references, in which the historical development is outlined. Our entire exposition largely draws from the survey articles [84,85] and the books [13,91].

Hereafter, $(\mathbf{v})_i = v_i$ and $(A)_{i,j} = a_{i,j}$ for a column vector $\mathbf{v} = [v_i]_i$ and a matrix $A = [a_{i,j}]_{i,j}$, log stands for $\log_2$ unless specified otherwise, "section.name" stands for Section "section.name," **ops** for arithmetic operations, W^T for the transpose of a matrix or vector W, and W^H for its Hermitian, that is, complex conjugate transpose.

18.2   Some Fundamental Transforms

Every transform discussed here can be thought of as a mapping of a vector into a vector. The transforms are fundamental for fast solution of a variety of general algebraic and numerical computational problems and for devising algorithms that support the record estimates for the asymptotic complexity of some fundamental algebraic and numerical computations (see Section 18.3).

Section 18.2.1 covers the **discrete Fourier transform (DFT)** and some closely related transforms, and outlines the main algorithm for computing DFT, namely, the FFT. The record complexity bounds for polynomial arithmetic, including multiplication, division, transformation under a shift of the variable, evaluation, **interpolation**, and approximating polynomial zeros, are based on the FFT. The FFT and fast polynomial algorithms are the basis for many other fast polynomial computations, performed both numerically and symbolically. Section 18.2.2 studies vector convolution and shows its equivalence to the DFT and also to generalized DFT. Section 18.2.3 recalls the sine, cosine, and some other transforms.

Abundant material and bibliography on transforms and convolution can be found in [10,12,13, 15,21,38,39,52,91,112,122,126].

18.2.1 The Discrete Fourier Transform and Its Inverse

The DFT of the coefficient vector $\mathbf{p} = [p_i]_{i=0}^n = [p_0, \ldots, p_n]^T$ of a polynomial

$$p(x) = p_0 + p_1 x + \cdots + p_n x^n \tag{18.1}$$

is the vector $[p(\omega^i)]_{i=0}^{K-1} = [p(1), \ldots, p(\omega^{K-1})]^T$, where ω is a primitive Kth root of 1, so that $\{1, \omega, \omega^2, \ldots, \omega^{K-1}\}$ is the set of all the Kth roots of unity, and $K = 2^k = n + 1$, for a natural k. For simplicity, the reader may assume that we study DFT in the complex field, but the study can be extended to other fields, rings, and even groups, where the Kth roots of unity are defined [27,34]. The problem of computing the DFT is solved by applying the FFT algorithm, which is an example of recursive algorithms based on the **divide-and-conquer** method. The FFT algorithm can be traced back to [116] and even to a work of C. F. Gauss of 1805, published posthumously in 1866, although its introduction in modern times has been credited to [36] (see [18,15,19,21,50,52,73,130] for details). To derive the FFT, write

$$p(x) = q(x^2) + xs(x^2) = q(y) + xs(y),$$

where
$y = x^2$
$q(y) = p_0 + p_2 y + \cdots + p_{n-1} y^{(n-1)/2}$
$s(y) = p_1 + p_3 y + \cdots + p_n y^{(n-1)/2}$
the polynomials $q(y)$ and $s(y)$ have degree of at most $(n-1)/2$

To evaluate $p(x)$ at $x = \omega^h$ for $h = 0, 1, \ldots, n$, we first compute $q(y)$ and $s(y)$ at $y = (\omega^2)^h$ and then $q(\omega^{2h}) + xs(\omega^{2h})$ at $x = \omega^h$. There exist only $K/2$ distinct values among all integer powers of ω^2, since ω^2 is a $(K/2)$nd root of unity. Half of the multiplications of ω^h by $s(\omega^{2h})$ can be saved because $\omega^i = -\omega^{i+K/2}$ for all i. By applying this method recursively, we compute the DFT in $1.5\,K \log K$ ops versus $2n^2$ ops in the classical algorithm, in both cases not counting the cost of computing the roots of unity.

For demonstration, we perform the FFT for the polynomial

$$p(x) = 3x^3 + 2x^2 - x + 5.$$

Here, $K = 4$, and the 4th roots of unity are $1, \omega = \sqrt{-1}, \omega^2 = -1$, and $\omega^3 = -\omega$. Write

$$p(x) = (5 + 2x^2) + x(-1 + 3x^2) = q(y) + xs(y), \quad y = x^2.$$

At the cost of performing two multiplications (by 1 and ω) and four additions/subtractions, this reduces DFT for $p(x)$ to the evaluation of $q(y)$ and $s(y)$ at the points 1 and $\omega^2 = -1$. Compute $q(1), q(\omega^2), s(1)$, and $s(\omega^2)$, by using two multiplications (of the values 2 and 3 by 1) and four additions and subtractions: $5 + 2$, $5 - 2$, $-1 + 3$, $-1 - 3$. Overall, four multiplications and eight additions are required, which is indeed $1.5\,K \log K = 12$ ops.

In numerical computations, one can use complex approximations to the roots of unity $\omega^i = \exp(2\pi i \sqrt{-1}/K)$, $i = 0, \ldots, K - 1$ [73]. If all the p_i are bounded integers, we can perform the computations over Z_m, the ring of integers modulo a sufficiently large integer m, or use other special techniques (cf. [129] and [13, Section 1.7]). For appropriate choices of the modulus m, the desired Kth roots of unity are readily available in Z_m (see [1, pp. 265–269], [18, pp. 86–87], or

[13, Section 3.3]). Performing the FFT over finite fields or rings is also required, e.g., in application to integer multiplication (see [1], [13, Chapter 1], [25]).

The FFT can be conveniently represented in the divide-and-conquer form of structured matrix factorization, and then the algorithm is called binary or radix split *FFT* [13], [91, Section 2.3]. It is still revised and ameliorated [48,49].

Among the variations of the FFT, distinct from the Cooley/Tukey best-known variant that we described, we recall the Prime-Factor (Good/Thomas), and the algorithms of Brunn, Rader, Bluestein, and Winograd [16,24,38,113,114,121,128].

The **inverse discrete Fourier transform** (IDFT) of a vector $\mathbf{r}$ of polynomial values $r_h = p(\omega^h)$, $h = 0, 1, \cdots, K-1$, on a set of all Kth roots of unity is the coefficient vector $\mathbf{p} = [p_0, \cdots, p_n]^T$ of $p(x)$. Computing the DFT and the IDFT is a special case of the more general problems of multipoint polynomial **evaluation** and interpolation (Section 18.3.2), equivalent to **structured matrix** computations (see Section 18.4.1).

The IDFT can be computed in at most $K + 1.5\,K \log K$ ops by means of applying the inverse *FFT* algorithm, which is again a divide-and-conquer procedure, reminiscent of the FFT (see [13, p. 12]). Alternatively, we reduce the DFT and the IDFT to each other and to matrix computations, based on the following vector equation: $\Omega\mathbf{p} = \mathbf{r}/\sqrt{K}$. Here, $\Omega = [\omega^{ij}/\sqrt{K}]$ is the Fourier matrix, $\mathbf{p} = [p_j]$ is the input vector of DFT, and $\mathbf{r} = [r_i]$, $i, j = 0, 1, \ldots, n$, where $\mathbf{r}/\sqrt{K}$ is the output vector $[p(\omega^h)]$ of DFT. It follows that $\Omega^{-1} = [\omega^{-ij}/\sqrt{K}]$ and $\mathbf{p} = \Omega^{-1}\mathbf{r}/\sqrt{K} = [\omega^{-ij}]\mathbf{r}/K$. Since ω^{-1} is also a primitive Kth root of 1, the latter matrix-by-vector product can be computed by means of an FFT. Then it remains to divide the resulting vector by K to obtain the IDFT.

The generalized DFT is the DFT relaxed from any restrictions on the number K of points and the basic value w, except that this value is required to have the reciprocal w^{-1}. In Section 18.2.2 we define the vector convolution, show that its computation is equivalent to computing generalized DFT, and apply the FFT to obtain the complexity bound of $O(K \log K)$ ops for both problems of a size K.

In Section 18.3.1 we extend these results to the evaluation of $p(x)$ on the set $\{ah^{2i} + fh^i + g, i = 0, 1, \ldots, n-1\}$ for any fixed 4-tuple of constants (a, f, g, h), by using $O(n \log n)$ ops [2].

The bound $O(n \log n)$ is a dramatic decrease of the order of n^2 ops required in the classical algorithms for the latter problem as well as for the IDFT, generalized DFT, and vector convolution. In practice, the DFT and the IDFT are frequently computed on K points for $K \geq 10,000$ or even $K \geq 100,000$, and so the practical impact of the FFT has been dramatic. Efficient implementations of the FFT, including some techniques that reduce the complexity of computing the DFT and the IDFT for specific smaller integers K, are covered in [10,126,129]. Public domain codes implementing the FFT and its natural and important extension to multivariate *FFT* can be freely accessed via netlib [3,4,46,47,123], and some libraries of arbitrary-precision integer arithmetic use FFT [4,9,55].

The DFT is a well-conditioned problem, and the FFT is a numerically stable algorithm if we consider both input and output as vectors and measure the errors in terms of vector norms. More formally, let $\mathbf{x}$ and $\mathbf{y}$ be a pair of K-dimensional complex vectors, for $K = 2^k$, such that $\mathbf{y} = DFT(\mathbf{x})$ is the DFT of $\mathbf{x}$. Let $FFT(\mathbf{x})$ denote the vector output by the matrix version of the FFT algorithm described earlier and performed in floating point arithmetic with d bits, $d \geq 10$, and let $e_K(x)$ express the error vector of dimension K. Then, according to [13, prop. 3.4.1], we have

$$FFT(x) = DFT(x + e_K(x));$$

$$\|e_K(x)\| \leq \left(\left(1 + \rho 2^{-d}\right)^k - 1 \right) \|x\|, \quad 0 < \rho < 4.83,$$

where $\| \cdot \| = \| \cdot \|_2$ denotes the Euclidean norm. Moreover,

$$\|e_K(x)\| \leq (5k)\, 2^{-d}\|x\| \quad \text{for any} \quad K \leq 2^{2^{d-6}}.$$

It immediately follows [13, cor. 3.4.1] that

$$\|FFT(x) - DFT(x)\| \leq 5\sqrt{K}(\log K)2^{-d}\|x\| \quad \text{if } K < 2^{2^{d-6}}.$$

Similar results hold for the IDFT and certain other computational problems reducible to computing the DFT. In particular, we can apply the FFT with a relatively low precision when we compute the product of two polynomials with integer coefficients (see Section 18.2.2). Then rounding off still gives us the output with no error. On the other hand, polynomial division and computing the greatest common divisor (GCD) and the zeros of polynomials (see Sections 18.3.1 and 18.3.3 and Chapter 17) are generally ill-conditioned problems [13,63,88].

The application of the FFT to the computation of the continuous Fourier transform is central to several engineering and numerical applications such as digital filters, image restoration, and numerical solution of PDEs [15,19,21,120].

18.2.2 Vector Convolution

The next computation is fundamental for computer algebra and signal and image processing. Given the coordinates $u_0, v_0, u_1, v_1, \ldots, u_n, v_n$ of two vectors $\mathbf{u} = (u_i)_{i=0}^n$ and $\mathbf{v} = (v_i)_{i=0}^n$, we seek the coordinates w_i, w_i^+ and/or w_i^- of the vectors $\mathbf{w} = (w_i)_{i=0}^{2n}$, $\mathbf{w}^+ (w_i^+)_{i=0}^n$, and $\mathbf{w}^- (w_i^-)_{i=0}^n$ of **convolution**, positive and negative wrapped convolution, respectively, where $w_i^+ = w_i + \hat{w}_i$, $w_i^- = w_i - \hat{w}_i$, $w_i = \sum_{j=0}^i u_j v_{i-j}$, $\hat{w}_i = w_{n+1+i} = \sum_{j=i+1}^n u_j v_{n+1+i-j}$, $i = 0, 1, \ldots, n$, $\hat{w}_n = 0$.

Now write

$$u(x) = \sum_{i=0}^n u_i x^i, \quad v(x) = \sum_{i=0}^n v_i x^i, \quad w(x) = \sum_{i=0}^{2n} w_i x^i,$$

$$w^+(x) = \sum_{i=0}^n w_i^+ x^i, \quad w_i^-(x) = \sum_{i=0}^n w_i^- x^i$$

and observe that $w(x) = u(x)v(x)$, so that the coefficient vector of the product of two polynomials is the convolution of their coefficient vectors, and furthermore $w^+(x) = w(x) \bmod (x^{n+1} - 1)$, $w^-(x) = w(x) \bmod (x^{n+1} + 1)$,

We reduce this problem essentially to the solution of three DFT problems. Let $n + 1 = K = 2^k$ for a natural k, for otherwise, we can pad $u(x)$ and $v(x)$ with l zero leading coefficients each, for $l < n$, to bring the degree values to the form $K - 1 = 2^k - 1$, $k = \lceil \log_2 n \rceil$. By applying Toom's evaluation–interpolation techniques [124], evaluate at first $u(x)$ and $v(x)$ at the $(2K)$th roots of unity (two FFTs), then multiply the $2K$ computed values pairwise, to obtain the values of $w(x)$ at the $(2K)$th roots of unity, and then compute the coefficients of $w(x)$ (via IDFT) at the overall cost of $9K \log K + 2K$ ops. By reducing $w(x)$ modulo $(x^{n+1} \pm 1)$, we can also obtain $w^+(x)$ and $w^-(x)$.

Let us compute w^+ by performing two DFTs and an IDFT at the Kth roots of unity. Consider the values of $w^+(x)$ at ω^h, where ω is a primitive $(n+1)$st root of unity and $h = 0, \ldots, n$. Letting $\hat{w}_n = 0$, we have

$$w^+ \left(\omega^h \right) = \sum_{i=0}^n (w_i + \hat{w}_i) \omega^{ih} = \sum_{i=0}^n \omega^{ih} \sum_{j=0}^i u_j v_{i-j} + \sum_{i=0}^{n-1} \omega^{ih} \sum_{j=i+1}^n u_j v_{n+1+i-j}.$$

Since $\omega^{n+1} = 1$ and $u_j = v_j = 0$ for $j < 0$, we rewrite the second summand as

$$\left(\omega^{n+1} \right)^h \sum_{i=0}^{n-1} \omega^{ih} \sum_{j=i+1}^n u_j v_{n+1+i-j} = \sum_{i=n+1}^{2n} \omega^{ih} \sum_{j=0}^i u_j v_{i-j}.$$

Let $w(x)$ be the polynomial $u(x)v(x)$. Then $w^+(\omega^h) = w(\omega^h)$, $h = 0, \ldots, n$, and we compute the positive wrapped convolution $\omega^+(x)$ at the overall complexity of $4.5 K \log K + K$ ops.

A similar argument applies to the negative wrapped convolution. Let ψ be a $(2n + 2)$nd primitive root of unity and let ω and h be as aforementioned. Then $w^-(x) = u(x)v(x)$, for $x = \psi \omega^h$, since $x^{n+1} = -1$, and it suffices to use three FFTs, each on $n + 1$ points $\psi\omega^h$, at the overall cost $4.5\,K\log K + K$.

Conversely, to reduce the generalized DFT problem to convolution, seek

$$r_h = \sum_{j=0}^{K-1} p_j\omega^{hj} = \omega^{-h^2/2}\sum_{j=0}^{K-1} p_j\omega^{(j+h)^2/2}\omega^{-j^2/2} \quad \text{for } h = 0,\ldots,K-1.$$

Let $w_i = r_h\omega^{h^2/2}, v_{i-j} = \omega^{(j+h)^2/2}$, and $u_j = p_j\omega^{-j^2/2}$, for $i = K-1,\ldots,2K-2$ and $j = 0,\ldots,K-1$. Essentially, we change indices by setting $i = 2K - 2 - h$. The undefined values of u_j are zero; in particular, $u_j = 0$ for $j \geq K$. Then

$$w_i = \sum_{j=0}^{K-1} u_j v_{i-j} + \sum_{j=K}^{i} u_j v_{i-j}, \quad i = K-1,\ldots,2K-2,$$

which is a part of the convolution problem $w_i = \sum_{j=0}^{i} u_j v_{i-j}$, where $i = 0,\ldots,2K-2$ and $u_s = v_s = 0$ for $s \geq K$. Hence, the generalized DFT is reduced to two wrapped convolutions, followed by $2K$ multiplications for computing the values u_j and recovering the values r_h from w_i. Thus the asymptotic complexity of generalized DFT is $O(K\log K)$, the same as DFT and (wrapped) convolution.

More generally, we only need $4.5K\log K + 2K$ ops to compute the **convolution of two vectors** of dimensions $m + 1$ and $n + 1$, respectively, which is the coefficient vector of the product of two polynomials of degrees at most m and n given with their coefficient vectors, where $K = 2^k$ and $k = \lceil\log(m + n + 1)\rceil$.

So far, we have assumed computations in the fields that support reduction to the FFT in $O(K\log K)$ ops. An arbitrary ring of constants with unity supports reduction to the FFT on a set of the Kth roots of unity only in $O(K\log K\log\log K)$ ops [27]. Thus in a ring we multiply our FFT-based complexity estimates for convolution by $\log\log K$, yielding an overall bound in $O(K\log K\log\log K)$.

The convolution problem and its solutions are immediately extended to multiplication of several polynomials [13].

For smaller m and n, the convolution and the associated polynomial product can be alternatively computed by means of the straightforward (classical) algorithm that uses $(m + 1)(n + 1)$ multiplications and mn additions.

For moderate m and n one may prefer the Karatsuba's algorithm in [72], where assuming for simplicity that $m = n$ is a power of two, we just recursively apply the following equations:

$$\begin{aligned}
u(x)v(x) &= \left(u_0(x) + x^{n/2}u_1(x)\right)\left(v_0(x) + x^{n/2}v_1(x)\right) \\
&= u_0(x)v_0(x)\left(1 - x^{n/2}\right) + (u_1(x) + u_0(x))(v_1(x) + v_0(x))\,x^{n/2} \\
&\quad + u_1(x)v_1(x)\left(x^n - x^{n/2}\right).
\end{aligned} \tag{18.2}$$

This means $O(n^{\log 3})$ ops, for $\log 3 = 1.5849\ldots$ and a small overhead constant.

Hereafter $\mu(n)$ denotes the number of ops sufficient for computing the convolution of two vectors of dimension $n + 1$ in a fixed field or ring by a fixed algorithm, so that

$$\mu(n) \leq c_{class}\,n^2, \quad \mu(n) \leq c_k n^{\log 3}, \quad \mu(n) \leq c_{ck} n\log n\log\log n$$

where c_{class}, c_k, and c_{ck} are the overhead constants that support the asymptotic complexity bounds based on the classical, Karatsuba's, and Cantor/Kaltofen's algorithms, respectively, $c_{class} < c_k < c_{ck}$.

18.2.3 The Sine and Cosine Transforms

Besides the FFT, other related transforms widely used in signal processing include sine, cosine, Hartley, and wavelet transforms. Given a vector $\mathbf{y} = [y_1, \ldots, y_n]$, its sine transform can be defined as the vector $\mathbf{x} = [x_1, \ldots, x_n]$, where

$$x_i = \sum_{j=1}^{n} y_j \sin \frac{\pi ij}{n+1}, \quad i = 1, \ldots, n,$$

or, equivalently, $\mathbf{x} = \sqrt{\frac{n+1}{2}} S y$, where $S = \left[\sqrt{\frac{2}{n+1}} \sin \frac{\pi ij}{n+1} \right]$, $i, j = 1, \ldots, n$, $S^T = S^{-1} = S$. The cosine transform of vector $[y_j]$ can be defined analogously by substituting sine by cosine. For further variants of sine and cosine transforms see [67,112] and [91, Section 3.11]. All these sine and cosine transforms can be performed in $O(n \log n)$ ops by means of FFT [67].

An important application of the sine transforms is to computations in the matrix algebra τ, introduced in [11]; further applications can be found in [64]. This algebra consists of all $n \times n$ matrices $A = [a_{ij}]$, such that

$$a_{i,j-1} + a_{i,j+1} = a_{i-1,j} + a_{i+1,j}, \quad i, j = 0, 1, \ldots, n - 1;$$
$$a_{i,j} = 0 \quad \text{if } i \in \{-1, n\}, \text{ or } j \in \{-1, n\}.$$

Then we write $A = \tau(\mathbf{a})$, where $\mathbf{a} = [a_{0,0}, a_{1,0}, \ldots, a_{n-1,0}]^T$ is the first column of the matrix A. This matrix algebra is related to Chebyshev-like polynomials and is studied in [13, Chapter 2].

For any matrices $A \in \tau$ and S mentioned earlier, $SAS = D$ is a diagonal matrix with nonzero entries $d_i = (S\mathbf{a})_i / \left(\sqrt{2/(n+1)} \sin \frac{\pi i}{n+1} \right)$, $i = 1, \ldots, n$. Furthermore, given two vectors $\mathbf{u}$ and $\mathbf{v}$, each having n components, we can apply sine transforms to compute the following vectors in $O(n \log n)$ ops: $\tau(\mathbf{u})\mathbf{v}$, the first column of $\tau(\mathbf{u})\tau(\mathbf{v})$, and $\tau(\mathbf{u})^{-1}\mathbf{v}$, if the matrix $\tau(\mathbf{u})$ is nonsingular.

The matrix algebra τ is important in the analysis of the spectral properties of band Toeplitz matrices, the computation of their tensor rank, and solving block Toeplitz linear systems (see Section 18.4.2).

18.3 Fast Polynomial and Integer Arithmetic

Computations with integers and polynomials, in one or more variables, is of fundamental importance in computational mathematics and computer science. Such operations lie at the core of every computer algebra system. In this section, we focus on univariate polynomials and show some extensions to integer arithmetic and multivariate polynomials. The connection between vector manipulation, on the one hand, and univariate polynomial and integer arithmetic, on the other hand, relies on representing a polynomial or an integer by a vector of coefficients or digits, respectively. A large number of problems in science and engineering can be solely expressed in terms of polynomials, which also serve as the basis for the study of more complex structures, such as rational functions, algebraic functions, power series, and transcendental functions.

We apply the transforms seen so far to yield the record asymptotic upper bounds on the complexity of several fundamental operations with univariate polynomials. We implicitly assume that the polynomials are represented by the vectors of their coefficients and are dense, that is, have most of their coefficients nonzero. Various representations of sparse polynomials (e.g., by the straight-line programs for their evaluation) and various complexity measures (e.g., in terms of the output size) can be found in [1,23,50,69,130].

For dense univariate polynomials we cover polynomial multiplication, division, multipoint evaluation, and interpolation, as well as further extensions of these basic operations. Due to their correlation to the DFT, the IDFT, and vector convolution, all of these problems have the same asymptotic complexity within a logarithmic factor.

Some algorithms for manipulating univariate polynomials can be adapted to performing the analogous operations over the integers and vice versa (see Section 18.3.4).

Lastly, Section 18.3.5 examines the case of polynomials in more than one indeterminates.

The fast computations with integers and polynomials rely on the FFT-based fast multiplication, where the two main alternatives are the classical and Karatsuba's algorithms. They prevail for the inputs of smaller sizes for multiplication and even more so for computations reduced to multiplication [8,50]. Moreover, the classical polynomial multiplication is a little more stable numerically. In current libraries and software packages, such as GMP, MAPLE, and Mathematica, one can find all three multiplication algorithms. Potentially competitive is the technique of binary segmentation (see Chapter 17), which is effective for polynomial computations in finite fields and in other cases where the input coefficients are represented by short binary integers.

18.3.1   Multiplication, Division, and Variable Shift

Multiplication of univariate polynomials is a basic operation, to which we ultimately reduce other fundamental operations. This section covers multiplication of polynomials, their division with remainder, computing the reciprocal of a polynomial, and the transformation of a polynomial under a shift of the variable. All these operations have the same asymptotic complexity as multiplication. Clearly, the "linear" operations of addition, subtraction, and multiplication by a constant for degree n polynomials can be performed in $n + 1$ ops.

Polynomial multiplication is the computation of the coefficients of the polynomial $w(x) = u(x)v(x)$ given the coefficients of two polynomials $u(x) = \sum_{i=0}^{m} u_i x^i$ and $v(x) = \sum_{i=0}^{n} v_i x^i$. We have already solved this problem in $\mu(K)$ ops, for $K = m + n + 1$, in Section 18.2.2 as the problem of computing the convolution of two vectors.

Given polynomials $u(x)$ and $v(x)$ of degrees m and n, respectively, polynomial division with a remainder is the computation of the coefficients of the unique pair of polynomials $q(x) = \sum_{g=0}^{m-n} q_g x^g$ (quotient) and $r(x) = \sum_{h=0}^{n-1} r_h x^h$ (remainder) such that

$$u(x) = q(x)v(x) + r(x), \quad \deg r(x) < n$$

$r(x)$, denoted $u(x) \bmod v(x)$, is also called the residue of $u(x)$ modulo $v(x)$.

A related problem is the computation of the reciprocal of a polynomial. For a natural n and a polynomial $u(x)$, $u(0) \neq 0$, compute $w(x) \bmod x^n$, that is, compute the first n coefficients of the formal power series $w(x)$ where $w(x)u(x) = 1$. Both polynomial division and computation of the reciprocal can be reduced to polynomial multiplication [1,13,18,91] and solved in $O(\mu(n))$ ops.

Another reduction of polynomial division to the FFT is direct, by means of Toom's techniques of evaluation–interpolation with no reduction to polynomial multiplication. This is easily done where it is known that $r(x) \equiv 0$, but see [95] on such a reduction to the FFT for approximate division of any pair of polynomials.

Over any ring of constants, multiplication, division, and the reciprocal computation have the same complexity bounds up to within constant factors, since each of the problems can be solved by means of a constant number of applications of any one of the others [1,13,91].

Shifting the variable, also known as a **Taylor shift**, is the problem of computing the coefficients $q_0(\Delta), q_1(\Delta), \ldots, q_n(\Delta)$ of the polynomial

$$q(y) = p(y + \Delta) = \sum_{h=0}^{n} p_h \left(y + \Delta\right)^h = \sum_{g=0}^{n} q_g(\Delta) y^g,$$

given a scalar Δ and the coefficients $p_0, p_1, \ldots, p_n$ of a polynomial $p(x)$. This is reduced to vector convolution and thus solved in $O(\mu(n))$ ops [2,13,91].

The practical impact of using FFT and the fast convolution on computer algebra is more limited than one may expect. A major reason is the growth of the precision required for representing the auxiliary parameters [43].

18.3.2 Evaluation, Interpolation, and Chinese Remainder Computations

Polynomial evaluation is a classical problem. Given the coefficients $p_0, \ldots, p_n$, compute the value of a polynomial

$$p(x) = p_0 + p_1 x + \cdots + p_n x^n,$$

at a point x_0. The so-called Horner's rule (invented by Ch'in Chiu-Shao in medieval China in the thirteenth century, and also published by Newton in 1669, 150 years prior to Horner in 1819) is the decomposition

$$p(x_0) = (\cdots ((p_n x_0 + p_{n-1}) x_0 + p_{n-2}) x_0 + \cdots + p_1) x_0 + p_0.$$

It uses n multiplications and n additions and is optimal [82], although about 50% of multiplications can be saved if we precondition the coefficients and count only operations depending on x [73].

More generally, given the coefficients p_i of a polynomial $p(x)$, one may need to evaluate $p(x)$ on a fixed set of points $\{x_0, \ldots, x_{K-1}\}$. The problem can be reduced to K applications of Horner's rule and solved in $2Kn$ ops. Alternative reduction to polynomial multiplications and divisions yields an $O(\mu(m) \log m)$ algorithm, where $m = \max\{n, K\}$. The cost turns into $O(m \log^2 m)$ provided $\mu(m) = O(m \log m)$, where polynomial arithmetic is implemented based on the FFT. We show the two stages of this approach referring the reader to [1,13,18,91] for further details.

Suppose $K \leq 2^k \leq n + 1$. Note that $p(x_i) = p(x) \bmod (x - x_i), i = 0, \ldots, K - 1$. The first stage is the fan-in process. Successively compute the polynomials $m_i^{(j)} = m_{2i}^{(j-1)} m_{2i+1}^{(j-1)}$, for $i = 0, \ldots, \frac{K}{2^j} - 1$ and $j = 1, \ldots, k - 1$, beginning with the moduli $m_i^{(0)} = x - x_i$ for all i. The degree of $m_i^{(j)}$ in x is 2^j. So, given j and $m_i^{(j-1)}$, we compute all "supermoduli" $m_i^{(j)}$ in $2^{k-j} \mu(2^j)$ ops.

The second stage is the fan-out process. Since $m_0^{(k-1)} = \prod_{i=0}^{K/2-1} m_i$ and $m_1^{(k-1)} = \prod_{i=K/2}^{K-1} m_i$, we have

$$p(x) \bmod (x - x_i) = \left(p(x) \bmod m_0^{(k-1)} \right) \bmod (x - x_i), \quad i = 0, \ldots, K/2 - 1,$$

$$p(x) \bmod (x - x_i) = \left(p(x) \bmod m_1^{(k-1)} \right) \bmod (x - x_i), \quad i = K/2, \ldots, K - 1.$$

Thus, two polynomial divisions reduce the evaluation of $p(x)$ to the evaluation of two polynomials of roughly half degree. This stage continues recursively and has the same asymptotic complexity as the fan-in stage, since polynomial division and multiplication have the same asymptotic complexity.

The overall complexity is $O(\mu(n) \log K)$. The algorithm can be adapted to the case $K > n + 1$ with complexity $O((K/n)\mu(n) \log K)$. Hence, the bound $O(\mu(m) \log m)$, stated earlier, for $m = \max\{n, K\}$.

In short, evaluation on a set of points reduces to multiplications, which can be implemented based on the DFT. The computation of the latter is, obviously, a special case of the evaluation problem. Thus, all basic problems seen so far have the same complexity within polylogarithmic factors. This is also the case for polynomial interpolation, which is the inverse of the polynomial evaluation problem and is stated as follows: Given two sets of scalars,

$$\{x_i : i = 0, \ldots, n; \; x_i \neq x_j \text{ for } i \neq j\} \quad \{r_i : i = 0, \ldots, n\},$$

evaluate the coefficients $p_0, p_1, \ldots, p_n$ of the polynomial $p(x)$ of Equation 18.1 satisfying $p(x_i) = r_i, i = 0, 1, \ldots, n$. The problem always has a unique solution. The classical interpolation algorithms compute it in $O(n^2)$ ops [35,58], but with FFT we only need $O(n \log^2 n)$ ops. This bound is optimal up to the factor $O(\log n)$ [18,25].

The fast algorithm [1,13,91] uses the Lagrange interpolation formula:

$$p(x) = L(x) \sum_{i=0}^{n} \frac{r_i}{L'(x_i)(x - x_i)}, \quad \text{where } L(x) = \prod_{i=0}^{n}(x - x_i), \tag{18.3}$$

$L'(x_i) = \prod_{k=1, \, k \neq i}^{n}(x_i - x_k)$ for $i = 0, 1, \ldots, n$, and $L'(x)$ is the formal derivative of $L(x)$. Interpolation then reduces to application of, at first, the fan-in method to computing the polynomials $L(x)$ and $L'(x)$, secondly, the evaluation algorithm to finding the values $L'(x_i)$ for all i, and, thirdly, polynomial multiplications to obtain $p(x)$ from (18.3). The overall cost is $O(\mu(n) \log n)$.

The Chinese remainder problem generalizes interpolation to its univariate polynomial version, though its name comes from its application to the integers (cf. Section 18.3.4), by Chinese mathematicians in the second century AD or even earlier [73]. Let us be given the coefficients of $2h$ polynomials $m_i(x), r_i(x)$, $i = 1, \ldots, h$, where $\deg m_i(x) > \deg r_i(x)$ and where the polynomials $m_i(x)$ are pairwise relatively prime, that is, pairwise have only constant common divisors or, equivalently, $\gcd(m_i, m_j) = 1$ when $i \neq j$. Then we are asked to compute the unique polynomial $p(x) = p(x) \bmod \prod_{i=1}^{h} m_i(x)$, such that $r_i(x) = p(x) \bmod m_i(x), i = 1, \ldots, h$. When every $m_i(x)$ is of the form $x - x_i$, we come back to the interpolation problem.

The Chinese remainder theorem states that there always exists a unique solution to this problem. The importance of this theorem cannot be overestimated, since it allows us to reduce computation with a high-degree polynomial to similar computations with several smaller-degree polynomials, namely, to computations modulo each $m_i(x)$. Then the **Chinese remainder algorithm** combines the results modulo each $m_i(x)$ to yield $p(x)$.

There are two approaches to recovering $p(x) = p(x) \bmod \prod_{i=1}^{h} m_i(x)$ from $r_i(x)$ and $m_i(x)$, $i = 1, \ldots, h$. The first uses essentially Lagrange's interpolation formula, simply by generalizing the polynomial values in formula (18.3) to polynomial remainders. The second approach is named after Newton and is incremental, in the sense that successive steps compute $p_{k-1}(x) = p(x) \bmod \prod_{i=1}^{k-1} m_i(x)$, for $k = 1, 2, \ldots, h$, and the last step gives the final result:

$$p_k(x) = p_{k-1}(x) + \left(\left(r_k(x) - p_{k-1}(x) \right) s_k(x) \bmod m_k(x) \right) \prod_{i=1}^{k-1} m_i(x),$$

$$\text{where } s_k(x) \prod_{i=1}^{k-1} m_i(x) \bmod m_k(x) = 1, \quad k = 1, 2, \ldots, h, \quad p_h(x) = p(x).$$

Let $N = \sum_{i=1}^{h} d_i$, $d_i = \deg m_i(x)$. Then the overall complexity of computing $p_h(x) = p(x)$ by this algorithm is $O(\mu(N)h + \sum_{i=1}^{h} \mu(d_i) \log d_i)$, which, based on the FFT, turns into $O(Nh \log N)$. The algorithm has a probabilistic version [22]. Further details can be found in [1,13,18,23,73,91,130].

18.3.3 GCD, LCM, and Padé Approximation

In this section, we study the computation of the **greatest common divisor (GCD)** and the **least common multiple (LCM) of two polynomials**, with applications to polynomial and rational computations.

The classical solution method is the Euclidean algorithm, which is a major general tool for many algebraic and numerical computations. Despite its ancient origins, the problem of computing greatest common divisors, with its numerous facets, is still an active area of research.

gcd$(u(x), v(x))$, the GCD of two polynomials

$$u(x) = \sum_{i=0}^{m} u_i x^i, \; v(x) = \sum_{j=0}^{n} v_j x^j, \quad m \ge n \ge 0, \quad u_m v_n \ne 0,$$

is their common divisor of the highest degree in x. It is unique if it is scaled to be monic. Otherwise it is unique up to within constant factors. For example,

$$\gcd\left(x^5 + x^4 + x^3 + x^2 + x + 1, x^4 - 2x^3 + 3x^2 - x - 7\right) = x + 1,$$

because $x + 1$ divides both polynomials and they have no common divisor of degree greater than or equal to two.

Algorithm 18.1 (Euclid's). *Set* $u_0(x) = u(x)$, $v_0(x) = v(x)$. *Compute*

$$u_{i+1}(x) = v_i(x),$$
$$v_{i+1}(x) = u_i(x) \bmod v_i(x) = u_i(x) - q_{i+1}(x)v_i(x), \quad i = 0, 1, \ldots, \ell - 1, \qquad (18.4)$$

where
$q_{i+1}(x)$ is the quotient polynomial
ℓ is such that $v_\ell(x) = 0$
At the end of this process, $u_\ell(x) = \gcd(u(x), v(x))$.

The correctness of the algorithm can be deduced from the equation

$$\gcd\left(u_i(x), v_i(x)\right) = \gcd\left(u_{i+1}(x), v_{i+1}(x)\right),$$

which holds for all i. Euclid's algorithm only involves arithmetic operations, so that the output coefficients of the GCD are rational functions in the input coefficients. The algorithm involves $O(m^2)$ ops, but using a matrix representation of the recurrence and the fan-in method yields an $O(\mu(m) \log m)$ bound [13,91].

The sequence $v_1(x), v_2(x), \ldots, v_l(x)$ is called a polynomial remainder sequence. It can be generalized to the sequence of remainder polynomials obtained if the division step (18.4) is substituted by

$$a_{i+1} v_{i+1}(x) = b_i u_i(x) \bmod v_i(x) = b_i u_i(x) - q_{i+1}(x)v_i(x), \quad i = 0, 1, \ldots, \ell - 1,$$

where the a_{i+1} and b_i, for $i = 0, \ldots, \ell - 1$, are scalars. When all scalars are units we recover the original equation. The polynomial pseudo-remainder sequence is obtained by setting $a_{i+1} = 1$ and $b_i = c_i^{d_i}$, for $i = 0, \ldots, \ell - 1$, where c_i is the leading coefficient of $v_i(x)$ and $d_i = \deg u_i(x) - \deg v_i(x) + 1$. There exists the third remainder sequence, called the subresultant sequence, which is closely related to the Sylvester resultant (see Chapter 17). These variations aim at palliating the swell of the intermediate coefficients. For details, refer to [1,5,13,23,37,50,130].

lcm$(u(x), v(x))$, the least common multiple (LCM) of the two polynomials $u(x)$ and $v(x)$, is their common multiple of the minimum degree. In the previous example, the LCM is $x^8 - 2x^7 + 4x^6 - 3x^5 - 3x^4 - 3x^3 - 4x^2 - x - 7$. Given $u(x)$, $v(x)$, and their GCD, we can immediately compute

$$\mathrm{lcm}(u(x), v(x)) = \frac{u(x)v(x)}{\gcd(u(x), v(x))}.$$

The cited references for the GCD also have alternative algorithms for the LCM that support the asymptotic complexity bound $O(\mu(m) \log m)$.

In the (m, n) Padé approximation problem, we are given two natural numbers m and n and the first $N + 1 = m + n + 1$ Taylor coefficients of an analytic function $V(x)$ decomposed at $x = 0$, and we seek two polynomials $R(x)$ and $T(x)$ satisfying the relations

$$R(x) - T(x)V(x) = 0 \bmod x^{N+1}, \quad N = m + n, \quad \deg T(x) \le n, \quad \deg R(x) \le m.$$

This is actually a special case of rational interpolation, also called rational function reconstruction (cf. [8,13,50,91] on both subjects). Its complexity is $O(\mu(N) \log N)$ ops.

Padé approximation is closely related to computing the minimum span for a linear recurrence, also known as the Berlekamp–Massey problem and having important applications to algebraic coding theory, sparse polynomial interpolation, and parallel matrix computations [7,13,59,71,91,130]. Given a natural s and $2s$ numbers $v_0, \ldots, v_{2s-1}$, compute the minimum natural $n \le s$ and n numbers $t_0, \ldots, t_{n-1}$ such that $v_i = t_{n-1}v_{i-1} + \cdots + t_0v_{i-n}$, for $i = n, n+1, \ldots, 2s - 1$. Its solution can be obtained by extending the solution of Padé approximation problem and requires $O(\mu(s) \log s)$ ops [7,20,90,91].

18.3.4   Integer Arithmetic

The algorithms for some basic operations on univariate polynomials can be applied to the integers and vice versa, and in many cases the complexity estimates do not change significantly when we shift from arithmetic operations with polynomials to binary operations with integers [1,13,73]. We illustrate this for the multiplication and GCD algorithms.

The basic premise of the integer/polynomial correlation, which goes into both directions, is that any binary integer can be thought of as a polynomial with coefficients in $\{0, 1\}$. For instance, the binary representation of 24 a 11000 corresponds to the univariate polynomial $u(x) = x^4 + x^3$ of degree 4 with the coefficient vector $[1, 1, 0, 0, 0]$. Operating with integers, however, we must take care of the carry bits, somewhat complicating the extension of effective algorithms from polynomials to integers. Nonetheless, historically, the fast algorithms for division, GCD, and Chinese Remaindering first appeared for integers and only later for polynomials.

Let $M(N)$ denote the complexity of integer multiplication modulo $2^N + 1$, Assume that for a pair of N-bit integers u and v,

$$u = \sum_{i=0}^{N-1} u_i 2^i, \quad v = \sum_{i=0}^{N-1} v_i 2^i, \quad u_i, v_i \in \{0, 1\},$$

we seek the integer uv. The classical algorithm has complexity $O(N^2)$. Karatsuba in [72] recursively applies the equation

$$uv = U_0 V_0 \left(1 - 2^{N-1}\right) + (U_1 + U_0)(V_1 + V_0) 2^{N-1} + U_1 V_1 \left(2^{2N-2} - 2^{N-1}\right),$$

where
$u = U_0 + 2^{N-1} U_1$
$v = V_0 + 2^{N-1} V_1$ (see Equation 18.2)

This algorithm uses $O(N^{\log 3})$ Boolean operations, where $\log 3 = 1.5849\ldots$, and has a small overhead constant. The Toom's evaluation–interpolation techniques demonstrated for polynomial multiplication enable us to reduce the complexity to $O(N^{1+\epsilon})$ for any positive ϵ. Finally, by exploiting fast convolution and the FFT over finite rings, Schönhage and Strassen reduced the asymptotic complexity bound to $O(N \log N \log \log N)$; see [1, pp. 270–274], [13, pp. 78–79], [118,119]. Only the (obvious) information lower estimate of order N is known for $M(N)$.

Integer division is the problem where, given two positive integers u, v with the bit sizes n, m, respectively, we seek the unique pair of integers q, r for which $u = qv + r$, where $0 \le r < v$. The classical algorithm has the Boolean complexity $O(mn)$, whereas the reduction to multiplication algorithm yields the $O(M(m))$ bound, that is, the asymptotic complexity of integer multiplication and division is the same [1,13,18,73].

Given two integers u and v, their GCD is the largest integer that divides both u and v, whereas their LCM is the smallest integer divisible by both u and v. For instance, $\gcd(16, 24) = 8$, whereas $\text{lcm}(16, 24) = 48$. The GCD of a pair of positive integers less than 2^n can be computed in $O(M(n) \log n) = O(n \log^2 n \log\log n)$ Boolean operations [1,13,60,73].

The algorithm and the Boolean complexity estimates have nontrivial extensions (using also the associated remainder sequences) to the reconstruction of a rational number with bounded numerator and bounded scaled denominator, both from the value of the ratio modulo a fixed large integer and from an approximation to the ratio within a fixed small error [50,60,77,107].

Given the integer residues r_i with respect to fixed integer moduli m_i, for $i = 0, 1, \ldots, k$, where $\gcd(m_i, m_j) = 1$ for $i \neq j$, the Chinese remainder algorithm computes an integer $p = p \mod \prod_{i=0}^{k} m_i$ such that $r_i = p \mod m_i$ for all i. The Chinese remainder theorem states that there exists a unique such integer. The Boolean complexity of computing it is $O(M(N) \log N)$ ops, where $N = \sum_i \lceil \log m_i \rceil$.

An important application of the Chinese remainder theorem is in reducing the bulk of an arbitrary-precision integer computation to computations with fixed-precision integers. We map the input integers into their residues moduli m_i, then perform the computation in the finite field or ring of integers modulo m_i for each i, that is, with the precision of $\lceil \log m_i \rceil$ bits, and finally, use the Chinese remainder algorithm to compute the exact answer (see [40] on extensions).

The aforementioned technique of modular arithmetic is one of the most efficient methods for conducting integer (as well as rational) arithmetic on modern computers with highly important impact on signal processing and cryptography. Typically, the moduli used are primes that fit in a computer word. The implementation of the Chinese remainder algorithm relies either on an extension of Lagrange's formula (18.3) or on Newton's incremental approach (cf. Section 18.3.2). For a comprehensive discussion as well as other alternatives, see [1,13,18,23,73,130].

18.3.5 Multivariate Polynomials

Multivariate polynomials generalize the univariate ones. They appear in a wide variety of scientific and engineering applications (see Chapter 17). This section outlines some basic results on multiplication, evaluation, and interpolation of multivariate polynomials.

As in the univariate case, we do not consider in depth the representation problem and just assume that polynomials are represented by a coefficient vector indexed by the corresponding monomials in some order. Other operations, such as addition, subtraction, taking integer powers and division, as well as the various representations are treated extensively in [23,130].

A polynomial in n variables is of the form

$$p(x_1, \ldots, x_n) = \sum_{i_1, \ldots, i_n} p_{i_1, i_2, \ldots, i_n} x_1^{i_1} x_2^{i_2} \cdots x_n^{i_n},$$

where the coefficients correspond to distinct monomials. Recall that the degree of p in x_j is the maximum i_j for which $p_{i_1, i_2, \ldots, i_n} \neq 0$, the total degree of a monomial $x_1^{i_1} x_2^{i_2} \cdots x_n^{i_n}$ is $i_1 + \cdots + i_n$, and the total degree of the polynomial p is the maximum total degree of any monomial. If d is the maximum degree in any variable, then the total number of terms is $O(d^n)$. A polynomial most of whose coefficients are nonzero is called dense.

To compute the product of two such polynomials, we can reduce the problem to the univariate case by means of Kronecker's substitution:

$$x_1 = y, \quad x_{k+1} = y^{D_1 \ldots D_k}, \quad k = 1, \ldots, n - 1.$$

Here, $D_j = 2d_j + 1$ exceeds the degree in x_j of the product polynomial provided both input polynomials have degrees of at most d_j in $x_j, j = 1, 2, \ldots, n$. Once the univariate product is computed, we can recover the product polynomial in the n variables x_i by inverting the Kronecker map. This method yields the complexity bound $O(N \log N \log\log N)$, where $N = \prod_{i=1}^{n} D_i$.

A certain improvement for polynomials with a large number of variables in [86] relies on the Toom's evaluation–interpolation method and (over any field of constants) yields the complexity bound

$$O\left(N \log N \log \log D\right) = O\left(nD^n \log D \log \log D\right), \qquad (18.5)$$

where $D = \max\{D_1, \ldots, D_n\}$ and the number of terms is of order $N = D^n$.

The algorithm in [26] for dense polynomials is applied over the fields of characteristic 0. It also relies on the Toom's evaluation–interpolation scheme. If the product polynomial has at most T terms, each of a total degree of at most T, then the algorithm uses $O(\mu(T) \log T) = O(T \log^2 T \log \log T)$ ops. This is inferior to the estimate (18.5) under the bound D on the degree of each variable but superior under the bound T on the total number of terms.

Alternatively, the extended algorithm of [72] yields complexity $O(D^{n \log 3})$.

Evaluation and interpolation of dense multivariate polynomials may use points on a grid (or lattice) in which each variable is assigned the values in a fixed set. Let $E(d)$ and $I(d)$ denote, respectively, the complexity of evaluating and interpolating a univariate polynomial of degree bounded by $(d - 1)/2$ on d points. Then, for grids of d_j values for each variable x_j, the complexity of evaluation and interpolation of a multivariate polynomial is $d^{n-1}E(d)$ and $nd^{n-1}I(d)$, respectively, where $d = \max\{d_1, d_2, \ldots, d_n\}$. By applying the FFT, we yield the bounds $O(d^n \log^2 d)$ and $O(nd^n \log^2 d)$, for evaluation and interpolation, respectively. For dense polynomials, these bounds are satisfactory. The approach of [26] adapts the algorithm of [6] and also relies on solving a transposed Vandermonde system fast (see the end of Section 18.4.2). It yields the bounds $O(T \log T \log \log T \log t)$ and $O(T \log^2 T \log \log T)$ for evaluation and interpolation, respectively, over the fields of characteristic zero. Here t is the actual number of nonzero terms and T is any available upper bound on t. One may use the total degree of the product polynomial in order to estimate T, when no other information is available.

With multivariate polynomials, sparsity considerations become more important. Typically, the critical computation is interpolation. In addition to multiplication, computing the GCD can also be reduced to interpolation [130]. So, in the rest of this section, we concentrate on sparse multivariate interpolation.

There are two main approaches with different advantages and drawbacks. One approach is Zippel's randomized algorithm. It requires a bound on the degree in each variable, but not on the number of nonzero terms. The computation involves Vandermonde matrices (cf. Section 18.4.1), and its complexity is $O^*(nd^2t)$, where d is the maximum degree in all variables, t denotes the number of nonzero terms of the input polynomial, and $O^*(\cdot)$ indicates that some polylogarithmic factors may have been omitted. There exists a deterministic version of this algorithm with higher, but still polynomial, complexity. For further information, see [59,70,130].

Another, historically the first, approach is due to [6]. The algorithm of [6] needs no degree bounds, but uses a bound on the actual number of terms t, on which both the algorithm and its estimated complexity, $O^*(ndt)$, depend. The algorithm applies in the fields of characteristic equal to zero or to a very large positive integer. It proceeds by finding the exponents of the nonzero monomials which is reduced to the solution of the Berlekamp–Massey problem (cf. Section 18.3.3). Then, at the cost $O^*(ndt)$, the algorithm computes the corresponding coefficients, by exploiting the structure of Toeplitz and Vandermonde matrices [6], [13, Section 1.9], [70], [91, Section 3.10].

18.4 Structured Matrices

Matrices with special structure are omnipresent in computations in sciences, engineering, control, and signal and image processing. They are used extensively in both computer algebra and numerical matrix computations, especially in computation of curves and surfaces, solution of PDEs, integral equations, singular integrals, conformal mappings, particle simulation, and Markov chains [17,29, 78–80,115]. Furthermore, they are involved in some fundamental parallel and sequential algorithms for computations with general matrices [13,93,94].

Typically, the n^2 entries of an $n \times n$ **structured matrix** are readily expressed via $O(n)$ parameters, and therefore linear (rather than quadratic) memory space is sufficient for its representation. This includes, e.g., the highly important classes of Vandermonde, Cauchy, circulant, Toeplitz, Hankel, Sylvester, Loewner, and Bézout matrices. Furthermore, the computations with such matrices are closely related to computations with polynomials, thus enabling the FFT-based dramatic acceleration versus the general matrices. In particular, the FFT-based multiplication of such a structured matrix by a vector as well as the solution of a nonsingular linear system of n equations with a structured matrix of coefficients takes $O(n \log n)$ or $O(n \log^2 n)$ ops, versus $2n^2 - n$ ops required for multiplying a general $n \times n$ matrix by a vector and versus cubic time in Gaussian elimination for general linear systems. Such an acceleration is our main subject to the end of the section.

The size limitation forced us to omit many important subjects, which would naturally fit into this section, but some relevant bibliography can be found in [13,68,85,91]. We also refer the reader to [13,78,91,94,127], on some other important classes of structured matrices, such as Frobenius (companion), multilevel Toeplitz, banded, and semiseparable matrices.

18.4.1 Vandermonde and Cauchy Matrices

An $m \times n$ Vandermonde matrix V has its entries $(V)_{i,j} = v_i^j$ for $i = 0, 1, \ldots, m, j = 0, 1, \ldots, n$ [91, Section 3.4]. If $m = n$, then this Vandermonde matrix V has the determinant

$$\det V = \prod_{0 \le i < k \le n} (v_k - v_i) \tag{18.6}$$

and therefore is nonsingular if and only if $v_i \ne v_k$ for $i \ne k$.

For example, for $m = 1$ and $n = 3$ we have

$$V = \begin{pmatrix} 1 & v_0 & v_0^2 & v_0^3 \\ 1 & v_1 & v_1^2 & v_1^3 \end{pmatrix}.$$

Some authors call V^T, rather than V, a Vandermonde matrix.

A special case of a Vandermonde matrix is the Fourier matrix $\Omega = (\omega^{ij}/\sqrt{K})_{i,j=0}^{K-1}$ of the DFT, where ω is a primitive Kth root of 1.

Multiplication of a Vandermonde matrix $V = (v_i^j)$ by a vector $\mathbf{p}$ is equivalent to the evaluation at the points $v_0, \ldots, v_m$ of the polynomial with the coefficient vector $\mathbf{p}$. The solution of a linear system $V\mathbf{x} = \mathbf{v}$, where $\mathbf{x}$ and $\mathbf{v}$ are vectors and V is a nonsingular square Vandermonde matrix, is equivalent to the interpolation from the vector $\mathbf{v}$ of the polynomial values at the points $v_0, \ldots, v_m$ to the coefficient vector $\mathbf{x}$. Due to the algorithms of Section 18.3.2 both of these operations with Vandermonde matrices can be performed in $O(N \log^2 N)$ ops for $N = m + n$. The Vandermonde-polynomial correlations have also been exploited in the reverse direction, in order to improve the known methods for polynomial evaluation and interpolation [104,111].

The next class of structured matrices is named after Cauchy [91, Section 3.6]. An $m \times n$ Cauchy matrix $C = C(\mathbf{s}, \mathbf{t})$ is defined by two vectors $\mathbf{s} = [s_i]$ and $\mathbf{t} = [t_j]$, such that $s_i \ne t_j$ for $i = 0, \ldots, m-1$, $j = 0, 1, \ldots, n - 1$, and $(C)_{ij} = 1/(t_i - s_j)$. For instance, if $m = 2, n = 3$, then

$$C = C(\mathbf{s}, \mathbf{t}) \begin{pmatrix} (t_0 - s_0)^{-1} & (t_0 - s_1)^{-1} & (t_0 - s_2)^{-1} \\ (t_1 - s_0)^{-1} & (t_1 - s_1)^{-1} & (t_1 - s_2)^{-1} \end{pmatrix}.$$

Cauchy matrices extend Hilbert matrices $(1/(i+j+1))_{i,j}$ and are extended to Loewner matrices $B = ((u_i - v_j)/(s_i - t_j))_{i,j}$ [44,45].

Both multiplication of an $n \times n$ Cauchy matrix by a vector and the solution (for vector $\mathbf{x}$) of a Cauchy's nonsingular linear system of n equations can be reduced to rational and polynomial evaluation and interpolation and thus take $O(n \log^2 n)$. The algorithm of [115] approximates the product $C\mathbf{v}$ at the cost $c_\epsilon n \log n$ ops where c_ϵ depends on the approximation error ϵ and the minimum distance $\min_{i,j} |t_i - s_j|$. Some typical applications of Cauchy matrices include the study of integral equations, conformal mappings, and singular integrals [80,115].

18.4.2   Circulant, Toeplitz, and Hankel Matrices

Toeplitz and Hankel matrices is our next subject [91, Chapter 2]. These are most used and most studied structured matrices.

T is a Toeplitz matrix if $(T)_{i,j} = (T)_{i+k,j+k}$, for all positive k, that is, if all its entries are invariant in their shifts in the diagonal direction. Such a matrix is defined by its first row and first column. A 4×3 Toeplitz matrix in the following text is defined by the coefficients of the polynomial $v(x) = v_2 x^2 + v_1 x + v_0$ whose multiplication by a polynomial $u(x) = u_2 x^2 + u_1 x + u_0$ is equivalent to multiplication of this matrix by the column vector of the coefficients of $u(x)$.

$$
\begin{pmatrix} v_2 & 0 & 0 \\ v_1 & v_2 & 0 \\ v_0 & v_1 & v_2 \\ 0 & v_0 & v_1 \\ 0 & 0 & v_0 \end{pmatrix} \begin{pmatrix} u_2 \\ u_1 \\ u_0 \end{pmatrix} = \begin{pmatrix} v_2 u_2 \\ v_1 u_2 + v_2 u_1 \\ v_0 u_2 + v_1 u_1 + v_2 u_0 \\ v_0 u_1 + v_1 u_0 \\ v_0 u_0 \end{pmatrix}. \tag{18.7}
$$

In general, the product of a Toeplitz matrix of such a form by a vector is precisely the vector convolution. Furthermore, a general Toeplitz matrix T can be embedded into a matrix of such a form as its middle block of rows. Then the product of the matrix T by a vector can be readily extracted from the associated polynomial product (convolution).

H is a Hankel matrix if $(H)_{i,j} = (H)_{i-k,j+k}$, that is, if all the entries of H are invariant in their shifts in the antidiagonal direction. H is defined by its first row and its last column. For example, a 4×3 Hankel matrix is

$$
H = \begin{pmatrix} v_0 & v_1 & v_2 & v_3 \\ v_1 & v_2 & v_3 & v_4 \\ v_2 & v_3 & v_4 & v_5 \end{pmatrix}.
$$

Hankel and Toeplitz matrices are related via the *reversion matrix*

$$
J = J^{-1} = \begin{pmatrix} 0 & 0 & \cdots & 1 \\ \vdots & \vdots & \cdots & \vdots \\ 0 & 1 & \cdots & 0 \\ 1 & 0 & \cdots & 0 \end{pmatrix}. \tag{18.8}
$$

Namely, TJ and JT are Hankel matrices for any Toeplitz matrix T, whereas HJ and JH are Toeplitz matrices for any Hankel matrix H. Consequently, many computations with Hankel matrices can be reduced to computations with Toeplitz matrices, and vice versa. Hereafter, we focus on the Toeplitz class.

$C_f = C_f(\mathbf{v})$, for a vector $\mathbf{v} = [v_0, \ldots, v_{m-1}]^T$ and a scalar f, is an f-circulant $m \times n$ matrix if $(C_f)_{i,j} = v_{i-j \bmod m}$ for $i \geq j$; $(C_f)_{i,j} = f v_{i-j \bmod m}$ for $i < j$. For example, for $m = n = 4$ we have

$$C_f(\mathbf{v}) = \begin{pmatrix} v_0 & fv_3 & fv_2 & fv_1 \\ v_1 & v_0 & fv_3 & fv_2 \\ v_2 & v_1 & v_0 & fv_3 \\ v_3 & v_2 & v_1 & v_0 \end{pmatrix}.$$

1-circulant and (-1)-circulant matrices are called circulant and skew-circulant or anticirculant, respectively. An f-circulant matrix is a special Toeplitz matrix, defined by its first column or row and the scalar f. It is possible to embed an $m \times n$ Toeplitz matrix T into an $m \times (m+n-1)$ circulant matrix $C_f(\mathbf{v})$ and then read the product $T\mathbf{w}$ from the product $C_f(\mathbf{v})\mathbf{w}$.

Circulant matrix manipulation is fast due to the FFT and the following result.

THEOREM 18.1   [33]. *Let C_f be an $n \times n$ f-circulant matrix, with complex $f \neq 0$, and let $\mathbf{c}_f^T$ denote its first row. Let Ω be the $n \times n$ Fourier matrix, $(\Omega)_{i,j} = \omega^{ij}/\sqrt{n}$, $i,j = 0, 1, \ldots, n-1$, where ω is a primitive n-th root of unity, $(\Omega^H)_{i,j} = \omega^{-ij}/\sqrt{n}$, and $\Omega^H \Omega = I$. Let $D_f = diag(1, g, g^2, \ldots, g^{n-1})$, $g^n = f$, and let D be another diagonal matrix with entries given by the vector $\sqrt{n} \, \Omega D_f \mathbf{c}_f$. Then $\Omega D_f C_f D_f^{-1} \Omega^H = D$, or equivalently, $C_f = D_f^{-1} \Omega^H D \Omega D_f$, and so if the matrix D is nonsingular, then $C_f^{-1} = D_f^{-1} \Omega^H D^{-1} \Omega D_f$ is a $(1/f)$-circulant matrix.*

The theorem reduces multiplication of an $n \times n$ f-circulant matrix by a vector to vector convolution, performed in $O(n \log n)$ ops. Because of the Toeplitz-into-circulant embedding, the same bound applies to Toeplitz, and therefore also to Hankel matrices. This alternative algorithm is slightly more effective than in the beginning of this subsection because the total size of the FFTs involved is smaller.

Theorem 18.1 implies that the inverse of an f-circulant matrix is completely defined by its first column. Now let $c_f(x)$ and $c_{1/f}(x)$ denote the two polynomials with the coefficient vectors given by the first columns of the circulant matrices C_f and C_f^{-1}, respectively. Then $c_{1/f}(x) = 1/c_f(x)$ mod $x^n - f$. For $f = 0$ this relates the reciprocal of a polynomial modulo x^n to the first column of the inverse of the associated triangular Toeplitz matrix. Therefore, we can reduce the problem of polynomial division to the inversion of circulant and/or triangular Toeplitz matrices and vice versa. In both ways we arrive at the solution in $O(n \log n)$ ops [13,91].

Computing Padé approximation is equivalent to the solution of a general nonsingular Toeplitz (or Hankel) linear system. In [20] this equivalence is the basis for solving such a linear system in $O(n \log^2 n)$ ops versus $(2/3)n^3 + O(n^2)$ ops in Gaussian elimination for a general nonsingular linear system. Later we describe alternative fast solutions of Toeplitz, Hankel, Cauchy, Vandermonde, and other structured linear systems of equations in nearly linear time.

We conclude this section by combining the results on Toeplitz and Vandermonde matrices and polynomials to compute the transpose of an $n \times n$ Vandermonde matrix V. The transpose V^T can be factorized into the product of the inverse matrix V^{-1}, an f-circulant matrix, the reversion matrix J, and a diagonal matrix [91, Equation(3.4.1)]. The factorization is essentially reduced to computing the coefficients of the polynomial $v(x) = \prod_i (x - v_i)$ and the values $v'(v_j)$ for $j = 0, 1, \ldots, m$. Therefore, for $m = n$ we can multiply the matrix V^T and its inverse by a vector in $O(n \log^2 n)$ ops.

18.4.3 Bézout Matrices

Bézout matrices arise in many fundamental algebraic operations. They support computation of the GCD of two univariate polynomials at a low Boolean complexity and numerically stable solution of a system of two univariate polynomials, generalized to the solution of a system of multivariate polynomial equations [78]. This section introduces Bézout's matrices, demonstrates their structure and correlations to structured matrices studied earlier, and discusses some applications and computational complexity issues [13, Section 2.9].

Consider two polynomials, $u(x) = \sum_{i=0}^{n} u_i x^i$ and $v(x) = \sum_{i=0}^{m} v_i x^i$, of degrees n and m, respectively, where $m \leq n$. The expression

$$\frac{u(z)v(w) - v(z)u(w)}{z - w} = \sum_{i,j=0}^{n-1} b_{i,j} z^i w^j \tag{18.9}$$

is easily verified to be a polynomial in w and z. This polynomial is called the generating function of the $n \times n$ Bézout matrix B or, simply, Bezoutian of $u(x)$ and $v(x)$, with $(B)_{i,j} = b_{i,j}$. The polynomial of (18.9) is sometimes also called the Bezoutian of $u(x)$ and $v(x)$.

By definition, $B(u, v) = -B(v, u)$. Observe, as particular cases, that

$$B(u, 1) = \begin{pmatrix} u_1 & \cdots & u_n \\ \vdots & \ddots & \\ u_n & & O \end{pmatrix}, \qquad B(u, x^n) = -\begin{pmatrix} O & & u_0 \\ & \ddots & \vdots \\ u_0 & \cdots & u_{n-1} \end{pmatrix}$$

are triangular Hankel matrices. It has been proved that $B(u, v)$, for general $u(x)$ and $v(x)$, can be written in terms of matrices of this form (see, for instance, [74]):

$$B(u, v) = B(v, 1)JB(u, x^n) - B(u, 1)JB(v, x^n), \tag{18.10}$$

where J is the reversion matrix of (18.8).

There is a deeper connection of Bezoutians to Hankel matrices [13, Section 2.9]. Given univariate polynomials $u(x), v(x)$, of respective degrees n, m where $n > m$, consider the infinite sequence of values $h_0, h_1, \ldots$, in the coefficient field of $u(x), v(x)$, known as the Markov parameters and defined as follows:

$$\frac{v(x^{-1})}{xu(x^{-1})} = \sum_{i=0}^{\infty} h_i x^i \qquad \text{or, equivalently,} \qquad \frac{v(x)}{u(x)} = \sum_{i=0}^{\infty} h_i x^{-i-1}.$$

The Markov parameters generate an $n \times n$ Hankel matrix $H(u, v)$ such that $(H)_{i,j} = h_{i+j}$ for $i, j = 0, 1, \ldots, n - 1$. Given $u(x), v(x)$, we can compute the entries of $H(u, v)$ by solving a lower triangular Toeplitz system of $2n - 1$ equations in $O(n \log n)$ ops. These equations are obtained by truncating the first formal power series mentioned earlier and expressing polynomial multiplication by a Toeplitz matrix as in Equation 18.7.

If $u(x)$ and $v(x)$ are relatively prime, then $H(u, v)$ is nonsingular. Moreover, for any nonsingular $n \times n$ Hankel matrix H, there exists a pair of relatively prime polynomials $u(x), v(x)$, of degrees n and $m < n$, respectively, such that $u(x)$ is monic and $H = H(u, v)$. Matrix $H(u, v)$, for polynomials $u(x)$ and $v(x)$ whose degrees m and n satisfy $n > m$, is related to the Bezoutian $B(u, v)$ of the same polynomials by

$$B(u, v) = B(u, 1)H(u, v)B(u, 1) \quad \text{and} \quad B(u, w)H(u, v) = I_n,$$

where

$w(x)$ is a univariate polynomial of degree less than n and such that $w(x)v(x) = 1 \bmod u(x)$

I_n is the $n \times n$ identity matrix

The second expression states, in words, that the inverse of any nonsingular Hankel matrix is a Bezoutian.

Bezoutians have been studied in classical elimination theory (starting with univariate polynomials and then extended to the multivariate case [78]), in stability theory, as well as in computing the reduction of a general matrix to the tridiagonal form. The reader may refer to [13, Chapter 2], for a comprehensive introduction, and [28,44,63,74,78] for further information.

18.4.4 Toward Unification of Structured Matrix Algorithms: Compress, Operate, Decompress

The structured matrices seen so far are

- Expressed via a linear number of parameters
- Multiplied by vectors in nearly linear time
- Algorithmically related to polynomials and the FFT

Next, by following [57,62,66,81,83,89,91], we unify the study of structured matrices M with such features. We express structured matrices via their smaller rank displacements, that is, the images $L(M)$ of appropriate linear displacement operators of the Sylvester type, $L = \nabla_{A,B}$:

$$L(M) = \nabla_{A,B}(M) = AM - MB,$$

or the Stein type, $L = \Delta_{A,B}$:

$$L(M) = \Delta_{A,B}(M) = M - AMB,$$

for fixed pairs $\{A, B\}$ of operator matrices.

The operators of both types are closely related to one another.

THEOREM 18.2 $\nabla_{A,B} = A\Delta_{A^{-1},B}$ *if the operator matrix A is nonsingular, and $\nabla_{A,B} = -\Delta_{A,B^{-1}}B$ if the operator matrix B is nonsingular.*

We choose a (generally nonunique) pair of operator matrices A and B for which the rank of the displacement (called the displacement rank) of the input matrix M is small. Then we say that the matrix M is $\nabla_{A,B}$-structured or $\Delta_{A,B}$-structured. Table 18.1, which is [91, Table 1.3], shows some sample choices for the structured matrices seen so far. For all of them we choose operator matrices among the unit f-circulant (shift) matrices $C_f = C_f(\mathbf{e_1})$ (which we defined in Section 18.4.2) for appropriate scalars f and the first unit coordinate vector $\mathbf{e_1} = (1, 0, \ldots, 0)^T$, their transposes C_f^T, and the (scaling) diagonal matrices $D(\mathbf{v}) = \mathrm{diag}\,(v_i)_{i=1}^n$ for appropriate vectors $\mathbf{v} = (v_i)_{i=1}^n$. Clearly, all Toeplitz matrices are ∇_{C_0,C_1}-structured, whereas a Cauchy matrix $C(\mathbf{s}, \mathbf{t})$ is $\nabla_{D(\mathbf{s}),D(\mathbf{t})}$-structured where the same pair of vectors $(\mathbf{s}, \mathbf{t})$ appears in the input and operator matrices. More details can be found in [91]. In particular all Toeplitz matrices are ∇_{C_e,C_f}-structured for any pair of scalars e and f such that $e \neq f$ and are Δ_{C_e,C_f}-structured for any pair of scalars e and f such that $ef \neq 1$.

We call a matrix M Toeplitz-like, Hankel-like, Vandermonde-like, or Cauchy-like if the displacement $L(M)$ has a smaller rank for the respective displacement operators L. This includes many important classes of structured matrices such as the products and inverses of Toeplitz, Hankel, Vandermonde, and Cauchy matrices, block matrices with Toeplitz and Hankel blocks, Frobenius,

TABLE 18.1 Pairs of Operators $\nabla_{A,B}$ and Structured Matrices

Operator Matrices		Class of Structured	Rank of
A	B	Matrices M	$\nabla_{A,B}(M)$
C_1	C_0	Toeplitz and its inverse	≤ 2
C_1	C_0^T	Hankel and Bezout	≤ 2
$C_0 + Z_0^T$	$C_0 + C_0^T$	Toeplitz+Hankel	≤ 4
$D(\mathbf{t})$	C_0	Vandermonde	≤ 1
C_0	$D(\mathbf{t})$	Inverse of Vandermonde	≤ 1
C_0^T	$D(\mathbf{t})$	Transposed Vandermonde	≤ 1
$D(\mathbf{s})$	$D(\mathbf{t})$	Cauchy	≤ 1
$D(\mathbf{t})$	$D(\mathbf{s})$	Inverse of Cauchy	≤ 1

Loewner, and Pick matrices (the latter matrices define the Nevanlinna/Pick celebrated problems of rational interpolation [81], [91, Section 3.8]).

To obtain displacements of smaller ranks for a block Toeplitz or block Hankel matrix M, one can either use block variations of the operator matrices C_f and C_f^T or apply displacement operators with the operator matrices C_f and C_f^T to appropriate permutations of the rows and columns of the matrix M that turn it into a block matrix with Toeplitz or Hankel blocks. We also note that the reflection matrix J relates the classes $\{\mathbf{T}\}$ and $\{\mathbf{H}\}$ of Toeplitz-like and Hankel-like matrices to one another so that $\{\mathbf{T}\}J = J\{\mathbf{T}\} = \{\mathbf{H}\}$ and $J\{\mathbf{H}\} = \{\mathbf{H}\}J = \{\mathbf{T}\}$.

Now recall that an $n \times n$ matrix M has a rank of at most r if and only if this matrix is the sum of r outer products of r pairs of vectors, $M = \sum_{j=1}^{r} \mathbf{g}_j \mathbf{h}_j^T = GH^T$ for $n \times r$ matrices $G = (\mathbf{g}_j)_{j=1}^{r}$ and $H = (\mathbf{h}_j)_{j=1}^{r}$. So the displacement $L(M)$ of a rank r can be expressed via $2nr$ parameters. This is strong compression versus using the n^2 input entries where $r \ll n$.

By inverting the operator L we express the original structured matrix M via these $2nr$ parameters as well. Many simple inversion expressions are specified in [57], [91, Examples 1.4.1 and 4.4.1–4.4.9], and [106]. In particular we have

$$M = \sum_{k=1}^{l} D(\mathbf{g}_k)C(\mathbf{s}, \mathbf{t})D(\mathbf{h}_k),$$

for $L(M) = D(\mathbf{s})M - MD(\mathbf{t}) = \sum_{i=1}^{l} \mathbf{g}_i \mathbf{h}_i^T = GH^T$, as well as

$$(e - f)M = \sum_{j=1}^{l} C_e(\mathbf{g}_j)C_f(J\mathbf{h}_j) \quad \text{so that} \quad (e - f)JM = \sum_{j=1}^{l} JC_e(\mathbf{g}_j)C_f(J\mathbf{h}_j),$$

for $L(M) = C_e M - MC_f = \sum_{i=1}^{l} \mathbf{g}_i \mathbf{h}_i^T, e \neq f$, and the reflection matrix J.

This yields memory efficient representations for Cauchy-like, Toeplitz-like, and Hankel-like matrices M, respectively, and reduces their multiplication by a vector essentially to at most r and at most $2r$ multiplications of Cauchy's and e- and f-circulant matrices by vectors, respectively. Thus, for smaller ranks r we can multiply such a matrix by a vector in nearly linear time. Similar representations of Vandermonde-like matrices have the same features.

To simplify the computations with structured matrices, we combine the latter techniques of their compression and decompression with some simple results on operating with the short displacement generators G and H instead of the entries of the structured input matrices.

THEOREM 18.3 *(see [91, Theorem 1.5.4]). For any 5-tuple $\{A, B, C, K, N\}$ of matrices of compatible sizes and a nonsingular matrix M, we have*

$$\nabla_{A,C}(KN) = \nabla_{A,B}(K)N + K\nabla_{B,C}(N),$$
$$\nabla_{B,A}(M^{-1}) = -M^{-1}\nabla_{A,B}(M)M^{-1}.$$

Similar results hold for the same operations where the Stein operator is used. Simple expressions in terms of displacement generators G and H can be readily obtained also for linear combinations of structured matrices, their transposes and blocks.

This suggests operating with short displacement generators instead of the input entries, thus saving computer time and memory and unifying the algorithms for various classes of structured matrices. At the end of the computations, we should decompress the output structured matrices from their computed short displacement generators.

18.4.5 Unified Solution of Structured Linear Systems and Extensions

Due to Theorem 18.3, we can devise algorithms for a structured nonsingular linear system of n equations $My = b$ by manipulating with short displacement generators of the matrix M and the auxiliary matrices. The algorithms can be unified for all classes of structured matrices expressed via their short displacement generators. To a large extent, our study applies to computations with banded and semiseparable (rank structured) matrices [127] and partly to sparse matrices, particularly to the ones whose associated graphs have small separator families [53,54,75,103].

Furthermore, the second statement of Theorem 18.3 reduces the inversion of an $n \times n$ matrix M having a displacement rank r to solving $2r$ linear systems of equations with matrices M and M^T, versus n linear systems of equations representing the inverse of general matrix.

Gaussian elimination uses $(2/3)n^3 + O(n^2)$ ops for a nonsingular linear system of n equations, but can be modified to run in $O(n^2)$ ops for structured systems whose coefficient matrix is given with a short displacement generator. For Toeplitz-like and Hankel-like systems this fast implementation is not compatible with pivoting (that is, row and column interchange) and thus is numerically unstable, but the problem can be overcome with the transition to Vandermonde-like or Cauchy-like matrices by means of "Displacement Transformation" (see the next subsection).

Recursive application of the 2×2 block Gaussian elimination yields the solution y (as well as the determinant of the input matrix) in $O(\mu(n) \log n)$ ops for μ defined in Section 18.2.2 (cf. [91, Chapter 5] and the bibliography therein). This divide-and-conquer algorithm was proposed by Morf in 1974 and 1980 and by Bitmead and Anderson in 1980 for Toeplitz-like matrices and extended to Cauchy-like matrices in 1996 in [110] and general structured matrices in [81,89] and [91, Chapter 5]. Stable numerical implementation of the algorithm is available for a symmetric positive definite Toeplitz input, but symmetrization of an unsymmetric matrix generally leads to numerical problems.

Iterative numerical algorithms are an alternative. The GMRES and the Conjugate Gradient algorithm with its variants [58, Chapter 9], [117] are highly effective for structured matrices as long as the iterations converge rapidly because every iteration step is essentially multiplication of an input matrix and its transpose by vectors. To yield rapid convergence, one should precondition the input matrix. Effective circulant preconditioners are well known for Toeplitz matrices and some classes of Toeplitz-like matrices [30,31,125]. More general randomized preconditioners are highly effective for both structured and general matrices [97,101,109].

The cited algorithms have linear convergence but can be accelerated when the approximations come closer to the solution. Namely, suppose we have an approximate inverse matrix X_0 such that the residual matrix $I - MX_0$ has norm, ν, noticeably less than one. Then we can apply Newton's iteration $X_{i+1} = 2X_i - X_i M X_i$, for which we have $I - MX_{i+1} = (I - MX_i)^2 = (I - MX_0)^{2^{i+1}}$ and therefore $\|I - MX_{i+1}\| \le \nu^{2^{i+1}}$ for $i = 0, 1, \ldots$. This means quadratic convergence, but one must nontrivially modify the algorithm to preserve both matrix structure and rapid convergence [91, Chapter 6], [98,102,105]. The iteration is closely related to the iterative refinement of the solutions of linear systems of equations extended to the refinement of approximate inverses [96].

Numerical iterative refinement [58, Section 3.5] of the solution to a linear system $Mx = b$ has a counterpart in symbolic (algebraic) computations. It is called Hensel's lifting, is applied where all input values are integers, and begins with computing the inverse $Q = M^{-1} \bmod p$ for a fixed basic prime p such that $(\det M) \bmod p \ne 0$. For smaller primes p such that $\log p = O(\log(n\|M\|))$, this computation uses lower precision and thus is not costly under the Boolean model. Then the solution x is recursively lifted from the initial vector $x \bmod p = Qb \bmod p$ to $x \bmod p^i$ for $i = 2, 3, \ldots, h$. Formally one writes $r^{(0)} = b$ and then for $i = 0, 1, \ldots, h - 1$ computes the vectors

$$\mathbf{u}^{(i)} = Q\mathbf{r}^{(i)} \bmod p, \quad \mathbf{r}^{(i+1)} = (p\mathbf{r}^{(i)} - M\mathbf{u}^{(i)})/p.$$

$\mathbf{x}^{(i+1)} = \sum_{j=0}^{i} \mathbf{u}^{(j)} s^j$. When h grows sufficiently large, the rational vector $\mathbf{x}$ is reconstructed from $\mathbf{x} \bmod p^h$ (see Section 18.3.4).

The cost of every lifting step is essentially the cost of multiplication of structured matrices M and Q by two vectors, performed with the precision of $O(\log p)$ bits. Thus, lifting is highly effective wherever the input matrix M and its inverse Q can be multiplied by a vector in fewer ops. (This includes structured matrices that we cover in this section as well as banded matrices and, more generally, sparse matrices whose associated graph has small separator families [53,54,75,103].) For linear systems with such matrices, lifting supports nearly optimal arithmetic and the Boolean complexity.

Hensel's lifting is also important in block Wiedemann's algorithm, which is highly effective for computing the determinant and Smith's factors of a sparse and/or structured integer matrix [93].

We refer the reader to the original papers [92,99,100,108] on the details and on the extension of this algorithm to lifting in the rings (rather than fields). By using this extension, one can begin lifting with a nonprime base 2^s and thus enjoy the practical benefits of computing with binary numbers modulo powers of two.

Polynomial systems of equations can be reduced to linear systems with so-called multilevel (or quasi) Toeplitz and Hankel matrices [17,41,78]. They can be multiplied by vectors fast by using the FFT, but not so for their inverses. This undermines the efficiency of lifting algorithms and Newton's iteration but not of the iterative algorithms in [17], which can be extended even to sparse linear systems whose associated graphs have no small separator families.

18.4.6 Displacement Transformation

A distinct path to the unification of structured matrix computations relies on the observation that according to Theorem 18.3 the two pairs (A, B) and (B, C) of the operator matrices for the matrices M and N define the pair (A, C) of the operator matrices for the product MN. If the matrices M and N have small displacement ranks for the structure associated with the respective pairs (A, B) and (B, C), then the matrix MN has small displacement rank for the structure associated with the pair (A, C). This suggests the Method of Displacement Transformation), proposed in [83] and to be outline next. Suppose we must solve a linear system $M\mathbf{y} = \mathbf{b}$ where the coefficient matrix M is $\nabla_{A,B}$-structured, but the known algorithms are more effective for linear systems with matrices having the $\nabla_{A,C}$-structures. Then we can proceed as follows:

- Fix a $\nabla_{B,C}$-structured matrix N andapply Theorem 18.3 to compute a short $\nabla_{A,C}$-generator for the matrix MN.
- Compute the solution $\mathbf{x}$ of the linear system $MN\mathbf{x} = \mathbf{b}$.
- Compute and output the solution $\mathbf{y} = N\mathbf{x}$ to the original linear system.

Likewise, suppose we must solve a linear system $N\mathbf{y} = \mathbf{b}$ where the coefficient matrix N is $\nabla_{B,C}$-structured, but the known algorithms are more effective for linear systems with matrices having the $\nabla_{A,C}$-structure. Then we can proceed as follows:

- Fix a $\nabla_{A,B}$-structured matrix M and apply Theorem 18.3 to compute a short $\nabla_{A,C}$-generator for the matrix MN.
- Compute the vector $M\mathbf{b}$.
- Compute and output the solution $\mathbf{y}$ to the original linear system by solving the equivalent linear system $MN\mathbf{y} = M\mathbf{b}$.

Clearly, the same method can be applied to matrix inversion and computing determinants, and it can be readily extended to the $\Delta_{A,B}$-structured input matrices M.

Furthermore, two successive displacement transformations of the two types aforementioned can transform a $\nabla_{A,B}$-structured input into a $\nabla_{C,D}$-structured matrix for any quadruple of matrices (A, B, C, D).

The following table shows some displacement transformations among three basic classes of structured matrices, for which we write HT, C, and V to denote the classes of Hankel-like and Toeplitz-like, Cauchy-like, and Vandermonde-like matrices, respectively.

A	B	AB
HT	V	V
V	V	C
V	V	HT
C	V	V
HT	HT	HT
C	C	C

Already the transition from the class C to itself enables useful changes of the operator matrices $D(\mathbf{s})$ and $D(\mathbf{t})$ (cf. [32]), but even more interesting is the effect of Vandermonde multipliers. The transition from the HT matrices to the C matrices allows us to avoid destroying matrix structure when we perform pivoting, that is, row/column interchange, required to support numerical stability of Gaussian elimination and other matrix computations. This application alone would have already made the Method of Displacement Transformation practically valuable.

A most natural choice for the Vandermonde multipliers V is the scaled Fourier matrix $c\Omega$, which is unitary up to the choice of a scalar c. The displacement rank increases by one when we multiply by this matrix. This small deficiency, however, has been overcome in [56,61] based on the following result.

THEOREM 18.4 *(Cf. Theorem 18.1 and [33]). $C = \Omega T D_{2n} \Omega^{-1}$ is a Cauchy-like matrix, having a r-generator $G^* = \Omega G$, $H^* = \Omega D(w) H$, for $F = F_{D_n, D_{-n}} = F(D_n, D_{-n})$, that is,*

$$F_{D_n, D_{-n}} C = D_n C - C D_{-n} = G^* (H^*)^T,$$

provided that T is a Toeplitz-like matrix with an F-generator G, H for $F = F_{Z_1, Z_{-1}} = F(Z_1, Z_{-1})$,

$$F_{Z_1, Z_{-1}} T = Z_1 T - T Z_{-1} = G H^T,$$

$D_k = diag(1, w_k, w_k^2, \ldots, w_k^{n-1})$, $w_k = exp(2\pi\sqrt{-1}/k)$, *and* $\Omega = 1/\sqrt{n}(w_n^{ij})$ *is a Fourier matrix,* $\Omega^{-1} = 1/\sqrt{n}(w_n^{-ij})$, $i, j = 0, 1, \ldots, n - 1$.

Due to this theorem, the transformation of a Toeplitz-like matrix into a Cauchy-like matrix is immediately reduced to the FFT, which leads to effective practical algorithms for solving Toeplitz-like and Hankel-like linear systems [56].

Finally, as another example of applications of the Method of Displacement Transformation, we recall fast algorithms for polynomial evaluation and interpolation based on manipulation with structured matrices of the classes HT, V, and C [104,111].

18.5 Research Issues and Summary

We have reviewed the known highly effective algorithms for the discrete Fourier transform (DFT) and its inverse, as well as for some related transforms, all of them based on the celebrated fast Fourier transform (FFT) algorithm. We have showed immediate application of the FFT to computing convolution of vectors (polynomial products), which is a fundamental operation of computer algebra and signal processing. We have also demonstrated further applications to other basic operations with integers, univariate and multivariate polynomials, power series, and to the fundamental computations with circulant, Toeplitz, Hankel, Vandermonde, Cauchy, and Bézout matrices, as well as to other structured matrices related to the aforementioned classes of matrices via the associated linear operators. We exemplified such relations and some basic techniques for extending the power of the FFT to numerous computational problems, and we supplied pointers to further bibliography.

Some of these techniques are quite recent, and further research is very promising, in particular, via the study of structured matrices and polynomial systems of equations see, e.g., [14,17,41,42,76,78].

18.6 Further Information

The main research journals in this area are (alphabetically) *Computers and Mathematics (with Applications), IEEE Transaction on Signal Processing, Journal of Symbolic Computation, Linear Algebra and Its Applications, Mathematics of Computation, SIAM Journal of Computing, SIAM Journal of Matrix Analysis and Applications,* and *Theoretical Computer Science.* New implementations are reported in the *ACM Transactions on Mathematical Software.*

The main annual conferences in this area are the *International Signal Processing and Expo,* the *ACM–SIGSAM International Symposium on Symbolic and Algebraic Computation,* the *ACM Symposium on Theory of Computing,* the *ACM–SIAM Symposium on Discrete Algorithms,* the *European Symposium on Algorithms,* the *IEEE Symposium on Foundations of Computer Science,* the *Symposium on Parallel Algorithms and Architectures,* and the *Workshop on Symbolic-Numeric Computation.* Many conferences on Matrix Computations, Applied Linear Algebra, Computational and Applied Math, and Networks and Systems (e.g., the SIAM, ICIAM, and MTNS conferences) regularly include Sessions on Computations with Structured Matrices.

The following books contain general information on the topics of this chapter [1,13,18,23,50, 73,91,130]. Some references on transforms and convolution have been listed at the beginning of Section 18.2.

Parallel algebraic computation is an important subject, which we have barely touched. We refer the interested reader to [13,65,85,87].

Defining Terms

Chinese remainder algorithm: The algorithm recovering a unique integer (or polynomial) p modulo M from the h residues of p modulo pairwise relatively prime integers (or, respectively, polynomials) $m_1, \cdots, m_h$, where M is the product $m_1 \cdots m_h$. (The residue of p modulo m is the remainder of the division of p by m.)

Convolution of two vectors: A vector that contains the coefficients of the product of two polynomials whose coefficients make up the given vectors. The positive and negative wrapped convolutions are the coefficient vectors of the two polynomials obtained via reduction of the polynomial product by $x^n + 1$ and $x^n - 1$, respectively.

Determinant (of an $n \times n$ matrix): A polynomial of degree n in the entries of the matrix with the property of being invariant in the elementary transformations of a matrix used in Gaussian elimination; the determinant of the product of matrices is the product of their determinants; the determinant of a triangular matrix is the product of its diagonal entries; the determinant of any matrix is nonzero if and only if the matrix is invertible (nonsingular).

Discrete Fourier transform (DFT): The vector of the values of a given polynomial at the set of all the Kth roots of unity. The inverse discrete Fourier transform (IDFT) of a vector **v:** The vector of the coefficients of a polynomial whose values at the Kth roots of unity form a given vector **v.**

Divide-and-conquer: A general algorithmic method of dividing a given problem into two (or more) subproblems of smaller sizes that are easier to solve, then synthesizing the overall solution from the solutions to the subproblems.

Fast Fourier transform (FFT): An algorithm that uses $1.5K \log K$ ops in order to compute the DFT at the Kth roots of unity, where $K = 2^k$, for a natural k. It also computes the IDFT in $K + 1.5K \log K$ ops.

Greatest common divisor (GCD) of two or several integers (or polynomials): The largest positive integer (or a polynomial of the largest degree) that divides both or all of the input integers (or polynomials).

Interpolation: The computation of the coefficients of a polynomial in one or more variables, given its values at a set of points. The inverse problem is the evaluation of a given polynomial on a set of points.

Least common multiple (LCM) of two or several integers (or polynomials): The smallest positive integer (or a polynomial of the smallest degree) divisible by both or by all of the input integers (or polynomials).

Ops: Arithmetic operations, i.e., additions, subtractions, multiplications, or divisions.

Structured matrix: A matrix whose every entry can be derived by a formula depending on a smaller number of parameters, typically on the order of at most $m + n$ parameters for an $m \times n$ matrix, as opposed to its mn entries. For instance, an $m \times n$ Cauchy matrix has $\frac{1}{s_i - t_j}$ as the entry in row i and column j for $m + n$ parameters s_i and t_j, $1, \ldots, m, j = 1, \ldots, n$.

Taylor shift of the variable: Recovery of the coefficient vector of a polynomial after a linear substitution of its variable, $y = x + \Delta$ for x, where Δ is a fixed shift value.

Acknowledgments

This material is based on work supported in part by IST Programme of the European Union as a Shared-cost RTD (FET Open) Project under Contract No IST-006413-2 (ACS-Algorithms for Complex Shapes) (first author) and by NSF Grant CCR 9732206 and PSC CUNY Awards 67297-0036 and 68291-0037 (second author).

References

1. Aho, A.V., Hopcroft, J.E., and Ullman, J.D., *The Design and Analysis of Computer Algorithms.* Addison-Wesley, Reading, MA, 1974.
2. Aho, A., Steiglitz, K., and Ullman, J., Evaluating polynomials at fixed set of points. *SIAM J. Comp.,* 4, 533–539, 1975.
3. Bailey, D.H., A high-performance FFT algorithm for the Cray-2. *J. Supercomp.,* 1(1), 43–60, 1987.
4. Bailey, D.H., A fortran-90 based multiprecision system. *ACM Trans. Math. Soft.,* 21(4), 379–387, 1995.
5. Basu, S., Pollack, R., and Roy, M.-F., *Algorithms in Real Algebraic Geometry,* volume 10 of *Algorithms and Computation in Mathematics.* Springer-Verlag, Berlin, Germany, 2003.

6. Ben-Or, M. and Tiwari, P., A deterministic algorithm for sparse multivariate polynomial interpolation. In *Proceedings of ACM Symposium Theory of Computing*, ACM Press, New York, 1988, pp. 301–309.

7. Berlekamp, E.R., *Algebraic Coding Theory*. McGraw-Hill, New York, 1968.

8. Bernstein, D.J., Fast multiplication and its applications. Preprint, 2003. Available from http://cr.yp.to/papers.html

9. Biehl, I., Buchmann, J., and Papanikolaou, T., LiDIA: A library for computational number theory. Technical Report SFB 124-C1, Universität des Saarlandes, Saarbrücken, Germany, 1995. http://www-jb.cs.uni-sb.de/linktop/LiDIA.

10. Bini, D. and Bozzo, E., Fast discrete transform by means of eigenpolynomials. *Comp. Math. (with Appl.)*, 26(9), 35–52, 1993.

11. Bini, D. and Capovani, M., Spectral and computational properties of band symmetric Toeplitz matrices. *Linear Algebra Appl.*, 52/53, 99–126, 1983.

12. Bini, D. and Favati, P., On a matrix algebra related to the Hartley transform. *SIAM J. Matrix Anal. Appl.*, 14(2), 500–507, 1993.

13. Bini, D. and Pan, V.Y., *Polynomial and Matrix Computations*, volume 1: *Fundamental Algorithms*. Birkhäuser, Boston, MA, 1994.

14. Bini, D.A., Pan, V.Y., and Verschelde, J., editors. Special Issue on Symbolic–Numerical Algorithms, *Theoretical Computer Science*, 409(2), 255–268, 2008.

15. Blahut, R.E., *Fast Algorithms for Digital Signal Processing*. Addison-Wesley, New York, 1984.

16. Bluestein, L.I., A linear filtering approach to the computation of the discrete Fourier transform. *Northeast Electron. Res. Eng. Meeting Record*, 10, 218–219, 1968.

17. Bondyfalat, D., Mourrain, B., and Pan, V.Y., Computation of a specified root of a polynomial system of equations using eigenvectors. *Linear Algebra Appl.*, 319, 193–209, 2000.

18. Borodin, A. and Munro, I., *The Computational Complexity of Algebraic and Numeric Problems*. American Elsevier, New York, 1975.

19. Bracewell, R., *The Fourier Transform and Its Applications, 3rd edition*. McGraw-Hill, New York, 1999.

20. Brent, R.P., Gustavson, F.G., and Yun, D.Y.Y., Fast solution of Toeplitz systems of equations and computation of Padé approximations. *J. Algorithms*, 1, 259–295, 1980.

21. Brigham, E.O., *The Fast Fourier Transform and Applications*. Prentice-Hall, Englewood Cliffs, NJ, 1988.

22. Brönnimann, H., Emiris, I.Z., Pan, V.Y., and Pion, S., Sign determination in residue number systems. *Theoretical Computer Science*, Special Issue on Real Numbers and Computers, 210, 173–197, 1999.

23. Buchberger, B., Collins, G.E., and Loos, R., editors, *Computer Algebra: Symbolic and Algebraic Computation, 2nd edition*. volume 4 of *Computing Supplementum*. Springer-Verlag, Wien, Austria, 1982.

24. Bruun, G., z-Transform DFT filters and FFTs. *IEEE Trans. Acoustics, Speech Signal Process. (ASSP)*, 26(1), 56–63, 1978.

25. Bürgisser, P., Clausen, M., and Shokrollahi, M.A., *Algebraic Complexity Theory*. Springer-Verlag, Berlin, Germany, 1997.

26. Canny, J., Kaltofen, E., and Lakshman, Y., Solving systems of non-linear polynomial equations faster. In *Proceedings of ACM International Symposium on Symbolic Algebraic Computations (ISSAC '89)*, ACM Press, New York, 1989, pp. 121–128.

27. Cantor, D.G. and Kaltofen, E., On fast multiplication of polynomials over arbitrary rings. *Acta Inform.*, 28(7), 697–701, 1991.

28. Cardinal, J.-P. and Mourrain, B., Algebraic approach of residues and applications. In *The Mathematics of Numerical Analysis*, Renegar, J., Shub, M., and Smale, S., editors, volume 32 of *Lectures in Applied Mathematics*, AMS, New York, 1996, pp. 189–210.

29. Chan, R., Scientific applications of iterative Toeplitz solvers. *Calcolo*, 33, 249–267, 1996.

30. Chan, R.H. and Ng, M.K., Conjugate gradient methods for Toeplitz systems. *SIAM Rev.*, 38, 427–482, 1996.

31. Chan, R.H. and Ng, M.K., Iterative methods for linear systems with matrix structures. *Fast Reliable Algorithms for Matrices with Structure*, Kailath, T. and Sayed, A.H., editors. SIAM, Philadelpha, PA, 1999, pp. 117–152.

32. Chen, Z. and Pan, V.Y., An efficient solution for Cauchy-like systems of linear equations. *Comp. Math. Appl.*, 48, 529-537, 2004.

33. Cline, R.E., Plemmons, R.J., and Worm, G., Generalized inverses of certain Toeplitz matrices. *Linear Algebra Appl.*, 8, 25–33, 1974.

34. Cole, R. and Hariharan, R., An $O(n \log n)$ algorithm for the maximum agreement subtree problem for binary trees. In *Proceedings of 7th ACM-SIAM Symposium on Discrete Algorithms*, SIAM, Philadelphia, PA, pp. 323–332, 1996.

35. Conte, C.D. and de Boor, C., *Elementary Numerical Analysis: An Algorithmic Approach*. McGraw-Hill, New York, 1980.

36. Cooley, J.W. and Tukey, J.W., An algorithm for the machine calculation of complex Fourier series. *Math. Comp.*, 19(90), 297–301, 1965.

37. Díaz, A. and Kaltofen, E., On computing greatest common divisors with polynomials given by black boxes for their evaluation. In *Proceedings ACM International Symposium on Symbolic Algebraic Computations (ISSAC '95)*, Levelt, A., editor. ACM Press, New York, 1995, pp. 232–239.

38. Duhamel, P. and Vetterli, M., Fast Fourier transforms: A tutorial review. *Signal Process.* 19, 259–299, 1990.

39. Elliott, D.F. and Rao, K.R., *Fast Transform Algorithms, Analyses, and Applications*. Academic Press, New York, 1982.

40. Emiris, I.Z., A complete implementation for computing general dimensional convex hulls. *Int. J. Comput. Geometry Appl.* (Special Issue on Geometric Software), 8(2), 223–253, 1998.

41. Emiris, I.Z. and Pan, V.Y., Symbolic and numerical methods for exploiting structure in constructing resultant matrices. *J. Symbolic Comp.*, 33, 393–413, 2002.

42. Emiris, I.Z., Mourrain, B., and Pan, V.Y., editors, *Theoretical Computer Science*, (Special Issue on Algebraic and Numerical Algorithms), 315 (2–3), 2004.

43. Fateman, R.J., Polynomial multiplication, powers and asymptotic analysis: some comments. *SIAM J. Comp.*, 3(3), 196–213, 1974.

44. Fiedler, M., Hankel and Loewner matrices. *Linear Algebra Appl.*, 58, 75–95, 1984.

45. Fiedler, M. and Ptak, V., Loewner and Bezout matrices. *Linear Algebra Appl.*, 101, 187–220, 1988.

46. Frigo, M. and Johnson, S.J., The fastest Fourier transform in the west. Available at http://theory.lcs.mit.edu/~fftw.

47. Frigo, M. and Johnson, S.J., FFTW, A free (GPL) C library for computing discrete Fourier transforms in one or more dimensions of arbitrary size. http://www.fftw.org/.

48. Frigo, M. and Johnson, S.J., The design and implementation of FFTW. *Proc. IEEE*, 93(2), 216–231, 2005.

49. Frigo, M. and Johnson, S.J., A modified split-radix FFT with fewer arithmetic operations. *IEEE Trans. Signal Process.*, 55(1), 111–119, 2007.

50. von zur Gathen, J. and Gerhard, J., *Modern Computer Algebra, 2nd edition*. Cambridge University Press, Cambridge, U.K., 2003.

51. Geddes, K.O., Czapor, S.R., and Labahn, G., *Algorithms for Computer Algebra*. Kluwer Academic Publishers, Norwell, MA, 1992.

52. Gentleman W.M. and Sande G., Fast Fourier transforms for fun and profit. *Proc. AFIPS*, 29, 563–578, 1966.

53. Gilbert, J.R. and Hafsteinsson, H., Parallel symbolic factorization of sparse linear Systems. *Parallel Comp.*, 14, 151–162, 1990.

54. Gilbert, J. R. and Schreiber, R., Highly parallel sparse Cholesky factorization. *SIAM J. Sci. Comp.*, 13, 1151–1172, 1992.

55. GNU multiple precision library. Free Software Foundation, 1996. ftp://prep.ai.mit.edu/pub/gnu/gmp-M.N.tar.gz.

56. Gohberg, I., Kailath, T., and Olshevsky, V., Fast Gaussian elimination with partial pivoting for matrices with displacement structure. *Math. Comp.*, 64(212), 1557–1576, 1995.

57. Gohberg, I. and Olshevsky, V., Complexity of multiplication with vectors for structured matrices. *Linear Algebra Appl.*, 202, 163–192, 1994.

58. Golub, G.H. and Van Loan, C.F., *Matrix Computations, 3rd edition*. The Johns Hopkins University Press, Baltimore, MD, 1996.

59. Grigoriev, D.Y., Karpinski, M., and Singer, M.F., Fast parallel algorithms for sparse multivariate polynomial interpolation over finite fields. *SIAM J. Comp.*, 19(6), 1059–1063, 1990.

60. Hardy, G.H. and Wright, E.M., *An Introduction to the Theory of Numbers, 5th edition*. Clarendon Press, Oxford, 1979.

61. Heinig, G., Inversion of generalized Cauchy matrices and the other classes of structured matrices. IMA Volume in Mathematics and Its Applications, *Linear Algebra for Signal Process.*, 69, 95–114, 1995.

62. Heinig, G. and Rost, K., *Algebraic Methods for Toeplitz-like Matrices and Operators*, volume 13 of *Operator Theory*. Birkhäuser, Basel, Switzerland, MA, 1984.

63. Householder, A.S., *The Numerical Treatment of a Single Nonlinear Equation*. McGraw-Hill, Boston, MA, 1970.

64. Huckle, T., Iterative methods for ill-conditioned Toeplitz matrices. *Calcolo*, 33, 1996.

65. Ja Ja, J., *An Introduction to Parallel Algorithms*. Addison-Wesley, Reading, MA, 1992.

66. Kailath, T., Kung, S.-Y., and Morf, M., Displacement ranks of matrices and linear equations. *J. Math. Anal. Appl.*, 68(2), 395–407, 1979.

67. Kailath, T. and Olshevsky, V., Displacement structure approach to discrete-trigonometric transform based preconditioners of G. Strang type and of T. Chan type. *Calcolo*, 33, 1996.

68. Kailath, T. and Sayed, A., editors, *SIAM Volume on Fast Reliable Algorithms for Matrices with Structure*, SIAM, Philadelphia, PA, 1999.

69. Kaltofen, E., Factorization of polynomials given by straight-line programs. In *Randomness and Computation*, Micali, S., editor, volume 5 of *Advances in Computing Research*. JAI Press, Greenwich, CT, 1989, pp. 375–412.

70. Kaltofen, E. and Lakshman, Y., Improved sparse multivariate polynomial interpolation algorithms. In *Proceedings of ACM International Symposium on Symbolic and Algebraic Computations (ISSAC '88)*, volume 358 of *Lecture Notes in Computer Science*, Springer-Verlag, London, U.K., 1988, pp. 467–474.

71. Kaltofen, E., Lakshman, Y.N., and Wiley, J.M., Modular rational sparse multivariate polynomial interpolation. In *Proceedings of ACM International Symposium on Symbolic Algebraic Computations (ISSAC '90)*. ACM Press, New York, 1990, pp. 135–139.

72. Karatsuba, A. and Ofman, Y., Multiplication of multidigit numbers on automata. *Soviet Physics Dokl.*, 7, 595–596, 1963.

73. Knuth, D.E., *The Art of Computer Programming: Seminumerical Algorithms*, volume 2. Addison-Wesley, Reading, MA, 1998.

74. Lancaster, P. and Tismenetsky, M., *The Theory of Matrices, Second Edition*. Academic Press, Orlando, FL, 1985.

75. Lipton, R.J., Rose, D., and Tarjan, R.E., Generalized nested dissection. *SIAM J. Numer. Anal.*, 16(2), 346–358, 1979.

76. Manocha, D. and Canny, J., Multipolynomial resultant algorithms. *J. Symbolic Comp.*, 15(2), 99–122, 1993.

77. Monagan, M., Maximal quotient rational reconstruction: an almost optimal algorithm for rational reconstruction. *Proceedings of ACM-SIGSAM International Symposium on Symbolic and Algebraic Computation (ISSAC'04)*. ACM Press, New York, 2004, pp. 243–249.

78. Mourrain, B. and Pan, V.Y., Multivariate polynomials, duality, and structured matrices. *J. Complexity*, 16(1), 110–180, 2000.

79. Neuts, M.F., *Structured Stochastic Matrices of M/G/1 Type and Their Applications*. Marcel Dekker, New York, 1989.

80. O'Donnell, S.T. and Rokhlin, V., A fast algorithm for numerical evaluation of conformal mappings. *SIAM J. Sci. Stat. Comp.*, 10(3), 475–487, 1989.

81. Olshevsky, V. and Pan, V.Y., A unified superfast algorithm for boundary rational tangential interpolation problem and for inversion and factorization of dense structured matrices. *Proceedings of 39th Annual IEEE Symposium on Foundation of Computer Science (FOCS '98)*. IEEE Computer Society Press, Los Alamitos, CA, 1998, pp. 192–201.

82. Pan, V.Y., On methods of computing the values of polynomials. *Uspekhi Matematicheskikh Nauk*, 21(1)(127), 103–134, 1966. (Transl. *Russian Mathematical Surveys*, 21(1)(127), 105–137, 1966.)

83. Pan, V.Y., Computations with dense structured matrices. *Math. Comp.*, 55(191), 179–190, 1990.

84. Pan, V.Y., Complexity of algorithms for linear systems of equations. In *Computer Algorithms for Solving Linear Algebraic Equations (State of the Art)*, Spedicato, E., editor, volume 77 of *NATO ASI Series, Series F: Computer and Systems Sciences*, pp. 27–56. Springer, Berlin, 1991, and Academic Press, Dordrecht, the Netherlands (1992).

85. Pan, V.Y., Complexity of computations with matrices and polynomials. *SIAM Rev.*, 34(2), 225–262, 1992.

86. Pan, V.Y., Simple multivariate polynomial multiplication. *J. Symb. Comp.*, 18, 183–186, 1994.

87. Pan, V.Y., Parallel computation of polynomial GCD and some related parallel computations over abstract fields. *Theoret. Comp. Sci.*, 162(2), 173–223, 1996.

88. Pan, V.Y., Solving a polynomial equation: Some history and recent progress. *SIAM Rev.*, 39(2), 187–220, 1997.

89. Pan, V.Y., Nearly optimal computations with structured matrices. *Proceedings of 11th Annual ACM-SIAM Symposium on Discrete Algorithms (SODA'00)*. ACM Press, New York, and SIAM Philadelphia, PA, 2000, pp. 953–962.

90. Pan, V.Y., New techniques for the computation of linear recurrence coefficients. *Finite Fields Their Appl.*, 6, 93–118, 2000.

91. Pan, V.Y., *Structured Matrices and Polynomials: Unified Superfast Algorithms*. Birkhäuser/Springer, Boston, MA/New York, 2001.

92. Pan, V.Y., Can we optimize Toeplitz/Hankel computations? *Proceedings 5th International Workshop on Computer Algebra in Scientific Computing (CASC '02)*, Mayr, E.W., Ganzha, V.G., and Vorozhtzov, E.V., editors. Technische Univ. München, Germany, 2002, pp. 253–264.

93. Pan, V.Y., On theoretical and practical acceleration of randomized computation of the determinant of an integer matrix. *Zapiski Nauchnykh Seminarov POMI (in English)*, 316, 163–187, St. Petersburg, Russia, 2004. Also available at http://comet.lehman.cuny.edu/vpan/

94. Pan, V.Y., Amended DSeSC power method for polynomial root-finding. *Comp. Math. (with Appl.)*, 49(9–10), 1515–1524, 2005.

95. Pan, V.Y., Landowne, E., and Sadikou, A., Univariate polynomial division with a remainder by means of evaluation and interpolation. *Info. Process. Lett.*, 44, 149–153, 1992.

96. Pan, V.Y., Branham, S., Rosholt, R., and Zheng, A., Newton's iteration for structured matrices and linear systems of equations. *SIAM Volume on Fast Reliable Algorithms for Matrices with Structure*, Kailath, T. and Sayed, A. editors. SIAM, Philadelphia, 1999, Chapter 7, pp. 189–210.

97. Pan, V.Y., Grady, D., Murphy, B., Qian, G., Rosholt, R.E., and Ruslanov, A., Schur aggregation for linear systems and determinants, Special Issue on Symbolic–Numerical Algorithms, Bini, D.A., Pan, V.Y., and Verschelde, J., editors, *Theoretical Computer Science*, 409(2), 255–268, 2008.

98. Pan, V.Y., Kunin, M., Rosholt, R.E. and Kodal, H., Homotopic residual correction processes. *Math. Comp.*, 75, 345–368, 2006.

99. Pan, V.Y., Murphy, B., and Rosholt, R.E., Unified nearly optimal algorithms for structured matrices. In *Numerical Methods for Structured Matrices and Applications: Georg Heinigs Memorial Volume.* Birkhauser Verlag, in press.

100. Pan, V.Y., Murphy, B., Rosholt, R.E., and Wang, X., Toeplitz and Hankel meet Hensel and Newton: Nearly optimal algorithms and their practical acceleration with saturated initialization. Technical Report 2004 013, Ph.D. Program in Computer Science, The Graduate Center, City University of New York, 2004.

101. Pan, V.Y. and Qian, G., Solving homogeneous linear systems with randomized preprocessing. *Linear Algebra Appl.*, accepted.

102. Pan, V.Y., Rami, Y., and Wang, X., Structured matrices and Newton's iteration: Unified approach. *Linear Algebra Appl.*, 343/344, 233–265, 2002.

103. Pan, V.Y. and Reif, J., Fast and efficient parallel solution of sparse linear systems. *SIAM J. Comp.*, 22(6), 1227–1250, 1993.

104. Pan, V.Y., Sadikou, A., Landowne, E., and Tiga, O., A new approach to fast polynomial interpolation and multipoint evaluation. *Comp. Math. (with Appl.)*, 25(9), 25–30, 1993.

105. Pan, V.Y., Van Barel, M., Wang, X., and Codevico, G., Iterative inversion of structured matrices. (Special Issue on Algebraic and Numerical Algorithms, Emiris, I.Z., Mourrain, B., and Pan, V.Y., editors), *Theoretical Computer Science*, 315(2–3), 581–592, 2004.

106. Pan, V.Y. and Wang, X., Inversion of displacement operators. *SIAM J. Matrix Anal. Appl.*, 24(3), 660–677, 2003.

107. Pan, V.Y. and Wang, X., On rational number reconstruction and approximation. *SIAM J. Comp.*, 33(2), 502–503, 2004.

108. Pan, V.Y. and Wang, X., Degeneration of integer matrices modulo an integer. *Linear Algebra Appl.*, 429, 2113–2130, 2008.

109. V.Y. Pan and Yan, X., Additive preconditioning, eigenspaces, and the inverse iteration. *Linear Algebra Appl.*, 430, 186–203, 2009.

110. Pan, V.Y. and Zheng, A.L., Superfast algorithms for Cauchy-like matrix computations and extensions. *Linear Algebra Appl.*, 310, 83-108, 2000.

111. Pan, V.Y., Zheng, A.L., Huang, X.H., and Yu, Y.Q., Fast multipoint polynomial evaluation and interpolation via computations with structured matrices. *Ann. Numerical Math.*, 4, 483–510, 1997.

112. Press, W., Flannery, B., Teukolsky, S., and Vetterling, W., *Numerical Recipes in C: The Art of Scientific Computing.* Cambridge University Press, Cambridge, 1988, and 2nd ed. 1992.

113. Rabiner, L., The chirp z-transform algorithm — a lesson in serendipity. *IEEE Signal Process. Mag.*, 24, 118–119, March 2004. (Historical commentary.)

114. Rader, C.M., Discrete Fourier transforms when the number of data samples is prime. *Proc. IEEE*, 56, 1107–1108, 1968.

115. Rokhlin, V., Rapid solution of integral equations of classical potential theory. *J. Comp. Phys.*, 60, 187–207, 1985.

116. Runge, C. and König, H., *Die Grundlehren der mathematischen Wissenshaften,* 11. Springer-Verlag, Berlin, Germany, 1924.

117. Saad, Y. and van der Vorst, H.A., Iterative solution of linear systems in the 20th century. *J. Comp. Appl. Math.*, 123, 1–33, 2000.

118. Schönhage, A., Grotefeld, A.F.W., and Vetter, E., *Fast Algorithms: A Multitape Turing Machine Implementation.* Wissenschaftsverlag, Mannheim, Germany, 1994.

119. Schönhage, A. and Strassen, V., Schnelle multiplikation großer zahlen. *Computing,* 7, 281–292, 1971. In German.

120. Smith, S.W., *The Scientist and Engineer's Guide to Digital Signal Processing, 2nd edition.* California Technical Publishing, San Diego, CA, 1999.

121. Storn, R., Some results in fixed point error analysis of the Bruun-FTT [sic] algorithm. *IEEE Trans. Signal Processing*, 41(7), 2371–2375, 1993.

122. Strang, G., Wavelet transforms versus Fourier transforms. *Bulletin (New Series) Am. Math. Soc.*, 28(2), 288–305, 1993.

123. Swarztrauber, P., FFT algorithms for vector computers. *Parallel Computing*, 1, 45–63, 1984. Implementation at http://www.psc.edu/general/software/packages/fftpack/fftpack.html or ftp://netlib.att.com/netlib.

124. Toom, A.L., The complexity of a scheme of functional elements realizing the multiplication of integers. *Soviet Math. Doklady*, 3, 714–716, 1963.

125. Trefethen, L.N. and Bau, D. III, *Numerical Linear Algebra*. SIAM, Philadelphia, PA, 1997.

126. Van Loan, C.F., *Computational Frameworks for the Fast Fourier Transform*. SIAM, Philadelphia, PA, 1992.

127. Vandebril, R., Van Barel, M., Golub, G., and Mastronardi, N., A bibliography on semiseparable matrices. *Calcolo*, 42(3–4), 249–270, 2005.

128. Winograd, S., On computing the discrete Fourier transform. *Math. Comp.*, 32, 175–199, 1978.

129. Winograd, S., *Arithmetic Complexity of Computations*. SIAM, Philadelphia, PA, 1980.

130. Zippel, R., *Effective Polynomial Computation*. Kluwer Academic, Boston, MA, 1993.

19

Basic Notions in Computational Complexity

Tao Jiang
University of California

Ming Li
University of Waterloo

Bala Ravikumar
Sonoma State University

19.1 Introduction

Computational complexity is aimed at measuring how complex a computational solution is. There are many ways to measure the complexity of a solution: how hard it is to understand it, how hard to express it, how long the solution process will take, and more. The last criterion—time—is most widely taken as the definition of complexity. This is because the agent that implements an **algorithm** is usually a computer—and from a user's point of view, the most important issue is how long one should wait to see the solution. However, there are other important measures of complexity, such as the amount of memory, hardware, information, communication, knowledge, or energy needed for the solution. Complexity theory is aimed at quantifying these resources precisely and studying the amounts of them required to accomplish computational tasks.

This technical sense of the word "complexity" did not take root until the mid-1960s. Today, complexity theory is not only a vibrant subfield of computer science, but also has direct or metaphoric impact on other fields such as dynamical systems and chaos theory. The seed of computational complexity is the formalization of the concept of an algorithm. Algorithms in turn must be planted in some computational model, ideally one that abstracts important features of real computing machines and processes. In this chapter we consider the **Turing machine** (TM), a computer model created and studied by the British mathematician Alan Turing in the 1930s [36]. Chapter 21 will discuss different (and equivalent) ways of formalizing algorithms.

With the advent of commercial computers in the 1950s and 1960s, when processor speed was much lower and memory cost unimaginably higher than today, it became critical to design efficient algorithms for solving large classes of problems. Just knowing that a problem was solvable, which had been the main concern of computability theory since the 1930s, was no longer enough.

TM provided a basis for Hartmanis and Stearns [13] to formally define the measures of **time complexity and space complexity** for computations. The latter refers to the amount of memory needed to execute the computation. They defined measures for other resources as well, and Blum [1] found a precise definition of complexity measure that was not tied to any particular resource or machine model. Together with earlier work by Rabin [28], these papers marked the beginning of computational complexity theory as an important new discipline. A closely related development was drawn together by Knuth, whose work on algorithms and data structures [18] created the field of algorithm design and analysis. All of this work has been recognized in annual Turing Awards given by the Association for Computing Machinery. Hartmanis' Turing Award lecture [12] recounts the origins of computational complexity theory and speculates on its future development.

The power of a computing machine depends on its internal structure as well as on the amount of time, space, or other resources it is allowed to consume. Restricting our model of choice, the TM, in various internal ways yields progressively simpler and weaker computing machines. These machines correspond to a natural hierarchy of grammars defined and advanced by Chomsky [5], which we describe in Chapter 20. In this chapter we present the most basic computational models, and use these models to classify problems first by solvability and then by complexity.

Central issues studied by researchers in computational complexity include the following:

- For a given amount of resources, or for a given type of resource, what problems can and cannot be solved?
- What is the relationship between problems requiring essentially the same amount of the resource or resources? May they be equivalent in some intrinsic sense?
- What connections are there between different kinds of resources? Given more time, can we reduce the demand on memory storage, or vice-versa?
- What general limits can be set on the kind of problems that can be solved when resources are limited?

The end thrust of Turing's famous paper [36] was to demonstrate rigorously that several fundamental problems in logic cannot be solved by algorithms at all. Complexity theory aims to show similar results for many more problems in the presence of resource limits. Such an unsolvability result, although "bad news" in most contexts (cryptographic security is an exception) can have practical benefits: it may lead to alternative models, goals, or solution strategies. As will come out in Chapters 22 and 23, complexity theory has so far been much more successful at drawing relative conclusions and relationships between problems than in proving absolute statements about (un)solvability.

19.2 Computational Models

Throughout this chapter, Σ denotes a finite alphabet of symbols; unless otherwise specified, $\Sigma = \{0, 1\}$. Then Σ^* denotes the set of all finite strings, including the empty string ϵ, over Σ. A **language** over Σ is simply a subset of Σ^*.

We use **regular expressions** in specifying some languages. In advance of the formal definition to come in Chapter 20, we define them as one kind of "patterns" for strings to "match." The basic patterns are ϵ and the characters in Σ, which match only themselves. The null pattern $\emptyset$ matches no strings. Two patterns joined by "$+$" match any string that matches either of them. Two or more patterns written in sequence match any string composed of substrings that match the respective patterns. A pattern followed by a "$*$" matches any string that can be divided into zero or more successive substrings, each of which matches the pattern—here the "zero" case applies to ϵ, which matches any starred pattern. For example, the pattern $0(0 + 1)^*1$ matches any string that begins with a 0 and ends with a 1, with zero or more binary characters in between. The pattern can be used as a name for the language of strings that match it. In like manner, the pattern $(0^*10^*1)^*0^*$ stands for the language of binary strings that have an even number of 1s. This pattern says that trailing 0s are immaterial to any such string, and the rest of the string can be broken into zero or more substrings, each of which ends in a 1 and has exactly one previous 1.

19.2.1 Finite Automata

The **finite automaton** (in its **deterministic** version) was introduced by McCulloch and Pitts in 1943 as a logical model for the behavior of neural systems [27]. Rabin and Scott introduced the nondeterministic finite automaton (NFA) in [29], and showed that NFAs are equivalent to deterministic finite automata, in the sense that they recognize the same class of languages. This class of languages, called the **regular languages**, had already been characterized by Kleene [17] and Chomsky and Miller [6] in terms of regular expressions and regular grammars, which will be described formally in Chapter 20.

In addition to their original role in the study of neural nets, finite automata have enjoyed great success in many fields such as the design and analysis of sequential circuits [20], asynchronous circuits [3], text-processing systems [22], and compilers. They also led to the design of more efficient algorithms. One excellent example is the development of linear-time string-matching algorithms, as described in [19]. Other applications of finite automata can be found in computational biology [32], natural language processing, and distributed computing [26].

A finite automaton, pictured in Figure 19.1, consists of an input tape and a finite control. The input tape contains a finite string of input symbols, and is read one symbol at a time from left to right. The finite control is connected to an input head that reads each symbol, and can be in one of a finite number of states. The input head is one-way, meaning that it cannot "backspace" to the left, and read-only, meaning that it cannot alter the tape. At each step, the finite control may change its state according to its current state and the symbol read, and the head advances to the next tape cell. In an NFA there may be more than one choice of next state in a step. Figure 19.1 also shows the second step of a computation on an input string beginning aabab. . . When the input head reaches the right end of the input tape, if the machine is in a state designated "final" (or "accepting"), we say that the input string is accepted; else we say it is rejected. The following is the formal definition.

DEFINITION 19.1 A nondeterministic finite automaton (NFA) is a quintuple $(Q, \Sigma, \delta, q_0, F)$, where

- Q is a finite set of states
- Σ is a finite set of input symbols
- δ, the state transition function, is a mapping from $Q \times \Sigma$ to subsets of Q
- $q_0 \in Q$ is the start state of the NFA
- $F \subseteq Q$ is the set of final states

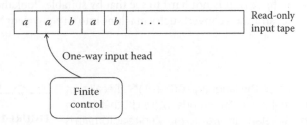

FIGURE 19.1 A finite automaton.

If δ maps $|Q| \times \Sigma$ to singleton subsets of Q, then we call such a machine a deterministic finite automaton (DFA).

Note that a DFA is treatable as a special case of an NFA, where the next state is always uniquely determined by the current state and the symbol read. On any input string $x \in \Sigma^*$, a DFA follows a unique computation path, starting in state q_0. If the final state in the path is in F, then x is accepted, and the path is an accepting path. An NFA, however, may have multiple computation paths on the same input x. It is useful to imagine that when an NFA has more than one next state, all options are taken in parallel, so that the "super-computation" is a tree of branching computation paths. Then the NFA is said to accept x if at least one of those paths is an accepting path.

REMARK 19.1   The concept of a nondeterministic automaton, and especially the notion of acceptance, can be nonintuitive and confusing at first. We can, however, explain them in terms that should be familiar, namely those of a solitaire game such as "Klondike." The game starts with a certain arrangement of cards, which we can regard as the input, and has a desired "final" position—in Klondike, when all the cards have been built up by suit from ace to king. At each step, the rules of the game dictate how a new arrangement of cards can be reached from the current one—and the element of nondeterminism is that there is often more than one choice of move (otherwise the game would be little fun!). Some positions have no possible move, and lose the game. Most crucially, some positions have moves that unavoidably lead to a loss, and other moves that keep open the possibility of winning. Now for a given position, the important analytical question is, "Is there a way to win?" The answer is "yes" so long as there is at least one sequence of moves that ends in a (or the) desired final position. For the starting position, this condition is much the same as for an input to be accepted by an NFA. (Practically speaking, some winnable starting positions may give so many chances to go wrong along the way that a player may have little chance to find a winning sequence of moves. That, however, is beside the point in defining which positions are winnable. If one can always (efficiently!) answer the yes/no question of whether a given position is winnable, then one can always avoid losing moves—and win—so long as the start position is winnable to begin with.)

In any event, the set of strings accepted by a (deterministic or nondeterministic) finite automaton M is denoted by $L(M)$. When we say that M accepts a language L, we mean that M accepts all strings in L and nothing else, i.e., $L = L(M)$. Two machines are equivalent if they accept the same language.

Nondeterminism is capable of modeling many important situations other than solitaire games. Concurrent computing offers some examples. Suppose a device or resource (such as a printer or a network interface) is controllable by more than one process. Each process could change the state of the device in a different way. Since there may be no way to predict the order in which processes may be given control in any step, the evolution of the device may best be regarded as nondeterministic.

Sometimes a state change can occur without input stimulus. This can be modeled by allowing an NFA to make ϵ-transitions, which change state without advancing the read head. Then the second argument of δ can be ϵ instead of a character in Σ, and these transitions can even be nondeterministic, for instance if $\delta(q_0, \epsilon) = \{q_1, q_2\}$. It is not hard to see that by suitable "lookahead" on states reachable by ϵ-transitions, one can always convert such a machine into an equivalent NFA that does not use them.

Example 19.1

We design an NFA to accept the language $0(0 + 1)^*1$. Recall that this regular expression defines those strings in $\{0, 1\}^*$ that begin with a 0 and end with a 1. A standard way to draw finite automata is exemplified by Figure 19.2. As a convention, each circle represents

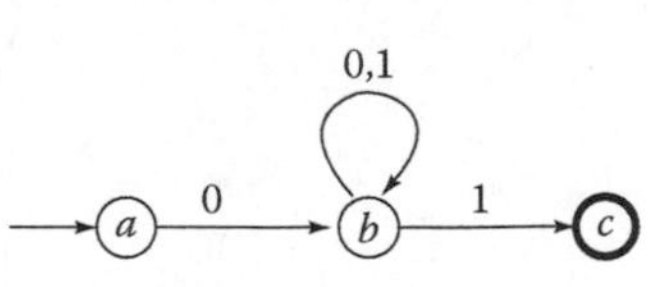

FIGURE 19.2   An NFA accepting $0(0 + 1)^*1$.

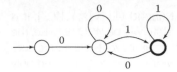

FIGURE 19.3 A DFA accepting $0(0+1)^*1$.

a state, and an unlabeled arrow points to the start state (here, state "a"). Final states such as "c" have darker (or double) circles. The transition function δ is represented by the labeled edges. For example, $\delta(a, 0) = \{b\}$, and $\delta(b, 1) = \{b, c\}$. When a transition is missing, such as on input 1 from state "a" and on both inputs from state "c," the transition is assumed to lead to an implicit nonaccepting "dead" state, which has transitions to itself on all inputs. In a DFA such a dead state must be included when counting the number of states, while in an NFA it can be left out.

The machine in Figure 19.2 is nondeterministic since from "b" on input 1 the machine has two choices: stay at "b" or go to "c." Figure 19.3 gives a DFA that accepts the same language. The DFA has four, not three, states, since a dead state reached by an initial "1" is not shown.

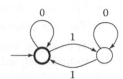

Example 19.2

The DFA in Figure 19.4 accepts the language of all strings in $\{0, 1\}^*$ with an even number of 1's, which has the regular expression $(0^*10^*1)^*0^*$.

FIGURE 19.4 A DFA accepting $(0^*10^*1)^*0^*$.

Example 19.3

For a final example of a regular language, we introduce a general "tiling problem" to be discussed further in Chapter 21, and then strip it down to a simpler problem. A tile is a unit-sized square divided into four quarters by joining two diagonals. Each quarter has a color chosen from a finite set C of colors. Suppose you are given a set T of different types of tiles, and have an unlimited supply of each type. A $k \times n$ rectangle is said to be tiled using the tiles in T if the rectangle can be filled with exactly kn unit tiles (with no overlaps) such that at every edge between two tiles, the quarters of the two tiles sharing that edge have the same color. The tile set T is said to tile an entire plane if the plane can be covered with tiles subject to the color compatibility stated above. As a standard application of König's infinity lemma (for which see [18, Chapter 2, p. 381]), it can be shown that the entire plane can be tiled with a tile set T if and only if all finite integer sided rectangles can be tiled with T. We will see in Chapter 21 that the problem of whether a given tile set T can tile the entire plane is unsolvable. Chapter 21 will also say more about the meaning and implications of this tiling problem.

Here we will consider a simpler problem: Let k be a fixed positive integer. Given a set T of unit tiles, we want to know whether T can tile an infinite strip of width k. The answer is yes if T can tile any $k \times n$ rectangle for all n. It turns out that this problem is solvable by an efficient algorithm. One-way to design such an algorithm is based on finite automata. We present the solution for $k = 1$ and leave the generalization for other values of k as an exercise. Number the quarters of each tile as in Figure 19.5. Given a tile set T, we want to know whether for all n, the $1 \times n$ rectangle can be tiled using T.

FIGURE 19.5 Numbering the quarters of a tile.

To use finite automata, we define a language that corresponds to valid tilings. Define Σ, the input alphabet, to be T, the tile set. Each tile in T can be described by a 4-tuple $[A, B, P, Q]$ where A, B, P, and Q are (possibly equal) members of the color set C. Next we define a language L over Σ to be the set of strings $T_1 T_2 \cdots T_n$ such that (1) each T_i is in Σ, and (2) for each i, $1 \le i \le n - 1$, T_i's third-quadrant color is the same as T_{i+1}'s first-quadrant color. These two conditions say that $T_1 T_2 \cdots T_n$ is a valid $1 \times n$ tiling.

We will now informally describe a DFA M_L that recognizes the language L. Basically, M_L "remembers" (using the current state as the memory) the relevant information—for this problem, it need only remember the third-quadrant color Q of the most recently seen tile. Suppose the DFA's current state is Q. If the next (input) tile is $[X,Y,W,Z]$, it is consistent with the last tile if and only if $Q = X$. In this case, the next state will be W. Otherwise, the tile sequence is inconsistent, so M_L enters a "dead state" from which it can never leave and rejects. All other states of M_L are accepting states. Then the infinite strip of width 1 can be tiled if and only if the language L accepted by M_L contains strings of all lengths. There are standard algorithms to determine this property for a given DFA.

The next three theorems show the satisfying result that all the following language classes are identical:

- The class of languages accepted by DFAs
- The class of languages accepted by NFAs
- The class of languages generated by regular expressions
- The class of languages generated by the right-linear, or Type-3, grammars, which are formally defined in Chapter 20 and informally used here

This class of languages is called the regular languages.

THEOREM 19.1   *For every NFA, there is an equivalent DFA.*

PROOF   An NFA might look more powerful since it can carry out its computation in parallel with its nondeterministic branches. But since we are working with a finite number of states, we can simulate an NFA $M = (Q, \Sigma, \delta, q_0, F)$ by a DFA $M' = (Q', \Sigma, \delta', q'_0, F')$, where

- $Q' = \{[S] : S \subseteq Q\}$
- $q'_0 = [\{q_0\}]$
- $\delta'([S], a) = [S'] = [\cup_{q_l \in S} \delta(q_l, a)]$
- F' is the set of all subsets of Q containing a state in F

Here square brackets have been placed around sets of states to help one view these sets as being states of the DFA M'. The idea is that whenever M' has read some initial segment y of its input x, its current state equals the set of states q such that M has a computation path reading y that leads to state q. When all of x is read, this means that M' is in an accepting state if and only if M has an accepting computation path. Hence $L(M) = L(M')$.

Example 19.4

Example 19.1 contains an NFA and an equivalent DFA accepting the same language. In fact the above proof provides an effective procedure for converting an NFA to a DFA. Although each NFA can be converted to an equivalent DFA, the resulting DFA may require exponentially many more states, since the above procedure may assign a state for every eligible subset of the states of the NFA. For any $k > 0$, consider the language $L_k = \{x \mid x \in \{0, 1\}^*$ and the kth letter from the right of x is a $1\}$. An NFA of $k + 1$ states (for $k = 3$) accepting L_k is given in Figure 19.6. Now we claim that any DFA M accepting L_k needs a separate state for every possible value $y \in \{0, 1\}^k$ of the last k bits read. Take any distinct $y_1, y_2 \in \{0, 1\}^k$ and let i be a position in which they differ. Let $z = 0^{k-i}$. Then M must accept one of the strings $y_1 z, y_2 z$ and reject the other. This is possible only if the state M is in after processing y_1 (with z to come) is different from that after y_2, and thus M needs a different state for each string in $\{0, 1\}^k$. The 2^k required states are also sufficient, as the reader may verify.

The remaining results of this section point forward to the **formal-language** models defined in Chapter 20. The point of including them here is to show the power of the finite automaton model.

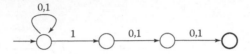

FIGURE 19.6 An NFA accepting L_3.

Regular expressions have been defined above, while a regular grammar over Σ consists of a set V of variable symbols, a starting variable $S \in V$, and a set P of substring-rewrite rules of the forms $A \rightarrow cB, A \rightarrow \epsilon$, or $A \rightarrow c$, where $A, B \in V$ and $c \in \Sigma$.

THEOREM 19.2 *A language L is generated by a regular grammar if and only if L is accepted by an NFA.*

PROOF   Let L be accepted by an NFA $M = (Q, \Sigma, \delta, q_0, F)$. We define an equivalent regular grammar $G = (\Sigma, V, S, P)$ by taking $V = Q$ with $S = q_0$, adding a rule $q_i \rightarrow cq_j$ whenever $q_j \in \delta(q_i, c)$, and adding rules $q_j \rightarrow \epsilon$ for all $q_j \in F$. Then the grammar simulates computations by the NFA in a direct manner, giving $L(G) = L(M)$.

Conversely, suppose L is the language of a regular grammar $G = (\Sigma, V, S, P)$. We design an NFA $M = (Q, \Sigma, \delta, q_0, F)$ by taking $Q = V \cup \{f\}$, $q_0 = S$, and $F = \{f\}$. To define the δ function, we have $B \in \delta(A, c)$ iff $A \rightarrow cB$. For rules $A \rightarrow c$, $\delta(A, c) = \{f\}$. Then $L(M) = L(G)$—if this is not clear already, the treatment of grammars in Chapter 20 will make it so.

THEOREM 19.3 *A language L is specified by a regular expression if and only if L is accepted by an NFA.*

PROOF   PROOF SKETCH. *Part 1.* We inductively convert a regular expression to an NFA that accepts the same language as follows:

- The pattern ϵ converts to the NFA $M_\epsilon = (\{q\}, \Sigma, \emptyset, q, \{q\})$, which accepts only the empty string.
- The pattern $\emptyset$ converts to the NFA $M_\emptyset = (\{q\}, \Sigma, \emptyset, q, \emptyset)$, which accepts no strings at all.
- For each $c \in \Sigma$, the pattern c converts to the NFA $M_c = (\{q, f\}, \Sigma, \delta(q, c) = \{f\}, q, \{f\})$, which accepts only the string c.
- A pattern of the form $\alpha + \beta$, where α and β are regular expressions that (by induction hypothesis) have corresponding NFAs M_α and M_β, converts to an NFA M that connects M_α and M_β in parallel: M includes all the states and transitions of M_α and M_β and has an extra start state q_0, which is connected by ϵ-transitions to the start states of M_α and M_β.
- A pattern of the form $\alpha \cdot \beta$, where α and β have the corresponding NFAs M_α and M_β, converts to an NFA M that connects M_α and M_β in series: M includes all the states and transitions of M_α and M_β and has extra ϵ-transitions from the final states of M_α to the start state of M_β. The start state of M is that of M_α, while the final states of M are those of M_β.
- A pattern of the form α^*, where α has the corresponding NFA M_α, converts to an NFA M that figuratively feeds M_α back into itself. M includes the states and transitions of M_α, plus ϵ-transitions from the final states of M_α back to its start state. This state is not only the start state of M but the only final state of M as well.

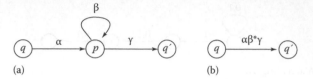

FIGURE 19.7 Converting an NFA to a regular expression.

Part 2. We now show how to convert an NFA to an equivalent regular expression. The idea used here is based on [2]; see also [3,38].

FIGURE   19.8   The reduced NFA.

Given an NFA M, add a new final state t, and add ϵ-transitions from each old final state of M to t. Also add a new start state s with an ϵ-transition to the old start state of M. The idea is to eliminate all states p other than s and t as follows. To eliminate a state p, we eliminate each arc coming in to p from some other state q as follows: For each triple of states q, p, q' as shown in Figure 19.7a, add the transition(s) shown in Figure 19.7b. (Note that if p does not have a transition leading back to p, then $\beta = \beta^* = \epsilon$.) After we have considered all such triples, we can delete state p and transitions related to p. Finally, we obtain Figure 19.8, and the final α is a regular expression for $L(M)$.

The last three theorems underline the importance of the class of regular languages, since it connects to notions of automata, grammars, patterns, and much else. However, regular languages and finite automata are not powerful enough to serve as our model for a modern computer. Many extremely simple languages cannot be accepted by DFAs. For example, $L = \{xx \mid x \in \{0,1\}^*\}$ cannot be accepted by a DFA. To show this, we can argue similarly to the "L_k" languages in Example 19.4 that for any two strings $y_1 = 0^m 1$ and $y_2 = 0^n 1$ with $n \neq m$, a hypothetical DFA M would need to be in a different state after processing y_1 from that after y_2, because with $z = 0^m 1$ it would need to accept $y_1 z$ and reject $y_2 z$. However, in this case we would conclude that M needs infinitely many states, contradicting the definition of a finite automaton. Hence the language L is not regular. Other ways to prove assertions of this kind include so-called "pumping lemmas" or a direct argument that some strings y contain more information than a finite state machine can remember [25]. We refer the interested readers to Chapter 20 and textbooks such as [9,10,14,38].

One can also try to generalize the DFA to allow the input head to move backward as well as forward, in order to review earlier parts of the input string, while keeping it read-only. However, such "two-way DFAs" are not more powerful—they can be simulated by normal DFAs. The point of departure for a more-powerful model is to allow the machine to write on its tape and later review what it has written. Then the tape becomes a storage medium, not just a sequence of events to react to. This ability to write down intermediate results for future reference makes DFAs into full-blown general-purpose computers.

19.2.2   Turing Machines

A TM, pictured in Figure 19.9, consists of a finite control, an infinite tape divided into cells, and a read/write head on the tape. The finite control can be in any one of a finite set Q of states. Each cell can contain one symbol from the tape alphabet Γ, which contains the input alphabet Σ and a special character B called the blank. Γ may contain other characters besides B and those in Σ, but most often $\Sigma = \{0, 1\}$ and $\Gamma = \{0, 1, B\}$. We refer to the two directions on the tape as left and right, and abbreviate them by L and R. At any time, the head is positioned over a particular cell that it is said to scan. Initially, the head scans a distinguished cell on the tape called the start cell, the finite control is in the start state q_0, and all cells contain B except for a contiguous finite sequence of cells,

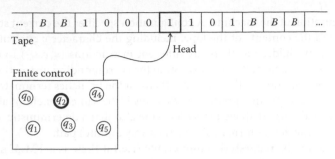

FIGURE 19.9 A turing machine.

extending from the start cell to the right, that contain characters in Σ. These cells hold the input string x; in case $x = \epsilon$ the whole tape is blank. The machine is said to begin its computation on input x at time 0, and computation unfolds in discrete time steps numbered $1, 2, \ldots$

In any step, contingent on its current state and the character being scanned, the device is allowed to perform one the following two basic operations:

1. Write a symbol from the tape alphabet Γ into the scanned cell
2. Shift the head one cell left or right

Then it may change its internal state in the same step. The allowed actions of a particular machine are specified by a finite set δ of instructions. Each instruction has the form (q, c, d, r) with $q, r \in Q$, $c \in \Gamma$, and either $d \in \Gamma$ or $d \in \{L, R\}$. This means that if the machine is in state q scanning character c on the tape, it may either change c to d (if $d \in \Gamma$) or move its head (if $d \in \{L, R\}$), and it enters state r. Either $c = d$ or $q = r$ is allowed. (Many texts use an alternate formalism in which both basic operations may be performed in the same step, so that instructions have the form (q, c, d, D, r) with $q, r \in Q$, $c, d \in \Gamma$, and $D \in \{L, R\}$, sometimes adding the option $D = S$ of keeping the head stationary. The differences do not matter for our purposes.)

If for every combination of q and c there is at most one instruction (q, c, d, r) that the machine can execute, then the machine is deterministic. Otherwise, the machine is nondeterministic. In order for computations to possibly halt, there must be some combinations q, c for which δ has no instruction (q, c, d, r). If a computation reaches such a state q while scanning c, the device is said to halt. Then if q is designated as a final state, we say the machine accepts its input string x; if q is not a final state, we say that the machine rejects the input. We adopt the convention that there is only one final state labeled q_f, and that q_f is also a halting state, meaning that there is no instruction (q, c, d, r) with source state $q = q_f$ at all.

DEFINITION 19.2 A TM is a 7-tuple $M = (Q, \Sigma, \Gamma, \delta, B, q_0, q_f)$, where each of the components has been described above.

Given an input x, a deterministic TM M carries out a uniquely determined succession of operations, which may or may not terminate in a finite number of steps. If it terminates, then the output $M(x)$ is determined to be the longest string of characters over Σ beginning in the cell in which the head halted and extending to the right. (If the scanned cell holds B or some other character in $\Gamma \backslash \Sigma$, then the output is ϵ.) A function $f : \Sigma^* \to \Sigma^*$ is **computable** if there is a TM M such that for all inputs $x \in \Sigma^*$, $M(x) = f(x)$.

A nondeterministic TM is analogous to an NFA. One may imagine that it carries out its computation in parallel. The computation may be viewed as a (possibly infinite) tree, each of whose nodes is labeled by a configuration of the machine. A configuration specifies the current state q, the position of the tape head, and the contents of the tape, including the character c currently being scanned. Each node has as many children as there are different instructions $(q, c, \cdot, \cdot)$ to execute, and each child is the configuration that results from executing the corresponding instruction. The root of the tree is the starting configuration of the machine. If any of the branches terminates in the final state q_f, we say the machine accepts the input. Note that this is the same "benefit of all doubt" criterion for acceptance that we discussed above for NFAs. Note also that a deterministic TM always defines a "tree" with a single branch that forms a simple (possibly infinite) path.

A language L is said to be **recursively enumerable (r.e.)** if there is a TM M such that $L = \{x : M \text{ accepts } x\}$. Furthermore, if M is total, i.e., if every computation of M on every input x halts, then $L(M)$ is recursive. These terms reflect an important correspondence between languages and functions. For any language $L \subseteq \Sigma^*$, define the characteristic function f_L by $f_L(x) = 1$ of $x \in L$, $f(x) = 0$ otherwise. Then a language L is recursive if and only if f_L is computable—and **computable functions** were originally defined with regard to formalisms that used recursion. Note that having L be acceptable by a TM M is *not* enough for f_L to be computable, because there may be inputs $x \notin L$ for which the computation of M on x never halts.

Conversely, given a function $f : \Sigma^* \to \Sigma^*$, and letting # be a new input symbol not in Σ, one can define the language $L_f = \{x\#y : y \text{ is an initial segment of } f(x)\}$. Recognizing L_f allows one to find successive bits of the value $f(x)$. Hence it is common in the field to identify function problems with language problems and concentrate on the latter. A language can also be identified with a property of strings and with the associated **decision problem** "given a string x, does x have the property?" For instance, the problem of deciding whether a given number is prime is identifiable with the language of (binary string encodings of) prime numbers. The problem is **decidable** if the associated language is recursive, and a total TM accepting the language is said to decide the problem. The term "decidable language" is a synonym for "**recursive language**," and "recursive function" is a synonym for "computable function." A TM M that does not halt on all inputs computes a **partial recursive function** (whose domain is a proper subset of Σ^*), and $L(M)$ is a **partially decidable** language (or problem, or property). Any problem or language that is not decidable by a TM is called **undecidable**, and any (partial or total) function that is not computable by a TM is called **uncomputable**.

Now when we say that a TM M' simulates another TM M, we usually mean more than saying they accept the same language or compute the same (partial) function. Usually there is some overt correspondence between computations of M and those of M'. This is so in the simulations claimed by the following theorem, which says that many variations in the basic machine model do not alter the notion of computability.

THEOREM 19.4   *All the following generalizations of TMs can be simulated by the one-tape deterministic TM model defined in Definition 19.2, with tape alphabet $\{0, 1, B\}$.*

- *Enlarging the tape alphabet Γ*
- *Adding more tapes*
- *Adding more read/write heads or other access points on each tape*
- *Having two- or higher-dimensional grids in place of tapes, where the head may move to any adjacent grid cell*
- *Allowing nondeterminism*

Extra tapes after the input tape are called worktapes provided they allow read-write access. A two-tape TM, or alternately a one-tape machine with two heads, has instructions with six components: current state, two characters read by the heads, two head actions, and next state. Although these generalizations do not make a TM compute more, they do make TMs more efficient and easier to **program**. Many more variants of TMs have been studied and used in the literature.

Of all the simulations in Theorem 19.4, the last one needs special comment. A nondeterministic computation branches like a tree. The easiest way for a deterministic TM to simulate it is by traversing the tree in a breadth-first manner, which is the same as trying all possibilities at any step with nondeterministic choices. However, even if there are at most two choices at any step, simulating n steps of the NTM could take on the order of 2^n steps by the DTM. Whether a more-efficient simulation is possible is bound up with the famous "P vs. NP" problem, to be discussed below and in Chapter 22.

Example 19.5

A DFA can be regarded as the special case of a TM in which every instruction moves the head right. TMs naturally accept more languages than DFAs can. For example, a TM can accept the nonregular language $L = \{xx \mid x \in \{0, 1\}^*\}$ as follows. Given an input string $w \in \{0, 1\}^*$:

- First find the middle point. A TM with two-tape heads can do this efficiently by moving one head twice for every move of the other, until the further-advanced head sees the blank that marks the end of w. This stage can also tell whether w has even or odd length and immediately reject in the latter case.

- Then check whether the scanned characters match while moving both heads one cell left until the leftmost head sees the blank to the left of the beginning of w. If all pairs match, accept, else reject.

For a TM with only one head the strategy is more cumbersome. One-way is to use "alias" characters a for 0 and b for 1, aliasing first the leftmost 0/1 character on the tape, then the rightmost, then the next-leftmost,... until finding the character just left of middle (if w has even length). Then "un-alias" it and check that the rightmost aliased character matches it, un-aliasing the latter as well. By looking for cases of an a or b immediately to the left of an un-aliased 0 or 1, the TM can repeat this check until all of the left half is compared with the right half. Whereas the two-head TM needs only $3n/2$ steps to decide whether a string w of length n belongs to L, the one-head TM takes about n^2 steps. It is known that n^2 steps are necessary (asymptotically) for any one-head TM to accept the language L (see, for instance, [14] or [25]).

Three restrictions of the notion of a TM tape merit special mention.

- A pushdown store (or stack) is a semiinfinite worktape with one head such that each time the head moves to the left, it erases the symbol scanned previously. This is a last-in first-out storage.

- A queue is a semiinfinite work tape with two heads that only move to the right, the leading head is write-only and the trailing head is read-only. This is a first-in first-out storage.

- A counter is a pushdown store with a single-letter alphabet, except for a special bottom-of-stack marker that allows testing the counter for zero. Then a push increments the counter by one, and a pop decrements it by one.

Example 19.6

A pushdown automaton (PDA) has one read-only input tape and one pushdown store. A PDA can be identified with a two-tape TM whose tape-1 head can never move left and whose tape-2 head can move left only while scanning a blank (combined with a previous step that writes a blank, this simulates popping the stack).

Pushdown automata have been thoroughly studied because nondeterministic PDAs accept precisely the class of context-free languages, to be defined in Chapter 20. Various types of PDAs have fundamental applications in compiler design.

The PDA has less power than a TM. For example, $L = \{xx \mid x \in \{0, 1\}^*\}$ cannot be accepted by a PDA, but it can be accepted by a TM as in Example 19.5. However, a PDA is more powerful than a DFA. For example, a PDA can accept the nonregular language $L' = \{0^n 1^n \mid n \geq 0\}$ easily: For each initial 0 read, push a 0 onto the stack; then pop a 0 for each 1 read, and accept if and only if a blank is read after the block of 1s exactly when the stack is empty. Indeed, this PDA is a counter machine with a single counter.

Two pushdown stores can easily be used to simulate a tape—let one stack represent the part of the tape currently to the left of the input head, and let the other stack represent the rightward portion. Much more subtle is the fact that two counters can simulate a tape; unlike the two-pushdown case this takes exponentially more time. Finally, a single queue can simulate a tape: send the lead head to the right end so that it can write the next-step update of the configuration that the trailing head is reading. This involves encoding the current state of the TM being simulated onto the tape of the queue machine that is simulating it. Hence a single-queue machine, with the input initially resting in the queue, is as powerful as a TM, although it may require the square of the running time. For comparisons of powers of pushdown stores, queues, counters, and tapes, see [25,37].

Much more important is the fact that there are single TMs that are capable of simulating any TM. Formally, a **universal TM** U takes an encoding $\langle M, w \rangle$ of a (deterministic) TM M and a string w as input, and simulates M on input w. U accepts $\langle M, w \rangle$ if and only if M accepts w. Intuitively, U models a general-purpose computer that takes a "program" M and "data" w, and executes M on input w. Universal TMs have many applications. For example, the definition of Kolmogorov complexity (see [25]) fundamentally relies on them.

Example 19.7

Let $L_u = \{\langle M, w \rangle \mid M$ accepts $w\}$. Then L_u is *the* language accepted by a universal TM, so it is recursively enumerable. We see Chapter 21, however, that L_u is not recursive. The same properties hold for the language $L_h = \{\langle M, w \rangle : M$ on input w halts$\}$, which is the language of the so-called **Halting Problem**.

19.2.3 Oracle Turing Machines

In order to study the comparative hardness of computational problems, we often need to extend the power of TMs by adding oracles to them.

Informally, a TM T with an oracle A operates similarly to a normal TM, with the exception that it can write down a string z and ask whether z is in the language A. The machine gets the correct yes/no answer from the oracle in one step, and can branch its computation accordingly. This feature can be used as often as desired. We now give the definition precisely.

DEFINITION 19.3   An **oracle TM** is a normal TM T with an extra oracle query tape, a special state $q_?$, and two distinguished states labeled q_y and q_n. Let A be any language over an alphabet Σ. Whenever T enters state $q_?$ with some string $z \in \Sigma^*$ on the query tape, control passes to state q_y if $z \in A$, or to q_n if $z \notin A$. The computation continues normally until the next time the machine enters $q_?$. The machine T with a given choice of oracle A is denoted by T^A.

Example 19.8

In Example 19.7, we know that the universal language $L_u = \{\langle M, w \rangle \mid M$ accepts $w\}$ is not Turing decidable. But if we can use L_u as the oracle set, there is a trivial oracle TM T such that T with oracle

L_u decides L_u. T simply copies its input x onto the query tape and enters $q_?$. If control passes to q_y, T accepts; otherwise from q_n it rejects.

For something less trivial, suppose we have $L_h = \{\langle M, w \rangle : M$ on input w halts$\}$ as the oracle set. Given a (nonoracle) TM M, there is a standard way to modify M to the code of an equivalent TM M' in which the accepting state q_f is the only place where M' can halt. This is done by making every other combination q, c where M might halt send control to an extra state that causes an infinite loop. Thus M' halts on w if and only if M accepts w. Now design an oracle TM T' that on any input x of the form $\langle M, w \rangle$ (rejecting if x does not have the form) writes $x' = \langle M', w \rangle$ on the query tape and enters $q_?$, accepting if control goes to q_y and rejecting from q_n. Then T' with oracle set L_h decides whether $x \in L_u$, since $x \in L_u \iff x' \in L_h$. This is a simple case where an oracle for one problem helps one decide a different problem.

A language A is **Turing-reducible** to a language B, written $A \leq_T B$, if there is an oracle TM that with oracle B decides A. For example, we have just shown that $L_u \leq_T L_h$. The important special case in which the oracle TM T makes exactly one query, accepting from q_y and rejecting from q_n, gets its own notation: $A \leq_m B$. Equivalently, $A \leq_m B$ if there is a computable function f such that for all $x \in \Sigma^*$, $x \in A \iff f(x) \in B$. The function f represents the computation done by T prior to making the sole query. This case is called a **many-one reduction** (hence the subscript "m") for the arcane historical reason that f need not be a one-to-one function. The term "many-one reduction" is standard now. The above example actually shows that $L_u \leq_m L_h$. It is not hard to show that also $L_h \leq_m L_u$, so that the Halting Problem and the membership problem for a universal TM are many-one equivalent.

19.2.4 Alternating Turing Machines

TMs can be naturally generalized to model parallel computation. A nondeterministic TM accepts an input if there exists a move sequence leading to acceptance. We can call any nondeterministic state entered along this sequence an existential state. We can naturally add another type of state, a universal state. When a machine enters a universal state, the machine will accept if and only if *all* moves from this state lead to acceptance. These machines are called **alternating TMs**.

Let us describe the computation of alternating TMs formally and precisely. An alternating TM is simply a nondeterministic TM with the extra power that some states can be universal. A configuration of an alternating TM A has the same form as was described for a deterministic TM, namely,

```
(current state, tape contents, head positions).
```

We write

$$\alpha \vdash \beta$$

if, in one step, A can move from configuration α to configuration β. A configuration with current state q is accepting if

- q is an accepting state (i.e., $q = q_f$); or
- q is existential, and there exists an accepting configuration β such that $\alpha \vdash \beta$; or
- q is universal, and or each configuration β such that $\alpha \vdash \beta$, β is an accepting configuration.

This definition may seem circular, but by working backwards from configurations with q_f in the current-state field, one may verify that it inductively defines the set of accepting configurations in a natural manner. Then A accepts an input x if its initial configuration (with current state q_0, x on the input tape, heads at initial positions) is accepting.

Alternating TMs were first proposed by [4] for the purpose of modeling parallel computation. In order to allow sublinear computation times, a random-access model is used to read the input.

When in a special "read" state, the alternating TM is allowed to write a number in binary which is then interpreted as the address of a location on the input tape, whose symbol is then read in unit time. By using universal states to relate different branches of the computation, one can effectively read the whole input in as little as logarithmic time.

Just as a nondeterministic TM is a model for solitaire games, an alternating TM is a model for general two-person games. Alternating TMs have been successfully used to provide a theoretical foundation of parallel computation as well as to establish the complexity of various two person games. For example, a chess position with White to move can be modeled from White's point of view as a configuration α whose first component is an existential state. The position is winning if there exists a move for White such that the resulting position β is winning. Here β with Black to move has a universal state q, and is a winning position (for White) if and only if either Black is checkmated (this is the base case $q = q_f$) or for all moves by Black to a position γ, γ is a winning position for White. Chapter 23, Sections 23.5 and 23.6, will demonstrate the significance of games for time and space complexity, to which we now turn.

19.3 Time and Space Complexity

With TMs, we can now formally define what we mean by time and space complexity. The formal investigation by [1,13] in the 1960s marked the beginning of the field of computational complexity.

An important point with space complexity is that a machine should be charged only for those cells it uses for calculation, and not for read-only input, which might be provided on cheaper nonwritable media such as CD-ROM or accessed piecemeal over a network. Hence we modify the TM of Figure 19.9 by making the tape containing the input read-only, and giving it one or more worktapes.

DEFINITION 19.4 Let M be a TM. If for all n, every sequence of legal moves on an input x of length n halts within $t(n)$ steps, we say that M is of time complexity $t(n)$. Similarly, if every such sequence uses at most $s(n)$ worktape cells, then M is of space complexity $s(n)$.

THEOREM 19.5 *Fix a number $c > 0$, a space bound $s(n)$, and a time bound $t(n)$.*

(a) *Any TM of $s(n)$ space complexity, using any number of tapes or grids of any dimension, can be simulated by a TM with a single (one-dimensional) worktape that has space complexity $s(n)/c$.*

(b) *Any TM of $t(n)$ time complexity can be simulated by a TM, with the same number and kinds of worktapes, that has time complexity $n + t(n)/c$.*

The proof of these so-called linear speed-up theorems involves enlarging the original TM's worktape alphabet Γ to an alphabet Γ' large enough that one character in Γ' can encode a block of c consecutive characters on a tape of the original machine (see [14]). The extra "$n+$" in the time for part (b) is needed to read and translate the input into the "compressed" alphabet Γ'. If we think of memory in units of bits the idea that this saves space and time is illusory, but if we regard the machine with Γ' as having a larger word size, the savings make sense. Definition 19.4 is phrased in a way that applies also to nondeterministic and alternating TMs, and the two statements in Theorem 19.5 hold for them as well.

In Theorem 19.5, if $s(n) \geq n$, then we do not need to separate the input tape from the worktape(s). For any TM M of linear space complexity, part (a) implies that we can simulate M by a one-tape TM

M' that on any input x uses only the cells initially occupied by x (except for one visit to the blank cell to the right of x to tell where x ends). Then M' is called a linear bounded automaton.

The main import and convenience of Theorem 19.5 is that one does not need to use "$O()$-notation" to define complexity classes: space complexity $O(s(n))$ is no different from space complexity $s(n)$, and similarly for time. As we shall see, it is not always possible to reduce the number of tapes and run in the same time complexity, so researchers have settled on TMs with any finite number of tapes as the bench model for time complexity.

DEFINITION 19.5

- DTIME$[t(n)]$ is the class of languages accepted by multitape deterministic TMs in time $t(n)$.
- NTIME$[t(n)]$ is the class of languages accepted by multitape nondeterministic TMs in time $t(n)$.
- DSPACE$[s(n)]$ is the class of languages accepted by deterministic TMs in space $s(n)$.
- NSPACE$[s(n)]$ is the class of languages accepted by multitape nondeterministic TMs in space $O(s(n))$.
- P is the complexity class $\bigcup_{c \in \mathcal{N}}$ DTIME$[n^c]$.
- NP is the complexity class $\bigcup_{c \in \mathcal{N}}$ NTIME$[n^c]$.
- PSPACE is the complexity class $\bigcup_{c \in \mathcal{N}}$ DSPACE$[n^c]$.
- ATIME$[s(n), t(n)]$ is the class of languages accepted by alternating TMs operating simultaneously in time $t(n)$ and space $s(n)$.

Example 19.9

In Example 19.5 we demonstrated how the language $L = \{xx | x \in \{0, 1\}^*\}$ can be decided by a TM. We gave a two-head, one-tape TM running in time $3n/2$, and it is easy to design a two-tape, one-head-per-tape TM that executes the same strategy in time $2n$. Theorem 19.5 says that by using a larger tape alphabet, one can push the time down to $(1 + \epsilon)n$ for any fixed $\epsilon > 0$. However, our basic one-tape, one-head TM model can do no better than time on the order of n^2.

Example 19.10

Any multitape, multihead TM, not just the one accepting L in the last example, can be simulated by our basic one-tape, one-head model in at most the square of the original's running time. For example, a two-tape machine M with tape alphabet Γ can be simulated by a one-tape machine M' with a Γ' large enough to encode all pairs of characters over Γ. Then M' can regard its single tape as having two "tracks," one for each tape of M. M' also needs to mark the locations of the two heads of M, one on each track—this can be facilitated by adding more characters to Γ'. Now to simulate one step of M, the one-tape machine M' must use its single head to update the computation at the two locations with the two head markers. If M runs in time $t(n)$, then the two head markers cannot be more than $t(n)$ cells apart. Thus to simulate each step by M, M' moves its head for at most $t(n)$ distance. Hence M' runs in time at most $t(n)^2$.

The simulation idea in Example 19.10 does not work if M uses two- or higher-dimensional tapes, or a more-general "random-access" storage (see the next section). However, a one-tape M' can be built to simulate it whose running time is still no worse than a polynomial in the time of M. It is important to note that our basic one-tape deterministic TM is known to simulate all of the extended models we offer above and below—except nondeterministic and alternating TMs—with at most polynomial slowdown. This is a key point in taking the class P, defined as above in terms of TM time, as the benchmark for which languages and functions are considered feasibly computable for general computation. (See Chapter 22 for more discussion of this point.) Polynomial time for nondeterministic TMs defines the class NP. Interestingly, polynomial space for nondeterministic TMs does equal polynomial space for deterministic TMs [31], and polynomial time for alternating TMs equals the same class, namely PSPACE.

Example 19.11

All of the basic arithmetical operations—plus, minus, times, and division—belong to P. Given two n-digit integers x and y, we can easily add or subtract them in $O(n)$ steps. We can multiply or divide them in $O(n^2)$ steps using the standard algorithms learned in school. Actually, by grouping blocks of digits in x and y and using some clever tricks one can bring the time down to $O(n^{1+\epsilon})$ bit-operations for any desired fixed $\epsilon > 0$, and the asymptotically fastest method known takes time $O(n \log n \log\log n)$ [34]. Computing x^y is technically not in P because the sheer length of the output may be exponential in n, but if we measure time as a function of output length as well as input length, it is in P. However, the operation of factoring a number into primes, which is a kind of inverse of multiplication, is commonly believed not to belong to P. The language associated to the factoring function (refer to "L_f" before Theorem 19.4) does belong to NP.

Example 19.12

There are many other important problems in NP that are not known to be in P. For example, consider the following "King Arthur" problem, which is equivalent to the problem called HAMILTONIAN CIRCUIT in Chapter 23. King Arthur plans to have a round table meeting. By one historical account he had 150 knights, so let $n = 150$. It is known that some pairs of knights hate each other, and some do not. King Arthur's problem is to arrange the knights around a round table so that no pair of knights who sit side by side hate each other. King Arthur can solve this problem by enumerating all possible permutations of n knights. But even at $n = 150$, there are 150! permutations. All the computers in the whole world, even if they started a thousand years ago and worked nonstop, would still be going on today, having examined only a tiny fraction of the 150! permutations. However, this problem is in NP because a nondeterministic TM can just guess an arrangement and verify the correctness of the solution—by checking if any two neighboring knights are enemies—in polynomial time. It is currently unknown if every problem in NP is also in P. This problem has a special property—namely, if it is in P then every problem in NP is also in P.

The following relationships are true:

$$P \subseteq NP \subseteq PSPACE.$$

Whether or not either of the above inclusions is proper is one of the most fundamental open questions in computer science and mathematics. Research in computational complexity theory centers around these questions. The first step in working on these questions is to identify the hardest problems in NP or PSPACE.

DEFINITION 19.6   Given two languages A and B over an alphabet Σ, a function $f : \Sigma^* \to \Sigma^*$ is called a **polynomial-time many-one reduction** from A to B if

(a) f is polynomial-time computable, and

(b) for all $x \in \Sigma^*$, $x \in A$ if and only if $f(x) \in B$.

One also writes $A \leq_m^p B$.

The only change from the definition of "many-one reduction" at the end of Section 19.2.3 is that we have inserted "polynomial-time" before "computable." There is also a polynomial-time version of Turing reducibility as defined there, which gets the notation $A \leq_T^p B$.

DEFINITION 19.7 A language B is called NP-complete if

1. B is in NP;
2. for every language $A \in$ NP, $A \leq_m^p B$.

In this definition, if only the second item holds, then we say the language B is NP-hard. (Curiously, while "NP-complete" is always taken to refer by default to polynomial-time many-one reductions, "NP-hard" is usually extended to refer to polynomial-time Turing reductions.) The upshot is that if a language B is NP-complete and B is in P, then NP $=$ P. An **NP-complete language** is in this sense a hardest language in the class NP. Working independently, Cook [7] and Levin [23] introduced NP-completeness, and Karp [15] further demonstrated its importance. PSPACE and other classes also have complete languages.

Chapters 22 and 23 develop the topics of this section in much greater detail. We also refer the interested reader to the textbooks [9,14,24,38,39].

19.4 Other Computing Models

Over the years, many alternative computing models have been proposed. Under reasonable definitions of running time for these models, they can all be simulated by TMs with at most a polynomial slow-down. The Ref. [37] provides a nice survey of various computing models other than TMs. We will discuss a few such alternatives very briefly and refer our readers to Chapter 23 and [37] for more information.

19.4.1 Random Access Machines

The random access machine [8] consists of a finite control where a program is stored, several arithmetic registers, and an infinite collection of memory registers $R[1], R[2], \ldots$. All registers have an unbounded word length. The basic instructions for the program are LOAD, STORE, ADD, MUL, GOTO, and conditional-branch instructions. The LOAD and STORE commands can use indirect addressing. Compared to TMs, this appears to be a closer but more complicated approximation of modern computers. There are two standard ways for measuring time complexity of the model:

- The unit-cost RAM: Here each instruction takes one unit of time, no matter how big the operands are. This measure is convenient for analyzing many algorithms.
- The log-cost RAM: Here each instruction is charged for the sum of the lengths of all data manipulated by the instruction. Equivalently, each use of an integer i is charged $\log i$ time units, since $\log i$ is approximately the length of i. This is a more realistic model, but sometimes less convenient to use.

Log-cost RAMs and TMs can simulate each other with polynomial overheads. The unit-cost assumption becomes unrealistic when the MUL instruction is used repeatedly to form exponentially large numbers. Taking MUL out, however, makes the unit-cost RAM polynomially equivalent to the TM as well.

19.4.2 Pointer Machines

Pointer machines were introduced by [21] in 1958 and in modified form by Schönhage in the 1970s. Schönhage called his form the "storage modification machine" [33], and both forms are sometimes named for their authors. We informally describe Schönhage's form here. A pointer machine is similar to a RAM, but instead of having an unbounded array of registers for its memory structure, it has modifiable pointer links that form a Δ-structure. A Δ-structure S, for a finite alphabet Δ of size k, is a finite directed graph in which each node has k out-edges labeled by the k symbols in Δ. Every node also has a cell holding an integer, as with a RAM. At every step of the computation, one node of S is distinguished as the center, which acts as a starting point for addressing. A word $w \in \Delta^*$ addresses the cell of the node formed by following the path of pointer links selected by the successive characters in w. Besides having all the RAM instructions, the pointer machine has various instructions and rules for moving its center and redirecting pointer links, thus modifying the storage structure. Under the log-cost criterion, pointer machines are polynomially equivalent to RAMs and TMs. There are many interesting studies on the precise efficiency of the simulations among these models, and we refer to the reader to the survey [37] as a center for further pointers on them.

19.4.3 Circuits and Nonuniform Models

A Boolean circuit is a finite, labeled, directed acyclic graph. Input nodes are nodes without ancestors; they are labeled with input variables $x_1, \ldots, x_n$. The internal nodes are labeled with functions from a finite set of Boolean operations such as {AND, OR, NOT} or {NAND}. The number of ancestors of an internal node is precisely the number of arguments of the Boolean function that the node is labeled with. A node without successors is an output node. The circuit is naturally evaluated from input to output: at each node the function labeling the node is evaluated using the results of its ancestors as arguments. Two cost measures for the circuit model are

- Depth: The length of a longest path from an input node to an output node
- Size: The number of nodes in the circuit

These measures are applied to a family $\{C_n \mid n \geq 1\}$ of circuits for a particular problem, where C_n solves the problem of size n. Subject to the uniformity condition that the layout of C_n be computable given n (in time polynomial in n), circuits are (polynomially) equivalent to TMs. Chapters 22 and 28 give full presentations of circuit complexity, while [16,30,37] have more details and pointers to the literature.

19.5   Further Information

The fundamentals of the theory of computation, automata theory, and formal languages can be found in Chapters 20 through 22 and in many text books including [9–11,14,24,35,38,39]. One central focus of research in this area is to understand the relationships between different resource complexity classes. This work is motivated in part by some major open questions about the relationships between resources (such as time and space) and the role of control mechanisms (such as nondeterminism or randomness). At the same time, new computational models are being introduced and studied. One recent model that has led to the resolution of a number of interesting problems is the interactive proofs (IP) model. IP is defined in terms of two Turing machines that communicate with each other. One of them has unlimited power and the other (called the verifier) is a probabilistic Turing machine whose time complexity is bounded by a polynomial. The study of IP has led to new ways to encrypt information as well as to the proof of some unexpected results about the difficulty of solving NP-hard problems (such as coloring, clique etc.) even approximately. See Chapter 24, Sections 24.3

and 24.5. Another new model is the quantum Turing machine, which can solve in polynomial time some problems such as factoring that are believed to require exponential time on any hardware that follows the laws of "classical" (prequantum) physics. There are also attempts to use molecular or cell-level interactions as the basic operations of a computer.

The following annual conferences present the leading research work in computation theory: ACM Annual Symposium on Theory of Computing (STOC), IEEE Symposium on the Foundations of Computer Science (FOCS), IEEE Conference on Computational Complexity (CCC, formerly Structure in Complexity Theory), International Colloquium on Automata, Languages and Programming (ICALP), Symposium on Theoretical Aspects of Computer Science (STACS), Mathematical Foundations of Computer Science (MFCS), and Fundamentals of Computation Theory (FCT). There are many related conferences in the following areas: computational learning theory, computational geometry, algorithms, principles of distributed computing, computational biology, and database theory. In each case, specialized computational models and concrete algorithms are studied for a specific application area. There are also conferences in both pure and applied mathematics that admit topics in computation theory and complexity. We conclude with a partial list of major journals whose primary focus is in theory of computation: *Journal of the ACM, SIAM Journal on Computing, Journal of Computer and System Sciences, Information and Computation, Theory of Computing Systems* (formerly *Mathematical Systems Theory*), *Theoretical Computer Science, Computational Complexity, Journal of Complexity, Information Processing Letters, International Journal of Foundations of Computer Science,* and *Acta Informatica.*

Defining Terms

Alternating turing machine: A generalization of a nondeterministic TM. In the latter, every state can be called an existential state since the machine accepts if one of the possible moves leads to acceptance. In an alternating TM there are also universal states, from which the machine accepts only if all possible moves out of that state lead to acceptance.

Algorithm: A finite sequence of instructions that is supposed to solve a particular problem.

Complexity class NP: The class of languages that can be accepted by a nondeterministic TM in polynomial time.

Complexity class P: The class of languages that can be accepted by a deterministic TM in polynomial time.

Complexity class PSPACE: The class of languages that can be accepted by a TM in polynomial space.

Computable function: A function that can be computed by an algorithm—equivalently, by a TM.

Decidable problem/language: A problem that can be decided by an algorithm—equivalently, whose associated language is accepted by a TM that halts for all inputs.

Deterministic: Permitting at most one next move at any step in a computation.

Finite automaton or finite-state machine: A restricted TM where the head is read-only and shifts only from left to right.

(Formal) language: A set of strings over some fixed alphabet.

Halting problem: The problem of deciding whether a given program (or TM) halts on a given input.

Many-one reduction: A reduction that maps an instance of one problem into an equivalent instance of another problem.

Nondeterministic: Permitting more than one choice of next move at some step in a computation.

NP-complete language: A language in NP such that every language in NP can be reduced to it in polynomial time.

Oracle Turing machine: A TM with an extra oracle tape and three extra states $q_?, q_y, q_n$. When the machine enters $q_?$, control goes to state q_y if the oracle tape content is in the oracle set; otherwise control goes to state q_n.

Partial recursive function: A partial function computed by a TM that need not halt for all inputs.

Partially decidable problem: One whose associated language is recursively enumerable. Equivalently, there exists a program that halts and outputs 1 for every instance having a yes answer, but is allowed not to halt or to halt and output 0 for every instance with a no answer.

Polynomial time reduction: A reduction computable in polynomial time.

Program: A sequence of instructions that can be executed, such as the code of a TM or a sequence of RAM instructions.

Pushdown automaton: A restricted TM where the tape acts as a pushdown store (or a stack), with an extra one-way read-only input tape.

Recursive language: A language accepted by a TM that halts for all inputs.

Recursively enumerable (r.e.) language: A language accepted by a TM.

Reduction: A computable transformation of one problem into another.

Regular language: A language which can be described by some right-linear/regular grammar (or equivalently by some regular expression).

Time/space complexity: A function describing the maximum time/space required by the machine on any input of length n.

Turing machine: A simplest formal model of computation consisting a finite-state control and a semi-infinite sequential tape with a read-write head. Depending on the current state and symbol read on the tape, the machine can change its state and move the head to the left or right. Unless otherwise specified, a TM is deterministic.

Turing reduction: A reduction computed by an oracle TM that halts for all inputs with the oracle used in the reduction.

Uncomputable function: A function that cannot be computed by any algorithm—equivalently, not by any TM.

Undecidable problem/language: A problem that cannot be decided by any algorithm—equivalently, whose associated language cannot be recognized by a TM that halts for all inputs.

Universal Turing machine: A TM that is capable of simulating any other Turing machine if the latter is properly encoded.

References

1. Blum, M., A machine independent theory of complexity of recursive functions. *J. Assoc. Comput. Mach.*, 14, 322–336, 1967.
2. Brzozowski, J. and McCluskey Jr., E., Signal flow graph techniques for sequential circuit state diagram. *IEEE Trans. Electron. Comput.*, EC-12(2), 67–76, 1963.
3. Brzozowski, J.A. and Seger, C.-J.H., *Asynchronous Circuits*, Springer-Verlag, New York, 1994.
4. Chandra, A.K., Kozen, D.C., and Stockmeyer, L.J., Alternation. *J. Assoc. Comput. Mach.*, 28, 114–133, 1981.
5. Chomsky, N., Three models for the description of language. *IRE Trans. Inform. Theory*, 2(2), 113–124, 1956.

6. Chomsky, N. and Miller, G., Finite-state languages. *Inform. Control,* 1, 91–112, 1958.
7. Cook, S., The complexity of theorem-proving procedures. *Proceedings of the Third ACM Symposium Theory of Computing,* Shaker Heights, Ohio, pp. 151–158, 1971.
8. Cook, S. and Reckhow, R., Time bounded random access machines. *J. Comput. Syst. Sci.,* 7, 354–375, 1973.
9. Floyd, R.W. and Beigel, R., *The Language of Machines: An Introduction to Computability and Formal Languages,* Computer Science Press, New York, 1994.
10. Gurari, E., *An Introduction to the Theory of Computation,* Computer Science Press, Rockville, MD, 1989.
11. Harel, D., *Algorithmics: The Spirit of Computing,* Addison-Wesley, Reading, MA, 1992.
12. Hartmanis, J., On computational complexity and the nature of computer science. *CACM,* 37(10), 37–43, 1994.
13. Hartmanis, J. and Stearns, R., On the computational complexity of algorithms. *Trans. Am. Math. Soc.,* 117, 285–306, 1965.
14. Hopcroft, J. and Ullman, J., *Introduction to Automata Theory, Languages and Computation,* Addison-Wesley, Reading, MA, 1979.
15. Karp, R.M., 1972. Reducibility among combinatorial problems. In *Complexity of Computer Computations,* Miller, R.E. and Thatcher, J.W. (Eds.), 4, Plenum Press, New York, 1972, pp. 85–104.
16. Karp, R.M. and Ramachandran, V., Parallel algorithms for shared-memory machines. In *Handbook of Theoretical Computer Science,* van Leeuwen, J. (Ed.), MIT Press, Cambridge, MA, 1990, pp. 869–941.
17. Kleene, S., Representation of events in nerve nets and finite automata. In *Automata Studies,* Princeton University Press, Princeton, NJ, 1956, pp. 3–41.
18. Knuth, D.E., *Fundamental Algorithms,* Vol. 1 of *The Art of Computer Programming,* Addison-Wesley, Reading, MA, 1969.
19. Knuth, D., Morris, J., and Pratt, V., Fast pattern matching in strings. *SIAM J. Comput.,* 6, 323–350, 1977.
20. Kohavi, Z., *Switching and Finite Automata Theory,* McGraw-Hill, New York, 1978.
21. Kolmogorov, A. and Uspenskii, V., On the definition of an algorithm. *Uspekhi Mat. Nauk.,* 13, 3–28, 1958.
22. Lesk, M., LEX–a lexical analyzer generator. Technical Report 39, Bell Laboratories, Murray Hill, NJ, 1975.
23. Levin, L., Universal sorting problems. *Problemi Peredachi Informatsii,* 9(3), 265–266, 1973. (In Russian.)
24. Lewis, H. and Papadimitriou, C.H., *Elements of the Theory of Computation,* Prentice-Hall, Englewood Cliffs, NJ, 1981.
25. Li, M. and Vitányi, P., *An Introduction to Kolmogorov Complexity and Its Applications,* Springer-Verlag, New York, 1993; 2nd edn., 1997.
26. Lynch, N., *Distributed Algorithms,* Morgan Kaufmann, San Francisco, CA 1996.
27. McCulloch, W. and Pitts, W., A logical calculus of ideas immanent in nervous activity. *B. Math. Biophys.,* 5, 115–133, 1943.
28. Rabin, M.O., Real time computation. *Israel J. Math.,* 1(4), 203–211, 1963.
29. Rabin, M.O. and Scott, D., Finite automata and their decision problems. *IBM J. Res. Dev.,* 3, 114–125, 1959.
30. Ruzzo, W.L., On uniform circuit complexity. *J. Comput. Syst. Sci.,* 22, 365–383, 1981.
31. Savitch, J., Relationships between nondeterministic and deterministic tape complexities. *J. Comput. Syst. Sci.,* 4(2), 177–192, 1970.
32. Searls, D., The computational linguistics of biological sequences. In *Artificial Intelligence and Molecular Biology,* Hunter, L. (Ed.), MIT Press, Cambridge, MA, 1993, pp. 47–120.
33. Schönhage, A., Storage modification machines. *SIAM J. Comput.,* 9, 490–508, 1980.
34. Schönhage, A. and Strassen, V., Schnelle Multiplikation grosser Zahlen. *Computing,* 7, 281–292, 1971.

35. Sipser, M., *Introduction to the Theory of Computation,* 1st edn. PWS Publishing Company, Boston, MA, 1997.
36. Turing, A., On computable numbers with an application to the Entscheidungsproblem. *Proc. London Math. Soc., Series 2,* 42, 230–265, 1936.
37. van Emde Boas, P., Machine models and simulations. In *Handbook of Theoretical Computer Science,* van Leeuwen, J. (Ed.), MIT Press, Cambridge, MA, 1990, pp. 1–66.
38. Wood, D., *Theory of Computation,* Harper and Row, New York, 1987.
39. Yap, C., *Introduction to Complexity Classes,* Oxford University Press, Oxford, UK, 2000.

20

Formal Grammars and Languages

Tao Jiang
University of California

Ming Li
University of Waterloo

Bala Ravikumar
Sonoma State University

Kenneth W. Regan
State University of New York at Buffalo

20.1   Introduction

Formal language theory as a discipline is generally regarded as growing from the work of linguist Noam Chomsky in the 1950s, when he attempted to give a precise characterization of the structure of natural languages. His goal was to define the syntax of languages using simple and precise mathematical rules. Later it was found that the syntax of programming languages can be described using one of Chomsky's grammatical models called **context-free grammars**. Much earlier, the Norwegian mathematician Axel Thue studied sequences of binary symbols subject to interesting mathematical properties, such as not having the same substring three times in a row. His work influenced Emil Post, Stephen Kleene, and others to study the mathematical properties of strings and collections of strings.

Soon after the advent of modern electronic computers, people realized that all forms of information—whether numbers, names, pictures, or sound waves—can be represented as strings. Then collections of strings known as languages became central to computer science. This chapter is concerned with fundamental mathematical properties of languages and language-generating systems, such as grammars. Every programming language from Fortran to Java can be precisely described by a grammar. Moreover, the grammar allows us to write a computer program (called the syntax analyzer in a compiler) to determine whether a string of statements is syntactically correct in the programming language. Many people would wish that natural languages such as English could be analyzed as precisely, that we could write computer programs to tell which English sentences are grammatically correct. Despite recent advances in natural language processing, many of which have been spurred by formal grammars and other theoretical tools, today's commercial products for grammar and style fall well short of that ideal. The main problem is that there is no common agreement on what are grammatically correct (English) sentences; nor has anyone yet been able to offer a grammar precise enough to propose as definitive. And style is a matter of taste(!) such as not beginning sentences with "and" or using interior exclamations. Formal languages and grammars have many applications in other fields, including molecular biology (see [17]) and symbolic dynamics (see [14]).

In this chapter, we will present some formal systems that define families of formal languages arising in many computer science applications. Our primary focus will be on **context-free languages** (CFL), since they are most widely used to describe the syntax of programming languages. In the rest of this section, we present some basic definitions and terminology.

DEFINITION 20.1   An alphabet is a finite nonempty set of symbols. Symbols are assumed to be indivisible.

For example, an alphabet for English can consist of as few as the 26 lower-case letters $a, b, \ldots, z$, adding some punctuation symbols if sentences rather than single words will be considered. Or it may include all of the symbols on a standard North American typewriter, which together with terminal control codes yields the 128-symbol ASCII alphabet, in which much of the world's communication takes place. The new world standard is an alphabet called UNICODE, which is intended to provide symbols for all the world's languages—as of this writing, over 38,000 symbols have been assigned. But most important aspects of formal languages can be modeled using the simple two-letter alphabet $\{0, 1\}$, over which ASCII and UNICODE are encoded to begin with. We usually use the symbol Σ to denote an alphabet.

DEFINITION 20.2   A string over an alphabet Σ is a finite sequence of symbols of Σ.

The number of symbols in a string x is called its length, denoted by $|x|$. It is convenient to introduce a notation ϵ for the empty string, which contains no symbols at all. The length of ϵ is 0.

DEFINITION 20.3   Let $x = a_1 a_2 \cdots a_n$ and $y = b_1 b_2 \cdots b_m$ be two strings. The concatenation of x and y, denoted by xy, is the string $a_1 a_2 \cdots a_n b_1 b_2 \cdots b_m$.

Then for any string x, $\epsilon x = x \epsilon = x$. For any string x and integer $n \geq 0$, we use x^n to denote the string formed by sequentially concatenating n copies of x.

DEFINITION 20.4   The set of all strings over an alphabet Σ is denoted by Σ^*, and the set of all nonempty strings over Σ is denoted by Σ^+. The empty set of strings is denoted by $\emptyset$.

DEFINITION 20.5   For any alphabet Σ, a language over Σ is a set of strings over Σ. The members of a language are also called the words of the language.

Example 20.1

The sets $L_1 = \{01, 11, 0110\}$ and $L_2 = \{0^n 1^n | n \geq 0\}$ are two languages over the binary alphabet $\{0, 1\}$. L_1 has three words, while L_2 is infinite. The string 01 is in both languages while 11 is in L_1 but not in L_2.

Since languages are just sets, standard set operations such as union, intersection, and complementation apply to languages. It is useful to introduce two more operations for languages: concatenation and Kleene closure.

DEFINITION 20.6   Let L_1 and L_2 be two languages over Σ. The concatenation of L_1 and L_2, denoted by $L_1 L_2$, is the language $\{xy | x \in L_1, y \in L_2\}$.

DEFINITION 20.7   Let L be a language over Σ. Define $L^0 = \{\epsilon\}$ and $L^i = LL^{i-1}$ for $i \geq 1$. The Kleene closure of L, denoted by L^*, is the language

$$L^* = \bigcup_{i \geq 0} L^i.$$

The positive closure of L, denoted by L^+, is the language

$$L^+ = \bigcup_{i \geq 1} L^i.$$

In other words, the Kleene closure of a language L consists of all strings that can be formed by concatenating zero or more words from L. For example, if $L = \{0, 01\}$, then $LL = \{00, 001, 010, 0101\}$, and L^* comprises all binary strings in which every 1 is preceded by a 0. Note that concatenating zero words always gives the empty string, and that a string with no 1s in it still makes the condition on "every 1" true. L^+ has the meaning "concatenate one or more words from L," and satisfies the properties $L^* = L^+ \cup \{\epsilon\}$ and $L^+ = LL^*$. Furthermore, for any language L, L^* always contains ϵ, and L^+ contains ϵ if and only if L does. Also note that Σ^* is in fact the Kleene closure of the alphabet Σ when Σ is viewed as a language of words of length 1, and Σ^+ is just the positive closure of Σ.

20.2   Representation of Languages

In general a language over an alphabet Σ is a subset of Σ^*. How can we describe a language rigorously so that we know whether a given string belongs to the language or not? As shown in Example 20.1, a finite language such as L_1 can be explicitly defined by enumerating its elements. An infinite language such as L_2 cannot be exhaustively enumerated, but in the case of L_2 we were able to give a simple rule characterizing all of its members. In English, the rule is, "some number of 0s followed by an equal number of 1s." Can we find systematic methods for defining rules that characterize a wide class of languages? In the following we will introduce three such methods: **regular expressions**, **pattern systems**, and **grammars**. Interestingly, only the last is capable of specifying the simple rule for L_2, although the first two work for many intricate languages. The term **formal languages** refers to languages that can be described by a body of systematic rules.

20.2.1   Regular Expressions and Languages

Let Σ be an alphabet.

DEFINITION 20.8   The regular expressions over Σ and the languages they represent are defined inductively as follows [5]:

1. The symbol $\emptyset$ is a regular expression, and represents the empty language.
2. The symbol ϵ is a regular expression, and represents the language whose only member is the empty string, namely $\{\epsilon\}$.
3. For each $c \in \Sigma$, c is a regular expression, and represents the language $\{c\}$, whose only member is the string consisting of the single character c.
4. If r and s are regular expressions representing the languages R and S, then $(r+s)$, (rs) and (r^*) are regular expressions that represent the languages $R \cup S$, RS, and R^*, respectively.

For example, $((0(0+1)^*) + ((0+1)^*0))$ is a regular expression over $\{0, 1\}$ that represents the language consisting of all binary strings that begin or end with a 0. Since the set operations union and concatenation are both associative, and since we can stipulate that Kleene closure takes precedence over concatenation and concatenation over union, many parentheses can be omitted from regular expressions. For example, the above regular expression can be written as $0(0 + 1)^* + (0 + 1)^*0$. We will also abbreviate the expression rr^* as r^+. Let us look at a few more examples of regular expressions and the languages they represent.

Example 20.2

The expression $0(0 + 1)^*1$ represents the set of all strings that begin with a 0 and end with a 1.

Example 20.3

The expression $0 + 1 + 0(0 + 1)^*0 + 1(0 + 1)^*1$ represents the set of all nonempty binary strings that begin and end with the same bit. Note the inclusion of the strings 0 and 1 as special cases.

Example 20.4

The expressions 0^*, 0^*10^*, and $0^*10^*10^*$ represent the languages consisting of strings that contain no 1, exactly one 1, and exactly two 1's, respectively.

Example 20.5

The expressions $(0+1)^*1(0+1)^*1(0+1)^*$, $(0+1)^*10^*1(0+1)^*$, $0^*10^*1(0+1)^*$, and $(0+1)^*10^*10^*$ all represent the same set of strings that contain at least two 1's.

Two or more regular expressions that represent the same language, as in Example 20.5, are called equivalent. It is possible to introduce algebraic identities for regular expressions in order to construct equivalent expressions. Two such identities are $r(s + t) = rs + rt$, which says that concatenation distributes over union the same way "times" distributes over "plus" in ordinary algebra (but taking care that concatenation isn't commutative), and $r^* = (r^*)^*$. These two identities are easy to prove; the reader seeking more detail may consult [16].

Example 20.6

Let us construct a regular expression for the set of all strings that contain no consecutive 0s. A string in this set may begin and end with a sequence of 1s. Since there are no consecutive 0s, every 0 that is not the last symbol of the string must be followed by a 1. This gives us the expression $1^*(01^+)^*1^*(\epsilon + 0)$. It is not hard to see that the second 1^* is redundant and thus the expression can in fact be simplified to $1^*(01^+)^*(\epsilon + 0)$.

Regular expressions were first introduced by [13] for studying the properties of neural nets. The above examples illustrate that regular expressions often give very clear and concise representations of languages. The languages represented by regular expressions are called the regular languages. Fortunately or unfortunately, not every language is regular. For example, there are no regular expressions that represent the languages $\{0^n1^n | n \geq 1\}$ or $\{xx | x \in \{0, 1\}^*\}$; the latter case is proved in Section 19.2.

20.2.2 Pattern Languages

Another way of representing languages is to use pattern systems [2] (see also [12]).

DEFINITION 20.9   A pattern system is a triple (Σ, V, p), where Σ is the alphabet, V is the set of variables with $\Sigma \cap V = \emptyset$, and p is a string over $\Sigma \cup V$ called the pattern.

DEFINITION 20.10   The language generated by a pattern system (Σ, V, p) consists of all strings over Σ that can be obtained from p by replacing each variable in p with a string over Σ.

An example pattern system is $(\{0, 1\}, \{v_1, v_2\}, v_1 v_1 0 v_2)$. The language it generates contains all words that begin with a 0 (since v_1 can be chosen as the empty string, and v_2 as an arbitrary string), and contains some words that begin with a 1, such as 110 (by taking $v_1 = 1$, $v_2 = \epsilon$) and 101001 (by taking $v_1 = 10$, $v_2 = 1$). However, it does not contain the strings $\epsilon, 1, 10, 11, 100, 101$, etc. The pattern system $(\{0, 1\}, \{v_1\}, v_1 v_1)$ generates the set of all strings that are the concatenation of two equal substrings, namely the set $\{xx | x \in \{0, 1\}^*\}$. The languages generated by pattern systems are called pattern languages.

Regular languages and pattern languages are really different. We have noted that the pattern language $\{xx | x \in \{0, 1\}^*\}$ is not a regular language, and one can prove that the set represented by the regular expression $0^* 1^*$ is not a pattern language. Although it is easy to write an algorithm to decide whether a given string is in the language generated by a given pattern system, such an algorithm would most likely have to be very inefficient [2].

20.2.3   General Grammars

Perhaps the most useful and general system for representing languages is based on the formal notion of a grammar.

DEFINITION 20.11   A grammar is a quadruple (Σ, V, S, P), where

1. Σ is a finite nonempty set called the terminal alphabet. The elements of Σ are called the terminals.
2. V is a finite nonempty set disjoint from Σ. The elements of V are called the nonterminals or variables.
3. $S \in V$ is a distinguished nonterminal called the start symbol.
4. P is a finite set of productions (or rules) of the form

$$\alpha \to \beta$$

where $\alpha \in (\Sigma \cup V)^* V (\Sigma \cup V)^*$ and $\beta \in (\Sigma \cup V)^*$, i.e., α is a string of terminals and nonterminals containing at least one nonterminal and β is a string of terminals and nonterminals.

Example 20.7

Let $G_1 = (\{0, 1\}, \{S, T, O, I\}, S, P)$, where P contains the following productions

$$S \to TO$$
$$S \to OI$$
$$T \to SI$$
$$O \to 0$$
$$I \to 1$$

As we shall see, the grammar G_1 can be used to describe the set $\{0^n 1^n | n \geq 1\}$.

Example 20.8

Let $G_2 = (\{0, 1, 2\}, \{S, A\}, S, P)$, where P contains the following productions

$$S \rightarrow 0SA2$$
$$S \rightarrow \epsilon$$
$$2A \rightarrow A2$$
$$0A \rightarrow 01$$
$$1A \rightarrow 11$$

This grammar G_2 can be used to describe the set $\{0^n 1^n 2^n \geq n \geq 0\}$.

Example 20.9

To construct a grammar G_3 to describe English sentences, one might let the alphabet Σ comprise all English words rather than letters. V would contain nonterminals that correspond to the structural components in an English sentence, such as <sentence>, <subject>, <predicate>, <noun>, <verb>, <article>, and so on. The start symbol would be <sentence>. Some typical productions are as follows:

$$<sentence> \rightarrow <subject><predicate>$$
$$<subject> \rightarrow <noun>$$
$$<predicate> \rightarrow <verb><article><noun>$$
$$<noun> \rightarrow mary$$
$$<noun> \rightarrow algorithm$$
$$<verb> \rightarrow wrote$$
$$<article> \rightarrow an$$

The rule <sentence> $\rightarrow$ <subject><predicate> models the fact that a sentence can consist of a subject phrase and a predicate phrase. The rules <noun> $\rightarrow$ mary and <noun> $\rightarrow$ algorithm mean that both "mary" and "algorithm" are possible nouns. This approach to grammar, stemming from Chomsky's work, has influenced even elementary-school teaching.

To explain how a grammar represents a language, we need the following concepts.

DEFINITION 20.12   Let (Σ, V, S, P) be a grammar. A *sentential form* of G is any string of terminals and nonterminals, i.e., a string over $\Sigma \cup V$.

DEFINITION 20.13   Let (Σ, V, S, P) be a grammar, and let γ_1, γ_2 be two sentential forms of G. We say that γ_1 *directly derives* γ_2, written $\gamma_1 \Rightarrow \gamma_2$, if $\gamma_1 = \sigma\alpha\tau$, $\gamma_2 = \sigma\beta\tau$, and $\alpha \rightarrow \beta$ is a production in P.

For example, the sentential form $00S11$ directly derives the sentential form $00OT11$ in grammar G_1, and $A2A2$ directly derives $AA22$ in grammar G_2.

DEFINITION 20.14   Let γ_1 and γ_2 be two sentential forms of a grammar G. We say that γ_1 **derives** γ_2, written $\gamma_1 \Rightarrow^* \gamma_2$, if there exists a sequence of (zero or more) sentential forms $\sigma_1, \ldots, \sigma_n$ such that

$$\gamma_1 \Rightarrow \sigma_1 \Rightarrow \cdots \Rightarrow \sigma_n \Rightarrow \gamma_2.$$

The sequence $\gamma_1 \Rightarrow \sigma_1 \Rightarrow \cdots \Rightarrow \sigma_n \Rightarrow \gamma_2$ is called a derivation of γ_2 from γ_1.

For example, in grammar G_1, $S \Rightarrow^* 0011$ because

$$S \Rightarrow \underline{QT} \Rightarrow 0\underline{T} \Rightarrow 0\underline{SI} \Rightarrow 0\underline{S}1 \Rightarrow 0\underline{QI}1 \Rightarrow 00\underline{I}1 \Rightarrow 0011$$

and in grammar G_2, $S \Rightarrow^* 001122$ because

$$S \Rightarrow 0\underline{S}A2 \Rightarrow 00\underline{S}A2A2 \Rightarrow 00\underline{A}2A2 \Rightarrow 0012\underline{A}2 \Rightarrow 0011\underline{A}22 \Rightarrow 001122.$$

Here the left-hand side of the relevant production in each derivation step is underlined for clarity.

DEFINITION 20.15   Let (Σ, V, S, P) be a grammar. The language generated by G, denoted by $L(G)$, is defined as

$$L(G) = \left\{ x \mid x \in \Sigma^*, S \Rightarrow^* x \right\}.$$

The words in $L(G)$ are also called the sentences of $L(G)$.

Clearly, $L(G_1)$ contains all strings of the form $0^n 1^n$, $n \geq 1$, and $L(G_2)$ contains all strings of the form $0^n 1^n 2^n$, $n \geq 0$. Although only a partial definition of G_3 is given, we know that $L(G_3)$ contains sentences like "mary wrote an algorithm" and "algorithm wrote an algorithm," but does not contain strings like "an wrote algorithm."

Formal grammars were introduced as such by [15], and had antecedents in work by Thue and others. However, the study of their rigorous use in describing formal (and natural) languages did not begin until the mid-1950s [3]. In the next section, we consider various restrictions on the form of productions in a grammar, and see how these restrictions can affect its power to represent languages. In particular, we will show that regular languages and pattern languages can all be generated by grammars under different restrictions.

20.3   Hierarchy of Grammars

Grammars can be divided into four classes by gradually increasing the restrictions on the form of the productions. Such a classification is due to Chomsky [3,4] and is called the Chomsky hierarchy.

DEFINITION 20.16   Let $G = (\Sigma, V, S, P)$ be a grammar.

1. G is also called a Type-0 grammar or an unrestricted grammar.
2. G is a Type-1 or context-sensitive grammar if each production $\alpha \to \beta$ in P satisfies $|\alpha| \leq |\beta|$. By "special dispensation," we also allow a Type-1 grammar to have the production $S \to \epsilon$, provided S does not appear on the right-hand side of any production.
3. G is a Type-2 or context-free grammar if each production $\alpha \to \beta$ in P satisfies $|\alpha| = 1$; i.e., α is a single nonterminal.
4. G is a Type-3 or right-linear or regular grammar if each production has one of the following three forms:

$$A \to cB, \quad A \to c, \quad A \to \epsilon,$$

where A, B are nonterminals (with $B = A$ allowed) and c is a terminal.

The language generated by a Type-i grammar is called a Type-i language, $i = 0, 1, 2, 3$. A Type-1 language is also called a **context-sensitive language** (CSL), and a Type-2 language is also called a

CFL. The "special dispensation" allows a CSL to contain ϵ, and thus allows one to say that every CFL is also a CSL. Many sources allow "right-linear" grammars to have productions of the form $A \rightarrow xB$, where x is any string of terminals, and/or exclude one of the forms $A \rightarrow c, A \rightarrow \epsilon$ from their definition of "regular" grammar (perhaps allowing $S \rightarrow \epsilon$ in the latter case). Regardless of the choice of definitions, every Type-3 grammar generates a regular language, and every regular language has a Type-3 grammar; we have proved this using finite automata in Chapter 19. Stated in other words:

THEOREM 20.1   *The class of Type-3 languages and the class of regular languages are equal.*

The grammars G_1 and G_3 given in the last section are context-free and the grammar G_2 is context-sensitive. Now we give some examples of unrestricted and right-linear grammars.

Example 20.10

Let $G_4 = (\{0, 1\}, \{S, A, O, I, T\}, S, P)$, where P contains

$$
\begin{array}{rcl rcl}
S & \rightarrow & AT & & & \\
A & \rightarrow & 0AO & A & \rightarrow & 1AI \\
00 & \rightarrow & 00 & 01 & \rightarrow & 10 \\
I0 & \rightarrow & 0I & I1 & \rightarrow & 1I \\
OT & \rightarrow & 0T & IT & \rightarrow & 1T \\
A & \rightarrow & \epsilon & T & \rightarrow & \epsilon
\end{array}
$$

Then G_4 generates the set $\{xx \mid x \in \{0, 1\}^*\}$. To understand how this grammar works, think of the nonterminal O as saying, "I must ensure that the right half gets a terminal 0 in the same place as the terminal 0 in the production $A \rightarrow 0AO$ that introduced me." The nonterminal I eventually forces the precise placement of a terminal 1 in the right-hand side in the same way. The nonterminal T makes sure that O and I place their 0 and 1 on the right-hand side rather than prematurely. Only after every O and I has moved right past any earlier-formed terminals 0 and 1 and been eliminated "in the context of" T, and the production $A \rightarrow \epsilon$ is used to signal that no additional O or I will be introduced, can the endmarker T be dispensed with via $T \rightarrow \epsilon$. For example, we can derive the word 0101 from S as follows:

$$S \Rightarrow \underline{A}T \Rightarrow 0\underline{A}OT \Rightarrow 01\underline{A}IOT \Rightarrow 01\underline{IO}T \Rightarrow 01\underline{IO}T \Rightarrow 010\underline{II}T \Rightarrow 0101\underline{T} \Rightarrow 0101.$$

Only the productions $A \rightarrow \epsilon$ and $T \rightarrow \epsilon$ prevent this grammar from being Type-1. The interested reader is challenged to write a Type-1 grammar for this language.

Example 20.11

We give a right-linear grammar G_5 to generate the language represented by the regular expression in Example 20.3, i.e., the set of all nonempty binary strings beginning and ending with the same bit. Let $G_5 = (\{0, 1\}, \{S, O, I\}, S, P)$, where P contains

$$
\begin{array}{rcl rcl}
S & \rightarrow & 0O & S & \rightarrow & 1I \\
S & \rightarrow & 0 & S & \rightarrow & 1 \\
O & \rightarrow & 0O & O & \rightarrow & 1O \\
I & \rightarrow & 0I & I & \rightarrow & 1I \\
O & \rightarrow & 0 & I & \rightarrow & 1
\end{array}
$$

Here O means to remember that the last bit must be a 0, and 1 similarly forces the last bit to be a 1. Note again how the grammar treats the words 0 and 1 as special cases.

Every regular grammar is a context-free grammar, but not every context-free grammar is context-sensitive. However, every context-free grammar G can be transformed into an equivalent one in which every production has the form $A \to BC$ or $A \to c$, where A, B, and C are (possibly identical) variables, and c is a terminal. If the empty string is in $L(G)$, then we can arrange to include $S \to \epsilon$ under the same "special dispensation" as for CSLs. This form is called **Chomsky normal form** [4], where it was used to prove the case $i = 1$ of the next theorem. The grammar G_1 in the last section is an example of a context-free grammar in Chomsky normal form.

THEOREM 20.2   *For each $i = 0, 1, 2$, the class of Type-i languages properly contains the class of Type-$(i + 1)$ languages.*

The containments are clear from the above remarks. For the proper containments, we have already seen that $\{0^n 1^n | n \geq 0\}$ is a Type-2 language that is not regular, and Chapter 21 will show that the language of the Halting Problem is Type-0 but not Type-1. One can prove by a technique called "pumping" that the Type-1 languages $\{0^n 1^n 2^n | n \geq 0\}$ and $\{xx | x \in \{0, 1\}^*\}$ are not Type-2. See [11] for this, and for a presentation of the algorithm for converting a context-free grammar into Chomsky normal form.

The four classes of languages in the Chomsky hierarchy have also been completely characterized in terms of Turing machines (see Chapter 19) and natural restrictions on them. We mention this here to make the point that these characterizations show that these classes capture fundamental properties of computation, not just of formal languages. A linear bounded automaton is a possibly nondeterministic Turing machine that on any input x uses only the cells initially occupied by x, except for one visit to the blank cell immediately to the right of x (which is the initially scanned cell if $x = \epsilon$). Pushdown automata may also be nondeterministic and were likewise introduced in Chapter 19.

THEOREM 20.3

 (a) *The class of Type-0 languages equals the class of languages accepted by Turing machines.*

 (b) *The class of Type-1 languages equals the class of languages accepted by linear bounded automata.*

 (c) *The class of Type-2 languages equals the class of languages accepted by pushdown automata.*

 (d) *The class of Type-3 languages equals the class of languages accepted by finite automata.*

PROOF   (a) Given a Type-0 grammar G, one can build a nondeterministic Turing machine M that accepts $L(G)$ by having M first write the start symbol S of G on a second tape. M always nondeterministically chooses a production and chooses a place (if any) on its second tape where it can be applied. If and when the second tape becomes an all-terminal string, M compares it to its input, and if they match, M accepts. Then $L(M) = L(G)$, and by Theorem 19.4, M can be converted into an equivalent deterministic single-tape Turing machine.

For the reverse simulation of a TM by a grammar we give full details. Given any TM M_0, we may modify M_0 into an equivalent TM $M = (Q, \Sigma, \Gamma, \delta, B, q_0, q_f)$ that has the following five properties: (1) M never writes a blank; (2) M when reading a blank always converts it to a nonblank symbol on the current step; (3) M begins with a transition from q_0 that overwrites the first input cell (remembering what it was) by a special symbol $\wedge$ that is never altered; (4) M never reenters state q_0 or moves left of $\wedge$; and (5) whenever M is about to accept, M moves left to the $\wedge$, where it executes an instruction that moves right and enters a distinguished state q_e. In state q_e it overwrites any nonblank character by a special new symbol # and moves right; when it hits the blank after having #-ed out the rightmost nonblank symbol on its tape, M finally goes to q_f and accepts.

Given M with these properties, take $V = \{S, A, \} \cup (Q \times \Gamma) \cup (\Gamma \setminus \Sigma)$. A single symbol in $Q \times \Gamma$ is written using square brackets; e.g., $[q, c]$ means that M is in state q scanning character c. The grammar G has the following productions, which intuitively can simulate any accepting computation by M in reverse:

(1) $S \to \wedge S_0$; $S_0 \to \#S_0 | [q_e, \#]$;

(2) $[r, d] \to [q, c]$, for all instructions $(q, c, d, r) \in \delta$ with $q, r \in Q$ and $c, d \in \Gamma$;

(3) $c[r, B] \to [q, c]A$, for all $(q, c, R, r) \in \delta$;

(4) $c[r, d] \to [q, c]d$, for all $(q, c, R, r) \in \delta$ and $d \in \Gamma$, $d \neq B$;

(5) $[r, d]c \to d[q, c]$, for all $(q, c, L, r) \in \delta$ and $d \in \Gamma$, $d \neq B$;

(6) $[q_0, c] \to c$ for all $c \in \Sigma$, and

(7) $A \to \epsilon$.

For all $x \in L(M)$, G can generate x by first using the productions in (1) to lay down a # for every cell used during the computation, using the productions (2–5) to simulate the computation in reverse, using (6) to restore the first bit of x (blank if $x = \epsilon$) one step after having eliminated the nonterminal $\wedge$, and using (7) to erase each A marking an initially-blank cell that M used. Conversely, the only way G can eliminate $\wedge$ and reach an all-terminal string is by winding back an accepting computation of M all the way to state q_0 scanning the first cell. Hence $L(G) = L(M)$.

(b) If the given TM M_0 is a linear bounded automaton, then we can patch the last construction to eliminate the productions in (3) and (7), yielding a context-sensitive grammar G. To do this, we need to make M postpone its one allowed visit to the blank cell after the input until the last step of an accepting computation. To do this, we make M nondeterministically guess which bit of its input x is the last one, and overwrite it by an immutable right endmarker $\$$ the same way it did with $\wedge$ on the left. Then we arrange that from state q_e, M will accept only if it sees a blank immediately to the right of the $\$$, meaning that its initial guess delimited exactly the true input x. (Technically this needs another state q_e'.) Now M never even scans a blank in the middle of an accepting computation, and we can delete the productions in (3) as well as (7). Moreover, if M_0 accepts ϵ, we can add the production $S \to \epsilon$ allowed by the "special dispensation" for context-sensitive grammars above.

Going the other way, if the grammar G in the first paragraph of this proof is context-sensitive, then the resulting TM M uses only $O(n)$ space, and can be converted to an equivalent linear bounded automaton by Theorem 19.5

(c) Given a context-free grammar G, we may assume that G is in Chomsky normal form. We can build a nondeterministic PDA M whose initial moves lay down a bottom-of-stack marker $\wedge$ and the start symbol S of G, and go to a "central" state q. For every production of the form $A \to BC$ in G, M has moves that pop the stack if A is uppermost and push C and then B, returning to state q. For every production of the form $A \to c$, M can pop an uppermost A from its stack if the currently-scanned input symbol is c; then it moves its input head right. If G has the production $S \to \epsilon$ as a special case, then M can pop the initial S. A computation path accepts if and only if the stack gets down to $\wedge$ precisely when M reaches the blank at the end of its input x. Then accepting paths of M on an input x are in 1–1 correspondence with **leftmost derivations** (see below) of x in G, so $L(M) = L(G)$.

Going from a PDA M to an equivalent CFG G is much trickier, and is covered well in [11].

(d) This has been proved in Theorem 19.2.

Since $\{xx | x \in \{0, 1\}^*\}$ is a pattern language, we know from discussions above that the class of pattern languages is not contained in the class of CFLs. It is contained in the class of CSLs, however.

THEOREM 20.4   *Every pattern language is context-sensitive.*

This was proved by showing that every pattern language is accepted by a linear bounded automaton [2], whereupon it is a corollary of Theorem 20.3b.

Given a class of languages, we are often interested in the so-called closure properties of the class.

DEFINITION 20.17   A class of languages is said to be closed under a particular operation (such as union, intersection, complementation, concatenation, or Kleene closure) if every application of the operation on language(s) of the class yields a language of the class.

Closure properties are often useful in constructing new languages from existing languages, and for proving many theoretical properties of languages and grammars. The closure properties of the four types of languages in the Chomsky hierarchy are summarized below. Proofs may be found in [8,10,11], the closure of the CSLs under complementation is the famous Immerman–Szelepcsényi Theorem, which is given as Theorem 22.4c.

THEOREM 20.5

1. *The class of Type-0 languages is closed under union, intersection, concatenation, and Kleene closure, but not under complementation.*

2. *The class of CFLs is closed under union, concatenation, and Kleene closure, but not under intersection or complementation.*

3. *The classes of context-sensitive and regular languages are closed under all of the five operations.*

For example, let $L_1 = \{0^m 1^n 2^p | m = n\}$, $L_2 = \{0^m 1^n 2^p | n = p\}$, and $L_3 = \{0^m 1^n 2^p | m = n \text{ or } n = p\}$. Now L_1 is the concatenation of the CFLs $\{0^n 1^n | n \geq 0\}$ and 2^*, so L_1 is context-free. Similarly L_2 is context-free. Since $L_3 = L_1 \cup L_2$, L_3 is context-free. However, intersecting L_1 with L_2 gives the language $\{0^m 1^n 2^p | m = n = p\}$, which is not context-free.

We will look at context-free grammars more closely in the next section and introduce the concepts of parsing and ambiguity.

20.4   Context-Free Grammars and Parsing

From a practical point of view, for each grammar $G = (\Sigma, V, S, P)$ representing some language, the following two problems are important:

1. Membership problem: Given a string over Σ, does it belong to $L(G)$?
2. Parsing problem: Given a string in $L(G)$, how can it be derived from S?

The importance of the membership problem is quite obvious—given an English sentence or computer program, we wish to know if it is grammatically correct or has the right format. Solving the membership problem for context-free grammars is an integral step in the lexical analysis of computer programs, namely the stage of decomposing each statement into tokens, prior to fully parsing the program. For this reason, the membership problem is also often referred to as lexical analysis (cf. [6]). Parsing is important because a derivation usually brings out the "meaning" of the string. For example, in the case of a Pascal program, a derivation of the program in the Pascal grammar tells the compiler how the program should be executed. The following theorem qualifies the decidability of the membership problem for the four classes of grammars in the Chomsky hierarchy.

Proofs of the first assertion can be found in [4,10,11], while the second assertion is treated below. Decidability and time complexity were defined in Chapter 19.

THEOREM 20.6   *The membership problem for Type-0 grammars is undecidable in general, but it is decidable given any context-sensitive grammar. For context-free grammars the problem is decidable in polynomial time, and for regular grammars, linear time.*

Since context-free grammars play a very important role in describing computer programming languages, we discuss the membership and parsing problems for context-free grammars in more detail in this and the next section. First, let us look at another example of a context-free grammar. For convenience, let us abbreviate a set of productions

$$A \rightarrow \alpha_1, \ldots, A \rightarrow \alpha_n$$

with the same left-hand side nonterminal as

$$A \rightarrow \alpha_1 | \cdots | \alpha_n.$$

Example 20.12

We construct a context-free grammar G_6 for the set of all valid real-number literals in Pascal. In general, a real constant in Pascal has one of the following forms:

$$m.n, \quad m\mathbf{e}q, \quad m.n\mathbf{e}q,$$

where m, q are signed or unsigned integers and n is an unsigned integer. Let Σ comprise the digits 0–9, the decimal point '.', the $+$ and $-$ signs, and the $\mathbf{e}$ for scientific notation. Let the set V of variables be $\{S, M, N, D\}$ and let the set P of the productions be

$$S \rightarrow M.N | M\mathbf{e}M | M.N\mathbf{e}M$$
$$M \rightarrow N | + N | - N$$
$$N \rightarrow DN | D$$
$$D \rightarrow 0 | 1 | 2 | 3 | 4 | 5 | 7 | 8 | 9$$

Then the grammar generates all valid Pascal real values (allowing redundant leading 0s). For instance, the value 12.3**e**-4 can be derived via

$$S \Rightarrow \underline{M}.N\mathbf{e}M \Rightarrow \underline{N}.N\mathbf{e}M \Rightarrow \underline{D}N.N\mathbf{e}M \Rightarrow 1\underline{N}.N\mathbf{e}M \Rightarrow 1\underline{D}.N\mathbf{e}M \Rightarrow$$
$$12.\underline{N}\mathbf{e}M \Rightarrow 12.\underline{D}\mathbf{e}M \Rightarrow 12.3\mathbf{e}\underline{M} \Rightarrow 12.3\mathbf{e} - \underline{N} \Rightarrow 12.3\mathbf{e}-\underline{D} \Rightarrow 12.3\mathbf{e}-4$$

Perhaps the most natural representation of derivations in a context-free grammar is a **derivation tree or a parse tree**. Every leaf of such a tree corresponds to a terminal (or to ϵ), and every internal node corresponds to a nonterminal. If A is an internal node with children $B_1, \ldots, B_n$, ordered from left to right, then $A \rightarrow B_1 \cdots B_n$ must be a production. The concatenation of all leaves from left to right yields the string being derived. For example, the derivation tree corresponding to the above derivation of 12.3e-4 is given in Figure 20.1. Such a tree also makes possible the extraction of the parts 12, 3, and -4, which are useful in the storage of the real value in a computer memory.

DEFINITION 20.18   A context-free grammar G is ambiguous if there is a string $x \in L(G)$ that has two distinct derivation trees. Otherwise G is unambiguous.

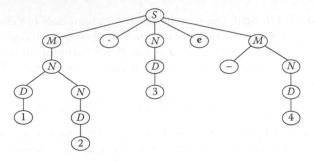

FIGURE 20.1 The derivation tree for 12.3e-4.

Unambiguity is a very desirable property because it promises a unique interpretation of each sentence in the language. It is not hard to see that the grammar G_6 for Pascal real values and the grammar G_1 defined in Example 20.7 are both unambiguous. The following example shows an ambiguous grammar.

Example 20.13

Consider a grammar G_7 for all valid arithmetic expressions that are composed of unsigned positive integers and symbols $+, *, (,)$. For convenience, let us use the symbol **n** to denote any unsigned positive integer—it is treated as a terminal. This grammar has the productions

$$S \to T + S | S + T | T$$
$$T \to F * T | T * F | F$$
$$F \to \mathbf{n} | (S)$$

Two possible different derivation trees for the expression $1 + 2 * 3 + 4$ are shown in Figure 20.2. Thus G_7 is ambiguous. The left tree means that the first addition should be done before the second addition, while the right tree says the opposite.

Although in the above example different derivations/interpretations of any expression always result in the same value because the operations addition and multiplication are associative, there are situations where the difference in the derivation can affect the final outcome. Actually, the grammar G_7 can be made unambiguous by removing the redundant productions $S \to T + S$ and $T \to F * T$. This corresponds to the convention that a sequence of consecutive additions or multiplications is

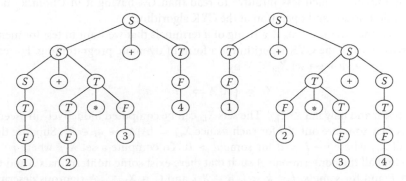

FIGURE 20.2 Different derivation trees for the expression $1 + 2 * 3 + 4$.

always evaluated from left to right. Deleting the two productions does not change the language of strings generated by G_7, but it does fix unique interpretations of those strings.

It is worth noting that there are CFLs that cannot be generated by any unambiguous context-free grammar. Such languages are said to be inherently ambiguous. An example taken from [11] (where this fact is proved) is

$$\left\{0^m 1^m 2^n 3^n \,\middle|\, m, n > 0\right\} \cup \left\{0^m 1^n 2^n 3^m \,\middle|\, m, n > 0\right\}.$$

The reason is that every context-free grammar G must yield two parse trees for some strings of the form $x = 0^n 1^n 2^n 3^n$, where one tree intuitively expresses that x is a member of the first set of the union, and the other tree expresses that x is in the second set.

We end this section by presenting an efficient algorithm for the membership problem for context-free grammars, following the treatment in [11]. The algorithm is due to Cocke, Younger, and Kasami, and is often called the CYK algorithm. Let $G = (\Sigma, V, S, P)$ be a context-free grammar in Chomsky normal form.

Example 20.14

If we use the algorithm in [11] to convert the grammar G_7 from Example 20.13 into Chomsky normal form, we are led to introduce new "alias variables" A, B, C, D for the operators and parentheses, and "helper variables" S_1, T_1, T_2, F_1, F_2 to break up the productions in G_7 with right-hand sides of length > 2 into length-2 pieces. The resulting grammar is:

$$S \rightarrow T_1 S | ST_2 | F_1 T | TF_2 | CS_1 | \mathbf{n}$$
$$T_1 \rightarrow TA$$
$$T_2 \rightarrow AT$$
$$T \rightarrow F_1 T | TF_2 | CS_1 | \mathbf{n}$$
$$F_1 \rightarrow FB$$
$$F_2 \rightarrow BF$$
$$F \rightarrow \mathbf{n} | CS_1$$
$$S_1 \rightarrow SD$$
$$A \rightarrow +$$
$$B \rightarrow *$$
$$C \rightarrow ($$
$$D \rightarrow)$$

While this grammar is much less intuitive to read than G_7, having it in Chomsky normal form facilitates the description and operation of the CYK algorithm.

Now suppose that $x = a_1 \cdots a_n$ is a string of n terminals that we want to test for membership in $L(G)$. The basic idea of the CYK algorithm is a form of dynamic programming. For each pair i, j, where $1 \le i \le j \le n$, define a set $X_{i,j} \subseteq V$ by

$$X_{i,j} = \left\{A \,\middle|\, A \Rightarrow^* a_i \cdots a_j\right\}.$$

Then $x \in L(G)$ if and only if $S \in X_{1,n}$. The sets $X_{i,j}$ can be computed inductively in ascending order of $j - i$. It is easy to figure out $X_{i,i}$ for each i since $X_{i,i} = \{A | A \rightarrow a_i \in P\}$. Suppose that we have computed all $X_{i,j}$ where $j - i < d$ for some $d > 0$. To compute a set $X_{i,j}$, where $j - i = d$, we just have to find all the nonterminals A such that there exist some nonterminals B and C satisfying $A \rightarrow BC \in P$ and for some k, $i \le k < j$, $B \in X_{i,k}$ and $C \in X_{k+1,j}$. A rigorous description of the algorithm in a Pascal-style pseudocode is given below.

TABLE 20.1 An Example Execution of the CYK Algorithm

		0	0	0	1	1	1
				$j \to$			
		1	2	3	4	5	6
	1	O					S
	2		O			S	T
i	3			O	S	T	
$\downarrow$	4				I		
	5					I	
	6						I

Algorithm CYK($x = a_1 \cdots a_n$)
1   **for** $i \leftarrow 1$ **to** n **do**
2      $X_{i,i} \leftarrow \{A | A \to a_i \in P\}$;
3   **for** $d \leftarrow 1$ **to** $n - 1$ **do**
4   **for** $i \leftarrow 1$ **to** $n - d$ **do**
5        $X_{i,i+d} \leftarrow \emptyset$;
6   **for** $t \leftarrow 0$ **to** $d - 1$ **do**
7          $X_{i,i+d} \leftarrow X_{i,i+d} \cup \{A | A \to BC \in P$ for some $B \in X_{i,i+t}$ and $C \in X_{i+t+1,i+d}\}$;

Table 20.1 shows the sets $X_{i,j}$ for the grammar G_1 and the string $x = 000111$. In this run it happens that every $X_{i,j}$ is either empty or a singleton. The computation proceeds from the main diagonal toward the upper-right corner.

We now analyze the asymptotic time complexity of the CYK algorithm. Step 2 is executed n times. Step 5 is executed $\sum_{d=1}^{n-1} n - d = (n-1)(n-1+n-(n-1))/2 = n(n-1)/2 = O(n^2)$ times. Step 7 is repeated for $\sum_{d=1}^{n-1} d(n - d) = O(n^3)$ times. Therefore, the algorithm requires asymptotically $O(n^3)$ time to decide the membership of a string length n in $L(G)$, for any grammar G in Chomsky normal form.

20.5 More Efficient Parsing for Context-Free Grammars

The CYK algorithm presented in the last section can be easily extended to solve the parsing problem for context-free grammars: In step 7, we also record a production $A \to BC$ and the corresponding value of t for any nonterminal A that gets added to $X_{i,i+d}$. Thus a derivation tree for x can be constructed by starting from the nonterminal S in $X_{1,n}$ and repeatedly applying the productions recorded for appropriate nonterminals in appropriate sets $X_{i,j}$. However, the cubic running time of this algorithm is generally too high for parsing applications. In practice, with compilation modules thousands of lines long, people seek grammars in other forms besides Chomsky's that permit parsing in linear or nearly linear time.

Before we present some of these forms, we discuss parsing strategies in general. Parsing algorithms fall into two basic types, called **top-down parsers** and **bottom-up parsers**. As indicated by their names, a top-down parser builds derivation trees from the top (root) to the bottom (leaves), while a bottom-up parser starts from the leaves and works up to the root. Although neither method is good for handling all context-free grammars, each provides efficient parsing for many important subclasses of the context-free grammars, including those used in most programming languages.

We will only consider unambiguous grammars. To simplify the description of the parsers, we will assume that each string to be parsed ends with a special delimiter $ that does not appear anywhere else in the string. This assumption makes the detection of the end of the string easy in a left-to-right scan. The assumption does not put any serious restriction on the range of languages that can be parsed—the $ is just like the end-of-file marker in a real input file. The following definition will be useful.

DEFINITION 20.19   A derivation from a sentential form to another is said to be leftmost (or rightmost) if at each step the leftmost (or rightmost, respectively) nonterminal is replaced.

For example, Example 20.14 gave a leftmost derivation of the word 12.3e-4 in the grammar G_6. For a given word x, leftmost derivations are in 1–1 correspondence with derivation trees, since we can find the leftmost derivation specified by a derivation tree by tracing the tree down from the root going from left to right. Rightmost derivations are likewise in 1–1 correspondence with derivation trees. Hence, in an unambiguous context-free grammar, every derivable string has a unique leftmost derivation and a unique rightmost derivation. The parsing methods considered next find one or the other.

20.5.1 Top-Down Parsing

An important member of the top-down parsers is the LL parser (see [1,6]). Here, the first "L" means scanning the input from left to right, and the second means leftmost derivation. In other words, for any input string x, the parser intends to find the sequence of productions used in the leftmost derivation of x.

Let $G = (\Sigma, V, S, P)$ be a context-free grammar. A parsing table T for G has rows indexed by members of V and columns indexed by members of Σ and $. Each entry $T[A, c]$ is either blank or contains one or more productions of the form $A \rightarrow \alpha$. Here we will suppose that G allows the construction of a parsing table T such that every non-blank entry $T[A, c]$ contains only one production. Then the LL parser for G is a device very similar to a pushdown automaton as described in Chapter 19. The parser has an input buffer, a pushdown stack, a parsing table, and an output stream. The input buffer contains the string to be parsed followed by the delimiter $. The stack contains a sequence of terminals or nonterminals, with another delimiter # that marks the bottom of the stack. Initially, the input pointer points to the first symbol of the input string, and the stack contains the start nonterminal S on top of #. Figure 20.3 illustrates schematically the components of the parser. As usual, the input pointer will only move to the right, while the stack pointer is allowed to move up and down.

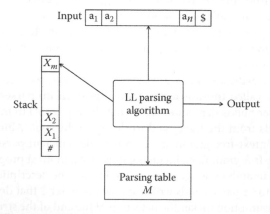

FIGURE 20.3 A schematic illustration of the LL parser.

The parser is controlled by an algorithm that behaves as follows. At any instant of time, the algorithm considers the symbol X on top of the stack and the current input symbol c pointed by the input pointer, and makes one of the following moves.

1. If X is a nonterminal, the algorithm consults the entry $T[X, a]$ of the parsing table T. If the entry is blank, the parser halts and states that the input string x is not in the language $L(G)$. If not, the entry is a production of the form $X \rightarrow Y_1 \cdots Y_k$. Then the algorithm replaces the top stack symbol X with the string $Y_1 \cdots Y_k$ (with Y_1 on top), and outputs the production.

2. If X is a terminal, X is compared with c. If $X = c$, the algorithm pops X off the stack and shifts the input pointer to the next input symbol. Otherwise, the algorithm halts and states that $x \notin L(G)$.

3. If $X =$ #, then provided $c =$ \$, the algorithm halts and declares the successful completion of parsing. Otherwise the algorithm halts and states that $x \notin L(G)$.

Intuitively, the parser reconstructs the derivation of a string $x = a_1 \cdots a_n$ as follows. Suppose that the leftmost derivation of x is

$$S = \gamma_0 \Rightarrow \gamma_1 \Rightarrow \cdots \Rightarrow \gamma_i \Rightarrow \gamma_{i+1} \Rightarrow \cdots \Rightarrow \gamma_m = x,$$

where each γ_j is a sentential form. Suppose, moreover, that the derivation step $\gamma_i \Rightarrow \gamma_{i+1}$ is the result of applying a production $X \rightarrow Y_1 \cdots Y_k$. This means that $\gamma_i = \alpha X \beta$ for some string α of terminals and sentential form β. Since no subsequent derivation will change α, this string must match a leading substring $a_1 \cdots a_j$ of x for some j. In other words, $\gamma_i = a_1 \cdots a_j X \beta$ and $\gamma_{i+1} = a_1 \cdots a_j Y_1 \cdots Y_k \beta$. Suppose that the parser has successfully reconstructed the derivation steps up to γ_i. To complete the derivation, the parser must transform the tail end of γ_i into $a_{j+1} \cdots a_n$. Thus, it keeps the string $X\beta$ on the stack and repeatedly replaces the top stack symbol (i.e., replaces the leftmost nonterminal) until a_{j+1} appears on top. At this point, a_{j+1} is removed from the stack, and the remainder of the stack must be transformed to match $a_{j+2} \cdots a_n$. The procedure is repeated until all the input symbols are matched.

The following example illustrates the parsing table for a simple context-free grammar, and how the parser operates.

Example 20.15

Consider again the language of valid arithmetic expressions from Example 20.13, where an ambiguous grammar G_7 was given that could be made unambiguous by removing two productions. Let us remove the ambiguity in a different way. The new grammar is called G_8 and has the following productions

$$S \rightarrow TS'$$
$$S' \rightarrow +S | \epsilon$$
$$T \rightarrow FT'$$
$$T' \rightarrow *T | \epsilon$$
$$F \rightarrow \mathbf{n} | (S)$$

It is easy to see that grammar G_8 is unambiguous. A parsing table for this grammar is shown in Table 20.2. We will discuss how such a table can be constructed shortly.

TABLE 20.2   An LL Parsing Table for Grammar G_8

| Nonterminal | Input Symbol | | | | | |
	n	+	*	(	)	$
S	$S \to TS'$			$S \to TS'$		
S'		$S' \to +S$			$S' \to \epsilon$	$S' \to \epsilon$
T	$T \to FT'$			$T \to FT'$		
T'		$T' \to \epsilon$	$T' \to *T$		$T' \to \epsilon$	$T' \to \epsilon$
F	$F \to n$			$F \to (S)$		

TABLE 20.3   The Steps in the LL Parsing of $(n + n) * n$

Stack	Input	Output
$S\#$	$(n + n) * n\$$	
$TS'\#$	$(n + n) * n\$$	$S \to TS'$
$FT'S'\#$	$(n + n) * n\$$	$T \to FT'$
$(S)T'S'\#$	$(n + n) * n\$$	$F \to (S)$
$S)T'S'\#$	$n + n) * n\$$	
$TS')T'S'\#$	$n + n) * n\$$	$S \to TS'$
$FT'S')T'S'\#$	$n + n) * n\$$	$T \to FT'$
$nT'S')T'S'\#$	$n + n) * n\$$	$F \to n$
$T'S')T'S'\#$	$+n) * n\$$	
$S')T'S'\#$	$+n) * n\$$	$T' \to \epsilon$
$+S)T'S'\#$	$+n) * n\$$	$S' \to +S$
$S)T'S'\#$	$n) * n\$$	
$TS')T'S'\#$	$n) * n\$$	$S \to TS'$
$FT'S')T'S'\#$	$n) * n\$$	$T \to FT'$
$nT'S')T'S'\#$	$n) * n\$$	$F \to n$
$T'S')T'S'\#$	$) * n\$$	
$S')T'S'\#$	$) * n\$$	$T' \to \epsilon$
$)T'S'\#$	$) * n\$$	$T' \to \epsilon$
$T'S'\#$	$*n\$$	
$*TS'\#$	$*n\$$	$T' \to *T$
$TS'\#$	$n\$$	
$FT'S'\#$	$n\$$	$T \to FT'$
$nT'S'\#$	$n\$$	$F \to n$
$T'S'\#$	$\$$	$T' \to \epsilon$
$S'\#$	$\$$	$S' \to \epsilon$
$\#$	$\$$	

To demonstrate how the parser operates, consider the input string $(n + n) * n$. Table 20.3 shows the content of the stack, the remaining input symbols, and the output after each step. If we trace the actions of the parser carefully, we see that the sequence of productions it outputs constitutes the leftmost derivation of $(n + n) * n$.

Now we turn to the question of how to construct an LL parser for a given grammar $G = (\Sigma, V, S, P)$. It suffices to show how to compute the entries $T[A, c]$, where $A \in V$ and $c \in \Sigma \cup \{\$\}$. We first need to introduce two functions $FIRST(\alpha)$ and $FOLLOW(A)$. The former maps a sentential form to a terminal or ϵ, and the latter maps a nonterminal to a terminal or $\$$.

DEFINITION 20.20   For each sentential form $\alpha \in \{\Sigma \cup V\}^*$, and for each nonterminal $A \in V$,

$$FIRST(\alpha) = \left\{ c \in \Sigma \mid \text{for some } \beta \in \{\Sigma \cup V\}^*, \alpha \Rightarrow^* c\beta \right\} \cup \left\{ \epsilon \mid \alpha \Rightarrow^* \epsilon \right\}$$

$$FOLLOW(A) = \left\{ c \in \Sigma \mid \text{for some } \alpha, \beta \in \{\Sigma \cup V\}^*, S \Rightarrow^* \alpha A c\beta \right\}$$
$$\cup \left\{ \$ \mid \text{for some } \alpha \in \{\Sigma \cup V\}^*, S \Rightarrow^* \alpha A \right\}.$$

Intuitively, for any sentential form α, $FIRST(\alpha)$ consists of all the terminals that appear as the first symbol of some sentential form derivable from α. The empty string ϵ is included in $FIRST(\alpha)$ as a special case if α derives ϵ. On the other hand, for any nonterminal A, $FOLLOW(A)$ consists of all the terminals that immediately follow an occurrence of A in some sentential form derivable from the start symbol S. The end delimiter $ is included in $FOLLOW(A)$ as a special case if A appears at the end of some sentential form derivable from S.

Algorithms for computing the $FIRST()$ and $FOLLOW()$ functions are fairly straightforward and can be found in [1,6]. It turns out that to construct the parsing table for a grammar G, we only need to know the values of $FIRST(\alpha)$ for those sentential forms α appearing on the right-hand sides of the productions in G.

Example 20.16

The following illustrate the functions $FIRST(\alpha)$ and $FOLLOW(A)$ for the grammar G_8 described in the above example. For the former, only those sentential forms appearing on the right-hand sides of the productions in G_8 are considered.

$$FIRST(TS') = \{(, \mathbf{n}\}$$
$$FIRST(+S) = \{+\}$$
$$FIRST(FT') = \{(, \mathbf{n}\}$$
$$FIRST(*T) = \{*\}$$
$$FIRST((S)) = \{(\}$$
$$FIRST(\mathbf{n}) = \{\mathbf{n}\}$$
$$FIRST(\epsilon = \{\epsilon\}$$
$$FIRST(S) = \{), \$\}$$
$$FIRST(S') = \{), \$\}$$
$$FIRST(T) = \{+,), \$\}$$
$$FIRST(T') = \{+,), \$\}$$
$$FIRST(F') = \{*, +,), \$\}$$

Given the functions $FIRST(\alpha)$ and $FOLLOW(A)$ for a grammar G, we can easily construct the **LL parsing table** $T[A, c]$ for G. The basic idea is as follows. Suppose that $A \rightarrow \alpha$ is a production and $c \in FIRST(\alpha)$. Then, the parser will replace A with α when A is on top of the stack and c is the current input symbol. The only complication occurs when α may derive ϵ. In this case, the parser should still replace A with α if the current input symbol is a member of $FOLLOW(A)$. The detailed algorithm is given below.

Algorithm LL-Parsing-Table($G = (\Sigma, V, S, P)$)
1   Initialize each entry of the table to blank.
2   **for** each production $A \rightarrow \alpha$ in P **do**
3   **for** each terminal $a \in FIRST(\alpha)$ **do**
4   add $A \rightarrow \alpha$ to $T[A, a]$;
5   **if** $\epsilon \in FIRST(\alpha)$ **then**
6   **for** each terminal or delimiter $a \in FOLLOW(A)$ **do**
7   add $A \rightarrow \alpha$ to $T[A, a]$;

The above algorithm can be applied to any context-free grammar to produce a parsing table. However, for some grammars the table may have entries containing multiple productions. Multiply defined entries in a parsing table, however, would present our parsing algorithm with an unwelcome choice. It would be possible for it to make a wrong choice and incorrectly report a string as not being derivable, and backtracking to the last choice to try another would blow up the running time unacceptably.

Example 20.17

Recall that we could make the grammar G_7 of Example 20.13 unambiguous by deleting two unnecessary productions. The resulting grammar, which we call G_9, has the following productions:

$$S \rightarrow S + T|T$$
$$T \rightarrow T * F|F$$
$$F \rightarrow \mathbf{n}|(S)$$

It is easy to see that both $FIRST(S + T)$ and $FIRST(T)$ contain the terminal $\mathbf{n}$. Hence, the entry $T[S, \mathbf{n}]$ of the parsing table is multiply defined, so this table is not well-conditioned for LL parsing.

A context-free grammar whose parsing table has no multiply defined entries is called an **LL(1) grammar**. Here, the "1" signifies the fact that the LL parser uses one input symbol of lookahead to decide its next move. For example, G_8 is an LL(1) grammar, while G_9 is not. It is easy to show that our LL parser runs in linear time for any LL(1) grammar.

What can we do for grammars that are not LL(1), such as G_9? The first idea is to extend the LL parser to use more input symbols of lookahead. In other words, we will allow the parser to see the next several input symbols before it makes a decision. For one more symbol of lookahead, this requires expanding the parsing table to have a column for every pair of symbols in Σ (plus $ as a possible second symbol), but so doing may separate and/or eliminate multiply defined entries in the original parsing table. The $FIRST()$ and $FOLLOW()$ functions have to be modified to take two (or more) lookahead symbols into consideration. For any constant $k > 1$, a grammar is said to be an **LL(k) grammar** if its parsing table using k lookahead symbols has no multiply defined entries. For example, the grammar G_1 given in Example 20.7 is not LL(1), but it is LL(2).

Although LL(k) grammars form a larger class than LL(1) grammars, there are still grammars that are not LL(k) for any constant k. The grammar G_7 and G_9 are examples. The texts [1,6] provide several techniques for dealing with non-LL(k) grammars, such as grammar transformations and backtracking. When backtracking is used, the parsing process is often called recursive-descent parsing, and can be very time consuming due to the use of many recursive calls.

20.5.2 Bottom-Up Parsing

The most popular bottom-up parsing technique is **LR parsing**. Here, the "L" again means scanning the input from left to right, while the "R" means constructing the rightmost derivation. For any input string x, the LR parser scans x from left to right and tries to find the reverse of the sequence of productions used in the rightmost derivation of x. It turns out that in bottom-up parsing rightmost derivations are easier to deal with than leftmost derivations. LR parsing is especially attractive in practice for many reasons summarized in [1]: (1) it can handle virtually all programming language constructs; (2) it has very efficient implementations; (3) it is more powerful than LL parsing; and (4) it detects syntactic errors quickly. The principal drawback of the method is that constructing an LR parser is very involved. Fortunately, there exist efficient algorithms that can automatically generate LR parsers from certain context-free grammars. Because of space limitations, we describe only the operation of an LR parser here, and refer the reader to [1] for the construction of such a parser.

Similar to an LL parser, an LR parser has an input buffer, a pushdown stack, a parsing table, and an output stream, and is controlled by an algorithm that is the same for all LR parsers. The input string is again assumed to have an end delimiter $. At any time during parsing, the stack stores a string of the form $q_m X_m q_{m-1} \cdots X_1 q_0$ (with q_0 at bottom), where each X_i is a grammar symbol (i.e., a terminal or nonterminal of the grammar involved) and q_i is a state symbol. The number of distinct states is finite, and each state symbol intends to summarize the information contained in the stack below it. The combination of the state on top of the stack and the current input symbol are used to index the parsing table and determine the move of the parser. It will be seen that the state symbols subsume all information in the grammar symbols, and a real parser omits the latter. However, we retain the grammar symbols $X_1, \ldots, X_m$ to make our illustration easier to follow, and for consistency with previous examples.

The parsing table consists of two parts: a parsing action function $ACTION(q, c)$, which maps a state and an input symbol to a move, and a function $GOTO(q, X)$, which maps a state and a grammar symbol to a state. For each state q and each input symbol c, the value of the function $ACTION(q, c)$ can be one of the following:

1. shift
2. reduce by $A \to \alpha$, where $A \to \alpha$ is a production in the grammar
3. accept
4. blank

The algorithm controlling the LR parser operates as follows. Suppose that the state on top of the stack is q and the current input symbol is c. It consults $ACTION(q, c)$ and makes one of the four types of moves as below.

1. If $ACTION(q, c) =$ shift, the parser pushes the string $GOTO(q, c)c$ on the stack and shifts its input pointer to the next input symbol.
2. If $ACTION(q, c) =$ reduce by $A \to \alpha$, the parser applies the production $A \to \alpha$ as follows. Let $k = |\alpha|$, and let the current stack content be $q_m X_m q_{m-1} \cdots X_1 q_0$. The parser first pops the top $2k$ symbols $q_m, X_m, \ldots, q_{m-k+1}, X_{m-k+1}$ off the stack. It then consults $GOTO(q_{m-k}, A)$ and pushes the string $GOTO(q_{m-k}, A)A$ onto the stack, resulting in a stack with content $GOTO(q_{m-k}, A)A q_{m-k} X_{m-k} \cdots X_1 q_0$. The parser also outputs the production $A \to \alpha$.
   It is always guaranteed in the above that $X_{m-k+1} \cdots X_m = \alpha$.
3. If $ACTION(q, a) =$ accept, the parser successfully terminates.
4. If $ACTION(q, a) =$ blank, the parser terminates and declares that the input string is not a member of the language.

Intuitively, the LR parser reconstructs the rightmost derivation of a string $x = a_1 \cdots a_n$ as follows. Suppose that the rightmost derivation of x is

$$S = \gamma_0 \Rightarrow \gamma_1 \Rightarrow \cdots \Rightarrow \gamma_i \Rightarrow \gamma_{i+1} \Rightarrow \cdots \Rightarrow \gamma_m = x,$$

where each γ_j is a sentential form. Furthermore, suppose that the derivation step $\gamma_i \Rightarrow \gamma_{i+1}$ is the result of applying a production $A \to Y_1 \cdots Y_k$. This means that $\gamma_i = \alpha A z$ for some sentential form $\alpha = X_1 \cdots X_t$ and string z of terminals, and $\gamma_{i+1} = \alpha Y_1 \cdots Y_k z = X_1 \cdots X_t Y_1 \cdots Y_k z$. Since no subsequent derivation will change z, this string must match a trailing substring $a_j \cdots a_n$ of x for some j. In other words, $\gamma_i = X_1 \cdots X_t A a_j \cdots a_n$ and $\gamma_{i+1} = X_1 \cdots X_t Y_1 \cdots Y_k a_j \cdots a_n$.

Suppose that the parser has successfully reconstructed the derivation steps in reverse from γ_m back to γ_{i+1}. At this point, the stack must be holding a string of the form

$$q_{t+h}Y_h \cdots q_{t+1}Y_1 q_t X_t \cdots q_1 X_1 q_0,$$

where $h \leq k$ and $q_0, q_1, \ldots q_{t+h}$ are some states, and the input pointer is pointing at a_{j+h-k}. Moreover, it must be that $Y_{h+1} = a_{j+h-k}, \ldots, Y_k = a_{j-1}$. To recover γ_i, the parser consults the state q_{t+h} on top of stack and the current input symbol a_{j+h-k}. It then shifts the $h - k$ input symbols $a_{j+h-k}, \ldots, a_{j-1}$ and $h - k$ appropriate state symbols onto the stack. It also advances the input pointer to a_j. Then, the parser reduces the string $Y_1 \cdots Y_k$ to the nonterminal A by replacing the top $2k$ stack symbols with A and an appropriate state symbol.

The above shift-and-reduce process is repeated until the sentential form $\gamma_0 = S$ is obtained. For this reason, the LR parser is sometimes called a shift-reduce parser.

Clearly, the state symbols stored on the stack play a key role in dictating the actions of the parser. Below we first give an example of LR parsing tables and show exactly how the parser operates on a specific input. Then we will briefly sketch how the states are chosen for a grammar and what they represent.

Example 20.18

Consider again the unambiguous grammar G_9 given in Example 20.17. For convenience, let us number the productions as follows.

$$(1) \quad S \rightarrow S + T$$
$$(2) \quad S \rightarrow T$$
$$(3) \quad T \rightarrow T * F$$
$$(4) \quad T \rightarrow F$$
$$(5) \quad F \rightarrow \mathbf{n}$$
$$(6) \quad F \rightarrow (S)$$

Tables 20.4 and 20.5 illustrate the functions $ACTION(q, c)$ and $GOTO(q, X)$ for the grammar. In the first table, shf means shift, pi means reduce by production i, acc means accept, and blank means reject. The states are numbered $0, 1, \ldots, 11$.

Now we demonstrate how the parser operates on the string $(\mathbf{n} + \mathbf{n}) * \mathbf{n}$. Table 20.6 shows the content of the stack, the remaining input symbols, and the output after each step. It is easy to see that the reverse sequence of the productions in the reduce steps constitute the rightmost derivation of $(\mathbf{n} + \mathbf{n}) * \mathbf{n}$.

TABLE 20.4   The Function $ACTION(q, c)$ for the Unambiguous Grammar G_9

State	n	+	*	(	)	$
0	shf			shf		
1		shf				acc
2		p2	shf		p2	p2
3		p4	p4		p4	p4
4	shf			shf		
5		p6	p6		p6	p6
6	shf			shf		
7	shf			shf		
8		shf			shf	
9		p1	shf		p1	p1
10		p3	p3		p3	p3
11		p5	p5		p5	p5

TABLE 20.5 The Function $GOTO(q, X)$ for the Unambiguous Grammar G_9

State	n	+	*	(	)	$	S	T	F
0	5			4			1	2	3
1		6							
2			7						
3									
4	5			4			8	2	3
5									
6	5			4				9	3
7	5			4					10
8		8			11				
9		7							
10									
11									

TABLE 20.6 The Steps in the LR Parsing of $(n + n) * n$

Stack	Input	Action
0	$(n + n) * n\$$	Shift
$4(0$	$n + n) * n\$$	Shift
$5n4(0$	$+n) * n\$$	Reduce by $F \rightarrow n$
$3F4(0$	$+n) * n\$$	Reduce by $T \rightarrow F$
$2T4(0$	$+n) * n\$$	Reduce by $S \rightarrow T$
$8S4(0$	$+n) * n\$$	Shift
$6 + 8S4(0$	$n) * n\$$	Shift
$5n6 + 8S4(0$	$) * n\$$	Reduce by $F \rightarrow n$
$3F6 + 8S4(0$	$) * n\$$	Reduce by $T \rightarrow F$
$9T6 + 8S4(0$	$) * n\$$	Reduce by $S \rightarrow S + T$
$8S4(0$	$) * n\$$	Shift
$11)8S4(0$	$*n\$$	Reduce by $F \rightarrow (S)$
$3F0$	$*n\$$	Reduce by $T \rightarrow F$
$2T0$	$*n\$$	Shift
$7 * 2T0$	$n\$$	Shift
$5n7 * 2T0$	$\$$	Reduce by $F \rightarrow n$
$10F7 * 2T0$	$\$$	Reduce by $T \rightarrow T * F$
$2T0$	$\$$	Reduce by $S \rightarrow T$
$1S0$	$\$$	Accept

There are several techniques for constructing an LR parsing table, such as simple-LR (SLR), canonical-LR, and lookahead-LR (LALR), as described by [1]. In general, these techniques all use states that are sets of items of the form $A \rightarrow \alpha \cdot \beta$, where $A \rightarrow \alpha\beta$ is a production and the $\cdot$ marks a place in the right-hand side. Such items are commonly known as the LR items. Each item expresses the assertion that the part α has already been obtained by previous shift/reduce steps and pushed on the stack, and the part β is expected to be obtainable from the next few input symbols by some shift/reduce steps. Since at any given time the parser may not be able to predict what input symbols should follow, it has to maintain a set of LR items to deal with all possibilities.

Again, not all context-free grammars have effective LR parsers. For example, the grammar with productions

$$S \rightarrow 0S0 \mid 1S1 \mid 0 \mid 1 \mid \epsilon$$

cannot be handled by LR parsing. This grammar generates the set of all palindromes. The grammars that have effective LR parsers are called LR grammars. In fact, there are context-free languages that cannot be represented by any LR grammars. The set of palindromes is one such language.

20.6   Further Information

The fundamentals of formal languages and grammars can be found in many text books including [6–11,18]. The central focus of research in this area has been to find formal grammatical representations of languages that are very expressive and are yet easy to parse. The research results have greatly benefited many fields of computer science, including programming languages, compiler design, and natural language processing. Chapter 19 presents the machine model counterparts of regular grammars, context-free grammars, context-sensitive grammars, and unrestricted grammars, and Chapter 21 introduces the concepts of decidability and undecidability, which has a close relation to formal grammars. The following annual conferences present the leading research work in formal languages and grammars: International Colloquium on Automata, Languages and Programming (ICALP), ACM Annual Symposium on Theory of Computing (STOC), IEEE Symposium on the Foundations of Computer Science (FOCS), ACM Symposium on Principles of Programming Languages (POPL), Symposium on Theoretical Aspects of Computer Science (STACS), Mathematical Foundations of Computer Science (MFCS), Fundamentals of Computation Theory (FCT), Foundation of Software Technology and Theoretical Computer Science (FSTTCS), and Conference on Developments in Language Theory (DLT). There are many related conferences, including Computational Learning Theory (COLT), Colloquium on Trees in Algebra and Programming (CAAP), and International Conference on Concurrency Theory (CONCUR), where either specific issues concerning formal grammars are considered or specialized grammatical systems are studied for a specific application area. We conclude with a list of major journals that publish papers in formal language theory: *Journal of the ACM, SIAM Journal on Computing, Journal of Computer and System Sciences, Information and Computation, Theory of Computing Systems* (formerly *Mathematical Systems Theory*), *Theoretical Computer Science, Information Processing Letters, International Journal of Foundations of Computer Science,* and *Acta Informatica.*

Defining Terms

Ambiguous context-free grammar: A context-free grammar in which some derivable terminal strings have two distinct derivation trees.

Bottom-up parsing: A process of building a derivation tree from the leaves up to the root.

Chomsky normal form: A form of context-free grammar in which every rule has the form $A \to BC$ or $A \to a$, where A, B, C are nonterminals and a is a terminal.

Context-free grammar: A grammar whose rules have the form $A \to \beta$, where A is a nonterminal and β is a string of nonterminals and terminals.

Context-free language: A language that can be described by some context-free grammar.

Context-sensitive grammar: A grammar whose rules have the form $\alpha \to \beta$, where α, β are strings of nonterminals and terminals, and $|\alpha| \leq |\beta|$.

Context-sensitive language: A language that can be described by some context-sensitive grammar.

Derivation or parsing: A sequence of applications of rules of a grammar that transforms the start symbol into a given terminal string or sentential form.

Derivation tree or parse tree: A rooted, ordered tree that describes a particular derivation of a string with respect to some context-free grammar.

(Formal) language: A set of strings over some fixed alphabet.

(Formal) grammar: A description of some language, typically consisting of a set of terminals, a set of nonterminals, a distinguished nonterminal called the start symbol, and a set of rules (or productions) of the form $\alpha \rightarrow \beta$, which determine which substrings α of a sentential form can be replaced by some another string β.

Leftmost (or rightmost) derivation: A derivation in which at each step, the leftmost (respectively, rightmost) nonterminal is rewritten.

LL parsing: A type of top-down parsing in which one reads the input from left to right in order to reconstruct a leftmost derivation.

LL(k) grammar: A context-free grammar whose LL(k) parsing table has no multiply defined entries.

LL(k) parsing: An LL parsing that uses k symbols of lookahead.

LR parsing: A type of bottom-up parsing in which one reads the input from left to right in order to reconstruct a rightmost derivation in reverse order of steps.

LR grammar: A context-free grammar that has an effective LR parser.

Membership problem (or lexical analysis): The problem or process of deciding whether a given string is generated by a given grammar.

Parsing problem: The problem of reconstructing a derivation of a given input string in a given grammar.

Regular expression: A description of some language using the operators union, concatenation, and Kleene closure.

Regular language: A language that can be described by some regular expression, or equivalently, by some right-linear/regular grammar.

Right-linear or regular grammar: A grammar whose rules have the form $A \rightarrow cB$, $A \rightarrow c$, or $A \rightarrow \epsilon$, where A, B are nonterminals, c is a terminal, and ϵ is the empty string.

Sentential form: A string of terminals and nonterminals obtained at some step of a derivation in a grammar.

Top-down parsing: A process of building derivation trees from the top (root) down to the bottom (leaves).

References

1. Aho, A.V., Ullman, J.D., and Sethi, I., *Compilers: Principles, Techniques, and Tools.* Addison-Wesley, Reading, MA, 1985.
2. Angluin, D., Finding patterns common to a set of strings. *J. Comput. Syst. Sci.* 21, 46–62, 1980.
3. Chomsky, N., Three models for the description of language. *IRE Trans. Info. Theory,* 2(2), 113–124, 1956.
4. Chomsky, N., Formal properties of grammars. In *Handbook of Mathematical Psychology,* Vol. 2. John Wiley & Sons, New York, 1963, pp. 323–418.
5. Chomsky, N. and Miller, G., Finite-state languages. *Inform. Control,* 1, 91–112, 1958.
6. Drobot, V., *Formal Languages and Automata Theory.* Computer Science Press, Rockville, MD, 1989.
7. Floyd, R.W. and Beigel, R., *The Language of Machines: An Introduction to Computability and Formal Languages.* Computer Science Press, New York, 1994.
8. Gurari, E., *An Introduction to the Theory of Computation.* Computer Science Press, Rockville, MD, 1989.
9. Harel, D., *Algorithmics: The Spirit of Computing.* Addison-Wesley, Reading, MA, 1992.
10. Harrison, M., *Introduction to Formal Language Theory.* Addison-Wesley, Reading, MA, 1978.
11. Hopcroft, J. and Ullman, J., *Introduction to Automata Theory, Languages and Computation.* Addison-Wesley, Reading, MA, 1979.

12. Jiang, T., Salomaa, A., Salomaa, K., and Yu, S., Decision problems for patterns. *J. Comput. Syst. Sci.*, 50(1), 53–63, 1995.
13. Kleene, S., Representation of events in nerve nets and finite automata. In *Automata Studies*. Princeton University Press, Princeton, NJ, 1956, pp. 3–41.
14. Lind, D. and Marcus, B., *An Introduction to Symbolic Dynamics and Coding*, Cambridge University Press, New York, 1995.
15. Post, E., Formal reductions of the general combinatorial decision problems. *Am. J. Math.*, 65, 197–215, 1943.
16. Salomaa, A., Two complete axiom systems for the algebra of regular events. *J. ACM.* 13(1), 158–169, 1966.
17. Searls, D., The computational linguistics of biological sequences. In *Artificial Intelligence and Molecular Biology*, L. Hunter (Ed.). MIT Press, Cambridge, MA, 1993, pp. 47–120.
18. Wood, D., *Theory of Computation*. Harper and Row, New York, 1987.

21
Computability

Tao Jiang
University of California

Ming Li
University of Waterloo

Bala Ravikumar
Sonoma State University

Kenneth W. Regan
State University of New York at Buffalo

21.1 Introduction

In the last two chapters, we have introduced several important computational models, including Turing machines and Chomsky's hierarchy of formal grammars. In this chapter, we will explore the limits of mechanical computation as defined by these models. We begin with a list of fundamental problems for which automatic computational solution would be very useful. One of these is the universal simulation problem: can one design a single algorithm that is capable of simulating any algorithm? Turing's demonstration that the answer is yes [21] supplied the proof for Babbage's dream of a single machine that could be programmed to carry out any computational task. We introduce a simple Turing machine programming language called "GOTO" in order to facilitate our own design of a universal machine. Next, we describe the schemes of primitive recursion and μ-recursion, which enable a concise, mathematical description of computable functions that is independent of any machine model. We show that the μ-recursive functions are the same as those computable on a Turing machine, and describe some computable functions, including one that solves a second problem on our list.

The success in solving ends there, however. We show in the last section of this chapter that all of the remaining problems on our list are unsolvable by Turing machines, and subject to the Church–Turing thesis, have no mechanical or human solver at all. That is to say, there is no Turing machine or physical device, no stand-alone product of human invention, that is capable of giving the correct answer to all—or even most—instances of these problems. The implication we draw is that in order to solve some important instances of these problems, human ingenuity is needed to guide powerful computers down the paths felt most likely to yield the answers. To cite Raymond Smullyan quoting P. Rosenbloom, the results on unsolvability imply that "man can never eliminate the necessity of using his own cleverness, no matter how cleverly he tries."

Almost the first consequence of formalizing computation was that we can formally establish its limits. Kurt Gödel showed that the process of proof in any formal axiomatic system of logic can be simulated by the basic arithmetical functions that computation is made of. Then he proved that any sound formal system that is capable of stating the grade-school rules of arithmetic can make statements that are neither provable nor disprovable in the system. Put another way, every sound formal system is incomplete in the sense that there are mathematical truths that cannot be proved in the system. Turing realized that Gödel's basic method could be applied to computational models themselves, and thus proved the first computational unsolvability results. Since then problems from many areas, including group theory, number theory, combinatorics, set theory, logic, cellular automata, dynamical systems, topology, and knot theory, have been shown to be unsolvable. In fact, proving unsolvability is now an accepted "solution" to a problem. It is just a way of saying that the problem is too general for a computer to handle—that supplementary information is needed to enable a mechanical solution.

Since Turing machines capture the power of mechanical computability, our study will be based on Turing's model. In the next section, we describe a Turing machine as a computer that can run programs written in a very simple language we call the "GOTO Language." This formalism is equivalent to Chapter 19's description of Turing machines using the standard 7-tuple notation. Our language provides an alternate way to write programs and makes proofs about Turing machines more intuitive.

21.2   Computability and a Universal Program

Turing's notion of mechanical computation was based on identifying the basic steps in any mechanical computation. He reasoned that an operation such as numerical multiplication is not primitive, because it can be divided into simpler steps such as using the times-table on individual pairs of digits, shifting, and adding. Addition itself can be broken down into simpler steps such as adding the lowest digits, computing the carry, and moving to the next digit. Turing concluded that the most basic features of mechanical computation are the ability to read and write on a storage medium, the ability to move about on that medium, and the ability to make simple logical decisions. Turing chose the storage medium to be a single linear tape divided into cells. He showed that such a tape could model spatial memory in three (or any number of) dimensions through the use of indexed coordinates. With much care he argued that human sensory input could be encoded by strings over a finite alphabet of cell symbols called the tape alphabet. (This bold discretization of sensory experience now seems a harbinger of the digital revolution that was to follow.) A decision step enables the computer to exert local control over the sequence of actions. Turing restricted the next action performed to be in a cell neighboring the one on which the current action occurred, and showed how nonlocal actions can be simulated by successions of steps of this kind. He also introduced an instruction to tell the computer to stop.

In summary, Turing proposed a model to characterize mechanical computation as being carried out as a sequence of instructions. Our "GOTO" formalism provides the following five kinds of instructions. Here i stands for a tape symbol and j stands for a line number.

> PRINT i
> MOVE RIGHT
> MOVE LEFT
> IF i IS SCANNED GOTO LINE j
> STOP

1 Print 0
2 Move left
3 If 1 is scanned goto line 2
4 Print 1
5 Move right
6 If 1 is scanned goto line 5
7 Print 1
8 Move right
9 If 1 is scanned goto line 1
10 Stop

FIGURE 21.1 The doubling program in the GOTO language.

When we speak about programs recognizing languages rather than computing functions, we replace STOP by statements ACCEPT and REJECT, each of which need occur only once in a program.

A program in this language is a sequence of instructions or "lines" numbered 1 to k. The input to the program is a string over a designated input alphabet Σ, which we take to be $\{0, 1\}$ throughout this chapter. The tape alphabet includes Σ and a special blank character B representing an empty cell, and may (but need not) contain other symbols. The input is stored on the tape, with the read head scanning the first symbol (or B if the input is empty), before the computation begins.

How much memory should we allow the computer to use? Rather than postulate that the tape is actually infinite—an unrealistic assumption—we prefer here to say that the tape has expandable boundaries. Initially the input defines the two boundaries of the tape. Whenever the machine moves left of the left boundary or right of the right boundary, a new memory cell containing the blank is attached. This convention clarifies what we mean by saying that if and when the machine halts by reaching the STOP instruction, the "result" of the computation is the entire content of the tape.

We present an example program written in the GOTO language. This program accomplishes the simple task of doubling the number of initial 1s in the input string. Informally, the program achieves its goal as follows: When it reads a 1, it changes the 1 to a 0, moves left looking for a new cell, and writes a 1 in that cell. Then it returns rightward to the 0 that marks where it had been, rewrites it as a 1, and moves right to look for more 1s. If it immediately finds another 1 it repeats the process from line 1, while if it doesn't, it halts right there. This program even has a "bug"—it "should" leave strings that do not begin with a 1 unchanged, but instead it alters them (Figure 21.1).

The main change from the traditional Turing machine formalism of Chapter 19 is that we have replaced "states" by line numbers. A Turing machine of the former kind can always be simulated in our GOTO language by making blocks of successive lines, themselves divided into subblocks (for each character) that are headed by "IF" statements, carry out the instructions for each state. Our formalism makes many programs more succinct and closer to programmers' experience, and highlights the role of (conditional) GOTO instructions in setting up loops and enabling statements to be repeated. Despite the popular scorn of goto statements, this feature is ultimately the most important aspect of programming and can be found in every imperative-style programming language—at least in the code produced by the compiler if the language has no goto instruction itself. Indeed, the above example could be rendered into a structured programming language such as C as follows,

```
do {
    do { PRINT 0; MOVE LEFT; } while (1 is scanned);
    PRINT 1;
    do MOVE RIGHT; while (1 is scanned);
    PRINT 1; MOVE RIGHT; }
while (1 is scanned);
```

and a C compiler (using a character array `tape[i]` and `++i`, `-i` for the moves) might plausibly convert this into something exactly like our GOTO program!

The simplicity of the GOTO language is rather deceptive. As the above example hints, any program in any known high-level programming language can be converted into an equivalent GOTO program, under suitable conventions on how inputs and outputs are represented on the tape. (If the program only reads from the standard input stream and writes to the standard output stream, then no such conventions are necessary.) There is strong reason to believe that any mechanical computation of any future kind can be expressed by a suitable GOTO program. Note, however, that a program written in the GOTO language need not always halt; i.e., on certain inputs the program may never reach a STOP instruction. On such inputs we say that the output of the program is undefined.

Now we can give a precise definition of what we mean by an algorithm, attempting to rule out this last situation. An algorithm is any program written in the GOTO language that has the additional property of halting on all inputs. Such programs will be called halting programs, and correspond to "total" deterministic Turing machines in Chapter 19. When we consider decision problems, which have yes/no answers, halting programs are required to end their computation with either an ACCEPT or a REJECT statement, on any input.

21.2.1 Some Computational Problems

We begin by listing a collection of computational problems for which a mechanical solution can be very helpful. By a mechanical solution, we mean a step-by-step process that takes into account all possible inputs, and that can be executed without any human assistance once a certain input is provided. An algorithm is required to work correctly on all instances.

We now list some problems that are fundamental either because they are inherently important or because they played a historical role in the development of computation theory [4,5]. For the first four, P stands for a program in our GOTO language, and x is a string over the input alphabet, which we fix to be $\{0, 1\}$.

1. *Universal simulation*: Given a program P and an input x to P, determine the output (if any) that P would produce on input x.

2. *Halting problem*: Given P and x, output 1 (for yes) if P would halt when given input x, and 0 (for no) if P would not halt.

3. *Type-0 grammar membership*: Given a type-0 grammar G and a string x, determine whether x can be derived from the start symbol of G.

4. *String compression*: Given a string x, find the shortest program P such that when P is started with empty tape, P eventually halts with x as its output. Here "shortest" means that the total number of symbols in the program's instructions is as small as possible.

5. *Tiling*: Given a finite set T of tile types, where all tiles of a type are unit squares with the same four colors on their four edges, determine whether every finite rectangle can be tiled by T. If k and n are the integer sides of the rectangle, being tiled means that kn tiles drawn from T can be arranged so that every two tiles that share an edge have the same color at that edge.

6. *Linear programming*: Given some number k of linear inequalities in n unknowns, determine whether there is an assignment of n values to the unknowns that satisfies all the inequalities.

7. *Integer equations*: Given k-many polynomial equations in n unknowns, determine whether there is an assignment of n integers to the unknowns that satisfies all the equations.

SOME REMARKS ABOUT THE ABOVE PROBLEMS: A solution to Problem 1 realizes Babbage's objective of a single program or machine capable of simulating all programs P. For cases where P run on x would never halt and produce output, we have left open whether we require the solution itself to halt and detect this fact—i.e., to be an algorithm. For any such algorithm to exist, there must be an algorithm to solve Problem 2, which is a yes/no decision problem. An algorithm for Problem 2 would be a boon to reliable software design, since it could be used to test whether a given block of code can cause infinite

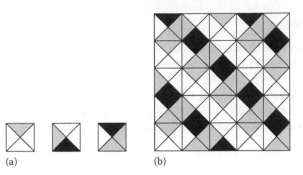

FIGURE 21.2 An example of tiling. (a) A set of tile types. (b) An example tiling of a 5 × 5 square area.

loops. Problem 3 is another decision problem; its solution would be useful for natural-language processing and much more. Problem 4 is a function-computation problem of central importance in information theory. For illustration, think of x as a large amount of scientific data for which we seek a concise theory P that can generate and hence explain it. A famous example is Kepler's laws, which explained Tycho Brahe's voluminous and meticulous observational data. Problem 4 thus asks whether the heart of science (to paraphrase Occam's Razor, "finding the simplest explanation that fits the facts") can be done automatically on a computer.

The tiles in Problem 5, which we introduced in detail in Chapter 19, are sometimes named after Hao Wang who wrote the first research paper about them [23]. Figure 21.2a shows an example of a set T of tile types, and Figure 21.2b shows how tiles drawn from T can be used to tile a 5 × 5 square area. The tiling problem is not merely an interesting puzzle. It has been an art form pursued by artists from many cultures for centuries. Tiling problems have deep significance in combinatorics, algebra, and formal languages. Note that our decision problem does not ask simply whether a given k × n rectangle can be tiled, but whether—given T—all k × n rectangles can be tiled via T. The full problem of linear programming adds to Problem 6 a clause saying: if there exist feasible solutions, i.e., assignments that satisfy all the so-called linear constraints, find one that maximizes (or minimizes) a given objective function (or cost function). This problem has central importance in economics, game theory, and operations research. Problem 7 is called Hilbert's tenth problem, and was 1 of 24 that David Hilbert posed as challenges for the new century at the International Congress of Mathematicians in 1900. It goes back 2000 years to the mathematician Diophantus' study of these so-called Diophantine equations. Actually, Hilbert posed the "meta-problem" of finding an algorithm that can solve any Diophantine equation, or at least tell whether it has a solution.

Recall from Chapter 19 that a decision problem is decidable if it has an algorithm, and undecidable otherwise. If our program correctly evaluates all instances for which the answer is "yes," but may fail to halt on some instances for which the answer is "no," then the program is a partial decision procedure, and the problem is partially decidable. A partially decidable problem, however, is undecidable—unless you can find an algorithm that removes the word "partially." Likewise, if our program correctly outputs $f(x)$ whenever $f(x)$ is defined, but may fail to halt when $f(x)$ is undefined, then the partial function f is partial computable.

In the remainder of the subsection, we present some simple algorithm design techniques and sketch how they make progress on solving some of these problems and special cases of them. These techniques may seem too obvious to warrant explicit description. However, we feel that such a description will help new readers to appreciate the limits on information processing that make certain problems undecidable.

21.2.1.1   Table-Lookup

For certain functions g it can be advantageous to create a table with one column for inputs x and one for values $g(x)$, looking up the value in the table whenever an evaluation $g(x)$ is needed. A function f that is defined on an infinite set such as Σ^* cannot have its values enumerated in a finite table in this manner, but sometimes the infinite table for f can be described in a finite way that constitutes an algorithm for f. Moreover, tables for other functions g may help the task of computing f, such as the digit-by-digit times-table used in multiplying integers of arbitrary size. These ideas come into play next.

21.2.1.2   Bounding the Search Domain

Many solutions to decision problems involve finding a witness that proves a "yes" or "no" answer for a given instance. The term reflects an analogy to a criminal trial where a key witness may determine the guilt or innocence of the defendant. Thus the first step in solving many decision problems is to identify the right kind of witness to look for. For example, consider the problem of determining whether a given number N is prime. Here a (counter-) witness would be a factor of N (other than 1 and N itself). If N is composite, it is easy to prove by simple division that the witness' claim is correct.

In cases where the given number is prime, a witness of a different kind needs to be searched for. This search may involve integers larger than N, and trying to summon every integer sequentially as a witness would violate the requirement of an algorithm to terminate in finite number of steps. This is often the main challenge in establishing decidability. The difficulty can be surmounted if, based on the structure of the problem, we can establish ahead of time an upper bound such that if any witness exists at all, one exists that meets the bound. Then a sufficient body of potential witnesses can be examined in a finite number of steps. In the case of composite N, the bound is N itself. For prime N, there is a known polynomial p such that a witness exists in the numbers between 1 and $2^{p(n)}$, where n is the number of digits in N, according to a certain witnessing scheme whose test for correct claims is easy to compute. This kind of "polynomial size-bounded witnessing scheme" characterizes the important complexity class NP, and is discussed much further in Chapter 22.

For another example, let us consider the special case of Problem 3 where the given G is a type-1 grammar, and we wish to determine whether a given string x can be generated from the start symbol S of G. A witness in this case can be a sequence of sentential forms starting from S and ending with x that forms a valid derivation in G. The length of x imposes a limit on the size of such sentential forms because G has no length-decreasing productions, and this in turn defines a (much larger) limit on the number of sequences that need be considered before all possibilities are exhausted. Readers may find the details in a standard text such as [10].

For one more example, consider the full version of the linear programming problem where one wishes to maximize a linear objective function f over the set of feasible solutions s. This set may be infinite, and so a table-lookup through all values $f(s)$ cannot be used. However, it is possible to reduce the search domain to a finite set as follows. The feasible solutions form a collection in n-dimensional space (where n is the number of variables plus the number of constraints), known as a convex polytope. Unless the polytope is empty or unbounded—cases that can be detected and resolved—the polytope has a finite number of "corner" points, which are similar to the vertices of a polygon, and which are easily computed. In this case, it is known that a linear objective function attains its maximum value at one (or more) of these corner points. Thus we know the problem is decidable via table-lookup of values at the corner points. In practice, there are intelligent algorithms that find a maximum-giving corner point after searching (usually) only a small part of this table.

21.2.1.3 Use of Subroutines

This is more a programming technique than an algorithm design tool. The idea is to use one program P as a single step in another program Q. Building programs from simpler programs is a natural way to deal with the complexity of the programmer's task. A simple example is using a lookup to the times table as a subroutine in multiplying two integers i and j. Let us examine this in the context of designing Turing machines, where i and j are represented on the tape by the string $1^i 0 1^j$ (namely, i 1s followed by a 0 and then by j 1s). The basic idea of our GOTO program is to duplicate the string of i 1s $j - 1$ times, meanwhile erasing the string 1^j bit-by-bit to count the iterations. A little thought reveals that our earlier GOTO program in Figure 21.1 can almost be used verbatim as a subroutine to call $j - 1$ times. The only hitch is that the first call would run $2i$-many 1s together so that further calls would duplicate too many 1s. To fix the problem, we introduce a new tape symbol 2, using two initial steps to convert the tape to $21^i 0 1^j$, and "patch" the subroutine so that it will not overwrite this 2. The new subroutine can be called by a line "k: IF 2 IS SCANNED GOTO m," where m is the number of the first line in the subroutine, and can return control to the point of call by replacing its STOP instruction by "IF 0 IS SCANNED GOTO $k + 1$." Careful writing will ensure that this latter 0 is the one initially separating 1^i from 1^j. The remaining details are left to the interested reader, while performing a similar patch without using a new symbol "2" is left to the obsessive reader. This subroutine mechanism is in fact no different from the one programmers in BASIC have used for decades.

21.2.2 A Universal Program

We will now solve Problem 1 by arguing the existence of a program U written in the GOTO language that takes as input a program P (also written in the GOTO language) and data x for P, and that produces the same output as P does on input x, if $P(x)$ halts and produces output at all. The last caveat hints that we shall only achieve a partial solution, formally showing only that the function $U(P, x) = P(x)$ is partial computable.

For convenience, we assume that all programs written in the GOTO language use the fixed alphabet $\{0, 1, B\}$. Since we have thus far used the full English alphabet for the notation of our GOTO programs, we must first address the issue of what the formal input to the program U will look like. This problem can be circumvented by encoding each instruction using only 0 and 1. The idea of such an encoding should not be mysterious—we could refer to the 0–1 encoding defined by the ASCII standard, which the terminal used to type this chapter has already carried out for these example programs. However, we prefer the more-succinct encoding defined by Table 21.1.

To encode an entire program, we simply write down in order (without the line numbers) the code for each instruction as given in the table. For example, here is the code for the doubling program shown in Figure 21.1:

$$000100101111011000110100111111011000110100111011100.$$

Note that the encoded string preserves all the information about the program, so that one can easily reverse the process to decode the string into a GOTO program. From now on, if P is a program in

TABLE 21.1 Encoding GOTO Instructions

Instruction	Code
PRINT i	0001^{i+1}
MOVE LEFT	001
MOVE RIGHT	010
IF i IS SCANNED GOTO j	$0111^j 01^{i+1}$
STOP	100

the GOTO language, *code* (*P*) will denote its binary encoding. When there is no confusion, we will identify *P* and *code* (*P*). We may also assume that all programs *P* have a unique STOP instruction that comes last. This convention ensures that a input string to *U* of the form *w* = *code* (*P*)*x* can be parsed into its *P* and *x* components. (When we consider decision problems we will use the code 100 for a unique final ACCEPT instruction, and assign some other code to REJECT.) Before proceeding further, readers may test their understanding of the encoding/decoding process by decoding the following string: 0100111011001001.

The basic idea behind the construction of a universal program is simple, although the details involved in actually constructing one are substantial. Turing in his original paper [21] exhibited a universal program in glorious gory detail, while simpler constructions may be found in more-recent sources such as [15]. Here we will content ourselves with a sketch that conveys the central ideas.

U has as its input a string *w* of the form *code* (*P*)*x*. (If *U* is given an input string not of this form, it can detect the flaw and immediately stop.) To simulate the computational steps of *P* on input *x*, *U* divides its tape into two segments, one containing the program *P*, and one modeling the contents of the tape of *P* as it changes with successive moves. The computation by *U* consists of a sequence of cycles, each of which simulates one step by *P* and is analogous to an *REW* cycle (for "read-evaluate-write") in many real computer systems.

To execute a cycle, *U* first needs to know the cell that the "virtual" tape head of *P* is currently scanning, and the instruction *P* is currently executing. We can assist *U* by extending its own work alphabet to include new "alias" symbols $0', 1', B'$ for the characters of *P*. *U* maintains the condition that there is exactly one aliased symbol in the "*P*" segment of its tape that marks the encoding of the current instruction, and exactly one in the other segment that marks the cell currently scanned by *P*. For example, suppose that after thirty-nine steps, *P* is reading the fourth symbol from the left on its tape containing 01001001. Then the second tape segment of *U* after 39 cycles consists of the string $0100'1001$. We can further assist *U* by adding a symbol $\wedge$ to divide the two segments, although the unique STOP instruction itself could serve as the divider. The computation by *U* on an input *w* = *code* (*P*)*x* can begin with some steps that prime the first symbol of *code* (*P*), insert a $\wedge$ before the first symbol of *x* (caterpillaring *x* one cell to the right), and prime the first symbol of *x*. We may suppose that each cycle by *U* begins with its own head scanning the $\wedge$.

At the beginning of a new cycle, *U* moves its head left to find the current instruction, and begins decoding it. The only information *U* needs to retain is which type of instruction it is, and in the case of a PRINT *i* or IF *i*... instruction, which character *i* is involved. To execute a PRINT *i*, MOVE RIGHT, or MOVE LEFT instruction, *P* unprimes the instruction, primes the next one, and marches down its tape to find the primed cell on its copy of *P*'s tape and execute the action. It is possible that a MOVE LEFT instruction may bump into the $\wedge$, in which case *U* makes another call to its "caterpillar" subroutine to move *P*'s tape over, and inserts a B' for the blank *P* would scan after that move. The only case that requires cumbersome action by *U* is an instruction IF *i* IS SCANNED GOTO *j*, when *U* finds that *P* really is scanning character *i*. Then *U* needs to find the *j*th instruction in the "*P*" part of its tape. Because we have used a unary encoding 1^j of the required line number *j*, it is not too difficult to write a subroutine that counts off the 1s in 1^j and advances an instruction marker each time beginning from line 1, knowing to stop when the *j*th instruction has been located. Finally, if the current instruction is STOP, *U* gleefully erases *P*, erases the $\wedge$, and unprimes the scanned symbol, leaving exactly the final output *P*(*x*).

One last refinement is needed to answer the objection that *U* is using extra tape symbols $0', 1', B', \wedge$ that we have expressly forbidden to GOTO programs. This use can be eliminated by one more level of encoding. Give each of the seven tape symbols its own three-bit code, and make *U* treat blocks of three cells as single cells in the simulation that was described above. *U* itself can be programmed to convert its input *code* (*P*)*x* to this encoding before the first cycle, and to invert it when restoring the final output *P*(*x*). Then *U* is a bona-fide GOTO program that meets all our requirements. It is even possible to run *U* on input *code* (*U*)*w* where *w* = *code* (*P*)*x*, producing (more slowly) the same output $P(x)$. It is important to note that the code of U itself is completely independent of any program P that might be simulated. The code of U itself is not long—a reader with good programming skill can make it shorter than the prose description we have just given.

Besides solving what was asked for in Problem 1, we have also shown that Problem 2 is partially decidable. Namely, for any "yes"-instance $w = code\ (P)x$ where P on input x halts, U on input w will eventually detect that fact—and the slight edit of changing U's own STOP instruction to ACCEPT will make U halt and accept w. However, on a "no"-instance where $P(x)$ does not halt, our U will blindly follow P and not halt either. The question is whether we can improve U so that it will detect every case in which $P(x)$ does not halt, and signal this by executing a REJECT instruction. We will see in Section 21.5 that all the programming skill in the world cannot produce such a U—the halting problem is undecidable.

Before presenting undecidability, however, we develop a fundamentally different way to formalize the notion of mechanical computability in the next section.

21.3 Recursive Function Theory

The main advantage of using the class of μ-recursive functions to define computation is their mathematical elegance. Proofs about this class can be presented in a rigorous and concise way, without long prose descriptions or complicated programs that are hard to verify. These functions need and make no reference to any computational machine model, so it is remarkable that they characterize "mechanical" computability.

An analogy to the two broad families of programming languages is in order. We have already discussed how Turing machines and our particular "GOTO" formalism abstract the essence of imperative programming languages, in which a program is a sequence of operational commands and the major program structures are subroutines and loops and other forms of iteration. By contrast, specifications in recursive function theory are declarative, and the major structures are forms of recursion. "Declarative" means that a function f is specified by a direct description of the value $f(x)$ on a general argument x, as opposed to giving steps to compute $f(x)$ on input x. Often this description is recursive, meaning that $f(x)$ is defined in terms of values $f(y)$ on other (usually smaller) arguments y. Programming languages built on declarative principles include Lisp, ML, and Haskell, which are known as functional languages. These languages have recursion syntax that is not greatly different from the recursion schemes presented here. They also draw upon Church's lambda calculus, which can be called the world's first general programming language. A formal proof of equivalence between lambda calculus and the Turing machine model (via a programming language called I) can be found in [11], which presents computability theory from a programming perspective.*

In this section, we will describe this functional approach to computation and code some simple functions using recursion. Owing to space limitation, we will not present a complete proof that the class of μ-recursive functions is the same as the class of (partial) computable functions on a Turing machine. The full proof can be found in standard texts such as [20]. All the functions we consider have one or more nonnegative integers as arguments, and produce a single nonnegative integer value.

---

* Turing created an addendum to his seminal paper [21] showing that his definition of a (partial) computable function was equivalent to the one proposed by Church. The lambda calculus uses essentially a single execution scheme called reduction to govern its computations, and by suitable conditioning one can make this scheme carry out recursion. Another declarative language, Prolog, also fixes a single execution scheme that tries to limit the operational decisions the programmer needs to make, and also relies upon recursion.

Before presenting formal definitions, we qualify the above ideas with a few examples. Consider first the simple definition of a two-variable linear function by

$$h(y, z) = z + 2 * y + 1. \tag{21.1}$$

Here $h(y, z)$ is defined with the aid of other functions (here, plus and times) and quantities (here, 2 and 1) that presumably have already been defined or given. This is an example of an explicit definition because all entities on the right-hand side are known—in particular, this definition does not involve recursion. If we rewrite the infix functions $+$ and $*$ in prefix-function style as "plus" and "times," the expression becomes

$$h(y, z) = plus\ (z, plus\ (times\ (2, y), 1)), \tag{21.2}$$

and we can glimpse another hallmark of functional languages: function names can be regarded as parameters the same way that variable names can. Now consider the somewhat-similar definition of a one-variable function by

$$f(x) = f(x - 1) + 2 * (x - 1) + 1, \tag{21.3}$$

together with a base case such as $f(0) = 0$. Here not every quantity on the right-hand side is known—one must first know $f(x - 1)$ to compute $f(x)$. However, this is still "declarative" insofar as $f(x)$ is defined in terms of known quantities and values $f(y)$ for other (smaller) arguments y. The reader may check that this is a recursive definition of the squaring function.

Why use recursion? One reason is that explicit definition by itself is known not to be powerful enough to capture the essence of mechanical computation. The next two sections define the two principal schemes of recursion in recursive function theory.

21.3.1 Primitive Recursive Functions

The class of primitive recursive functions is built up from the following set of basic functions, which are the only ones we need to presuppose are "known":

1. The successor function S is defined for all x by $S(x) = x + 1$.
2. The zero function Z is defined for all x by $Z(x) = 0$. The constant 0 is also provided here.
3. For all fixed numbers i and n with $1 \leq i \leq n$, the projection function p_i^n is defined for all n-tuples $(x_1, x_2, \ldots, x_n)$ by $p_i^n(x_1, x_2, \ldots, x_n) = x_i$.

The primitive recursive functions are constructed from the basic functions by applications of the following two operations. The case $n = 0$ is allowed in them; a zero-variable function is the same as a constant, and a 0-tuple is the empty list.

1. *Functional composition:* Given k-many functions $g_1, \ldots, g_k$ that each take n variables, and a function h that takes k variables, one can define a function f of n variables by

$$f(x_1, \ldots, x_n) = h\left(g_1(x_1, \ldots, x_n), g_2(x_1, \ldots, x_n), \ldots, g_k(x_1, \ldots, x_n)\right). \tag{21.4}$$

   If $g_1, \ldots, g_k$ and h are primitive recursive, then f is defined to be primitive recursive.
2. *Primitive recursion:* Given a function g that takes n variables, and a function h that takes $n + 2$ variables, one can define a function f of $n + 1$ variables by

$$f(x_1, \ldots, x_n, 0) = g(x_1, \ldots, x_n); \tag{21.5}$$
$$f(x_1, \ldots, x_n, S(y)) = h(x_1, \ldots, x_n, y, f(x_1, \ldots, x_n, y)). \tag{21.6}$$

   If g and h are primitive recursive, then f is defined to be primitive recursive.

Here Equation 21.5 is the basis and Equation 21.6 is the recursion step. It is conventional to call $x_1, \ldots, x_n$ the parameters and y the recursion variable. From a computational viewpoint, the scheme is easy to interpret. Given integer values for variables $x_1, \ldots, x_n$ and z, how can we evaluate $f(x_1, \ldots, x_n, z)$? We start building a table T in which each row y contains the value of $f(x_1, \ldots, x_n, y)$. The basis step gives us the top row via $T[0] = f(x_1, \ldots, x_n, 0) = g(x_1, \ldots, x_n)$. Whenever we have filled a row y, we can fill the next row via the recursion step, via $T[y + 1] = f(x_1, \ldots, x_n, S(y)) = h(x_1, \ldots, x_n, y, T[y])$. As soon as row z is filled, using y such that $z = S(y)$, we are done. The point is that provided g and h are computable, the function f is also computable. Functional composition likewise preserves computability. Moreover, since the basic functions are all total and produce nonnegative values, every function that we can build up in this manner is also total and produces nonnegative values.

DEFINITION 21.1   A function is said to be primitive recursive if it can be built up from the successor, zero and projection functions by a finite number of applications of composition and primitive recursion.

Example 21.1

To show how the scheme of primitive recursion models the informal recursion defining the function f of one variable (so we have $n = 0$) in Equation 21.3, take "$g()$" to be the constant 0, and take h to be the two-variable function $h(y, z) = z + 2y + 1$, which happens to be our example of "explicit definition" in Equation 21.1. Then we have $f(0) = 0$ and

$$f(S(y)) = h(y, f(y)) = f(y) + 2 * y + 1.$$

With "$x - 1$" in place of "y" and "x" in place of "$S(y)$," this is the same as Equation 21.3. We will return to this notational difference later.

As the prefix form Equation 21.2 indicates, h itself can be built up via functional composition from the plus and times functions. It is interesting to see how the usual functions of arithmetic can themselves be constructed from the rather Spartan basis we have been given. To begin with, the constants $1, 2, \ldots$ are formally introduced by functional composition, with "$g_1()$" as the constant 0 and "h" as the successor function, via $1 = S(0), 2 = S(1) = S(S(0)), 3 = S(2)$, and so on.

Example 21.2

Addition. Take $g(x) = x$ and $h(x, y, z) = S(z)$. Formally, g is the basis function p_1^1, and h is the functional composition of the successor function with p_3^3. Then primitive recursion gives us $plus\,(x, 0) = g(x) = x$ and

$$plus\,(x, S(y)) = h(x, y, plus\,(x, y)) = S(plus\,(x, y)) = S(x + y) = x + y + 1,$$

as we would demand. Hence this formal definition of plus correctly computes addition, and we may use the standard "$+$" notation in the formal examples that follow.

Example 21.3

Multiplication. Take $g(x) = 0$ and $h(x, y, z) = x + z$. Formally, g is the zero function (of one variable rather than the constant zero), and h is the functional composition of plus with the two functions p_1^3 and p_3^3 (so $k = 2$ here). Then primitive recursion gives us $times\,(x, 0) = g(x) = 0$ and

$$times\,(x, S(y)) = h(x, y, times\,(x, y)) = x + times\,(x, y) = x * (y + 1),$$

again as we would demand. Hence this formal definition of times correctly computes multiplication. Note that we had to go to some length (of making h a function of three variables) so that our definition exactly agrees with the formal requirements in Equation 21.6.

Example 21.4

Exponentiation. Take $g(x) = 1$ and $h(x, y, z) = x * z$. Formally, g is the one-variable function that always outputs 1 and is defined by composing S and the zero function Z, while h is the same as in Example 21.3 but with times in place of plus. Then primitive recursion gives us $exp\,(x, 0) = g(x) = 1$ (note that even 0^0 equals 1) and

$$exp\,(x, S(y)) = h(x, y, exp\,(x, y)) = x * exp\,(x, y) = x^{y+1}.$$

Once again the correctness of this definition for all values of x and y is easy to verify, via a simple proof by induction that follows the recursion.

It is now straightforward to omit some of the formal apparatus and write the definitions more succinctly. For instance, the last example becomes

$$exp\,(x, 0) = 1$$
$$exp\,(x, y + 1) = x * exp\,(x, y).$$

This resembles a program one would actually write, especially in a language like C that does not provide exponentiation as a built-in operator.

At this point the alert reader, noting the way our schemes all involve nonnegative numbers, will first wonder how on earth we can ever define subtraction this way. The key is that the syntax of primitive recursion allows us to define a function $P(y)$ that computes "proper subtraction by 1," and then use P to define proper subtraction itself. The word "proper" here means that any negative value is replaced by 0, in order to maintain our restriction to the nonnegative numbers. The definitions are

$$P(0) = 0$$
$$P(S(y)) = y$$
$$sub\,(x, 0) = x$$
$$sub\,(x, S(y)) = P(sub\,(x, y)).$$

For P we took $h(y, z) = y$, i.e., $h = p_1^2$, and for sub we took $h(y, z) = P(z)$. To trace this out, $sub\,(3, 2) = P(sub\,(3, 1)) = P(P(sub\,(3, 0))) = P(P(3)) = P(2) = 1$, and $sub\,(2, 3) = P(P(P(2))) = P(0) = 0$, which is the "proper" value.

Second, the reader may have felt uncomfortable defining functions in terms of "$S(y)$" rather than "y." For example, the primitive recursion for the factorial function, with 0! standardly defined to be 1, gives us

$$\texttt{fact(0)} = 1 | \texttt{fact(y + 1)} = (\texttt{y} + 1) * \texttt{fact(y)}; @$$

here "$|$" separates the base and recursion cases. This would actually be valid syntax in the programming language ML except that "$\texttt{fact(y+1)}$" is an illegal function header. The syntax of ML forces one to write it this way:

$$\texttt{fact(0)} = 1 | \texttt{fact(y)} = \texttt{y} * \texttt{fact(y - 1)}; @$$

this is literally the example used in many texts. To make the formal Equation 21.6 for primitive recursion reflect the syntax of programming languages, we can use P in place of S to change it to

$$f\,(x_1, \ldots, x_n, y) = h\,(x_1, \ldots, x_n, y, f\,(x_1, \ldots, x_n P(y))), \qquad (21.7)$$

and alternately make the middle argument of h be $P(y)$ instead of y. Either way, one might then expect to be able to recover the function S by defining it in terms of P and the other two basis functions, just as we defined P in terms of S above. However, this is impossible—one could never define any increasing functions at all. This curious asymmetry partly explains why primitive recursion was defined the way it is. Nevertheless, if S as well as P is provided in the basis, then one can use the modified definition and obtain exactly the same class of primitive recursive functions. For instance, addition is definable by $plus\,(x, 0) = x \mid plus\,(x, y) = S(plus\,(x, P(y)))$, and so on. Hence primitive recursion is for the most part exactly what ML and other functional languages do.*

Finally, the reader may wonder what has become of functions defined on strings. A string over an alphabet Σ can always be identified with its number in the standard lexicographic enumeration of Σ^*, with ϵ corresponding to 0. Then a string function $f : \Sigma^* \to \Sigma^*$ can be called primitive recursive if the corresponding numerical function (of one variable) is primitive recursive. For instance, the function that appends a '1' to a binary string x corresponds to $2x + 2$. Cutting the other way, under some transparent encoding of negative and rational and complex numbers (etc.) by strings, one can extend the concept of primitive recursion to define addition and multiplication and nearly all familiar mathematical functions in their full generality. The meaning and proof of the following statement should now be clear; full detail can be found in [20].

THEOREM 21.1   *Every primitive recursive function is computable by a Turing machine.*

The converse is false, however. A famous example of a computable total function that is not primitive recursive is Ackermann's function; this and other examples may be found in [13]. To obtain all computable functions we need to introduce one more scheme of recursion—at the inevitable cost, however, of opening a Pandora's box of functions that are no longer total.

21.3.2   μ-Recursive Functions

We will add a new operation called minimalization that does not preserve totality. Again we restrict numerical arguments to be nonnegative integers.

DEFINITION 21.2   A possibly-partial function f of n variables is defined by **μ-recursion** from a function g of $n + 1$ variables, written

$$f(x_1, \ldots, x_n) = \mu y \cdot g(x_1, \ldots, x_n, y),$$

if whenever $f(x_1, \ldots, x_n)$ is defined, it equals the least number y such that $g(x_1, \ldots, x_n, y) = 1$. If $f(x_1, \ldots, x_n)$ is undefined, there must be no y such that $g(x_1, \ldots, x_n, y) = 1$. The class of μ-recursive functions is the class of all functions that can be built up from the successor, zero, and projection functions by the operations of composition, primitive recursion, and μ-recursion.

The computation of $f(x_1, \ldots, x_n)$ that is implicit in Definition 21.2 can be described by building a table as before. First fill in the row $T[0] = g(x_1, \ldots, x_n, 0)$, then $T[1] = g(x_1, \ldots, x_n, 1)$, and so on. If and when one finds a y whose value $T[y]$ equals 1, halt and output y. The "if" is the big difference from the algorithm for primitive recursion, because if $g(x_1, \ldots, x_n, y)$ never takes

---

* Primitive recursion has its counterpart in imperative languages as well, aside from the fact that most of them support recursion directly. The "table $T[y]$" computation above shows how primitive recursion can be simulated by a simple for-loop for $y = 0$ to z do... end that fixes its bounds and never alters y in the loop body. A theorem [12] in programming languages states that the primitive recursive functions are exactly the total functions computable by programs that use only if-then-else and simple nested for-loops.

the value 1, this procedure will never halt. This procedure is called an unbounded search. Compared another way to primitive recursion, μ-recursion increments its recursion variable rather than decrement it.

There is nothing special about "$=1$" here: zero or any other constant could be used instead. Our use of 1 suggests the special case in which g is a total function that takes on only the values 0 and 1. Then we can regard its output as a Boolean truth value, with $1 =$ true and $0 =$ false, and call g a predicate. The class of μ-recursive functions is not changed under the restriction that g be a predicate. Then we can read the syntax "$\mu y.g(x_1, \ldots, x_n, y)$" in English as "the least y such that $g(x_1, \ldots, x_n, y)$ is true." From all this we can see that whereas primitive recursion corresponds to a for loop, μ-recursion corresponds to a while loop, with $g(\ldots)$ as the test condition.

Example 21.5

Partial square-root function. Define the predicate $g(x, y)$ to hold if and only if $x = y^2$. Then the function f defined for all x by $f(x) = \mu y.g(x, y)$ computes the square root of x when x is a perfect square. When x is not a perfect square, however, the recursion is undefined, so f is a partial recursive function.

Example 21.6

Linear programming. The standard simplex algorithm uses a while loop that executes a basic pivot step until a predicate expressing optimality holds. Hence the function that embodies the solution to a linear programming problem is μ-recursive. In point of fact, because a bound on the number of polytope corner points is explicitly definable from the problem instance, the same function can be computed via a simple for loop, so it is primitive recursive. However, the former method is usually much faster.

Part (a) of Theorem 21.2 expresses the fact that while loops, together with if-then-else, suffice to make a general-purpose programming language. Part (b) is the gist of the famous theorem, credited in various forms to various sources, that at most one while loop is needed in any program.

THEOREM 21.2

  (a) *A (partial) function is μ-recursive if and only if it is a Turing-computable (partial) function.*

  (b) *Moreover, given any Turing machine T, we can find a primitive recursive function u and a primitive recursive predicate t such that for all x, $T(x) = u(\mu y \cdot t(x, y))$.*

In the standard proof of part (b), the predicate $t(x, y)$ is designed to hold if and only if y encodes the sequence of configurations of a halting computation of T on input x, and the function u picks off the output from the final configuration. To complete the proof of (a), all one needs to show is that given a Turing machine that computes g in Definition 21.2, one can build a Turing machine that computes f. This is done by following the unbounded-search procedure sketched above.

The corresponding theorem for formal languages also merits mention here. In Chapter 19 we defined the characteristic function of a language L to be the function f_L defined for all x by $f_L(x) = 1$ if $x \in L$, $f_L(x) = 0$ if $x \notin L$. This is simply the predicate corresponding to membership in L. The partial characteristic function still takes the value 1 when $x \in L$, but is undefined when $x \notin L$.

THEOREM 21.3

  (a) *A language is recursive if and only if its characteristic function is μ-recursive.*

  (b) *A language is r.e. if and only if its partial characteristic function is μ-recursive.*

Part (a) explains how the term "recursive" became applied to languages and predicates as a synonym for "decidable." It is important to recall that not all languages L accepted by Turing machines have computable characteristic functions (i.e., are decidable); unless we find a Turing machine accepting L that halts for all inputs, all we know is that the partial characteristic function of L is (partial-) computable. Before proceeding to undecidable languages, we take time to interpret these two theorems and others presented in Chapters 19 and 20.

21.4 Equivalence of Computational Models and the Church–Turing Thesis

In Chapter 19 we introduced various machine models, the most important of which is the Turing machine. In Chapter 20 we introduced the grammar hierarchy of Chomsky [2,3], of which the most powerful was the type-0 grammar. Here we have presented the purely mathematical model of μ-recursive functions. Although these models were defined over different domains for different purposes, they are all equivalent in a precise technical sense—they all define the same class of computable functions and decidable languages, and the same class of partial computable functions and partially decidable languages. We can summarize all this by saying that Turing machines, type-0 grammars, and μ-recursive functions have the same problem-solving power.

This equivalence extends to vastly many other computational models, of which we mention a few:

1. *Cellular Automata:* Cellular automata are intended to model the evolution of a colony of microorganisms. Each cell is a deterministic finite automaton that receives its input in discrete time steps from neighboring cells, so that its current state is defined by its own previous state and the previous states of its neighbors. All the cells execute the same deterministic finite automaton (DFA). There are different schemes for specifying the representation of the input to a cellular automaton and its output. But under any reasonable scheme, the largest class of problems that can be solved on cellular automata coincides with the class of **solvable problems** on a Turing machine.

2. *String-Rewriting Systems:* A string-rewriting system is similar to a grammar. The main difference is that there are no nonterminals. Let the input alphabet be Σ. The production rules of a rewriting system T will be of the form $\alpha \rightarrow \beta$ where α and β are strings over Σ. One can apply such a rule by replacing any occurrence of α in a string by β. T is defined as a finite set of rewrite rules, along with a finite set of initial strings. The language generated by T is defined as the set of strings that can be obtained from an initial string by applying the rewrite rules a finite number of times. The systems proposed before 1930 by Thue and Post fall roughly into this category. It turns out that the class of string-rewriting languages is the same as the r.e. languages (see [1]).

3. *Tree-Rewriting Systems:* These are similar to string-rewriting systems except that the local edits are done on subtrees of a tree, and rules may have more than one argument. The subtrees typically represent terms in algebraic or logical expressions that are being operated on. Under reasonable schemes for encoding numbers or strings by trees, all known tree-rewriting systems generate r.e. languages or compute partial recursive functions. Church's λ-calculus and most formal systems of logic fall into this category.

4. *Extensions of Turing's Model:* As mentioned in Chapter 19, one can also create numerous modifications to the basic Turing machine model, such as having multidimensional tapes or binary trees with MOVE UP, MOVE DOWN LEFT, and MOVE DOWN RIGHT instructions (the latter are tantamount to having random-access to stored values), allowing nondeterminism or alternation, making computation probabilistic (see Section 24.2),

and so on. All of these machines compute the same functions as the simple one-tape Turing machine.

5. *Random-Access Machines and High-Level Programming Languages:* These can be mentioned in tandem because a RAM, as described in Chapter 19, is just an idealization of assembly or machine language. Every high-level language yet devised can be compiled into some machine language. Even the standard Java Virtual Machine is little more than a RAM, with some added handling of class objects via pointers that is not unlike the workings of a pointer machine, and some hooks to enable the host system to control physical devices and network communications. Without excessive effort one can extend the construction of a **universal Turing machine** in Section 21.2 to handle the case where P is a RAM program rather than a GOTO program. The registers of P can be simulated on the tape by adding one more tape symbol # and using strings of the form #i#j#, where i is the register's number and j is its contents. As stated in Section 19.4, this simulation is even fairly efficient. Hence all these high-level languages have the same problem-solving power as the lowly one-tape Turing machine.

The convergence of so many disparate formal models on the same class of languages or functions is the main evidence for the assertion that they all exactly capture the informal notion of what is mechanically or humanly computable. This assertion is called the Church–Turing thesis. In one form, it asserts that every problem that is humanly solvable is solvable by a Turing machine. Put more precisely, any cognitive process that a human being could or will ever use to distinguish certain numbers or strings as "good" defines an r.e. language—and if it also would determine that any other given number or string is "bad," it defines a **recursive language**. An extension of the thesis claims that no one will ever design a physical device to compute functions that are not μ-recursive. The Church–Turing thesis is not a mathematical conjecture and is not subject to mathematical proof; it is not even clear whether the extension is resolvable scientifically.

21.5   Undecidability

The Church–Turing thesis implies that if a language is undecidable in the formal sense defined above, then the problem it represents is really, humanly, physically undecidable. The existence of languages that are not even partially decidable can be established by a counting argument: Turing machines can be counted 1, 2, 3, . . ., but the mathematician Georg Cantor proved that the totality of all sets of integers cannot be so counted. Hence there are sets left over that are not accepted, let alone decided, by any program. This argument, however, does not apply to languages or problems that one can state, since these are also countable. The remarkable fact is that many easily-stated problems of high practical relevance are undecidable. This section shows that the five remaining problems on our list in Section 21.2.1, namely 2 through 5 and 7, are all unsolvable.

21.5.1   Diagonalization and Self-Reference

Undecidability is inextricably tied to the concept of self-reference, and so we begin by looking at this perplexing and sometimes paradoxical concept. The simplest examples of self-referential paradox are statements such as "This statement is false" and "Right now I am lying." If the former statement is true, then by what it says, it is false; and if false, it is true. . . . The idea and effects of self-reference go back to antiquity; a version of the latter "liar" paradox ascribed to the Cretan poet Epimenides even found its way into the New Testament, Titus 1:12–13. For a more colorful example, picture a barber of Seville hanging out an advertisement reading, "I shave those who do not shave themselves." When the statement is applied to the barber himself, we need to ask: Does he shave himself? If yes, then he is one of those who do shave themselves, which are not the people his statement says he shaves. The contrary answer no is equally untenable. Hence the statement can be neither true nor false (it may be good ad copy), and this is the essence of the paradox. Such paradoxes have made entry into modern mathematics in various forms. We will present some more examples in the next few paragraphs. Many variations on the theme of self-reference can be found in the books of the logician and puzzlist Raymond Smullyan, including [18,19].

Berry's paradox concerns English descriptions of natural numbers. For example, the number 24 can be described by many different phrases: "twenty-four," "six times four," "four factorial," etc. We are interested in the shortest of such descriptions, namely one(s) having the fewest letters. Here, "two dozen" beats all of the above. Clearly there are (infinitely) many positive integers whose shortest descriptions require one hundred letters or more. (A simple counting argument can be used to show this. The set of positive integers is infinite, but the set of positive integers with English descriptions of fewer than one hundred letters is finite.) Let D denote the set of positive integers that do not have English descriptions of fewer than one hundred letters. Thus D is not empty. It is a well-known fact in set theory that any nonempty subset of positive integers has a smallest integer. Let x be the smallest integer in D. Does x have an English description of fewer than one hundred letters? By the definition of the set D and x, the answer is yes: such a description of x is, "the smallest positive integer that cannot be described in English in fewer than one hundred letters." This is an absurdity, because the quoted part of the last sentence is clearly a description of x, and it contains fewer than one hundred letters.

Russell's paradox similarly turns on issues in defining sets. In formal mathematics, we can perfectly easily describe "the set of all sets that do not include themselves as elements" by the definition $S = \{x | x \notin x\}$. The question "Is $S \in S$?" leads to a real conundrum. This also resembles the barber paradox, with "$\notin$" read as "does not shave." This paradox forced the realization that the formal notion of a *set,* and importantly the formal rules that apply to sets, do not and cannot apply to everything that we informally regard as being a "set."

Our last example is a charming paradox named for the mathematician William Zwicker. Consider the collection of all two-person games that are normal in the sense that every play of the game must end after a finite number of moves. Tic-tac-toe is normal since it always ends within nine moves, while chess is normal because the official "fifty move rule" prevents games from going on forever. Now here is hypergame. In the first move of hypergame, the first player calls out a normal game—and then the two players go on to play that game, with the second player making the first move. Now we need to ask, "Is hypergame normal?" If yes, then it is legal for the first player to call out "hypergame!"—since it is a normal game. By the rules, the second player must then play the first move of hypergame—and this move can be calling out "hypergame!" Thus the players can keep saying "hypergame" without end, but this contradicts the definition of a normal game. On the other hand, suppose hypergame is not normal. Then in the first move, player 1 cannot call out hypergame and must call a normal game instead—so that the infinite move sequence given above is not possible and hypergame is normal after all!

Let us try to implement Zwicker's paradox. To play hypergame, we need a way of formalizing and encoding the rules of a game as a string x, and we need a decision procedure `isNormal(x)` to tell if the game is normal. Then the rules of hypergame are easily formalized: pick a string x, verify `isNormal(x)`, and play game x. Let h be the string encoding of these rules. Now we get a real contradiction when `isNormal(h)` is run. We must conclude that either (1) our formalization of games is inadequate or inconsistent, or (2) a decision procedure `isNormal` simply cannot exist. Now (1) is the way out for Russell's paradox with "sets" in place of "games." For computation, however, we know that our formalization is adequate and consistent—and hence we will be faced with conclusions of (2), namely that our corresponding computational problems are unsolvable.

Before showing how the above paradoxes can be modified and ingrained into our problems, we need to review the 0–1 encoding of GOTO programs from Section 21.2.2, including the conventions that ACCEPT has the same code 100 as STOP for programs that accept languages, and that such an ACCEPT statement be last and unique. We may assign the code 101 to REJECT, which may appear anywhere. If a binary string x encodes a program P, it is easy to decode x into P, and we may identify x with P. If x does not encode a legal GOTO program, this fact is easy to detect. Then we may choose to treat x as an alternate code for the trivial GOTO program that consists of a single REJECT statement.

Now we can define the so-called "diagonal language" L_d as follows:

$$L_d = \{x \mid x \text{ is a GOTO program that does not accept the string } x\} \tag{21.8}$$

This language consists of all programs in the GOTO language that do not halt in the ACCEPT statement when given their own encoding as input—they may either REJECT or not halt at all on that input. For example, consider $x = 01111101101100$, which encodes a program that accepts any string beginning with 1 and rejects any string beginning with 0. Then $x \in L_d$ since the program does not accept 01111101101100. Note the self-reference in Equation 21.8. Although the definition of L_d seems artificial, its importance will become clear when we use it to show the undecidability of other problems. First we prove that L_d is not even accepted by any Turing machine, let alone decided by one.

THEOREM 21.4 L_d *is not recursively enumerable.*

PROOF   Suppose for the sake of contradiction that L_d is r.e. Then there is a GOTO program that accepts L_d—call it P. Now what does P do on input $x = code\,(P)$? If P accepts x, then x is not in L_d, but this contradicts $L(P) = L_d$. But if P does not accept x, then x *is* in L_d, and this also contradicts $L(P) = L_d$. Hence a program P such that $L(P) = L_d$ cannot exist, and so L_d is not r.e.

The definition of L_d is motivated by Russell's paradox, reading "$\notin$" as "does not accept." Whereas in Russell's paradox we had to conclude that S is not a set, here we conclude that L_d is not a Turing-acceptable set.

We can similarly carry over Zwicker's paradox by treating a given string x as formally defining "Game-x" as follows: The first player decodes x into a GOTO program P, and then tries to choose some string x' in the language $L(P)$. If $L(P)$ is empty, in particular if x decodes to the trivial program "1. REJECT" as stipulated above, then the game ends then and there. But if the first player finds such an x', then the second player must play the same way with x'. Then we can say that x is normal if every play of Game-x must terminate (by reaching a GOTO program that accepts the empty language) in a finite number of steps. Finally define L_Z to be the set of normal strings. By applying the reasoning from Zwicker's paradox, one can imitate the above proof to show that L_Z is not recursively enumerable.

21.5.2   Reductions and More Undecidable Problems

Recall from Chapter 19 (Section 19.3) the notion of Turing reducibility. Basically, a language L_1 is Turing reducible to L_2 if there is a halting Turing machine for language L_1 using an oracle for language L_2. If L_1 is reducible to L_2 and L_2 is decidable, then so is L_1. This is because one can replace queries to oracles by executing a halting computation for L_2. The contrapositive of this statement can be used to show undecidability. If L_1 is undecidable, then so is L_2. We will first express Problem (2) as a language:

$$L_U = \{code(P)111x \mid P \text{ accepts the string } x\}.$$

Thus L_U takes as input a program in GOTO, and a binary string x, and accepts the encoded pair (P, x) if and only if P accepts x. (Note 111 is used as a separator between P and x.) The universal program presented in "A Universal Program" accepts the language L_U hence it is recursively enumerable. We will show that L_U is not recursive. First, we will show a simple fact about recursive languages.

THEOREM 21.5 *Recursive languages are closed under complement.*

PROOF Let P be a GOTO program for language L. The program P' obtained by interchanging the ACCEPT and REJECT instructions is easily seen to accept the language $\overline{L}$. This standard trick works to complement the computations of most of the deterministic devices (such as DFA).

Now we show that L_U is not recursive.

THEOREM 21.6 L_U *is not recursive.*

PROOF Consider the language $L_U' = \{x \mid x$ when interpreted as a GOTO program accepts its own encoding$\}$. Obviously, $L_U' = \overline{L_d}$. Since L_d is not recursively enumerable, it is not recursive. (Recall that the set of recursive languages is a subset of **recursively enumerable languages**.) By the above theorem, L_U' is not recursive. Finally, note that L_U' can be reduced to L_U as follows. Given an algorithm for L_U, we can construct an algorithm for L_U' as follows: Let P be an algorithm for L_U. To construct an algorithm for L_U' simply note the connection between the two problems. An input string x belongs to L_U' if and only if $x111x$ belongs to L_U. Thus, a simple copy program (similar to one presented in Section 21.2) can be first used to convert the input x into $x111x$. Move the scanning head back to the leftmost character of the first copy of x. Now simply run the program P. Note that the program P' described above is being constructed using only P, not x. This reduction shows that L_U is not recursive.

Next we consider problem (3) in our list. Earlier we showed that a special case of this problem (when the input is restricted to type-1 grammar) is totally solvable. It is not hard to see that the general problem is partially solvable. (To see this, suppose there is a derivation for a string x starting from S, the start symbol of the grammar. Suppose the length of one such derivation is k. A program can try all possible derivations of length 1, 2, etc., until it succeeds. Such a program will always halt on strings x generated by the grammar G. Thus the language

$$L_0 = \{G\#x \mid G \text{ is a type-0 grammar and } x \text{ can be generated by } G\}$$

is recursively enumerable. A standard result from formal language theory [10] is that for every Turing machine M, there is a type-0 grammar G such that $L(M) = T(G)$. This conversion from M to G is the reduction that shows that the language is not recursive.

The string compression problem, numbered 4 on our list, is not a decision problem, but reduction techniques can still be used to show that it is unsolvable. We refer the reader to [17] for details.

By a fairly elaborate reduction (from L_d), it can be shown that the tiling problem (5) in our list is also not partially decidable. We will not do it here and refer the interested reader to [8]. But we would like to point out how the undecidability result can be used to infer a result about aperiodic tilings. This deduction is interesting because the result appears to have some deep implications and is hard to deduce directly. We need the following discussion before we can state the result. A different way to pose the tiling problem is whether a given set of tiles can tile an entire plane in such a way that all the adjacent tiles have the same color on the meeting quarter. (Note that this question is different from the way we originally posed it: Can a given set of tiles tile any finite rectangular region? Interestingly, the two problems are identical in the sense that the answer to one version is "yes" if and only if it is "yes" for the other version.) Call a tiling of the plane periodic if one can identify a $k \times k$ square such that the entire tiling is made by repeating this $k \times k$ square tile. Otherwise, call it aperiodic. Consider the question: Is there a (finite) set of unit tiles that can tile the plane, but only aperiodically? The answer is "yes" and it can be shown from the total undecidability of the tiling problem. Suppose the answer is "no." Then, for any given set of tiles, the entire plane can be tiled if and only if the plane can be tiled periodically. But such a periodic tiling can be found, if one exists, by trying to tile a $k \times k$ region for successively increasing values of k. This process will eventually succeed (in a finite number of steps) if the tiling exists. This would make the tiling problem partially decidable, which contradicts the total undecidability of the problem. This means that the assumption that the entire plane can be tiled if and only if some $k \times k$ region can be tiled is wrong. Thus there must exist a finite set of tiles that can tile the entire plane, but only aperiodically.

We conclude with a brief remark about problem 7 in our list. After many years of effort by several mathematicians and computer scientists (including Davis and Robinson), Matiyasevich found an effective way to transform a given Turing machine T into a set of equations in variables $x, y_1, \ldots, y_m$ such that for any x, T on input x halts if and only if the other m variables can be set to solve the equations. This reduction shows that Hilbert's tenth problem is undecidable. Details behind this reduction can be found in [6].

21.6 Further Information

The fundamentals of computability can be found in many books including the classic texts [4,14,16]. More-recent books on automata and formal languages have also devoted at least a few chapters to computability [6–10,17,22]. Early work on computability was motivated by a quest to address profound questions about the basis of logical reasoning, mathematical proofs and automatic computation. Various formalisms discussed in this chapter were proposed at around the same time, and soon thereafter, their equivalence was tested. Thus, in a short time, the Church-Turing thesis took deep roots. Subsequent work has focused on whether specific problems are decidable or not. Another direction of research has been to make finer distinctions among unsolvable problems by introducing degrees of unsolvability. Recursive function theory and lambda calculus also led to the development of functional programming languages such as Lisp, Scheme, Haskell, and ML. Computability theory is also closely related to logic, formal deductive systems, and complexity theory. Logic and deductive systems are of interest to philosophers and researchers in artificial intelligence, as well as to computation theorists. Although there are no journals devoted exclusively to computability, many theory journals (such as those listed at the end of Chapters 19 and 20) publish papers on this topic. In addition, *Annals of Pure and Applied Logic* publishes papers on logic and computability.

Defining Terms

Decision problem: A computational problem with a yes/no answer. Equivalently, a function whose range consists of two values $\{0, 1\}$.

Decidable problem: A decision problem that can be solved by a GOTO program that halts on all inputs in a finite number of steps. For emphasis, the equivalent term totally decidable problem is used. The associated language is called recursive.

Partially decidable problem: A decision problem that can be solved by a GOTO program that halts (and outputs ACCEPT) on all yes-instances. The program may or may not halt on no-instances. Equivalently, the collection of yes-instance strings forms a type-0 language (see Chapter 20).

Recursively enumerable language: Same as partially decidable language.

μ-recursive function: A function that is a basic function (Zero, Successor or Projection), or one that can be obtained from other μ-recursive functions using composition and μ-recursion.

Recursive language: A language that can be accepted by a GOTO program that halts on all inputs. The associated problem is called decidable.

Solvable problem: A computational problem that can be solved by a halting GOTO program. The problem may have a nonbinary output.

Totally undecidable problem: A problem that cannot be solved by a GOTO program. Equivalently, one for which the set of yes-instance strings is not a type-0 language.

Undecidable problem: A decision problem that is not (totally) decidable. It could be partially decidable or totally undecidable.

Universal Turing machine: A Turing machine that can simulate any other Turing machine.

Unsolvable problem: A computational problem that is not solvable. The associated function is called an uncomputable function.

References

1. Book, R. and Otto, F., *String Rewriting Systems,* Springer-Verlag, Berlin, Germany, 1993.
2. Chomsky, N., Three models for the description of language, *IRE Trans. Inform. Theory,* 2(2), 113–124, 1956.
3. Chomsky, N., Formal properties of grammars. In *Handbook of Mathematical Psycholgy,* Vol. 2. John Wiley & Sons, New York, 1963, pp. 323–418.
4. Davis, M., *Computability and Unsolvability,* McGraw-Hill, New York, 1958.
5. Davis, M., What is computation? In *Mathematics Today–Twelve Informal Essays,* Steen, L. (Ed.), Random House, Inc., New York, 1980, pp. 241–259.
6. Floyd, R.W. and Beigel, R., *The Language of Machines: An Introduction to Computability and Formal Languages,* Computer Science Press, New York, 1994.
7. Gurari, E., *An Introduction to the Theory of Computation,* Computer Science Press, Rockville, MD, 1989.
8. Harel, D., 1992. *Algorithmics: The Spirit of Computing,* Addison-Wesley, Reading, MA, 1992.
9. Harrison, M., *Introduction to Formal Language Theory,* Addison-Wesley, Reading, MA, 1978.
10. Hopcroft, J. and Ullman, J., *Introduction to Automata Theory, Languages and Computation.* Addison-Wesley, Reading, MA, 1979.
11. Jones, N.D., *Computability and Complexity from a Programming Perspective,* MIT Press, Cambridge, MA, 1997.
12. Meyer, A. and Ritchie, D., The complexity of LOOP programs. *Proceedings of the 22nd ACM National Conference,* 1967, pp. 465–469.
13. McNaughton, R., *Elementary Computability, Formal Languages and Automata,* ZB Publishing Industries, Lawrence, KS, 1993,
14. Minsky, M., *Computation: Finite and Infinite Machines,* Prentice Hall, Englewood Cliffs, NJ, 1967.
15. Robinson, Minsky's small universal Turing machine. *Int. J. Math.,* 2(5), 551–562, 1991.
16. Rogers, H., *Theory of Recursive Functions and Effective Computability,* MIT Press, Cambridge, MA, 1967.
17. Sipser, M., *Introduction to the Theory of Computation,* 1st edn., PWS, Boston, MA, 1996.

18. Smullyan, R., *What is the Name of this Book?* Prentice-Hall, Englewood Cliffs, NJ, 1978.
19. Smullyan, R., *Satan, Cantor and Infinity.* Alfred A. Knopf, New York, 1992.
20. Sudkamp, T., *Languages and Machines An Introduction to the Theory of Computer Science,* Addison-Wesley Longman, Reading, MA, 1997.
21. Turing, A., On computable numbers with an application to the Entscheidungsproblem. *Proc. London Math. Soc., Series 2,* 42, 230–265, 1936.
22. Wood, D., *Theory of Computation,* Harper and Row, New York, 1987.
23. Wang, H., Proving theorems by pattern recognition. *Bell Syst. Tech. J.,* 40, 1–42, 1961.

22

Complexity Classes

Eric Allender
Rutgers University

Michael C. Loui
University of Illinois at Urbana-Champaign

Kenneth W. Regan
State University of New York at Buffalo

22.1 Introduction

The purposes of complexity theory are to ascertain the amount of computational resources required to solve important computational problems, and to classify problems according to their difficulty. The resource most often discussed is computational time, although memory (space) and circuitry (or hardware) have also been studied. The main challenge of the theory is to prove lower bounds, i.e., that certain problems cannot be solved without expending large amounts of resources. Although it is easy to prove that inherently difficult problems exist, it has turned out to be much more difficult to prove that any *interesting* problems are hard to solve. There has been much more success in providing strong evidence of intractability, based on plausible, widely held conjectures. In both cases, the mathematical arguments of intractability rely on the notions of *reducibility* and *completeness*—which are the topics of Chapter 23. Before one can understand reducibility and completeness, however, one must grasp the notion of a complexity class—and that is the topic of this chapter.

First, however, we want to demonstrate that complexity theory really can prove—to even the most skeptical practitioner—that it is hopeless to try to build hardware or programs that solve certain problems. As our example, we consider the manufacture and testing of logic circuits and communication protocols. Many problems in these domains are solved by building a logical formula over a certain vocabulary, and then determining whether the formula is logically valid, or whether counterexamples (i.e., bugs) exist. The choice of vocabulary for the logic is important here, as the next paragraph illustrates.

One particular logic that was studied by Stockmeyer and Meyer [42] is called WS1S. (We need not be concerned with any details of this logic.) They showed that any circuit that takes as input a formula with up to 610 symbols and produces as output a correct answer saying whether the formula is valid, requires at least 10^{125} gates. The 610 symbols can be encoded within 458 bytes, i.e., fewer than 6 lines of ASCII text. According to Stockmeyer [43], "Even if gates were the size of a proton and were connected by infinitely thin wires, the network would densely fill the known universe."

Of course, Stockmeyer's theorem holds for one particular sort of circuitry, but the awesome size of the lower bound makes it evident that, no matter how innovative the architecture, no matter how clever the software, no computational machinery will enable us to solve the validity problem in this logic. For the practitioner testing validity of logical formulas, the lessons are (1) be careful with the choice of the logic, (2) use small formulas, and often (3) be satisfied with something less than full validity testing.

In contrast to this result of Stockmeyer, most lower bounds in complexity theory are stated asymptotically. For example, one might show that a particular problem requires time $\Omega(t(n))$ to solve on a Turing machine, for some rapidly growing function t. For the Turing machine model, no other type of lower bound is possible, because Turing machines have the linear-speed-up property (see Theorem 19.5). This property makes Turing machines mathematically convenient to work with, since constant factors become irrelevant, but it has the by-product—which some find disturbing— that for any n there is a Turing machine that handles inputs of length n in just n steps by looking up answers in a big table. Nonetheless, these asymptotic lower bounds essentially always can be translated into concrete lower bounds on, say, the number of components of a particular technology, or the number of clock cycles on a particular vendor's machine, that are required to compute a given function on a certain input size.*

Sadly, to date, few general complexity–theoretic lower bounds are known that are interesting enough to translate into concrete lower bounds in this sense. Even worse, for the vast majority of important problems that are believed to be difficult, no nontrivial lower bound on complexity is known today. Instead, complexity theory has contributed (1) a way of dividing the computational world up into complexity classes, and (2) evidence suggesting that these complexity classes are probably distinct. If this evidence can be replaced by mathematical proof, then we will have an abundance of interesting lower bounds.

22.1.1 What Is a Complexity Class?

Typically, a complexity class is defined by (1) a model of computation, (2) a resource (or collection of resources), and (3) a function known as the complexity bound for each resource.

The models used to define complexity classes fall into two main categories: (1) machine-based models, and (2) circuit-based models. Turing machines (TMs) and random-access machines (RAMs) are the two principal families of machine models; they were described in Chapter 19. We describe circuit-based models later, in Section 22.3. Other kinds of (Turing) machines were also introduced in Chapter 19, including deterministic, nondeterministic, alternating, and oracle machines.

When we wish to model real computations, deterministic machines and circuits are our closest links to reality. Then why consider the other kinds of machines? There are two main reasons.

The most potent reason comes from the computational problems whose complexity we are trying to understand. The most notorious examples are the hundreds of natural NP-complete

---

* The skeptical practitioner can still argue that these lower bounds hold only for the worst-case behavior of an algorithm, and that these bounds are irrelevant if the worst case arises very rarely in practice. There is a complexity theory of problems that are hard on average (as a counterpoint to the average case analysis of algorithms considered in Chapter 11), but to date only a small number of natural problems have been shown to be hard in this sense, and this theory is beyond the scope of this volume. See Section 22.5 at the end of this chapter.

problems [15]. To the extent that we understand anything about the complexity of these problems, it is because of the model of *nondeterministic* Turing machines. Nondeterministic machines do not model physical computation devices, but they do model real computational problems. There are many other examples where a particular model of computation has been introduced in order to capture some well-known computational problem in a complexity class. This phenomenon is discussed at greater length in Chapter 24.

The second reason is related to the first. Our desire to understand real computational problems has forced upon us a repertoire of models of computation and resource bounds. In order to understand the relationships between these models and bounds, we combine and mix them and attempt to discover their relative power. Consider, for example, nondeterminism. By considering the complements of languages accepted by nondeterministic machines, researchers were naturally led to the notion of alternating machines. When alternating machines and deterministic machines were compared, a surprising virtual identity of deterministic space and alternating time emerged. Subsequently, alternation was found to be a useful way to model efficient parallel computation. (See Sections 22.2.8 and 22.3.4) This phenomenon, whereby models of computation are generalized and modified in order to clarify their relative complexity, has occurred often through the brief history of complexity theory, and has generated some of the most important new insights.

Other underlying principles in complexity theory emerge from the major theorems showing relationships between complexity classes. These theorems fall into two broad categories. **Simulation theorems** show that computations in one class can be simulated by computations that meet the defining resource bounds of another class. The containment of nondeterministic logarithmic space (NL) in polynomial time (P), and the equality of the class P with alternating logarithmic space, are simulation theorems. **Separation theorems** show that certain complexity classes are distinct. Complexity theory currently has precious few of these. The main tool used in those separation theorems we have is called **diagonalization**. We illustrate this tool by giving proofs of some separation theorems in this chapter. In the next chapter, however, we show some apparently severe limitations of this tool. This ties in to the general feeling in computer science that **lower bounds are hard to prove**. Our current inability to separate many complexity classes from each other is perhaps the greatest challenge posed by computational complexity theory.

22.2 Time and Space Complexity Classes

We begin by emphasizing the fundamental resources of time and space for deterministic and nondeterministic Turing machines. We concentrate on resource bounds between logarithmic and exponential, because those bounds have proved to be the most useful for understanding problems that arise in practice.

Time complexity and space complexity were defined in Definition 19.4. We repeat Definition 19.5 of that chapter to define the following fundamental time classes and fundamental space classes, given functions $t(n)$ and $s(n)$:

- DTIME[$t(n)$] is the class of languages decided by deterministic Turing machines of time complexity $t(n)$.
- NTIME[$t(n)$] is the class of languages decided by nondeterministic Turing machines of time complexity $t(n)$.
- DSPACE[$s(n)$] is the class of languages decided by deterministic Turing machines of space complexity $s(n)$.
- NSPACE[$s(n)$] is the class of languages decided by nondeterministic Turing machines of space complexity $s(n)$.

We sometimes abbreviate DTIME[$t(n)$] to DTIME[t] (and so on) when t is understood to be a function, and when no reference is made to the input length n.

22.2.1 Canonical Complexity Classes

The following are the canonical complexity classes:

- L $=$ DSPACE[$\log n$] (deterministic log space)
- NL $=$ NSPACE[$\log n$] (nondeterministic log space)
- P $=$ DTIME[$n^{O(1)}$] $= \bigcup_{k \geq 1}$ DTIME[n^k] (polynomial time)
- NP $=$ NTIME[$n^{O(1)}$] $= \bigcup_{k \geq 1}$ NTIME[n^k] (nondeterministic polynomial time)
- PSPACE $=$ DSPACE[$n^{O(1)}$] $= \bigcup_{k \geq 1}$ DSPACE[n^k] (polynomial space)
- E $=$ DTIME[$2^{O(n)}$] $= \bigcup_{k \geq 1}$ DTIME[k^n]
- NE $=$ NTIME[$2^{O(n)}$] $= \bigcup_{k \geq 1}$ NTIME[k^n]
- EXP $=$ DTIME[$2^{n^{O(1)}}$] $= \bigcup_{k \geq 1}$ DTIME[2^{n^k}] (deterministic exponential time)
- NEXP $=$ NTIME[$2^{n^{O(1)}}$] $= \bigcup_{k \geq 1}$ NTIME[2^{n^k}] (nondeterministic exponential time)
- EXPSPACE $=$ DSPACE[$2^{n^{O(1)}}$] $= \bigcup_{k \geq 1}$ DSPACE[2^{n^k}] (exponential space)

It is worth emphasizing that NP does *not* stand for "non-polynomial time"; the class P is a subclass of NP, and NP $\neq$ P is not known.

The space classes PSPACE and EXPSPACE are defined in terms of the DSPACE complexity measure. By Savitch's Theorem (see Theorem 22.3) the NSPACE measure with polynomial bounds also yields PSPACE, and with $2^{n^{O(1)}}$ bounds yields EXPSPACE.

22.2.2 Why Focus on These Classes?

The class P contains many familiar problems that can be solved efficiently, such as finding shortest paths in networks, parsing context-free grammars, sorting, matrix multiplication, and linear programming. By definition, in fact, P contains all problems that can be solved by (deterministic) programs of reasonable worst-case time complexity.

But P also contains problems whose best algorithms have time complexity $n^{10^{500}}$. It seems ridiculous to say that such problems are computationally feasible. Nevertheless, there are four important reasons to include these problems:

1. For the main goal of proving lower bounds, it is sensible to have an overly generous notion of the class of feasible problems. That is, if we show that a problem is not in P, then we have shown in a very strong way that solution via deterministic algorithms is impractical.

2. The theory of complexity-bounded reducibility (Chapter 23) is predicated on the simple notion that if functions f and g are both easy to compute, then the composition of f and g should also be easy to compute. If we want to allow algorithms of time complexity n^2 to be considered feasible (and certainly many algorithms of this complexity are used daily), then we are immediately led to regard running times $n^4, n^8, \ldots$ as also being feasible. Put another way, the choice is either to lay down an arbitrary and artificial limit on feasibility (and to forgo the desired property that the composition of easy functions be easy), or to go with the natural and overly generous notion given by P.

3. Polynomial time has served well as the intellectual boundary between feasible and infeasible problems. Empirically, problems of time complexity $n^{10^{500}}$ do not arise in practice, while problems of time complexity $O(n^4)$, and those proved or believed to be $\Omega(2^n)$, occur often. Moreover, once a polynomial-time algorithm for a problem is found, the foot is in the door, and an armada of mathematical and algorithmic techniques can be used to improve the algorithm. Linear programming may be the best known example. The breakthrough $O(n^8)$ time algorithm of Khachiyan [30], for $n \times n$ instances, was impractical in itself, but it prompted an innovation by Karmarkar [28] that produced an algorithm whose running time of about $O(n^3)$ on all cases competes well commercially with the simplex method, which runs in $O(n^3)$ time in most cases but takes 2^n time in some. Of course, if it should turn out that the Hamiltonian circuit problem (or some other natural and interesting NP-complete problem) has complexity $n^{10^{500}}$, then the theory would need to be overhauled. For the time being, this seems unlikely.

4. We would like our fundamental notions to be independent of arbitrary choices we have made in formalizing our definitions. There is much that is arbitrary and historically accidental in the prevalent choice of the Turing machine as the standard model of computation. This choice does not affect the class P itself, however, because the natural notions of polynomial time for essentially all models of sequential computation that have been devised yield the same class. The random-access and pointer machine models described in Section 19.4 can be simulated by Turing machines with at most a cubic increase in time. Many feel that our "true" experience of running time on real sequential computers falls midway between Turing machines and these more-powerful models, but this only bolsters our conviction that the class P gives the "true" notion of polynomial time.

By analogy to the famous *Church–Turing thesis* (see Section 21.4), which states that the definition of a (partial) recursive function captures the intuitive notion of a computable process, several authorities have proposed the following:

Polynomial-Time Church-Turing Thesis   The class P captures the true notion of those problems that are computable in polynomial time by sequential machines, and is the same for any physically relevant model and minimally reasonable measure of overall computational effort that will ever be devised.

This thesis is not universally accepted. Many people consider probabilistic and quantum models of computation to be "physically relevant." These models and their possibly-greater capabilities are discussed in more detail in Chapter 24. Whether the quantum models' time measures adequately reflect the physical effort needed to realize quantum computation is also controversial. This thesis extends also to parallel computing models if the technologically important notion of parallel *work* (see Chapter 28) is taken as the measure of effort.

Another way in which the concept of P is robust is that P is characterized by many concepts from logic and mathematics that do not mention machines or time. Some of these characterizations are surveyed in Chapter 24.

The class NP can also be defined by means other than nondeterministic Turing machines. NP equals the class of problems whose solutions can be *verified* quickly, by deterministic machines in polynomial time. Equivalently, NP comprises those languages whose membership proofs can be checked quickly.

For example, one language in NP is the set of composite numbers, written in binary. A proof that a number z is composite can consist of two factors $z_1 \geq 2$ and $z_2 \geq 2$ whose product $z_1 z_2$ equals z. This proof is quick to check if z_1 and z_2 are given, or guessed. Correspondingly, one can design a nondeterministic Turing machine N that on input z branches to write down "guesses" for z_1 and z_2, and then deterministically multiplies them to test whether $z_1 z_2 = z$. Then $L(N)$, the language accepted by N, equals the set of composite numbers, since there exists an accepting computation path if and only if z really is composite. Note that N does not really solve the problem—it just checks the candidate solution proposed by each branch of the computation.

Another important language in NP is the set of satisfiable Boolean formulas, called *SAT*. A Boolean formula ϕ is satisfiable if there exists a way of assigning `true` or `false` to each variable so that the resulting value of ϕ is `true`. For example, the formula $x \wedge (\overline{x} \vee y)$ is satisfiable, but $x \wedge \overline{y} \wedge (\overline{x} \vee y)$ is not satisfiable. A nondeterministic Turing machine N, after checking the syntax of ϕ and counting the number n of variables, can nondeterministically write down an n-bit 0-1 string a on its tape, and then deterministically (and easily) evaluate ϕ for the truth assignment denoted by a. The computation path corresponding to each individual a accepts if and only if $\phi(a) = $ `true`, and so N itself accepts ϕ if and only if ϕ is satisfiable; i.e., $L(N) = SAT$. Again, this *checking* of given assignments differs significantly from trying to *find* an accepting assignment.

The above characterization of NP as the set of problems with easily verified solutions is formalized as follows: $A \in$ NP if and only if there exist a language $A' \in$ P and a polynomial p such that for every x, $x \in A$ if and only if there exists a y such that $|y| \leq p(|x|)$ and $(x, y) \in A'$. Here, whenever x belongs to A, y is interpreted as a positive solution to the problem represented by x, or equivalently, as a proof that x belongs to A. The difference between P and NP is that between solving and checking, or between finding a proof of a mathematical theorem and testing whether a candidate proof is correct. In essence, NP represents all sets of theorems with proofs that are short (i.e., of polynomial length), while P represents those statements that can proved or refuted quickly from scratch.

Having an intuitive NP-style characterization does not make a problem hard. Indeed, the set of composite numbers is now known to belong to P, owing to the efficient deterministic primality test of Agrawal et al. [1]! NP-completeness, which is treated in depth in the next chapter, is the requisite notion of hardness, for languages such as SAT. This notion, together with the many known NP-complete problems, is perhaps the best justification for interest in the classes P and NP. All of the other canonical complexity classes listed above have natural and important problems that are complete for them—under various kinds of reducibility, the next chapter's other main theme.

Further motivation for studying L, NL, and PSPACE, comes from their relationships to P and NP. Namely, L and NL are the largest space-bounded classes known to be contained in P, and PSPACE is the smallest space-bounded class known to contain NP. Similarly, EXP is of interest primarily because it is the smallest deterministic time class known to contain NP. The closely-related class E is not known to contain NP. We will see in Section 22.2.7 an important reason for interest in E, and another reason is mentioned toward the end of Section 24.3.

22.2.3 Constructibility

Before we go further, we need to introduce the notion of *constructibility*. Without attending to it, no meaningful theory of complexity is possible.

The most basic theorem that one should expect from complexity theory would say, "If you have more resources, you can do more." Unfortunately, if we aren't careful with our definitions, then this claim is false:

THEOREM 22.1 **(Gap Theorem).** *There is a computable, strictly increasing time bound* $t(n)$ *such that* $\text{DTIME}[t(n)] = \text{DTIME}[2^{2^{t(n)}}]$.

That is, there is an empty gap between time $t(n)$ and time doubly exponentially greater than $t(n)$, in the sense that anything that can be computed in the larger time bound can already be computed in the smaller time bound. That is, even with much more time, you can't compute more. This gap can be made much larger than doubly exponential; for any computable r, there is a computable time bound t such that $\text{DTIME}[t(n)] = \text{DTIME}[r(t(n))]$. Exactly analogous statements hold for the $\text{NTIME}, \text{DSPACE}$, and NSPACE measures.

Fortunately, the gap phenomenon cannot happen for time bounds t that anyone would ever be interested in. Indeed, the proof of the Gap Theorem proceeds by showing that one can define a time bound t such that no machine has a running time that is between $t(n)$ and $2^{2^{t(n)}}$. This theorem indicates the need for formulating only those time bounds that actually describe the complexity of some machine.

A function $t(n)$ is **time-constructible** if there exists a deterministic Turing machine that halts after exactly $t(n)$ steps for every input of length n. A function $s(n)$ is **space-constructible** if there exists a deterministic Turing machine that uses exactly $s(n)$ worktape cells for every input of length n. (Most authors consider only functions $t(n) \geq n + 1$ to be time-constructible, and many limit attention to $s(n) \geq \log n$ for space bounds. There do exist sublogarithmic space-constructible functions, but we prefer to avoid the tricky theory of $o(\log n)$ space bounds.)

For example, $t(n) = n + 1$ is time-constructible. Furthermore, if $t_1(n)$ and $t_2(n)$ are time-constructible, then so are the functions $t_1 + t_2, t_1 t_2, t_1^{t_2}$, and c^{t_1} for every integer $c > 1$. Consequently, if $p(n)$ is a polynomial, then $p(n) = \Theta(t(n))$ for some time-constructible polynomial function $t(n)$. Similarly, $s(n) = \log n$ is space-constructible, and if $s_1(n)$ and $s_2(n)$ are space-constructible, then so are the functions $s_1 + s_2, s_1 s_2, s_1^{s_2}$, and c^{s_1} for every integer $c > 1$. Many common functions are space-constructible: e.g., $n \log n, n^3, 2^n, n!$.

Constructibility helps eliminate an arbitrary choice in the definition of the basic time and space classes. For general time functions t, the classes $\text{DTIME}[t]$ and $\text{NTIME}[t]$ may vary depending on whether machines are required to halt within t steps on all computation paths, or just on those paths that accept. If t is time-constructible and s is space-constructible, however, then $\text{DTIME}[t]$, $\text{NTIME}[t], \text{DSPACE}[s]$, and $\text{NSPACE}[s]$ can be defined without loss of generality in terms of Turing machines that always halt.

As a general rule, any function $t(n) \geq n + 1$ and any function $s(n) \geq \log n$ that one is interested in as a time or space bound, is time- or space-constructible, respectively. As we have seen, little of interest can be proved without restricting attention to constructible functions. This restriction still leaves a rich class of resource bounds.

The Gap Theorem is not the only case where intuitions about complexity are false. Most people also expect that a goal of algorithm design should be to arrive at an optimal algorithm for a given problem. In some cases, however, no algorithm is remotely close to optimal.

THEOREM 22.2   (Speed-Up Theorem). *There is a decidable language A such that for every machine M that decides A, with running time $u(n)$, there is another machine M' that decides A much faster: its running time $t(n)$ satisfies $2^{2^{t(n)}} \leq u(n)$, for all but finitely many n.*

This statement, too, holds with any computable function $r(t)$ in place of 2^{2^t}. Put intuitively, the program M' running on an old IBM PC is better than the program M running on the fastest hardware to date. Hence A has no best algorithm, and no well-defined time-complexity function. Unlike the case of the Gap Theorem, the speed-up phenomenon may hold for languages and time bounds of interest. For instance, a problem of time complexity bounded by $t(n) = n^{\log n}$, which is just above polynomial time, may have arbitrary polynomial speed-up—i.e., may have algorithms of time complexity $t(n)^{1/k}$ for all $k > 0$.

One implication of the Speed-Up Theorem is that the complexities of some problems need to be sandwiched between upper and lower bounds. Actually, there is a sense in which every

problem has a well defined lower bound on time. For every language A there is a
computable function t_0 such that for every time-constructible function t, there is
some machine that accepts A within time t if and only if $t = \Omega(t_0)$ [31]. A catch,
however, is that t_0 itself may not be time-constructible.

22.2.4 Basic Relationships

Clearly, for all time functions $t(n)$ and space functions $s(n)$, $\mathrm{DTIME}[t(n)] \subseteq$
$\mathrm{NTIME}[t(n)]$ and $\mathrm{DSPACE}[s(n)] \subseteq \mathrm{NSPACE}[s(n)]$, because a deterministic machine
is a special case of a nondeterministic machine. Furthermore, $\mathrm{DTIME}[t(n)] \subseteq$
$\mathrm{DSPACE}[t(n)]$ and $\mathrm{NTIME}[t(n)] \subseteq \mathrm{NSPACE}[t(n)]$, because at each step, a k-tape
Turing machine can write on at most $k = O(1)$ previously unwritten cells. The next
theorem presents additional important relationships between classes.

THEOREM 22.3 *Let $t(n)$ be a time-constructible function, and let $s(n)$ be a space-constructible function, $s(n) \geq \log n$.*

    (a) $\mathrm{NTIME}[t(n)] \subseteq \mathrm{DTIME}[2^{O(t(n))}]$

    (b) $\mathrm{NSPACE}[s(n)] \subseteq \mathrm{DTIME}[2^{O(s(n))}]$

    (c) $\mathrm{NTIME}[t(n)] \subseteq \mathrm{DSPACE}[t(n)]$

    (d) **(Savitch's Theorem)** $\mathrm{NSPACE}[s(n)] \subseteq \mathrm{DSPACE}[s(n)^2]$

As a consequence of the first part of this theorem, $\mathrm{NP} \subseteq \mathrm{EXP}$. No better general
upper bound on deterministic time is known for languages in NP, however. See
Figure 22.1 for other known inclusion relationships between canonical complexity
classes. (The classes AC^0, TC^0, and NC^1 are defined in Section 22.3.4.)

Although we do not know whether allowing nondeterminism strictly increases the class of
languages decided in polynomial time, Savitch's Theorem says that for space classes, nondeterminism does not help by more than a polynomial amount.

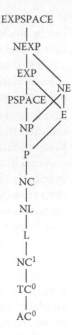

FIGURE 22.1
Inclusion relationships among the canonical complexity classes.

22.2.5 Complementation

For a language A over an alphabet Σ, define $\overline{A}$ to be the **complement** of A in the set of words over Σ:
$\overline{A} = \Sigma^* - A$. For a class of languages $\mathcal{C}$, define $\mathrm{co}\text{-}\mathcal{C} = \{\overline{A} : A \in \mathcal{C}\}$. If $\mathcal{C} = \mathrm{co}\text{-}\mathcal{C}$, then $\mathcal{C}$ is **closed**
under complementation.

In particular, $\mathrm{co}\text{-}\mathrm{NP}$ is the class of languages that are complements of languages in NP. For the
language SAT of satisfiable Boolean formulas, $\overline{\mathrm{SAT}}$ is the set of unsatisfiable formulas, whose value
is `false` for every truth assignment, together with the syntactically incorrect formulas. A closely
related language in $\mathrm{co}\text{-}\mathrm{NP}$ is the set of Boolean tautologies, namely, those formulas whose value
is `true` for every truth assignment. The question of whether NP equals $\mathrm{co}\text{-}\mathrm{NP}$ comes down to
whether every tautology has a short (i.e., polynomial-sized) proof. The only obvious general way to
prove a tautology ϕ in m variables is to verify all 2^m rows of the truth table for ϕ, and this proof has
exponential size. Most complexity theorists believe that there is no general way to reduce this proof
size to a polynomial, hence that $\mathrm{NP} \neq \mathrm{co}\text{-}\mathrm{NP}$.

Questions about complementation bear directly on the P vs. NP question. It is easy to show that P
is closed under complementation (see Theorem 22.4). Consequently, if $\mathrm{NP} \neq \mathrm{co}\text{-}\mathrm{NP}$, then $\mathrm{P} \neq \mathrm{NP}$.

THEOREM 22.4 (**Complementation Theorems**). *Let t be a time-constructible function, and let s be a space-constructible function, with $s(n) \geq \log n$ for all n. Then*

(a) DTIME[t] *is closed under complementation.*

(b) DSPACE[s] *is closed under complementation.*

(c) (**Immerman–Szelepcsényi Theorem**) NSPACE[s] *is closed under complementation.*

The Complementation Theorems are used to prove the Hierarchy Theorems (Theorem 22.5) in Section 22.2.6.

22.2.6 Hierarchy Theorems and Diagonalization

Diagonalization is the most useful technique for proving the existence of computationally difficult problems. In this section, we will see examples of two rather different types of arguments, both of which can be called "diagonalization," and we will see how these are used to prove hierarchy theorems in complexity theory.

A hierarchy theorem is a theorem that says "If you have more resources, you can compute more." As we saw in Section 22.2.3, this theorem is possible only if we restrict attention to constructible time and space bounds. Next, we state hierarchy theorems for deterministic and nondeterministic time and space classes. In the following, $\subset$ denotes strict inclusion between complexity classes.

THEOREM 22.5 (**Hierarchy Theorems**). *Let t_1 and t_2 be time-constructible functions, and let s_1 and s_2 be space-constructible functions, with $s_1(n), s_2(n) \geq \log n$ for all n.*

(a) *If $t_1(n) \log t_1(n) = o(t_2(n))$, then* DTIME[$t_1$] $\subset$ DTIME[t_2].

(b) *If $t_1(n + 1) = o(t_2(n))$, then* NTIME[t_1] $\subset$ NTIME[t_2].

(c) *If $s_1(n) = o(s_2(n))$, then* DSPACE[s_1] $\subset$ DSPACE[s_2].

(d) *If $s_1(n) = o(s_2(n))$, then* NSPACE[s_1] $\subset$ NSPACE[s_2].

As a corollary of the Hierarchy Theorem for DTIME,

$$\text{P} \subseteq \text{DTIME}[n^{\log n}] \subset \text{DTIME}[2^n] \subseteq \text{E};$$

hence we have the strict inclusion P $\subset$ E. Although we do not know whether P $\subset$ NP, there exists a problem in E that cannot be solved in polynomial time. Other consequences of the Hierarchy Theorems are NE $\subset$ NEXP and NL $\subset$ PSPACE.

In the Hierarchy Theorem for DTIME, the hypothesis on t_1 and t_2 is $t_1(n) \log t_1(n) = o(t_2(n))$, instead of $t_1(n) = o(t_2(n))$, for technical reasons related to the simulation of machines with multiple worktapes by a single universal Turing machine with a fixed number of worktapes. Other computational models, such as random access machines, enjoy tighter time hierarchy theorems.

All proofs of the Hierarchy Theorems use the technique of **diagonalization**. For example, the proof for DTIME constructs a Turing machine M of time complexity t_2 that considers all machines $M_1, M_2, \ldots$ whose time complexity is t_1; for each i, the proof finds a word x_i that is accepted by M if and only if $x_i \notin L(M_i)$, the language decided by M_i. Consequently, $L(M)$, the language decided by M, differs from each $L(M_i)$, hence $L(M) \notin$ DTIME[t_1]. The diagonalization technique resembles the classic method used to prove that the real numbers are uncountable, by constructing a number whose jth digit differs from the jth digit of the jth number on the list. To illustrate the diagonalization technique, we outline proofs of the Hierarchy Theorems for DSPACE and for NTIME. In this subsection, $\langle i, x \rangle$ stands for the string $0^i 1x$, and *zeroes*(y) stands for the number of 0's that a given string y starts with. Note that *zeroes*$(\langle i, x \rangle) = i$.

PROOF   (of the DSPACE Hierarchy Theorem) We construct a deterministic Turing machine M that decides a language A such that $A \in \text{DSPACE}[s_2] - \text{DSPACE}[s_1]$.

Let U be a deterministic universal Turing machine, as described in Section 21.2. On input x of length n, machine M performs the following:

1. Using the space-constructibility of s_2, lay out $s_2(n)$ cells on a worktape.

2. Let $i = zeroes(x)$.

3. Simulate the universal machine U on input $\langle i, x \rangle$. Accept x if U tries to use more than s_2 worktape cells. (We omit some technical details, such as interleaving multiple worktapes onto the fixed number of worktapes of M, and the way in which the constructibility of s_2 is used to ensure that this process halts.)

4. If U accepts $\langle i, x \rangle$, then reject; if U rejects $\langle i, x \rangle$, then accept.

Clearly, M always halts and uses space $O(s_2(n))$. Let $A = L(M)$.

Suppose $A \in \text{DSPACE}[s_1(n)]$. Then there is some Turing machine M_j accepting A using space at most $s_1(n)$. The universal Turing machine U can easily be given the property that its space needed to simulate a given Turing machine M_j is at most a constant factor higher than the space used by M_j itself. More precisely, there is a constant k depending only on j (in fact, we can take $k = |j|$), such that U, on inputs z of the form $z = \langle j, x \rangle$, uses at most $k s_1(|x|)$ space.

Since $s_1(n) = o(s_2(n))$, there is an n_0 such that $k s_1(n) \leq s_2(n)$ for all $n \geq n_0$. Let x be a string of length greater than n_0 such that the first $j + 1$ symbols of x are $0^j 1$. Note that the universal Turing machine U, on input $\langle j, x \rangle$, simulates M_j on input x and uses space at most $k s_1(n) \leq s_2(n)$. Thus, when we consider the machine M defining A, we see that on input x the simulation does not stop in Step 3, but continues on to Step 4, and thus $x \in A$ if and only if U rejects $\langle j, x \rangle$. Consequently, M_j does not accept A, contrary to our assumption. Thus $A \notin \text{DSPACE}[s_1(n)]$.   □

A more sophisticated argument is required to prove the Hierarchy Theorem for NTIME. To see why, note that it is necessary to diagonalize against nondeterministic machines, and thus it is necessary to use a nondeterministic universal Turing machine as well. In the deterministic case, when we simulated an accepting computation of the universal machine, we would reject, and if we simulated a rejecting computation of the universal machine, we would accept. That is, we would do exactly the opposite of what the universal machine does, in order to "fool" each simulated machine M_i. If the machines under consideration are nondeterministic, then M_i can have both an accepting path and a rejecting path on input x, in which case the universal nondeterministic machine would accept input $\langle i, x \rangle$. If we simulate the universal machine on an input and accept upon reaching a rejecting leaf and reject if upon reaching an accepting leaf, then this simulation would still accept (because the simulation that follows the rejecting path now accepts). Thus, we would fail to do the opposite of what M_i does.

The following careful argument guarantees that each machine M_i is fooled on some input. It draws on a result of Book et al. [6] that every language in NTIME$[t(n)]$ is accepted by a two-tape nondeterministic Turing machine that runs in time $t(n)$.

PROOF   (of the NTIME Hierarchy Theorem) Let $M_1, M_2, \ldots$ be an enumeration of two-tape nondeterministic Turing machines running in time $t_1(n)$. Let f be a rapidly growing function such that time $f(i, n, s)$ is enough time for a deterministic machine to compute the function

$$(i, n, s) \mapsto \begin{cases} 1 & \text{if } M_i \text{ accepts } 1^n \text{ in } \leq s \text{ steps} \\ 0 & \text{otherwise} \end{cases}$$

Letting $f(i, n, s)$ be greater than $2^{2^{i+n+s}}$ is sufficient.

Now divide Σ^* into regions, so that in region $j = \langle i, y \rangle$, we try to "fool" machine M_i. Note that each M_i is considered infinitely often. The regions are defined by functions $start(j)$ and $end(j)$, defined as follows: $start(1) = 1$, $start(j+1) = end(j) + 1$, where taking $i = zeroes(j)$, we have $end(j) = f(i, start(j), t_2(start(j)))$. The important point is that, on input $1^{end(j)}$, a deterministic machine can, in time $t_2(end(j))$, determine whether M_i accepts $1^{start(j)}$ in at most $t_2(start(j))$ steps.

By picking f appropriately easy to invert, we can guarantee that, on input 1^n, we can in time $t_2(n)$ determine which region j contains n.

Now it is easy to verify that the following routine can be performed in time $t_2(n)$ by a nondeterministic machine. (In the pseudo-code below, U is a "universal" nondeterministic machine with four tapes, which is therefore able to simulate one step of machine M_i in $O(i^3)$ steps.)

1. On input 1^n, determine which region j contains n. Let $j = \langle i, y \rangle$.

2. If $n = end(j)$, then accept if and only if M_i does *not* accept $1^{start(j)}$ within $t_2(start(j))$ steps.

3. Otherwise, accept if and only if U accepts $\langle i, 1^{n+1} \rangle$ within $t_2(n)$ steps. (Here, it is important that we are talking about $t_2(n)$ steps of U, which may be only about $t_2(n)/i^3$ steps of M_i.)

Let us call the language accepted by this procedure A. Clearly $A \in \text{NTIME}[t_2(n)]$. We now claim that $A \notin \text{NTIME}[t_1(n)]$.

Assume otherwise, and let M_i be the nondeterministic machine accepting A in time $t_1(n)$. Recall that M_i has only two tapes. Let c be a constant such that $i^3 t_1(n+1) < t_2(n)$ for all $n \geq c$. Let y be a string such that $|y| \geq c$, and consider stage $j = \langle i, y \rangle$. Then for all n such that $start(j) \leq n < end(j)$, we have $1^n \in A$ if and only if $1^{n+1} \in A$. However this contradicts the fact that $1^{start(j)} \in A$ if and only if $1^{end(j)} \notin A$. $\qquad \square$

Although the diagonalization technique successfully separates some pairs of complexity classes, diagonalization does not seem strong enough to separate P from NP. (See Theorem 24.25.)

22.2.7 Padding Arguments

A useful technique for establishing relationships between complexity classes is the *padding argument*. Let A be a language over alphabet Σ, and let # be a symbol not in Σ. Let f be a numeric function. The f-*padded version* of A is the language

$$A' = \left\{ x \#^{f(n)} : x \in A \text{ and } n = |x| \right\}$$

That is, each word of A' is a word in A concatenated with $f(n)$ consecutive # symbols. The padded version A' has the same information content as A, but because each word is longer, the computational complexity of A' is lower!

The proof of Theorem 22.6 illustrates the use of a padding argument.

THEOREM 22.6   *If* P = NP, *then* E = NE.

PROOF   Since E $\subseteq$ NE, we prove that NE $\subseteq$ E.

Let $A \in$ NE be decided by a nondeterministic Turing machine M in at most $t(n) = k^n$ time for some constant integer k. Let A' be the $t(n)$-padded version of A. From M, we construct a nondeterministic Turing machine M' that decides A' in linear time: M' checks that its input has the correct format, using the time-constructibility of t; then M' runs M on the prefix of the input preceding the first # symbol. Thus, $A' \in$ NP.

If P $=$ NP, then there is a deterministic Turing machine D' that decides A' in at most $p'(n)$ time for some polynomial p'. From D', we construct a deterministic Turing machine D that decides A, as follows. On input x of length n, since $t(n)$ is time-constructible, machine D constructs $x\#^{t(n)}$, whose length is $n + t(n)$, in $O(t(n))$ time. Then D runs D' on this input word. The time complexity of D is at most $O(t(n)) + p'(n + t(n)) = 2^{O(n)}$. Therefore, NE $\subseteq$ E. $\square$

A similar argument shows that the E $=$ NE question is equivalent to the question of whether NP $-$ P contains a subset of 1^*, that is, a language over a single-letter alphabet.

Padding arguments sometimes can be used to give tighter hierarchies than can be obtained by straightforward diagonalization. For instance, Theorem 22.5 leaves open the question of whether, say, DTIME$[n^3 \log^{1/2} n] =$ DTIME$[n^3]$. We can show that these classes are not equal, by using a padding argument. We will need the following lemma, whose proof is similar to that of Theorem 22.6.

LEMMA 22.1 [Translational Lemma] Let t_1, t_2, and f be time-constructible functions. If we have DTIME$[t_1(n)] =$ DTIME$[t_2(n)]$, then DTIME$[t_1(f(n))] =$ DTIME$[t_2(f(n))]$.

THEOREM 22.7 *For any real number $a > 0$ and integer $k \geq 1$, DTIME$[n^k] \subset$ DTIME$[n^k \log^a n]$.*

PROOF   Suppose for contradiction that DTIME$[n^k] =$ DTIME$[n^k \log^a n]$. For now let us also suppose that $a > 1/2$. Taking $f(n) = 2^{n/k}$, and using the linear speed-up property, we obtain from the Translational Lemma the identity DTIME$[2^n n^a] =$ DTIME$[2^n]$. This does not yet give the desired contradiction to the DTIME Hierarchy Theorem—but it is close. We will need to use the Translational Lemma twice more.

Assume that DTIME$[2^n n^a] =$ DTIME$[2^n]$. Using the Translational Lemma with $f(n) = 2^n$ yields DTIME$[2^{2^n} 2^{an}] =$ DTIME$[2^{2^n}]$. Applying the Lemma once again on the classes DTIME$[2^n n^a] =$ DTIME$[2^n]$, this time using $f(n) = 2^n + an$, we obtain DTIME$[2^{2^n} 2^{an} f(n)^a] =$ DTIME$[2^{2^n} 2^{an}]$. Combining these two equalities yields DTIME$[2^{2^n} 2^{an} f(n)^a] =$ DTIME$[2^{2^n}]$. Since $f(n)^a > 2^{an}$, we have that $2^{an} f(n)^a > 2^{2an} = 2^n 2^{bn}$ for some $b > 0$ (since $a > 1/2$). Thus DTIME$[2^{2^n} 2^n 2^{bn}] =$ DTIME$[2^{2^n}]$, and this contradicts the DTIME Hierarchy Theorem, since $2^{2^n} \log 2^{2^n} = o(2^{2^n} 2^n 2^{bn})$.

Finally, for any fixed $a > 0$, not just $a > 1/2$, we need to apply the Translational Lemma several more times. $\square$

One consequence of this theorem is that within P, there can be no "complexity gaps" of size $(\log n)^{\Omega(1)}$.

22.2.8   Alternating Complexity Classes

In this section, we define time and space complexity classes for alternating Turing machines, and we show how these classes are related to the classes introduced already. Alternating Turing machines and their configurations are defined in Chapter 19.

The possible computations of an alternating Turing machine M on an input word x can be represented by a tree T_x in which the root is the initial configuration, and the children of a non-terminal node C are the configurations reachable from C by one step of M. For a word x in $L(M)$, define an **accepting subtree** S of T_x as follows:

- S is finite.
- The root of S is the initial configuration with input word x.
- If S has an existential configuration C, then S has exactly one child of C in T_x; if S has a universal configuration C, then S has all children of C in T_x.
- Every leaf is a configuration whose state is the accepting state q_A.

Observe that each node in S is an accepting configuration.

We consider only alternating Turing machines that always halt. For $x \in L(M)$, define the time taken by M to be the height of the shortest accepting tree for x, and the space to be the maximum number of nonblank worktape cells among configurations in the accepting tree that minimizes this number. For $x \notin L(M)$, define the time to be the height of T_x, and the space to be the maximum number of nonblank worktape cells among configurations in T_x.

Let $t(n)$ be a time-constructible function, and let $s(n)$ be a space-constructible function. Define the following complexity classes:

- ATIME$[t(n)]$ is the class of languages decided by alternating Turing machines of time complexity $O(t(n))$.

- ASPACE$[s(n)]$ is the class of languages decided by alternating Turing machines of space complexity $O(s(n))$.

Because a nondeterministic Turing machine is a special case of an alternating Turing machine, for every $t(n)$ and $s(n)$, NTIME$(t) \subseteq$ ATIME(t) and NSPACE$(s) \subseteq$ ASPACE(s). The next theorem states further relationships between computational resources used by alternating Turing machines, and resources used by deterministic and nondeterministic Turing machines.

THEOREM 22.8 **(Alternation Theorems).** *Let $t(n)$ be a time-constructible function, and let $s(n)$ be a space-constructible function, $s(n) \geq \log n$.*

(a) NSPACE$[s(n)] \subseteq$ ATIME$[s(n)^2]$

(b) ATIME$[t(n)] \subseteq$ DSPACE$[t(n)]$

(c) ASPACE$[s(n)] \subseteq$ DTIME$[2^{O(s(n))}]$

(d) DTIME$[t(n)] \subseteq$ ASPACE$[\log t(n)]$

In other words, space on deterministic and nondeterministic Turing machines is polynomially related to time on alternating Turing machines. Space on alternating Turing machines is exponentially related to time on deterministic Turing machines. The following corollary is immediate.

THEOREM 22.9

(a) ASPACE$[O(\log n)] =$ P.

(b) ATIME$[n^{O(1)}] =$ PSPACE.

(c) ASPACE$[n^{O(1)}] =$ EXP.

Note that Theorem 22.8(a) says, for instance, that NL is contained in ATIME$(\log^2(n))$. For this to make sense, it is necessary to modify the definition of alternating Turing machines to allow them to read individual bits of the input in constant time, rather than requiring n time units to traverse the entire input tape. This has become the standard definition of alternating Turing machines, because it is useful in establishing relationships between Turing machine complexity and circuit complexity, as explained in the upcoming section.

22.3 Circuit Complexity

Up to now, this chapter has been concerned only with complexity classes that were defined in order to understand the nature of sequential computation. Although we called them "machines," the models discussed here and in Chapter 19 are closer in spirit to software, namely to sequential algorithms or to single-processor machine-language programs. Circuits were originally studied to model hardware. The hardware of electronic digital computers is based on digital gates, connected into combinational and sequential networks. Whereas a software program can branch and even modify itself while running, hardware components on today's typical machines are fixed and cannot reconfigure themselves. Also, circuits capture well the notion of nonbranching, straight-line computation.

Furthermore, circuits provide a good model of parallel computation. Many machine models, complexity measures, and classes for parallel computation have been devised, but the circuit complexity classes defined here coincide with most of them. Chapter 28 surveys parallel models and their relation to circuits in more detail.

22.3.1 Kinds of Circuits

A circuit can be formalized as a directed graph with some number n of sources, called *input nodes* and labeled $x_1, \ldots, x_n$, and one sink, called the *output node*. The edges of the graph are called *wires*. Every non-input node v is called a *gate*, and has an associated *gate function* g_v that takes as many arguments as there are wires coming into v. In this survey we limit attention to Boolean circuits, meaning that each argument is 0 or 1, although arithmetical circuits with numeric arguments and $+, *$ (etc.) gates have also been studied in complexity theory. Formally g_v is a function from $\{0,1\}^r$ to $\{0,1\}$, where r is the *fan-in* of v. The value of the gate is transmitted along each wire that goes out of v. The *size* of a circuit is the number of wires in it.

We restrict attention to circuits C in which the graph is acyclic, so that there is no "feedback." Then every Boolean assignment $x \in \{0,1\}^n$ of values to the input nodes determines a unique value for every gate and wire, and the value of the output gate is the output $C(x)$ of the circuit. The circuit accepts x if $C(x) = 1$.

The sequential view of a circuit is obtained by numbering the gates in a manner that respects the edge relation, meaning that for all edges (u, v), g_u has a lower number than g_v. Then the gate functions in that order become a sequence of basic instructions in a straight-line program that computes $C(x)$. The size of the circuit becomes the number of steps in the program. However, this view presumes a single processing unit that evaluates the instructions in sequence, and ignores information that the graphical layout provides. A more powerful view regards the gates as simple processing units that can act in parallel. Every gate whose incoming wires all come from input nodes can act and compute its value at Step 1, and every other gate can act and transmit its value at the first step after all gates on its incoming wires have computed their values. The number of steps for this process is the *depth* of the circuit. Depth is a notion of *parallel time complexity*. A circuit with small depth is a fast circuit. The circuit size in this view is the amount of hardware needed. Chapter 28 gives much more information on the correspondence between circuits and parallel machines, and gives formal definitions of size and depth.

A *circuit family* **C** consists of a sequence of circuits $\{C_1, C_2, \ldots\}$, where each C_n has n input nodes. The language accepted by the family is $L(\mathbf{C}) = \{x : C_{|x|} \text{ accepts } x\}$. (Circuit families computing functions $f : \{0,1\}^* \to \{0,1\}^*$ are defined in Chapter 29.)

The *size complexity* of the family is the function $z(n)$ giving the number of nodes in C_n. The *depth complexity* is the function $d(n)$ giving the depth of C_n.

Another aspect of circuits that must be specified in order to define complexity classes is the underlying technology. By technology we mean the types of gates that are used as components in the circuits. Three types of technology are considered in this chapter:

(1) Bounded fan-in gates, usually taken to be the "standard basis" of binary $\wedge$ (AND), binary $\vee$ (OR), and unary $\neg$ (NOT) gates. A notable alternative is to use *NAND* gates.

(2) Unbounded fan-in $\wedge$ and $\vee$ gates (together with unary $\neg$ gates).

(3) Threshold gates. For our purposes, it suffices to consider the simplest kind of threshold gate, called the MAJORITY gate, which also uses the Boolean domain. A MAJORITY gate outputs 1 if and only if at least $r/2$ of its r incoming wires have value 1. These gates can simulate unbounded fan-in $\wedge$ and $\vee$ with the help of "dummy wires." Threshold circuits also have unary $\neg$ gates.

The difference between (1) and (2) corresponds to general technological issues about high-bandwidth connections, whether they are feasible and how powerful they are. Circuits of type (1) can be converted to equivalent circuits that also have bounded fan-out, with only a constant-factor penalty in size and depth. Thus the difference also raises issues about one-to-many broadcast and all-to-one reception.

Threshold gates model the technology of *neural networks*, which were formalized in the 1940s. The kind of threshold gate studied most often in neural networks uses Boolean arguments and values, with "1" for "firing" and "0" for "off." It has numerical *weights* $w_1, \ldots, w_r$ for each of the r incoming wires and a threshold t. Letting $a_1, \ldots, a_r$ stand for the incoming 0-1 values, the gate outputs 1 if $\Sigma_{i=1}^{r} a_i w_i \geq t$, 0 otherwise. Thus the MAJORITY gate is the special case with $w_1 = \cdots = w_r = 1$ and $t = r/2$. A depth-2 (sub-)circuit of MAJORITY gates can simulate this general threshold gate.

22.3.2 Uniformity and Circuit Classes

One tricky aspect of circuit complexity is the fact that many functions that are *not computable* have trivial circuit complexity! For instance, let K be a noncomputable set of numbers, such as the indices of halting Turing machines, and let A be the language $\{x : |x| \in K\}$. For each n, if $n \in K$, then define C_n by attaching a $\neg$ gate to input x_1 and an OR gate whose two wires come from the $\neg$ gate and x_1 itself. If $n \notin K$, then define C_n similarly but with an AND gate in place of the OR. The circuit family $[C_n]$ so defined accepts A and has size and depth 2. The rub, however, is that there is no algorithm to tell *which* choice for C_n to define for each n. A related anomaly is that there are uncountably many circuit families. Indeed, every language is accepted by some circuit family $[C_n]$ with size complexity $2^{O(n)}$ and depth complexity 3 (unbounded fan-in) or $O(n)$ (bounded fan-in). Consequently, for general circuits, size complexity is at most exponential, and depth complexity is at most linear.

The notion of *uniform circuit complexity* avoids both anomalies. A circuit family $[C_n]$ is uniform if there is an easy algorithm Q that, given n, outputs an encoding of C_n. Either the adjacency-matrix or the edge-list representation of the graphs of the circuits C_n, together with the gate type of each node, may serve for our purposes as the *standard encoding scheme* for circuit families. If Q runs in polynomial time, then the circuit family is P-*uniform*, and so on.

P-uniformity is natural because it defines those families of circuits that are feasible to construct. However, we most often use circuits to model computation in *subclasses* of P. Allowing powerful computation to be incorporated into the step of *building* $C_{|x|}$ may overshadow the computation done by the circuit $C_{|x|}$ itself. The following much more stringent condition has proved to be most useful for characterizing these subclasses, and also works well for circuit classes at the level of polynomial time.

DEFINITION 22.1 A circuit family $[C_n]$ is **DLOGTIME-uniform** if there is a Turing machine M that can answer questions of the forms "Is there a path of edges from node u to node v in C_n?" and "What gate type does node u have?" in $O(\log n)$ time.

This uniformity condition is sufficient to build an encoding of C_n in sequential time roughly proportional to the size of C_n, and even much faster in parallel time. We will not try to define DLOGTIME as a complexity class, but note that since the inputs u, v to M can be presented by strings of length $O(\log n)$, the computation by M takes linear time in the (scaled down) input length. This definition presupposes that the size complexity $z(n)$ of the family is polynomial, which will be our chief interest here. The definition can be modified for $z(n)$ more than polynomial by changing the time limit on M to $O(\log z(n))$. Many central results originally proved using L-uniformity extend without change to DLOGTIME-uniformity, as explained later in this section. Unless otherwise stated, "uniform" means DLOGTIME-uniform throughout this and the next two chapters. We define the following circuit complexity classes:

DEFINITION 22.2   Given complexity functions $z(n)$ and $d(n)$,

- SIZE[$z(n)$] is the class of all languages accepted by DLOGTIME-uniform bounded fan-in circuit families whose size complexity is at most $z(n)$;
- DEPTH[$d(n)$] is the class of all languages accepted by DLOGTIME-uniform bounded fan-in circuit families whose depth complexity is at most $d(n)$;
- SIZE, DEPTH[$z(n), d(n)$] is the class of all languages accepted by DLOGTIME-uniform bounded fan-in circuit families whose size complexity is at most $z(n)$ and whose depth complexity is at most $d(n)$.

Nonuniform circuit classes can be approached by an alternative view introduced by Karp and Lipton [29], by counting the number of bits of information needed to set up the preprocessing. For integer-valued functions t, a, define DTIME[$t(n)$]/ADV[$a(n)$] to be the class of languages accepted by Turing machines M as follows: for all n there is a word y_n of length at most $a(n)$ such that for all x of length n, on input (x, y_n), M accepts if and only if $x \in L$, and M halts within $t(n)$ steps. Here y_n is regarded as "advice" on how to accept strings of length n. The class DTIME[$n^{O(1)}$]/ADV[$n^{O(1)}$] is called P/poly. Karp and Lipton observed that P/poly is equal to the class of languages accepted by polynomial-sized circuits. Indeed, P/poly is now the standard name for this class.

22.3.3   Circuits and Sequential Classes

The importance of P/poly and uniformity is shown by the following basic theorem. We give the proof since it is used often in the next chapter.

THEOREM 22.10   *Every language in* P *is accepted by a family of polynomial-sized circuits that is* DLOGTIME-*uniform. Conversely, every language with* P-*uniform polynomial-sized circuits belongs to* P.

PROOF   Let $A \in$ P. By Example 19.10, A is accepted by a Turing machine M with just one tape and tape head that runs in polynomial time $p(n)$. Let δ be the transition function of M, whereby for all states q of M and characters c in the worktape alphabet Γ of M, $\delta(q, c)$ specifies the character written to the current cell, the movement of the head, and the next state of M. We build a circuit of "δ-gates" that simulates M on inputs x of a given length n as follows, and then show how to simulate δ-gates by Boolean gates.

Lay out a $p(n) \times p(n)$ array of cells. Each cell (i, j) $(0 \leq i, j \leq p(n))$ is intended to hold the character on tape cell j after step i of the computation of M, and if the tape head of M is in that cell, also the state of M after step i. Cells $(0, 0)$ through $(0, n - 1)$ are the input nodes of C_n, while cells $(0, n)$ through $(0, p(n))$ can be treated as "dummy wires" whose value is the blank B in the alphabet Γ. The key idea is that the value in cell (i, j) for $i \geq 1$ depends only on the values in cells $(i - 1, j - 1)$, $(i - 1, j)$, and $(i - 1, j + 1)$. Cell $(i - 1, j - 1)$ is relevant in case its value includes the component for the tape head being there, and the head moves right at step i; cell $(i - 1, j + 1)$ similarly for a left move.

When the boundary cases $j = 0$ or $j = p(n)$ are handled properly, each cell value is computed by the same finite function of the three cells above, and this function defines a "δ-gate" for each cell (see Figure 22.2). Finally, we may suppose that M is coded to signal acceptance by moving its tape head to the left end and staying in a special state q_a. Thus node $(i, 0)$ becomes the output gate of the circuit, and the accepting output values are those with q_a in the state component. Since in $p(n)$ steps

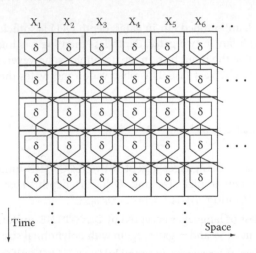

FIGURE 22.2 Conversion from Turing machine to Boolean circuits.

M can visit at most $p(n)$ tape cells, the array is large enough to hold all the computations of M on inputs of length n.

Since each argument and value of a δ-gate comes from a finite domain, we may take an (arbitrary) binary encoding of the domain, and replace all δ-gates by identical fixed-size subcircuits of Boolean gates that compute δ under the encoding. If the alphabet Σ over which A is defined is $\{0, 1\}$, then no recoding need be done at the inputs; otherwise, we similarly adopt a binary encoding of Σ. The Boolean circuits C_n thus obtained accept A. They also are DLOGTIME-uniform, intuitively by the very regular structure of the identical δ-gates.

Conversely, given a P-uniform family **C**, a Turing machine can accept $L(\mathbf{C})$ in polynomial time given any input x by first constructing $C_{|x|}$ in polynomial time, and then evaluating $C_{|x|}(x)$. $\quad\square$

A caching strategy that works for Turing machines with any fixed number of tapes yields the following improvement:

THEOREM 22.11   *If $t(n)$ is a time-constructible function, then* DTIME$[t] \subseteq$ SIZE$[t \log t]$.

Connections between space complexity and circuit depth are shown by the next result.

THEOREM 22.12

(a) *If $d(n) \geq \log n$, then* DEPTH$[d(n)] \subseteq$ DSPACE$[d(n)]$.

(b) *If $s(n)$ is a space-constructible function, $s(n) \geq \log n$, then* NSPACE$[s(n)] \subseteq$ DEPTH$[s(n)^2]$.

22.3.4 Circuits and Parallel Classes

Since the 1970s, research on circuit complexity has focused on problems that can be solved quickly in parallel, with feasible amounts of hardware—circuit families of polynomial size and depth as small as possible. Note, however, that the meaning of the phrase "as small as possible" depends on the technology used. With unbounded fan-in gates, depth $O(1)$ is sufficient to carry out interesting computation, whereas with fan-in two gates, depth less than $\log n$ is impossible if the value at the output gate depends on all of the input bits. In any technology, however, a circuit with depth nearly logarithmic is considered to be very fast. This observation motivates the following definitions. Let $\log^k n$ stand for $(\log n)^k$.

DEFINITION 22.3   For all $k \geq 0$,

(a) NC^k denotes the class of languages accepted by DLOGTIME-uniform bounded fan-in circuit families of polynomial size and $O(\log^k n)$ depth. That is, NC^k abbreviates SIZE, DEPTH$[n^{O(1)}, O(\log^k n)]$. NC denotes $\cup_{k \geq 0} NC^k$.

(b) AC^k denotes the class of languages accepted by DLOGTIME-uniform families of circuits of unbounded fan-in $\wedge$, $\vee$, and $\neg$ gates, again with polynomial size and $O(\log^k n)$ depth.

(c) TC^k denotes the class of languages accepted by DLOGTIME-uniform families of circuits of MAJORITY and $\neg$ gates, again with polynomial size and $O(\log^k n)$ depth.

The case $k = 0$ in these definitions gives constant-depth circuit families. A function f is said to belong to one of these classes if the language $A_f = \{\langle x, i, b\rangle : 1 \leq i \leq |f(x)| \text{ and bit } i \text{ of } f(x) \text{ is } b\}$ belongs to the class. NC^0 is not studied as a language class in general, since the output gate can depend on only a constant number of input bits, but NC^0 is interesting as a function class.

Some notes on the nomenclature are in order. Nicholas Pippenger was one of the first to study polynomial-size, polylog-depth circuits in the late 1970s, and NC was dubbed "Nick's Class." There is no connotation of nondeterminism in NC. The "A" in AC^k connotes both alternating circuits and alternating Turing machines for reasons described below. The "T" in TC^k stands for the presence of threshold gates.

The following theorem expresses the relationships at each level of the hierarchies defined by these classes.

THEOREM 22.13   *For each $k \geq 0$,*

$$NC^k \subseteq AC^k \subseteq TC^k \subseteq NC^{k+1} .$$

PROOF   The first inclusion is immediate (for each k), and the second conclusion follows from the observation noted above that MAJORITY gates can simulate unbounded fan-in AND and OR gates. The interesting case is $TC^k \subseteq NC^{k+1}$. For this, it suffices to show how to simulate a single MAJORITY gate with a fan-in two circuit of logarithmic depth. To simulate MAJORITY$(w_1, \ldots, w_r)$, we add up the one-bit numbers $w_1, \ldots, w_r$ and test whether the sum is at least $r/2$. We may suppose for simplicity that the fan-in r is a power of 2, $r = 2^m$. The circuit has m distinguished nodes that represent the sum written as an m-bit binary number. Then the sum is at least $r/2 = 2^{m-1}$ if and only if the node representing the most significant bit of the sum has value 1.

To compute the sum efficiently, we use a standard "carry-save" technique: There is a simple $O(1)$ depth fan-in two circuit that takes as input three b-bit binary numbers a_1, a_2, a_3 and produces as output two $(b + 1)$-bit numbers b_1, b_2 such that $a_1 + a_2 + a_3 = b_1 + b_2$. Thus in one phase, the original sum of r bits is reduced to taking the sum of $\frac{2}{3}r$ numbers, and after $O(\log r)$ additional phases, the problem is reduced to taking the sum of two $\log r$-bit numbers, and this sum can be produced by a full carry-lookahead adder circuit of $O(\log r)$ depth. Finally, since the circuits have polynomial size, r is polynomial in n, and so $O(\log r) = O(\log n)$.

Thus in particular, $\cup_k \text{AC}^k = \cup_k \text{TC}^k = \text{NC}$. The only proper inclusion known, besides the trivial case $\text{NC}^0 \subset \text{AC}^0$, is $\text{AC}^0 \subset \text{TC}^0$, discussed below. For all we know at this time, TC^0 may be equal not only to NC, but even to NP!

Several relationships between complexity classes based on circuits and classes based on Turing machines are known:

THEOREM 22.14 $\quad \text{NC}^1 \subseteq \text{L} \subseteq \text{NL} \subseteq \text{AC}^1$.

In fact, the connection with Turing machines is much closer than this theorem suggests. Using alternating Turing machines, we define the following complexity classes:

- ASPACE, TIME$[s(n), t(n)]$ is the class of languages recognized by alternating Turing machines that use space at most $s(n)$ and also run in time at most $t(n)$.
- ASPACE, ALTS$[s(n), a(n)]$ is the class of languages recognized by alternating Turing machines that use space at most $s(n)$ and make at most $a(n)$ alternations between existential and universal states.
- ATIME, ALTS$[s(n), a(n)]$ is the class of languages recognized by alternating Turing machines that run in time $t(n)$ and make at most $a(n)$ alternations between existential and universal states.

THEOREM 22.15

(a) For all $k \geq 1$, $\text{NC}^k = $ ASPACE, TIME$[O(\log n), O(\log^k n)]$

(b) For all $k \geq 1$, $\text{AC}^k = $ ASPACE, ALTS$[O(\log n), O(\log^k n)]$

(c) $\text{NC}^1 = $ ATIME$[O(\log n)]$

(d) $\text{AC}^0 = $ ATIME, ALTS$[O(\log n), O(1)]$

For AC^1 and the higher circuit classes, changing the uniformity condition to L-uniformity does not change the class of languages. However, it is not known whether L-uniform NC^1 differs from NC^1, or L-uniform AC^0 from AC^0. Thus the natural extension (c,d) of the results in (a,b) is another advantage of DLOGTIME-uniformity—noting for (c) in particular that Definition 22.1 adopts the "extended connection language" of [38]. Insofar as the containment of NC^1 in L is believed to be proper by many researchers, the definition of L-uniform NC^1 may allow more computing power to the "preprocessing stage" than to the circuits themselves. Avoiding this anomaly is a reason to adopt DLOGTIME-uniformity.

As discussed in Chapter 28, many other models of parallel computation can be used to define NC. This robustness of NC supports the belief that NC is not merely an artifact of some arbitrary choices made in formulating the definitions, but instead captures a fundamental aspect of parallel computation. The criticism has been made that NC is overly generous in allowing polynomial size. Again, the justification in complexity theory is that the ultimate goal is to prove lower bounds, and a lower bound proved against a generous upper-bound notion is impervious to this criticism.

22.3.5 Why Focus on These Circuit Classes?

The class AC^0 is particularly important for the following reasons:

- It captures the complexity of important basic operations such as integer addition and subtraction.
- It corresponds closely to first-order logic, as described in Section 24.5.

- Most important, it is one of the few complexity classes for which lower bounds are actually known, instead of merely being conjectured.

It is known that AC^0 circuits, even nonuniform ones, cannot recognize the language PARITY of strings that have an odd number of 1's. Consequently, constant depth unbounded fan-in AND/OR/NOT circuits for PARITY must have super-polynomial size. However, PARITY does have constant-depth polynomial-size threshold circuits; indeed, it belongs to TC^0.

Note that this also implies that AC^0 is somehow "finer" than the notion of constant space, because the class of regular languages, which includes PARITY, can be decided in constant space. There are a number of important papers on proving lower bounds for classes of constant-depth circuits. Still, the fact that TC^0 is not known to differ from NP is a wide gulf in our knowledge. Separating NC from P, or L from P, or L from NP would imply separating TC^0 from NP.

TC^0 is important because it captures the complexity of important basic operations such as sorting, and integer multiplication and division. Also, TC^0 is a good complexity-theoretic counterpart to popular models of neural networks.

NC^1 is important because it captures the complexity of the basic operation of evaluating a Boolean formula on a given assignment. The problem of whether NC^1 equals TC^0 thus captures the question of whether basic calculations in logic are harder than basic operations in arithmetic, or harder than basic neural processes. Several other characterizations of NC^1 besides the one given for $ATIME[O(\log n)]$ are known. NC^1 equals the class of languages definable by polynomial-size Boolean formulas (as opposed to polynomial-sized circuits; a formula is equivalent to a circuit of fan-out 1). Also, NC^1 equals the class of languages recognized by bounded-width branching programs [4]. The programs are representable as compositions of permutations of five elements, and encoding the compositions as words over a finite alphabet reduces the task to membership in a regular language. Thus NC^1 has a complete language under AC^0 reductions (which are studied in the next chapter) that is regular, and since all regular languages belong to NC^1, this gives a sense in which NC^1 captures the circuit complexity of regular expressions.

22.4 Research Issues and Summary

The complexity class is the fundamental notion of complexity theory. What makes a complexity class useful to the practitioner is the close relationship between complexity classes and real computational problems. The strongest such relationship comes from the concept of completeness, which is a chief subject of the next chapter. Even in the absence of lower bounds separating complexity classes, the apparent fundamental difference between models such as deterministic and nondeterministic Turing machines, for example, provides insight into the nature of problem solving on computers.

The initial goal when trying to solve a computational problem is to find an efficient polynomial-time algorithm. If this attempt fails, then one could attempt to prove that no efficient algorithm exists, but to date nobody has succeeded doing this for any problem in PSPACE. With the notion of a complexity class to guide us, however, we can attempt to discover the class that exactly captures our current problem. A main theme of the next chapter is the surprising fact that most natural computational problems are *complete* for one of the canonical complexity classes. When viewed in the abstract setting provided by the model that defines the complexity class, the aspects of a problem that make an efficient algorithm difficult to achieve are easier to identify. Often this perspective leads to a redefinition of the problem in a way that is more amenable to solution.

Figure 22.1 shows the known inclusion relationships between canonical classes. Perhaps even more significant is what is currently not known. Although AC^0 differs from TC^0, TC^0 (let alone P!) is not known to differ from NP, nor NP from EXP, nor EXP from EXPSPACE. The only other proper inclusions known are (immediate consequences of) NL $\subset$ PSPACE $\subset$ EXPSPACE, P $\subset$ E $\subset$ EXP, and NP $\subset$ NE $\subset$ NEXP—and these follow simply from the simulation and hierarchy theorems proved in this chapter.

We have given two examples of diagonalization arguments. Diagonalization is still the main tool for showing the existence of hard-to-compute problems inside a complexity class. Unfortunately, the languages constructed by diagonalization arguments rarely correspond to computational problems that arise in practice. In some cases, however, one can show that there is an efficient reduction from a difficult problem (shown to exist by diagonalization) to a more natural problem—with the consequence that the natural problem is also difficult to solve. Thus diagonalization inside a complexity class (the topic of this chapter) can work hand in hand with reducibility (the topic of the next chapter) to produce intractability results for natural computational problems.

22.5 Further Information

Primary sources for the results presented in this chapter are Theorem 22.1 [9,45]; Theorem 22.2 [5]; Theorems 22.3 and 22.4 [17,25,32,39,44]; Theorem 22.5 [17,23,40]; Theorem 22.6 [7]; Lemma 22.1 [37]; Theorems 22.8 and 22.9 [11]; Theorem 22.10 [39]; Theorem 22.11 [35]; Theorem 22.12 [10]; and Theorem 22.15 [3,12,38,41]. Theorems 22.13 and 22.14 are a combination of results in the last four papers; see also the influential survey by Cook [13]. Our proof of Theorem 22.5(b) follows [49].

For Section 22.3.1, a comparison of arithmetical circuits with Boolean circuits may be found in [47], the result that bounded fan-in circuits can be given bounded fan-out is due to [21], and the sharpest simulation of general weighted threshold gates by MAJORITY gates is due to [20]. The theorem in Section 22.3.5 that *PARITY* is not in AC^0 is due to [2,14], and the strongest lower bounds known on the size of constant-depth circuits for *PARITY* are those in [18]. The results mentioned for TC^0 may be found in [3,19,24,36].

The texts [22,33] present many of these results in greater technical detail. Three chapters of the *Handbook of Theoretical Computer Science*, [27,46,8] respectively, describe more complexity classes, compare complexity measures for more machine models, and present more information on circuit complexity. Relationships between circuits and parallel and neural models are covered very accessibly in [34]. Average-case complexity is discussed by [16,26,48]. See also Chapter 24 and Section 24.13 for further sources.

Defining Terms

Canonical complexity classes: The classes defined by logarithmic, polynomial, and exponential bounds on time and space, for deterministic and nondeterministic machines. These are the most central to the field, and classify most of the important computational problems.

Circuit: A network of input, output, and logic gates, contrasted with a Turing machine in that its hardware is static and fixed.

Circuit complexity: The study of the size, depth, and other attributes of circuits that decide specified languages or compute specified functions.

Diagonalization: A proof technique for showing that a given language does not belong to a given complexity class, used in many separation theorems.

Padding argument: A method for transferring results about one complexity bound to another complexity bound, by padding extra dummy characters onto the inputs of the machines involved.

Polynomial-time Church–Turing Thesis: An analogue of the classical Church–Turing Thesis, for which see Chapter 22, stating that the class P captures the true notion of feasible (polynomial time) sequential computation.

Separation theorems: Theorems showing that two complexity classes are distinct. Most known separation theorems have been proved by diagonalization.

Simulation theorems: Theorems showing that one kind of computation can be simulated by another kind within stated complexity bounds. Most known containment or equality relationships between complexity classes have been proved this way.

Space-constructible function: A function $s(n)$ that gives the actual space used by some Turing machine on all inputs of length n, for all n.

Time-constructible function: A function $t(n)$ that is the actual running time of some Turing machine on all inputs of length n, for all n.

Uniform circuit complexity: The study of complexity classes defined by uniform circuit families.

Uniform circuit family: A sequence of circuits, one for each input length n, that can be efficiently generated by a Turing machine.

Acknowledgments

The first author was supported by the National Science Foundation under Grants CCF 0830133 and 0132787; portions of this work were performed while a visiting scholar at the Institute of Mathematical Sciences, Madras, India. The second author was supported by the National Science Foundation under Grants DUE-0618589 and EEC-0628814. The third author was supported by the National Science Foundation under Grant CCR-9409104.

References

1. Agrawal, M., Kayal, K., and Saxena, T., PRIMES is in P. *Ann. Math.,* 160, 781–793, 2004.
2. Ajtai, M., Σ_1^1 formulae on finite structures. *Ann. Pure Appl. Logic,* 24, 1–48, 1983.
3. Barrington, D.M., Immerman, N., and Straubing, H., On uniformity within NC^1. *J. Comp. Sys. Sci.,* 41, 274–306, 1990.
4. Barrington, D.M., Bounded-width polynomial-size branching programs recognize exactly those languages in NC^1. *J. Comp. Sys. Sci.,* 38, 150–164, 1989.
5. Blum, M., A machine-independent theory of the complexity of recursive functions. *J. Assn. Comp. Mach.,* 14, 322–336, 1967.
6. Book, R., Greibach, S., and Wegbreit, B., Time- and tape-bounded Turing acceptors and AFLs. *J. Comp. Sys. Sci.,* 4, 606–621, 1970.
7. Book, R., Comparing complexity classes. *J. Comp. Sys. Sci.,* 9, 213–229, 1974.
8. Boppana, R. and Sipser, M., The complexity of finite functions. In *Handbook of Theoretical Computer Science,* J. Van Leeuwen, Ed., Vol. A, pp. 757–804. Elsevier, Amsterdam, the Netherlands, and MIT Press, Cambridge, MA, 1990.
9. Borodin, A., Computational complexity and the existence of complexity gaps. *J. Assn. Comp. Mach.,* 19, 158–174, 1972.
10. Borodin, A., On relating time and space to size and depth. *SIAM J. Comp.,* 6, 733–744, 1977.
11. Chandra, A., Kozen, D., and Stockmeyer, L., Alternation. *J. Assn. Comp. Mach.,* 28, 114–133, 1981.
12. Chandra, A., Stockmeyer, L., and Vishkin, U., Constant-depth reducibility. *SIAM J. Comp.,* 13, 423–439, 1984.

13. Cook, S., A taxonomy of problems with fast parallel algorithms. *Inform. Control,* 64, 2–22, 1985.

14. Furst, M., Saxe, J., and Sipser, M., Parity, circuits, and the polynomial-time hierarchy. *Math. Sys. Thy.,* 17, 13–27, 1984.

15. Garey, M. and Johnson, D.S., *Computers and Intractability: A Guide to the Theory of NP-Completeness.* W.H. Freeman, San Francisco, CA 1988. First edition was 1979.

16. Gurevich, Y., Average case completeness. *J. Comp. Sys. Sci.,* 42, 346–398, 1991.

17. Hartmanis, J. and Stearns, R., On the computational complexity of algorithms. *Trans. AMS,* 117, 285–306, 1965.

18. Håstad, J., Almost optimal lower bounds for small-depth circuits. In *Randomness and Computation,* S. Micali, Ed., Vol. 5, *Advances in Computing Research,* pp. 143–170. JAI Press, Greenwich, CT, 1989.

19. Hesse, W., Allender, E., and Barrington, D.M., Uniform constant-depth threshold circuits for division and iterated Multiplication. *J. Comp. Sys. Sci.,* 65, 695–716, 2002.

20. Hofmeister, T., A note on the simulation of exponential threshold weights. In *Proceedings of 2nd International Computing and Combinatorics Conference,* Vol. 1090, *Lecture Notes in Computer Science,* pp. 136–141. Springer-Verlag, Berlin, 1996.

21. Hoover, H., Klawe, M., and Pippenger, N., Bounding fan-out in logical networks. *J. Assn. Comp. Mach.,* 31, 13–18, 1984.

22. Hopcroft, J. and Ullman, J., *Introduction to Automata Theory, Languages, and Computation.* Addison-Wesley, Reading, MA, 1979.

23. Ibarra, O., A note concerning nondeterministic tape complexities. *J. Assn. Comp. Mach.,* 19, 608–612, 1972.

24. Immerman, N. and Landau, S., The complexity of iterated multiplication. *Inform. Control,* 116, 103–116, 1995.

25. Immerman, N., Nondeterministic space is closed under complementation. *SIAM J. Comp.,* 17, 935–938, 1988.

26. Impagliazzo, R., A personal view of average-case complexity. In *Proceedings of 10th Annual IEEE Conference on Structure in Complexity Theory,* pp. 134–147, 1995.

27. Johnson, D.S., A catalog of complexity classes. In *Handbook of Theoretical Computer Science,* J. Van Leeuwen, Ed., Vol. A, pp. 67–161. Elsevier, Amsterdam, the Netherlands and MIT Press, Cambridge, MA, 1990.

28. Karmarkar, N., A new polynomial-time algorithm for linear programming. *Combinatorica,* 4, 373–395, 1984.

29. Karp, R. and Lipton, R., Turing machines that take advice. *L'Enseignement Mathématique,* 28, 191–210, 1982.

30. Khachiyan, L., A polynomial algorithm in linear programming. *Soviet Mathematics Doklady,* 20(1), 191–194, 1979. English translation.

31. Levin, L.A., Computational complexity of functions. *Theor. Comp. Sci.,* 157(2), 267–271, 1996.

32. Lewis II, P., Stearns, R., and Hartmanis, J., Memory bounds for recognition of context-free and context-sensitive languages. In *Proceedings of 6th Annual IEEE Symposium on Switching Circuit Theory and Logical Design,* pp. 191–202, 1965.

33. Papadimitriou, C., *Computational Complexity.* Addison-Wesley, Reading, MA, 1994.

34. Parberry, I., *Circuit Complexity and Neural Networks.* MIT Press, Cambridge, MA, 1994.

35. Pippenger, N. and Fischer, M., Relations among complexity measures. *J. Assn. Comp. Mach.,* 26, 361–381, 1979.

36. Reif, J. and Tate, S., On threshold circuits and polynomial computation. *SIAM J. Comp.,* 21, 896–908, 1992.

37. Ruby, S. and Fischer, P., Translational methods and computational complexity. In *Proceedings of 6th Annual IEEE Symposium on Switching Circuit Theory and Logical Design,* pp. 173–178, 1965.

38. Ruzzo, W., On uniform circuit complexity. *J. Comp. Sys. Sci.,* 22, 365–383, 1981.

39. Savitch, W., Relationship between nondeterministic and deterministic tape complexities. *J. Comp. Sys. Sci.,* 4, 177–192, 1970.

40. Seiferas, J., Fischer, M., and Meyer, A., Separating nondeterministic time complexity classes. *J. Assn. Comp. Mach.,* 25, 146–167, 1978.

41. Sipser, M., Borel sets and circuit complexity. In *Proceedings of 15th Annual ACM Symposium on the Theory of Computing,* pp. 61–69, 1983.

42. Stockmeyer, L. and Meyer, A.R., Cosmological lower bound on the circuit complexity of a small problem in logic. *J. ACM* 49: 753–784, 2002.

43. Stockmeyer, L., Classifying the computational complexity of problems. *J. Symb. Logic,* 52, 1–43, 1987.

44. Szelepcsényi, R., The method of forced enumeration for nondeterministic automata. *Acta Informatica,* 26, 279–284, 1988.

45. Trakhtenbrot, B., Turing computations with logarithmic delay. *Algebra i Logika,* 3, 33–48, 1964.

46. van Emde Boas, P., Machine models and simulations. In *Handbook of Theoretical Computer Science,* J. Van Leeuwen, Ed., Vol. A, pp. 1–66. Elsevier, Amsterdam, the Netherlands and MIT Press, Cambridge, MA, 1990.

47. von zur Gathen, J., Efficient exponentiation in finite fields. In *Proceedings of 32nd Annual IEEE Symposium on Foundations of Computer Science,* pp. 384–391, 1991.

48. Wang, J., Average-case computational complexity theory. In *Complexity Theory Retrospective II,* L. Hemaspaandra and A. Selman, Eds., pp. 295–328. Springer-Verlag, Berlin, 1997.

49. Zak, S., A Turing machine time hierarchy. *Theor. Comp. Sci.,* 26, 327–333, 1983.

23

Reducibility and Completeness

Eric Allender
Rutgers University

Michael C. Loui
*University of Illinois at
Urbana-Champaign*

Kenneth W. Regan
State University of New York at Buffalo

23.1 Introduction

There is little doubt that the notion of reducibility is the most useful tool that complexity theory has delivered to the rest of the computer science community.

For most computational problems that arise in real-world applications, such as the Traveling Salesperson Problem, we still know only a little about their deterministic time or space complexity. We cannot now tell whether classes such as P and NP are distinct. And yet, even without such hard knowledge, it has been useful in practice to take some new problem *A* whose complexity needs to be analyzed, and announce that *A* has roughly the same complexity as the Traveling Salesperson Problem, by exhibiting efficient ways of reducing each problem to the other. Thus, we can say a lot about problems being equivalent in complexity to each other, even if we cannot pinpoint what that complexity is.

One reason for this success is that when one partitions the many thousands of real-world computational problems into equivalence classes according to the reducibility relation, there are surprisingly few classes in this partition. Thus, the complexity of almost any problem arising in practice can be classified by showing that it is equivalent to one of the short list of representative problems. It was not originally expected that this would be the case.

Even more amazingly, these "representative problems" correspond in a natural way to abstract models of computation—that is, they correspond to complexity classes. These classes were defined in Chapter 22 using a small set of abstract machine concepts: Turing machines, nondeterminism, alternation, time, space, circuits. With this and a few simple functions that define time and space bounds, we are able to characterize the complexity of the overwhelming majority of natural computational problems—most of which bear no topical resemblance to any question about Turing machines. This tool has been much more successful than we had any right to expect it would be.

All this leads us to believe that it is no mere accident that problems easily lend themselves to being placed in one class or another. That is, we are disposed to think that these classes really *are* distinct, the classification is real, and the mathematics developed to deal with them really does describe some important aspect of nature. Nondeterministic Turing machines, with their magic ability to soar through immense search spaces, seem to be much more powerful than our mundane deterministic machines, and this reinforces our belief. However, until P versus NP and similar long-standing questions of complexity theory are completely resolved, our best method of understanding the complexity of real-world problems is to use the classification provided by reducibility, and to trust in a few plausible conjectures.

In this chapter, we discuss reducibility. We will learn about the different types of reducibility, and the related notion of completeness. It is especially useful to understand NP-completeness. We define NP-completeness precisely, and give examples of NP-complete problems. We show how to prove that a problem is NP-complete, and give some help for coping with NP-completeness. After that, we describe problems that are complete for other complexity classes, under the most efficient reducibility relations. Finally, we cover several important topics that supplement reducibility in appraising the status and structure of the P versus NP question: relativization, "natural proofs," and sparse languages.

23.2 Reducibility Relations

In mathematics, as in everyday life, a typical way to solve a new problem is to reduce it to a previously solved problem. Frequently, an instance of the new problem is expressed completely in terms of an instance of the prior problem, and the solution is then interpreted in the terms of the new problem. This kind of **reduction** is called **many-one reducibility**, and is defined below.

A different way to solve the new problem is to use a subroutine that solves the prior problem. For example, we can solve an optimization problem whose solution is feasible and maximizes the value of an objective function g by repeatedly calling a subroutine that solves the corresponding decision problem of whether there exists a feasible solution x whose value $g(x)$ satisfies $g(x) \geq k$. This kind of reduction is called **Turing reducibility**, and is also defined below.

Let A_1 and A_2 be languages. A_1 is many-one reducible to A_2, written $A_1 \leq_m A_2$, if there exists a computable function f such that for all x, $x \in A_1$ if and only if $f(x) \in A_2$. The function f is called the **transformation function**. A_1 is Turing reducible to A_2, written $A_1 \leq_T A_2$, if A_1 can be decided by a deterministic oracle Turing machine M using A_2 as its oracle, i.e., $A_1 = L(M^{A_2})$. (Computable functions and oracle machines are defined in Chapter 19. The oracle for A_2 models a hypothetical efficient subroutine for A_2.)

If f or M above consumes too much time or space, the reductions they compute are not helpful. To study complexity classes defined by bounds on time and space resources, it is natural to consider resource-bounded reducibilities. Let A_1 and A_2 be languages.

- A_1 is **Karp reducible** to A_2, written $A_1 \leq_m^p A_2$, if A_1 is many-one reducible to A_2 via a transformation function that is computable deterministically in polynomial time.
- A_1 is **Cook reducible** to A_2, written $A_1 \leq_T^p A_2$, if A_1 is Turing reducible to A_2 via a deterministic **oracle Turing machine** of polynomial-time complexity.

The term "polynomial-time reducibility" usually refers to Karp reducibility. If $A_1 \leq_m^p A_2$ and $A_2 \leq_m^p A_1$, then A_1 and A_2 are equivalent under Karp reducibility. Equivalence under Cook reducibility is defined similarly.

Karp and Cook reductions are useful for finding relationships between languages of high complexity, but they are not at all useful for distinguishing between problems in P, because all problems in P are equivalent under Karp (and hence Cook) reductions. (Here and later we ignore the special cases $A_1 = \emptyset$ and $A_1 = \Sigma^*$, and consider them to reduce to any language.) To investigate the many interesting complexity classes inside P, we will want to define more restrictive reducibilities, and we do this in the beginning in Section 23.5. Now, however, we focus on Cook and Karp reducibility.

The key property of Cook and Karp reductions is that they preserve polynomial-time feasibility. Suppose $A_1 \leq_m^p A_2$ via a transformation f. If M_2 decides A_2, and M_f computes f, then to decide whether an input word x is in A_1, we may use M_f to compute $f(x)$, and then run M_2 on input $f(x)$. If the time complexities of M_2 and M_f are bounded by polynomials t_2 and t_f, respectively, then on inputs x of length $n = |x|$, the time taken by this method to decide A_1 is at most $t_f(n) + t_2(t_f(n))$, which is also a polynomial in n. In summary, if A_2 is feasible, and there is an efficient reduction from A_1 to A_2, then A_1 is feasible. Although this is a simple observation, this fact is important enough to state as a theorem. First, though, we need the concept of "closure."

A class of languages $\mathcal{C}$ is **closed under a reducibility** $\leq_r$, if for all languages A_1 and A_2, whenever $A_1 \leq_r A_2$ and $A_2 \in \mathcal{C}$, necessarily $A_1 \in \mathcal{C}$.

THEOREM 23.1   P *is closed under both Cook and Karp reducibility.*

Note that this is an instance of an idea that motivated our identification of P with the class of "feasible" problems in Chapter 22, namely that the composition of two feasible functions should be feasible. Similar considerations give us the following theorem.

THEOREM 23.2   *Karp reducibility and Cook reducibility are transitive; i.e.,*

1. *If $A_1 \leq_m^p A_2$ and $A_2 \leq_m^p A_3$, then $A_1 \leq_m^p A_3$.*
2. *If $A_1 \leq_T^p A_2$ and $A_2 \leq_T^p A_3$, then $A_1 \leq_T^p A_3$.*

We shall see the importance of closure under a reducibility in conjunction with the concept of completeness, which we define in the next section.

23.3   Complete Languages and the Cook–Levin Theorem

Let $\mathcal{C}$ be a class of languages that represent computational problems. A language A_0 is $\mathcal{C}$-hard under a reducibility $\leq_r$ if for all A in $\mathcal{C}$, $A \leq_r A_0$. A language A_0 is $\mathcal{C}$-complete under $\leq_r$ if A_0 is $\mathcal{C}$-hard, and $A_0 \in \mathcal{C}$. Informally, if A_0 is $\mathcal{C}$-hard, then A_0 represents a problem that is at least as difficult to solve as any problem in $\mathcal{C}$. If A_0 is $\mathcal{C}$-complete, then in a sense, A_0 is one of the most difficult problems in $\mathcal{C}$.

There is another way to view completeness. Completeness provides us with tight relative lower bounds on the complexity of problems. If a language A is complete for complexity class $\mathcal{C}$, then we have a lower bound on its complexity. Namely, A is as hard as the most difficult problem in $\mathcal{C}$, assuming that the complexity of the reduction itself is small enough not to matter. The lower bound is tight because A is in $\mathcal{C}$; that is, the upper bound matches the lower bound.

In the case $\mathcal{C} = $ NP, the reducibility $\leq_r$ is usually taken to be Karp reducibility unless otherwise stated. Thus, we say:

- A language A_0 is NP-**hard** if A_0 is NP-hard under Karp reducibility.
- A_0 is NP-**complete** if A_0 is NP-complete under Karp reducibility.

However, some sources take the term "NP-hard" to refer to Cook reducibility.

Many important languages are now known to be NP-complete. Before we get to them, let us discuss some implications of the statement "A_0 is NP-complete," and also some things this statement does not mean.

The first implication is that if there exists a deterministic Turing machine that decides A_0 in polynomial time—that is, if $A_0 \in$ P—then because P is closed under Karp reducibility (Theorem 23.1 in Section 23.2), it would follow that NP $\subseteq$ P, hence P $=$ NP. In essence, the question of whether P is the same as NP comes down to the question of whether any particular NP-complete language is in P. Put another way, all of the NP-complete languages stand or fall together: if one is in P, then all are in P; if one is not, then all are not. Another implication, which follows by a similar closure argument applied to co-NP, is that if $A_0 \in$ co-NP then NP $=$ co-NP. It is also believed unlikely that NP $=$ co-NP, as was noted in Chapter 22 regarding whether all tautologies have short proofs.

A common misconception is that the above property of NP-complete languages is actually their definition, namely: if $A \in$ NP, and $A \in$ P implies P $=$ NP, then A is NP-complete. This "definition" is wrong. A theorem due to Ladner [22] shows that P $\neq$ NP if and only if there exists a language A' in NP $-$ P such that A' is not NP-complete. Thus, if P $\neq$ NP, then A' is a counterexample to the "definition."

Another common misconception arises from a misunderstanding of the statement "If A_0 is NP-complete, then A_0 is one of the most difficult problems in NP." This statement is true on one level: if there is any problem at all in NP that is not in P, then the NP-complete language A_0 is one such problem. However, note that there are NP-complete problems in NTIME[n]—and these problems are, in some sense, much simpler than many problems in NTIME[$n^{10^{500}}$]. We discuss the difficulty of NP-complete problems in more detail after studying several examples.

We now prove the **Cook–Levin Theorem**, which established the first important NP-complete problem. Recall the definition of SAT, the language of satisfiable Boolean formulas, from Section 22.2.2. In this and later Karp-reduction proofs, we highlight the construction of the transformation f, check if the complexity of f is a polynomial, and verify the correctness of the reduction.

THEOREM 23.3 (**Cook–Levin Theorem**). *SAT is NP-complete.*

PROOF   Let $A \in$ NP. Without loss of generality we may assume that $A \subseteq \{0, 1\}^*$. There is a polynomial q and a polynomial-time computable relation R such that for all x,

$$x \in A \iff (\exists y : |y| = q(|x|))\, R(x, y).$$

By the construction of the proof of Theorem 22.11, there is a polynomial p such that for all n, we can build in time $p(n)$ a Boolean circuit C_n, using only binary NAND gates, that decides R on inputs of length $n + q(n)$. C_n has n input nodes labeled $x_1, \ldots, x_n$ and $q = q(n)$ more input nodes labeled $y_1, \ldots, y_q$. C_n has at most $p(n)$ wires, which we label $w_1, \ldots, w_m$, where $m \leq p(n)$ and w_m is a special wire leading out of the output gate.

Construction. We first write a boolean formula ϕ_n in the x, y, and w variables to express that every gate in C_n functions correctly and C_n outputs 1. For every NAND gate in C_n with incoming wires u, v, and for each outgoing wire w of the gate, we add to ϕ_n the following conjunction ψ_w of three clauses

$$\psi_w = (u \vee w) \wedge (v \vee w) \wedge (\bar{u} \vee \bar{v} \vee \bar{w}).$$

These clauses are satisfied by those assignments to u, v, w such that $w = \neg(u \wedge v)$. Intuitively, ψ_w asserts that the given NAND gate functions correctly for wire w. Note that ψ_w is defined for each wire w except those coming from the inputs—those wires appear only as "u" or "v" and furnish occurrences of the x and y variables. Finally, for the output wire, ϕ_n has the singleton clause (w_m). So ϕ_n has fewer than $3p(n) + 1$ clauses in all.

Now given x, we form the desired formula $f(x) = \phi_x$ by building ϕ_n, where $n = |x|$, and simply appending n singleton clauses that force the corresponding assignment to the $x_1, \ldots, x_n$ variables. (For example, if $x = 1001$, append $x_1 \wedge \bar{x}_2 \wedge \bar{x}_3 \wedge x_4$.)

Complexity. C_n is built up in roughly $O(p(n))$ time. Building ϕ_n from C_n, and appending the singleton clauses for x, takes a similar polynomial amount of time.

Correctness. Formally, we need to show that for all x, $x \in A \iff f(x) \in SAT$. By construction, for all x, $x \in A$ if and only if there exists an assignment to the y variables and to the w variables that satisfies ϕ_x. Hence, the reduction is correct. $\square$

A glance at the proof shows that ϕ_x is always a boolean formula in conjunctive normal form (CNF) with clauses of one, two, or three literals each. By introducing some new "dummy" variables, we can arrange that each clause has exactly three literals. Thus, we have actually shown that the following restricted form of the satisfiability problem is NP-complete:

3-SATISFIABILITY (*3SAT*)

Instance: A Boolean expression ϕ in conjunctive normal form with three literals per clause.

Question: Is ϕ satisfiable?

One concrete implication of the Cook–Levin Theorem is that if deciding SAT is easy (i.e., in polynomial time), then factoring integers is likewise easy, because the decision version of factoring belongs to NP. (See Section 24.7.1) This is a surprising connection between the two ostensibly unrelated problems.

The main impact, however, is that once one language has been proved complete for a class such as NP, others can be proved complete by constructing transformations. If A_0 is NP-complete, then to prove that another language A_1 is NP-complete, it suffices to prove that $A_1 \in$ NP, and to construct a polynomial-time transformation that establishes $A_0 \leq^p_m A_1$. Since A_0 is NP-complete, for every language A in NP, $A \leq^p_m A_0$, hence by transitivity (Theorem 23.2 in Section 23.2), $A \leq^p_m A_1$.

Hundreds of computational problems in many fields of science and engineering have been proved to be NP-complete, almost always by reduction from a problem that was previously known to be NP-complete. We give some practically motivated examples of these reductions, and also some advice on how to cope with NP-completeness.

23.4   NP-Complete Problems and Completeness Proofs

This and the next two sections are directed toward practitioners who have a computational problem, do not know how to solve it, and want to know how hard it is—specifically, is it NP-complete, or NP-hard? The following step-by-step procedure will help in answering these questions, and may help in identifying cases of the problem that are tractable even if the problem is NP-hard for general cases. In brief, the steps are

1. State the problem in general mathematical terms, and formalize the statement.
2. Ascertain whether the problem belongs to NP.
3. If so, try to find it in a compendium of known NP-complete problems.
4. If you cannot find it, try to construct a reduction from a related problem that is known to be NP-complete or NP-hard.
5. Try to identify special cases of your problem that are (a) hard, (b) easy, and/or (c) the ones you need. Your work in Steps 1–4 may help you here.
6. Even if your cases are NP-hard, they may still be amenable to direct attack by sophisticated methods on high-powered hardware.

These steps are interspersed with a traditional "theorem–proof" presentation and several long examples, but the same sequence is maintained. We emphasize that trying to do the formalization and proofs asked for in these steps may give you useful positive information about your problem.

Step 1. Give a formal statement of the problem. State it without using terms that are specific to your own particular discipline. Use common terms from mathematics and data objects in computer science, e.g., graphs, trees, lists, vectors, matrices, alphabets, strings, logical formulas, mathematical equations. For example, a problem in evolutionary biology that a phylogenist would state in terms of "species" and "characters" and "cladograms" can be stated in terms of trees and strings, using an alphabet that represents the taxonomic characters. Standard notions of size, depth, and distance in trees can express the objectives of the problem.

If your problem involves computing a function that produces a lot of output, look for associated yes/no decision problems, because decision problems have been easier to characterize and classify. For instance, if you need to compute matrices of a certain kind, see whether the essence of your problem can be captured by yes/no questions about the matrices, perhaps about individual entries of them. Many optimization problems looking for a solution of a certain minimum cost or maximum value can be turned into decision problems by including a target cost/value "k" as an input parameter, and phrasing the question of whether a solution exists of cost less than (or value greater than) the target k. Several problems given in the examples below have this form.

It may also help to simplify, even oversimplify, your problem by removing or ignoring some particular elements of it. Doing so may make it easier to ascertain what general category of decision problem yours is in or closest to. In the process, you may learn useful information about the problem that tells you what the effects of those specific elements are. We say more about this under Section 23.4.2.

Step 2. When you have an adequate formalization, ask first, does your decision problem belong to NP? This is true if and only if candidate solutions that would bring about a "yes" answer can be tested in polynomial time—see the extended discussion in Section 22.2.2. If it does belong to NP, that is good news for now! Even if not, you may still proceed to determine whether it is NP-hard. The problem may be complete for a class such as PSPACE that contains NP. Examples of such problems are given later in this chapter.

Step 3. See whether your problem is already listed in a compendium of (NP-)complete problems. The book [14] lists several hundred NP-complete problems arranged by category. The following are intended as a small representative sample. The first five (together with 3SAT) receive extended treatment in [14], while the last four—plus a problem stated before Theorem 23.6—receive comparable treatment here. (The language corresponding to each problem is the set of instances whose answers are "yes.")

Vertex Cover

Instance: A graph G and an integer k.

Question: Does G have a set W of k vertices such that every edge in G is incident on a vertex in W?

Clique

Instance: A graph G and an integer k.

Question: Does G have a set K of k vertices such that every two vertices in K are adjacent in G?

Hamiltonian Circuit

Instance: A graph G.

Question: Does G have a circuit that includes every vertex exactly once?

3-Dimensional Matching

Instance: Sets W, X, Y with $|W| = |X| = |Y| = q$ and a subset $S \subseteq W \times X \times Y$.

Question: Is there a subset $S' \subseteq S$ of size q such that no two triples in S' agree in any coordinate?

Partition

Instance: A set S of positive integers.

Question: Is there a subset $S' \subseteq S$ such that the sum of the elements of S' equals the sum of the elements of $S - S'$?

Independent Set

Instance: A graph G and an integer k.

Question: Does G have a set U of k vertices such that no two vertices in U are adjacent in G?

Graph Colorability

Instance: A graph G and an integer k.

Question: Is there an assignment of colors to the vertices of G so that no two adjacent vertices receive the same color, and at most k colors are used overall?

Traveling Salesperson (TSP)

Instance: A set of m "cities" $C_1, \ldots, C_m$, with a distance $d(i, j)$ between every pair of cities C_i and C_j, and an integer D.

Question: Is there a tour of the cities whose total length is at most D, i.e., a permutation $c_1, \ldots, c_m$ of $\{1, \ldots, m\}$, such that $d(c_1, c_2) + \cdots + d(c_{m-1}, c_m) + d(c_m, c_1) \leq D$?

Knapsack

Instance: A set $U = \{u_1, \ldots, u_m\}$ of objects, each with an integer size $size(u_i)$ and an integer profit $profit(u_i)$, a target size s_0, and a target profit p_0.

Question: Is there a subset $U' \subseteq U$ whose total cost and total profit satisfy

$$\sum_{u_i \in U'} size(u_i) \leq s_0 \quad \text{and} \quad \sum_{u_i \in U'} profit(u_i) \geq p_0?$$

The languages of all of these problems are easily seen to belong to NP. For example, to show that TSP is in NP, one can build a nondeterministic Turing machine that simply guesses a tour and checks that the tour's total length is at most D.

Some comments on the last two problems are relevant to Steps 1 and 2. TSP provides a single abstract form for many concrete problems about sequencing a series of test examples so as to minimize the variation between successive items. The KNAPSACK problem models the filling of a knapsack with items of various sizes, with the goal of maximizing the total value (profit) of the items. Many scheduling problems for multiprocessor computers can be expressed in the form of KNAPSACK instances, where the "size" of an item represents the length of time a job takes to run, and the size of the knapsack represents an available block of machine time.

If yours is on the list of NP-complete problems, you may skip Step 4, and the compendium may give you further information for Steps 5 and 6. You may still wish to pursue Step 4 if you need more detailed study of particular transformations to and from your problem.

If your problem is not on the list, it may still be close enough to one or more problems on the list to help with the next step.

Step 4. Construct a reduction from an already-known NP-complete problem. Broadly speaking, Karp reductions come in three kinds:

- A restriction from your problem to a special case that is already known to be NP-complete
- A minor adjustment of an already-known problem
- A combinatorial transformation

The first two kinds of reductions are usually quite easy to do, and we give several examples before proceeding to the third kind.

Example 23.1

PARTITION $\leq_m^p$ KNAPSACK, by restriction: Given a PARTITION instance with integers s_i, the corresponding instance of KNAPSACK takes *size* $(u_i) = profit$ $(u_i) = s_i$ (for all i), and sets the targets s_0 and p_0 both equal to $(\sum_i s_i)/2$. The condition in the definition of the KNAPSACK problem of not exceeding s_0 nor being less than p_0 requires that the sum of the selected items meet the target $(\sum_i s_i)/2$ exactly, which is possible if and only if the original instance of PARTITION is solvable.

In this way, the PARTITION problem can be regarded as a restriction or special case of the KNAPSACK problem. Note that the reduction itself goes from the more-special problem *to* the more-general problem, even though one thinks of the more-general problem as the one being restricted. The implication is that if the restricted problem is NP-hard, then the more-general problem is NP-hard as well, not vice versa.

Example 23.2

HAMILTONIAN CIRCUIT $\leq_m^p$ TSP by restriction: Let a graph G be given as an instance of the HAMILTONIAN CIRCUIT problem, and let G have m vertices $v_1, \ldots, v_m$. These vertices become the "cities" of the TSP instance that we build. Now define a distance function d as follows:

$$d(i,j) = \begin{cases} 1 & \text{if } (v_i, v_j) \text{ is an edge in } G \\ m+1 & \text{otherwise} . \end{cases}$$

Set $D = m$. Clearly, d and D can be computed in polynomial time from G. If G has a Hamiltonian circuit, then the length of the tour that corresponds to this circuit is exactly m. Conversely, if there is a tour whose length is at most m, then each step of the tour must have distance 1, not $m + 1$. Then each step corresponds to an edge of G, so the corresponding sequence of vertices forms a Hamiltonian circuit in G. Thus, the function f defined by $f(G) = (\{d(i,j) : 1 \leq i, j \leq m\}, D)$ is a polynomial-time transformation from HAMILTONIAN CIRCUIT to TSP.*

Minor Adjustments. Here, we consider cases where two problems look different but are really closely connected. Consider CLIQUE, INDEPENDENT SET, and VERTEX COVER. A graph G has a clique of size k if and only if its complementary graph G' has an independent set of size k. It follows that the function f defined by $f(G, k) = (G', k)$ is a Karp reduction from INDEPENDENT SET to CLIQUE. To forge a link to the VERTEX COVER problem, note that all vertices *not* in a given vertex cover form an independent set, and vice versa. Thus, a graph G on n vertices has a vertex cover of size at most k if and only if G has an independent set of size at least $n - k$. Hence, the function $g(G, k) = (G, n - k)$ is a Karp reduction from INDEPENDENT SET to VERTEX COVER. (Note that the same f and g also provide

---

* Technically we need f to be a function from Σ^* to Σ^*. However, given a string x we can decide in polynomial time whether x encodes a graph G that can be given as an instance of HAMILTONIAN CIRCUIT. If x does not encode a well-formed instance, then define $f(x)$ to be a fixed instance I_0 of TSP for which the answer is "no." Because this sort of thing can generally always be done, we are free to regard the domain of a reduction function f to be the set of "well-formed instances" of the problem we are reducing from. Henceforth, we try to ignore such encoding details.

reductions from CLIQUE to INDEPENDENT SET and from VERTEX COVER to INDEPENDENT SET, respectively. This does not happen for all reductions, and gives a sense in which these three problems are unusually close to each other.)

23.4.1 NP-Completeness by Combinatorial Transformation

The following examples show how the combinatorial mechanism of one problem (here, 3SAT) can be transformed by a reduction into the seemingly much different mechanism of another problem.

THEOREM 23.4   INDEPENDENT *Set is NP-complete. Hence also* CLIQUE *and* VERTEX COVER *are NP-complete.*

PROOF   We have remarked already that the languages of these three problems belong to NP, and shown already that INDEPENDENT SET $\leq_m^p$ CLIQUE and INDEPENDENT SET $\leq_m^p$ VERTEX COVER. It suffices to show that $3SAT \leq_m^p$ INDEPENDENT SET.

Construction. Let the boolean formula ϕ be a given instance of 3SAT with variables $x_1, \ldots, x_n$ and clauses $C_1, \ldots, C_m$. The graph G_ϕ we build consists of a "ladder" on $2n$ vertices labeled $x_1, \bar{x}_1, \ldots, x_n, \bar{x}_n$, with edges $(x_i, \bar{x}_i)$ for $1 \leq i \leq n$ forming the "rungs," and m "clause components." Here the component for each clause C_j has one vertex for each literal x_i or $\bar{x}_i$ in the clause, and all pairs of vertices within each clause component are joined by an edge. Finally, each clause-component node with a label x_i is connected by a "crossing edge" to the node with the opposite label $\bar{x}_i$ in the ith "rung," and similarly each occurrence of $\bar{x}_i$ in a clause is joined to the rung node x_i. This finishes the construction of G_ϕ. See Figure 23.1.

Also set $k = n + m$. Then the reduction function f is defined for all arguments ϕ by $f(\phi) = (G_\phi, k)$.

Complexity. It is not hard to see that f is computable in polynomial time given (a straightforward encoding of) ϕ.

Correctness. To complete the proof, we need to argue that ϕ is satisfiable if and only if G_ϕ has an independent set of size $n + m$. To see this, first note that any independent set I of that size must contain exactly one of the two nodes from each "rung," and exactly one node from each clause component—because the edges in the rungs and the clause component prevent any more nodes from being added. And if I selects a node labeled x_i in a clause component, then I must also select x_i in the ith rung. If I selects $\bar{x}_j$ in a clause component, then I must also select $\bar{x}_j$ in the rung. In this manner I induces a truth assignment in which $x_i = $ true and $x_j = $ false, and so on for

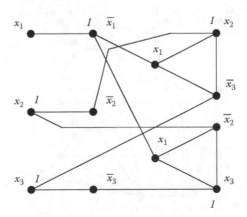

FIGURE 23.1 Construction in the proof of NP-completeness of INDEPENDENT SET for the formula $(x_1 \vee x_2 \vee \bar{x}_3) \wedge (x_1 \vee \bar{x}_2 \vee x_3)$. The independent set of size 5 corresponding to the satisfying assignment $x_1 = $ false, $x_2 = $ true, and $x_3 = $ true is shown by nodes marked I.

all variables. This assignment satisfies φ, because the node selected from each clause component tells how the corresponding clause is satisfied by the assignment. Going the other way, if φ has a satisfying assignment, then that assignment yields an independent set I of size $n + m$ in like manner.   □

Since the φ in this proof is a $3SAT$ instance, every clause component is a triangle. The idea, however, also works for CNF formulas with any number of variables in a clause, such as the $φ_x$ in the proof of the Cook–Levin Theorem.

Now we modify the above idea to give another example of an NP-completeness proof by combinatorial transformation.

THEOREM 23.5   *GRAPH COLORABILITY is NP-complete*

PROOF   *Construction.* Given the $3SAT$ instance φ, we build $G_φ$ similarly to the last proof, but with several changes. See Figure 23.2. On the left, we add a special node labeled "B" and connect it to all $2n$ rung nodes. On the right, we add a special node "G" with an edge to B. In any possible 3-coloring of $G_φ$, without loss of generality B will be colored "blue" and the adjacent G will be colored "green." The third color "red" stands for literals made true, whereas green stands for falsity.

Now for each occurrence of a positive literal x_i in a clause, the corresponding clause component has two nodes labeled x_i and x'_i with an edge between them; and similarly an occurrence of a negated literal $\bar{x}_j$ gives nodes $\bar{x}_j$ and $\bar{x}'_j$ with an edge between them. The primed ("inner") nodes in each component are connected by edges into a triangle, but the unprimed ("outer") nodes are not. Each outer node of each clause component is instead connected by an edge to G. Finally, each outer node x_i is connected by a "crossing edge" to the rung node $\bar{x}_i$, and each outer node $\bar{x}_j$ to rung node x_j, exactly as in the INDEPENDENT SET reduction. This finishes the construction of $G_φ$.

Complexity. The function f such that given any φ outputs $G_φ$, also fixing $k = 3$, is clearly computable in polynomial time.

Correctness. The key idea is that every three-coloring of B, G, and the rung nodes, which corresponds to a truth assignment to the variables of φ, can be extended to a 3-coloring of a clause component if and only if at least one of the three crossing edges from the component goes to a green rung node. If all three of these edges go to red nodes, then the links to G force each outer node in the component to be colored blue—but then it is impossible to three-color the inner triangle since

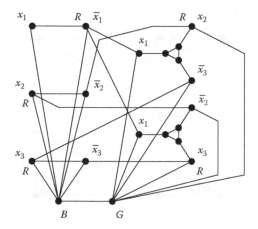

FIGURE 23.2 Construction in the proof of NP-completeness of GRAPH COLORABILITY for the formula $(x_1 \lor x_2 \lor \bar{x}_3) \land (x_1 \lor \bar{x}_2 \lor x_3)$. The nodes shown colored R correspond to the satisfying assignment $x_1 = $ false, $x_2 = $ true, and $x_3 = $ true, and these together with G and B essentially force a 3-coloring of the graph, which the reader may complete. Note the structural resemblance to Figure 23.1.

blue cannot be used. Conversely, any crossing edge to a green node allows the outer node x_i or $\bar{x}_j$ to be colored red, so that one red and two blues can be used for the outer nodes, and this allows the inner triangle to be colored as well. Hence, G_ϕ is 3-colorable if and only if ϕ is satisfiable. $\square$

Note that we have also shown that the restricted form of GRAPH COLORABILITY with k fixed to be 3 (i.e., given a graph G, is G 3-colorable?) is NP-complete. Had we stated the problem this way originally, we would now conclude instead that the more-general graph-colorability problem is NP-complete, similarly to the KNAPSACK and TSP examples above.

Many other reductions from 3SAT use the same basic pattern of a truth-assignment selection component for the variables, components for the clauses (whose behavior depends on whether a variable in the clause is satisfied), and links between these components that make the reduction work correctly. For another example of this pattern, a standard proof that HAMILTONIAN CIRCUIT is NP-complete uses subgraphs V_i for each pair $x_i, \bar{x}_i$ and C_j for each clause. There are two possible ways a circuit can enter V_i, and these correspond to the choices of $x_i = \texttt{true}$ or $x_i = \texttt{false}$ in an assignment. The whole graph is built so that if the circuit enters V_i on the "$x_i = \texttt{true}$" side, then the circuit has the opportunity to visit all the nodes in the C_j components for all clauses in which x_i occurs positively, and similarly for occurrences of $\bar{x}_i$ if the circuit enters on the negative side. Hence, the circuit can run through every C_j if and only if ϕ is satisfiable. Full details may be found in the text by Papadimitriou [25]. For our last fully worked-out example, we show a somewhat different pattern in which the individual variables as well as the clauses correspond to top-level components of the following problem.

DISJOINT CONNECTING PATHS

Instance: A graph G with two disjoint sets of distinguished vertices $s_1, \ldots, s_k$ and $t_1, \ldots, t_k$, where $k \geq 1$.

Question: Does G contain paths $P_1, \ldots, P_k$, with each P_i going from s_i to t_i, such that no two paths share a vertex?

THEOREM 23.6 *DISJOINT CONNECTING PATHS is NP-complete.*

PROOF First, it is easy to see that DISJOINT CONNECTING PATHS belongs to NP: one can design a polynomial-time nondeterministic Turing machine that simply guesses k paths and then deterministically checks that no two of these paths share a vertex. Now let ϕ be a given instance of $3SAT$ with n variables and m clauses. Take $k = n + m$.

Construction and complexity. The graph G_ϕ we build has distinguished path-origin vertices $s_1, \ldots, s_n$ for the variables and $S_1, \ldots, S_m$ for the clauses of ϕ. G_ϕ also has corresponding sets of path-destination nodes $t_1, \ldots, t_n$ and $T_1, \ldots, T_m$. The other vertices in G_ϕ are nodes u_{ij} for each occurrence of a positive literal x_i in a clause C_j, and nodes v_{ij} for each occurrences of a negated literal $\bar{x}_i$ in C_j. For each i, $1 \leq i \leq n$, G_ϕ is given the edges for a directed path from s_i through all u_{ij} nodes to t_i, and another from s_i through all v_{ij} nodes to t_i. (If there are no occurrences of the positive literal x_i in any clause then the former path is just an edge from s_i right to t_i, and likewise for the latter path if the negated literal $\bar{x}_i$ does not appear in any clause.) Finally, for each j, $1 \leq j \leq m$, G_ϕ has an edge from S_j to every node u_{ij} or v_{ij} for the jth clause, and edges from those nodes to T_j. Clearly these instructions can be carried out to build G_ϕ in polynomial time given ϕ. (See Figure 23.3.)

Correctness. The first point is that for each i, no path from s_i to t_i can go through both a "u-node" and a "v-node." Setting x_i true corresponds to avoiding u-nodes, and setting x_i false entails avoiding v-nodes. Thus, the choices of such paths for all i represent a truth assignment. The key point is that for each j, one of the three nodes between S_j and T_j will be free for the taking if and only if

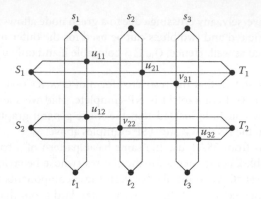

FIGURE 23.3 Construction in the proof of NP-completeness of DISJOINT CONNECTING PATHS for the formula $(x_1 \vee x_2 \vee \bar{x}_3) \wedge (x_1 \vee \bar{x}_2 \vee x_3)$.

the corresponding positive or negative literal was made true in the assignment, thus satisfying the clause. Hence, G_ϕ has the $n + m$ required paths if and only if ϕ is satisfiable.

23.4.2   Significance of NP-Completeness

Suppose that you have proved that your problem is NP-complete. What does this mean, and how should you approach the problem now?

Exactly what it means is that your problem does not have a polynomial-time algorithm, unless every problem in NP has a polynomial-time algorithm; i.e., unless NP ≠ P. We have discussed above the reasons for believing that NP ≠ P. In practical terms, you can draw one definite conclusion: Do not bother looking for a "magic bullet" to solve the problem. A simple formula or an easily tested deciding condition will not be available; otherwise it probably would have been spotted already during the thousands of person-years that have been spent trying to solve similar problems. For example, the NP-completeness of GRAPH 3-COLORABILITY effectively ended hopes that an efficient mathematical formula for deciding the problem would pop out of research on "chromatic polynomials" associated to graphs. Notice that NP-hardness does not say that one needs to be "extra clever" to find a feasible solving algorithm—it says that one probably does not exist at all.

The proof itself means that the combinatorial mechanism of the problem is rich enough to simulate boolean logic. The proof, however, may also unlock the door to finding saving graces in Steps 5 and 6.

Step 5. Analyze the instances of your problem that are in the range of the reduction. You may tentatively think of these as "hard cases" of the problem. If these differ markedly from the kinds of instances that you expect to see, then this difference may help you refine the statement and conditions of your problem in ways that may actually define a problem in P after all.

To be sure, avoiding the range of one reduction still leaves wide open the possibility that another reduction will map into your instances of interest. However, it often happens that special cases of NP-complete problems belong to P—and often, the boundary between these and the NP-complete cases is sudden and sharp. For one example, consider SAT. The restricted case of three variables per clause is NP-complete, but the case of two variables per clause belongs to P.

For another example, note that the proof of NP-completeness for DISJOINT CONNECTING PATHS given above uses instances in which $k = n + m$; i.e., in which k depends on the number of variables.

The case $k = 2$, where you are given G and s_1, s_2, t_1, t_2 and need to decide whether there are vertex-disjoint paths from s_1 to t_1 and from s_2 to t_2, belongs to P. (The polynomial-time algorithm for this case is nontrivial and was not discovered until 1978 by Garey and Johnson [14].)

However, one must also be careful in one's expectations. Suppose we alter the statement of DISJOINT CONNECTING PATHS by requiring also that no two vertices in two different paths may have an edge between them. Then the case $k = 2$ of the new problem is NP-complete. (Showing this is a nice exercise; the idea is to make one path climb the "variable ladder" and send the other path through all the clause components.)

23.4.3 Strong NP-Completeness for Numerical Problems

An important difference between hard and easy cases applies to certain NP-complete problems that involve numbers. For example, above we stated that the PARTITION problem is NP-complete; thus, it is unlikely to be solvable by an efficient algorithm. Clearly, however, we can solve the PARTITION problem by a simple dynamic programming algorithm, as follows.

For an instance of PARTITION, let S be a set of positive integers $\{s_1, \ldots, s_m\}$, and let s^* be the total, i.e., $s^* = \sum_{i=1}^{m} s_i$. Initialize a linear array B of Boolean values so that $B[0] = \texttt{true}$, and each other entry of B is $\texttt{false}$. For $i = 1$ to m, and for $t = s^*$ down to 0, if $B[t] = \texttt{true}$, then set $B[t + s_i]$ to $\texttt{true}$. After the ith iteration, $B[t]$ is $\texttt{true}$ if and only if a subset of $\{s_1, \ldots, s_i\}$ sums to t. The answer to this instance of PARTITION is "yes," if $B[s^*/2]$ is ever set to $\texttt{true}$.

The running time of this algorithm depends critically on the representation of S. If each integer in S is represented in binary notation, then the running time is exponential in the total length of the representation. If each integer is represented in unary—that is, each s_i is represented by s_i consecutive occurrences of the same symbol—then total length of the representation would be greater than s^*, and the running time would be only a polynomial in the length. Put another way, if the magnitudes of the numbers involved are bounded by a polynomial in m, then the above algorithm runs in time bounded by a polynomial in m. Since the length of the encoding of such a low-magnitude instance is $O(m \log m)$, the running time is polynomial in the length of the input. The bottom line is that these cases of the PARTITION problem are feasible to solve completely.

A problem is NP-complete in the strong sense, if there is a fixed polynomial p such that for each instance x of the problem, the value of the largest number encoded in x is at most $p(|x|)$. That is, the integer values are polynomial in the length of the standard representation of the problem. By definition, the 3SAT, VERTEX COVER, CLIQUE, HAMILTONIAN CIRCUIT, and 3-DIMENSIONAL MATCHING problems defined in Section 23.4 are NP-complete in the strong sense, but PARTITION and KNAPSACK are not. The PARTITION and KNAPSACK problems can be solved in polynomial time, if the integers in their statements are bounded by a polynomial in n—for instance, if numbers are written in unary rather than binary notation.

The concept of strong NP-completeness reminds us that the representation of information can have a major impact on the computational complexity of a problem.

23.4.4 Coping with NP-Hardness

Step 6. Even if you cannot escape NP-hardness, the cases you need to solve may still respond to sophisticated algorithmic methods, possibly needing high-powered hardware.

There are two broad families of direct attack that have been made on hard problems. Exact solvers typically take exponential time in the worst case, but provide feasible runs in certain concrete cases. Whenever they halt, they output a correct answer—and some exact solvers also output a proof that their answer is correct. Solvers for SAT are surveyed by Zhang and Malik [32]. Heuristic algorithms typically run in polynomial time in all cases, and often aim to be correct only most of the time, or to find approximate solutions (see Sections 23.6.1 and 23.6.2. They are more common. Popular heuristic methods include genetic algorithms, simulated annealing, neural networks, relaxation to linear programming, and stochastic (Markov) process simulation. Experimental systems dedicated to certain NP-complete problems have yielded some interesting results. For surveys of solvers for TSP see Johnson and McGeogh [18] and Applegate et al. [3].

There are two ways to attempt to use this research. One is to find a problem close to yours for which people have produced solvers, and try to carry over their methods and heuristics to the specific features of your problem. The other (much more speculative) is to construct a Karp reduction from your problem to their problem, ask to run their program or machine itself on the transformed instance, and then try to map the answer obtained back to a solution of your problem. The hitches are (1) that the currently-known Karp reductions f tend to lose much of the potentially helpful structure of the source instance x when they form $f(x)$, and (2) that approximate solutions for $f(x)$ may map back to terribly suboptimal or even infeasible answers to x. (See, however, the notion of L-reductions in Section 24.9) All of this indicates that there is much scope for further research on important practical features of relationships between NP-complete problems. See also Chapter 35.

23.4.5 Beyond NP-Hardness

If your problem belongs to NP and you cannot prove that it is NP-hard, it may be an "NP-intermediate" problem; i.e., neither in P nor NP-complete. The theorem of Ladner mentioned in Section 23.3 shows that NP-intermediate problems exist, assuming NP $\neq$ P. However, very few natural problems are currently counted as good candidates for such intermediate status: factoring, discrete logarithm, graph isomorphism, and several problems relating to lattice bases form a very representative list. For the first two, see Chapter 16. The vast majority of natural problems in NP have resolved themselves as being either in P or NP-complete. Unless you uncover a specific connection to one of those four intermediate problems, it is more likely offhand that your problem simply needs more work.

The observed tendency of natural problems in NP to "cluster" as either being in P or NP-complete, with little in between, reinforces the arguments made early in this chapter that P is really different from NP.

Finally, if your problem seems not to be in NP, or alternatively if some more stringent notion of feasibility than polynomial time is at issue, then you may desire to know whether your problem is complete for some other complexity class. We now turn to this question.

23.5 Complete Problems for NL, P, and PSPACE

We first investigate the log-space analog of the P versus NP question, namely whether NL = L. We show that there are natural computational problems that are NL-complete. The question is, under which reducibility? Polynomial-time reducibility is too blunt an instrument here, because NL is contained in P, and so all languages in NL are technically complete for NL under both $\leq_m^p$ and $\leq_T^p$ reductions. We need a reducibility that is fine enough to preserve the distinction between deterministic and nondeterministic log space that we are attempting to establish and study. The simplest way is to replace the polynomial-time bound in Karp reductions by a log-space bound.

- A language A_1 is **log-space reducible** to a language A_2, written $A_1 \leq_m^{log} A_2$, if A_1 is many-one reducible to A_2 via a transformation function that is computable by a deterministic Turing machine in $O(\log n)$ space.

There is a log-space analog of Cook reducibility, but we do not use it here. Now we observe that $\leq_m^{log}$ reductions have the properties we desire:

THEOREM 23.7

(a) *(Closure) If $A_1 \leq_m^{log} A_2$ and $A_2 \in$ L, then $A_1 \in$ L.*

(b) *(Transitivity) If $A_1 \leq_m^{log} A_2$ and $A_2 \leq_m^{log} A_3$, then $A_1 \leq_m^{log} A_3$.*

(c) *(Refinement of $\leq_m^p$ reductions) If $A_1 \leq_m^{log} A_2$, then $A_1 \leq_m^p A_2$.*

The proof of (a) and (b) is somewhat tricky and rests on the fact that if two functions f and g from strings to strings are computable in log space, then so is the function h defined by $h(x) = g(f(x))$. The hitch is that a log-space Turing machine M_f computing f can output the characters of $f(x)$ serially but does not have space to store them. This becomes a problem whenever the machine M_g computing g, whose input is the output from M_f, requests the ith character of $f(x)$, where i may be less than the index j of the previous request. The solution is that since only space and not time is constrained, we may restart the computation of $M_f(x)$ from scratch upon the request, and let M_g count the characters that M_f outputs serially until it sees the ith one. Such a counter, and similar ones tracking the movements of M_f's actual input head and M_g's "virtual" input head, can be maintained in $O(\log n)$ space. Thus, we need no physical output tape for M_f or input tape for M_g, and we obtain a tandem machine that computes $g(f(x))$ in log space. Part (c) is immediate by the function-class counterpart of the inclusion L $\subseteq$ P.

The definition of "NL-complete" is an instance of the general definition of completeness at the beginning of Section 23.3: A language A_1 is NL-complete ("under $\leq_m^{log}$ reductions" is assumed), if $A_1 \in$ NL and for every language $A_2 \in$ NL, $A_2 \leq_m^{log} A_1$. One defines "P-complete" in a similar manner—again with $\leq_m^{log}$ reductions assumed. Together with the observation that whenever $A_1, A_2 \in$ L we have $A_1 \leq_m^{log} A_2$ (ignoring technicalities for A_1 or A_2 equal to $\emptyset$ or Σ^*), we obtain a similar state of affairs to what is known about NP-completeness and the P versus NP question:

THEOREM 23.8   *Let A be NL-complete. Then the following statements are equivalent:*

- NL $=$ L.
- $A \in$ L.
- *Some NL-complete language belongs to* L.
- *All NL-complete languages belong to* L.
- *All languages in* L *are NL-complete.*

Substitute "P" for "NL" and the same equivalence holds. Note that here we are applying completeness to a class, namely P itself, whose definition does not involve nondeterminism.

The Cook–Levin Theorem provides two significant inferences: evidence of intractability, and a connection between computation and boolean logic. NL-completeness is not to any comparable degree a notion of intractability, but does provide a fundamental link between computations and graphs, via the following important problem.

GRAPH ACCESSIBILITY PROBLEM (GAP)

Instance: A directed graph G, and two nodes s and t of G.

Question: Does G have a directed path from node s to node t?

Other names are the s–t connectivity problem and the reachability problem.

The link involves the concept of an instantaneous description (ID) of a Turing machine M. Let us suppose for simplicity that M has just two tapes: one read-only input tape that holds the input x, and one work tape with alphabet $\{0, 1, B\}$, where B is the blank character. Let us also suppose that M never writes a B. Then any step of a computation of M on the fixed input x is describable by giving

- The current state q of M
- The contents y of the work tape
- The position i of the input tape head
- The position j of the work tape head

Then the 4-tuple (q, y, i, j) is called an ID of M on input x. The restriction on writing B allows us to identify y with a string in $\{0, 1\}^*$. Without loss of generality, we always have $1 \leq i \leq n + 1$, where $n = |x|$, and if $s(n)$ is a space bound on M, also $1 \leq j \leq s(n)$. An ID is also called a **configuration**.

Now define G_x to be the graph whose nodes are all possible IDs of M on input x, and whose directed edges comprise all pairs (I, J) such that M, if set up in the configuration I, has a transition that takes it to the configuration J in one step. If M is deterministic, then every node in G_x has at most one outgoing arc. Nondeterministic TMs, however, give rise to directed graphs G_x of out-degree more than one. Note that G_x does depend on x, since the step(s) taken from an ID (q, y, i, j) may depend on bit x_i of x.

THEOREM 23.9   *GAP is* NL-*complete.*

PROOF   *GAP* belongs to NL because guessing successive edges in a path from 1 to R (when one exists) needs only $O(\log n)$ space to store the label of the current node, and to locate where on the input tape the adjacency information for the current node is stored. To show NL-hardness, let $A \in$ NL. Then A is accepted by a nondeterministic $O(\log n)$ space bounded Turing machine M. We prove that $A \leq_m^{log}$ GAP.

Construction. It is easy to modify M to have the properties supposed in the above discussion of IDs and still run in $O(\log n)$ space. We may also code M so that any accepting computation has a final phase that blanks out all used work tape cells, leaves the input head in cell $n + 1$, and halts in a special accepting state q_a. This ensures that every accepting computation (if any) ends in the unique ID $I_t = (q_a, \lambda, n + 1, 1)$, where λ stands for the empty string.

Now given any x, define G_x as above. Let node s be the unique starting ID $I_s = (q_0, \lambda, 1, 1)$, and let node t be I_t. Note that the size of G_x is polynomial—if M runs in $k \log n$ space, then the size is $O(n^{k+2})$.

Complexity. We show that the transformation f that takes a string x as input and produces the list of edges in G_x as output can be computed by a machine M_f that uses $O(\log n)$ space. For each node $I = (q, y, i, j)$ in turn, M_f reads the ith symbol of x and then produces an edge (I, J) for each J such that M can move from I to J when reading that symbol. The only memory space that M_f needs is the space to step through each I in turn, and count up to the ith input position, and produce each J. $O(\log n)$ space is sufficient for all of this.

Correctness. By the construction, paths in G_x correspond to valid sequences of transitions by M on input x. Hence, there exists a path from s to t in G_x if and only if M has an accepting computation on input x.   □

To see an example of a reduction between two NL-complete problems, consider the related problem *SC* of whether a given directed graph is strongly connected, meaning that there is a path from every node u to every other node v. Then *GAP* $\leq_m^{log}$ *SC*: Take an instance graph G with distinguished nodes s and t and add an edge from every node to s and from t to every node. Computing this transformation needs only $O(\log n)$ space to store the labels of nodes s and t and find their adjacency information on the input tape, changing '0' for "non-edge" to '1' for "edge" as appropriate while writing to the output tape. This transformation is correct because the new edges cannot cause a path from s to t to exist when there was not one beforehand, but do allow any such path to be extended to and from any other pair of nodes. SC belongs to NL since a log-space machine can cycle through all pairs (u, v) and nondeterministically guess a path in each case, so SC is NL-complete.

Further variations of the connectivity theme and many other problems are NL-complete. For an interesting contrast to the current situation with NP-completeness, the complements of all these problems are also NL-complete under $\leq_m^{log}$ reductions! This is true because NL is closed under complementation (Theorem 24.4(c)). The next problem, however, is apparently harder than the NL-complete problems.

CIRCUIT VALUE PROBLEM

Instance: A boolean circuit C, and an assignment I to the inputs of C.

Question: Does $C(I)$ evaluate to `true`?

THEOREM 23.10   *CVP is P-complete under $\leq_m^{log}$ reductions.*

PROOF   That $CVP \in P$ is clear, and completeness is essentially proved by the construction in the proof of Theorem 22.11, which gives a polynomial-size circuit family $\{C_n\}$ that accepts any given language in P. $\qquad\square$

Thus, CVP belongs to L if and only if P $=$ L, to NL if and only if P $=$ NL, and to NC if and only if P $=$ NC. The third statement follows from NC likewise being closed under $\leq_m^{log}$ reductions. For more P-complete problems, more detail, and discussion of the kind of "intractability" that P-completeness is evidence for, see Chapter 28.

The last problem we consider here is a generalization of SAT. Quantified boolean formulas may use the quantifiers $\forall$ and $\exists$ as well as $\{\wedge, \vee, \neg\}$. The formula is *closed* if every variable is quantified. For example, an instance $\phi(x_1, \ldots, x_n)$ of SAT is satisfiable if and only if the closed quantified boolean formula $\phi' = (\exists x_1)(\exists x_2) \ldots (\exists x_n)\phi$ is true. Another example of a quantified boolean formula is $\forall x \forall y \exists z(x \wedge (\bar{y} \vee z))$, and this one happens to be false.

QUANTIFIED BOOLEAN FORMULAS (QBF)

Instance: A closed quantified boolean formula ϕ.

Question: Is ϕ true?

THEOREM 23.11   *QBF is PSPACE-complete under $\leq_m^{log}$ reductions, hence also under $\leq_m^{p}$ reductions.*

PROOF   Given an n-variable instance ϕ, $O(n)$ space suffices to maintain a stack with current assignments to each variable, and with this stack one can evaluate ϕ by unwinding one quantifier at a time. So $QBF \in$ PSPACE. For hardness, let $A \in$ PSPACE, and let M be a Turing machine that accepts A in polynomial space. We prove that $A \leq_m^{log} QBF$.

Construction. Given an input x to M, define the ID graph of G_x as before for the proof of Theorem 23.9, but for a polynomial rather than logarithmic space bound. The size of G_x is bounded by $2^{s(n)}$ for some polynomial s, where $n = |x|$. We first define, by induction on r, formulas $\Phi_r(I, J)$

expressing that M started in configuration I can reach configuration J in at most 2^r transitions. The base case formula $\Phi_0(I, J)$ asserts that (I, J) is an edge of G_x.

The idea of the induction is to assert that there is an ID K that is "halfway between" I and J. The straightforward definition $\Phi_r(I, J) := (\exists K)[\Phi_{r-1}(I, K) \wedge \Phi_{r-1}(K, J)]$, however, blows Φ_r up to size exponential in r because of the two occurrences of "Φ_{r-1}" on the right-hand side. The trick is to define, for $r \geq 1$,

$$\Phi_r(I, J) := (\exists K) \left(\forall I', J'\right) : \left[\left(I' = I \wedge J' = K\right) \vee \left(I' = K \wedge J' = J\right)\right] \to \Phi_{r-1}\left(I', J'\right).$$

The single occurrence of "Φ_{r-1}" makes the size of $\Phi_r(I, J)$ roughly proportional to r. Now let I_s be the starting ID of M on input x, and I_t the unique accepting ID, from the proof of Theorem 23.9. Then M accepts x if and only if $\Phi_{s(n)}(I_s, I_t)$ is true.

To convert $\Phi_{s(n)}(I_s, I_t)$ into an equivalent instance ϕ_x of *QBF*, we can represent IDs by blocks of $s(n)$ boolean variables. All we need to do is code up polynomially-many instances of the predicates "$I - J$" and "M has a transition from I to J." This is similar to the coding done in the proof of Theorem 23.3, since levels of the circuits C_n in that proof are essentially IDs.

Complexity and Correctness. The only real care needed in the straightforward buildup of $\Phi_{s(n)}(I_1, I_R)$ and then ϕ_x is keeping track of the variables. Since there are only polynomially many of them, each has a tag of length $O(\log n)$, and the housekeeping can be done by a deterministic log-space machine. The reduction is correct since $x \in A$ if and only if $\phi_x \in QBF$.   □

Remarks. This construction is unaffected if M is nondeterministic, and works for any constructible space bound by $s(n) \geq \log n$, producing a formula ϕ_x with $O(s(n)^2)$ boolean variables. Since ϕ_x can be evaluated deterministically in $O(s(n)^2)$ space, we have also proved Savitch's Theorem (Theorem 22.3(d)), namely that NSPACE$[s(n)] \subseteq$ DSPACE$[s(n)^2]$. We have also essentially proved that PSPACE equals alternating polynomial time (see Theorem 22.9(b)), since ϕ_x can be evaluated in $O(s(n)^2)$ *time* by an ATM M that makes existential and universal moves corresponding to the leading "$\exists$" and "$\forall$" quantifier blocks in ϕ_x. This in turn yields an alternative reduction from any language A in PSPACE to QBF, since the proof method for the Cook–Levin Theorem (Theorem 23.3) extends to convert this M directly into a quantified boolean formula.

One family of PSPACE-complete problems consists of connectivity problems for graphs that, although of exponential size, are specified by a "hierarchical," recursive, or some other scheme that enables one to test whether (u, v) is an edge in time polynomial in the length of the labels of u and v. The graph G_x in the last proof is of this kind, since its edge relation $\Phi_1(I, J)$ is polynomial-time decidable. Another family comprises many two-player combinatorial games, where the question is whether the player to move has a winning strategy. A reduction from QBF to the game question transforms a formula such as $(\exists x)(\forall y)(\exists z) \ldots B$ into reasoning of the form "there exists a move for Black such that for all moves by White, there exists a move for Black such that . . . Black wins," starting from a carefully constructed position. Decision problems that exhibit this kind of "there exists. . . for all. . . " alternation (with polynomially many turns) are often PSPACE-complete.

All PSPACE-complete problems are NP-hard, since NP $\subseteq$ PSPACE; hence, PSPACE $=$ P if and only if any one of them belongs to P. The only definite lower bound that follows from the results in this section is that no problem that is PSPACE-complete under $\leq_m^{log}$ reductions belongs to L or even NL, because these classes are closed under $\leq_m^{log}$ reductions and PSPACE $\neq$ NL. It is, however, still possible to have a problem that is PSPACE-complete under $\leq_m^p$ reductions belong to L, since if PSPACE $=$ P then all languages in P are PSPACE-complete under $\leq_m^p$ reductions.

To investigate problems and classes within L, we need even finer reducibility relations than $\leq_m^{log}$. Amazingly, these new reductions, which are based on our tiniest canonical complexity class, are effective not just within L but for natural problems in all the complexity classes in these chapters, including all the problems defined above.

23.6 AC⁰ **Reducibilities**

A function f is in AC^0 if and only if the language $\left\{ \langle x, i, b \rangle : i \le |f(x)| \wedge \text{bit } i \text{ of } f(x) \text{ equals } b \right\}$ belongs to AC^0.

- A language A_1 is AC^0 reducible to a language A_2, written $A_1 \le_m^{AC^0} A_2$, if A_1 is many-one reducible to A_2 via a transformation in AC^0.

- A_1 is AC^0-Turing reducible to A_2, written $A_1 \le_T^{AC^0} A_2$, if A_1 is recognized by a DLOGTIME-uniform family of circuits of polynomial size and constant depth, consisting of $\neg$ gates, unbounded fan-in $\wedge$ and $\vee$ gates, and oracle gates for A_2. (An oracle gate for A_2 takes m inputs $x_1, \ldots, x_m$ and outputs 1 if $x_1 \ldots x_m$ is in A_2, and outputs 0 otherwise.)

The next theorem summarizes basic relationships among the five reducibility relations defined thus far.

THEOREM 23.12 *For any languages A_1, A_2, and A_3:*
(Transitivity)

(a) *If $A_1 \le_m^{AC^0} A_2$ and $A_2 \le_m^{AC^0} A_3$, then $A_1 \le_m^{AC^0} A_3$.*

(b) *If $A_1 \le_T^{AC^0} A_2$ and $A_2 \le_T^{AC^0} A_3$, then $A_1 \le_T^{AC^0} A_3$.*

(Refinement)

(c) *$A_1 \le_m^{AC^0} A_2 \implies A_1 \le_m^{log} A_2 \implies A_1 \le_m^{p} A_2 \implies A_1 \le_T^{p} A_2$.*

(d) *$A_1 \le_m^{AC^0} A_2 \implies A_1 \le_T^{AC^0} A_2 \implies A_1 \le_T^{p} A_2$.*

In prose, (c) says that AC^0 reducibility implies log-space reducibility, which implies Karp reducibility, which implies Cook reducibility; and (d) says that AC^0 reducibility implies AC^0-Turing reducibility, which implies Cook reducibility. However, AC^0-Turing reducibility is known not to imply log-space reducibility or even Karp reducibility—any language that does not many-one reduce to its complement shows this.

Next, we list which of our canonical complexity classes are closed under which reducibilities. Note that the subclasses of P are not known to be closed under the more powerful reducibilities, and that the nondeterministic time classes are not known to be closed under the Turing reducibilities—mainly because they are not known to be closed under complementation.

THEOREM 23.13

(a) P, PSPACE, EXP, *and* EXPSPACE *are closed under Cook reducibility, and hence under AC^0 reducibility, AC^0-Turing reducibility, log-space reducibility, and Karp reducibility as well.*

(b) NP *and* NEXP *are closed under Karp reducibility, hence also under AC^0 reducibility and log-space reducibility.*

(c) L, NL, *and* NC *are closed under both log-space reducibility and AC^0-Turing reducibility, hence also under AC^0 reducibility.*

(d) NC^1, TC^0, *and* AC^0 *are closed under AC^0-Turing reducibility, hence also under AC^0 reducibility.*

For contrast, note that E and NE are *not* closed under AC^0 reducibility—hence they are not closed under any of the other reducibilities, either. To see this, let A be any language in EXP − E; such languages exist by the time hierarchy theorem (Theorem 22.5). Then for some $k > 0$, the language $A_k = \left\{ x10^{|x|^k} : x \in A \right\}$ belongs to E, and it is easy to see that $A \leq_m^{AC^0} A_k$. If E were closed under $\leq_m^{AC^0}$ reductions, A would be in E, a contradiction. The same can be done for NE using NEXP in place of EXP.

Note that this gives an easy proof that E $\neq$ NP, because NP has a closure property that E does not share. On the other hand, although this inequality tells us that exactly one of the following must hold,

- NP $\subset$ E
- E $\subset$ NP
- NP $\not\subseteq$ E and E $\not\subseteq$ NP,

it is not known which of these is true.

23.6.1 Why Have So Many Kinds of Reducibility?

We have already discussed one reason to consider different kinds of reducibility: in order to explore the important subclasses of P, more restrictive notions such as $\leq_m^{AC^0}$ and $\leq_m^{log}$ are required. However, that does not explain why we study both Karp and Cook reducibility, or both AC^0 and AC^0-Turing reductions. It is worth taking a moment to explain this.

If our goal was merely to classify the deterministic complexity of problems, then Cook reducibility ($\leq_T^p$) would be the most natural notion to study. The class of problems that is Cook reducible to A (usually denoted by P^A) characterizes what can be computed quickly if A is easy to compute. Note in particular that A is Cook-reducible to its complement $\overline{A}$, and A and $\overline{A}$ have the same deterministic complexity.

However, if A is an NP-complete language, then A probably does not have the same nondeterministic complexity as $\overline{A}$. That is, $\overline{A} \in$ NP only if NP = co−NP. It is worth emphasizing that, if we know only that A is complete for NP under $\leq_T^p$ reductions, the hypothesis $\overline{A} \in$ NP does not allow us to conclude NP = co−NP. That is, we get stronger evidence that $\overline{A}$ has high nondeterministic complexity, if we know that A is complete for NP under the more restrictive kind of reducibility.

This is a general phenomenon: if we know that a language is complete under more restrictive kind of reducibility, then we know more about its complexity. We have already seen one other example: knowing a language A is complete for PSPACE under $\leq_m^{log}$ reductions tells you that $A \notin$ NL, whereas completeness under $\leq_m^p$ reductions does not even entail that $A \notin$ L. In the latter case, we must appeal to the unproven conjecture that P $\neq$ PSPACE to infer that A is not in L. For another example, if A is complete for NP under $\leq_m^{AC^0}$ reductions, then we know that $A \notin AC^0$, whereas if we know only that A is NP-complete under $\leq_m^{log}$ reductions, then we cannot conclude anything about the complexity of A, because we cannot yet rule out the possibility that L = NP.

23.6.2 Canonical Classes and Complete Problems

It is an amazing and surprising fact that most computational problems that arise in practice turn out to be complete for some natural complexity class—and complete under some extremely restrictive reducibility such as $\leq_m^{AC^0}$. Indeed, the lion's share of those natural problems known to be complete for NP under $\leq_m^p$ reductions, and for P under $\leq_m^{log}$ reductions, etc., is in fact complete under $\leq_m^{AC^0}$ reductions, too. We observe this after filling out our spectrum of complexity classes with some more decision problems.

INTEGER MULTIPLICATION

Instance: The binary representation of integers x and y, and a number i.

Question: Is the ith bit of the binary representation of $x \cdot y$ equal to 1?

BOOLEAN FORMULA VALUE PROBLEM (BFVP)

Instance: A boolean formula ϕ and a 0-1 assignment I to the variables in ϕ.

Question: Does $\phi(I)$ evaluate to `true`?

DEGREE-ONE CONNECTIVITY (GAP$_1$)

Instance: A directed graph in which each node has at most one outgoing edge, and nodes s, t of G.

Question: Is there a path from node s to node t in G?

REGULAR EXPRESSIONS WITH ($\cup, \cdot, ^*$)

Instance: A regular expression α with the standard union, concatenation, and Kleene closure operations (see Chapter 20).

Question: Is there a string that does not match α?

REGULAR EXPRESSIONS WITH ($\cup, \cdot, ^2$)

Instance: A regular expression α with union, concatenation, and "squaring" operators (where α^2 denotes $\alpha \cdot \alpha$).

Question: Is there a string that does not match α?

REGULAR EXPRESSIONS WITH ($\cup, \cdot, ^*, ^2$)

Instance: A regular expression α composed of the union, concatenation, "squaring," and Kleene star operators.

Question: Is there a string that does not match α?

N × N CHECKERS

Instance: A position in checkers played on an $N \times N$ board, with Black to move.

Question: Is this a winning position for Black?

THEOREM 23.14 *The following problems are complete for the given complexity classes under $\leq_m^{AC^0}$ reductions, except that INTEGER MULTIPLICATION is only known to be complete under $\leq_T^{AC^0}$ reductions.*

- TC^0 : *INTEGER MULTIPLICATION.*
- NC^1 : *BFVP.*
- L: *GAP$_1$.*
- NL: *GAP.*
- P: *CVP.*
- NP: *SAT, CLIQUE, VERTEX COVER, and so on.*
- PSPACE : *QBF, REGULAR EXPRESSIONS WITH ($\cup, \cdot, ^*$).*
- EXP : *N × N CHECKERS.*
- NEXP : *REGULAR EXPRESSIONS WITH ($\cup, \cdot, ^2$).*
- EXPSPACE : *REGULAR EXPRESSIONS WITH ($\cup, \cdot, ^*, ^2$).*

The last three problems also belong to E, NE, and DSPACE[$2^{O(n)}$], respectively (under suitable encodings), and so they are complete for these respective classes as well. Note that a "tiny" reducibility still gives complete problems for a big class! However, TC^0 is not known to have any complete problems under $\leq_m^{AC^0}$ reductions.

Note that the class NC does not appear anywhere in the list above. NC is not known (or generally believed) to have any complete language under log-space reductions. In fact, if NC does have a language that is complete under $\leq_m^{log}$ reducibility, then there is some k such that $NC^k = NC^{k+1} = \cdots = NC$. This is considered unlikely. This behavior is typical of certain "hierarchy classes," and the polynomial hierarchy class PH (defined in the next chapter) behaves similarly with regard to $\leq_m^p$ reductions.

To (im)prove the claim about $\leq_m^{AC^0}$ reductions in Theorem 23.14, we can show that all the reductions in this chapter are computable by uniform AC^0 circuits without any $\wedge$ *or* $\vee$ gates at all! The circuits have only the constants 0 and 1, the inputs $x_1, \ldots, x_n$, and $\neg$ gates. A function $f : \{0,1\}^* \to \{0,1\}^*$ computed by circuits of this kind is called a projection. In a projection, every bit j of the output depends on at most one bit i of the input: it is either always 0, always 1, always x_i, or always the complement of x_i. An AC^0 projection f can be defined without reference to circuits: there is a deterministic Turing machine M that, given binary numbers n and j, decides in $O(\log n)$ time which of these four cases holds, also computing i in the latter two cases. (Note that $m = |f(x)|$ depends only on $n = |x|$; M must also test whether $j > m$ in $O(\log n)$ time.) Now observe:

- Most of the construction of ϕ in our proof of the Cook–Levin Theorem depended only on the length n of the argument x. The only dependence on x itself was in the very last piece of ϕ, and under the encoding, x was basically copied bit by bit as the signs of the literals. The construction for P-completeness of CVP has the same property.

- In the reductions shown from SAT to graph problems, each edge of the target graph depended on only one bit of information of the form, "is variable x_i in clause C_j?" (To meet the technical requirements for projections, we must use a convoluted encoding of formulas and graphs, but this is the essential idea.)

- In the construction for NL-completeness of GAP, each edge of G_x depended on what the machine N could do while reading just one bit of x.

- Even in the trickiest proof in this chapter, for PSPACE-completeness of QBF, the only ID whose dependence on x needs to be made explicit is the starting ID I_s, and for this x is just copied bit by bit.

Similar ideas work for BFVP, and GAP$_1$; the reader is invited to investigate the remaining problems.

Our point is not to emphasize projections at the expense of other reductions, but to show that the reductions themselves can be incredibly easy to compute. Thus, the complexity levels shown in these completeness results are entirely intrinsic to the target problems. In practice, with classes around P or NP, finding and proving a $\leq_m^p$ or $\leq_m^{log}$ reduction is usually easier, free of encoding fuss, and sufficient for one's purposes.

The list in Theorem 23.14 only begins to illustrate the phenomenon of completeness. Take a computational problem from the practical literature, and chances are it is complete for one of our short list of canonical classes. Here are some more examples, all complete under AC^0 reductions:

- Does a given deterministic finite automaton M accept a given input x? This is L-complete. For certain fixed M, however, the problem is NC^1-complete, and since every regular language belongs to NC^1, this gives a sense in which NC^1 characterizes the complexity of regular languages.

- Do $N - 1$ pairs (i, j) of numbers in $\{1, \ldots, N\}$ form one linked list starting from 1 and ending at N? This is L-complete. Various related problems about permutations, list-ranking, depth-first search, and breadth-first search are also L-complete.

- Satisfiability for 2CNF formulas? This is NL-complete.

- Is $L(G) = \emptyset$ for a given context-free grammar G? This is P-complete.
- REGULAR EXPRESSIONS WITH $(\cup, \cdot)$ only? This is NP-complete.
- Can a multithreaded finite-state program avert deadlock? This is PSPACE-complete.

Game problems in which play must halt after polynomially many moves tend to be PSPACE-complete, but if exponentially long games are possible, as with suitably generalized versions of Chess and Go as well as Checkers to arbitrarily large boards, they tend to be EXP-complete—and hence intractable to solve!

There are some exceptional problems: matrix determinant, matrix permanent (see Section 24.10), and the "NP-intermediate" problems mentioned at the end of the Section 23.4. But overall, few would have expected 40 years ago that so many well-studied problems would quantize into so few complexity levels under efficient reductions.

Changing the conditions on a problem also often makes it jump into a new canonical completeness level. The regular-expression problems amply show this. Special cases of NP-complete problems overwhelmingly tend either to remain NP-hard or jump all the way down to P. BFVP is the special case of CVP where every gate in the circuit has fanout 1. Even SAT itself is a restricted case of QBF.

Problems complete for a given class share an underlying mathematical structure that is brought out by the reductions between them. Note that the transformations map tiny local features of one instance x to tiny local features of $f(x)$—particularly when f is a projection! How such local transformations can propagate global decision properties between widely varying problems is a scientific phenomenon that has been studied for itself.

The main significance of completeness, however, is the evidence of intractability it provides. Although in many cases this evidence is based on an unproven conjecture, sometimes it is absolute. Consider the problem REGULAR EXPRESSIONS WITH $(\cup,^2, \cdot)$, which is complete for NEXP. If this problem were in P, then by closure under Karp reducibility (Theorem 23.1 in Section 23.2), we would have NEXP $\subseteq$ P, a contradiction of the Hierarchy Theorems (Theorem 22.5). Therefore, this decision problem is infeasible: it has no polynomial-time algorithm. In contrast, decision problems in NEXP $-$ P that have been constructed by diagonalization are artificial problems that nobody would want to solve anyway. It is an important point that although diagonalization produces unnatural problems by itself, the combination of diagonalization and completeness shows that natural problems are intractable.

However, the next section points out some limitations of current diagonalization techniques.

23.7   Relativization of P versus NP and "Natural Proofs"

Two obstacles to proving relationships between complexity classes are posed by relativization and the "Natural Proofs" mechanism. We cover the former first. Let A be a language. Define P^A (respectively, NP^A) to be the class of languages accepted in polynomial time by deterministic (nondeterministic) oracle Turing machines with oracle A.

Proofs that use the diagonalization technique on Turing machines without oracles generally carry over to oracle Turing machines. Thus, for instance, the proof of DTIME hierarchy theorem also shows that, for any oracle A, DTIME$^A[n^2]$ is properly contained in DTIME$^A[n^3]$. This can be seen as a strength of the diagonalization technique, since it allows an argument to "relativize" to computation carried out relative to an oracle. In fact, there are examples of lower bounds (for deterministic, "unrelativized" circuit models) that make crucial use of the fact that the time hierarchies relativize in this sense.

But it can also be seen as a weakness of the diagonalization technique. The following important theorem demonstrates why.

THEOREM 23.15 *There exist languages A and B such that* $P^A = NP^A$, *and* $P^B \neq NP^B$.

PROOF Take $A = QBF$. Then $NP^A \subseteq PSPACE$ since all branches of a polynomial-time non-deterministic machine with an oracle in PSPACE can be evaluated sequentially, reusing the space. But $PSPACE = P^{QBF}$, so all three classes are equal. Now consider a machine N that on inputs of the form 0^n guesses a string y of length n, queries y, and accepts if and only if the oracle answers "yes." Strings not of the form 0^n are rejected forthwith. Then N runs in linear time, so for all oracle languages B, $L(N^B) \in NP^B$. Deterministic oracle machines M that run in less than 2^n time, however, cannot query all strings of length n. If M has a finite oracle B_0 of shorter-length strings, and M^{B_0} on input 0^n leaves some string w_n of length n unqueried, then by taking $B_1 = B_0$ if M^{B_0} accepts 0^n, $B_1 = B_0 \cup \{w_n\}$ otherwise, we arrange that $M^{B_1}(0^n) = M^{B_0}(0^n) \neq N^{B_1}(0^n)$, so $L(M^{B_1}) \neq L(N^{B_1})$. Doing this for each sub-exponential time-bounded M in turn, also taking n large enough to leave previously considered computations unaffected, builds B so that in particular $L(N^B) \notin P^B$. $\square$

This shows that resolving the P versus NP question requires techniques that do not relativize, i.e., that do not apply equally to oracle Turing machines. Thus, diagonalization as we currently know is unlikely to succeed in separating P from NP, because the diagonalization arguments we know (and in fact most of the arguments we know) relativize. Results analogous to Theorem 23.15 hold for many other pairs of relativizable complexity classes, thus limiting our ability to separate them or prove them equal.

The response beginning in the 1970s was to employ "concrete" characterizations of complexity classes that do not relativize, such as by families of boolean circuits or arithmetical functions. The latter figures into the proof of IP = PSPACE, even though there is an oracle A such that $IP^A \neq PSPACE^A$ (see Theorem 24.8, Section 24.6, and remarks before it). Success began with the proof by Furst et al. [13] that families of polynomial-size, constant-depth unbounded fan-in circuits cannot compute the parity function. Subsequent lower bounds were proved against more-powerful circuit families. This raised hope for proving NP $\neq$ P by showing that polynomial-size circuits (without the constant-depth restriction) cannot solve SAT, since such circuits solve all problems in P (see Section 22.3).

Razborov and Rudich [26] blunted this optimism by discerning a common feature in lower-bound proofs to date against target classes $\mathcal{C}$ of "easy" circuits, and giving tangible evidence that lower bounds against general polynomial-size circuits cannot have this feature. The evidence rests on the majority belief in the existence of polynomial-time computable functions that circuits of less than exponential size fail to invert in most cases. Such functions are called strongly one-way and are defined formally in Section 24.7.1. The feature is that the proofs define properties Π of n-ary boolean functions f_n such that:

1. The functions f_n defining the problem being shown hard have property Π, for all n.
2. For all sequences g_n of boolean functions computed by circuits in $\mathcal{C}$, there are infinitely many n such that g_n does not have property Π.
3. For some constant $c > 0$ and all n, the proportion of n-ary boolean functions having property Π is at least $1/2^{cn}$.
4. Whether a given f_n has property Π is decidable in time $2^{n^{O(1)}}$.

Items 1 and 2 imply that the f_n does not have easy circuits, while item 3 says that a fairly large number of boolean functions have property Π, and item 4 says that Π is not too complex—hence by inference not too hard for humans to understand. Then Π is called a natural property against $\mathcal{C}$, and a lower-bound proof employing Π is a natural proof. Since f_n has length $N = 2^n$ when presented as a truth table, the time bound in item 4 is quasi-polynomial in N. Razborov and Rudich actually stated a polynomial complexity bound $N^{O(1)} = 2^{O(n)}$, but the above suffices for their main theorem.

THEOREM 23.16 *If there exists a natural property against all polynomial-size circuits, then strongly one-way functions do not exist. In particular, then there are circuits of size $2^{n^{o(1)}}$ that factor n-bit whole numbers.*

Such circuits would beat the best-known algorithms for factoring and thereby diminish faith in the security of RSA and other cryptosystems based on the hardness of factoring, as covered in any standard reference on cryptography. Hence, Theorem 23.16 is widely taken as evidence that efforts to prove NP $\neq$ P must either employ hardness properties Π of complexity greater than singly-exponential in n, or that hold for very few functions besides the one being lower-bounded, or use fundamentally different approaches. The belief that certain functions in TC^0 can achieve a property related to strong one-wayness suggests that lower bounds against constant-depth circuits with threshold gates will be equally difficult to obtain.

Theorems 23.15 and 23.16 are also taken by many to suggest that the P versus NP problem may be independent of important formal systems of logic, in the manner of Gödel's Theorems. This evidence of difficulty contributed to P versus NP being listed as one of the seven "Millennium Prize Problems" by Harvard's Clay Mathematical Institute. For more see Aaronson [1].

23.8 Isomorphism and Sparse Languages

Despite their variety, the known NP-complete languages are similar in the following sense. Two languages A and B are P-**isomorphic** if there exists a function h such that

- For all x, $x \in A$ if and only if $h(x) \in B$.
- h is bijective (i.e., one-to-one and onto).
- Both h and its inverse h^{-1} are computable in polynomial time.

All known NP-complete languages are P-isomorphic. Thus, in some sense, they are merely different encodings of the same problem. This is yet another example of the "amazing fact" alluded to in Section 23.6.2 that natural complete languages exhibit unexpected similarities.

Because of this and other considerations, Berman and Hartmanis [7] conjectured that *all* NP-complete languages are P-isomorphic. This conjecture implies that P $\neq$ NP, because if P $=$ NP, then there are finite NP-complete languages, and no infinite language (such as SAT) can be isomorphic to a finite language.

Between the finite languages and the infinite languages lie the sparse languages, which are defined as follows. For a language A over an alphabet Σ, the **census function** of A, denoted $c_A(n)$, is the number of words x in A such that $|x| \leq n$. Clearly, $c_A(n) < |\Sigma|^{n+1}$. If $c_A(n)$ is bounded by a polynomial in n, then A is **sparse**. From the definitions, it follows that if A is sparse, and A is P-isomorphic to B, then B is sparse.

If a sparse NP-complete language S exists, then we could use S to solve NP-complete problems efficiently, by the following method. Let A be a language in NP, and let f be a transformation function that reduces A to S in polynomial time $t_f(n)$. To quickly decide membership in A for every word x whose length is at most n, deterministically, compute $f(x)$ and check whether $f(x) \in S$ by looking up for $f(x)$ in a table. The table would consist of all words in S whose length is at most $t_f(n)$. The number of entries in this table would be $c_S(t_f(n))$, which is a polynomial in n, and hence the total space occupied by the table would be bounded by a polynomial in n.

The Berman–Hartmanis conjecture implies that there is no sparse NP-complete language, because SAT is not sparse. A stronger reason for believing there are no such languages is:

THEOREM 23.17   **(Mahaney's Theorem)** *If there is a sparse NP-complete language, then* $P = NP$.

Subsequently other complexity classes were shown unlikely to possess sparse complete languages.

THEOREM 23.18   *If there is a sparse language that is complete for* P *(respectively, for* NL*) under log-space reducibility, then* $L = P$ *(respectively,* $L = NL$*).*

23.9   Advice, Circuits, and Sparse Oracles

In the computation of an oracle Turing machine, the oracle language provides assistance in the form of answers to queries, which may depend on the input word. A different kind of assistance, called "advice," depends only on the length of the input word.

Recall the discussion of "advice" in Section 22.3.2: A function α is an advice function if for every nonnegative integer n, $\alpha(n)$ is a binary word whose length is bounded by a polynomial in n. (An advice function need not be computable.)

- P/poly is the class of languages $A = \big\{ x : \langle x, \alpha(|x|) \rangle \in A' \big\}$ for some advice function α and some language A' in P.

As expounded in that section, P/poly comprises those problems that can be solved by nonuniform circuit families of polynomial size.

Berman and Hartmanis also pointed out the following connection between sparse languages and circuit complexity.

THEOREM 23.19   *A language A belongs to* P/poly *if and only if* $A \in P^S$ *for some sparse language S, i.e., A is Cook reducible to a sparse language.*

Note that whereas having a sparse language complete for NP under Karp reductions gave NP = P in Section 23.8, having one complete under Cook reductions is only known to give NP $\subseteq$ P/poly. The reasons for disbelieving NP $\subseteq$ P/poly are nearly as strong as those for NP $\neq$ P; for one, NP $\subseteq$ P/poly implies that there are lookup tables to help one efficiently solve instances of SAT or factor integers. There is other evidence against containment in P/poly that we present in Section 24.2, but the contrast to the uniform class P remains. These connections to circuit complexity and to the isomorphism conjecture have motivated research into the complexity of sparse languages under various types of reducibility.

All of the concepts and classes considered in this chapter, however, combine to reveal the full state of our current ignorance about complexity. The oracle set B and the language $L(N^B) \in NP^B - P^B$ in Theorem 23.15 are both sparse. While this suggests that exponential time lower bounds may apply even to sparse languages in NP, this also gives $L(N^B) \in P/poly$. And proving NP $\not\subseteq$ P/poly is obstructed by the "Natural Proofs" mechanism, Theorem 23.16. Worst of all, similar considerations currently block proving NP $\neq$ TC^0, and even proving that NEXP is not contained in non-uniform TC^0! Nor is a super-linear deterministic time lower bound known for SAT, nor a super-linear circuit size lower bound for any language in NP or in E. Lower bounds of the kind described at the outset of Chapter 23 are known only for problems in classes higher than these. Thus, reducibility and completeness give sharp relative structure to a landscape whose absolute face is almost completely in shadows.

23.10 Research Issues and Summary

Thanks to the notions of reducibility and completeness, it is possible to give "tight lower bounds" on the complexity of many natural problems, even without yet knowing whether $P = NP$. In this chapter, we have seen some examples showing how to prove that problems are NP-complete. We have also explored some of the other notions to which reducibility gives rise, including the notions of relativized computation, P-isomorphism, and the complexity of sparse languages. The most challenging research issue is proving absolute lower bounds.

There are many natural and important problems that are complete for complexity classes that do not appear in our list of "canonical" complexity classes. In order that these problems can be better understood, it is necessary to introduce some additional complexity classes. That is the topic of Chapter 24.

23.11 Further Information

The Cook–Levin Theorem was originally stated and proved for Cook reductions [11], and later for Karp reductions [20]. Independently, Levin [23] proved an equivalent theorem using a variant of Karp reductions. Our circuit-based proof stems from Schnorr [28]. Karp [20] showed the large-scale impact of NP-completeness, and Theorems 23.4 through 23.6 come from there.

A much more extensive discussion of NP-completeness and techniques for proving problems to be NP-complete may be found in [14]. This classic reference also contains a list of hundreds of NP-complete problems. An analogous treatment of problems complete for P can be found in [15]—see also Chapter 28. Most textbooks on algorithm design or complexity theory also contain a discussion of NP-completeness.

Primary sources for other completeness results in this chapter include: Theorem 23.9 [27]; Theorem 23.10 [21]; Theorem 23.11 [29]. Jones [19] studied log-space reductions in detail, and also introduced $\leq_m^{AC^0}$ reductions under the name "log-bounded rudimentary reductions." An important paper for the theory of AC^0 is [6], which also discusses the complete problems in TC^0 and NC^1 in Theorem 23.14. The remaining problems in Theorem 23.14 may be found in [31]. The texts by Hopcroft and Ullman [17] and Papadimitriou [25] give more examples of problems complete for other classes.

There are many other notions of reducibility, including truth table, randomized, and truth-table reductions. A good treatment of this material can be found in the two volumes of [5].

The first relativization results, including Theorem 23.15, are due to Baker et al. [4], and many papers have proved more of them. The role of relativization in complexity theory (and even the question of what constitutes a nonrelativizing proof technique) is fraught with controversy. Longer discussions of the issues involved may be found in [2,12,16]. Theorem 23.16 is from [26].

Sparse languages came to prominence in connection with the Berman–Hartmanis conjecture [7], where Theorem 23.19 is ascribed to Albert Meyer. Theorem 23.17 is from [24], and Theorem 23.18 from [9,10]. Two new surveys of sparse languages and their impact on complexity theory are [8,30].

Information about practical efforts to solve instances of NP-complete and other hard problems is fairly easy to find on the World Wide Web, by searches on problem names such as `Traveling Salesman` (note that variants such as "Salesperson" and the British "Travelling" are also used). Four helpful sites with further links are the Center for Discrete Mathematics and Computer Science (DIMACS), the TSP Library (TSPLIB), the International SAT Competitions Web page, and the Genetic Algorithms Archive;

http://dimacs.rutgers.edu/
http://www.iwr.uni-heidelberg.de/groups/comopt/software/TSPLIB95/

http://www.satcompetition.org/
http://kmh.yeungnam-c.ac.kr/comScience/general/algorithm/galist.html

are the current URLs. TSPLIB has downloadable test instances of the *TSP* problem drawn mostly from practical sources. There is also an extensive bibliography on the *TSP* problem called TSPBIB, maintained by P. Moscato at: http://www.ing.unlp.edu.ar/cetad/mos/TSPBIB_home.html.

Parameterized complexity provides another approach to devise workable algorithms for special cases of NP-completeness, by identifying a suitable "parameter" that may be bounded on instances that arise in practice. For pointers, consult the Section 24.13.

Defining Terms

Configuration: For a Turing machine, synonymous with instantaneous description.

Cook reduction ($\leq_T^p$)**:** A reduction computed by a deterministic polynomial-time oracle Turing machine.

Cook–Levin theorem: The theorem that the language SAT of satisfiable boolean formulas (defined in Chapter 22) is NP-complete.

Instantaneous description (ID): A string that encodes the current state, head position, and (work) tape contents at one step of a Turing machine computation.

Karp reduction ($\leq_m^p$)**:** A reduction given by a polynomial-time computable transformation function.

Natural proof: A lower-bound proof that employs a hardness property of Boolean functions, of a kind itemized in Section 23.7 that appeals to human intuition.

NP: The class of languages accepted by **N**ondeterministic **P**olynomial-time Turing machines. The acronym does not stand for "non-polynomial"—every problem in P belongs to NP.

NP-complete: A language A is NP-complete if A belongs to NP and every language in NP reduces to A. Usually this term refers to Karp reducibility.

NP-hard: A language A is NP-hard if every language in NP reduces to A. Either Cook reducibility or Karp reducibility may be intended.

Oracle Turing machine: A Turing machine that may write "query strings" y on a special tape and learn instantly whether y belongs to a language A_2 given as its *oracle*. These are defined in more detail in Chapter 19.

P: The class of languages accepted by (equivalently, decision problems solved by) deterministic polynomial-time Turing machines. Less technically, the class of feasibly solvable problems.

Reduction: A function or algorithm that maps a given instance of a (decision) problem A_1 into one or more instances of another problem A_2, such that an efficient solver for A_2 could be plugged in to yield an efficient solver for A_1.

Sparse language: A language with a polynomially bounded number of strings of any given length.

Transformation function: A function f that maps instances x of one decision problem A_1 to those of another problem A_2 such that for all such x, $x \in A_1 \iff f(x) \in A_2$. (Here we identify a decision problem with the language of inputs for which the answer is "yes.")

Acknowledgments

The first author was supported by the National Science Foundation under Grants CCF 0830133 and 0132787; portions of this work were performed while being a visiting scholar at the Institute of

Mathematical Sciences, Madras, India. The second author was supported by the National Science Foundation under Grants DUE-0618589 and EEC-0628814. The third author was supported by the National Science Foundation under Grant CCR-9409104.

References

1. Aaronson, S., Is P versus NP formally independent? *Bull. Eur. Assn. Theor. Comp. Sci.* 81, 109–136, 2003.
2. Allender, E., Oracles versus proof techniques that do not relativize. In *Proceedings of 1st Annual International Symposium on Algorithms and Computation,* 1990.
3. Applegate, D.L., Bixby, R.M., Chvátal, V., and Cook, W.J., *The Traveling Salesman Problem,* Princeton University Press, Princeton, NJ, 2006.
4. Baker, T., Gill, J., and Solovay, R., Relativizations of the P=NP? question. *SIAM J. Comp.,* 4, 431–442, 1975.
5. Balcázar, J., Díaz, J., and Gabarró, J., *Structural Complexity I,II,* Springer-Verlag, Berlin, Germany, 1990. Part I published in 1988.
6. Barrington, D.M., Immerman, N., and Straubing, H., On uniformity within NC^1. *J. Comp. Sys. Sci.,* 41, 274–306, 1990.
7. Berman, L. and Hartmanis, J., On isomorphisms and density of NP and other complete sets. *SIAM J. Comp.,* 6, 305–321, 1977.
8. Cai, J.-Y. and Ogihara, M., Sparse hard sets. In *Complexity Theory Retrospective II,* L. Hemaspaandra and A. Selman, Eds., pp. 53–80, Springer-Verlag, New York, 1997.
9. Cai, J. and Sivakumar, D., Sparse hard sets for P: Resolution of a conjecture of Hartmanis. *J. Comp. Sys. Sci.,* 58(2), 280–296, 1999.
10. Cai, J., Naik, A., and Sivakumar, D., On the existence of hard sparse sets under weak reductions. In *Proceedings of 13th Annual Symposium on Theoretical Aspects of Computer Science,* Vol. 1046, *Lecture Notes in Computer Science,* pp. 307–318, Springer-Verlag, Berlin, Germany, 1996.
11. Cook, S., The complexity of theorem-proving procedures. In *Proceedings of 3rd Annual ACM Symposium on the Theory of Computing,* pp. 151–158, 1971.
12. Fortnow, L., The role of relativization in complexity theory. *Bull. EATCS,* 52, 229–244, 1994.
13. Furst, M., Saxe, J., and Sipser, M., Parity, circuits, and the polynomial-time hierarchy. *Math. Sys. Thy.,* 17, 13–27, 1984.
14. Garey, M. and Johnson, D.S., *Computers and Intractability: A Guide to the Theory of NP-Completeness,* Freeman, San Francisco, CA, 1988. First edition was 1979.
15. Greenlaw, R., Hoover, J., and Ruzzo, W.L., *Limits to Parallel Computation: P-Completeness Theory,* Oxford University Press, Oxford, U.K., 1995.
16. Hartmanis, J., New directions in structural complexity theory. In *Proceedings of 15th Annual International Conference on Automata, Languages, and Programming,* Vol. 317, *Lecture Notes in Computer Science,* pp. 271–286, Springer-Verlag, Berlin, Germany, 1988.
17. Hopcroft, J. and Ullman, J., *Introduction to Automata Theory, Languages, and Computation,* Addison-Wesley, Reading, MA, 1979.
18. Johnson, D.S. and McGeogh, L., The traveling salesman problem: A case study in local optimization. In *Local Search in Combinatorial Optimization,* E.H.L. Aarts and J.K. Lenstra, Eds., John Wiley & Sons, New York, 1997.
19. Jones, N., Space-bounded reducibility among combinatorial problems. *J. Comp. Sys. Sci.,* 11, 68–85, 1975. Corrigendum: *J. Comp. Sys. Sci.,* 15, 241, 1977.
20. Karp, R., Reducibility among combinatorial problems. In *Complexity of Computer Computations,* R.E. Miller and J.W. Thatcher, Eds., pp. 85–104, Plenum Press, New York, 1972.
21. Ladner, R., The circuit value problem is log-space complete for P. *SIGACT News,* 7, 18–20, 1975.

22. Ladner, R., On the structure of polynomial-time reducibility. *J. Assn. Comp. Mach.*, 22, 155–171, 1975.

23. Levin, L., Universal sequential search problems. *Problems of Information Transmission*, 9, 265–266, 1973.

24. Mahaney, S., Sparse complete sets for NP: Solution of a conjecture of Berman and Hartmanis. *J. Comp. Sys. Sci.*, 25, 130–143, 1982.

25. Papadimitriou, C., *Computational Complexity*, Addison-Wesley, Reading, MA, 1994.

26. Razborov, A. and Rudich, S., Natural proofs. *J. Comp. Sys. Sci.*, 55, 24–35, 1997.

27. Savitch, W., Relationship between nondeterministic and deterministic tape complexities. *J. Comp. Sys. Sci.*, 4, 177–192, 1970.

28. Schnorr, C., Satisfiability is quasilinear complete in NQL. *J. Assn. Comp. Mach.*, 25, 136–145, 1978.

29. Stockmeyer, L., The complexity of decision problems in automata theory and logic. Technical Report MAC-TR-133, Project MAC, M.I.T., Cambridge, MA, 1974.

30. van Melkebeek, D. and Ogihara, M., Sparse hard sets for P. In *Advances in Algorithms, Languages, and Complexity*, Du, D. and Ko, K., Eds., Kluwer Academic Press, Dordrecht, the Netherlands, 1997.

31. Wagner, K. and Wechsung, G., *Computational Complexity*, D. Reidel, Boston, MA, 1986.

32. Zhang, L and Malik, S., The quest for efficient Boolean satisfiability solvers. In *Proceedings of 14th International Conference on Computer Aided Verification*, Vol. 2404, *Lecture Notes in Computer Science*, pp. 17–36, Springer-Verlag, Berlin, Germany, 2002.

24

Other Complexity Classes and Measures

Eric Allender
Rutgers University

Michael C. Loui
University of Illinois at Urbana-Champaign

Kenneth W. Regan
State University of New York at Buffalo

24.1 Introduction

In the previous two chapters, we have

- Introduced the basic complexity classes
- Summarized the known relationships between these classes
- Seen how reducibility and completeness can be used to establish tight links between natural computational problems and complexity classes

Some natural problems do not seem to be complete for any of the complexity classes we have seen so far. For example, consider the problem of taking as input a graph G and a number k, and deciding whether k is exactly the length of the shortest traveling salesperson's tour. This is clearly related to the Traveling Salesperson Problem (TSP) discussed in Chapter 23, but in contrast to TSP, it seems not to belong to NP, and also seems not to belong to co-NP.

To classify and understand this and other problems, we will introduce a few more complexity classes. We cannot discuss all of the classes that have been studied—there are further pointers to the literature at the end of this chapter. Our goal is to describe some of the most important classes, such as those defined by probabilistic and interactive computation.

A common theme is that the new classes arise from the interaction of complexity theory with other fields, such as randomized algorithms, quantum mechanics, formal logic, combinatorial optimization, and matrix algebra. Complexity theory provides a common formal language for analyzing computational performance in these areas. Other examples can be found in other chapters of this Handbook.

24.2 The Polynomial Hierarchy

Recall from Theorem 22.9(b) that PSPACE is equal to the class of languages that can be recognized in polynomial time on an alternating Turing machine, and that NP corresponds to polynomial time on a nondeterministic Turing machine, which is just an alternating Turing machine that uses only existential states. Thus, in some sense, NP sits near the very "bottom" of PSPACE, and as we allow more use of the power of alternation, we slowly climb up toward PSPACE.

Many natural and important problems reside near the bottom of PSPACE in this sense, but are neither known nor believed to be in NP. (We shall see some examples later in this chapter.) Most of these problems can be accepted quickly by alternating Turing machines that make only two or three alternations between existential and universal states. This observation motivates the definition in the next paragraph.

With reference to Chapter 19, define a k-**alternating Turing machine** to be a machine such that on every computation path, the number of changes from an existential state to universal state, or from a universal state to an existential state, is at most $k - 1$. Thus, a nondeterministic Turing machine, which stays in existential states, is a 1-alternating Turing machine.

It turns out that the class of languages recognized in polynomial time by 2-alternating Turing machines that start out in existential states is precisely NP^{SAT}. This is a manifestation of something more general, and it leads us to the following definitions.

Let $\mathcal{C}$ be a class of languages. Define

- $NP^{\mathcal{C}} = \bigcup_{A \in \mathcal{C}} NP^A$,
- $\Sigma_0^P = \Pi_0^P = P$;

and for $k \geq 0$, define

- $\Sigma_{k+1}^P = NP^{\Sigma_k^P}$,
- $\Pi_{k+1}^P = \text{co-}\Sigma_{k+1}^P$.

Observe that $\Sigma_1^P = NP^P = NP$, because each of polynomially many queries to an oracle language in P can be answered directly by a (nondeterministic) Turing machine in polynomial time. Consequently, $\Pi_1^P = \text{co-NP}$. For each k, $\Sigma_k^P \subseteq \Sigma_{k+1}^P$, and $\Pi_k^P \subseteq \Sigma_{k+1}^P$, but these inclusions are not known to be strict. See Figure 24.1.

The classes Σ_k^P and Π_k^P constitute the **polynomial hierarchy**. Define

$$PH = \bigcup_{k \geq 0} \Sigma_k^P .$$

It is straightforward to prove that $PH \subseteq PSPACE$, but it is not known whether the inclusion is strict. In fact, if $PH = PSPACE$, then the polynomial hierarchy collapses to some level, i.e., $PH = \Sigma_m^P$ for some m.

We have already hinted that the levels of the polynomial hierarchy correspond to k-alternating Turing machines. The next theorem makes this correspondence explicit, and also gives us a third equivalent characterization.

THEOREM 24.1   *For any language A, the following are equivalent:*

1. *$A \in \Sigma_k^P$.*
2. *A is decided in polynomial time by a k-alternating Turing machine that starts in an existential state.*
3. *There exists a language $B \in P$ and a polynomial p such that for all x, $x \in A$ if and only if*

$$(\exists y_1 : |y_1| \leq p(|x|)) \, (\forall y_2 : |y_2| \leq p(|x|)) \cdots$$
$$(Q y_k : |y_k| \leq p(|x|)) \, \big[(x, y_1, \ldots, y_k) \in B \big],$$

where the quantifier Q is $\exists$ if k is odd, $\forall$ if k is even.

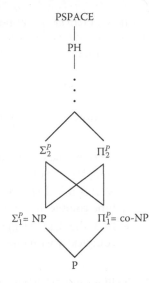

FIGURE 24.1   The polynomial hierarchy.

In Section 23.9, we discussed some of the startling consequences that would follow, if NP were included in P/poly, but observed that this inclusion was not known to imply P = NP. It is known, however, that if NP $\subseteq$ P/poly, then PH collapses to its second level, Σ_2^P [43]. It is generally considered likely that PH does not collapse to any level, and hence that all of its levels are distinct. Hence, this result is considered as a strong evidence that NP is not a subset of P/poly.

Also inside the polynomial hierarchy is the important class BPP of problems that can be solved efficiently and reliably by probabilistic algorithms, to which we now turn.

24.3   Probabilistic Complexity Classes

Since the 1970s, with the development of randomized algorithms for computational problems (see Chapter 12), complexity theorists have placed randomized algorithms on a firm intellectual foundation. In this section, we outline some basic concepts in this area.

A **probabilistic Turing machine** M can be formalized as a nondeterministic Turing machine with exactly two choices at each step. During a computation, M chooses each possible next step with independent probability $1/2$. Intuitively, at each step, M flips a fair coin to decide what to do next. The probability of a computation path of t steps is $1/2^t$. The probability that M accepts an input string x, denoted by $p_M(x)$, is the sum of the probabilities of the accepting computation paths.

Throughout this section, we consider only machines whose time complexity $t(n)$ is time-constructible. Without loss of generality, we may assume that every computation path of such a machine halts in exactly t steps.

Let A be a language. A probabilistic Turing machine M decides A with

	for all $x \in A$	for all $x \notin A$
unbounded two-sided error	if $p_M(x) > 1/2$	$p_M(x) \leq 1/2$
bounded two-sided error	if $p_M(x) > 1/2 + \epsilon$	$p_M(x) < 1/2 - \epsilon$
	for some constant ϵ	
one-sided error	if $p_M(x) > 1/2$	$p_M(x) = 0$

Many practical and important probabilistic algorithms make one-sided errors. For example, in the Solovay–Strassen primality testing algorithm covered in Chapter 16, when the input x is a prime number, the algorithm always says "prime"; when x is composite, the algorithm *usually* says "composite," but may occasionally say "prime." Using the definitions above, this means that the Solovay–Strassen algorithm is a one-sided error algorithm for the set A of composite numbers. It is also a bounded two-sided error algorithm for $\overline{A}$, the set of prime numbers.

These three kinds of errors suggest three complexity classes:

- PP is the class of languages decided by probabilistic Turing machines of polynomial time complexity with unbounded two-sided error.
- BPP is the class of languages decided by probabilistic Turing machines of polynomial time complexity with bounded two-sided error.
- RP is the class of languages decided by probabilistic Turing machines of polynomial time complexity with one-sided error.

In the literature, RP is sometimes also called R.

A probabilistic Turing machine M is a PP-machine (respectively, a BPP-machine, an RP-machine), if M has polynomial time complexity, and M decides with two-sided error (bounded two-sided error, one-sided error).

Through repeated Bernoulli trials, we can make the error probabilities of BPP-machines and RP-machines arbitrarily small, as stated in the following theorem. (Among other things, this theorem implies that RP $\subseteq$ BPP.)

THEOREM 24.2   *If $L \in$ BPP, then for every polynomial $q(n)$, there exists a BPP-machine M such that $p_M(x) > 1 - 1/2^{q(n)}$ for every $x \in L$, and $p_M(x) < 1/2^{q(n)}$ for every $x \notin L$.*

If $L \in$ RP, then for every polynomial $q(n)$, there exists an RP-machine M such that $p_M(x) > 1 - 1/2^{q(n)}$ for every x in L.

It is important to note just how minuscule the probability of error is (provided that the coin flips are truly random). If the probability of error is less than $1/2^{5000}$, then it is less likely that the algorithm produces an incorrect answer than that the computer will be struck by a meteor. An algorithm whose probability of error is $1/2^{5000}$ is essentially as good as an algorithm that makes no errors. For this reason, many computer scientists consider BPP to be the class of practically feasible computational problems.

Next, we define a class of problems that have probabilistic algorithms that make no errors. Define

- ZPP $=$ RP $\cap$ co-RP.

The letter Z in ZPP is for zero probability of error, as we now demonstrate. Suppose $A \in$ ZPP. Here is an algorithm that checks membership in A. Let M be an RP-machine that decides A, and let M' be an RP-machine that decides $\overline{A}$. For an input string x, alternately run M and M' on x, repeatedly, until a computation path of one machine accepts x. If M accepts x, then accept x; if M' accepts x, then reject x. This algorithm works correctly because when an RP-machine accepts its input, it does not make a mistake. This algorithm might not terminate, but with very high probability, the algorithm terminates after a few iterations.

The next theorem expresses some known relationships between probabilistic complexity classes and other complexity classes, such as classes in the polynomial hierarchy (see Section 24.2).

THEOREM 24.3

(a) $P \subseteq ZPP \subseteq RP \subseteq BPP \subseteq PP \subseteq PSPACE$.

(b) $RP \subseteq NP \subseteq PP$.

(c) $BPP \subseteq \Sigma_2^P \cap \Pi_2^P$.

(d) $PH \subseteq P^{PP}$.

(e) $TC^0 \subset PP$.

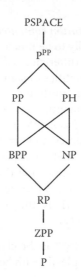

FIGURE 24.2 Probabilistic complexity classes. Many researchers believe $BPP = P$.

(Note that the last inclusion is strict! TC^0 is not known to be different from NP, but it is a proper subset of PP.) Figure 24.2 illustrates many of these relationships. PP is not considered to be a feasible class because it contains NP.

Whether various inexpensive sources of coin-flip bits for probabilistic algorithms meet the randomness conditions for these classes is still controversial, but efficient pseudorandom generators (PRGs) of the kind covered in Chapter 17 of *Algorithms and Theory of Computation Handbook, Second Edition: Special Topics and Techniques* seem to work well in practice. Using such a PRG converts a probabilistic algorithm into a similarly efficient deterministic one, and is said to de-randomize both the algorithm and the problem it solves.

There is a simple sense in which any probabilistic algorithm with small error probability can be de-randomized. If the error probability is brought below $1/2^n$, then there is one sequence r_n of coin flips that gives the right answer on all inputs of length n, and r_n can be hard-wired into the algorithm to yield a deterministic (but nonuniform) circuit family. More formally:

THEOREM 24.4 $BPP \subseteq P/poly$.

This does not imply that $BPP = P$ because r_n may not be constructible in polynomial time. However, if there is any problem in the exponential time class E that requires circuits of size $2^{\Omega(n)}$, a hypothesis supported by many researchers, then $BPP = P$ follows [41].

Another important way in which BPP, RP, and ZPP differ from PP, NP, and most other complexity classes discussed thus far is that they are not known to have any complete languages. The standard approach to construct a complete set for BPP fails because there is no computable way to weed out those polynomial-time probabilistic Turing machines that are not BPP-machines from those that are. The same goes for RP and ZPP—for discussion see [7,60]. However, if $BPP = P$ then all of these classes have the same complete sets that P has.

Log-space analogs of these probabilistic classes have also been studied, of which the most important is RL, defined by probabilistic TMs with one-sided error that run in log space and may use polynomially many random bits in any computation. For a quarter century, the problem of testing whether an undirected graph is connected was an important example of a problem in RL that was not known to be in L. In another watershed for de-randomization, this problem was shown to be in L [57], and many people now conjecture that $RL = L$.

24.4 Quantum Computation

A probabilistic computer enables one to sample efficiently over exponentially many possible computation paths. However, only one path is active at a time, it is unclear whether true random sampling can be achieved, and if $BPP = P$ then the added capability may be too weak to matter anyway.

A *quantum computer*, however, can harness parallelism and randomness in ways that appear theoretically to be much more powerful, ways that are able to solve problems believed not to lie in BPP. A key difference is that by quantum interference, opening up new computation paths may cause the probabilities of some other paths to vanish, whereas for a BPP-machine those probabilities would remain positive.

A quantum computer controls some number m of *qubits*, and is able to maintain up to 2^m-many *basis states* in *superposition*. Subject to mathematical limitations from the theory of quantum mechanics, and practical obstacles to maintaining *coherence* of the superposition, the machine gives effects that are explainable as results of running up to 2^m-many computations in parallel, all while taking sequential time that is polynomial in m. An *observation* of the system then yields a basis state with a probability distribution that depends on the answer to the problem. From enough observations, the answer can be inferred, with a small error probability (and in some cases with certainty).

- BQP is the class of languages decided by quantum machines of polynomial time complexity with bounded two-sided error.

A prime motivation for studying BQP is that it includes language versions of the integer factoring and discrete logarithm problems, which are defined in Section 24.7.1 and further detailed in Chapter 13 of *Algorithms and Theory of Computation Handbook, Second Edition: Special Topics and Techniques*. The public-key cryptosystems in common use today rely on the presumed intractability of these problems, and are theoretically breakable by eavesdroppers armed with quantum computers. This is one of the several reasons why the interplay of quantum mechanics and computational complexity is important to cryptographers.

In terms of the complexity classes depicted in Figure 24.2, the only inclusions that are known involving BQP are

$$\text{BPP} \subseteq \text{BQP} \subseteq \text{PP}.$$

In particular, it is important to emphasize that it is not generally believed that quantum computers can solve NP-complete problems quickly. Chapter 13 of *Algorithms and Theory of Computation Handbook, Second Edition: Special Topics and Techniques* gives further details on how quantum computers work and their connections to cryptography.

24.5 Formal Logic and Complexity Classes

There is a surprisingly close connection between important complexity classes and natural notions that arise in the study of formal logic. This connection has led to important applications of complexity theory to logic, and vice versa. Below, we present some basic notions from formal logic, and then we show some of the connections between logic and complexity theory.

Descriptive complexity refers to the ability to describe and characterize individual problems and whole complexity classes by certain kinds of formulas in formal logic. These descriptions do not depend on an underlying machine model—they are machine-independent. Furthermore, computational problems can be described in terms of their native data structures, rather than under ad hoc string encodings.

A **relational structure** consists of a set V (called the universe), a tuple $E_1, \ldots, E_k$ of relations on V, and a tuple $c_1, \ldots, c_\ell$ of elements of V ($k, \ell \geq 0$). Its *type* τ is given by the tuple $(a_1, \ldots, a_k)$ of arities of the respective relations, together with ℓ. In this chapter, V is always finite. For example, directed graphs $G = (V, E)$ are relational structures with the one binary relation E, and their type has $k = 1$, $a_1 = 2$, and $\ell = 0$, the last since there are no distinguished vertices. For another example, instances of the GRAPH ACCESSIBILITY PROBLEM (GAP) from Section 23.5 consist of a directed graph $G = (V, E)$ along with two distinguished vertices $s, t \in V$, so that they have $\ell = 2$.

An ordinary binary string x can be regarded as a structure $(V, X, \leq)$, where $\leq$ is a total order on V that sequences the bits, and for all i $(1 \leq i \leq |x|)$, $x_i = 1$ if and only if $X(u_i)$ holds. Here, u_i is the ith element of V under the total order, and x_i is the ith bit of x. It is often desirable to regard the ordering $\leq$ as fixed, and focus attention on the single unary relation $X(\cdot)$ as the essence of the string.

24.5.1 Systems of Logic

For our purposes, a system of logic (or logic language) $\mathcal{L}$ consists of the following:

1. A tuple $(E_1, \ldots, E_k)$ of relation symbols, with corresponding arities $a_1, \ldots, a_k \geq 1$, and a tuple $(c_1, \ldots, c_\ell)$ of constant symbols $(k, \ell \geq 0)$. These symbols constitute the vocabulary of $\mathcal{L}$, and can be identified with the corresponding type τ of relational structures.

2. Optionally, a further finite collection of relation and constant symbols whose interpretations are fixed in all universes V is under consideration. By default this collection contains the symbol $=$, which is interpreted as the equality relation on V.

3. An unbounded supply of variable symbols $u, v, w, \ldots$ ranging over elements of V, and optionally, an unbounded supply of variable relation symbols $R_1, R_2, R_3, \ldots$, each with an associated arity and ranging over relations on V.

4. A complete set of Boolean connectives, for which we use $\wedge$, $\vee$, $\neg$, $\rightarrow$, and $\leftrightarrow$, and the quantifiers $\forall$, $\exists$. Additional kinds of operators for building up formulas are discussed later.

The well-formed formulas of $\mathcal{L}$, and the free, bound, positive, and negative occurrences of the symbols in a formula, are defined in the usual inductive manner. A sentence is a formula ϕ with no free variables. A formula, or a whole system, is called first-order, if it has no relation variables R_i; otherwise it is second-order.

Just as machines of a particular type define complexity classes, so also do logical formulas of a particular type which define important classes of languages. The most common nomenclature for these classes begins with a prefix such as FO or F_1 for first-order systems, and SO or F_2 for second-order. SO$\exists$ denotes systems whose second-order formulas are restricted to the form $(\exists R_1)(\exists R_2) \ldots (\exists R_k)\psi$ with ψ first order. After this prefix, in parentheses, we list the vocabulary, and any extra fixed-interpretation symbols or additions to formulas. For instance, SO$\exists$(*Graphs*, $\leq$) stands for the second-order existential theory of graphs whose nodes are labeled and ordered. (The predicate $=$ is always available in the logics we study, and thus it is not explicitly listed with the other fixed-interpretation symbols such as $\leq$.)

The fixed-interpretation symbols deserve special mention. Many authorities treat them as part of the vocabulary. A finite universe V may without loss of generality be identified with the set $\{1, \ldots, n\}$, where $n \in \mathbb{N}$. Important fixed-interpretation symbols for these sets, besides $=$ and $\leq$, are *suc*, $+$, and $*$, respectively, standing for the successor, addition, and multiplication relations. (Here $+(i, j, k)$ stands for $i + j = k$, etc.) Insofar as they deal with the numeric coding of V and do not depend on any structures that are being built on V, such fixed-interpretation symbols are commonly called numerical predicates.

24.5.2 Languages, Logics, and Complexity Classes

Let us see how a logical formula describes a language, just as a Turing machine or a program does. A formal inductive definition of the following key notion, and much further information on systems of logic, may be found in the standard text [23].

DEFINITION 24.1 Let ϕ be a sentence in a system $\mathcal{L}$ with vocabulary τ. A relational structure $\mathcal{R}$ of type τ **satisfies** (or **models**) ϕ, written $\mathcal{R} \models \phi$, if ϕ becomes a true statement about $\mathcal{R}$ when the elements of $\mathcal{R}$ are substituted for the corresponding vocabulary symbols of ϕ. The language of ϕ is $L_\phi = \{\mathcal{R} : \mathcal{R} \models \phi\}$.

We say that ϕ describes L_ϕ, or describes the property of belonging to L_ϕ. Finally, given a system $\mathcal{L}$ of vocabulary τ, $\mathcal{L}$ itself stands for the class of structures of type τ that are described by formulas in $\mathcal{L}$. If τ is the vocabulary Strings of binary strings, then L_ϕ is a language in the familiar sense of a subset of $\{0, 1\}^*$, and systems $\mathcal{L}$ over τ define ordinary classes of languages. Thus, defining sets of structures over τ generalizes the notion of defining languages over an alphabet.

For example, the formula $(\forall u)X(u)$, using the bit-predicate X over binary strings, describes the language 1^*, while $(\forall v, w)[v \neq w \leftrightarrow E(v, w)]$ defines complete (loop-free) graphs. The formula

$$Undir = (\forall v, w)[E(v, w) \rightarrow E(w, v)] \wedge (\forall u)\neg E(u, u)$$

describes the property of being an undirected simple graph, treating an undirected edge as a pair of directed edges, and ruling out "self-loops." Given unary relation symbols $X_1, \ldots, X_k$, the formula

$$Uniq_{X_1,\ldots,X_k} = (\forall v)\left[\bigvee_{1 \leq i \leq k} X_i(v) \wedge \bigwedge_{1 \leq i < j \leq k} \neg\left(X_i(v) \wedge X_j(v)\right) \right]$$

expresses that every element v is assigned exactly one i such that $X_i(v)$ holds. Given an arbitrary finite alphabet $\Sigma = \{c_1, \ldots, c_k\}$, the vocabulary $\{X_1, \ldots, X_k\}$, together with this formula, enables us to define languages of strings over Σ. (Since the presence of $Uniq$ does not affect any of the syntactic characterizations that follow, we may now regard Strings as a vocabulary over any Σ.) Given a unary relation symbol R and the numerical predicate suc on V, the formula

$$Alts_R = (\exists s, t)(\forall u, v)[\neg Suc(u, s) \wedge \neg Suc(t, u) \wedge R(s) \wedge \neg R(t) \wedge (Suc(u, v) \rightarrow (R(u) \leftrightarrow \neg R(v)))]$$

says that R is true of the first element s, false of the last element t, and alternates true and false in-between. This requires $|V|$ to be even. The following examples are used again below.

1. The regular language $(10)^*$ is described by the first-order formula $\phi_1 = Alts_X$.
2. $(11)^*$ is described by the second-order formula $\phi_2 = (\exists R)(\forall u)[X(u) \wedge Alts_R]$.
3. GRAPH THREE-COLORABILITY:

$$\phi_3 = Undir \wedge (\exists R_1, R_2, R_3)\left[Uniq_{R_1,R_2,R_3} \wedge (\forall v, w)(E(v, w) \rightarrow \bigvee_{1 \leq i \leq 3} R_i(v) \wedge \neg R_i(w)) \right].$$

4. GAP (i.e., s-t connectivity for directed graphs):

$$\phi_4 = (\forall R)\neg(\forall u, v)[R(s) \wedge \neg R(t) \wedge (R(u) \wedge E(u, v) \rightarrow R(v))].$$

Formula ϕ_4 says that there is no set $R \subseteq V$ that is closed under the edge relation and contains s but does not contain t, and this is equivalent to the existence of a path from s to t. Much trickier is the fact that deleting "$Uniq_{R_1,R_2,R_3}$" from ϕ_3 leaves a formula that still defines exactly the set of undirected 3-colorable graphs. This fact hints at the delicacy of complexity issues in logic.

Much of this study originated in research on database systems, because database query languages correspond to logics. First-order logic is notoriously limited in expressive power, and this limitation has motivated the study of extensions of first-order logic, such as the following first-order operators.

DEFINITION 24.2

(a) *Transitive closure* (TC): Let ϕ be a formula in which the first-order variables $u_1, \ldots, u_k$ and $v_1, \ldots, v_k$ occur freely, and regard ϕ as implicitly defining a binary relation S on V^k. That is, S is the set of pairs $(\vec{u}, \vec{v})$ such that $\phi(\vec{u}, \vec{v})$ holds. Then $\text{TC}_{(u_1, \ldots, u_k, v_1, \ldots, v_k)} \phi$ is a formula, and its semantics is the reflexive-transitive closure of S.

(b) *Least fixed point* (LFP): Let ϕ be a formula with free first-order variables $u_1, \ldots, u_k$ and a free k-ary relation symbol R that occurs only positively in ϕ. In this case, for any relational structure $\mathcal{R}$ and $S \subseteq V^k$, the mapping $f_\phi(S) = \{(e_1, \ldots, e_k) : \mathcal{R} \models \phi(S, e_1, \ldots, e_k)\}$ is monotone. That is, if $S \subseteq T$, then for every tuple of domain elements $(e_1, \ldots, e_k)$, if $\phi(R, u_1, \ldots, u_k)$ evaluates to `true` when R is set to S and each u_i is set to e_i, then ϕ also evaluates to `true` when R is set to T, because R appears positively. Thus, the mapping f_ϕ has a least fixed point in V^k. Then $\text{LFP}_{(R, u_1, \ldots, u_k)} \phi$ is a formula, and its semantics is the least fixed point of f_ϕ, i.e., the smallest S such that $f_\phi(S) = S$.

(c) *Partial fixed point* (PFP): Even if f_ϕ above is not monotone, $\text{PFP}_{(R, u_1, \ldots, u_k)} \phi$ is a formula whose semantics is the first fixed point found in the sequence $f_\phi(\emptyset), f_\phi(f_\phi(\emptyset)), \ldots$, if it exists, $\emptyset$ otherwise.

The first-order variables $u_1, \ldots, u_k$ remain free in these formulas. The relation symbol R is bound in $\text{LFP}_{(R, u_1, \ldots, u_k)} \phi$, but since this formula is fixing R uniquely rather than quantifying over it, the formula $\text{LFP}_{(R, u_1, \ldots, u_k)} \phi$ is still regarded as first-order (provided ϕ is first-order).

A somewhat less natural but still useful operation is the "deterministic transitive closure" operator. We write "DTC" for the restriction of (a) above to cases where the implicitly defined binary relation S is a partial function. The DTC restriction is enforcible syntactically by replacing any (sub)-formula ϕ to which TC is applied by $\phi'' = \phi \wedge (\forall w_1, \ldots, w_k)[\phi' \to \wedge_{i=1}^{k} v_i = w_i]$, where ϕ' is the result of replacing each v_i in ϕ by w_i, $1 \leq i \leq k$.

For example, s–t connectivity is definable by the FO(TC) and FO(LFP) formulas

$$\phi_4' = (\exists u, v) \left[u = s \wedge v = t \wedge \text{TC}_{(u,v)} E(u, v) \right],$$

$$\phi_4'' = (\exists u, v) \left[u = s \wedge v = t \wedge \text{LFP}_{(R, u, v)} \psi \right],$$

where $\psi = (u = v \vee E(u, v) \vee (\exists w)[R(u, w) \wedge R(w, v)])$. To understand how ϕ_4'' works, starting with S as the empty binary relation and substituting the current S for R at each turn, the first iteration yields $S = \{(u, v) : u = v \vee E(u, v)\}$, the second iteration gives pairs of vertices connected by a path of length at most 2, then 4, $\ldots$, and the fixed-point is the reflexive-transitive closure E^* of E. Then ϕ_4'' is read as if it were $(\exists u, v)(u = s \wedge v = t \wedge E^*(u, v))$, or more simply, as if it were $E^*(s, t)$.

Note however, that writing DTC$\ldots$ in place of TC$\ldots$ in ϕ_4' changes the property defined by restricting it to directed graphs in which each non-sink vertex has out-degree 1. It is not known whether s–t connectivity can be expressed using the DTC operator. This question is equivalent to whether $\text{L} = \text{NL}$.

24.5.3   Logical Characterizations of Complexity Classes

As discussed by [25], there is a uniform encoding method *Enc* such that for any vocabulary τ and (finite) relational structure $\mathcal{R}$ of type τ, $Enc(\mathcal{R})$ is a standard string encoding of $\mathcal{R}$. For instance with $\tau = Graphs$, an n-vertex graph becomes the size-n^2 binary string that lists the entries of its adjacency matrix in row-major order. Thus, one can say that a language L_ϕ over any vocabulary belongs to a complexity class $\mathcal{C}$ if the string language $Enc(L_\phi) = \{Enc(\mathcal{R}) : \mathcal{R} \models \phi\}$ is in $\mathcal{C}$.

The following theorems of the form "$\mathcal{C} = \mathcal{L}$" all hold in the following strong sense: for every vocabulary τ and $\mathcal{L}(\tau)$-formula ϕ, $Enc(L_\phi) \in \mathcal{C}$; and for every language $A \in \mathcal{C}$, there is a $\mathcal{L}(Strings)$-formula ϕ such that $L_\phi = A$. Although going to strings via *Enc* may seem counter to the motivation expressed in the first paragraph of this whole section, the generality and strength of these results have a powerful impact in the desired direction: they define the right notion of complexity class $\mathcal{C}$ for any vocabulary τ. Hence, we omit the vocabulary τ in the following statements.

THEOREM 24.5

   (a) $\text{PSPACE} = \text{FO}(\text{PFP}, \leq)$.

   (b) $\text{PH} = \text{SO}$.

   (c) **(Fagin's Theorem)** $\text{NP} = \text{SO}\exists$.

   (d) $\text{P} = \text{FO}(\text{LFP}, \leq)$.

   (e) $\text{NL} = \text{FO}(\text{TC}, \leq)$.

   (f) $\text{L} = \text{FO}(\text{DTC}, \leq)$.

   (g) $\text{AC}^0 = \text{FO}(+, *)$.

One other result should be mentioned with the above. Define the spectrum of a formula ϕ by $S_\phi = \{n : \text{for some } \mathcal{R} \text{ with } n \text{ elements}, \mathcal{R} \models \phi\}$. Jones and Selman [42] proved that a language A belongs to NE if and only if there is a vocabulary τ and a sentence $\phi \in \text{FO}(\tau)$ such that $A = S_\phi$ (identifying numbers and strings). Thus, spectra characterize NE.

The ordering $\leq$ is needed in results (a), (d), (e), and (f). Chandra and Harel [17] proved that $\text{FO}(\text{LFP})$ without $\leq$ cannot even define $(11)^*$ (and their proofs work also for $\text{FO}(\text{PFP})$). Put another way, without an (ad-hoc) ordering on the full database, one cannot express queries of the kind "Is the number of widgets in Toledo even?" even in the powerful system of first-order logic with PFP. Note that, as a consequence of what we know about complexity classes, it follows that $\text{FO}(\text{PFP}, \leq)$ is more expressive than $\text{FO}(\text{TC}, \leq)$. This result is an example of an application of complexity theory to logic. In contrast, when the ordering is not present, it is much easier to show directly that $\text{FO}(\text{PFP})$ is more powerful than $\text{FO}(\text{TC})$ than to use the tools of complexity theory. Furthermore, the hypotheses $\text{FO}(\text{LFP}) = \text{FO}(\text{PFP})$ and $\text{FO}(\text{LFP}, \leq) = \text{FO}(\text{PFP}, \leq)$ are both equivalent to $\text{P} = \text{PSPACE}$ [2]. This shows how logic can apply to complexity theory.

24.5.4 A Short Digression: Logic and Formal Languages

There are two more logical characterizations that seem at first to have little to do with complexity theory. Characterizations such as these have been important in circuit complexity, but those considerations are beyond the scope of this chapter.

Let SF stand for the class of star-free regular languages, which are defined by regular expressions without Kleene stars, but with $\emptyset$ as an atom and complementation ($\sim$) as an operator. For example, $(10)^* \in \text{SF}$ via the equivalent expression $\sim[(\sim \emptyset)(00 + 11)(\sim \emptyset) + 0(\sim \emptyset) + (\sim \emptyset)1]$.

A formula is monadic if each of its relation symbols is unary. A second-order system is monadic if every relation variable is unary. Let mSO denote the monadic second-order formulas. The formula ϕ_2 above defines $(11)^*$ in $\text{mSO}\exists(suc)$. The following results are specific to the vocabulary of strings.

THEOREM 24.6

   (a) $\text{REG} = \text{mSO}(\textit{Strings}, \leq) = \text{mSO}\exists(\textit{Strings,suc})$.

   (b) $\text{SF} = \text{FO}(\textit{Strings}, \leq)$.

Theorem 24.6, combined with Theorem 24.5(b) and (c), shows that SO is much more expressive than mSO, and $\text{SO}\exists(\leq)$ is similarly more expressive than $\text{mSO}\exists(\leq)$. A seemingly smaller change to mSO$\exists$ also results in a leap of expressiveness from the regular languages to the level of NP.

Lynch [51] showed that if we consider $\mathrm{mSO\exists}(+)$ instead of $\mathrm{mSO\exists}(\leq)$ (for strings), then the resulting class contains $\mathrm{NTIME}[n]$, and hence contains many NP-complete languages, such as GRAPH THREE-COLORABILITY.

24.6 Interactive Models and Complexity Classes

24.6.1 Interactive Proofs

In Section 22.2, we characterized NP as the set of languages whose membership proofs can be checked quickly, by a deterministic Turing machine M of polynomial time complexity. A different notion of proof involves interaction between two parties, a prover P and a verifier V, who exchange messages. In an **interactive proof system**, the prover is an all-powerful machine, with unlimited computational resources, analogous to a teacher. The verifier is a computationally limited machine, analogous to a student. Interactive proof systems are also called "Arthur–Merlin games:" the wizard Merlin corresponds to P, and the impatient Arthur corresponds to V.

Formally, an interactive proof system comprises the following:

- A read-only input tape on which an input string x is written.
- A prover P, whose behavior is not restricted.
- A verifier V, which is a probabilistic Turing machine augmented with the capability to send and receive messages. The running time of V is bounded by a polynomial in $|x|$.
- A tape on which V writes messages to send to P, and a tape on which P writes messages to send to V. The length of every message is bounded by a polynomial in $|x|$.

A computation of an interactive proof system (P, V) proceeds in rounds, as follows. For $j = 1, 2, \ldots$, in round j, V performs some steps, writes a message m_j, and temporarily stops. Then P reads m_j and responds with a message m'_j, which V reads in round $j + 1$. An interactive proof system (P, V) accepts an input string x, if the probability of acceptance by V satisfies $p_V(x) > 1/2$.

In an interactive proof system, a prover can convince the verifier about the truth of a statement without exhibiting an entire proof, as the following example illustrates.

Example 24.1

Consider the graph nonisomorphism problem: the input consists of two graphs G and H, and the decision is "yes" if and only if G is not isomorphic to H. Although there is a short proof that two graphs are isomorphic (namely: the proof consists of the isomorphism mapping G onto H), nobody has found a general way of proving that two graphs are not isomorphic that is significantly shorter than listing all $n!$ permutations and showing that each fails to be an isomorphism. (That is, the graph nonisomorphism problem is in **co-NP**, but is not known to be in **NP**.) In contrast, the verifier V in an interactive proof system is able to take statistical evidence into account, and determine "beyond all reasonable doubt" that two graphs are nonisomorphic, using the following protocol.

In each round, V randomly chooses either G or H with equal probability; if V chooses G, then V computes a random permutation G' of G, presents G' to P, and asks P whether G' came from G or from H (and similarly if V chooses H). If P gave an erroneous answer on the first round, and G is isomorphic to H, then after k subsequent rounds, the probability that P answers all the subsequent queries correctly is $1/2^k$. (To see this, it is important to understand that the prover P does not see the coins that V flips in making its random choices; P sees only the graphs G' and H' that V sends as messages.) V accepts the interaction with P as "proof" that G and H are nonisomorphic if P is able to pick the correct graph for 100 consecutive rounds. Note that V has ample grounds to accept this as a convincing demonstration: if the graphs are indeed isomorphic, the prover P would have to have an incredible streak of luck to fool V.

The complexity class IP comprises the languages A for which there exists a verifier V and an ϵ such that

- There exists a prover $\hat{P}$ such that for all x in A, the interactive proof system $(\hat{P}, V)$ accepts x with probability greater than $1/2 + \epsilon$
- For every prover P and every $x \notin A$, the interactive proof system (P, V) rejects x with probability greater than $1/2 + \epsilon$

It is straightforward to show that IP $\subseteq$ PSPACE. It was originally believed likely that IP was a small subclass of PSPACE. Evidence supporting this belief was the construction by Fortnow and Sipser [29] of an oracle language B for which co-NP^B $-$ IP$^B \neq \emptyset$, so that IP^B is strictly included in PSPACE^B. Using a proof technique that does not relativize, however, Shamir [59] (building on the work of Lund et al. [49]) proved that in fact, IP and PSPACE are the same class.

THEOREM 24.7   IP = PSPACE.

If NP is a proper subset of PSPACE, as is widely believed, then Theorem 24.7 says that interactive proof systems can decide a larger class of languages than NP.

Notice that in the interactive proof for graph nonisomorphism, if the input graphs G and H are isomorphic, then the verifier V learns only the fact of isomorphism; V gains no additional knowledge about the relationship between G and H. This property of the proof is called **zero-knowledge**; we define the property formally below. The zero-knowledge property is particularly useful for authentication. A user might wish to convince a server that he is authorized to use its service by proving that he has the correct password, but he does not want to reveal the password to the server. A zero-knowledge proof would authenticate the user but provide no additional knowledge to the server. Zero-knowledge proofs can also be applied to secret sharing between two parties who do not initially trust each other to follow a secure communication protocol. Instead of asking a trusted third party to confirm that they are following the protocol, the two parties could use a zero-knowledge proof to convince each other that they are complying with the protocol.

Consider a k-round interactive proof with messages $m_1, m_1', m_2, \ldots, m_k'$ as described above, and for each j, let β_j be the binary sequence of random choices by V during its computation in round j. Define the history h to be the concatenation of the messages and binary sequences, i.e., $h = m_1 m_1' \cdots m_k' \beta_1 \cdots \beta_k$. Each interactive proof system (P, V) generates for each input x a probability distribution $D_{P,V,x}$ on the histories, so that $D_{P,V,x}(h)$ is the probability that h occurs. A probabilistic Turing machine M (with output) also generates a probability distribution: for each input x, let $D_{M,x}'(y)$ be the probability that M on input x produces output y. An interactive proof system (P, V) has the perfect zero-knowledge property if there exists a polynomial-time probabilistic Turing machine M that on each input x generates a probability distribution $D_{M,x}'$ that is identical to $D_{P,V,x}$. This definition captures the meaning of zero-knowledge because the verifier V gains no new ability to compute something beyond what it could already have computed (represented by M). From an information-theoretic point of view, every history h provides no new information to V because the probability of h could have been determined a priori. Weaker but practical versions of the zero-knowledge property include statistical indistinguishability and computational indistinguishability [31].

24.6.2   Probabilistically Checkable Proofs

In an interactive proof system, the verifier does not need a complete conventional proof to become convinced about the membership of a word in a language, but uses random choices to query parts of a proof that the prover may know. This interpretation inspired another notion of "proof": a proof consists of a (potentially) large amount of information that the verifier need only inspect in a few places in order to become convinced. The following definition makes this idea more precise.

A language L has a **probabilistically checkable proof**, if there exists an oracle BPP-machine M such that

- For all $x \in L$, there is an oracle language B_x such that M^{B_x} accepts x.
- For all $x \notin L$, and for every language B, machine M^B rejects x.

Intuitively, the oracle language B_x represents a proof of membership of x in L. Notice that B_x can be finite since the length of each possible query during a computation of M^{B_x} on x is bounded by the running time of M. The oracle language takes the role of the prover in an interactive proof system—but in contrast to an interactive proof system, the prover cannot change strategy adaptively in response to the questions that the verifier poses. This change results in a potentially stronger system, since a machine M that has bounded error probability relative to all languages B might not have bounded error probability relative to some adaptive prover. Although this change to the proof system framework may seem modest, it leads to a characterization of a class that seems to be much larger than PSPACE.

THEOREM 24.8 *A language A has a probabilistically checkable proof if and only if $A \in$ NEXP.*

Although the notion of probabilistically checkable proofs seems to lead us away from feasible complexity classes, by considering natural restrictions on how the proof is accessed, we can obtain important insights into familiar complexity classes.

Let $PCP[r(n), q(n)]$ denote the class of languages with probabilistically checkable proofs in which the probabilistic oracle Turing machine M makes $r(n)$ random binary choices, and queries its oracle $q(n)$ times. (For this definition, we assume that M has either one or two choices for each step.) It follows from the definitions that $BPP = PCP[n^{O(1)}, 0]$, and $NP = PCP[0, n^{O(1)}]$.

THEOREM 24.9 $NP = PCP[O(\log n), O(1)]$.

Theorem 24.9 asserts that for every language L in NP, a proof that $x \in L$ can be encoded so that the verifier can be convinced of the correctness of the proof (or detect an incorrect proof) by using only $O(\log n)$ random choices, and inspecting only a constant number of bits of the proof!

This surprising characterization of NP has important applications to the complexity of finding approximate solutions to optimization problems, as discussed in the next section.

24.7 Classifying the Complexity of Functions

Up to now, we have considered only complexity for languages and decision problems, for which the output is "yes" or "no," nothing else. Most of the functions that we actually compute are functions that produce more than one bit of output. For example, instead of merely deciding whether a graph has a clique of size m, we often want to find such a clique, if one exists. Problems in NP are naturally associated with this kind of search problem.

One reason for emphasis on languages is that search problems can be reduced to decision problems in several ways. Given a function f, define the languages:

$$A_f = \left\{ \langle x, i \rangle : 1 \le i \le |f(x)| \text{ and bit } i \text{ of } f(x) \text{ is a } 1 \right\},$$
$$G_f = \left\{ \langle x, w \rangle : w \text{ is an initial substring of } f(x) \right\}.$$

Say f is polynomially length bounded, if there is a polynomial p such that for all x, $|f(x)| \leq p(|x|)$. If so, then $f \leq_T^p A_f$ and $f \leq_T^p G_f$, making both languages equivalent to f under Cook reductions. However, several kinds of problems embrace aspects of functions not well captured by these associated decision problems: inversion, optimization, approximation, and counting. In turning to these, we define classes and notions of reductions specific to functions themselves, beginning with the class that corresponds to P for languages.

- FP is the set of functions computable in polynomial time by deterministic Turing machines.

In an analogous way, we define FL, FNC^k, etc., to be the set of functions computable by deterministic log-space machines, by NC^k circuits, etc. We also define FPSPACE to be the class of polynomial length-bounded functions f computable by deterministic machines in polynomial space.

To study functions that appear to be difficult to compute, we again use the notions of reducibility and completeness. Analogous to Cook reducibility to oracle languages, we consider Cook reducibility to a function given as an oracle. For a polynomial length-bounded function f, we say that a language A is Cook reducible to f if there is a polynomial-time oracle Turing machine M that accepts A, where the oracle is accessed as follows: M writes a string y on the query tape, and in the next step y is replaced by $f(y)$. As usual, we let P^f and FP^f denote the class of languages and functions computable in polynomial time with oracle f, respectively.

Let $\mathcal{C}$ be a class of functions. When $\mathcal{C}$ is at least as big as FP, then we will use Cook reducibility to define completeness. That is, a function f is $\mathcal{C}$-complete, if f is in $\mathcal{C}$ and $\mathcal{C} \subseteq$ FP^f. When we are discussing classes $\mathcal{C}$ within FP, for which polynomial-time is too powerful to give a meaningful notion of reducibility), we refer to hardness and completeness under AC^0-Turing reducibility, which was defined in Section 23.6. Although many other kinds of reducibility have been studied for functions just as with languages, these two suffice for our purposes in this chapter.

24.7.1 Inversion and One-Way Functions

Say that a program D inverts a function f if for all $y \in Ran(f)$, $D(y)$ outputs x such that $f(x) = y$. The behavior of $D(y)$ for $y \notin Ran(f)$ need not be specified, and f need not be 1-1. We presume that f is polynomially honest, meaning that for some polynomial p and all x, $p(|f(x)|) \geq |x|$. This ensures that any solution x can be written down in time polynomial in $|y|$.

- A polynomial-time computable function is **weakly one-way** if no polynomial-time deterministic Turing machine inverts f.

For cryptographic purposes one needs assurance that prospective inverters fail on most inputs, not just some. A function f is $s(n)$-hard to invert if for all sufficiently large n, and all $s(n)$-sized circuits D_n,

$$\mathrm{Prob}_{x \in \Sigma^n}[D_n(f(x)) \text{ inverts } f(x)] \leq 1/s(n).$$

- A polynomial-time computable function is **one-way**, if it is n^k-hard to invert for all k, and **strongly one-way** if it is 2^{n^ϵ}-hard to invert, for some $\epsilon > 0$.

Strongly one-way functions are conjectured to exist, in particular ones based on the suspected hardness of integer factoring or the discrete logarithm problem, which are covered in Chapters 16 of this book and Chapters 18 and 19 of *Algorithms and Theory of Computation Handbook, Second Edition: Special Topics and Techniques.* Here, we observe that a complexity class of languages captures some aspects of these problems.

- A language A belongs to UP if there is a polynomial-time nondeterministic Turing machine N such that $L(N) = A$, and for all $x \in A$, $N(x)$ has exactly one accepting computation path.

The "U" stands for "unique," and one may also think of a polynomial-time decidable witness predicate R for A, such that whenever $y \in A$, there is a unique z such that $R(y, z)$ holds. Clearly P $\subseteq$ UP $\subseteq$ NP. An important example of a language in UP is

$$G_{fact} = \left\{ \langle n, w \rangle : w \text{ is an initial substring of the prime factorization of } n \right\}.$$

Note that the prime factorization can be written in a unique way, and verified in polynomial time even with n in binary notation. The complement of G_{fact} also belongs to UP. Thus, by the above remarks on search reducing to decision, integer factorization Cook-reduces to a language in UP $\cap$ co-UP. In all we have:

THEOREM 24.10

(a) Weakly one-way functions exist if and only if P $\neq$ NP.

(b) Weakly one-way functions that are 1-1 exist if and only if P $\neq$ UP.

(c) Weakly one-way permutations of Σ^* *exist if and only if* UP $\cap$ co-UP $\neq$ P.

(d) Integer factorization is NP-*hard only if* NP $=$ UP $=$ co-UP $=$ co-NP.

Since cracking the RSA public-key cryptosystem reduces to factoring, Part (d) is the evidence that this problem is not NP-hard. Nevertheless, belief that the deterministic time or circuit complexity of factoring is 2^{n^ϵ} for some $\epsilon > 0$ (the current best-known upper bound has ϵ approaching $1/3$) has remained steady for 30 years. For more on these topics, see Chapters 18 and 19 of *Algorithms and Theory of Computation Handbook, Second Edition: Special Topics and Techniques*.

24.8 Optimization Classes

Given an optimization (minimization) problem, we most often study the following associated decision problem:

"Is the optimal value at most k"?

Alternatively, we could formulate the decision problem as the following:

"Is the optimal value exactly k?"

For example, consider the TSP again. TSP asks whether the length of the optimal tour is at most d_0. Define EXACT TSP to be the decision problem that asks whether the length of the optimal tour is exactly d_0. It is not clear that EXACT TSP is in NP or in co-NP, but EXACT TSP can be expressed as the intersection of TSP and its complement $\overline{\text{TSP}}$: the length of the optimal tour is d_0, if there is a tour whose length is at most d_0, and no tour whose length is at most $d_0 - 1$. Similar remarks apply to the optimization problem MAX CLIQUE: given an undirected graph G, find the maximum size of a clique in G.

Exact versions of many optimization problems can be expressed as the intersection of a language in NP and a language in co-NP. This observation motivates the definition of a new complexity class:

- D^P is the class of languages A such that $A = A_1 \cap A_2$ for some languages A_1 in NP and A_2 in co-NP.

The letter D in D^P means difference: $A \in D^P$ if and only if A is the difference of two languages in NP, i.e., $A = A_1 - A_3$ for some $A_1, A_3 \in$ NP.

Not only is EXACT TSP in D^P, but also EXACT TSP is D^P-complete. Exact versions of many other NP-complete problems, including CLIQUE, are also D^P-complete [54].

Although it is not known whether D^P is contained in NP, it is straightforward to prove that

$$\text{NP} \subseteq D^P \subseteq P^{\text{NP}} \subseteq \Sigma_2^P \cap \Pi_2^P .$$

Thus, D^P lies between the first two levels of the polynomial hierarchy.

We have characterized the complexity of computing the optimal value of an instance of an optimization problem, but we have not yet characterized the complexity of computing the optimal solution itself. An optimization algorithm produces not only a "yes" or "no" answer, but also, when feasible solutions exist, an optimal solution.

First, for a maximization problem, suppose that we have a subroutine that solves the decision problem "Is the optimal value at least k?" Sometimes, with repeated calls to the subroutine, we can construct an optimal solution. For example, suppose subroutine S solves the CLIQUE problem; for an input graph G and integer k, the subroutine outputs "yes" if G has a clique of k (or more) vertices. To construct the largest clique in an input graph, first, determine the size K of the largest clique by binary search on $\{1, \ldots, n\}$ with $\log_2 n$ calls to S. Next, for each vertex v, in sequence, determine whether deleting v produces a graph whose largest clique has size K by calling S. If so, then delete v and continue with the remaining graph. If not, then look for a clique of size $K - 1$ among the neighbors of v.

The method outlined in the last paragraph uses S in the same way as an oracle Turing machine queries an oracle language in NP. With this observation, we define the following classes:

- FP^{NP} is the set of functions computable in polynomial time by deterministic oracle Turing machines with oracle languages in NP.
- $FP^{\text{NP}[\log n]}$ is the set of functions computable in polynomial time by deterministic oracle Turing machines with oracle languages in NP that make $O(\log n)$ queries during computations on inputs of length n

FP^{NP} and $FP^{\text{NP}[\log n]}$ contain many well-studied optimization problems [44]. The problem of producing the optimal tour in the TSP is FP^{NP}-complete. The problem of determining the size of the largest clique subgraph in a graph is $FP^{\text{NP}[\log n]}$-complete.

24.9 Approximability and Complexity

As discussed in Chapter 34, because polynomial-time algorithms for NP-hard optimization problems are unlikely to exist, we ask whether a polynomial-time algorithm can produce a feasible solution that is close to optimal.

Fix an optimization problem Π with a positive integer-valued objective function g. For each problem instance x, let $\text{OPT}(x)$ be the optimal value, that is, $g(z)$, where z is a feasible solution to x that achieves the best possible value of g. Let M be a deterministic Turing machine that on input x produces as output a feasible solution $M(x)$ for Π. We say M is an ϵ-approximation algorithm if for all x,

$$\frac{|g(M(x)) - \text{OPT}(x)|}{\max\{g(M(x)), \text{OPT}(x)\}} \le \epsilon .$$

(This definition handles both minimization and maximization problems.) The problem Π has a **polynomial-time approximation scheme** if for every fixed ϵ, there is a polynomial-time ϵ-approximation algorithm. Although the running time is polynomial in $|x|$, the time may be exponential in $1/\epsilon$.

Several NP-complete problems, including KNAPSACK, have polynomial-time approximation schemes. It is natural to ask whether all NP-complete optimization problems have polynomial-time approximation schemes. We define an important class of optimization problems, MAX–SNP, whose complete problems apparently do not.

First, we define a reducibility between optimization problems that preserves the quality of solutions. Let Π_1 and Π_2 be optimization problems with objective functions g_1 and g_2, respectively. An **L-reduction** from Π_1 to Π_2 is defined by a pair of polynomial-time computable functions f and f' that satisfy the following properties:

1. If x is an instance of Π_1 with optimal value $\text{OPT}(x)$, then $f(x)$ is an instance of Π_2 whose optimal value satisfies $\text{OPT}(f(x)) \leq c \cdot \text{OPT}(x)$ for some constant c.

2. If z is a feasible solution of $f(x)$, then $f'(z)$ is a feasible solution of x, such that

$$\left| \text{OPT}(x) - g_1(f'(z)) \right| \leq c' \left| \text{OPT}(f(x)) - g_2(z) \right|$$

for some constant c'.

The second property implies that if z is an optimal solution to $f(x)$, then $f'(z)$ is an optimal solution to x. From the definitions, it follows that if there is an L-reduction from Π_1 to Π_2, and there is a polynomial-time approximation scheme for Π_2, then there is a polynomial-time approximation scheme for Π_1.

To define MAX–SNP, it will help to recall the characterization of NP as SO$\exists$ in Section 25.5.3. This characterization says that for any A in NP, there is a first-order formula ψ such that $x \in A$ if and only if

$$\exists S_1 \dots \exists S_l \, \psi \, (x, S_1, \dots, S_l) .$$

For many important NP-complete problems, it is sufficient to consider having only a single second-order variable S, and to consider formulas ψ having only universal quantifiers. Thus, for such a language A we have a quantifier-free formula ϕ such that $x \in A$ if and only if

$$\exists S \forall u_1 \dots \forall u_k \, \phi \, (S, u_1, \dots, u_k) .$$

Now define MAX–SNP$_0$ to be the class of optimization problems mapping input x to

$$\max_S \left| \{ (y_1, \dots, y_k) : \phi \, (S, y_1, \dots, y_k) \} \right| .$$

For example, we can express in this form the MAX CUT problem, the problem of finding the largest cut in an input graph $G = (V, E)$ with vertex set V and edge set E. A set of vertices S is the optimal solution if it maximizes

$$\left| \{ (v, w) : E(v, w) \wedge S(v) \wedge \neg S(w) \} \right| .$$

That is, the optimal solution S maximizes the number of edges (v, w) between vertices v in S and vertices w in $V - S$.

Define MAX–SNP to be the class of all optimization problems that are L-reducible to a problem in MAX–SNP$_0$. MAX–SNP contains many natural optimization problems. MAX CUT is MAX–SNP-complete, and MAX CLIQUE is MAX–SNP-hard, under L-reductions.

A surprising connection between the existence of probabilistically checkable proofs (Section 24.6) and the existence of approximation algorithms comes out in the next major theorem.

THEOREM 24.11 *If there is a polynomial-time approximation scheme for some MAX–SNP-hard problem, then* P $=$ NP.

In particular, unless $P = NP$, there is no polynomial-time approximation scheme for MAX CUT or MAX CLIQUE. To prove this theorem, all we need to do is show its statement for a particular problem that is MAX–SNP-complete under L-reductions. However, we prefer to show the *idea* of the proof for the MAX CLIQUE problem, which although MAX–SNP-hard is not known to belong to MAX–SNP. It gives a strikingly different kind of reduction from an arbitrary language A in NP to CLIQUE over the reduction from A to SAT to CLIQUE in Section 24.4, and its discovery by Feige et al. [26] stimulated the whole area.

PROOF   Let $A \in NP$. By Theorem 24.9, namely $NP = PCP[O(\log n), O(1)]$, there is a probabilistic oracle Turing machine M constrained to use $r(n) = O(\log n)$ random bits and make at most a constant number ℓ of queries in any computation path, such that:

- For all $x \in A$, there exists an oracle language B_x such that $\text{Prob}_{s\in\{0,1\}^{r(n)}}[M^{B_x}(x, s) = 1] > 3/4$;
- For all $x \notin A$, and for every language B, $\text{Prob}_{s\in\{0,1\}^{r(n)}}[M^B(x, s) = 1] < 1/4$.

Now define a transcript of M on input x to consist of a string $s \in \{0,1\}^{r(n)}$ together with a sequence of ℓ pairs (w_i, a_i), where w_i is an oracle query and $a_i \in \{0, 1\}$ is a possible yes/no answer. In addition, a transcript must be valid, meaning that for all i, $0 \le i < \ell$, on input x with random bits s, having made queries $w_1, \ldots, w_i$ to its (unspecified) oracle and received answers $a_1, \ldots, a_i$, machine M writes w_{i+1} as its next query string. Thus, a transcript provides enough information to determine a full computation path of M on input x, and the transcript is accepting if and only if this computation path accepts. Finally, call two transcripts consistent, if whenever a string w appears as "w_i" in one transcript and "w_j" in the other, the corresponding answer bits a_i and a_j are the same.

Construction: Let G_x be the undirected graph whose node set V_x is the set of all accepting transcripts, and whose edges connect pairs of transcripts that are consistent.

Complexity: Since $r(n) + \ell = O(\log n)$, there are only polynomially many transcripts, and since consistency is easy to check, G_x is constructed in polynomial time.

Correctness: If $x \in A$, then take the oracle B_x specified above and let C be the set of accepting transcripts whose answer bits are given by B_x. These transcripts are consistent with each other, and there are at least $(3/4)2^{r(n)}$ such accepting transcripts, so C forms a clique of size at least $(3/4)2^{r(n)}$ in G_x. Now suppose $x \notin A$, and suppose C' is a clique of size greater than $(1/4)2^{r(n)}$ in G_x. Because the transcripts in C' are mutually consistent, there exists a single oracle B that produces all the answer bits to queries in transcripts in C'. But then $\text{Prob}_s[M^B(x, s) = 1] > 1/4$, contradicting the PCP condition on M.

Thus, we have proved the statement of the theorem for MAX CLIQUE. The proof of the general statement is similar. $\square$

Note that the cases $x \in A$ and $x \notin A$ in this proof lead to a "$(3/4, 1/4)$ gap" in the maximum clique size ω of G_x. If there were a polynomial-time algorithm guaranteed to determine ω within a factor better than 3, then this algorithm could tell the "3/4" case apart from the "1/4" case, and hence decide whether $x \in A$. Since G_x can be constructed in polynomial time (in particular, G_x has size at most $2^{r(n)+\ell} = n^{O(1)}$), $P = NP$ would follow. Hence, we can say that CLIQUE is NP-hard to approximate within a factor of 3. A long sequence of improvements to this basic construction has pushed the hardness of approximation not only to any fixed constant factor, but also to factors that increase with n. Moreover, approximation-preserving reductions have extended this kind of hardness result to many other optimization problems.

24.10 Counting

Two other important classes of functions deserve special mention:

- #P is the class of functions f such that there exists a nondeterministic polynomial-time Turing machine M with the property that $f(x)$ is the number of accepting computation paths of M on input x.
- #L is the class of functions f such that there exists a nondeterministic log-space Turing machine M with the property that $f(x)$ is the number of accepting computation paths of M on input x.

Some functions in #P are clearly at least as difficult to compute as some NP-complete problems are to decide. For instance, consider the following problem.

NUMBER OF SATISFYING ASSIGNMENTS TO A **3CNF** FORMULA (#3CNF)

Instance: A boolean formula in conjunctive normal form with at most three variables per clause.

Question: The number of distinct assignments to the variables that cause the formula to evaluate to true.

Note that #3CNF is in #P, and note also that the NP-complete problem of determining whether $x \in 3SAT$ is merely the question of whether #3CNF(x)= 0.

In apparent contrast to #P, all functions in #L can be computed by NC circuits.

THEOREM 24.12   *Relationships between counting classes.*

- $\text{FP} \subseteq \#\text{P} \subseteq \text{FPSPACE}$,
- $\text{P}^{\text{PP}} = \text{P}^{\#\text{P}}$ *(and thus also* $\text{FP}^{\text{NP}} \subseteq \text{FP}^{\#\text{P}}$),
- $\text{FL} \subseteq \#\text{L} \subseteq \text{FNC}^2$.

It is not surprising that #P and #L capture the complexity of various functions that involve counting, but as the following examples illustrate, it sometimes is surprising which functions are difficult to compute.

The proof of the Cook–Levin Theorem that appears in Chapter 23 also proves that #3*CNF* is complete for #P, because it shows that for every nondeterministic polynomial-time machine M and every input x, one can efficiently construct a formula with the property that each accepting computation of M on input x corresponds to a distinct satisfying assignment, and vice versa. Thus, the number of satisfying assignments equals the number of accepting computation paths. A reduction with this property is called parsimonious.

Most NP-complete languages that one encounters in practice are known to be complete under parsimonious reductions. (The reader may wish to check which of the reductions presented in Chapter 23 are parsimonious.) For any such complete language, it is clear how to define a corresponding complete function in #P.

Similarly, for the GRAPH ACCESSIBILITY PROBLEM (GAP), which is complete for NL, we can define the function that counts the number of paths from the start vertex s to the terminal vertex t. For reasons that will become clear soon, we consider two versions of this problem: one for general directed graphs, and one for directed acyclic graphs. (The restriction of GAP to acyclic graphs remains NL-complete.)

NUMBER OF PATHS IN A GRAPH (#Paths)

Instance: A directed graph on n vertices, with two distinguished vertices s and t.

Question: The number of simple paths from s to t. (A path is a simple path, if it visits no vertex more than once.)

NUMBER OF PATHS IN A DIRECTED ACYCLIC GRAPH (#DAG-Paths)

Instance: A directed acyclic graph on n vertices, with two distinguished vertices s and t.

Question: The number of paths from s to t. (In an acyclic graph, all paths are simple.)

As one might expect, the function #DAG-Paths is complete for #L, but it may come as a surprise that #Paths is complete for #P [67]! That is, although it is easy to decide whether there is a path between two vertices, it seems quite difficult to count the number of distinct paths, unless the underlying graph is acyclic.

As another example of this phenomenon, consider the problem 2SAT, which is the same as 3SAT except that each clause has at most two literals. Then 2SAT is complete for NL, but the problem of counting the number of satisfying assignments for these formulas is complete for #P.

A striking illustration of the relationship between #P and #L is provided by the following two important problems from linear algebra. Recall that the determinant and permanent of a matrix M with entries $M_{i,j}$ are respectively given by

$$\sum_{\pi} \text{sign}(\pi) \prod_{i=1}^{n} M_{i,\pi(i)} \quad \text{and} \quad \sum_{\pi} \prod_{i=1}^{n} M_{i,\pi(i)},$$

where the sum is over all permutations π on $\{1, \dots, n\}$, $\text{sign}(\pi)$ is -1, if π can be written as the composition of an odd number of transpositions, and $\text{sign}(\pi)$ is 1 otherwise.

DETERMINANT

Instance: An integer matrix.

Question: The determinant of the matrix.

PERMANENT

Instance: An integer matrix.

Question: The permanent of the matrix.

The reader is probably familiar with the determinant function, which can be computed efficiently by Gaussian elimination. The permanent may be less familiar, although its definition is formally simpler. Nobody has ever found an efficient way to compute the permanent, however. If M is the adjacency matrix of a bipartite graph G with two sets of nodes of equal size, then the permanent of M is the number of perfect matchings in G.

We need to introduce slight modification of our function classes to classify these problems, however, because #P and #L consist of functions that take only nonnegative values, whereas both the permanent and determinant can be negative.

Define GapL to be the class of functions that can be expressed as the difference of two #L functions, and define GapP to be the difference of two #P functions.

THEOREM 24.13

    (a) PERMANENT *is complete for* GapP.

    (b) DETERMINANT *is complete for* GapL.

The class of problems that are AC^0-Turing reducible to DETERMINANT is one of the most important subclasses of NC, and in fact contains most of the natural problems for which NC algorithms are known.

24.11   Kolmogorov Complexity

Until now, we have considered only dynamic complexity measures, namely, the time and space used by Turing machines. Kolmogorov complexity is a static complexity measure that captures the difficulty of describing a string. For example, the string consisting of three million zeroes can be described with fewer than three million symbols (as in this sentence). In contrast, for a string consisting of three million randomly generated bits, with high probability there is no shorter description than the string itself.

Let U be a universal Turing machine (see Section 21.2). Let ϵ denote the empty string. The **Kolmogorov complexity** of a binary string y with respect to U, denoted by $K_U(y)$, is the length of the shortest binary string i such that on input $\langle i, \epsilon \rangle$, machine U outputs y. In essence, i is a description of y, for it tells U how to generate y.

The next theorem states that different choices for the universal Turing machine affect the definition of Kolmogorov complexity in only a small way.

THEOREM 24.14   (Invariance Theorem). *There exists a universal Turing machine U such that for every universal Turing machine U', there is a constant c such that for all y, $K_U(y) \leq K_{U'}(y) + c$.*

Henceforth, let K be defined by the universal Turing machine of Theorem 24.14. For every integer n and every binary string y of length n, because y can be described by giving itself explicitly, $K(y) \leq n + c'$ for a constant c'. Call y **incompressible** if $K(y) \geq n$. Since there are 2^n binary strings of length n, and only $2^n - 1$ possible shorter descriptions, there exists an **incompressible string** for every length n.

Kolmogorov complexity gives a precise mathematical meaning to the intuitive notion of "randomness." If someone flips a coin 50 times and it comes up "heads" each time, then intuitively, the sequence of flips is not random—although from the standpoint of probability theory the all-heads sequence is precisely as likely as any other sequence. Probability theory does not provide the tools for calling one sequence "more random" than another; Kolmogorov complexity theory does.

Kolmogorov complexity provides a useful framework for presenting combinatorial arguments. For example, when one wants to prove that an object with some property P exists, then it is sufficient to show that any object that does not have property P has a short description; thus, any incompressible (or "random") object must have property P. This sort of argument has been useful in proving lower bounds in complexity theory. For example, Dietzfelbinger et al. [19] use Kolmogorov complexity to show that no Turing machine with a single worktape can compute the transpose of a matrix in less than time $\Omega(n^{3/2}/\sqrt{\log n})$.

24.12   Research Issues and Summary

As stated in the introduction to Chapter 22, the goals of complexity theory are (1) to ascertain the amount of computational resources required to solve important computational problems, and (2) to classify problems according to their difficulty. The preceding two chapters have explained how complexity theory has devised a classification scheme in order to meet the second goal. The present chapter has presented a few of the additional notions of complexity that have been devised in order to capture more problems in this scheme. Progress toward the first goal, namely proving lower bounds, depends on knowing that levels in this classification scheme are in fact distinct. Thus, the core research questions in complexity theory are expressed in terms of separating complexity classes:

- Is L different from NL?
- Is P different from RP, or from BPP?

- Is P different from BQP, or from NP?
- Is NP different from PSPACE?

Motivated by these questions, much current research is devoted to efforts to understand the power of nondeterminism, randomization, and interaction. In these studies, researchers have gone well beyond the theory presented in Chapters 22 through 24:

- Beyond Turing machines and Boolean circuits, to restricted and specialized models in which nontrivial lower bounds on complexity can be proved;
- Beyond deterministic reducibilities, to nondeterministic and probabilistic reducibilities, and refined versions of the reducibilities considered here;
- Beyond worst-case complexity, to average-case complexity.

We have illustrated how research in complexity theory has had direct applications to other areas of computer science and mathematics. Probabilistically checkable proofs were used to show that obtaining approximate solutions to some optimization problems is as difficult as solving them exactly. Complexity theory provides new tools for studying questions in finite model theory, a branch of mathematical logic. NP-completeness and related notions of computational intractability have proved to be very useful in physics, chemistry, economics, and other fields. Fundamental questions in complexity theory are intimately linked to practical questions about the use of cryptography for computer security, such as the existence of one-way functions and the strength of public key cryptosystems.

This last point illustrates the urgent practical need for progress in computational complexity theory. Many popular cryptographic systems in current use are based on unproven assumptions about the difficulty of certain computational tasks, such as integer factoring and computing discrete logarithms—see Chapters 18 through 19 of *Algorithms and Computation Handbook, Second Edition: Special Topics and Techniques* for more background on cryptography. All of these systems are thus based on wishful thinking and conjecture. The need to resolve these open questions and replace conjecture with mathematical certainty should be self-evident. In the brief history of complexity theory, we have learned that many popular expectations, such as co-NL being different from NL or co-NP not having efficient interactive proofs, turn out to be incorrect.

With precisely defined models and mathematically rigorous proofs, research in complexity theory will continue to provide sound insights into the difficulty of solving real computational problems.

24.13 Further Information

Primary sources for major theorems presented in this chapter include Theorem 24.1 [18,64,72]; Theorem 24.3(a,b) [30], (c) [46,61], (d) [66], (e) [5]; Theorem 24.4 [3]; Theorem 24.5(a) [1], (b) [64], (c) [24], (d,e,f) [39], (g) ([48], cf. [13]); Theorem 24.6(a) [16], (b) [52,58]; Theorem 24.7 [59]; Theorem 24.8 [11]; Theorem 24.9 [8] (see also [20]; Theorem 24.10 various including [32,38]; Theorem 24.11 [9]; Theorem 24.13(a) [67], (b) [69]. The operators in Definition 24.2 are from [1,39]. Interactive proof systems were defined by Goldwasser et al. [31], and in the "Arthur-Merlin" formulation, by Babai and Moran [10]. A large compendium of optimization problems and hardness results collected By P. Crescenzi and V. Kann is available at: http://www.nada.kth.se/ viggo/problemlist/ compendium.html

The class #P was introduced by Valiant [67], and #L by Alvarez and Jenner [6]. Li and Vitányi [47] give a far-reaching and comprehensive scholarly treatment of Kolmogorov complexity, with many applications, as well as the source of Theorem 24.14.

Readers seeking more information on quantum computing can consult survey articles such as [4,14,27,36,56] or texts such as [37,53] (among others). Some people have proposed *DNA* computing as a quite different source of massive parallelism; for a survey, see [45].

Five contemporary textbooks on complexity theory are [12,15,22,35,55]. Wagner and Wechsung [70] provide is an exhaustive survey of complexity theory that covers work published before 1986. Another perspective of some of the issues covered in these three chapters may be found in the survey [65].

A good general reference is the *Handbook of Theoretical Computer Science* [68], volume A. The following chapters in the handbook are particularly relevant: "Machine models and simulations," by P. van Emde Boas, pp. 1–66; "A catalog of complexity classes," by D.S. Johnson, pp. 67–161; "Machine-independent complexity theory," by J.I. Seiferas, pp. 163–186; "Kolmogorov complexity and its applications," by M. Li and P.M.B. Vitányi, pp. 187–254; and "The complexity of finite functions," by R.B. Boppana and M. Sipser, pp. 757–804, which covers circuit complexity.

A collection of articles edited by Hartmanis [33] includes an overview of complexity theory, and chapters on sparse complete languages, on relativizations, on interactive proof systems, and on applications of complexity theory to cryptography. For historical perspectives on complexity theory, see [34,62,63].

There are many areas of complexity theory that we have not been able to cover in these chapters. Some of them cross-pollinate with other fields of computer science and are reflected in other chapters of this handbook. Three others are average-case complexity, resource-bounded measure theory, and parameterized complexity. Surveys on the first two are by Lutz [50] and Wang [71], while the third stems from Downey and Fellows [21] (see also the recent textbook [28]).

An excellent on-line resource for complexity theory is the Electronic Colloquium on Computational Complexity: http://eccc.hpi-web.de/eccc/. Here you can find recent research papers, as well as pointers to books, lecture notes, and surveys that are available on-line. Additional material can be found on Wikipedia, and in the "Complexity Zoo," which is a guide to the bewildering menagerie of complexity classes: http://qwiki.caltech.edu/wiki/Complexity_Zoo

Research papers on complexity theory are presented at several annual conferences, including the annual ACM Symposium on Theory of Computing; the annual International Colloquium on Automata, Languages, and Programming, sponsored by the European Association for Theoretical Computer Science (EATCS); and the annual Symposium on Foundations of Computer Science, sponsored by the IEEE. The annual Conference on Computational Complexity (formerly Structure in Complexity Theory), also sponsored by the IEEE, is entirely devoted to complexity theory. Research articles on complexity theory regularly appear in the following journals, among others: *Chicago Journal on Theoretical Computer Science, Computational Complexity, Information and Computation, Journal of the ACM, Journal of Computer and System Sciences, SIAM Journal on Computing, Theoretical Computer Science,* and *Theory of Computing Systems* (formerly *Mathematical Systems Theory).* Each issue of *ACM SIGACT News* and *Bulletin of the EATCS* contains a column on complexity theory.

Defining Terms

Descriptive complexity: The study of classes of languages described by formulas in certain systems of logic.

Incompressible string: A string whose **Kolmogorov complexity** equals its length, so that it has no shorter encodings.

Interactive proof system: A protocol in which one or more provers try to convince another party called the verifier that the prover(s) possess(es) certain true knowledge, such as the membership of a string x in a given language, often with the goal of revealing no further details about this knowledge. The prover(s) and verifier are formally defined as probabilistic Turing machines with special "interaction tapes" for exchanging messages.

Kolmogorov complexity: The minimum number of bits into which a string can be compressed without losing information. This is defined with respect to a fixed but universal decompression scheme, given by a universal Turing machine.

L-reduction: A Karp reduction that preserves approximation properties of optimization problems.

One-way function: A polynomial-time computable function f for which given y in the range of f, it is difficult to find x in the domain of f such that $f(x) = y$. This property has implications for cryptography.

Optimization problem: A computational problem in which the object is not to decide some yes/no property, as with a decision problem, but to find the best solution in those "yes" cases where a solution exists.

Polynomial hierarchy: The collection of classes of languages accepted by k-alternating Turing machines, over all $k \geq 0$ and with initial state existential or universal. The bottom level ($k = 0$) is the class P, and the next level ($k = 1$) comprises NP and co-NP.

Polynomial-time approximation scheme (PTAS): A meta-algorithm that for every $\epsilon > 0$ produces a polynomial time ϵ-approximation algorithm for a given optimization problem.

Probabilistic Turing machine: A Turing machine in which some transitions are random choices among finitely many alternatives.

Probabilistically checkable proof: An interactive proof system in which provers follow a fixed strategy, one not affected by any messages from the verifier. The prover's strategy for a given instance x of a decision problem can be represented by a finite oracle language B_x, which constitutes a proof of the correct answer for x.

Relational structure: The counterpart in formal logic of a data structure or class instance in the object-oriented sense. Examples are strings, directed graphs, and undirected graphs. Sets of relational structures generalize the notion of languages as sets of strings.

Acknowledgments

The first author was supported by the National Science Foundation under Grants CCF 0830133 and 0132787; portions of this work were performed while a visiting scholar at the Institute of Mathematical Sciences, Madras, India. The second author was supported by the National Science Foundation under Grants DUE-0618589 and EEC-0628814. The third author was supported by the National Science Foundation under Grant CCR-9409104.

References

1. Abiteboul, S. and Vianu, V., Datalog extensions for database queries and updates. *J. Comp. Sys. Sci.*, 43, 62–124, 1991.
2. Abiteboul, S. and Vianu, V., Computing with first-order logic. *J. Comp. Sys. Sci.*, 50, 309–335, 1995.
3. Adleman, L., Two theorems on random polynomial time. In *Proceedings of 19th Annual IEEE Symposium on Foundations of Computer Science*, pp. 75–83, 1978.
4. Aharonov, D., Quantum computation. In *Annual Review of Computational Physics VI*, D. Stauffer, Ed., pp. 259–346. World Scientific Press, Singapore, 1999.
5. Allender, E., The permanent requires large uniform threshold circuits. *Chicago J. Theoret. Comp. Sci.*, article 7, 1999.

6. Alvarez, C. and Jenner, B., A very hard log-space counting class. *Theor. Comp. Sci.*, 107, 3–30, 1993.
7. Ambos-Spies, K., A note on complete problems for complexity classes. *Info. Proc. Lett.*, 23, 227–230, 1986.
8. Arora, S. and Safra, S., Probabilistic checking of proofs: A new characterization of NP. *J. ACM*, 45, 70–122, 1998.
9. Arora, S., Lund, C., Motwani, R., Sudan, M., and Szegedy, M., Proof verification and hardness of approximation problems. *J. ACM*, 45, 501–555, 1998.
10. Babai, L. and Moran, S., Arthur-Merlin games: A randomized proof system, and a hierarchy of complexity classes. *J. Comp. Sys. Sci.*, 36, 254–276, 1988.
11. Babai, L., Fortnow, L., and Lund, C., Nondeterministic exponential time has two-prover interactive protocols. *Comp. Complexity*, 1, 3–40, 1991. Addendum in Vol. 2 of same journal.
12. Balcázar, J., Díaz, J., and Gabarró, J., *Structural Complexity I, II.* Springer-Verlag, Berlin, Germany, 1990. Part I published in 1988.
13. Barrington, D.M., Immerman, N., and Straubing, H., On uniformity within NC^1. *J. Comp. Sys. Sci.*, 41, 274–306, 1990.
14. Berthiaume, A., Quantum computation. In *Complexity Theory Retrospective II*, L. Hemaspaandra and A. Selman, Eds., pp. 23–51. Springer-Verlag, New York, 1997.
15. Bovet, D. and Crescenzi, P., *Introduction to the Theory of Complexity.* Prentice Hall International (U.K.) Limited, Hertfordshire, U.K., 1994.
16. Büchi, J., Weak second-order arithmetic and finite automata. *Zeitschrift für Mathematische Logik und Grundlagen der Mathematik*, 6, 66–92, 1960.
17. Chandra, A. and Harel, D., Structure and complexity of relational queries. *J. Comp. Sys. Sci.*, 25, 99–128, 1982.
18. Chandra, A., Kozen, D., and Stockmeyer, L., Alternation. *J. Assn. Comp. Mach.*, 28, 114–133, 1981.
19. Dietzfelbinger, M., Maass, W., and Schnitger, G., The complexity of matrix transposition on one-tape off-line Turing machines. *Theor. Comp. Sci.*, 82, 113–129, 1991.
20. Dinur, I., The PCP theorem by gap amplification. *J. ACM*, 54, article 12, 2007.
21. Downey, R. and Fellows, M., Fixed-parameter tractability and completeness I: Basic theory. *SIAM J. Comp.*, 24, 873–921, 1995.
22. Du, D-Z. and Ko, K-I., *Theory of Computational Complexity.* Wiley, New York, 2000.
23. Enderton, H.B., *A Mathematical Introduction to Logic.* Academic Press, New York, 1972.
24. Fagin, R., Generalized first-order spectra and polynomial-time recognizable sets. In *Complexity of Computation: Proceedings of Symposium in Applied Mathematics of the American Mathematical Society and the Society for Industrial and Applied Mathematics*, Vol. VII, R. Karp, Ed., pp. 43–73. SIAM-AMS, 1974.
25. Fagin, R., Finite model theory—a personal perspective. *Theor. Comp. Sci.*, 116, 3–31, 1993.
26. Feige, U., Goldwasser, S., Lovász, L., Safra, S., and Szegedy, M., Interactive proofs and the hardness of approximating cliques. *J. ACM*, 43, 268–292, 1996.
27. Fenner, S., Counting complexity and quantum computation. In *Mathematics of Quantum Computation*, R.K. Brylinski and G. Chen, Eds., pp. 171–219. Chapter 8. CRC Press, Boca Raton, FL, 2002.
28. Flum, J. and Grohe, M., *Parameterized Complexity Theory.* Springer Verlag, Berlin, Germany, 2006.
29. Fortnow, L. and Sipser, M., Are there interactive protocols for co-NP languages? *Info. Proc. Lett.*, 28, 249–251, 1988.
30. Gill, J., Computational complexity of probabilistic Turing machines. *SIAM J. Comp.*, 6, 675–695, 1977.
31. Goldwasser, S., Micali, S., and Rackoff, C., The knowledge complexity of interactive proof systems. *SIAM J. Comp.*, 18, 186–208, 1989.

32. Grollmann, J. and Selman, A.L., Complexity measures for public-key cryptosystems. *SIAM J. Comp.*, 17, 309–335, 1988.

33. Hartmanis, J., Ed., *Computational Complexity Theory*. American Mathematical Society, Providence, RI, 1989.

34. Hartmanis, J., On computational complexity and the nature of computer science. *Comm. Assn. Comp. Mach.*, 37, 37–43, 1994.

35. Hemaspaandra, L.A. and Ogihara, M., *The Complexity Theory Companion*. Springer, Berlin, Germany, 2002.

36. Hirvensalo, M., An introduction to quantum computing. In *Current Trends in Theoretical Computer Science: Entering the 21st Century*, G. Păun, G. Rozenberg, and A. Salomaa, Eds., pp. 643–663. World Scientific Press, Singapore, 2001.

37. Hirvensalo, M., *Quantum Computing, Springer Series on Natural Computing*, Springer-Verlag, Berlin, 2001.

38. Homan, C.M. and Thakur, M., One-way permutations and self-witnessing languages. *J. Comp. Sys. Sci.*, 67, 608–622, 2003.

39. Immerman, N., Descriptive and computational complexity. In *Computational Complexity Theory*, volume 38 of *Proceedings of the Symposium in Applied Mathematics*, J. Hartmanis, Ed., pp. 75–91. American Mathematical Society, Providence, RI, 1989.

40. Immerman, N., *Descriptive Complexity*, Springer Graduate Texts in Computer Science, Springer-Verlag, New York, 1999.

41. Impagliazzo, R. and Wigderson, A., P = BPP if E requires exponential circuits: Derandomizing the XOR Lemma. In *Proceedings of 29th Annual ACM Symposium on the Theory of Computing*, pp. 220–229, 1997.

42. Jones, N. and Selman, A., Turing machines and the spectra of first-order formulas. *J. Assn. Comp. Mach.*, 39, 139–150, 1974.

43. Karp, R. and Lipton, R., Turing machines that take advice. *L'Enseignement Mathématique*, 28, 191–210, 1982.

44. Krentel, M., The complexity of optimization problems. *J. Comp. Sys. Sci.*, 36, 490–509, 1988.

45. Kurtz, S., Mahaney, S., Royer, J., and Simon, J., Biological computing. In *Complexity Theory Retrospective II*, L. Hemaspaandra and A. Selman, Eds., pp. 179–195. Springer-Verlag, New York, 1997.

46. Lautemann, C., BPP and the polynomial hierarchy. *Info. Proc. Lett.*, 17, 215–217, 1983.

47. Li, M. and Vitányi, P., *An Introduction to Kolmogorov Complexity and Its Applications*. 2nd ed., Springer-Verlag, New York, 1997.

48. Lindell, S., How to define exponentiation from addition and multiplication in first-order logic on finite structures, 1994. Unpublished manuscript. See [40] for details.

49. Lund, C., Fortnow, L., Karloff, H., and Nisan, N., Algebraic methods for interactive proof systems. *J. Assn. Comp. Mach.*, 39, 859–868, 1992.

50. Lutz, J., The quantitative structure of exponential time. In *Complexity Theory Retrospective II*, L. Hemaspaandra and A. Selman, Eds., pp. 225–260. Springer-Verlag, New York, 1997.

51. Lynch, J., Complexity classes and theories of finite models. *Math. Sys. Thy.*, 15, 127–144, 1982.

52. McNaughton, R. and Papert, S., *Counter-Free Automata*. MIT Press, Cambridge, MA, 1971.

53. Nielsen, M. and Chuang, I., *Quantum Computation and Quantum Information*. Cambridge University Press, New York, 2000.

54. Papadimitriou, C. and Yannakakis, M., The complexity of facets (and some facets of complexity). *J. Comp. Sys. Sci.*, 28, 244–259, 1984.

55. Papadimitriou, C., *Computational Complexity*. Addison-Wesley, Reading, MA, 1994.

56. Reiffel, E. and Polak, W., An introduction to quantum computing for non-physicists. *Comp. Surveys*, 32, 300–335, 2000.

57. Reingold, O., Undirected ST-connectivity in log-space. In *Proceedings of ACM Symposium on Theory of Computing*, 376–385, 2005.

58. Schützenberger, M.P., On finite monoids having only trivial subgroups. *Info. Control,* 8, 190–194, 1965.

59. Shamir, A., IP = PSPACE. *J. Assn. Comp. Mach.,* 39, 869–877, 1992.

60. Sipser, M., On relativization and the existence of complete sets. In *Proceedings of 9th Annual International Conference on Automata, Languages, and Programming,* volume 140 of *Lecture Notes in Computer Science,* pp. 523–531. Springer-Verlag, New York, 1982.

61. Sipser, M., Borel sets and circuit complexity. In *Proceedings of 15th Annual ACM Symposium on the Theory of Computing,* pp. 61–69, 1983.

62. Sipser, M., The history and status of the P versus NP question. In *Proceedings of 24th Annual ACM Symposium on the Theory of Computing,* pp. 603–618, 1992.

63. Stearns, R., Juris Hartmanis: the beginnings of computational complexity. In *Complexity Theory Retrospective,* A. Selman, Ed., pp. 5–18. Springer-Verlag, New York, 1990.

64. Stockmeyer, L., The polynomial time hierarchy. *Theor. Comp. Sci.,* 3, 1–22, 1976.

65. Stockmeyer, L., Classifying the computational complexity of problems. *J. Symb. Logic,* 52, 1–43, 1987.

66. Toda, S., PP is as hard as the polynomial-time hierarchy. *SIAM J. Comp.,* 20, 865–877, 1991.

67. Valiant, L., The complexity of computing the permanent. *Theor. Comp. Sci.,* 8, 189–201, 1979.

68. van Leeuwen, J., Ed., *Handbook of Theoretical Computer Science,* volume A. Elsevier, Amsterdam, the Netherlands, and MIT Press, Cambridge, MA, 1990.

69. Vinay, V., Counting auxiliary pushdown automata and semi-unbounded arithmetic circuits. In *Proceedings of 6th Annual IEEE Conference on Structure in Complexity Theory,* pp. 270–284, 1991.

70. Wagner, K. and Wechsung, G., *Computational Complexity.* D. Reidel, Dordrecht, the Netherlands, 1986.

71. Wang, J., Average-case computational complexity theory. In *Complexity Theory Retrospective II,* L. Hemaspaandra and A. Selman, Eds., pp. 295–328. Springer-Verlag, Berlin, 1997.

72. Wrathall, C., Complete sets and the polynomial-time hierarchy. *Theor. Comp. Sci.,* 3, 23–33, 1976.

25

Parameterized Algorithms

Rodney G. Downey
Victoria University

Catherine McCartin
Massey University

25.1 Introduction

Since the advent of classical complexity theory in the early 1970s, the twin notions of *NP*-hardness and *NP*-completeness have been accepted as concrete measures of computational intractability. However, a verdict of *NP*-hardness does not do away with the need for solving hard computational problems, since the bulk of these problems are of both theoretical and practical importance.

The field of parameterized complexity theory and parameterized computation has developed rapidly over the past 20 years as a robust approach to dealing with hard computational problems arising from applications in diverse areas of science and industry. The parameterized paradigm augments classical complexity theory and computation, providing, on the one hand, systematic and practical algorithm design techniques for hard problems, and, on the other hand, more finely grained complexity analysis and stronger computational lower bounds for natural computational problems.

The theory is based on the simple observation that many hard computational problems have certain aspects of their input, or expected solution, that vary only within a moderate range, at least for instances that are of practical importance. By exploiting such small associated parameters, many classically intractable problems can be efficiently solved.

Apparent parameterized intractability is established via a completeness program, which parallels the traditional paradigm, but allows for stratification of problems into a far more richly-structured hierarchy of complexity classes.

A number of approaches have been proposed to deal with the central issue of computational intractability, including polynomial time approximation, randomization, and heuristic algorithms. The parameterized paradigm is orthogonal to each of these earlier approaches, yet a range of fundamental connections has begun to emerge.

The aim of this chapter is to survey the current state of the art in the field of parameterized complexity, canvassing main techniques and important results. We concentrate on the distinctive algorithmic techniques that have emerged in the field, in particular those that lead to practical and useful algorithms for classically intractable problems.

While there are a large number of applications of these ideas in many diverse arenas, our plan is to present the ideas concentrating mainly on a small number of problems, particularly VERTEX COVER and some variants (defined in Section 25.2) Our intention is to facilitate understanding of the principal techniques without the need for lengthy explanations of a diverse range of problems. Even so, space limitations mean that we cannot canvass all algorithmic techniques used in the field. The use of bounded variable integer linear programming and general graph modification techniques are two important omissions. Finally, we mention that implementations of many of the techniques that we will introduce have performed much better in practice than one might reasonably expect, but we have left out experimental discussions. We refer the reader to the recently published special issue of *The Computer Journal* [36], the monographs of Niedermeier [55] and Fernau [45], and the *ACM SIGACT News* article by Guo and Niedermeier [46] for discussions on these points and for a tour of the wide range of applications.

25.2   The Main Idea

It is generally accepted that solving an *NP*-hard problem will necessarily entail a combinatorial explosion of the search space. However, it is not necessarily the case that all instances of an *NP*-hard problem are equally hard to solve, hardness sometimes depends on the particular structure of a given instance, or of the expected solution. Instances of *NP*-hard problems arising from "real life" often exhibit more regular structure than the general problem description might, at first, suggest.

For example, suppose that we are concerned with solving computational problems to do with relational databases. Typically, a real-life database will be huge, and the queries made to it will be relatively small. Moreover, real-life queries will be questions that people actually ask. Hence, such queries will tend to be of low logical complexity. Thus, an algorithm that works very efficiently for small formulae with low logical depth might well be perfectly acceptable in practice. Alternatively, suppose that we are concerned with computational problems where the focus is to recognize a particular substructure in the problem input. If the size of the expected substructure is small, then an algorithm that works very efficiently for small solutions may be acceptable in practice.

The main idea of parameterized complexity is to develop a framework that addresses complexity issues in this situation, where we know in advance that certain parameters of the problem at hand are likely to be bounded, and that this might significantly affect the complexity.

The basic insight that underpins parameterized complexity and parameterized computation arose from consideration of two well-known *NP*-complete problems for simple undirected graphs.

A vertex cover of $G = (V, E)$ is a set of vertices $V' \subseteq V$ that covers all edges: that is $V' = \{v_1, \ldots, v_k\}$ is a vertex cover for G iff, for every edge $(u, v) \in E$, either $u \in V'$ or $v \in V'$. A dominating set of $G = (V, E)$ is a set of vertices $V' \subseteq V$ that covers all vertices: that is $V' = \{v_1, \ldots, v_k\}$ is a dominating set for G iff, for every vertex $v \in V$, either $v \in V'$ or there is some $u \in V'$ such that $(u, v) \in E$.

VERTEX COVER

Instance: A graph $G = (V, E)$ and a positive integer k.

Question: Does G have a vertex cover of size at most k?

DOMINATING SET

Instance: A graph $G = (V, E)$ and a positive integer k.

Question: Does G have a dominating set of size at most k?

Although both of these problems are, classically, *NP*-complete, the parameter k contributes to the complexity of these two problems in two qualitatively different ways.

DOMINATING SET: Essentially the only known algorithm for this problem is to try all possibilities. The brute force algorithm of trying all k-subsets runs in time $O(n^{k+1})$ (we use n to denote $|V|$ and m to denote $|E|$).

VERTEX COVER: After many rounds of improvement, there is now an algorithm running in time $O(1.286^k + kn)$ [26] for determining if a graph $G = (V, E)$ has a vertex cover of size k. This has been implemented and is practical for n of unlimited size and k up to around 400 [39,66].

The table below shows the contrast between these two kinds of complexity.

These observations are formalized in the framework of parameterized complexity theory [33,34]. In classical complexity, a decision problem is specified by two items of information:

1. The input to the problem
2. The question to be answered

	$n = 50$	$n = 100$	$n = 150$
$k = 2$	625	2,500	5,625
$k = 3$	15,625	125,000	421,875
$k = 5$	390,625	6,250,000	31,640,625
$k = 10$	1.9×10^{12}	9.8×10^{14}	3.7×10^{16}
$k = 20$	1.8×10^{26}	9.5×10^{31}	2.1×10^{35}

The ratio $\frac{n^{k+1}}{2^k n}$ for various values of n and k

In parameterized complexity, there are three parts to a problem specification:

1. The input to the problem
2. The aspects of the input that constitute the parameter
3. The question to be answered

The notion of fixed-parameter tractability (FPT) is the central concept of the theory. Intuitively, a problem is fixed-parameter tractable if we can somehow confine any "bad" complexity behavior to some limited aspect of the problem, the parameter.

More formally, we consider a parameterized language to be a subset $L \subseteq \Sigma^* \times \Sigma^*$. If L is a parameterized language and $(I, k) \in L$ then we refer to I as the *main part* and k as the *parameter*.

DEFINITION 25.1 **(Fixed Parameter Tractability (FPT)).** A parameterized language $L \subseteq \Sigma^* \times \Sigma^*$ is *fixed-parameter tractable* if there is an algorithm (or a k-indexed collection of algorithms) that correctly decides, for input $(I, k) \in \Sigma^* \times \Sigma^*$, whether $(I, k) \in L$ in time $f(k) \cdot n^c$, where n is the size of the main part of the input I, k is the parameter, c is a constant (independent of both n and k), and f is an arbitrary function dependent only on k.

Usually, the parameter k will be a positive integer, but it could be, for instance, a graph or an algebraic structure, or a combination of integer values bounding various aspects of the problem. The parameter will often bound the size of some part of the input instance or the solution. Alternatively, it can bound the complexity of the input instance in some well-defined sense. For example, in Sections 25.4.2 and 25.4.3 we introduce width metrics for graphs which precisely capture various notions of complexity in graphs. A single classical problem can often be parameterized in several natural ways, each leading to a separate parameterized problem.

In this chapter, we will concentrate mainly on the techniques for demonstrating parameterized tractability. There is also a very well-developed theory of parameterized intractability, used to address problems like DOMINATING SET, which we introduce in Section 25.5, but for which space limitations preclude a deeper treatment. As we see in Section 25.5, there is a completeness and hardness theory, akin to that of NP-completeness, that can be used to demonstrate parameterized intractability, based around a parameterized analog of NONDETERMINISTIC TURING MACHINE ACCEPTANCE.

25.3   Practical FPT Algorithms

In this section we introduce the main practical techniques that have emerged in the field of FPT algorithm design. We focus first on two simple algorithmic strategies that are not part of the usual toolkit of polynomial algorithm design, but which have lead, in many important cases, to practical and useful algorithms for natural parameterized versions of *NP*-hard problems. These two techniques, (1) Kernelization and (2) Depth-Bounded Search Trees, have formed the backbone of practical FPT algorithm design. In Section 25.3.3 we show how these two techniques can be profitably combined using the concept of interleaving. We also introduce iterative compression, a relatively new technique that has been successfully applied to a range of parameterized minimization problems, where the parameter is the size of the solution set.

25.3.1   Kernelization

Kernelization is based on an old idea, that of preprocessing, or reducing, the input data of a computational problem. It often makes sense to try to eliminate those parts of the input data that are relatively easy to cope with, shrinking the given instance to some hard core that must be dealt with using a computationally expensive algorithm. In fact, this is the basis of many heuristic algorithms for *NP*-hard problems, in a variety of areas, that seem to work reasonably well in practice. In other words, it is something that many practitioners, faced with a real-world *NP*-hard problem, already do.

A compelling example of the effectiveness of data reduction, for a classically-posed *NP*-hard problem, is given by Weihe [67]. He considered the following problem in the context of the European railroad network: given a set of trains, select a set of stations such that every train passes through at least one of those stations and such that the number of selected stations is minimum. Weihe modeled this problem as a path cover by vertices in an undirected graph. Here, we formulate the problem as domination of one set of vertices by another in a bipartite graph.

TRAIN COVERING BY STATIONS

Instance: A bipartite graph $G = (V_S \cup V_T, E)$, where the set of vertices V_S represents railway stations and the set of vertices V_T represents trains. E contains an edge $(s, t), s \in V_s, t \in V_T$, iff the train t stops at the station s.

Problem: Find a minimum set $V' \subseteq V_S$ such that V' covers V_T, that is, for every vertex $t \in V_T$, there is some $s \in V'$ such that $(s, t) \in E$.

Weihe employed two simple data reduction rules for this problem. For our problem formulation they translate to the following:

REDUCTION RULE TCS1

Let $N(t)$ denote the neighbors of t in V_S. If $N(t) \subseteq N(t')$ then remove t' and all adjacent edges of t' from G. If there is a station that covers t, then this station also covers t'.

REDUCTION RULE TCS2

Let $N(s)$ denote the neighbors of s in V_T. If $N(s) \subseteq N(s')$ then remove s and all adjacent edges of s from G. If there is a train covered by s, then this train is also covered by s'.

In practice, exhaustive application of these two simple data reduction rules allowed for the problem to be solved in minutes, for a graph modeling the whole European train schedule, consisting of around 1.6×10^5 vertices and 1.6×10^6 edges.

This impressive performance begs the question: Why should data reduction be of more concrete use in the parameterized paradigm? The answer comes from the observation that a data reduction scheme for a parameterized problem can often give upper bounds on the size of the reduced instance in terms solely of the parameter. Once such a reduction scheme is established, a trivial FPT algorithm manifests as a brute force search of the reduced instance. Thus, in the parameterized context, data reduction can often lead directly to an FPT algorithm to solve the problem. This contrasts with the classical context, where data reduction can clearly lead to a useful heuristic, but without any provable performance guarantee.

To illustrate the kernelization concept, we start with a simple data reduction scheme for the standard parameterized version of the NP-hard VERTEX COVER problem (introduced in Section 25.2.) As for subsequent examples given in this section, we paraphrase the treatment given in [49].

K-VERTEX COVER

Instance: A graph $G = (V, E)$.

Parameter: A positive integer k.

Question: Does G have a vertex cover of size $\leq k$?

Vertices with no adjacent edges are irrelevant, both to the problem instance and to any solution. This leads to

REDUCTION RULE VC1

Remove all isolated vertices.

In order to cover an edge in E, one of its endpoints must be in the solution set. If one of these endpoints is a degree one vertex, then the other endpoint has the potential to cover more edges than the degree one vertex, leading to:

REDUCTION RULE VC2

For any degree one vertex v, add its single neighbor u to the solution set and remove u and all of its incident edges from the graph.

The reduced instance thus consists of both a smaller graph and a smaller parameter, $(G, k) \rightarrow (G', k - 1)$.

These two data reduction rules are applicable in any problem-solving context. However, in the parameterized setting, where we are looking only for a small solution, with size bounded by parameter k, we can do more. Buss [21] originally observed that, for a simple graph G, any vertex of degree greater than k must belong to every k-element vertex cover of G (otherwise all the neighbors of the vertex must be included, and there are more than k of these).

REDUCTION RULE VC3

If there is a vertex v of degree at least $k + 1$, add v to the solution set and remove v and all of its incident edges from the graph.

The reduced instance again consists of both a smaller graph and a smaller parameter, $(G, k) \rightarrow (G', k - 1)$.

After exhaustively applying these three rules, we have a new, reduced, instance, (G', k'), where no vertex in the reduced graph has degree greater than k', or less than two. Thus, any vertex remaining can cover at most k' edges in this reduced instance. Since the solution set can contain at most k' vertices, if the reduced graph is a YES instance, then it must have at most k'^2 edges, and consequently at most k'^2 vertices.

Thus, in a polynomial amount of time, we have reached a situation where we can either declare our original instance to be a NO instance, or, by means of a brute force search of the reduced instance, in time $O(k'^{2k'})$, $k' \le k$, decide whether our original instance admits a vertex cover of size at most k.

The important point is that the reduced instance can either be immediately declared a NO instance or, otherwise, has size bounded by a function of the parameter. We formalize this idea in terms of a reduction to a problem kernel, or kernelization.

DEFINITION 25.2 (Kernelization). Let $L \subseteq \Sigma^* \times \Sigma^*$ be a parameterized language. Let $\mathcal{L}$ be the corresponding parameterized problem, that is, $\mathcal{L}$ consists of input pairs (I, k), where I is the main part of the input and k is the parameter. A reduction to a problem kernel, or kernelization, comprises replacing an instance (I, k) by a reduced instance (I', k'), called a problem kernel, such that

(i) $k' \le k$,

(ii) $|I'| \le g(k)$, for some function g depending only on k, and

(iii) $(I, k) \in L$ if and only if $(I', k') \in L$.

The reduction from (I, k) to (I', k') must be computable in time polynomial in $|I|$.

The kernelization for K-VERTEX COVER described above uses rules that examine only local substructures of the input (a vertex and its neighborhood). For a range of problems, this approach proves adequate for producing a reasonably-sized problem kernel. Another possibility is to consider the global properties of a problem instance.

Chen et al. [26] have used this second approach in exploiting a well-known theorem of Nemhauser and Trotter [54] to construct a problem kernel for VERTEX COVER having at most $2k$ vertices. This seems to be the best that one could hope for, since a problem kernel of size $(2 - \epsilon) \cdot k$, with constant $\epsilon > 0$, would imply a factor $2 - \epsilon$ polynomial-time approximation algorithm for VERTEX COVER. The existence of such an algorithm is a long-standing open question in the area of approximation algorithms for *NP*-hard problems.

THEOREM 25.1 [54] *For an n-vertex graph $G = (V, E)$ with m edges, we can compute two disjoint sets $C' \subseteq V$ and $V' \subseteq V$, in $O(\sqrt{n} \cdot m)$ time, such that the following three properties hold:*

(i) *There is a minimum size vertex cover of G that contains C'.*

(ii) *A minimum vertex cover for the induced subgraph $G[V']$ has size at least $|V'|/2$.*

(iii) *If $D \subseteq V'$ is a vertex cover of the induced subgraph $G[V']$, then $C = D \cup C'$ is a vertex cover of G.*

THEOREM 25.2 [26] *Let $(G = (V, E), k)$ be an instance of K-VERTEX COVER. In $O(k \cdot |V| + k^3)$ time we can reduce this instance to a problem kernel $(G = (V', E'), k')$ with $|V'| \le 2k$.*

The kernelization begins by applying the three reduction rules described above, VC1, VC2, and VC3, to produce a reduced instance (G', k'), where G' contains at most $O(k'^2)$ vertices and edges, and $k' \le k$. This reduction takes $O(k \cdot |V|)$ time.

For the resulting reduced instance (G', k') we compute the two sets C' and V' as described in Theorem 25.1. Determining the two sets C' and V' involves computation of a maximum matching on a graph constructed from G' and can be achieved in time $O(\sqrt{k^2} \cdot k^2) = O(k^3)$.

The set C' contains vertices that have to be in the vertex cover, so we define a new parameter $k'' = k' - |C'|$. Due to Theorem 25.1, we know that if $|V'| > 2k''$ then there is no vertex cover of size k for the original graph G. Otherwise, we let the induced subgraph $G[V']$ be the problem kernel, having size at most $2k'' \leq 2k$. By Theorem 25.1, the remaining vertices for a minimum vertex cover of G can be found by searching for a minimum vertex cover in $G[V']$.

Recently, a third alternative to both local and global data reduction schemes has been explored. In this third case, local rules are generalized to examine arbitrarily large substructures. Continuing with our running example, K-VERTEX COVER, we show that the local rule **VC2**, which entails the deletion of any degree-1 vertex and the admission of its sole neighbor into the vertex cover, can be generalized to the crown reduction rule.

A crown in a graph $G = (V, E)$ consists of an independent set $I \subseteq V$ (no two vertices in I are connected by an edge) and a set H containing all vertices in V adjacent to I. A crown in G is formed by $I \cup H$ iff there exists a size $|H|$ maximum matching in the bipartite graph induced by the edges between I and H, that is, every vertex of H is matched. It is clear that degree-1 vertices in V, coupled with their sole neighbors, can be viewed as the most simple crowns in G.

If we find a crown $I \cup H$ in G, then we need at least $|H|$ vertices to cover all edges in the crown. Since all edges in the crown can be covered by admitting at most $|H|$ vertices into the vertex cover, there is a minimum size vertex cover that contains all vertices in H and no vertices in I. These observations lead to the following reduction rule.

REDUCTION RULE CR
For any crown $I \cup H$ in G, add the set of vertices H to the solution set and remove $I \cup H$ and all of the incident edges of $I \cup H$ from G.
The reduced instance thus consists of a smaller graph and a smaller parameter, $(G, k) \rightarrow (G', k-|H|)$. For both instance and parameter the reduction may be significant.

We are now faced with two issues: How to find crowns efficiently? and how to bound the size of the problem kernel that eventuates?

In [3] it is shown that finding a crown in a graph G can be achieved in polynomial time by computing maximum matchings in G. The size of the reduced instance that results is bounded above via the following theorem.

THEOREM 25.3 [3] *A graph that is crown-free and has a vertex cover of size at most k can contain at most $3k$ vertices.*

Another strategy for employment of crown reductions makes use of the following lemma from [29].

LEMMA 25.1 [29] *If a graph $G = (V, E)$ has an independent set $V' \subset V$ such that $|N(V')| < |V'|$, then a crown $I \cup H$ with $I \subseteq V'$ can be found in G in time $O(n + m)$.*

The following simple crown kernelization algorithm, given in [65], uses this strategy to produce either a correct NO answer, or a problem kernel of size at most $4k$, for K-VERTEX-COVER.

We start by computing a maximal matching M in G. Since we have to pick one vertex for each edge in the matching it follows that the size of a minimum vertex cover of G is at least $|M|/2$. Thus, if $|V(M)| > 2k$, then we output NO. Otherwise, if $|V(M)| \leq 2k$, then there are two possibilities: Since M is a maximal matching it must be the case that $V(G) - V(M)$ is an independent set in G. If we assume that G does not contain any isolated vertices then each vertex in $V(G) - V(M)$ must be adjacent to some vertex in $V(M)$. Thus, if $|V(G) - V(M)| > 2k$ then, by Lemma 25.1, we can find a crown $I \cup H$ in G in time $O(n + m)$. The reduced instance is $(G[V - (I \cup H)], k - |H|)$. If $|V(G) - V(M)| \leq 2k$ then $|V(G)| = |V(M)| + |V(G) - V(M)| \leq 2k + 2k = 4k$ so G is the required problem kernel of size at most $4k$.

The three kernelizations given here for K-Vertex-Cover make for a compelling argument in support of data reduction in the parameterized context. In comparison with polynomial approximation, kernelization achieves the conjectured best possible result for this particular problem. In [3,4] Abu-Khzam et al. report on experiments solving large instances of the K-Vertex Cover problem in the context of computational biology applications. A common problem in many of these applications involves finding the maximum clique in a graph. However, the K-Clique problem is $W[1]$-hard and so not directly amenable to an FPT algorithmic approach. A graph G has a maximum clique of size k iff its complement graph $\bar{G}$ has a minimum vertex cover of size $n - k$. Thus, one approach to the clique problem is to solve the vertex cover problem on the complement graph with parameter $n - k$. Results from [3,4] show that the kernelization techniques presented here for K-Vertex Cover perform far better in practice than is suggested by the theory, both in terms of running time and in terms of the size of the kernels that can be achieved. Implementations combining kernelization with a depth-bounded search tree approach (see Sections 25.3.2 and 25.3.3) work effectively on real data for k up to around 1000 [3].

However, since K-Vertex Cover is considered to be the "success story" of parameterized computation, it is fair to ask whether or not the program works so well in general. There are by now a plethora of kernelization algorithms in the literature, solving a wide variety of problems with practical applications in diverse areas of science and industry. Many of these yield sufficiently small problem kernels to be of concrete practical use. In this regard, the benchmark is a linear kernel, where the size of the fully reduced instance is a (small) linear function of the parameter. Examples of parameterized problems having linear problem kernels include K-Dominating Set and K-Connected Vertex Cover restricted to planar graphs (for general graphs K-Connected Vertex Cover has so far only been shown to have an exponentially bounded problem kernel and K-Dominating Set is $W[2]$-hard so has no problem kernel bounded by any function of the parameter) and K-Tree Bisection and Reconnection (see Section 25.6 for details of this problem from computational biology). In some documented cases, for example see [50], even though the provable bound on the size of the kernel might be large, even an exponential function of the parameter, the underlying data reduction still performs very well in practice.

We conclude this section by noting the following two caveats regarding kernelization:

For some problems obtaining a problem kernel is trivial. The following example is given in [65]. We consider the K-Dominating Set problem for cubic graphs, where all vertices have degree three. No vertex in such a graph can dominate more than four vertices, itself and three neighbours. Thus, we can safely answer NO whenever the input graph has more than $4k$ vertices. Note that this problem kernel of $4k$ vertices and at most $6k$ edges is obtained without the application of any reduction rule at all. However, by the same argument we see that no cubic graph has a dominating set of size less than $n/4$. Thus, for any nontrivial problem instance, we have $k \geq n/4$ and $4k \geq n$. The bound obtained for the size of the kernel is at least as large as the size of the instance itself.

In this case, a more sensible problem to consider arises from the idea of bounding above the guarantee, first introduced in [52]. Given a cubic graph G and parameter k, it makes more sense to ask if there is a dominating set of size $n/4 + k$ for G. Now, the parameter contributes to the problem in a nontrival fashion, since it is has become a bound on the distance of the solution from some guaranteed minimum.

Two of the three kernelizations given here for K-Vertex-Cover result in problem kernels that we commonly call linear kernels, since the number of vertices in the fully reduced instance is a linear function of the parameter. A more accurate description is to say that they are polynomial kernels, since the number of graph edges in the reduced instance may be quadratic in the size of the parameter. Recent results [15] suggest that some parameterized problems likely won't admit any polynomial kernel (i.e any problem kernel whose size is an arbitrary polynomial function of the parameter), under reasonable complexity-theoretic hypotheses, even though they can be shown to be FPT using some of the not-quite-practical FPT methods we introduce later in Section 25.4. This suggests that, for such problems to have FPT algorithms with provably fast running times, these must be of rather unusual types.

25.3.2 Depth-Bounded Search Trees

Many parameterized problems can be solved by the construction of a search tree whose depth depends only upon the parameter. The total size of the tree will necessarily be an exponential function of the parameter, to keep the size of the tree manageable the trick is to find efficient branching rules to successively apply to each node in the search tree.

Continuing with our running example, consider the following simple algorithm for the K-VERTEX COVER problem.

We construct a binary tree of height k. We begin by labeling the root of the tree with the empty set and the graph $G = (V, E)$. Now we pick any edge $(u, v) \in E$. In any vertex cover of G we must have either u or v, in order to cover the edge (u, v), so we create children of the root node corresponding to these two possibilities. The first child is labeled with $\{u\}$ and $G - u$, the second with $\{v\}$ and $G - v$. The set of vertices labeling a node represents a possible vertex cover, and the graph labeling a node represents what remains to be covered in G. In the case of the first child we have determined that u will be in our possible vertex cover, so we delete u from G, together with all its incident edges, as these are all now covered by a vertex in our possible vertex cover.

In general, for a node labeled with a set S of vertices and subgraph H of G, we arbitrarily choose an edge $(u, v) \in E(H)$ and create the two child nodes labeled, respectively, $S \cup \{u\}$, $H - u$, and $S \cup \{v\}$, $H - v$. At each level in the search tree the size of the vertex sets that label nodes will increase by one. Any node that is labeled with a subgraph having no edges must also be labeled with a vertex set that covers all edges in G. Thus, if we create a node at height at most k in the tree that is labeled with a subgraph having no edges, then a vertex cover of size at most k has been found.

There is no need to explore the tree beyond height k, so this algorithm runs in time $O(2^k \cdot n)$.

In many cases, it is possible to significantly improve the $f(k)$, the function of the parameter that contributes exponentially to the running time, by shrinking the search tree. In the case of K-VERTEX COVER, Balasubramanian et al. [11] observed that, if G has no vertex of degree three or more, then G consists of a collection of cycles. If such a G is sufficiently large, then this graph cannot have a size k vertex cover. Thus, at the expense of an additive constant factor (to be invoked when we encounter any subgraph in the search tree containing only vertices of degree at most two), we need consider only graphs containing vertices of degree three or greater.

We again construct a binary tree of height at most k. We begin by labeling the root of the tree with the empty set and the graph G. Now we pick any vertex $v \in V$ of degree three or greater. In any vertex cover of G we must have either v or all of its neighbors, so we create children of the root node corresponding to these two possibilities. The first child is labeled with $\{v\}$ and $G - v$, the second with $\{w_1, w_2, \ldots, w_p\}$, the neighbors of v, and $G - \{w_1, w_2, \ldots, w_p\}$. In the case of the first child, we are still looking for a size $k - 1$ vertex cover, but in the case of the second child we need only look for a vertex cover of size $k - p$, where p is at least 3. Thus, the bound on the size of the search tree is now somewhat smaller than 2^k.

Using a recurrence relation to determine a bound on the number of nodes in this new search tree, it can be shown that this algorithm runs in time $O(5^{k\backslash 4} \cdot n)$.

A third search tree algorithm for K-VERTEX COVER, given in [49] uses the following three branching rules:

BRANCHING RULE VC1

If there is a degree one vertex v in G, with single neighbor u, then there is a minimum size cover that contains u (by the argument given for rule VC2 in Section 25.3.1) Thus, we create a single child node labeled with $\{u\}$ and $G - u$.

BRANCHING RULE VC2

If there is a degree two vertex v in G, with neighbors w_1 and w_2, then either both w_1 and w_2 are in a minimum size cover, or v together with all other neighbors of w_1 and w_2 are in a minimum size cover.

To see that this rule is correct, assume that there is a minimum size cover containing v and only one of its neighbors. Replacing v with its second neighbor would then also yield a minimum size cover and this is the cover that will be constructed in the first branching case. Thus, if there is a cover that is smaller than the cover containing both w_1 and w_2 then this cover must contain v and neither w_1 nor w_2. This implies that all other neighbors of w_1 and w_2 must be in this cover.

BRANCHING RULE VC3

If there is a degree three vertex v in G, then either v or all of its neighbors are in a minimum size cover.

Using a recurrence relation, it can be shown that if there is a solution of size at most k then the size of the corresponding search tree has size bounded above by $O(1.47^k)$.

The basic method of finding efficient branching rules is to look for a structure in the problem input which gives rise to only a few alternatives, one of which must lead to an acceptable solution, if such a solution exists. In all of the examples given here for K-VERTEX COVER, this structure consists of a vertex and its one or two hop neighborhood. The smallest search tree found so far for K-VERTEX COVER has size $O(1.286^k)$ [26] and is achieved by more complex case analysis than that described here, although the structures identified still consist simply of small local neighborhoods in the input graph.

We now briefly canvas two quite different examples of such structures which give rise to efficient branching rules for two unrelated parameterized problems. Space limitations mean that most of the problem details are left out; the intention is merely to demonstrate the nature of the possibilities for search tree algorithms.

For the CLOSEST STRING problem [51], we are given a set $S = \{s_1, s_2, \ldots, s_k\}$ of k strings, each of length l, over an alphabet Σ, and the task is to find a string whose Hamming distance is at most d from each of the $s_i \in S$. The structure that we identify is a candidate string, $\widehat{s}$. At the root node of the search tree, $\widehat{s}$ is simply one of the input strings. If any other string $s_i \in S$ differs from $\widehat{s}$ in more than $2d$ positions, then there is no solution for the problem. At each step we look for an input string s_i that differs from $\widehat{s}$ in more than d positions but less than $2d$ positions. Choosing $d + 1$ of these positions, we branch into $d + 1$ subcases, in each subcase modifying one position in $\widehat{s}$ to match s_i.

For the MAXIMUM AGREEMENT FOREST problem [47], we are given two phylogenetic X-trees, T_1 and T_2, each an unrooted binary tree with leaves labeled by a common set of species X and (unlabeled) interior nodes having degree exactly three. The (labeled) topologies of T_1 and T_2 may differ. The task is to find at most k edges that can be cut from T_1 so that the resulting forest agrees with the topologies of both T_1 and T_2. One structure that we can identify here is called a minimal incompatible quartet, essentially a set of four leaf labels, $Q = \{l_1, l_2, l_3, l_4\}$, that gives rise to two different topologies in the restriction of each of the trees to those leaves labeled by Q. Given any solution set of edges E from T_1 that gives rise to an agreement forest F, we can obtain an equivalent set of edges E' that produces F by cutting at least one of a certain set of four edges induced by Q in T_1. Thus, after finding an incompatible quartet, we branch into four subcases, in each subcase cutting one of these edges.

Finally, we note that search trees inherently allow for a parallel implementation: when branching into subcases, each branch can be further explored with no reference to other branches. This has proven of concrete use in practice for VERTEX COVER [28]. Along with the idea introduced in the next section, this is one of two powerful arguments in support of the use of the depth-bounded search tree approach to obtain really practical FPT algorithms.

25.3.3 Interleaving

It is often possible to combine the two methods outlined above. For example, for the K-VERTEX COVER problem, we can first reduce any instance to a problem kernel and then apply a search tree method to the kernel itself.

Niedermeier and Rossmanith [56] have developed the technique of interleaving depth-bounded search trees and kernelization. They show that applying kernelization repeatedly during the course of a search tree algorithm can significantly improve the overall time complexity in many cases.

Suppose we take any fixed-parameter algorithm that satisfies the following conditions: The algorithm works by first reducing an instance to a problem kernel, and then applying a depth-bounded search tree method to the kernel. Reducing any given instance to a problem kernel takes at most $P(|I|)$ steps and results in a kernel of size at most $q(k)$, where both P and q are polynomially bounded. The expansion of a node in the search tree takes $R(|I|)$ steps, where R is also bounded by some polynomial. The size of the search tree is bounded by $O(\alpha^k)$. The overall time complexity of such an algorithm running on instance (I, k) is

$$O(P(|I|) + R(q(k))\alpha^k).$$

The strategy developed in [56] is basically to apply kernelization at any step of the search tree algorithm where this will result in a significantly smaller problem instance. To expand a node in the search tree labeled by instance (I, k) we first check whether or not $|I| > c \cdot q(k)$, where $c \geq 1$ is a constant whose optimal value will depend on the implementation details of the algorithm. If $|I| > c \cdot q(k)$ then we apply the kernelization procedure to obtain a new instance (I', k'), with $|I'| \leq q(k)$, which is then expanded in place of (I, k). A careful analysis of this approach shows that the overall time complexity is reduced to

$$O(P(|I|) + \alpha^k).$$

This really does make a difference. In [55] the 3-HITTING SET problem is given as an example. An instance (I, k) of this problem can be reduced to a kernel of size k^3 in time $O(|I|)$, and the problem can be solved by employing a search tree of size 2.27^k. Compare a running time of $O(2.27^k \cdot k^3 + |I|)$ (without interleaving) with a running time of $O(2.27^k + |I|)$ (with interleaving).

Note that, although the techniques of kernelization and depth-bounded search tree are simple algorithmic strategies, they are not part of the classical toolkit of polynomial-time algorithm design since they both involve costs that are exponential in the parameter.

25.3.4 Iterative Compression

Iterative compression is a relatively new technique for obtaining FPT algorithms, first introduced in a paper by Reed et al. [59]. Although currently only a small number of results are known, it seems to be applicable to a range of parameterized minimization problems, where the parameter is the size of the solution set. Most of the currently known iterative compression algorithms solve feedback set problems in graphs, problems where the task is to destroy certain cycles in the graph by deleting at most k vertices or edges. In particular, the K-GRAPH BIPARTISATION problem, where the task is to find a set of at most k vertices whose deletion destroys all odd-length cycles, has been shown to be FPT by means of iterative compression [59]. This had been a long-standing open problem in parameterized complexity theory.

To illustrate the concept, we again paraphrase the treatment given in [49]. The central idea is to employ a compression routine.

DEFINITION 25.3   **(Compression Routine).** A *compression routine* is an algorithm that, given a problem instance I and a solution of size k, either calculates a smaller solution or proves that the given solution is of minimum size.

Using such a routine we can find an optimal solution for a parameterized problem by inductively building up the problem structure and iteratively compressing intermediate solutions. The idea is that, if the compression routine is an FPT algorithm, then so is the whole algorithm. The manner by which the problem structure is inductively produced will normally be straightforward, the trick is in finding an efficient compression routine. Continuing with our running example, we now describe an algorithm for K-VERTEX COVER based on iterative compression.

Given a problem instance $(G = (V, E), k)$, we build the graph G vertex by vertex. We start with an initial set of vertices $V' = \emptyset$ and an initial solution $C = \emptyset$. At each step, we add a new vertex v to both V' and C, $V' \leftarrow V' \cup \{v\}$, $C \leftarrow C \cup \{v\}$. We then call the compression routine on the pair $(G[V'], C)$, where $G[V']$ is the subgraph induced by V' in G, to obtain a new solution C'. If $|C'| > k$ then we output NO, otherwise we set $C \leftarrow C'$.

If we successfully complete the nth step where $V' = V$, we output C with $|C| \leq k$. Note that C will be an optimal solution for G.

The compression routine takes a graph G and a vertex cover C for G and returns a smaller vertex cover for G if there is one, otherwise, it returns C unchanged. Each time the compression routine is used it is provided with an intermediate solution of size at most $k + 1$.

The implementation of the compression routine proceeds as follows. We consider a smaller vertex cover C' as a modification of the larger vertex cover C. This modification retains some vertices $Y \subseteq C$ while the other vertices $S = C \setminus Y$ are replaced with $|S| - 1$ new vertices from $V \setminus C$.

The idea is to try by brute force all $2^{|C|}$ partitions of C into such sets Y and S. For each such partition, the vertices from Y along with all of their adjacent edges are deleted. In the resulting instance $G' = G[V \setminus Y]$, it remains to find an optimal vertex cover that is disjoint from S. Since we have decided to take no vertex from S into the vertex cover, we have to take that endpoint of each edge that is not in S. At least one endpoint of each edge in G' is in S, since S is a vertex cover for G'. If both endpoints of some edge in G' are in S, then this choice of S cannot lead to a vertex cover C' with $S \cap C' = \emptyset$. We can quickly find an optimal vertex cover for G' that is disjoint from S by taking every vertex that is not in S and has degree greater than zero. Together with Y, this gives a new vertex cover C' for G. For each choice of Y and S, this can be done in time $O(m)$, leading to $O(2^{|C|}m) = O(2^k m)$ time overall for one call of the compression routine. With at most n iterations of the compression algorithm, we get an algorithm for K-VERTEX COVER running in time $O(2^k mn)$.

Note that, in general, a compression routine will have running time exponential in the size of the solution provided to it, it is therefore important that each intermediate solution considered has size bounded by some $k' = f(k)$, where k is the parameter value for the original problem.

The employment of an FPT compression routine in the manner described here for K-VERTEX COVER will work effectively for any parameterized minimization problem which is monotone in the sense that NO instances are closed under element addition. That is, given a problem instance (I, k) that is a NO instance, any problem instance (I', k) with $I \subseteq I'$ is also a NO instance. If a problem is monotone in this sense then we can immediately answer NO if we encounter an intermediate solution that cannot be compressed to meet the original parameter bound. Note that many minimization problems are not monotone in this sense. For example, a NO instance $(G = (V, E), k)$ for K-DOMINATING SET can be changed to a YES instance by means of the addition of a single vertex that is adjacent to all vertices in V.

Finally, as noted in [49], the employment of compression routines is not restricted to the mode detailed here. For example, we could start with a suboptimal solution, perhaps provided by some type of parameterized approximation algorithm as detailed in Section 25.6, and then repeatedly compress this solution until it is either good enough or we are not willing to invest more calculation time.

25.4 Not-Quite-Practical FPT Algorithms

In this section we introduce two techniques that lead to not-quite-practical FPT algorithms, color-coding and dynamic programming on bounded width graph decompositions. Both of these techniques have potential for practical application and have been extensively studied from a theoretical point of view. However, in contrast to the methods introduced in the previous section, these approaches have so far lead to only isolated practical implementations and experimental results.

We also introduce a series of algorithmic meta-theorems. These are based on results from descriptive complexity theory and topological graph theory and provide us with general FPT algorithms, sometimes nonconstructive, pertaining to large classes of problems. We view these theorems not as an end in themselves, but as being useful signposts in the search for practically efficient fixed-parameter algorithms.

25.4.1 Color-Coding

This technique is useful for problems that involve finding small subgraphs in a graph, such as paths and cycles. Introduced by Alon et al. [10], it can be used to derive seemingly efficient randomized FPT algorithms for several subgraph isomorphism problems.

We formulate a parameterized version of the SUBGRAPH ISOMORPHISM problem as follows:

K-SUBGRAPH ISOMORPHISM
Instance: $G = (V, E)$ and a graph $H = (V^H, E^H)$ with $|V^H| = k$.
Parameter: A positive integer k.
Question: Is H isomorphic to a subgraph in G?

The idea is that, in order to find the desired set of vertices V' in G, such that $G[V']$ is isomorphic to H, we randomly color all the vertices of G with k colors and expect that, with some high degree of probability, all vertices in V' will obtain different colors. In some special cases of the SUBGRAPH ISOMORPHISM problem, dependent on the nature of H, this will simplify the task of finding V'.

If we color G uniformly at random with k colors, a set of k distinct vertices will obtain different colors with probability $(k!)/k^k$. This probability is lower-bounded by e^{-k}, so we need to repeat the process e^k times to have sufficiently high probability of obtaining the required coloring.

We can derandomize this kind of algorithm using hashing, but at the cost of extending the running time. We need a list of colorings of the vertices in G such that, for *each* subset $V' \subseteq V$ with $|V'| = k$ there is at least one coloring in the list by which all vertices in V' obtain different colors. Formally, we require a k-perfect family of hash functions from $\{1, 2, ..., |V|\}$, the set of vertices in G, onto $\{1, 2, ..., k\}$, the set of colors.

DEFINITION 25.4 (*k*-**Perfect Hash Functions**). A k-perfect family of hash functions is a family $\mathcal{H}$ of functions from $\{1, 2, ..., n\}$ onto $\{1, 2, ..., k\}$ such that, for each $S \subset \{1, 2, ..., n\}$ with $|S| = k$, there exists an $h \in \mathcal{H}$ such that h is bijective when restricted to S.

By a variety of sophisticated methods, Alon et al. [10] have proved the following:

THEOREM 25.4 [10] *Families of k-perfect hash functions from $\{1, 2, ..., n\}$ onto $\{1, 2, ..., k\}$ can be constructed which consist of $2^{O(k)} \cdot \log n$ hash functions. For such a hash function, h, the value $h(i)$, $1 \leq i \leq n$, can be computed in linear time.*

We can color G using each of the hash functions from our k-perfect family in turn. If the desired set of vertices V' exists in G, then, for at least one of these colorings, all vertices in V' will obtain different colors as we require.

We now give a very simple example of this technique. The subgraph that we will look for is a cycle of length k. We use k colors. If a k-cycle exists in the graph, then there must be a coloring that assigns a different color to each vertex in the cycle.

For each coloring h, we check every ordering $c_1, c_2, \ldots, c_k$ of the k colors to decide whether or not it realizes a k-cycle. We first construct a directed graph G' as follows:

For each edge $(u, v) \in E$, if $h(u) = c_i$ and $h(v) = c_{i+1 \pmod k}$ for some i, then replace (u, v) with arc $\langle u, v \rangle$, otherwise delete (u, v).

In G', for each v with $h(v) = c_1$, we use a breadth first search to check for a cycle C from v to v of length k.

A deterministic algorithm will need to check $2^{O(k)} \cdot \log |V|$ colorings, and, for each of these, $k!$ orderings. We can decide if an ordering of colors realizes the k-cycle in time $O(k \cdot |V|^2)$. Thus, our algorithm is FPT, but, arguably, not practically efficient. The main drawback is that the $2^{O(k)} \cdot \log |V|$ bound on the size of the family of hash functions hides a large constant in the $O(k)$ exponent.

More interesting examples of applications of color-coding to subgraph isomorphism problems, based on dynamic programming, can be found in [10].

25.4.2   Bounded Width Metrics

Faced with intractable graph problems, many authors have turned to the study of various restricted classes of graphs for which such problems can be solved efficiently. A number of graph width metrics naturally arise in this context which restrict the inherent complexity of a graph in various senses.

The idea here is that a useful width metric should admit efficient algorithms for many (generally) intractable problems on the class of graphs for which the width is small. This leads to consideration of these measures from a parameterized point of view. The corresponding naturally parameterized problem has the following form:

Let $w(G)$ denote any measure of graph width.

Instance: A graph $G = (V, E)$.
Parameter: A positive integer k.
Question: Is $w(G) \leq k$?

One of the most successful measures in this context is the notion of treewidth which arose from the seminal work of Robertson and Seymour on graph minors and immersions [61–63]. Treewidth measures, in a precisely defined way, how tree-like a graph is. The fundamental idea is that we can lift many results from trees to graphs that are tree-like. Related to treewidth is the notion of pathwidth which measures, in the same way, how path-like a graph is.

Many generally intractable problems become fixed-parameter tractable for the class of graphs that have bounded treewidth or bounded pathwidth, with the parameter being the treewidth or pathwidth of the input graph. Furthermore, treewidth and pathwidth generalize many other well-studied graph properties. For example, planar graphs with radius k have treewidth at most $3k$, series parallel multigraphs have treewidth two, chordal graphs (graphs having no induced cycles of length four or more) with maximum clique size k have treewidth at most $k - 1$, interval graphs G' with maximum clique size k have pathwidth at most $k - 1$.

A graph G has treewidth at most k if we can associate a tree T with G in which each node represents a subgraph of G having at most $k + 1$ vertices, such that all vertices and edges of G are represented in at least one of the nodes of T, and for each vertex v in G, the nodes of T where v is represented from a subtree of T. Such a tree is called a tree decomposition of G, of width k.

DEFINITION 25.5 (Tree Decomposition and Treewidth). Let $G = (V, E)$ be a graph. A *tree decomposition*, *TD*, of G is a pair $(T, \mathcal{X})$ where

1. $T = (I, F)$ is a tree, and
2. $\mathcal{X} = \{X_i \mid i \in I\}$ is a family of subsets of V, one for each node of T, such that

   (i) $\bigcup_{i \in I} X_i = V$,

   (ii) for every edge $\{v, w\} \in E$, there is an $i \in I$ with $v \in X_i$ and $w \in X_i$, and

   (iii) for all $i, j, k \in I$, if j is on the path from i to k in T, then $X_i \cap X_k \subseteq X_j$.

The *width* of a tree decomposition $((I, F), \{X_i \mid i \in I\})$ is $\max_{i \in I} |X_i| - 1$. The treewidth of a graph G, denoted by $tw(G)$, is the minimum width over all possible tree decompositions of G.

DEFINITION 25.6 (Path Decomposition and Pathwidth). A *path decomposition*, *PD*, of a graph G is a tree decomposition $(P, \mathcal{X})$ of G where P is simply a path (i.e., the nodes of P have degree at most two). The *pathwidth* of G, denoted by $pw(G)$, is the minimum width over all possible path decompositions of G.

Any path decomposition of G is also a tree decomposition of G, so the pathwidth of G is at least equal to the treewidth of G. For many graphs, the pathwidth will be somewhat larger than the treewidth. For example, let B_k denote the complete binary tree of height k, having $2^k - 1$ vertices, then $tw(B_k) = 1$, but $pw(B_k) = k$.

Graphs of treewidth and pathwidth at most k are also called partial k-trees and partial k-paths, respectively, as they are exactly the subgraphs of k-trees and k-paths. There are a number of other variations equivalent to the notions of treewidth and pathwidth (see, e.g., [14]). For algorithmic purposes, the characterizations provided by the definitions given above tend to be the most useful.

The typical method employed to produce FPT algorithms for problems restricted to graphs of bounded treewidth (pathwidth) proceeds in two stages.

1. Find a bounded-width tree (path) decomposition of the input graph that exhibits the underlying tree (path) structure.
2. Perform dynamic programming on this decomposition to solve the problem.

The following lemma encapsulates the two properties of tree decompositions on which the dynamic programming approach relies.

LEMMA 25.2 (**Connected Subtrees**). Let $G = (V, E)$ be a graph and $TD = (T, \mathcal{X})$ a tree decomposition of G.

   (i) For all $v \in V$, the set of nodes $\{i \in I \mid v \in X_i\}$ forms a connected subtree of T.
   (ii) For each connected subgraph G' of G, the nodes in T which contain a vertex of G' induce a connected subtree of T.

In order for this approach to produce practically efficient FPT algorithms, as opposed to proving that problems are theoretically tractable, it is important to be able to produce the necessary decomposition reasonably efficiently.

Determining the treewidth or pathwidth of a graph is an *NP*-hard problem. However, polynomial time approximation algorithms have been found [16]. There is a polynomial time algorithm that, given a graph *G* with treewidth *k*, finds a tree decomposition of width at most $O(k \cdot \log n)$ for *G*. There is a polynomial time algorithm that, given a graph *G* with pathwidth *k*, finds a path decomposition of width at most $O(k \cdot \log^2 n)$ for *G*.

Bodlaender [13] gave the first linear-time FPT algorithms (i.e., linear in |*G*|) for the constructive versions of both K-TREEWIDTH and K-PATHWIDTH. Perkovic and Reed [58] have improved upon Bodlaender's work, although the $f(k)$'s involved mean that the algorithms given in both [13] and [58] are not workable in practice. However, there are far simpler FPT algorithms that produce tree and path decompositions having width at most a constant factor larger than the optimum [60,63].

For some graph classes, the optimal treewidth and pathwidth, or good approximations of these, can be found using practically efficient polynomial time algorithms. Examples are chordal bipartite graphs, interval graphs, permutation graphs, circle graphs [18], and co-graphs [19].

For planar graphs, Alber et al. [5–7] have introduced the notion of a layerwise separation property pertaining to the underlying parameterized problem that one might hope to solve via a small-width tree decomposition. The layerwise separation property holds for all problems on planar graphs for which a linear problem kernel can be constructed.

For problems having this property, we can exploit the layer structure of planar graphs, along with knowledge about the structure of "YES" instances of the problem, in order to find small separators in the input graph such that each of the resulting components has small treewidth. Tree decompositions for each of the components are then merged with the separators to produce a small-width tree decomposition of the complete graph.

This approach leads to, for example, algorithms that solve K-VERTEX COVER and, more interestingly, K-DOMINATING SET, on planar graphs in time $2^{O(\sqrt{k})} \cdot n$. The algorithms work by constructing a tree decomposition of width $O(\sqrt{k})$ for the kernelized graph, and then performing dynamic programming on this decomposition.

An algorithm that uses dynamic programming on a tree works by computing some value, or table of values, for each node in the tree. The important point is that the value for a node can be computed using only information directly associated with the node itself, along with values already computed for the children of the node.

Extending the idea of dynamic programming on trees to dynamic programming on bounded-width tree decompositions is really just a matter of having to construct more complicated tables of values. Instead of considering a single vertex at each node and how it interacts with the vertices at its child nodes, we now need to consider a reasonably small *set* of vertices represented at each node, and how this small set of vertices can interact with each of the small sets of vertices represented at its child nodes.

The most important factor in dynamic programming on tree decompositions is the size of the tables produced. The table size is usually $O(c^k)$, where *k* is the width of the tree decomposition and *c* depends on the combinatorics of the underlying problem that we are trying to solve. We can trade off different factors in the design of such algorithms. For example, a fast approximation algorithm that produces a tree decomposition of width 3*k*, or even k^2, for a graph of treewidth *k* could be quite acceptable if *c* is small.

Cliquewidth, first introduced in [31], is another graph width metric that has more recently gained prominence in algorithm design. A graph that has cliquewidth *k* can be recursively constructed from single vertices with labels in $[k] = \{1, 2, \ldots, k\}$ using only the composition operations of graph union $G = G_1 \cup G_2$, vertex relabeling $G = (G_1)_{i \to j}$, and cross-product edge insertion between label sets $G = (G_1)_{i \times j}$.

The series of composition operations (called a k-expression) that produces such a cliquewidth-k graph G gives rise to a decomposition of G into a tree of subgraphs of G. This decomposition then leads to a linear-time dynamic programming algorithm for many problems restricted to cliquewidth-k graphs. However, in contrast with treewidth and pathwidth, there is no known FPT algorithm that constructs such a decomposition for a given cliquewidth-k graph. A polynomial time approximation algorithm has recently been presented in [57].

We note that, even though finding exact bounded width graph decompositions, for graphs with small width, does not yet appear to be feasible, in practice heuristics and approximations have proven to be quite effective. In addition, many graphs derived in practical applications are themselves constructed inductively, making them prime candidates for these methods. For examples of dynamic programming algorithms on various bounded width graph decompositions see, for example, [20].

25.4.3 Algorithmic Meta-Theorems

Descriptive complexity theory relates the logical complexity of a problem description to its computational complexity. In this context there are some useful results that relate to fixed-parameter tractability. We can view these results not as an end in themselves, but as being useful signposts in the search for practically efficient fixed-parameter algorithms.

We will consider graph properties that can be defined in first-order logic and monadic second-order logic.

In first order logic we have an (unlimited) supply of individual variables, one for each vertex in the graph. Formulas of first-order logic (FO) are formed by the following rules:

1. *Atomic formulas:* $x = y$ and $R(x_1, ..., x_k)$, where R is a k-ary relation symbol and $x, y, x_1, ..., x_k$ are individual variables, are FO-formulas.
2. *Conjunction, disjunction:* If ϕ and ψ are FO-formulas, then $\phi \wedge \psi$ is an FO-formula and $\phi \vee \psi$ is an FO-formula.
3. *Negation:* If ϕ is an FO-formula, then $\neg\phi$ is an FO-formula.
4. *Quantification:* If ϕ is an FO-formula and x is an individual variable, then $\exists x\ \phi$ is an FO-formula and $\forall x\ \phi$ is an FO-formula.

We can state that a graph has a clique of size k using an FO-formula. Here, the binary relation $E(x_i, x_j)$ indicates the existence of an edge between vertices x_i and x_j.

$$\exists x_1 \ldots x_k \bigwedge_{1 \leq i \leq j \leq k} E(x_i, x_j)$$

We can state that a graph has a dominating set of size k using an FO-formula,

$$\exists x_1 \ldots x_k \, \forall y \bigvee_{1 \leq i \leq k} \left(E(x_i, y) \vee (x_i = y) \right)$$

In monadic second-order logic we have an (unlimited) supply of both individual variables, one for each vertex in the graph, and set variables, one for each subset of vertices in the graph. Formulas of monadic second-order logic (MSO) are formed by the rules for FO and the following additional rules:

1. *Additional atomic formulas:* For all set variables X and individual variables y, Xy is an MSO-formula.

2. *Set quantification:* If φ is an MSO-formula and X is a set variable, then $\exists X$ φ is an MSO-formula, and $\forall X$ φ is an MSO-formula.

We can state that a graph is k-colorable using an MSO-formula,

$$\exists X_1 \ldots \exists X_k \left(\forall x \bigvee_{i=1}^{k} X_i x \ \wedge \ \forall x \forall y \Big(E(x,y) \rightarrow \bigwedge_{i=1}^{k} \neg (X_i x \wedge X_i y) \Big) \right)$$

The problems that we are interested in are special cases of the model-checking problem.

Let Φ be a class of formulas (logic), and let $\mathcal{D}$ be a class of finite relational structures. The model-checking problem for Φ on $\mathcal{D}$ is the following problem.

Instance: A structure $\mathcal{A} \in \mathcal{D}$, and a sentence (no free variables) φ $\in \Phi$.
Question: Does $\mathcal{A}$ satisfy φ?

The model-checking problems for FO and MSO are PSPACE-complete in general. However, as the following results show, if we restrict the class of input structures then in some cases these model-checking problems become tractable.

The most well-known result, paraphrased here, is due to Courcelle [30].

THEOREM 25.5   [30] *The model-checking problem for MSO restricted to graphs of bounded treewidth is linear-time fixed-parameter tractable.*

Seese [64] has proved a converse to Courcelle's theorem.

THEOREM 25.6   [64] *Suppose that $\mathcal{F}$ is any family of graphs for which the model-checking problem for MSO is decidable, then there is a number n such that, for all $G \in \mathcal{F}$, the treewidth of G is less than n.*

Courcelle's theorem tells us that if we can define the problem that we are trying to solve as a model-checking problem, and we can define the particular graph property that we are interested in as an MSO-formula, then there is an FPT algorithm that solves the problem for input graphs of bounded treewidth. The theorem by itself doesn't tell us how the algorithm works.

The automata-theoretic proof of Courcelle's theorem given by Abrahamson and Fellows [2] provides a generic algorithm that relies on dynamic programming over labeled trees (see [35] for details of this approach). However, this generic algorithm is really just further proof of theoretical tractability. The importance of the theorem is that it provides a powerful engine for demonstrating that a large class of problems is FPT. If we can couch a problem in the correct manner then we know that it is worth looking for an efficient FPT algorithm that works on graphs of bounded treewidth.

The next result concerns classes of graphs having bounded local treewidth. The local tree width of a graph G is defined via the following function.

$$ltw(G, r) = \max \{ tw(N_r(v)) \mid v \in V(G) \}$$

where $N_r(v)$ is the neighborhood of radius r about v (including v).

A class of graphs $\mathcal{F}$ has bounded local treewidth if there is a function $f : \mathbb{N} \rightarrow \mathbb{N}$ such that, for all $G \in \mathcal{F}$ and $r \geq 1$, $ltw(G, r) \leq f(r)$. The concept is a relaxation of bounded treewidth for classes of graphs. Instead of requiring that the treewidth of a graph overall is bounded by some constant, we require that, for each vertex in the graph, the treewidth of each neighborhood of radius r about that vertex is bounded by some uniform function of r.

Examples of classes of graphs that have bounded local treewidth include graphs of bounded treewidth (naturally), graphs of bounded degree, planar graphs, and graphs of bounded genus. Frick and Grohe [44] have proved the following theorem.

THEOREM 25.7 [44] *Parameterized problems that can be described as model-checking problems for FO are fixed-parameter tractable on classes of graphs of bounded local treewidth.*

This theorem tells us, for example, that parameterized versions of problems such as DOMINATING SET, INDEPENDENT SET, or SUBGRAPH ISOMORPHISM are FPT on planar graphs, or on graphs of bounded degree. As with Courcelle's theorem, it provides us with a powerful engine for demonstrating that a large class of problems is FPT, but leaves us with the job of finding practically efficient FPT algorithms for these problems.

The last meta-theorem that we will present has a somewhat different flavor. We first need to introduce some ideas from topological graph theory.

A graph H is a minor of a graph G iff there exists some subgraph, G^H of G, such that H can be obtained from G^H by a series of edge contractions.

We define an edge contraction as follows. Let $e = (u, v)$ be an edge of the graph G^H. By G^H/e we denote the graph obtained from G^H by contracting the edge e into a new vertex v_e which becomes adjacent to all former neighbors of u and of v. H can be obtained from G^H by a series of edge contractions iff there are graphs $G_0, ..., G_n$ and edges $e_i \in G_i$ such that $G_0 = G^H$, $G_n \simeq H$, and $G_{i+1} = G_i/e_i$ for all $i < n$.

Note that every subgraph of a graph G is also a minor of G. In particular, every graph is its own minor.

A class of graphs $\mathcal{F}$ is minor-closed if, for every graph $G \in \mathcal{F}$, every minor of G is also contained in $\mathcal{F}$. A very simple example is the class of graphs with no edges. Another example is the class of acyclic graphs. A more interesting example is the following:

Let us say that a graph $G = (V, E)$ is within k vertices of a class of graphs $\mathcal{F}$ if there is a set $V' \subseteq V$, with $|V'| \leq k$, such that $G[V - V'] \in \mathcal{F}$. If $\mathcal{F}$ is any minor-closed class of graphs, then, for every $k \geq 1$, the class of graphs within k vertices of $\mathcal{F}$, $\mathcal{W}_k(\mathcal{F})$, is also minor-closed.

Note that for each integer $k \geq 1$, the class of graphs of treewidth or pathwidth at most k is minor-closed.

Let $\mathcal{F}$ be a class of graphs which is closed under taking of minors, and let H be a graph that is not in $\mathcal{F}$. Each graph G which has H as a minor is not in $\mathcal{F}$, otherwise H would be in $\mathcal{F}$. We call H a forbidden minor of $\mathcal{F}$. A minimal forbidden minor of $\mathcal{F}$ is a forbidden minor of $\mathcal{F}$ for which each proper minor is in $\mathcal{F}$. The set of all minimal forbidden minors of $\mathcal{F}$ is called the obstruction set of $\mathcal{F}$.

In a long series of papers, collectively entitled Graph Minors, Robertson and Seymour [63] have essentially proved that any minor-closed class of graphs $\mathcal{F}$ must have a finite obstruction set. Robertson and Seymour have also shown that, for a fixed graph H, it can be determined whether H is a minor of a graph G in time $O(f(|H|) \cdot |G|^3)$.

We can now derive the following theorem.

THEOREM 25.8 (Minor-Closed Membership). *If $\mathcal{F}$ is a minor-closed class of graphs then membership of a graph G in $\mathcal{F}$ can be determined in time $O(f(k) \cdot |G|^3)$, where k is the collective size of the graphs in the obstruction set for $\mathcal{F}$.*

This meta-theorem tells us that if we can define a graph problem via membership in a minor-closed class of graphs $\mathcal{F}$, then the problem is FPT, with the parameter being the collective size of the graphs in the obstruction set for $\mathcal{F}$. However, it is important to note two major difficulties that we now face.

First, for a given minor-closed class we have a proof of the existence of a finite obstruction set, but no effective method for obtaining the obstruction set. Second, the minor testing algorithm has very large hidden constants (around 2^{500}), and the sizes of obstruction sets in many cases are known to be very large.

Thus, again we have a theorem that provides us with a powerful engine for demonstrating that a large class of problems is, in fact, FPT, but the problem of finding practically efficient FPT algorithms for such problems remains open.

25.5 Parameterized Intractability

The question arizes: What do we do with a problem for which we know of no FPT algorithm? Good examples are the problems DOMINATING SET or INDENDENT SET for which we know of no algorithm significantly better than trying all possibilities. For a fixed k, this takes time $\Omega(n^{k+1})$. Of course, we would like to prove that there is *no* FPT algorithm for such a problem, but, as with classical complexity, the best we can do is to formulate some sort of completeness/hardness program. Showing that K-DOMINATING SET is not FPT would also show, as a corollary, that $P \neq NP$.

Any completeness programme needs three things. First, it needs a notion of easiness, which we have—FPT. Second, it needs a notion of reduction, and third, it needs some core problem which we believe to be intractable.

Following naturally from the concept of fixed-parameter tractability is an appropriate notion of reducibility that expresses the fact that two parameterized problems have compatible parameterized complexity. That is, if problem (language) A reduces to problem (language) B, and problem B is fixed-parameter tractable, then so too is problem A.

DEFINITION 25.7 (Parameterized Transformation). A *parameterized transformation** from a parameterized language L to a parameterized language L' (symbolically $L \leq_{FPT} L'$) is an algorithm that computes, from input consisting of a pair (I, k), a pair $\langle I', k' \rangle$ such that:

1. $(I, k) \in L$ if and only if $\langle I', k' \rangle \in L'$,
2. $k' = g(k)$ is a computable function only of k, and
3. the computation is accomplished in time $f(k)n^c$, where n is the size of the main part of the input I, k is the parameter, c is a constant (independent of both n and k), and f is an arbitrary function dependent only on k.

If $A \leq_{FPT} B$ and $B \leq_{FPT} A$, then we say that A and B are *FPT-equivalent*.

Now we have two ingredients: easiness and reductions. We need the final component for our program to establish the apparent parameterized intractability of computational problems: the identification of a core problem to reduce from.

In classical NP-completeness this is the heart of the Cook–Levin theorem: the argument that a nondeterministic Turing machine is such an opaque object that it does not seem reasonable that we can determine in polynomial time if it has an accepting path from amongst the exponentially many possible paths. The idea of Downey and Fellows, introduced in the fundamental papers [33,34], was to look at the following parameterized version of nondeterministic Turing machine acceptance.

---

* Strictly speaking, this is a parameterized many–one reduction as an analog of the classical Karp reduction. Other variations such as parameterized Turing reductions are possible. The function g can be arbitrary, rather than computable, for other nonuniform versions. We give the reduction most commonly met in practice.

SHORT NONDETERMINISTIC TURING MACHINE ACCEPTANCE

Instance: A nondeterministic Turing machine (of arbitrary fanout) M.

Parameter: A positive integer k.

Question: Does M have a computation path accepting the empty string in at most k steps?

In the same sense that NP-completeness of the $q(n)$-STEP NONDETERMINISTIC TURING MACHINE ACCEPTANCE, where $q(n)$ is a polynomial in the size of the input, provides us with very strong evidence that no NP-complete problem is likely to be solvable in polynomial time, using SHORT NONDETERMINISTIC TURING MACHINE ACCEPTANCE as a hardness core provides us with very strong evidence that no parameterized language L, for which SHORT NONDETERMINISTIC TURING MACHINE ACCEPTANCE $\leq_{FPT} L$, is likely to be fixed-parameter tractable. That is, if we accept the idea for the basis of NP-completeness, then we should also accept that the SHORT NONDETERMINISTIC TURING MACHINE ACCEPTANCE problem is not solvable in time $O(|M|^c)$ for any fixed c. Our intuition would again be that all paths would need to be tried.

We remark that the hypothesis "SHORT NONDETERMINISTIC TURING MACHINE ACCEPTANCE is not FPT" is somewhat stronger than P$\neq$NP. Furthermore, connections between this hypothesis and classical complexity have recently become apparent. If SHORT NONDETERMINISTIC TURING MACHINE ACCEPTANCE *is* FPT then we know that the EXPONENTIAL TIME HYPOTHESIS, which states that n-variable 3SAT is not in subexponential time (DTIME($2^{o(n)}$)), fails. See Impagliazzo et al. [48], Cai and Juedes [24], and Estivill-Castro et al. [41].

The class of problems FPT-equivalent to SHORT NONDETERMINISTIC TURING MACHINE ACCEPTANCE is called $W[1]$, for reasons discussed below. The parameterized analog of the classical Cook–Levin theorem (that CNFSAT is NP-complete) uses the following parameterized version of 3SAT:

WEIGHTED CNF SAT

Instance: A CNF formula X.

Parameter: A positive integer k.

Question: Does X have a satisfying assignment of weight k? Here the weight of an assignment
is the Hamming weight, that is, the number of literals set to be true.

Similarly, we can define WEIGHTED nCNF SAT, where the clauses have only n variables and n is some number fixed in advance. WEIGHTED nCNF SAT, for any fixed $n \geq 2$, is complete for $W[1]$.

THEOREM 25.9 [22,34] *WEIGHTED nCNF SAT* $\equiv_{FPT}$ *SHORT NONDETERMINISTIC TURING MACHINE ACCEPTANCE.*

The original theorems and classes were first characterized in terms of boolean circuits of a certain structure. These characterizations lend themselves to easier membership proofs, we define them here for completeness.*

We consider a 3CNF formula as a circuit consisting of one input (of unbounded fanout) for each variable, possibly inverters below the variable, and structurally a large *and* of small *or*'s (of size 3) with a single output line. We can similarly consider a 4CNF formula to be a large *and* of small *or*'s where small is defined to be 4. More generally, it is convenient to consider the model of a decision circuit. This is a circuit consisting of large and small gates with a single output line, and no restriction on the fanout of gates. For such a circuit, the depth is the maximum number of gates on any path from the input variables to the output line, and the weft is the large gate depth. More precisely, the

---

* Other approaches to characterization of parameterized hardness classes have been proposed, notably that of Flum and
 Grohe [43] using finite model theory. We refer the reader to [43].

weft is defined to be the maximum number of large gates on any path from the input variables to the output line, where a gate is called large if it's fanin exceeds some predetermined bound.

The weight of an assignment to the input variables of a decision circuit is the Hamming weight, the number of variables made true by the assignment.

Let $\mathcal{F} = \{C_1, ..., C_n, ...\}$ be a family of decision circuits. Associated with $\mathcal{F}$ is a basic parameterized language

$$L_{\mathcal{F}} = \{\langle C_i, k \rangle : C_i \text{ has a weight } k \text{ satisfying assignment}\}.$$

We will denote by $L_{\mathcal{F}(t,h)}$ the parameterized language associated with the family of weft t, depth h, decision circuits.

DEFINITION 25.8   (W[1] [34]). We define a language L to be in the class $W[1]$ iff there is a parameterized transformation from L to $L_{\mathcal{F}(1,h)}$, for some h.

We remark that, since publication of the original papers of Downey and Fellows, hundreds of problems have been shown to be $W[1]$-complete and many have been shown to be $W[1]$ hard. We refer the reader to Downey and Fellows [35] for a list of examples, as of 1998, and to Flum and Grohe [43] for some more recent examples.

Notice that, in Theorem 25.9, we did not say that WEIGHTED CNF SAT is $W[1]$-complete. The reason for this is that we do not believe that it is!

Classically, using a padding argument, we know that CNF SAT $\equiv_m^P$ 3CNF SAT. However, the classical reduction *doesn't* define a parameterized transformation from WEIGHTED CNF SAT to WEIGHTED 3CNF SAT, it is not structure-preserving enough to ensure that parameters map to parameters. In fact, it is conjectured [33] that there is no parameterized transformation at all from WEIGHTED CNF SAT to WEIGHTED 3CNF SAT. If the conjecture is correct, then WEIGHTED CNF SAT is not in the class $W[1]$.

The point here is that parameterized reductions are more refined than classical ones, and hence we believe that we get a wider variety of apparent hardness behavior when intractable problems are classified according to this more finely-grained analysis.

We can view an input formula X for WEIGHTED CNF SAT as a product of sums. Extending this reasoning, we can define WEIGHTED t-NORMALIZED SAT as the weighted satisfiability problem for a formula X where X is a product of sums of products of sums ... with t alternations. We can define WEIGHTED SAT to be the weighted satisfiability problem for a formula X that is unrestricted.

DEFINITION 25.9   [W[t]].   For each $t \geq 1$, we define a language L to be in the class $W[t]$ iff there is a parameterized transformation from L to $L_{\mathcal{F}(t,h)}$ for some h.

In [33] Downey and Fellows show that, for all $t \geq 1$, WEIGHTED t-NORMALIZED SAT is complete for $W[t]$. Thus, $W[1]$ is the collection of parameterized languages FPT-equivalent to WEIGHTED 3CNF SAT, $W[2]$ is the collection of parameterized languages FPT-equivalent to WEIGHTED CNF SAT, and for each $t \geq 2$, $W[t]$ is the collection of parameterized languages FPT-equivalent to WEIGHTED t-NORMALIZED SAT.

These classes form part of the basic hierarchy of parameterized problems below.

$$FPT \subseteq W[1] \subseteq W[2] \subseteq \cdots \subseteq W[t] \subseteq W[SAT] \subseteq W[P] \subseteq AW[*] = AW[t] \subseteq AW[P] \subseteq XP$$

This sequence is commonly termed the W-hierarchy. The complexity class $W[1]$ can be viewed as the parameterized analog of NP, since it suffices to establish likely parameterized intracability.

We remark that a number of natural problems have been found at various levels of this hierarchy. For example, DOMINATING SET is complete for the level $W[2]$.

The classes $W[SAT]$, $W[P]$, and the AW classes were introduced by Abrahamson et al. in [1]. The class $W[SAT]$ is the collection of parameterized languages FPT-equivalent to WEIGHTED SAT. The class $W[P]$ is the collection of parameterized languages FPT-equivalent to WEIGHTED CIRCUIT SAT, the weighted satisfiability problem for a decision circuit C that is unrestricted. A standard translation of Turing machines into circuits shows that k-WEIGHTED CIRCUIT SAT is the same as the problem of deciding whether or not a deterministic Turing machine accepts an input of weight k. It is conjectured that the containment $W[SAT] \subseteq W[P]$ is proper [35].

$AW[P]$ captures the notion of alternation. $AW[*]$ and $AW[t]$ capture alternation with circuits of various logical depths. $AW[t]$ has the following core problem:

PARAMETERIZED QUANTIFIED CIRCUIT SAT$_t$

Instance: A decision circuit C of weft t whose inputs correspond to a sequence $s_1, \ldots, s_r$ of pairwise disjoint sets of variables.

Parameter: r, $k_1, \ldots, k_n$.

Question: Is it the case that there exists a size k_1 subset t_1 of s_1, such that for every size k_2 subset t_2 of s_2, there exists a size k_3 subset t_3 of s_3, such that ... (alternating quantifiers) such that, when $t_1 \cup t_2 \cup \cdots \cup t_r$ are set to true, and all other variables are set to false, C is satisfied?

In fact $AW[1]$ is the same as $AW[t]$ for all $t \geq 1$ (see [35]). To emphasize this fact, the class $AW[1]$ is renamed $AW[*]$.

The class $AW[P]$ is the collection of parameterized languages FPT-equivalent to the weighted satisfiability problem for an unrestricted decision circuit that applies alternating quantifiers to the inputs. That is, $AW[P]$ is defined in the same way as $AW[t]$ but using arbitrary circuits in the place of weft t circuits in the core problem.

Many parameterized analogs of game problems are complete for the AW classes, such as the parameterized analog of GEOGRAPHY, which is complete for $AW[*]$. Instead of asking whether or not player one has a winning strategy, the parameterized analog of GEOGRAPHY asks whether or not player one has a winning strategy using at most k moves.

The class XP, introduced in [35], is the collection of parameterized languages L such that the kth slice of L (the instances of L having parameter k) is complete for $DTIME(n^k)$. XP is provably distinct from FPT and seems to be the parameterized class corresponding to the classical class EXP (exponential time).

It is conjectured that all of the containments of the W-hierarchy are proper, but all that is currently known is that FPT is a proper subset of XP.

If we compare classical and parameterized complexity it is evident that the framework provided by parameterized complexity theory allows for more finely grained complexity analysis of computational problems. It is deeply connected with algorithmic heuristics and exact algorithms in practice. We refer the reader to either the survey by Flum and Grohe [42], or those in two recent issues of *The Computer Journal* [36] for further insight.

We can consider many different parameterizations of a single classical problem, each of which leads to either a tractable, or (likely) intractable, version in the parameterized setting. This allows for an extended dialog with the problem at hand. This idea toward the solution of algorithmic problems is explored in, for example, [38].

The reader may note that parameterized complexity is addressing intractability within polynomial time. In this vein, the parameterized framework can be used to demonstrate that many classical problems that admit a PTAS don't, in fact, admit any PTAS with a practical running time, unless $W[1] = FPT$ (see end of Section 25.6). It has been used to show that resolution is not automizable unless $W[P] = FPT$ [8,40]. It can also be used to show that the large hidden constants (various towers of two's) in the running times of generic algorithms obtained though the use of algorithmic metatheorems (Section 25.4.3) cannot be improved upon (see [43]).

We finally note that there are alternative useful parameterized complexity hierarchies, such as the A and M-hierarchies, see, e.g., Flum and Grohe [43].

Rather than further pursuing parameterized intractability and the rich area of parameterized structural complexity theory, we have concentrated this survey on the collection of distinctive techniques that has been developed for fixed-parameter tractable algorithm design. It would take a survey of comparable length to comprehensively tackle the area of parameterized intractability. We will simply refer the reader to Downey and Fellows [35], Flum and Grohe [43], or to survey articles such as Downey [32] for further details.

25.6 Parameterized Approximation Algorithms

We close this chapter with the discussion of a relatively new topic, parameterized approximation, introduced independently by three papers presented at IWPEC 2006 (The 3rd International Workshop on Parameterized Complexity and Exact Algorithms) [23,27,37].

There are various ways in which parameterized complexity and parameterized computation can interact with approximation algorithms. The interplay between the two fields is covered comprehensively in [53]. Here, we will consider only the most natural extension of parameterized complexity in this direction. We first need to introduce some definitions, we follow those given in [53].

Each input instance to an NP-optimization problem has associated with it a set of feasible solutions. A cost measure is defined for each of these feasible solutions. The task in solving the optimization problem is to find a feasible solution where the measure is as good as possible.

Formally, we define an NP-optimization problem as a 4-tuple $(I, sol, cost, goal)$ where

- I is the set of instances.
- For an instance $x \in I$, $sol(x)$ is the set of feasible solutions for x. The length of each $y \in sol(x)$ is polynomially bounded in $|x|$, and it can be decided in time polynomial in $|x|$ whether $y \in sol(x)$ holds for given x and y.
- Given an instance x and a feasible solution y, $cost(x, y)$ is a polynomial-time computable positive integer.
- $goal$ is either max or min.

The cost of an optimum solution for instance x is denoted by $opt(x)$.

$$opt(x) = goal\{cost(x, y) \mid y \in sol(x)\}$$

If y is a solution for instance x then the performance ratio of y is defined as

$$R(x, y) = \begin{cases} cost(x, y)/opt(x) & \text{if } goal = min, \\ opt(x)/cost(x, y) & \text{if } goal = max. \end{cases}$$

For a real number $c > 1$, we say that an algorithm is a c-approximation algorithm if it always produces a solution with performance ratio at most c.

One obvious parameter of an optimization problem instance is the optimum cost. This leads to a standard parameterization of an optimization problem X, where we define the corresponding parameterized decision problem $X_\leq$ as

$X_\leq$
Instance: An instance x of X.
Parameter: A positive integer k.
Question: Is $opt(x) \leq k$?

We can define $X_\geq$ analogously. For many problems, if we can solve $X_\leq$, or $X_\geq$, via an FPT algorithm, then we can also actually find an optimum solution for any instance x by repeatedly applying the algorithm to slightly modified versions of x. This strategy is known as polynomial-time self-reducibility. In some cases, an FPT algorithm that solves $X_\leq$, or $X_\geq$, will construct an optimal solution simply as a side effect of deciding the answer to the question.

Many of the standard problems in the parameterized complexity literature are standard parameterizations of optimization problems, for example, K-VERTEX COVER, K-CLIQUE, K-DOMINATING SET, and K-INDEPENDENT SET. If such a standard parameterization is fixed-parameter tractable then this means that we have an efficient algorithm for exactly determining the optimum for those instances of the corresponding optimization problem where the optimum is small. A $W[1]$-hardness result for a standard parameterization of an optimization problem shows that such an algorithm is unlikely to exist. In this case, we can ask the question: Is it possible to efficiently approximate the optimum as long as it is small?

We use the following definition for an FPT-approximation algorithm proposed by Chen et al. [27].

DEFINITION 25.10 (Standard FPT-approximation Algorithm). Let $X = (I, sol, cost, goal)$ be an optimization problem. A *standard FPT-approximation algorithm with performance ratio c* for X is an algorithm that, given input (x, k) satisfying

$$\begin{cases} opt(x) \leq k & \text{if } goal = min, \\ opt(x) \geq k & \text{if } goal = max, \end{cases} \tag{25.1}$$

computes a $y \in sol(x)$ in time $f(k) \cdot |x|^{O(1)}$ such that

$$\begin{cases} cost(x, y) \leq k \cdot c & \text{if } goal = min, \\ cost(x, y) \geq k/c & \text{if } goal = max. \end{cases} \tag{25.2}$$

For inputs not satisfying (25.1) the output can be arbitrary.

One example of this type of approximability is given by the K-SUBTREE PRUNE AND REGRAFT problem. The input to this problem is a pair of phylogenetic X-trees, T_1 and T_2, each an unrooted binary tree with leaves labeled by a common set of species X and (unlabeled) interior nodes having degree exactly three. The (labeled) topologies of T_1 and T_2 may differ. The task is to find at most k subtree prune and regraft operations that will transform T_1 into T_2.

A single subtree prune and regraft (SPR) operation on T_1 begins by pruning a subtree of T_1 by detaching an edge $e = (u, v)$ in T_1 from one of its (non-leaf) endpoints, say u. The vertex u and its two remaining adjacent edges, (u, w) and (u, x) are then contracted into a single edge, (w, x). Let T_u be the phylogenetic tree, previously containing u, obtained from T_1 by this process. We create a new vertex u' by subdividing an edge in T_u. We then create a new tree T' by adding an edge f between u' and v. We say that T' has been obtained from T_u by a single subtree prune and regraft (SPR) operation.

A related operation on the phylogenetic tree T_1 is the tree bisection and reconnection (TBR) operation. A single tree bisection and reconnection (TBR) operation begins by detaching an edge

$e = (u, v)$ in T_1 from both of its endpoints, u and v. Contractions are then applied to either one or both of u and v to create two new phylogenetic trees (note that a contraction is necessary only in the case of a non-leaf vertex.) Let T_u and T_v be the phylogenetic trees, previously containing, respectively, u and v, obtained from T_1 by this process. We create new vertices u' and v' by subdividing edges in T_u and T_v respectively. We then create a new tree T' by adding an edge f between u' and v'. We say that T' has been obtained from T_1 by a single tree bisection and reconnection (TBR) operation. Note that the effect of a single TBR operation can be achieved by the application of either one or two SPR operations and that every SPR operation is also a TBR operation.

In [47] it is shown that the k-TREE BISECTION AND RECONNECTION problem can be solved via an FPT algorithm for the equivalent k-MAXIMUM AGREEMENT FOREST problem, running in time $O(k^4 \cdot n^5)$. The algorithm employed in [47] uses a kernelization strategy from [9] as a preprocessing step, followed by a search tree strategy. Neither of the techniques used in [47] is applicable in the case of the k-SUBTREE PRUNE AND REGRAFT problem. It has long been conjectured that the k-SUBTREE PRUNE AND REGRAFT problem is fixed-parameter tractable, however, proof of this conjecture is a long-standing open problem in the parameterized complexity community.

The algorithm given in [47] can easily be adapted to give either a NO answer to the k-TREE BISECTION AND RECONNECTION problem or, otherwise, a minimal set of TBR operations that transforms T_1 into T_2. Given two phylogenetic trees, T_1 and T_2, if there is a set S of at most k SPR operations that transforms T_1 into T_2 then S is also a set of at most k TBR operations transforming T_1 into T_2. Thus, in this case, the algorithm given in [47] will return a solution S' consisting of at most k TBR operations. This set S' will translate into a set of at most $2k$ SPR operations transforming T_1 into T_2. If there is no size k set of SPR operations that transforms T_1 into T_2, then the algorithm given in [47] might still return a solution S' consisting of at most k TBR operations and again, this set S' will translate into a set of at most $2k$ SPR operations transforming T_1 into T_2. However, in this case, there is no guarantee of success.

This is currently the only example (known to us) of a problem that is not proven to be FPT but that does have a standard FPT-approximation algorithm, although there are other examples of FPT-approximation algorithms appearing in the literature. For example, as recorded in [53], k-TREEWIDTH is FPT, in fact for every k, if a tree decomposition of a given graph $G = (V, E)$ of width k exists, then this can be computed in time linear in $|V|$ [13]. The algorithm given in [13] is rather complex and not at all workable in practice. However, there are far simpler FPT algorithms that produce tree decompositions having width at most a constant factor larger than the optimum [60,63].

In the negative, for some parameterized problems, it is possible to show that there is no standard FPT-approximation algorithm for any performance ratio, be it a constant ratio, or, otherwise, some ratio that is a function of the parameter k.

Given a graph $G = (V, E)$, an independent dominating set V' is an independent set of vertices $V' \subseteq V$ such that V' is a dominating set for G. The corresponding optimization problem is the MINIMUM INDEPENDENT DOMINATING SET problem, where the goal is to minimize $|V'|$. Downey et al. [37] prove that the standard parameterization of this problem is completely inapproximable.

THEOREM 25.10 [37] *If* k-INDEPENDENT DOMINATING SET *has a standard fpt-approximation algorithm with performance ratio function $f(k)$, for some computable function $f(k)$, then $W[2] = FPT$.*

This particular problem is not monotone, where monotone here means that, if we extend a feasible solution with additional vertices, then it remains feasible. Clearly, we can arbitrarily add vertices to V' such that V' remains a dominating set, but such a V' may no longer be independent. Thus, we can manufacture an instance where the the optimum is k and where every feasible solution also has size k. It would be more interesting to obtain inapproximability results for the monotone problem K-DOMINATING SET.

Important progress on parameterized approximation, at the level of $W[P]$, has recently been achieved by Eikmeyer et al. [40]. They looked at *multiplicative* approximation ratios, proving the following.

THEOREM 25.11 [40] *All known "natural" W[P] complete problems (including all the ones from [35,1]) have no FPT approximation algorithms with approximation ratio exp(log$^\gamma$ k) for some constant $\gamma \in (0, 1)$ (γ depending on the problem) unless W[P] = FPT.*

One illustration of an application of Theorem 25.11 is the problem MIN LINEAR INEQUALITY BY DELETION which has parameter k and asks "Does the deletion of k elements from a system of linear inequalities result in a system that is solvable?"

We end this discussion with a final negative result, which illustrates a direct connection between parameterized complexity and classical approximation algorithms.

We say that a problem X admits a polynomial-time approximation scheme (PTAS) if there is an algorithm A such that, for every instance x of X and every $\epsilon > 0$, A produces a $(1 + \epsilon)$-approximate solution, $y \in sol(x)$, in time $|x|^{f(1/\epsilon)}$, for some arbitrary computable function f.

Such an algorithm A runs in polynomial time for every fixed value of ϵ, but if ϵ is small then the exponent of the polynomial $|x|^{f(1/\epsilon)}$ can be very large. Two restricted classes of approximation schemes have been defined that avoid this problem. An efficient polynomial-time approximation scheme (EPTAS) is a PTAS with running time of the form $f(1/\epsilon) \cdot |x|^{O(1)}$. A fully polynomial-time approximation scheme (FPTAS) is a PTAS with running time of the form $(1/\epsilon)^{O(1)} \cdot |x|^{O(1)}$.

Parameterized complexity affords evidence to show that, in some cases, an EPTAS will not be forthcoming. The following theorem has been proposed independently by Bazgan [12] and Cesati and Trevisan [25].

THEOREM 25.12 [12,25] *If an optimization problem X admits an EPTAS, then the standard parameterization of X is FPT.*

We can use the contrapositive of Theorem 25.12 to show that an EPTAS likely does not exist for a particular *NP*-optimization problem.

COROLLARY 25.1 If the standard parameterization of an optimization problem is $W[1]$-hard, then the optimization problem does not have an EPTAS (unless *FPT = W[1]*).

Note that the converse of Theorem 25.12 is not true. For example, K-VERTEX COVER is FPT, as we have repeatedly shown throughout this article, but MINIMUM VERTEX COVER is *APX*-hard, implying that it doesn't have any type of PTAS at all (unless $P = NP$).

25.7 Conclusions

Our aim in this chapter has been to introduce the reader to the distinctive algorithmic techniques of parameterized complexity, in the hope that these might find a useful place in the repertoire of a widening circle of algorithm designers. We have endeavored to strike a balance between high-level, generalized, descriptions and technical details. Space limitations have inevitably meant that many technical details be omitted. There are many aspects of the parameterized paradigm which we have not canvassed at all. In this regard, we enthusiastically refer the reader to the recently published collection of parameterized complexity survey papers in the *Computer Journal* [36], the monographs [43,45,55], as well as the original parameterized complexity text [35], for more extensive coverage of parameterized complexity theory and parameterized algorithms.

References

1. K. A. Abrahamson, R. Downey, and M. Fellows: Fixed parameter tractability and completeness IV: On completeness for W[P] and PSPACE analogs. *Annals of Pure and Applied Logic* 73, 235–276, 1995.

2. K. A. Abrahamson and M. R. Fellows: Finite automata, bounded treewidth, and well-quasi-ordering. *Graph Structure Theory*, N. Robertson and P. Seymour (Eds.), *Contempory Mathematics* Vol 147, American Mathematical Society, Providence, RI, pp 539–564, 1993.

3. F. N. Abu-Khzam, R. L. Collins, M. R. Fellows, M. A. Langston, W. II. Suters, and C. T. Symons: Kernelization algorithms for the vertex cover problem: Theory and experiments. *Proceedings of the 6th ALENEX* 2004, SIAM, New Orleans, LA, pp. 62–69, 2004.

4. F. N. Abu-Khzam, M. A. Langston, P. Shanbhag, and C. T. Symons: Scalable parallel algorithms for FPT problems. *Algorithmica* 45, 269–284, 2006.

5. J. Alber: Exact algorithms for NP-hard problems on planar and related graphs: Design, analysis, and implementation. PhD thesis, Universität Tübingen, Germany, January 2003.

6. J. Alber, H. L. Bodlaender, H. Fernau, T. Kloks, and R. Niedermeier: Fixed parameter algorithms for dominating set and related problems on planar graphs. *Algorithmica* 33, 461–493, 2002.

7. J. Alber, H. Fernau, and R. Niedermeier: Parameterized complexity: Exponential speed-up for planar graph problems. *Proceedings of 28th ICALP*, Crete, Greece. LNCS 2076, Springer-Verlag, Berlin, Germany, pp. 261–272, 2001.

8. M. Alekhnovich and A. Razborov: Resolution is not automatizable unless W[P] is tractable. *Proceedings of the 42nd IEEE FOCS*, Las Vegas, NV, pp 210–219, 2001.

9. B. Allen and M. Steel: Subtree transfer operations and their induced metrics on evolutionary trees. *Annals of Combinatorics* 5, 1–13, 2000.

10. N. Alon, R. Yuster, and U. Zwick: Color-coding. *Journal of the ACM* 42 (4), 844–856, 1995.

11. R. Balasubramanian, M. Fellows, and V. Raman: An improved fixed parameter algorithm for vertex cover. *Information Processing Letters* 65 (3), 163–168, 1998.

12. C. Bazgan: Schémas d'approximation et complexité paramétrée. Rapport de stage de DEA d'Informatique à Orsay, 1995.

13. H. L. Bodlaender: A linear time algorithm for finding tree decompositions of small treewidth. *SIAM Journal of Computing* 25, 1305–1317, 1996.

14. H. L. Bodlaender: A partial *k*-arboretum of graphs with bounded treewidth. Technical Report UU-CS-1996-02, Department of Computer Science, Utrecht University, Utrecht, the Netherlands, 1996.

15. H. L. Bodlaender, R. G. Downey, M. R. Fellows, D. Hermelin: On problems without polynomial kernels (extended abstract). *Proceedings of the 35th International Colloquium on Automata, Languages and Programming*, Reykjavik, Iceland, L. Aceto, I. Damgård, L. A. Glodberg, M. M. Halldórsson, A. Ingólfsdóttir, and I. Walukiewicz (Eds.), *LNCS* 5215, pp. 563–574, 2008.

16. H. L. Bodlaender, J. R. Gilbert, H. Hafsteinsson, and T. Kloks: Approximating treewidth, pathwidth, and minimum elimination tree height. *Journal of Algorithms* 18, 238–255, 1995.

17. H. L. Bodlaender and T. Kloks: Efficient and constructive algorithms for the pathwidth and treewidth of graphs. *Journal of Algorithms* 21, 358–402, 1996.

18. H. L. Bodlaender, T. Kloks, and D. Kratsch: Treewidth and pathwidth of permutation graphs. *Proceedings of the 20th International Colloquium on Automata, Langauges and Programming*, A. Lingas,

R. Karlsson, and S. Carlsson (Eds.), Lund, Sweden. *LNCS* 700, Springer-Verlag, Berlin, Germany, pp 114–125, 1993.

19. H. L. Bodlaender and R. H. Möhring: The pathwidth and treewdith of cographs. *SIAM Journal of Discrete Mathematics* 6, 181–188, 1993.

20. R. B. Borie, R. G. Parker and C. A Tovey: *Solving Problems on Recursively Constructed Graphs. ACM Computing Surveys*, 41(1), 1–51, 2008.

21. S. Buss: private communication, 1989.

22. L. Cai, J. Chen, R. G. Downey, and M. R. Fellows: On the parameterized complexity of short computation and factorization. *Archive for Mathematical Logic* 36, 321–337, 1997.

23. L. Cai and X. Huang: Fixed-parameter approximation: conceptual framework and approximability results. *Proceedings of IWPEC 2006*, Zurich, Switzerland. *LNCS* 4169, pp 96–108, 2006.

24. L. Cai and D. Juedes: Subexponential parameterized algorithms collapse the *W*-hierarchy. *Proceedings of ICALP 2001*, Crete, Greece, *LNCS* 2076, Springer-Verlag, 2001.

25. M. Cesati and l. Trevisan: On the efficiency of polynomial time approximation schemes. *Information Processing Letters* 64 (4), 165–171, 1997.

26. J. Chen, I.A. Kanj, and W. Jia: Vertex cover: Further observations and further improvements. *Journal of Algorithms* 41, 280–301, 2001.

27. Y. Chen, M. Grohe, and M. Gruber: On parameterized approximability. *Proceedings of IWPEC 2006*, Zurich, Switzerland. *LNCS* 4169, pp 109–120, 2006.

28. J. Cheetham, F. Dehne, A. Rau-Chaplin, U. Stege, and P. Taillon: Solving large FPT problems on coarse-grained parallel machines. *Journal of Computer and System Sciences* 67, 691–706, 2003.

29. B. Chor, M. R. Fellows, and D. W. Juedes: Linear kernels in linear time, or how to save k colors in $O(n^2)$ steps. *Proceedings of the 30th WG*, Bonn, Germany. *LNCS* 3353, pp 257–269, 2004.

30. B. Courcelle: The monadic second-order logic of graphs I: Recognizable sets of finite graphs. *Information and Computation* 85, 12–75, 1990.

31. B. Courcelle, J. Engelfriet, and G. Rozenberg: Handle-rewriting hypergraph grammars. *Journal of Computing and Systems Sciences* 46 (2), 218–270, 1993.

32. R. G. Downey, Parameterized complexity for the skeptic. *18th Annual IEEE Conference on Computational Complexity*, Aarhus, Denmark, pp 147–169, 2003.

33. R. G. Downey and M. R. Fellows: Fixed parameter tractability and completeness I: Basic theory. *SIAM Journal of Computing* 24, 873–921, 1995.

34. R. G. Downey and M. R. Fellows: Fixed parameter tractability and completeness II: Completeness for W[1]. *Theoretical Computer Science A* 141, 109–131, 1995.

35. R. G. Downey and M. R. Fellows: *Parameterized Complexity*. Springer-Verlag, Berlin, Germany, 1999.

36. R. G. Downey, M. R. Fellows, and M. Langston: *The Computer Journal Special Issue on Parameterized Complexity*, 51(1 and 3), 2008.

37. R. G. Downey, M. R. Fellows, and C. McCartin: Parameterized approximation algorithms. *Proceedings of IWPEC 2006*, Zurich, Switzerland, *LNCS* 4169, pp 121–129, 2006.

38. R. Downey, M. Fellows, and U. Stege: Computational tractability: The view from Mars. *Bulletin of the European Association for Theoretical Computer Science*, 69, 73–97, 1999.

39. F. Dehne, A. Rau-Chaplin, U. Stege, and P. Taillon: Solving large FPT problems on coarse grained parallel machines. *Journal of Computer and System Sciences* 67 (4), 691–706, 2003.

40. K. Eickmeyer, M. Grohe, and M. Grüber, Approximation of natural W[P]-complete minimisation problems is hard. *Proceedings of the 23rd IEEE Conference on Computational Complexity (CCC'08)*, College Park, MD, pp. 8–18, 2008.

41. V. Estivill-Castro, R. Downey, M. R. Fellows, E. Prieto-Rodriguez, and F. A. Rosamond: Cutting up is hard to do: The parameterized complexity of k-cut and related problems. *Electronic Notes in Theoretical Computer Science* 78, pp 205–218, 2003.

42. J. Flum and M. Grohe: Parameterized complexity and subexponential time, *Bulletin of the European Association for Theoretical Computer Science* 84, 71–100, 2004.

43. J. Flum and M. Grohe: *Parameterized Complexity Theory*. Springer-Verlag, Berlin, Germany, 2006.

44. M. Frick and M. Grohe: Deciding first-order properties of locally tree-decomposable graphs. *Proceedings of the 26th International Colloquium on Automata, Languages and Programming*, Prague, Czech Republic. *LNCS* 1644, pp 331–340, Springer-Verlag, Berlin, Germany, 1999.

45. Henning Fernau: *Parameterized Algorithmics: A Graph-Theoretic Approach.* Habilitationsschrift, Universitat Tubingen, Tubingen, Germany, 2005.

46. J. Guo and R. Niedermeier: Invitation to data reduction and problem kernelization. *ACM SIGACT News* 38(1), 31–45, 2007.

47. M. Hallett and C. McCartin: A faster FPT algorithm for the maximum agreement forest problem. *Theory of Computing Systems* 41 (3), 539–550, 2007.

48. R. Impagliazzo, R. Paturi, and F. Zane: Which problems have strongly exponential complexity? *JCSS* 63(4), 512–530, 2001.

49. F. Huffner, R. Niedermeier, and S. Wernicke: Techniques for practical fixed-parameter algorithms. *The Computer Journal*, 51(1), 7–25, 2008.

50. J. Gramm, J. Guo, F. Huffner, and R. Niedermeier: Data reduction, exact and heurisitc algorithms for clique cover. *Proceedings of 8th ALENEX*, Miami, FL, pp. 86–94, 2006.

51. J. Gramm, R. Niedermeier, and P. Rossmanith: Fixed-parameter algorithms for closest string and related problems. *Algorithmica* 37, 25–42, 2003.

52. M. Mahajan and V. Raman: Parameterizing above the guarantee: MAXSAT and MAXCUT. *Journal of Algorithms* 31, 335–354, 1999.

53. D. Marx: Parameterized complexity and approximation algorithms. *The Computer Journal*, 51(1), 60–78, 2008.

54. G. L. Nemhauser and L. E. Trotter Jr: Vertex packings: Structural properties and algorithms. *Mathematical Programming* 8, 232–248, 1975.

55. R. Niedermeier: *Invitation to Fixed-Parameter Algorithms.* Oxford University Press, New York, 2006.

56. R. Niedermeier and P. Rossmanith: A general method to speed up fixed-parameter tractable algorithms. *Information Processing Letters* 73, 125–129, 2000.

57. S.-I. Oum: Approximating rank-width and clique-width quickly. *Proceedings of 31st International Workshop on Graph-Theoretic Concepts in Computer Science*, Metz, France, pp. 49–58, 2005.

58. L. Perkovic and B. Reed: An improved algorithm for finding tree decompositions of small width. *International Journal of Foundations of Computer Science* 11 (3), 365–371, 2000.

59. B. Reed, K. Smith, and A. Vetta: Finding odd cycle transversals. *Operations Research Letters* 32, 299–301, 2004.

60. B. Reed: Finding approximate separators and computing treewidth quickly. *STOC '92: Proceedings of the 24th Annual ACM Symposium on Theory of Computing*, Victoria, British Columbia, Canada, pp. 221–228, 1992.

61. N. Robertson and P. D. Seymour: *Graph Minors—A Survey.* Surveys in Combinatorics, I. Anderson (Ed.), Cambridge University Press, Cambridge, pp. 153–171, 1985.

62. N. Robertson and P. D. Seymour: Graph minors II. Algorithmic aspects of tree-width. *Journal of Algorithms* 7, 309–322, 1986.

63. N. Robertson and P. D. Seymour: Graph minors I–XV. Appearing in *Journal of Combinatorial Theory Series B*, 1983–1996.

64. D. Seese: The structure of models of decidable monadic theories of graphs. *Annals of Pure and Applied Logic* 53, 169–195, 1991.

65. C. Sloper and J. A. Telle: An overview of techniques for designing parameterized algorithms. *The Computer Journal*, 51(1), 122–136, 2008.

66. U. Stege: Resolving conflicts in problems in computational biochemistry. PhD dissertation, ETH, 2000.

67. K. Weihe: Covering trains by stations or the power of data reduction. *Proceedings of Algorithms and Experiments (ALEX98)*, R. Battiti and A. A. Bertossi (Eds.), Trento, Italy, pp. 1–8, 1998.

26

Computational Learning Theory*

Sally A. Goldman
Washington University

26.1 Introduction

Since the late 1950s, computer scientists (particularly those working in the area of artificial intelligence) have been trying to understand how to construct computer programs that perform tasks we normally think of as requiring human intelligence, and which can improve their performance over time by modifying their behavior in response to experience. In other words, one objective has been to design computer programs that can learn. For example, Samuels designed a program to play checkers in the early 1960s that could improve its performance as it gained experience playing against human opponents. More recently, research on artificial neural networks has stimulated interest in the design of systems capable of performing tasks that are difficult to describe algorithmically (such as recognizing a spoken word or identifying an object in a complex scene), by exposure to many examples.

As a concrete example consider the task of handwritten character recognition. A learning algorithm is given a set of examples where each contains a handwritten character as specified by a set of attributes (e.g., the height of the letter) along with the name (label) for the intended character. This set of examples is often called the training data. The goal of the learner is to efficiently construct a rule (often referred to as a hypothesis or classifier) that can be used to take some previously unseen character and with high accuracy determine the proper label.

---

* This chapter is unchanged from the first edition in 1988, and thus does not discuss the many recent developments in the field since that time.

Computational learning theory is a branch of theoretical computer science that formally studies how to design computer programs that are capable of learning, and identifies the computational limits of learning by machines. Historically, researchers in the artificial intelligence community have judged learning algorithms empirically, according to their performance on sample problems. While such evaluations provide much useful information and insight, often it is hard using such evaluations to make meaningful comparisons among competing learning algorithms.

Computational learning theory provides a formal framework in which to precisely formulate and address questions regarding the performance of different learning algorithms so that careful comparisons of both the predictive power and the computational efficiency of alternative learning algorithms can be made. Three key aspects that must be formalized are the way in which the learner interacts with its environment, the definition of successfully completing the learning task, and a formal definition of efficiency of both data usage (sample complexity) and processing time (time complexity). It is important to remember that the theoretical learning models studied are abstractions from real-life problems. Thus close connections with experimentalists are useful to help validate or modify these abstractions so that the theoretical results help to explain or predict empirical performance. In this direction, computational learning theory research has close connections to machine learning research. In addition to its predictive capability, some other important features of a good theoretical model are simplicity, robustness to variations in the learning scenario, and an ability to create insights to empirically observed phenomena.

The first theoretical studies of machine learning were performed by inductive inference researchers (see [10]) beginning with the introduction of the first formal definition of learning given by Gold [24]. In Gold's model, the learner receives a sequence of examples and is required to make a sequence of guesses as to the underlying rule (concept) such that this sequence converges at some finite point to a single guess that correctly names the unknown rule. A key distinguishing characteristic is that Gold's model does not attempt to capture any notion of the efficiency of the learning process, whereas the field of computational learning theory emphasizes the computational feasibility of the learning algorithm.

Another closely related field is that of pattern recognition (see [21] and [20]). Much of the theory developed by learning theory researchers to evaluate the amount of data needed to learn directly adapts results from the fields of statistics and pattern recognition. A key distinguishing factor is that learning theory researchers study both the data (information) requirements for learning and the time complexity of the learning algorithm. (In contrast, pattern recognition researchers tend to focus on issues related to the data requirements.) Finally, there are a lot of close relations to work on artificial neural networks. While much of neural network research is empirical, there is also a good amount of theoretical work (see [20]).

This chapter is structured as follows. In Section 26.2 we describe the basic framework of concept learning and give notation that we use throughout the chapter. Next, in Section 26.3 we describe the PAC (distribution-free) model that began the field of computational learning theory. An important early result in the field is the demonstration of the relationships between the VC-dimension, a combinatorial measure, and the data requirements for **PAC learning**. We then discuss some commonly studied noise models and general techniques for PAC learning from noisy data.

In Section 26.4 we cover some of the other commonly studied formal learning models. First we study an **online learning model**. Unlike the PAC model in which there is a training period, in the online learning model the learner must improve the quality of its predictions as it functions in the world. Next we study the query model which is very similar to the online model except that the learner plays a more active role.

As in other theoretical fields, along with having techniques to prove positive results, another important component of learning theory research is to develop and apply methods to prove when a learning problem is hard. In Section 26.5 we describe some techniques used to show that learning problems are hard. Next, in Section 26.6, we explore a variation of the PAC model called the weak learning model, and study techniques for boosting the performance of a mediocre learning algorithm. Finally, we close with a brief overview of some of the many current research issues being studied within the field of computational learning theory.

26.2 General Framework

For ease of exposition, we initially focus on concept learning in which the learner's goal is to infer how an unknown target function classifies (as positive or negative) examples from a given domain. In the character recognition example, one possible task of the learner would be to classify each character as a numeral or nonnumeral. Most of the definitions given here naturally extend to the general setting of learning functions with multiple-valued or real-valued outputs. Later in Section 26.4 we briefly discuss the more general problem of learning a real-valued function.

26.2.1 Notation

The instance space (domain) $\mathcal{X}$ is the set of all possible objects (instances) to be classified. Two very common domains are the Boolean hypercube $\{0, 1\}^n$ and continuous n-dimensional space $\mathfrak{R}^n$. For example, if there are n Boolean attributes then each example can be expressed as an element of $\{0, 1\}^n$. Likewise, if there are n real-valued attributes, then $\mathfrak{R}^n$ can be used. A concept f is a Boolean function over domain $\mathcal{X}$. Each $x \in \mathcal{X}$ is referred to as an example (point).

A **concept class** $\mathcal{C}$ is a collection of subsets of $\mathcal{X}$. That is $\mathcal{C} \subseteq 2^{\mathcal{X}}$. Typically (but not always), it is assumed that the learner has prior knowledge of $\mathcal{C}$. Examples are classified according to membership of a target concept $f \in \mathcal{C}$. Often, instead of using this set theoretic view of $\mathcal{C}$, a functional view is used in which $f(x)$ gives the classification of concept $f \in \mathcal{C}$ for each example $x \in \mathcal{X}$. An example $x \in \mathcal{X}$ is a positive example of f if $f(x) = 1$ (equivalently $x \in f$), or a negative example of f if $f(x) = 0$ (or equivalently $x \notin f$). Often $\mathcal{C}$ is decomposed into subclasses $\mathcal{C}_n$ according to some natural size measure n for encoding an example. For example, in the Boolean domain, n is the number of Boolean attributes. Let $\mathcal{X}_n$ denote the set of examples to be classified for each problem of size n, $\mathcal{X} = \bigcup_{n \geq 1} \mathcal{X}_n$.

We use the class $\mathcal{C}_{\text{halfspace}}$, the set of halfspaces in $\mathfrak{R}^n$, and the class $\mathcal{C}_{\text{halfspace}}^{\cap_s}$, the set of all the intersections of up to s halfspaces in $\mathfrak{R}^d$, to illustrate some of the basic techniques for designing learning algorithms* (see Figure 26.1). We now formally define a halfspace in $\mathfrak{R}^n$ and describe how the points in $\mathfrak{R}^n$ are classified by it. Let $\vec{x} = (x_1, \ldots, x_n)$ denote an element of $\mathfrak{R}^n$. A halfspace defined by $\vec{a} \in \mathfrak{R}^n$ and $b \in \mathfrak{R}$ classifies as positive the set of points $\{\vec{x} \mid \vec{a} \cdot \vec{x} \geq b\}$. For example, in two dimensions (i.e., $n = 2$), $5x_1 - 2x_2 \geq -3$ defines a halfspace. The point $(0, 0)$ is a positive example and $(2, 10)$ is a negative example. For $\mathcal{C}_{\text{halfspace}}^{\cap_s}$, an example is positive exactly when it is classified as positive by each of the halfspaces forming the intersection. Thus the set of positive points forms a (possibly open) convex polytope in $\mathfrak{R}^n$.

We also study some Boolean concepts. For these concepts the domain is $\{0, 1\}^n$. Let $x_1, \ldots, x_n$ denote the n Boolean attributes. A literal is either x_i or $\overline{x_i}$ where $i = 1, \ldots, n$. A term is a conjunction of literals. Finally, a DNF formula is a disjunction of terms. One of the biggest open problems of computational learning theory is whether or not the concept class of DNF formulas is efficiently PAC learnable. Since the problem of learning general DNF formulas is a long-standing open problem, several subclasses have been studied. A monotone DNF formula is a DNF formula in which there are no negated variables. A read-once DNF formula is a DNF formula in which each variable appears at most once. In addition to adding a restriction that the formula be monotone and/or read-once, one

---

* As a convention, when the number of dimensions can be arbitrary we use n, and when the number of dimensions must be a constant we use d.

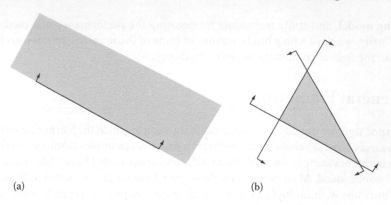

(a)   (b)

FIGURE 26.1 (a) Shows (in $\Re^2$) a concept from $\mathcal{C}_{\text{halfspace}}$ and (b) shows a concept for $\mathcal{C}_{\text{halfspace}}^{\cap_s}$ for $s = 3$. The points from $\Re^2$ that are classified as positive are shaded. The unshaded points are classified as negative.

can also limit either the size of each term or the number of terms. A k-term DNF formula is a DNF formula in which there are at most k terms. Finally, a k-DNF formula is a DNF formula in which at most k literals are used by each term.

As one would expect, the time and sample complexity of the learning algorithm depends on the complexity of the underlying target concept. For example, the complexity for learning a monotone DNF formula is likely to depend on the number of terms (conjuncts) in the formula. To give a careful definition of the size of a concept $f \in \mathcal{C}$, we associate a language $\mathcal{R}_\mathcal{C}$ with each concept class $\mathcal{C}$ that is used for representing concepts in $\mathcal{C}$. Each $r \in \mathcal{R}_\mathcal{C}$ denotes some $f \in \mathcal{C}$, and every $f \in \mathcal{C}$ has at least one representation $r \in \mathcal{R}_\mathcal{C}$. Each concept $f \in \mathcal{C}_n$ has a size denoted by $|f|$, which is the representation length of the shortest $r \in \mathcal{R}_\mathcal{C}$ that denotes f. For ease of exposition, in the remainder of this chapter we use $\mathcal{C}$ and $\mathcal{R}_\mathcal{C}$ interchangeably.

To appreciate the significance of the choice of the representation class, consider the problem of learning a regular language.* The question as to whether an algorithm is efficient depends heavily on the representation class. As defined more formally below, an efficient learning algorithm must have time and sample complexity polynomial in $|f|$ where f is the target concept. The target regular language could be represented as a deterministic finite-state automaton (DFA) or as a nondeterministic finite-state automaton (NFA). However, the length of the representation as a NFA can be exponentially smaller than the shortest representation as a DFA. Thus, the learner may be allowed exponentially more time when learning the class of regular languages as represented by DFAs versus when learning the class of regular languages as represented by NFAs. Often to make the representation class clear, the concept class names the representation. Thus instead of talking about learning a regular language, we talk about learning a DFA or an NFA. Thus the problem of learning a DFA is easier than learning an NFA. Similar issues arise when learning Boolean functions. For example, whether the function is represented as a decision tree, a DNF formula, or a Boolean circuit greatly affects the representation size.

26.3   PAC Learning Model

The field of computational learning theory began with Valiant's seminal work [48] in which he defined the PAC† (distribution-free) learning model. In the PAC model examples are generated

---

   * See the Section 26.2.1 and Section 30.2 in this handbook for background on regular languages, DFAs and NFAs.

   † PAC is an acronym coined by Dana Angluin for probably approximately correct.

according to an unknown probability distribution $\mathcal{D}$, and the goal of a learning algorithm is to classify with high accuracy (with respect to the distribution $\mathcal{D}$) any further (unclassified) examples.

We now formally define the PAC model. To obtain information about an unknown target function $f \in \mathcal{C}_n$, the learner is provided access to labeled (positive and negative) examples of f, drawn randomly according to some unknown target distribution $\mathcal{D}$ over $\mathcal{X}_n$. The learner is also given as input ϵ and δ such that $0 < \epsilon, \delta < 1$, and an upper bound k on $|f|$. The learner's goal is to output, with probability at least $1 - \delta$, a hypothesis $h \in \mathcal{C}_n$ that has probability at most ϵ of disagreeing with f on a randomly drawn example from $\mathcal{D}$ (thus, h has error at most ϵ). If such a learning algorithm $\mathcal{A}$ exists (that is, an algorithm $\mathcal{A}$ meeting the goal for any $n \geq 1$, any target concept $f \in \mathcal{C}_n$, any target distribution $\mathcal{D}$, any $\epsilon, \delta > 0$, and any $k \geq |f|$), then $\mathcal{C}$ is PAC learnable. A PAC learning algorithm is a polynomial-time (efficient) algorithm if the number of examples drawn and the computation time are polynomial in $n, k, 1/\epsilon$, and $1/\delta$. We note that most learning algorithms are really functions from samples to hypotheses (i.e., given a sample S the learning algorithm produces a hypothesis h). It is only for the analysis in which we say that for a given ϵ and δ, the learning algorithm is guaranteed with probability at least $1 - \delta$ to output a hypothesis with error at most ϵ given a sample whose size is a function of ϵ, δ, n, and k. Thus, empirically, one can generally run a PAC algorithm on provided data and then empirically measure the error of the final hypothesis. One exception is when trying to empirically use statistical query algorithms since, for most of these algorithms, the algorithm uses ϵ for more than just determining the desired sample size. (See Section 26.3.3 for a discussion of statistical query algorithms, and Goldman and Scott [27] for a discussion of their empirical use).

As originally formulated, PAC learnability also required the hypothesis to be a member of the concept class $\mathcal{C}_n$. We refer to this more stringent learning model as proper PAC-learnability. The work of Pitt and Valiant [42] shows that a prerequisite for proper PAC-learning is the ability to solve the consistent hypothesis problem, which is the problem of finding a concept $f \in \mathcal{C}$ that is consistent with a provided sample. Their result implies that if the consistent hypothesis problem is NP-hard for a given concept class (as they show is the case for the class of k-term DNF formulas) and NP $\neq$ RP, then the learning problem is hard.* A more general form of learning in which the goal is to find any polynomial-time algorithm that classifies instances accurately in the PAC sense is commonly called prediction. In this less stringent variation of the PAC model, the algorithm need not output a hypothesis from the concept class $\mathcal{C}$ but instead is just required to make its prediction in polynomial time. This idea of prediction in the PAC model originated in the paper of Haussler et al. [30], and is discussed in Pitt and Warmuth [43]. Throughout the remainder of this chapter, when referring to the PAC learning model we allow the learner to output any hypothesis that can be evaluated in polynomial time. That is, given a hypothesis h and an example x, we require that $h(x)$ can be computed in polynomial time. We refer to the model in which the learner is required to output a hypothesis $h \in \mathcal{C}_n$ as the proper PAC learning model.

Many variations of the PAC model are known to be equivalent (in terms of what concept classes are efficiently learnable) to the model defined above. We now briefly discuss one of these variations. In the above definition of the PAC model, along with receiving ϵ, δ, and n as input, the learner also receives k, an upperbound on $|f|$. The learner must be given ϵ and δ. Further, by looking at just one example, the value of n is known. Yet giving the learner knowledge of k may appear to make the problem easier. However, if the demand of polynomial-time computation is replaced with expected polynomial-time computation, then the learning algorithm need not be given the parameter k, but could "guess" it instead. We now briefly review the standard doubling technique used to convert an algorithm A designed to have k as input to an algorithm B that has no prior knowledge of k. Algorithm B begins with an estimate, say 1, for its upperbound on k and runs algorithm A using this

---

* For background on complexity classes see Chapter 32 (NP defined) of this handbook and Chapter 20 (RP defined) of *Algorithm and Theory of Computation Handbook, Second Edition: Special Topics and Techniques.*

estimate to obtain hypothesis h. Then algorithm B uses a hypothesis testing procedure to determine if the error of h is at most ϵ. Since the learner can only gather a random sample of examples, it is not possible to distinguish a hypothesis with error ϵ from one with error just greater than ϵ. However, by drawing a sufficiently large sample and looking at the empirical performance of h on that sample, we can distinguish a hypothesis with error at most $\epsilon/2$ from one with error more than ϵ. In particular, given a sample* of size $m = \lceil \frac{32}{\epsilon} \ln \frac{2}{\delta} \rceil$, if h misclassifies at most $\frac{3\epsilon}{4} \cdot m$ examples then B accepts h. Otherwise, algorithm B doubles its estimate for k and repeats the process. For the technical details of the above argument as well as for a discussion the other equivalences, see [29].

26.3.1 Sample Complexity Bounds, the VC-Dimension, and Occam's Razor

Although we are concerned with the time complexity of the learning algorithm, a fundamental quantity to first compute is the sample complexity (data) requirements. Blumer et al. [16] identified a combinatorial parameter of a class of functions defined by Vapnik and Chervonenkis [50]. They give strong results relating this parameter, called the **VC-dimension**, to information-theoretic bounds on the sample size needed to have accurate generalization. Note that given a sufficiently large sample there is still the computational problem of finding a "good" hypothesis.

We now define the VC-dimension. We say that a finite set $S \subseteq \mathcal{X}$ is shattered by the concept class $\mathcal{C}$ if for each of the $2^{|S|}$ subsets $S' \subseteq S$, there is a concept $f \in \mathcal{C}$ that contains all of S' and none of $S - S'$. In other words, for any of the $2^{|S|}$ possible labelings of S (where each example $s \in S$ is either positive or negative), there is some $f \in \mathcal{C}$ that realizes the desired labeling (see Figure 26.2). We can now define the VC-dimension of a concept class $\mathcal{C}$. The VC-dimension of $\mathcal{C}$, denoted VCD($\mathcal{C}$), is the smallest d for which no set of $d + 1$ examples is shattered by $\mathcal{C}$. Equivalently, VCD($\mathcal{C}$) is the cardinality of the largest finite set of points $S \subseteq X$ that is shattered by $\mathcal{C}$.

So to prove that VCD($\mathcal{C}$) $\geq d$ it suffices to give d examples that can be shattered. However, to prove VCD($\mathcal{C}$) $\leq d$ one must show that no set of $d + 1$ examples can be shattered. Since the VC-dimension is so fundamental in determining the sample complexity required for PAC learning, we now go through several sample computations of the VC-dimension.

Axis-parallel rectangles in $\mathfrak{R}^2$. For this concept class the points lying on or inside the target rectangle are positive, and points lying outside the target rectangle are negative. First, it is easily seen that there is a set of four points (e.g., $\{(0, 1), (0, -1), (1, 0), (-1, 0)\}$) that

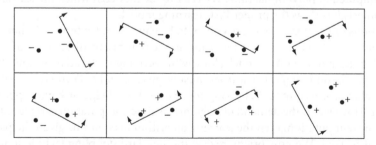

FIGURE 26.2 A demonstration that there are three points that are shattered by the class of two-dimensional halfspaces (i.e., $\mathcal{C}_{\text{halfspace}}$ with $n = 2$). Notice that all eight possible ways that the points can be classified as positive or negative can be realized.

---

* Chernoff bounds are used to compute the sample size so that the hypothesis testing procedure gives the desired output with high probability.

can be shattered. Thus $VCD(\mathcal{C}) \geq 4$. We now argue that no set of five points can be shattered. The smallest bounding axis-parallel rectangle defined by the five points is in fact defined by at most four of the points. For p a nondefining point in the set, we see that the set cannot be shattered since it is not possible for p to be classified as negative while also classifying the others as positive. Thus $VCD(\mathcal{C}) = 4$.

Halfspaces in $\Re^2$. Points lying in or on the halfspace are positive, and the remaining points are negative. It is easily shown that any three noncollinear points (e.g., $(0, 1), (0, 0), (1, 0)$) are shattered by $\mathcal{C}$ (recall Figure 26.2). Thus $VCD(\mathcal{C}) \geq 3$. We now show that no set of size four can be shattered by $\mathcal{C}$. If at least three of the points are collinear then there is no halfspace that contains the two extreme points but does not contain the middle points. Thus the four points cannot be shattered if any three are collinear. Next, suppose that the points form a quadrilateral. There is no halfspace which labels one pair of diagonally opposite points positive and the other pair of diagonally opposite points negative. The final case is that one point p is in the triangle defined by the other three. In this case there is no halfspace which labels p differently from the other three. Thus clearly the four points cannot be shattered. Therefore we have demonstrated that $VCD(\mathcal{C}) = 3$. Generalizing to halfspaces in $\Re^n$, it can be shown that $VCD(\mathcal{C}_{halfspace}) = n + 1$ (see [16]).

Closed sets in $\Re^2$. All points lying in the set or on the boundary of the set are positive, and all points lying outside the set are negative. Any set can be shattered by $\mathcal{C}$, since a closed set can assume any shape in $\Re^2$. Thus, the largest set that can be shattered by $\mathcal{C}$ is infinite, and hence $VCD(\mathcal{C}) = \infty$.

We now briefly discuss techniques that aid in computing the VC-dimension of more complex concept classes. Suppose we wanted to compute the VC-dimension of $\mathcal{C}_{halfspace}^{\cap_s}$, the class of intersections of up to s halfspaces over $\Re^d$. We would like to make use of our knowledge of the VC-dimension of $\mathcal{C}_{halfspace}$, the class of halfspaces over $\Re^d$. Blumer et al. [16] gave the following result: let $\mathcal{C}$ be a concept class with $VCD(\mathcal{C}) \leq D$. Then the class defined by the intersection of up to s concepts from $\mathcal{C}$ has VC-dimension* at most $2Ds\lg(3s)$. In fact, the above result applies when replacing intersection with any Boolean function. Thus the concept class $\mathcal{C}_{halfspace}^{\cap_s}$ (where each halfspace is defined over $\Re^d$) has VC-dimension at most $2(d + 1)s\lg(3s)$.

We now discuss the significance of the VC-dimension to the PAC model. One important property is that for $D = VCD(\mathcal{C})$, the number of different labelings that can be realized (also referred to as behaviors) using $\mathcal{C}$ for a set S is at most $\left(\frac{e|S|}{D}\right)^D$. Thus for a constant D we have polynomial growth in the number of labelings versus the exponential growth of $2^{|S|}$. A key result in the PAC model [16] is an upperbound on the sample complexity needed to PAC learn in terms of ϵ, δ, and the VC-dimension of the hypothesis class. They proved that one can design a learning algorithm A for concept class $\mathcal{C}$ using hypothesis space $\mathcal{H}$ in the following manner. Any concept $h \in \mathcal{H}$ consistent with a sample of size $\max\left(\frac{4}{\epsilon}\lg\frac{2}{\delta}, \frac{8\,VCD(\mathcal{H})}{\epsilon}\lg\frac{13}{\epsilon}\right)$ has error at most ϵ with probability at least $1 - \delta$. To obtain a polynomial-time PAC learning algorithm what remains is to solve the algorithmic problem of finding a hypothesis from $\mathcal{H}$ consistent with the labeled sample. As an example application, see Figure 26.3. Furthermore, Ehrenfencht and Haussler [22] proved an information-theoretic lower bound that learning any concept class $\mathcal{C}$ requires $\Omega\left(\frac{1}{\epsilon}\log\frac{1}{\delta} + \frac{VCD(\mathcal{C})}{\epsilon}\right)$ examples in the worst case.

A key limitation of this technique to design a PAC learning algorithm is that the hypothesis must be drawn from some fixed hypothesis class $\mathcal{H}$. In particular, the complexity of the hypothesis class

---

* Note that throughout this chapter, lg is used for the base-2 logarithm. When the base of the logarithm is not significant (such as when using asymptotic notation), we use log.

PAC-learn-halfspace(n, ϵ, δ)

Draw a labeled sample S of size max $\left(\frac{4}{\epsilon} \lg \frac{2}{\delta}, \frac{8(n+1)}{\epsilon} \lg \frac{13}{\epsilon} \right)$

Use a polynomial-time linear programming algorithm to find a halfspace h
consistent with S

Output h

FIGURE 26.3 A PAC algorithm to learn a concept from $\mathcal{C}_{\text{halfspace}}$. Recall that $\text{VCD}(\mathcal{C}_{\text{halfspace}}) = n + 1$.

must be independent of the sample size. However, often the algorithmic problem of finding such a hypothesis from the desired class is NP-hard.

As an example suppose we tried to modify PAC-learn-halfspace from Figure 26.3 to efficiently PAC learn the class of intersections of at most s halfspaces in $\mathfrak{R}^d$ for d the number of dimensions a constant. Suppose we are given a sample S of m example points labeled according to some $f \in \mathcal{C}_{\text{halfspace}}^{\cap s}$. The algorithmic problem of finding a hypothesis from $\mathcal{C}_{\text{halfspace}}^{\cap s}$ can be formulated as a set covering problem in the following manner. An instance of the set covering problem is a set $\mathcal{U}$ and a family $\mathcal{T}$ of subsets of $\mathcal{U}$ such that $\left(\bigcup_{t \in \mathcal{T}} t \right) = \mathcal{U}$. A solution is a smallest cardinality subset $\mathcal{G}$ of $\mathcal{T}$ such that $\left(\bigcup_{g \in \mathcal{G}} g \right) = \mathcal{U}$. Consider any halfspace g that correctly classifies all positive examples from S. We say that g *covers* all of the negative examples from S that are also classified as negative by g. Thus $\mathcal{U}$ is the set of negative examples from S, and $\mathcal{T}$ is the set of representative halfspaces (one for each behavior with respect to S) that correctly classify all positive examples from S.

So finding a minimum sized hypothesis from $\mathcal{C}_{\text{halfspace}}^{\cap s}$ consistent with the sample S is exactly the problem of finding the minimum number of halfspaces that are required to cover all negative points in S. Given that the set covering problem is NP-complete, how can we proceed? We can apply the greedy approximation algorithm that has a ratio bound of $\ln |\mathcal{U}| + 1$ to find a hypothesis consistent with S that is the intersection of at most $\ln |S| + 1$ halfspaces. However, since the VC-dimension of the hypothesis grows with the size of the sample, the basic technique described above cannot be applied. In general, when using a set covering approach, the size of the hypothesis often depends on the size of the sample.

Blumer et al. [15,16] extended this basic technique by showing that finding a hypothesis h consistent with a sample S for which the size of h is sublinear in $|S|$ is sufficient to guarantee PAC learnability. In other words, by obtaining sufficient data compression one obtains good generalization. More formally, let $\mathcal{H}_{k,n,m}^A$ be the hypothesis space used by algorithm A when each example has size n, the target concept has size k, and the sample size is m. We say that algorithm A is an Occam algorithm for concept class $\mathcal{C}$ if there exists a polynomial $p(k, n)$ and a constant β, $0 \leq \beta < 1$, such that for any sample S with $|S| = m \geq 1$ and any k, n, A outputs a hypothesis[*] h consistent with S such that $|h| \leq p(k, n)m^{\beta}$. Let A be an Occam algorithm for concept class $\mathcal{C}$ that has hypothesis space $\mathcal{H}_{k,n,m}^A$. If $\text{VCD}(\mathcal{H}_{k,n,m}^A) \leq p(k, n)(\lg m)^{\ell}$ (so $|h| \leq p(k, n)(\lg m)^{\ell}$) for some polynomial $p(k, n) \geq 2$ and $\ell \geq 1$, then A is a PAC learning algorithm for $\mathcal{C}$ using sample size

$$ m = \max \left(\frac{4}{\epsilon} \lg \frac{2}{\delta}, \frac{2^{\ell+4}p(k, n)}{\epsilon} \left(\lg \frac{8(2\ell + 2)^{\ell+1}p(k, n)}{\epsilon} \right)^{\ell+1} \right). $$

Figure 26.4 presents an **Occam algorithm** to PAC learn a concept from $\mathcal{C}_{\text{halfspace}}^{\cap s}$ (i.e., the intersection of at most s halfspaces over $\mathfrak{R}^d$). Since $\text{VCD}(\mathcal{C}_{\text{halfspace}}^{\cap s}) \leq 2(d + 1)s \lg(3s)$, and the hypothesis consists of the intersection of at most $s(1 + \ln m)$ halfspaces, it follows that $\text{VCD}(\mathcal{H}) \leq 2(d + 1)s(1 + \ln m) \lg(3s(1 + \ln m)) \leq 3ds \ln^2 m$. The size of the sample S then follows by noting that

---

[*] Recall that we use $|h|$ to denote the number of bits required to represent h.

> PAC-learn-intersection-of-halfspaces (d, s, ϵ, δ)
> Draw a labeled sample S of size $m = \max\left(\frac{4}{\epsilon}\lg\frac{2}{\delta}, \frac{192ds}{\epsilon}\left(6 + 4\lg3 + \lg\frac{ds}{\epsilon}\right)^3\right)$
> Form a set $\mathcal{T}$ of halfspaces that are consistent with the positive examples
> Let N be the negative examples from S
> Let h be the always true hypothesis
> Repeat until $N = \emptyset$
>   Find a halfspace $t \in \mathcal{T}$ consistent with the max. number of exs. from N
>   Let $h = h \cap t$
>   Remove the examples from N that were covered by t
> Output h

FIGURE 26.4 An Occam algorithm to PAC learn a concept from $\mathcal{C}_{\text{halfspace}}^{\cap_s}$.

$p(s, d) = 3ds$ and $\ell = 3$. By combining the fact that $\text{VCD}(\mathcal{C}_{\text{halfspace}}) = d + 1$ and the general upper bound of $\left(\frac{e|S|}{D}\right)^D$ on the number of behaviors on a sample of size $|S|$ for a class of VC-dimension D, we get that $|\mathcal{T}| \leq \left(\frac{em}{d+1}\right)^{d+1} = O(m^{d+1})$. (For d constant this quantity is polynomial in the size of the sample.) We can construct $\mathcal{T}$ in polynomial time by either using simple geometric arguments or the more general technique of Blumer et al. [16]. Finally, the time complexity of the greedy set covering algorithm is $O(\sum_{t \in \mathcal{T}} |t|) = O(m \cdot m^{d+1})$, which is polynomial in s (the number of halfspaces), ϵ, and δ for d (the number of dimensions) constant.

For the problem of learning the intersection of halfspaces the greedy covering technique provided substantial data compression. Namely, the size of our hypothesis only had a logarithmic dependence on the size of the sample. In general, only a sublinear dependence is required as given by the following result of Blumer et al. [15]. Let A be an Occam algorithm for concept class $\mathcal{C}$ that has hypothesis space $\mathcal{H}_{k,n,m}^A$. If $\text{VCD}(\mathcal{H}_{k,n,m}^A) \leq p(k, n)m^\beta$ (so $|h| \leq p(k, n)m^\beta$) for some polynomial $p(k, n) \geq 2$ and $\beta < 1$, then A is a PAC learning algorithm for $\mathcal{C}$ using sample size

$$m = \max\left(\frac{2}{\epsilon}\ln\frac{1}{\delta}, \left(\frac{2\ln 2}{\epsilon} \cdot p(k, n)\right)^{\frac{1}{1-\beta}}\right).$$

26.3.2 Models of Noise

The basic definition of PAC learning assumes that the data received is drawn randomly from $\mathcal{D}$ and properly labeled according to the target concept. Clearly, for learning algorithms to be of practical use they must be robust to noise in the training data. In order to theoretically study an algorithm's tolerance to noise, several formal models of noise have been studied. In the model of random classification noise [9], with probability $1 - \eta$, the learner receives the uncorrupted example (x, ℓ). However, with probability η, the learner receives the example $(x, \bar{\ell})$. So in this noise model, learner usually gets a correct example, but some small fraction η of the time the learner receives an example in which the label has been inverted. In the model of malicious classification noise [47], with probability $1 - \eta$, the learner receives the uncorrupted example (x, ℓ). However, with probability η, the learner receives the example (x, ℓ') in which x is unchanged, but the label ℓ' is selected by an adversary who has infinite computing power and has knowledge of the learning algorithm, the target concept, and the distribution $\mathcal{D}$. In the previous two models, only the labels are corrupted. Another noise model is that of malicious noise [49]. In this model, with probability $1 - \eta$, the learner receives the uncorrupted example (x, ℓ). However, with probability η, the learner receives an example (x', ℓ') about which no assumptions whatsoever may be made. In particular, this example (and its label) may be maliciously selected by an adversary. Thus in this model, the learner usually gets a correct example, but some small fraction η of the time the learner gets noisy examples and the nature of the noise is unknown.

We now define two noise models that are only defined when the instance space is $\{0, 1\}^n$. In the model of uniform random attribute noise [47], the example $(b_1 \cdots b_n, \ell)$ is corrupted by a random process that independently flips each bit b_i to $\bar{b}_i$ with probability η for $1 \leq i \leq n$. Note that the label of the "true" example is never altered. In this noise model, the attributes' values are subject to noise, but that noise is as benign as possible. For example, the attributes' values might be sent over a noisy channel, but the label is not. Finally, in the model of product random attribute noise [26], the example $(b_1 \cdots b_n, \ell)$ is corrupted by a random process of independently flipping each bit b_i to $\bar{b}_i$ with some fixed probability $\eta_i \leq \eta$ for each $1 \leq i \leq n$. Thus unlike the model of uniform random attribute noise, the noise rate associated with each bit of the example may be different.

26.3.3 Gaining Noise Tolerance in the PAC Model

Some of the first work on designing noise-tolerant PAC algorithms was done by Angluin and Laird [9]. They gave an algorithm for learning Boolean conjunctions that tolerates random classification noise of a rate approaching the information-theoretic barrier of $1/2$. Furthermore, they proved that the general technique of finding a hypothesis that minimizes disagreements with a sufficiently large sample allows one to handle random classification noise of any rate approaching $1/2$. However, they showed that even the very simple problem of minimizing disagreements (when there are no assumptions about the noise) is NP-hard. Until recently, there have been a small number of efficient noise-tolerant PAC algorithms, but no general techniques were available to design such algorithms, and there was little work to characterize which concept classes could be efficiently learned in the presence of noise.

The first (computationally feasible) tool to design noise-tolerant PAC algorithms was provided by the **statistical query model**, first introduced by Kearns [31], and since improved by Aslam and Decatur [12]. In this model, rather than sampling labeled examples, the learner requests the value of various statistics. A relative-error statistical query [12] takes the form $\mathbf{SQ}(\chi, \mu, \theta)$ where χ is a predicate over labeled examples, μ is a relative error bound, and θ is a threshold. As an example, let χ to be "$(h(x) = 1) \land (\ell = 0)$" which is true when x is a negative example but the hypothesis classifies x as positive. So the probability that χ is true for a random example is the false positive error of hypothesis h. For target concept f, we define $P_\chi = \mathrm{Pr}_{\mathcal{D}}[\chi(x, f(x))) = 1]$ where $\mathrm{Pr}_{\mathcal{D}}$ is used to denote that x is drawn at random from distribution $\mathcal{D}$. If $P_\chi < \theta$ then $\mathbf{SQ}(\chi, \mu, \theta)$ may return $\perp$. If $\perp$ is not returned, then $\mathbf{SQ}(\chi, \mu, \theta)$ must return an estimate $\hat{P}_\chi$ such that $P_\chi(1 - \mu) \leq \hat{P}_\chi \leq P_\chi(1 + \mu)$. The learner may also request unlabeled examples (since we are only concerned about classification noise).

Let's take our algorithm, `PAC-learn-intersection-of-halfspaces` and reformulate it as a relative-error statistical query algorithm. First we draw an unlabeled sample S_u that we use to generate the set $\mathcal{T}$ of halfspaces to use for our covering. Similar to before, we place a halfspace in $\mathcal{T}$ corresponding to each possible way in which the points of S_u can be divided. Recall that before, we only place a halfspace in $\mathcal{T}$ if it properly labeled the positive examples. Since we have an unlabeled sample we cannot use such an approach. Instead, we use a statistical query (for each potential halfspace) to check if a given halfspace is consistent with most of the positive examples. Finally, when performing the greedy covering step we cannot pick the halfspace that covers the most negative examples, but rather the one that covers the "most" (based on our empirical estimate) negative weight. Figure 26.5 shows our algorithm in more depth.

We now cover the key ideas in proving that `SQ-learn-intersection-of-halfspaces` is correct. First, we pick S_u so that any hypothesis consistent with S_u (if we knew the labels) would have error at most $\frac{\epsilon}{6r}$ with probability $1 - \frac{\delta}{2}$. Since our hypothesis class is $\mathcal{C}_{\text{halfspace}}^{\cap r}$, and $\mathrm{VCD}(\mathcal{C}_{\text{halfspace}}^{\cap r}) \leq 2(d + 1)r\lg(3r)$, we obtain the sample size for S_u.

```
SQ-learn-intersection-of-halfspaces(d, s, ε)
    Let r = 2s ln(3/ε)
    Draw an unlabeled sample S_u = max { (24r/ε) lg (4/δ), (96(d+1)r²lg(3r)/ε) lg (78r/ε) }
    Form the set T of halfspaces - one for each linear separation of S_u into two sets
    T_good = ∅
    For each t ∈ T
        P̂_X = SQ (t(x) = 0 ∧ ℓ = 1, 1/2, ε/2r)
        If P̂_X ≤ ε/4r or ⊥ Then T_good = T_good ∪ t

    Let h be the always true hypothesis
    While SQ (h(x) = 1 ∧ ℓ = 0, 1/9, ε/2) > 4ε/9 and the loop has been executed < r times
        t_max = argmax_{t∈T_good} SQ (h(x) = 1 ∧ ℓ = 0 ∧ t(x) = 0, 1/3, 7ε/30s)
        h = h ∩ t_max
    Output h
```

FIGURE 26.5 A relative-error statistical query algorithm to learn a concept from $C_{halfspace}^{\cap_s}$ over $\Re^d$.

For each of the s halfspaces that form the target concept, there is some halfspace in T consistent with that halfspace over S_u, and thus that has error at most $\epsilon/(6r)$ on the positive region. For each such halfspace t our estimate $\hat{P}_X$ (for the probability that $t(x) = 0$ and $\ell = 1$) is at most $\frac{\epsilon}{6r} \cdot \frac{3}{2} = \frac{\epsilon}{4r}$. Thus, we place s halfspaces in T_{good} that produce a hypothesis with false negative error $\leq \epsilon/(6r)$. Let $e_i = \Pr[h(x) = 1 \wedge \ell = 0]$ (i.e., the false positive error) after i rounds of the while loop. Since e_i is our current error and $\epsilon/(6r)$ is a lower bound on the error we could achieve, by an averaging argument it follows that there is a $t \in T_{good}$ for which

$$P = \Pr[(h(x) = 1) \wedge (\ell = 0) \wedge (t(x) = 0)] \geq \left(e_i - \frac{\epsilon}{6r}\right) \cdot \frac{1}{s} \geq \left(e_i - \frac{\epsilon}{6}\right) \cdot \frac{1}{s}.$$

Thus for the halfspace t_{max} selected to add to h we know that the statistical query returns an estimate $\hat{P} \geq \frac{2}{3} \cdot P \geq \frac{2}{3s}\left(e_i - \frac{\epsilon}{6}\right)$. Thus for t_{max} we know that $P \geq \frac{3}{4} \cdot \frac{2}{3s}\left(e_i - \frac{\epsilon}{6}\right) = \frac{1}{2s}\left(e_i - \frac{\epsilon}{6}\right)$. Solving the recurrence $e_{i+1} \leq e_i - \frac{1}{2s}\left(e_i - \frac{\epsilon}{6}\right)$ (with $e_i \leq 1$) yields that $e_i \leq \frac{\epsilon}{6} + \left(1 - \frac{1}{2s}\right)^i$. Picking $r = i = 2s \ln \frac{3}{\epsilon}$ suffices to ensure that $e_i \leq \epsilon/2$ as desired.

Once $e_i \leq 2\epsilon/5$, we exit the while loop (since $\frac{2\epsilon}{5} \cdot \frac{10}{9} = \frac{4\epsilon}{9}$). Thus we enter the loop only when $e_i > 2\epsilon/5$, and hence for t_{max}, $P \geq \frac{1}{s}(e_i - \frac{\epsilon}{6})$. So we choose $\theta = \frac{1}{s}(\frac{2\epsilon}{5} - \frac{\epsilon}{6}) = \frac{7\epsilon}{30s}$. Finally, when we exit the loop $\Pr[h(x) = 1 \wedge \ell = 0] \leq \frac{4\epsilon}{9} \cdot \frac{9}{8} = \frac{\epsilon}{2}$. Thus the total error is at most ϵ as desired.

Notice that in the statistical query model there is not a confidence parameter δ. This is because the SQ oracle guarantees that its estimates meet the given requirements. However, when we use random labeled examples to simulate the statistical queries, we can only guarantee that with high probability the estimates meet their requirements. Thus the results on converting an SQ algorithm into a PAC algorithm reintroduce the confidence parameter δ.

By applying uniform convergence results and Chernoff bounds it can be shown that if one draws a sufficiently large sample then a statistical query can be estimated by the fraction of the sample that satisfies the predicate. For example, in the relative-error SQ model with a set Q of possible queries, Aslam and Decatur [12] show that a sample of size $O\left(\frac{1}{\mu^2\theta} \log \frac{|Q|}{\delta}\right)$ suffices to appropriately answer $[\chi, \mu, \theta]$ for every $\chi \in Q$ with probability at least $1 - \delta$. (If VCD(Q) $= q$ then a sample of size $O\left(\frac{q}{\mu^2\theta} \log \frac{1}{\mu\theta} + \frac{1}{\mu^2\theta} \log \frac{1}{\delta}\right)$ suffices.)

To handle random classification noise of any rate approaching $1/2$ more complex methods are used for answering the statistical queries. Roughly, using knowledge of the noise process and a sufficiently accurate estimate of the noise rate (which must itself be determined by the algorithm), the noise process can be "inverted." The total sample complexity required to simulate an SQ algorithm in the presence of classification noise of rate $\eta \leq \eta_b$ is $\tilde{O}\left(\frac{\log(|\mathcal{Q}|/\delta)}{\mu_*^2 \theta_* \rho(1-2\eta_b)^2}\right)$ where μ_* (respectively, θ_*) is the minimum value of μ (respectively, θ) across all queries and $\rho \in [\theta_*, 1]$, The soft-oh notation ($\tilde{O}$) is similar to the standard big-oh notation except log factors are also removed. Alternatively, for the query space $\mathcal{Q}$, the sample complexity is $O\left(\frac{\text{VCD}(\mathcal{Q})}{\mu_*^2 \theta_* \rho_*(1-2\eta_b)^2}\left[\log\left(\frac{1}{\mu_* \theta_* \rho_*(1-2\eta_b)}\right) + \log\frac{1}{\delta}\right]\right)$. Notice that the amount by which η_b is less than $1/2$ is just $\frac{1}{1/2-\eta_b} = \frac{2}{1-2\eta_b}$. Thus the above are polynomial as long as $\frac{1}{2} - \eta_b$ is at least one over a polynomial.

26.4 Exact and Mistake Bounded Learning Models

The PAC learning model is a batch model—there is a separation between the training phase and the performance phase. In the training phase the learner is presented with labeled examples—no predictions are expected. Then at the end of the training phase the learner must output a hypothesis h to classify unseen examples. Also, since the learner never finds out the true classification for the unlabeled instances, all learning occurs in the training phase. In many settings, the learner does not have the luxury of a training phase but rather must learn as it performs. We now study two closely related learning models designed for such a setting.

26.4.1 Online Learning Model

To motivate the online learning model (also known as the mistake-bounded learning model), suppose that when arriving at work (in Boston) you may either park in the street or park in a garage. In fact, between your office building and the garage there is a street on which you can always find a spot. On most days, street parking is preferable since you avoid paying the $15 garage fee. Unfortunately, when parking on the street you risk being towed ($75) due to street cleaning, snow emergency, special events, etc. When calling the city to find out when they tow, you are unable to get any reasonable guidance and decide the best thing to do is just learn from experience. There are many pieces of information that you might consider in making your prediction, e.g., the date, the day of the week, the weather. We make the following assumption: after you commit yourself to one choice or the other you learn of the right decision. In this example, the city has rules dictating when they tow; you just don't know them. If you park on the street, at the end of the day you know if your car was towed; otherwise when walking to the garage you see if the street is clear (i.e., you learn if you would have been towed). The online model is designed to study algorithms for learning to make accurate predictions in circumstances such as these. Note that unlike the problems addressed by many techniques from reinforcement learning, here there is immediate (versus delayed) feedback.

We now define the online learning model for the general setting in which the target function has a real-valued output (without loss of generality, scaled to be between 0 and 1). An online learning algorithm for C is an algorithm that runs under the following scenario. A learning session consists of a set of trials. In each trial, the learner is given an unlabeled instance $x \in \mathcal{X}$. The learner uses its current hypothesis to predict a value $p(x)$ for the unknown (real-valued) target concept $f \in C$ and then the learner is told the correct value for $f(x)$. Several loss functions have been considered to measure the quality of the learner's predictions. Three commonly used loss functions are the following: the square loss defined by $\ell_2(p(x), f(x)) = (f(x) - p(x))^2$, the log loss defined by $\ell_{\log}(p(x), f(x)) = -f(x)\log p(x) - (1 - f(x))\log(1 - p(x))$, and the absolute loss defined by $\ell_1(p(x), f(x)) = |f(x) - p(x)|$.

The goal of the learner is to make predictions so that total loss over all predictions is minimized. In this learning model, most often a worst-case model for the environment is assumed. There is some known concept class from which the target concept is selected. An adversary (with unlimited computing power and knowledge of the learner's algorithm) then selects both the target function and the presentation order for the instances. In this model there is no training phase. Instead, the learner receives unlabeled instances throughout the entire learning session. However, after each prediction the learner "discovers" the correct value. This feedback can then be used by the learner to improve its hypothesis.

We now discuss the special case when the target function is Boolean and correspondingly, predictions must be either 0 or 1. In this special case the loss function most commonly used is the absolute loss. Notice that if the prediction is correct then the value of the loss function is 0, and if the prediction is incorrect then the value of the loss function is 1. Thus the total loss of the learner is exactly the number of prediction mistakes. Thus, in the worst-case model we assume that an adversary selects the order in which the instances are presented to the learner and we evaluate the learner by the maximum number of mistakes made during the learning session. Our goal is to minimize the worst-case number of mistakes using an efficient learning algorithm (i.e., each prediction is made in polynomial time). Observe that such mistake bounds are quite strong in that the order in which examples are presented does not matter; however, it is impossible to tell how early the mistakes will occur. Littlestone [37] has shown that in this learning model $VCD(\mathcal{C})$ is a lower bound on the number of prediction mistakes.

26.4.1.1 Handling Irrelevant Attributes

Here we consider the common scenario in which there are many attributes the learner could consider, yet the target concept depends on a small number of them. Thus most of the attributes are irrelevant to the target concept. We now briefly describe one early algorithm, Winnow of [37], that handles a large number of irrelevant attributes. More specifically, for Winnow, the number of mistakes only grows logarithmically with the number of irrelevant attributes. Winnow (or modifications of it) have many nice features such as noise tolerance and the ability to handle the situation in which the target concept is changing. Also, Winnow can directly be applied to the agnostic learning model [35]. In the agnostic learning model no assumptions are made about the target concept. In particular, the learner is unaware of any concept class that contains the target concept. Instead, we compare the performance of an agnostic algorithm (typically in terms of the number of mistakes or based on some other loss function) to the performance of the best hypothesis selected from a comparison or "touchstone" class where the best hypothesis from the touchstone class is the one that incurs the minimum loss over all functions in the touchstone class.

Winnow is similar to the classical perceptron algorithm [45], except that it uses a multiplicative weight-update scheme that permits it to perform much better than classical perceptron training algorithms when many attributes are irrelevant. The basic version of Winnow is designed to learn the concept class of a linearly separable Boolean function, which is a map $f : \{0, 1\}^n \to \{0, 1\}$ such that there exists a hyperplane in $\mathfrak{R}^n$ that separates the inverse images $f^{-1}(0)$ and $f^{-1}(1)$ (i.e., the hyperplane separates the points on which the function is 1 from those on which it is 0). An example of a linearly separable function is any monotone disjunction: if $f(x_1, \ldots, x_n) = x_{i_1} \vee \cdots \vee x_{i_k}$, then the hyperplane $x_{i_1} + \cdots + x_{i_k} = 1/2$ is a separating hyperplane. For each attribute x_i there is an associated weight w_i. There are two parameters, θ which determines the threshold for predicting 1 (positive), and α which determines the adjustment made to the weight of an attribute that was partly responsible for a wrong prediction. The pseudocode for Winnow is shown in Figure 26.6.

Winnow has been successfully applied to many learning problems. It has the very nice property that one can prove that the number of mistakes only grows logarithmically in the number of variables (and linearly in the number of relevant variables). For example, it can be shown that for the learning of monotone disjunctions of at most k literals, if Winnow is run with $\alpha > 1$ and $\theta \geq 1/\alpha$, then the total number of mistakes is at most $\alpha k(\log_\alpha \theta + 1) + n/\theta$. Observe that Winnow need not have

```
Winnow(θ, α)
    For i = 1,...,n, initialize w_i = 1
    To predict the value of x = (x_1,...,x_n) ∈ X:
        If ∑_{i=1}^{n} w_i x_i > θ, then predict 1
        Else predict 0
    Let ρ be the correct prediction
    If the prediction was 1 and ρ = 0 then
        If x_i = 1 then let w_i = w_i/α
    If the prediction was 0 and ρ = 1 then
        If x_i = 1 then let w_i = w_i · α
```

FIGURE 26.6 The algorithm `winnow`.

prior knowledge of k although the number of mistakes depends on k. Littlestone [37] showed how to optimally choose θ and α if an upperbound on k is known a priori. Also, while `Winnow` is designed to learn a linearly separable class, reductions (discussed in "Prediction-Preserving Reductions") can be used to apply `Winnow` to classes for which the positive and negative points are not linearly separable, e.g., k-DNF.

26.4.1.2 The Halving Algorithm and Weighted Majority Algorithm

We now discuss some of the key techniques for designing good online learning algorithms for the special case of concept learning (i.e., learning Boolean-valued functions). If one momentarily ignores the issue of computation time, then the halving algorithm [37] performs very well. It works as follows. Initially all concepts in the concept class $\mathcal{C}$ are candidates for the target concept. To make a prediction for instance x, the learner takes a majority vote based on all remaining candidates (breaking a tie arbitrarily). Then when the feedback is received, all concepts that disagree are removed from the set of candidates. It can be shown that at each step the number of candidates is reduced by a factor of at least 2. Thus, the number of prediction mistakes made by the halving algorithm is at most $\lg|\mathcal{C}|$.

Clearly, the halving algorithm will perform poorly if the data is noisy. We now briefly discuss the weighted majority algorithm [38], which is one of several multiplicative weight-update schemes for generalizing the halving algorithm to tolerate noise. Also, the weighted majority algorithm provides a simple and effective method for constructing a learning algorithm A that is provided with a pool of "experts," one of which is known to perform well, but A does not know which one. Associated with each expert is a weight that gives A's confidence in the accuracy of that expert. When asked to make a prediction, A predicts by combining the votes from its experts based on their associated weights. When an expert suggests the wrong prediction, A passes that information to the given expert and reduces its associated weight using a multiplicative weight-updating scheme. Namely, the weight associated with each expert that mispredicts is multiplied by some weight $0 \leq \beta < 1$. By selecting $\beta > 0$ this algorithm is robust against noise in the data. Figure 26.7 shows the weighted majority algorithm in more depth.

We now briefly discuss some learning problems in which weighted majority is applicable. Suppose one knows that the correct prediction comes from some target concept selected from a known concept class. Then one can apply the weighted majority algorithm where each concept in the class is one of the algorithms in the pool. For such situations, the weighted majority algorithm is a robust generalization of the halving algorithm. (In fact, the halving algorithm corresponds to the special case where $\beta = 0$.) As another example, the weighted majority algorithm can often be applied to help in situations in which the prediction algorithm has a parameter that must be selected and the best choice for the parameter depends on the target. In such cases one can build the pool of algorithms by choosing various values for the parameter.

```
Weighted Majority Algorithm (WM)
    Let wᵢ be the weight associated with algorithm Aᵢ
    For 1 ≤ i ≤ n
        Let wᵢ = 1
    To make a prediction for an instance x ∈ X:
        Let 𝒜₀ be the set of algorithms that predicted 0
        Let 𝒜₁ be the set of algorithms that predicted 1
        Let q₀ = Σ_{Aᵢ∈𝒜₀} wᵢ
        Let q₁ = Σ_{Aᵢ∈𝒜₁} wᵢ
        Predict 0 iff q₀ ≥ q₁
    If a mistake is made:
        Let γ be the prediction
        For 1 ≤ i ≤ n
            If Aᵢ's prediction for x is γ
                Let wᵢ = wᵢ · β (where 0 ≤ β < 1)
                Inform Aᵢ of the mistake
```

FIGURE 26.7 The weighted majority algorithm.

We now describe some of the results known about the performance of the weighted majority algorithm. If the best algorithm in the pool $\mathcal{A}$ makes at most m mistakes, then the worst-case number of mistakes made by the weighted majority algorithm is $O(\log|\mathcal{A}| + m)$ where the constant hidden within the big-oh notation depends on β. Specifically, the number of mistakes made by the weighted majority algorithm is at most $\dfrac{\log|\mathcal{A}| + m\log\frac{1}{\beta}}{\log\frac{2}{1+\beta}}$ if one algorithm in $\mathcal{A}$ makes at most m mistakes, $\dfrac{\log\frac{|\mathcal{A}|}{k} + m\log\frac{1}{\beta}}{\log\frac{2}{1+\beta}}$ if there exists a set of k algorithms in $\mathcal{A}$ such that each algorithm makes at most m mistakes, and $\dfrac{\log\frac{|\mathcal{A}|}{k} + \frac{m}{k}\log\frac{1}{\beta}}{\log\frac{2}{1+\beta}}$ if the total number of mistakes made by a set of k algorithms in $\mathcal{A}$ is m.

When $|\mathcal{A}|$ is not polynomial, the weighted majority algorithm (when directly implemented), is not computationally feasible. Recently, Mass and Warmuth [40] introduced what they call the virtual weight technique to implicitly maintain the exponentially large set of weights so that the time to compute a prediction and then update the "virtual" weights is polynomial. More specifically, the basic idea is to simulate Winnow by grouping concepts that "behave alike" on seen examples into blocks. For each block only one weight has to be computed and one constructs the blocks so that the number of concepts combined in each block as well as the weight for the block can be efficiently computed. While the number of blocks increases as new **counterexamples** are received, the total number of blocks is polynomial in the number of mistakes. Thus all predictions and updates can be performed in time polynomial in the number of blocks, which is in turn polynomial in the number of prediction mistakes of Winnow. Many variations of the basic weighted majority algorithm have also been studied. The results of Cesa-Bianchi et al. [19] demonstrate how to tune β as a function of an upper bound on the noise rate.

26.4.2 Query Learning Model

A very well-studied formal learning model is the membership and equivalence query model developed by Angluin [3]. In this model (often called the exact learning model) the learner's goal is to learn exactly how an unknown (Boolean) target function f, taken from some known concept class $\mathcal{C}$, classifies all instances from the domain. This goal is commonly referred to as exact identification. The learner has available two types of queries to find out about f: one is a **membership query**, in which the learner supplies an instance x from the domain and is told $f(x)$. The other query provided is an **equivalence query** in which the learner presents a candidate function h and either is told that $h \equiv f$ (in which case learning is complete), or else is given a counterexample x for which $h(x) \neq f(x)$. There is a very close relationship between this learning model and the online learning model (supplemented with membership queries) when applied to a classification problem. Using a standard transformation [3,37], algorithms that use membership and equivalence queries can easily be converted to online learning algorithms that use membership queries. Under such a transformation the number of counterexamples provided to the learner in response to the learner's equivalence queries directly corresponds to the number of mistakes made by the online algorithm.

In this model a number of interesting polynomial time algorithms are known for learning deterministic finite automata [2], Horn sentences [5], read-once formulas [6], read-twice DNF formulas [1], decision trees [18], and many others. It is easily shown that membership queries alone are not sufficient for efficient learning of these classes, and Angluin has developed a technique of "approximate fingerprints" to show that equivalence queries alone are also not enough [4]. (In both cases the arguments are information theoretic, and hold even when the computation time is unbounded.) The work of Bshouty et al. [17] extended Angluin's results to establish tight bounds on how many equivalence queries are required for a number of these classes. Maass and Turán studied upper and lower bounds on the number of equivalence queries required for learning (when computation time is unbounded), both with and without membership queries [39].

It is known that any class learnable exactly from equivalence queries can be learned in the PAC setting [3]. At a high level the exact learning algorithm is transformed to a PAC algorithm by having the learner use random examples to "search" for a counterexample to the hypothesis of the exact learner. If a counterexample is found, it is given as a response to the equivalence query. Furthermore, if a sufficiently large sample was drawn and no counterexample was found then the hypothesis has error at most ϵ (with probability at least $1 - \delta$). The converse does not hold [13]. That is, there are concept classes that are efficiently PAC learnable but cannot be efficiently learned in the exact model.

We now describe Angluin's algorithm for learning monotone DNF formulas (see Figure 26.8). A prime implicant t of a formula f is a satisfiable conjunction of literals such that t implies f but no proper subterm of t implies f. For example, $f = (a \wedge c) \vee (b \wedge \bar{c})$ has prime implicants $a \wedge c$, $b \wedge \bar{c}$, and $a \wedge b$. The number of prime implicants of a general DNF formula may be exponentially larger than the number of terms in the formula. However, for monotone DNF the number of prime implicants is no greater than the number of terms in the formula. The key to the analysis is to show that at each iteration, the term t is a new prime implicant of f, the target concept. Since the loop iterates exactly once for each prime implicant there are at most m counterexamples where m is the number of terms in the target formula. Since at most n membership queries are performed during each iteration there are at most nm membership queries overall.

```
Learn-Monotone-DNF
    h ← false
    Do forever
        Make equivalence query with h
        If "yes," output h and halt
        Else let x = b₁b₂ ··· bₙ be the counterexample
            Let t = ⋀_{bᵢ=1} yᵢ
            For i = 1,...,n
                If bᵢ = 1 perform membership query on x with iᵗʰ bit flipped
                If "yes," t = t \ {yᵢ} and x = b₁ ··· b̄ᵢ ··· bₙ
            Let h = h ∨ t
```

FIGURE 26.8 An algorithm that uses membership and equivalence queries to exactly learn an unknown monotone DNF formula over the domain $\{0, 1\}^n$. Let $\{y_1, y_2, \ldots, y_n\}$ be the Boolean variables and let h be the learner's hypothesis.

26.5 Hardness Results

In order to understand what concept classes are learnable, it is essential to develop techniques to prove when a learning problem is hard. Within the learning theory community, there are two basic type of hardness results that apply to all of the models discussed here. There are **representation-dependent hardness results** in which one proves that one cannot efficiently learn C using a hypothesis class of H. These hardness results typically rely on some complexity theory assumption* such as RP $\neq$ NP. For example, given that RP $\neq$ NP, Pitt and Warmuth [43] showed that k-term-DNF is not learnable using the hypothesis class of k-term-DNF. While such results provide some information, what one would really like to obtain is a hardness result for learning a concept class using any reasonable (i.e., polynomially evaluatable) hypothesis class. (For example, while we have a representation-dependent hardness result for learning k-term-DNF, there is a simple algorithm to PAC learn the class of k-term-DNF formulas using a hypothesis class of k-CNF formulas.) **Representation-independent hardness results** meet this more stringent goal. However, they depend on cryptographic (versus complexity theoretic) assumptions. Kearns and Valiant [32] gave representation-independent hardness results for learning several concept classes such as Boolean formulas, deterministic finite automata, and constant-depth threshold circuits (a simplified form of "neural networks"). These hardness results are based on assumptions regarding the intractability of various cryptographic schemes such as factoring Blum integers and breaking the RSA function.

26.5.1 Prediction-Preserving Reductions

Given that we have some representation-independent hardness result (assuming the security of various cryptographic schemes) one would like a "simple" way to prove that other problems are hard in a similar fashion as one proves a desired algorithm is intractable by reducing a known NP-complete problem to it.[†] Such a complexity theory for predictability has been provided by Pitt and Warmuth [43]. They formally define a prediction preserving reduction from concept class C over domain X to concept class C' over domain X' (denoted by $C \leq C'$) that allows an efficient learning algorithm for C' to be used to obtain any efficient learning algorithm for C. The requirements for such a prediction-preserving reduction are (1) an efficient instance transformation g from X to X', and (2) the existence of an image concept. The instance transformation g must be polynomial time computable. Hence, if $g(x) = x'$ then the size of x' must be polynomially related to the size of x. So for $x \in X_n$, $g(x) \in X_{p(n)}$ where $p(n)$ is some polynomial function of n. It is also important that g be independent of the target function. We now define what is meant by the existence of an image concept. For every $f \in C_n$ there must exist some $f' \in C'_{p(n)}$ such that for all $x \in X_n$, $f(x) = f'(g(x))$ and the number of bits to represent f' is polynomially related to the number of bits to represent f.

As an example, let C be the class of DNF formulas over $X = \{0, 1\}^n$, and C' be the class of monotone DNF formulas over $X' = \{0, 1\}^{2n}$. We now show that $C \leq C'$. Let $y_1, \ldots, y_n$ be the variables for each concept from C. Let $y'_1, \ldots, y'_{2n}$ be the variables for C'. The intuition behind the reduction is that variable y_i for the DNF problem is associated with variable y_{2i-1} for the monotone DNF problem. And variable $\overline{y}_i$ for the DNF problem is associated with variable y_{2i} for the monotone DNF problem. So for example $x = b_1 b_2 \cdots b_n$ where each $b_i \in \{0, 1\}$ and $g(x) = b_1(1 - b_1)b_2(1 - b_2) \cdots b_n(1 - b_n)$. Given a target concept $f \in C$, the image concept f' is obtained by replacing each nonnegated variable y_i from f with y'_{2i-1} and each negated variable $\overline{y}_i$ from f with y'_{2i}. It is easily confirmed that all required conditions are met.

---

* For background on complexity classes see Chapter 32 (NP defined) of this handbook and Chapter 20 (RP defined) of *Algorithms and Theory of Computation Handbook, Second Edition: Special Topics and Techniques.*

† See Chapter 34 of this handbook for background on reducibility and completeness.

If $C \leq C'$, what implications are there with respect to learnability? Observe that if we are given a polynomial prediction algorithm A' for C', one can use A' to obtain a polynomial prediction algorithm A for C as follows. If A' requests a labeled example then A can obtain a labeled example $x \in \mathcal{X}$ from its oracle and give $g(x)$ to A'. Finally, when A' outputs hypothesis h', A can make a prediction for $x \in \mathcal{X}$ using $h(g(x))$. Thus if C is known not to be learnable then C' also is not learnable. Thus the reduction given above implies that the class of monotone DNF formulas is just as hard to learn in the PAC model as the class of arbitrary DNF formulas. Equivalently, if there were an efficient PAC algorithm for the class of monotone DNF formulas, then there would also be an efficient PAC algorithm for arbitrary DNF formulas.

Pitt and Warmuth [43] gave a prediction preserving reduction from the class of Boolean formulas to class of DFAs. Thus since Kearns and Valiant [32] proved that Boolean formula are not predictable (under cryptographic assumptions), it immediately follows that DFAs are not predictable. In other words, DFAs cannot be efficiently learned from random examples alone. Recall that any algorithm to exactly learn using only equivalence queries can be converted into an efficient PAC algorithm. Thus if DFAs are not efficiently PAC learnable (under cryptographic assumptions), it immediately follows that DFAs are not efficiently learnable from only equivalence queries (under cryptographic assumptions). Contrasting this negative result, recall that DFAs are exactly learnable from membership and equivalence queries [2], and thus are PAC learnable with membership queries.

Notice that for $C \leq C'$, the result that an efficient learning algorithm for C' also provides an efficient algorithm for C relies heavily on the fact that membership queries are NOT allowed. The problem created by membership queries (whether in the PAC or exact model) is that algorithm A' for C' may make a membership query on an example $x' \in \mathcal{X}'$ for which $g^{-1}(x')$ is not in $\mathcal{X}$. For example, the reduction described earlier demonstrates that if there is an efficient algorithm to PAC learn monotone DNF formulas then there is an efficient algorithm to PAC learn DNF formulas. Notice that we already have an algorithm Learn-Monotone-DNF that exactly learns this class with membership and equivalence queries (and thus can PAC learn the class when provided with membership queries). Yet, the question of whether or not there is an algorithm with access to a membership oracle to PAC learn DNF formulas remains one of the biggest open questions within the field. We now describe the problem with using the algorithm Learn-Monotone-DNF to obtain an algorithm for general DNF formulas. Suppose that we have four Boolean variables (thus the domain is $\{0, 1\}^4$). The algorithm Learn-Monotone-DNF could perform a membership query on the example 00100111. While this example is in the domain $\{0, 1\}^8$ there is no $x \in \{0, 1\}^4$ for which $g(x) = 00100111$. Thus the provided membership oracle for the DNF problem cannot be used to respond to the membership query posed by Learn-Monotone-DNF.

We now define a more restricted type of reduction $C \leq_{wmq} C'$ that yields results even when membership queries are allowed [7]. For these reductions, we just add the following third condition to the two conditions already described: for all $x' \in \mathcal{X}'$, if x' is not in the image of g (i.e., there is no $x \in \mathcal{X}$ such that $g(x) = x'$), then the classification of x' for the image concept must always be positive or always be negative. As an example, we show that $C \leq_{wmq} C'$ where C is the class of DNF formulas and C' is the class of read-thrice DNF formulas (meaning that each literal can appear at most three times). Thus learning a read-thrice DNF formula (even with membership queries) is as hard as learning general DNF formula (with membership queries). Let s be the number of literals in f. (If s is not known a priori, the standard doubling technique can be applied.) The mapping g maps from an $x \in \{0, 1\}^n$ to an $x' \in \{0, 1\}^{sn}$. More specifically, $g(x)$ simply repeats, s times, each bit of x. To see that there is an image concept, note that we have s variables for each concept in C' associated with each variable for a concept in C. Thus we can rewrite $f \in C$ as a formula $f' \in C'$ in which each variable only appears once. At this point we have shown $C \leq C'$ but still need to do more to satisfy the new third condition. We want to ensure that the s variables (call them $\ell'_1, \ldots, \ell'_s$) for f' associated with one literal ℓ_i of f all take the same value. We do this by defining our final image concept as: $f' \wedge \Gamma_1 \wedge \cdots \wedge \Gamma_n$ where n is the number of variables and Γ_i is of the form

$(\ell'_1 \to \ell'_2) \wedge \cdots \wedge (\ell'_{s-1} \to \ell'_s) \wedge (\ell'_s \to \ell'_1)$. This formula evaluates to true exactly when $\ell'_1, \ldots, \ell'_s$ have the same value. Thus an x' for which no $g(x) = x'$ will not satisfy some Γ_i and thus we can respond "no" to a membership query on x'. Finally, f' is a read-thrice DNF formula. This reduction also proves that Boolean formulas $\leq_{wmq}$ read-thrice Boolean formulas.

26.6 Weak Learning and Hypothesis Boosting

As originally defined, the PAC model requires the learner to produce, with arbitrarily high probability, a hypothesis that is arbitrarily close to the target concept. While for many problems it is easy to find simple algorithms ("rules-of-thumb") that are often correct, it seems much harder to find a single hypothesis that is highly accurate. Informally, a weak learning algorithm is one that outputs a hypothesis that has some advantage over random guessing. (To contrast this sometimes a PAC learning algorithm is called a strong learning algorithm.) This motivates the question: are there concept classes for which there is an efficient weak learner, but there is no efficient PAC learner? Somewhat surprisingly the answer is no [46]. The technique used to prove this result is to take a weak learner and transform (boost) it into a PAC learner. The general method of converting a rough rule-of-thumb (weak learner) into a highly accurate prediction rule (PAC learner) is referred to as **hypothesis boosting**.*

We now formally define the weak learning model [32,46]. As in the PAC model, there is the instance space $\mathcal{X}$ and the concept class $\mathcal{C}$. Also the examples are drawn randomly and independently according to a fixed but unknown distribution $\mathcal{D}$ on $\mathcal{X}$. The learner's hypothesis h must be a polynomial time function that given an $x \in \mathcal{X}$ returns a prediction of $f(x)$ for $f \in \mathcal{C}$, the unknown target concept. The accuracy requirements for the hypothesis of a weak learner are as follows. There is a polynomial function $p(n)$ and algorithm A such that, for all $n \geq 1, f \in \mathcal{C}_n$, for all distributions $\mathcal{D}$, and for all $0 < \delta \leq 1$, algorithm A, given n, δ, and access to labeled examples from $\mathcal{D}$, outputs a hypothesis h such that, with probability $\geq 1 - \delta$, $error_\mathcal{D}(h)$ is at most $(1/2 - 1/p(n))$. Algorithm A should run in time polynomial in n and $1/\delta$.

If $\mathcal{C}$ is strongly learnable, then it is weakly learnable—just fix $\epsilon = 1/4$ (or any constant less than $1/2$). The converse (weak learnability implying strong learnability) is not at all obvious. In fact, if one restricts the distributions under which the weak learning algorithm runs then weak learnability does not imply strong learnability. Namely, Kearns and Valiant [32] have shown that under a uniform distribution, monotone Boolean functions are weakly, but not strongly, learnable. Thus it is important to take advantage of the requirement that the weak learning algorithm must work for all distributions. Using this property, Schapire [46] proved the converse result: if concept class $\mathcal{C}$ is weakly learnable, then it is strongly learnable.

Proving that weak learnability implies strong learnability has also been called the hypothesis boosting problem, because a way must be found to boost the accuracy of slightly-better-than-half hypothesis to be arbitrarily close to 1. There have been several boosting algorithms proposed since the original work of Schapire [46]. Figure 26.9 describes AdaBoost [23] that has shown promise in empirical use. The key to forcing the weak learner to output hypothesis that can be combined to create a highly accurate hypothesis is to create different distributions on which the weak learner is trained.

---

* We note that one can easily boost the confidence δ by first designing an algorithm A that works for say $\delta = 1/2$ and then running A several times taking a majority vote. For an arbitrary $\delta > 0$ the number of runs of A needed are polynomial in $\lg 1/\delta$.

```
AdaBoost
    For i = 1,..., m
        D₁(i) = 1/m
    For t = 1,..., T
```
Call WeakLearn providing it with the distribution D_t to obtain the hypothesis h_t

Calculate the error of $h_t : \epsilon_t = \sum_{i:h_t(x_i) \neq y_i} D_t(i)$

If $\epsilon_t > 1/2$, set $T = t - 1$ and abort the loop

Let $\beta_t = \epsilon_t / (1 - \epsilon_t)$

Update distribution D_t:

    Let Z_t be a normalization constant so D_{t+1} is a valid probability distribution

    If $h_t(x_i) = y_i$

        Let $D_{t+1} = \frac{D_t(i)}{Z_t} \cdot \beta_t$

    Else

        Let $D_{t+1} = \frac{D_t(i)}{Z_t}$

Output the final hypothesis:

$$h_{fin}(x) = \arg\max_{y \in Y} \sum_{t:h_t(x)=y} \log \frac{1}{\beta_t}$$

FIGURE 26.9 The procedure AdaBoost to boost the accuracy of a mediocre hypothesis (created by WeakLearn) to a very accurate hypothesis. The input is a sequence $\langle (x_1, y_1), \ldots, (x_m, y_m) \rangle$ of labeled examples where each label comes from the set $Y = \{1, \ldots, k\}$.

Freund and Schapire [23] have done some experiments showing that by applying AdaBoost to some simple rules of thumb or using C4.5 [44] as the weak learner they can perform better than previous algorithms on some standard benchmarks. Also, Freund and Schapire have presented a more general version for AdaBoost for the situation in which the predictions are real-valued (versus binary). Some of the practical advantages of AdaBoost are that is fast, simple and easy to program, requires no parameters to tune (besides T, the number of rounds), no prior knowledge is needed about the weak learner, it is provably effective, and it is flexible, since you can combine it with any classifier that finds weak hypothesis.

26.7 Research Issues and Summary

In this chapter, we have described the fundamental models and techniques of computational learning theory. There are many interesting research directions besides those discussed here. One general direction of research is in defining new, more realistic models, and models that capture different learning scenarios. Here are a few examples. Often, as in the problem of weather prediction, the target concept is probabilistic in nature. Thus Kearns and Schapire [34] introduced the p-concepts model. Also there has been a lot of work in extending the VC theory to real-valued domains (e.g., [28]). In both the PAC and online models, many algorithms use membership queries. While most work has assumed that the answers provided to the membership queries are reliable, in reality a learner must be able to handle inconclusive and noisy results from the membership queries. See Blum et al. [14] for a summary of several models introduced to address this situation. As one last example, there has been much work recently in exploring models of a "helpful teacher," since teaching is often used to assist human learning (e.g., [8,25]).

Finally, there has been work to bridge the computational learning research with the research from other fields such as neural networks, natural language processing, DNA analysis, inductive logic programming, information retrieval, expert systems, and many others.

26.8   Further Information

Good introductions to computational learning theory (along with pointers to relevant literature) can be found in such textbooks as [33,41], and [11]. Many recent results can be found in the proceedings from the following conferences: ACM Conference on Computational Learning Theory (COLT), European Conference on Computational Learning Theory (EuroCOLT), International Workshop on Algorithmic Learning Theory (ALT), International Conference on Machine Learning (ICML), IEEE Symposium on Foundations of Computer Science (FOCS), ACM Symposium on Theoretical Computing (STOC), and Neural Information Processing Conference (NIPS).

Some major journals in which many learning theory papers can be found are *Machine Learning, Journal of the ACM, SIAM Journal of Computing, Information and Computation,* and *Journal of Computer and System Sciences.* See Kodratoff and Michalski [36] for background information on machine learning, and see Angluin and Smith [10] for a discussion of inductive inference models and research.

Defining Terms

Classification noise: A model of noise in which the label may be reported incorrectly. In the random classification noise model, with probability η the label is inverted. In the malicious classification noise model, with probability η an adversary can choose the label. In both models with probability $1 - \eta$ the example is not corrupted. The quantity η is referred to as the noise rate.

Concept class: A set of rules from which the target function is selected.

Counterexample: A counterexample x to a hypothesis h (where the target concept is f) is either an example for which $f(x) = 1$ and $h(x) = 0$ (a positive counterexample) or for which $f(x) = 0$ and $h(x) = 1$ (a negative counterexample).

Equivalence query: A query to an oracle in which the learner provides a hypothesis h and is either told that h is logically equivalent to the target or otherwise given a counterexample.

Hypothesis boosting: The process of taking a weak learner that predicts correctly just over half of the time and transforming (boosting) it into a PAC (strong) learner whose predictions are as accurate as desired.

Hypothesis class: The class of functions from which the learner's hypothesis is selected.

Membership query: The learner supplies the membership oracle with an instance x from the domain and is told the value of $f(x)$.

Occam algorithm: An algorithm that draws a sample S and then outputs a hypothesis consistent with the examples in S such that the size of the hypothesis is sublinear in S (i.e., it performs some data compression).

Online learning model: The learning session consists of a set of trials where in each trial, the learner is given an unlabeled instance $x \in \mathcal{X}$. The learner uses its current hypothesis to predict a value $p(x)$ for the unknown (real-valued) target and is then told the correct value for $f(x)$. The performance of the learner is measured in terms of the sum of the loss over all predictions.

PAC learning: This is a batch model in which first there is a training phase in which the learner sees examples drawn randomly from an unknown distribution $\mathcal{D}$, and labeled according to the unknown target concept f drawn from a known concept class $\mathcal{C}$. The learner's goal is to output a hypothesis that has error at most ϵ with probability at least $1 - \delta$. (The learner receives ϵ and δ as inputs.) In the proper PAC learning model the learner must output a hypothesis from $\mathcal{C}$. In the nonproper PAC learning model the learner can output any hypothesis h for which $h(x)$ can be computed in polynomial time.

Representation-dependent hardness result: A hardness result in which one proves that one cannot efficiently learn C using a hypothesis class of $\mathcal{H}$. These hardness results typically rely on some complexity theory assumption such as RP $\neq$ NP.

Representation-independent hardness result: A hardness result in which no assumption is made about the hypothesis class used by the learner (except the hypothesis can be evaluated in polynomial time). These hardness results typically rely on cryptographic assumptions such as the difficulty of breaking RSA.

Statistical query model: A learning model in which the learner does not have access to labeled examples, but rather only can ask queries about statistics about the labeled examples and receive unlabeled examples. Any statistical query algorithm can be converted to a PAC algorithm that can handle random classification noise (as well as some other types of noise).

Vapnik–Chervonenkis (VC) dimension: A finite set $S \subseteq \mathcal{X}$ is shattered by C if for each of the $2^{|S|}$ subsets $S' \subseteq S$, there is a concept $f \in C$ that contains all of S' and none of $S - S'$. In other words, for any of the $2^{|S|}$ possible labelings of S (where each example $s \in S$ is either positive or negative), there is some $f \in C$ that realizes the desired labeling (see Figure 26.2). The Vapnik–Chervonenkis dimension of C, denoted as VCD(C), is the smallest d for which no set of $d + 1$ examples is shattered by C. Equivalently, VCD(C) is the cardinality of the largest finite set of points $S \subseteq X$ that is shattered by C.

Weak learning model: A variant of the PAC model in which the learner is only required to output a hypothesis that with probability $\geq 1 - \delta$, has error at most $(1/2 - 1/p(n))$. That is, it does noticeably better than random guessing. Recall in the standard PAC (strong learning) model the learner must output a hypothesis with arbitrarily low error.

References

1. Aizenstein, H. and Pitt, L., Exact learning of read-twice DNF formulas. In *Proceedings of the 32nd Annual IEEE Symposium Foundations Computer Science*. IEEE Computer Society Press, Washington, DC, pp. 170–179, 1991.

2. Angluin, D., Learning regular sets from queries and counterexamples. *Inform. Comput.*, 75(2), 87–106, November 1987.

3. Angluin, D., Queries and concept learning. *Mach. Learn.*, 2(4), 319–342, April 1988.

4. Angluin, D., Negative results for equivalence queries. *Mach. Learn.*, 5, 121–150, 1990.

5. Angluin, D., Frazier, M., and Pitt, L., Learning conjunctions of Horn clauses. *Mach. Learn.*, 9, 147–164, 1992.

6. Angluin, D., Hellerstein, L., and Karpinski, M., Learning read-once formulas with queries. *J. ACM*, 40, 185–210, 1993.

7. Angluin, D. and Kharitonov, M. When won't membershipt queries help? *J. Comput. Syst. Sci.*, 50(2), 336–355, 1995.

8. Angluin, D. and Kriķis, M., Teachers, learners and black boxes. In *Proceedings of the 10th Annual Conference on Computational Learning Theory*. ACM Press, New York, pp. 285–297, 1997.

9. Angluin, D. and Laird, P., Learning from noisy examples. *Mach. Learn.*, 2(4), 343–370, 1988.

10. Angluin, D. and Smith, C., Inductive inference. In *Encyclopedia of Artificial Intelligence*, John Wiley & Sons, New York, pp. 409–418, 1987.

11. Anthony, M. and Biggs, N., *Computational Learning Theory, Cambridge Tracts in Theoretical Computer Science (30)*. Cambridge University Press, Cambridge, U.K., 1992.

12. Aslam, J.A. and Decatur, S.E., Specification and simulation of statistical query algorithms for efficiency and noise tolerance. *J. Comp. Syst. Sci.*, 1998. To appear.

13. Blum, A., Separating distribution-free and mistake-bound learning models over the Boolean domain. *SIAM J. Comput.*, 23(5), 990–1000, 1994.

14. Blum, A., Chalasani, P., Goldman, S.A., and Slonim, D.K., Learning with unreliable boundary queries. In *Proceedings of the Eighth Annual Conference on Computational Learning Theory*. 1995. ACM Press, New York, pp. 98–107, 1995. To appear in COLT '95 Special Issue of *J. Comput. Syst. Sci.*

15. Blumer, A., Ehrenfeucht, A., Haussler, D., and Warmuth, M.K., Occam's razor. *Inform. Proc. Lett.*, 24, 377–380, 1987.

16. Blumer, A., Ehrenfeucht, A., Haussler, D., and Warmuth, M.K., Learnability and the Vapnik-Chervonenkis dimension. *J. ACM*, 36(4), 929–965, 1989.

17. Bshouty, N., Goldman, S., Hancock, T., and Matar, S., Asking questions to minimize errors. *J. Comp. Syst. Sci.*, 52(2), 268–286, April, 1996.

18. Bshouty, N.H. and Mansour, Y., Simple learning algorithms for decision trees and multivariate polynomials. In *Proceedings of the 36th Annual Symposium on Foundations of Computer Science,* 1995. IEEE Computer Society Press, Los Alamitos, CA, pp. 304–311, 1995.

19. Cesa-Bianchi, N., Freund, Y., Helmbold, D.P., Haussler, D., Schapire, R.E., and Warmuth, M.K., How to use expert advice. *J. ACM*, 44(3), 427–485, 1997.

20. Devroye, L., Györfi, L., and Lugosi, G., *A Probabilistic Theory of Pattern Recognition, Applications of Mathematics*. Springer-Verlag, New York, 1996.

21. Duda, R.O. and Hart, P.E., *Pattern Classification and Scene Analysis*. John Wiley & Sons, New York, 1973.

22. Ehrenfeucht, A. and Haussler, D., A general lower bound on the number of examples needed for learning. *Inform. Comput.*, 82(3), 247–251, September 1989.

23. Freund, Y. and Schapire, R., Experiments with a new boosting algorithm. In *Proceedings of the 13th International Conference on Machine Learning*. Morgan Kaufmann, San Francisco, CA, pp. 148–156, 1996.

24. Gold, E.M., Language identification in the limit. *Inform. Control,* 10, 447–474, 1967.

25. Goldman, S. and Mathias, D., *J. Comput. Syst. Sci.,* (52)2, 255–267, April 1996.

26. Goldman, S. and Sloan, R., Can PAC learning algorithms tolerate random attribute noise? *Algorithmica*, 14(1), 70–84, July 1995.

27. Goldman, S.A. and Scott, S.D., A theoretical and empirical study of a noise-tolerant algorithm to learn geometric patterns. In *Proceedings of the 13th International Conference on Machine Learning*. Morgan Kaufmann, San Francisco, CA, pp. 191–199, 1996. To appear in *Mach. Learn.*

28. Haussler, D., Decision theoretic generalizations of the PAC model for neural net and other learning applications. *Inform. Comput.,* 100(1), 78–150, September 1992.

29. Haussler, D., Kearns, M., Littlestone, N., and Warmuth, M.K., Equivalence of models for polynomial learnability. *Inform. Comput.,* 95(2), 129–161, December 1991.

30. Haussler, D., Littlestone, N., and Warmuth, M.K., Predicting {0, 1} functions on randomly drawn points. *Inform. Comput.,* 115(2), 284–293, 1994.

31. Kearns, M., Efficient noise-tolerant learning from statistical queries. In *Proceedings of the 25th Annual ACM Symposium Theory of Computing*. ACM Press, New York, pp. 392–401, 1993.

32. Kearns, M. and Valiant, L.G., Cryptographic limitations on learning Boolean formulae and finite automata. In *Proceedings of the 21st Symposium on Theory of Computing,* 1989. ACM Press, New York, pp. 433–444, 1989. To appear in *J. ACM*.

33. Kearns, M. and Vazirani, U., *An Introduction to Computational Learning Theory*. MIT Press, Cambridge, MA, 1994.

34. Kearns, M.J. and Schapire, R.E., Efficient distribution-free learning of probabilistic concepts. *J. Comput. Syst. Sci.,* 48(3), 464–497, 1994.

35. Kearns, M.J., Schapire, R.E., and Sellie, L.M., Toward efficient agnostic learning. *Mach. Learn.*, 17(2/3), 115–142, 1994.

36. Kodratoff, Y. and Michalski, R.S., Eds., *Machine Learning: An Artificial Intelligence Approach*, Vol. III. Morgan Kaufmann, Los Altos, CA, 1990.
37. Littlestone, N., Learning when irrelevant attributes abound: A new linear-threshold algorithm. *Mach. Learn.*, 2, 285–318, 1988.
38. Littlestone, N. and Warmuth, M.K., The weighted majority algorithm. *Inform. Comput.*, 108(2), 212–261, 1994.
39. Maass, W. and Turán, G., Lower bound methods and separation results for on-line learning models. *Mach. Learn.*, 9, 107–145, 1992.
40. Maass, W. and Warmuth, M.K., Efficient learning with virtual threshold gates. In *Proceedings of the 12th International Conference on Machine Learning*. Morgan Kaufmann, San Francisco, CA, pp. 378–386, 1995. To appear in *Inform. Comput.*
41. Natarajan, B.K., *Machine Learning: A Theoretical Approach*. Morgan Kaufmann, San Mateo, CA, 1991.
42. Pitt, L. and Valiant, L., Computational limitations on learning from examples. *J. ACM*, 35, 965–984, 1988.
43. Pitt, L. and Warmuth, M.K., Prediction preserving reducibility. *J. Comput. Syst. Sci.*, 41(3), 430–467, December 1990. Special issue for the *Third Annual Conference of Structure in Complexity Theory*, Washington, DC, June 19–88.
44. Quinlan, J.R., *C45: Programs for Machine Learning*. Morgan Kaufmann, San Francisco, CA, 1993.
45. Rosenblatt, F., The perceptron: A probabilistic model for information storage and organization in the brain. *Psych. Rev.*, 65, 386–407, 1958. Reprinted in *Neurocomputing* (MIT Press, Cambridge, MA, 1988).
46. Schapire, R.E., The strength of weak learnability. *Mach. Learn.*, 5(2), 197–227, 1990.
47. Sloan, R., Four types of noise in data for PAC learning. *Inform. Proc. Lett.*, 54, 157–162, 1995.
48. Valiant, L.G., A theory of the learnable. *Commun. ACM*, 27(11), 1134–1142, November 1984.
49. Valiant, L.G., Learning disjunctions of conjunctions. In *Proceedings of the Ninth International Joint Conference on Artificial Intelligence*, Vol. 1 (Los Angeles, CA, 1985.) International Joint Committee for Artificial Intelligence, pp. 560–566, 1985.
50. Vapnik, V.N. and Chervonenkis, A.Y., On the uniform convergence of relative frequencies of events to their probabilities. *Theor. Probab. Appl.*, 16(2), 264–280, 1971.

27

Algorithmic Coding Theory

Atri Rudra
State University of New York at Buffalo

27.1 Introduction

Error-correcting codes (or just **codes**) are clever ways of representing data so that one can recover the original information even if parts of it are corrupted. The basic idea is to judiciously introduce redundancy so that the original information can be recovered when parts of the (redundant) data have been corrupted.

Perhaps the most natural and common application of error correcting codes is for communication. For example, when packets are transmitted over the Internet, some of the packets get corrupted or dropped. To deal with this, multiple layers of the TCP/IP stack use a form of error correction called CRC Checksum (Peterson and Davis, 1996). Codes are used when transmitting data over the telephone line or via cell phones. They are also used in deep space communication and in satellite broadcast. Codes also have applications in areas not directly related to communication. For example, codes are used heavily in data storage. CDs and DVDs can work even in the presence of scratches precisely because they use codes. Codes are used in Redundant Array of Inexpensive Disks (RAID) (Chen et al., 1994) and error correcting memory (Chen and Hsiao, 1984). Codes are also deployed in other applications such as paper bar codes, for example, the bar code used by UPS called MaxiCode (Chandler et al., 1989).

In addition to their widespread practical use, codes have also been used in theoretical computer science; especially in the last 15 years or so (though there are a few notable exceptions). They have found numerous applications in computational complexity and cryptography. Doing justice to these connections is beyond the scope of this chapter: we refer the reader to some of the surveys on these connections (Guruswami, 2004a, 2006c; Sudan, 2000; Trevisan, 2004).

In this chapter, we will think of codes in the communication scenario. In this framework, there is a sender who wants to send k message symbols over a noisy channel. The sender first encodes the k message symbols into n symbols (called a **codeword**) and then sends it over the channel. As codes introduce redundancy, $n \geq k$. The receiver gets a **received word** consisting of n symbols. The receiver then tries to decode and recover the original k message symbols. We assume that the sender and receiver only communicate via the channel: specifically the receiver has no side information about the contents of the message. The main challenge in algorithmic coding theory is to come up with good codes along with efficient **encoding** and **decoding** algorithms. Next, we elaborate on some of the core issues in meeting the aforementioned challenge. The first issue is combinatorial while the second is algorithmic.

The combinatorial issue is the inherent tension between the amount of redundancy used and the number of errors such codes can tolerate. Let us illustrate this issue with two examples of codes whose symbols are bits: such codes are also called **binary codes**. In our first code (also called the repetition code), every message bit is repeated a fixed number of times (for concreteness say 100 times). Intuitively, this code should be able to tolerate many errors. In fact, here is a natural decoding procedure for such a code—for every contiguous 100 bits in the received word, declare the corresponding message bit to be the majority bit. In a typical scenario such a decoding procedure will recover the original message. However, the problem with the repetition code is that it uses too much redundancy. In our example, for every 100 bits that are transmitted, there is only one bit of information. On the other extreme is the so-called parity code. In a parity code, the parity of all the bits in the message is tagged at the end of the code and then sent over the channel. For example, for the message 0000001, 00000011 is sent over the channel. Parity codes use the minimum amount of redundancy. However, for such a code it is not possible to even detect two errors. Using our last example, suppose that when 00000011 is transmitted, the first two bits are flipped. That is, the received word is 11000011. Note that 11000011 is a valid codeword for the message 1100001. Now if the decoding procedure gets the received word 11000011 it is justified in outputting 1100001 as the transmitted message, which of course is not desirable.

The problem with the parity code is that the codewords corresponding to two different messages do not differ by much. In other words, we need the codewords corresponding to different messages to be far apart. However, in order to make sure that any two codewords corresponding to any pair of different messages are far apart, one needs to map messages into a large space (i.e., n needs to be much larger than k). This in turn implies that the code uses more redundancy. In a nutshell, the combinatorial challenge is to design codes that balance these two contradictory goals well. We will formalize this inherent trade-off more formally in Section 27.2.

However, meeting the combinatorial challenge is just one side of the coin. To use codes in practice, one also needs efficient encoding and decoding algorithms. For this chapter, by an efficient algorithm we mean an algorithm whose running time is polynomial in n (we will also talk about extremely efficient algorithms that run in time that is linear in n). Typically, the definition of the code gives an efficient encoding procedure for free.* The decoding procedure is generally the more challenging algorithmic task. As we will see in this chapter, the combinatorial and the algorithmic challenges are intertwined. Indeed, all the algorithms that we will consider exploit the specifics of the code (which in turn will be crucial to show that the codes have good combinatorial property).

A **random code** generally has the required good combinatorial properties with high probability. However, to transmit messages over the channel, one needs an explicit code that has good properties. This is not a great concern in some cases as one can search for a good code in a brute-force fashion (given that a code picked randomly from an appropriate ensemble has the required property). The bigger problem with random codes is that they do not have any inherent structure that can be exploited to design efficient decoding algorithms. Thus, the challenge in algorithmic coding theory is to design explicit codes with good combinatorial properties along with efficient encoding and decoding algorithms. This will be the underlying theme for the rest of this chapter.

---

* This will not be the case when we discuss linear-time encoding.

Before we move on, we would like to point out an important aspect that we have ignored till now: How do we model the noise in the channel? The origins of coding theory can be traced to the seminal works of Shannon (1948) and Hamming (1950). These lead to different ways of modeling noise. Shannon pioneered the practice of modeling the channel as a stochastic process while Hamming modeled the channel as an adversary. The analysis of algorithms in these two models are quite different and these two different schools of modeling the noise seem to be divergent.

Needless to say, it is impossible to do justice to the numerous facets of algorithmic coding theory in the confines of this chapter. Instead of briefly touching upon a long list of research themes, we will focus on the following three that have seen a spurt of research activity in the last decade or so:

- Low-density parity-check (or LDPC) codes were defined in the remarkable thesis of Gallager (1963). These codes were more or less forgotten for a few decades. Interestingly, there has been a resurgence in research activity in such codes (along with iterative, message-passing decoding algorithms that were also first designed by Gallager) which has led to codes with good combinatorial properties along with extremely efficient encoding and decoding algorithms for stochastic channels.

- Expander codes are LDPC codes with some extra properties. These codes have given rise to codes with linear-time encoding and decoding algorithms for the adversarial noise model.

- List decoding is a relaxation of the usual decoding procedure in which the decoding algorithm is allowed to output a list of possible transmitted messages. This allows for error correction from more adversarial errors than the usual decoding procedure. As another benefit list decoding bridges the gap between the Shannon and Hamming schools of coding theory.

The progress in some of the aforementioned themes are related: we will see glimpses of these connections in this chapter. In Section 27.2, we will first define the basic notions of codes. Then we will review some early results in coding theory that will also set the stage for the results in Section 27.3. Interestingly, some of the techniques developed earlier are still crucial to many of the current developments.

27.2 Basics of Codes and Classical Results

We first fix some notations that will be used frequently in this chapter.

For any integer $m \geq 1$, we will use $[m]$ to denote the set $\{0, 1, \ldots, m-1\}$. Given positive integers n and m, we will denote the set of all length n vectors over $[m]$ by $[m]^n$. By default, all vectors in this chapter will be row vectors. The logarithm of x in base 2 will be denoted by $\log x$. For bases other than 2, we will specify the base of the logarithm explicitly: for example logarithm of x in base q will be denoted by $\log_q x$. A finite field with q elements will be denoted by $\mathbb{F}_q$. For any real value x in the range $0 \leq x \leq 1$, we will use $H_q(x) = x \log_q(q-1) - x \log_q x - (1-x) \log_q(1-x)$ to denote the q-ary entropy function. For the special case of $q = 2$, we will simply use $H(x)$ for $H_2(x)$. For any finite set S, we will use $|S|$ to denote the size of the set.

We now move on to the definitions of the basic notions of error correcting codes.

27.2.1 Basic Definitions for Codes

Let $q \geq 2$ be an integer. An error correcting code (or simply a code) C is a subset of $[q]^n$ for positive integers q and n. The elements of C are called codewords. The parameter q is called the alphabet size of C. In this case, we will also refer to C as a q-ary code. When $q = 2$, we will refer to C as a binary code. The parameter n is called the **block length** of the code. For a q-ary code C, the quantity

$k = \log_q |C|$ is called the **dimension** of the code (this terminology makes more sense for certain classes of codes called **linear codes**, which we will discuss shortly). For a q-ary code C with block length n, its **rate** is defined as the ratio $R = \frac{\log_q |C|}{n}$.

Often it will be useful to use the following alternate way of looking at a code. We will think of a q-ary code C with block length n and $|C| = M$ as a function $[M] \to [q]^n$. Every element x in $[M]$ is called a message and $C(x)$ is its associated codeword. If M is a power of q, then we will think of the message as length-k vector in $[q]^k$.

Given any two vectors $\mathbf{v} = \langle v_1, \ldots, v_n \rangle$ and $\mathbf{u} = \langle u_1, \ldots, u_n \rangle$ in $[q]^n$, their **Hamming distance** (or simply distance), denoted by $\Delta(\mathbf{v}, \mathbf{u})$, is the number of positions that they differ in. The (**minimum**) **distance** of a code C is the minimum Hamming distance between any two codewords in the code. More formally $\mathrm{dist}(C) = \min_{\substack{c_1, c_2 \in C \\ c_1 \neq c_2}} \Delta(c_1, c_2)$. The relative distance of a code C of block length n is defined as $\delta = \frac{\mathrm{dist}(C)}{n}$.

27.2.1.1 Code Families

The focus of this chapter will be on the asymptotic performance of algorithms. For such analysis to make sense, we need to work with an infinite family of codes instead of a single code. In particular, an infinite family of q-ary codes $\mathcal{C}$ is a collection $\{C_i | i \in \mathbb{Z}\}$, where for every i, C_i is a q-ary code of block length n_i and $n_i > n_{i-1}$. The rate of the family $\mathcal{C}$ is defined as

$$R(\mathcal{C}) = \liminf_i \left\{ \frac{\log_q |C_i|}{n_i} \right\}.$$

The relative distance of such a family is defined as

$$\delta(\mathcal{C}) = \liminf_i \left\{ \frac{\mathrm{dist}(C_i)}{n_i} \right\}.$$

From this point on, we will overload notation by referring to an infinite family of codes simply as a code. In particular, from now on, whenever we talk a code C of length n, rate R, and relative distance δ, we will implicitly assume the following. We will think of n as large enough so that its rate R and relative distance δ are (essentially) same as the rate and the relative distance of the corresponding infinite family of codes.

Given this implicit understanding, we can talk about the asymptotics of different algorithms. In particular, we will say that an algorithm that works with a code of block length n is efficient if its running time is $O(n^c)$ for some fixed constant c.

27.2.2 Linear Codes

We will now consider an important subclass of codes called linear codes. Let q be a prime power. A q-ary code C of block length n is said to be linear if it is a linear subspace (over some field $\mathbb{F}_q$) of the vector space $\mathbb{F}_q^n$.

The size of a q-ary linear code is obviously q^k for some integer k. In fact, k is the dimension of the corresponding subspace in $\mathbb{F}_q^n$. Thus, the dimension of the subspace is the same as the dimension of the code. (This is the reason behind the terminology of dimension of a code.) We will denote a q-ary linear code of dimension k, length n, and distance d as an $[n, k, d]_q$ code. Most of the time, we will drop the distance part and just refer to the code as an $[n, k]_q$ code.

Any $[n, k]_q$ code C can be defined in the following two ways.

- C can be defined as a set $\{\mathbf{x}\mathbf{G} | \mathbf{x} \in \mathbb{F}_q^k\}$, where $\mathbf{G}$ is an $k \times n$ matrix over $\mathbb{F}_q$. $\mathbf{G}$ is called a generator matrix of C. Given the generator matrix $\mathbf{G}$ and a message $\mathbf{x} \in \mathbb{F}_q^k$, one can compute $C(x)$ using $O(nk)$ field operations (by multiplying $\mathbf{x}$ with $\mathbf{G}$). As an example, the $[7,4]_2$ Hamming code has the following generator matrix:

$$
G_{\text{Ham}} = \begin{pmatrix} 1 & 0 & 0 & 0 & 0 & 1 & 1 \\ 0 & 1 & 0 & 0 & 1 & 1 & 0 \\ 0 & 0 & 1 & 0 & 1 & 0 & 1 \\ 0 & 0 & 0 & 1 & 1 & 1 & 1 \end{pmatrix}.
$$

- C can also be characterized by the subspace $\{\mathbf{c} | \mathbf{c} \in \mathbb{F}_q^n \text{ and } H\mathbf{c}^T = \mathbf{0}\}$, where H is an $(n-k) \times n$ matrix over $\mathbb{F}_q$. H is called the parity check matrix of C. The following matrix is the parity check matrix of the $[7,4]_2$ Hamming code:

$$
H_{\text{ham}} = \begin{pmatrix} 0 & 0 & 0 & 1 & 1 & 1 & 1 \\ 0 & 1 & 1 & 0 & 0 & 1 & 1 \\ 1 & 0 & 1 & 0 & 1 & 0 & 1 \end{pmatrix}. \tag{27.1}
$$

It can be verified that for every $\mathbf{x} \in \mathbb{F}_2^4$, $H_{\text{Ham}} (\mathbf{x}G_{\text{Ham}})^T = \mathbf{0}$.

27.2.3 Reed–Solomon Codes

In this subsection, we review a family of codes that we will encounter in multiple places in this chapter. A Reed–Solomon code (named after its inventors, Reed and Solomon (1960)) is a linear code that is based on univariate polynomials over finite fields. More formally, an $[n, k+1]_q$ Reed–Solomon code with $k < n$ and $q \geq n$ is defined as follows. Let $\alpha_1, \ldots, \alpha_n$ be distinct elements from $\mathbb{F}_q$ (which is why we needed $q \geq n$). Every message $\mathbf{m} = \langle m_0, \ldots, m_k \rangle \in \mathbb{F}_q^{k+1}$ is thought of as a degree k polynomial over $\mathbb{F}_q$ by assigning the $k+1$ symbols to the $k+1$ coefficients of a degree k polynomial. In other words, $P_{\mathbf{m}}(X) = m_0 + m_1 X + \cdots + m_k X^k$. The codeword corresponding to $\mathbf{m}$ is defined as $RS(\mathbf{m}) = \langle P_{\mathbf{m}}(\alpha_1), \ldots, P_{\mathbf{m}}(\alpha_n) \rangle$. Now a degree k polynomial can have at most k roots in any field. This implies that any two distinct degree k polynomials can agree in at most k places. In other words, an $[n, k+1]_q$ Reed–Solomon code is an $[n, k+1, d = n-k]_q$ code.

By the Singleton bound (cf. Theorem 27.5), the distance of any code of dimension $k+1$ and length n is at most $n-k$. Thus, the Reed–Solomon codes have optimal distance. The optimal distance property along with its nice algebraic structure has the made Reed–Solomon codes the center of much research in coding theory. In addition to their nice theoretical applications, the Reed–Solomon codes have found widespread use in practical applications. We refer the reader to Wicker and Bhargava (1999) for more details on some of the applications of the Reed–Solomon codes.

Next, we turn to the question of how we will model the noise in the channel.

27.2.4 Modeling the Channel Noise

As was mentioned earlier, Shannon modeled the channel noise probabilistically. The channel is assumed to be memory-less, that is, the noise acts independently on each symbol. The channel has an input alphabet $\mathcal{X}$ and an output alphabet $\mathcal{Y}$. The behavior of the channel is modeled by a probability transition matrix M, where for any $x \in \mathcal{X}$ and $y \in \mathcal{Y}$, $M(x,y)$ denotes the probability that when x is transmitted through the channel, the receiver gets y. We now look at some specific examples of such stochastic channels which we will encounter later in this chapter.

DEFINITION 27.1 (Binary Symmetric Channel). Given a real $0 \leq p \leq 1/2$, the *Binary Symmetric Channel* with *cross-over probability* p, denoted by BSC_p is defined as follows. The input and output alphabets are the same: $\mathcal{X} = \mathcal{Y} = \{0, 1\}$. For any pair $(x, y) \in \{0, 1\} \times \{0, 1\}$, $M(x,y) = p$ if $x \neq y$ and $M(x,y) = 1 - p$ otherwise.

In other words, BSC_p channel flips each bit with probability p. There is a generalization of this channel to a q-ary alphabet.

DEFINITION 27.2 (q-ary Symmetric Channel). Let $q \geq 2$ be an integer and let $0 \leq p \leq 1 - 1/q$ be a real. The q-ary *Symmetric Channel* with *cross-over probability* p, denoted by $\text{QSC}_{q,p}$ is defined as follows. The input and output alphabets are the same: $\mathcal{X} = \mathcal{Y} = [q]$. For any pair $(x, y) \in [q] \times [q]$, $M(x, y) = \frac{p}{q-1}$ if $x \neq y$ and $M(x, y) = 1 - p$ otherwise.

In other words, in $\text{QSC}_{q,p}$ every symbol from $[q]$ remains untouched with probability $1 - p$ while it is changed to each of the other symbols in $[q]$ with probability $\frac{p}{q-1}$. We will also look at the following noise model.

DEFINITION 27.3 (Binary Erasure Channel). Given a real $0 \leq \alpha \leq 1$, the *Binary Erasure Channel* with *erasure probability* α, denoted by BEC_α is defined as follows. The input alphabet is $\mathcal{X} = \{0, 1\}$ while the output channel is given by $\mathcal{Y} = \{0, 1, ?\}$ where ? denotes an *erasure*. For any $x \in \{0, 1\}$, $M(x, x) = 1 - \alpha$ and $M(x, ?) = \alpha$ (the rest of the entries are 0).

BEC_p is a strictly weaker noise model than BSC_p. The intuitive reason is that in BEC_p, the receiver knows which symbols have errors while this is not the case in BSC_p. In Section 27.2.5.1, we will formally see this difference.

Finally, we turn to the adversarial noise model pioneered by Hamming.

DEFINITION 27.4 (Worst-Case Noise Model). Let $q \geq 2$ be an integer and let $0 \leq p \leq 1 - 1/q$ be a real. Then the q-ary *Hamming Channel* with *maximum fraction of errors* p, denoted by $\text{HAM}_{q,p}$ is defined as follows. The input and output alphabets of the channel are both $[q]$. When a codeword of length n is transmitted over $\text{HAM}_{q,p}$ any arbitrary set of pn positions is in error. Moreover, for each such position, the corresponding symbol $x \in [q]$ is mapped to a completely arbitrary symbol in $[q] \setminus \{x\}$.

Two remarks are in order. First, we note that while the notations in Definitions 27.1 through 27.3 are standard, the notation $\text{HAM}_{q,p}$ in Definition 27.4 is not standard. We use this notation in this chapter for uniformity with the standard definitions of the stochastic noise models. Second, the worst-case noise model is stronger than the stochastic noise model in the following sense (for concreteness we use $q = 2$ below). Given a decoding algorithm D for a code C of block length n that can tolerate p fraction of errors (i.e., it recovers the original message for any error pattern with at most pn errors), one can use D for reliable communication over $\text{BSC}_{p-\varepsilon}$ for any $\varepsilon > 0$ (i.e., D fails to recover the transmitted message with a probability exponentially small in n). This claim follows from the simple argument. Note that the expected number of errors in $\text{BSC}_{p-\varepsilon}$ is $(p - \varepsilon)n$. Now as the errors for each position are independent in $\text{BSC}_{p-\varepsilon}$, by the Chernoff bound, with all but an exponentially small probability the actual number of errors will be at most pn in which case D can recover the transmitted message.

27.2.5 The Classical Results

27.2.5.1 Shannon's Result

The main contribution of Shannon's work was a precise characterization of when error correction can and cannot be achieved on stochastic channels. Let us make this statement more precise for the BSC_p channel. First we note that one cannot hope for perfect decoding. For example, there is some chance (albeit very tiny) that all the bits of the transmitted codeword may be flipped during transmission. In such a case there is no hope for any decoder to recover the transmitted message. Given a decoding function (or a decoding algorithm), we define the decoding error probability to be the maximum, over all transmitted messages, of the probability that the decoding function outputs a message different from the transmitted message. We would like the decoding error probability of our decoding algorithm to vanish with the block length. Ideally, we would like this error probability to be exponentially small in n.

Shannon proved a general theorem that pinpoints the trade-off between the rate of the code and reliable communication for stochastic channels. In this section, we will instantiate Shannon's theorem for the stochastic channels we saw in Section 27.2.4.

Shannon's theorem implies the following for BSC_p:

THEOREM 27.1   *For every real $0 \le p < 1/2$ and $0 < \varepsilon < 1/2 - p$, there exists δ in $\Theta(\varepsilon^2)$ such that the following holds for large enough n. There exists an encoding function $E : \{0,1\}^k \to \{0,1\}^n$ and a decoding function $D : \{0,1\}^n \to \{0,1\}^k$ for $k = \lfloor(1 - H(p + \varepsilon))n\rfloor$ such that the decoding error probability is at most $2^{-\delta n}$.*

The theorem aforementioned states that as long as the rate is at most $1 - H(p + \varepsilon)$, one can have reliable communication over BSC_p. In particular, the decoding error probability can be made as small as desired by picking a large enough block length n. The proof of the theorem is one of the early uses of the probabilistic method (cf. Alon and Spencer, 1992). The proof of Theorem 27.1 proceeds as follows. First the encoding function E is picked at random. That is, for every message $m \in \{0,1\}^k$, its corresponding codeword $E(m)$ is picked uniformly at randomly from $\{0,1\}^n$. Further, this choice is independent for every distinct message. The decoding function D performs maximum likelihood decoding (or MLD), that is, for every received word $y \in \{0,1\}^n$, $D(y) = \arg\min_{m \in \{0,1\}^k} \Delta(y, E(m))$. In fact, the analysis shows that with high probability over the random choice of E, the functions E and (the corresponding MLD function) D satisfy Theorem 27.1. Shannon also showed the following converse result.

THEOREM 27.2   *For every real $0 \le p < 1/2$ and $0 < \varepsilon < 1/2 - p$ the following holds. For large enough n and $k = \lfloor(1 - H(p) + \varepsilon)n\rfloor$ there does not any exist any pair of encoding and decoding functions $E : \{0,1\}^k \to \{0,1\}^n$ and $D : \{0,1\}^n \to \{0,1\}^k$ such that D has small decoding error probability.*

We note that unlike Theorem 27.1, in Theorem 27.2, the decoding error probability can be as large as a constant (recall that in Theorem 27.1 the decoding error probability was exponentially small). In other words, for any code with rate at least $1 - H(p) + \varepsilon$, the decoding error probability for any decoder cannot be made arbitrarily small, no matter large the block length of the code (unlike the situation in Theorem 27.1). The reader might be puzzled by the appearance of the entropy function in these two theorems. Without going into details of the analysis, we point out the following fact that is used crucially in the analysis: $2^{H(p)n}$ is a very good estimate of the number of vectors in $\{0,1\}^n$ that are within a Hamming distance of $\lfloor pn \rfloor$ from any fixed vector in $\{0,1\}^n$ (cf. MacWilliams and Sloane, 1981).

Theorems 27.1 and 27.2 pin down the best possible rate with which reliable communication over BSC_p can be achieved to $1 - H(p)$. This quantity $1 - H(p)$ is called the capacity of BSC_p.

Shannon's general result also implies the following special cases.

THEOREM 27.3 *Let $0 \leq \alpha \leq 1$. The capacity of BEC_α is $1 - \alpha$.*

THEOREM 27.4 *Let $q \geq 2$ be an integer and let $0 \leq p \leq 1 - 1/q$ be a real. The capacity of $\mathrm{QSC}_{q,p}$ is $1 - H_q(p)$.*

The proof of Theorem 27.4 is a straightforward generalization of the proof of Theorems 27.1 and 27.2. The positive part of Theorem 27.3 (i.e., showing the existence of of a code of rate around $1 - \alpha - \varepsilon$, $\varepsilon > 0$, that can be used for reliable communication over BEC_α) is also proved by picking a random encoding function (as in the proof of Theorem 27.1). The decoding function is as follows: Declare an error if more than one codeword agrees with the received word on the unerased positions, otherwise output the (unique) codeword that agrees with the unerased positions.

27.2.5.1.1 Problems Left Open by Shannon's Result

Even though Shannon's theorem pinpoints the best possible rate for reliable communication over various stochastic channels, there are two unsatisfactory aspects of Shannon's proofs.

First, the encoding functions are chosen completely at random. Note that a general encoding function has no succinct representation. So, even if one found a good encoding function as guaranteed by Shannon's result, the lack of a succinct representation would seem to preclude any efficient encoding algorithm. However, as we saw in Section 27.2.2, linear codes do have a succinct representation and can be encoded in time quadratic in the block length. A natural question to ask is whether we can prove Theorem 27.1 and (the positive parts of) Theorems 27.3 and 27.4 for random linear codes*? The answer is yes (and in fact, the analysis of Theorem 27.1 becomes somewhat easier for random linear codes). However, these codes are still not explicit. This is the right time to clarify what we mean by an explicit code. We say that a code is explicit if a succinct representation of such a code can be computed by an algorithm that runs in time polynomial in the block length of the code.

As was mentioned in the introduction, even though having an explicit encoding function that satisfies (for example) the conditions of Theorem 27.1 would be nice, in some applications one could use a brute force algorithm that runs in exponential time to find a good encoding function. The real problem with random (linear) codes is that they do not seem to have any structural property that can be exploited to design efficient decoding algorithms. Recall that Shannon's proof used the maximum likelihood decoding function. This is a notoriously hard decoding function to implement in polynomial time. In fact, there exist linear codes for which MLD is intractable (Berlekamp et al., 1978). Further, for any nontrivial code, the only known implementation of the MLD function is the brute force exponential time algorithm. Thus, the grand challenge in algorithmic coding theory after Shanon's work was the following:

Question 27.1 *Can one design an explicit code with rate that achieves the capacity of BSC_p (and other stochastic channels)? Further, can one come up with efficient decoders with negligible decoding error probability for such codes?*

We will see a positive resolution of the question aforementioned in Section 27.3.1 for BSC_p. In fact to get within ε of capacity for codes of block length n, results in Section 27.3.1 achieve a decoding time complexity of $n2^{1/\varepsilon^c}$, where $c \geq 2$ is a constant (the encoding complexity is $n/\varepsilon^{O(1)}$). Now if we think of ε as constant and n as growing (as we will do in most of this chapter), then this implies linear-time encoding and decoding. However, codes in practice typically have moderate block lengths and the

---

* A random linear code with encoding function that maps k bits to n bits can be chosen by picking a random $k \times n$ matrix over the appropriate alphabet as the generator matrix.

$2^{1/\varepsilon^c}$ factor becomes prohibitive. For example, even to get within 10% of capacity, this factor is at least 2^{100}! This leads to the following question.

Question 27.2 *Can one design an explicit code with rate that is within ε of the capacity of BSC_p along with encoders and decoders (with negligible decoding error probability for such codes) that have a time complexity of $n/\varepsilon^{O(1)}$?*

In Section 27.3.2, we will review some work from the turn of the century that answers the aforementioned question in the affirmative for the weaker BEC_α model.

27.2.6 Hamming's Work

As was mentioned earlier, Hamming studied worst-case noise. In such a scenario, the distance of the code becomes an important parameter: the larger the distance of the code, the larger the number of errors that can be corrected. Before we make this notion more precise, let us briefly look at what it means to do error correction in this noise model. In Hamming's model, we will insist on perfect decoding. Note that if all the symbols can be in error then we cannot hope for such a decoding. Hence, we will also need an upper bound on the number of errors that can be tolerated. More precisely, we will examine unique decoding. Under unique decoding, given an upper bound on the total number of errors $p_U n$ (so p_U is the fraction of errors), the decoding algorithm has to output the transmitted message for any error pattern with at most $p_U n$ many errors. Given this setup, the natural question to ask is how large can p_U be? Proposition 27.1 relates this question to the distance of the code.

PROPOSITION 27.1 For any q-ary code C of block length n and minimum distance d, there does not exist any unique decoding algorithm for $\mathrm{HAM}_{q,p}$ for any $p \geq \frac{1}{n}\lceil\frac{d}{2}\rceil$. Further for any $p \leq \frac{1}{n}\lfloor\frac{d-1}{2}\rfloor$, unique decoding can be done on $\mathrm{HAM}_{q,p}$.

The negative result follows from the following argument. Consider two codewords $c_1, c_2 \in C$ such that $\Delta(c_1, c_2) = d$, where for simplicity assume that d is even. Now consider a received word $\mathbf{y}$ such that $\Delta(\mathbf{y}, c_1) = \Delta(\mathbf{y}, c_2) = d/2$. Note that this is a possible received word under $\mathrm{HAM}_{q,d/(2n)}$. Now the decoder has no way of knowing whether c_1 or c_2 is the transmitted codeword.[*] Thus, unique decoding is not possible. For the positive side of the argument, using triangle inequality, one can show that every received word in the channel $\mathrm{HAM}_{q,(d-1)/(2n)}$ has a unique closest by codeword and thus, for example, an MLD algorithm can recover the transmitted codeword.

The aforementioned result pinpoints the maximum number of errors that can be corrected using unique decoding to be $d/2$. Thus, in order to study the trade-off between the rate of the code and the fraction of errors that can be tolerated (as was done in the Shannon's work) it is enough to study the trade-off between the rate and the (relative) distances of codes. Further, we mention that for specific families of codes (such as the Reed–Solomon codes from Section 27.2.3 among others) polynomial time unique decoding algorithms are known. There is a huge body of work that deals with the trade-off between rate and distance (and polynomial time algorithms to correct up to half the distance). This body of work is discussed in detail in any standard coding theory text book such as MacWilliams and Sloane (1981) and van Lint (1999). For this chapter, we will need the following trade-off.

THEOREM 27.5 *(Singleton Bound) Any code of dimension k and block length n has distance $d \leq n - k + 1$.*

---

[*] Recall we are assuming that the sender and the receiver only communicate via the channel.

Let us now return to the trade-off between the rate of a code and the fraction of errors that can be corrected. By Proposition 27.1 and Theorem 27.5, to correct p fraction of errors via unique decoding, the code can have a rate of at most $1 - 2p$. This trade-off can also be achieved by explicit codes and efficient encoding and decoding algorithms. In particular, a Reed–Solomon code with relative distance of $\delta = 2p$ has a rate (slightly more than) $1 - 2p$. Further, as was mentioned earlier, there are polynomial time unique decoding algorithms that can correct up to $\delta/2 = p$ fraction of errors. Recall that as a Reed–Solomon code is a linear code, a quadratic time encoding algorithm is immediate. However, if we are interested in extremely efficient algorithms, that is, linear time-encoding and-decoding algorithms, then algorithms for the Reed–Solomon codes do not suffice. This leads to the following question:

Question 27.3 *Let $0 < p < 1$ and $\varepsilon > 0$ be arbitrary reals. Do explicit codes of rate $1 - 2p - \varepsilon$ exist such that they can be encoded as well as uniquely decoded from a p fraction of errors in time linear in the block length of the code?*

In Section 27.3.3, we will see a positive resolution of the above question.

Let us pause for a bit and see how the bound of $1 - 2p$ compares with corresponding capacity result for $QSC_{q,p}$. Theorem 27.4 states that it is possible to have reliable communication for rates at most $1 - H_q(p)$. Now for large enough q, $1 - H_q(p)$ is almost $1 - p$ (see, e.g., Rudra 2007, Chapter 2). Thus, there is a gap in how much redundancy one needs to use to achieve reliable communication in Shannon's stochastic noise model and Hamming's worst-case noise model. Another way to look at this gap is the following question. Given a code of rate R (say over large alphabets), how much error can we hope to tolerate? In the q-ary Symmetric Channel, we can tolerate close to $1 - R$ fraction of errors, while in the worst-case noise model we can only tolerate half as much, that is, $(1 - R)/2$ fraction of errors. In Section 27.2.7, we will look at a relaxation of unique decoding called list decoding that can bridge this gap.

Before we wrap up this subsection, we would like to mention that no q-ary code with fixed q can achieve the Singleton bound (this follows, e.g., from the so called Plotkin bound (cf. MacWilliams and Sloane 1981)). In other words, for any q-ary code of rate R that has relative distance δ close to $1 - R$, q must depend on n (e.g., the Reed–Solomon codes require $q \geq n$.). For small fixed values of q, the following is the best known existential result:

THEOREM 27.6 *(Gilbert–Varshamov Bound) For any $q \geq 2$, there exist q-ary linear codes with rate R and relative distance δ such that $\delta \geq H_q^{-1}(1 - R)$,*

Theorem 27.6 is proved by picking a random linear code of large enough block length n and dimension $k = Rn$ and then showing that with high probability, the relative distance of such codes is at least $H_q^{-1}(1 - R)$.

27.2.7 List Decoding

Recall the argument for the upper bound on the fraction of errors that can be corrected via unique decoding: For any code of relative distance δ, there exists a received word that has two codewords within a fractional Hamming distance of at most $\delta/2$ from it. However, this is an overly pessimistic estimate of the maximum fraction of errors that can be corrected, since the way Hamming spheres pack in space, for most choices of the received word there will be at most one codeword within distance p from it even for p much greater than $\delta/2$. Therefore, always insisting on a unique answer will preclude decoding most such received words owing to a few pathological received words that have more than one codeword within distance roughly $\delta/2$ from them.

A notion called list decoding provides a clean way to get around this predicament and yet deal with worst-case error patterns. Under list decoding, the decoder is required to output a list of all codewords within distance p from the received word. The notion of list decoding itself is quite old and dates back to work in 1950s by Elias (1957) and Wozencraft (1958). However, the algorithmic aspects of list decoding were not revived until the more recent works (Goldreich and Levin, 1989; Sudan, 1997) which studied the problem for complexity-theoretic motivations.

We now state the definition of list decoding that we will use in this chapter.

DEFINITION 27.5 (List-Decodable Codes). Let C be a q-ary code of block length n. Let $L \geq 1$ be an integer and $0 < \rho < 1$ be a real. Then C is called (ρ, L)-list decodable if every Hamming ball of radius ρn has at most L codewords in it. That is, for every $\mathbf{y} \in \mathbb{F}_q^n$, $|\{c \in C | \Delta(c, \mathbf{y}) \leq \rho n\}| \leq L$.

27.2.7.1 List-Decoding Capacity

In Section 27.2.7, we informally argued that list decoding has the potential to correct more errors than unique decoding. We will now make this statement more precise. The following results were implicit in Zyablov and Pinsker (1982) but were formally stated and proved in Elias (1991).

THEOREM 27.7 *Let $q \geq 2$ be an integer and $0 < \delta \leq 1$ be a real. For any integer $L \geq 1$ and any real $0 < \rho < 1 - 1/q$, there exists a (ρ, L)-list decodable q-ary code with rate at least $1 - H_q(\rho) - \frac{1}{L+1} - \frac{1}{n^\delta}$.*

THEOREM 27.8 *Let $q \geq 2$ be an integer and $0 < \rho \leq 1 - 1/q$. For every $\varepsilon > 0$, there do not exist any q-ary code with rate $1 - H_q(\rho) + \varepsilon$ that is $(\rho, L(n))$-list decodable for any function $L(n)$ that is polynomially bounded in n.*

Theorems 27.7 and 27.8 show that to be able to list decode with small lists on $\text{HAM}_{q,p}$ the best rate possible is $1 - H_q(p)$. Another way to interpret Theorem 27.7 is the following. One can have codes of rate $1 - H_q(p) - \varepsilon$ that can tolerate p fraction of adversarial errors under list decoding with a worst-case list size of $O(1/\varepsilon)$. In other words, one can have the same trade-off between rate and fraction of errors as in $\text{QSC}_{q,p}$ if one is willing to deal with a small list of possible answers. Due to this similarity, we will call the quantity $1 - H_q(p)$ the list-decoding capacity of the $\text{HAM}_{q,p}$ channel.

The proof of Theorem 27.7 is similar to the proof in Shannon's work: it can be shown that with high probability a random code has the desired property. Theorem 27.8 is proved by showing that for any code of rate $1 - H_q(p) + \varepsilon$, there exists a received word with superpolynomially many codewords with a relative Hamming distance of p. Note that as a list-decoding algorithm must output all the near by codewords, this precludes the existence of a polynomial time list-decoding algorithm. Now, the only list-decoding algorithm known for random codes used in the proofs of Theorem 27.7 and 27.8 is the brute-force list-decoding algorithm that runs in exponential time. Thus, the grand challenge for list decoding is the following.

Question 27.4 *Can one design an explicit code with rate that achieves the list-decoding capacity of $\text{HAM}_{q,p}$? Further, can one come up with efficient list-decoding algorithms for such codes?*

In Section 27.3.4, we will look at some recent work that answer the question aforementioned in the positive for large enough q.

27.3 Explicit Constructions and Efficient Algorithms

27.3.1 Code Concatenation

In this subsection, we return to Question 27.1. Forney answered the question in the affirmative by using a code composition method called code concatenation. Before we show how Forney used concatenated codes to design binary codes that achieve the capacity of BSC_p, we digress a bit to talk about code concatenation.

27.3.1.1 Background and Definition

We start by formally defining concatenated codes. Say we are interested in a code over $[q]$. Then the outer code C_{out} is defined over $[Q]$, where $Q = q^k$ for some positive integer k. The second code, called the inner code is defined over $[q]$ and is of dimension k. (Note that the message space of C_{in} and the alphabet of C_{out} have the same size.) The concatenated code, denoted by $C = C_{out} \circ C_{in}$, is defined as follows. Let the rate of C_{out} be R and let the block lengths of C_{out} and C_{in} be N and n respectively. Define $K = RN$ and $r = k/n$. (In what follows, it be might be instructive to think of the codes as the following maps: $C_{out} : [Q]^K \rightarrow [Q]^N$ and $C_{in} : [Q] \rightarrow [q]^n$.) The input to C is a vector $\mathbf{m} = \langle m_1, \ldots, m_K \rangle \in ([q]^k)^K$. Let $C_{out}(\mathbf{m}) = \langle x_1, \ldots, x_N \rangle \in [Q]^N$. The codeword in C corresponding to $\mathbf{m}$ is defined as follows

$$C(\mathbf{m}) = \langle C_{in}(x_1), C_{in}(x_2), \ldots, C_{in}(x_N) \rangle.$$

It is easy to check that C has rate rR, dimension kK, and block length nN. While the concatenated construction still requires an inner q-ary code, this is a small/short code with block length n, which is typically logarithmic or smaller in the length of the outer code. A good choice for the inner code can therefore be found efficiently by a brute-force search, leading to a polynomial time construction of the final concatenated code.

Ever since its discovery and initial use in Forney (1966), code concatenation has been a powerful tool for constructing error-correcting codes. The popularity of code concatenation arises because it often difficult to give a direct construction of codes over small alphabets. On the other hand, over large alphabets, an array of powerful algebraic constructions (such as the Reed–Solomon codes) with excellent parameters are available. This paradigm draws its power from the fact that a concatenated code, roughly speaking, inherits the good features of both the outer and inner codes. For example, the rate of the concatenated code is the product of the rates of the outer and inner codes, and the minimum distance is at least the product of the distances of the outer and inner codes. The alphabet size of the concatenated code equals that of the inner code.

27.3.1.2 Achieving Capacity on BSC_p

We now return to the question of achieving the capacity of BSC_p for some $0 \le p < 1/2$. More precisely, say we want to construct the codes of rate $1 - H(p) - \varepsilon$ that allow for reliable communication over BSC_p. As mentioned earlier, we will use a concatenated code to achieve this. We now spell out the details.

Let b be an integer parameter we will fix later and say we want to construct a binary code C. We will pick $C = C_{out} \circ C_{in}$, where the outer and inner codes have the following properties:

- C_{out} is a code over $\mathbb{F}_{2^b}$ of length n with rate $1 - \varepsilon/2$. Further, let D_{out} be a unique decoding algorithm that can correct a small fraction $\gamma = \gamma(\varepsilon)$ fraction of worst-case errors.
- C_{in} is a binary code of dimension b with rate $1 - H(p) - \varepsilon/2$. Further, let D_{in} be a decoding algorithm that can recover the transmitted message over BSC_p with probability at least $1 - \gamma/2$.

Let us defer for a bit how we obtain the codes C_{out} and C_{in} (and their decoding algorithms). Assuming we have the requisite outer and inner codes, we analyze the parameters of $C = C_{out} \circ C_{in}$ and present a natural decoding algorithm for C. Note that C has rate $(1 - \varepsilon/2) \cdot (1 - H(p) - \varepsilon/2) \geq 1 - H(p) - \varepsilon$ as required (its block length is $N = nb/(1 - H(p) - \varepsilon/2)$).

The decoding algorithm for C is fairly natural. For notational convenience define $b' = b/(1 - H(p) - \varepsilon/2)$. The decoding algorithm has the following steps:

1. Given a received word $\mathbf{y} \in \mathbb{F}_2^N$, divide it up into n contiguous blocks, each consisting of b' bits– denote the ith block by y_i. Note that b' is the block length of C_{in}.

2. For every i, decode y_i, using D_{in} to get y'_i. Let $\mathbf{y}' = (y'_1, \ldots, y'_n) \in \mathbb{F}_{2^b}^n$ be the intermediate result. Note that $\mathbf{y}'$ is a valid received word for D_{out}.

3. Decode $\mathbf{y}'$ using D_{out} to get a message m and output that has the transmitted message.

We now briefly argue that the aforementioned decoder recovers the transmitted message with high probability. Note that as the noise in BSC_p acts on each bit independently, in step 2 for any block i, y_i is a valid received word given a codeword from C_{in} was transmitted over BSC_p. Thus, by the stated property of D_{in}, for any block D_{in} makes an error with probability at most $\gamma/2$. Further, as the noise on each block is independent, by the Chernoff bound, except with exponentially small probability, at most γ fraction of the blocks are decoded incorrectly. In other words, at the beginning of step 3, $\mathbf{y}'$ is a received word with at most γ fraction of errors. Thus, by the stated property of D_{out}, step 3 will output the transmitted message for C with high probability as desired.

We now return to the task of getting our hands on appropriate outer and inner codes. We start with C_{in}. Note that by Theorem 27.1, there exists codes of block length b' that achieve the BSC_p capacity with decoding error probability $2^{-\Theta(b'\varepsilon^2)}$ (which we want to be at most $\gamma/2$). Picking C_{in} to be such a code with b' (and hence b) in $\Theta(\log(1/\gamma)/\varepsilon^2)$ will satisfy the required properties. Further, by the discussion in Section 27.2.5.1, C_{in} is also a linear code. This implies that such a code can be found by a brute-force search (which will imply constant time complexity that depends only on ε). Further, C_{in} can be encoded in time $O(b^2)$ (since any linear code has quadratic encoding time complexity). D_{in} is the MLD algorithm that runs in time $2^{O(b)}$.

We now turn to the outer codes. Interestingly, C_{out} is also a binary code. We think of C_{out} as a code over $\mathbb{F}_{2^b}$ by simply grouping together the blocks of b bits. Note that if a decoding algorithm can decode from γ fraction of worst-case errors over $\mathbb{F}_2$, then it can also decode from γ fraction of worst-case errors over $\mathbb{F}_{2^b}$. In Forney (1966), the code C_{out} was in turn another concatenated code (where the outer code is the Reed–Solomon code and the inner code was chosen from the so called Wozencraft ensemble, Justesen (1972)). However, for our purposes we will use binary codes from Theorem 27.13 (with γ in $O(\varepsilon^2)$). Finally, we estimate the decoding time complexity for C. The decoding time is dominated by step 2 of the algorithm, which takes time $n2^{O(b)}$. Recalling that $n = N(1 - H(p) - \varepsilon/2)/b$ and the instantiations of γ and b, we have the following:

THEOREM 27.9   *There exist explicit codes that get within ε of capacity of BSC_p that can be decoded and encoded in time $N2^{O(\log(1/\varepsilon)/\varepsilon^2)}$ and $N/\varepsilon^{O(1)}$ respectively (where N is the block length of the code).*

Thus, we have answered Question 27.1. In fact, encoding and decoding can be carried out in linear time (assuming that ε is fixed while N is increasing).

27.3.2 LDPC Codes

In this subsection, we return to Question 27.2. Unfortunately, a positive answer to this question is not known to date. In this subsection we will look at a family of codes called Low Density Parity

Check (or LDPC) codes along with iterative message passing decoding algorithms that experimentally seem to answer Question 27.2 in the affirmative, though no theoretical guarantees are known. However, for the weaker model of BEC_α, the corresponding question can be answered in the affirmative. We will focus mostly on BEC in this subsection, though we will also discuss results in BSC. We start with the definitions related to LDPC codes and a high level overview of iterative message passing decoding algorithms. Both LDPC codes and message passing algorithms were introduced and studied in the remarkable thesis (Gallager, 1963) which was the way ahead of its time.

27.3.2.1 Definitions

The LDPC codes are linear binary codes with sparse parity check matrices. In particular, each row and column of the parity check matrix has at most a fixed number of 1s. A useful way to think about an LDPC code is by its corresponding factor graph. Given a binary code of dimension k and block length n, its factor graph is a bipartite graph where the left side has n vertices called variable nodes (each of which corresponds to a position in a codeword). The right side has $n - k$ vertices called check nodes (corresponding to a parity check or a row of the parity check matrix). Every check node is connected to variable nodes whose corresponding codeword symbol appears in the associated parity check. In other words, the incidence matrix of the factor graph is exactly the parity check matrix. Figure 27.1 illustrates the factor graph for the $[7,4]_2$ Hamming code, whose corresponding parity check matrix is the one given by Equation 27.1.

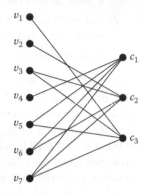

FIGURE 27.1 Factor graph of the $[7,4]_2$ Hamming code.

Gallager considered regular LDPC codes for which the corresponding factor graph is (d_v, d_c)-regular (i.e., every left vertex has degree d_v and every right vertex has degree d_c). Later on in the subsection, we will briefly look at irregular LDPC codes for which the corresponding factor graph is not regular. For the rest of the subsection we will exclusively think of the LDPC codes in terms of their factor graphs. Finally, for the remainder of the subsection, for notational convenience we will think of the bits to take values form $\{-1, 1\}$. -1 and 1 correspond to the conventional 1 and 0 respectively. Note that in this notation, parity of two bits is just their product.

27.3.2.2 Iterative Message Passing Decoding Algorithms for LDPC Codes

As the name suggests, iterative message passing decoding algorithms occur in rounds. In particular, in alternating rounds, check nodes pass messages to their neighboring variable nodes and vice versa. Initially, every variable node v_j ($1 \leq j \leq n$) has its corresponding symbol in the received word y_j (note that for BSC and BEC channels this is a random variable). In the first round, every variable node sends a message to its neighboring check nodes (which typically is just y_j for v_j). A check node after receiving messages from its neighboring variable nodes, computes a message using a predetermined function on the received messages and sends it back to its neighboring variable nodes. The variable node v_j upon receiving messages from its neighboring check nodes computes another message using another predetermined function on the received messages and y_j and sends it to its neighboring check nodes. Messages are passed back and forth in this manner till a predetermined number of rounds is completed.

Three remarks are in order. First, the functions used by variable and check nodes to compute messages can depend on the iteration number. However, typically these functions have the same structure over different rounds. Second, the message sent to a neighboring variable (resp. check) node v (resp. c) by a check (resp. variable) node is independent of the message sent to it by v (resp. c) in the previous round. In other words only extrinsic information is used to compute new messages.

This is a very important restriction that is useful in the analysis of the algorithm. Third, there is an intuitive interpretation of the messages in the algorithm. In particular, they are supposed to be votes on the value of codeword bits. If the messages take values in $\{-1, 1\}$ then they correspond to the actual bit value. One can add 0 to denote an erasure or an absentee vote. If a message takes a real value then the sign will denote the vote, while the absolute value denotes the confidence in the vote.

We now state some of the results in Gallager (1963), which among other things will give concrete examples of the general paradigm discussed earlier.

27.3.2.3   Gallager's Work

To present the main ideas in Gallager's work we apply his methods to BEC even though Gallager (1963) did not explicitly consider this channel. The first step is to design a (d_v, d_c)-regular factor graph with n variable nodes such that it has no cycle of sub-logarithmic size (i.e., the girth of the graph is $\Omega(\log n)$).

In the second step we need to specify the functions that variable and check nodes use to compute the messages. For the BEC the following is the natural choice. When a variable node needs to send a message to its neighboring check node it sends the corresponding codeword bit if it is known (either from the received word or as a message from a check node in an earlier round), otherwise it sends an erasure. On the other hand when a check node c needs to send a message to a variable node v, it sends an erasure if at least one neighbor other than v sent it an erasure in the previous round. Otherwise it sends v the parity of the messages it received in the previous round. (In other words, let $m_1, \ldots, m_{d_c-1}$ be the messages received by c in the previous round from all its neighbors other than v (recall the messages are in $\{-1, 0, 1\}$, where 0 denotes an erasure). Then c sends the message $\prod_{i=1}^{d_c-1} m_i$ to v.) It can be shown by induction that if the message to v is not an erasure, then the message is the correct codeword bit for v. This algorithm can be implemented as a peeling decoder, where each edge of the factor graph is used to pass a message only once. This implies that the decoding algorithm is linear in the number of edges (which in turn is linear in n).

The analysis of the aforementioned decoding algorithm proceeds as follows. The first step of the analysis is to obtain a recursive relation on the fraction of messages that are erasures in a particular round (in terms of the fraction in the previous round). This part crucially uses the facts that only extrinsic information is used to compute new messages and that the factor graph has logarithmic girth. The second step of the analysis is to come up with a threshold α^* on the erasure probability such that for any $\alpha < \alpha^*$, under BEC_α, the decoder in its last iteration would have a negligible fraction of messages as erasures. Using the number of iterations is logarithmic in n (since the girth is logarithmic), it can be shown that the expected fraction of messages that are erasures vanishes as n increases. This implies that except with a negligible probability, the decoder outputs the transmitted codeword. Making the earlier discussion formal results in the following (though this does not achieve the capacity of the BEC).

THEOREM 27.10   *For integers $3 \leq d_v < d_c$, there exists an explicit family of codes of rate $1 - \frac{d_v}{d_c}$ that can be reliably decoded in linear time on BEC_α, provided $\alpha < \alpha^*$. The threshold α^* is given by the expression $\frac{1-\gamma}{(1-\gamma^{d_c-1})^{d_v-1}}$, where γ is the unique positive root of the polynomial $((d_v - 1)(d_c - 1) - 1)x^{d_c-2} - \sum_{j=0}^{d_c-3} x^j$.*

We briefly mention how Gallager's result for BSC "extends" the aforementioned techniques. The first main difference is in the maps used to compute new messages. The check nodes use the parity of the all incoming messages as their function. For the variable nodes, Gallager proposed two functions. In the first one, which leads to the so-called Gallager's Algorithm A, the variable node sends its received bit to a check node unless messages coming from all other check nodes in the previous node indicate otherwise. In that case, it sends the complement of the received value. In the second function that leads to Gallager's algorithm B, the variable node sends the complement of its received value if more than a pre-fixed number of messages from check nodes in the previous round say otherwise.

27.3.2.4 Irregular LDPC Codes

We now briefly look at the some more recent work that builds on Gallager's work and achieves the capacity of BEC with extremely efficient encoding and decoding algorithms. For the rest of the subsection, we will concentrate mostly on the BEC.

The work of Luby et al. (2001a) introduced the study of LDPC codes based on irregular factor graphs. We start with some intuition as to why having an irregular factor graph might help while running Gallager's decoding algorithms. From the perspective of the variable node, it is beneficial to have more adjacent check nodes as the variable node would obtain more information from the check nodes which should intuitively help the variable node in computing its bit value. On the other side, a check node would prefer to have fewer adjacent variable nodes because the parity function becomes more unpredictable with more inputs. However, for the rate to be positive, the number of check nodes has to be fewer than the number of variable nodes. Meeting these contradictory goals is difficult. Due to their less stringent conditions on vertex degrees irregular graphs provide more flexibility in meeting the competing degree requirements discussed earlier. The motivation of having a spread of degrees is that variable nodes with high degree could be expected to converge to their correct value faster than their regular factor graph counterpart. This would in turn lead to the neighboring check nodes getting better information, which would then be relayed to variable nodes with smaller degrees. Thus, the hope is that this cascading effect would lead to better algorithms. Of course, making this intuition formal requires some effort, which leads to the following result.

THEOREM 27.11   *There exist codes that get within ε of capacity of* BEC$_\alpha$. *Further, these codes can be decoded and encoded in time* $O(n \log(1/\varepsilon))$.

We wrap up this section with two remarks regarding the aforementioned result. First, unlike Gallager's result designing an explicit irregular factor graph with a large girth seems like a difficult task. The results of Luby et al. (2001a) instead work on ensembles of irregular factor graphs. By an ensemble of codes, we mean a family of codes that are parametrized by two distributions $\{\lambda_i\}$ and $\{\rho_i\}$. Here λ_i (resp. ρ_i) is the fraction of edges that are incident on variable (resp. check) nodes of degree i. It is shown that there exist appropriate choices of these distributions for which if one samples from the corresponding ensemble then with high probability the resulting factor graph will have the required properties. Thus, we no longer have the explicit construction of codes. Second, given a low density parity check matrix (or the equivalent sparse factor graph), it is nontrivial to get linear-time encoding schemes. Luby et al. (2001a) used a cascade of so called low-density generator matrix code to obtain linear-time encoding. An alternate approach from Richardson and Urbanke (2001b) is to find an approximate lower triangulation of the parity check matrix that is still sparse, which suffices for a linear-time encoding.

Thus, Theorem 27.11 answers Question 27.2, though for the weaker BEC. The techniques mentioned earlier have been extended to the other stochastic channels such as the BSC. These techniques perform very well experimentally, though rigorous theoretical guarantees have so far been elusive.

27.3.3 Expander Codes

In this subsection, we will look at codes that are constructed from certain combinatorial objects called expanders. Expanders are sparse graphs that are still well-connected. There are two main ways in which expanders are used to define codes: (1) using the graph as a factor graph for LDPC codes, which we discussed in Section 27.3.2 and (2) using the edges of the graph to move symbols around during encoding. Both techniques will be required to obtain a positive answer to Question 27.3. Before we delve into the techniques mentioned earlier, we first define the version of expanders graphs that we will need in this subsection.

DEFINITION 27.6 (Expander Graphs). A bipartite graph $G = (L, R, E)$ is said to be an $(\ell, n, d_L, \alpha, \gamma)$-*expander* if the following holds. $|L| = \ell$, $|R| = n$ and the degree of each node in L is d_L. More importantly, every subset of nodes $A \subseteq L$ with $|A| \leq \alpha\ell$ has at least $\gamma|A|$ neighbors in Y.

We will think degree parameter d_L as a constant. Ideally, we would like γ to be as large as possible. Note that $\gamma \leq d_L$. It can be shown that random bipartite graph with the left-degrees being d_L with high probability have $\gamma = d(1 - \varepsilon)$ for any $\varepsilon > 0$. Such expanders are called loss-less expanders.

27.3.3.1 The Basic Construction

As was mentioned earlier, one way to use an $(n, m, d_L, \alpha, \gamma)$-expander is as a factor graph for an LDPC code. Note that such a code will have a rate at least $1 - m/n$. The decoding algorithm for such a code follows by several rounds of bit-flipping. (Given an assignment to the variable nodes, a check node or the corresponding parity check is unsatisfied if the parity of the values of the variable nodes in its neighborhood is not 0. Note that if all the check nodes are satisfied then the vector corresponding to the variable nodes is indeed a codeword.) More precisely, in every round, every variable node flips its value in parallel if the number of neighboring checks nodes with unsatisfied parity checks is at least $2d_L/3$. Otherwise it does not change its value. We make some remarks on this decoding algorithm. First, the proof of correctness of this algorithm proceeds by showing that if the fraction of errors to begin with is bounded then in each round the number of bits that are in error decrease by a factor of $2/3$. This claim crucially uses the connectivity and the sparsity properties of the underlying expander graph to argue that most of the unsatisfied check nodes are adjacent to a single variable node that has an erroneous bit (and thus, flipping the value corresponding to that variable node would satisfy the parity check). There is a danger that too many variable nodes with correct values can also flip their bits. However, again using the connectivity property of the underlying expander one can show that this is not the case. Second, a careful accounting for the nodes that need to be flipped in each round leads to a linear-time implementation. Picking parameters correctly, one can make the aforementioned argument rigorous and obtain the following result.

THEOREM 27.12 *Let* $0 < \varepsilon < \frac{1}{12}$ *and C be the LDPC code corresponding to an* $(n, m, d, \alpha, d(1-\varepsilon))$-*expander. C has rate at least* $1 - m/n$ *and can be decoded from a* $\alpha(1 - 3\varepsilon)$ *fraction of errors in* $O(n)$ *time.*

Note that for the aforementioned code to be explicit, one needs explicit constructions of loss-less expanders. Such a construction was recently obtained (Capalbo et al., 2002).

27.3.3.2 Linear-Time Encoding

We now briefly discuss the code construction in Spielman (1996) that in addition to the linear decoding complexity as guaranteed by Theorem 27.12 also has linear encoding time complexity. These codes are the only binary codes known that have provable linear-time encoding and decoding for worst-case errors. Unlike the construction of Theorem 27.12 where expanders are used to define the parity check matrix, in Spielman's construction, the expander graph is used to define the generator matrix. In particular, Spielman's construction first defines what he called error reduction codes. These are systematic codes, that is, the codeword consists of the message bits followed by some parity check bits. The structure of the expander graphs is used to compute the parity checks from the message bits. In this construction, the variable nodes only correspond to the message bits. Further, the expander graphs are (d_v, d_c)-regular graphs (as was the case with Gallager's LDPC codes from Section 27.3.2.3). Thus, such error reduction codes have a trivial linear-time encoding algorithm. Unfortunately, these codes by themselves cannot be decoded from a large number of errors. However, an algorithm similar to the one used in Theorem 27.12 can be used to obtain an intermediate received word that has at most half the number of errors in the original received word. Such error reduction codes can be used recursively to obtain the final linear-time encodable and decodable codes.

Here we just state a special case of the general result in Spielman (1996).

THEOREM 27.13   *For every small enough $\gamma > 0$, there exists explicit binary codes of rate $1/(1+\gamma)$ that can be encoded in linear time and decoded in linear time from up to $\Omega(\gamma^2/\log^2(1/\gamma))$ fraction of errors.*

27.3.3.3   Approaching Half the Singleton Bound

Unfortunately the result of Spielman (1996) only works for about 10^{-6} fraction of errors. We will now see another general technique introduced in Alon et al. (1995) that uses expanders to improve the fraction of errors that can be corrected. We start with an informal description of the construction. The code uses two codes C_{out} and C_{in} and an expander graph G. The final code C^* is constructed as follows. Let $C' = C_{\text{out}} \circ C_{\text{in}}$ be the code concatenation of C_{out} and C_{in} (see Section 27.3.1 for more details on code concatenation). For our purposes C' will be a binary code. Now symbols in codewords from C' are redistributed in the following manner to obtain codewords in C^*. Let G be a (d_L, d_R)-regular bipartite expander. The symbols from a codeword in C' are blocked into d_L bits (each corresponding to a codeword in C_{in}) and then placed on the left vertices of G. These symbols are then pushed along the edges in some predetermined order. For example, the ℓ'th bit in the r'th C_{in} encoding can be sent to the ℓ'th right vertex neighbor of the r'th left vertex. A right vertex then collects the bits on its incident edges and then juxtaposes them to form symbols in $\{0,1\}^{d_R}$. These juxtaposed symbols on the right vertices form the codeword in C^*. Note that as G is used to redistribute the symbols, the rates of C^* and C' are the same.

We now briefly discuss how the aforementioned technique was used in Guruswami and Indyk (2005). The code C_{in} will be a Reed–Solomon code over constant-sized alphabet. C_{out} will be the binary code from Theorem 27.13, which we will think of as a code over a suitable larger alphabet by simply grouping together the bits of appropriate length (recall we did a similar thing with the outer code in Section 27.3.1.2). Since C_{out} can be encoded in linear time, C^* can also be encoded in linear time. The decoding algorithm is very natural. First, given the received word $\mathbf{y}$, we invert the symbol redistribution using G in the encoding procedure to get an intermediate received word $\mathbf{y}'$. In the next stage, $\mathbf{y}'$ is then decoded using the natural decoder for the concatenated code C' (as discussed in Section 27.3.1.2): C_{in} can be decoded using the polynomial time unique decoding algorithm for the Reed–Solomon codes while C_{out} can be decoded using the linear-time decoding algorithm from Theorem 27.13. Note that the resulting decoding algorithm also runs in linear time. The proof of correctness of this algorithm proceeds by showing that a certain pseudorandomness property of G smoothens out the errors when passing from $\mathbf{y}$ to $\mathbf{y}'$. More precisely, for most of the inner blocks corresponding to received words for C_{in}, the fraction of errors is roughly the fraction of errors in $\mathbf{y}'$. For each such block, the decoding algorithm for C_{in} corrects all the errors. The few remaining errors are then corrected by the decoding algorithm for C_{out}. Selecting parameters carefully and formalizing the argument earlier leads to the following result, which answers Question 27.3 in the affirmative.

THEOREM 27.14 *For every $0 < R < 1$, and all $\varepsilon > 0$, there is an explicit family of codes of rate at least $1 - R - \varepsilon$ over an alphabet of size $2^{O(\log(1/\varepsilon)/(\varepsilon^4 R))}$, that can encoded in linear time and decoded from a $(1 - R - \varepsilon)/2$ fraction of errors in linear time.*

27.3.4 List Decoding

In this section we return to Question 27.4. The answer to this question is known in the affirmative only for codes over large alphabets. In this subsection we review the sequence of work that has led to this partial positive answer. Reed–Solomon codes will play a crucial role in our discussions.

27.3.4.1 List Decoding of Reed–Solomon Codes

Consider the $[n, k + 1]_q$ the Reed–Solomon codes with the set of evaluation points as the nonzero elements of $\mathbb{F}$, which is denoted by $\mathbb{F}_q^* = \{1, \gamma, \gamma^2, \ldots, \gamma^{n-1}\}$ where $n = q - 1$ and γ is the generator of the cyclic group $\mathbb{F}_q^*$. Under list decoding of such a Reed–Solomon code, given the received word $\mathbf{y} = \langle y_0, \ldots, y_{n-1} \rangle$, we are interested in all degree k polynomials $f(X)$ such that for at least $(1+\delta)\sqrt{R}$ fraction of positions $0 \le i \le n - 1$, $f(\gamma^i) = y_i$. We now sketch the main ideas of the algorithms in Sudan (1997) and Guruswami and Sudan (1999). The algorithms have two main steps: the first is an interpolation step and the second one is a root finding step. In the interpolation step, the list-decoding algorithm finds a bivariate polynomial $Q(X, Y)$ that fits the input. That is,

$$\text{for every position } i, Q(\gamma^i, y_i) = 0.$$

Such a polynomial $Q(\cdot, \cdot)$ can be found in polynomial time if we search for one with large enough total degree. This amounts to solving a system of linear equations. After the interpolation step, the root finding step finds all factors of $Q(X, Y)$ of the form $Y - f(X)$. The crux of the analysis is to show that

$$\text{for every degree } k \text{ polynomial } f(X) \text{ that satisfies } f(\gamma^i) = y_i \text{ for at least } (1 + \delta)\sqrt{R}$$
$$\text{fraction of positions } i, Y - f(X) \text{ is indeed a factor of } Q(X, Y).$$

However, the aforementioned analysis is not true for every bivariate polynomial $Q(X, Y)$ that satisfies $Q(\gamma^i, y_i) = 0$ for all positions i. The main ideas in Sudan (1997) and Guruswami and Sudan (1999) were to introduce more constraints on $Q(X, Y)$. In particular, Sudan (1997) added the constraint that a certain weighted degree of $Q(X, Y)$ is below a fixed upper bound. Specifically, $Q(X, Y)$ was restricted to have a nontrivially bounded $(1, k)$-weighted degree. The $(1, k)$-weighted degree of a monomial $X^i Y^j$ is $i + jk$ and the $(1, k)$-weighted degree of a bivariate polynomial $Q(X, Y)$ is the maximum $(1, k)$-weighted degree among its monomials. The intuition behind defining such a weighted degree is that given $Q(X, Y)$ with the $(1, k)$-weighted degree of D, the univariate polynomial $Q(X, f(X))$, where $f(X)$ is some degree k polynomial, has total degree at most D. The upper bound D is chosen carefully such that if $f(X)$ is a codeword that needs to be output, then $Q(X, f(X))$ has more than D zeroes and thus $Q(X, f(X)) \equiv 0$, which in turn implies that $Y - f(X)$ divides $Q(X, Y)$. To get to the bound of $1 - (1 + \delta)\sqrt{R}$, Guruswami and Sudan (1999) added a further constraint on $Q(X, Y)$ that required it to have r roots at (γ^i, y_i), where r is some parameter (in Sudan (1997) $r = 1$ while in Guruswami and Sudan (1999), r is roughly $1/\delta$). Choosing parameters carefully leads to the following result.

THEOREM 27.15 *Let $0 < R < 1$. Then any Reed–Solomon code of rate at least R can be list-decoded from a $1 - \sqrt{R}$ fraction of errors.*

We note that the aforementioned result holds for any Reed–Solomon code and not just the code where the set of evaluation points is $\mathbb{F}_q^*$.

27.3.4.2 List Decoding of Reed–Solomon Like Codes

We now discuss the recent developments in Parvaresh and Vardy (2005) and Guruswami and Rudra (2008). These consider the variants of the Reed–Solomon codes that are no longer linear. The codes considered in Guruswami and Rudra (2008) are a strict subset of those considered in Parvaresh and Vardy (2005). For the ease of presentation, we will present the ideas of Parvaresh and Vardy (2005) using these smaller subset of codes.

Folded Reed–Solomon code with folding parameter $m \geq 1$, is exactly the Reed–Solomon code considered in Section 27.3.4.1, but viewed as a code over $(\mathbb{F}_q)^m$ by bundling together m consecutive symbols of codewords in the Reed–Solomon code. For example with $m = 2$ (and n even), the Reed–Solomon codeword $\langle f(1), f(\gamma), f(\gamma^2), f(\gamma^3), \ldots, f(\gamma^{n-2}), f(\gamma^{n-1}) \rangle$ will correspond to the following codeword in the folded Reed–Solomon code: $\langle (f(1), f(\gamma)), (f(\gamma^2), f(\gamma^3)), \ldots, (f(\gamma^{n-2}), f(\gamma^{n-1})) \rangle$. We will now briefly present the ideas which can be used to show that folded Reed–Solomon codes with folding parameter m can be list decoded up to a $1 - (1 + \delta) \left(\frac{m}{m-s+1} \right) R^{s/(s+1)}$ fraction of errors for any $1 \leq s \leq m$. Note that Theorem 27.15 handles the $m = s = 1$ case.

We now consider the next nontrivial case of $m = s = 2$. The ideas for this case can be easily extended to the general $m = s$ case. Note that now given the received word $\langle (y_0, y_1), (y_2, y_3), \ldots, (y_{n-2}, y_{n-1}) \rangle$ we want to find all degree k polynomials $f(X)$ such that for at least $2(1 + \delta) \sqrt[3]{R^2}$ fraction of positions $0 \leq i \leq n/2 - 1$, $f(\gamma^{2i}) = y_{2i}$ and $f(\gamma^{2i+1}) = y_{2i+1}$. As in the Reed–Solomon case, we will have an interpolation and a root finding step. The interpolation step is a straightforward generalization of the Reed–Solomon case: we find a trivariate polynomial $Q(X, Y, Z)$ that fits the received word, that is, for every $0 \leq i \leq n/2 - 1$, $Q(\gamma^{2i}, y_{2i}, y_{2i+1}) = 0$. Further, $Q(X, Y, Z)$ has an upper bound on its $(1, k, k)$-weighted degree (which is a straightforward generalization of the $(1, k)$-weighted degree for the bivariate case) and has a multiplicity of r at every point. For the root finding step, it suffices to show that for every degree k polynomial $f(X)$ that needs to be output, $Q(X, f(X), f(\gamma X)) \equiv 0$. This, however does not follow from weighted degree and multiple root properties of $Q(X, Y, Z)$. Here we will need two new ideas, the first of which (due to Guruswami and Rudra (2008)) is to show that with a suitable choice of an irreducible polynomial* $E(X)$ of degree $q - 1$, $f(X)^q \equiv f(\gamma X) \mod (E(X))$.[†] The second idea, due to Parvaresh and Vardy (2005) is the following. We first obtain the bivariate polynomial (over an appropriate extension field) $T(Y, Z) \equiv Q(X, Y, Z) \mod (E(X))$. Note that by the first idea, we are looking for solutions on the curve $Z = Y^q$ (Y corresponds to $f(X)$ and Z corresponds to $f(\gamma X)$ in the extension field). The crux of the argument is to show that all the polynomials $f(X)$ that need to be output correspond to (in the extension field) some root of the equation $T(Y, Y^q) = 0$.

To go from $s = m$ to any $s \leq m$ requires the following idea due to Guruswami and Rudra (2008): First, the problem of list decoding folded Reed–Solomon code with folding parameter m is reduced to the problem of list decoding folded Reed–Solomon code with folding parameter s. Then the algorithm outlined in the previous paragraph is used for the folded Reed–Solomon code with folding parameter s. A careful tracking of the agreement parameter in the reduction brings down the final agreement fraction that is required for the original folded Reed–Solomon code with folding parameter m from $m(1 + \delta) \sqrt[m+1]{R^m}$ (which can be obtained without the reduction) to $(1 + \delta) \left(\frac{m}{m-s+1} \right) \sqrt[s+1]{R^s}$. Choosing parameters carefully leads to the following result:

---

* An irreducible polynomial $E(X)$ over $\mathbb{F}_q$ has no nontrivial polynomial factors.

[†] This idea shows that folded Reed–Solomon codes are special cases of the codes considered in Parvaresh and Vardy (2005).

THEOREM 27.16   *For every $0 < R < 1$ and $0 < \varepsilon \leq R$, there is an explicit family of folded Reed–Solomon codes that have rate at least R and which can be list decoded up to a $1 - R - \varepsilon$ fraction of errors in time (and outputs a list of size at most) $(N/\varepsilon^2)^{O(\varepsilon^{-1}\log(1/R))}$ where N is the block length of the code. The alphabet size of the code as a function of the block length N is $(N/\varepsilon^2)^{O(1/\varepsilon^2)}$.*

One drawback of the aforementioned result is that the alphabet size of the code increases with the block length. However, it is shown in Guruswami and Rudra (2008) that the code concatenation and expander-based techniques from Section 27.3.3.3 can be used to obtain the following result (which resolves Question 27.4 for large enough alphabets).

THEOREM 27.17   *For every R, $0 < R < 1$, every $\varepsilon > 0$, there is a polynomial time constructible family of codes over an alphabet of size $2^{O(\varepsilon^{-4}\log(1/\varepsilon))}$ that have rate at least R and which can be list decoded up to a fraction $(1 - R - \varepsilon)$ of errors in polynomial time.*

We remark that the best known explicit construction of codes with list decoding algorithms for binary codes use code concatenation (Section 27.3.1). Expander codes (Section 27.3.3) have been used to obtain explicit codes with linear time list decoding algorithms.

27.4   Research Issues and Summary

The goal of algorithmic coding theory is to design (explicit) codes along with efficient encoding and decoding algorithms such that the best possible combinatorial trade-off between the rate of the code and the fraction of errors that can be corrected is achieved. (This trade-off is generally captured by the notion of the capacity of a channel.) This is generally achieved via two steps. First, the combinatorial trade-off is established. Generally random codes achieve this trade-off. The next and more challenging step is to achieve the capacity of the channel (possibly with explicit codes and) with efficient encoding and decoding algorithms. In the first part of this chapter, we presented results on the capacity of some classical channels. In the second part, we presented mostly recent work on the progress toward capacity-achieving explicit codes along with efficient encoding and decoding algorithms.

Almost 60 years since the birth of coding theory, codes over large alphabets are better understood. However, challenging open questions related to such codes still remain. For example, as we saw in Section 27.3.4, Question 27.4 has been answered for codes over large alphabets. Although these codes do achieve the best possible trade-off between rate and fraction of errors, the guarantee on the worst-case list size if far from satisfactory. In particular, to get within ε of capacity the worst-case size guaranteed Theorem 27.17 grows as $n^{1/\varepsilon}$. This should be compared to the worst-case list size of $O(1/\varepsilon)$ achieved by random codes (Theorem 27.7).

As was alluded to the earlier, most of the outstanding open questions in algorithmic coding theory relate to codes over small or fixed alphabets, in particular binary codes. For the stochastic noise models, getting to within ε of capacity (e.g., the BSC) with encoding and decoding time complexities that have polynomial dependence on $1/\varepsilon$ and linear dependence on the block length is an important challenge. One promising avenue is to prove that irregular LDPC codes from Section 27.3.2.4 meet this challenge. The experimental results suggest that this might be true.

One of the biggest open questions in coding theory is to design explicit binary codes that have the same rate vs. distance trade-off as random binary codes, that is, meet the Gilbert–Varshamov (GV) bound (Theorem 27.6). In fact achieving this bound of any constant rate is wide open. Another classical open question in this vein is to determine the optimal trade-off between rate and distance for binary codes: the best lower bound on rate for given distance bound is the GV bound while the best upper bound is achieved by the so-called MRRW bound (McEliece et al., 1977). There is a gap between the two bounds. Closing the gap between these two bounds remains an important open problem. Another challenge is to achieve a positive resolution of Question 27.4 for codes over fixed alphabets in general and binary codes in particular. One of the obstacles in meeting challenges related to binary codes is that not many explicit constructions of such codes with some constant rate and constant relative distance are known. In fact, all known constructions of such codes either use code concatenation or expander graphs.

27.5 Further Information

Given the space limitations of this chapter, many important research themes in algorithmic coding theory have regretfully been omitted. The results from classical work in coding theory can be found in the standard coding textbooks such as MacWilliams and Sloane (1981) and van Lint (1999). Another great resource for such results are the two volumes of the *Handbook of Coding theory* (Pless and Huffman, 1998). An excellent source for most of the material covered in this chapter is Sudan's notes from his course on Coding Theory (Sudan, 2001). A conspicuous absence from this chapter is any discussion of algebraic geometric codes (cf. Høholdt et al. (1998)). These are codes based on some deep mathematics involving function fields that can be thought of as a generalization of the Reed–Solomon codes. These techniques give codes with excellent rate vs. distance properties over small alphabets. For alphabets of size greater than 49 these codes beat the Gilbert–Varshamov bound.

For more details on code concatenation, the reader can consult any of the aforementioned references (there is a dedicated chapter in the handbook of coding theory, (Dumer, 1998)). Guruswami's introductory survey on LDPC codes (Guruswami, 2006b) is a good place to get more details regarding the material presented in Section 27.3.2. The book Richardson and Urbanke (2008) has a more comprehensive treatment. Another valuable resource is the February 2001 issue of Volume 47 of the journal *IEEE Transactions on Information Theory*: this was a special issue dedicated to iterative decoding and specifically contains the sequence of papers Luby et al. (2001a,b); Richardson and Urbanke (2001a,b); and Richardson et al. (2001). This series of papers perhaps constituted the most important post-Gallager work on LDPC codes and laid the foundation for the recent spurt in research activity in LDPC codes. Guruswami's survey (Guruswami, 2004a) is a good starting point for more details on expander codes. The material presented in Section 27.3.3 appeared in Sipser and Spielman (1996); Spielman (1996); Alon et al. (1995); and Guruswami and Indyk (2005). Sudan (2000) is a nice introductory survey on list decoding and its applications in complexity theory. The authoritative text on list decoding is Guruswami (2004b). For more details on the recent developments see the survey (Guruswami, 2006a) or the author's thesis (Rudra, 2007).

As was mentioned in the introduction, codes have found numerous applications in theoretical science. The reader is referred to these surveys (and the references within) for more details: Trevisan (2004); Sudan (2000); Guruswami (2004a, 2006c).

Many, if not most, papers on algorithmic coding theory are published in the *IEEE Transactions on Information Theory*. A number of research papers are presented at the annual International Symposium on Information Theory (ISIT). Papers are also presented at the IEEE Information Theory workshop and the Allerton Conference on Communication, Control and Computing.

Defining Terms

Alphabet: The set of symbols over which codewords are defined.

Binary codes: Codes defined over an alphabet of size two.

Block length: Length of the codewords in a code.

Capacity: Threshold on the rate of a code for which reliable communication is possible on a channel.

Code: A collection of vectors of the same length defined over a fixed set.

Codeword: A vector in a code.

Decoding: Function that when given the received word outputs what it thinks was the transmitted message. It can also report a failure.

Dimension: The number $\log_q(|C|)$, where the code C is defined over an alphabet of size q.

Encoding: Function that converts the original message to a codeword.

Hamming distance: Number of positions in which two vectors of the same length differ.

Linear code: A code over $\mathbb{F}_q^n$ and block length n that is a linear subspace of $\mathbb{F}_q^n$.

Minimum distance: The minimum Hamming distance between any two distinct codewords in a code.

Random codes: Code that maps the k message symbols to n symbols, each of which is picked uniformly and independently at random.

Rate: Ratio of the dimension and the block length of a code.

Received word: The corrupted codeword obtained after transmission.

References

Alon, N. and Spencer, J. (1992). *The Probabilistic Method.* John Wiley & Sons, Inc., New York.

Alon, N., Edmonds, J., and Luby, M. (1995). Linear time erasure codes with nearly optimal recovery (extended abstract). In *Proceedings of the 36th Annual Symposium on Foundations of Computer Science (FOCS)*, pp. 512–519.

Berlekamp, E. R., McEliece, R. J., and van Tilborg, H. C. A. (1978). On the inherent intractability of certain coding problems. *IEEE Transactions on Information Theory*, 24:384–386.

Capalbo, M. R., Reingold, O., Vadhan, S. P., and Wigderson, A. (2002). Randomness conductors and constant-degree lossless expanders. In *Proceedings of the 34th annual ACM symposium on Theory of computing (STOC)*, pp. 659–668.

Chandler, D. G., Batterman, E. P., and Shah, G. (1989). Hexagonal, information encoding article, process and system. U.S. Patent Number 4,874,936.

Chen, C. L. and Hsiao, M. Y. (1984). Error-correcting codes for semiconductor memory applications: A state-of-the-art review. *IBM Journal of Research and Development*, 28(2):124–134.

Chen, P. M., Lee, E. K., Gibson, G. A., Katz, R. H., and Patterson, D. A. (1994). RAID: High-performance, reliable secondary storage. *ACM Computing Surveys*, 26(2):145–185.

Dumer, I. I. (1998). Concatenated codes and their multilevel generalizations. In V. S. Pless and W. C. Huffman, editors, *Handbook of Coding Theory*, volume 2. pp. 1911–1988. North Holland, Amsterdam, the Netherlands.

Elias, P. (1957). List decoding for noisy channels. Technical Report 335, Research Laboratory of Electronics, MIT.

Elias, P. (1991). Error-correcting codes for list decoding. *IEEE Transactions on Information Theory*, 37:5–12.

Forney, G. D. (1966). *Concatenated Codes.* MIT Press, Cambridge, MA.

Gallager, R. G. (1963). *Low-Density Parity-Check Codes.* MIT Press, Cambridge, MA.

Goldreich, O. and Levin, L. (1989). A hard-core predicate for all one-way functions. In *Proceedings of the 21st Annual ACM Symposium on Theory of Computing*, pp. 25–32.

Guruswami, V. (2004a). Error-correcting codes and expander graphs. *SIGACT News*, 25–41.

Guruswami, V. (2004b). List decoding of error-correcting codes. Number 3282 in *Lecture Notes in Computer Science*. Springer (Winning Thesis of the 2002 ACM Doctoral Dissertation Competition).

Guruswami, V. (2006a). Algorithmic results in list decoding. In *Foundations and Trends in Theoretical Computer Science (FnT-TCS)*, volume 2. NOW publishers, Hanover, MA.

Guruswami, V. (2006b). Iterative decoding of low-density parity check codes. *Bulletin of the EATCS*, 90:53–88.

Guruswami, V. (2006c). List decoding in pseudorandomness and average-case complexity. In *IEEE Information Theory Workshop*, 13–17 March, pp. 32–36.

Guruswami, V. and Indyk, P. (2005). Linear-time encodable/decodable codes with near-optimal rate. *IEEE Transactions on Information Theory*, 51(10):3393–3400.

Guruswami, V. and Rudra, A. (2008). Explicit codes achieving list decoding capacity: Error-correction with optimal redundancy. *IEEE Transactions on Information Theory*, 54(1):135–150.

Guruswami, V. and Sudan, M. (1999). Improved decoding of Reed–Solomon and algebraic–geometric codes. *IEEE Transactions on Information Theory*, 45(6):1757–1767.

Hamming, R. W. (1950). Error detecting and error correcting codes. *Bell System Technical Journal*, 29:147–160.

Høholdt, T., van Lint, J. H., and Pellikaan, R. (1998). Algebraic geometry codes. In V. S. Pless, W. C. H. and A.Brualdi, R., editors, *Handbook of Coding Theory*. North Holland, Amsterdam, the Netherlands.

Justesen, J. (1972). A class of constructive asymptotically good algebraic codes. *IEEE Transactions on Information Theory*, 18:652–656.

Luby, M., Mitzenmacher, M., Shokrollahi, M. A., and Spielman, D. A. (2001a). Efficient erasure correcting codes. *IEEE Transactions on Information Theory*, 47(2):569–584.

Luby, M., Mitzenmacher, M., Shokrollahi, M. A., and Spielman, D. A. (2001b). Improved low-density parity-check codes using irregular graphs. *IEEE Transactions on Information Theory*, 47(2): 585–598.

MacWilliams, F. J. and Sloane, N. J. A. (1981). *The Theory of Error-Correcting Codes*. Elsevier/North-Holland, Amsterdam, the Netherlands.

McEliece, R. J., Rodemich, E. R., Rumsey Jr., H., and Welch, L. R. (1977). New upper bounds on the rate of a code via the Delsarte-Macwilliams inequalities. *IEEE Transactions on Information Theory*, 23:157–166.

Parvaresh, F. and Vardy, A. (2005). Correcting errors beyond the Guruswami–Sudan radius in polynomial time. In *Proceedings of the 46th Annual IEEE Symposium on Foundations of Computer Science (FOCS)*, pp. 285–294.

Peterson, L. L. and Davis, B. S. (1996). *Computer Networks: A Systems Approach*. Morgan Kaufmann Publishers, San Francisco, CA.

Pless, V. S. and Huffman, W. C., editors (1998). *Handbook of Coding Theory*. North Holland, Amsterdam, the Netherlands.

Reed, I. S. and Solomon, G. (1960). Polynomial codes over certain finite fields. *SIAM Journal on Applied Mathematics*, 8:300–304.

Richardson, T. J. and Urbanke, R. L. (2001a). The capacity of low-density parity-check codes under message-passing decoding. *IEEE Transactions on Information Theory*, 47(2):599–618.

Richardson, T. J. and Urbanke, R. L. (2001b). Efficient encoding of low-density parity-check codes. *IEEE Transactions on Information Theory*, 47(2):638–656.

Richardson, T. and Urbanke, R. (2008). *Modern Coding Theory*. Cambridge University Press, Cambridge, U.K.

Richardson, T. J., Shokrollahi, M. A., and Urbanke, R. L. (2001). Design of capacity-approaching irregular low-density parity-check codes. *IEEE Transactions on Information Theory*, 47(2):619–637.

Rudra, A. (2007). *List Decoding and Property Testing of Error Correcting Codes*. PhD thesis, University of Washington, Seattle, WA.

Shannon, C. E. (1948). A mathematical theory of communication. *Bell System Technical Journal*, 27:379–423, 623–656.

Sipser, M. and Spielman, D. (1996). Expander codes. *IEEE Transactions on Information Theory*, 42(6):1710–1722.

Spielman, D. (1996). Linear-time encodable and decodable error-correcting codes. *IEEE Transactions on Information Theory*, 42(6):1723–1732.

Sudan, M. (1997). Decoding of Reed–Solomon codes beyond the error-correction bound. *Journal of Complexity*, 13(1):180–193.

Sudan, M. (2000). List decoding: Algorithms and applications. *SIGACT News*, 31:16–27.

Sudan, M. (2001). Lecture notes on algorithmic introduction to coding theory. Available at http://people.csail.mit.edu/madhu/FTOI

Trevisan, L. (2004). Some applications of coding theory in computational complexity. *Quaderni di Matematica*, 13:347–424.

van Lint, J. H. (1999). *Introduction to Coding Theory*. Graduate Texts in Mathematics 86 (Third Edition) Springer-Verlag, Berlin, Germany.

Wicker, S. B. and Bhargava, V. K., editors (1999). *Reed–Solomon Codes and Their Applications*. John Wiley & Sons, Inc, Piscataway, NJ.

Wozencraft, J. M. (1958). List decoding. Quarterly Progress Report, Research Laboratory of Electronics, MIT, 48:90–95.

Zyablov, V. V. and Pinsker, M. S. (1981 (in Russian); pp. 236-240 (in English), 1982). List cascade decoding. *Problems of Information Transmission*, 17(4):29–34.

28

Parallel Computation: Models and Complexity Issues

Raymond Greenlaw
Armstrong Atlantic State University

H. James Hoover
University of Alberta

28.1 Introduction

The theory of **parallel computation** is concerned with the development and analysis of parallel computing models, the techniques for solving and classifying problems on such models, and the implications of that work.

Even though we have seen dramatic increases in computer performance, humans are adept at inventing problems for which a single processor is simply too slow. Problem complexity has increased significantly over the past 20 years. The number of computational operations required to solve some of these problems is mind boggling, yet problems and their data sizes continue to grow rapidly. Only by using many processors in parallel is there any hope of quickly solving such problems. And, as a result of increased problem complexity, inherent technological limitations to individual processor speeds, and economic factors, the first decade of the twenty-first century has seen a resurgence in parallel computing [1]. In principle, having more processors working on a problem means it can be solved much faster. Ideally, we expect a speedup of the following form:

$$\text{Parallel Time} = \frac{\text{Sequential Time}}{\text{Number of Processors}}$$

In practice, parallel algorithms seldom attain ideal speedup,* are more complex to design, and are more difficult to implement than single-processor algorithms. The designer of a parallel algorithm must grapple with these fundamental issues:

1. What parallel resources are available?
2. What kind of speedup is desired?
3. What machine architecture should be used?
4. How can the problem be decomposed to exploit parallelism?
5. Whether the problem is even amenable to a parallel attack?

The dilemma of parallel computation is that at some point every problem begins to resist speedup. The reality is that as one adds processors one must devote (disproportionately) more resources to interprocessor communication, and one must deal with more problems caused by processors waiting for others to finish. Furthermore, some problems appear to be so resistant to speedup via parallelism that they have earned the name inherently sequential.

28.1.1 Pragmatic versus Asymptotic Parallelism

The practice of parallel computation can be loosely divided into the pragmatic and the asymptotic.

The goal of pragmatic parallel computation is simply to speed up the computation as much as possible using whatever parallel techniques and equipment are available. For example, doing arithmetic on 32 bits in parallel, overlapping independent I/O operations on a storage subsystem, fetching instructions and precomputing branch conditions, or using four processors interconnected on a common bus to share workload are all pragmatic uses of parallelism. Pragmatic parallelism is tricky, very problem-specific, and highly effective at obtaining the modest factor of 5 or 10 speedup that can suddenly make a problem reasonable to solve. This chapter is not directly concerned with pragmatic parallel computation as defined above, although the models and techniques we explain in this chapter are.

This chapter focuses on asymptotic parallel computation, which in contrast to pragmatic parallelism, is more concerned with the architecture for general parallel computation; parallel algorithms for solving broadly applicable, fundamental problems; and the ultimate limitations of parallel computation. For example, are shared memory multiprocessors more powerful than mesh-connected parallel machines? Given an unlimited number of processors, just how fast can one do n-bit arithmetic? Or, can every polynomial time sequential problem be solved in polylogarithmic time on a parallel machine? Asymptotic parallel computation does provide tools for the pragmatic person too, for example, the algorithm designer with many processors available can make use of the results from this field. The field of asymptotic parallel computation can be subdivided into the following main areas:

---

* The speedup is simply the ratio (Sequential Time/Parallel Time). In general, the speedup is less than the number of processors.

1. Models: comparing and evaluating different parallel machine architectures
2. Algorithm design: using a particular architecture to solve a specific problem
3. Computational complexity: classifying problems according to their intrinsic difficulty of solution

Many books are devoted to the design of parallel algorithms, for example [18,24,38]. The focus of this chapter is on models of parallel computation and complexity. When discussing parallel computation it is common to adopt the following informal definitions:

- A problem is feasible if it can be solved by a parallel algorithm with worst-case time and processor complexity $n^{O(1)}$.

- A problem is feasible and feasibly highly parallel if it can be solved by an algorithm with worst-case time complexity $(\log n)^{O(1)}$ and processor complexity $n^{O(1)}$.

- A problem is inherently sequential if it is feasible but has no feasible highly parallel algorithm for its solution.

28.1.2 Chapter Overview

Section 28.2—two fundamental models of parallel computation: the parallel random access machines and uniform Boolean circuit families

Section 28.3—the fundamental parallel complexity classes P and NC

Section 28.4—other important parallel models and mutual simulations

Section 28.5—P-completeness, the theory of inherently sequential problems

Section 28.6—various examples of P-complete problems

Section 28.7—research issues and summary

Section 28.8—Further Information, references for further reading.

defining terms

28.2 Two Fundamental Models of Parallel Computation

An important part of parallel computation involves the description, classification, and comparison of abstract models of parallel machines. In this section we examine two important models.

28.2.1 Introduction

One of the prerequisites to studying a problem's parallel complexity is specifying the resources that we are concerned about consuming, for example, time and processors, and the abstract machine model on which we want to compute our solutions. For practicing programmers, these two items are usually dictated by the real machines at their disposal. For example, having only a small number of processors, each with a very large memory, is a much different situation from having tens of thousands of processors each with a very small memory. In the first situation we wish to minimize the number of processors required by our solution, and will choose a model that ignores memory consumption. In the second situation, where minimizing local memory may be more important, we choose a model that provides many processors (although not extravagantly many) and in which all basic operations are assumed to take the same amount of time to execute. The single biggest issue to be faced from the standpoint of developing a parallel algorithm is the granularity of the parallel operations. Should the problem be broken up into very small units of computations, which can be done in parallel but then may incur costly communication overhead, or should the problem be

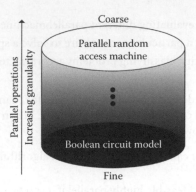

FIGURE 28.1 Parallel models at the opposite ends of the parallel operations granularity scale.

divided into relatively large chunks that can be done in parallel but where each chunk is processed sequentially?

The opposite ends of our granularity spectrum are captured by the two main models of parallel computation: parallel random access machines (see Section 28.2.2) and uniform Boolean circuit families (see Section 28.2.3). For high-level, coarse-granularity situations the preferred model is the PRAM, while for more detailed questions of implementability and small resource bounds, the low-level, finer granularity uniform Boolean circuit model is used. Figure 28.1 illustrates the parallel granularity spectrum.

Although there are many differences among parallel models, for feasible, highly parallel computations, most models are equivalent to within a polynomial in both time and hardware resources, simultaneously. By this we mean that if the size-n instances of some problem can be solved in time $T(n) = (\log n)^{O(1)}$ and processors $P(n) = n^{O(1)}$ on a machine from model M_1, then there exists a machine from model M_2 that can solve the size n instances of the problem in time $T(n)^{O(1)}$ and $P(n)^{O(1)}$ processors. Thus, if a problem is feasibly highly parallel on one model, the problem is feasibility highly parallel on all other equivalent models (see Section 28.4 for more details). This statement says that the class of problems solvable by feasibly highly parallel computations is robust and insensitive to minor variations in the underlying computational model.

28.2.2 Parallel Random Access Machines

One of the natural ways of modeling parallel computation is the generalization of the single processor to a shared memory multiprocessor, as illustrated in Figure 28.2. This machine consists of many powerful independent machines (usually viewed as being of the same type) that share a large common memory which is used for computation and communication. The parallel machines envisioned in such a model contain very large numbers of processors, on the order of millions. Such enormous machines do not exist yet, but machines with hundreds of thousands of processors can be built now, and economics rather than technology is the main factor that limits their size.

The formal model of the shared memory multiprocessor is called a parallel random access machine (PRAM), and was introduced independently in [16,19,20]. The PRAM model consists of a collection of random access machine (RAM) processors that run in parallel and communicate via a common memory. The basic PRAM model consists of an unbounded collection of numbered RAM processors $P_0, P_1, P_2, \ldots$ and an unbounded collection of shared memory cells $C_0, C_1, C_2, \ldots$ (see Figure 28.2). Inputs and outputs to the PRAM computation are placed in shared memory to allow concurrent access. Each instruction is executed in unit time, synchronized over all active processors.

Each RAM processor P_i has its own local memory, knows its own index i, and has instructions for direct and indirect read/write access to the shared memory. Local memory is unbounded, and

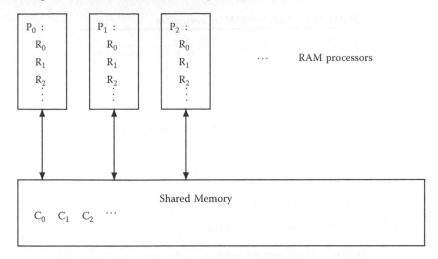

FIGURE 28.2 The parallel random access machine (PRAM). A PRAM consists of a large number of processors connected to a common memory. Each processor is quite powerful by itself, and operates independently of the others. Processors communicate by passing information through the shared memory. Arbitration may be required when more than one processor attempts to access a given memory cell at the same time.

consists of memory cells $R_0, R_1, R_2, \ldots$, with each cell capable of holding an integer of unbounded size. The usual complexity measures for each individual processor's RAM computation are time, in the form of the number of instructions executed, and space, in the form of the number of memory cells accessed.

A typical PRAM instruction set, with addressing modes, is given in Table 28.1. In this simple machine, local memory cell R_0 serves as an accumulator so that at most one read and one write to shared memory occurs for each instruction. The multiply and divide instructions take only a constant operand in order to prevent the rapid generation and testing of very large numbers. These restrictions also prevent the consumption of an exponential amount of space in polynomial time.

Two important technical issues must be dealt with by the model. The first is the manner in which a finite number of the processors from the potentially infinite pool are activated for a computation. A common way is for processor P_0 to have a special activation register that specifies the maximum index of an active processor. Any non-halted processor with an index smaller than the value in the register can execute its program. Initially only processor P_0 is active, and all others are suspended waiting to execute their first instruction. P_0 then computes the number of processors required for the computation and loads this value into the special register. Computation proceeds until P_0 halts, at which point all active processors halt. The SIMDAG model is an example of a PRAM using such a convention [20]. Another common approach is to have active processors explicitly activate new ones via FORK instructions [16].

The second technical issue concerns the way in which simultaneous access to shared memory is arbitrated. In all models, it is assumed that the basic instruction cycle separates shared memory read operations from write operations. Each PRAM instruction is executed in a cycle with three phases. First the read operation (if any) from shared memory is performed, then the computation associated with the instruction (if any) is done, and finally the write operation (if any) to shared memory is performed. This cycle eliminates read/write conflicts to shared memory, but does not eliminate all access conflicts. These conflicts are dealt with in a number of ways as described in Table 28.2 (see [15,48] for more details). All of these variants of the PRAM are deterministic, except for the ARBITRARY CRCW-PRAM, for which it is possible that repeated executions on identical inputs result in different outputs.

TABLE 28.1   Sample PRAM Instruction Set

Instruction	Description
$\alpha \leftarrow \beta$	Move data to cell with address α from cell with address β

Address	Description
R_i	Local cell R_i
R_{R_i}	Local cell with address given by contents of R_i
C_i	Shared cell C_i
C_{R_i}	Shared cell with address given by contents of R_i

IDENT	Load the processor number into R_0
CONST c	Load the constant c into R_0
ADD α	Add contents of α to R_0
SUB α	Subtract contents of α from R_0
MULT c	Multiply contents of R_0 by constant c
DIV c	Divide contents of R_0 by constant c and truncate
GOTO i	Branch to instruction i
IFZERO i	Branch to instruction i if contents of R_0 is 0
HALT	Stop execution of this processor

Any given PRAM computation will use some specific time and hardware resources. The complexity measure corresponding to time is simply the time taken by the longest running processor. The measure corresponding to hardware is the maximum number of active processors during the computation.

Our standard PRAM model will be the CREW-PRAM with a processor activation register in processor P_0. This means that processor P_0 is guaranteed to have run for the duration of the computation, and the largest value in the activation register is an upper bound on the number of processors used. We also need to specify how the inputs are provided to a PRAM computation, how the outputs are extracted, and how the cost of the computation is accounted. Definition 28.1 formalizes these ideas.

DEFINITION 28.1   Let M be a PRAM. The input/output conventions for M are as follows. An input $x \in \{0, 1\}^n$ is presented to M by placing the integer n in shared memory cell C_0, and the bits $x_1, \ldots, x_n$ of x in shared memory cells $C_1, \ldots, C_n$. M displays its output $y \in \{0, 1\}^m$ similarly: integer m in shared memory cell C_0, and the bits $y_1, \ldots, y_m$ of y in shared memory cells $C_1, \ldots, C_m$.

M computes in parallel time $T(n)$ and processors $P(n)$ if and only if for every input $x \in \{0, 1\}^n$, machine M halts within at most $T(n)$ time steps, activates at most $P(n)$ processors, and presents some output $y \in \{0, 1\}^*$.

M computes in sequential time $T(n)$ if and only if it computes in parallel time $T(n)$ using 1 processor.

TABLE 28.2 How Different PRAM Models Resolve Access Conflicts

CRCW	Concurrent-Read Concurrent-Write
	Allows simultaneous reads and writes to the same memory cell with a mechanism for arbitrating simultaneous writes to the same cell
	PRIORITY—only the write by the lowest numbered contending processor succeeds
	COMMON—the write succeeds only if all processors are writing the same value
	ARBITRARY—any one of the writes succeeds
CREW	Concurrent-Read Exclusive-Write
	Allows simultaneous reads of the same memory cell, but only one processor may attempt to write to a cell
CROW	Concurrent-Read Owner-Write
	A common restriction of the CREW-PRAM that preassigns an owner to each common memory cell. Simultaneous reads of the same memory cell are allowed, but only the owner can write to the cell, thus ensuring exclusive-write access (see [14])
EREW	Exclusive-Read Exclusive-Write
	Requires that no two processors simultaneously access any given memory cell

With these conventions in place, and having decided on one version of the PRAM model to be used for all computations, we can talk about a function being computed in parallel time $T(n)$ and processors $P(n)$.

DEFINITION 28.2 Let f be a function from $\{0, 1\}^*$ to $\{0, 1\}^*$. The function f is computable in parallel time $T(n)$ and processors $P(n)$ if and only if there is a PRAM M that on input x outputs $f(x)$ in time $T(n)$ and processors $P(n)$.

Note that no explicit accounting is made of the local or shared memory used by the computation. Since the PRAM is prevented from generating large numbers, that is, for $T \geq \log n$ no number may exceed $O(T)$ bits in T steps, a computation of time T with P processors cannot store more than $O(PT^2)$ bits of information. Hence, for our purposes P and T together adequately characterize the memory requirement of the computation, and there is no need to parameterize memory separately.

All of the various PRAM models are polynomially equivalent with respect to feasible, highly parallel computations, and so any one of them is suitable for defining the complexity classes P and NC that we present in Section 28.3. The final important point to note about the PRAM model is that it is generally not difficult to see (in principle) how to translate an informally described parallel algorithm into a PRAM algorithm.

28.2.3 Uniform Boolean Circuit Families

Boolean circuits are designed to capture very fine-grained parallel computation at the resolution of a single bit—they are a formal model of the combinational logic circuit. Circuits are basic technology, consisting of very simple logical gates connected by bit-carrying wires. They have no memory and no notion of state. Circuits avoid almost all issues of machine organization and instruction repertoire. Their computational components correspond directly with devices that we can actually fabricate, although the circuit model is still an idealization of real electronic devices. The circuit model ignores a host of important practical considerations such as circuit area, volume, pin limitations, power dissipation, packaging, and signal propagation delay. Such issues are addressed more accurately by more complex VLSI models (see [35]). But for much of the study of parallel computation, the Boolean circuit model provides a good compromise between simplicity and realism.

Each circuit is an acyclic directed graph in which the edges carry unidirectional logical signals, and the vertices compute elementary Boolean logical functions. Formally, we denote the logical functions by the sets $B_k = \{f \mid f : \{0, 1\}^k \to \{0, 1\}\}$. That is, each B_k is the set of all k-ary Boolean functions. We refer informally to such functions by strings "1," "0," "¬," "∧," "∨," "NOT," "AND," "OR," and so on. The entire graph computes a Boolean function from the inputs to the outputs in a natural way.

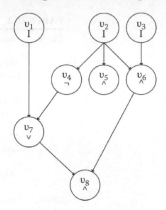

FIGURE 28.3 A sample Boolean circuit. A Boolean circuit with inputs $\langle v_1, v_2, v_3 \rangle$ and outputs $\langle v_5, v_8 \rangle$. The circuit has size 8, depth 3, and width 2. Input v_2 has fanout 3. Gate v_4 has fanin 1 and fanout 1.

DEFINITION 28.3   A Boolean circuit α is a labeled finite oriented directed acyclic graph. Each vertex v has a type $\tau(v) \in \{I\} \cup B_0 \cup B_1 \cup B_2$. A vertex v with $\tau(v) = I$ has indegree 0 and is called an input. The inputs of α are given by a tuple $\langle x_1, \dots, x_n \rangle$ of distinct vertices. A vertex v with outdegree 0 is called an output. The outputs of α are given by a tuple $\langle y_1, \dots, y_m \rangle$ of distinct vertices. A vertex v with $\tau(v) \in B_i$ must have indegree i and is called a gate.

Note that fanin is less than or equal to two, but fanout is unrestricted. Inputs and gates can also be outputs. See Figure 28.3 for an example.

DEFINITION 28.4   A Boolean circuit α with inputs $\langle x_1, \dots, x_n \rangle$ and outputs $\langle y_1, \dots, y_m \rangle$ computes a function $f : \{0, 1\}^n \to \{0, 1\}^m$ in the following way: input x_i is assigned a value $\nu(x_i)$ from $\{0, 1\}$ representing the ith bit of the argument to the function. Every other vertex v is assigned the unique value $\nu(v) \in \{0, 1\}$ obtained by applying $\tau(v)$ to the value(s) of the vertices incoming to v. The value of the function is the tuple $\langle \nu(y_1), \dots, \nu(y_m) \rangle$ in which output y_j contributes the jth bit of the output.

The most common resource measures of interest for a circuit are its size and depth.

DEFINITION 28.5   The size of α, denoted size(α), is the number of vertices in α. The depth of α, denoted depth(α), is the length of the longest path in α from an input to an output.

A less common measure is width, which intuitively corresponds to the maximum number of gate values, not counting inputs, that need to be preserved when the circuit is evaluated level-by-level.

DEFINITION 28.6   The width of α, denoted width(α), is

$$\max_{0 < i < \text{depth}(\alpha)} \left| \left\{ v : \begin{array}{l} 0 < d(v) \le i \text{ and} \\ \text{there is an edge from vertex } v \text{ to a vertex } w, d(w) > i \end{array} \right\} \right|$$

where $d(w)$, the depth of w, is the length of the longest path from any input to vertex w.

Each circuit α is described by a binary string denoted by $\overline{\alpha}$. This description can be thought of as a blueprint for that circuit, or alternatively as a parallel program executed by a universal circuit simulator. In any case, although we speak of circuits, we actually generate and manipulate circuit descriptions (exactly as we manipulate programs and not Turing machines). One common description is the standard encoding, the precise details of which are not important to this chapter (see [39]). The main point is that circuit descriptions are simple objects to generate and manipulate.

An individual circuit with n inputs and m outputs is a finite object computing a function from binary strings of length n to binary strings of length m. In contrast to a PRAM computation in which one algorithm handles all possible lengths of inputs, different circuits are required for different length inputs. The collection of different circuits for the various input lengths is called a circuit family.

The simplest kind of circuit family is used for computing some function f whose output length m is a function, possibly constant, only of the length of the input. That is, the length of $f(x)$ on n-bit inputs x is some function $m(n)$. In this case we can represent the function $f : \{0,1\}^* \to \{0,1\}^*$ by an infinite sequence of circuits, $\{\alpha_n\}$, where circuit α_n computes f restricted to inputs of length n. Such a sequence of circuits is called a Boolean circuit family.

DEFINITION 28.7 A Boolean circuit family $\{\alpha_n\}$ is a collection of circuits, each α_n computing a function $f^n : \{0,1\}^n \to \{0,1\}^{m(n)}$. The function computed by $\{\alpha_n\}$, denoted f_α, is the function $f_\alpha : \{0,1\}^* \to \{0,1\}^*$, defined by $f_\alpha(x) \equiv f^{|x|}(x)$.

Functions where the output length varies based on the value of x as well as x's length can be handled by special encodings (similar to the output convention used for PRAMs). The case where the length of the output is always 1 is particularly important for defining formal languages.

DEFINITION 28.8 Let $\{\alpha_n\}$ be a Boolean circuit family that computes the function $f_\alpha : \{0,1\}^* \to \{0,1\}$. The language decided by $\{\alpha_n\}$, denoted L_α, is the set $L_\alpha = \{x \in \{0,1\}^* \mid f_\alpha(x) = 1\}$.

Defining an infinite collection of circuits with no computational constraints gives nonuniform circuit families. Nonuniform circuit families are unexpectedly powerful in that they can compute non-computable functions. These circuit families are widely used as objects of lower bound proofs, where their power merely serves to strengthen the significance of the lower bounds. However, nonuniform circuits are a somewhat unsatisfactory model in which to consider upper bounds. In particular, there may be no effective way, given n, to obtain a description of the nth circuit α_n.

The uniform approach gives an explicit algorithm for constructing the elements of the circuit family. Each circuit family is defined by a program in some computational model that takes n as input and then outputs the encoding $\overline{\alpha_n}$ of the nth member. In doing so, an infinite object, the family, is effectively described by a finite object, the program. The question then becomes, how much computational power is permitted in producing the description $\overline{\alpha_n}$?

Borodin, arguing that the circuit constructor should have no more computational power than the object it constructs, introduced the notion of uniformity [2]. One such example of a weak constructor is a Turing machine that is limited to only $O(\log n)$ work space on inputs of length n. Such machines have limited computing power but can still describe a wide class of useful circuit families.

DEFINITION 28.9 A family $\{\alpha_n\}$ of Boolean circuits is logarithmic space uniform if the transformation $1^n \to \overline{\alpha_n}$ can be computed in $O(\log(\text{size}(\alpha_n)))$ space on a deterministic Turing machine.

Note how the complexity of producing the description of α_n is expressed in terms of the size of the resulting circuit. Logarithmic space uniformity is sometimes called Borodin–Cook uniformity, and was first mentioned in [6]. This notion of uniformity has the desirable property that the description $\overline{\alpha_n}$ can be produced in polynomial time sequentially, or in polylogarithmic time in parallel with a polynomial number of processors. Thus, the circuit constructor is reasonable from both sequential and parallel perspectives.

28.2.4 Equivalence of PRAMs and Uniform Boolean Circuit Families

We have remarked that many parallel models are equivalent with respect to feasible highly parallel algorithms. That is, if a problem has a feasible highly parallel solution on one model, then it also has such a solution on any equivalent model. Originally, the notion of feasible and highly parallel came from the observation that certain problems had polylogarithmic time and polynomial processor solutions on many different models. This ability to support feasibly highly parallel algorithms has now become the defining characteristic of all reasonable parallel models. In order for any new parallel model to be considered reasonable, it must be able to simulate relatively efficiently some existing reasonable model and vice versa. (See the remarks in Section 28.7 about more-realistic parallel models.)

In the sense just described, PRAMs and uniform Boolean circuits are both reasonable parallel models, and can simulate each other in a way that maintains feasible highly parallel computations. This result, as other simulation results, is quite technical, and sensitive to the precise details of each model involved. The basic ideas are relatively simple, and easily worked to completion by those with a penchant for detail (see [22] for more details). To simulate a circuit operating on input x of length n, the PRAM first uses the uniformity condition to compute the description of the circuit from the family that handles the inputs of length n. Then it initializes the inputs and does a gate-by-gate simulation of the circuit, evaluating in parallel all the gates at the same level. The simulation of a PRAM by a circuit is a bit more complex. The circuit for inputs of length n has to account for the worst-case time and processor consumption of the PRAM, and the product of these two gives the size of the circuit. The circuit is then organized as a sequence of layers, each layer of which simulates one time step of each processor of the PRAM.

THEOREM 28.1 *A function f from $\{0, 1\}^*$ to $\{0, 1\}^*$ can be computed by a logarithmic space uniform Boolean circuit family $\{\alpha_n\}$ with $depth(\alpha_n) = (\log n)^{O(1)}$ and $size(\alpha_n) = n^{O(1)}$ if and only if f can be computed by a CREW-PRAM M on inputs of length n in time $T(n) = (\log n)^{O(1)}$ and processors $P(n) = n^{O(1)}$.*

Similar simulation results among the various models of parallel computation (see Section 28.4) allow us to observe that if a problem is inherently sequential on one reasonable parallel model, then it is inherently sequential on all other reasonable parallel models.

28.3 Fundamental Parallel Complexity Classes

28.3.1 Introduction

The most important complexity classes in parallel computation are P, NC, and the class of P-complete problems. See Figure 28.4 for the context of these classes. The problems in P are considered to be easy to solve on a single processor; they are tractable or feasible. Such problems may still take unacceptable amounts of sequential time to solve and so are ideal candidates for parallel computation. NC consists of those problems in P that can be solved very fast in parallel. The P-complete problems appear to be outside of NC. The classes NC and P are defined formally in the next section, and the P-complete problems, which require more technical background, are defined formally in Section 28.5.

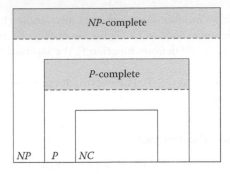

FIGURE 28.4 The classes *NC* and *P*.

28.3.2 Nick's Class (*NC*) and Polynomial Time (*P*)

Because of their simplicity, language recognition and decision problems are the standard mechanisms for defining the computational classes of complexity theory. We refer the reader to Chapters 22 through 24 for background and more in-depth discussion. The following definitions aid in defining the classes* *NC* and *P*.

DEFINITION 28.10   Let *L* be a language over $\{0,1\}^*$. The characteristic function of *L* is the function f_L defined on all $x \in \{0,1\}^*$ such that $f_L(x) = 1$ if $x \in L$, and $f_L(x) = 0$ if $x \notin L$.

DEFINITION 28.11   A language $L \subseteq \{0,1\}^*$ is decidable in sequential time $T(n)$ if and only if the characteristic function of *L* can be computed in sequential time $T(n)$.

DEFINITION 28.12   A language $L \subseteq \{0,1\}^*$ is decidable in parallel time $T(n)$ with $P(n)$ processors if and only if the characteristic function of *L* is computable in parallel time $T(n)$ and processors $P(n)$.

A single sequential processor running in polynomial time can easily simulate a polynomial number of processors running in polynomial time, and conversely. These observations mean that we can ignore sequential computation and restrict our attention to PRAMs, as per the following lemma.

LEMMA 28.1   A language *L* is decidable in sequential time $n^{O(1)}$ if and only if *L* is decidable in parallel time $n^{O(1)}$ with $n^{O(1)}$ processors.

We can now define the class of feasible highly parallel problems, *NC*, and the class of polynomial time sequential problems, *P*.

DEFINITION 28.13   The class *NC* is the set of all languages *L* that are decidable in parallel time $(\log n)^{O(1)}$ and processors $n^{O(1)}$.

DEFINITION 28.14   The class *P* is the set of all languages *L* that are decidable in parallel time $n^{O(1)}$ and processors $n^{O(1)}$.

---

* *NC* was named after Nick Pippenger. *P* stands for polynomial.

Many complexity classes have functional analogs, typically denoted by prefixing the class with the letter F. For example, FNC denotes function NC — the class of functions computable in the resource bounds of the class NC. Similarly, FP denotes function P. We assume the reader has an intuition for what these classes are.

From Lemma 28.1, we know that $NC \subseteq P$. The important open question for parallel computation is whether this inclusion is proper.

28.3.2.1 A Basic Example—Parallel Sums

Many problems in P have truly dramatic speed improvements when solved in parallel. Here we consider a simple but important problem, Parallel Sums, defined as follows:

Given: Natural numbers $a_1, a_2, \ldots, a_n$, and t.

Problem: Is $t = a_1 + a_2 + \cdots + a_n$?

We can solve this problem sequentially by adding numbers together two at a time. Such a procedure requires $n - 1$ additions to compute the total, and the total can be checked against t to solve the problem. Parallel Sums is an example of a problem that is sequentially feasible. Now consider solving Parallel Sums using a number of processors operating in parallel.

With limited parallelism, which means using a fixed number of processors, we can never achieve more than a constant factor of improvement over the best-known sequential time. For example, with two processors we can never be more than twice as fast as with one. But suppose that processors were so plentiful and inexpensive that we could consider using thousands or more in parallel. This assumption introduces a qualitative change in the way we approach parallel computation. What kind of speedup could be achieved if we had essentially an unlimited number of processors?

With polynomially bounded parallelism, we have the potential to achieve more than a constant factor speedup. For example, suppose that we use $P(n) = n/2$ processors to solve an instance of Parallel Sums consisting of n numbers and a value t. That is, we have one processor for each pair of numbers. The computation can then be organized as a binary tree in which we add as many pairs of numbers as possible in each time step. Assuming a unit cost for addition, the time to compute the addition of n numbers is the height of the tree, so the computation can be done in $O(\log n)$ elapsed time using $n/2$ processors.

We must emphasize the importance of this example. Using only a relatively small number of processors (about half the size of the problem instance), we have achieved an exponential* improvement in the solution of parallel sums by going from $O(n)$ sequential time to $O(\log n)$ parallel time. Problems that exhibit this kind of improvement are exactly those that belong to the class NC.

28.3.3 Does *NC* Equal *P*?

Many of the problems in P have highly parallel solutions similar to Parallel Sums, and one wonders if perhaps every problem in P can be solved fast on a parallel machine.

> Does every problem with a feasible sequential solution also have a feasible highly parallel solution? That is, does NC equal P?

Unfortunately, it seems that some problems in P do not lie in NC, and one of the main tasks of complexity theory is to identify these suspected inherently sequential problems. The theory of P-completeness described in Section 28.5 provides evidence suggesting NC and P are indeed different classes.

---

* By exponential we mean going from a polynomial function to a polylogarithmic function. Therefore, an improvement from n^3 to $(\log n)^2$ is considered an exponential improvement.

28.4   Parallel Models and Simulation Results

28.4.1   Introduction

Parallel models of computation vary widely in modes of communication, instruction sets, and in what constitutes a processor. There is no clearly superior architecture for a general purpose parallel computer, and the same is true for the theoretical models of parallel computation. Thus, there is room for inventing new architectures and computational features. These new models are evaluated by two basic methods: assessing how well they speed up existing sequential computation classes, and how well they do in mutual simulations with existing parallel models. The comparison with existing parallel models also provides information about how easy the model is to use and how close the model is to a real machine. The ultimate aim of this exercise is to identify the key attributes that must be present in any reasonable parallel model.

In this section we introduce a number of synchronous parallel models and some simulation results. A parallel machine is synchronous if all processors must complete the execution of their current instruction before any processor begins execution of its next instruction. Brief descriptions, providing only a taste of the models, are given in Sections 28.4.3 and 28.4.4. Many of the important technical issues concerning the models can be extrapolated from the PRAM and Boolean circuit models given in Sections 28.2.2 and 28.2.3, respectively. The meanings of the names of complexity classes used in this section are easily inferred, and are mostly historical. For example, UAG-TIME stands for uniform aggregate time and ATIME, SPACE represents alternating Turing machines with simultaneous time and space bounds.

The simulation results that relate two seemingly different parallel models M_1 and M_2 follow a basic pattern. If any problem Π can be solved on a machine of type M_1 using T_{M_1} time and P_{M_1} processors, we want to know whether Π can be solved on an M_2 machine in some related bounds? For example, using $f_T(T_{M_1})$ time and $f_P(P_{M_1})$ processors, where f_T and f_P are two well-behaved functions relating M_1's resources of time and processors respectively, to M_2's. What about vice versa? Do time and processor bounds for problem Π on model M_2 imply time and processor bounds on M_1?

The simulation results that relate a parallel model and a sequential model have a different purpose. Here the focus is on how parallel resources relate to sequential resources. This approach provides an understanding of what kinds of sequential problems can be automatically parallelized. A useful framework for this direction was proposed by Cook in [7]; we follow his discussion and build upon it in the next section.

28.4.2   Cook's Classification Scheme for Parallel Models

The various parallel models, which are defined in the following sections, are grouped into two classes: the fixed structure models and the modifiable structure models (see Table 28.3). "The fixed structure models correspond to sequential machines with a fixed storage structure such as Turing machine tape." They consist of alternating Turing machines, bounded fanin uniform Boolean circuit families, conglomerates, k-PRAMs, and uniform aggregates. This class represents parallel models whose interconnection pattern is fixed throughout a computation.

"The modifiable structure models correspond to the modifiable sequential Storage Modification Machines [41] and RAMs." They consist of hardware modification machines, MRAMs (standard RAMs with a multiplication instruction [23]), PRAMs, SIMDAGs, unbounded fanin uniform Boolean circuit families, and vector machines. This class represents parallel machines whose communication links are allowed to vary during a computation.

The above mentioned parallel machines can be grouped according to their time resource usage as related to sequential space on a Turing machine. For a fixed structure model X, the following relationship typically holds:

TABLE 28.3 Classification Scheme of Parallel Models

Fixed Structure Models
    Alternating Turing machine
    Bounded fanin uniform Boolean circuit family
    Conglomerate
    k-Parallel random access machine
    Uniform aggregate

Modifiable Structure Models
    Hardware modification machine
    Multiplication random access machine
    Parallel random access machine
    Single instruction stream multiple data Stream global Memory machine
    Unbounded fanin uniform Boolean circuit family
    Vector machine

$$\text{X-TIME}(S(n)) \subseteq \text{DSPACE}(S(n)) \subseteq \text{NSPACE}(S(n)) \subseteq \text{X-TIME}(S(n)^2) \tag{28.1}$$

whereas a modifiable structure model Y usually satisfies

$$\text{Y-TIME}(S(n)) \subseteq \text{DSPACE}(S(n)^2) \subseteq \text{Y-TIME}(S(n)^2) \tag{28.2}$$

For more on DSPACE (deterministic Turing machine space) and NSPACE (nondeterministic Turing machine space) see Chapter 22. Equation* 28.2 is not known to be true for some of the models in the modifiable class. For example, it is not known how to simulate $S(n)$ space bounded Turing machines by vector machines or MRAMs running in time $S(n)$. The SIMDAG can simulate NSPACE($S(n)$) in linear time.

28.4.3 The Fixed Structure Models

28.4.3.1 Uniform Families of Bounded Fanin Boolean Circuits

As presented in Section 28.2.3, the resources of interest for circuits are depth, size, and width. Circuit depth is a useful measure of parallel time because a lower bound on the running time for a problem Π using the circuit model can be applied to parallel machines that are implemented using circuit technology, or to parallel models that are equivalent to circuits with respect to the time resource. The fixed structure of circuits as opposed to the programmable nature of some other parallel models facilitates the proof of lower bounds as noted in [43].

A shortcoming of the circuit model is that it does not provide a very good measure of hardware. The size of the circuit provides an upper bound on the hardware size, but intuitively, this measure is a poor bound because a circuit is not allowed to reuse any of its gates. Width seems to provide a reasonable measure of hardware size for synchronous circuits (see Theorem 28.4). Uniform circuit depth corresponds to sequential space in the following way:

THEOREM 28.2 *[2] For $T(n) \geq \log n$ UDEPTH($T(n)$) $\subseteq$ DSPACE($T(n)$) $\subseteq$ NSPACE($T(n)$) $\subseteq$ UDEPTH($T(n)^2$).*

The time bound $T(n)$ has to be computable within the same resource constraints established by the uniformity conditions of the model. This requirement is in practice only a technical annoyance, and

---

* As is customary, we refer to the relations in 28.1 and 28.2 as Equations 28.1 and 28.2, respectively, even though they are not mathematical equations.

so from now on, unless otherwise stated, we assume that $T(n) \geq \log n$, and that $T(n)$ is constructible in $O(\log n)$ space.

28.4.3.2 Uniform Aggregates

Aggregates, defined in [12], are circuits with feedback which have a special input convention that does not count the hardware required to hold the input. The feedback allows gates to be reused during a computation and thus reflects more accurately the amount of hardware required to solve problems. The resources of interest for an aggregate are hardware and time. Aggregates, like circuits, work for only one fixed input length per aggregate, and thus they require the same notions of family and uniformity as for circuits. The unusual input convention for aggregates allows a computation to use less than linear hardware. Aggregates like circuits provide a reasonable measure of parallel time. In fact, the following theorem shows that they provide the same measure of time as circuits. (All of these are results from [12,13].)

THEOREM 28.3 $UAG\text{-}TIME(T(n)) = UDEPTH(T(n)).$

As a corollary to Theorems 28.2 and 28.3, we get the following result.

COROLLARY 28.1 $UAG\text{-}TIME(T(n)) \subseteq DSPACE(T(n)) \subseteq NSPACE(T(n)) \subseteq UAG\text{-}TIME$ $(T(n)^2).$

Theorem 28.2 and Corollary 28.1 provide evidence for the Parallel Computation Thesis that is discussed later in Section 28.4.5. The following result relates uniform synchronous circuit width to uniform aggregate hardware. A circuit is synchronous if for all gates in the circuit all inputs to a gate v come from gates at depth $(d(v) - 1)$ (see Definition 28.5).

THEOREM 28.4 $DSPACE(S(n)) = UAG\text{-}HARDWARE(S(n)) = USWIDTH(S(n)).$

Width represents the maximum number of gates that have to be active at any one time during a level-by-level evaluation of a circuit, and thus circuit width gives a reasonable measure for hardware size.

28.4.3.3 Conglomerates

A conglomerate is defined as a collection of interconnected finite state machines $M_0, M_1, M_2, \ldots$ [19]. Each M_i is deterministic and has $r \geq 1$ inputs and 1 output. The conglomerate is unlike a circuit in that it may have loops, and unlike an aggregate in that its input convention implies hardware size $\Omega(n)$. A uniformity requirement similar to what one does for circuits and aggregates is imposed on the connection function of conglomerates. On a given input of size n, the first n machines each contain one bit of the input. All machines begin in a designated initial state. The computation halts if M_0 enters a special final state.

THEOREM 28.5 *[20] If C is a conglomerate whose connection function f can be computed in space n on a Turing machine then* $CONG\text{-}TIME(T(n)) \subseteq DSPACE(T(n)) \subseteq NSPACE(T(n)) \subseteq$ $CONG\text{-}TIME(T(n)^2).$

Time is the only resource that has been formally studied for the conglomerate. The obvious definition of hardware for a conglomerate is the number of finite state machines that are used during a computation. For values of hardware size $\Omega(n)$, conglomerates provide the same measure of hardware as aggregates do.

28.4.3.4 k-PRAMs

The k-PRAM [40] does not have any global memory structure. Instead a processor is allowed to communicate with another processor through a call by passing parameters or via a return through the channel registers. A processor is also allowed the following instruction: if child processor returned then $label_1$ else $label_2$. This mechanism provides an additional way to communicate because the return time can affect the parent's computation. Communication among processors in a k-PRAM is much more restricted than in PRAM models because their processors share a global memory. The constant k indicates the branching factor out of a processor. The inability of a processor to communicate with an unbounded number of processors is the motivation for placing the model in the fixed structure class. Theorem 28.6 shows that time on this nondeterministic model is polynomially related to sequential space, although this result is not as tight as for some of the other parallel models.

THEOREM 28.6 *[40] If $T(n) \geq n$ is k-PRAM-countable then $NSPACE(T(n)) \subseteq k$-NPRAM-TIME$(T(n)^2)$. If $T(n) \geq n$ is RAM-constructible then k-NPRAM-TIME$(T(n)) \subseteq NSPACE(T(n)^3)$.*

28.4.3.5 Alternating Turing Machines

An alternating Turing machine [3] is a generalization of a nondeterministic Turing machine to have universal acceptance states as well as existential acceptance states (see Chapter 23). ATM resources are time, space, and alternation, where alternation counts the maximum number of changes between existential and universal acceptance states, in an accepting subtree of an ATM computation.

ATMs are particularly suited for problems that can be modeled as two-person games. One drawback of the ATM as a parallel model is that there does not seem to be a resource measure for the ATM that corresponds to hardware [7].

The result stated next demonstrates the close relationship between ATMs and uniform families of bounded fanin Boolean circuits.

THEOREM 28.7 *[39] If S and T are computable in deterministic time $O(T(n))$ then ATIME, SPACE$(T(n), S(n)) = UDEPTH, SIZE(T(n), 2^{O(S(n))})$.*

The next theorem of Ruzzo and Tompa (see [43] for a proof) shows that ATMs with alternation considered as a resource are a good model of unbounded fanin, parallel computation (see Section 28.4.4.1).

THEOREM 28.8 *If $T(n)$ and $S(n)$ are suitable functions then ATM-ALT, SPACE$(T(n), S(n)) = CRCW$-PRAM-TIME, PROC$(T(n), 2^{O(S(n))})$.*

28.4.4 The Modifiable Structure Models

28.4.4.1 SIMDAGs and CRCW-PRAMs

The SIMDAG was formally defined in [19]. SIMDAG is an acronym for Single Instruction Stream, Multiple Data Stream, Global Memory. The model is more commonly known as the CRCW-PRAM. Goldschlager proposed a fair cost per instruction version of the SIMDAG, called the charged

SIMDAG. The charged SIMDAG satisfies Equation 28.1, whereas the following theorem holds for the SIMDAG.

THEOREM 28.9 *[19] If $T(n) \geq \log n$ is SIMDAG-countable then NSPACE $(T(n)) \subseteq$ SIMDAG-TIME$(T(n)) \subseteq$ DSPACE$(T(n)^2)$.*

28.4.4.2 Uniform Families of Unbounded Fanin Boolean Circuits

Lower bounds for unbounded fanin Boolean circuits were proved in [17]. The following theorem by Stockmeyer and Vishkin [43] relates CRCW-PRAMs to uniform families of unbounded fanin Boolean circuits in such a way that these lower bounds also hold for the CRCW-PRAM.

THEOREM 28.10 *[43] There is a constant c and a function $q(P, T, n)$ bounded above by a polynomial in P, T, and n such that the following holds. Let M be a CRCW-PRAM with processor bound $P(n)$ that operates in time $T(n)$. There is a constant d_M and, for each n, a circuit C_n of size $d_M q(P(n), T(n), n)$ and depth $cT(n)$ such that C_n realizes the input–output behavior of M on inputs of size n.*

Combining Theorem 28.10 with results from [17], we get the following corollary.

COROLLARY 28.2 [43] A CRCW-PRAM with a polynomially bounded number of processors that operates in constant time cannot compute parity, multiply integers, find the transitive closure of a graph, determine whether a graph has a perfect matching, or sort integers.

This corollary illustrates why Theorem 28.10 is a useful step toward unifying the modifiable parallel models. It allows the lower bounds already known for one model to be transferred over to another. It is possible to simulate unbounded fanin Boolean circuits via PRAMs [43]. See Section 28.2.4 for the bounded fanin case of this simulation.

28.4.4.3 Hardware Modification Machines

The hardware modification machine (HMM) (see [11–13]) was defined to measure hardware more effectively. A hardware modification machine is made up of a finite collection of finite state machines. The machines are connected together in the same manner as in the conglomerate, the main difference being that the connections may be changed locally during a computation. A single HMM works for all input lengths so there is no need to define uniformity. The resources of interest for HMMs are time and hardware. The input convention for the HMM is such that we can consider hardware values which are sublinear. The following theorem shows that the HMM satisfies Equation 28.2.

THEOREM 28.11 *[12,13] DSPACE$(T(n)) \subseteq$ HMM-TIME$(T(n)) \subseteq$ DSPACE$(T(n)^2)$.*

This theorem illustrates that despite the HMM's processors being allowed only a bounded number of connections at any step during the computation, the HMM can still simulate DSPACE linearly. The following theorem shows the relationship between DSPACE and HMM hardware.

THEOREM 28.12 *[12,13] HMM-HARDWARE$(H(n)) \subseteq$ DSPACE $(H(n)(\log n + \log H(n)))$.*

Other interesting results for parallel pointer machines are given in [33].

28.4.5 Parallel Computation Thesis

The Parallel Computation Thesis is that time-bounded parallel machines are polynomially related to space-bounded sequential machines [19]. That is, for any function $T(n)$,

$$\text{PARALLEL-TIME}(T(n)^{O(1)}) = \text{SEQUENTIAL-SPACE}(T(n)^{O(1)}). \tag{28.3}$$

The Parallel Computation Thesis, by constructively relating sequential space to parallel time, provides an automatic mechanism for obtaining a fast parallel algorithm from a highly space-efficient sequential algorithm.

Any model satisfying Equation 28.1 or 28.2 gives support to the Parallel Computation Thesis. For example, Theorem 28.2, Corollary 28.1, Theorem 28.5, and Theorem 28.11 are results of the form described in Equation (28.1) for uniform bounded fanin circuits, uniform aggregates, conglomerates, and HMMs, respectively. The next few results provide additional evidence supporting the Parallel Computation Thesis for vector machines, CREW-PRAMs, and ATMs.

THEOREM 28.13   *[36] If $T(n)$ is VM-countable and $T(n) \geq \log n$ then NSPACE$(T(n)) \subseteq$ VM-TIME$(T(n)^2)$. If $T(n)$ is RAM-constructible and $T(n) \geq \log n$ then VM-TIME$(T(n)) \subseteq$ DSPACE$(T(n)^2)$.*

THEOREM 28.14   *[16] CREW-PRAM-TIME$(T(n)) \subseteq$ DSPACE$(T(n)^2) \subseteq$ CREW-PRAM-TIME $(T(n)^2)$.*

THEOREM 28.15   *[3] ATIME$(T(n)) \subseteq$ DSPACE$(T(n)) \subseteq$ NSPACE$(T(n)) \subseteq$ ATIME$(T(n)^2)$.*

The proofs of the simulation of sequential space by parallel time are similar to, and motivated by, Savitch's Theorem (see Chapter 22). The general structure of such proofs is as follows:

1. Let M be an $S(n)$ space bounded Turing machine (deterministic or nondeterministic) with initial configuration C_0 and unique final configuration C_f.
2. Observe that M has at most $2^{O(S(n))}$ possible configurations.
3. Construct the state transition matrix or transition graph for M.
4. Compute the transitive closure by repeated squaring of the transition matrix or perform path doubling in the transition graph.
5. Accept if and only if the entry at position $C_0 C_f$ is 1 in the transition matrix, or if and only if there is a path from C_0 to C_f in the transition graph.

For the other direction, of simulating parallel time by sequential space, the basic technique is to perform a depth-first search on the instructions executed by the machine being simulated. This search is done using a recursive procedure. The recursive procedure's initial call often has the form VERIFY(q, x, t), where q is a final condition, x is the input, and t is the time bound. The procedure VERIFY checks to see that the final state or accept instruction is reached on input x in time t. The recursion depth is the same as the running time, say $T(n)$, of the machine being simulated. In situations where the recursion requires only constant space at each level, the whole simulation can be performed using linear space. If the recursion needs space $T(n)$ in each call to store parameters and return values, then the simulation takes quadratic space.

28.5 *P*-Completeness Theory

The following is a basic presentation of *P*-completeness theory. For an in-depth study, see [22]. We begin by looking at reducibility, completeness, and proof methodologies. In Section 28.6 we present a number of examples of *P*-complete problems.

28.5.1 Reducibility

The theory of *P*-completeness is useful for categorizing problems that are likely to be inherently sequential; this theory parallels the theory of *NP*-completeness. One of the key ideas required in the development of *P*-completeness theory is the ability to relate one problem to another, via the notion of reducibility (see Chapter 23 for more on this subject). Here we focus on one of the most basic forms.

DEFINITION 28.15   A language L is *NC* many–one reducible or *NC* **reducible** to L', written $L \leq_m^{NC} L'$, if and only if there is a function f in *FNC* such that $x \in L$ if and only if $f(x) \in L'$.

Since *NC* reducibility is transitive, we can use a series of individual reductions to relate two problems to each other.

LEMMA 28.2   *NC* reducibility is transitive. That is, whenever $L \leq_m^{NC} L'$ and $L' \leq_m^{NC} L''$, then $L \leq_m^{NC} L''$.

In many reductions from a language L to a language L', the exact complexity of L' is unknown. Although this reduction gives us no absolute information about the computational complexity of L, it still provides useful information about the relative difficulties of the two languages. In particular, assuming the reduction is not too powerful, it implies that L is no more difficult to decide than L'. It is important to note that if the reduction is allowed too much power, the reduction can mask the complexity of L'. The following lemma shows that *NC* reducibility has the appropriate level of power, since it preserves membership in *NC*.

LEMMA 28.3   If $L' \in NC$ and $L \leq_m^{NC} L'$ then $L \in NC$.

A less powerful notion called AC^0 reducibility is discussed at length in Chapter 23.

28.5.2 Completeness

The idea of *P*-completeness is to identify those problems in *P* that are the hardest. The following definition formalizes this notion.

DEFINITION 28.16   A language L is *P*-hard under *NC* reducibility if and only if $L' \leq_m^{NC} L$ for every $L' \in P$. A language L is *P*-complete under *NC* reducibility if and only if $L \in P$ and L is *P*-hard.

We now state an important theory involving *P*-complete problems and the class *NC*.

THEOREM 28.16   *If any P-complete language is in NC then NC equals P.*

Theorem 28.16 tells us that the fundamental question of whether *NC* equals *P* has the same answer as the question of whether any *P*-complete problem is in *NC*. Much evidence has accumulated showing that it is unlikely that *NC* equals *P*, and so the *P*-complete problems are likely to be inherently sequential. Thus, when trying to design a highly parallel algorithm, one should avoid solving *P*-complete problems. That is, do not make use of a subroutine call to solve a *P*-complete problem since this call will create a parallel bottleneck.

28.5.3 Proof Methodology for *P*-Completeness

The steps required to show that a language *L* is *P*-complete are as follows:

1. Demonstrate that *L* is in *P* as follows

   a. Provide an algorithm *A* for *L*

   b. Prove that algorithm *A* is correct

   c. Prove that algorithm *A* runs in polynomial time

2. Prove that *L* is *P*-hard as follows

   a. For all $L' \in P$, prove that $L' \leq_m^{NC} L$ (this proof is usually done by providing a function *f* reducing a known *P*-complete problem to *L*)

   b. Prove that *f* is a valid reduction

   c. Prove that *f* is in *FNC*

In the next section we examine the most fundamental *P*-complete problems. These are the problems that, by application of Lemma 28.2, are most frequently used to demonstrate a new problem is *P*-complete.

28.6 Examples of *P*-Complete Problems

Our goal in this section is to acquaint the reader with a number of *P*-complete problems. The full proofs of *P*-completeness may be found in [22].

28.6.1 Generic Machine Simulation

The canonical device for performing sequential computations is the Turing machine, with its single processor and serial access to memory. Of course, the usual machines that we call sequential are not nearly so primitive, but fundamentally they all suffer from the same bottleneck created by having just one processor. So, to say that a problem is inherently sequential, is to say that solving the problem on a parallel machine is not substantially better than solving it on a Turing machine. What could be more sequential than the problem of simulating a Turing machine computation? If we could just discover how to simulate efficiently, in parallel, every Turing machine that uses polynomial time, then every feasible sequential computation could be translated automatically into a highly parallel form. Thus, we are interested in the following problem.

DEFINITION 28.17   Generic Machine Simulation Problem (GMSP)
Given: A string *x*, a description $\overline{M}$ of a Turing machine *M*, and a non-negative integer *t* coded in unary. The input is coded as $x\#\overline{M}\#^t$, where # is a delimiter character not otherwise present in the string and $\#^t$ is the unary encoding of *t*.
Problem: Does *M* accept *x* within *t* steps?

Intuitively at least, it is easy to see that GMSP is solvable in polynomial time sequentially—just interpret M's program step-by-step on input x until either M accepts or t steps have been simulated, whichever comes first. Such a step-by-step simulation of an arbitrary Turing machine by a fixed one is the essence of the fundamental result that universal Turing machines exist. Given a reasonable encoding $\overline{M}$ of M, the simulation of it by the universal machine will take time polynomial in t and the lengths of x and $\overline{M}$, which in turn is polynomial in the length of the universal machine's input. (This is why we insist that t be encoded in unary.)

It is easy to *NC* reduce an arbitrary language L in P to the generic machine simulation problem. Let M_L be a Turing machine recognizing L in polynomial time and let $r(n) = n^{O(1)}$ be an easy-to-compute upper bound on that running time. To accomplish the reduction, given a string x, simply generate the string $f(x) = x\#\overline{M_L}\#^{r(|x|)}$. Then $f(x)$ will be a "yes" instance of the generic machine simulation problem if and only if x is in L. This transformation is easily performed in *NC* (see [22] for details).

THEOREM 28.17 *The generic machine simulation problem is P-complete.*

GMSP is in fact a bit too generic to be very useful for proving other problems are P-complete. A problem that is better suited for this purpose is described in Section 28.6.2.

28.6.2   The Circuit Value Problem and Variants

The fundamental P-complete problem is the circuit value problem, defined as follows:

DEFINITION 28.18   Circuit Value Problem (CVP)
Given: An encoding $\overline{\alpha}$ of a Boolean circuit α, inputs $x_1, \ldots, x_n$, and a designated output y.
Problem: Is output y of α TRUE on input $x_1, \ldots x_n$?

THEOREM 28.18   *[32] The circuit value problem is P-complete.*

The circuit value problem plays the same role in P-completeness theory that satisfiability (SAT) [4] does in NP-completeness theory. Like SAT, CVP is the fundamental P-complete problem in the sense that it is the problem most frequently used to show that other problems are P-complete. Also like SAT, CVP has many restricted variants that are P-complete and can often simplify the construction of reductions. Table 28.4 summarizes these variants of CVP.

There are two key ideas in the proof of Theorem 28.18. The first is that for any circuit there is a simple sequential algorithm that given any input to the circuit evaluates individual gates of the circuit in a fixed order, evaluating each exactly once, and arriving at the circuit's designated output value in polynomial time. Thus CVP is in P.

The second key idea (see Chapter 22 for full details) is used to show that CVP is P-complete. Every language $L \in$ P has an associated Turing machine that decides membership of a length n string x in L in time $T(n)$, polynomial in n. Any polynomially time-bounded Turing machine computation can only use a polynomial amount of tape, and this fact means that the state of the tape at any instant in time can be simulated by a constant-depth polynomially sized circuit slice. On an input of length n, the Turing machine runs for at most $T(n)$ steps, and so at most $T(n)$ tape simulation slices are required. Thus there is a circuit family associated with L such that the nth circuit of the family decides membership of all strings of length n. This observation reduces the question of $x \in L$ to whether a specific polynomially sized circuit with input x outputs the value 1; and therefore, it shows that CVP is P-hard.

TABLE 28.4   Other *P*-Complete Variants of the Circuit Value Problem

Topologically, Ordered CVP	A **topological ordering** is a vertex numbering so that the source vertex of each edge is less than the sink vertex. All CVP variants here remain *P*-complete even if we require that the vertices be presented in topological order in the circuit encoding (and thus the evaluation order is provided in the CVP instance)
NORCVP	CVP is restricted to contain only NOR gates. Reductions are often simpler when only one gate type needs to be simulated
Monotone CVP (MCVP)	CVP restricted to **monotone** gates, that is, AND's and OR's. Useful in the situation where negations are hard to simulate
Alternating, Monotone CVP (AMCVP)	A special case of MCVP. A monotone circuit is **alternating** if on any path from an input to an output the gates alternate between OR and AND. For AMCVP, inputs must connect only to OR gates, and outputs must be OR gates. Reductions often replace individual gates by certain small gadgets. The alternating property reduces the number and kinds of interactions between gadgets that must be considered
Fanin 2, Fanout 2 AMCVP (AM2CVP)	A restriction of AMCVP, where all gates are restricted to having fanin and fanout two, with the obvious exception of inputs and outputs. Having a fixed fanout often simplifies reductions
Synchronous, AM2CVP (SAM2CVP)	Restriction of AM2CVP to synchronous circuits (defined in Section 28.4.3.2). All output vertices are required to be on the highest level of the circuit, so that it can be partitioned into layers, with all edges going from one layer to the next higher one, and all outputs on the last layer. Note that in a circuit that is both alternating and synchronous, all gates on any given level must be of the same type. The fanin two and fanout two restrictions further imply that every level contains exactly the same number of vertices. This structural regularity simplifies some reductions

28.6.3   Additional *P*-Complete Problems

There are hundreds of *P*-complete problems (see [22]), from many areas of computer science: circuit complexity, graph theory, graph searching, combinatorial optimization and flow, local optimality, logic, formal languages, algebra, geometry, real analysis, games, and miscellaneous topics. We next present three of the more interesting *P*-complete problems. This first problem is from graph theory and was proved *P*-complete in [8].

DEFINITION 28.19   Lexicographically First Maximal Independent Set (LFMIS)
Given: An undirected graph $G = (V, E)$ with an ordering on the vertices and a designated vertex v.
Problem: Is vertex v in the lexicographically first maximal independent set of G?

We now prove LFMIS is *P*-complete.

THEOREM 28.19   *[8] The lexicographically first maximal independent set problem is P-complete.*

PROOF   Membership in *P* follows from the standard greedy algorithm: vertices are processed in numerical order and added to the independent set if they do not have an edge to any vertex already contained in the independent set.

Completeness follows by reducing the NOR circuit value problem (NORCVP) to LFMIS using the construction given in [22]. Without loss of generality, we assume that the instance α of NORCVP has its gates numbered (starting from 1) in topological order with inputs numbered first and outputs last. Suppose y is the designated output gate in the instance of NORCVP. We construct from α an instance of LFMIS, namely an undirected graph G. The graph G will be exactly the same as the graph underlying the circuit α, except that we add a new vertex, numbered 0, that is adjacent to all 0-inputs of α. It is easy to verify by induction that a vertex i in G is included in the lexicographically first maximal independent set if and only if either i equals 0 (the new vertex), or gate i in α has value TRUE. A choice of v equal to y completes the reduction.

The proof that the reduction can be performed in NC amounts to showing that the required edges from the new vertex 0 to the 0-inputs can be produced easily. The remainder of the circuit connections are easy to output directly. □

The next problem is from graph searching and was proved P-complete in [37]. The reader is referred there or to [22] for details.

DEFINITION 28.20 Lexicographically First Depth-First Search Ordering
Given: An undirected graph $G = (V, E)$ with fixed ordered adjacency lists, and two designated vertices u and v.
Problem: Is vertex u visited before vertex v in the depth-first search of G induced by the order of the adjacency lists?

The final problem is from formal languages and was proved P-complete in [27]. The reader is referred there or to [22] for details.

DEFINITION 28.21 Context-Free Grammar Membership
Given: A context-free grammar $G = (N, T, P, S)$ and a string $x \in T^*$.
Problem: Is $x \in L(G)$?

It is likely that all these problems are inherently sequential. However, each of the problems has a related variant that is feasibly highly parallel. For example, each of these are in NC: finding some maximal independent set, finding the lexicographically first depth-first numbering for directed acyclic graphs, and context-free grammar membership for ϵ-free (ϵ denotes the empty string) grammars. References to these algorithms can be found in [22]. In practice, when the problem that you want to solve is P-complete, there is often a closely related variant that is feasible and highly parallel. Whether this variant is useful to you is of course another matter.

28.6.4 Open Problems

There are a few simple problems for which it is unknown whether the problem is in NC or is P-complete. We give a sampling of these here. The original references for these, and other, problems may be found in [22].

DEFINITION 28.22 Integer Greatest Common Divisor (IntegerGCD)
Given: Two n-bit positive integers a and b.
Problem: Compute the greatest common divisor of a and b.

DEFINITION 28.23 Lexicographically First Maximal Matching (LFMM)
Given: An undirected graph $G = (V, E)$ with an ordering on its edges and distinguished edge $e \in E$.
Problem: Is e in the lexicographically first maximal matching of G? A matching is maximal if it cannot be extended.

DEFINITION 28.24   Directed or Undirected Depth-First Search (DFS)
Given: A graph $G = (V, E)$ and a vertex s.
Problem: Construct a depth-first search numbering of G starting from vertex s.

DEFINITION 28.25   Maximum Matching (MM)
Given: An undirected graph $G = (V, E)$.
Problem: Find a maximum matching of G. A matching is a subset of edges $E' \subseteq E$ such that no two edges in E' share a common endpoint. A matching is maximum if no matching of larger cardinality exists.

DEFINITION 28.26   Subtree Isomorphism (STI)
Given: Two unrooted trees $T = (V, E)$ and $T' = (V', E')$.
Problem: Is T *isomorphic* to a subtree of T'?

It is notable that LFMM is CC-complete (circuits composed of comparator gates only), and that DFS, MM, and STI have randomized feasible highly parallel algorithms (i.e., they are in RNC, which is the class random NC).

28.7   Research Issues and Summary

Many issues in parallel computation remain unresolved. The big question of asymptotic complexity theory is whether every feasible sequential problem has a feasible highly parallel solution, that is, does NC equal P? Although we have strong evidence* suggesting that there are fundamental limits to the speedup attainable through parallel computation, this evidence does not constitute proof. It seems that deciding this question will require a major breakthrough in complexity theory. Another important complexity question is whether randomization helps in highly parallel computation. It is clear that NC is a subset of RNC (the randomized version of NC), but not known if the two classes are equal. Finally, there are many problems that have feasible highly parallel algorithms but whose algorithms are not optimal. That is, their work (time-processor product) is more than a constant factor greater than the running time of the best known sequential algorithm. Attempting to improve the time or processor bounds for such problems is one way of understanding the relationship between sequential and parallel computation.

From the standpoint of algorithm design, the most important issue is finding a parallel model that is more realistic. That is, a model that reflects the kinds of parallel machines that will actually be built in the foreseeable future, and on which algorithms will be implemented. The works [9,10,44,45] are some first steps in this direction. The main assumption of these models is that all parallel machines will essentially consist of a moderate number (thousands not millions) of highly powerful individual processors each with substantial memory, and interconnected by a high (but not infinite) capacity network.

Each model is parameterized so that any given instance of these kinds of parallel machines can be modeled so as to account for the architectural features which dominate the design of a parallel algorithm for that machine. The parameters include the number of processors, the communication bandwidth on the interconnection network, the latency or delay in transmitting over the network, granularity, and the overhead of initiating communication over the network.

---

* We have never explicitly stated the evidence in this chapter due to its technical nature. Chapter 5 of [22] goes into the evidence in detail.

The presence of these parameters forces the designer of the parallel algorithm not only to account for the computational aspects of the problem (deciding what is to be done and by what processor), but also to consider the communication aspects of the problem (where should data be placed and how should accesses be scheduled). By putting equal emphasis on computation and communication, such models more accurately capture reality, while avoiding architectural details such as the kind of interconnection network. The task is now to develop parallel algorithms that can be used for as wide a range of parameter values as possible, and thus be robust over a much broader class of real machines.

As noted in the introduction to this chapter, due to limitations that have been encountered in the design of faster individual processors, we expect to see a resurgence of interest in parallel computing. Many more systems are being constructed now having a handful to a few dozens processors, and this trend is likely to continue.

28.8 Further Information

Much of our material is adapted from *Limits to Parallel Computation* [22], and many more details can be found there. The relationships among various parallel models can be found in the excellent surveys [7,15,28]. Additional papers surveying other aspects of parallel models and parallel computing include [25,30,42,46,48].

The book *Efficient Parallel Algorithms* [18] has a brief discussion of parallel models of computation followed by substantial material on parallel algorithms. The text *An Introduction to Parallel Algorithms* [24] devotes a chapter to discussing parallel models and then extensively delves into parallel algorithms. The text *Introduction to Parallel Algorithms and Architectures: Arrays, Trees, Hypercubes* [34] contains a detailed discussion of many different types of parallel models and algorithms. The collection *Synthesis of Parallel Algorithms* [38] contains about 20 chapters organized around parallel algorithms for particular types of problems, together with an introductory chapter on *P*-completeness [21], and one surveying PRAM models [15]. The chapter "Parallel algorithms for shared-memory machines" [28] in the *Handbook of Theoretical Computer Science* [47] describes a variety of highly parallel algorithms. In the same handbook, the chapter "A catalog of complexity classes" [26] is a thorough overview of basic complexity theory and of the current state of knowledge about most complexity classes. It is an excellent reference for establishing the context of each class and its established relationships to others. The papers [29–31] provide extensive bibliographies of papers about parallel algorithms and parallel algorithm development for combinatorial optimization problems.

Defining Terms

Circuit Value Problem: [32]
Given: An encoding $\bar{\alpha}$ of a Boolean circuit α, inputs $x_1, \ldots, x_n$, and a designated output y.
Problem: Is output y of α TRUE on input $x_1, \ldots x_n$?

NC: The set of all languages L that are decidable in parallel time $(\log n)^{O(1)}$ and processors $n^{O(1)}$.

NC **many–one reducibility:** A language L is *NC* **many–one reducible** or *NC* **reducible** to L', written $L \leq_m^{NC} L'$, if there is a function f in *FNC* such that $x \in L$ if and only if $f(x) \in L'$.

P: The set of all languages L that are decidable in sequential time $n^{O(1)}$.

Parallel Computation Thesis: Sequential space is polynomially related to parallel time.

Path Systems: [5]
Given: A path system $P = (X, R, S, T)$, where $S \subseteq X$, $T \subseteq X$, and $R \subseteq X \times X \times X$.

Problem: Is there an admissible vertex in S? A vertex x is **admissible** if and only if $x \in T$, or there exists admissible $y, z \in X$ such that $(x, y, z) \in R$.

P-**complete**: A language L is *P*-**hard under** *NC* **reducibility** if $L' \leq_m^{NC} L$ for every $L' \in P$. A language L is *P*-**complete under** *NC* **reducibility** if $L \in P$ and L is *P*-hard.

Acknowledgments

A warm thanks to Mike Atallah for inviting us to work with him on this exciting project. A special thanks to Larry Ruzzo for many enlightening discussions about the material in this chapter. Thanks to the anonymous referees for carefully reading the chapter and providing us with many helpful suggestions. The original version of this chapter was written while Ray was on sabbatical at the Universitat Politècnica de Catalunya in Barcelona. The department's hospitality is greatly appreciated. Thanks to Chiang Mai University where Ray completed the revision.

This research was partially supported by National Science Foundation grant CCR-9209184; a Fulbright Scholarship, Senior Research Award; and a Spanish Fellowship for Scientific and Technical Investigation (Raymond Greenlaw). This research was partially supported by the Natural Sciences and Engineering Research Council of Canada grant OGP 38937 (H. James Hoover).

References

1. Asanovic, K., Bodik, R., Catanzaro, B.C., Gabis, J.J., Husbands, P., Keutzen, K., Patterson, D.A., Plishker, W.L., Shalf, J., Williams, S.W., and Yelick, K.A., The landscape of parallel computing research: A view from Berkeley, Technical Report No. UCB/EECS-2006-186, 2006.

2. Borodin, A., On relating time and space to size and depth. *SIAM Journal on Computing*, 6(4), 733–744, 1977.

3. Chandra, A.K., Kozen, D.C., and Stockmeyer, L.J., Alternation. *Journal of the ACM*, 28(1), 114–133, 1981.

4. Cook, S.A., The complexity of theorem proving procedures. In *Conference Record of 3rd Annual ACM Symposium on Theory of Computing*, pp. 151–158, Shaker Heights, OH, 1971.

5. Cook, S.A., An observation on time-storage trade off. *Journal of Computer and System Sciences*, 9(3), 308–316, 1974.

6. Cook, S.A., Deterministic CFL's are accepted simultaneously in polynomial time and log squared space. In *Conference Record of the 11th Annual ACM Symposium on Theory of Computing*, 338–345, Atlanta, GA, 1979. See also [49].

7. Cook, S.A., Towards a complexity theory of synchronous parallel computation. *L'Enseignement Mathématique*, XXVII(1–2), 99–124, 1981. Also in [36, pp. 75–100].

8. Cook, S.A., A taxonomy of problems with fast parallel algorithms. *Information and Control*, 64(1–3), 2–22, 1985.

9. Culler, D., Karp, R., Patterson, D., Sahay, A., Schauser, K.E., Santos, E., Subramonian, R., and von Eicken, T., Logp: Towards a realistic model of parallel computation. In *Proceedings of the 5th Annual ACM SIGPLAN Symposium on Principles and Practices of Parallel Programming*, pp. 1–12, ACM, 1993.

10. de la Torre, P. and Kruskal, C.,Towards a single model of efficient computation in real parallel machines. *Future Generations Computer Systems*, 8, 395–408, 1992. Preliminary version in *PARLE'91: Parallel Architectures and Languages Europe, Lecture Notes in Computer Science*, Aars, van Leeuwen and Rems, Eds., Springer-Verlag, pp. 6–24, July 1991.

11. Dymond, P.W., Simultaneous resource bounds and parallel computation, PhD Thesis, University of Toronto, Toronto, ON, Department of Computer Science Technical Report 145/80, 1980.

12. Dymond, P.W. and Cook, S.A., Hardware complexity and parallel computation. In *21st Annual Symposium on Foundations of Computer Science*, pp. 360–372, IEEE, Syracuse, NY, 1980.

13. Dymond, P.W. and Cook, S.A., Complexity theory of parallel time and hardware. *Information and Computation*, 80(3), 205–226, 1989.

14. Dymond, P.W. and Ruzzo, W.L.,Parallel random access machines with owned global memory and deterministic context-free language recognition. In *Automata, Languages, and Programming: 13th International Colloquium*, Kott, L., Ed., Vol. 226: *Lecture Notes in Computer Science*, pp. 95–104, Springer-Verlag, Rennes, France, 1986.

15. Fich, F.E., The complexity of computation on the parallel random access machine. In [38], chapter 20, pp. 843–899, 1993.

16. Fortune, S. and Wyllie, J.C., Parallelism in random access machines. In *Conference Record of the Tenth Annual ACM Symposium on Theory of Computing*, pp. 114–118, San Diego, CA, 1978.

17. Furst, M.L., Saxe, J.B., and Sipser, M., Parity, circuits, and the polynomial-time hierarchy. *Mathematical Systems Theory*, 17(1), 13–27, 1984.

18. Gibbons, A.M. and Rytter, W., *Efficient Parallel Algorithms*, Cambridge University Press, Cambridge, U.K., 1988.

19. Goldschlager, L.M., Synchronous parallel computation, PhD Thesis, University of Toronto, Toronto, ON, Computer Science Department Technical Report 114, 1977.

20. Goldschlager, L.M., A universal interconnection pattern for parallel computers. *Journal of the ACM*, 29(4), 1073–1086, 1982.

21. Greenlaw, R., Polynomial completeness and parallel computation. In [38], chapter 21, pp. 901–953, 1993.

22. Greenlaw, R., Hoover, H.J., and Ruzzo, W.L., *Limits to Parallel Computation: P-Completeness Theory*, Oxford University Press, New York, 1995.

23. Hartmanis, J. and Simon, J., On the power of multiplication in random access machines. In *15th Annual Symposium on Switching and Automata Theory*, pp. 13–23, 1974.

24. JáJá, J., *An Introduction to Parallel Algorithms*, Addison-Wesley, Reading, MA, 1992.

25. Johnson, D.S., The *NP*-completeness column: An ongoing guide (7th). *Journal of Algorithms*, 4(2), 189–203, 1983.

26. Johnson, D.S., A catalog of complexity classes. In [47], chapter 2, pp. 67–161, 1990.

27. Jones, N.D. and Laaser, W.T., Complete problems for deterministic polynomial time. In *Conference Record of Sixth Annual ACM Symposium on Theory of Computing*, 40–46, Seattle, WA, 1974.

28. Karp, R.M. and Ramachandran, V., Parallel algorithms for shared-memory machines. In [47], chapter 17, pp. 869–941, 1990.

29. Kindervater, G.A.P. and Lenstra, J.K., An introduction to parallelism in combinatorial optimization. In *Parallel Computers and Computation*, van Leeuwen, J. and Lenstra, J.K., Eds., Vol. 9: *CWI Syllabus*, 163–184, Center for Mathematics and Computer Science, Amsterdam, the Netherlands, 1985.

30. Kindervater, G.A.P. and Lenstra, J.K., Parallel algorithms. In *Combinatorial Optimization: Annotated Bibliographies*, O'hEigeartaigh, M., Lenstra, J.K., and Rinnooy Kan, A.H.G., Eds., chapter 8, pp. 106–128, John Wiley & Sons, Chichester, U.K., 1985.

31. Kindervater, G.A.P. and Trienekens, H.W.J.M., Experiments with parallel algorithms for combinatorial problems, Technical Report 8550/A, Erasmus University Rotterdam, Econometric Inst., 1985.

32. Ladner, R.E., The circuit value problem is log space complete for *P*. *SIGACT News*, 7(1), 18–20, 1975.

33. Lam, T.W. and Ruzzo, W.L., The power of parallel pointer manipulation. In *Proceedings of the 1989 ACM Symposium on Parallel Algorithms and Architectures*, pp. 92–102, Santa Fe, NM, 1989.

34. Leighton, F.T., *Introduction to Parallel Algorithms and Architectures: Arrays, Trees, Hypercubes*, Morgan Kaufmann, San Mateo, CA, 1992.

35. Lengauer, T., VLSI theory. In [47], chapter 16, pp. 837–868, 1990.

36. Pratt, V.R. and Stockmeyer, L.J., A characterization of the power of vector machines. *Journal of Computer and System Sciences*, 12(2), 198–221, 1976.

37. Reif, J.H., Depth-first search is inherently sequential. *Information Processing Letters*, 20(5), 229–234, 1985.

38. Reif, J.H., Ed., *Synthesis of Parallel Algorithms*, Morgan Kaufmann, San Francisco, CA, 1993.

39. Ruzzo, W.L., On uniform circuit complexity. *Journal of Computer and System Sciences*, 22(3), 365–383, 1981.

40. Savitch, W.J. and Stimson, M.J., Time bounded random access machines with parallel processing. *Journal of the ACM*, 26(1), 103–118, 1979.

41. Schönhage, A., Storage modification machines. *SIAM Journal on Computing*, 9(3), 490–508, 1980.

42. Spirakis, P.G., Fast parallel algorithms and the complexity of parallelism (basic issues and recent advances). In *Parcella'88. Fourth International Workshop on Parallel Processing by Cellular Automata and Arrays Proceedings*, Wolf, G., Legendi, T., and Schendel, U., Eds., Vol. 342: *Lecture Notes in Computer Science*, pp. 177–189, Springer-Verlag, Berlin, East Germany, 1988 (published 1989).

43. Stockmeyer, L.J. and Vishkin, U., Simulation of parallel random access machines by circuits. *SIAM Journal on Computing*, 13(2), 409–422, 1984.

44. Valiant, L.G., A bridging model for parallel computation. *Communications of the ACM*, 33(8), 103–111, 1990.

45. Valiant, L.G., General purpose parallel architectures. In [47], chapter 18, pp. 943–971, 1990.

46. van Emde Boas, P., The second machine class: Models of parallelism. In *Parallel Computers and Computation*, van Leeuwen, J. and Lenstra, J.K., Eds., Vol. 9: *CWI Syllabus*, pp. 133–161, 1985, Center for Mathematics and Computer Science, Amsterdam, the Netherlands.

47. van Leeuwen, J., Ed., *Handbook of Theoretical Computer Science, Vol. A: Algorithms and Complexity*, M.I.T. Press/Elsevier, Amsterdam, the Netherlands, 1990.

48. Vishkin, U., Synchronous parallel computation—a survey. Preprint. Courant Institute, New York University, New York, 1983.

49. von Braunmühl, B., Cook, S.A., Mehlhorn, K., and Verbeek, R., The recognition of deterministic CFL's in small time and space. *Information and Control*, 56(1–2), 34–51, 1983.

29

Distributed Computing: A Glimmer of a Theory

Eli Gafni
University of California

29.1 Introduction

29.1.1 What Is the Area About?

Distributed computing theory develops computability and complexity theories for models whose computation involves many processors interacting in certain limited unpredictable manner through some communication objects, where a processor is shorthand for a sequential piece of code that includes instructions, some of which involve access to the communication objects.

One of the defining characteristics of the models tackled by distributed computing is a nondeterminism due to run-time events whose possibility of occurrence or the order in which they might occur is unpredictable. Moreover, distributed computing takes as its primary domain models whose nondeterminism gives their processors differing views of the world. Thus, for example, distributed computing will examine a model in which the failure of a processor is noticeable to some of its processors but not to others, but it will not examine a model in which the outcome of a flip of a coin (i.e., a randomized outcome) is made available to all processors simultaneously. In the parlance of knowledge [20], distributed computing studies systems in which the current state of a processor is not common knowledge, even when the input is known to all of the processors.

Research in distributed computing has focused on a number of models distinguished by the different communication objects and different timing constraints. Among those investigated most extensively are the asynchronous message-passing model, the asynchronous shared-memory model, asynchronous models in which at most t processors may fail-stop, and synchronous models with Byzantine, omission, or fail-stop failures.

29.1.2   What Is This Chapter About?

Here I argue that in the chaos that has characterized the field of distributed computing theory, we have begun to see signs of an underlying order. In the 1980s, as Lynch and Lamport [25] have observed, "the field seemed to consist of a collection of largely unrelated results about individual models." They were right at the time. Researchers concentrated on obtaining efficient solutions to problems, and although this was and is an important and productive line of research, most of the ideas leading to efficiency seem ad hoc and are closely tied to a specific model and, at times, to a specific problem instance. Few of the ideas leading to efficient solutions supply a methodology that can be applied in other contexts.

Furthermore, the arguments that have led to the most fundamental impossibility results of the field seem to share no common thread. One fundamental negative result, FLP impossibility [16], states that no agreement can be reached in an asynchronous system in which even a single processor may fail-stop unannounced. The original arguments that led to this result were procedural and very much tied to the properties of the operations specific to the model. Concurrently, the impossibility of reaching Byzantine agreement in time equal to the number of faults was proven [15] through the use of a chain of compatible views that led from a run in which one decision has to be taken to a run in which another decision has to be taken. Consequently, the argument went, there must be a link in the chain at which incompatible decisions are taken by different processors in the same run. At about the same time, the procedural inductive proof of the nonexistence of a solution to the so-called two-armies problem was replaced by a slick proof using a newly developed knowledge theory [18,20]. Although the connection between the latter two impossibility results, one in the context of synchronous systems and the other in the context of asynchronous systems, was at the time vaguely understood, the connection between them and FLP impossibly seemed nonexistent.

In recent years tools have been developed, and research has been fruitful in making connections between results and models in a way that parallels what has been done in complexity theory. Where complexity theory is not much help in supplying a receipt to determine whether a specific problem at hand is $\Theta(n^2)$ or $\Theta(n^3)$ other then referral to some generic problems that have been thoroughly investigated, distributed computing can similarly provide only hints in the form of pointers to similar problems, in determining time and message complexity of a specific problem. Where complexity theory has tool and techniques to categorize problems in a "broad-brush," e.g., P vs. NP, distributed computing theory has over recent years built a formidable machinery to classify problems by the models in which they are solvable.

Problems have been identified that are analogous to **complete problems**. They characterize a model. The problem is solvable in the characterized model, and any model in which the problem is solvable, has a set of solvable problems that is a superset of the problems solvable in the characterized model. Collections of models, whose set of solvable problems is identical, have been identified, where in some models the level of the unpredictability of the possible order of events is lesser than in others. A problem is then checked for solvability in the model that exhibit the least unpredictability, if one exists. It provides, on the one hand, short uncluttered impossibility argument, or on the other hand, it provides a succinct protocol. Such a protocol may be viewed as written in a high-level programming language, with a guaranteed automatic compilation of the protocol to a protocol in the target model.

Some starts that stalled, and some partial successes in the direction indicated above occurred in the 1980s. Knowledge theory [20] was developed, showed some promise for certain problems, but eventually led, in my opinion, to no major breakthrough, in the understanding of distributed computing. A way of coaching 1-resilient models within graph theory led to a complete characterization of these models [9], providing an automatic impossibility proof on one hand or a protocol on the other, to the question of a problem solvability in the model. This raised the hope that protocol derivation, or the realization of its nonexistence, can in general be automated. And finally, a helpful instance of the "high-level-programming," mentioned above, was discovered with the realization of the equivalence between message-passing systems and shared-memory systems on the one hand, and between shared memory and atomic-snapshot shared-memory on the other. These instances of model equivalence allowed the investigation of the power of message-passing systems, within the much less detailed model of atomic-snapshot shared-memory.

The breakthrough was enabled when researchers started to zooming on complete problems. Herlihy [19] showed that the number processor that can achieve consensus characterizes and differentiates certain models. While consensus takes n input values and allows a single output, Chaudhuri proposed to check for models that allow several but less than n outputs. In 1993 three teams were able to show that her problems are also complete for certain models. The techniques employed resulted in the revelation of a connection between distributed computing and topology, be it algebraic [22], combinatorial [5], or point-set topology [29].

Extensive research effort followed. If by "theory" one means a body of tightly related results, building one upon the other, and readily providing explanations to many observable phenomena, then, as I will attempt to show in this chapter, a theory of distributed computing has emerged.

The survey in this chapter has a narrow focus. The advances outlined above are presented. The utility of the "modern tools" that result in are demonstrated in deriving past results. Section 29.2, on models, discusses distributed systems in general and defines the various notions of emulation and tasks which are central to this chapter. We then embark upon a sequence of emulations. We start with the asynchronous shared-memory model for two and three processors, characterizing them for fail-stop and Byzantine failures and, via transformation, we apply all these results to the message-passing model. Then we investigate the shared-memory model in the synchronous domain, and we show how an impossibility result in the asynchronous model translates into lower bound on complexity in the synchronous model. The next step is to link the two models through the notion of failure detectors (FDs), and the new notion of an iterated model, and to explicate the utility of characterizing a synchronous system as an asynchronous one with a (weak) FD [12]. The concluding section raises some of the major questions that, in my opinion, are facing this emerging theory of distributed computing. Obviously, the chapter leaves unaddressed many of the very large number of techniques developed in distributed computing; in most cases, how they may be incorporated into the framework reported here is not yet clear. The reader is encouraged to look up this wealth of techniques in the excellent textbook by Lynch [26], and two of the surveys by Lamport and Lynch [25] and Attiya [6].

29.2 Models

An asynchronous distributed computation model is the set of all sequential interleaving of communication actions performed by sequential processes or processors on shared communication objects. A communication action may be thought of as the invocation of a remote procedure call by a processor at an object. An object may be thought of as a processor that executes the remote procedure call, changes its state, and responds to the invoking processor by returning a value. The returned value causes the invoking processor to change its state, which in turn determines the next parameter for the next access to a communication object. In this chapter we assume that processors never halt, and therefore after each return of an invocation, they enter a state in which a new invocation is enabled. In the network model, processors access unidirectional point-to-point communication channels.

A single communication object is associated with two processors, called sender and receiver, respectively. The sender can invoke an action send(m) on the object. The effect of the action is to place a message m in the buffer of the object. After placing the message in its buffer the object responds by returning an ok to the sender. The receiver invokes an action receive which moves a message from the buffer of the object to the receiver, if such a message exists, or the object responds by notifying exception otherwise. In the shared-memory model, communication objects are read/write registers on which the action of read and write can be invoked.

In this chapter we assume that communication objects do not fail. Yet, in light of the view of a communication object as a "restricted" processor, it is not surprising that when communication failures are taken into account [3], they give rise to results reminiscent of processor failures.

Given a **protocol**—the instantiation of processors with codes—and the initial conditions of processors and objects, we define a space R of **runs** to be a subset of the infinite sequences of processors names. Since we assume that a processor has a single enabled invocation at a time, such a sequence when interpreted as the order in which enabled invocations were executed completely determines the evolution of the computation. Before the system starts all runs in which the processor has its current input are possible. As the system evolves, the local state of the processor excludes some runs. Thus with a local state of a processor we associate a **view**—the set of all runs in R that are not excluded by the local state. By making processors maintain their history in local memory we may assume that consecutive views of a processor are monotonically nondecreasing. Thus, with each run $r \in R$ of a protocol p we can associate a limit view $\lim(V_i(r, p))$ of processor P_i. A protocol f is **full-information** if for all i, r, and p we have $\lim(V_i(r, f)) \subseteq \lim(V_i(r, p))$. Intuitively, a full-information protocol does not economize on the size of its local state, or the size of the parameter to its object invocation. Models which are **oblivious**, that is, the sequence of communication objects a processor will access is the same for all protocols, possess a full-information protocol. All the models in this chapter do. In the rest of this chapter, a protocol stands for the full-information one, and correspondingly a model is associated with a single protocol—its full-information protocol. One can define the notion of full-information protocol with respect to a specific protocol in a nonoblivious model, but we will not need this notion here. A sequence of runs $r_1, r_2, \ldots$ converges to a run r, if r_k and r share a longer and longer prefix as k increases.

It can be observed from the definition of a view, that two views of the same processor, are either disjoint, or related by containment. Given an intermediate view $V_i(r)$ of a processor P_i in run r, we say that a processor **outputs** its view in r if for all P_j which have infinitely many distinct views in r, $\lim(V_j(r)) \subseteq V_i(r)$. Processor P_i is **faulty** in r if it outputs finitely many views. Processor P_i is **participating** in r if it outputs any nontrivial view. Otherwise it is **sleeping** in r. A model A with n processors with communication object O_A **wait-free emulates** a model B with n processors and communication objects O_B if there is a map m from runs R_A in A to runs R_B in B such that

1. The sets of sleeping processors and faulty processors in r and $m(r)$ are identical.
2. The map m is continuous with respect to prefixes. That is, if $r_1, r_2, \ldots$ in A converges to r, then $m(r_1), m(r_2), \ldots$ in B converges to $m(r)$. This captures the idea that the map does not predict the future.
3. The map m does not utilize detailed information about the past of a run, if this detailed information is not available through processors' views. Formally, for all P_j nonfaulty and for all r in A, $m(\lim(V_j(r))) \subseteq \lim(V_j(m(r)))$.

We say that A **nonblocking emulates** B if we relax the first condition by allowing the mapping m to fail any nonfaulty processor as long as an infinite sequence is mapped into an infinite sequence. Two models are wait-free (nonblocking) equivalent if they wait-free (nonblocking) emulate each other.

A specification of a problem Π on n processors, is a relation from runs to sets of "output-sequences." Each output in the sequence is associate with a unique processor. A model with n processors wait-free solves Π if there exists a map from views to outputs, such the map of the projection of a run on views that are output in the run, is an output-sequence that relates to the run. A problem Π on n processors is nonblocking solvable in a model, if the relaxed problem $\bar{\Pi}$ is wait-free solvable, where $\bar{\Pi}$ takes each element in Π and closes the output-sequence set with respect to removal of infinite suffixes of processors' outputs (as long as the sequence remains infinite).

A **task** is a relaxation of a problem Π such that only bounded prefixes of output-sequences matter. That is to say that past some number of outputs any output is acceptable. Since the notion of participating set is invariant over models, the runs that are distinguished by different output requirements in a task are those that differ in their participating set. Thus in this chapter we employ the notion of task in this restricted sense. In the consensus task a processor first outputs its private value, which is either 0 or 1, and then outputs a consensus value. Consensus values agree, and match the input of at least one of the participating processors. In the **election** task a processor outputs its ID and then outputs an election value which is an ID of a participating processor. All election values agree. A run with a single participating processor is a **solo-execution** of that processor.

A model is t-resilient if we require that it solves a problem only over runs in which at most t processors are faulty.

A synchronous model is one which progresses in rounds. In each round all the communication actions enabled by the beginning of the round are executed by the end of the round.

29.3 Asynchronous Models

29.3.1 Two-Processor Shared-Memory Model

Consider a two-processor single-writer/multireader (SWMR) shared-memory system. In such a system, there are two processors P_1 and P_0, and two shared-memory cells C_1 and C_0. Processor P_i writes exclusively to C_i, but it can read the other cell. Both shared-memory cells are initialized to $\perp$. W.l.o.g. computation proceeds with each processor alternately writing to its cell and reading the cell of the other processor.

Can this two-processor system 1-resiliently (wait-free in this case, since for $n = 2$, $n - 1 = 1$) elect one of the processors as the leader? No one-step full-information protocol, and consequently no one-step protocol at all, for solving this problem exists. Consider the state of processor P_1 after writing and reading. It could have read what processor P_0 wrote (denoted by $P_1 : w_0$), or it could have missed what processor P_0 wrote (denoted by $P_1 : \perp$). Thus, we have four possible views, two for each processor, after one step.

In the graph whose nodes are these views, two views are connected by an undirected edge if there is an execution that gives rise to the two views. The resulting graph appears in Figure 29.1. Since a processor has a single view in an execution, edges connect nodes labeled by distinct processor IDs. The two nodes of distinct IDs which do not share an edge are $P_1 : \perp$ and $P_0 : \perp$. This follows from the fact that in shared memory in which processors first write and then read, the processor that writes second must read the value of the processor that writes first.

The edge $\{P_1 : \perp, P_0 : w_1\}$ corresponds to the execution: P_1 writes, P_1 reads, P_0 writes, P_0 reads. If we could map the view of a processor after one step into an output, then processor P_1 in this edge is bound to elect P_1, since the possibility of a solo execution by the processor has not yet been eliminated. Similarly, in the edge $\{P_0 : \perp, P_1 : w_0\}$, processor P_0 is bound to elect P_0. Thus, no matter

FIGURE 29.1 One-step view graph.

what processor is elected by P_1 and P_0 in the views $P_1 : w_0$ and $P_0 : w_1$, respectively, we are bound to create an edge where at one end P_1 is elected and at the other end P_0 is elected. Because there is an execution in which both processors are elected, we must conclude that there is no one-step 1-resilient protocol for election with two processors.

To confirm that no k-step full-information protocol exists, we could draw the graph of the views after k steps, and observe that the graph contains a path connecting the views $\{P_1 : \bot\}$ and $\{P_0 : \bot\}$.

It is not easy to see that indeed the observation above holds. Given that our goal is an argument that will generalize to more than two processors, we have to be able to get a handle on the general explicit structure of the shared-memory model for any number of processors. This has been an elusive challenge. Instead, we turn to iterated shared memory, a model in which the structure of the graph of a k-step two-processor full-information protocol is easily verified to be a path. We then argue that for two processors, the shared-memory model and the iterated shared-memory model are nonblocking equivalent. We then show that this line of argumentation generalizes to any number of processors.

29.3.2 Two-Processor Iterated Shared-Memory Model

For any model M in which the notion of one-shot use of the model exists, one can define the iterated counterpart $\bar{M}$ of M. In $\bar{M}$, the processors go through a sequence of stages of one-shot use of M in which the output of the $(k-1)$th stage is in turn the input to the kth stage.

To iterate the two-processor SWMR shared-memory model, we take two sequences of cells: $C_{1,1}, C_{1,2}, C_{1,3}, \ldots,$ and $C_{0,1}, C_{0,2}, C_{0,3}, \ldots,.$ Processor P_1 writes its input to $C_{1,1}$ and then reads $C_{0,1}$. Inductively, P_1 then takes its view after reading $C_{0,(k-1)}$, writes this view into $C_{1,k}$, reads $C_{0,k}$, and so on.

The iterated model is related to the notion of a communication-closed layer [14]. This accounts for why algorithms in the iterated model are easy both to understand and to prove correct: one may imagine that there is a barrier synchronization after each stage such that no processor proceeded to the current stage until all processors had executed (asynchronously) the previous stage.

In an execution, if the view of P_1 after reading $C_{0,(k-1)}$ is X and the corresponding view for P_0 is Y, then the graph of the views after the kth stage, given the view after the $(k-1)$th stage, appears in Figure 29.2. This graph is the same as the graph for the one-shot SWMR shared-memory model when X is the input to P_1 (and stands therefore for w_1) and Y is the input to P_0 (and stands therefore for w_0). To get the graph of all the possible views after the kth stage, we inductively take the graph after the $(k-1)$th stage, replace each edge with a path of three edges, and label the nodes appropriately. Thus, after k stages, we get a path of 3^k edges. At one end of this path is a view that is bound to output P_1, at the other a view bound to be output P_0, which leads to the conclusion that there is no k-stage protocol in the model for any k.

It is easy to see that the shared-memory model nonblocking implements its iterated version by dividing cell C_i into a sequence of on-demand virtual cells $C_{i,1}, C_{i,2}, C_{i,3}, \ldots,.$ Processors then "pretend" to read only the appropriate cell. To the see that a nonblocking emulation in the reverse direction is also possible we consider processors to WriteRead sequence numbers. Processor P_1 keeps an estimate v_{p_0} of the last sequence number P_0 wrote. To $WriteRead_1(v)$, processor P_1 writes the pair v_{p_0}, v into the next cell in the sequence. It then read the other. If it contains $\bot$, or it contains the same pair it has written, then the operation terminates, and P_1 returns the pair it wrote to its cell.

$$P_1 : \bot \qquad\qquad P_0 : X \qquad\qquad P_1 : Y \qquad\qquad P_0 : \bot$$

FIGURE 29.2 One-step view graph after the kth stage with input X, Y.

Otherwise, it updates v_{p_0} to the maximum between the value it held and the value it read, and continues [8].

29.3.3 Characterization of Solvability for Two Processors

What tasks can two processors in shared memory solve 1-resiliently? They can solve, for instance, the following task. Processor P_1 in a solo execution outputs 1, and processor P_0 in a solo execution outputs 10. In every case, the two processors must output values between 1 and 10 whose absolute difference is exactly 1. This task can be solved easily, since in the iterated shared-memory model, after three stages the path contains 10 nodes and these nodes can be associated one-to-one with the integers 1 through 10. The task may be represented using domino pieces. There is an infinite number of pieces, each labeled P_1, x on one side and P_0, y on the other side, where all x and y are real numbers and $|x - y| = 1$, $1 \leq x \leq 10$, $1 \leq y \leq 10$. The task is solvable iff one can create a domino path with the pieces such that one side of the path is labeled $P_1, 1$ and the other side $P_0, 10$. It is easy to see that if processor P_0 had to output 11 (rather than 10) in a solo execution, the problem would not be solvable.

A generalized version is solvable if processors in solo executions output their integer inputs (which come, for the moment, from a bounded domain of integers) and the tuples of inputs are such that one input value is odd and the other is even. To see that the input (1,8) is solvable, take the output from the second stage of the iterated shared-memory model and fold three consecutive view edges on a single output edge. In algebraic and combinatorial topology, an edge is called a 1-simplex, a node is called a 0-simplex, an edge that is subdivided into a path is called a one-dimensional subdivided simplex, and a graph is called a **complex**. Thus, for a problem to be solvable 1-resiliently by two processors, the output complex must contain a subdivided simplex in which the labels on the two boundary nodes are the processors with their corresponding solo-execution outputs and the ID labels on the path alternate (colored subdivided simplex).

What if we want to solve the infinite version of the task where the possible integer inputs are not bounded? In this case, the difficulty is that we cannot place an a priori upper bound on the number of iterated steps we need to take in the iterated model. One solution is to map the infinite line to a semicircle, then do the appropriate convergence on the semicircle, and map back. Another solution, denoted by Π, proceeds as follows. Processor P_1 with input k_1 takes k_1 steps if it reads $\bot$ continuously. Otherwise, after P_1 reads a value for P_0, it stops once it reads $\bot$ or reads that P_0 has read a value from P_1. Clearly, if the input values are k_1 and k_2, then the view complex is a path of length $k_1 + k_2$ that can be folded into the interval $[k_1, k_2]$ with enough views to cover all the integers (see [4]).

In the case of shared memory (not iterated), if a processor halts once it takes some number of steps or once it observes the other processor take one step, we say that the processor halts within a unit of time. Consider the full-information version of Π (a protocol in the iterated model). An execution of the full-information protocol can be interpreted as an execution of a nonblocking emulation of the atomic-snapshot shared-memory model. If we take this view, then Π translates into a unit-time algorithm.

This conclusion, in fact, holds true for any task that is solvable 1-resiliently by two processors. It is easy to see that by defining the unit as any desirable $\epsilon > 0$, we can get two processors to output real numbers that are within an ϵ-ball (ϵ-agreement). To solve any problem, we fix an embedding of a path that may account for the solvability of the task. Consequently, there exists an ϵ such that for any interval I of length ϵ, all the simplexes that overlap the interval have a common intersection. Processors then conduct ϵ-agreement on the path, and each processor adopts as an output the node of its label which is closest to its ϵ-agreement output.

This view of convergence does not generalize easily to more than two processors. Therefore, we propose another interpretation for the two-processor convergence process [8]. After an $\epsilon/2$-agreement as above, P_i observes the largest common intersection s_i of the simplexes overlapping the

ϵ-length interval that is centered around P_i's ϵ-agreement value. It must be that $s_1 \cup s_2$ is a simplex and that $s_1 \cap s_2 \neq \emptyset$. P_i posts s_i in shared memory. It then takes the intersection of the s_j's it observes posted. If a node labeled by P_i's ID is in the intersection, P_i outputs that node. Otherwise, P_i sees only one node, v, in the intersection, and v has a label different from P_i's ID. In this case, P_i outputs one of the nodes labeled by its own ID which appear in a simplex along with v (these nodes are said to be "in the **link** of v").

Thus, since solvability amounts to ϵ-agreement, and ϵ-agreement can be achieved 1-resiliently within a unit of time, we conclude that any task 1-resiliently solvable by two processors can be solved within a unit of time.

29.3.4   Three-Processor 2-Resilient Model

We consider now the three-processor 2-resilient SWMR shared-memory model. W.l.o.g. processors alternate between writing and reading the other two cells one by one. Obviously, we cannot elect a leader (since two processors cannot), but perhaps we can solve the $(3,2)$ set-consensus problem in which processors elect at most two leaders (i.e., each processor outputs an ID of a participating processor, and the union of the outputs is of cardinality at most 2).

The structure of the full-information protocol of the one-shot shared-memory model for three processors is not as easy to identify as that for two processors. Two-processor executions have many hidden properties that are lost when we have three processors. For example, in a two-processor execution, when a processor reads the value of the other processor's cell, then in conjunction with the value it has last written to its own cell, the processor has an instantaneous view of how the memory looks, as if it read both cell in a single atomic operation. Such an instantaneous copy is called an atomic snapshot (or, for short, a snapshot) [1]. In a three-processor system, this property is lost; we cannot interpret a read operation as returning a snapshot.

To get the effect of processor P_1 reading cells C_2 and C_0 instantaneously, we have P_1 read C_2 and C_0 repeatedly until the values the processor reads do not change over two consecutive repetitions. If all values written are distinct, then the values the processor reads reside simultaneously in the memory in an instant that is after the first of the two consecutive repetitions and before the second of the repetitions. Thus, P_1 may safely return these values.

Clearly, three processors can 2-resiliently nonblocking implement one-shot snapshot memory— a memory in which a processor writes its cell and then obtains a vector of values for all cells, and this vector is a snapshot. Yet, a one-shot snapshot may give processors the following views: $P_1 : w_1, w_2, w_0, P_2 : w_1, w_2, w_0, P_0 : \bot, w_2, w_0$. This is the result of the execution: P_2 writes, P_0 writes, P_0 takes a snapshot, P_1 writes, P_1 and P_2 take a snapshot. Can we require that the set of processors S_i that return at most i values return snapshots only of values written by processors from S_i? In the example above, we have $S_2 = \{P_0\}$, but P_0 returns a value from P_2 which is not in S_2. A snapshot whose values are restricted in this way is called an immediate snapshot [7,29]. A recursive distributed procedure will return immediate snapshots (program for P_i):

1. Procedure-immediate-snapshot $ISN(P_i, k)$
2. Write input to cell $C_{k,i}$
3. For $j = 1$ to n, read cell $C_{k,j}$
4. If the number of values ($\neq \bot$) is k, then return everything read; else, call immediate snapshot $ISN(P_i, k - 1)$

We assume that all cells are initialized to $\bot$ and that in a system with n processors, P_i starts by calling immediate snapshot $ISN(P_i, n)$.

It is not difficult to prove that the view complex of a one-shot immediate snapshot is a subdivided simplex. Simple extension of the algorithm outlined for two processors shows that, in general, the

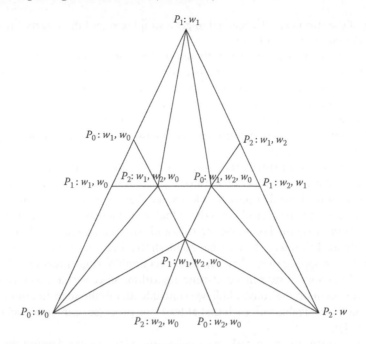

FIGURE 29.3 Three-processor one-shot immediate-snapshot view complex.

iterated immediate-snapshot model nonblocking implements the shared-memory atomic-snapshot model [8]. The view complex for three processors is shown in Figure 29.3.

We now argue by way of example that after the first stage in the iterated immediate-snapshot model, processors cannot elect two leaders. By definition, the view $P_i : w_i$ has to be mapped to P_i. The views $P_i : w_i, w_j$ are mapped to P_i or P_j. The rest are mapped to any processor ID. Such a mapping of views to processor IDs constitutes a Sperner coloring of the subdivided simplex [28]. The Sperner Lemma then says that there must be a triangle colored by the three colors. Since a triangle is at least one execution, we have proven that no election of two leaders is possible. The argument we made about the coloring of a path by two processors' IDs is just the one-dimensional instance of the Sperner Lemma. By the recursive properties of iterated models, we conclude that the structure of a k-step three-processor 2-resilient iterated immediate snapshot is the structure of a subdivided triangle.

What nontrivial tasks can we solve? For one, the task of producing a one-shot immediate snapshot is far from trivial. In general, we can solve anything that the view complex of a sufficiently large number of iterations of one-shot immediate snapshots can map to, color and boundary preserving, simplicially. This includes any subdivided triangle A [23].

We show algorithmically how any three processors converge to a triangle on a colored subdivided triangle A. Embedding a reasonably large-enough complex of the iterated immediate snapshots embedded over A yields two-dimensional ϵ-agreement over A, since triangles of the view complex can be inscribed in smaller and smaller circles as a function of the number of iterations we take. Thus, as we did for two processors, we may argue now that an $\epsilon > 0$ exists such that for any ϵ-ball in A, all simplexes that overlap the ball have a common intersection. P_i then conducts $\epsilon/2$-agreement and posts the largest such intersection s_i of the simplexes overlapping the $\epsilon/2$-radius ball whose center is P_i's $\epsilon/2$-agreement value. P_i takes the intersection of the s_i's posted and, if P_i's color is present, adopts the value of that A node. Otherwise, P_i removes a node of P_i's color, if one exists, from the union of the simplexes P_i observed to get a simplex x_i and then starts a new ϵ-agreement from a node of P_i's color in the link of x_i.

As we argued for two processors, when there are three processors, at least one processor will terminate after the first ϵ-agreement. If two processors show up for the second agreement, they have identified the processor that will not proceed; the link is at worst a closed path. The convergence of the remaining two processors is interpreted to take place on one side of the closed path rather than on the other, and the decision about on which side convergence takes place is made according to a predetermined rule that is a function of the starting nodes. The convergence of two processors on a path was outlined in Section 29.3.3.

To see that ϵ-agreement for three processors is solvable 2-resiliently on a triangle within a unit of time, we notice that if we again take an iterated algorithm with the stopping rule that processors halt once they reach some bound or once they learn that they read from each other, as in the case of two processors, it can easily be argued inductively that we get a subdivided simplex "growing from the center." As we increase the bound on the solo and pairs executions, we "add a layer" around the previous subdivided simplex. Thus, we have a mesh that becomes as fine as we need and results in ϵ-agreement. Our stopping rule, when converted via nonblocking emulation of shared memory by iterated shared memory, results in a unit-time algorithm. Since we have seen that solving a task amounts to ϵ-agreement on the convex hull, we conclude that in the snapshot model, an algorithm exists for any wait-free solvable task such that at least one processor can obtain an output within a unit of time (see [4]).

Now, if we view three-processor tasks as a collection of triangular domino pieces that can be reshaped into triangles of any size we want, we see that again the question of solvability amounts to a tiling problem. Where in the two-processor case we had two distinct domino ends that we had to tile in between, we now have three distinct domino corners, corresponding to solo executions, and three distinct domino sides, corresponding to executions in which only two processors participated. To solve a three-processor task, we have to pack the domino pieces together to form a subdivided triangle that complies with the boundary conditions on the sides. Unfortunately, this two-dimensional tiling problem is undecidable in general [17].

29.3.5 Three-Processor 1-Resilient Model

What kind of three-processor tasks are 1-resiliently solvable [9]? It stands to reason that such systems are more powerful than two-processor 1-resiliency. Perhaps the two nonfaulty processors can gang up on the faulty one and decide.

In light of our past reasoning, the natural way to answer this question is to ask what the analogue to the one-shot immediate snapshot is in this situation. A little thought shows that the analogue comprises two stages of the wait-free one-shot immediate snapshot where the nodes that correspond to solo executions have been removed. (We need to two stages because one stage with solo executions removed is a task that is not solvable 1-resiliently by three processors.) This structure can be solved as a task by the three-processor 1-resilient model, and the iteration of this structure is a model that can nonblocking implement a shared-memory model of the three-processor 1-resilient model.

An inductive argument shows that this structure is connected, and consequently consensus is impossible. Furthermore, the link of any node is connected. If we have a task satisfying these conditions, the convergence argument of Section 29.3.4 shows how to solve the task. We start at possibly only two nodes because one processor may wait on one of the other two. By simple connectivity, we can converge so that at least one processor terminates. The other two can converge on the link of the terminating processor since this link is connected.

Thus a task is solvable 1-resiliently in a system with three processors if it contains paths with connected links, connecting solo executions of two processors. Checking whether this holds amounts to a reachability problem and therefore is decidable.

29.3.6   Models with Byzantine Failure

What if a processor not only may fail to take further steps but also may write anything to its cell in the asynchronous SWMR memory? Such a processor is said to fail in a Byzantine fashion. Byzantine failures have been dealt with in the asynchronous model within the message-passing system [10]. Here we define it for the shared-memory environment and show that the essential difficulty is actually in transforming a message-passing system to a shared-memory one with write-once cells.

Why a Byzantine failure is not more harmful than a regular fail-stop failure when in a write-once cells shared-memory environment? If we assume that cell C_i is further subdivided into subcells $C_{i1}, C_{i2}, \ldots$ and that these subcells are write-once only, we can nonblocking emulate iterated snapshot shared memory in which a processor writes and then reads all cells in a snapshot. All processors now have to comply with writing in the form of snapshots—namely, they must write a set of values from the previous stage, otherwise what they write will be trivially discarded. Yet, faulty processors may post snapshots that are inconsistent with the snapshots posted by other, nonfaulty processors. We observe that we can resolve inconsistent snapshots by letting their owners to revise these snapshots to snapshots that are the union of the inconsistent ones. This allows processors to affirm some snapshots and not affirm others. A processor that observes another snapshot that is inconsistent with its own snapshot will suggests the union snapshot if its snapshot has not been affirmed yet. This processor waits with a snapshot consistent with the others until one of its snapshots has been affirmed. It then writes one of its affirmed snapshots as a final one. Thus, all other processors may check that a processor's final snapshot have been affirmed.

A processor affirms a snapshot by observing it to be consistent with the rest and writing an affirmation for it. If we wait for at least $n/2 + 1$ affirmations (discard a processor which affirms two inconsistent snapshots as faulty), then it can be seen that no inconsistent snapshots will be affirmed. In other words, we have transformed Byzantine failure to fail-stop at the cost of nonblocking emulating the original algorithm if it was not written in the iterated style.

The next problem to be addressed is how to nonblocking emulate subdivision to a write-once subcell of a cell C_i when a processor may overwrite the values previously written to the cell. The following procedure nonblocking emulates a read of $C_{i,k}$:

1. If read $C_{i,k} \neq \perp$ or if read $f + 1$ processors claiming a value $v \neq \perp$ for $C_{i,k}$, claim v for $C_{i,k}$
2. If read $2f + 1$ processors claiming v for $C_{i,k}$ or if read $f + 1$ processors wrote Confirm(v) for $C_{i,k}$, write Confirm(v) for $C_{i,k}$
3. If read $2f + 1$ processors claiming Confirm(v) for $C_{i,k}$, accept v for $C_{i,k}$

Clearly, once a processor accepts a value, all processors will accept that value eventually and so a value may not change.

Since the availability of write-once cells allows Byzantine agreement we conclude that like Byzantine agreement the implementation of write-once cells requires that less of a third of the processors fail. In hindsight, we recognize that Bracha discovered in 1987 that a shared-memory system with Byzantine failure and write-once cells can be nonblocking implemented on a message-passing system with the same failures as the shared-memory system provided that $3f < n$ [10]. It took another three years for researchers to independently realize the simpler transformation from message passing to shared memory for fail-stop faults [2].

29.3.7   Message-Passing Model

Obviously, shared memory can nonblocking emulate message passing. The conceptual breakthrough made by Attiya, Bar-Noy, and Dolev (ABD) [2] was to realize that for $2f < n$, message passing can f-resiliently wait-free emulate f-resilient shared memory, as follows. To emulate a write, a processor takes the latest value it is to write, sends the value to all of the processors, and waits for acknowledgment from a majority. To read, a processor asks for the latest value for the cell from a majority of the processors and then writes it. A processor keeps an estimate of the latest value for a cell; this estimate is the value with the highest sequence number that the processor has ever observed for the cell. Since two majorities intersect, the emulation works.

All results of shared memory now apply verbatim, by either the fail-stop or the Byzantine transformation, to message passing. To derive all the above directly in message passing is complex, because, in one way or the other, hidden ABD or Bracha transformations sit there and obscure the real issues. The ABD and Bracha transformations have clarified why shared memory is a much cleaner model than message passing to think and argue about.

29.3.8 From Safe-Bits to Atomic Snapshots

We now show that in retrospect the result of the research conducted during the second part of the 1980s on the power of shared-memory made out of safe-bits is not surprising.

We started this section with a SWMR shared-memory model. Can such a model be nonblocking implemented from the most basic primitives? Obviously, this is a question about the power of models, specifically about relaxing the atomicity of read-write shared-memory registers. This problem was raised by Lamport [24] in 1986, and quite a few researchers have addressed it since then. Here we show that the nonblocking version of the problem can be answered trivially.

The primitive Lamport considers is a single-writer/single-reader safe-bit. A safe-bit is an object to which the writer writes a bit and from which the reader can read provided that the interval of operation execution of reading does not overlap with writing.

We now show how a stage of the iterated immediate-snapshot model can be nonblocking implemented by safe-bits. We assume an unlimited number of safe-bits, all initialized to 0, per pair of processors. To write a value, a processor writes it in unary, starting at a location known to the reader. To read, a processor counts the number of 1s it encounters until it meets a 0.

We observe that the recursive program for one-shot immediate snapshot itself consists of stages. At most k processors arrive at the immediate-snapshot stage called k (notice that these stages run from n down to 1). All we need is for at least one out of the k processors to remain trapped at stage k. The principle used to achieve this is the flag principle, namely, if k processors raise a flag and then count the number of flags raised, at least one processor will see k flags. For this principle to hold, we do not need atomic variables. Processors first write to all the other processors that they are at stage k and only then start the process of reading, so the last processor to finish writing will always read k flags.

Given that a processor writes its input to all other processors before it starts the immediate-snapshot stage, when one processor encounters another in the immediate snapshot and needs to know the other processor's input, the input is already written.

This shows that any task solvable by the most powerful read-write objects can be solved by single-writer/single-reader safe-bits. Can one nonblocking emulate by safe-bits any object that can be read-write emulated? The ingenious transformations cited in the introduction have made it clear that the answer to this question is "yes." Nonetheless, the field still awaits a theory that will allow us to deal with wait-free emulation in the same vein as we have dealt with nonblocking emulation here.

29.3.9 Geometric Protocols

What if we draw a one-dimensional immediate snapshot complex on a plan and repeat subdivide it forever? We can then argue that each point in our drawing corresponds to an infinite run with respect to views that are outputed. Obviously such a construction can be done for any dimension.

We obtain an embedding of the space of runs in the Euclidean unit simplex. Such an embedding was the quest that eluded the authors of [29]. An embedding gives rise to "geometric-protocols."

Consider the problem of 2-processors election when we are given that the infinite symmetric run will not happen. One may work out an explicit protocol (which is not trivial). Geometrically, we take an embedding, and a processor waits until its view is completely on one side of the symmetric run. It then decides according to the solo execution that side contains.

29.4 Synchronous Systems

29.4.1 Shared-Memory Model

In retrospect, given the iterated-snapshots model in which conceptually processors go in lock-step, and its equivalence to the asynchronous shared memory, it is hard to understand the dichotomy between synchrony and asynchrony. In one case, the synchronous, we consider processors that are not completely coordinated because of various type of failures. In the other case, the asynchronous, processors are not coordinated because of speed mismatch. Why is this distinction of such fundamental importance? In fact, we argue that it is not. To exemplify this we show how one can derive a result in one model from a result in the other.

We consider the SWMR shared-memory model where its computation evolves in rounds. In other words, all communication events in the different processors and communication objects proceed in lockstep. At the beginning of a round, processors write their cells and then read all cells in any order. Anything written in the round by a processor is read by all processors. This model may be viewed as a simple variant of the parallel RAM (PRAM) model, in which processors read all cells, rather than just one cell, in a single round.

With no failures, asynchronous systems can emulate synchronous ones. What is the essential difference between synchronous and asynchronous systems when failures are involved? A 1-resilient asynchronous system can "almost" emulate a round of the synchronous system. It falls short because one processor (say, if we are dealing with the synchronous shared-memory system below) misbehaves—but this processor does not really misbehave: what happens is that some of the other processors miss it because they read its cell too early, and they cannot wait on it, since one processor is allowed to fail-stop. If we call this processor faulty, we have the situation where we have at most one fault in a round, but the fault will shift all over the place and any processor may look faulty sooner or later. In contrast, in a synchronous system, a faulty behavior is attributed to some processor, and we always assume that the number of faults is less than n, the number of processors in the system.

We now introduce the possibility of faults into the synchronous system, and we will examine three types of faults. The first type is the analogue of fail-stop: a processor dies in the middle of writing, and the last value it wrote may have been read by some processors and not read by the others.

The algorithm to achieve consensus is quite easy. Each processor writes an agreement proposal at each round and then, at the next round, proposes the plurality of values it has read (with a tie-breaking rule common to all processors). At the first round, processors propose their initial value. After $t + 1$ rounds, where t is an upper bound on the number of faults, a processor decides on the value it would have proposed at round $t + 2$. The algorithm works because, by the pigeon principle, there is a clean round, namely, a round at which no processor dies. At the clean round, all processors take the plurality of common information, which results in a unanimous proposal at the next round. This proposal will be sustained to the end. Various techniques exist that can lead to early termination in executions with fewer faults than expected.

The next type of fault is omission [27]: a processor may be resurrected from fail-stop and continue to behave correctly, only to later die again to be perhaps resurrected again, and so on.

We reduce omission failure to fail-stop. A processor P_i that fails to read a value of P_j at round k goes into a protocol introduced in Section 29.5 to commit P_j as faulty. Such a protocol has the property that if P_i succeeds, then all processors will consider P_j faulty in the next round. Otherwise P_i obtain a value for P_j. We see that if a correct processor P_i fails to read a value from P_j, all processors will stop reading cell C_j at the next round, which is exactly the effective behavior of a fail-stop processor.

The last type of fault is Byzantine. We assume $n = 3f + 1$. Here, in addition to omitting a value, a processor might not obey the protocol and instead write anything (but not different values ($\neq \perp$) to different processors). We want to achieve the effect of true shared memory, in which if a correct processor reads a value, then all other processors can read the same value after the correct processor has done so.

We encountered the same difficulty in the asynchronous case, with the difference that in the asynchronous case things happen "eventually." If we adopt the same algorithm, whatever must happen eventually (asynchronous case) translates into happening in some finite number of rounds (synchronous case). Moreover, if something that is supposed to happen for a correct processor in the asynchronous case does not happen for a processor in the synchronous case within a prescribed number of rounds, the other processors infer that the processor involved is faulty and announce it as faulty. Thus, the asynchronous algorithm for reading a value which we gave in Section 29.3 translates to (code for processor P_i):

1. Round 1: $v := read(cell)$.
2. Round 2: Write v, read values for v from all nonfaulty processors.
3. Round 3: If $2f + 1$ for value $v \neq \perp$ in Round 2, write $confirm(v)$; else, write $v := faulty$, read cells.
4. Round 4: If $2f + 1$ for v in Round 3 or if $f + 1$ for $confirm(v)$, write $confirm(v)$; else, $v := faulty$. If $2f + 1$ $confirm(v)$ up to now, then $accept(v)$. If $v = faulty$ and $accept(v)$, consider the processor that wrote v faulty.

A little thought shows that the "eventual" of the asynchronous case translates into a single round in the synchronous case. If a value is accepted, it will be accepted by all correct processors in the next round. If a value is not accepted by the end of the third round, then all correct processors propose $v := faulty$. In any case, at the end of the fourth round, either a real value or $v = faulty$ or both will be accepted. After a processor P_i is accepted as faulty, at the next round (which may be at the next phase) all processors will accept it as faulty and will ignore it. Thus, ignoring for the moment the problem that a faulty processor may write incorrect values, we have achieved the effect of write-once shared memory with fail-stop.

To deal with the issue of a faulty processor writing values a correct processor would not write, we can check on previous writes and see whether the processor observes its protocol. Here we do not face the difficulty we faced in the asynchronous case: a processor cannot avoid reading a value of a correct processor, and a processor may or may not (either is possible) read a value from a processor that failed. Nevertheless, a processor's value may be inconsistent with the values of correct processors. We notice that correct processors are consistent among themselves. Processors then can draw a "conflict" graph and, by eliminating edges to remove conflicts, declare the corresponding processors faulty. Since an edge contains at least one faulty processor, we can afford the cost of failing a correct processor.

We now argue a lower bound of $f + 1$ rounds for consensus for any of our three failure modes. It suffices to prove this bound for fail-stop failure, because fail-stop is a special case of omission failure and of Byzantine failure. Suppose an $(m < t + 1)$-rounds consensus algorithm exists for the fail-stop type of failure. We emulate the synchronous system by a 1-resilient asynchronous system in iterated atomic-snapshot shared memory. At a round, there is a unique single processor whose value other processors may fail to read. We consider such a processor to be faulty. We have seen that, without any extra cost, if a correct processor fails to read a value of processor P_j, then all correct processors will consider P_j faulty in the next round, which amounts to fail-stop. Thus, the asynchronous system emulates m rounds of the synchronous system. At each simulated round, there is at most one fault, for a total of at most f faults. This means that the emulated algorithm should result in consensus in the 1-resilient asynchronous system, which is impossible.

We can apply the same logic for set consensus and show that with f faults and k-set consensus, we need at least $\lfloor f/k \rfloor + 1$ rounds. This involves simulating the algorithm in a k-resilient asynchronous atomic-snapshot shared-memory system in which k-set consensus is impossible. (The first algorithm in this section automatically solves k-set consensus in the prescribed number of rounds; see [11]).

Notice that the above impossibility result for the asynchronous model translated into a lower bound on round complexity in the synchronous model. We now present the failure-detector framework, a framework in which speed mismatch is "transformed" into failure, and thus unifying synchrony and asynchrony.

29.5 Failure Detectors: A Bridge between Synchronous and Asynchronous Systems

In section 29.4, we have seen that a 1-resilient asynchronous system looks like a synchronous system in which at each round a single but arbitrary processor may fail. Thus, synchronous systems can achieve consensus because in this setting when one processor considers another processor faulty, it is indeed the case, and one processor is never faulty. In the asynchronous setting, the processor considered faulty in a round is not faulty, only slow; we make a mistake in declaring it faulty.

Chandra and Toueg [13] have investigated the power of systems via the properties of their of augmenting subsystem called FD that issues faulty declarations. The most interesting FDs are those with properties called S and $\Diamond S$.

FD S can be described as follows:

- All processors that take finitely many steps in the underlying computation (faulty processors) will eventually be declared forever faulty by all the local FDs of processors that took infinitely many steps (correct processors).
- Some correct processor is never declared faulty by the local FDs of correct processors.

In $\Diamond S$, these properties hold eventually. Chandra et al. [12] then showed that of all the FDs that provide for consensus, $\Diamond S$ is the weakest: any FD that provides for consensus can nonblocking emulate $\Diamond S$.

Thus, in a sense, if we take a system with timing constraints and call it synchronous if the constraints provide for consensus, then we have an alternative way of checking on whether the constraints provide for nonblocking emulation of $\Diamond S$.

In many systems, this alternative technique is natural. For instance, consider a system in which the pending operations stay pending for at most some bounded but otherwise unknown real time and the processors can delay themselves for times that grow longer and longer without bound (of course, a delay operation is not pending until the prescribed delay time has elapsed). It is easy to see that this system can nonblocking implement $\Diamond S$ and, as a result, achieve consensus. Thus, we have a variety of consensus algorithms from which to choose. In fact, we can transform any synchronous consensus algorithm for omission to an algorithm for $\Diamond S$. The transformation proceeds in two stages. We first transform a synchronous omission algorithm A to an algorithm B for consensus in S. We accomplish this by nonblocking emulating A round by round, where each processor waits on the others until either the expected value is observed or the processor waited on is declared faulty by the FD. Such failure can be considered omission failure. Since in S, one of the processors is never declared faulty, we get that the number of omission faults is less than n, and the emulation results in consensus.

To transform a consensus algorithm A in S to a consensus algorithm B for $\Diamond S$, we use a layer algorithm called Eventual. We assume A is safe in the sense that when running A in $\Diamond S$, the liveness conditions that result in consensus are weakened but the safety conditions that make processors in A over S agree or preserve validity (output has to be one of the inputs) are maintained.

Algorithm Eventual has the property that a processor's output consists of either committing to some input value or adopting one.

1. If all processors start with the same input, then all commit to that input.
2. If one processor commits to an input value, then all other processors commit to or adopt that value.

The consensus algorithm B for shared memory with $\Diamond S$ is to run alternately with A followed by Eventual. The output of A is the input to Eventual, and the output of Eventual is the input to A. When a processor commits to a value in Eventual, it outputs it as the output of B. Algorithm Eventual is simple. Processors repeatedly post their input values and take snapshots until the number of postings exceeds n. If a processor observes only a single value, it commits to that value; otherwise, it adopts a plurality value.

The notion of FDs is very appealing. It unifies synchronous and asynchronous systems. In the FD framework, all systems are of the same type but each system possesses FDs with distinctive properties. Research into the question of how this unified view can be exploited in distributed computing—for example, it might enable understanding of the "topology" of FDs—has hardly begun.

Another direction of possible research is to enrich the semantics of FD. If we take the iterated snapshot system we may consider attaching a separate FD subsystem to each layer. A processor which is "late" arriving to a layer is declared faulty by that layer subsystem. We may now investigate the power of such system as a function of the properties of their FDs. We can for instance model a t-resilient system as a system in which at most t processors will be declared faulty at a layer. When one considers such a system, the dichotomy of systems between synchronous and asynchronous is completely blurred. The traditional FD of Chandra and Toueg cannot capture such a property.

29.6 Research Issues and Summary

In this chapter, we have considered a body of fairly recent research in distributed computing whose results are related and draw upon one another. We have argued that a unified view of these results derives from the synergy between the application of results from topology and the use of the interpretive power of distributed computing to derive transformations that make topology apply.

Many interesting and important questions, aside from matters of complexity, remain. The most practical of these questions concern computations that are amenable to algorithms whose complexity is a function of the concurrency rather than the size of the system. Such algorithms are usually referred to as fast. Examples of tasks that can be solved by application of fast algorithms are numerous, but a fundamental understanding of exactly what, why, and how computations are amenable to such algorithms is lacking. Extension of the theory presented in this chapter to nonterminating tasks and to long-lived objects is next on the list of questions.

29.7 Further Information

Current research on the theoretical aspects of distributed computing is reported in the proceedings of the annual ACM Symposium on Principles of Distributed Computing (PODC) and the annual International Workshop on Distributed Algorithms on Graphs (WDAG). Relatively recent books and surveys are [6,21,26].

Defining Terms

t-**Resilient system:** A system in which at most *t* processors are faulty.

0-Simplex: A singleton set. item[One-dimensional subdivided simplex:] An embedding of 1-simplex that is partitioned into 1-simplexes (a path).

1-Simplex: A set consisting of two elements.

Atomic snapshot: An atomic read operation that returns the entire shared memory.

Chain-of-runs: A sequence of runs in which two consecutive runs are indistinguishable to a processor.

Cleanround: A round in which no new faulty behavior is exhibited.

Communication-closed-layers: A distributed program partitioned to layers that communicate remotely only among themselves and communicate locally unidirectionally.

Complete problem: A problem in a model that characterizes the set of all other problems in the model in the sense that they are reducible to it.

Complex: A set of simplexes that is closed under subset.

Consensus problem: A decision task in which all processors agree on a single input value.

Election: A consensus over the IDs as inputs.

Failure-detector: An oracle that updates a processor on the operational status of the rest of the processors.

Fast solution: A solution to a problem whose complexity depends on the number of participating processors rather than the size of the entire system.

Faulty processor: A processor whose view is output finitely many times in a run.

Full-information protocol: A protocol that induces the finest partition of the set of runs, compared to any other protocol in the model.

Immediate snapshots: A restriction of the atomic snapshots that achieves a certain closure property that atomic snapshots do not have.

Link (of a simplex in a complex): The set of all simplexes in a complex that are contained in a simplex with the given simplex.

Nonblocking (emulation): An emulation that may increase the set of faulty processors.

Oblivious (to a parameter): Not a function of that parameter.

Outputs: Map over views that are eventually known to all nonfaulty processors.

Participating (set of processors): Processors in a run that are not sleeping.

Processor: A sequential piece of code.

Protocol: A set of processors whose codes refer to common communication objects.

Run: An infinite sequence of global "instantaneous-description" of a system, such that one element is the preceding one after the application of a pending operation.

Safe-bit: A single-writer single-reader register bit whose read is defined and returns the last value written only if does not overlap a write operation to the register.

Sleeping (in a run): A processor whose state in a run does not change.

Solo-execution: A run in which only a single processor is not sleeping.

Task: A relation from inputs and set of participating processors to outputs.

View: A set of runs compatible with a processor's local state.

Wait-free (solution): A solution to a nontermination problem in which a processor that is not faulty outputs infinitely many output values.

Acknowledgment

I am grateful to Hagit Attiya for detailed illuminating comments on an earlier version of this chapter.

References

1. Afek, Y., Attiya, H., Dolev, D., Gafni, E., Merrit, M., and Shavit, N., Atomic snapshots of shared memory. In *Proceedings of the 9th ACM Symposium on Principles of Distributed Computing*, pp. 1–13, Quebec, Canada, 1990.
2. Attiya, H., Bar-Noy, A., and Dolev, D., Sharing memory robustly in message-passing systems. *Journal of the ACM*, 42(1), 124–142, January 1995.
3. Afek, Y., Greenberg, D.S., Merritt, M., and Taubenfeld, G., Computing with faulty shared objects. *Journal of the Association of the Computing Machinery*, 42, 1231–1274, 1995.
4. Attiya, H., Lynch, N., and Shavit, N., Are wait-free algorithms fast? *Journal of the ACM*, 41(4), 725–763, July 1994.
5. Attiya, H. and Rajsbaum, S., The combinatorial structure of wait-free solvable tasks. In *WDAG: International Workshop on Distributed Algorithms*, Springer-Verlag, Berlin, Germany, 1996.
6. Attiya, H., Distributed computing theory. In *Handbook of Parallel and Distributed Computing*, A.Y. Zomaya (Ed.), McGraw-Hill, New York, 1995.
7. Borowsky, E. and Gafni, E., Immediate atomic snapshots and fast renaming. In *Proceedings of the 12th ACM Symposium on Principles of Distributed Computing*, pp. 41–51, Ithaca, NY, 1993.
8. Borowsky, E. and Gafni, E., A simple algorithmically reason characterization of wait-free computations. In *Proceedings of the 16th ACM Symposium on Principles of Distributed Computing*, pp. 189–198, Santa Barbara, CA, 1997.
9. Biran, O., Moran, S., and Zaks, S., A combinatorial characterization of the distributed tasks which are solvable in the presence of one faulty processor. In *Proceedings of the 7th ACM Symposium on Principles of Distributed Computing*, pp. 263–275, Toronto, ON, Canada, 1988.
10. Bracha, G., Asynchronous byzantine agreement protocols. *Information and Computation*, 75(2), 130–143, November 1987.
11. Chaudhuri, S., Herlihy, M., Lynch, N.A., and Tuttle, M.R., A tight lower bound for k-set agreement. In *34th Annual Symposium on Foundations of Computer Science*, pp. 206–215, Palo Alto, CA, IEEE, November 3–5, 1993.
12. Chandra, T.D., Hadzilacos, V., and Toueg, S., The weakest failure detector for solving consensus. *Journal of the ACM*, 43(4), 685–722, July 1996.
13. Chandra, T.D. and Toueg, S., Unreliable failure detectors for reliable distributed systems. *Journal of the ACM*, 43(2), 225–267, March 1996.
14. Elrad, T.E. and Francez, N., Decomposition of distributed programs into communication closed layers. *Science of Computer Programming*, 2(3), 155–173, 1982.
15. Fischer, M.J. and Lynch, N.A., A lower bound on the time to assure interactive consistency. *Information Processing Letters*, 14(4), 183–186, 1982.
16. Fischer, M., Lynch, N., and Paterson, M., Impossibility of distributed consensus with one faulty process. *Journal of the ACM*, 32(2), 374–382, 1985.
17. Gafni, E. and Koutsoupias, E., 3-processor tasks are undecidable. In *Proceedings of the 14th Annual ACM Symposium on Principles of Distributed Computing*, 271, ACM, New York, August 1995.

18. Gray, J.N., Notes on data base operating systems. In *LNCS, Operating Systems, an Advanced Course,* R. Bayer, R.M. Graham, and G. Seegmuller (Eds.), Vol. 60, Springer-Verlag, Heidelberg, Germany, 1978.
19. Herlihy, M.P., Wait-free synchronization. *ACM Transactions on Programming Languages and Systems,* 11(1), 124–149, January 1991. [Supersedes 1988 PODC version.]
20. Halpern, J.Y. and Moses, Y., Knowledge and common knowledge in a distributed environment. *Journal of the ACM,* 37(3), 549–587, July 1990.
21. Herlihy, M. and Rajsbaum, S., Algebraic topology and distributed computing—A primer. *Lecture Notes in Computer Science,* 1000, 203, 1995.
22. Herlihy, M. and Shavit, N., The asynchronous computability theorem for *t*-resilient tasks. In *Proceedings of the 25th ACM Symposium on the Theory of Computing,* pp. 111–120, San Diego, CA, 1993.
23. Herlihy, M. and Shavit, N., A simple constructive computability theorem for wait-free computation. In *Proceedings of the 26th ACM Symposium on the Theory of Computing,* Montreal, QC, Canada, 1994.
24. Lamport, L., On interprocess communication. *Distributed Computing,* 1, 77–101, 1986.
25. Lamport, L. and Lynch, N., Distributed computing: Models and methods. In *Handbook of Theoretical Computer Science,* J. van Leewen (Ed.), Vol. B: *Formal Models and Semantics,* Chapter 19, pp. 1157–1199, MIT Press, New York, 1990.
26. Lynch, N., *Distributed Algorithms,* Morgan Kaufmann, San Francisco, CA, 1996.
27. Neiger, G. and Toueg, S., Automatically increasing the fault-tolerance of distributed algorithms. *Journal of Algorithms,* 11(3), 374–419, September 1990.
28. Spanier, E.H., *Algebraic Topology,* Springer-Verlag, New York, 1966.
29. Saks, M. and Zaharoglou, F., Wait-free *k*-set agreement is impossible: The topology of public knowledge. In *Proceedings of the 25th ACM Symposium on the Theory of Computing,* pp. 101–110, San Diego, CA, 1993.

30

Linear Programming*

Vijay Chandru
National Institute of Advanced Studies

M.R. Rao
Indian School of Business

Linear programming has been a fundamental topic in the development of computational sciences. The subject has its origins in the early work of L. B. J. Fourier on solving systems of linear inequalities, dating back to the 1820s. More recently, a healthy competition between the simplex and interior point methods has led to rapid improvements in the technologies of linear programming. This combined with remarkable advances in computing hardware and software have brought linear programming tools to the desktop, in a variety of application software for decision support. Linear programming has provided a fertile ground for the development of various algorithmic paradigms. Diverse topics such as symbolic computation, numerical analysis, computational complexity, computational geometry, combinatorial optimization, and randomized algorithms all have some linear programming connection. This chapter reviews this universal role played by linear programming in the science of algorithms.

---

* Dedicated to George Dantzig on this the 50th Anniversary of the Simplex Algorithm.

30.1   Introduction

Linear programming has been a fundamental topic in the development of the computational sciences [50]. The subject has its origins in the early work of Fourier [30] on solving systems of linear inequalities, dating back to the 1820s. The revival of interest in the subject in the 1940s was spearheaded by Dantzig [19] in the United States and Kantorovich [45] in the former USSR. They were both motivated by the use of linear optimization for optimal resource utilization and economic planning. Linear programming, along with classical methods in the calculus of variations, provided the foundations of the field of mathematical programming, which is largely concerned with the theory and computational methods of mathematical optimization. The 1950s and 1960s marked the period when linear programming fundamentals (**duality**, **decomposition** theorems, network flow theory, **matrix factorizations**) were worked out in conjunction with the advancing capabilities of computing machinery [20].

The 1970s saw the realization of the commercial benefits of this huge investment of intellectual effort. Many large-scale **linear programs** were formulated and solved on mainframe computers to support applications in the industry (for example, oil, airlines) and for the state (for example, energy planning, military logistics). The 1980s were an exciting period for linear programmers. The polynomial time-complexity of linear programming was established. A healthy competition between the simplex and interior point methods ensued that finally led to rapid improvements in the technologies of linear programming. This combined with remarkable advances in computing hardware and software have brought linear programming tools to the desktop, in a variety of application software (including spreadsheets) for decision support.

The fundamental nature of linear programming in the context of algorithmics is borne out by a few examples.

- Linear programming is at the starting point for variable elimination techniques on algebraic constraints [12], which in turn forms the core of algebraic and symbolic computation.
- Numerical linear algebra and particularly sparse matrix technology was largely driven in its early development by the need to solve large-scale linear programs [37,62].
- The complexity of linear programming played an important role in the 1970s in the early stages of the development of the polynomial hierarchy and particularly in the $\mathcal{N}P$-completeness and $\mathcal{P}$-completeness in the theory of computation [68,69].
- Linear-time algorithms based on prune and search techniques for low-dimensional linear programs have been used extensively in the development of computational geometry [25].
- Linear programming has been the testing ground for very exciting new developments in randomized algorithms [61].
- **Relaxation** strategies based on linear programming have played a unifying role in the construction of approximation algorithms for a wide variety of combinatorial optimization problems [17,33,77].

In this chapter we will encounter the basic algorithmic paradigms that have been invoked in the solution of linear programs. An attempt has been made to provide intuition about some fairly deep and technical ideas without getting bogged down in details. However, the details are important and the interested reader is invited to explore further through the references cited. Fortunately, there are many excellent texts, monographs, and expository papers on linear programming [5,9,16,20,63, 66,68,70,72,73] that the reader can choose from, to dig deeper into the fascinating world of linear programming.

30.2   Geometry of Linear Inequalities

Two of the many ways in which linear inequalities can be understood are through algorithms or through nonconstructive geometric arguments. Each approach has its own merits (aesthetic and otherwise). Since the rest of this chapter will emphasize the algorithmic approaches, in this section we have chosen the geometric version. Also, by starting with the geometric version, we hope to hone the reader's intuition about a convex **polyhedron**, the set of solutions to a finite system of linear inequalities.* We begin with the study of linear, homogeneous inequalities. This involves the geometry of (convex) **polyhedral cones**.

30.2.1   Polyhedral Cones

A homogeneous linear equation in n variables defines a hyperplane of dimension $(n - 1)$ which contains the origin and is therefore a linear subspace. A homogeneous linear inequality defines a halfspace on one side of the hyperplane, defined by converting the inequality into an equation. A system of linear homogeneous inequalities therefore defines an object which is the intersection of finitely many halfspaces, each of which contains the origin in its boundary. A simple example of such an object is the nonnegative orthant. Clearly the objects in this class resemble cones with the apex defined at the origin and with a prismatic boundary surface. We call them convex polyhedral cones.

A convex polyhedral cone is the set of the form

$$\mathcal{K} = \{x | Ax \leq 0\}$$

Here A is assumed to be an $m \times n$ matrix of real numbers. A set is convex if it contains the line segment connecting any pair of points in the set. A convex set is called a convex cone if it contains all nonnegative scalar multiples of points in it. A convex set is polyhedral if it is represented by a finite system of linear inequalities. As we shall deal exclusively with cones that are both convex and polyhedral, we shall refer to them as cones.

The representation of a cone as the solutions to a finite system of homogeneous linear inequalities is sometimes referred to as the constraint or implicit description. It is implicit because it takes an algorithm to generate points in the cone. An explicit or edge description can also be derived for any cone.

THEOREM 30.1   *Every cone $\mathcal{K} = \{x : Ax \leq 0\}$ has an edge representation of the following form.*
$\mathcal{K} = \{x : x = \sum_{j=1}^{L} e^j \mu_j, \ \ \mu_j \geq 0 \ \forall j\}$ *where each distinct edge of $\mathcal{K}$ is represented by a point e^j.*

Thus, for any cone we have two representations:

- Constraint representation: $\mathcal{K} = \{x : Ax \leq 0\}$
- Edge representation: $\mathcal{K} = \{x : x = E\mu, \ \ \mu \geq 0\}$

The matrix E is a representation of the edges of $\mathcal{K}$. Each column $E_{i.}$ of E contains the coordinates of a point on a distinct edge. Since positive scaling of the columns is permitted, we fix the representation by scaling each column so that the last non-zero entry is either 1 or -1. This scaled matrix E is called the Canonical Matrix of Generators of the cone $\mathcal{K}$.

---

* For the study of infinite systems of linear inequalities see Chapter 32 of this handbook.

Every point in a cone can be expressed as a positive combination of the columns of E. Since the number of columns of E can be huge, the edge representation does not seem very useful. Fortunately, the following tidy result helps us out.

THEOREM 30.2   [11]. *For any cone $\mathcal{K} \in \mathfrak{R}^n$, every $\bar{x} \in \mathcal{K}$ can be expressed as the positive combination of at most n edge points.*

30.2.1.1 Conic Duality

The representation theory for convex polyhedral cones exhibits a remarkable duality relation. This duality relation is a central concept in the theory of linear inequalities and linear programming as we shall see later.

Let $\mathcal{K}$ be an arbitrary cone. The dual of $\mathcal{K}$ is given by

$$\mathcal{K}^* = \left\{ u : x^T u \leq 0, \ \forall x \in \mathcal{K} \right\}$$

THEOREM 30.3   *The representations of a cone and its dual are related by*
$\mathcal{K} = \{x : Ax \leq 0\} = \{x : x = E\mu, \ \mu \geq 0\}$ *and*
$\mathcal{K}^* = \{u : E^T u \leq 0\} = \{u : u = A^T \lambda, \ \lambda \geq 0\}.$

COROLLARY 30.1 $\mathcal{K}^*$ *is a convex polyhedral cone and duality is involutory (i.e., $(\mathcal{K}^*)^*). = \mathcal{K}$.*

As we shall see, there is not much to linear inequalities or linear programming, once we have understood convex polyhedral cones.

30.2.2 Convex Polyhedra

The transition from cones to polyhedra may be conceived of, algebraically, as a process of dehomogenization. This is to be expected, of course, since polyhedra are represented by systems of (possibly inhomogeneous) linear inequalities and cones by systems of homogeneous linear inequalities. Geometrically, this process of dehomogenization corresponds to realizing that a polyhedron is the Minkowski or set sum of a cone and a **polytope** (bounded polyhedron). But before we establish this identity, we need an algebraic characterization of polytopes. Just as cones in $\mathfrak{R}^n$ are generated by taking positive linear combinations of a finite set of points in $\mathfrak{R}^n$, polytopes are generated by taking convex linear combinations of a finite set of (generator) points.

DEFINITION 30.1   Given K points $\{x^1, x^2, \cdots, x^K\}$ in $\mathfrak{R}^n$ the Convex Hull of these points is given by

$$C.H. \left(\{x^i\} \right) = \left\{ \bar{x} : \bar{x} = \sum_{i=1}^{K} \alpha_i x^i, \ \sum_{i=1}^{K} \alpha_i = 1, \ \alpha \geq 0 \right\}$$

i.e., the convex hull of a set of points in $\mathfrak{R}^n$ is the object generated in $\mathfrak{R}^n$ by taking all convex linear combinations of the given set of points. Clearly, the convex hull of a finite list of points is always bounded.

THEOREM 30.4   ([84]). *P is a polytope if and only if it is the convex hull of a finite set of points.*

DEFINITION 30.2 An **extreme point** of a convex set S is a point $x \in S$ satisfying

$$x = \alpha \bar{x} + (1 - \alpha)\tilde{x}, \quad \bar{x}, \tilde{x} \in S, \quad \alpha \in (0, 1) \quad \rightarrow \quad x = \bar{x} = \tilde{x}$$

Equivalently, an extreme point of a convex set S is one that cannot be expressed as a convex linear combination of some other points in S. When S is a polyhedron, extreme points of S correspond to the geometric notion of corner points. This correspondence is formalized in the corollary below.

COROLLARY 30.2 A polytope P is the convex hull of its extreme points.

Now we go on to discuss the representation of (possibly unbounded) convex polyhedra.

THEOREM 30.5 *Any convex polyhedron P represented by a linear inequality system $\{y : yA \leq c\}$ can be also represented as the set addition of a convex cone R and a convex polytope Q.*

$$P = Q + R = \{x : x = \bar{x} + \bar{r}, \ \bar{x} \in Q, \ \bar{r} \in R\}$$

$$Q = \left\{ \bar{x} : \bar{x} = \sum_{i=1}^{K} \alpha_i x^i, \ \sum_{i=1}^{K} \alpha_i = 1, \ \alpha \geq 0 \right\}$$

$$R = \left\{ \bar{r} : \bar{r} = \sum_{j=1}^{L} \mu_j r^j, \ \mu \geq 0 \right\}$$

It follows from the statement of the theorem that P is nonempty if and only if the polytope Q is nonempty. We proceed now to discuss the representations of R and Q, respectively.

The cone R associated with the polyhedron P is called the recession or characteristic cone of P. A hyperplane representation of R is also readily available. It is easy to show that

$$R = \{r : Ar \leq 0\}$$

An obvious implication of Theorem 30.5 is that P equals the polytope Q if and only if $R = \{0\}$. In this form, the vectors $\{r^j\}$ are called the extreme rays of P.

The polytope Q associated with the polyhedron P is the convex hull of a finite collection $\{x^i\}$ of points in P. It is not difficult to see that the minimal set $\{x^i\}$ is precisely the set of extreme points of P. A nonempty pointed polyhedron P, it follows, must have at least one extreme point.

The affine hull of P is given by

$$A.H.\{P\} = \left\{ x : x = \sum \alpha_i x^i \right\}$$

$$x^i \in P \ \forall i \quad \text{and} \quad \sum \alpha_i = 1$$

Clearly, the x^i can be restricted to the set of extreme points of P in the definition above. Furthermore, $A.H.\{P\}$ is the smallest affine set that contains P. A hyperplane representation of $A.H.\{P\}$ is also possible. First let us define the implicit linear equality system of P to be

$$\{A^= x = b^=\} = \{A_{i.} x = b_i \ \forall x \in P\}$$

Let the remaining inequalities of P be defined as

$$A^+ x \leq b^+$$

It follows that P must contain at least one point $\bar{x}$ satisfying

$$A^=\bar{x} = b^= \quad \text{and} \quad A^+\bar{x} < b^+$$

LEMMA 30.1 $A.H.\{P\} = \{x : A^=x = b^=\}$.

The dimension of a polyhedron P in $\Re^n$ is defined to be the dimension of the affine hull of P, which equals the maximum number of affinely independent points, in $A.H.\{P\}$, minus one. P is said to be full-dimensional if its dimension equals n or, equivalently, if the affine hull of P is all of $\Re^n$.

A supporting hyperplane of the polyhedron P is a hyperplane H

$$H = \left\{x : b^Tx = z^*\right\}$$

satisfying

$$b^Tx \le z^* \ \forall x \in P$$
$$b^T\hat{x} = z^* \text{ for some } \hat{x} \in P$$

A supporting hyperplane H of P is one that touches P such that all of P is contained in a halfspace of H. Note that a supporting plane can touch P at more than one point.

A face of a nonempty polyhedron P is a subset of P that is either P itself or is the intersection of P with a supporting hyperplane of P. It follows that a face of P is itself a nonempty polyhedron. A face of dimension, one less than the dimension of P, is called a facet. A face of dimension one is called an edge (note that extreme rays of P are also edges of P). A face of dimension zero is called a vertex of P (the vertices of P are precisely the extreme points of P). Two vertices of P are said to be adjacent if they are both contained in an edge of P. Two facets are said to be adjacent if they both contain a common face of dimension one less than that of a facet. Many interesting aspects of the facial structure of polyhedra can be derived from the following representation lemma.

LEMMA 30.2 F is a face of $P = \{x : Ax \le b\}$ if and only if F is nonempty and $F = P \cap \{x : \tilde{A}x = \tilde{b}\}$, where $\tilde{A}x \le \tilde{b}$ is a subsystem of $Ax \le b$.

As a consequence of the lemma, we have an algebraic characterization of extreme points of polyhedra.

THEOREM 30.6   *Given a polyhedron P, defined by $\{x : Ax \le b\}$, x^i is an extreme point of P if and only if it is a face of P satisfying $A^ix^i = b^i$ where $((A^i), (b^i))$ is a submatrix of (A, b) and the rank of A^i equals n.*

Now we come to Farkas' Lemma, which says that a linear inequality system has a solution if and only if a related (polynomial size) linear inequality system has no solution. This lemma is representative of a large body of theorems in mathematical programming known as *theorems of the alternative*.

LEMMA 30.3   [27]. Exactly one of the alternatives

$$I. \ \exists \ x : Ax \le b; \quad II. \ \exists \ y \ge 0 : A^Ty = 0, \ b^Ty < 0$$

is true for any given real matrices A, b.

30.2.3 Optimization and Dual Linear Programs

The two fundamental problems of linear programming (which are polynomially equivalent) are

- **Solvability:** This is the problem of checking if a system of linear constraints on real (rational) variables is solvable or not. Geometrically, we have to check if a polyhedron, defined by such constraints, is nonempty.

- **Optimization:** This is the problem (LP) of optimizing a linear objective function over a polyhedron described by a system of linear constraints.

Building on polarity in cones and polyhedra, duality in linear programming is a fundamental concept which is related to both the complexity of linear programming and to the design of algorithms for solvability and optimization. We will encounter the solvability version of duality (called Farkas' Lemma) while discussing the Fourier elimination technique below. Here we will state the main duality results for optimization. If we take the primal linear program to be

$$(P) \quad \min_{x \in \Re^n} \{cx : Ax \geq b\}$$

there is an associated dual linear program

$$(D) \quad \max_{y \in \Re^m} \left\{ b^T y : A^T y = c^T, \, y \geq 0 \right\}$$

and the two problems satisfy

1. For any $\hat{x}$ and $\hat{y}$ feasible in (P) and (D) (i.e., they satisfy the respective constraints), we have $c\hat{x} \geq b^T \hat{y}$ (weak duality).

2. (P) has a finite optimal solution if and only if (D) does.

3. x^* and y^* are a pair of optimal solutions for (P) and (D), respectively, if and only if x^* and y^* are feasible in (P) and (D) (i.e., they satisfy the respective constraints) and $cx^* = b^T y^*$ (strong duality).

4. x^* and y^* are a pair of optimal solutions for (P) and (D), respectively, if and only if x^* and y^* are feasible in (P) and (D) (i.e., they satisfy the respective constraints) and $(Ax^* - b)^T y^* = 0$ (complementary slackness).

The strong duality condition above gives us a good stopping criterion for optimization algorithms. The complementary slackness condition, on the other hand gives us a constructive tool for moving from dual to primal solutions and vice versa. The weak duality condition gives us a technique for obtaining lower bounds for minimization problems and upper bounds for maximization problems.

Note that the properties above have been stated for linear programs in a particular form. The reader should be able to check that if, for example, the primal is of the form

$$\left(P'\right) \quad \min_{x \in \Re^n} \{cx : Ax = b, \, x \geq 0\}$$

then the corresponding dual will have the form

$$\left(D'\right) \quad \max_{y \in \Re^m} \left\{ b^T y : A^T y \leq c^T \right\}$$

The tricks needed for seeing this is that any equation can be written as two inequalities, an unrestricted variable can be substituted by the difference of two nonnegatively constrained variables, and an inequality can be treated as an equality by adding a nonnegatively constrained variable to the lesser side. Using these tricks, the reader could also check that dual construction in linear programming is involutory (i.e., the dual of the dual is the primal).

30.2.4   Complexity of Linear Equations and Inequalities

30.2.4.1   Complexity of Linear Algebra

Let us restate the fundamental problem of linear algebra as a decision problem.

$$CLS = \left\{(A, b) : \exists\, x \in Q^n, Ax = b\right\} \tag{30.1}$$

In order to solve the decision problem on CLS it is useful to recall homogeneous linear equations. A basic result in linear algebra is that any linear subspace of Q^n has two representations, one from hyperplanes and the other from a vector basis.

$$\mathcal{L} = \left\{x \in Q^n : Ax = 0\right\}$$
$$\mathcal{L} = \left\{x \in Q^n : x = Cy, \ y \in Q^k\right\}$$

Corresponding to a linear subspace $\mathcal{L}$ there exists a dual (orthogonal complementary) subspace $\mathcal{L}^*$ with the roles of the hyperplanes and basis vectors of $\mathcal{L}$ exchanged.

$$\mathcal{L}^* = \{z : Cz = 0\}$$
$$\mathcal{L}^* = \{z : z = Ax\}$$
$$dimension\mathcal{L} + dimension\mathcal{L}^* = n$$

Using these representation results it is quite easy to establish the Fundamental Theorem of Linear Algebra.

THEOREM 30.7   *Either $Ax = b$ for some x or $yA = 0$, $yb \neq 0$ for some y.*

Along with the basic theoretical constructs outlined above, let us also assume knowledge of the Gaussian Elimination Method for solving a system of linear equations. It is easily verified that on a system of size m by n, this method uses $O(m^2 n)$ elementary arithmetic operations. However we also need some bound on the size of numbers handled by this method. By the size of a rational number we mean the length of binary string encoding the number. And similarly for a matrix of numbers.

LEMMA 30.4   For any square matrix S of rational numbers, the size of the determinant of S is polynomially related to the size of S itself.

Since all the numbers in a basic solution (i.e., basis-generated) of $Ax = b$ are bounded in size by subdeterminants of the input matrix (A, b), we can conclude that CLS is a member of $\mathcal{NP}$. The Fundamental Theorem of Linear Algebra further establishes that CLS is in $\mathcal{NP} \cap co\mathcal{NP}$. And finally the polynomiality of Gaussian Elimination establishes that CLS is in $\mathcal{P}$.

30.2.4.2   Complexity of Linear Inequalities

From our earlier discussion of polyhedra, we have the following algebraic characterization of extreme points of polyhedra.

THEOREM 30.8   *Given a polyhedron P, defined by $\{x : Ax \leq b\}$, x^i is an extreme point of P if and only if it is a face of P satisfying $A^i x^i = b^i$ where $((A^i), (b^i))$ is a submatrix of (A, b) and the rank of A^i equals n.*

COROLLARY 30.3 The decision problem of verifying the membership of an input string (A, b) in the language $\mathcal{L}_I = \{(A, b) : \exists x \text{ such that } Ax \leq b\}$ belongs to $\mathcal{NP}$.

PROOF It follows from the theorem that every extreme point of the polyhedron $P = \{x : Ax \leq b\}$ is the solution of an $(n \times n)$ linear system whose coefficients come from (A, b). Therefore we can guess a polynomial length string representing an extreme point and check its membership in P in polynomial time.

A consequence of Farkas' Lemma is that the decision problem of testing membership of input (A, b) in the language

$$\mathcal{L}_I = \{(A, b) : \exists x \text{ such that } Ax \leq b\}$$

is in $\mathcal{NP} \cap co\mathcal{NP}$. That $\mathcal{L}_I$ can be recognized in polynomial time follows from algorithms for linear programming that we now discuss.

We are now ready for a tour of some algorithms for linear programming. We start with the classical technique of Fourier, which is interesting because of its simple syntactic specification. It leads to simple proofs of the duality principle of linear programming that was alluded to above. We will then review the simplex method of linear programming [19], a method that uses the vertex-edge structure of a convex polyhedron to execute an optimization march. The simplex method has been finely honed over almost five decades now. We will spend some time with the ellipsoid method and in particular with the polynomial equivalence of solvability (optimization) and **separation** problems. This aspect of the ellipsoid method [35] has had a major impact on the identification of many tractable classes of combinatorial optimization problems. We conclude the tour of the basic methods with a description of Karmarkars [47] breakthrough in 1984, which was an important landmark in the brief history of linear programming. A noteworthy role of interior point methods has been to make practical the theoretical demonstrations of tractability of various aspects of linear programming, including solvability and optimization, that were provided via the ellipsoid method.

In later sections we will review the more sophisticated (and naturally esoteric) aspects of linear programming algorithms. This will include **strongly polynomial** algorithms for special cases, randomized algorithms, and specialized methods for large-scale linear programming. Some readers may notice that we do not have a special section devoted to the discussion of parallel computation in the context of linear programming. This is partly because we are not aware of a well-developed framework for such a discussion. We have instead introduced discussion and remarks about the effects of parallelism in the appropriate sections of this chapter.

30.3 Fourier's Projection Method

Linear programming is at the starting point for variable elimination techniques on algebraic constraints [12], which in turn forms the core of algebraic and symbolic computation. Constraint systems of linear inequalities of the form $Ax \leq b$, where A is an $m \times n$ matrix of real numbers, are widely used in mathematical models. Testing the solvability of such a system is equivalent to linear programming. We now describe the elegant syntactic variable elimination technique due to Fourier [30] (Figure 30.1).

Suppose we wish to eliminate the first variable x_1 from the system $Ax \leq b$. Let us denote

$$I^+ = \{i : A_{i1} > 0\}; \quad I^- = \{i : A_{i1} < 0\}; \quad I^0 = \{i : A_{i1} = 0\}$$

Our goal is to create an equivalent system of linear inequalities $\tilde{A}\tilde{x} \leq \tilde{b}$ defined on the variables $\tilde{x} = (x_2, x_3, \ldots, x_n)$.

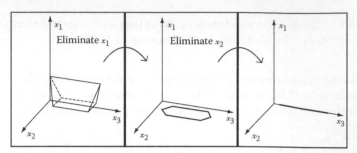

FIGURE 30.1 Variable elimination and projection.

- If I^+ is empty then we can simply delete all the inequalities with indices in I^-, since they can be trivially satisfied by choosing a large enough value for x_1. Similarly, if I^- is empty we can discard all inequalities in I^+.

- For each $k \in I^+$, $l \in I^-$ we add $-A_{l1}$ times the inequality $A_k x \leq b_k$ to A_{k1} times $A_l x \leq b_l$. In these new inequalities the coefficient of x_1 is wiped out, i.e., x_1 is eliminated. Add these new inequalities to those already in I^0.

- The inequalities $\{\tilde{A}_{i1}\tilde{x} \leq \tilde{b}_i\}$ for all $i \in I^0$ represent the equivalent system on the variables $\tilde{x} = (x_2, x_3, \ldots, x_n)$.

Repeat this construction with $\tilde{A}\tilde{x} \leq \tilde{b}$ to eliminate x_2 and so on until all variables are eliminated. If the resulting $\tilde{b}$ (after eliminating x_n) is nonnegative we declare the original (and intermediate) inequality systems as being consistent. Otherwise* $\tilde{b} \not\geq 0$ and we declare the system inconsistent.

As an illustration of the power of elimination as a tool for theorem proving, we show now that Farkas' Lemma is a simple consequence of the correctness of Fourier elimination. The lemma gives a direct proof that solvability of linear inequalities is in $\mathcal{NP} \cap co\mathcal{NP}$.

Farkas' Lemma   Exactly one of the alternatives

$$I. \ \exists \ x \in \mathfrak{R}^n : Ax \leq b; \quad II. \ \exists \ y \in \mathfrak{R}^m_+ : y^t A = 0, \ y^t b < 0$$

is true for any given real matrices $A, \ b$.

PROOF   Let us analyze the case when Fourier Elimination provides a proof of the inconsistency of a given linear inequality system $Ax \leq b$. The method clearly converts the given system into $RAx \leq Rb$ where RA is zero and Rb has at least one negative component. Therefore, there is some row of R, say r, such that $rA = 0$ and $rb < 0$. Thus $\neg I$ implies II. It is easy to see that I and II cannot both be true for fixed A, b.

In general, the Fourier elimination method is quite inefficient. Let k be any positive integer and n the number of variables be $2^k + k + 2$. If the input inequalities have left-hand sides of the form $\pm x_r \pm x_s \pm x_t$ for all possible $1 \leq r < s < t \leq n$, it is easy to prove by induction that

---

*Note that the final $\tilde{b}$ may not be defined if all the inequalities are deleted by the monotone sign condition of the first step of the construction described above. In such a situation we declare the system $Ax \leq b$ strongly consistent, since it is consistent for any choice of b in $\mathfrak{R}^m$. In order to avoid making repeated references to this exceptional situation, let us simply assume that it does not occur. The reader is urged to verify that this assumption is indeed benign.

after k variables are eliminated, by Fourier's method, we would have at least $2^{\frac{n}{2}}$ inequalities. The method is therefore exponential in the worst case and the explosion in the number of inequalities has been noted, in practice as well, on a wide variety of problems. We will discuss the central idea of minimal generators of the projection cone that results in a much improved elimination method [41].

First let us identify the set of variables to be eliminated. Let the input system be of the form

$$P = \left\{ (x, u) \in \Re^{n_1 + n_2} \mid Ax + Bu \leq b \right\}$$

where u is the set to be eliminated. The projection of P onto x or equivalently the effect of eliminating the u variables is

$$P_x = \left\{ x \in \Re^{n_1} \mid \exists u \in \Re^{n_2} \text{ such that } Ax + Bu \leq b \right\}$$

Now W, the projection cone of P, is given by

$$W = \left\{ w \in \Re^m \mid wB = 0, \ w \geq 0 \right\}$$

A simple application of Farkas' Lemma yields a description of P_x in terms of W.

Projection Lemma Let G be any set of generators (e.g., the set of extreme rays) of the cone W. Then $P_x = \{x \in \Re^{n_1} \mid (gA)x \leq gb \ \ \forall g \in G\}$.

The lemma, sometimes attributed to Černikov [10], reduces the computation of P_x to enumerating the extreme rays of the cone W or equivalently the extreme points of the polytope $W \cap \{w \in \Re^m \mid \sum_{i=1}^m w_i = 1\}$.

30.4 The Simplex Method

Consider a polyhedron $\mathcal{K} = \{x \in \Re^n : Ax = b, \ x \geq 0\}$. Now $\mathcal{K}$ cannot contain an infinite (in both directions) line, since it is lying within the nonnegative orthant of $\Re^n$. Such a polyhedron is called a pointed polyhedron (Figure 30.2). Given a pointed polyhedron $\mathcal{K}$ we observe that

- If $\mathcal{K} \neq \emptyset$ then $\mathcal{K}$ has at least one extreme point.
- If $\min\{cx : Ax = b, \ x \geq 0\}$ has an optimal solution, then it has an optimal extreme point solution.

These observations together are sometimes called the fundamental theorem of linear programming since they suggest simple finite tests for both solvability and optimization. To generate all extreme points of $\mathcal{K}$, in order to find an optimal solution, is an impractical idea. However, we may

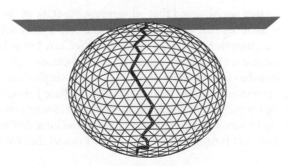

FIGURE 30.2 The simplex path.

try to run a partial search of the space of extreme points for an optimal solution. A simple local improvement search strategy of moving from extreme point to adjacent extreme point until we get to a local optimum is nothing but the simplex method of linear programming [19,20]. The local optimum also turns out to be a global optimum, because of the convexity of the polyhedron $\mathcal{K}$ and the objective function cx.

Procedure: **Primal Simplex($\mathcal{K}$,c)**

0. **Initialize**

   - $x_0 :=$ an extreme point of $\mathcal{K}$

   - $k := 0$

1. **Iterative Step**

   do

   If for all edge directions $\mathcal{D}_k$ at x_k, the objective function is nondecreasing, i.e.,

   $$cd \geq 0 \ \ \forall\, d \in \mathcal{D}_k$$

   then exit and return optimal x_k.

   Else pick some d_k in $\mathcal{D}_k$ such that $cd_k < 0$.

   If $d_k \geq 0$ **then** declare the linear program unbounded in

   objective value and exit.

   Else $x_{k+1} := x_k + \theta_k * d_k$, where

   $$\theta_k = \max\{\theta : x_k + \theta * d_k \geq 0\}$$

   $k := k + 1$

   od

2. **End**

Remarks

1. In the initialization step we assumed that an extreme point x_0 of the polyhedron $\mathcal{K}$ is available. This also assumes that the solvability of the constraints defining $\mathcal{K}$ has been established. These assumptions are reasonable since we can formulate the solvability problem as an optimization problem, with a self-evident extreme point, whose optimal solution either establishes unsolvability of $Ax = b$, $x \geq 0$, or provides an extreme point of $\mathcal{K}$. Such an optimization problem is usually called a Phase I model. The point is, of course, that the simplex method as described above can be invoked on the Phase I model and if successful, can be invoked once again to carry out the intended minimization of cx. There are several different formulations of the Phase I model that have been advocated. Here is one.

$$\min \{v_0 : Ax + bv_0 = b, \ x \geq 0, \ v_0 \geq 0\}$$

The solution $(x, v_0)^T = (0, \ldots, 0, 1)$ is a self-evident extreme point, and $v_0 = 0$ at an optimal solution of this model is a necessary and sufficient condition for the solvability of $Ax = b$, $x \geq 0$.

2. The scheme for generating improving edge directions uses an algebraic representation of the extreme points as certain bases, called feasible bases, of the vector space generated by the columns of the matrix A. It is possible to have linear programs for which an extreme point is geometrically overdetermined (degenerate), i.e., there are more than d facets of $\mathcal{K}$ that contain the extreme point, where d is the dimension of $\mathcal{K}$. In such a situation, there would be several feasible bases corresponding to the same extreme point. When this happens, the linear program is said to be primal degenerate.

3. There are two sources of nondeterminism in the primal simplex procedure. The first involves the choice of edge direction d_k made in Step 1. At a typical iteration there may be many edge directions that are improving in the sense that $cd_k < 0$. Dantzig's Rule, Maximum Improvement Rule, and Steepest Descent Rule are some of the many rules that have been used to make the choice of edge direction in the simplex method. There is, unfortunately, no clearly dominant rule, and successful codes exploit the empirical and analytic insights that have been gained over the years to resolve the edge selection nondeterminism in the simplex method.

   The second source of nondeterminism arises from degeneracy. When there are multiple feasible bases corresponding to an extreme point, the simplex method has to pivot from basis to adjacent basis by picking an entering basic variable (a pseudo edge direction) and by dropping one of the old ones. A wrong choice of the leaving variables may lead to cycling in the sequence of feasible bases generated at this extreme point. Cycling is a serious problem when linear programs are highly degenerate, as in the case of linear relaxations of many combinatorial optimization problems. The Lexicographic Rule (Perturbation Rule) for choice of leaving variables in the simplex method is a provably finite method (i.e., all cycles are broken).

   A clever method proposed by Bland (cf. [72]) preorders the rows and columns of the matrix A. In case of nondeterminism in either entering or leaving variable choices, Bland's Rule just picks the lowest index candidate. All cycles are avoided by this rule also.

30.4.1 Implementation Issues: Basis Representations

The enormous success of the simplex method has been primarily due to its ability to solve large size problems that arise in practice. A distinguishing feature of many of the linear problems that are solved routinely in practice is the sparsity of the constraint matrix. So from a computational point of view, it is desirable to take advantage of the sparseness of the constraint matrix. Another important consideration in the implementation of the simplex method is to control the accumulation of round off errors that arise because the arithmetic operations are performed with only a fixed number of digits and the simplex method is an iterative procedure.

An algebraic representation of the simplex method in matrix notation is as follows:

```
0: Find an initial feasible extreme point x⁰, and the corresponding
   feasible basis B (of the vector space generated by the columns
   of the constraint matrix A). If no such x₀ exists, stop, there
   is no feasible solution. Otherwise, let t = 0, and go to Step 1.
1: Partition the matrix A as A = (B,N), the solution vector x as
   x = (xB,xN), and the objective function vector c as c = (cB,cN),
   corresponding to the columns in B.
```

2: The extreme point x^t is given by $x^t = (x_B, 0)$, where $Bx_B = b$

3: Solve the system $\pi_B B = c_B$ and calculate $r = c_N - \pi_B N$. If $r \geq 0$, stop, the current solution $x^t = (x_B, 0)$ is optimal. Otherwise, let $r_k = \min_j\{r_j\}$, where r_j is the jth component of r (actually one may pick any $r_j < 0$ as r_k).

4: Let a_k denote the kth column of N corresponding to r_k. Find y_k such that $B y_k = a_k$

5: Find $x_B(p)/y_{pk} = \min_i\{x_B(i)/y_{ik} : y_{ik} > 0\}$ where $x_B(i)$ and y_{ik} denote the ith component of x_B and y_k, respectively.

6: The new basis $\hat{B}$ is obtained from B by replacing the pth column of B by the kth column of N. Let the new feasible basis $\hat{B}$ be denoted as B. Return to Step 1.

30.4.2 LU Factorization

At each iteration, the simplex method requires the solution of the following systems:

$$Bx_B = b; \pi_B B = c_B \quad \text{and} \quad By_k = a_k.$$

After row interchanges, if necessary, any basis B can be factorized as $B = LU$ where L is a lower triangular matrix and U is an upper triangular matrix. So solving $LUx_B = b$ is equivalent to solving the triangular systems $Lv = b$ and $Ux_B = v$. Similarly, for $By_k = a_k$, we solve $Lw = a_k$ and $Uy_k = w$. Finally, for $\pi_B B = c_B$, we solve $\pi_B L = \lambda$ and $\lambda U = c_B$.

Let the current basis B and the updated basis $\hat{B}$ be represented as

$$B = (a_1, a_2, \ldots, a_{p-1}, a_p, a_{p+1}, \ldots, a_m) \quad \text{and} \quad \hat{B} = (a_1, a_2, \ldots, a_{p-1}, a_{p+1}, a_{p+2}, \ldots, a_m, a_k).$$

An efficient implementation of the simplex method requires the updating of the triangular matrices L and U as triangular matrices $\hat{L}$ and $\hat{U}$ where $B = LU$ and $\hat{B} = \hat{L}\hat{U}$. This is done by first obtaining $H = (u_1, u_2, \ldots, u_{p-1}, u_{p+1}, \ldots, u_m, w)$ where u_i is the ith column of U and $w = L^{-1}a_k$. The matrix H has zeros below the main diagonal in the first $p - 1$ columns and zeros below the element immediately under the diagonal in the remaining columns. The matrix H can be reduced to an upper triangular matrix by Gaussian elimination which is equivalent to multiplying H on the left by matrices M_i, $i = p, p + 1, \ldots, m - 1$, where M_j differs from an identity matrix in column j which is given by $(0, \ldots, 0, 1, m_j, 0 \ldots 0)^T$, where m_j is in position $j + 1$. Now $\hat{U}$ is given by $\hat{U} = M_{m-1}, M_{m-2}, \ldots, M_p H$ and $\hat{L}$ is given by $\hat{L} = LM_p^{-1}, \ldots, M_{m-1}^{-1}$. Note that M_j^{-1} is M_j with the sign of the off-diagonal term m_j reversed.

The LU factorization preserves the sparsity of the basis B, in that the number of nonzero entries in L and U is typically not much larger than the number of nonzero entries in B. Furthermore, this approach effectively controls the accumulation of round off errors and maintains good numerical accuracy. In practice, the LU factorization is periodically recomputed for the matrix $\hat{B}$ instead of updating the factorization available at the previous iteration. This computation of $\hat{B} = \hat{L}\hat{U}$ is achieved by Gaussian elimination to reduce $\hat{B}$ to an upper triangular matrix (for details, see for instance [37,62,63]). There are several variations of the basic idea of factorization of the basis matrix B, as described here, to preserve sparsity and control round off errors.

REMARK 30.1   The simplex method is not easily amenable to parallelization. However, some steps such as identification of the entering variable and periodic refactorization can be efficiently parallelized.

30.4.3   Geometry and Complexity of the Simplex Method

An elegant geometric interpretation of the simplex method can be obtained by using a column space representation [20], i.e., $\Re^{m+1}$ coordinatized by the rows of the $(m + 1) \times n$ matrix $\left(\begin{smallmatrix} c \\ A \end{smallmatrix} \right)$. In fact it is this interpretation that explains why it is called the simplex method. The bases of A correspond to an arrangement of simplicial cones in this geometry, and the pivoting operation corresponds to a physical pivot from one cone to an adjacent one in the arrangement. An interesting insight that can be gained from the column space perspective is that Karmarkar's interior point method can be seen as a natural generalization of the simplex method [14,81].

However, the geometry of linear programming, and of the simplex method, has been largely developed in the space of the x variables, i.e., in $\Re^n$. The simplex method walks along edge paths on the combinatorial graph structure defined by the boundary of convex polyhedra. These graphs are quite dense (Balinski's Theorem [87] states that the graph of d-dimensional polyhedron must be d-connected). A polyhedral graph can also have a huge number of vertices since the Upper Bound Theorem of McMullen, see [87], states that the number of vertices can be as large as $O(k^{\lfloor d/2 \rfloor})$ for a polytope in d dimensions defined by k constraints. Even a polynomial bound on the diameter of polyhedral graphs is not known. The best bound obtained to date is $O(k^{1+\log d})$ of a polytope in d dimensions defined by k constraints. Hence it is no surprise that there is no known variant of the simplex method with a worst-case polynomial guarantee on the number of iterations.

Klee and Minty [49] exploited the sensitivity of the original simplex method of Dantzig, to projective scaling of the data, and constructed exponential examples for it. These examples were simple projective distortions of the hypercube to embed long isotonic (improving objective value) paths in the graph. Scaling is used in the Klee–Minty construction, to trick the choice of entering variable (based on most negative reduced cost) in the simplex method and thus keep it on an exponential path. Later, several variants of the entering variable choice (best improvement, steepest descent, etc.) were all shown to be susceptible to similar constructions of exponential examples (cf. [72]).

Despite its worst-case behavior, the simplex method has been the veritable workhorse of linear programming for five decades now. This is because both empirical [7,20] and probabilistic [9,39] analyses indicate that the expected number of iterations of the simplex method is just slightly more than linear in the dimension of the primal polyhedron.

Probabilistic analyses includes various average-case analyses of the simplex and the self-dual simplex algorithm [9,59,79]. Recently Spielman and Teng [80] have introduced the concept of smoothed analysis of algorithms. In the case study of the simplex method, the smoothed complexity is defined as the maximum of the expected running time of the simplex method under small independent random perturbations where the maximum assumes an adversarial choice of the randomly perturbed input coefficients, i.e., of the matrix A and the vector b. If the perturbation is Gaussian with mean 0 and a small standard deviation, Spielman and Teng [80] show that the simplex method has a smoothed complexity which is polynomial in the input size and in $1/\sigma$. This provides a plausible explanation of why the simplex method performs well on typical inputs where the data is inherently a little noisy.

The ellipsoid method of Shor [78] was devised to overcome poor scaling in convex programming problems and therefore turned out to be the natural choice of an algorithm to first establish polynomial-time solvability of linear programming. Later Karmarkar [47] took care of both projection and scaling simultaneously and arrived at a superior algorithm.

30.5　The Ellipsoid Method

The Ellipsoid Algorithm of Shor [78] gained prominence in the late 1970s when Hačijan (pronounced Khachiyan) [38] showed that this convex programming method specializes to a polynomial-time algorithm for linear programming problems. This theoretical breakthrough naturally led to intense study of this method and its properties. The survey paper by Bland et al. [8] and the monograph by Akgül [2] attest to this fact. The direct theoretical consequences for combinatorial optimization problems was independently documented by Padberg and Rao [67], Karp and Papadimitriou [48], and Grötschel et al. [34]. The ability of this method to implicitly handle linear programs with an exponential list of constraints and maintain polynomial-time convergence is a characteristic that is the key to its applications in combinatorial optimization. For an elegant treatment of the many deep theoretical consequences of the ellipsoid algorithm, the reader is directed to the monograph by Lovász [51], and the book by Grötschel et al. [35].

Computational experience with the ellipsoid algorithm, however, showed a disappointing gap between the theoretical promise and practical efficiency of this method in the solution of linear programming problems. Dense matrix computations as well as the slow average-case convergence properties are the reasons most often cited for this behavior of the ellipsoid algorithm. On the positive side though, it has been noted (cf. [24]) that the ellipsoid method is competitive with the best known algorithms for (nonlinear) convex programming problems.

Let us consider the problem of testing if a polyhedron $Q \in \Re^d$, defined by linear inequalities, is nonempty. For technical reasons let us assume that Q is rational, i.e., all extreme points and rays of Q are rational vectors or equivalently that all inequalities in some description of Q involve only rational coefficients. The ellipsoid method does not require the linear inequalities describing Q to be explicitly specified. It suffices to have an oracle representation of Q. Several different types of oracles can be used in conjunction with the ellipsoid method [35,48,67]. We will use the strong separation oracle described below.

> Oracle: **Strong Separation** (Q,y)
> Given a vector $y \in \Re^d$, decide whether $y \in Q$, and if not find a hyperplane that separates y from Q; more precisely, find a vector $c \in \Re^d$ such that $c^T y < \min\{c^T x : x \in Q\}$.

The ellipsoid algorithm initially chooses an ellipsoid large enough to contain a part of the polyhedron Q if it is nonempty. This is easily accomplished because we know that if Q is nonempty then it has a rational solution whose (binary encoding) length is bounded by a polynomial function of the length of the largest coefficient in the linear program and the dimension of the space (Figure 30.3).

The center of the ellipsoid is a feasible point if the separation oracle tells us so. In this case, the algorithm terminates with the coordinates of the center as a solution. Otherwise, the separation oracle outputs an inequality that separates the center point of the ellipsoid from the polyhedron Q. We translate the hyperplane defined by this inequality to the center point. The hyperplane slices the ellipsoid into two halves, one of which can be discarded. The algorithm now creates a new ellipsoid that is the minimum volume ellipsoid containing the remaining half of the old one. The algorithm questions if the new center is feasible and so on. The key is that the new ellipsoid has substantially smaller volume than the previous one. When the volume of the current ellipsoid shrinks

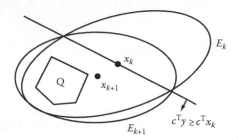

FIGURE 30.3 Shrinking ellipsoids.

to a sufficiently small value, we are able to conclude that Q is empty. This fact is used to show the polynomial time convergence of the algorithm.

Ellipsoids in $\Re^d$ are denoted as $E(A, y)$ where A is a $d \times d$ positive definite matrix and $y \in \Re^d$ is the center of the ellipsoid $E(A, y)$.

$$E(A, y) = \left\{ x \in \Re^d \mid (x - y)^T A^{-1} (x - y) \leq 1 \right\}.$$

The ellipsoid algorithm is described on the iterated values, A_k and x^k, which specify the underlying ellipsoids $E_k(A_k, x^k)$.

Procedure: **Ellipsoid** (Q)

0. **Initialize**

   - $N := N(Q)$ (comment: iteration bound)

   - $R := R(Q)$ (comment: radius of the initial ellipsoid/sphere E_0)

   - $A_0 := R^2 I$

   - $x_0 := 0$ (comment: center of E_0)

   - $k := 0$

1. **Iterative Step**

   while $k < N$

   call Strong Separation (Q, x^k)

   if $x^k \in Q$ **halt**

   else hyperplane $\{x \in \Re^d \mid c^T x = c_0\}$ separates x^k from Q

   Update

   $$b := \frac{1}{\sqrt{c^T A_k c}} A_k c$$

   $$x^{k+1} := x^k - \frac{1}{d+1} b$$

   $$A_{k+1} := \frac{d^2}{d^2 - 1} \left(A_k - \frac{2}{d+1} b b^T \right)$$

   $$k := k + 1$$

   endwhile

2. **Empty Polyhedron**

- **halt** and **declare** "$\mathcal{Q}$ is empty"

3. **End**

The crux of the complexity analysis of the algorithm is on the a priori determination of the iteration bound. This in turn depends on three factors. The volume of the initial ellipsoid E_0, the rate of volume shrinkage ($\frac{vol(E_{k+1})}{vol(E_k)} < e^{-\frac{1}{(2d)}}$), and the volume threshold at which we can safely conclude that $\mathcal{Q}$ must be empty. The assumption of $\mathcal{Q}$ being a rational polyhedron is used to argue that $\mathcal{Q}$ can be modified into a full-dimensional polytope without affecting the decision question (Is $\mathcal{Q}$ nonempty?). After careful accounting for all these technical details and some others (e.g., compensating for the round-off errors caused by the square root computation in the algorithm), it is possible to establish the following fundamental result.

THEOREM 30.9 *There exists a polynomial $g(d, \phi)$ such that the* ellipsoid method *runs in time bounded by $T g(d, \phi)$ where ϕ is an upper bound on the size of linear inequalities in some description of $\mathcal{Q}$, and T is the maximum time required by the oracle* Strong Separation$(\mathcal{Q}, y)$ *on inputs y of size at most $g(d, \phi)$.*

The size of a linear inequality is just the length of the encoding of all the coefficients needed to describe the inequality. A direct implication of the theorem is that solvability of linear inequalities can be checked in polynomial time if strong separation can be solved in polynomial time. This implies that the standard linear programming solvability question has a polynomial-time algorithm (since separation can be effected by simply checking all the constraints). Happily, this approach provides polynomial-time algorithms for much more than just the standard case of linear programming solvability. The theorem can be extended to show that the optimization of a linear objective function over $\mathcal{Q}$ also reduces to a polynomial number of calls to the strong separation oracle on $\mathcal{Q}$. A converse to this theorem also holds, namely separation can be solved by a polynomial number of calls to a solvability/optimization oracle [35]. Thus, optimization and separation are polynomially equivalent. This provides a very powerful technique for identifying tractable classes of optimization problems. Semi-definite programming and submodular function minimization are two important classes of optimization problems that can be solved in polynomial time using this property of the ellipsoid method.

30.5.1 Semidefinite Programming

The following optimization problem, defined on symmetric ($n \times n$) real matrices,

$$(SDP) \quad \min_{X \in \Re^{n \times n}} \left\{ \sum_{ij} C \bullet X : A \bullet X = B, X \succeq 0 \right\},$$

is called a semidefinite program. Note that $X \succeq 0$ denotes the requirement that X is a positive semidefinite matrix, and $F \bullet G$ for $n \times n$ matrices F and G denotes the product matrix ($F_{ij} * G_{ij}$). From the definition of positive semi-definite matrices, $X \succeq 0$ is equivalent to

$$q^T X q \geq 0 \quad \text{for every} \quad q \in \Re^n$$

Thus (SDP) is really a linear program on $O(n^2)$ variables with an (uncountably) infinite number of linear inequality constraints. Fortunately, the strong separation oracle is easily realized for these constraints. For a given symmetric X we use Cholesky factorization to identify the minimum eigenvalue λ_{min}. If λ_{min} is nonnegative then $X \succeq 0$ and if, on the other hand, λ_{min} is negative we have a separating inequality

$$\gamma_{min}^T X \gamma_{min} \geq 0$$

where γ_{min} is the eigenvector corresponding to λ_{min}. Since the Cholesky factorization can be computed by an $O(n^3)$ algorithm, we have a polynomial-time separation oracle and an efficient algorithm for (SDP) via the Ellipsoid method. Alizadeh [3] has shown that interior point methods can also be adapted to solving (SDP) to within an additive error ϵ in time polynomial in the size of the input and $\log 1/\epsilon$.

This result has been used to construct efficient approximation algorithms for maximum stable sets and cuts of graphs [33], Shannon capacity of graphs, and minimum colorings of graphs. It has been used to define hierarchies of relaxations for integer linear programs that strictly improve on known exponential-size linear programming relaxations [52].

30.5.2 Minimizing Submodular Set Functions

The minimization of submodular set functions is a generic optimization problem that contains a large class of important optimization problems as special cases [26]. Here we will see why the ellipsoid algorithm provides a polynomial-time solution method for submodular minimization.

DEFINITION 30.3 Let N be a finite set. A real valued set function f defined on the subsets of N is

$$\text{submodular if } f(X \cup Y) + f(X \cap Y) \leq f(X) + f(Y) \text{ for } X, Y \subseteq N$$

Example 30.1

Let $G = (V, E)$ be an undirected graph with V as the node set and E as the edge set. Let $c_{ij} \geq 0$ be the weight or capacity associated with edge $(ij) \in E$. For $S \subseteq V$, define the cut function $c(S) = \Sigma_{i \in S, j \in V \setminus S} c_{ij}$. The cut function defined on the subsets of V is submodular since $c(X) + c(Y) - c(X \cup Y) - c(X \cap Y) = \Sigma_{i \in X \setminus Y, j \in Y \setminus X} 2c_{ij} \geq 0$.

The optimization problem of interest is

$$\min\{f(X) : X \subseteq N\}$$

The following remarkable construction that connects submodular function minimization with convex function minimization is due to Lovász (cf. [35]).

DEFINITION 30.4 The Lovász extension $\hat{f}(\cdot)$ of a submodular function $f(\cdot)$ satisfies

- $\hat{f} : [0, 1]^N \to \Re$.
- $\hat{f}(x) = \Sigma_{I \in \mathcal{I}} \lambda_I f(x_I)$ where $x = \Sigma_{I \in \mathcal{I}} \lambda_I x_I$, $x \in [0, 1]^N$, x_I is the incidence vector of I for each $I \in \mathcal{I}$, $\lambda_I > 0$ for each I in $\mathcal{I}$, and $\mathcal{I} = \{I_1, I_2, \cdots, I_k\}$ with $\emptyset \neq I_1 \subset I_2 \subset \cdots \subset I_k \subseteq N\}$. Note that the representation $x = \Sigma_{I \in \mathcal{I}} \lambda_I x_I$ is unique given that the $\lambda_I > 0$ and that the sets in $\mathcal{I}$ are nested.

It is easy to check that $\hat{f}(\cdot)$ is a convex function. Lovász also showed that the minimization of the submodular function $f(\cdot)$ is a special case of convex programming by proving

$$\min\{f(X) : X \subseteq N\} = \min\left\{\hat{f}(x) : x \in [0,1]^N\right\}$$

Further, if x^* is an optimal solution to the convex program and

$$x^* = \sum_{I \in \mathcal{I}} \lambda_I x_I$$

then for each $\lambda_I > 0$, it can be shown that $I \in \mathcal{I}$ minimizes f. The ellipsoid method can be used to solve this convex program (and hence submodular minimization) using a polynomial number of calls to an oracle for f (this oracle returns the value of $f(X)$ when input X).

30.6 Interior Point Methods

The announcement of the polynomial solvability of linear programming followed by the probabilistic analyses of the simplex method in the early 1980s left researchers in linear programming with a dilemma. We had one method that was good in a theoretical sense but poor in practice and another that was good in practice (and on average) but poor in a theoretical worst-case sense. This left the door wide open for a method that was good in both senses. Narendra Karmarkar closed this gap with a breathtaking new projective scaling algorithm. In retrospect, the new algorithm has been identified with a class of nonlinear programming methods known as logarithmic barrier methods. Implementations of a primal-dual variant of the logarithmic barrier method have proven to be the best approach at present. The recent monograph by Wright [85] is dedicated to primal-dual interior point methods. It is a variant of this method that we describe below.

It is well known that moving through the interior of the feasible region of linear program using the negative of the gradient of the objective function, as the movement direction, runs into trouble because of getting jammed into corners (in high dimensions, corners make up most of the interior of a polyhedron). This jamming can be overcome if the negative gradient is balanced with a centering direction. The centering direction in Karmarkar's algorithm is based on the **analytic center** y_c of a full dimensional polyhedron $\mathcal{D} = \{x : A^T y \leq c\}$ which is the unique optimal solution to

$$\max\left\{\sum_{j=1}^{n} \ln(z_j) : A^T y + z = c\right\}$$

Recall that the primal and dual forms of a linear program may be taken as

$$(P) \quad \min\{cx : Ax = b, x \geq 0\}$$
$$(D) \quad \max\left\{b^T y : A^T y \leq c\right\}$$

The logarithmic barrier formulation of the dual (D) is

$$(D_\mu) \quad \max\left\{b^T y + \mu \sum_{j=1}^{n} \ln(z_j) : A^T y + z = c\right\}$$

Notice that (D_μ) is equivalent to (D) as $\mu \to 0^+$. The optimality (Karush–Kuhn–Tucker) conditions for (D_μ) are given by

$$D_x D_z e = \mu e$$
$$Ax = b$$
$$A^T y + z = c$$

where D_x and D_z denote $n \times n$ diagonal matrices whose diagonals are x and z, respectively. Notice that if we set μ to 0, the above conditions are precisely the primal-dual optimality conditions; complementary slackness, primal and dual feasibility of a pair of optimal (P) and (D) solutions. The problem has been reduced to solving the above equations in x, y, z. The classical technique for solving equations is Newton's method which prescribes the directions

$$\Delta y = -\left(AD_xD_z^{-1}A^{\mathrm{T}}\right)^{-1} AD_z^{-1} (\mu e - D_xD_ze)$$

$$\Delta z = -A^{\mathrm{T}}\Delta y \qquad\qquad (30.2)$$

$$\Delta x = D_z^{-1} (\mu e - D_xD_ze) - D_xD_z^{-1}\Delta z$$

The strategy is to take one Newton step, reduce μ, and iterate until the optimization is complete. The criterion for stopping can be determined by checking for feasibility $(x, z \geq 0)$ and if the duality gap $(x^t z)$ is close enough to 0. We are now ready to describe the algorithm.

Procedure: **Primal-Dual Interior**

0. **Initialize**

   - $x_0 > 0, \ y_0 \in \Re^m, \ z_0 > 0, \mu_0 > 0, \epsilon > 0, \rho > 0$
   - $k := 0$

1. **Iterative Step**

   do

   Stop if $Ax_k = b, \ A^{\mathrm{T}}y_k + z_k = c$ and $x_k^{\mathrm{T}}z_k \leq \epsilon$.

   $x_{k+1} \leftarrow x_k + \alpha_k^P\Delta x_k$

   $y_{k+1} \leftarrow y_k + \alpha_k^D\Delta y_k$

   $z_{k+1} \leftarrow z_k + \alpha_k^D\Delta z_k$

   ```
   /* Δx_k, Δy_k, Δz_k are the Newton directions from (30.2) */
   ```

   $\mu_{k+1} \leftarrow \rho\mu_k$

   $k := k + 1$

   od

2. **End**

Remarks

1. The primal-dual algorithm has been used in several large-scale implementations. For appropriate choice of parameters, it can be shown that the number of iterations in the worst case is $O(\sqrt{n}\log(\epsilon_0/\epsilon))$ to reduce the duality gap from ϵ_0 to ϵ [70,85]. While this is sufficient to show that the algorithm is polynomial time, it has been observed that the average number of iterations is more like $O(\log n\log(\epsilon_0/\epsilon))$. However, unlike the simplex method we do not have a satisfactory theoretical analysis to explain this observed behavior.

2. The stepsizes α_k^P and α_k^D are chosen to keep x_{k+1} and z_{k+1} strictly positive. The ability in the primal–dual scheme to choose separate stepsizes for the primal and dual variables is a major computational advantage that this method has over the pure primal or dual methods. Empirically this advantage translates to a significant reduction in the number of iterations.

3. The stopping condition essentially checks for primal and dual feasibility and near complementary slackness. Exact complementary slackness is not possible with interior solutions. It is possible to maintain primal and dual feasibility through the algorithm, but this would require a Phase I construction via artificial variables. Empirically, this feasible variant has not been found to be worthwhile. In any case, when the algorithm terminates with an interior solution, a post-processing step is usually invoked to obtain optimal extreme point solutions for the primal and dual. This is usually called the purification of solutions and is based on a clever scheme described by Megiddo [56].

4. Instead of using Newton steps to drive the solutions to satisfy the optimality conditions of (D_μ), Mehrotra [60] suggested a predictor–corrector approach based on power series approximations. This approach has the added advantage of providing a rational scheme for reducing the value of μ. It is the predictor-corrector based primal–dual interior method that is considered the current winner in interior point methods. The OB1 code of Lustig et al. [53] is based on this scheme. CPLEX 4.0 [18], a general purpose linear (and integer) programming solver, also contains implementations of interior point methods. Saltzman [71] describes a parallelization of the OB1 method to run on shared-memory vector multiprocessor architectures. Recent computational studies of parallel implementations of simplex and interior point methods on the SGI Power Challenge (SGI R8000) platform indicate that on all but a few small linear programs in the NETLIB linear programming benchmark problem set, interior point methods dominate the simplex method in run times. New advances in handling Cholesky factorizations in parallel are apparently the reason for this exceptional performance of interior point methods.

    As in the case of the simplex method, there are a number of special structures in the matrix A that can be exploited by interior point methods to obtain improved efficiencies. Network flow constraints, generalized upper bounds (GUB), and variable upper bounds (VUB) are structures that often come up in practice and which can be useful in this context [15,83].

5. Interior-point methods, like ellipsoid methods, do not directly exploit the linearity in the problem description. Hence they generalize quite naturally to algorithms for semidefinite and convex programming problems. More details of these generalizations are given in Chapter 32 of this handbook. Kamath and Karmarkar [46] have proposed an interior-point approach for integer programming problems. The main idea is to reformulate an integer program as the minimization of a quadratic energy function over linear constraints on continuous variables. Interior-point methods are applied to this formulation to find local optima.

30.7 Strongly Polynomial Methods

The number of iterations and hence the number of elementary arithmetic operations required for the ellipsoid method as well as the interior point method is bounded by a polynomial function of the number of bits required for the binary representation of the input data. Recall that the size of a rational number a/b is defined as the total number of bits required in the binary representation of a and b. The dimension of the input is the number of data items in the input. An algorithm is said to be strongly polynomial if it consists of only elementary arithmetic operations (performed on rationals of size bounded by a polynomial in the size of the input) and the number of such elementary arithmetic operations is bounded by a polynomial in the dimension of the input.

It is an open question as to whether there exists a strongly polynomial algorithm for the general linear programming problem. However, there are some interesting partial results:

- Tardos [82] has devised an algorithm for which the number of elementary arithmetic operations is bounded by a polynomial function of n, m, and the number of bits required for the binary representation of the elements of the constraint matrix A which is $m \times n$. The number of elementary operations does not depend upon the right-hand side or the cost coefficients.

- Megiddo [57] described a strongly polynomial algorithm for checking the solvability of linear constraints with at most two non-zero coefficients per inequality. Later, Hochbaum and Naor [40] showed that Fourier elimination can be specialized to provide a strongly polynomial algorithm for this class.

- Megiddo [58] and Dyer [22] have independently designed strongly polynomial (linear-time) algorithms for linear programming in fixed dimensions. The number of operations for these algorithms is linear in the number of constraints and independent of the coefficients but doubly exponential in the number of variables.

The rest of this section details these three results and some of their consequences.

30.7.1 Combinatorial Linear Programming

Consider the linear program, *(LP) Max*$\{cx : Ax = b, x \geq 0\}$, where A is a $m \times n$ integer matrix. The associated dual linear program is *Min* $\{yb : y A \geq c\}$. Let L be the maximum absolute value in the matrix and let $\Delta = (nL)^n$. We now describe Tardos' algorithm for solving *(LP)* which permits the number of elementary operations to be free of the magnitudes of the rim coefficients b and c.

The algorithm uses Procedure 1 to solve a system of linear inequalities. Procedure 1, in turn, calls Procedure 2 with any polynomial-time linear programming algorithm as the required subroutine. Procedure 2 finds the optimal objective function value of a linear program and a set of variables which are zero in some optimal solution, if the optimum is finite. Note that Procedure 2 only finds the optimal objective value and not an optimal solution. The main algorithm also calls Procedure 2 directly with Subroutine 1 as the required subroutine. For a given linear program, Subroutine 1 finds the optimal objective function value and a dual solution, if one exists. Subroutine 1, in turn, calls Procedure 2 along with any polynomial-time linear programming algorithm as the required subroutine. We omit the detailed descriptions of the Procedures 1 and 2 and Subroutine 1 and instead only give their input/output specifications.

Algorithm: **Tardos**

> INPUT: A linear programming problem max$\{cx : Ax = b, x \geq 0\}$
> OUTPUT: An optimal solution, if it exists and the optimal objective function value.
>
> 1. Call Procedure 1 to test whether $\{Ax = b, x \geq 0\}$ is feasible. If the system is not feasible, the optimal objective function value $= -\infty$, stop.
> 2. Call Procedure 1 to test whether $\{yA \geq c\}$ is feasible. If the system is not feasible, the optimal objective function value $= +\infty$, stop.
> 3. Call Procedure 2 with the inputs as the linear program *Max*$\{cx : Ax = b, x \geq 0\}$ and Subroutine 1 as the required subroutine. Let $x_i = 0$, $i \in K$ be the set of equalities identified.

4. Call Procedure 1 to find a feasible solution x^* to $\{Ax = b, x \geq 0, x_i = 0, i \in K\}$. The solution x^* is optimal and the optimal objective function value is cx^*.

5. **End**

Specification of Procedure 1

    INPUT: A linear system $Ax \leq b$, where A is a $m \times n$ matrix .

    OUTPUT: Either $Ax \leq b$ is infeasible or $\hat{x}$ is a feasible solution.

Specification of Procedure 2

    INPUT: Linear program $Max\{cx : Ax = b, x \geq 0\}$, which has a feasible solution and a subroutine which for a given integer vector $\bar{c}$ with $\| \bar{c} \|_\infty \leq n^2 \Delta$ and a set K of indices, determines $\max\{\bar{c}x : Ax = b, x \geq 0, x_i = 0, i \in K\}$ and if the maximum is finite, finds an optimal dual solution.

    OUTPUT: The maximum objective function value z^* of the input linear program $\max\{c x : A x = b, x \geq 0,\}$ and the set K of indices such $x_i = 0, i \in K$ for some optimum solution to the input linear program.

Specification of Subroutine 1

    INPUT: A Linear program $\max\{\bar{c}x : Ax = b, x \geq 0, x_i = 0, i \in K\}$, which is feasible and $\| \bar{c} \|_\infty \leq n^2 \Delta$.

    OUTPUT: The Optimal objective function value z^* and an optimal dual solution y^*, if it exists.

The validity of the algorithm and the analysis of the number of elementary arithmetic operations required are in the paper by Tardos [82]. This result may be viewed as an application of techniques from diophantine approximation to linear programming. A scholarly account of these connections is given in the book by Schrijver [72].

REMARK 30.2   Linear programs with $\{0, \pm 1\}$ elements in the constraint matrix A arise in many applications of polyhedral methods in combinatorial optimization. Network flow problems (shortest path, maximum flow, and transshipment) [1] are examples of problems in this class. Such linear programs, and more generally linear programs with the matrix A made up of integer coefficients of bounded magnitude, are known as combinatorial linear programs. The algorithm described shows that combinatorial linear programs can be solved by strongly polynomial methods.

30.7.2   Fourier Elimination and $LI(2)$

We now describe a special case of the linear programming solvability problem for which Fourier elimination yields a very efficient (strongly polynomial) algorithm. This is the case $LI(2)$ of linear inequalities with at most two variables per inequality. Nelson [64] observed that Fourier elimination is subexponential in this case. He showed that the number of inequalities generated never exceeds $O(mn^{\lceil \log n \rceil} \log n)$. Later Aspvall and Shiloach [4] obtained the first polynomial-time algorithm for solving $LI(2)$ using a graph representation of the inequalities. We give a high-level description of the technique of Hochbaum and Naor [40] that combines Fourier elimination and a graph reasoning technique to obtain the best known sequential complexity bounds for $LI(2)$.

An interesting property of $LI(2)$ systems is that they are closed under Fourier elimination. Therefore, the projection of an $LI(2)$ system on to a subspace whose coordinates are a subset of the variables is also an $LI(2)$ system. Note that $LI(3)$ does not have this closure property. Indeed $LI(3)$ is unlikely to have any special property, since any system of linear inequalities can be reduced to an instance of $LI(3)$ with $0, \pm 1$ coefficients [43].

Given an instance of $LI(2)$ of the form $Ax \leq b$ with each row of A containing at most two nonzero coefficients we construct a graph $G(V,E)$ as follows. The vertices V are $x_0, x_1, \ldots, x_n$ corresponding to the variables of the constraints (x_0 is an extra dummy variable). The edges E of G are composed of pairs (x_i, x_j) if x_i and x_j are two variables with nonzero coefficients of at least one inequality in the system. There are also edges of the form $x_0, x_k)$ if x_k is the only variable with a nonzero coefficient in some constraint. Let us also assume that each edge is labelled with all the inequalities associated with its existence.

Aspvall and Shiloach [4] describe a grapevine algorithm that takes as input a rumour $x_j = \alpha$ and checks its authenticity, i.e., checks if α is too small, too large, or within the range of feasible values of x_j. The idea is simply to start at node x_j and set $x_j = \alpha$. Obviously, each variable x_k that is a neighbor of x_j in G gets an implied lower bound or upper bound (or both), depending on the sign of the coefficient of x_k in the inequality shared with x_j. These bounds get propagated further to neighbors of the x_k and so on. If this propagation is carried out in a breadth-first fashion, it is not hard to argue that the implications of setting $x_j = \alpha$ are completely revealed in $3n - 2$ stages. Proofs of inconsistency can be traced back to delineate if α was either too high or too low a value for x_j.

The grapevine algorithm is similar to Bellman and Ford's classical shortest path algorithm for graphs which also takes $O(mn)$ effort. This subroutine provides the classification test for being able to execute binary search in choosing values for variables. The specialization of Fourier's algorithm for $LI(2)$ can be described now.

Algorithm Fourier $LI(2)$
```
For j = 1, 2, ⋯ , n
```

1. The inequalities of each edge (x_j, x_k) define a convex polygon Q_{jk} in x_j, x_k-space. Compute J_k the sorted collection of x_j coordinates of the corner (extreme) points of Q_{jk}. Let J denote the sorted union (merge) of the J_k (x_k a neighbor of x_j in G).

2. Perform a binary search on the sequence J for a feasible value of x_j. If we succeed in finding a feasible value for x_j among the values in J we fix x_j at that value and contract vertex x_j with x_0. Move to the next variable $j \leftarrow j+1$ and repeat.

3. Else we know that the sequence is too coarse and that all feasible values lie in the strict interior of some interval $[x_j^1, x_j^2]$ defined by consecutive values in J. In this latter case we prune all but the two essential inequalities, defining the edges of the polygon Q_{jk}, for each of the endpoints x_j^1 and x_j^2.

4. Eliminate x_j using standard Fourier elimination.

5. End

Notice that at most four new inequalities are created for each variable elimination. Also note that the size of J is always $O(m)$. The complexity is dominated by the search over J. Each search step requires a call to the grapevine procedure and there are at most $n \log m$ calls. Therefore the overall time-complexity is $O(mn^2 \log m)$, which is strongly polynomial in that it is polynomial and independent of the size of the input coefficients.

An open problem related to $LI(2)$ is the design of a strongly polynomial algorithm for optimization of an arbitrary linear function over $LI(2)$ constraints. This would imply, via duality, a strongly polynomial algorithm for generalized network flows (flows with gains and losses).

30.7.3 Fixed Dimensional LPs: Prune and Search

Consider the linear program $\max\{cx : Ax \leq b\}$ where A is an $m \times n$ matrix that includes the nonnegativity constraints. Clearly, for fixed dimension n, there is a polynomial-time algorithm because there are at most $\binom{m}{n}$ system of linear equations to be solved, to generate all extreme points of the feasible region. However, Megiddo [56] and Dyer [22] have shown that for the above linear program with fixed n, there is a linear-time algorithm that requires $O(m)$ elementary arithmetic operations on numbers of size bounded by a polynomial in the input data. The algorithm is highly recursive. Before we give an outline of the algorithm, some definitions are required.

DEFINITION 30.5   Given a linear program $\max\{cx : Ax \leq b\}$ and a linear equality $fx = q$,

(i) The inequality $fx < q$ is said to hold for the optimum if either

   (a) We know that $Ax \leq b$ is feasible and
   $$\max\{cx : Ax \leq b, fx \leq q\} > \max\{cx : Ax \leq b, fx = q\}$$

   or

   (b) We know a row vector $y \geq 0$ such that $yA = f$ and $yb < q$.

(ii) The inequality $fx > q$ is said to hold for the optimum if either

   (a) We know that $Ax \leq b$ is feasible and
   $$\max\{cx : Ax \leq b, fx \geq q\} > \max\{cx : Ax \leq b, fx = q\}$$

   or

   (b) We know a vector $y \geq 0$ such that $yA = -f$ and $yb < -q$.

DEFINITION 30.6   For a given linear program $\max\{cx : Ax \leq b\}$ and a given linear equation $fx = q$, the position of the optimum of the linear program relative to the linear equation is said to be known if either we know that $fx < q$ holds for an optimum or $fx > q$ holds for an optimum.

An outline of the algorithm is presented below. The algorithm requires an oracle, denoted as Procedure 1, with inputs as the linear program $\max\{cx : Ax \leq b\}$ where A is an $m \times n$ matrix and a system of p linear equations $Fx = d$ with the rank of F being r. The output of Procedure 1 is either a solution to the linear program (possibly unbounded or infeasible) or a set of $\lceil p/2^{2^{r-1}} \rceil$ equations in $Fx = d$ relative to each of which we know the position of the optimum of the linear program.

Algorithm Sketch: **Prune & Search**

Call Procedure 1 with inputs as the linear program $\max\{cx : Ax \leq b\}$ and the system of m equations $Ax = b$. Procedure 1 either solves the linear program or identifies $k = \lceil m/2^{2^{n-1}} \rceil$ equations in $Ax = b$ relative to each of which we know the position of the optimum of the linear program. The identified equations are then omitted from the system $Ax = b$. The resulting subsystem, $A_1 x = b_1$ has $m_1 = m - k$ equations. Procedure 1 is applied again with the original given linear program and the system of equations $A_1 x = b_1$ as the inputs. This process is

repeated until either the linear program is solved or we know the
position of the optimum with respect to each of the equations in
$Ax = b$. In the latter case the system $Ax \leq b$ is infeasible.

We next describe the input/output specification of Procedure 1. The procedure is highly recursive
and splits into a lot of cases. This procedure requires a linear-time algorithm for the identification
of the median of a given set of rationals in linear time.

Specification of Procedure 1

INPUT: Linear program $\max\{cx : Ax \leq b\}$ where A is an $m \times n$ matrix
and a system of p equations $Fx = d$ where rank of F is r.

OUTPUT: A solution to the linear program or a set of $\lceil p/2^{2^{r-1}} \rceil$
equations in $Fx = d$ relative to each of which we know the
position of the optimum as in Definition 30.6.

For fixed n, Procedure 1 requires $O(p + m)$ elementary arithmetic operations on numbers of size
bounded by a polynomial in the size of the input data. Since at the outset $p = m$, algorithm Prune &
Search solves linear programs with fixed n in linear time. Details of the validity of the algorithm
as well as analysis of its linear time complexity for fixed n are given by Megiddo [56], Dyer [22], and
in the book by Schrijver [72]. As might be expected, the linear-time solvability of linear programs
in fixed dimension has important implications in the field of computational geometry which deals
largely with two-and three-dimensional geometry. The book by Edelsbrunner [25] documents these
connections.

The linear programming problem is known to be $\mathcal{P}$-complete and therefore we do not expect to
find a parallel algorithm that achieves polylog run time. However, for fixed n, there are simple polylog
algorithms [23]. In a recent paper, Sen [74] shows that linear programming in fixed dimension n
can be solved in $O(\log\log^{n+1} m)$ steps using m processors in a *CRCW PRAM*.

30.8 Randomized Methods for Linear Programming

The advertising slogan for randomized algorithms has been simplicity and speed [61]. In the case of
fixed-dimensional linear programming there appears to be some truth in the advertisement. In stark
contrast with the very technical deterministic algorithm outlined in the last section, we will see that
an almost trivial randomized algorithm will achieve comparable performance (but of course at the
cost of determinism).

Consider a linear programming problem of the form

$$\min\{cx : Ax \leq b\}$$

with the following properties:

- The feasible region $\{x : Ax \leq b\}$ is nonempty and bounded.
- The objective function c has the form $(1, 0, \ldots, 0)$.
- The minimum to the linear program is unique and occurs at an extreme point of the
  feasible region.
- Each vertex of $\{x : Ax \leq b\}$ is defined by exactly n constraints where A is $m \times n$.

Note that none of these assumptions compromise the generality of the linear programming problem
that we are considering.

Let S denote the constraints $Ax \leq b$. A feasible $B \subseteq S$ is called optimal if it defines the uniquely optimal extreme point of the feasible region. The following randomized algorithm due to Sharir and Welzl [76] uses an incremental technique to obtain the optimal basis of the input linear program.

Algorithm: **ShW**

> INPUT: The constraint set S and a feasible basis T.
> OUTPUT: The optimal basis for the linear program.
>
> 1. If S equals T, return T;
> 2. Pick a random constraint $s \in S$. Now define
> $\bar{T} = \mathbf{ShW}(S \setminus \{s\}, T)$;
> 3. If the point defined by $\bar{T}$ satisfies s, output $\bar{T}$;
>   Else output $\mathbf{ShW}(S, \mathbf{opt}(\{s\} \cup \bar{T}))$
> 4. **End**

The subroutine **opt** when given an input of $n + 1$ or less constraints $H \subseteq S$ returns an optimal basis for the linear program with constraints defined by H (and objective function cx). It has been shown [55] that algorithm **ShW** has an expected running time of $O(\min\{mn \exp \sqrt[4]{n \ln(m+1)},$ $n^4 2^n m\})$. Thus algorithm **ShW** is certainly linear expected time for fixed n but has a lower complexity than Prune & Search if n is allowed to vary.

30.9 Large-Scale Linear Programming

Linear programming problems with thousands of rows and columns are routinely solved either by variants of simplex method or by interior point methods. However, for several linear programs that arise in combinatorial optimization, the number of columns (or rows in the dual) are too numerous to be enumerated explicitly. The columns, however, often have a structure which is exploited to generate the columns as and when required in the simplex method. Such an approach, which is referred to as **column generation**, is illustrated next on the cutting stock problem [32], which is also known as the bin packing problem in the computer science literature.

30.9.1 Cutting Stock Problem

Rolls of sheet metal of standard length L are used to cut required lengths $l_i, i = 1, 2 \ldots, m$. The jth cutting pattern should be such that a_{ij}, the number of sheets length l_i cut from one roll of standard length L must satisfy $\sum_{i=1}^{m} a_{ij} l_i \leq L$. Suppose $n_i, i = 1, 2, \ldots, m$ sheets of length l_i are required. The problem is to find cutting patterns so as to minimize the number of rolls of standard length L that are used to meet the requirements. A linear programming formulation of the problem is as follows:

Let $x_j, j = 1, 2, \ldots, n$ denote the number of times the jth cutting pattern is used. In general, $x_j, j = 1, 2, \ldots, n$ should be an integer but in the formulation below the variables are permitted to be fractional.

$$
\begin{aligned}
(P1) \quad &\min \sum_{j=1}^{n} x_j \\
\text{Subject to} \quad &\sum_{j=1}^{n} a_{ij} x_j \geq n_i \quad i = 1, 2, \ldots, m \\
&x_j \geq 0 \qquad\qquad j = 1, 2, \ldots, n \\
\text{where} \quad &\sum_{i=1}^{m} l_i a_{ij} \leq L \quad j = 1, 2, \ldots, n
\end{aligned}
$$

The formulation can easily be extended to allow for the possibility of p standard lengths, L_k, $k = 1, 2, \ldots, p$ from which the n_i units of length $l_i, i = 1, 2, \ldots, m$ are to be cut.

The cutting stock problem can also be viewed as a bin packing problem. Several bins, each of standard capacity L are to be packed with n_i units of item i, each of which uses up capacity of l_i in a bin. The problem is to minimize the number of bins used.

30.9.1.1 Column Generation

In general, the number of columns in (P1) is too large to enumerate all the columns explicitly. The simplex method, however, does not require all the columns to be explicitly written down. Given a basic feasible solution and the corresponding simplex multipliers $w_i, i = 1, 2, \ldots, m$, the column to enter the basis is determined by applying dynamic programming to solve the following knapsack problem:

$$(P2) \quad z = \max \quad \sum_{i=1}^{m} w_i a_i$$

$$\text{Subject to} \quad \sum_{i=1}^{m} l_i a_i \leq L$$

$$a_i \geq 0 \text{ and integer } i = 1, 2, \ldots, m$$

Let $a_i^*, i = 1, 2, \ldots, m$ denote an optimal solution to (P2). If $z > 1$, the kth column to enter the basis has coefficients $a_{ik} = a_i^*, i = 1, 2, \ldots, m$.

Using the identified columns, a new improved (in terms of the objective function value) basis is obtained and the column generation procedure is repeated. A major iteration is one in which (P2) is solved to identify, if there is one, a column to enter the basis. Between two major iterations, several minor iterations may be performed to optimize the linear program using only the available (generated) columns.

If $z \leq 1$ the current basic feasible solution is optimal to (P1). From a computational point of view, alternative strategies are possible. For instance, instead of solving (P2) to optimality, a column to enter the basis can be identified as soon as a feasible solution to (P2) with an objective function value greater than 1 has been found. Such an approach would reduce the time required to solve (P2) but may increase the number of iterations required to solve (P1).

A column, once generated may be retained, even if it comes out of the basis at a subsequent iteration, so as to avoid generating the same column again later on. However, at a particular iteration some columns that appear unattractive in terms of their reduced costs may be discarded in order to avoid having to store a large number of columns. Such columns can always be generated again subsequently, if necessary. The rationale for this approach is that such unattractive columns will rarely be required subsequently.

The dual of (P1) has a large number of rows. Hence column generation may be viewed as row generation in the dual. In other words, in the dual we start with only a few constraints explicitly written down. Given an optimal solution w to the current dual problem (i.e., with only a few constraints which have been explicitly written down) find a constraint that is violated by w or conclude that no such constraint exists. The problem to be solved for identifying a violated constraint, if any, is exactly the separation problem that we encountered in Section 30.5.

30.9.2 Decomposition

Large-scale linear programming problems sometimes have a block diagonal structure. Consider for instance, the following linear program:

$$(P3) \quad \max \sum_{j=1}^{p} c^j x^j \tag{30.3}$$

$$\text{Subject to} \quad \sum_{j=1}^{p} A^j x^j = b \tag{30.4}$$

$$D^j x^j = d^j \quad j = 2, 3, \ldots, p \tag{30.5}$$

$$x^j \geq 0 \quad j = 1, 2, \ldots, p \tag{30.6}$$

where

A^j is an $m \times n_j$ matrix

D^j is an $m_j \times n_j$ matrix

x_j is an $n_j \times 1$ column vector

c^j is an $1 \times n_j$ row vector

b is an $m \times 1$ column vector

d^j is an $m_j \times 1$ column vector

The constraints 30.4 are referred to as the linking master constraints. The p sets of constraints 30.5 and 30.6 are referred to as subproblem constraints. Without the constraints 30.4, the problem decomposes into p separate problems which can be handled in parallel. The Dantzig–Wolfe [21] decomposition approach to solving ($P3$) is described next.

Clearly, any feasible solution to ($P3$) must satisfy constraints 30.5 and 30.6. Now consider the polyhedron P_j, $j = 2, 3, \ldots, p$ defined by the constraints 30.5 and 30.6. By the representation theorem of polyhedra (see Section 30.2.1) a point $x^j \in P_j$ can be written as

$$x^j = \sum_{k=1}^{h_j} x^{jk} \rho_{jk} + \sum_{k=1}^{g_j} y^{jk} \mu_{jk}$$

$$\sum_{k=1}^{h_j} \rho_{jk} = 1$$

$$\rho_{jk} \geq 0 \ \ k = 1, 2, \ldots, h_j$$

$$\mu_{jk} \geq 0 \ \ k = 1, 2, \ldots, g_j$$

where

x^{jk} ($k = 1, 2, \ldots, h_j$) are the extreme points

y^{jk} ($k = 1, 2, \ldots, g_j$) are the extreme rays of P_j

Now substituting for x^j, $j = 2, 3, \ldots, p$ in (30.3) and (30.4), ($P3$) is written as

$$(P4) \quad \max \left\{ c^1 x^1 + \sum_{j=2}^{p} \left[\sum_{k=1}^{h_j} \left(c^j x^{jk} \right) \rho_{jk} + \sum_{k=1}^{g_j} \left(c^j y^{jk} \right) \mu_{jk} \right] \right\}$$

$$\text{Subject to} \quad A^1 x^1 + \sum_{j=2}^{p} \left[\sum_{k=1}^{h_j} \left(A^j x^{jk} \right) \rho_{jk} + \sum_{k=1}^{g_j} \left(A^j y^{jk} \right) \mu_{jk} \right] = b \qquad (30.7)$$

$$\sum_{k=1}^{h_j} \rho_{jk} = 1 \ \ j = 2, 3, \ldots, p \qquad (30.8)$$

$$\rho_{jk} \geq 0 \ \ j = 2, 3, \ldots, p; \ \ k = 1, 2, \ldots, h_j$$

$$\mu_{jk} \geq 0 \ \ j = 2, 3, \ldots, p; \ \ k = 1, 2, \ldots, g_j$$

In general, the number of variables in ($P4$) is an exponential function of the number of variables $n_j, j = 1, 2, \ldots, p$ in ($P3$). However, if the simplex method is adapted to solve ($P4$), the extreme points or the extreme rays of $P_j, j = 2, 3, \ldots, p$, and consequently the columns in ($P4$) can be generated, as and when required, by solving the linear programs associated with the p subproblems. This column generation is described next.

Given a basic feasible solution to ($P4$), let w and u be row vectors denoting the simplex multipliers associated with constraints 30.7 and 30.8, respectively. For $j = 2, \ldots, p$, solve the following linear programming subproblems:

$$(S_j) \qquad z_j = \min \left(wA^j - c^j \right) x^j$$
$$D^j x^j = d^j$$
$$x^j \geq 0$$

Suppose z_j is finite. An extreme point solution x^{jt} is then identified. If $z_j + u_j < 0$, a candidate column to enter the basis is given by $\binom{A^j x^{jt}}{1}$. On the other hand if $z_j + u_j \geq 0$, there is no extreme point of P_j that gives a column to enter the basis at this iteration. Suppose the optimal solution to S_j is unbounded. An extreme ray y^{jt} of P_j is then identified and a candidate column to enter the basis is given by $\binom{A^j y^{jt}}{0}$. If the simplex method is used to solve S_j, the extreme point or the extreme ray is identified automatically. If a column to enter the basis is identified from any of the subproblems, a new improved basis is obtained and the column generation procedure is repeated. If none of the subproblems identify a column to enter the basis, the current basic solution to ($P4$) is optimal.

As in Section 30.9.1 a major iteration is when the subproblems are solved. Instead of solving all the p subproblems at each major iteration, one option is to update the basis as soon as a column to enter the basis has been identified in any of the subproblems. If this option is used at each major iteration, the subproblems that are solved first are typically the ones that were not solved at the previous major iteration.

The decomposition approach is particularly appealing if the subproblems have a special structure that can be exploited. Note that only the objective functions for the subproblems change from one major iteration to another. Given the current state of the art, ($P4$) can be solved in polynomial time (polynomial in the problem parameters of ($P3$)) by the ellipsoid method but not by the simplex method or interior point methods. However ($P3$) can be solved in polynomial time by interior point methods.

It is interesting to note that the reverse of decomposition is also possible. In other words, suppose we start with a statement of a problem and an associated linear programming formulation with a large number columns (or rows in the dual). If the column generation (or row generation in the dual) can be accomplished by solving a compact linear program, then a compact formulation of the original problem can be obtained. Here compact refers to the number of rows and columns being bounded by a polynomial function of the parameters (not the number of the columns in the original linear programming formulation) in the statement of the original problem. This result due to Martin [54] enables one to solve the problem in the polynomial time by solving the compact formulation directly using interior point methods.

30.9.3   Compact Representation

Consider the following linear program:

$$(P5) \qquad \min \ cx$$
$$\text{Subject to} \qquad Ax \geq b$$
$$x \geq 0$$
$$x \in \bigcap_{j=1}^{p} P_j$$

where

A is an $m \times n$ matrix

x is an $n \times 1$ vector

c is an $1 \times n$ vector

b is an $m \times 1$ vector

P_j, $j = 1, 2, \ldots, p$ is a polyhedron and p is bounded by a polynomial in m and n

Without loss of generality, it is assumed that $P_j, j = 1, 2, \ldots, p$ is nonempty and ($P5$) is feasible. Given $\bar{x}$ such that $A\bar{x} \geq b$, the constraint identification or separation problem $S_j(\bar{x})$ with respect to P_j is to either (a) conclude that $\bar{x} \in P_j$, or (b) find a valid inequality $D_i^j x \leq d_i^j$ that is satisfied by $x \in P_j$ but $D_i^j \bar{x} > d_i^j$.

Suppose the separation problem $S_j(\bar{x})$ can be solved by the following linear program:

$$
\begin{aligned}
S_j(\bar{x}) : z_j &= \max \left(\bar{x}^T G^j + g^j \right) y \\
F^j y &\leq f^j \\
y &\geq 0
\end{aligned}
$$

where

G^j is an $n \times k$ matrix

F^j is an $r \times k$ matrix

g^j is a $1 \times k$ vector

f^j is an $r \times 1$ vector

y is a $k \times 1$ vector

r and k are bounded by a polynomial in m and n

$\bar{x} \in P_j \bigcap \{x : Ax \geq b\}$ if and only if $z_j \leq h^j \bar{x} + k_j$ where h^j is a $1 \times n$ vector and k_j is a scalar.

Now, if w^j denotes the dual variables associated with the $F^j y \leq f^j$ constraints in S_j, it follows from the duality theorem of linear programming that a **compact representation** of ($P5$) is given by

$$
\begin{aligned}
\min \ & cx \\
\text{Subject to} \quad Ax &\geq b \\
(F^j)^T w^j - (G^j)^T x &\geq (g^j)^T \quad j = 1, 2, \ldots, p \\
(f^j)^T w^j - h^j x &\leq k_j \quad j = 1, 2, \ldots, p \\
x &\geq 0 \\
w^j &\geq 0 \quad j = 1, 2, \ldots, p
\end{aligned}
$$

Note that this approach to obtaining a compact formulation is predicated on being able to formulate the separation problem as a compact linear program. This may not always be possible. In fact, Yannakakis [86] shows that for a b-matching problem under a symmetry assumption, no compact formulation is possible. This despite the fact that b-matching can be solved in polynomial time using a polynomial-time separation oracle.

30.9.3.1 An Application: Neural Net Loading

The decision version of the Hopfield neural net loading problem (cf. [65]) is as follows.

Given y^i ($i = 1, 2, \ldots, p$) where each y^i is an n-dimensional training vector whose components are $(+1, -1)$, construct a symmetric $n \times n$ synaptic weight matrix W, such that for every n-dimensional vector v whose components are $(+1, -1)$, the following holds:

$$
\text{If } d\left(y^i, v\right) \leq k \text{ then } y^i = sgn(Wv) \text{ for } i = 1, 2, \ldots, p
$$

A feasible W would represent a Hopfield net with a radius of direct attraction of at least k around each training vector, i.e., a robust network of associative memory. Here k is specified and $d(y^i, v)$ is the Hamming distance between y^i and v. For $t = 1, 2, \ldots, p$, let $v^{tq}, q = 1, 2, \ldots, m_t$ be all the vectors whose components are $(+1, -1)$ and $d(y^t, v^{tq}) \leq k$. The Hopfield neural net loading problem is to find a matrix W such that

$$(P6) \quad \sum_{j=1}^{n} y_i^t w_{ij} v_j^{tq} \geq 1 \quad \text{for } i = 1, 2, \ldots, n; \ t = 1, 2, \ldots, p; \text{ and } q = 1, 2, \ldots, m_t$$

This is a linear program with the number of inequalities equal to $pn\binom{n}{k}$, which is huge. The synaptic weights have to be symmetric, so in addition to the inequalities given above, (P6) would include the constraints $w_{ij} = w_{ji}$ for $i = 1, 2, \ldots, n$ and $j = 1, 2, \ldots, n$.

Given the weights $\bar{w}_{uv}, u = 1, 2, \ldots, n, v = 1, 2, \ldots, n$, the separation problem, $S_{it}(\bar{w})$ for specified i and t, where $1 \leq i \leq n$ and $1 \leq t \leq p$, can be formulated as follows [13]:

Let

$$u_{ij}^{tq} = \begin{cases} 1 & \text{if } v_j^{tq} = -y_j^t \\ 0 & \text{if } v_j^{tq} = y_j^t \end{cases}$$

Since $d(y^t, v^{tq}) \leq k$ it follows that

$$\sum_{j=1}^{n} u_{ij}^{tq} \leq k \quad q = 1, 2, \ldots, m_t$$

Note that for specified i, t, and q, the inequality in (P6) is equivalent to

$$\sum_{j=1}^{n} y_i^t w_{ij} \left[y_j^t \left(-u_{ij}^{tq} \right) + y_j^t \left(1 - u_{ij}^{tq} \right) \right] \geq 1$$

which reduces to

$$-\sum_{j=1}^{n} 2 y_i^t y_j^t w_{ij} u_{ij}^{tq} + \sum_{j=1}^{n} y_i^t y_j^t w_{ij} \geq 1$$

Consequently, the separation problem after dropping the superscript q is

$$S_{it}(\bar{w}) : z_{it} = \max \sum_{j=1}^{n} 2 y_i^t y_j^t \bar{w}_{ij} u_{ij}^t$$

$$\text{Subject to} \quad \sum_{j=1}^{n} u_{ij}^t \leq k \tag{30.9}$$

$$0 \leq u_{ij}^t \leq 1 \quad j = 1, 2, \ldots, n \tag{30.10}$$

Note that $S_{it}(\bar{w})$ is trivial to solve and always has a solution such that $u_{ij}^t = 0$ or 1. It follows that for given i and t, $\bar{w}$ satisfies the inequalities in (P6) corresponding to $q = 1, 2, \ldots, m_t$ if and only if

$$z_{it} \leq \sum_{j=1}^{n} y_i^t y_j^t \bar{w}_{ij} - 1$$

Let θ_{it} and $\beta_{ij}^t, j = 1, 2, \ldots, n$, denote the dual variables associated with constraints 30.9 and 30.10, respectively. Now applying the compact representation result stated above, a compact formulation of the neural net loading problem (*P6*) is as follows:

$$\theta_{it} + \beta ij^t - 2y_i^t y_j^t w_{ij} \geq 0 \quad i = 1, 2, \ldots, n \ \ t = 1, 2, \ldots, p \ \ j = 1, 2, \ldots, n$$

$$k\,\theta_{it} + \sum_{j=1}^{n} \beta ij^t - \sum_{j=1}^{n} y_i^t y_j^t w_{ij} \leq -1, \quad i = 1, 2, \ldots, n \ \ t = 1, 2, \ldots, p$$

$$\theta_{it} \geq 0 \quad i = 1, 2, \ldots, n; \ \ t = 1, 2, \ldots, p$$

$$\beta_{ij}^t \geq 0 \quad i = 1, 2, \ldots, n; \ \ t = 1, 2, \ldots, p; \ \ j = 1, 2, \ldots, n$$

With the symmetry condition $w_{ij} = w_{ji}$ for $i = 1, 2, \ldots, n$ and $j = 1, 2, \ldots, n$ added in, we now have a linear programming formulation of the Hopfield net loading problem that is compact in the size of the network $n \times n$ and the size of the training set $n \times p$.

30.10 Linear Programming: A User's Perspective

This chapter has been written for readers interested in learning about the algorithmics of linear programming. However, for someone who is primarily a user of linear programming software, there are a few important concerns that we address briefly here.

1. *The expression of linear programming models.* The data sources from which the coefficients of a linear programming model are generated may be organized in a format that is incompatible with the linear programming software in use. Tools to facilitate this translation have come to be known as matrix generators [28]. Over the years such tools have evolved into more complete modeling languages (e.g., AMPL [29]), and GAMS [6].

2. *The viewing, reporting, and analysis of results.* This issue is similar to that of model expression. The results of a linear programming problem when presented as raw numerical output are often difficult for a user to digest. Report writers and modeling languages like the ones mentioned above usually provide useful features for processing the output into a user-friendly form. Because of the widespread use of linear programming in decision support, user interfaces based on spreadsheets have become popular with software vendors [75].

3. *Tools for error diagnosis and model correction.* Many modeling exercises using linear programming involve a large amount of data and are prone to numerical as well as logical errors. Some sophisticated tools [36] are now available for helping users in this regard.

4. *Software support for linear programming model management.* Proliferation of linear programming models can occur in practice because of several reasons. The first is that when a model is being developed, it is likely that several versions are iterated on before converging to a suitable model. Scenario analysis is the second type of source for model proliferation. Finally, iterative schemes such as those based on column generation or stochastic linear programming may require the user to develop a large number of models. The software support in optimization systems for helping users in all such situations have come to be known as tools for model management [28,31].

5. *The choice of a linear programming solution procedure.* For linear programs with special structure (e.g., network flows [1]) it pays to use specialized versions of linear programming algorithms. In the case of general linear optimization software, the user may be provided a choice of a solution method from a suite of algorithms. While the right choice

of an algorithm is a difficult decision, we hope that insights gained from reading this chapter would help the user.

30.11 Further Information

Research publications in linear programming are dispersed over a large range of journals. The following is a partial list which emphasizes the algorithmic aspects: *Mathematical Programming, Mathematics of Operations Research, Operations Research, INFORMS Journal on Computing, Operations Research Letters, SIAM Journal on Optimization, SIAM Journal on Computing, SIAM Journal on Discrete Mathematics, Algorithmica, Combinatorica.*

Linear programming professionals frequently use the following newsletters:

- *INFORMS Today* (earlier OR/MS Today) published by The Institute for Operations Research and Management Science
- *INFORMS CSTS Newsletter* published by the INFORMS computer science technical section
- *Optima* published by the Mathematical Programming Society

The linear programming FAQ (frequently asked questions) facility is maintained at http://www.mcs.anl.gov/home/otc/Guide/faq/

To have a linear program solved over the Internet check the following locations: http://www.mcs.anl.gov/home/otc/Server/; http://www.mcs.anl.gov/home/otc/Server/lp/

The universal input standard for linear programs is the MPS format [62].

Defining Terms

Analytic center: The interior point of a polytope at which the product of the distances to the facets is uniquely maximized.

Column generation: A scheme for solving linear programs with a huge number of columns.

Compact representation: A polynomial size linear programming representation of an optimization problem.

Decomposition: A strategy of divide and conquer applied to large scale linear programs.

Duality: The relationship between a linear program and its dual.

Extreme point: A corner point of a polyhedron.

Linear program: Optimization of a linear function subject to linear equality and inequality constraints.

Matrix factorization: Representation of a matrix as a product of matrices of simple form.

Polyhedral cone: The set of solutions to a finite system of homogeneous linear inequalities on real-valued variables.

Polyhedron: The set of solutions to a finite system of linear inequalities on real-valued variables. Equivalently, the intersection of a finite number of linear half-spaces in $\Re^n$.

Polytope: A bounded polyhedron.

Relaxation: An enlargement of the feasible region of an optimization problem. Typically, the relaxation is considerably easier to solve than the original optimization problem.

Separation: Test if a given point belongs to a convex set and if it does not, identify a separating hyperplane.

Strongly polynomial: Polynomial algorithms with the number of elementary arithmetic operations bounded by a polynomial function of the dimensions of the linear program.

References

1. Ahuja, R.K., Magnanti, T.L., and Orlin, J.B., *Network Flows,* Prentice Hall, Englewood Cliffs, NJ, 1993.
2. Akgül, M., *Topics in Relaxation and Ellipsoidal Methods, Research Notes in Mathematics,* Pitman Publishing Ltd., London, 1984.
3. Alizadeh, F., Interior point methods in semidefinite programming with applications to combinatorial optimization, *SIAM Journal on Optimization,* 5(1), 13–51, Feb. 1995.
4. Aspvall, B.I. and Shiloach, Y., A polynomial time algorithm for solving systems of linear inequalities with two variables per inequality, *FOCS,* 205–217, 1979.
5. Bertsimas, D. and Tsitsiklis, J.N., *Introduction to Linear Optimization,* Athena Scientific, Belmont, MA, 1997.
6. Bisschop, J. and Meerhaus, A., On the development of a General Algebraic Modeling System (GAMS) in a strategic planning environment, *Mathematical Programming Study,* 20, 1–29, 1982.
7. Bixby, R.E., Progress in linear programming, *ORSA Journal on Computing,* 6(1), 15–22, 1994.
8. Bland, R., Goldfarb, D., and Todd, M.J., The ellipsoid method: A survey, *Operations Research,* 29, 1039–1091, 1981.
9. Borgwardt, K.H., *The Simplex Method: A Probabilistic Analysis,* Springer-Verlag, Berlin, Germany, 1987.
10. Černikov, R.N., The solution of linear programming problems by elimination of unknowns, *Doklady Akademii Nauk,* 139, 1314–1317, 1961. Translation in *Soviet Mathematics Doklady,* 2, 1099–1103, 1961.
11. Caratheodory, C., Über den Variabiltätsbereich der Fourierschen Konstanten von positiven harmonischen Funktionen, *Rendiconto del Circolo Matematico di Palermo,* 32, 193–217, 1911.
12. Chandru, V., Variable elimination in linear constraints, *The Computer Journal,* 36(5), 463–472, Aug. 1993.
13. Chandru, V., Dattasharma, A., Keerthi, S.S., Sancheti, N.K., and Vinay, V., Algorithms for the design of optimal discrete neural networks. In *Proceedings of the Sixth ACM/SIAM Symposium on Discrete Algorithms,* SIAM Press, San Franciso, CA, Jan. 1995.
14. Chandru. V. and Kochar, B.S., A class of algorithms for linear programming, *Research Memorandum 85-14,* School of Industrial Engineering, Purdue University, West Lafayette, IN, Nov. 1985.
15. Chandru, V. and Kochar, B.S., Exploiting special structures using a variant of Karmarkar's algorithm, *Research Memorandum 86-10,* School of Industrial Engineering, Purdue University, West Lafayette, IN, Jun. 1986.
16. Chvatal, V., *Linear Programming,* Freeman Press, New York, 1983.
17. Cook, W., Lovász, L., and Seymour, P., Eds., *Combinatorial Optimization: Papers from the DIMACS Special Year,* Series in Discrete Mathematics and Theoretical Computer Science, Volume 20, AMS, 1995.
18. CPLEX, *Using the CPLEX Callable Library and CPLEX Mixed Integer Library,* CPLEX Optimization, 1993.
19. Dantzig, G.B., Maximization of a linear function of variables subject to linear inequalities. In *Activity Analysis of Production and Allocation,* C. Koopmans, Ed., John Wiley & Sons, New York, pp. 339–347, 1951.
20. Dantzig, G.B., *Linear Programming and Extensions,* Princeton University Press, Princeton, NJ, 1963.

21. Dantzig, G.B. and Wolfe, P., The decomposition algorithm for linear programming, *Econometrica*, 29, 767–778, 1961.

22. Dyer, M.E., Linear time algorithms for two- and three-variable linear programs, *SIAM Journal on Computing*, 13, 31–45, 1984.

23. Dyer, M.E., A parallel algorithm for linear programming in fixed dimension. In *Proceedings of the 11th Annual ACM Symposium on Computational Geometry*, ACM Press, New York, pp. 345–349, 1995.

24. Ecker, J.G. and Kupferschmid, M., An ellipsoid algorithm for nonlinear programming, *Mathematical Programming*, 27, 1983.

25. Edelsbrunner, M., *Algorithms in Combinatorial Geometry*, Springer-Verlag, Berlin, Germany, 1987.

26. Edmonds, J., Submodular functions, matroids and certain polyhedra. In *Combinatorial Structures and Their Applications*, R. Guy et al., Eds., Gordon Breach, New York, pp. 69–87, 1970.

27. Farkas, Gy., A Fourier-féle mechanikai elv alkalmazásai (in Hungarian), *Mathematikai és Természettudományi Értesítö*, 12, 457–472, 1894.

28. Fourer, R., Software for Optimization: A Buyer's Guide (Parts I and II), *INFORMS Computer Science Technical Section Newsletter*, 17(1/2), 1996.

29. Fourer, R., Gay, D.M., and Kernighian, B.W., *AMPL: A Modeling Language for Mathematical Programming*, Scientific Press, San Francisco, CA, 1993.

30. Fourier, L.B.J., reported in Analyse des travaux de l'Academie Royale des Sciences, pendant l'annee 1824, Partie mathematique, *Histoire de l'Academie Royale des Sciences de l'Institut de France 7*, 1827, xlvii-lv (Partial English Translation in D.A. Kohler, Translation of a Report by Fourier on his Work on Linear Inequalities, *Opsearch*, 10, 38–42, 1973).

31. Geoffrion, A.M., An introduction to structured modeling, *Management Science*, 33, 547–588, 1987.

32. Gilmore, P. and Gomory, R.E., A linear programming approach to the cutting stock problem, Part I, *Operations Research*, 9, 849–854; Part II, *Operations Research*, 11, 863–887, 1963.

33. Goemans, M.X. and Williamson, D.P., 878 approximation algorithms MAX CUT and MAX 2SAT, *Proceedings of ACM STOC*, 422–431, 1994.

34. Grötschel, M., Lovász, L., and Schrijver, A., The ellipsoid method and its consequences in combinatorial optimization, *Combinatorica*, 1, 169–197, 1982.

35. Grötschel, M., Lovász, L., and Schrijver, A., *Geometric Algorithms and Combinatorial Optimization*, Springer-Verlag, New York, 1988.

36. Greenberg, H.J., *A Computer-Assisted Analysis System for Mathematical Programming Models and Solutions: A User's Guide for ANALYZE*, Kluwer Academic Publishers, Boston, MA, 1993.

37. Golub, G.B. and van Loan, C.F., *Matrix Computations*, The Johns Hopkins University Press, Baltimore, MD, 1983.

38. Hačijan, L.G., A polynomial algorithm in linear programming, *Soviet Mathematics Doklady*, 20, 191–194, 1979.

39. Haimovich, M., The simplex method is very good! On the expected number of pivot steps and related properties of random linear programs, Unpublished Manuscript, 1983.

40. Hochbaum, D. and Naor, J., Simple and fast algorithms for linear and integer programs with two variables per inequality. In *Proceedings of the Symposium on Discrete Algorithms (SODA)*, 1992. SIAM Press (also in the *Proceedings of the Second Conference on Integer Programming and Combinatorial Optimization IPCO*, Pittsburgh, PA, Jun. 1992).

41. Huynh, T., Lassez, C., and Lassez, J.-L., Practical issues on the projection of polyhedral sets, *Annals of Mathematics and Artificial Intelligence*, 6, 295–316, 1992.

42. IBM, *Optimization Subroutine Library—Guide and Reference (Release 2)*, 3rd ed., 1991.

43. Itai, A., Two-Commodity Flow, *Journal of the ACM*, 25, 596–611, 1978.

44. Kalai, G. and Kleitman, D.J., A quasi-polynomial bound for the diameter of graphs of polyhedra, *Bulletin of the American Mathematical Society*, 26, 315–316, 1992.

45. Kantorovich, L.V., Mathematical methods of organizing and planning production (in Russian), Publication House of the Leningrad State University, Leningrad, 1939; English translation in *Management Science*, 6, 366–422, 1959.

46. Kamath, A. and Karmarkar, N.K., A continuous method for computing bounds in integer quadratic optimization problems, *Journal of Global Optimization*, 2, 229–241, 1992.

47. Karmarkar, N.K., A new polynomial-time algorithm for linear programming, *Combinatorica*, 4, 373–395, 1984.

48. Karp, R.M. and Papadimitriou, C.H., On linear characterizations of combinatorial optimization problems, *SIAM Journal on Computing*, 11, 620–632, 1982.

49. Klee, V. and Minty, G.J., How good is the simplex algorithm? In *Inequalities III*, O. Shisha, Ed., Academic Press, New York, 1972.

50. Lenstra, J.K., Rinnooy Kan, A.H.G., and Schrijver, A., Eds., *History of Mathematical Programming: A Collection of Personal Reminiscences*, North Holland, Amsterdam, the Netherlands, 1991.

51. Lovász, L., *An Algorithmic Theory of Numbers, Graphs and Convexity*, SIAM Press, Philadelphia, PA, 1986.

52. Lovász, L. and Schrijver, A., Cones of matrices and setfunctions, *SIAM Journal on Optimization*, 1, 166–190, 1991.

53. Lustig, I.J., Marsten, R.E., and Shanno, D.F., Interior point methods for linear programming: Computational state of the art, *ORSA Journal on Computing*, 6(1), 1–14, 1994.

54. Martin, R.K., Using separation algorithms to generate mixed integer model reformulations, *Operations Research Letters*, 10, 119–128, 1991.

55. Matousek, J., Sharir, M., and Welzl, E., A subexponential bound for linear programming. In *Proceedings of the Eighth Annual ACM Symposium on Computational Geometry*, ACM Press, New York, pp. 1–8, 1992.

56. Megiddo, N., On finding primal- and dual-optimal bases, *ORSA Journal on Computing*, 3, 63–65, 1991.

57. Megiddo, N., Towards a genuinely polynomial algorithm for linear programming, *SIAM Journal on Computing*, 12, 347–353, 1983.

58. Megiddo, N., Linear programming in linear time when the dimension is fixed, *JACM*, 31, 114–127, 1984.

59. Megiddo, N., Improved asymptotic analysis of the average number of steps performed by the self-dual simplex method, *Mathematical Programming*, 35(2), 140–172, 1986.

60. Mehrotra, S., On the implementation of a primal-dual interior point method, *SIAM Journal on Optimization*, 2(4), 575–601, 1992.

61. Motwani, R. and Raghavan, P., *Randomized Algorithms*, Cambridge University Press, New York, 1996.

62. Murtagh, B.A., *Advanced Linear Programming: Computation and Practice*, McGraw-Hill, New York, 1981.

63. Murty, K.G., *Linear Programming*, John Wiley & Sons, New York, 1983.

64. Nelson, C.G., An $O(n^{\log n})$ algorithm for the two-variable-per-constraint linear programming satisfiablity problem, Technical Report AIM-319, Department of Computer Science, Stanford University, Palo Alto, CA, 1978.

65. Orponen, P., Neural networks and complexity theory. In *Proceedings of the 17th International Symposium on Mathematical Foundations of Computer Science*, I.M. Havel and V. Koubek, Eds., *Lecture Notes in Computer Science* 629, Springer-Verlag, London, U.K., pp. 50–61, 1992.

66. Padberg, N.W., *Linear Optimization and Extensions*, Springer-Verlag, Berlin, Germany, 1995.

67. Padberg, M.W. and Rao, M.R., The Russian method for linear inequalities, Part III, Bounded integer programming, Preprint, New York University, New York, 1981.

68. Papadimitriou, C.H. and Steiglitz, K., *Combinatorial Optimization: Algorithms and Complexity*, Prentice-Hall, Englewood Cliffs, NJ, 1982.

69. Papadimitriou, C.H. and Yannakakis, M., Optimization, approximation, and complexity classes, *Journal of Computer and Systems Sciences*, 43, 425–440, 1991.
70. Saigal, R., *Linear Programming: A Modern Integrated Analysis*, Kluwer Press, Boston, MA, 1995.
71. Saltzman, M.J., Implementation of an interior point LP algorithm on a shared-memory vector multiprocessor. In *Computer Science and Operations Research*, O. Balci, R. Sharda and S.A. Zenios, Eds., Pergamon Press, Oxford, U.K., pp. 87–104, 1992.
72. Schrijver, A., *Theory of Linear and Integer Programming*, John Wiley & Sons, Chichester, U.K., 1986.
73. Schrijver, A., *Combinatorial Optimization: Polyhedra and Efficiency*, Springer, Berlin, Germany, 2003.
74. Sen, S., Parallel multidimensional search using approximation algorithms: with applications to linear-programming and related problems, *Proceedings of SPAA*, ACM, New York, 1996.
75. Sharda, R., Linear programming solver software for personal computers: 1995 report, *OR/MS Today*, 22(5), 49–57, 1995.
76. Sharir, M. and Welzl, E., A combinatorial bound for linear programming and related problems. In *Proceedings of the Ninth Symposium on Theoretical Aspects of Computer Science, Lecture Notes in Computer Science 577*, Springer-Verlag, Berlin, Germany, pp. 569–579, 1992.
77. Shmoys, D.B., Computing near-optimal solutions to combinatorial optimization problems, in [17], cited above, 355–398, 1995.
78. Shor, N.Z., Convergence rate of the gradient descent method with dilation of the space, *Cybernetics*, 6, 1970.
79. Smale, S., On the average number of steps in the simplex method of linear programming, *Mathematical Programming*, 27, 241–262, 1983.
80. Spielman, D.A. and Teng, S-H., Smoothed analysis of algorithms: Why the simplex method usually takes polynomial time. In *Proceedings of the 33rd Annual ACM Symposium on Theory of Computing*, 296-305, 2001.
81. Stone, R.E. and Tovey, C.A., The simplex and projective scaling as iterated reweighting least squares, *SIAM Review*, 33, 220–237, 1991.
82. Tardos, E., A strongly polynomial algorithm to solve combinatorial linear programs, *Operations Research*, 34, 250–256, 1986.
83. Todd, M.J., Exploiting special structure in Karmarkar's algorithm for linear programming, Technical Report 707, School of Operations Research and Industrial Engineering, Cornell University, Ithaca, NY, Jul. 1986.
84. Weyl, H., Elemetere Theorie der konvexen polyerer, *Comm. Math. Helv.*, 1, 3–18, 1935. (English translation in *Annals of Mathematics Studies*, 24, 1950.)
85. Wright, S.J., *Primal-Dual Interior-Point Methods*, SIAM Press, Philadelphia, PA, 1997.
86. Yannakakis, M., Expressing combinatorial optimization problems by linear programs. In *Proceedings of ACM Symposium of Theory of Computing*, pp. 223–228, 1988.
87. Ziegler, M., *Convex Polytopes*, Springer-Verlag, New York, 1995.

31

Integer Programming*

Vijay Chandru
National Institute of Advanced Studies

M.R. Rao
Indian School of Business

Integer programming is an expressive framework for modeling and solving discrete optimization problems that arise in a variety of contexts in the engineering sciences. Integer programming representations work with implicit algebraic constraints (linear equations and inequalities on integer valued variables) to capture the feasible set of alternatives, and linear objective functions (to minimize or maximize over the feasible set) that specify the criterion for defining optimality. This algebraic approach permits certain natural extensions of the powerful methodologies of linear programming to be brought to bear on combinatorial optimization and on fundamental algorithmic questions in the geometry of numbers.

---

* Readers unfamiliar with linear programming methodology are strongly encouraged to consult Chapter 30 on linear programming in this handbook.

31.1   Introduction

In 1957 the Higgins lecturer of mathematics at Princeton, Ralph Gomory, announced that he would lecture on solving **linear programs** in integers. The immediate reaction he received was "But that's impossible!" This was his first indication that others had thought about the problem [58]. Gomory went on to work on the foundations of the subject of integer programming as a scientist at IBM from 1959 until 1970 (when he took over as Director for Research). From **cutting planes** and the polyhedral combinatorics of corner polyhedra to group knapsack **relaxations**, the approaches that Gomory developed remain striking to researchers even today.

There were other pioneers in integer programming who collectively played a similar role in developing techniques for linear programming in Boolean or 0–1 variables. These efforts were directed at combinatorial optimization problems such as routing, scheduling, layout, and network design. These are generic examples of combinatorial optimization problems that often arise in computer engineering and decision support.

Unfortunately, almost all interesting generic classes of integer programming problems are $\mathcal{NP}$-hard. The scale at which these problems arise in applications and the explosive exponential complexity of the search spaces preclude the use of simplistic enumeration and search techniques. Despite the worst-case intractability of integer programming, in practice we are able to solve many large problems and often enough with off-the-shelf software. Effective software for integer programming is usually problem specific and based on sophisticated algorithms that combine approximation methods with search schemes and that exploit mathematical (and not just syntactic) structure in the problem at hand.

An abstract formulation of combinatorial optimization is

$$(CO) \quad \min\{f(I) : I \in \mathcal{I}\}$$

where
  $\mathcal{I}$ is a collection of subsets of a finite ground set $E = \{e_1, e_2, \ldots, e_n\}$
  f is a criterion (objective) function that maps 2^E (the power set of E) to the reals

The most general form of an integer linear program is

$$(MILP) \quad \min_{\mathbf{x} \in \Re^n} \left\{ \mathbf{cx} : A\mathbf{x} \geq \mathbf{b},\ x_j \text{ integer } \forall j \in J \right\}$$

which seeks to minimize a linear function of the decision vector $\mathbf{x}$ subject to linear inequality constraints and the requirement that a subset of the decision variables are integer valued. This model captures many variants. If $J = \{1, 2, \ldots, n\}$, we say that the integer program is pure, and mixed otherwise. Linear equations and bounds on the variables can be easily accommodated in the inequality constraints. Notice that by adding in inequalities of the form $0 \leq x_j \leq 1$ for a $j \in J$ we have forced x_j to take value 0 or 1. It is such Boolean variables that help capture combinatorial optimization problems as special cases of (MILP).

Section 31.2 contains preliminaries on linear inequalities, polyhedra, linear programming, and an overview of the complexity of integer programming. These are the tools we will need to analyze and solve integer programs. Section 31.3 is the testimony on how integer programs model combinatorial optimization problems. In addition to working a number of examples of such integer programming formulations, we shall also review formal representation theories of (Boolean) **mixed integer linear programs**.

With any mixed integer program we associate a `linear programming relaxation` obtained by simply ignoring the integrality restrictions on the variables. The point being, of course, that we have polynomial-time (and practical) algorithms for solving linear programs (see Chapter 30 of this handbook). Thus the linear programming relaxation of (MILP) is given by

$$(LP) \quad \min_{\mathbf{x} \in \Re^n} \{\mathbf{cx} : A\mathbf{x} \geq \mathbf{b}\}$$

The thesis underlying the integer linear programming approaches is that this linear programming relaxation retains enough of the structure of the combinatorial optimization problem to be a useful weak representation. In Section 31.4 we shall take a closer look at this thesis in that we shall encounter special structures for which this relaxation is tight. For general integer programs, there are several alternate schemes for generating linear programming relaxations with varying qualities of approximation. A general technique for improving the quality of the linear programming relaxation is through the generation of valid inequalities or cutting planes.

The computational art of integer programming rests on useful interplays between search methodologies and algebraic relaxations. The paradigms of branch and bound and branch and cut are the two enormously effective partial enumeration schemes that have evolved at this interface. These will be discussed in Section 31.5. It may be noted that all general purpose integer programming software available today use one or both of these paradigms.

A general principle is that we often need to disaggregate integer formulations to obtain higher quality linear programming relaxations. To solve such huge linear programs we need specialized techniques of large-scale linear programming. These aspects are described in Chapter 30 in this handbook. The reader should note that the focus in this chapter is on solving hard combinatorial optimization problems. We catalog several special structures in integer programs that lead to tight linear programming relaxations (Section 31.6) and hence to polynomial-time algorithms. These include structures such as network flows, matching, and matroid optimization problems. Many hard problems actually have pieces of these nice structures embedded in them. Successful implementations of combinatorial optimization have always used insights from special structures to devise strategies for hard problems.

The inherent complexity of integer linear programming has led to a long-standing research program in approximation methods for these problems. Linear programming relaxation and Lagrangean relaxation (Section 31.6) are two general approximation schemes that have been the real workhorses of computational practice. Primal–dual strategies and semidefinite relaxations (Section 31.7) are two recent entrants that appear to be very promising.

Pure integer programming with variables that take arbitrary integer values is a natural extension of diophantine equations in number theory. Such problems arise in the context of cryptography, dependence analysis in programs, the geometry of numbers, and Presburgher arithmetic. Section 31.8 covers this aspect of integer programming.

We conclude the chapter with brief comments on future prospects in combinatorial optimization from the algebraic modeling perspective.

31.2 Preliminaries

31.2.1 Polyhedral Preliminaries

Polyhedral combinatorics is the study of embeddings of combinatorial structures in Euclidean space and their algebraic representations. We will make extensive use of some standard terminology from polyhedral theory. Definitions of terms not given in the brief review below can be found in [98,127].

A (convex) **polyhedron** in $\mathfrak{R}^n$ can be algebraically defined in two ways. The first and more straightforward definition is the implicit representation of a polyhedron in $\mathfrak{R}^n$ as the solution set to a finite system of linear inequalities in n variables. A single linear inequality $\mathbf{a}\mathbf{x} \leq a_0$; $a \neq 0$ defines a half-space of $\mathfrak{R}^n$. Therefore geometrically a polyhedron is the intersection set of a finite number of half-spaces.

A polytope is a bounded polyhedron. Every polytope is the convex closure of a finite set of points. Given a set of points whose convex combinations generate a polytope we have an explicit or parametric algebraic representation of it. A polyhedral cone is the solution set of a system of homogeneous linear inequalities. Every (polyhedral) cone is the conical or positive closure of a finite set of vectors. These generators of the cone provide a parametric representation of the cone. And finally a polyhedron can be alternately defined as the Minkowski sum of a polytope and a cone. Moving from one representation of any of these polyhedral objects to another defines the essence of the computational burden of polyhedral combinatorics. This is particularly true if we are interested in minimal representations.

A set of points $\mathbf{x}^1, \ldots, \mathbf{x}^m$ is affinely independent if the unique solution of $\Sigma_{i=1}^m \lambda_i \mathbf{x}^i = \mathbf{0}$, $\Sigma_{i=1}^m \lambda_i = 0$ is $\lambda_i = 0$ for $i = 1, \ldots, m$. Note that the maximum number of affinely independent points in $\mathfrak{R}^n$ is $n + 1$. A polyhedron P is of dimension k, $\dim P = k$, if the maximum number of affinely independent points in P is $k + 1$. A polyhedron $P \subseteq \mathfrak{R}^n$ of dimension n is called full-dimensional.

An inequality $\mathbf{a}\mathbf{x} \leq a_0$ is called valid for a polyhedron P if it is satisfied by all $\mathbf{x}$ in P. It is called supporting if in addition there is an $\tilde{\mathbf{x}}$ in P that satisfies $\mathbf{a}\tilde{\mathbf{x}} = a_0$. A face of the polyhedron is the set of all $\mathbf{x}$ in P that also satisfy a valid inequality as an equality. In general, many valid inequalities might represent the same face. Faces other than P itself are called proper. A facet of P is a maximal nonempty and proper face. A facet is then a face of P with a dimension of $\dim P - 1$. A face of dimension zero, i.e., a point $\mathbf{v}$ in P that is a face by itself, is called an **extreme point** of P. The extreme points are the elements of P that cannot be expressed as a strict convex combination of two distinct points in P. For a full-dimensional polyhedron, the valid inequality representing a facet is unique up to multiplication by a positive scalar, and facet-inducing inequalities give a minimal implicit representation of the polyhedron. Extreme points, on the other hand, give rise to minimal parametric representations of polytopes.

The two fundamental problems of linear programming (which are polynomially equivalent) are

- **Solvability.** This is the problem of checking if a system of linear constraints on real (rational) variables is solvable or not. Geometrically, we have to check if a polyhedron, defined by such constraints, is nonempty.
- **Optimization.** This is the problem (LP) of optimizing a linear objective function over a polyhedron described by a system of linear constraints.

Building on polarity in cones and polyhedra, duality in linear programming is a fundamental concept that is related to both the complexity of linear programming and to the design of algorithms for solvability and optimization. Here we will state the main duality results for optimization. If we take the primal linear program to be

$$(P) \quad \min_{\mathbf{x} \in \mathfrak{R}^n} \{\mathbf{c}\mathbf{x} : A\mathbf{x} \geq \mathbf{b}\}$$

there is an associated dual linear program

$$(D) \quad \max_{\mathbf{y} \in \mathfrak{R}^m} \{\mathbf{b}^T \mathbf{y} : A^T y = \mathbf{c}^T, \mathbf{y} \geq \mathbf{0}\}$$

and the two problems satisfy the following:

1. For any $\hat{\mathbf{x}}$ and $\hat{\mathbf{y}}$ feasible in (P) and (D) (i.e., they satisfy the respective constraints), we have $\mathbf{c}\hat{\mathbf{x}} \geq \mathbf{b}^T\hat{\mathbf{y}}$ (weak duality). Consequently, (P) has a finite optimal solution if and only if (D) does.

2. $\mathbf{x}^*$ and $\mathbf{y}^*$ are a pair of optimal solutions for (P) and (D), respectively, if and only if $\mathbf{x}^*$ and $\mathbf{y}^*$ are feasible in (P) and (D) (i.e., they satisfy the respective constraints) **and** $\mathbf{c}\mathbf{x}^* = \mathbf{b}^T\mathbf{y}^*$ (strong duality).

3. $\mathbf{x}^*$ and $\mathbf{y}^*$ are a pair of optimal solutions for (P) and (D), respectively, if and only if $\mathbf{x}^*$ and $\mathbf{y}^*$ are feasible in (P) and (D) (i.e., they satisfy the respective constraints) and $(A\mathbf{x}^* - \mathbf{b})^T\mathbf{y}^* = 0$ (complementary slackness).

The strong duality condition gives us a good stopping criterion for optimization algorithms. The complementary slackness condition, on the other hand gives us a constructive tool for moving from dual to primal solutions and vice versa. The weak duality condition gives us a technique for obtaining lower bounds for minimization problems and upper bounds for maximization problems.

Note that the properties above have been stated for linear programs in a particular form. The reader should be able to check, that if for example the primal is of the form

$$(P') \quad \min_{\mathbf{x} \in \Re^n}\{\mathbf{c}\mathbf{x} : A\mathbf{x} = \mathbf{b}, \mathbf{x} \geq 0\}$$

then the corresponding dual will have the form

$$(D') \quad \max_{\mathbf{y} \in \Re^m}\left\{\mathbf{b}^T\mathbf{y} : A^T\mathbf{y} \leq \mathbf{c}^T\right\}$$

The tricks needed for seeing this is that any equation can be written as two inequalities, an unrestricted variable can be substituted by the difference of two nonnegatively constrained variables and an inequality can be treated as an equality by adding a nonnegatively constrained variable to the lesser side. Using these tricks, the reader could also check that duality in linear programming is involutory (i.e., the dual of the dual is the primal).

31.2.2 Linear Diophantine Systems

Let us first examine the simple case of solving a single equation with integer (rational) coefficients and integer variables.

$$a_1 x_1 + a_2 x_2 + \cdots + a_n x_n = b \tag{31.1}$$

A classical technique is to use the Euclidean algorithm to eliminate variables. Consider the first iteration in which we compute a_{12}, and the integers δ_1 and δ_2 where

$$a_{12} = gcd(a_1, a_2) = \delta_1 a_1 + \delta_2 a_2$$

Now we have a reduced equation that is equivalent to Equation 31.1.

$$a_{12}x_{12} + a_3 x_3 + \cdots + a_n x_n = b \tag{31.2}$$

It is apparent that integer solutions to Equation 31.2 are linear projections of integer solutions to Equation 31.1. However, it is not a simple elimination of a variable as happens in the case of equations over reals. It is a projection to a space whose dimension is one less than the dimension we began with. The solution scheme reduces the equation to a univariate equation and then inverts the projection maps. All of this can be accomplished in polynomial time since the Euclidean algorithm is good.

Solving a system of linear diophantine equations $A\mathbf{x} = \mathbf{b}$, $\mathbf{x} \in Z^n$ now only requires a matrix version of the simple scheme described above. An integer matrix, of full row rank, is said to be in Hermite normal form if it has the appearance $[L|0]$ where L is nonsingular and lower triangular with nonnegative entries satisfying the condition that the largest entry of each row is on the main diagonal. A classical result (cf. [116]) is that any integer matrix A with full row rank has a unique Hermite normal form $\mathrm{HNF}(A) = AK = [L|0]$ where K is a square unimodular matrix (an integer matrix with determinant ± 1).

The matrix K encodes the composite effect of the elementary column operations on the matrix A needed to bring it to normal form. The elementary operations are largely defined by repeated invocation of the Euclidean algorithm in addition to column swaps and subtractions. Polynomial-time computability of Hermite normal forms of integer matrices was first proved by Kannan and Bachem [78] using delicate and complicated analysis of the problems of intermediate swell. Subsequently, a much easier argument based on modulo arithmetic was given by Domich et al. [37]. As consequences, we have that

- Linear Diophantine systems can be solved in polynomial time. Assuming A has been preprocessed to have full row rank, to solve $A\mathbf{x} = b, \mathbf{x} \in Z^n$ we first obtain $\mathrm{HNF}(A) = AK = [L|0]$. The input system has a solution if and only if $L^{-1}\mathbf{b}$ is integral and if so, a solution is given by $\mathbf{x} = K\left(\begin{smallmatrix} L^{-1}\mathbf{b} \\ 0 \end{smallmatrix}\right)$.

- KRONECKER: (cf. [116]) $A\mathbf{x} = \mathbf{b}, \mathbf{x} \in Z^n$ has no solution if and only if there exists a $\mathbf{y} \in \Re^m$ such that $\mathbf{y}^t A$ is integral and $\mathbf{y}^t \mathbf{b}$ is not. A certificate of unsolvability is always available from the construction described above.

- The solutions to a linear Diophantine system are finitely generated. In fact, a set of generators can be found in polynomial time. $\{\mathbf{x} \in Z^n | A\mathbf{x} = \mathbf{b}\} = \{\mathbf{x}^0 + \Sigma_{i=1}^m \lambda_j \mathbf{x}^j | \lambda_j \in Z\}$, $\mathbf{x}^0 = K\left(\begin{smallmatrix} L^{-1}\mathbf{b} \\ 0 \end{smallmatrix}\right)$ and $[\mathbf{x}^1, \mathbf{x}^2, \ldots, \mathbf{x}^m] = K(0, I)^t$.

In summary, linear Diophantine systems are a lot like linear systems over the reals (rationals). The basic theory and complexity results for variable elimination in both constraint domains are similar. This comfortable situation changes when we move on to linear inequalities over the integers or equivalently to nonnegative solutions to linear Diophantine systems.

31.2.3 Computational Complexity of Integer Programming

Any algorithm for integer programming is a universal polynomial-time decision procedure on a nondeterministic Turing machine. This statement is credible not only because the solvable systems of linear inequalities (with rational coefficients) over integer-valued variables describe an $\mathcal{NP}$-complete language but also because integer programs are expressive enough to capture most decision problems in $\mathcal{NP}$ via straightforward reductions. This expressiveness derives from our ability to embed sentential inference and combinatorial structures with equal ease in integer programs. We will see ample demonstration of this expressiveness in the next section. This relationship of integer programming with $\mathcal{NP}$ is akin to the relationship of linear programming with $\mathcal{P}$.

31.2.3.1 Complexity of Linear Inequalities

From our earlier discussion of polyhedra, we have the following algebraic characterization of extreme points of polyhedra.

THEOREM 31.1 *Given a polyhedron P, defined by $\{\mathbf{x} : A\mathbf{x} \leq \mathbf{b}\}$, where A is $m \times n$, $\mathbf{x}^i$ is an extreme point of P if and only if it is a face of P satisfying $A^i\mathbf{x}^i = \mathbf{b}^i$ where $((A^i), (\mathbf{b}^i))$ is a submatrix of $(A, \mathbf{b})$ and the rank of A^i equals n.*

COROLLARY 31.1 The decision problem of verifying the membership of an input string $(A, \mathbf{b})$ in the language $\mathcal{L}_I = \{(A, \mathbf{b}) : \exists \mathbf{x} \text{ such that } A\mathbf{x} \leq \mathbf{b}\}$ belongs to $\mathcal{NP}$.

PROOF It follows from the theorem that every extreme point of the polyhedron $Q = \{\mathbf{x} : A\mathbf{x} \leq \mathbf{b}\}$ is the solution of an $(n \times n)$ linear system whose coefficients come from $(A, \mathbf{b})$. Therefore we can guess a polynomial length string representing an extreme point and check its membership in Q in polynomial time.

A consequence of Farkas Lemma [46] (duality in linear programming) is that the decision problem of testing membership of input $(A, \mathbf{b})$ in the language

$$\mathcal{L}_I = \{(A, \mathbf{b}) : \exists \mathbf{x} \text{ such that } A\mathbf{x} \leq \mathbf{b}\}$$

is in $\mathcal{NP} \bigcap co\mathcal{NP}$. That $\mathcal{L}_I \in \mathcal{P}$, follows from algorithms for linear programming [80,84]. This is as far down the polynomial hierarchy that we can go, since $\mathcal{L}_I$ is known to be $\mathcal{P}$-complete (that is, complete for $\mathcal{P}$ with respect to log-space reductions).

31.2.3.2 Complexity of Linear Inequalities in Integer Variables

The complexity of integer programming is polynomially equivalent to the complexity of the language

$$\mathcal{L}_{IP} = \{(A, \mathbf{b}) : \exists \mathbf{x} \in Z^n \text{ such that } A\mathbf{x} \leq \mathbf{b}\}$$

We are assuming that the input coefficients in $A, \mathbf{b}$ are integers (rationals). It is very easy to encode Boolean satisfiability as a special case of $\mathcal{L}_{IP}$ with appropriate choice of $A, \mathbf{b}$ as we shall see in the next section. Hence $\mathcal{L}_{IP}$ is $\mathcal{NP}$-hard.

It remains to argue that the decision problem of verifying the membership of an input string $(A, \mathbf{b})$ in the language $\mathcal{L}_{IP}$ belongs to $\mathcal{NP}$. We will have to work a little harder since integer programs may have solutions involving numbers that are large in magnitude. Unlike the case of linear programming there is no extreme point characterization for integer programs. However, since linear diophantine systems are well behaved, we are able to salvage the following technical result that plays a similar role. Let α denote the largest integer in the integer matrices $A, \mathbf{b}$ and let $q = \max\{m, n\}$ where A is $m \times n$.

31.2.3.2.1 Finite Precision Lemma [106]
If B is an $r \times n$ submatrix of A with rank $B < n$ then $\exists$ a nonzero integral vector $\mathbf{z} = (z_1, z_2, \cdots z_n)$ in the null space of B such that $|z_j| \leq (\alpha q)^{q+1} \forall j$.

Repeated use of this lemma shows that if $\{\mathbf{x} \in Z^n : A\mathbf{x} \leq \mathbf{b}\}$ is solvable then there is a polynomial size certificate of solvability and hence that $\mathcal{L}_{IP}$ belongs to $\mathcal{NP}$.

As with any $\mathcal{NP}$-hard problem, researchers have looked for special cases that are polynomial-time solvable. Table 31.1 is a summary of the important complexity classification results in integer programming that have been obtained to date.

TABLE 31.1   Summary of Complexity Results

	Input Data	Generic Problem	Complexity
	Solvability	Does $\exists$ an $\mathbf{x}$ satisfying:	
1.	$n, m, A, \mathbf{b}$	$A\mathbf{x} = \mathbf{b}; \mathbf{x} \geq 0$, integer	$\mathcal{NP}$-complete [17,55,82]
2.	$n, m, A, \mathbf{b}$	$A\mathbf{x} \leq \mathbf{b}; \mathbf{x} \in \{0,1\}^n$	$\mathcal{NP}$-complete [82,113]
3.	$n, m, A, \mathbf{b}, d(> 0)$	$A\mathbf{x} \equiv \mathbf{b} \pmod{d}; \mathbf{x} \geq 0$, integer	$\mathcal{P}$ [6,51]
4.	$n, m, A, \mathbf{b}$	$A\mathbf{x} = \mathbf{b};\mathbf{x}$ integer	$\mathcal{P}$ [6,51]
5.	$n, m, A, \mathbf{b}$	$A\mathbf{x} = \mathbf{b}; \mathbf{x} \geq 0$	$\mathcal{P}$ [80,84]
6.	$n, (m = 1), (A \geq 0), (\mathbf{b} \geq 0)$	$A\mathbf{x} = \mathbf{b}; \mathbf{x} \geq 0$, integer	$\mathcal{NP}$-complete [113]
7.	$n, (m = 2), (A \geq 0), (\mathbf{b} \geq 0)$	$a^1\mathbf{x} \geq b_1; a^2\mathbf{x} \leq b_2; \mathbf{x} \geq 0$, integer	$\mathcal{NP}$-complete [79]
8.	$(n = k), m, A, \mathbf{b}$	$A\mathbf{x} = \mathbf{b}; \mathbf{x} \geq 0$, integer	$\mathcal{P}$ [90]
9.	$n, (m = k), A, \mathbf{b}$	$A\mathbf{x} \leq \mathbf{b};\mathbf{x}$ integer	$\mathcal{P}$ [90]
		Find an $\mathbf{x}$ that	
	Optimization	maximizes $\mathbf{cx}$ subject to:	
10.	$n, m, A, \mathbf{b}, \mathbf{c}$	$A\mathbf{x} = \mathbf{b}; \mathbf{x} \geq 0$, integer	$\mathcal{NP}$-hard [55,82]
11.	$n, m, A, \mathbf{b}, \mathbf{c}$	$A\mathbf{x} \leq \mathbf{b}; \mathbf{x} \in \{0,1\}^n$	$\mathcal{NP}$-hard [82,113]
12.	$n, (m = 1), (A \geq 0), (\mathbf{b} \geq 0), \mathbf{c}$	$A\mathbf{x} = \mathbf{b}; \mathbf{x} \geq 0$, integer	$\mathcal{NP}$-hard [113]
13.	$n, (m = 1), (A \geq 0), (\mathbf{b} \geq 0), \mathbf{c}$	$A\mathbf{x} \leq \mathbf{b}; \mathbf{x} \in \{0,1\}^n$	$\mathcal{NP}$-hard [113]
14.	$n, m, A, \mathbf{b}, \mathbf{c}, (d_i \geq 2\forall i)$	$A\mathbf{x} \equiv \mathbf{b} \pmod{\mathbf{d}}; \mathbf{x} \geq 0$, integer	$\mathcal{NP}$-hard [21]
15.	$(n = k), m, A, \mathbf{b}, \mathbf{c}$	$A\mathbf{x} = \mathbf{b}; \mathbf{x} \geq 0$, integer	$\mathcal{P}$ [90]
16.	$n, (m = k), A, \mathbf{b}, \mathbf{c}$	$A\mathbf{x} \leq \mathbf{b};\mathbf{x}$ integer	$\mathcal{P}$ [90]
17.	$n, m, A, \mathbf{b}, \mathbf{c}$	$A\mathbf{x} = \mathbf{b}; \mathbf{x} \geq 0$	$\mathcal{P}$ [80,84]
18.	$n, m, (A \text{ is graphic})^a, \mathbf{b}_1, \mathbf{b}_2, \mathbf{d}_1, \mathbf{d}_2$	$\mathbf{d}_1 \leq \mathbf{x} \leq \mathbf{d}_2, \mathbf{b}_1 \leq A\mathbf{x} \leq \mathbf{b}_2, \mathbf{x} \in Z^n$	$\mathcal{P}$ [45,104]

^a A is a graphic matrix if it has entries from $\{0, \pm1, \pm2\}$ such that the sum of absolute values of the entries in any column is at most 2.

31.3   Integer Programming Representations

We will first discuss several examples of combinatorial optimization problems and their formulation as integer programs. Then we will review a general representation theory for integer programs that gives a formal measure of the expressiveness of this algebraic approach. We conclude this section with a representation theorem due to Benders [11] that has been very useful in solving certain large-scale combinatorial optimization problems in practice.

31.3.1   Formulations

Formulating decision problems as integer or mixed integer programs is often considered an art form. However, there are a few basic principles which can be used by a novice to get started. As in all art forms though, principles can be violated to creative effect. We list below a number of example formulations, the first few of which may be viewed as principles for translating logical conditions into models.

1. Discrete Choice

CONDITION   MODEL

$X \in \{s_1, s_2, \ldots, s_p\}$ $X = \sum_{j=1}^{p} s_j \, \delta_j$

$\qquad\qquad\qquad\qquad\quad \sum_{j=1}^{p} \delta_j = 1, \ \delta_j = 0 \text{ or } 1 \ \forall j$

2. Dichotomy

CONDITION   MODEL

$g(\mathbf{x}) \geq 0$ $g(\mathbf{x}) \geq \delta\, \beta g$ βg and βh are
or   finite lower
$h(\mathbf{x}) \geq 0$ $h(\mathbf{x}) \geq (1-\delta)\, \beta h$ bounds on g, h,
or both $\delta = 0$ or 1 respectively.

3. κ-Fold Alternatives

condition   model

at least k of $g_i(\mathbf{x}) \geq \delta_i\, \beta g_i \quad i = 1, \dots, m$
$g_i(\mathbf{x}) \geq 0, i = 1, \dots, m$ $\sum_{i=1}^{m} \delta_i \leq m - k$
must hold $\delta_i = 0$ or 1

4. Conditional Constraints

condition   model

$(f(\mathbf{x}) > 0 \;\Rightarrow\; g(\mathbf{x}) \geq 0)$ $g(\mathbf{x}) \geq \delta\beta g$ $\bar{f}/\beta g$ is an
$\Updownarrow$ $f(\mathbf{x}) \leq (1-\delta)\bar{f}$ upper/lower bound
$(f(\mathbf{x}) \leq 0$ or $g(\mathbf{x}) \geq 0$ or *both*$)$ $\delta = 0$ or 1 on f/g

5. Fixed Charge Models

condition   model

$f(x) = 0$ if $x = 0$ $f(x) = Ky + cx$
$f(x) = K + cx$ if $x > 0$ $x \leq Uy$ U an upper bound
                                 $x \geq 0$ on x
                                 $y = 0$ or 1

6. Piecewise Linear Models (Figure 31.1)

condition   model

                                 $f(x) = 5\delta_1 + \delta_2 + 3\delta_3$
$\delta_2 > 0 \Rightarrow \delta_1 = 4$ $4W_1 \leq \delta_1 \leq 4$
$\delta_3 > 0 \Rightarrow \delta_2 = 6$ $6W_2 \leq \delta_2 \leq 6W_1$
                                 $0 \leq \delta_3 \leq 5W_2$
                                 $W_1, W_2 = 0$ or 1

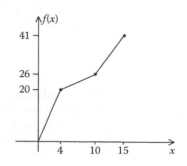

FIGURE 31.1   A polyline.

7. Capacitated Plant Location Model

$i = \{1, 2, \dots, m\}$ possible locations for
                                plants
$j = \{1, 2, \dots, n\}$ demand sites
$k_i = $ capacity of plant i, if opened
$f_i = $ fixed cost of opening plant i
$c_{ij} = $ per unit production cost at i plus
         transportion cost from i to j
$d_j = $ demand at location j

Choose plant locations so as to minimize total cost and meet all demands.
Formulation:

$$\min \sum_i \sum_j c_{ij} x_{ij} + \sum_i f_i y_i$$
$$s.t. \quad \sum_i x_{ij} \geq d_j \ \forall j$$
$$\sum_j x_{ij} \leq k_i y_i \ \forall i$$
$$x_{ij} \geq 0 \quad \forall i, j$$
$$y_i = 0 \text{ or } 1 \ \forall i$$

If the demand d_j is less than the capacity k_i for some ij combination, it is useful to add the constraint $x_{ij} \leq d_j y_i$ to improve the quality of the linear programming relaxation.

8. Traveling Salesman Problem (Alternate Formulations): A recurring theme in integer programming is that the same decision problem can be formulated in several different ways. Principles for sorting out the better ones have been the subject of some discourse [124]. We now illustrate this with the well-known traveling salesman problem. Given a complete directed graph D(N,A) with distance c_{ij} of arc (i, j), we are to find the minimum length tour beginning at node 1 and visiting each node of D(N,A) exactly once before returning to the start node 1.

   Formulation 1

$$\min \sum_{i=1}^{n} \sum_{j=1}^{n} c_{ij} x_{ij}$$
$$s.t. \quad \sum_{j=1}^{n} x_{ij} = 1 \ \forall i$$
$$\sum_{i=1}^{n} x_{ij} = 1 \ \forall j$$
$$\sum_{i \in \phi} \sum_{j \notin \phi} x_{ij} \geq 1 \ \forall \ \phi \subset N$$
$$x_{ij} = 0 \text{ or } 1 \quad \forall \ ij$$

   Formulation 2

$$\min \sum_{i=1}^{n} \sum_{j=1}^{n} c_{ij} x_{ij}$$
$$s.t. \quad \sum_{j=1}^{n} x_{ij} = 1 \ \forall i$$
$$\sum_{i=1}^{n} x_{ij} = 1 \ \forall j$$
$$\sum_{j=1}^{n} y_{ji} - \sum_{j=2}^{n} y_{ij} = 1 \ \forall i \neq 1$$
$$y_{ij} \leq (n-1) x_{ij} \ \ i = 1, 2, \ldots, n$$
$$j = 2, \ldots, n$$
$$x_{ij} = 0 \text{ or } 1 \quad \forall ij$$
$$y_{ij} \geq 0 \quad \forall ij$$

9. Nonlinear Formulations: If we allow arbitrary nonlinear objective and constraint functions, the general integer programming problem is closely related to Hilbert's tenth problem and is undecidable [72]. However, when the integer variables are restricted to 0–1, the problem is of the form

$$(NIP) \quad \min \left\{ f(\mathbf{x}) \mid g_i(\mathbf{x}) \leq 0, \ i = 1, \ldots m, \ \mathbf{x} \in \{0, 1\}^n \right\}$$

and we can capture a rich variety of decision problems (modular design, cluster analysis, capital budgeting under uncertainty, and production planning in flexible manufacturing systems, to name a few). Restricting the constraints to linear assignment constraints while allowing quadratic cost functions yields the quadratic assignment problem (QAP)

$$\min \sum_{i \neq j} \sum_{k \neq l} c_{ijkl} x_{ik} x_{jl}$$
$$s.t. \quad \sum_i x_{ik} = 1 \ \forall k$$
$$\sum_k x_{ik} = 1 \ \forall i$$
$$x_{ik} = 0 \text{ or } 1 \ \forall ik$$

which includes the traveling salesman problem and plant location problems as special cases. Karmarkar [81] has advocated solving integer programs, by first formulating them as minimizing an indefinite quadratic function over a polyhedral region, and then solving the continuous model using interior point methods.

The other side of the coin is that these problems are extremely hard to solve, and the most successful strategies to date for these problems are via linearization techniques and semidefinite relaxations.

10. Covering and Packing Problems: A wide variety of location and scheduling problems can be formulated as set covering or set packing or set partitioning problems. The three different types of covering and packing problems can be succinctly stated as follows: Given

   (a) A finite set of elements $\mathcal{M} = \{1, 2, \ldots, m\}$

   (b) A family F of subsets of $\mathcal{M}$ with each member F_j, $j = 1, 2, \ldots, n$ having a profit (or cost) c_j associated with it

find a collection, S, of the members of F that maximizes the profit (or minimizes the cost) while ensuring that every element of $\mathcal{M}$ is in

(P1): At most one member of S (set packing problem),

(P2): At least one member of S (set covering problem), and

(P3): Exactly one member of S (set partitioning problem)

The three problems (P1), (P2), and (P3) can be formulated as integer linear programs as follows:

Let A denote the $m \times n$ matrix where

$$A_{ij} = \begin{cases} 1 & \text{if element } i \in F_j \\ 0 & \text{otherwise} \end{cases}$$

The decision variables are x_j, $j = 1, 2, \ldots, n$ where

$$x_j = \begin{cases} 1 & \text{if } F_j \text{ is chosen} \\ 0 & \text{otherwise} \end{cases}$$

The set packing problem is

$$(P1) \qquad \text{Max} \qquad c\mathbf{x}$$
$$\text{Subject to} \qquad A\mathbf{x} \leq \mathbf{e}_m$$
$$x_j = 0 \ or 1, \quad j = 1, 2, \ldots, n$$

where $\mathbf{e}_m$ is an m-dimensional column vector of 1s.

The set covering problem (P2) is (P1) with less than or equal to constraints replaced by greater than or equal to constraints and the objective is to minimize rather than maximize. The set partitioning problem (P3) is (P1) with the constraints written as equalities. The set partitioning problem can be converted to a set packing problem or a set covering problem (see [101]) using standard transformations. If the right-hand side vector $\mathbf{e}_m$ is replaced by a nonnegative integer vector b, (P1) is referred to as the generalized set packing problem.

The airline crew scheduling problem is a classic example of the set partitioning or the set covering problem. Each element of $\mathcal{M}$ corresponds to a flight segment. Each subset

F_j corresponds to an acceptable set of flight segments of a crew. The problem is to cover, at minimum cost, each flight segment exactly once. This is a set partitioning problem. If dead heading of crew is permitted, we have the set covering problem.

11. Packing and Covering Problems in a Graph: Suppose A is the node-edge incidence matrix of a graph. Now, (P1) is a weighted matching problem. If in addition, the right-hand side vector e_m is replaced by a nonnegative integer vector b, (P1) is referred to as a weighted b-matching problem. In this case, each variable x_j that is restricted to be an integer may have a positive upper bound of u_j. Problem (P2) is now referred to as the weighted edge covering problem. Note that by substituting for $x_j = 1 - y_j$, where $y_j = 0 \ or \ 1$, the weighted edge covering problem is transformed to a weighted b-matching problem in which the variables are restricted to be 0 or 1.

    Suppose A is the edge-node incidence matrix of a graph. Now, (P1) is referred to as the weighted vertex packing problem and (P2) is referred to as the weighted vertex covering problem. It is easy to see that the weighted vertex packing problem and the weighted vertex covering problem are equivalent in the sense that the complement of an optimal solution to one problem defines an optimal solution to the other. The set packing problem can be transformed to a weighted vertex packing problem in a graph G as follows:

> G contains a node for each x_j and an edge between nodes j and k exists if and only if the columns $A_{\cdot j}$ and $A_{\cdot k}$ are not orthogonal. G is called the intersection graph of A. Given G, the complement graph $\bar{G}$ has the same node set as G, and there is an edge between nodes j and k in $\bar{G}$ if and only if there is no such corresponding edge in G. A clique in a graph is a subset, k, of nodes of G such that the subgraph induced by k is complete. Clearly, the weighted vertex packing problem in G is equivalent to finding a maximum weighted clique in $\bar{G}$.

12. Satisfiability and Inference Problems: In propositional logic, a truth assignment is an assignment of true or false to each atomic proposition $x_1, x_2, \ldots, x_n$. A literal is an atomic proposition x_j or its negation $\neg x_j$. For propositions in conjunctive normal form, a clause is a disjunction of literals and the proposition is a conjunction of clauses. A clause is obviously satisfied by a given truth assignment if at least one of its literals is true. The satisfiability problem consists of determining whether there exists a truth assignment to atomic propositions such that a set S of clauses is satisfied.

    Let T_i denote the set of atomic propositions such that if any one of them is assigned true, the clause $i \in S$ is satisfied. Similarly let F_i denote the set of atomic propositions such that if any one of them is assigned false, the clause $i \in S$ is satisfied. The decision variables are

$$x_j = \begin{cases} 1 & \text{if atomic proposition } j \text{ is assigned true} \\ 0 & \text{if atomic proposition } j \text{ is assigned false} \end{cases}$$

The satisfiability problem is to find a feasible solution to

$$\text{(P4)} \quad \sum_{j \in T_i} x_j - \sum_{j \in F_i} x_j \geq 1 - |F_i| \quad i \in S$$

$$x_j = 0 \text{ or } 1 \quad \text{for } j = 1, 2, \ldots, n$$

By substituting $x_j = 1 - y_j$, where $y_j = 0$ or 1, for $j \in F_i$, (P4) is equivalent to the set covering problem

$$(P5) \qquad \text{Min} \qquad \sum_{j=1}^{n} (x_j + y_j)$$

$$\text{subject to} \quad \sum_{j \in T_i} x_j + \sum_{j \in F_i} y_j \geq 1, \quad i \in S$$

$$x_j + y_j \geq 1, \quad j = 1, 2, \ldots, n$$

$$x_j, y_j = 0 \text{ or } 1, \quad j = 1, 2, \ldots, n$$

Clearly (P4) is feasible if and only if (P5) has an optimal objective function value equal to n.

Given a set S of clauses and an additional clause $k \notin S$, the logical inference problem is to find out whether every truth assignment that satisfies all the clauses in S also satisfies the clause k. The logical inference problem is

$$(P6) \qquad \text{Min} \qquad \sum_{j \in T_k} x_j - \sum_{j \in F_k} x_j$$

$$\text{subject to} \quad \sum_{j \in T_i} x_j - \sum_{j \in F_i} x_j \geq 1 - \mid F_i \mid, \quad i \in S$$

$$x_j = 0 \text{ or } 1, \quad j = 1, 2, \ldots, n$$

The clause k is implied by the set of clauses S, if and only if (P6) has an optimal objective function value greater than $-\mid F_k \mid$. It is also straightforward to express the MAX-SAT problem (i.e., find a truth assignment that maximizes the number of satisfied clauses in a given set S) as an integer linear program.

13. Multiprocessor Scheduling: Given n jobs and m processors, the problem is to allocate each job to one and only one of the processors so as to minimize the make span time, i.e., minimize the completion time of all the jobs. The processors may not be identical and hence job j if allocated to processor i requires p_{ij} units of time. The multiprocessor scheduling problem is

$$(P7) \qquad \text{Min} \qquad T$$

$$\text{subject to} \quad \sum_{i=1}^{m} x_{ij} = 1, \quad j = 1, 2, \ldots, n$$

$$\sum_{j=1}^{n} p_{ij} x_{ij} - T \leq 0, \quad i = 1, 2, \ldots, m$$

$$x_{ij} = 0 \text{ or } 1 \quad \forall \, ij$$

Note that if all p_{ij} are integers, the optimal solution will be such that T is an integer.

31.3.2 Jeroslow's Representability Theorem

Jeroslow [73], building on joint work with Lowe [74], characterized subsets of n-space that can be represented as the feasible region of a mixed integer (Boolean) program. They proved that a set is the feasible region of some mixed integer/linear programming problem (MILP) if and only if it is the union of finitely many polyhedra having the same recession cone (defined below). Although this result is not widely known, it might well be regarded as the fundamental theorem of mixed integer modeling.

The basic idea of Jeroslow's results is that any set that can be represented in a mixed integer model can be represented in a disjunctive programming problem (i.e., a problem with either/or constraints). A recession direction for a set S in n-space is a vector $\mathbf{x}$ such that $\mathbf{s} + \alpha\mathbf{x} \in S$ for all $\mathbf{s} \in S$ and all $\alpha \geq 0$. The set of recession directions is denoted $rec(S)$. Consider the general mixed integer constraint set below.

$$f(\mathbf{x}, \mathbf{y}, \lambda) \leq \mathbf{b}$$
$$\mathbf{x} \in \Re^n, \quad \mathbf{y} \in \Re^p \tag{31.3}$$
$$\lambda = (\lambda_1, \ldots, \lambda_k), \quad \text{with} \quad \lambda_j \in \{0, 1\} \quad \text{for } j = 1, \ldots, k$$

Here $\mathbf{f}$ is vector-valued function, so that $\mathbf{f}(\mathbf{x}, \mathbf{y}, \lambda) \leq \mathbf{b}$ represents a set of constraints. We say that a set $S \subset \Re^n$ is represented by 31.3 if

$$\mathbf{x} \in S \text{ if and only if } (\mathbf{x}, \mathbf{y}, \lambda) \text{ satisfies 31.3 for some } \mathbf{y}, \lambda$$

If $\mathbf{f}$ is a linear transformation, so that 31.3 is a MILP constraint set, we will say that S is *MILP representable*. The main result can now be stated.

THEOREM 31.2   [73,74] *A set in n-space is MILP representable if and only if it is the union of finitely many polyhedra having the same set of recession directions.*

31.3.3   Benders Representation

Any mixed integer linear program (MILP) can be reformulated so that there is only one continuous variable. This reformulation, due to Benders [11], will in general have an exponential number of constraints. Benders representation suggests an algorithm for mixed integer programming (known as Benders Decomposition in the literature because of its similarity to Dantzig–Wolfe Decomposition, cf. [98]) that uses dynamic activation of these rows (constraints) as and when required.

Consider the (MILP)

$$\max\{\mathbf{cx} + \mathbf{dy} : A\mathbf{x} + G\mathbf{y} \leq \mathbf{b}, \mathbf{x} \geq \mathbf{0}, y \geq \mathbf{0} \text{ and integer}\}$$

Suppose the integer variables $\mathbf{y}$ are fixed at some values, then the associated linear program is

$$(LP) \quad \max\{\mathbf{cx} : \mathbf{x} \in \mathcal{P} = \{\mathbf{x} : A\mathbf{x} \leq \mathbf{b} - G\mathbf{y}, \mathbf{x} \geq \mathbf{0}\}\}$$

and its dual is

$$(DLP) \quad \min\{\mathbf{w}(\mathbf{b} - G\mathbf{y}) : \mathbf{w} \in Q = \{\mathbf{w} : \mathbf{w}A \geq \mathbf{c}, \mathbf{w} \geq \mathbf{0}\}\}$$

Let $\{\mathbf{w}^k\}$, $k = 1, 2, \ldots, K$ be the extreme points of Q and $\{\mathbf{u}^j\}$, $j = 1, 2, \ldots, J$ be the extreme rays of the recession cone of Q, $\mathcal{C}_Q = \{\mathbf{u} : \mathbf{u}A \geq \mathbf{0}, \mathbf{u} \geq \mathbf{0}\}$. Note that if Q is nonempty, the $\{\mathbf{u}^j\}$ are all the extreme rays of Q.

From linear programming duality, we know that if Q is empty and $\mathbf{u}^j(\mathbf{b} - G\mathbf{y}) \geq 0, j = 1, 2, \ldots, J$ for some $\mathbf{y} \geq \mathbf{0}$ and integer then (LP) and consequently $(MILP)$ has an unbounded solution. If Q is nonempty and $\mathbf{u}^j(\mathbf{b} - G\mathbf{y}) \geq 0, j = 1, 2, \ldots, J$ for some $\mathbf{y} \geq \mathbf{0}$ and integer, then (LP) has a finite optimum given by

$$\min_k \left\{\mathbf{w}^k(\mathbf{b} - G\mathbf{y})\right\}$$

Hence, an equivalent formulation of (*MILP*) is

$$\max \; \alpha$$
$$\alpha \leq \mathbf{dy} + \mathbf{w}^k(\mathbf{b} - \mathbf{Gy}), \quad k = 1, 2, \cdots, K$$
$$\mathbf{u}^j(\mathbf{b} - \mathbf{Gy}) \geq 0, \quad j = 1, 2, \cdots, J$$
$$\mathbf{y} \geq \mathbf{0} \; \text{ and integer}$$
$$\alpha \quad \text{unrestricted}$$

which has only one continuous variable α as promised.

31.3.4 Aggregation

An integer linear programming problem with only one constraint, other than upper bounds on the variables, is referred to as a **knapsack problem**. An integer linear programming problem with m constraints can be represented as a knapsack problem. In this section, we show this representation for an integer linear program with bounded variables [54,99]. We show how two constraints can be aggregated into a single constraint. By repeated application of this aggregation, an integer linear program with m constraints can be represented as a knapsack problem.

Consider the feasible set S defined by two constraints with integer coefficients and m nonnegative integer variables with upper bounds, i.e.,

$$S = \left\{ \mathbf{x} : \sum_{j=1}^n a_j x_j = d_1; \; \sum_{j=1}^n b_j x_j = d_2; \; 0 \leq x_j \leq u_j \text{ and } x_j \text{ integer} \right\}$$

Consider now the problem

$$(P) \quad z = \max \left\{ \left| \sum_{j=1}^n a_j x_j - d_1 \right| \; : \; 0 \leq x_j \leq u_j \text{ and } x_j \text{ integer}, \; j = 1, 2, \ldots, n \right\}$$

This problem is easily solved by considering two solutions: one in which the variables x_j with positive coefficients are set to u_j, while the other variables are set to zero; and the other in which the variables x_j with negative coefficients are set to u_j, while the other variables are set to zero.

Let α be an integer greater than z, the maximum objective value of (*P*).

It is easy to show that S is equivalent to

$$K = \left\{ \mathbf{x} : \sum_{j=1}^n \left(a_j + \alpha b_j\right) x_j = d_1 + \alpha d_2; \; 0 \leq x_j \leq u_j \text{ and } x_j \text{ integer} \right\}$$

Note that if $\mathbf{x}^* \in S$, then clearly $\mathbf{x}^* \in K$. Suppose $\mathbf{x}^* \in K$. Now we show that $\sum_{j=1}^n b_j x_j^* = d_2$. Suppose not and that

$$\sum_{j=1}^n b_j x_j^* = d_2 + k \tag{31.4}$$

where k is some arbitrary integer, positive or negative. Now multiplying 31.4 by α, and subtracting it from the equality constraint defining K, we have $\sum_{j=1}^n a_j x_j^* = d_1 - \alpha k$. But since $|\sum_{j=1}^n a_j x_j^* - d_1| < \alpha$, it follows that $k = 0$ and $\mathbf{x}^* \in S$.

31.4   Polyhedral Combinatorics

One of the main purposes of writing down an algebraic formulation of a combinatorial optimization problem as an integer program is to then examine the linear programming relaxation and understand how well it represents the discrete integer program [110]. There are somewhat special but rich classes of such formulations for which the linear programming relaxation is sharp or tight. These special structures are presented next.

31.4.1   Special Structures and Integral Polyhedra

A natural question of interest is whether the LP associated with an ILP has only integral extreme points. For instance, the linear programs associated with matching and edge covering polytopes in a bipartite graph have only integral vertices. Clearly, in such a situation, the ILP can be solved as LP. A polyhedron or a polytope is referred to as being integral if it is either empty or has only integral vertices.

DEFINITION 31.1   A $0, \pm 1$ matrix is totally unimodular if the determinant of every square submatrix is 0 or ± 1.

THEOREM 31.3   [65] *Let $A = \begin{pmatrix} A_1 \\ A_2 \\ A_3 \end{pmatrix}$ be a $0, \pm 1$ matrix and $\mathbf{b} = \begin{pmatrix} \mathbf{b}_1 \\ \mathbf{b}_2 \\ \mathbf{b}_3 \end{pmatrix}$ be a vector of appropriate dimensions. Then A is totally unimodular if and only if the polyhedron*

$$P(A, \mathbf{b}) = \{\mathbf{x} \, : \, A_1 \mathbf{x} \leq \mathbf{b}_1; A_2 \mathbf{x} \geq \mathbf{b}_2; A_3 \mathbf{x} = \mathbf{b}_3; \mathbf{x} \geq \mathbf{0}\}$$

is integral for all integral vectors $\mathbf{b}$.

The constraint matrix associated with a network flow problem (see for instance [1]) is totally unimodular. Note that for a given integral $\mathbf{b}$, $P(A, \mathbf{b})$ may be integral even if A is not totally unimodular.

DEFINITION 31.2   A polyhedron defined by a system of linear constraints is totally dual integral (TDI) if for each objective function with integral coefficients, the dual linear program has an integral optimal solution whenever an optimal solution exists.

THEOREM 31.4   [44] *If $P(A) = \{\mathbf{x} \, : \, A\mathbf{x} \leq \mathbf{b}\}$ is TDI and $\mathbf{b}$ is integral, then $P(A)$ is integral.*

Hoffman and Kruskal [65] have in fact shown that the polyhedron $P(A, \mathbf{b})$ defined in Theorem 31.3 is TDI. This follows from Theorem 31.3 and the fact that A is totally unimodular if and only if A^T is totally unimodular.

Balanced matrices, first introduced by Berge [14] have important implications for **packing and covering** problems (see also [15]).

DEFINITION 31.3   A $0, 1$ matrix is balanced if it does not contain a square submatrix of odd order with two ones per row and column.

THEOREM 31.5   [14,53] *Let A be a balanced $0, 1$ matrix. Then the set packing, set covering, and set partitioning polytopes associated with A are integral, i.e., the polytopes*

$P(A) = \{\mathbf{x} : \mathbf{x} \geq \mathbf{0}; A\mathbf{x} \leq \mathbf{1}\}$
$Q(A) = \{\mathbf{x} : \mathbf{0} \leq \mathbf{x} \leq \mathbf{1}; A\mathbf{x} \geq \mathbf{1}\}$ *and*
$R(A) = \{\mathbf{x} : \mathbf{x} \geq \mathbf{0}; A\mathbf{x} = \mathbf{1}\}$
are integral.

Let $A = \begin{pmatrix} A_1 \\ A_2 \\ A_3 \end{pmatrix}$ be a balanced $0, 1$ matrix. Fulkerson et al. [53] have shown that the polytope $P(A) = \{\mathbf{x} : A_1\mathbf{x} \leq \mathbf{1}; A_2\mathbf{x} \geq \mathbf{1}; A_3\mathbf{x} = \mathbf{1}; \mathbf{x} \geq \mathbf{0}\}$ is TDI and by the theorem of Edmonds and Giles [44] it follows that $P(A)$ is integral.

Truemper [122] has extended the definition of balanced matrices to include $0, \pm 1$ matrices.

DEFINITION 31.4   A $0, \pm 1$ matrix is balanced if for every square submatrix with exactly two nonzero entries in each row and each column, the sum of the entries is a multiple of 4.

THEOREM 31.6   [27] *Suppose A is a balanced $0, \pm 1$ matrix. Let $\mathbf{n}(A)$ denote the column vector whose ith component is the number of -1s in the ith row of A. Then the polytopes*
$P(A) = \{\mathbf{x} : A\mathbf{x} \leq \mathbf{1} - \mathbf{n}(A); \mathbf{0} \leq \mathbf{x} \leq \mathbf{1}\}$
$Q(A) = \{\mathbf{x} : A\mathbf{x} \geq \mathbf{1} - \mathbf{n}(A); \mathbf{0} \leq \mathbf{x} \leq \mathbf{1}\}$
$R(A) = \{\mathbf{x} : A\mathbf{x} = \mathbf{1} - \mathbf{n}(A); \mathbf{0} \leq \mathbf{x} \leq \mathbf{1}\}$
are integral.

Note that a $0, \pm 1$ matrix A is balanced if and only if A^{T} is balanced. Moreover A is balanced (totally unimodular) if and only if every submatrix of A is balanced (totally unimodular). Thus if A is balanced (totally unimodular) it follows that Theorem 31.6 (Theorem 31.3) holds for every submatrix of A.

Totally unimodular matrices constitute a subclass of balanced matrices, i.e., a totally unimodular $0, \pm 1$ matrix is always balanced. This follows from a theorem of Camion [18], which states that a $0, \pm 1$ is totally unimodular if and only if for every square submatrix with an even number of nonzeros entries in each row and in each column, the sum of the entries equals a multiple of 4. The 4×4 matrix in Figure 31.2 illustrates the fact that a balanced matrix is not necessarily totally unimodular. Balanced $0, \pm 1$ matrices have implications for solving the satisfiability problem. If the given set of clauses defines a balanced $0, \pm 1$ matrix, then as shown by Conforti and Cornuejols [27], the satisfiability problem is trivial to solve and the associated MAXSAT problem is solvable in polynomial time by linear programming. A survey of balanced matrices is in Conforti et al. [30].

DEFINITION 31.5   A $0, 1$ matrix A is perfect if the set packing polytope $P(A) = \{\mathbf{x} : A\mathbf{x} \leq \mathbf{1}; \mathbf{x} \geq \mathbf{0}\}$ is integral.

The chromatic number of a graph is the minimum number of colors required to color the vertices of the graph so that no two vertices with the same color have an edge incident between them. A graph

$$A = \begin{bmatrix} 1 & 1 & 0 & 0 \\ 1 & 1 & 1 & 1 \\ 1 & 0 & 1 & 0 \\ 1 & 0 & 0 & 1 \end{bmatrix} \qquad A = \begin{bmatrix} 1 & 1 & 0 \\ 0 & 1 & 1 \\ 1 & 0 & 1 \\ 1 & 1 & 1 \end{bmatrix}$$

FIGURE 31.2 A balanced matrix and a perfect matrix.

G is perfect if for every node induced subgraph *H*, the chromatic number of *H* equals the number of nodes in the maximum clique of *H*. The connections between the integrality of the set packing polytope and the notion of a perfect graph, as defined by Berge [12,13], are given in Fulkerson [52], Lovász [92], Padberg [100], and Chvátal [26].

THEOREM 31.7 [52,92,26] *Let A be* 0, 1 *matrix whose columns correspond to the nodes of a graph G and whose rows are the incidence vectors of the maximal cliques of G. The graph G is perfect if and only if A is perfect.*

Let G_A denote the intersection graph associated with a given 0, 1 matrix *A* (see Section 31.3.1). Clearly, a row of *A* is the incidence vector of a clique in G_A. In order for *A* to be perfect, every maximal clique of G_A must be represented as a row of *A* because inequalities defined by maximal cliques are facet defining. Thus by Theorem 31.7, it follows that a 0, 1 matrix *A* is perfect if and only if the undominated (a row of *A* is dominated if its support is contained in the support of another row of *A*) rows of *A* form the clique-node incidence matrix of a perfect graph.

Balanced matrices with 0, 1 entries, constitute a subclass of 0, 1 perfect matrices, i.e., if a 0, 1 matrix *A* is balanced, then *A* is perfect. The 4 × 3 matrix in Figure 31.2 is an example of a matrix that is perfect but not balanced.

DEFINITION 31.6   A 0, 1 matrix *A* is ideal if the set covering polytope

$$Q(A) = \{\mathbf{x} \ : \ A\mathbf{x} \geq \mathbf{1}; \mathbf{0} \leq \mathbf{x} \leq \mathbf{1}\}$$

is integral.

Properties of ideal matrices are described by Lehman [88], Padberg [102], and Cornuejols and Novick [33]. The notion of a 0, 1 perfect (ideal) matrix has a natural extension to a 0, ±1 perfect (ideal) matrix. Some results pertaining to 0, ±1 ideal matrices are contained in Hooker [66], while some results pertaining to 0, ±1 perfect matrices are given in Conforti et al. [28].

An interesting combinatorial problem is to check whether a given 0, ±1 matrix is totally unimodular, balanced, or perfect. Seymour's [119] characterization of totally unimodular matrices provides a polynomial time algorithm to test whether a given 0, 1 matrix is totally unimodular. Conforti et al. [31] give a polynomial time algorithm to check whether a 0, 1 matrix is balanced. This has been extended by Conforti et al. [29] to check in polynomial time whether a 0, ±1 matrix is balanced. An open problem is that of checking in polynomial time whether a 0, 1 matrix is perfect. For linear matrices (a matrix is linear if it does not contain a 2 × 2 submatrix of all ones), this problem has been solved by Fonlupt and Zemirline [48] and Conforti and Rao [32].

31.4.2 Matroids

Matroids and submodular functions have been studied extensively, especially from the point of view of combinatorial optimization (see for instance Nemhauser and Wolsey [98]). Matroids have nice properties that lead to efficient algorithms for the associated optimization problems. One of the interesting examples of matroid optimization is the problem of finding a maximum or minimum weight spanning tree in a graph. Two different but equivalent definitions of a matroid are given first. A greedy algorithm to solve a linear optimization problem over a matroid is presented. The matroid intersection problem is then discussed briefly.

DEFINITION 31.7 Let $N = \{1, 2, \cdot, n\}$ be a finite set and let $\mathcal{F}$ be a set of subsets of N. Then $I = (N, \mathcal{F})$ is an independence system if $S_1 \in \mathcal{F}$ implies that $S_2 \in \mathcal{F}$ for all $S_2 \subseteq S_1$. Elements of $\mathcal{F}$ are called independent sets. A set $S \in \mathcal{F}$ is a maximal independent set if $S \cup \{j\} \notin \mathcal{F}$ for all $j \in N \backslash S$. A maximal independent set T is a maximum if $|T| \geq |S|$ for all $S \in \mathcal{F}$.

DEFINITION 31.8 The rank $r(Y)$ of a subset $Y \subseteq N$ is the cardinality of the maximum independent subset $X \subseteq Y$. Note that $r(\phi) = 0$, $r(X) \leq |X|$ for $X \subseteq N$, and the rank function is nondecreasing, i.e., $r(X) \leq r(Y)$ for $X \subseteq Y \subseteq N$.

DEFINITION 31.9 A matroid $M = (N, \mathcal{F})$ is an independence system in which every maximal independent set is a maximum.

Example 31.1

Let $G = (V, E)$ be an undirected connected graph with V as the node set and E as the edge set.
  (i) Let $I = (E, \mathcal{F})$ where $F \in \mathcal{F}$ if $F \subseteq E$ is such that at most one edge in F is incident to each node of V, i.e., $F \in \mathcal{F}$ if F is a matching in G. Then $I = (E, \mathcal{F})$ is an independence system but not a matroid.
  (ii) Let $M = (E, \mathcal{F})$ where $F \in \mathcal{F}$ if $F \subseteq E$ is such that $G_F = (V, F)$ is a forest, i.e., G_F contains no cycles. Then $M = (E, \mathcal{F})$ is a matroid and maximal independent sets of M are spanning trees.

An alternate but equivalent definition of matroids is in terms of submodular functions.

DEFINITION 31.10 Let N be a finite set. A real valued set function f defined on the subsets of N is submodular if $f(X \cup Y) + f(X \cap Y) \leq f(X) + f(Y)$ for $X, Y \subseteq N$.

Example 31.2

Let $G = (V, E)$ be an undirected graph with V as the node set and E as the edge set. Let $c_{ij} \geq 0$ be the weight or capacity associated with edge $(ij) \in E$. For $S \subseteq V$, define the cut function $c(S) = \Sigma_{i \in S, j \in V \backslash S} c_{ij}$. The cut function defined on the subsets of V is submodular since $c(X) + c(Y) - c(X \cup Y) - c(X \cap Y) = \Sigma_{i \in X \backslash Y, j \in Y \backslash X} 2c_{ij} \geq 0$.

DEFINITION 31.11 A nondecreasing integer valued submodular function r defined on the subsets of N is called a matroid rank function if $r(\phi) = 0$ and $r(\{j\}) \leq 1$ for $j \in N$. The pair (N, r) is called a matroid.

DEFINITION 31.12 A nondecreasing, integer-valued, submodular function f defined on the subsets of N is called a polymatroid function if $f(\phi) = 0$. The pair (N, r) is called a polymatroid.

31.4.2.1 Matroid Optimization

In order to decide whether an optimization problem over a matroid is polynomially solvable or not, we need to first address the issue of representation of a matroid. If the matroid is given either by listing the independent sets or by its rank function, many of the associated linear optimization problems are trivial to solve. However, matroids associated with graphs are completely described by the graph and the condition for independence. For instance, the matroid in which the maximal independent sets are spanning trees, the graph $G = (V, E)$ and the independence condition of no cycles describes the matroid.

Most of the algorithms for matroid optimization problems require a test to determine whether a specified subset is independent or not. We assume the existence of an oracle or subroutine to do this checking in running time which is a polynomial function of $|N| = n$.

31.4.2.1.1 *Maximum Weight Independent Set*

Given a matroid $M = (N, \mathcal{F})$ and weights w_j for $j \in N$, the problem of finding a maximum weight independent set is $\max_{F \in \mathcal{F}} \left\{ \Sigma_{j \in F} w_j \right\}$. The greedy algorithm to solve this problem is as follows:

Procedure: **Greedy**

0. **Initialize:** Order the elements of N so that $w_i \geq w_{i+1}$, $i = 1, 2, \cdots, n-1$. Let $T = \phi$, $i = 1$.

1. **If** $w_i \leq 0$ or $i > n$, **stop** T is optimal, i.e., $x_j = 1$ for $j \in T$ and $x_j = 0$ for $j \notin T$. If $w_i > 0$ and $T \cup \{i\} \in \mathcal{F}$, add element i to T.

2. **Increment** i by 1 and return to Step 1.

Edmonds [41,42] derived a complete description of the matroid polytope, the convex hull of the characteristic vectors of independent sets of a matroid. While this description has a large (exponential) number of constraints, it permits the treatment of linear optimization problems on independent sets of matroids as linear programs. Cunningham [36] describes a polynomial algorithm to solve the separation problem* for the matroid polytope. The matroid polytope and the associated greedy algorithm have been extended to polymatroids [41,96].

The separation problem for a polymatroid is equivalent to the problem of minimizing a submodular function defined over the subsets of N, see Nemhauser and Wolsey [98]. A class of submodular functions that have some additional properties can be minimized in polynomial time by solving a maximum flow problem [109,112]. The general submodular function can be minimized in polynomial time by the ellipsoid algorithm [61].

The uncapacitated plant location problem (see Section 31.6.1) can be reduced to maximizing a submodular function. Hence it follows that maximizing a submodular function is $\mathcal{NP}$-hard.

31.4.2.2　Matroid Intersection

A matroid intersection problem involves finding an independent set contained in two or more matroids defined on the same set of elements.

Let $G = (V_1, V_2, E)$ be a bipartite graph. Let $M_i = (E, \mathcal{F}_i)$, $i = 1, 2$ where $F \in \mathcal{F}_i$ if $F \subseteq E$ is such that no more than one edge of F is incident to each node in V_i. The set of matchings in G constitute the intersection of the two matroids $M_i, i = 1, 2$. The problem of finding a maximum weight independent set in the intersection of two matroids can be solved in polynomial time [41,43,50,87]. The two (poly) matroid intersection polytope has been studied by Edmonds [43].

The problem of testing whether a graph contains a Hamiltonian path is $\mathcal{NP}$-complete. Since this problem can be reduced to the problem of finding a maximum cardinality independent set in the intersection of three matroids, it follows that the matroid intersection problem involving three or more matroids is $\mathcal{NP}$-hard.

---

* The separation problem for a convex body K is to test if an input point **x** belongs to K and if it does not, to produce a linear halfspace that separates **x** from K. It is known [61,83,103] that linear optimization over K is polynomially equivalent to separation from K.

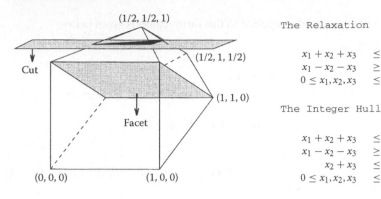

FIGURE 31.3 Relaxation, cuts, and facets.

31.4.3 Valid Inequalities, Facets, and Cutting Plane Methods

In Section 31.4.1 we were concerned with conditions under which the packing and covering polytopes are integral. But in general these polytopes are not integral and additional inequalities are required to have a complete linear description of the convex hull of integer solutions. The existence of finitely many such linear inequalities is guaranteed by Weyl's Theorem [123].

Consider the feasible region of an ILP given by

$$P_I = \{\mathbf{x} \; : \; A\mathbf{x} \leq \mathbf{b}; \mathbf{x} \geq 0 \text{ and integer}\} \tag{31.5}$$

Recall that an inequality $\mathbf{fx} \leq f_0$ is referred to as a valid inequality for P_I if $\mathbf{fx}^* \leq f_0$ for all $\mathbf{x}^* \in P_I$. A valid linear inequality for $P_I(A, \mathbf{b})$ is said to be facet defining if it intersects $P_I(A, \mathbf{b})$ in a face of dimension one less than the dimension of $P_I(A, \mathbf{b})$. In the example shown in Figure 31.3, the inequality $x_2 + x_3 \leq 1$ is a facet defining inequality of the integer hull.

Let $\mathbf{u} \geq 0$ be a row vector of appropriate size. Clearly $\mathbf{u}A\mathbf{x} \leq \mathbf{ub}$ holds for every $\mathbf{x}$ in P_I. Let $(\mathbf{u}A)_j$ denote the jth component of the row vector $\mathbf{u}A$ and $\lfloor(\mathbf{u}A)_j\rfloor$ denote the largest integer less than or equal to $(\mathbf{u}A)_j$. Now, since $\mathbf{x} \in P_I$ is a vector of nonnegative integers, it follows that $\Sigma_j \lfloor(\mathbf{u}A)_j\rfloor x_j \leq \lfloor\mathbf{ub}\rfloor$ is a valid inequality for P_I. This scheme can be used to generate many valid inequalities by using different $\mathbf{u} \geq 0$. Any set of generated valid inequalities may be added to the constraints in 31.5 and the process of generating them may be repeated with the enhanced set of inequalities. This iterative procedure of generating valid inequalities is called Gomory–Chvátal (GC) rounding. It is remarkable that this simple scheme is complete, i.e., every valid inequality of P_I can be generated by finite application of GC rounding [25,116].

The number of inequalities needed to describe the convex hull of P_I is usually exponential in the size of A. But to solve an optimization problem on P_I, one is only interested in obtaining a partial description of P_I that facilitates the identification of an integer solution and prove its optimality. This is the underlying basis of any cutting plane approach to combinatorial problems.

31.4.3.1 The Cutting Plane Method

Consider the optimization problem

$$\max\{\mathbf{cx} \; : \; \mathbf{x} \in P_I\{\mathbf{x} : A\mathbf{x} \leq \mathbf{b}; \mathbf{x} \geq 0 \text{ and integer}\}\}$$

The generic cutting plane method as applied to this formulation is given below.

Procedure: **Cutting Plane**

1. Initialize $A' \leftarrow A$ and $\mathbf{b}' \leftarrow \mathbf{b}$.
2. Find an optimal solution $\bar{\mathbf{x}}$ to the linear program

$$\max\{\mathbf{cx} : A'\mathbf{x} \leq \mathbf{b}'; \mathbf{x} \geq \mathbf{0}\}$$

   If $\bar{\mathbf{x}} \in P_I$, stop and return $\bar{\mathbf{x}}$.
3. Generate a valid inequality $\mathbf{fx} \leq f_0$ for P_I such that $f\bar{\mathbf{x}} > f_0$ (the inequality "cuts" $\bar{\mathbf{x}}$).
4. Add the inequality to the constraint system, update

$$A' \leftarrow \begin{pmatrix} A' \\ \mathbf{f} \end{pmatrix}, \quad \mathbf{b}' \leftarrow \begin{pmatrix} \mathbf{b}' \\ f_0 \end{pmatrix}$$

   Go to Step 2.

In Step 3 of the cutting plane method, we require a suitable application of the GC rounding scheme (or some alternate method of identifying a cutting plane). Notice that while the GC rounding scheme will generate valid inequalities, the identification of one that cuts off the current solution to the linear programming relaxation is all that is needed. Gomory [57] provided just such a specialization of the rounding scheme that generates a cutting plane. While this met the theoretical challenge of designing a sound and complete cutting plane method for integer linear programming, it turned out to be a weak method in practice. Successful cutting plane methods, in use today, use considerable additional insights into the structure of facet-defining cutting planes. Using facet cuts makes a huge difference in the speed of convergence of these methods. Also, the idea of combining cutting plane methods with search methods has been found to have a lot of merit. These branch and cut methods will be discussed in the next section.

31.4.3.2 The b-Matching Problem

Consider the **b**-matching problem:

$$\max\{\mathbf{cx} : A\mathbf{x} \leq \mathbf{b}, \mathbf{x} \geq \mathbf{0} \text{ and integer}\} \tag{31.6}$$

where
  A is the node-edge incidence matrix of an undirected graph
  $\mathbf{b}$ is a vector of positive integers

Let G be the undirected graph whose node-edge incidence matrix is given by A and let $W \subseteq V$ be any subset of nodes of G (i.e., subset of rows of A) such that $\mathbf{b}(W) = \Sigma_{i \in W} b_i$ is odd. Then the inequality

$$\mathbf{x}(W) = \sum_{e \in E(W)} x_e \leq \frac{1}{2}(\mathbf{b}(W) - 1) \tag{31.7}$$

is a valid inequality for integer solutions to 31.6 where $E(W) \subseteq E$ is the set of edges of G having both ends in W. Edmonds [40] has shown that the inequalities 31.6 and 31.7 define the integral **b**-matching polytope (see also [45]). Note that the number of inequalities 31.7 is exponential in the number of nodes of G. An instance of the successful application of the idea of using only a partial description of P_I is in the blossom algorithm for the matching problem, due to Edmonds [40].

An implication of the ellipsoid method for linear programming is that the linear program over P_I can be solved in polynomial time if and only if the associated separation problem can be solved in polynomial time (see Grötschel et al. [61], Karp and Papadimitriou [83], and Padberg and Rao [103]). The separation problem for the **b**-matching problem with or without upper bounds was shown by Padberg and Rao [104], to be solvable in polynomial time. The procedure involves a minor modification of the algorithm of Gomory and Hu [60] for multiterminal networks. However, no polynomial (in the number of nodes of the graph) linear programming formulation of this separation problem is known. Martin [95] has shown that if the separation problem can be expressed as a compact linear program then so can the optimization problem. Hence an unresolved issue is whether there exists a polynomial size (compact) formulation for the **b**-matching problem. Yannakakis [126] has shown that, under a symmetry assumption, such a formulation is impossible.

31.4.3.3 Other Combinatorial Problems

Besides the matching problem several other combinatorial problems and their associated polytopes have been well studied and some families of facet defining inequalities have been identified. For instance the set packing, graph partitioning, plant location, maximum cut, traveling salesman, and Steiner tree problems have been extensively studied from a polyhedral point of view (see for instance Nemhauser and Wolsey [98]).

These combinatorial problems belong to the class of $\mathcal{NP}$-complete problems. In terms of a worst-case analysis, no polynomial time algorithms are known for these problems. Nevertheless, using a cutting plane approach with branch and bound or branch and cut (see Section 31.5), large instances of these problems have been successfully solved, see Crowder et al. [35], for general $0 - 1$ problems, Barahona et al. [7] for the maximum cut problem, Padberg and Rinaldi [105] for the traveling salesman problem and Chopra et al. [24] for the Steiner tree problem.

31.5 Partial Enumeration Methods

In many instances, to find an optimal solution to an integer linear programing problem (ILP), the structure of the problem is exploited together with some sort of partial enumeration. In this section, we review the branch and bound (B & B) and branch and cut (B & C) methods for solving an ILP.

31.5.1 Branch and Bound

The branch and bound (B & B) method is a systematic scheme for implicitly enumerating the finitely many feasible solutions to an ILP. Although, theoretically the size of the enumeration tree is exponential in the problem parameters, in most cases, the method eliminates a large number of feasible solutions. The key features of branch and bound method are

1. Selection/Removal of one or more problems from a candidate list of problems.
2. Relaxation of the selected problem so as to obtain a lower bound (on a minimization problem) on the optimal objective function value for the selected problem.
3. **Fathoming**, if possible, of the selected problem.
4. Branching Strategy: If the selected problem is not fathomed, branching creates subproblems which are added to the candidate list of problems.

The above four steps are repeated until the candidate list is empty. The B & B method sequentially examines problems that are added and removed from a candidate list of problems.

31.5.1.1   Initialization

Initially the candidate list contains only the original ILP which is denoted as

$$(P)\qquad \min\{\mathbf{cx} : A\mathbf{x} \le \mathbf{b}, \mathbf{x} \ge 0 \text{ and integer}\}$$

Let $F(P)$ denote the feasible region of (P) and $z(P)$ denote the optimal objective function value of (P). For any $\bar{\mathbf{x}}$ in $F(P)$, let $z_P(\bar{\mathbf{x}}) = \mathbf{c\bar{x}}$.

Frequently, heuristic procedures are first applied to get a good feasible solution to (P). The best solution known for (P) is referred to as the current incumbent solution. The corresponding objective function value is denoted as z_I. In most instances, the initial heuristic solution is not either optimal or at least not immediately certified to be optimal. So further analysis is required to ensure that an optimal solution to (P) is obtained. If no feasible solution to (P) is known, z_I is set to ∞.

31.5.1.2   Selection/Removal

In each iterative step of B & B, a problem is selected and removed from the candidate list for further analysis. The selected problem is henceforth referred to as the candidate problem (CP). The algorithm terminates if there is no problem to select from the candidate list. Initially there is no issue of selection since the candidate list contains only the problem (P). However, as the algorithm proceeds, there would be many problems on the candidate list and a selection rule is required. Appropriate selection rules, also referred to as branching strategies, are discussed later. Conceptually, several problems may be simultaneously selected and removed from the candidate list. However, most sequential implementations of B & B select only one problem from the candidate list and this is assumed henceforth. Parallel aspects of B & B on $0 - 1$ integer linear programs are discussed in Cannon and Hoffman [19] and for the case of traveling salesman problems in [4].

The computational time required for the B & B algorithm depends crucially on the order in which the problems in the candidate list are examined. A number of clever heuristic rules may be employed in devising such strategies. Two general purpose selection strategies that are commonly used are

A. Choose the problem that was added last to the candidate list. This last-in-first-out rule (LIFO) is also called depth-first search (DFS), since the selected candidate problem increases the depth of the active enumeration tree.

B. Choose the problem on the candidate list that has the least lower bound. Ties may be broken by choosing the problem that was added last to the candidate list. This rule would require that a lower bound be obtained for each of the problems on the candidate list. In other words, when a problem is added to the candidate list, an associated lower bound should also be stored. This may be accomplished by using ad hoc rules or by solving a relaxation of each problem before it is added to the candidate list.

Rule (A) is known to empirically dominate the rule (B) when storage requirements for candidate list and computation time to solve (P) are taken into account. However, some analysis indicates that rule (B) can be shown to be superior if minimizing the number of candidate problems to be solved is the criterion (see Parker and Rardin [108]).

31.5.1.3   Relaxation

In order to analyze the selected candidate problem, (CP), a relaxation (CP_R) of (CP) is solved to obtain a lower bound $z(CP_R) \le z(CP)$. (CP_R) is a relaxation of (CP) if

1. $F(CP) \subseteq F(CP_R)$
2. for $\bar{\mathbf{x}} \in F(CP)$, $z_{CP_R}(\bar{\mathbf{x}}) \le z_{CP}(\bar{\mathbf{x}})$
3. for $\bar{\mathbf{x}}, \hat{\mathbf{x}} \in F(CP)$, $z_{CP_R}(\bar{\mathbf{x}}) \le z_{CP_R}(\hat{\mathbf{x}})$ implies that $z_{CP}(\bar{\mathbf{x}}) \le z_{CP}(\hat{\mathbf{x}})$

Relaxations are needed because the candidate problems are typically hard to solve. The relaxations used most often are either linear programming or Lagrangean relaxations of (CP); see Section 31.6 for details. Sometimes, instead of solving a relaxation of (CP), a lower bound is obtained by using some ad hoc rules such as penalty functions.

31.5.1.4 Fathoming

A candidate problem is fathomed if

(FC1) Analysis of (CP_R) reveals that (CP) is infeasible. For instance if $F(CP_R) = \phi$, then $F(CP) = \phi$.

(FC2) Analysis of (CP_R) reveals that (CP) has no feasible solution better than the current incumbent solution. For instance if $z(CP_R) \geq z_I$, then $z(CP) \geq z(CP_R) \geq z_I$.

(FC3) Analysis of (CP_R) reveals an optimal solution of (CP). For instance, if $\mathbf{x}_R$ is optimal for (CP_R) and is feasible in (CP), then $(\mathbf{x}_R)$ is an optimal solution to (CP) and $z(CP) = z_{CP}(\mathbf{x}_R)$.

(FC4) Analysis of (CP_R) reveals that (CP) is dominated by some other problem, say CP^*, in the candidate list. For instance if it can be shown that $z(CP^*) \leq z(CP)$, then there is no need to analyze (CP) further.

If a candidate problem (CP) is fathomed using any of the criteria above, then further examination of (CP) or its descendants (subproblems) obtained by separation is not required. If (FC3) holds, and $z(CP) < z_I$, the incumbent is updated as $\mathbf{x}_R$ and z_I is updated as $z(CP)$.

31.5.1.5 Separation/Branching

If the candidate problem (CP) is not fathomed, then CP is separated into several problems, say $(CP_1), (CP_2), \ldots, (CP_q)$ where $\bigcup_{t=1}^{q} F(CP_t) = F(CP)$ and typically

$$F(CP_i) \cap F(CP_j) = \phi \quad \forall i \neq j$$

For instance a separation of (CP) into $(CP_i), i = 1, 2, \ldots, q$ is obtained by fixing a single variable, say x_j, to one of the q possible values of x_j in an optimal solution to (CP). The choice of the variable to fix depends upon the separation strategy which is also part of the branching strategy. After separation, the subproblems are added to the candidate list. Each subproblem (CP_t) is a restriction of (CP) since $F(CP_t) \subseteq F(CP)$. Consequently $z(CP) \leq z(CP_t)$ and $z(CP) = \min_t z(CP_t)$.

The various steps in the B & B algorithm are outlined below.

Procedure: **B & B**

0. **Initialize:** Given the problem (P), the incumbent value z_I is obtained by applying some heuristic (if a feasible solution to (P) is not available, set $z_I = +\infty$). Initialize the candidate list $C \leftarrow \{(P)\}$.

1. **Optimality:** If $C = \emptyset$ and $z_I = +\infty$, then (P) is infeasible, stop. Stop also if $C = \emptyset$ and $z_I < +\infty$, the incumbent is an optimal solution to (P).

2. **Selection:** Using some candidate selection rule, select, and remove a problem $(CP) \in C$.

3. **Bound:** Obtain a lower bound for (CP) by either solving a relaxation (CP_R) of (CP) or by applying some ad hoc rules. If

(CP_R) is infeasible, return to Step 1. Else, let $\mathbf{x}_R$ be an
optimal solution of (CP_R).

4. **Fathom:** If $z(CP_R) \geq z_I$, return to Step 1. Else if $\mathbf{x}_R$ is feasible
   in (CP) and $z(CP) < z_I$, set $z_I \leftarrow z(CP)$, update the incumbent as
   $\mathbf{x}_R$ and return to Step 1. Finally, if $\mathbf{x}_R$ is feasible in (CP) but
   $z(CP) \geq z_I$, return to Step 1.

5. **Separation:** Using some separation or branching rule, separate
   (CP) into $(CP_i), i = 1, 2, \ldots, q$ and set $C \leftarrow C \cup \{CP_1), (CP_2), \ldots, (CP_q)\}$
   and return to Step 1.

6. **End Procedure.**

Although the B & B method is easy to understand, the implementation of this scheme for a particular ILP is a nontrivial task requiring

A. A relaxation strategy with efficient procedures for solving these relaxations
B. Efficient data structures for handling the rather complicated book-keeping of the candidate list
C. Clever strategies for selecting promising candidate problems
D. Separation or branching strategies that could effectively prune the enumeration tree

A key problem is that of devising a relaxation strategy (A), i.e., to find good relaxations which are significantly easier to solve than the original problems and tend to give sharp lower bounds. Since these two are conflicting, one has to find a reasonable trade-off.

31.5.2 Branch and Cut

In the last few years, the branch and cut (B & C) method has become popular for solving combinatorial optimization problems. As the name suggests, the B & C method incorporates the features of both the branch and bound method presented above and the cutting plane method presented in the previous section. The main difference between the B & C method and the general B & B scheme is in the bound step (Step 3).

A distinguishing feature of the B & C method is that the relaxation (CP_R) of the candidate problem (CP) is a linear programming problem and instead of merely solving (CP_R), an attempt is made to solve (CP) by using cutting planes to tighten the relaxation. If (CP_R) contains inequalities that are valid for (CP) but not for the given ILP, then the GC rounding procedure may generate inequalities that are valid for (CP) but not for the ILP. In the B & C method, the inequalities that are generated are always valid for the ILP, and hence can be used globally in the enumeration tree.

Another feature of the B & C method is that often heuristic methods are used to convert some of the fractional solutions, encountered during the cutting plane phase, into feasible solutions of the (CP) or more generally of the given ILP. Such feasible solutions naturally provide upper bounds for the ILP. Some of these upper bounds may be better than the previously identified best upper bound and if so, the current incumbent is updated accordingly.

We thus obtain the B & C method by replacing the bound step (Step 3) of the B & B method by Steps 3(a) and 3(b), and also by replacing the fathom step (Step 4) by Steps 4(a) and 4(b) given below.

3(a) **Bound:** Let (CP_R) be the LP relaxation of (CP). Attempt to solve
     (CP) by a cutting plane method which generates valid
     inequalities for (P). Update the constraint System of (P) and
     the incumbent as appropriate.

Let $Fx \leq f$ denote all the valid inequalities generated during this phase. Update the constraint system of (P) to include all the generated inequalities, i.e., set $A^T \leftarrow (A^T, F^T)$ and $\mathbf{b}^T \leftarrow (\mathbf{b}^T, \mathbf{f}^T)$. The constraints for all the problems in the candidate list are also to be updated.

During the cutting plane phase, apply heuristic methods to convert some of the identified fractional solutions into feasible solutions to (P). If a feasible solution, $\bar{\mathbf{x}}$, to (P), is obtained such that $\mathbf{c}\bar{\mathbf{x}} < z_I$, update the incumbent to $\bar{\mathbf{x}}$ and z_I to $\mathbf{c}\bar{\mathbf{x}}$. Hence the remaining changes to B & B are as follows:

3 (b) **If** (CP) is solved go to Step 4 (a). **Else,** let $\hat{\mathbf{x}}$ be the solution obtained when the cutting plane phase is terminated (we are unable to identify a valid inequality of (P) that is violated by $\hat{\mathbf{x}}$.) go to Step 4 (b).

4 (a) **Fathom by Optimality:** Let $\mathbf{x}^*$ be an optimal solution to (CP). If $z(CP) < z_I$, set $\mathbf{x}_I \leftarrow z(CP)$ and update the incumbent as $\mathbf{x}^*$. Return to Step 1.

4 (b) **Fathom by Bound:** If $\mathbf{c}\hat{\mathbf{x}} \geq z_I$, return to Step 1. Else go to Step 5.

The incorporation of a cutting plane phase into the B & B scheme involves several technicalities which require careful design and implementation of the B & C algorithm. Details of the state of the art in cutting plane algorithms including the B & C algorithm are reviewed in Jünger et al. [75].

31.6 Relaxations

The effectiveness of a partial enumeration strategy such as branch and bound is closely related to the quality of the relaxations used to generate the bounds and incumbents. We describe four general relaxation methods that together cover the most successful computational techniques for bound evaluation in the practice of partial enumeration for integer programming. These are linear programming relaxation, Lagrangean relaxation, group relaxation, and semidefinite relaxation methods.

31.6.1 LP Relaxation

A linear programming relaxation is derived from an integer programming formulation by relaxing the integrality constraints. When there are alternate integer programming formulations, modeling the same decision problem, it becomes necessary to have some criteria for selecting from among the candidate relaxations. We illustrate these ideas on the plant location model, a prototypical integer programming example.

31.6.1.1 Plant Location Problems

Given a set of customer locations $N = \{1, 2, \ldots, n\}$ and a set of potential sites for plants $M = \{1, 2, \ldots, m\}$, the plant location problem is to identify the sites where the plants are to be located so that the customers are served at a minimum cost. There is a fixed cost f_i of locating the plant at site i and the cost of serving customer j from site i is c_{ij}. The decision variables are

y_i is set to 1 if a plant is located at site i and to 0 otherwise.

x_{ij} is set to 1 if site i serves customer j and to 0 otherwise.

A formulation of the problem is

$$(P8) \quad \text{Min} \quad \sum_{i=1}^{m}\sum_{j=1}^{n} c_{ij}x_{ij} + \sum_{i=1}^{m} f_i y_i$$

$$\text{subject to} \quad \sum_{i=1}^{m} x_{ij} = 1 \quad j = 1, 2, \ldots, n$$

$$x_{ij} - y_i \leq 0 \quad i = 1, 2, \ldots, m; \; j = 1, 2, \ldots, n$$

$$y_i = 0 \text{ or } 1 \quad i = 1, 2, \ldots, m$$

$$x_{ij} = 0 \text{ or } 1 \quad i = 1, 2, \ldots, m; \; j = 1, 2, \ldots, n$$

Note that the constraints $x_{ij} - y_i \leq 0$ are required to ensure that customer j may be served from site i only if a plant is located at site i. Note that the constraints $y_i = 0$ or 1, force an optimal solution in which $x_{ij} = 0$ or 1. Consequently the $x_{ij} = 0$ or 1 constraints may be replaced by nonnegativity constraints $x_{ij} \geq 0$.

The linear programming relaxation associated with (P8) is obtained by replacing constraints $y_i = 0$ or 1, and $x_{ij} = 0$ or 1 by nonnegativity constraints on x_{ij} and y_i. The upper bound constraints on y_i are not required provided $f_i \geq 0$, $i = 1, 2, \ldots, m$. The upper bound constraints on x_{ij} are not required in view of constraints $\sum_{i=1}^{m} x_{ij} = 1$.

Remark: It is frequently possible to formulate the same combinatorial problem as two or more different ILPs. Suppose we have two ILP formulations (F1) and (F2) of the given combinatorial problem with both (F1) and (F2) being minimizing problems. Formulation (F1) is said to be stronger than (F2) if (LP1), the linear programming relaxation of (F1), always has an optimal objective function value which is greater than or equal to the optimal objective function value of (LP2) which is the linear programming relaxation of (F2).

It is possible to reduce the number of constraints in (P8) by replacing the constraints $x_{ij} - y_i \leq 0$ by an aggregate:

$$\sum_{j=1}^{n} x_{ij} - n y_i \leq 0, \quad i = 1, 2, \ldots, m$$

However, the disaggregated (P8) is a stronger formulation than the formulation obtained by aggregating the constraints as above. By using standard transformations, (P8) can also be converted into a set packing problem.

31.6.2 Lagrangean Relaxation

This approach has been widely used for about two decades now in many practical applications. Lagrangean relaxation, like linear programming relaxation, provides bounds on the combinatorial optimization problem being relaxed (i.e., lower bounds for minimization problems).

Lagrangean relaxation has been so successful because of a couple of distinctive features. As was noted earlier, in many hard combinatorial optimization problems, we usually have some nice tractable embedded subproblems which admit efficient algorithms. Lagrangean relaxation gives us a framework to jerry-rig an approximation scheme that uses these efficient algorithms for the subproblems as subroutines. A second observation is that it has been empirically observed that well-chosen Lagrangean relaxation strategies usually provide very tight bounds on the optimal objective value of integer programs. This is often used to great advantage within partial enumeration schemes to get very effective pruning tests for the search trees.

Practitioners also have found considerable success with designing heuristics for combinatorial optimization by starting with solutions from Lagrangean relaxations and constructing good feasible solutions via so-called dual ascent strategies. This may be thought of as the analogue of rounding strategies for linear programming relaxations (but with no performance guarantees—other than empirical ones).

Consider a representation of our combinatorial optimization problem in the form

$$(P) \quad z = \min \left\{ \mathbf{cx} : A\mathbf{x} \geq \mathbf{b}, \ \mathbf{x} \in X \subseteq \mathfrak{R}^n \right\}$$

Implicit in this representation is the assumption that the explicit constraints ($A\mathbf{x} \geq \mathbf{b}$) are small in number. For convenience let us also assume that X can be replaced by a finite list $\{\mathbf{x}^1, \mathbf{x}^2, \ldots, \mathbf{x}^T\}$.

The following definitions are with respect to (P):

- Lagrangean $L(\mathbf{u},\mathbf{x}) = \mathbf{u}(A\mathbf{x} - \mathbf{b}) + \mathbf{cx}$ where $\mathbf{u}$ are the Lagrange multipliers.
- Lagrangean Subproblem $\min_{\mathbf{x} \in X}\{L(\mathbf{u},\mathbf{x})\}$
- Lagrangean-Dual Function $\mathcal{L}(\mathbf{u}) = \min_{\mathbf{x} \in X}\{L(\mathbf{u},\mathbf{x})\}$
- Lagrangean-Dual Problem   (D) $d = \max_{\mathbf{u} \geq 0}\{\mathcal{L}(\mathbf{u})\}$

It is easily shown that (D) satisfies a weak duality relationship with respect to (P), i.e., $z \geq d$. The discreteness of X also implies that $\mathcal{L}(\mathbf{u})$ is a piecewise linear and concave function (see Shapiro [118]). In practice, the constraints X are chosen such that the evaluation of the Lagrangean dual function $\mathcal{L}(\mathbf{u})$ is easily made.

An Example: Traveling Salesman Problem (TSP)

For an undirected graph G, with costs on each edge, the TSP is to find a minimum cost set H of edges of G such that it forms a Hamiltonian cycle of the graph. H is a Hamiltonian cycle of G if it is a simple cycle that spans all the vertices of G. Alternately H must satisfy

1. Exactly two edges of H are adjacent to each node.
2. H forms a connected, spanning subgraph of G.

Held and Karp [64] used these observations to formulate a Lagrangean relaxation approach for TSP, that relaxes the degree constraints (condition 1 above). Notice that the resulting subproblems are minimum spanning tree problems that can be easily solved.

The most commonly used general method of finding the optimal multipliers in Lagrangean relaxation is subgradient optimization (cf. Held et al. [63]). Subgradient optimization is the non-differentiable counterpart of steepest descent methods. Given a dual vector $\mathbf{u}^k$, the iterative rule for creating a sequence of solutions is given by

$$\mathbf{u}^{k+1} = \mathbf{u}^k + t_k \gamma \left(\mathbf{u}^k \right)$$

where t_k is an appropriately chosen step size, and $\gamma(\mathbf{u}^k)$ is a subgradient of the dual function $\mathcal{L}$ at $\mathbf{u}^k$. Such a subgradient is easily generated by

$$\gamma \left(\mathbf{u}^k \right) = A\mathbf{x}^k - \mathbf{b}$$

where $\mathbf{x}^k$ is an optimal solution of $\min_{\mathbf{x} \in X}\{L(\mathbf{u}^k, \mathbf{x})\}$.

Subgradient optimization has proven effective in practice for a variety of problems. It is possible to choose the step sizes $\{t_k\}$ to guarantee convergence to the optimal solution. Unfortunately, the method is not finite, in that the optimal solution is attained only in the limit. Further, it is not a pure descent method. In practice, the method is heuristically terminated and the best solution in the generated sequence is recorded. In the context of nondifferentiable optimization, the ellipsoid algorithm was devised by Shor [121] to overcome precisely some of these difficulties with the subgradient method.

The ellipsoid algorithm may be viewed as a scaled subgradient method in much the same way as variable metric methods may be viewed as scaled steepest descent methods (cf. [2]). And if we use the ellipsoid method to solve the Lagrangean dual problem, we obtain the following as a consequence of the polynomial-time equivalence of optimization and separation.

THEOREM 31.8   *The Lagrangean dual problem is polynomial-time solvable if and only if the Lagrangean subproblem is. Consequently, the Lagrangean dual problem is $\mathcal{NP}$-hard if and only if the Lagrangean subproblem is.*

The theorem suggests that in practice, if we set up the Lagrangean relaxation so that the subproblem is tractable, then the search for optimal Lagrangean multipliers is also tractable.

31.6.3   Group Relaxations

A relaxation of the integer programming problem is obtained by dropping the nonnegativity restrictions on some variables. Consider the integer linear program

$$(ILP) \quad \max\{\mathbf{cx} \ : \ A\mathbf{x} = \mathbf{b}, \mathbf{x} \geq \mathbf{0} \text{ and integer}\}$$

where the elements of A, b are integral.

Suppose the linear programming relaxation of (ILP) has a finite optimum, then an extreme-point optimal solution $\mathbf{x}^* = \begin{pmatrix} B^{-1}\mathbf{b} \\ 0 \end{pmatrix}$ where B is a nonsingular submatrix of A. Let $A = (B, N), \mathbf{x} = \begin{pmatrix} \mathbf{x}_B \\ \mathbf{x}_N \end{pmatrix}$, and $\mathbf{c} = (\mathbf{c}_B, \mathbf{c}_N)$. By dropping the nonnegativity constraint on $\mathbf{x}_B$, substituting $\mathbf{x}_B = B^{-1}(\mathbf{b} - N\mathbf{x}_N)$, ignoring the constant term $\mathbf{c}_B B^{-1} b$ in the objective function, and changing the objective function from maximize to minimize, we obtain the following relaxation of (ILP):

$$(ILP_B) \quad \min\left\{\left(\mathbf{c}_B B^{-1}N - \mathbf{c}_N\right)\mathbf{x}_N \ : \ \mathbf{x}_B \right.$$
$$= B^{-1}\left(\mathbf{b} - N\mathbf{x}_N\right), \ \mathbf{x}_B \text{ integer}, \ \mathbf{x}_N \geq \mathbf{0} \text{ and integer}\right\}$$

Now $\mathbf{x}_B = B^{-1}(\mathbf{b} - N\mathbf{x}_N)$ and $\mathbf{x}_B$ integer is equivalent to requiring

$$B^{-1}N\mathbf{x}_N \equiv B^{-1}\mathbf{b}(mod \ 1)$$

where the congruence is with respect to each element of the vector $B^{-1}\mathbf{b}$ taken modulo 1.

Hence, (ILP_B) can be written as

$$(ILP_G) \quad \min\left\{\left(\mathbf{c}_B B^{-1}N - \mathbf{c}_N\right)\mathbf{x}_N \ : \ B^{-1}N\mathbf{x}_N\right.$$
$$\equiv B^{-1}\mathbf{b}(mod \ 1), \ \mathbf{x}_N \geq \mathbf{0} \text{ and integer}\right\}$$

Since A and $\mathbf{b}$ are integer matrices, the fractional parts of elements of $B^{-1}N$ and $B^{-1}\mathbf{b}$ are of the form $\left(\frac{k}{detB}\right)$ where $k \in \{0, 1, \ldots, |detB| - 1\}$. The congruences in (ILP_G) are equivalent to working in a product of cyclic groups. The structure of the product group is revealed by the Hermite normal form of B (see Section 31.2.2). Hence, (ILP_G) is referred to as a group (knapsack) problem and is solved by a dynamic programming algorithm [59].

31.6.4 Semidefinite Relaxation

Semidefinite programs are linear optimization problems defined over a cone of positive semidefinite matrices. These are models that generalize linear programs and are specializations of convex programming models. There are theoretical and practical algorithms for solving semidefinite programs in polynomial time [3]. Lovász and Schrijver [94] suggest a general relaxation strategy for $0 - 1$ integer programming problems that obtains semidefinite relaxations. The first step is to consider a homogenized version of a $0 - 1$ integer program (solvability version):

$$F_I = \left\{ \mathbf{x} \in \Re^{n+1} : A\mathbf{x} \geq 0,\ x_0 = 1,\ x_i \in \{0, 1\} \text{ for } i = 1, 2, \dots, n \right\}$$

The problem is to check if F_I is nonempty. Note that any $0 - 1$ integer program can be put in this form by absorbing a general right-hand side b as the negative of x_0 column of A. Now a linear programming relaxation of this integer program is given by

$$\left\{ \mathbf{x} \in \Re^{n+1} : A\mathbf{x} \geq 0,\ x_0 = 1,\ 0 \leq x_i \leq x_0 \text{ for } i = 1, 2, \dots, n \right\}$$

Next, we define two polyhedral cones.

$$\begin{aligned} \mathcal{K} &= \left\{ \mathbf{x} \in \Re^{n+1} : A\mathbf{x} \geq 0,\ 0 \leq x_i \leq x_0 \text{ for } i = 0, 1, \dots, n \right\} \\ \mathcal{K}_I &= \text{Cone generated by } 0 - 1 \text{ vectors in } P_I \end{aligned}$$

Lovász and Schrijver [94] show how we might construct a family of convex cones $\{\mathcal{C}\}$ such that $\mathcal{K}_I \subseteq \mathcal{C} \subseteq \mathcal{K}$ for each $\mathcal{C}$.

  i. Partition the cone constraints of $\mathcal{K}$ into $T_1 = \{A_1\mathbf{x} \geq 0\}$ and $T_2 = \{A_2\mathbf{x} \geq 0\}$, with the constraints $\{0 \leq x_i \leq x_0 \text{ for } i = 0, 1, \cdots, n\}$ repeated in both (the overlap can be larger).

  ii. Multiply each constraint in T_1 with each constraint in T_2 to obtain a quadratic homogeneous constraint.

  iii. Replace each occurrence of a quadratic term $x_i x_j$ by a new variable X_{ij}. The quadratic constraints are now linear homogeneous constraints in X_{ij}.

  iv. Add the requirement that the $(n + 1) \times (n + 1)$ matrix X is symmetric and positive semidefinite.

  v. Add the constraints $X_{0i} = X_{ii}$ for $i = 1, 2, \dots, n$.

The system of constraints on the X_{ij} constructed in steps (iii), (iv), and (v) above define a cone $M_+(T_1, T_2)$ parametrized by the partition T_1, T_2. We finally project the cone $M_+(T_1, T_2)$ to $\Re^{n+1}$ as follows.

$$\mathcal{C}_+(T_1, T_2) = \text{Diagonals of matrices in } M_+(T_1, T_2)$$

These resulting cones $\{\mathcal{C}_+(T_1, T_2)\}$ satisfy

$$\mathcal{K}_I \subseteq \mathcal{C}_+(T_1, T_2) \subseteq \mathcal{K}$$

where the X_{ii} in $\mathcal{C}_+(T_1, T_2)$ are interpreted as the original x_i in $\mathcal{K}$ and $\mathcal{K}_I$.

One semidefinite relaxation of the $0 - 1$ integer program F_I is just $C_+(T_1, T_2)$ along with the normalizing constraint $X_{00} = x_0 = 1$. For optimization versions of integer programming, we simply carry over the objective function. An amazing result obtained by Lovász and Schrijver [94] is that this relaxation when applied to the vertex packing polytope (see Section 31.3.4) is at least as good as one obtained by adding "clique, odd hole, odd antihole, and odd wheel" valid inequalities (see [62] for the definitions) to the linear programming relaxation of the set packing formulation. This illustrates the power of this approach to reveal structure and yet obtain a tractable relaxation. In particular, it also implies a polynomial-time algorithm for the vertex packing problem on perfect graphs (cf. [62]). Another remarkable success that semidefinite relaxations have registered is the recent result on finding an approximately maximum weight edge cutset of a graph. This result of Goemans and Williamson [56] will be described in the next section. While the jury is still out on the efficacy of semidefinite relaxation as a general strategy for integer programming, there is little doubt that it provides an exciting new weapon in the arsenal of integer programming methodologies.

31.7 Approximation with Performance Guarantees

The relaxation techniques we encountered in the previous section are designed with the intent of obtaining a good lower (upper) bound on the optimum objective value for a minimization (maximization) problem. If in addition, we are able to construct a good feasible solution using a heuristic (possibly based on a relaxation technique) we can use the bound to quantify the suboptimality of the incumbent and hence the quality of the approximation.

In the past few years, there has been significant progress in our understanding of performance guarantees for approximation of $\mathcal{NP}$-hard combinatorial optimization problems (cf. [120]). A ρ-approximate algorithm for an optimization problem is an approximation algorithm that delivers a feasible solution with objective value within a factor of ρ of optimal (think of minimization problems and $\rho \geq 1$). For some combinatorial optimization problems, it is possible to efficiently find solutions that are arbitrarily close to optimal even though finding the true optimal is hard. If this were true of most of the problems of interest we would be in good shape. However, the recent results of Arora et al. [5] indicate exactly the opposite conclusion.

A PTAS or polynomial-time approximation scheme for an optimization problem is a family of algorithms A_ρ, such that for each $\rho > 1$, A_ρ is a polynomial-time ρ-approximate algorithm. Despite concentrated effort spanning about two decades, the situation in the early 1990s was that for many combinatorial optimization problems, we had no PTAS and no evidence to suggest the nonexistence of such schemes either. This led Papadimitriou and Yannakakis [107] to define a new complexity class (using reductions that preserve approximate solutions) called MAXSNP, and they identified several complete languages in this class. The work of Arora and colleagues completed this agenda by showing that, assuming $\mathcal{P} \neq \mathcal{NP}$, there is no PTAS for a MAXSNP-complete problem.

An implication of these theoretical developments is that for most combinatorial optimization problems, we have to be quite satisfied with performance guarantee factors $\rho \geq 1$ that are of some small fixed value. (There are problems, like the general traveling salesman problem, for which there are no ρ-approximate algorithms for any finite value of ρ—of course assuming $\mathcal{P} \neq \mathcal{NP}$.) Thus one avenue of research is to go problem by problem and knock ρ down to its smallest possible value.

A different approach would be to look for other notions of good approximations based on probabilistic guarantees or empirical validation. A good example of the benefit to be gained from randomization is the problem of computing the volume of a convex body. Dyer and Frieze [38] have shown that this problem is $\#\mathcal{P}$-hard. Barany and Furedi [8] provide evidence that no polynomial-time deterministic approximation method with relative error less than $(cn)^{\frac{n}{2}}$, where c is a constant, is likely to exist. However, Dyer et al. [39] have designed a fully polynomial randomized approximation scheme (FPRAS) for this problem. The FPRAS of Dyer et al. [39] uses techniques from integer programming and the geometry of numbers (see Section 31.8).

31.7.1 LP Relaxation and Rounding

Consider the well-known problem of finding the smallest weight vertex cover in a graph (see Section 31.3.1). So we are given a graph $G(V, E)$ and a nonnegative weight $w(v)$ for each vertex $v \in V$. We want to find the smallest total weight subset of vertices S such that each edge of G has at least one end in S (This problem is known to be MAXSNP-hard.). An integer programming formulation of this problem is given by

$$\min \left\{ \sum_{v \in V} w(v)x(v) : x(u) + x(v) \geq 1, \ \forall (u, v) \in E, \quad x(v) \in \{0, 1\} \ \forall v \in V \right\}$$

To obtain the linear programming relaxation, we substitute the $x(v) \in \{0, 1\}$ constraint with $x(v) \geq 0$ for each $v \in V$. Let $\mathbf{x}^*$ denote an optimal solution to this relaxation. Now let us round the fractional parts of $\mathbf{x}^*$ in the usual way, that is, values of 0.5 and up are rounded to 1 and smaller values to 0. Let $\hat{\mathbf{x}}$ be the 0–1 solution obtained. First note that $\hat{x}(v) \leq 2x^*(v)$ for each $v \in V$. Also, for each $(u, v) \in E$, since $x^*(u) + x^*(v) \geq 1$, at least one of $\hat{x}(u)$ and $\hat{x}(v)$ must be set to 1. Hence $\hat{\mathbf{x}}$ is the incidence vector of a vertex cover of G whose total weight is within twice the total weight of the linear programming relaxation (which is a lower bound on the weight of the optimal vertex cover). Thus we have a 2-approximate algorithm for this problem which solves a linear programming relaxation and uses rounding to obtain a feasible solution.

The deterministic rounding of the fractional solution worked quite well for the vertex cover problem. One gets a lot more power from this approach by adding in randomization to the rounding step. Raghavan and Thompson [111] proposed the following obvious randomized rounding scheme. Given a 0 − 1 integer program, solve its linear programming relaxation to obtain an optimal $\mathbf{x}^*$. Treat the $x_j^* \in [0, 1]$ as probabilities, i.e., let Probability$\{x_j = 1\}= x_j^*$, to randomly round the fractional solution to a 0 − 1 solution. Using Chernoff bounds on the tails of the binomial distribution, they were able to show, for specific problems, that with high probability, this scheme produces integer solutions which are close to optimal. In certain problems, this rounding method may not always produce a feasible solution. In such cases, the expected values have to be computed as conditioned on feasible solutions produced by rounding. More complex (nonlinear) randomized rounding schemes have been recently studied and have been found to be extremely effective. We will see an example of nonlinear rounding in the context of semidefinite relaxations of the max-cut problem below.

31.7.2 Primal–Dual Approximation

The linear programming relaxation of the vertex cover problem, we saw above, is given by

$$(P_{VC}) \quad \min \left\{ \sum_{v \in V} w(v)x(v) : x(u) + x(v) \geq 1, \ \forall (u, v) \in E, \quad x(v) \geq 0 \ \forall v \in V \right\}$$

and its dual is

$$(D_{VC}) \quad \max \left\{ \sum_{(u,v) \in E} y(u, v) : \sum_{u | (u,v) \in E} y(u, v) \leq w(v), \ \forall v \in V, \quad y(u, v) \geq 0 \ \forall (u, v) \in E \right\}$$

The primal–dual approximation approach would first obtain an optimal solution y^* to the dual problem (D_{VC}). Let $\hat{V} \subseteq V$ denote the set of vertices for which the dual constraints are tight, i.e.,

$$\hat{V} = \left\{ v \in V : \sum_{u \mid (u,v) \in E} y^*(u, v) = w(v) \right\}$$

The approximate vertex cover is taken to be $\hat{V}$. It follows from complementary slackness that $\hat{V}$ is a vertex cover. Using the fact that each edge (u, v) is in the star of at most two vertices (u and v), it also follows that $\hat{V}$ is a 2-approximate solution to the minimum weight vertex cover problem.

In general, the primal–dual approximation strategy is to use a dual solution, to the linear programming relaxation, along with complementary slackness conditions as a heuristic to generate an integer (primal) feasible solution which for many problems turns out to be a good approximation of the optimal solution to the original integer program.

It is a remarkable property of the vertex covering (and packing) problem that all extreme points of the linear programming relaxation have values 0, $\frac{1}{2}$, or 1 [97]. It follows that the deterministic rounding of the linear programming solution to (P_{VC}) constructs exactly the same approximate vertex cover as the primal–dual scheme described above. However, this is not true in general.

31.7.3 Semidefinite Relaxation and Rounding

The idea of using semidefinite programming to approximately solve combinatorial optimization problems appears to have originated in the work of Lovász [93] on the Shannon capacity of graphs. Grötschel et al. [62] later used the same technique to compute a maximum stable set of vertices in perfect graphs via the ellipsoid method. As we saw in Section 31.6.4, Lovász and Schrijver [94] have devised a general technique of semidefinite relaxations for general $0 - 1$ integer linear programs. We will present a lovely application of this methodology to approximate the maximum weight cut of a graph (the maximum sum of weights of edges connecting across all strict partitions of the vertex set). This application of semidefinite relaxation for approximating MAXCUT is due to Goemans and Williamson [56].

We begin with a quadratic Boolean formulation of MAXCUT

$$\max \left\{ \frac{1}{2} \sum_{(u,v) \in E} w(u, v)(1 - x(u)x(v)) : \quad x(v) \in \{-1, 1\} \, \forall \, v \in V \right\}$$

where $G(V, E)$ is the graph and $w(u, v)$ is the nonnegative weight on edge (u, v). Any $\{-1, 1\}$ vector of **x** values provides a bipartition of the vertex set of G. The expression $(1 - x(u)x(v))$ evaluates to 0 if u and v are on the same side of the bipartition and 2 otherwise. Thus, the optimization problem does indeed represent exactly the MAXCUT problem.

Next we reformulate the problem in the following way:

- We square the number of variables by allowing each **x**(v) to denote an n-vector of variables (where n is the number of vertices of the graph).
- The quadratic term $x(u)x(v)$ is replaced by **x**$(u) \cdot$ **x**(v), which is the inner product of the vectors.
- Instead of the $\{-1, 1\}$ restriction on the $x(v)$, we use the Euclidean normalization $\|\mathbf{x}(v)\| = 1$ on the **x**(v).

So we now have a problem

$$\max\left\{\frac{1}{2}\sum_{(u,v)\in E} w(u,v)(1-\mathbf{x}(u)\cdot\mathbf{x}(v)):\quad \|\mathbf{x}(v)\|=1\ \forall\ v\in V\right\}$$

which is a relaxation of the MAXCUT problem (Note that if we force only the first component of the $\mathbf{x}(v)$'s to have nonzero value, we would just have the old formulation as a special case.).

The final step is in noting that this reformulation is nothing but a semidefinite program. To see this we introduce $n\times n$ Gram matrix Y of the unit vectors $\mathbf{x}(v)$. So $Y=X^TX$ where $X=(\mathbf{x}(v):\ v\in V)$. So the relaxation of MAXCUT can now be stated as a semidefinite program.

$$\max\left\{\frac{1}{2}\sum_{(u,v)\in E} w(u,v)(1-Y_{(u,v)}):\quad Y\succeq 0,\ \ Y_{(v,v)}=1\ \forall\ v\in V\right\}$$

Note that we are able to solve such semidefinite programs to an additive error ϵ in time polynomial in the input length and $\log\frac{1}{\epsilon}$ using either the ellipsoid method or interior point method (see [3] and Chapters 30 and 34 of this handbook).

Let $\mathbf{x}^*$ denote the near-optimal solution to the semidefinite programming relaxation of MAXCUT (Convince yourself that $\mathbf{x}^*$ can be reconstructed from an optimal Y^* solution.). Now we encounter the final trick of Goemans and Williamson. The approximate maximum weight cut is extracted from $\mathbf{x}^*$ by randomized rounding. We simply pick a random hyperplane H passing through the origin. All the $v\in V$ lying to one side of H get assigned to one side of the cut and the rest to the other. Goemans and Williamson observed the following inequality.

LEMMA 31.1   For $\mathbf{x}_1$ and $\mathbf{x}_2$, two random n-vectors of unit norm, let $x(1)$ and $x(2)$ be ± 1 values with opposing sign if H separates the two vectors and with same signs otherwise. Then $\tilde{E}(1-\mathbf{x}_1^T\mathbf{x}_2)\leq 1.1393\cdot\tilde{E}(1-x(1)x(2))$ where $\tilde{E}$ denotes the expected value.

By linearity of expectation, the lemma implies that the expected value of the cut produced by the rounding is at least 0.878 times the expected value of the semidefinite program. Using standard conditional probability techniques for derandomizing, Goemans and Williamson show that a deterministic polynomial-time approximation algorithm with the same margin of approximation can be realized. Hence we have a cut with value at least 0.878 of the maximum value.

31.8   Geometry of Numbers and Integer Programming

Given an integer program with a fixed number of variables (k) we seek a polynomial-time algorithm for solving them. Note that the complexity is allowed to be exponential in k, which is independent of the input length. Clearly if the integer program has all $(0,1)$ integer variables this is a trivial task, since complete enumeration works. However if we are given an "unbounded" integer program to begin with, the problem is no longer trivial.

31.8.1   Lattices, Short Vectors, and Reduced Bases

Euclidean **lattices** are a simple generalization of the regular integer lattice $\mathcal{Z}^n$. A (point) lattice is specified by $\{\mathbf{b}_1,\cdots,\mathbf{b}_n\}$ a *basis* (where $\mathbf{b}_i$ are linearly independent n-dimensional rational vectors).

The lattice L is given by

$$L = \left\{ \mathbf{x} : \mathbf{x} = \sum_{i=1}^{n} z_i \mathbf{b}_i; \ z_i \in Z \ \forall i \right\}$$

$$B = \left(\mathbf{b}_1 \vdots \mathbf{b}_2 \vdots \cdots \vdots \mathbf{b}_n \right) \text{ a basis matrix of } L$$

THEOREM 31.9 $|detB|$ *is an invariant property of the lattice L (i.e., for every basis matrix B_i of L we have invariant $d(L) = |detB_i|$).*

Note that

$$d(L) = \prod_{i=1}^{n} |\mathbf{b}_i| \text{ where } |\mathbf{b}_i| \text{ denotes the Euclidean length of } \mathbf{b}_i$$

if and only if the basis vectors $\{\mathbf{b}_i\}$ are mutually orthogonal. A sufficiently orthogonal basis B (called a reduced basis) is one that satisfies a weaker relation

$$d(L) \leq c_n \prod_{i=1}^{n} |\mathbf{b}_i| \text{ where } c_n \text{ is a constant that depends only on } n$$

One important use (from our perspective) of a reduced basis is that one of the basis vectors has to be short. Note that there is substantive evidence that finding the shortest lattice vector is $\mathcal{NP}$-hard [62]. Minkowski proved that in every lattice L there exists a vector of length no larger than $c\sqrt{n} \sqrt[n]{d(L)}$ (with c no larger than 0.32). This follows from the celebrated Minkowski's Convex Body Theorem which forms the centerpiece of the geometry of numbers [20].

THEOREM 31.10 **(Minkowski's Convex Body Theorem).** *If $K \subseteq \mathfrak{R}^n$ is a convex body that is centrally symmetric with respect to the origin, and $L \subseteq \mathfrak{R}^n$ is a lattice such that $vol(K) \geq 2^n d(L)$, then K contains a lattice point different from the origin.*

However, no one has been successful thus far in designing an efficient algorithm (polynomial-time) for constructing the short lattice vector guaranteed by Minkowski. This is where the concept of a reduced basis comes to the rescue. We are able to construct a reduced basis in polynomial time and extract a short vector from it. To illustrate the concepts we will now discuss an algorithm due to Gauss (cf. [6]) that proves the theorem for the special case of planar lattices ($n = 2$). It is called the 60° algorithm because it produces a reduced basis $\{\mathbf{b}_1, \mathbf{b}_2\}$ such that the acute angle between the two basis vectors is at least 60°.

Procedure: 60° Algorithm

Input: Basis vectors $\mathbf{b}_1$ and $\mathbf{b}_2$ with $|\mathbf{b}_1| \geq |\mathbf{b}_2|$.
Output: A reduced basis $\{\mathbf{b}_1, \mathbf{b}_2\}$ with at least 60° angle between the basis vectors.

   0. repeat until $|\mathbf{b}_1| < |\mathbf{b}_2|$
   1. swap $\mathbf{b}_1$ and $\mathbf{b}_2$
   2. $\mathbf{b}_2 \leftarrow (\mathbf{b}_2 - m\mathbf{b}_1)$ and $m = \left[\frac{\mathbf{b}_2^T \mathbf{b}_1}{\mathbf{b}_1^T \mathbf{b}_1} \right] \in Z$.
      Here $[\alpha]$ denotes the integer nearest to α.
   3. end

Remarks

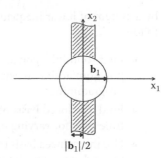

1. In each iteration the projection of $(\mathbf{b}_2 - m\mathbf{b}_1)$ onto the direction of $\mathbf{b}_1$ is of length atmost $|\mathbf{b}_1|/2$.

2. When the algorithm stops, $\mathbf{b}_2$ must lie in one of the shaded areas at the top or at the bottom of Figure 31.4 (since $|\mathbf{b}_2| > |\mathbf{b}_1|$ and since the projection of $\mathbf{b}_2$ must fall within the two vertical lines about $\mathbf{b}_1$).

3. The length of $\mathbf{b}_1$ strictly decreases in each iteration. Hence the algorithm is finite. That it is polynomial time takes a little more argument [85].

FIGURE 31.4 A reduced basis.

4. The short vector $\mathbf{b}_1$ produced by the algorithm satisfies
$$|\mathbf{b}_1| \le (1.075)(d(L))^{\frac{1}{2}}).$$

The only known polynomial-time algorithm for constructing a reduced basis in an arbitrary dimensional lattice [89] is not much more complicated than the 60° algorithm. However, the proof of polynomiality is quite technical (see [6] for an exposition).

31.8.2 Lattice Points in a Triangle

Consider a triangle T in the Euclidean plane defined by the inequalities

$$
\begin{aligned}
a_{11}x_1 + a_{12}x_2 &\le d_1 \\
a_{21}x_1 + a_{22}x_2 &\le d_2 \\
a_{31}x_1 + a_{32}x_2 &\le d_3
\end{aligned}
$$

The problem is to check if there is a lattice point $(x_1, x_2) \in Z^2$ satisfying the three inequalities. The reader should verify that checking all possible lattice points within some bounding box of T leads to an exponential algorithm for skinny and long triangles. There are many ways of realizing a polynomial-time search algorithm [47,76] for two-variable integer programs. Following Lenstra [91] we describe one that is a specialization of his [90] powerful algorithm for integer programming (searching for a lattice point in a polyhedron).

First we use a nonsingular linear transform τ that makes T equilateral (round). The transform sends the integer lattice Z^2 to a generic two-dimensional lattice L. Next we construct a reduced basis $\{\mathbf{b}_1, \mathbf{b}_2\}$ for L using the 60° algorithm. Let β_1 denote the length of the short vector $\mathbf{b}_1$. If l denotes the length of a side of the equilateral triangle $\tau(T)$ we can conclude that T is guaranteed to contain a lattice point if $\frac{l}{\beta_1} \ge \sqrt{6}$. Else $l < \sqrt{6}\beta_1$ and we can argue that just a few (no more than three) of the lattice lines $\{$Affine Hull $\{\mathbf{b}_1\} + k\mathbf{b}_2\}_{k \in Z}$ can intersect T. We recurse to the lower dimensional segments (lattice lines intersecting with T) and search each segment for a lattice point. Hence this scheme provides a simple polynomial-time algorithm for searching for a lattice point in a triangle.

31.8.3 Lattice Points in Polyhedra

In 1979, H.W. Lenstra, Jr., announced that he had a polynomial-time algorithm for integer programming problems with a fixed number of variables. The final published version [90] of his algorithm resembles the procedure described above for the case of a triangle and a planar lattice. As we have noted before, integer programming is equivalent to searching a polyhedron specified by a system of linear inequalities for an integer lattice point. Lenstra's algorithm then proceeds as follows.

- First round the polyhedron using a linear transformation. This step can be executed in polynomial time (even for varying n) using any polynomial-time algorithm for linear programming.
- Find a reduced basis of the transformed lattice. This again can be done in polynomial time even for varying n.
- Use the reduced basis to conclude that a lattice point must lie inside or recurse to several lower dimensional integer programs by slicing up the polyhedron using a long vector of the basis to ensure that the number of slices depends only on n.

A slightly different approach based on Minkowski's convex body theorem was designed by Kannan [77] to obtain an $O(n^{\frac{9n}{2}}L)$ algorithm for integer programming (where L is the length of the input). The following two theorems are straightforward consequences of these polynomial-time algorithms for integer programming with a fixed number of variables.

THEOREM 31.11   (Fixed Number of Constraints). *Checking the solvability of*

$$A\mathbf{x} \le \mathbf{b}; \quad \mathbf{x} \in Z^n$$

where A is ($m \times n$) is solvable in polynomial time if m is held fixed.

THEOREM 31.12   (Mixed Integer Programming). *Checking the solvability of*

$$A\mathbf{x} \le \mathbf{b}; \quad x_j \in Z \text{ for } j = 1, \ldots, k; \quad x_j \in \Re \text{ for } j = k+1, \ldots, n$$

where A is ($m \times n$) is solvable in polynomial time if $\min\{m, k\}$ is held fixed.

A related but more difficult question is that of counting the number of feasible solutions to an integer program or equivalently counting the number of lattice points in a polytope. Building on results described above, Barvinok [9] was able to show the following.

THEOREM 31.13   (Counting Lattice Points). *Counting the size of the set*

$$\left\{ \mathbf{x} \in Z^k \ : \ A\mathbf{x} \le \mathbf{b} \right\}$$

is solvable in polynomial time if k is held fixed.

31.8.4   An Application in Cryptography

In 1982, Shamir [117] pointed out that Lenstra's algorithm could be used to devise a polynomial-time algorithm for cracking the basic Merkle–Hellman Cryptosystem (a knapsack-based public-key cryptosystem). In such a cryptosystem the message to be sent is a (0-1) string $\bar{\mathbf{x}} \in \{0, 1\}^n$. The message is sent as an instance of the (0,1) knapsack problem, which asks for an $\mathbf{x} \in \{0, 1\}^n$ satisfying $\Sigma_{i=1}^n a_i x_i = a_0$. The knapsack problem is an $\mathcal{NP}$-hard optimization problem (we saw in Section 31.3.4 that any integer program can be aggregated to a knapsack). However, if $\{a_i\}$ form a super-increasing sequence, i.e., $a_i > \sum_{j=1}^{i-1} a_j$ $\forall i$, the knapsack problem can be solved in time $O(n)$ by a simple algorithm:

Procedure: **Best-Fit**

```
0. i ← n
1. If aᵢ ≤ a₀, x̄ᵢ ← 1, a₀ ← a₀ − aᵢ
2. i ← (i − 1)
3. If a₀ = 0 stop and return the solution x
4. If i = 1 stop the knapsack is infeasible.
5. repeat 1.
```

However, an eavesdropper could solve this problem as well and hence the $\{a_i\}$ have to be encrypted. The disguise is chosen through two secret numbers M and U such that $M > \sum_{i=1}^{n} a_i$ and U is relatively prime to M (i.e., $gcd(U, M) = 1$). Instead of $\{a_i\}$, the sequence $\{\tilde{a}_i = Ua_i(\text{mod } M)\}_{i=1,2,\cdots,n}$ is published and $\tilde{a}_0 = Ua_0(\text{mod } M)$ is transmitted.

Any receiver who knows U and M can easily reconvert $\{\tilde{a}_i\}$ to $\{a_i\}$ and apply the best fit algorithm to obtain the message $\bar{x}$. To find a_i given $\tilde{a}_i$, U and M, he runs the Euclidean algorithm on (U, M) to obtain $1 = PU + QM$. Hence, P is the inverse multiplier of U since $PU \equiv 1 \pmod{M}$. Using the identity $a_i \equiv P\tilde{a}_i \pmod{M}$ $\forall i = 0, 1, \ldots, n$, the intended receiver can now use the best fit method to decode the message. An eavesdropper knows only the $\{\tilde{a}_i\}$ and is therefore supposedly unable to decrypt the message.

The objective of Shamir's cryptanalyst (code breaker) is to find a $\hat{P}$ and in $\hat{M}$ (positive integer) such that

$$\hat{a}_i \equiv \hat{P}\tilde{a}_i \text{mod} \hat{M} \; \forall i = 0, 1, \cdots, n$$
(*) $\quad \{\hat{a}_i\}_{i=1,\cdots,n}$ is a super increasing sequence.
$$\sum_{i=1}^{n} \hat{a}_i < \hat{M}$$

It can be shown that for all pairs $(\hat{P}, \hat{M})$ such that $(\hat{P}/\hat{M})$ is sufficiently close to (P/M) will satisfy (*). Using standard techniques from diophantine approximation it would be possible to guess P and M from the estimate $(\hat{P}/\hat{M})$. However, the first problem is to get the estimate of (P/M). This is where integer programming helps.

Since $a_i \equiv P\tilde{a}_i \pmod{M}$ for $i = 1, 2, \ldots, n$ we have $a_i P\tilde{a}_i - y_i M$ for some integers $y_1, y_2, \ldots, y_n$. Dividing by $\tilde{a}_i M$ we have $\left(\frac{a_i}{\tilde{a}_i M}\right) = \frac{P}{M} - \frac{y_i}{\tilde{a}_i}$ for $i = 1, 2, \ldots, n$. For small $i(1, 2, 3, \ldots, t)$ the LHS ≈ 0 since $\sum_{i=1}^{n} a_i < M$ and the $\{a_i\}_{i=1,\ldots,n}$ are super-increasing. Thus $(y_1/\tilde{a}_1), (y_2/\tilde{a}_2), \ldots, (y_t/\tilde{a}_t)$ are close to (P/M) and hence to each other. Therefore a natural approach to estimating (P/M) would be to solve the integer program:

$$
\begin{aligned}
&\text{Find } y_1, y_2, \cdots, y_t \in \mathcal{Z} \text{ such that :}\\
\text{(IP)} \quad &\epsilon_i \le \tilde{a}_i y_1 - \tilde{a}_1 y_i \le \bar{\epsilon}_i \text{ for } i = 2, 3, \cdots, t\\
&0 < y_i < \tilde{a}_i \text{ for } i = 1, 2, \cdots, t
\end{aligned}
$$

This t variable integer program provides an estimate of (P/M), whence Diophantine approximation methods can be used to find $\hat{P}$ and $\hat{M}$. If we denote by d a density parameter of the instance where, $d = \frac{\log a_0}{n \log 2}$, the above scheme works correctly (with probability one as $n \to \infty$) if t is chosen to be $(\lfloor d \rfloor + 2)$ [86,117]. Moreover, the scheme is polynomial-time in n for fixed d. The probabilistic performance is not a handicap, since as Shamir points out, "... a cryptosystem becomes useless when most of its keys can be efficiently cryptanalyzed" [117].

31.9 Integrated Methods for Integer Programming

The integration of logic-based methodologies and mathematical programming approaches is evidenced in the recent emergence of constraint logic programming (CLP)systems [16,115] and logico-mathematical programming [22,73]. In CLP, we see a structure of Prolog-like programming language in which some of the predicates are constraint predicates whose truth values are determined by the solvability of constraints in a wide range of algebraic and combinatorial settings. The solution scheme is simply a clever orchestration of constraint solvers in these various domains and the role of conductor is played by SLD-resolution. The clean semantics of logic programming is preserved in CLP. A bonus is that the output language is symbolic and expressive.

An orthogonal approach to CLP is to use constraint programming methods to solve inference problems in logic. Imbeddings of logics in mixed integer programming sets were proposed by Williams [125] and Jeroslow [73]. Efficient algorithms have been developed for inference problems in many types and fragments of logic, ranging from Boolean to predicate to belief logics [23]. Embeddings of the satisfiability problem in first-order (predicate) logic in infinite dimensional integer programs have been used to provide new proofs of classical theorems in logic via fairly rudimentary techniques in topology [10,22].

The research agenda to bring the techniques of search, inference, and relaxation into an integrated framework for addressing optimization has been active for about a decade now. The two books by Hooker [68,69] bring out the complementarity and consequent cross fertilization of these techniques.

31.10 Prospects in Integer Programming

The current emphasis in software design for integer programming is in the development of shells (e.g., CPLEX [34], MINTO [114], and OSL [71]) wherein a general-purpose solver like Branch & Cut is the driving engine. Problem-specific code for generation of cuts and facets can be easily interfaced with the engine. We believe that this trend will eventually lead to the creation of general purpose problem solving languages for combinatorial optimization akin to AMPL [49] for linear and nonlinear programming.

A promising line of research is the development of an empirical science of algorithms for combinatorial optimization [67]. Computational testing has always been an important aspect of research on the efficiency of algorithms for integer programming. However, the standards of test designs and empirical analysis have not been uniformly applied. We believe that there will be important strides in this aspect of integer programming, and more generally of algorithms of all kinds. It may be useful to stop looking at algorithmics as purely a deductive science, and start looking for advances through repeated application of hypothesize and test paradigms [70], i.e., through empirical science.

A persistent theme in the integer programming approach to combinatorial optimization, as we have seen, is that the representation (formulation), of the problem, deeply affects the efficacy of the solution methodology. A proper choice of formulation can therefore make the difference between a successful solution of an optimization problem and the more common perception that the problem is insoluble, and one must be satisfied with the best that heuristics can provide. Formulation of integer programs has been treated more as an art form than a science by the mathematical programming community (exceptions are Jeroslow [73] and Williams [124]). We believe that progress in representation theory can have an important influence on the future of integer programming as a broad-based problem solving methodology.

31.11   Further Information

Research publications in integer programming are dispersed over a large range of journals. The following is a partial list which emphasize the algorithmic aspects: *Mathematical Programming, Mathematics of Operations Research, Operations Research, Discrete Mathematics, Discrete Applied Mathematics, Journal of Combinatorial Theory (Series B), INFORMS Journal on Computing, Operations Research Letters, SIAM Journal on Computing, SIAM Journal on Discrete Mathematics, Journal of Algorithms, Algorithmica, Combinatorica.*

Integer programming professionals frequently use the following newsletters to communicate with each other:

- *INFORMS Today* (earlier *OR/MS Today*) published by the Institute for Operations Research and Management Science (INFORMS).
- *INFORMS CSTS Newsletter* published by the INFORMS computer science technical section.
- *Optima* published by the Mathematical Programming Society.

The International Symposium on Mathematical Programming (ISMP) is held once every 3 years and is sponsored by the Mathematical Programming Society. The most recent ISMP was held in August 1997 in Lausanne, Switzerland. A conference on Integer Programming and Combinatorial Optimization (IPCO) is held on years when the symposium is not held. Some important results in integer programming are also announced in the general conferences on algorithms and complexity (e.g., SODA (SIAM), STOC (ACM), and FOCS (IEEE)). The annual meeting of the Computer Science Technical Section (CSTS) of the INFORMS held each January (partial proceedings published by Kluwer Press) is an important source for recent results in the computational aspects of integer programming.

Defining Terms

ρ-**Approximation:** An approximation method that delivers a feasible solution with objective value within a factor ρ of the optimal value of a combinatorial optimization problem.

Cutting plane: A valid inequality for an **integer polyhedron** that separates the polyhedron from a given point outside it.

Extreme point: A corner point of a polyhedron.

Fathoming: Pruning a search tree.

Integer polyhedron: A polyhedron, all of whose extreme points are integer valued.

Knapsack problem: An integer linear program with a single linear constraint other than the trivial bounds and integrality constraints on the variables.

Lattice: A point lattice generated by taking integer linear combinations of a set of basis vectors.

Linear program: Optimization of a linear function subject to linear equality and inequality constraints.

Mixed integer linear program: A linear program with the added constraint that some of the decision variables are integer valued.

Packing and covering: Given a finite collection of subsets of a finite ground set, to find an optimal subcollection that is pairwise disjoint (packing) or whose union covers the ground set (covering).

Polyhedron: The set of solutions to a finite system of linear inequalities on real-valued variables. Equivalently, the intersection of a finite number of linear half-spaces in $\Re^n$.

Reduced basis: A basis for a lattice that is nearly orthogonal.

Relaxation: An enlargement of the feasible region of an optimization problem. Typically, the relaxation is considerably easier to solve than the original optimization problem.

References

1. Ahuja, R.K., Magnati, T.L., and Orlin, J.B., *Network Flows: Theory, Algorithms and Applications,* Prentice Hall, Englewood Cliffs, NJ, 1993.
2. Akgul, M., *Topics in Relaxation and Ellipsoidal Methods, Research Notes in Mathematics,* Pitman Publishing Ltd., Boston, MA, 1984.
3. Alizadeh, F., Interior point methods in semidefinite programming with applications to combinatorial optimization, *SIAM Journal on Optimization,* 5, 13–51, 1995.
4. Applegate, D., Bixby, R.E., Chvátal, V., and Cook, W., Finding cuts in large TSP's, Technical Report, AT&T Bell Laboratories, Aug. 1994.
5. Arora, S., Lund, C., Motwani, R., Sudan, M., and Szegedy, M., Proof verification and hardness of approximation problems, in *Proceedings of the 33rd IEEE Symposium on Foundations of Computer Science,* pp. 14–23, 1992.
6. Bachem, A. and Kannan, R., Lattices and the basis reduction algorithm, Technical Report, Computer Science, Carnegie Mellon University, Pittsburgh, PA, 1984.
7. Barahona, F., Jünger, M., and Reinelt, G., Experiments in quadratic 0−1 programming, *Mathematical Programming,* 44, 127–137, 1989.
8. Barany I. and Furedi, Z., Computing the volume is difficult, in *Proceedings of 18th Annual ACM Symposium on Theory of Computing,* ACM Press, Berkeley, CA, pp. 442–447, 1986.
9. Barvinok, A., A polynomial time algorithm for counting integral points in polyhedra when the dimension is fixed. in *Proceedings of the 34th IEEE Conference on the Foundations of Computer Science (FOCS),* IEEE Press, Palo Alto, CA, pp. 566–572, 1993.
10. Borkar, V.S., Chandru, V., and Mitter, S.K., Mathematical programming embeddings of logic, *Journal of Automed Reasoning,* 29(1), 91–106, 2002.
11. Benders, J.F., Partitioning procedures for solving mixed-variables programming problems, *Numerische Mathematik,* 4, 238–252, 1962.
12. Berge, C., Farbung von Graphen deren samtliche bzw. deren ungerade Kreise starr sind (Zusammen-fassung), Wissenschaftliche Zeitschrift, Martin Luther Universitat Halle-Wittenberg, Mathematisch-Naturwiseenschaftliche Reihe, pp. 114–115, 1961.
13. Berge, C., Sur certains hypergraphes generalisant les graphes bipartites, in *Combinatorial Theory and Its Applications I,* Erdos, P., Renyi, A., and Sos, V., Eds., Colloq. Math. Soc. Janos Bolyai, 4, North Holland, Amsterdam, the Netherlands, pp. 119–133, 1970.
14. Berge, C., Balanced matrices, *Mathematical Programming,* 2, 19–31, 1972.
15. Berge, C. and Las Vergnas, M., Sur un theoreme du type Konig pour hypergraphes, International Conference on Combinatorial Mathematics, *Annals of the New York Academy of Sciences,* 175, 32–40, 1970.
16. Borning, A., Ed., *Principles and Practice of Constraint Programming, Lecture Notes in Computer Science* Volume 874, Springer-Verlag, Berlin, Germany, 1994.
17. Borosh, I. and Treybig, L.B., Bounds on positive solutions of linear diophantine equations, *Proceedings of the American Mathematical Society,* 55, 299, 1976.
18. Camion, P., Characterization of totally unimodular matrices, *Proceedings of the American Mathematical Society,* 16, 1068–1073, 1965.

19. Cannon, T.L. and Hoffman, K.L., Large-scale zero-one linear programming on distributed workstations, *Annals of Operations Research*, 22, 181–217, 1990.

20. Cassels, J.W.S., *An Introduction to the Geometry of Numbers*, Springer-Verlag, Berlin, Germany, 1971.

21. Chandru, V., Complexity of the supergroup approach to integer programming, PhD Thesis, Operations Research Center, MIT, Cambridge, MA, 1982.

22. Chandru., V. and Hooker, J.N., Extended Horn sets in propositional logic, *JACM*, 38, 205–221, 1991.

23. Chandru, V. and Hooker, J.N., *Optimization Methods for Logical Inference*, Wiley Interscience, New York, 1999.

24. Chopra, S., Gorres, E.R., and Rao, M.R., Solving Steiner tree problems by Branch and Cut, *ORSA Journal of Computing*, 3, 149–156, 1992.

25. Chvátal, V., Edmonds polytopes and a hierarchy of combinatorial problems, *Discrete Mathematics*, 4, 305–337, 1973.

26. Chvátal, V., On certain polytopes associated with graphs, *Journal of Combinatorial Theory*, B, 18, 138–154, 1975.

27. Conforti, M. and Cornuejols, G., A class of logical inference problems solvable by linear programming, *FOCS*, 33, 670–675, 1992.

28. Conforti, M., Cornuejols, G., and De Francesco, C., Perfect 0, ±1 matrices, preprint, Carnegie Mellon University, Pittsburgh, PA, 1993.

29. Conforti, M., Cornuejols, G., Kapoor, A., and Vuskovic, K., Balanced 0, ±1 matrices, Parts I–II, preprints, Carnegie Mellon University, Pittsburgh, PA, 1994.

30. Conforti, M., Cornuejols, G., Kapoor, A., Vuskovic, K., and Rao, M.R., Balanced matrices, in *Mathematical Programming, State of the Art 1994*, Birge, J.R. and Murty, K.G., Eds., University of Michigan, Ann Arbor, MI, 1994.

31. Conforti, M., Cornuejols, G., and Rao, M.R., Decomposition of balanced 0,1 matrices, Parts I–VII, preprints, Carneigie Mellon University, Pittsburgh, PA, 1991.

32. Conforti, M. and Rao, M.R., Testing balancedness and perfection of linear matrices, *Mathematical Programming*, 61, 1–18, 1993.

33. Cornuejols, G. and Novick, B., Ideal 0,1 matrices, *Journal of Combinatorial Theory*, 60, 145–157, 1994.

34. CPLEX, *Using the CPLEX Callable Library and CPLEX Mixed Integer Library*, CPLEX Optimization, 1993.

35. Crowder, H., Johnson, E.L., and Padberg, M.W., Solving large scale 0–1 linear programming problems, *Operations Research*, 31, 803–832, 1983.

36. Cunningham, W.H., Testing membership in matroid polyhedra, *Journal of Combinatorial Theory*, 36B, 161–188, 1984.

37. Domich, P.D., Kannan, R., and Trotter, L.E., Hermite normal form computation using modulo determinant arithmetic, *Mathematics of Operations Research*, 12, 50–59, 1987.

38. Dyer, M. and Frieze, A., On the complexity of computing the volume of a polyhedron, *SIAM Journal of Computing*, 17, 967–974, 1988.

39. Dyer, M., Frieze, A., and Kannan, R., A random polynomial time algorithm for approximating the volume of convex bodies, in *Proceedings of 21st Symposium on Theory of Computing*, ACM Press, New York, pp. 375–381, 1989.

40. Edmonds, J., Maximum matching and a polyhedron with 0–1 vertices, *Journal of Research of the National Bureau of Standards*, 69B, 125–130, 1965.

41. Edmonds, J., Submodular functions, matroids and certain polyhedra, in *Combinatorial Structures and Their Applications*, Guy, R. et al., Eds., Gordon Breach, New York, pp. 69–87, 1970.

42. Edmonds, J., Matroids and the greedy algorithm, *Mathematical Programming*, 1, 127–136, 1971.

43. Edmonds, J., Matroid intersection, *Annals of Discrete Mathematics*, 4, 39–49, 1979.

44. Edmonds, J. and Giles, R., A min-max relation for submodular functions on graphs, *Annals of Discrete Mathematics*, 1, 185–204, 1977.

45. Edmonds, J. and Johnson, E.L., Matching well solved class of integer linear programs, in *Combinatorial Structure and Their Applications*, Guy, R. et al., Eds., Gordon Breach, New York, pp. 89–92, 1970.

46. Farkas, Gy., A Fourier-féle mechanikai elv alkalmazásai (in Hungarian), *Mathematikai és Természettudományi Értesitö*, 12, 457–472, 1894.

47. Feit, S.D., A fast algorithm for the two-variable integer programming problem, *JACM*, 31, 99–113, 1984.

48. Fonlupt, J. and Zemirline, A., A polynomial recognition algorithm for K_4 \ e-free perfect graphs, Research Report, University of Grenoble, Grenoble, France, 1981.

49. Fourer, R., Gay, D.M., and Kernighan, B.W., *AMPL: A Modeling Language for Mathematical Programming*, Scientific Press, San Francisco, CA, 1993.

50. Frank, A., A weighted matroid intersection theorem, *Journal of Algorithms*, 2, 328–336, 1981.

51. Frumkin, M.A., Polynomial-time algorithms in the theory of linear diophentine equations, in *Fundamentals of Computation Theory*, Karpinski, M., Ed., *Lecture Notes in Computer Science*, Springer-Verlag, Berlin, Germany, 56, 1977.

52. Fulkerson, D.R., The perfect graph conjecture and the pluperfect graph theorem, in *Proceedings of the Second Chapel Hill Conference on Combinatorial Mathematics and Its Applications*, Bose, R.C. et al., Eds., pp. 171–175, 1970.

53. Fulkerson, D.R., Hoffman, A., and Oppenheim, R., On balanced matrices, *Mathematical Programming Study*, 1, 120–132, 1974.

54. Garfinkel, R. and Nemhauser, G.L., *Integer Programming*, John Wiley & Sons, New York, 1972.

55. Gathen, J.V.Z. and Sieveking, M., Linear integer inequalities are NP-complete, *SIAM Journal of Computing*, 1976.

56. Goemans, M.X. and Williamson, D.P., .878 approximation algorithms MAX CUT and MAX 2SAT, in *Proceedings of ACM STOC*, pp. 422–431, 1994.

57. Gomory, R.E., Outline of an algorithm for integer solutions to linear programs, *Bulletin of the American Mathematical Society*, 64, 275–278, 1958.

58. Gomory, R.E., Early integer programming, in *History of Mathematical Programming*, Lenstra, J.K. et al., Eds., North Holland, Amsterdam, the Netherlands, 1991.

59. Gomory, R.E., On the relation between integer and noninteger solutions to linear programs, *Proceedings of the National Academy of Sciences of the United States of America*, 53, 260–265, 1965.

60. Gomory, R.E. and Hu, T.C., Multi-terminal network flows, *SIAM Journal of Applied Mathematics*, 9, 551–556, 1961.

61. Grötschel, M., Lovász L., and Schrijver, A., The ellipsoid method and its consequences in combinatorial optimization, *Combinatorica*, 1, 169–197, 1982.

62. Grötschel, M., Lovász, L., and Schrijver, A., *Geometric Algorithms and Combinatorial Optimization*, Springer-Verlag, New York, 1988.

63. Held, M., Wolfe, P., and Crowder, H.P., Validation of subgradient optimization, *Mathematical Programming*, 6, 62–88, 1974.

64. Held, M. and Karp, R.M., The travelling-salesman problem and minimum spanning trees, *Operations Research*, 18, 1138–1162, Part II, 1970. *Mathematical Programming*, 1, 6–25, 1971.

65. Hoffman, A.J. and Kruskal, J.K., Integral boundary points of convex polyhedra, *Linear Inequalities and Related Systems*, Kuhn, H.W. and Tucker, A.W., Eds., Princeton University Press, Princeton, NJ, 1, 223–246, 1956.

66. Hooker, J.N., Resolution and the integrality of satisfiability polytopes, preprint, GSIA, Carnegie Mellon University, Pittsburgh, PA, 1992.

67. Hooker, J.N., Towards and empirical science of algorithms, *Operations Research*, 1993.

68. Hooker, J.N., *Logic Based Methods for Optimization: Combining Optimization and Constraint Satisfaction*, Wiley, New York, 2000.
69. Hooker, J.N., *Integrated Methods for Optimization*, Springer Science + Business Media, LC, New York, 2007.
70. Hooker, J.N. and Vinay, V., Branching rules for satisfiability, *Journal of Automated Reasoning*, 15(3), 359–383, 1995.
71. Wilson, D.G and Rudin, B.D., Introduction to the IBM optimization subroutine library, *IBM Systems Journal*, 31(1), 4–10, 1992.
72. Jeroslow, R.G., There cannot be any algorithm for integer programming with quadratic constraints, *Operations Research*, 21, 221–224, 1973.
73. Jeroslow, R.G., *Logic-Based Decision Support: Mixed Integer Model Formulation*, Annals of Discrete Mathematics, Vol. 40, North Holland, Amsterdam, the Netherlands, 1989.
74. Jeroslow, R.G. and Lowe, J.K., Modeling with integer variables, *Mathematical Programming Studies*, 22, 167–184, 1984.
75. Jünger, M., Reinelt, G., and Thienel, S., Practical problem solving with cutting plane algorithms, in *Combinatorial Optimization: Papers from the DIMACS Special Year*, Cook, W., Lovász, L., and Seymour, P., Eds., *Series in Discrete Mathematics and Theoretical Computer Science*, Vol. 20, AMS, pp. 111–152, 1995.
76. Kannan, R., A polynomial algorithm for the two-variable integer programming problem, *JACM*, 27, 1980.
77. Kannan, R., Minkowski's convex body theorem and integer programming, *Mathematics of Operations Research*, 12, 415–440, 1987.
78. Kannan, R. and Bachem, A., Polynomial algorithms for computing the Smith and Hermite normal forms of an integer matrix, *SIAM Journal of Computing*, 8, 1979.
79. Kannan, R. and Monma, C.L., On the computational complexity of integer programming problems, in *Lecture Notes in Economics and Mathematical Systems* 157, Henn, R., Korte, B., and Oettle, W., Eds., Springer-Verlag, Berlin, Germany, 1978.
80. Karmarkar, N.K., A new polynomial-time algorithm for linear programming, *Combinatorica*, 4, 373–395, 1984.
81. Karmarkar, N.K., An interior-point approach to NP-complete problems—Part I, *Contemporary Mathematics*, 114, 297–308, 1990.
82. Karp, R., Reducibilities among combinatorial problems, in *Complexity of Computer Computations*, Miller, R.E. and Thatcher, J.W., Eds., Plenum Press, New York, pp. 85–103, 1972.
83. Karp, R.M. and Papadimitriou, C.H., On linear characterizations of combinatorial optimization problems, *SIAM Journal of Computing*, 11, 620–632, 1982.
84. Khachiyan, L.G., A polynomial algorithm in linear programming, *Doklady Akademiia Nauk SSSR*, 244(5), 1093–1096, 1979, translated into English in *Soviet Mathematics Doklady*, 20(1), 191–194, 1979.
85. Lagarias, J.C., Worst-case complexity bounds for algorithms in the theory of integral quadratic forms, *Journal of Algorithms*, 1, 142–186, 1980.
86. Lagarias, J.C., Knapsack public key cryptosystems and diophantine approximation, *Advances in Cryptology*, Proceedings of CRYPTO 83, Plenum Press, New York, pp. 3–23, 1983.
87. Lawler, E.L., Matroid intersection algorithms, *Mathematical Programming*, 9, 31–56, 1975.
88. Lehman, A., On the width-length inequality, mimeographic notes, 1965, *Mathematical Programming*, 17, 403–417, 1979.
89. Lenstra, A.K., Lenstra, Jr., H.W., and Lovász, L., Factoring Polynomials with Rational Coefficients, Report 82-05, University of Amsterdam, Amsterdam, the Netherlands, 1982.
90. Lenstra, Jr., H.W., Integer programming with a fixed number of variables, *Mathematics of Operations Research*, 8, 538–548, 1983.
91. Lenstra, Jr., H.W., Integer programming and cryptography, *The Mathematical Intelligencer*, 6, 1984.

92. Lovász, L., Normal hypergraphs and the perfect graph conjecture, *Discrete Mathematics,* 2, 253–267, 1972.

93. Lovász, L., On the Shannon capacity of a graph, *IEEE Transactions on Information Theory,* 25, 1–7, 1979.

94. Lovász, L. and Schrijver, A., Cones of matrices and set functions, *SIAM Journal on Optimization,* 1, 166–190, 1991.

95. Martin, R.K., Using separation algorithms to generate mixed integer model reformulations, *Operations Research Letters,* 10, 119–128, 1991.

96. McDiarmid, C.J.H., Rado's theorem for polymatroids, *Proceedings of the Cambridge Philosophical Society,* 78, 263–281, 1975.

97. Nemhauser, G.L. and Trotter, Jr., L.E., Properties of vertex packing and independence system polyhedra, *Mathematical Programming,* 6, 48–61, 1974.

98. Nemhauser, G.L. and Wolsey, L.A., *Integer and Combinatorial Optimization,* John Wiley & Sons, New York, 1988.

99. Padberg, M.W., Equivalent knapsack-type formulations of bounded integer linear programs: An alternative approach, *Naval Research Logistics Quarterly,* 19, 699–708, 1972.

100. Padberg, M.W., Perfect zero-one matrices, *Mathematical Programming,* 6, 180–196, 1974.

101. Padberg, M.W., Covering, packing and knapsack problems, *Annals of Discrete Mathematics,* 4, 265–287, 1979.

102. Padberg, M.W., Lehman's forbidden minor characterization of ideal 0,1 matrices, *Discrete Mathematics,* 111, 409–420, 1993.

103. Padberg, M.W. and Rao, M.R., The Russian method for linear inequalities, Part III, Bounded integer programming, Preprint, New York University, New York, 1981.

104. Padberg, M.W. and Rao, M.R., Odd minimum cut-sets and *b*-matching, *Mathematics of Operations Research,* 7, 67–80, 1982.

105. Padberg, M.W. and Rinaldi, G., A branch and cut algorithm for the resolution of large scale symmetric travelling salesman problems, *SIAM Review,* 33, 60–100, 1991.

106. Papadimitriou, C.H. and Steiglitz, K., *Combinatorial Optimization: Algorithms and Complexity,* Prentice Hall, Englewood Cliffs, NJ, 1982.

107. Papadimitriou, C.H. and Yannakakis, M., Optimization, approximation, and complexity classes, *Journal of Computer and Systems Sciences,* 43, 425–440, 1991.

108. Parker, G. and Rardin, R.L., *Discrete Optimization,* Academic Press, London, U.K., 1988.

109. Picard, J.C. and Ratliff, H.D., Minimum cuts and related problems, *Networks,* 5, 357–370, 1975.

110. Pulleyblank, W.R., Polyhedral combinatorics, in *Handbooks in Operations Research and Management Science (Volume 1: Optimization),* Nemhauser, G.L, Rinnoy Kan, A.H.G., and Todd, M.J., Eds., North Holland, Amsterdam, the Netherlands, pp. 371–446, 1989.

111. Raghavan, P. and Thompson, C.D., Randomized rounding: A technique for provably good algorithms and algorithmic proofs, *Combinatorica,* 7, 365–374.

112. Rhys, J.M.W., A selection problem of shared fixed costs and network flows, *Management Science,* 17, 200–207, 1970.

113. Sahni, S., Computationally related problems, *SIAM Journal of Computing,* 3, 1974.

114. Savelsbergh, M.W.P., Sigosmondi, G.S., and Nemhauser, G.L., MINTO, a Mixed INTeger Optimizer, *Operations Research Letters,* 15, 47–58, 1994.

115. Saraswat, V. and Van Hentenryck, P., Eds., *Principles and Practice of Constraint Programming,* MIT Press, Cambridge, MA, 1995.

116. Schrijver, A., *Theory of Linear and Integer Programming,* John Wiley & Sons, New York, 1986.

117. Shamir, A., A polynomial-time algorithm for breaking the basic Merkle-Hellman cryptosystem, in *Proceedings of the Symposium on the Foundations of Computer Science,* IEEE Press, Chicago, IL, pp. 145–152, 1982.

118. Shapiro, J.F., A survey of Lagrangean techniques for discrete optimization, *Annals of Discrete Mathematics*, 5, 113–138, 1979.

119. Seymour, P., Decompositions of regular matroids, *Journal of Combinatorial Theory*, B, 28, 305–359, 1980.

120. Shmoys, D.B., Computing near-optimal solutions to combinatorial optimization problems, in *Combinatorial Optimization: Papers from the DIMACS Special Year*, Cook, W., Lovász, L., and Seymour, P., Eds., *Series in Discrete Mathematics and Theoretical Computer Science*, Vol. 20, AMS, pp. 355–398, 1995.

121. Shor, N.Z., Convergence rate of the gradient descent method with dilation of the space, *Cybernetics*, 6, 1970.

122. Truemper, K., Alpha-balanced graphs and matrices and GF(3)-representability of matroids, *Journal of Combinatorial Theory*, B, 55, 302–335, 1992.

123. Weyl, H., Elemetere Theorie der konvexen polyerer, *Commentari Mathematici Helvetici*, 1, 3–18, 1935. (English translation in *Annals of Mathematics Studies*, 24, Princeton, 1950.)

124. Williams, H.P., Experiments in the formulation of integer programming problems, *Mathematical Programming Study*, 2, 1974.

125. Williams, H.P., Linear and integer programming applied to the propositional calculus, *International Journal of Systems Research and Information Science*, 2, 81–100, 1987.

126. Yannakakis, M., Expressing Combinatorial optimization problems by linear programs, in *Proceedings of ACM Symposium of Theory of Computing*, pp. 223–228, 1988.

127. Ziegler, M., *Convex Polytopes*, Springer-Verlag, New York, 1995.

32

Convex Optimization

Florian Jarre
Universität Düsseldorf

Stephen A. Vavasis
University of Waterloo

32.1 Introduction

Nonlinear constrained **optimization** refers to the problem of minimizing $f(\mathbf{x})$ subject to $\mathbf{x} \in D$, where D is a subset of $\mathbf{R}^n$ and f is a continuous function from D to $\mathbf{R}$. Thus, an input instance is a specification of D, called the **feasible set**, and f, called the objective function. The output from an optimization algorithm is either a point $\mathbf{x}^* \in D$, called an **optimizer** or **global optimizer**, such that $f(\mathbf{x}^*) \leq f(\mathbf{x})$ for all $\mathbf{x} \in D$, or else a statement (preferably accompanied by a certificate) that no such $\mathbf{x}^*$ exists. The terms **minimizer** and global minimizer are also used.

Stated in this manner, nonlinear optimization encompasses a huge range of scientific and engineering problems. In fact, this statement of the problem is too general: there are undecidable problems that can naturally fit into the framework of the last paragraph. Thus, most optimization algorithms are limited to some subclass of the general case. Furthermore, most algorithms compute something easier to find than a true global minimizer, like a local minimizer.

This chapter focuses on **convex programming**, the subclass of nonlinear optimization in which the set D is convex and the function f is also convex. The term convex is defined in Section 32.2. Convex programming is interesting for three reasons: (a) convex problems arise in many applications, some of which are described in subsequent sections, (b) mathematicians have developed a rich body of theory on the topic of convexity, and (c) there are powerful algorithms for efficiently solving convex problems, which is the topic of most of this chapter. The development of **self-concordant barrier functions** by Nesterov and Nemirovskii [44] has led to much new research in this field. The purpose of this chapter is to describe some of the fundamentals of convex optimization and sketch some of the recent developments based on self-concordance. At the end of this chapter, some issues concerning general (nonconvex) nonlinear optimization are raised. This chapter does not cover low dimensional convex programming in which the number of unknowns is very small, e.g., less than 10. Such problems arise in computational geometry and are covered in that chapter.

The remainder of this chapter is organized as follows. In Section 32.2 we cover the general definitions and the principles of convex optimization. In Section 32.3 we cover the **ellipsoid method** for convex optimization. In Sections 32.4 and 32.5 we describe an interior-point method for convex programming, based on self-concordant barrier functions. In Section 32.6 the **interior-point method** is specialized to semidefinite programming and Section 32.7 is further specialized to **linear programming**. In Section 32.8 we discuss some Turing-machine complexity issues for convex optimization. Finally, in Section 32.10 we make some brief remarks on the applications to nonconvex optimization.

32.2   Underlying Principles

32.2.1   Convexity

The main theme of this chapter is convexity. A set $D \subset \mathbf{R}^n$ is said to be **convex** if, for any $\mathbf{x}, \mathbf{y} \in D$, and for any $\lambda \in [0, 1]$, $\lambda \mathbf{x} + (1 - \lambda)\mathbf{y} \in D$. In other words, for any $\mathbf{x}, \mathbf{y} \in D$, the segment joining $\mathbf{x}, \mathbf{y}$ also lies in D. Some useful properties of **convex sets** are as follows:

- A convex set is connected.
- The intersection of two convex sets is convex. In fact, an infinite intersection of convex sets is convex.
- An affine transformation applied to a convex set yields a convex set. An affine transformation from $\mathbf{R}^n$ to $\mathbf{R}^m$ is a mapping of the form $\mathbf{x} \mapsto A\mathbf{x} + \mathbf{b}$, where A is an $m \times n$ matrix and $\mathbf{b}$ is an m-vector.

Let D be a convex subset of $\mathbf{R}^n$. A function $f : D \to \mathbf{R}$ is said to be convex if for all $\mathbf{x}, \mathbf{y} \in D$, and for all $\lambda \in [0, 1]$,

$$f(\lambda \mathbf{x} + (1 - \lambda)\mathbf{y}) \leq \lambda f(\mathbf{x}) + (1 - \lambda)f(\mathbf{y}). \tag{32.1}$$

This definition may be stated geometrically as follows. If $\mathbf{x}, \mathbf{y} \in D$, then the segment in $\mathbf{R}^{n+1}$ joining $(\mathbf{x}, f(\mathbf{x}))$ to $(\mathbf{y}, f(\mathbf{y}))$ lies above the graph of f. Some useful properties of the **convex functions** are as follows:

- A convex function composed with an affine transformation is convex.
- The pointwise maximum of two convex functions is convex.
- If f, g are convex functions, so is $\alpha f + \beta g$ for any nonnegative scalars α and β.
- If f is a convex function on D, then the set $\{\mathbf{x} \in D : f(\mathbf{x}) \leq \alpha\}$ is convex for any choice of $\alpha \in \mathbf{R}$.

Some examples of convex functions are as follows:

- A **linear function** $f(\mathbf{x}) = \mathbf{a}^T \mathbf{x} + c$, where $\mathbf{a} \in \mathbf{R}^n$ and $c \in \mathbf{R}$ are given. (A constant function is a special case of a linear function.)
- More generally, a **positive semidefinite** quadratic function. Recall that a quadratic function has the form $f(\mathbf{x}) = \frac{1}{2}\mathbf{x}^T H \mathbf{x} + \mathbf{a}^T \mathbf{x} + c$, where H is a given $n \times n$ **symmetric matrix**,

a is a given vector, and c a given scalar. Recall that an $n \times n$ real matrix H is **symmetric** if $H^T = H$, where superscript T denotes matrix transpose. This function is convex provided that H is a **positive semidefinite** matrix. Recall that a real symmetric matrix H is said to be positive semidefinite if either of the following two equivalent conditions holds:

1. For every $\mathbf{w} \in \mathbf{R}^n$, $\mathbf{w}^T H \mathbf{w} \geq 0$.

2. Every eigenvalue of H is nonnegative.

If every eigenvalue of H is positive (or, equivalently, if $\mathbf{w}^T H \mathbf{w} > 0$ for every nonzero vector $\mathbf{w}$), we say H is **positive definite**.

- The function $f(x) = -\ln x$ defined on the positive real numbers.

Feasible sets D for optimization problems are typically defined as the set of points $\mathbf{x}$ satisfying a list of equality constraints, that is, **constraints** of the form $g(\mathbf{x}) = 0$ where g is a real-valued function, and inequality constraints, that is, constraints of the form $h(\mathbf{x}) \leq 0$, where h is a real-valued function. The entries of $\mathbf{x}$ are called the variables or decision variables or unknowns of the optimization problem. Some commonly occurring constraints are as follows.

- A linear equation $\mathbf{a}^T \mathbf{x} = b$, where $\mathbf{a}$ is a fixed vector in $\mathbf{R}^n$ and b is a scalar.
- A linear inequality $\mathbf{a}^T \mathbf{x} \leq b$.
- A p-norm constraint of the form $\|\mathbf{x} - \mathbf{x}_0\|_p \leq r$ where $\mathbf{x}_0 \in \mathbf{R}^n, p \geq 1$ and $r \geq 0$ are given.
- A constraint that X must be symmetric and positive semidefinite, where X is a matrix whose entries are variables.

  The constraint that X is symmetric amounts to $n(n-1)/2$ linear equality constraints among off-diagonal elements of X. The semidefinite constraint can be written in the form $-\pi(X) \leq 0$ where

$$\pi(X) = \min_{\|\mathbf{w}\|_2 = 1} \mathbf{w}^T X \mathbf{w}.$$

  The function $-\pi$ is a convex function of matrices. In the case that X is symmetric, $\pi(X)$ is the minimum eigenvalue of X.

A convex constraint is defined to be either a linear equality or inequality constraint, or is a constraint of the form $h(\mathbf{x}) \leq \mathbf{0}$ where h is a convex function. In fact, all of the aforementioned constraints are convex constraints. Because the intersection of convex sets is convex, any arbitrary conjunction of constraints of any type chosen from the aforementioned list, or affine transformations of these constraints, yields a convex set.

A convex set defined by linear constraints (i.e., the constraints of the first two types in the aforementioned list) is said to be a **polyhedron**.

Finally, we come to the definition of the main topic of this chapter. A convex optimization or **convex programming** instance is an optimization instance in which the feasible set D is defined by convex constraints, and the objective function f is also convex.

An alternative (more general) definition is that convex programming means minimizing a convex function $f(\mathbf{x})$ over a convex set D. The distinction (compared to the definition in the last paragraph) is that a convex set D is sometimes represented by means other than convex constraints. For example, the set $D = \{(x, y) \in \mathbf{R}^2 : x \geq 0, xy \geq 1\}$ is convex even though $xy \geq 1$ is not a convex constraint. The interior-point method described in the following text can sometimes be applied in this more general setting.

Some special cases of convex programming problems include the following:

1. **Linear programming** (LP) refers to optimization problems in which $f(\mathbf{x})$ is a linear function of $\mathbf{x}$ and D is a polyhedron, that is, it is defined by the conjunction of linear constraints.

2. **Quadratic programming** (QP) refers to optimization problems in which $f(\mathbf{x})$ is a quadratic function and D is a polyhedron. Such problems will be convex if the matrix in the quadratic function is positive semidefinite.

3. **Semidefinite programming** (SDP) refers to optimization problems in which $f(\mathbf{x})$ is a linear function and D is defined by linear and semidefinite constraints.

The word programming in these contexts is not directly related to programming a computer in the usual sense. This terminology arises for historical reasons.

One reason to be interested in convex optimization is that these classes of problems occur in many applications. Linear programming [6] is the problem that launched optimization as a discipline of study and is widely used for scheduling and planning problems. Quadratic programming arises in applications in which the objective function involves distance or energy (which are typically quadratic functions). Quadratic programming also arises as a subproblem of more general optimization algorithms; see, Nash and Sofer, [39] for details.

Finally, semidefinite programming is more general than both linear and convex programming, and arises in a number of unexpected contexts. The recent survey [64] presents many applications in which, at first glance, the optimization problem seemingly has nothing to do with positive semidefinite matrices. One recently discovered interesting class of applications is approximation algorithms for combinatorial problems. See the chapter on approximation algorithms.

32.2.2 Derivatives

Derivatives play a crucial role in optimization for two related reasons: necessary conditions for optimality usually involve derivatives, and most optimization algorithms involve derivatives. Both of these points are developed in more detail in the following text. For a function $f : \mathbf{R}^n \to \mathbf{R}$, we denote its first derivative (the **gradient**) by ∇f and its second derivative (the **Hessian** matrix) as $\nabla^2 f$. Recall that, under the assumption that f is C^2, the Hessian matrix is always symmetric.

A convex function is not necessarily differentiable: for example, the convex function $f(x) = |x|$ is not differentiable at the origin. Convex functions, however, have the special property that even when they fail to be differentiable, they possess a **subdifferential**, which is a useful generalization of derivative. Let D be a nonempty convex subset of $\mathbf{R}^n$. Let f be a convex function on a convex set D, and let $\mathbf{x}$ be a point in the interior of D. This definition can also be generalized to the case when D has no interior, that is, its affine hull has dimension less than n. (The affine hull of a set X is the minimal affine set containing X. An affine set is a subset of $\mathbf{R}^n$ of the form $\{A\mathbf{x} + \mathbf{b} : \mathbf{x} \in \mathbf{R}^p\}$ where A is an $n \times p$ matrix and $\mathbf{b}$ is an n-vector.) The subdifferential of f at $\mathbf{x}$ is defined to be the set of vectors $\mathbf{v} \in \mathbf{R}^n$ such that $f(\mathbf{y}) \geq f(\mathbf{x}) + \mathbf{v}^T(\mathbf{y} - \mathbf{x})$ for all $\mathbf{y} \in D$. For example, the subdifferential of the function $f(x) = |x|$ at $x = 0$ is the closed interval $[-1, 1]$. Some useful properties of the subdifferential are

- The subdifferential is a nonempty, closed, convex set.
- When f is differentiable at a point $\mathbf{x}$ in the interior of D, the subdifferential is a singleton set whose unique member is the ordinary derivative $\nabla f(\mathbf{x})$.

An element of the subdifferential is called a subgradient of f.

Since a convex function is not necessarily differentiable, a fortiori it is not necessarily twice-differentiable. When f is twice-continuously differentiable, there is a simple characterization of convexity. Let f be a C^2 function defined on a convex set D with a nonempty interior. Then f is convex if and only if the second derivative of f, that is, its Hessian matrix $\nabla^2 f(\mathbf{x})$, is positive semidefinite for all $\mathbf{x} \in D$.

32.2.3 Optimality Conditions

Recall that $\mathbf{x}^* \in D$ is the optimizer if $f(\mathbf{x}^*) \leq f(\mathbf{x})$ for all $\mathbf{x} \in D$. Verifying optimality thus apparently requires a global knowledge of the function f. Given a feasible point $\mathbf{x}^*$, it is desirable to be able to check whether $\mathbf{x}^*$ is an optimizer using information about D and f only in a small neighborhood of $\mathbf{x}^*$. For general optimization problems, such a local characterization is not possible because general optimization problems can have local minimizers that are not globally optimal. We define this term as follows: $\mathbf{x}^* \in D$ is a local minimizer of f if there exists an open set $N \subset \mathbf{R}^n$ containing $\mathbf{x}^*$ such that $f(\mathbf{x}) \geq f(\mathbf{x}^*)$ for all $\mathbf{x} \in N \cap D$.

In the case of convex optimization, local minimizers are always global minimizers and we state this as a theorem.

THEOREM 32.1 *Let f be a convex function defined on a convex domain D. Let $\mathbf{x}^* \in D$ be a local minimizer of f. Then $\mathbf{x}^*$ is a global minimizer of f.*

PROOF Suppose $\mathbf{x}^*$ is a local minimizer, and let $\mathbf{x}$ be any other feasible point. Consider the sequence of points $\mathbf{x}_1, \mathbf{x}_2, \ldots$ converging to $\mathbf{x}^*$ given by $\mathbf{x}_i = (1 - 1/i)\mathbf{x}^* + (1/i)\mathbf{x}$; clearly these points lie in D since D is convex. By the convexity of f, $f(\mathbf{x}_i) \leq (1 - 1/i)f(\mathbf{x}^*) + (1/i)f(\mathbf{x})$. On the other hand, by the local minimality of $\mathbf{x}^*$, there is an i sufficiently large such that $f(\mathbf{x}_i) \geq f(\mathbf{x}^*)$. Combining these inequalities, we have $f(\mathbf{x}^*) \leq (1 - 1/i)f(\mathbf{x}^*) + (1/i)f(\mathbf{x})$, which simplifies to $f(\mathbf{x}^*) \leq f(\mathbf{x})$.

For general optimization problems, **local optimizers** have local characterizations in terms of derivatives. Usually derivative characterizations are either necessary or sufficient for local minimality, but rarely is there a single condition that is both necessary and sufficient. In the case of convex optimization, there exist derivative conditions that are both necessary and sufficient for global optimality, provided that a certain **constraint qualification** holds, and provided that the functions in question are differentiable.

To set the stage, let us first consider the general nonconvex case. Given a point $\mathbf{x}^* \in D$, we will say that a nonzero vector $\mathbf{v}$ is a **feasible direction** at $\mathbf{x}^*$ if there exists a sequence of points $\mathbf{x}_1, \mathbf{x}_2, \ldots \in D$ converging to $\mathbf{x}^*$ and a sequence of positive scalars $\alpha_1, \alpha_2, \ldots$ converging to zero such that $\mathbf{x}_k - \mathbf{x}^* = \alpha_k \mathbf{v} + o(\alpha_k)$. Assuming the objective function f is differentiable at $\mathbf{x}^*$, it is easily seen that a necessary condition for local minimality is that $\nabla f(\mathbf{x}^*)^{\mathrm{T}} \mathbf{v} \geq 0$ for any feasible direction $\mathbf{v}$.

In order to apply the condition in the previous paragraph, one must be able to determine the set of feasible directions at $\mathbf{x}^*$. Assume that D is defined via a sequence of equality constraints $g_1(\mathbf{x}) = \cdots = g_p(\mathbf{x}) = 0$ and inequality constraints $h_1(\mathbf{x}) \leq 0, \ldots, h_q(\mathbf{x}) \leq 0$ with $g_1, \ldots, g_p$ and $h_1, \ldots, h_q$ continuous. We say that an inequality constraint is active at $\mathbf{x}^*$ if it is satisfied as an equality. Clearly the inactive constraints can be ignored in determining local optimality: if $h_i(\mathbf{x}^*) < 0$, then by continuity, $h_i(\mathbf{x}) < 0$ for all $\mathbf{x}$ sufficiently close to $\mathbf{x}^*$ and hence this constraint has no effect on the problem locally. Thus, the feasible directions at $\mathbf{x}^*$ are determined by equality and active inequality constraints. Let $A \subset \{1, \ldots, q\}$ index the inequality constraints that are active at $\mathbf{x}^*$, that is, $h_i(\mathbf{x}^*) = 0$ for all $i \in A$. For uniform terminology, we will regard equality constraints as always being active.

Assume further that all active constraints at $\mathbf{x}^*$ are C^1 in a neighborhood of $\mathbf{x}^*$. Consider replacing the active constraints locally at $\mathbf{x}^*$ by linearizations. For an equality constraint, say $g_i(\mathbf{x}) = 0$, we know that $g_i(\mathbf{x}^* + \mathbf{v}) = g_i(\mathbf{x}^*) + \mathbf{v}^{\mathrm{T}} \nabla g_i(\mathbf{x}^*) + o(\|\mathbf{v}\|) = \mathbf{v}^{\mathrm{T}} \nabla g_i(\mathbf{x}^*) + o(\|\mathbf{v}\|)$. This means that a feasible direction $\mathbf{v}$ for this one constraint in isolation must satisfy $\mathbf{v}^{\mathrm{T}} \nabla g_i(\mathbf{x}^*) = 0$. Similarly a feasible direction $\mathbf{v}$ for an inequality constraint $h_i(\mathbf{x}) \leq 0$ active at $\mathbf{x}^*$ must satisfy $\mathbf{v}^T \nabla h_i(\mathbf{x}^*) \leq 0$. This leads to the following definition. A nonzero vector $\mathbf{v}$ is said to be a **linearized feasible direction** at $\mathbf{x}^*$ if $\mathbf{v}^T \nabla g_i(\mathbf{x}^*) = 0$ for all $i = 1, \ldots, p$, and $\mathbf{v}^T \nabla h_i(\mathbf{x}^*) \leq 0$ for all $i \in A$. Observe that the linearized feasible directions at $\mathbf{x}^*$ form a polyhedral cone, and assuming we have reasonably explicit representations of the constraints, membership in this cone is easy to check.

Under the assumption of differentiability, the set of feasible directions is a subset of the linearized feasible directions, but in general, it could be a proper subset. A **constraint qualification** is a condition on the constraints that guarantees that every linearized feasible direction is also a feasible direction. There are many constraint qualifications proposed in the literature. Two simple ones are as follows: (1) All the active constraints are linear, and (2) all the constraints are C^1 and their gradient vectors are linearly independent. If either (1) or (2) holds at a feasible point $\mathbf{x}^*$, then every linearized feasible direction is a feasible direction.

Assuming a constraint qualification holds, we can now state necessary conditions for local minimality, known as the **Karush–Kuhn–Tucker** (KKT) conditions. The conditions state that (1) $\mathbf{x}^*$ must be feasible and (2) for every linearized feasible direction $\mathbf{v}$ at $\mathbf{x}^*$, $\mathbf{v}^T \nabla f(\mathbf{x}^*) \geq 0$.

Using Farkas' lemma (a result, i.e., equivalent to linear programming duality; see, e.g., [46]), we can reformulate condition (2) as the following equivalent statement: The gradient $\nabla f(\mathbf{x}^*)$ can be written in the form

$$\nabla f\left(\mathbf{x}^*\right) = \sum_{i=1}^{p} \lambda_i \nabla g_i\left(\mathbf{x}^*\right) - \sum_{i \in A} \mu_i \nabla h_i\left(\mathbf{x}^*\right) \tag{32.2}$$

where $\mu_i \geq 0$ for each $i \in A$. The variables λ_i, μ_i are called multipliers. In the case of equality constraints, we have only the first summation present on the right-hand side. In this special case, the KKT conditions are called the Lagrange multiplier conditions.

The third (and final) way to write the KKT conditions is as follows. We introduce multipliers μ_i for constraints in $\{1, \ldots, q\} - A$ and force them to be zero. This allows us to write the conditions without explicit reference to A. We state this as a theorem.

THEOREM 32.2 *Consider minimizing $f(\mathbf{x})$ over a domain D, where*

$$D = \left\{ \mathbf{x} \in \mathbf{R}^n : g_1(\mathbf{x}) = \cdots = g_p(\mathbf{x}) = 0; \ h_1(\mathbf{x}) \leq 0, \ldots, h_q(\mathbf{x}) \leq 0 \right\}. \tag{32.3}$$

Let $\mathbf{x}^* \in D$ be a local minimizer of f, and assume that a constraint qualification holds at $\mathbf{x}^*$. Assume that $g_1, \ldots, g_p, h_1, \ldots, h_q, f$ are all C^1 in a neighborhood of $\mathbf{x}^*$. Then there exist parameters $\lambda_1, \ldots, \lambda_p$ and $\mu_1, \ldots, \mu_q$ satisfying

$$\nabla f\left(\mathbf{x}^*\right) = \sum_{i=1}^{p} \lambda_i \nabla g_i\left(\mathbf{x}^*\right) - \sum_{i=1}^{q} \mu_i \nabla h_i\left(\mathbf{x}^*\right),$$
$$h_i\left(\mathbf{x}^*\right) \mu_i = 0 \quad \text{for } i = 1, \ldots, q,$$
$$\mu_i \geq 0 \quad \text{for } i = 1, \ldots, q,$$
$$g_i(\mathbf{x}^*) = 0 \quad \text{for } i = 1, \ldots, p,$$
$$h_i(\mathbf{x}^*) \leq 0 \quad \text{for } i = 1, \ldots, q. \tag{32.4}$$

The first condition is the same as before, except we have inserted dummy multipliers for inactive inequality constraints. The second condition expresses the fact that μ_i must be zero for every inactive constraint. This condition is known as the complementarity condition. The last two relations express the feasibility of $\mathbf{x}^*$.

A point that satisfies all of these conditions is said to be a KKT point or stationary point. In general, the KKT conditions are not sufficient for local minimality. To see this, consider the case of an unconstrained problem, in which case the KKT conditions reduce to the single requirement that $\nabla f(\mathbf{x}^*) = \mathbf{0}$. This condition is satisfied at local maxima as well as at local minima. To address this shortcoming, the KKT conditions are often accompanied by second-order conditions about the second derivative of the objective function. We omit second-order conditions from this discussion.

Let us now specialize this discussion to convex programming, where everything becomes simpler. Assuming a constraint qualification, the KKT conditions are necessary for local and hence global minimality. It turns out that they are also sufficient for local (and hence global) minimality, provided that the constraints are convex and differentiable. (The differentiability assumption can be dropped because one can use subgradients in place of derivatives in the KKT conditions.) Thus, the second-order optimality conditions are extraneous in the case of convex programming.

Unfortunately, the constraint qualification cannot be discarded for convex programming. For example, the convex optimization problem of minimizing $x + y$ subject to $x^2 + y^2 \leq 1$ and $x \geq 1$ has a single feasible point $(1, 0)$ and hence trivially this point is optimal. However, $(1, 0)$ is not a KKT point; the failure of the KKT conditions at $(1, 0)$ arises from the fact that the linearized feasible directions are not feasible directions. (There are no feasible directions!) In the case of convex programming, there is a specialized constraint qualification called the **Slater condition**, which applies to **feasible regions** defined by convex constraints. The condition states that there exists a point $\mathbf{x} \in D$ such that all the nonlinear constraints (i.e., all the h_i that are not linear functions) are inactive at $\mathbf{x}$. Note that the existence of a single such point means that the linearized feasible directions are the same as the feasible directions at every point in the domain.

32.3   The Ellipsoid Algorithm

There are many general purpose algorithms in the literature for nonlinear optimization. There is nothing to prevent such algorithms from being applied to convex problems. In this case, convergence to a KKT point means convergence to a global minimizer, so convexity already buys us something even with an algorithm that knows nothing about convexity. Rather than covering general nonlinear optimization algorithms, we will cover two specific algorithms for convex programming in this chapter: the ellipsoid method and an interior-point method. The ellipsoid method is easier to understand and is easier to set up than interior-point methods, but it can be inefficient in practice. Interior-point methods have come to be the preferred approach for convex programming.

The ellipsoid method is due to Yudin and Nemirovskii [76], although some of the ideas appeared earlier in Shor [54]. The treatment that follows is based on a section of the book by Nemirovskii and Yudin [39]. In that book, the ellipsoid method is called the modified method of centers of gravity. It is a variant of a method called method of centers of gravity also described in [39]. The method of centers of gravity requires at each step the computation of the centroid of a region defined by convex constraints. The method of centers of gravity is optimal in an information-theoretic sense (i.e., compared to other convex optimization methods, it uses the fewest number of function evaluations to compute an approximate optimizer within a prespecified tolerance of optimal) but is computationally intractable because there is no known efficient way to compute the required center of gravity.

For this section we are considering the convex programming problem

$$
\begin{aligned}
\text{minimize} \quad & f(\mathbf{x}) \\
\text{subject to} \quad & h_1(\mathbf{x}) \leq 0 \\
& \vdots \\
& h_q(\mathbf{x}) \leq 0
\end{aligned}
\tag{32.5}
$$

where $f, h_1, \ldots, h_q$ are convex. In other words, any linear equality constraints have been removed by solving for some of the variables and reducing to a lower dimension.

Recall that an ellipsoid is a subset of $\mathbf{R}^n$ defined as follows. Let B be a symmetric positive definite $n \times n$ matrix. Let $\mathbf{c}$ be an n-vector. Then the set

$$E = \left\{ \mathbf{x} \in \mathbf{R}^n : (\mathbf{x} - \mathbf{c})^{\mathrm{T}} B^{-1} (\mathbf{x} - \mathbf{c}) \leq 1 \right\} \tag{32.6}$$

is said to be an ellipsoid. Point $\mathbf{c}$ is said to be the center of the ellipsoid. An ellipsoid is easily seen to be a convex set because it is the image of the unit ball $\{\mathbf{x} \in \mathbf{R}^n : \|\mathbf{x}\| \leq 1\}$ under an affine transformation. From now on, $\| \cdot \|$ denotes the ordinary Euclidean norm unless otherwise noted.

In order to explain the ellipsoid method, we need the following preliminary result.

THEOREM 32.3 *Let D be a convex subset of $\mathbf{R}^n$, and let f be a convex function from $\mathbf{R}^n$ to $\mathbf{R}$. Suppose f is differentiable at a point $\mathbf{x} \in D$. Let $\mathbf{y} \in D$ be another point. If $f(\mathbf{y}) \leq f(\mathbf{x})$, then*

$$\nabla f(\mathbf{x})^{\mathrm{T}}(\mathbf{y} - \mathbf{x}) \leq 0. \tag{32.7}$$

We omit the proof of this result, which follows from the definition of derivative and standard convexity arguments. In the case when f is not differentiable at $\mathbf{x}$, a similar result holds, where in place $\nabla f(\mathbf{x})$ we can use any subgradient of f at $\mathbf{x}$.

The ellipsoid method is initialized with an ellipsoid E_0 containing the optimal solution $\mathbf{x}^*$. There is no general-purpose method for computing E_0; it must come from some additional knowledge about the convex programming problem. In Section 32.8 we discuss the issue of determining E_0 for the special case of linear programming.

The method now constructs a sequence of ellipsoids $E_1, E_2, \ldots$ with the following two properties: each E_i contains the optimal solution $\mathbf{x}^*$, and the volume of the E_i's decreases at a fixed rate. The algorithm terminates when it determines that, for some p, E_p is sufficiently small so that its center is a good approximation to the optimizer.

The procedure for obtaining E_{i+1} from E_i is as follows. We first check whether $\mathbf{c}_i$, the center of E_i, is feasible. Case 1 is that it is infeasible. Then we select a violated constraint, say h_j such that $h_j(\mathbf{c}_i) > 0$. Let H_i be the halfspace defined by the single linear inequality

$$H_i = \left\{ \mathbf{y} : \nabla h_j(\mathbf{c}_i)^{\mathrm{T}}(\mathbf{y} - \mathbf{c}_i) \leq 0 \right\}. \tag{32.8}$$

(When h_j is not differentiable at $\mathbf{c}_i$, we instead use any subgradient.) Note that the boundary of this halfspace passes through $\mathbf{c}_i$. Observe now that the entire feasible region is contained in H_i. The reason is that for any feasible point $\mathbf{x}$, we have $h_j(\mathbf{x}) \leq 0 \leq h_j(\mathbf{c}_i)$, hence $\mathbf{x} \in H_i$ by 32.7. In particular, the optimizer is in H_i.

We must also rule out the degenerate case when $\nabla h_j(\mathbf{c}_i) = \mathbf{0}$, because in this case 32.8 does not specify a halfspace. In this case, $\mathbf{c}_i$ is the global minimizer of h_j by the KKT conditions. Since $h_j(\mathbf{c}_i) > 0$ and $\mathbf{c}_i$ is a global minimizer, there does not exist an $\mathbf{x}$ such that $h_j(\mathbf{x}) \leq 0$. This means that the feasible region is empty and the problem has no solution.

Case 2 is that the center $\mathbf{c}_i$ of E_i is feasible. In this case, the ellipsoid algorithm takes H_i to be the halfspace

$$H_i = \left\{ \mathbf{y} : \nabla f(\mathbf{c}_i)^{\mathrm{T}}(\mathbf{y} - \mathbf{c}_i) \leq 0 \right\}. \tag{32.9}$$

(When f is not differentiable at $\mathbf{c}_i$, we instead use a subgradient. When $\nabla f(\mathbf{c}_i) = \mathbf{0}$, we are done, i.e., the ellipsoid method has found the exact optimizer.) Observe that by 32.7, every point $\mathbf{x}$ with a lower objective function value than $\mathbf{c}_i$ (and in particular, the optimizer) lies in H_i.

Thus, in both cases, we compute a halfspace H_i whose boundary passes through $\mathbf{c}_i$ such that the optimizer is guaranteed to lie in H_i. By induction, the optimizer also lies in E_i. The ellipsoid algorithm now computes a new ellipsoid E_{i+1} that is a superset of $E_i \cap H_i$. By induction, we know that the optimizer must thus lie in E_{i+1}.

The formulas for E_{i+1} are as follows. Let us suppose $E_i = \{\mathbf{x} : (\mathbf{x} - \mathbf{c}_i)^{\mathrm{T}} B_i^{-1} (\mathbf{x} - \mathbf{c}_i) \leq 1\}$ and $H_i = \{\mathbf{a}_i^{\mathrm{T}} (\mathbf{x} - \mathbf{c}_i) \leq 0\}$ where $\mathbf{a}_i$ is a gradient/subgradient of either a constraint or the objective function. We define

$$E_{i+1} = \left\{ \mathbf{x} : (\mathbf{x} - \mathbf{c}_{i+1})^{\mathrm{T}} B_{i+1}^{-1} (\mathbf{x} - \mathbf{c}_{i+1}) \leq 1 \right\} \tag{32.10}$$

where

$$B_{i+1} = \frac{n^2}{n^2 - 1} \cdot \left(B_i - \frac{2 B_i \mathbf{a}_i \mathbf{a}_i^{\mathrm{T}} B_i}{(n+1) \mathbf{a}_i^{\mathrm{T}} B_i \mathbf{a}_i} \right) \tag{32.11}$$

and

$$\mathbf{c}_{i+1} = \mathbf{c}_i - \frac{B_i \mathbf{a}_i}{(n+1)\sqrt{\mathbf{a}_i^{\mathrm{T}} B_i \mathbf{a}_i}}. \tag{32.12}$$

Properties of these formulas are summarized in the following theorem.

THEOREM 32.4 *With B_{i+1} defined by 32.11, $\mathbf{c}_{i+1}$ defined by 32.12, and E_{i+1} defined by 32.10, we have the following properties:*

- *Matrix B_{i+1} is symmetric and positive definite.*
- *$E_i \cap H_i \subset E_{i+1}$.*
- *If we let* vol *denote volume, then*

$$\frac{vol\,(E_{i+1})}{vol\,(E_i)} = \frac{n}{n+1} \left(\frac{n^2}{n^2 - 1} \right)^{(n-1)/2}. \tag{32.13}$$

Each of these properties is verified by substituting definitions and then algebraically simplifying. See, Bland et al. [4] for details.

Using the standard inequality $1 + x \leq e^x$, one can show that the volume decrease factor on the right-hand side of (32.13) is bounded above by $\exp(-1/(2n+2))$. Thus, the volumes of the ellipsoids form a decreasing geometric sequence.

The algorithm is terminated at an ellipsoid E_p that is judged to be sufficiently small. Once E_p is reached, the ellipsoid algorithm returns as an approximate optimizer a point $\mathbf{c}^*$, where $\mathbf{c}^*$ is the center of one of $E_0, \dots, E_p$, and is chosen as follows. Among all of $\{\mathbf{c}_0, \dots, \mathbf{c}_p\}$, consider the subset of feasible $\mathbf{c}_i$'s, that is, consider the set $\{\mathbf{c}_0, \dots, \mathbf{c}_p\} \cap D$. Choose $\mathbf{c}^*$ from this subset to have the smallest objective function value. This is the approximate minimizer. If $\{\mathbf{c}_0, \dots, \mathbf{c}_p\} \cap D$ is empty, then one of the following is true: either E_0 was chosen incorrectly, p has not been chosen sufficiently large, or $vol(D) = 0$.

Thus, the ellipsoid method is summarized as follows.

Ellipsoid Algorithm for 32.5
Choose E_0 centered at $\mathbf{c}_0$ so that $\mathbf{x}^* \in E_0$.
$i := 0$.
while $vol(E_i)$ too large do
  if $\mathbf{c}_i \notin D$
    choose a j such that $h_j(\mathbf{c}_i) > 0$.

       $\mathbf{a}_i := \nabla h_j(\mathbf{c}_i).$
    else
       $\mathbf{a}_i := \nabla f(\mathbf{c}_i).$
    end if
    define $E_{i+1}, \mathbf{c}_{i+1}$ by 32.10 through 32.12.
    $i := i + 1.$
  end while
  Let $\mathbf{c}^*$ minimize f over $\{\mathbf{c}_0, \ldots, \mathbf{c}_i\} \cap D.$

We now analyze the relationship between the number of steps p and the degree to which $\mathbf{c}^*$ approximates the optimizer. This complexity analysis requires some additional assumptions. Let the volume of the feasible region D be v_D. We assume that v_D is a positive real number. There are two cases when this assumption fails. The first case is that $v_D = \infty$. In this case, there is no difficulty with the ellipsoid method itself, but the upcoming complexity analysis no longer holds. The assumption also fails if $v_D = 0$. This happens when the dimension of the affine hull of D is less than n, including the case when $D = \emptyset$.

Since v_D is finite and positive, there is an ellipsoid that contains D. Let us assume that in fact E_0 contains D. There is no way to detect whether E_0 contains D or that $0 < v_D < \infty$ within the framework of the ellipsoid algorithm itself; instead, this must be discerned from additional information about the problem. There are special classes of convex programming problems, notably linear and quadratic programming, in which it is possible to initialize and terminate the ellipsoid algorithm without prior information (other than the specification of the problem) even when $v_D = \infty$ or $v_D = 0$.

Under the assumption that v_D is finite and positive and that E_0 contains D, let us now analyze the running time of the ellipsoid algorithm, following the analysis in [39]. Let us assume that the last iteration of the algorithm is the pth and generates an ellipsoid E_p whose volume is less than $v_D \gamma^n$, where $\gamma \in (0, 1)$ will be defined in the following text. Define the bijective affine map $\phi : \mathbf{R}^n \to \mathbf{R}^n$ by the formula $\phi(\mathbf{x}) = (1 - \gamma)\mathbf{x}^* + \gamma\mathbf{x}$. Note that this mapping shrinks $\mathbf{R}^n$ toward $\mathbf{x}^*$. Let $D' = \phi(D)$. Observe that D' contains $\mathbf{x}^*$, and by convexity, $D' \subset D$. Finally, observe that $\text{vol}(D') = v_D \gamma^n$. Since $\text{vol}(E_p) < v_D \gamma^n$, there exists at least one point $\mathbf{y}' \in D'$ that is not in E_p. Since $\mathbf{y} \in D' \subset D \subset E_0$ and $\mathbf{y}' \notin E_p$, there must have been an iteration, say m, such that $\mathbf{y}' \notin E_m$ but $\mathbf{y}' \in E_{m-1}$. This means that $\mathbf{y}' \notin H_{m-1}$. Recall that H_{m-1} was chosen according to one of the two rules depending on whether $\mathbf{c}_{m-1}$ was feasible or not. For the step under consideration, $\mathbf{c}_{m-1}$ cannot be infeasible as the following argument shows. Recall that when $\mathbf{c}_i$ is infeasible for some i, H_i contains the entire feasible region. Since $\mathbf{y}'$ is feasible and not in H_{m-1}, this means that $\mathbf{c}_{m-1}$ must have been feasible. In particular, this means that the ellipsoid method will be able to return a feasible $\mathbf{c}^*$.

Furthermore, we can use the existence of $\mathbf{c}_{m-1}$ to get bounds on the distance to optimality of $\mathbf{c}^*$. Recall that, because step $m - 1$ falls into case 2, halfspace H_{m-1} contains all feasible points whose objective function value is less than $\mathbf{c}_{m-1}$. Thus, $f(\mathbf{y}') \geq f(\mathbf{c}_{m-1})$. Further, $f(\mathbf{c}_{m-1}) \geq f(\mathbf{c}^*)$. Let $\mathbf{y} = \phi^{-1}(\mathbf{y}')$, that is, $\mathbf{y}' = (1 - \gamma)\mathbf{x}^* + \gamma\mathbf{y}$. Observe that $\mathbf{y} \in D$ since $\mathbf{y}' \in D'$. By convexity, $f(\mathbf{y}') \leq (1 - \gamma)f(\mathbf{x}^*) + \gamma f(\mathbf{y})$, that is,

$$f\left(\mathbf{y}'\right) - f\left(\mathbf{x}^*\right) \leq \gamma\left(f(\mathbf{y}) - f\left(\mathbf{x}^*\right)\right) \tag{32.14}$$

so

$$f\left(\mathbf{c}^*\right) - f\left(\mathbf{x}^*\right) \leq \gamma\left(f(\mathbf{y}) - f\left(\mathbf{x}^*\right)\right). \tag{32.15}$$

Let us say that a feasible point $\mathbf{x}$ is an ϵ-approximate optimizer if

$$\frac{f(\mathbf{x}) - f\left(\mathbf{x}^*\right)}{f\left(\mathbf{x}^\#\right) - f\left(\mathbf{x}^*\right)} \leq \epsilon \tag{32.16}$$

where $\mathbf{x}^{\#}$ is the feasible point with the worst objective function value. Thus, assuming f is bounded, a 0-approximate optimizer is the true optimizer, and every feasible point is a 1-approximate optimizer. We see from (32.15) that the ellipsoid method computes a γ-approximate optimizer.

Recall we defined $\gamma \in (0, 1)$ as a free parameter. Thus, the ellipsoid method can return arbitrarily good approximate optima. The price paid for a better approximation is a higher running time. Let the volume of the initial ellipsoid be denoted v_0. Recall that we run the method until $\text{vol}(E_p) < v_D \gamma^n$. Recall also that the ellipsoids decrease by a multiplicative factor less than $\exp(-1/(2n+2))$ per step. Thus, we must pick p sufficiently large so that $\exp(-p/(2n+2))v_0 < v_D \gamma^n$, that is,

$$p > (2n+2)(n \ln(1/\gamma) + \ln v_0 - \ln v_D). \tag{32.17}$$

Thus, we see that the running time depends logarithmically on the desired degree of accuracy. The running time also grows in the case when E_0 is much larger than D.

This raises the question: Is there a case when $\ln v_0$ would have to be much larger than $\ln v_D$? The answer is no: For any convex set D satisfying $0 < \text{vol}(D) < \infty$, there exists an ellipsoid E such that $D \subset E$ and such that $\text{vol}(E) \leq n^n \text{vol}(D)$. Thus, if we had complete information about D, we could ensure that the term $\ln v_0 - \ln v_D$ in 32.17 is never more than $n \ln n$.

32.4 A Primal Interior-Point Method

Interior-point methods for linear programming were introduced by Karmarkar [26]. It was discovered later that Karmarkar's interior-point method is related to log-barrier methods introduced earlier in the literature (see, Fiacco and McCormick [11] for details) and also to an affine scaling method due to Dikin [7]. Soon after Karmarkar's method was published, several papers, for example, Kapoor and Vaidya [25] and Ye and Tse [75] extended the method to quadratic programming. A further extension of interior-point methods to some classes of nonlinear constraints came in the work of Jarre [19,20] and Mehrotra and Sun [35]. An extension to the general case of convex programming came later with the landmark monograph by Nesterov and Nemirovskii [43] and the introduction of self-concordant functions. In this chapter, we reverse the historical development by first presenting the general convex programming case, and then specializing to linear programming. (It should be pointed out that lecture notes containing the main ideas of self-concordance had circulated for some years previous to publication of the monograph in 1994. In addition, many of the journal citations in this chapter have dates three or four years after the initial circulation of the result via preprints.)

Interior-point methods are generally believed to be the best approach for large-scaled structured linear programming and more general convex programming. As for complexity theory, interior-point methods represent only a slight improvement on the ellipsoid method. The difference is that interior-point methods work extremely well in practice (much better than their known worst-case bounds), whereas the running time of the ellipsoid method tends to be equal to its worst-case bound.

Interior-point methods are generally classified into three categories: path-following, potential reduction, and projective. Karmarkar's original method was a projective method, but projective methods have faded in importance since 1984. Both of the remaining two classes are used in practice, though at the time of this writing, path-following methods seem to be preferred. Because of space limitations, we cover only path-following methods in this section. Potential reduction is briefly described in the context of semidefinite programming in Section 32.6.

A second way of classifying interior-point methods is whether they are primal or primal-dual. The algorithm presented in this section is a primal algorithm. When we specialize to semidefinite and linear programming in upcoming sections, we present primal-dual algorithms, which are generally considered to be superior to primal algorithms.

In this section we closely follow the treatment of self-concordance due to Jarre [21,22] rather than [43]. The foundation of all interior-point methods is Newton's method. Let $g : \mathbf{R}^n \to \mathbf{R}$ be a three-times differentiable function, and suppose we want to minimize g over $\mathbf{R}^n$ (no constraints). Suppose we have a current point $\mathbf{x}_c$, called an iterate, and we would like to compute a better point $\mathbf{x}^+$. Newton's method for optimization is based on expanding g as a Taylor series around $\mathbf{x}_c$:

$$g(\mathbf{x}) = g(\mathbf{x}_c) + \nabla g(\mathbf{x}_c)^{\mathrm{T}} (\mathbf{x} - \mathbf{x}_c) + \frac{1}{2} (\mathbf{x} - \mathbf{x}_c)^{\mathrm{T}} \nabla^2 g(\mathbf{x}_c) (\mathbf{x} - \mathbf{x}_c) + O\left(\|\mathbf{x} - \mathbf{x}_c\|^3\right). \tag{32.18}$$

A reasonable approach to minimizing g might be to minimize $q(\mathbf{x})$, where $q(\mathbf{x})$ is the quadratic function that arises from dropping high-order terms from 32.18:

$$q(\mathbf{x}) = g(\mathbf{x}_c) + \nabla g(\mathbf{x}_c)^{\mathrm{T}} (\mathbf{x} - \mathbf{x}_c) + \frac{1}{2} (\mathbf{x} - \mathbf{x}_c)^{\mathrm{T}} \nabla^2 g(\mathbf{x}_c) (\mathbf{x} - \mathbf{x}_c). \tag{32.19}$$

This function $q(\mathbf{x})$ is called a quadratic model. In the case when $\nabla^2 g(\mathbf{x}_c)$ is a positive definite matrix, a standard linear algebra result tells us that the minimizer of $q(\mathbf{x})$ is

$$\mathbf{x}^+ = \mathbf{x}_c - \left(\nabla^2 g(\mathbf{x}_c)\right)^{-1} \nabla g(\mathbf{x}_c) \tag{32.20}$$

where the exponent -1 denotes matrix inversion. In a pure Newton's method, $\mathbf{x}^+$ would be taken as the next iterate, and a new quadratic model would be formed. In the case when $\nabla^2 g(\mathbf{x}_c)$ is not positive definite, the pure Newton's method has to be modified in some manner.

In the case of convex programming, however, we have already seen that $\nabla^2 g(\mathbf{x}_c)$ is always at least positive semidefinite, so indefiniteness is not as severe a difficulty as in the general case. Suppose g is C^2 and $\nabla^2 g(\mathbf{x})$ is positive definite for all $\mathbf{x}$; then the Newton step is always well-defined. This assumption of positive definiteness implies that g is strictly convex. A function $g(\mathbf{x})$ is said to be strictly convex if for all points $\mathbf{x},\mathbf{y}$ in the domain such that $\mathbf{x} \neq \mathbf{y}$, and for all $\lambda \in (0, 1)$,

$$g(\lambda \mathbf{x} + (1 - \lambda)\mathbf{y}) < \lambda g(\mathbf{x}) + (1 - \lambda)g(\mathbf{y}). \tag{32.21}$$

Strict convexity has the following useful consequence. A strictly convex function defined on a convex domain D has at most one point satisfying the KKT conditions, and hence at most one minimizer. (But a strictly convex function need not have a minimizer: consider the function $g(x) = e^x$.)

One interesting feature of Newton's method is its invariance under affine transformation. In particular, for any fixed nonsingular square matrix A and fixed vector $\mathbf{b}$, Newton's method applied to $g(\mathbf{x})$ starting at $\mathbf{x}_c$ is equivalent to Newton's method applied to $h(\mathbf{x}) = g(A\mathbf{x} + \mathbf{b})$ starting at a point $\mathbf{x}_c'$ defined by $A\mathbf{x}_c' + \mathbf{b} = \mathbf{x}_c$. (Note that other classical methods for optimization, such as steepest descent, do not have this desirable property.) Suppose we want to measure progress in Newton's method, that is, determine whether the current iterate $\mathbf{x}$ is near the minimizer $\mathbf{x}^*$. The obvious distance measure $\|\mathbf{x} - \mathbf{x}^*\|$ is not useful, since $\mathbf{x}^*$ is not known a priori; the measure $g(\mathbf{x}) - g(\mathbf{x}^*)$ suffers from the same flaw. The gradient norm $\|\nabla g(\mathbf{x})\|$ seems like a better choice since we know that $\nabla g(\mathbf{x}^*) = \mathbf{0}$ at optimum, except that this measure of nearness is not preserved under affine transformation. Nesterov and Nemirovskii propose the metric $\left(\nabla g(\mathbf{x})(\nabla^2 g(\mathbf{x}))^{-1}\nabla g(\mathbf{x})\right)^{1/2}$ to measure the degree of optimality. This measure has the advantage of invariance under affine transformation.

Newton's method as described so far does not handle constraints. A classical approach to incorporate constraints into Newton's method is a barrier-function approach. Assume from now on that D is a convex, closed set with an interior point.

Let $F(\mathbf{x})$ be a function defined on the interior of D with the following two properties:

- F is strictly convex.
- Let $\mathbf{x}$ be an arbitrary point on the boundary of D, and let $\mathbf{x}_1, \mathbf{x}_2, \ldots$ be a sequence of interior points converging to $\mathbf{x}$. Then $F(\mathbf{x}_k) \to \infty$ as $k \to \infty$.

In this case, F is said to be a **barrier function** for D. For example, the function $F(x) = -\ln x$ is a barrier function for the nonnegative real numbers $\{x \in \mathbf{R} : x \geq 0\}$. Note that a set D admits many different barrier functions. Barrier functions are a classical technique in optimization—see for instance [11]—although their prominence in the literature had waned during the 1970s and 1980s until the advent of interior-point methods.

For the rest of this section we will make the assumption that the objective function is a linear function $\mathbf{c}^{\mathrm{T}}\mathbf{x}$. Thus, our problem is written:

$$
\begin{aligned}
\text{minimize} \quad & \mathbf{c}^{\mathrm{T}}\mathbf{x} \\
\text{subject to} \quad & h_1(\mathbf{x}) \leq 0, \\
& \quad \vdots \\
& h_q(\mathbf{x}) \leq 0.
\end{aligned}
\tag{32.22}
$$

This assumption is without loss of generality for the following reason. Given a general convex programming problem with a nonlinear objective function of the form 32.5, we can introduce a new variable z and a new constraint $f(\mathbf{x}) \leq z$ (which is convex, since $f(\mathbf{x}) - z$ is a convex function), and then we replace the objective function with minimize z.

The barrier-function method for minimizing $\mathbf{c}^{\mathrm{T}}\mathbf{x}$ on D is as follows. We choose a sequence of positive parameters $\mu_0, \mu_1, \mu_2, \ldots$ strictly decreasing to zero. On the kth iteration of the barrier-function method, we minimize $\mathbf{c}^{\mathrm{T}}\mathbf{x} + \mu_k F(\mathbf{x})$ on the interior of D using Newton's method. The starting point for Newton's method is taken to be the (approximately-computed) minimizer of $\mathbf{c}^{\mathrm{T}}\mathbf{x} + \mu_{k-1}F(\mathbf{x})$ from the previous iteration. The kth iteration continues until a sufficiently good approximate minimizer of $\mathbf{c}^{\mathrm{T}}\mathbf{x} + \mu_k F(\mathbf{x})$ is found, at which point we move to iteration $k + 1$. Since $\mu_k \to 0$ as $k \to \infty$, in the limit we are minimizing the original objective function $\mathbf{c}^{\mathrm{T}}\mathbf{x}$.

The reason this method makes progress is as follows. Because the barrier function blows up near the boundaries of D, the minimizer of the sum $\mathbf{c}^{\mathrm{T}}\mathbf{x} + \mu_k F(\mathbf{x})$ will be bounded away from the boundaries of D. As we let μ_k approach zero, the importance of the barrier function is diminished, and the original optimization problem dominates. The intuition behind the barrier function method is as follows. Newton's method works very well in the unconstrained case, but has no notion of inequality constraints. The barrier function has information about the constraints encoded in its Hessian in a way that is useful for Newton's method. In particular, the barrier term prevents Newton's method from coming close to the boundary too soon.

Consider $\mu > 0$ fixed for a moment. Because of the assumption that F is strictly convex, the objective function $\mathbf{c}^{\mathrm{T}}\mathbf{x} + \mu F(\mathbf{x})$ has at most one minimizer. Let this minimizer be denoted $\mathbf{x}(\mu)$. By the KKT conditions, $\mathbf{c} + \mu \nabla F(\mathbf{x}(\mu)) = \mathbf{0}$. The **central path** is defined to be the collection of points $\{\mathbf{x}(\mu) : 0 < \mu < \infty\}$. The case when $\mathbf{c}^{\mathrm{T}}\mathbf{x} + \mu F(\mathbf{x})$ has no minimizer at all can occur if the domain D is unbounded and $\mathbf{c}^{\mathrm{T}}\mathbf{x}$ has no lower bound over the domain. We do not consider that case here. An example of the central path is presented in Figure 32.1.

Everything presented so far concerning barrier function methods was known in the 1960s. The important recent progress has been the discovery of the self-concordance conditions, due to Nesterov and Nemirovskii. The following conditions are called the finite-difference self-concordance conditions. Let int(D) denote the interior of the feasible region D. Suppose $\mathbf{x} \in \text{int}(D)$ and let E be the open ellipsoid

$$
E = \left\{ \mathbf{y} : (\mathbf{y} - \mathbf{x})^{\mathrm{T}} \nabla^2 F(\mathbf{x})(\mathbf{y} - \mathbf{x}) < 1 \right\}.
\tag{32.23}
$$

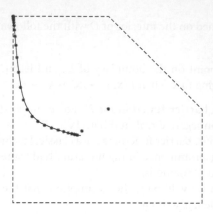

FIGURE 32.1 The central path, which is the solid trajectory, and iterates in a small-step path-following interior-point method, which are asterisks. The dashed segments indicate the boundary of the feasible region. The example problem is linear programming with two variables and five constraints. The iterates lie close to but not exactly on the central path, and their spacing decreases geometrically.

The first finite-difference self-concordance condition is that

$$E \subset \text{int}(D). \tag{32.24}$$

Next, suppose $\mathbf{y} \in E$ and let

$$r = \left((\mathbf{y} - \mathbf{x})^{\mathrm{T}} \nabla^2 F(\mathbf{x})(\mathbf{y} - \mathbf{x})\right)^{1/2} \tag{32.25}$$

[so that, by assumption, $r < 1$ and $\mathbf{y} \in \text{int}(D)$]. The second condition is that for all $\mathbf{h} \in \mathbf{R}^n$,

$$(1 - r)^2 \mathbf{h}^{\mathrm{T}} \nabla^2 F(\mathbf{x}) \mathbf{h} \leq \mathbf{h}^{\mathrm{T}} \nabla^2 F(\mathbf{y}) \mathbf{h} \leq \frac{1}{(1 - r)^2} \mathbf{h}^{\mathrm{T}} \nabla^2 F(\mathbf{x}) \mathbf{h}. \tag{32.26}$$

We assume from now on that F satisfies 32.24 and 32.26. The finite-difference self-concordance conditions are the easiest to use in the analysis of interior-point methods, but they are difficult to verify because of multitude of free parameters in 32.26. In the next section we will state the differential self-concordance condition, which implies 32.24 and 32.26 but is easier to verify analytically.

The rationale behind 32.26 is as follows. The condition that $r < 1$ means that $\mathbf{y}$ lies in an ellipsoidal open set around $\mathbf{x}$, where the ellipsoid is defined by the symmetric positive definite matrix $\nabla^2 F(\mathbf{x})$. It is very natural to define a neighborhood in this manner, because other simpler definitions of a neighborhood around $\mathbf{x}$ would not be invariant under affine transformations. An ellipsoid like 32.23 is sometimes known as a Dikin ellipsoid. Condition 32.26 says that the Hessian of F does not vary too much over this natural neighborhood. Thus, we see the significance of the term self-concordant: variation in the Hessian of F is small with respect to the neighborhood defined by the Hessian itself. Subtracting $\mathbf{h}^{\mathrm{T}} \nabla^2 F(\mathbf{x}) \mathbf{h}$ from all three quantities of 32.26 and simplifying yields

$$\left| \mathbf{h}^{\mathrm{T}} \left(\nabla^2 F(\mathbf{y}) - \nabla^2 F(\mathbf{x}) \right) \mathbf{h} \right| \leq \left(\frac{1}{(1 - r)^2} - 1 \right) \mathbf{h}^{\mathrm{T}} \nabla^2 F(\mathbf{x}) \mathbf{h}. \tag{32.27}$$

The self-concordance condition is not sufficient by itself to yield a good algorithm. The difficulty is that the minimizer of F could be very close to a boundary of F, causing a very steep gradient in a small neighborhood and slowing down convergence. An example of this undesirable behavior, due to Jarre, is the self-concordant barrier function $-(\ln x)/\epsilon - \ln(1 - x)$ for the interval $[0, 1]$ where

$\epsilon > 0$ is small. This barrier function has its minimizer arbitrarily close to 1, and very large derivatives near the minimizer. To prevent this behavior, we impose a second condition, called the self-limiting condition by Jarre, that for all $\mathbf{x} \in \text{int}(D)$,

$$\nabla F(\mathbf{x})^{\mathrm{T}} (\nabla^2 F(\mathbf{x}))^{-1} \nabla F(\mathbf{x}) \leq \theta \tag{32.28}$$

where $\theta \geq 1$ is a parameter that measures the quality of the self-concordant function. This parameter θ is called the **self-concordance parameter**. One way to interpret 32.28 is that it is the simplest possible upper bound on the first derivative of F that is invariant under affine transformations.

The parameter θ plays a crucial role in the analysis of interior-point methods. Thus, the question arises, what are typical values for θ? In subsequent sections we present specific classes of self-concordant barrier functions arising in applications, along with their accompanying values of θ. For this section the reader should imagine that $\theta = O(q)$, where q is the number of inequality constraints.

One kind of path-following chooses the parameter sequence $\mu_0, \mu_1, \mu_2, \ldots$ according to the rule

$$\mu_k = (1 - \sigma)\mu_{k-1}, \tag{32.29}$$

where σ is a problem-dependent scalar that is constant across all iterations. This is known as small-step path-following in the literature, and is the easiest rule to analyze, though not very efficient in practice. Because of the small changes in μ per step, there is a need for only one Newton step at each iteration.

Let us fix the iteration number k and cease to write it as a subscript. Note that $\nabla^2 g = \mu \nabla^2 F$ since the second derivative of the linear term is zero. From iterate $\mathbf{x}$ the next iterate $\mathbf{x}^+$ is computed by applying one step of Newton's method to g, that is,

$$\mathbf{x}^+ := \mathbf{x} - \mu^{-1} \left(\nabla^2 F(\mathbf{x}) \right)^{-1} (\mathbf{c} + \mu \nabla F(\mathbf{x})) \tag{32.30}$$

where we have used the fact that $\nabla g(\mathbf{x}) = \mathbf{c} + \mu \nabla F(\mathbf{x})$. Now we define $\mu^+ = (1 - \sigma)\mu$ by 32.29 and repeat the iteration.

In order to carry out a complexity analysis, we would like to claim that if the current iterate $\mathbf{x}$ is close to the central path for parameter μ, then the next iterate $\mathbf{x}^+$ is also close to the central path for parameter μ^+. This requires a notion of proximity to the central path. Here we use the metric mentioned earlier. For each $(\mathbf{y}, \mu)$, we let

$$\lambda(\mathbf{y}, \mu) = \mu^{-1} \left[(\mathbf{c} + \mu \nabla F(\mathbf{y}))^{\mathrm{T}} (\nabla^2 F(\mathbf{y}))^{-1} (\mathbf{c} + \mu \nabla F(\mathbf{y})) \right]^{1/2}. \tag{32.31}$$

Observe that if $\mathbf{y}$ is the point on the central path for parameter μ, then $\lambda(\mathbf{y}, \mu) = 0$.

Returning to our complexity analysis, we claim that if $\lambda(\mathbf{x}, \mu) \leq 1/5$, then $\lambda(\mathbf{x}^+, \mu^+) \leq 1/5$. This claim will require the correct choice of σ in 32.29. To prove this claim, we first analyze the intermediate quantity $\lambda(\mathbf{x}^+, \mu)$. In other words, we would like to know how much progress is made during one step of Newton's method applied to g. To preserve generality, let us state this as a theorem.

THEOREM 32.5   *Let $\mu > 0$ and $\mathbf{x} \in \text{int}(D)$ be given. Let λ denote $\lambda(\mathbf{x}, \mu)$, and suppose $\lambda < 1$. Let $\mathbf{x}^+$ be chosen by 32.30. Then*

$$\lambda\left(\mathbf{x}^+, \mu\right) \leq \frac{\lambda^2}{(1 - \lambda)^2}. \tag{32.32}$$

PROOF   Let $\mathbf{v} = -\mu^{-1}(\nabla^2 F(\mathbf{x}))^{-1}(\mathbf{c} + \mu \nabla F(\mathbf{x}))$, which is the update in 32.30, and let $\mathbf{y} = \mathbf{x} + s\mathbf{v}$ for some $s \in [0, 1]$ to be chosen later. Note that if $s = 1$, then $\mathbf{y} = \mathbf{x}^+$. We see from the definition 32.31 of $\lambda(\cdot, \cdot)$ that r in 32.25 is λs and hence satisfies $r < 1$ by assumption. Therefore, by 32.27, for any $\mathbf{h} \in \mathbf{R}^n$,

$$\left| \mathbf{h}^T \left(\nabla^2 F(\mathbf{x}) - \nabla^2 F(\mathbf{y}) \right) \mathbf{h} \right| \leq \left(\frac{1}{(1-s\lambda)^2} - 1 \right) \mathbf{h}^T \nabla^2 F(\mathbf{x}) \mathbf{h}. \tag{32.33}$$

We now apply the generalized Cauchy–Schwarz inequality from Jarre [22], which is the following. Let A, M be two symmetric matrices such that A is positive definite and such that $|\mathbf{z}^T M \mathbf{z}| \leq \mathbf{z}^T A \mathbf{z}$ for all vectors $\mathbf{z}$. Then for all $\mathbf{a}, \mathbf{b}$, $(\mathbf{a}^T M \mathbf{b})^2 \leq (\mathbf{a}^T A \mathbf{a})(\mathbf{b}^T A \mathbf{b})$. We omit the proof here. Observe that if we define $A = \gamma \nabla^2 F(\mathbf{x})$, where γ is the scalar on the right-hand side of 32.33, and $M = \nabla^2 F(\mathbf{x}) - \nabla^2 F(\mathbf{y})$, then the hypothesis of the generalized Cauchy–Schwarz inequality holds by 32.33. Therefore, we can apply the conclusion, with "a" taken to be an indeterminate vector $\mathbf{h}$ and "b" taken to be $\mathbf{v}$ to obtain

$$\left| \mathbf{v}^T \left(\nabla^2 F(\mathbf{x}) - \nabla^2 F(\mathbf{y}) \right) \mathbf{h} \right| \leq \left(\frac{1}{(1-s\lambda)^2} - 1 \right) \sqrt{\mathbf{v}^T \nabla^2 F(\mathbf{x}) \mathbf{v}} \sqrt{\mathbf{h}^T \nabla^2 F(\mathbf{x}) \mathbf{h}} \tag{32.34}$$

$$= \left(\frac{1}{(1-s\lambda)^2} - 1 \right) \lambda \sqrt{\mathbf{h}^T \nabla^2 F(\mathbf{x}) \mathbf{h}}. \tag{32.35}$$

Let us now come up with new expressions for the two terms on the left-hand side of 32.34. Recall by choice of $\mathbf{v}$ that the first term $\mathbf{v}^T \nabla^2 F(\mathbf{x}) \mathbf{h}$ evaluates to $-(\mathbf{c}/\mu + \nabla F(\mathbf{x}))^T \mathbf{h}$. For the second term, recall that $\mathbf{y} = \mathbf{x} + s\mathbf{v}$. Thinking of $\mathbf{y}$ as a function of s, observe that $\nabla^2 F(\mathbf{y})\mathbf{v} = \nabla^2 F(\mathbf{x} + s\mathbf{v})\mathbf{v} = d(\nabla F(\mathbf{x} + s\mathbf{v}))/ds = d(\mathbf{c}/\mu + \nabla F(\mathbf{x} + s\mathbf{v}))/ds$. Thus, we have

$$\left| -(\mathbf{c}/\mu + \nabla F(\mathbf{x}))^T \mathbf{h} - \frac{d(\mathbf{c}/\mu + \nabla F(\mathbf{x} + s\mathbf{v}))^T}{ds} \mathbf{h} \right|$$

$$\leq \left(\frac{1}{(1-s\lambda)^2} - 1 \right) \lambda \sqrt{\mathbf{h}^T \nabla^2 F(\mathbf{x}) \mathbf{h}}. \tag{32.36}$$

Let us define

$$p(s) = (1-s)(\mathbf{c}/\mu + \nabla F(\mathbf{x}))^T \mathbf{h} - (\mathbf{c}/\mu + \nabla F(\mathbf{x} + s\mathbf{v}))^T \mathbf{h}. \tag{32.37}$$

Observe that $p(0) = 0$ because the two terms cancel. Observe also that 32.36 is equivalent to

$$|p'(s)| \leq \left(\frac{1}{(1-s\lambda)^2} - 1 \right) \lambda \sqrt{\mathbf{h}^T \nabla^2 F(\mathbf{x}) \mathbf{h}}. \tag{32.38}$$

Therefore, by integrating both sides of 32.38 for s from 0 to 1, we conclude that

$$|p(1)| \leq \frac{\lambda^2}{1-\lambda} \sqrt{\mathbf{h}^T \nabla^2 F(\mathbf{x}) \mathbf{h}}. \tag{32.39}$$

On the other hand, directly from 32.37 we have $p(1) = -(\mathbf{c}/\mu + \nabla F(\mathbf{x} + \mathbf{v}))^T \mathbf{h}$. We combine this with 32.39, using the fact that $\mathbf{x} + \mathbf{v} = \mathbf{x}^+$, to obtain

$$\left| (\mathbf{c}/\mu + \nabla F(\mathbf{x}^+))^T \mathbf{h} \right| \leq \frac{\lambda^2}{1-\lambda} \sqrt{\mathbf{h}^T \nabla^2 F(\mathbf{x}) \mathbf{h}}. \tag{32.40}$$

We can now put an upper bound on the right-hand side of 32.40 using self-concordance to obtain

$$\left| (\mathbf{c}/\mu + \nabla F(\mathbf{x}^+))^T \mathbf{h} \right| \leq \frac{\lambda^2}{(1-\lambda)^2} \sqrt{\mathbf{h}^T \nabla^2 F(\mathbf{x}^+) \mathbf{h}}. \tag{32.41}$$

Now finally we will define the hitherto indeterminate $\mathbf{h}$ to be $\mathbf{h} = (\nabla^2 F(\mathbf{x}^+))^{-1}(\mathbf{c}/\mu + \nabla F(\mathbf{x}^+))$. If we substitute this into 32.41 and use the definition of $\lambda(\cdot, \cdot)$, we obtain

$$\lambda\left(\mathbf{x}^+, \mu\right)^2 \leq \frac{\lambda^2}{(1-\lambda)^2}\lambda\left(\mathbf{x}^+, \mu\right). \tag{32.42}$$

Dividing both sides by $\lambda(\mathbf{x}^+, \mu)$ proves 32.32.

In the specific case $\lambda \leq 1/5$, we see that $\lambda(\mathbf{x}^+, \mu) \leq 1/16$. Next, we want to show that $\lambda(\mathbf{x}^+, \mu^+) \leq 1/5$. Observe that the function $\mathbf{y} \mapsto (\mathbf{y}^\mathsf{T}\nabla^2 F(\mathbf{x}^+)^{-1}\mathbf{y})^{1/2}$ is a norm on $\mathbf{R}^n$, and hence obeys the triangle inequality. Thus, we can apply this triangle inequality to $\mu^+\lambda(\mathbf{x}^+, \mu^+) - \mu\lambda(\mathbf{x}^+, \mu)$ plugging in 32.31 to obtain

$$\begin{aligned}
\mu^+\lambda\left(\mathbf{x}^+, \mu^+\right) - \mu\lambda\left(\mathbf{x}^+, \mu\right) &\leq \left|\mu - \mu^+\right| \\
&\quad \cdot \left[\nabla F\left(\mathbf{x}^+\right)^\mathsf{T}\left(\nabla^2 F\left(\mathbf{x}^+\right)\right)^{-1}\nabla F\left(\mathbf{x}^+\right)\right]^{1/2} \tag{32.43} \\
&\leq \left|\mu - \mu^+\right| \cdot \sqrt{\theta} \tag{32.44}
\end{aligned}$$

where recall that θ was defined in 32.28. If we substitute 32.29 for μ^+ and simplify, we obtain

$$\lambda\left(\mathbf{x}^+, \mu^+\right) \leq \frac{\sigma\sqrt{\theta} + \lambda\left(\mathbf{x}^+, \mu\right)}{1 - \sigma}. \tag{32.45}$$

If we choose

$$\sigma = \frac{1}{9\sqrt{\theta}} \tag{32.46}$$

and use the inequality $\lambda(\mathbf{x}^+, \mu) \leq 1/16$, we conclude finally that $\lambda(\mathbf{x}^+, \mu^+) \leq 25/128 < 1/5$.

Observe that 32.46 controls how fast μ can decrease in 32.29. Thus, the rate of convergence of the interior-point method is determined by the value of the self-concordance parameter.

The next step in our analysis is to show that as μ is driven to zero, the interior-point method converges to the true minimizer of 32.22. Let $(\mathbf{x}, \mu)$ be an iterate of the aforementioned algorithm, so that $\lambda(\mathbf{x}, \mu) \leq 1/5$. Let $\mathbf{x}^*$ denote the global minimizer for the original problem 32.22. We now show that $\mathbf{c}^\mathsf{T}\mathbf{x} - \mathbf{c}^\mathsf{T}\mathbf{x}^* \leq O(\mu)$, so the difference of the iterates from optimality tends to zero. This argument is broken up as two separate theorems. First, we show that $\mathbf{c}^\mathsf{T}\mathbf{x} \leq \mathbf{c}^\mathsf{T}\hat{\mathbf{x}} + O(\mu)$, where $\hat{\mathbf{x}}$ is the point on the central path for μ. Second, we show $\mathbf{c}^\mathsf{T}\hat{\mathbf{x}} \leq \mathbf{c}^\mathsf{T}\mathbf{x}^* + O(\mu)$.

THEOREM 32.6 *Let $\mathbf{x} \in int(D)$ and $\mu > 0$ satisfy $\lambda(\mathbf{x}, \mu) \leq 1/5$. Let $\hat{\mathbf{x}} \in D$ be the central path point for μ, i.e., $\mathbf{c} + \mu\nabla F(\hat{\mathbf{x}}) = \mathbf{0}$. Then*

$$\mathbf{c}^\mathsf{T}\mathbf{x} \leq \mathbf{c}^\mathsf{T}\hat{\mathbf{x}} + \mu\left(0.27\sqrt{\theta} + 0.044\right). \tag{32.47}$$

PROOF Let us consider applying Newton's method for minimizing $\mathbf{c}^\mathsf{T}\mathbf{x} + \mu F(\mathbf{x})$ starting from $\mathbf{x}$, keeping μ fixed. Observe that 32.32 says that Newton's method converges quadratically, and it must converge to $\hat{\mathbf{x}}$. Let the Newton iterates be $\mathbf{y}^{(0)}(=\mathbf{x}), \mathbf{y}^{(1)}, \mathbf{y}^{(2)}, \ldots$ and sequence of proximities be denoted $\lambda_0, \lambda_1, \ldots$, so $\lambda_0 \leq 1/5$ by assumption. Let us consider two consecutive iterates in this sequence $\mathbf{y}^{(k)}, \mathbf{y}^{(k+1)}$ that we denote $\mathbf{y}$ and $\mathbf{y}^+$. By the definition of Newton's method, we have

$$\mathbf{y}^+ - \mathbf{y} = -\mu^{-1}\left(\nabla^2 F(\mathbf{y})\right)^{-1}(\mathbf{c} + \mu\nabla F(\mathbf{y})) \tag{32.48}$$

so

$$\mu \cdot \left| \mathbf{c}^T \mathbf{y}^+ - \mathbf{c}^T \mathbf{y} \right| = \left| \mathbf{c}^T \left(\nabla^2 F(\mathbf{y}) \right)^{-1} (\mathbf{c} + \mu \nabla F(\mathbf{y})) \right| \tag{32.49}$$

$$= \left| (\mathbf{c} + \mu \nabla F(\mathbf{y}))^T \left(\nabla^2 F(\mathbf{y}) \right)^{-1} (\mathbf{c} + \mu \nabla F(\mathbf{y})) \right.$$

$$\left. - \mu \nabla F(\mathbf{y})^T \left(\nabla^2 F(\mathbf{y}) \right)^{-1} (\mathbf{c} + \mu \nabla F(\mathbf{y})) \right| \tag{32.50}$$

$$\leq t + \mu t^{1/2} s^{1/2} \tag{32.51}$$

where

$$t = (\mathbf{c} + \mu \nabla F(\mathbf{y}))^T \left(\nabla^2 F(\mathbf{y}) \right)^{-1} (\mathbf{c} + \mu \nabla F(\mathbf{y})) \tag{32.52}$$

and

$$s = \nabla F(\mathbf{y})^T \left(\nabla^2 F(\mathbf{y}) \right)^{-1} \nabla F(\mathbf{y}). \tag{32.53}$$

We obtained 32.51 by applying first the triangle inequality to split up the two terms of 32.50, and then the Cauchy–Schwarz inequality to the second term. Observe that $t = (\lambda_k \mu)^2$ by definition of λ_k. More over, observe that $s \leq \theta$ by 32.28. Substituting these bounds into 32.51 and dividing by μ yields

$$\left| \mathbf{c}^T \mathbf{y}^+ - \mathbf{c}^T \mathbf{y} \right| \leq \mu \left(\lambda_k^2 + \lambda_k \sqrt{\theta} \right). \tag{32.54}$$

Thus,

$$\left| \mathbf{c}^T \hat{\mathbf{x}} - \mathbf{c}^T \mathbf{x} \right| \leq \mu \sum_{k=0}^{\infty} \left(\lambda_k^2 + \lambda_k \sqrt{\theta} \right). \tag{32.55}$$

Now we use the fact that the λ's are a decreasing quadratic series starting from $1/5$ and bounded by 32.32 to find that the sum of the λ_k's is at most 0.27 and the sum of λ_k^2 is at most 0.044. This proves 32.47.

THEOREM 32.7   *Let $(\hat{\mathbf{x}}, \mu)$ be a point on the central path. Let $\mathbf{x}^*$ be the optimal solution. Then*

$$\mathbf{c}^T \hat{\mathbf{x}} \leq \mathbf{c}^T \mathbf{x}^* + \mu \theta. \tag{32.56}$$

PROOF   Let $\mathbf{h} = \mathbf{x}^* - \hat{\mathbf{x}}$. Then $\mathbf{c}^T \hat{\mathbf{x}} - \mathbf{c}^T \mathbf{x}^* = -\mathbf{c}^T \mathbf{h}$. To prove the theorem, we must show that $-\mathbf{c}^T \mathbf{h} \leq \mu \theta$. If $\mathbf{c}^T \mathbf{h} \geq 0$ then this result is trivial, so suppose $\mathbf{c}^T \mathbf{h} < 0$. Let $\psi(t) = F(\hat{\mathbf{x}} + t\mathbf{h})$. Observe that ψ is a convex function of t satisfying $\psi'(t) = \nabla F(\hat{\mathbf{x}} + t\mathbf{h})^T \mathbf{h}$ and $\psi''(t) = \mathbf{h}^T \nabla^2 F(\hat{\mathbf{x}} + t\mathbf{h}) \mathbf{h}$. Observe that

$$\left| \psi'(t) \right| = \left| \nabla F \left(\hat{\mathbf{x}} + t\mathbf{h} \right)^T \mathbf{h} \right| \tag{32.57}$$

$$= \left| \nabla F \left(\hat{\mathbf{x}} + t\mathbf{h} \right)^T \left(\nabla^2 F \left(\hat{\mathbf{x}} + t\mathbf{h} \right) \right)^{-1/2} \left(\nabla^2 F \left(\hat{\mathbf{x}} + t\mathbf{h} \right) \right)^{1/2} \mathbf{h} \right| \tag{32.58}$$

$$\leq \left(\nabla F \left(\hat{\mathbf{x}} + t\mathbf{h} \right)^T \left(\nabla^2 F \left(\hat{\mathbf{x}} + t\mathbf{h} \right) \right)^{-1} \nabla F \left(\hat{\mathbf{x}} + t\mathbf{h} \right) \right)^{1/2} \left(\mathbf{h}^T \nabla^2 F \left(\hat{\mathbf{x}} + t\mathbf{h} \right) \mathbf{h} \right)^{1/2} \tag{32.59}$$

$$\leq \theta^{1/2} \psi''(t)^{1/2} \tag{32.60}$$

where 32.59 follows from 32.58 by the Cauchy–Schwarz inequality, and 32.60 follows from 32.28. Thus, $\psi''(t) \geq \psi'(t)^2 / \theta$. Observe also that $\psi'(0) = \nabla F(\hat{\mathbf{x}})^T \mathbf{h} = -\mathbf{c}^T \mathbf{h} / \mu$ since $\mathbf{c} + \mu \nabla F(\hat{\mathbf{x}}) = \mathbf{0}$ (by the optimality of $\hat{\mathbf{x}}$).

Now, consider the function χ defined by $\chi(t) = (-\mu/(\mathbf{c}^T\mathbf{h}) - t/\theta)^{-1}$. Observe that $\chi(0) = -\mathbf{c}^T\mathbf{h}/\mu$, so $\chi(0) = \psi'(0)$. Also, note that $\chi'(t) = \theta^{-1}(-\mu/(\mathbf{c}^T\mathbf{h}) - t/\theta)^{-2} = \chi(t)^2/\theta$.

Thus, χ is the solution to the initial value problem $u(0) = -\mathbf{c}^T\mathbf{h}/\mu$, $u'(t) = u(t)^2/\theta$. On the other hand, ψ' is a solution to the differential inequality $u(0) = -\mathbf{c}^T\mathbf{h}/\mu$, $u'(t) \geq u(t)^2/\theta$. Since u^2/θ is an increasing function of u for positive u, a theorem about differential inequalities tells us that ψ' must dominate χ for $t \geq 0$. But notice that χ blows up to ∞ at $t = -\mu\theta/(\mathbf{c}^T\mathbf{h})$ (recall we are assuming $\mathbf{c}^T\mathbf{h} < 0$). Thus, ψ' must blow up at some t_0 satisfying $t_0 \leq -\mu\theta/(\mathbf{c}^T\mathbf{h})$. On the other hand, we already know that ψ does not blow up on $[0, 1)$ because by convexity $\hat{\mathbf{x}} + t\mathbf{h} \in \text{int}(D)$ for $t < 1$. Thus, $-\mu\theta/(\mathbf{c}^T\mathbf{h}) \geq 1$, i.e., $-\mathbf{c}^T\mathbf{h} \leq \mu\theta$. This proves 32.56.

We can now summarize the interior-point method presented in this section.

Interior-point algorithm for 32.22

Start with $\mu_0 > 0$ and $\mathbf{x}_0 \in \text{int}(D)$ satisfying $\lambda(\mathbf{x}_0, \mu_0) \leq 1/5$.

$i := 0$.

while $\mu_i(\theta + 0.27\sqrt{\theta} + 0.044) > \epsilon$

    $\mathbf{x}_{i+1} := \mathbf{x}_i - \mu_i^{-1}(\nabla^2 F(\mathbf{x}_i))^{-1}(\mathbf{c} + \mu_i\nabla F(\mathbf{x}_i))$.

    $\mu_{i+1} := (1 - 1/(9\sqrt{\theta}))\mu_i$.

    $i := i + 1$.

end while

return $\mathbf{x}_i$.

The preceding algorithm lacks an initialization procedure, that is, a method to construct μ_0 and $\mathbf{x}_0$. In many situations it is easy to find $\mathbf{x}_0 \in \text{int}(D)$, but not so easy to find an $\mathbf{x}_0$ that is close to the central path (i.e., that satisfies $\lambda(\mathbf{x}_0, \mu_0) \leq 1/5$ for some μ_0). In this case, there is a general-purpose iterative procedure that starts at interior point $\mathbf{x}_0$ and eventually produces a point near the central path. This iterative procedure is based on the following observation. For any $\mathbf{x}_0 \in \text{int}(D)$, $\mathbf{x}_0$ is the minimizer of the artificial objective function $g(\mathbf{x}) = \mathbf{c}_0^T\mathbf{x} + \mu_0 F(\mathbf{x})$, where $\mathbf{c}_0 = -\nabla F(\mathbf{x}_0)$ and $\mu_0 = 1$: this claim is easily verified by checking that $\nabla g(\mathbf{x}_0) = \mathbf{0}$. Therefore, we can use a path-following method on the sequence of objective functions $\mathbf{c}_0^T\mathbf{x} + \mu F(\mathbf{x})$ and try to drive μ toward ∞ instead of 0. Once μ is sufficiently large, the first term of the objective function no longer matters, and can be replaced by $\mathbf{c}^T\mathbf{x}$ where $\mathbf{c}$ is the actual gradient of the linear functional that is under consideration. We omit the details.

The running time of this initialization procedure is determined by how close $\mathbf{x}_0$ is to the boundary of D. If $\mathbf{x}_0$ is close to the boundary, then $\nabla F(\mathbf{x}_0)$ is large, and the initialization procedure requires more steps to make μ sufficiently large. Thus, a good starting point for an interior-point method would be near the analytic center of D. The analytic center is the point $\mathbf{x}_a$ that minimizes the barrier function $F(\mathbf{x})$, i.e., the point satisfying $\nabla F(\mathbf{x}_a) = \mathbf{0}$. The analytic center is the limit of the central path as $\mu \to \infty$.

This approach to initialize an interior-point method, in which one works with an artificial objective function first and then the actual objective function, is commonly called phase I–phase II in the literature. Similar approaches have also been used for initializing the simplex method; see, Dantzig [6] for more details.

If no interior feasible point is known, then there is no general-purpose procedure to initialize the interior-point algorithm, and an initial point must be constructed from additional information about the problem. In the special cases of linear, quadratic, and semidefinite programming, there are initialization techniques that do not need any other additional *a priori* information.

We would like to compare the complexity of the ellipsoid method with the interior-point method. A complete comparison is not possible because of the incompatible assumptions made concerning the initialization procedure. Therefore, we compare only the rates of convergence. Recall that the ellipsoid method guarantees error tolerance of γ after $O(n^2 \log(1/\gamma))$ iterations. Each iteration requires a rank-one update to an $n \times n$ matrix given by 32.11, which requires $O(n^2)$ operations.

The convergence rate of the interior-point method is determined as follows. Convergence to tolerance ϵ is achieved after $\mu \leq O(\epsilon/\theta)$ as seen from the algorithm. On each step μ is decreased by a factor of $1 - \text{const} \cdot \theta^{-1/2}$. Thus, the number of steps p to reduce the error tolerance to ϵ must satisfy $(1 - \text{const} \cdot \theta^{-1/2})^p \leq \epsilon/\theta$. By taking logarithms, using the approximation $\ln(1 + \delta) \approx \delta$, and dropping the low-order term, we conclude that the number of iterations is $O(\sqrt{\theta} \log(1/\epsilon))$. Each iteration requires the solution of a system of $n \times n$ linear equations given by 32.30, which takes $O(n^3)$ operations.

Let us assume for now that $\theta = O(n)$, where n is the dimension of the problem. Actual values for barrier parameters are provided in Section 32.5. In this case, we conclude that the interior-point method improves on the number of iterations over the ellipsoid method by a factor of $n^{1.5}$. On the other hand, an interior-point iteration is a factor of $O(n)$ more expensive than an ellipsoid iteration, so the total savings comes out to a factor of $O(\sqrt{n})$. Karmarkar [26] showed how to save another factor of $\sqrt{n}$ in an interior-point algorithm for LP by approximately solving the linear systems using information from the previous iteration. The amortized cost of this technique is $O(n^{2.5})$ per iteration rather than $O(n^3)$, but this technique is not often used in practice.

In fact, as mentioned in the introduction, interior-point methods in practice are far more efficient than the ellipsoid method. The reason for the improvement in practice is not so much the theoretical factor of $O(\sqrt{n})$ or $O(n)$ mentioned in the last paragraph. Rather, interior-point methods are efficient in practice because the number of iterations is usually much less than $O(\sqrt{\theta} \log(1/\epsilon))$: in particular, usually it is possible to decrease μ at a much faster rate than 32.29 and still maintain approximate centrality. In such an algorithm, the decrease in μ is chosen adaptively rather than according to a fixed formula like 32.29. Nonetheless, even for these algorithms, there is no known upper bound better than $O(\sqrt{\theta} \log(1/\epsilon))$ iterations. The reason for the mismatch between the upper bound and practical behavior is not completely understood. Todd and Ye [61] showed a lower bound of $\Omega(n^{1/3} \log(1/\epsilon))$ for linear programming (where $n = \theta$) that holds for a large variety of interior-point methods.

A second reason why interior-point methods outperform the ellipsoid method is that the linear system given by 32.30 in practice is often sparse, meaning that most entries of $\nabla^2 F(\mathbf{x})$ are zeros. Many special methods have been developed for solving sparse systems of linear equations [15]; the running time for solving sparse equations can be much better than the worst-case estimate of $O(n^3)$. Sparsity considerations are the primary reason that the amortized $O(n^{2.5})$ linear system solver mentioned earlier is not used in practice. The ellipsoid method as presented in Section 32.3 is not able to take advantage of sparsity because 32.11 implies that B_{i+1} in general will be a completely dense $n \times n$ matrix. On the other hand, Todd [58] and Burrell and Todd [5] show that a different way of representing B_{i+1} leads to better preservation of sparsity in the ellipsoid method, in which case the sparsity properties of the two algorithms may be comparable.

32.5   Additional Remarks on Self-Concordance

In this section we give some additional theoretical background on self-concordance and an example.

The first question is, for what convex sets do self-concordant barrier functions exist? It turns out that *every* closed, convex subset D of $\mathbf{R}^n$ with a nonempty interior has a self-concordant barrier function with parameter $\theta = O(n)$. Given such a set D, Nesterov and Nemirovskii prove that the function

$$F(\mathbf{x}) = \text{const} \cdot \ln \text{vol}\left(\left\{\mathbf{a} \in \mathbf{R}^n : \mathbf{a}^T(\mathbf{y} - \mathbf{x}) \leq 1 \quad \forall \mathbf{y} \in D\right\}\right) \tag{32.61}$$

is self-concordant with parameter $O(n)$. This function is called the **universal** barrier. This barrier is not useful in practice because there is evidently no algorithm to evaluate 32.61, let alone its derivatives, for general convex sets. In practice, one constructs self-concordant barrier functions for certain special cases of commonly occurring forms of constraints.

Another question is, what is the best possible value of θ? In the last paragraph it was claimed that it is theoretically possible to always choose $\theta \leq \text{const} \cdot n$. On the other hand, it can be proved that $\theta \geq 1$ in all cases. This lower bound of 1 is tight. The barrier function

$$F(\mathbf{x}) = \ln \left(1 - x_1^2 - \cdots - x_n^2\right) \tag{32.62}$$

when D is the unit ball in $\mathbf{R}^n$ has self-concordance parameter 1. This result generalizes to any ellipsoidal constraint, since self-concordance is preserved by affine transformations.

How does one verify that 32.62 is indeed self-concordant? The two conditions 32.24 and 32.26 in the last section appear nontrivial to check, even for a simple function written down in closed form like 32.62.

It turns out that there is a simpler definition of self-concordance, which is as follows. A function F is self-concordant if for all $\mathbf{x} \in \text{int}(D)$ and for all $\mathbf{h} \in \mathbf{R}^n$

$$\left|\nabla^3 F(\mathbf{x})[\mathbf{h}, \mathbf{h}, \mathbf{h}]\right| \leq 2 \left(\mathbf{h}^\mathsf{T} \nabla^2 F(\mathbf{x})\mathbf{h}\right)^{3/2}. \tag{32.63}$$

The notation on the left-hand side means the application of the trilinear form $\nabla^3 F(\mathbf{x})$ to the three vectors $\mathbf{h}, \mathbf{h}, \mathbf{h}$. (We could have used analogous notation for the right-hand side: $2(\nabla^2 F(\mathbf{x})[\mathbf{h}, \mathbf{h}])^{3/2}$.) It should be apparent at an intuitive level why 32.63 and 32.26 are related: a bound on the third derivative in terms of the second means that the second derivative cannot vary too much over a neighborhood defined in terms of the second derivative.

The proof that 32.63 implies 32.26 is based on such an argument. The trickiest part of the proof is an inequality involving trilinear forms and related to the generalized Cauchy–Schwarz inequality aforementioned. The trilinear inequality is proved in [43] and has been simplified by Jarre [22]. Furthermore, under an assumption of sufficient differentiability, the other direction holds: 32.26 implies 32.63 as noted by Todd [59].

We now present one particularly simple barrier function, easily analyzed by 32.63. We consider the following barrier function for the positive orthant $O_n = \{\mathbf{x} \in \mathbf{R}^n : x_1 \geq 0, \ldots, x_n \geq 0\}$:

$$F(\mathbf{x}) = -\sum_{i=1}^{n} \ln x_i. \tag{32.64}$$

This barrier is the key to linear programming. The gradient $\nabla F(\mathbf{x})$ is seen to be $(-1/x_1, \ldots, -1/x_n)$ and the Hessian $\text{diag}(1/x_1^2, \ldots, 1/x_n^2)$, where $\text{diag}(\cdot)$ denotes a diagonal matrix with the specified entries. This matrix is positive definite so we see that 32.64 is indeed a strictly convex function that tends to ∞ at the boundaries of the orthant. The third derivative is $\text{diag}(-2/x_1^3, \ldots, -2/x_n^3)$, where diag denotes a diagonal tensor. Thus,

$$\nabla^3 F(\mathbf{x})[\mathbf{h}, \mathbf{h}, \mathbf{h}] = -2h_1^3/x_1^3 - \cdots - 2h_n^3/x_n^3 \tag{32.65}$$

compared to

$$\mathbf{h}^\mathsf{T} \nabla^2 F(\mathbf{x})\mathbf{h} = h_1^2/x_1^2 + \cdots + h_n^2/x_n^2. \tag{32.66}$$

It is now obvious that 32.63 holds. Furthermore, 32.28 is also easily checked; one finds that $\theta = n$ for this barrier. Nesterov and Nemirovskii show that n is also a lower bound for the barrier parameter on the positive orthant, and hence this barrier is optimal.

32.6 Semidefinite Programming and Primal-Dual Methods

In this section we specialize the interior-point framework to semidefinite programming. Interior-point methods for semidefinite programming were developed independently by Nesterov and Nemirovskii [43] and Alizadeh [1]. In this section we follow Alizadeh's treatment.

The following optimization problem is said to be primal standard form **semidefinite programming** (SDP):

$$
\begin{aligned}
\text{minimize} \quad & C \cdot X \\
\text{subject to} \quad & A_1 \cdot X = b_1, \\
& \quad \vdots \\
& A_m \cdot X = b_m, \\
& X \succeq 0.
\end{aligned}
\tag{32.67}
$$

The notation here is as follows. All of $C, A_1, \ldots, A_m, X$ are symmetric $n \times n$ matrices. Matrices $A_1, \ldots, A_m$ and C are given as data, as are scalars $b_1, \ldots, b_m$. Matrix X is the unknown. The notation $Y \cdot Z$ means elementwise inner product, that is, $Y \cdot Z = \Sigma_{i,j} y_{ij} z_{ij}$. This is equal to trace XY if X, Y are symmetric. The constraint $X \succeq 0$ means that X is positive semidefinite. More generally, the notation $X \succeq Y$ means $X - Y$ is positive semidefinite. This order is known as the Löwner partial order.

The set S of symmetric positive semidefinite matrices is convex as noted in Section 32.2. Furthermore, the function $F(X) = -\ln \det(X)$ is a self-concordant barrier function on S whose parameter of self-concordance is n. We omit the proof; see Nesterov and Nemirovskii [43]. This barrier is optimal for S in the sense that there is no self-concordant barrier on S with parameter less than n. (A set like the feasible region of 32.67, which is defined as the conjunction of a semidefinite constraint and equality constraints, may admit a better barrier. See remarks on this issue in Section 32.7.)

Given this specification of the barrier, it is now straightforward to set up the interior-point method of Section 32.4 and solve 32.67. The only difficulty is the inclusion of linear equality constraints. This is handled by regarding the barrier function F as restricted to the affine set given by $\{X : A_i \cdot X = b_i \text{ for each } i\}$. Restricting $F(X)$ to this affine set means that the Newton step 32.30 now must be modified with linear projections on the Hessian and gradient. An example of a projected Newton step for LP is presented in Section 32.7.

In this section we introduce primal-dual interior-point methods. In practice, primal-dual methods are preferred for SDP and LP over the primal method of Section 32.4. The framework of self-concordant barrier functions can be extended to primal-dual interior-point methods at a fairly general level [43], but we treat only the two special cases of primal-dual SDP and LP.

The dual problem for a convex optimization can be obtained from the KKT conditions 32.4: the multipliers in the KKT conditions turn out to be the solution to another convex optimization problem called the dual. An alternative way to derive the dual is by considering the KKT conditions of the barrier problem rather than the original problem. The SDP barrier problem is to minimize $C \cdot X - \mu \ln \det(X)$ subject to $X \succ 0$, $A_i \cdot X = b_i$ for $i = 1, \ldots, m$. The notation $X \succ 0$ means X is positive definite. Consider the KKT conditions for this barrier problem. Observe that the relation $X \succ 0$ drops out of the KKT conditions because it is inactive at the optimizer (because the barrier function blows up at the boundary of the feasible region, since $\det X = 0$ if X is semidefinite but not positive definite). Note also that the gradient of $\ln \det(X)$ is the function $X \mapsto X^{-1}$. Therefore, the KKT condition for the barrier problem is

$$
C - \mu X^{-1} = y_1 A_1 + \cdots + y_m A_m
\tag{32.68}
$$

and feasibility is

$$
A_1 \cdot X = b_1, \ldots, A_m \cdot X = b_m.
\tag{32.69}
$$

Here, $y_1, \ldots, y_m$, are unconstrained Lagrange multipliers.

We claim that these are also the KKT conditions for another SDP barrier problem. Consider the SDP

$$
\begin{aligned}
\text{maximize} \quad & b_1 y_1 + \cdots + b_m y_m \\
\text{subject to} \quad & C - S = y_1 A_1 + \cdots + y_m A_m, \\
& S \succeq 0,
\end{aligned}
\tag{32.70}
$$

where $S, y_1, \ldots, y_m$ are the variables. Attaching a barrier changes the objective to maximizing $b_1 y_1 + \cdots + b_m y_m + \mu \ln \det(S)$. The KKT optimality condition is that $b_1 = A_1 \cdot X, \ldots, b_m = A_m \cdot X$, and $X = \mu S^{-1}$. Here X is a multiplier for the equality constraint of 32.70. But notice that the identification $X = \mu S^{-1}$ means that 32.68 and 32.69 are satisfied. Thus, the two barrier problems have the same KKT conditions. We say that 32.70 is the dual to 32.67. One can check that the dual of the dual, after some simplification, is again the primal.

Let us assume that a constraint qualification such as the Slater condition holds. Then the KKT conditions are necessary and sufficient for both primal and dual optimality. Furthermore, the primal and dual barrier objective function optimal values satisfy the following relation:

$$
C \cdot X = \left(S + y_1 A_1 + \cdots + y_m A_m \right) \cdot X
\tag{32.71}
$$

$$
= S \cdot X + y_1 A_1 \cdot X + \cdots + y_m A_m \cdot X
\tag{32.72}
$$

$$
= S \cdot X + b_1 y_1 + \cdots + b_m y_m
\tag{32.73}
$$

$$
= n\mu + b_1 y_1 + \cdots + b_m y_m.
\tag{32.74}
$$

The last line was obtained by noting that since $SX = \mu I$, $S \cdot X = n\mu$. This follows because $S \cdot X = \text{trace } SX = \text{trace } \mu I$.

Thus, we see that for the barrier optimizers, the original primal function minimum value is exactly $n\mu$ greater than the dual objective function maximum value. Since we drive μ to zero in an interior-point method, we conclude that the two SDP's 32.67 and 32.70 have the same optimum value. Because of relationship 32.74, the parameter μ in an interior-point method is sometimes called the duality gap.

A primal-dual interior-point method for 32.67 and 32.70 is a method that simultaneously updates the primal and dual iterates, attempting to shrink the duality gap to zero. Alizadeh proposes a potential-reduction primal-dual algorithm for SDP, which is a generalization of Ye's [71] potential-reduction algorithm for linear programming. We define $q = n + \sqrt{n}$ and define

$$
\psi(X, S) = q \ln(X \cdot S) - \ln \det(XS).
\tag{32.75}
$$

The motivation for this definition is as follows. The first term decreases to $-\infty$ as the duality gap tends to 0. On the other hand, the second term is a penalty for loss of centrality. Unlike the method of Section 32.4, a potential-reduction method does not explicitly maintain proximity to the central path. Each iteration reduces the potential by a fixed constant amount, thus driving the potential to $-\infty$. The step taken is related to Newton's method for decreasing the potential function. To gain further insight, consider the case when X, S are on the central path, so that $XS = \mu I$: then the value of the potential function is $q \ln n + \sqrt{n} \ln \mu$, which tends to $-\infty$ as $\mu \to 0$. This algorithm has the same theoretical running time as the algorithm in Section 32.4, namely $O(\sqrt{n} \log(1/\epsilon))$ iterations to reduce the duality gap to $O(\epsilon)$.

We conclude this section with some applications of SDP, drawn from Vandenberghe and Boyd [63]. The most direct application is to eigenvalue optimization problems. For instance, if we want to minimize the maximum eigenvalue of an unknown symmetric matrix A that is parameterized by n free variables $x_1, \ldots, x_n$ according to the formula $A = A_0 + x_1 A_1 + \cdots + x_n A_n$, where $A_0, \ldots, A_n$ are given symmetric matrices. This can be stated as the following semidefinite program: find the minimum t such that $tI - A_0 - x_1 A_1 - \cdots - x_n A_n \succeq 0$. Some applications and further references for eigenvalue optimization are in [63]. Another application is quadratically constrained quadratic programming, which is minimizing a convex quadratic function subject to ellipsoidal and linear constraints. An ellipsoidal constraint $(\mathbf{x} - \mathbf{c})^T A(\mathbf{x} - \mathbf{c}) \leq 1$ can be transformed into an SDP constraint of the form:

$$\begin{bmatrix} I & R(\mathbf{x} - \mathbf{c}) \\ (\mathbf{x} - \mathbf{c})^T R^T & 1 \end{bmatrix} \succeq 0 \tag{32.76}$$

where R is a matrix such that $R^T R = A$ (e.g., the Cholesky factor of A). Note that a collection of SDP constraints can be concatenated along the diagonal to make a single large SDP constraint in order to put the problem in standard form. (These transformations do not necessarily lead to the most efficient algorithm for ellipsoidal constraints.) Another geometric problem that can be transformed to SDP is finding the smallest ellipsoid that contains the union of a set of given ellipsoids. See Vandenberghe and Boyd [63]. There are also applications of SDP to combinatorial optimization and control theory described in [1,63]. It should be noted that some of the combinatorial implications of semidefinite programming were observed before the advent of interior-point methods because the ellipsoid method can also solve semidefinite problems. See Grötschel et al. [17,18].

32.7 Linear Programming

A special case of SDP is linear programming. If we assume that each matrix A_i is diagonal and C is diagonal, then one can show that the solution X is also diagonal without loss of generality. In this case, the inequality $X \succeq 0$ means that each diagonal entry of X is nonnegative. Thus, if we let $\mathbf{x}$ be the vector of diagonal entries of X, and similarly for S, A_i's and C, then we obtain the primal and dual forms of linear programming:

$$\begin{aligned} \text{minimize} \quad & \mathbf{c}^T\mathbf{x} \\ \text{subject to} \quad & A\mathbf{x} = \mathbf{b}, \\ & \mathbf{x} \geq 0 \end{aligned} \tag{32.77}$$

and

$$\begin{aligned} \text{maximize} \quad & \mathbf{b}^T\mathbf{y} \\ \text{subject to} \quad & A^T\mathbf{y} + \mathbf{s} = \mathbf{c}, \\ & \mathbf{s} \geq 0. \end{aligned} \tag{32.78}$$

In these equations, A, $\mathbf{b}$,$\mathbf{c}$ are given: A is an $m \times n$ matrix assumed to have rank m (and hence $m \leq n$), $\mathbf{b}$ is an m-vector, and $\mathbf{c}$ is an n-vector. The variables are $\mathbf{x}$ for the primal and $(\mathbf{y},\mathbf{s})$ for the dual.

The barrier function for linear programming is the specialization of the SDP barrier to the case of diagonal matrices, which turns out to be the barrier for the positive orthant 32.64. Recall that this barrier has parameter n. Although this barrier is optimal for the orthant, it is not necessarily optimal for the orthant in conjunction with the linear constraints. The combination of the two types of constraints means that the feasible region is actually a polytope of dimension $n - m$ in the primal case, and m in the dual case. Since the noncomputable universal barrier has parameter $O(n - m)$ in the one case and $O(m)$ in the other case, the question arises whether there is a computable barrier with a better parameter than n for linear programming. Partial progress on this barrier was made by Vaidya and is described in [43]: Vaidya's barrier has parameter $O(\sqrt{mn})$ for the dual problem.

Using the standard barrier 32.64, we can write down LP with the barrier functions and derive the KKT conditions. This yields the following system of equations that describe the central path for both the primal and dual. These equations are a special case of 32.68 and 32.69 for SDP derived in Section 32.6.

$$Ax = b, \tag{32.79}$$

$$A^T y + s = c, \tag{32.80}$$

$$x_i s_i = \mu \quad \text{for } i = 1, \ldots, n, \tag{32.81}$$

$$x_i, s_i > 0 \quad \text{for } i = 1, \ldots, n. \tag{32.82}$$

We have introduced the LP central path as a special case of SDP, which is in turn a special case of general convex programming, but, the LP central path was discovered and analyzed before the others. This discovery is usually credited to [2,32,34,55], and several of these authors considered algorithms that follow this path. The first interior-point method for LP was Karmarkar's [26] projective method, which requires $O(n \log(1/\epsilon))$ iterations. Gill et al. [16] spotted the connection between Newton's method and Karmarkar's interior-point method. Renegar [49] also related interior-point methods to Newton's method and used this analysis to reduce the running time to $O(\sqrt{n} \log(1/\epsilon))$ iterations. Note that the $\sqrt{n}$ factor is the specialization of the $\sqrt{\theta}$ factor from the general convex case since $\theta = n$ for 32.64.

Kojima et al. [30] and Monteiro and Adler [36] introduced a primal-dual path-following method for 32.79–32.82. The idea is similar to what we have seen before: an iterate is a solution to 32.79 through 32.82, except 32.81 is satisfied only approximately. We then decrease μ and take a step of Newton's method to regain centrality. The iteration bound is also $O(\sqrt{n} \log(1/\epsilon))$. Because of the simple form, we can write down Newton's method explicitly. First, we need a measure of proximity to the central path. It turns out that the standard proximity measure $\lambda(\cdot, \cdot)$ for the primal or dual alone, after some manipulation and substitution of primal-dual approximations, has the form $\|XSe/\mu - e\|$, where $X = \text{diag}(x)$, $S = \text{diag}(s)$, and e is the vector of all 1's.

Assuming we have a primal-dual point (x,y,s) satisfying 32.79, 32.80 and 32.82, and a parameter μ such that $\|XSe/\mu - e\| \leq \lambda_0$ (where $\lambda_0 = 1/5$ for example), we can take a step to solve 32.79 through 32.82 for a smaller value of μ, say $\bar{\mu}$. The Newton step $(\Delta x, \Delta y, \Delta s)$ linearizes 32.81 and is defined by

$$A\Delta x = 0, \tag{32.83}$$

$$A^T \Delta y + \Delta s = 0, \tag{32.84}$$

$$X\Delta s + S\Delta x = -XSe + \bar{\mu}e. \tag{32.85}$$

Many of the best current interior-point methods are based on 32.83 through 32.85. Observe that 32.83 through 32.85 define a square system of linear equations with special structure; currently there is significant research on how to solve these equations efficiently and accurately. See Wright [70] for more information on this matter.

It should be pointed out that Newton's method for solving 32.79 through 32.82 as a system of nonlinear equations, which is the way 32.83 through 32.85 were obtained, is not the same as Newton's method for minimizing the barrier in either the primal or dual. Newton's method for minimizing the barrier in the primal is equivalent to Newton's method for solving a system of nonlinear equations similar to 32.79 through 32.82 except with the third equation replaced by $s_i = \mu/x_i$. The equation $s_i = \mu/x_i$ is of course equivalent to $x_i s_i = \mu$, but these two equations induce different Newton steps. See El-Bakry et al. [10] for more information on this matter.

32.8 Complexity of Convex Programming

In this section we consider complexity issues for linear, semidefinite, and convex programming. Complexity means establishing bounds on the asymptotic running time for large problems.

An immediate issue we must confront is the problem of modeling the numerical computations in the ellipsoid and interior-point method. Actual computers have floating-point arithmetic, which provides a fixed number of significant digits (about 16 decimal digits in IEEE standard double precision) for arithmetic. Restricting arithmetic to a fixed number of digits precludes the solution of poorly conditioned problems, and hence is not desirable for a complexity theory.

One way of modeling the notion that the precision of arithmetic should not have an a priori limit is the Turing machine model of computation, covered in detail in another chapter. For the purpose of numerical algorithms, the Turing machine model can be regarded as a computational device that carries out operations on arbitrary-precision rational numbers, and the running time of a rational operation is polynomially dependent on the number of bits in the numerators and denominators. The Turing machine model is also called the bit complexity model in the literature.

In this model, the data for a linear programming instance is presented as a triple $(A, \mathbf{b}, \mathbf{c})$ where the entries of these items must be rational numbers. In fact, without loss of generality, let us assume the data is integral since we can clear common denominators in a rational representation, causing at most a polynomial increase in the size of the problem.

Let L be the total number of bits to write this data. We assume that L is at least $mn + m + n$, that is, each coordinate entry of $(A, \mathbf{b}, \mathbf{c})$ requires at least one bit.

Khachiyan [27] showed that if an LP instance has a feasible solution, then it must have a feasible solution lying in a ball of radius $O(2^{cL})$ of the origin, where c is a universal constant. This ball can thus be used as the initial ellipsoid for the ellipsoid method. Furthermore, when the solution is known to accuracy $O(2^{-dL})$, then a rounding procedure can determine the exact answer.

A word is needed here concerning the exact answer. For a linear programming problem, it can be shown that the exact answer is the solution to a square system of linear equations whose coefficients are all rational and are derived from the initial data $(A, \mathbf{b}, \mathbf{c})$. This is because the optimal solution in a linear programming problem is always attained at a vertex (provided the feasible region has a vertex). The vertices of a polytope are determined by which constraints are active at that vertex. Solving a system of equations over the rationals can be done exactly with only a polynomial increase in bit-length, a result due to Edmonds [9].

Thus, in the case of linear programming with integer data, it is possible to deduce from the problem data itself valid initialization and termination procedures without additional knowledge. These procedures are also guaranteed to correctly diagnose problems with no feasible points, feasible sets with volume 0, and problems with an unbounded objective function. Khachiyan's analysis shows that the number of arithmetic operations for the ellipsoid method is bounded by $O(n^4 L)$. (The actual Turing machine running time is higher by a factor of L^c to account for the time to do each arithmetic operation.) This bound is a polynomial in the length of the input, which is L. Thus, Khachiyan's linear programming algorithm is polynomial time. This result generalizes to quadratic programming.

A similar analysis can be carried out for interior-point methods; see Vavasis [65] for the details. One reaches the conclusion that Karmarkar's method requires $O(nL)$ iterations to find the optimal solution, whereas Renegar's requires $O(\sqrt{n}L)$ iterations, where each iteration involves solving a system of equations. In the Turing-machine polynomial-time analysis of both the ellipsoid and interior-point methods, it is necessary to truncate the numerical data after each major iteration to $O(L)$ digits (else the number of digits grows exponentially). This truncation requires additional analysis to make sure that the invariants of the two algorithms are not violated.

Until 1979, the question of polynomiality of LP was a well-known open question. The reason is that the dominant algorithm in practice for solving LP problems until recently has been the simplex method of Dantzig [6]. The simplex method has many variants because many rules are possible for the selection of the pivot column. Klee and Minty [28] showed that some common variants of the simplex method are exponential time in the worst case, even though they are observed to run in polynomial time in practice. Exponential means that the running time could be as large as $\Omega(2^n)$ operations. Kalai [24] proposed a variant of simplex that has worst-case running time lower than exponential but still far above polynomial. There is no known variant of simplex that is polynomial-time, but on the other hand, there is also no proof that such a variant does not exist.

The ellipsoid algorithm, while settling the question of polynomiality, is not completely satisfactory for the following reason. Observe that the number of arithmetic operations of the simplex method depends on the dimensions of the relevant matrices and vectors but not on the data in the matrix. This is common for other numerical computations that have finite algorithms, such as solving systems of equations, finding maximum flows, and so on. An algorithm whose number of arithmetic operations is polynomial in the dimension of the problem, and whose intermediate numerical data require a polynomially-bounded number of bits, is said to be strongly polynomial time. Notice that the ellipsoid method is not strongly polynomial time because the number of arithmetic operations depends on the number of bits L in the data rather than on n.

The question of whether there is a strongly polynomial time algorithm for LP remains open. Some partial progress was made by Megiddo [33], who showed that linear programming in a low, fixed dimension is strongly polynomial. Low dimensional linear programming arises in computational geometry. Partial progress in another direction was made by Tardos [57], who proposed an LP algorithm based on the ellipsoid method such that the number of arithmetic operations depends only on the number of bits in A, and is independent of the number of bits in **b** and **c**. This algorithm uses rounding to the nearest integer as a main operation. Tardos's result is important because many network optimization problems can be posed as LP instances in which the entries of A have all small integer data, but **b** and **c** have complicated numerical data. Tardos's result was generalized by Vavasis and Ye as described in the following text.

We should add a word on how interior-point methods are initialized and terminated in practice. Three common techniques for initialization are phase I–phase II methods, big-M methods, and infeasible methods. Phase I–phase II methods were described at the end of Section 32.4. A big-M method appends a dummy variable to the LP instance, whose entry in the objective function vector is some very large number M. Because of the new variable, a feasible interior point is easily found. Because of the large weight on the new variable, it is driven to zero at optimum. A final class of initialization methods are infeasible interior-point methods. In these methods, an iterate satisfying 32.82 and approximately satisfying 32.81 is maintained, but the iterate does not necessarily satisfy 32.79 or 32.80. A trajectory is followed that simultaneously achieves feasibility and optimality in the limit.

All three initialization methods are guaranteed to work correctly and in polynomial time if the parameters are chosen correctly. For instance, M in the big-M method should be at least 2^{cL} for a theoretical guarantee. Infeasible interior-point methods are currently the best in practice but are the most difficult to analyze. See Wright [70] for details.

As for termination, many practical interior-point methods simply stop when μ is sufficiently small and output the iterate at that step as the optimizer. Khachiyan's solution for termination at an exact optimizer is generally not used in practiced. Ye [73] proposed a termination procedure based on projection that also gives the exact solution, but possibly at a much earlier stage than the theoretical worst-case $\sqrt{n}L$ iterations.

Finally, we mention a recent LP interior-point algorithm by Vavasis and Ye [67] that is based on decomposing the system of Equations 32.83 through 32.85 into layers. This algorithm has the property that its running time bound to find the exact optimizer is polynomial in a function of A that does not depend on **b** and **c**. If A contains integer data, then the bound by Vavasis and Ye depends polynomially on the number of bits in A. Thus, the layered algorithm gives a new proof of Tardos's result, but it is more general because no integrality assumption is made. In the case of generic real data, the running time of the algorithm depends on a parameter $\bar{\chi}_A$ that was discovered independently by a number of authors including Stewart [56] and Todd [62]. The earliest discoverer of the parameter is apparently Dikin [8].

Vavasis and Ye use a big-M initialization, but they show that the running time is independent of how big M is (i.e., the running time bound does not degrade if M is chosen too large). Furthermore, finite termination to an exact optimizer is built into the layered algorithm. Another consequence of the algorithm is a characterization of the LP central path as being composed of at most $O(n^2)$ alternating straight and curved segments such that the straight segments can be followed with a line search in a single step. There are still some obstacles preventing this algorithm from being completely practical; in particular, it is not known how to efficiently estimate the parameter $\bar{\chi}_A$ in a practical setting. Moreover, it is not known whether the approach generalizes beyond linear programming since there is no longer a direct relation between the layered step of [67] and Newton's method.

The complexity results in this section are not known to carry over to SDP. Consider the following SDP feasibility question: Given $A_1, \ldots, A_m$, each a symmetric $n \times n$ matrix with integer entries, and given m integer scalars $b_1, \ldots, b_m$, does there exist a real symmetric $n \times n$ matrix X satisfying $A_1 \cdot X = b_1, \ldots, A_m \cdot X = b_m, X \succeq 0$? This problem is not known to be solvable in polynomial time; in fact, it is not known even to lie in NP. The best Turing machine complexity bound, due to Porkolab and Khachiyan [47], is $O(mL2^{p(n)})$ operations, where $p(n)$ is a polynomial function of n, and L is the number of bits to write $A_1, \ldots, A_m, b_1, \ldots, b_m$. The technique uses a decision procedure for problems over the reals due to Renegar [50]. Note that the Turing machine is not required to compute such an X, but merely produce a yes/no answer. Such a problem is called a decision problem or language recognition problem.

There are two hurdles to generalize Khachiyan's LP analysis to this SDP decision problem. The first is that the exact solution to an SDP is irrational even if the original problem data is all integral. This hurdle by itself is apparently not such a significant difficulty. For instance, in [68] it is shown that a certain optimization problem (the trust-region problem) with an irrational optimizer nonetheless has an associated decision problem solvable in polynomial time. The trust-region problem is much simpler than SDP, but the arguments of [68] might generalize to SDP.

The second hurdle, apparently more daunting, is that points in the feasible region of an SDP problem may be double-exponentially large in the size of the data (as opposed to LP, where they are at most single-exponentially large). For instance, consider the example of minimizing x_n subject to $x_0 \geq 2$, and $x_1 \geq x_0^2, x_2 \geq x_1^2, \ldots, x_n \geq x_{n-1}^2$. (Recall that, as shown at the end of Section 32.6, convex quadratic constraints can be expressed in the SDP framework.) Clearly any feasible point satisfies $x_n \geq 2^{2^n}$. Note that even writing down 2^{2^n} as a binary number requires an exponential number of digits. This example is from Alizadeh [1], who attributes it to M. Ramana.

Since problems beyond linear and quadratic programming do not have rational solutions even when the data is rational, many researchers have abandoned the Turing machine model of complexity and adopted a real-number model of computation, in which the algorithm manipulates real numbers with unit cost per arithmetic operation. The preceding complexity bounds for the simplex method and layered-step method are both valid in the real number model. Other LP algorithms, as well as interior-point and ellipsoid algorithms for more general convex problems, need some new notion to replace the factor L. Several recent papers, for example [13,51,72], have defined a condition number for convex programming problems and have bounded the complexity in terms of this condition number.

32.9 First-Order Methods for Convex Optimization*

As indicated at the beginning of this chapter, convex programming is a part of nonlinear optimization for which there exists a rich theory and efficient algorithms. Moreover, many important problems

---

* This section was written by F. Jarre

arising in applications from engineering and other areas can be formulated as convex optimization problems.

This section concentrates on applications resulting in convex optimization problems

$$\text{minimize } f(\mathbf{x}) \mid \mathbf{x} \in D,$$

for which the number of unknowns, that is the dimension n of $\mathbf{x}$ is so large that interior-point methods cannot be applied in the fashion as outlined in Sections 32.4 and 32.5.

This section is thus a further restriction of the general class of nonlinear optimization. The aim of this restriction is to preserve the structure—typically sparsity of the data—throughout the transformations used in the algorithms for solving the problem. We consider two aspects. First, we briefly review complexity estimates of first-order methods, and subsequently some aspects of current implementations are discussed.

We always assume that D is convex and f is convex, and when f is (twice) differentiable at $\mathbf{x}$ we denote the gradient (column vector) by $\nabla f(\mathbf{x})$ and the Hessian matrix by $\nabla^2 f(\mathbf{x})$. For simplicity, subgradients are also denoted by $\nabla f(\mathbf{x})$. As before, the minimizer of the aforementioned problem is denoted by $\mathbf{x}^* \in D$.

32.9.1 Complexity of First-Order Methods

For the discussion of complexity we assume that D is a closed and bounded convex set, that is, $D \subset \{\mathbf{x} \in \mathbf{R}^n \mid \|\mathbf{x}\| \leq r\}$ for some finite $r \in \mathbf{R}$. We further assume that D has a simple structure in the sense that computations with regard to D can be neglected; we only consider evaluations of f later. Here, the values of $f(\mathbf{x})$ and $\nabla f(\mathbf{x})$ shall be given by a subroutine (oracle). Note that the information given by a call of the oracle is purely local.

Leaning on the excellent discussion of complexity issues of convex programs [42] by Nesterov we distinguish two cases. When the dimension n is moderate, and more than $O(n)$ evaluations of the oracle are affordable, the ellipsoid method of Section 32.3 is known to be optimal—if no other information of the problem is used apart from the oracle.

A surprising result is that even in situations where n is very large and less than $O(n)$ evaluations of the oracle are affordable we can state optimal algorithms that generate approximate solutions. These optimal algorithms depend on further assumptions on f: we either assume Lipschitz continuity of f (LC) or a bounded Hessian (BH),

$$
\begin{array}{lll}
(LC) & \exists L < \infty : & \|\nabla f(\mathbf{x})\| \leq L \quad \forall \mathbf{x} \in D, \\
(BH) & \exists M < \infty : & \|\nabla^2 f(\mathbf{x})\| \leq M \; \forall \mathbf{x} \in D.
\end{array}
$$

Under assumption (LC) the subgradient algorithm is optimal with $O(L^2 r^2 \epsilon^{-2})$ iterations for finding a point $\mathbf{x} \in D$ such that $f(\mathbf{x}) \leq f(\mathbf{x}^*) + \epsilon$.

Under the stronger assumption (BH), Nesterov [40] has proposed an accelerated version of descent algorithm that generates an ϵ-approximate solution in $O(M^{1/2} r^2 \epsilon^{-1/2})$ iterations. Note that this is a huge improvement; when $\epsilon = 0.01$ for example, then $\epsilon^{-1/2} = 10$ compared to $\epsilon^{-2} = 10,000$. In a simplified form (from [42]), this algorithm for the class (BH) is given by the scheme:

Start: Let $\mathbf{y}_0 \in D$ be given. Set $\mathbf{x}_{-1} := \mathbf{y}_0$.
For $k = 0, 1, 2, \ldots$ iterate

1. $\mathbf{x}_k := \arg\min\limits_{\mathbf{x} \in D} \; \nabla f(\mathbf{y}_k)^{\mathsf{T}} \mathbf{x} + \dfrac{M}{2} \|\mathbf{x} - \mathbf{y}_k\|^2$.

2. $\mathbf{y}_{k+1} := \mathbf{x}_k + \dfrac{k}{k+3} (\mathbf{x}_k - \mathbf{x}_{k-1})$.

The key observation made by Nesterov [41] is that a black box approach based on the model (LC) does not exploit the structure that is given for many practical optimization problems. If, for example, D is the standard simplex in $\mathbf{R}^n$ and we know (in addition to the availability of the oracle) that f is given by $f(\mathbf{x}) := \max_{1 \le j \le m} \mathbf{a}_j^\mathsf{T} \mathbf{x}$ then it is possible to extract global information from the values of f and its subgradients at certain given points—and to use this global information to accelerate the algorithm beyond the best possible rate of convergence of $O(L^2 r^2 / \epsilon^2)$ iterations (which holds in the absence of knowledge of the structure of f).

A systematic way of extracting global information is by approximating the nonsmooth function f with a smooth function f_μ for a small parameter $\mu > 0$,

$$f(\mathbf{x}) = \max_{u \mid u \ge 0, \ \sum_j u^{(j)} = 1} \sum_{j=1}^m u^{(j)} \mathbf{a}_j^\mathsf{T} \mathbf{x}$$

$$\approx f_\mu(\mathbf{x}) := \max_{u \mid u \ge 0, \ \sum_j u^{(j)} = 1} \left(\sum_{j=1}^m u^{(j)} \mathbf{a}_j^\mathsf{T} \mathbf{x} - \mu u^{(j)} \ln u^{(j)} \right) - \mu \ln m$$

$$= \mu \ln \left(\frac{1}{m} \sum_{j=1}^m e^{\mathbf{a}_j^\mathsf{T} \mathbf{x} / \mu} \right).$$

The smooth function f_μ can be minimized by algorithms for the class (BH) reducing the complexity of finding an ϵ-approximate minimizer for f from $O(\epsilon^{-2})$ iterations to $O(\epsilon^{-1})$ iterations. (In both cases, an ϵ-minimizer of f is generated!)

As detailed in [42] and the references therein, this acceleration is generalizable for functions f of the form

$$f(\mathbf{x}) = \max_{u \in D_u} (A\mathbf{x} - b)^\mathsf{T} u - \phi(u)$$

where ϕ is a concave function and D_u is a convex domain for u.

We now turn to numerical aspects, and from now on we assume that f is differentiable.

32.9.2 Ill-Conditioning

We begin the discussion of numerical aspects with a word of caution: A strong advantage of the interior-point approach in Section 32.4 is the fact that the theoretical rate of convergence is independent of any condition number. In contrast to this, cheaper algorithms used for solving large-scale problems are quite sensitive to condition numbers. We address two types of approach:

- A class of methods that is nearly perfect in exploiting and preserving sparsity are so-called direct methods such as genetic algorithms or simulated annealing. For low and moderate dimensions there are efficient implementations of such algorithms that tend to generate fairly good approximations to a global minimizer, even when the problem under consideration is not convex. (As mentioned, in the convex case, local minimizers are always global minimizers.) For large-scale problems or for poorly conditioned problems this is no longer the case.

  To explain this statement we note that a convex problem is poorly conditioned, if, for example, there exist feasible points $\mathbf{x} \in D$ such that $\|\mathbf{x} - \mathbf{x}^*\|$ is large, but $|f(\mathbf{x}) - f(\mathbf{x}^*)|$ is very small. (There are also other forms of ill-conditioning; see e.g. [13].) If the location of the exact minimizer $\mathbf{x}^*$ is to be approximated, direct methods tend to be very very slow in such situations. If, however, only the optimal value $f(\mathbf{x}^*)$ is to be approximated by some $f(\mathbf{x})$ with $\mathbf{x} \in D$ and the set D is reasonable, then direct methods may indeed be suitable.

- A class of algorithms that is better suitable for high dimensional problems are first-order (gradient based) methods. Similarly as for the direct methods the computational effort for a single step of a first-order method depends to a much lesser extent on the dimension of the problem than for a single step of an interior point method; first-order methods also avoid the costly evaluation of second derivatives, as well as the need to solve large scale linear systems.

However, first-order methods are also sensitive to ill-conditioning.

32.9.3 First-Order Approaches

In the unconstrained case, $D = \mathbf{R}^n$, the most efficient way to carry out the minimization of the convex function f may be by a limited memory BFGS-approach (LBFGS) for solving $\nabla f(\mathbf{x}) = 0$, see Nocedal [44] for details.

In the constrained case it is not advisable to solve the barrier problems arising in Section 32.4 by first-order methods as these barrier problems typically get systematically more and more ill-conditioned when the iterates approach $\mathbf{x}^*$. The effect of this type of ill-conditioning is less severe, when the Newton systems are solved to full machine precision by using direct methods. However, when this is not possible, it is more efficient to solve suitable reformulations of the optimality conditions by first-order methods.

In Section 32.9.3.1 we discuss such reformulations by means of semidefinite programs.

32.9.3.1 First-Order Approaches for Semidefinite Programs

In [29] Kocvara and Stingl describe a modified barrier method for the solution of large-scale semidefinite programming problems. The modification is such that the systematic ill-conditioning of the standard interior-point methods is avoided and a preconditioned conjugate gradient method can be used for solving the linear systems arising at each iteration.

The approaches by Malick et al. [31] and Povh et al. [48] use an augmented Lagrangian approach which also generates subproblems with bounded condition numbers under standard assumptions. A similar augmented Lagrangian approach is combined with a Newton cg-method in Zhao et al. [77]. All approaches present very promising numerical results when the desired final accuracy is moderate or when the semidefinite programs to be solved are well-conditioned in a certain sense.

32.9.3.2 An Augmented Primal-Dual Approach

A very general framework for convex optimization is given by the following convex conic problem:

$$(P) \qquad \text{minimize } \mathbf{c}^{\mathsf{T}}\mathbf{x} \mid \mathbf{x} \in K \cap (L + \mathbf{b}),$$

where

K is a closed convex cone

$L + \mathbf{b}$ is some affine space

The data of problem (P) consists of the two vectors $\mathbf{b}, \mathbf{c} \in \mathbf{R}^n$, and the sets K, L. For simplicity we assume that $L + \mathbf{b} = \{\mathbf{x} \mid A\mathbf{x} = \bar{\mathbf{b}}\}$ with an $m \times n$-matrix A and a vector $\bar{\mathbf{b}} \in \mathbf{R}^m$. By $L^{\perp} = \{\mathbf{s} \mid \exists \mathbf{y} : \mathbf{s} = A^{\mathsf{T}}\mathbf{y}\}$ we denote the orthogonal complement of L. Moreover, we assume that

$$\mathbf{c} \in L \quad \text{and} \quad \mathbf{b} \in L^{\perp}.$$

This assumption simplifies our derivations in the following text; it can easily be met by simply projecting $\mathbf{b}$ and $\mathbf{c}$ onto $L^{\perp}$ and L – without changing the feasible set or the optimal value of (P)!

When K is the nonnegative orthant $\mathbf{R}^n_+$ of $\mathbf{R}^n$, problem (P) is a linear program, when K is the semidefinite cone (and $\mathbf{x}$ is obtained from stacking the columns of the lower triangular part of the unknown matrix X on top of each other), then (P) is a semidefinite program, and when K is a cartesian product of second order cones, (P) is a second order cone program. More generally, any convex optimization problem can be transformed to the format (P).

We assume that K does not contain a straight line and has a nonempty interior. Then, the same is true for its dual cone

$$K^D := \{\, \mathbf{s} \mid \mathbf{x}^T \mathbf{s} \geq 0 \; \forall \, \mathbf{x} \in K \,\}.$$

For many important cases, K^D is known explicitly, in fact, for linear, second order cone programs and semidefinite programs, $K = K^D$.

Problem (P) has a dual problem which takes the form

(D) minimize $\mathbf{b}^T \mathbf{s} \mid \mathbf{s} \in K^D \cap (L^\perp + \mathbf{c})$.

If there exist a feasible point $\mathbf{x}$ for (P) in the interior of K and a feasible point $\mathbf{s}$ for (D) in the interior of K^D, then both problems have finite optimal solutions $\mathbf{x}^*$ and $\mathbf{s}^*$, and the optimal values satisfy

$$\mathbf{c}^T \mathbf{x}^* + \mathbf{c}^T \mathbf{s}^* = 0, \tag{32.86}$$

see Nesterov and Nemirovskii [43] for details. Conversely, if two points $\mathbf{x}$ and $\mathbf{s}$ are feasible for (P) and (D) and satisfy Equation 32.86, then $\mathbf{x}$ and $\mathbf{s}$ are optimal solutions for (P) and (D).

This observation can be used for solving (P) and (D). Let $\mathbf{z}$ be the vector $\mathbf{z} = (\mathbf{x}^T, \mathbf{s}^T)^T \in \mathbf{R}^{2n}$. We define an affine set $\mathbf{A} \subset \mathbf{R}^{2n}$ by Equation 32.86 and the conditions $\mathbf{x} \in L + \mathbf{b}$, $\mathbf{s} \in L^\perp + \mathbf{c}$. We also define a convex cone $\mathbf{K} \subset \mathbf{R}^{2n}$ by $\mathbf{K} = K \times K^D$. By the construction of $\mathbf{A}$ and $\mathbf{K}$, any point $\mathbf{z} \in \mathbf{A} \cap \mathbf{K}$ is optimal for (P) and (D).

As detailed in [23] the projections onto $\mathbf{A}$ are easy to compute, and for many applications such as linear, semidefinite, or second order cone programming, the projections onto $\mathbf{K}$ are easy to compute as well. If we are able to compute the projection onto the intersection of $\mathbf{A}$ and $\mathbf{K}$ then problems (P) and (D) can be solved.

For a given point $\mathbf{z} \in \mathbf{R}^{2n}$ let $\Phi(\mathbf{z})$ be the sum of the squares of the distances to $\mathbf{A}$ and $\mathbf{K}$. The function Φ is convex and differentiable and its minimizer provides a solution to (P) and (D). The values of the function Φ and its derivative directly follow once the projections onto $\mathbf{A}$ and $\mathbf{K}$ are known. The idea in [23] is thus to minimize Φ by a limited memory BFGS-approach.

Unfortunately, the function Φ is generally very ill-conditioned near its minimizer $\mathbf{z}^*$. The authors of [23] derive a regularizing function for semidefinite programs that is added to the function Φ and does not change the minimizer $\mathbf{z}^*$ of Φ but drastically improves the condition number near $\mathbf{z}^*$. Numerical examples for large semidefinite programs underline the importance of both, the regularization term and the LBFGS acceleration as compared to a simpler steepest descent or nonlinear conjugate gradient approaches.

Summarizing, large-scale convex programs force to use simpler first-order (or direct) methods which are much more sensitive to ill-conditioning than Newton-type methods as analyzed in Section 32.5. The approaches as described in [23,29,31,48,77] all aim at modifying the optimality conditions in a way that is well-conditioned under standard assumptions. The approaches outlined earlier for solving semidefinite programs can be modified to more general classes of convex programs.

Acknowledgment

The author likes to thank Steve Vavasis for his suggestions that helped to improve this section.

32.10   Nonconvex Optimization

The powerful techniques of convex programming can sometimes carry over to nonconvex problems, although generally none of the accompanying theory carries over to the nonconvex setting. For instance, the barrier function method was originally invented for general inequality-constrained nonconvex optimization—see, for example, Wright [69]—although it is does not necessarily converge to a global or even local optimizer in that setting. More recently, El-Bakry et al. [10] have looked at following the primal-dual central path for nonconvex optimization (which, as noted earlier, is not the same as following the barrier trajectory). In this section we will briefly summarize some known complexity results and approaches to nonconvex problems.

Perhaps the simplest class of nonconvex continuous optimization problems is nonconvex quadratic programming, that is, minimizing $\mathbf{x}^T H \mathbf{x}/2 + \mathbf{c}^T \mathbf{x}$ subject to linear constraints, where H is a symmetric but not positive semidefinite matrix. This problem is NP-hard [53], and NP-complete when posed as a decision problem [64]. Even when H has only one negative eigenvalue, the problem is still NP-hard [46]. Testing whether a point is a local minimizer is NP-hard [37]. If a polynomial time algorithm existed for computing an approximate answer guaranteed to be within a constant factor of optimal, this would imply that $\tilde{P} = \tilde{NP}$, where $\tilde{P}$ and $\tilde{NP}$ are complexity classes believed to be unequal [3].

In spite of all these negative results, some positive results are possible for nonconvex optimization. First, there are classes of nonconvex problems that can be transformed to SDP by nonlinear changes of variables; see [63] for examples. Sometimes the transformation from nonconvex to convex is not an exact transformation. For instance, if the feasible region is strictly enlarged, then the convex problem is said to be a relaxation of the nonconvex problem. The relaxed problem, while not equivalent to the original nonconvex problem, nonetheless yields a lower bound on the optimum value which may be useful for other purposes. See Vandenberghe and Boyd [63] for examples of semidefinite relaxations of nonconvex problems.

Another approach is to use convex optimization as a subroutine for nonconvex optimization. There are many examples of this in the nonlinear optimization literature; see Nash and Sofer [38] for details. Sometimes this can be done in a way that gives provable complexity bounds for nonconvex problems, such as the result by Vavasis [66] for nonconvex quadratic programming.

Finally, sometimes interior-point techniques, while not yielding global minima or approximations to global minima, can still deliver interesting results in polynomial time. For instance, Ye [74] shows that a primal-dual potential reduction algorithm applied to nonconvex quadratic programming can approximate a KKT point in polynomial time.

32.11   Research Issues and Summary

We have shown that convex optimization can be solved via two efficient algorithms, the ellipsoid method and interior-point methods. These methods can be derived in the general setting of convex programming and can also be specialized to semidefinite and linear programming. Perhaps the most outstanding research issue in the field is a better understanding of the complexity of semidefinite programming. Is this problem solvable in polynomial time? Is there a guaranteed efficient method for finding a starting point, and for determining whether a feasible region is empty? Another topic of great interest recently is to understand primal-dual path-following methods for SDP. The difficulty is that there are several different, inequivalent ways to generalize Equations 32.83 through 32.85 (which define the primal-dual path-following step for linear programming) to semidefinite programming, and it is not clear which generalization to prefer. See Todd [60] for more information. Another issue not completely understood for SDP is the correct handling of sparsity, since the matrices $A_1, \ldots, A_m$ are often sparse [14].

Linear programming is known to be polynomial time but not known to be strongly polynomial time. This is another notorious open problem.

32.12   Further Information

Interior-point methods for linear programming are covered in-depth by Wright [70]. Interior-point methods for semidefinite programming are the topic of the review article by Vandenberghe and Boyd [63]. Interior-point methods for general convex programming were introduced by Nesterov and Nemirovskii [43]. General theory of convexity is covered by Rockafellar [52]. General theory and algorithms for nonlinear optimization is covered by Fletcher [12] and Nash and Sofer [38]. Vavasis's book [65] covers complexity issues in optimization. The web page http://www.mcs.anl.gov/otc/ surveys optimization algorithms and software and includes the archive of interior-point preprints.

Defining Terms

Barrier function: Let D be a convex feasible region whose interior is nonempty. A barrier function $F : \text{int}(D) \rightarrow \mathbf{R}$ is a strictly convex function that tends to infinity as the boundary of D is approached.

Central path: Given a convex programming problem, the central path point for some $\mu > 0$ is the minimizer of $f(\mathbf{x}) + \mu F(\mathbf{x})$, where $f(\mathbf{x})$ is the objective function and $F(\mathbf{x})$ is a barrier function for D. The central path is the trajectory of all such points as μ varies from 0 to ∞. This curve joins the analytic center ($\mu = \infty$) of the region to the optimizer ($\mu = 0$). Many interior-point methods select iterates that are close to the central path according to a proximity measure.

Constraint: A feasible region $D \subset \mathbf{R}^n$ in an optimization problem is often represented as the set of points $\mathbf{x}$ satisfying a sequence of constraints involving $\mathbf{x}$. There are two main kinds of constraints: equality constraints, which have the form $g(\mathbf{x}) = 0$, and inequality constraints, which have the form $h(\mathbf{x}) \leq 0$.

Some specific types of constraints are as follows: A linear equality constraint has the form $\mathbf{a}^T\mathbf{x} = b$, where $\mathbf{a} \in \mathbf{R}^n$, $b \in \mathbf{R}$ are specified. A linear inequality constraint has the form $\mathbf{a}^T\mathbf{x} \leq b$. An ellipsoidal constraint has the form $(\mathbf{x} - \mathbf{c})^T A(\mathbf{x} - \mathbf{c}) \leq 1$, where $\mathbf{c}$ is a vector and A is a symmetric $n \times n$ positive definite matrix. A semidefinite constraint has the form

X is symmetric positive semidefinite

where X is a matrix composed of variables.

For a point $\mathbf{x}_1 \in D$, an inequality constraint $h(\mathbf{x}) \leq 0$ is active at $\mathbf{x}_1$ if $h(\mathbf{x}_1) = 0$, else the constraint is inactive. Equality constraints by default are always active.

A convex constraint is either a linear equality constraint defined above, or is an inequality constraint $h(\mathbf{x}) \leq 0$, where h is a convex function. The feasible region defined by one or more convex constraints is convex.

Constraint qualification: A constraint qualification is a condition imposed on the constraints at some feasible point that ensures that every linearized feasible direction is also a feasible direction.

Convex function: Let f be a real-valued function defined on a convex set D. This function is convex if for all $\mathbf{x}, \mathbf{y} \in D$ and for all $\lambda \in [0, 1]$, $f(\lambda \mathbf{x} + (1 - \lambda)\mathbf{y}) \geq \lambda f(\mathbf{x}) + (1 - \lambda)f(\mathbf{y})$. This function is strictly convex if the preceding inequality is strict whenever $\mathbf{x} \neq \mathbf{y}$ and $\lambda \in (0, 1)$.

Convex programming: An optimization problem in which the objective function is convex and the feasible region is specified via a sequence of convex constraints is called convex programming.

Convex set: A set $D \subset \mathbf{R}^n$ is convex if for any $\mathbf{x}, \mathbf{y} \in D$ and for any $\lambda \in [0,1]$, $\lambda\mathbf{x} + (1 - \lambda)\mathbf{y} \in D$.

Ellipsoid: Let A be an $n \times n$ symmetric positive definite matrix and let $\mathbf{c}$ be an n-vector. The set $E = \{\mathbf{x} \in \mathbf{R}^n : (\mathbf{x} - \mathbf{c})^T A(\mathbf{x} - \mathbf{c}) \leq 1\}$ is an ellipsoid.

Ellipsoid method: The ellipsoid method, due to Yudin and Nemirovskii [76], is a general-purpose algorithm for solving convex programming problems. It constructs a sequence of ellipsoids with shrinking volume each of which contains the optimizer. See Section 32.3.

Feasible direction: Let $D \subset \mathbf{R}^n$ be a feasible region and say $\mathbf{x} \in D$. A nonzero vector $\mathbf{v}$ is a feasible direction at $\mathbf{x}$ if there exists a sequence of points $\mathbf{x}_1, \mathbf{x}_2, \ldots$ all in D converging to $\mathbf{x}$ and a sequence of positive scalars $\alpha_1, \alpha_2, \ldots$ converging to zero such that $\mathbf{x}_k - \mathbf{x} = \alpha_k \mathbf{v} + o(\alpha_k)$.

Feasible region or feasible set: Defined under *optimization*.

Global optimizer: See *optimizer*.

Gradient: The gradient of a function $f : \mathbf{R}^n \to \mathbf{R}$ is its first derivative and is denoted ∇f. Note that $\nabla f(\mathbf{x})$ is an n-vector.

Hessian: The Hessian of a function $f : \mathbf{R}^n \to \mathbf{R}$ is its second derivative and is denoted $\nabla^2 f$. The Hessian is an $n \times n$ matrix which, under the assumption that f is C^2, is symmetric.

Interior-point method: An interior-point method for convex programming is an algorithm that constructs iterates interior to the feasible region and approaching the optimizer along some trajectory (usually the central path). See Sections 32.4 through 32.7.

Karush–Kuhn–Tucker (KKT) conditions: Consider the optimization problem $\min\{f(\mathbf{x}) : \mathbf{x} \in D\}$ in which D is specified by equality and inequality constraints and f is differentiable. A point $\mathbf{x}$ satisfies the KKT conditions if $\mathbf{x}$ is feasible and if for every linearized feasible direction $\mathbf{v}$ at $\mathbf{x}$, $\mathbf{v}^T \nabla f(\mathbf{x}) \geq 0$. If $\mathbf{x}$ is a local optimizer and a constraint qualification holds at $\mathbf{x}$, then $\mathbf{x}$ must satisfy the KKT conditions. The standard form of the KKT conditions is given by 32.4 earlier.

Linear function: The function $\mathbf{x} \mapsto \mathbf{a}^T \mathbf{x} + b$, where $\mathbf{a}, b$ are given, is linear.

Linear programming: Linear programming is an optimization problem in which $f(\mathbf{x})$ is a linear function and D is a polyhedron. The standard primal-dual format for linear programming is 32.77 and 32.78.

Linearized feasible direction: Let $D \subset \mathbf{R}^n$ be a feasible region specified as sequence of p equality constraints $g_1(\mathbf{x}) = \cdots = g_p(\mathbf{x}) = 0$ and q inequality constraints $h_1(\mathbf{x}) \leq 0, \ldots, h_q(\mathbf{x}) \leq 0$. Say $\mathbf{x} \in D$, and let $A \subset \{1, \ldots, q\}$ be the constraints active at D. Assume all these constraints are C^1 in a neighborhood of $\mathbf{x}$. A nonzero vector $\mathbf{v}$ is a linearized feasible direction at $\mathbf{x}$ if $\mathbf{v}^T \nabla g_i(\mathbf{x}) = 0$ for all $i = 1, \ldots, p$ and $\mathbf{v}^T \nabla h_i(\mathbf{x}) \leq 0$ for all $i \in A$.

Local optimizer: For the optimization problem $\min\{f(\mathbf{x}) : \mathbf{x} \in D\}$, $D \subset \mathbf{R}^n$, a local optimizer or local $N \subset \mathbf{R}^n$ containing $\mathbf{x}^*$ such that $f(\mathbf{x}) \geq f(\mathbf{x}^*)$ for all $\mathbf{x} \in D \cap N$. In the case of convex programming problems, a local optimizer is also a global optimizer.

Minimizer: See *optimizer*.

Objective function: Defined under *optimization*.

Optimization: Optimization refers to solving problems of the form: $\min\{f(\mathbf{x}) : \mathbf{x} \in D\}$ where D (the feasible region) is a subset of $\mathbf{R}^n$ and f (the objective function) is a real-valued function. The vector $\mathbf{x}$ is called the vector of variables or decision variables or unknowns. Set D is often described via constraints.

Optimizer: For the optimization problem of minimizing $f(\mathbf{x})$ subject to $\mathbf{x} \in D$, the optimizer or minimizer is the point $\mathbf{x}^* \in D$ such that $f(\mathbf{x}^*) \leq f(\mathbf{x})$ for all $\mathbf{x} \in D$. Also called global optimizer. Some authors use the term optimizer to mean local optimizer.

Polyhedron: A polyhedron is the set defined by a sequence of linear constraints.

Positive definite: A square matrix A is positive definite if $x^T A x > 0$ for all nonzero vectors x. This term is applied mainly to symmetric matrices.

Positive semidefinite: A square matrix A is positive semidefinite if $x^T A x \geq 0$ for all vectors x. This term is applied mainly to symmetric matrices.

Quadratic programming: Quadratic programming is the optimization problem of minimizing a quadratic function $f(x) = x^T H x/2 + c^T x$ subject to $x \in D$, where D is a polyhedron.

Self-concordant barrier function: A barrier function F is said to be self-concordant if it satisfies 32.63 for all $x \in \text{int}(D)$ and all $h \in \mathbf{R}^n$. The parameter of self-concordance is the smallest θ such that F satisfies 32.28 for all $x \in \text{int}(D)$.

Semidefinite programming: Semidefinite programming is the problem of minimizing a linear function subject to linear and semidefinite constraints. The standard primal-dual form for SDP is given by 32.67 and 32.70.

Slater condition: The Slater condition is a constraint qualification for convex programming. Let D be a convex feasible region specified by convex constraints. The Slater condition is that there exists a point $x_0 \in D$ such that all the nonlinear constraints are inactive at x_0. This condition serves as a constraint qualification for all of D, that is, it implies that for every point $x \in D$, the linearized feasible directions at x coincide with the feasible directions at x.

Stationary point: For an optimization problem, a stationary point is a point satisfying the KKT conditions.

Subdifferential: For a convex function f defined on a convex set D, the subdifferential of f at x is defined to be the set of vectors $v \in \mathbf{R}^n$ such that $f(y) \geq f(x) + v^T(y - x)$ for all $y \in D$. The subdifferential is a nonempty closed convex set that coincides with the ordinary derivative in the case that f is differentiable at x and x is in the interior of D.

Subgradient: A subgradient of f at x is an element of the subdifferential.

Symmetric matrix: A square matrix A is symmetric if $A = A^T$, where the superscript T indicates matrix transpose.

Acknowledgments

The author received helpful comments on this chapter from Florian Jarre, Michael Todd, and the anonymous referee. This work has been supported in part by NSF grant CCR-9619489. Research supported in part by NSF through grant DMS-9505155 and ONR through grant N00014-96-1-0050. Support was also received from the Mathematical, Information, and Computational Sciences Division subprogram of the Office of Computational and Technology Research, U.S. Dept. of Energy, under Contract W-31-109-Eng-38 through Argonne National Laboratory. Support was also received from the J.S. Guggenheim Foundation.

References

1. Alizadeh, F., Interior point methods in semidefinite programming with applications to combinatorial optimization. *SIAM Journal on Optimization*, 5, 13–51, 1995.
2. Bayer, D.A. and Lagarias, J.C., The nonlinear geometry of linear programming. I. Affine and projective scaling trajectories. *Transactions of the AMS*, 314, 499–526, 1989.
3. Bellare, M. and Rogaway, P., The complexity of approximating a nonlinear program. In *Complexity in Numerical Optimization*, Pardalos, P.M., Ed., World Scientific, Singapore, 1993.

4. Bland, R., Goldfarb, D., and Todd, M., The ellipsoid method: A survey. *Operations Research,* 29, 1039–1091, 1981.

5. Burrell, B. and Todd, M., The ellipsoid method generates dual variables. *Mathematics of Operations Research,* 10, 688–700, 1985.

6. Dantzig, G., *Linear Programming and Extensions.* Princeton University Press, Princeton, NJ, 1963.

7. Dikin, I.I., Iterative solution of problems of linear and quadratic programming. *Soviet Mathematics Doklady,* 8, 674–675, 1967.

8. Dikin, I.I., On the speed of an iterative process. *Upravlyaemye Sistemi,* 12, 54–60, 1974.

9. Edmonds, J., Systems of distinct representatives and linear algebra. *Journal of Research of National Bureau of Standards,* 71B, 241–245, 1967.

10. El-Bakry, A.S., Tapia, R.A., Tsuchiya, T., and Zhang, Y., On the formulation and theory of the Newton interior point method for nonlinear programming. *Journal of Optimization Theory and Applications,* 89(3), 507–541, 1996.

11. Fiacco, A.V. and McCormick, G.P., *Nonlinear Programming: Sequential Unconstrained Minimization Techniques.* John Wiley & Sons, Chichester, U.K., 1968.

12. Fletcher, R., *Practical Methods of Optimization,* 2nd ed. John Wiley & Sons, Chichester, U.K., 1987.

13. Freund, R. and Vera, J., Some characterizations and properties of the "distance to ill-posedness" and the condition measure of a conic linear system. Technical Report 3862-95-MSA, MIT Sloan School of Management, Cambridge, MA, 1995.

14. Fujisawa, K., Kojima, M., and Nakata, K., Exploiting sparsity in primal-dual interior-point methods for semidefinite programming. Technical Report B-324, Mathematical and Computing Sciences, Tokyo Institute of Technology, Tokyo, Japan, 1997.

15. George, A. and Liu, J.W.H., *Computer Solution of Large Sparse Positive Definite Systems.* Prentice-Hall, Englewood Cliffs, NJ, 1981.

16. Gill, P., Murray, W., Saunders, M., Tomlin, J., and Wright, M., On projected Newton barrier methods for linear programming and an equivalence to Karmarkar's projective method. *Mathematical Programming,* 36, 183–209, 1986.

17. Grötschel, M., Lovász, L., and Schrijver, A., The ellipsoid method and its consequences in combinatorial optimization. *Combinatorica,* 1, 169–197, 1981.

18. Grötschel, M., Lovász, L., and Schrijver, A., *Geometric Algorithms and Combinatorial Optimization.* Springer-Verlag, New York, 1988.

19. Jarre, F., On the method of analytic centers for solving smooth convex programs. In *Optimization,* Dolecki, S., Ed., volume 1405 of *Lecture Notes in Mathematics,* pp. 69–85. Springer-Verlag, Berlin, Germany, 1989.

20. Jarre, F., On the convergence of the method of analytic centers when applied to convex quadratic programs. *Mathematical Programming,* 49, 341–358, 1991.

21. Jarre, F., Interior-point methods for convex programming. *Applied Mathematics and Optimization,* 26, 287–311, 1992.

22. Jarre, F., Interior-point methods via self-concordance or relative Lipschitz condition. Habilitationsschrift, Bayerische Julius-Maximilians-Universität Würzburg, Würzburg, Germany, 1994.

23. Jarre, F. and Rendl, F., An augmented primal-dual method for linear conic programs. *SIAM Journal on Optimization* 19 (2), 808–823, 2008.

24. Kalai, G., A subexponential randomized simplex algorithm. In *Proceedings of the 24th ACM Symposium on the Theory of Computing,* pp. 475–482, 1992.

25. Kapoor, S. and Vaidya, P.M., Fast algorithms for convex quadratic programming and multicommodity flows. In *Proceedings of 18th Annual ACM Symposium on Theory of Computing,* pp. 147–159, 1986.

26. Karmarkar, N., A new polynomial-time algorithm for linear programming. *Combinatorica,* 4, 373–395, 1984.

27. Khachiyan, L.G., A polynomial algorithm in linear programming. *Doklady Akademii Nauk SSSR,* 244, 1093–1086, 1979. Translated in *Soviet Mathematics Doklady* 20, 191–194, 1979.

28. Klee, V. and Minty, G.J., How good is the simplex algorithm? In *Inequalities III*, Shisha, O., Ed., Academic Press, New York, 1972.

29. Kocvara, M. and Stingl, M., On the solution of large-scale sdp problems by the modified barrier method using iterative solvers. *Mathematical Programming* (Series B), 109(2–3):413–444, 2007.

30. Kojima, M., Mizuno, S., and Yoshise, A., A polynomial-time algorithm for a class of linear complementarity problems. *Mathematical Programming*, 44, 1–26, 1989.

31. Malick, J., Povh, J., Rendl, F., and Wiegele, A., Regularization methods for semidefinite programming. Optimization online, http://www.optimization-online.org/db_html/2007/10/1800.html, 2007.

32. McLinden, L., An analogue of Moreau's proximation theorem, with applications to the nonlinear complementarity problem. *Pacific Journal of Mathematics*, 88, 101–161, 1980.

33. Megiddo, N., Linear programming in linear time when the dimension is fixed. *Journal of the ACM*, 31, 114–127, 1984.

34. Megiddo, N., Pathways to the optimal set in linear programming. In *Progress in Mathematical Programming: Interior Point and Related Method*, Megiddo, N., Ed., pp. 131–158. Springer-Verlag, New York, 1989.

35. Mehrotra, S. and Sun, J., A method of analytic centers for quadratically constrained convex quadratic programs. *SIAM Journal of Numerical Analysis*, 28, 529–544, 1991.

36. Monteiro, R.C. and Adler, I., Interior path following primal-dual algorithm. Part I: linear programming. *Mathematical Programming*, 44, 27–41, 1989.

37. Murty, K.G. and Santosh, N.K., Some NP-complete problems in quadratic and nonlinear programming. *Mathematical Programming*, 39, 117–129, 1987.

38. Nash, S. and Sofer, A., *Linear and Nonlinear Programming*. McGraw-Hill, New York, 1996.

39. Nemirovsky, A.S. and Yudin, D.B., *Problem Complexity and Method Efficiency in Optimization*. John Wiley & Sons, Chichester, U.K., 1983. Translated by E. R. Dawson from *Slozhnost' Zadach i Effektivnost' Metodov Optimizatsii*, 1979, Glavnaya redaktsiya fiziko-matematicheskoi literatury, Izdatelstva "Nauka."

40. Nesterov, Y., A method for unconstrained convex minimization problems with the rate of convergence $O(1/k^2)$. *Soviet Mathematics Doklady*, 269(3), 543–547, 1983.

41. Nesterov, Y., Smooth minimization of non-smooth functions. *Mathematical Programming*, 103(1), 127–152, 2005.

42. Nesterov, Y., How to advance in Structural Convex Optimization *OPTIMA, MPS Newsletter*, 78, 2–5, 2008.

43. Nesterov, Y. and Nemirovskii, A., *Interior Point Polynomial Methods in Convex Programming: Theory and Algorithms*, volume 13 of *SIAM Studies in Applied Mathematics*. SIAM Press, Philadelphia, PA, 1994.

44. Nocedal, J., Updating Quasi-Newton matrices with limited storage. *Mathematics of Computation*, 35, 773–782, 1980.

45. Papadimitriou, C.H. and Steiglitz, K., *Combinatorial Optimization: Algorithms and Complexity*. Prentice Hall, Englewood Cliffs, NJ, 1982.

46. Pardalos, P.M. and Vavasis, S.A., Quadratic programming with one negative eigenvalue is NP-hard. *Journal of Global Optimization*, 1, 15–22, 1991.

47. Porkolab, L. and Khachiyan, L., On the complexity of semidefinite programs. Technical Report RRR 50-95, Rutgers Center for Operations Research (RUTCOR), Rutgers University, Piscataway, NJ, 1995.

48. Povh, J., Rendl, F., and Wiegele, A., A boundary point method to solve semidefinite programs. *Computing*, 78(3), 277–286, 2006.

49. Renegar, J., A polynomial-time algorithm based on Newton's method for linear programming. *Mathematical Programming*, 40, 59–94, 1988.

50. Renegar, J., On the computational complexity and geometry of the first-order theory of the reals. Part I: introduction, preliminaries; the geometry of semi-algebraic sets; the decision problem for the existential theory of the reals. *Journal of Symbolic Computing*, 13, 255–299, 1992.

51. Renegar, J., Linear programming, complexity theory, and elementary functional analysis. Unpublished manuscript, Department of Operations Research and Industrial Engineering, Cornell University, Ithaca, NY, 1993.

52. Rockafellar, R.T., *Convex Analysis*. Princeton University Press, Princeton, NJ, 1970.

53. Sahni, S., Computationally related problems. *SIAM Journal on Computing*, 3, 262–279, 1974.

54. Shor, N.Z., Convergence rate of the gradient descent method with dilatation of the space. *Kibernetika*, 6(2), 80–85, 1970. Translated in *Cybernetics*, 6(2), 102–108.

55. Sonnevend, G., An "analytic center" for polyhedrons and new classes of global algorithms for linear (smooth, convex) programming. In *System Modelling and Optimization: Proceedings of the 12th IFIP–Conference*, Budapest, Hungary, Sep. 1985, Prekopa, A., Szelezsan, J., and Strazicky, B., Eds., volume 84 of *Lecture Notes in Control and Information Sciences*, pp. 866–876. Springer-Verlag, Berlin, Germany, 1986.

56. Stewart, G.W., On scaled projections and pseudoinverses. *Linear Algebra and Its Applications*, 112, 189–193, 1989.

57. Tardos, É., A strongly polynomial algorithm to solve combinatorial linear programs. *Operations Research*, 34, 250–256, 1986.

58. Todd, M., On minimum volume ellipsoids containing part of a given ellipsoid. *Mathematics of Operations Research*, 7, 253–261, 1982.

59. Todd, M., Unpublished lecture notes for OR635: Interior point methods. 1995.

60. Todd, M., On search directions in interior-point methods for semidefinite programming. Technical Report TR1205, School of Operations Research and Industrial Engineering, Cornell University, Ithaca, NY, 1997.

61. Todd, M. and Ye, Y., A lower bound on the number of iterations of long-step and polynomial interior-point linear programming algorithms. Technical Report 1082, School of Operations Research and Industrial Engineering, Cornell University, Ithaca, NY, 1994. To appear in *Annals of Operations Research*.

62. Todd, M.J., A Dantzig-Wolfe-like variant of Karmarkar's interior-point linear programming algorithm. *Operations Research*, 38, 1006–1018, 1990.

63. Vandenberghe, L. and Boyd, S., Semidefinite programming. *SIAM Review*, 38, 49–95, 1996.

64. Vavasis, S.A., Quadratic programming is in NP. *Information Processing Letters*, 36, 73–77, 1990.

65. Vavasis, S.A., *Nonlinear Optimization: Complexity Issues*. Oxford University Press, New York, 1991.

66. Vavasis, S.A., Approximation algorithms for indefinite quadratic programming. *Mathematical Programming*, 57, 279–311, 1992.

67. Vavasis, S.A. and Ye, Y., A primal-dual interior point method whose running time depends only on the constraint matrix. *Mathematical Programming*, 74, 79–120, 1996.

68. Vavasis, S.A. and Zippel, R., Proving polynomial-time for sphere-constrained quadratic programming. Technical Report 90-1182, Department of Computer Science, Cornell University, Ithaca, New York, 1990.

69. Wright, M.H., *Numerical Methods for Nonlinearly Constrained Optimization*. PhD Thesis, Stanford University, Palo Alto, CA, 1976.

70. Wright, S.J., *Primal-Dual Interior-Point Methods*. SIAM Press, Philadelphia, PA, 1997.

71. Ye, Y., An $O(n^3 L)$ potential reduction algorithm for linear programming. *Mathematical Programming*, 50, 239–258, 1991.

72. Ye, Y., Toward probabilistic analysis of interior-point algorithms for linear programming. *Mathematics of Operations Research*, 1993, to appear, 1991.

73. Ye, Y., On the finite convergence of interior-point algorithms for linear programming. *Mathematical Programming*, 57, 325–336, 1992.

74. Ye, Y., On the complexity of approximating a KKT point of quadratic programming. Preprint, 1996.

75. Ye, Y. and Tse, E., An extension of Karmarkar's projective algorithm for convex quadratic programming. *Mathematical Programming*, 44, 157–179, 1989.

76. Yudin, D.B. and Nemirovskii, A.S., Informational complexity and efficient methods for solving complex extremal problems. *Ekonomika i Matematicheskie Metody,* 12, 357–369, 1976. Translated in *Matekon: Translations of Russian and East European Math. Economics,* 13, 25–45, Spring 1977.
77. Zhao, X., Sun, D., and Toh, K.C., A Newton-CG augmented Lagrangian method for semidefinite programming. Optimization online, http://www.optimization-online.org/db_html/2008/03/1930.html, 2008.

33

Simulated Annealing Techniques

Albert Y. Zomaya
The University of Sydney

Rick Kazman
University of Hawaii and
Carnegie Mellon University

33.1 Introduction

This chapter will present the essential components of the simulated annealing (SA) algorithm and review its origins and potential for solving a wide range of optimization problems, in a manner that is accessible to the widest possible audience. Some historical perspective and description of recent research results will also be provided. During the course of this review bibliographical references will be provided to guide the interested reader to sources that contain additional theoretical results and complete details of individual applications.

Many problems in a variety of disciplines can be formulated as optimization problems; and most of these can be solved by adopting one of two "popular" approaches: divide-and-conquer or hill-climbing techniques (other approaches can be adopted, see for example, [31], and see also Chapters 30–32). In the first approach, the solution is problem-dependent, and typically detailed information about the problem is required in order to develop a solution strategy. Also, not many problems can be subdivided into smaller parts that can be solved separately and then recombined. In the second approach, most hill-climbing algorithms are based on gradient descent methods. These methods suffer from a major drawback of getting trapped in a local minimum. That is, the algorithm may get "trapped" in a valley, from which all paths lead to locally worse solutions, and will never get to an optimal solution that lies outside the valley. The SA algorithm avoids local minima by introducing an element of randomness into the search process [36].

Over the last few years the SA algorithm and its many extensions and refinements have been extensively employed to solve a wide range of application domains, especially in combinatorial optimization problems [2,5,8,13,15,47,54,57,58]. An important characteristic of the SA algorithm is that it does not require specialist knowledge about how to solve a particular problem. This makes the algorithm generic in the sense that it can be used in a variety of optimization problems without the need to change the basic structure of the computations.

The versatility of SA has attracted many researchers over a number years, and has recently spawned a number of variations to the original algorithm, including parallel versions to speed up the rate of computations [18,19,25,42,51,62].

33.2   The Basic Idea

The SA is a stochastic optimization method modeled on the behavior of condensed matter at low temperatures. It borrows techniques from statistical mechanics to find global optima of systems with large numbers of degrees of freedom. The method is analogous to the way that liquids freeze and crystallize or metals cool and anneal. The core of the process is slow cooling, allowing enough time for the redistribution of atoms as they lose mobility until a minimum energy state is reached [57].

During the physical annealing process a solid is placed in a heat bath and the temperature is continually raised until the solid has melted and the particles of the solid are physically disarranged or positioned in random order. The orientation of the particles are referred to as the spins. From such a high energy level, the heat bath is cooled slowly by lowering the temperature to allow the particles to align themselves in a regular crystalline lattice structure. This final structure corresponds to a stable low energy state.

The temperature must be lowered slowly to allow the solid to reach equilibrium after each temperature drop. Otherwise, irregular alignments may occur, resulting in defects that get frozen into the solid. Of course, this can result in a metastable (highly unstable) structure rather than the required stable low energy structure.

The idea of annealing was combined with the well-known Monte Carlo algorithm [40], which was originally used to perform numerical averaging over large systems in statistical mechanics. The SA algorithm maintains the speed and reliability of gradient descent algorithms while at the same time avoiding local minima [36].

33.3   Global Optimization Problems

There are two classes of algorithms that can be used to solve global optimization problems [48]. The first are random or Monte Carlo type of algorithms that make use of pseudo random variables. The second are deterministic algorithms that do not take advantage of randomization; the earliest global optimization methods belong to this class. Global optimization techniques aim at solving the following general class of problems [31]:

$$\text{maximize} \quad f(x)$$

$$\text{subject to:} \quad h_i(x) \leq 0 \quad i = 1, 2, ..., n \tag{33.1}$$

$$x \in X$$

where f and h_i (for $i = 1, 2, \ldots, n$) are real-valued functions defined on a domain containing $X \subseteq \Re^m$, where m is the number of variables. If $n = 0$ and $X = [a_1, b_1] \times [a_2, b_2] \times \cdots \times [a_m, b_m]$ defines a hyperrectangle of $\Re^m$, the problem is called unconstrained, and if $m = 1$, the problem is univariate.

An assumption that is usually made to enable the solution of Equation 33.1 is that the functions f and $h_i(i = 1, 2, \ldots, n)$ can be evaluated at all points of X. This means that there are methods by which given the values of variables, the corresponding evaluations of functions can be provided. This is true in a number of practical situations in which measurements of functions can be made through empirical means, but no analytical (or closed-form) expressions are available [41,43,46].

If no other assumptions are made on f and $h_i (i = 1, 2, \ldots, n)$, the problem is intractable. It is important to note that no matter how many function evaluations are performed, there is no guarantee that the minimum can be obtained [29]. To help in finding a solution, another simple assumption that is usually made is that the slopes of the functions f and h_i $(i = 1, 2, \ldots, n)$ are bounded. In such case, these function are said to be Lipschitz. A real-valued function defined on a compact set $X \subseteq \mathcal{R}^m$ said to be Lipschitz if it satisfies the condition

$$\forall x \in X \quad \forall y \in X \quad | \alpha(x) - \alpha(y) | \le L \| x - y \| \tag{33.2}$$

where

L is a constant (Lipschitz constant)

$\| \bullet \|$ is the Euclidean norm (other norms can also be used)

In general, deterministic algorithms evaluate a given cost function at points on a grid. A major problem with such deterministic algorithms is that they require knowledge about the problem, or in other words, the cost function (such as the Lipchitz constant L) that needs to be evaluated.

Most global optimization algorithms are of the random type and are related to the so-called multistart algorithm [7]. In this case, a local optimization algorithm is executed from different initial or starting points that are chosen at random, usually from a uniform distribution on the domain of the cost function. However, a multistart algorithm is still inefficient because it will inevitably find each local extremum (maximum or minimum) several times. In addition, local search are the most time consuming part of any procedure (the multistart algorithm in this case). A typical multistart algorithm that can take advantage of a local search procedure Ω is shown below.

PROCEDURE MULTISTART

1 $k \leftarrow 1$
2 $\prod(0) = -\infty$
3 **Generate** a point x from the uniform
    distribution over X
4 **Apply** Ω to x to get $\tilde{x}$
5 **If** $f(\tilde{x}) > \prod(k - 1)$ **then**
6      $\prod(k) \leftarrow f(\tilde{x})$
7      $x_k = \tilde{x}$
8 **Else**
9      $\prod(k) \leftarrow \prod(k - 1)$
10     $x_k = x_{k-1}$
11   **Increment** k; **Return** to 3

From a computational point of view, a local search procedure should not be called more than once in every region of attraction. In this case, the region of attraction of the local maximum $\bar{x}_k$ is defined as the set of points in X from which the local search procedure (Ω) will converge to $\bar{x}_k$.

The SA algorithm can also be classified as a random search technique, but the algorithm avoids getting trapped in local minima by accepting, in addition to movements corresponding to improvement in cost function value, also movements corresponding to a deterioration in cost function value. However, the deterioration movements are accepted with a finite probability. These two different types of movements allow the algorithm to move away from local minima and traverse more states in the region X (Figure 33.1).

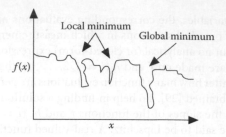

FIGURE 33.1 Local and global minima.

33.3.1 The Metropolis Algorithm

The SA algorithm is based on a procedure introduced by [40] to simulate the equilibrium states of a multibody system at a given finite temperature. The Metropolis procedure consisted of a number of simple principles. At each step, a small perturbation of the configuration (of the given system) is chosen at random and the resulting change in the energy of the system, Δ, is calculated. The new configuration is accepted with probability 1 if $\Delta \leq 0$, and with probability $e^{-\Delta/k_B T}$ if $\Delta > 0$. To temperature (T) can be viewed as a control parameter. The procedure is given below.

> **PROCEDURE METROPOLIS**
> 1 **Generate** some random initial configuration s
> 2 **Repeat**
> 3 $s' \leftarrow$ Some random neighboring configuration of s
> 4 $\Delta \leftarrow E(s') - E(s)$
> 5 $\rho_T(\Delta) \leftarrow \min(1, e^{-\Delta/k_B T})$
> 6   **If** $random(0.1) \leq \rho_T(\Delta)$ **then** $s \leftarrow s'$
> 7 **Until false**
> 8 **End**

In the above algorithm, $E(s)$ stands for the energy associated with state s and k_B is the Boltzmann constant ($k_B = 1.38 \times 10^{-16}$ ergs/K(Kelvin)). The energy function $E(\bullet)$ is usually replaced by a cost function; then the above procedure can be used to simulate the behavior of the optimization problem with the given cost function.

The essence of the Metropolis procedure can be summarized as follows. If Φ_s is the set of states $\varphi \in S$ reachable in exactly one move (perturbation) from s. Each move must be reversible ($\varphi \in \Phi_s \Rightarrow s \in \Phi_\varphi$). The number of possible moves (i.e., $\omega = |\Phi_s|$) must be the same from any state s. It also must be possible to reach any state in Φ from any other in a finite number of moves.

The function $\rho_T(\Delta) = \min(1, e^{-\Delta/k_B T})$ is used to choose a random move, and each move has a probability of $1/\omega$ of being accepted. Following the sequence of accept and reject moves given in the above algorithm, leads to a Markov process with transition function [7,12]:

$$
\Phi_T(s'|s) = \begin{bmatrix} \dfrac{1}{\omega}\rho_T(E(s') - E(s)) & s' \in \Phi_s \\[2ex] 1 - \displaystyle\sum_{\varphi \in \Phi_s} \dfrac{1}{\omega}\rho_T(E(\varphi) - E(s)) & s' = s \\[2ex] 0 & \text{otherwise} \end{bmatrix}
$$

If T is positive, the above process is irreducible, which means that for any two states $s, \varphi \in \Phi$, there is a nonzero probability that state φ will be reached from s. Defining π_T as the stationary (equilibrium) distribution of this process, it can be found that

$$\pi_T(s) = \frac{e^{(-E(s)/T)}}{\sum_{\varphi \in \Phi} e^{(-E(\varphi)/T)}} \quad \forall s \in \Phi$$

The above follows from the principle of detailed reversibility [7,12],

$$\pi_T(s)\Omega_T(s'|s) = \pi_T(s')\Omega_T(s|s') \quad \forall s, s' \in \Phi$$

where π_T is the Boltzmann distribution. Simply stated, detailed reversibility means that the likelihood of any transition equals that of the opposite transition (i.e., in the opposite direction).

33.4   Simulated Annealing

As mentioned earlier, the SA algorithm is an optimization technique based on the behavior of condensed matter at low temperatures. The procedure employs methods that originated from statistical mechanics to find global minima of systems with very large degrees of freedom. The correspondence between combinatorial optimization problems and the way natural systems search for the ground state (lowest energy state) was first realized by [10,36]. Based on the analogy given in Figure 33.2, Kirkpatrick et al. [36] applied Monte Carlo methods, which are usually found in statistical mechanics, to the solution of global optimization problems, and it was shown that better solutions can be obtained by simulating the annealing process that takes place in natural systems.

Further, Kirkpatrick et al. [36] generalized the basic concept that was introduced in [40] by using a multitemperature approach in which the temperature is lowered slowly in stages. At each stage the system is simulated by the Metropolis procedure until the system reaches equilibrium.

At the outset, the system starts with a high T, then a cooling (or annealing) scheme is applied by slowly decreasing T according to some given procedure. At each T a series of random new states are generated. States that improve the cost function are accepted. Now, rather than always rejecting states that do not improve the cost function, these states can be accepted with some finite probability depending on the amount of increase and T. This process randomizes the iterative improvement phase and also allows occasional uphill moves (i.e., moves that do not improve the solution) in an attempt to reduce the probability of falling into a local minimum.

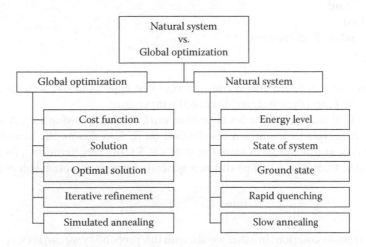

FIGURE 33.2 A one-to-one analogy between a natural process and global optimization.

As T decreases, configurations that increase the cost function are more likely to be rejected and the process eventually terminates with a configuration (solution) that has the lowest cost. This whole procedure has been proved to lead to a solution that is arbitrary close to the global minimum [1,38,57].

In addition to having a well-formulated cost function, the design of an efficient annealing algorithm requires three other ingredients [1,43]:

- The problem to be optimized must be represented by a set of data structures that provide a concise description of the nature of that problem. These data structures must allow for the fast generation and assessment of the different iterations as well as the efficient computation of the overall cost function.

- A large number of random rearrangements (i.e., perturbations, moves, or iterations) must be generated to adequately explore the search space.

- An annealing schedule should be devised to control the temperature during the annealing process. This procedure should specify the initial and final temperatures and the rate at which the temperature is lowered.

A general description of the SA algorithm is given below. After the cost function E is selected the algorithm can be described as follows:

PROCEDURE SIMULATED ANNEALING
1   **Set** $S \leftarrow S_0$ (random initial state)
2   **Set** $T \leftarrow T_0$ (initial temperature)
3   **While** (stopping criterion is not satisfied) **do**
4   **While** (required number of states is not generated) **do**
5   **Generate** a new state (S') by perturbing S.
6   **Evaluate** E
7   **Compute** $\Delta E = E(S') - E(S)$
8   **If** $(\Delta E \leq 0)$ **then**
9                  $S \leftarrow S'$
10   **Else**
11   **Generate** a random variable $\alpha, 0 \leq \alpha \leq 1$
12   **If** $\alpha \leq e^{(-\Delta E)/T}$ **then** $S \leftarrow S'$
13   **End**
14   **End**
15   **Update** T (decrement)
16 **End**

In the above procedure the Boltzmann constant (k_B) that appears in the Metropolis procedure is combined with the T, and the whole term is called temperature.

From a theoretical standpoint, the SA algorithm works in the following way. Assume a random walk on S that converges to a uniform distribution on S. Also denote the transition probability distribution by $\Lambda(s, \bullet)$ (the Markov chain is in state $s \in S$). In every iteration, given state s_i, a new state s' is generated from $\Lambda(s_i, \bullet)$. Then this new generated state is accepted with probability

$$\min \left\{ 1, e^{(f(s_{i+1}) - f(s_i))/T} \right\}$$

which is the Metropolis criterion. In other words, with this probability we can set $s_{i+1} = s'$, otherwise, $s_{i+1} = s_i$. Now, if the Markov chain given by Λ is filtered using the procedure, the sequence of states generated will converge to the Boltzmann distribution π_T (where T is the temperature). Therefore, a sequence of states can be generated $\{S_i(T)\}_{i=0}^{\infty}$ with the property that for every $\varepsilon > 0$,

$$\lim_{i \to \infty} \Pr\left(S_i(T) \in S_\varepsilon\right) = \pi_T\left(S_\varepsilon\right) \tag{33.3}$$

where S_ε is any level set ($S_\varepsilon \in S$). It is agreed upon that any adaptive search algorithm (e.g., SA) is based on the same property of the family of Boltzmann distributions, which can be stated as [20,27]

$$\lim_{T \to 0} \pi_T\left(S_\varepsilon\right) = 1 \quad \forall \varepsilon > 0 \tag{33.4}$$

By using Equations 33.3 and 33.4, it can be concluded that

$$\lim_{T \to 0} \lim_{i \to \infty} \Pr\left(S_i(T) \in S_\varepsilon\right) = 1 \quad \forall \varepsilon > 0 \tag{33.5}$$

Basically, Equation 33.5 governs the behavior of the SA algorithm. A number of states $S_0, S_1, \ldots$ are generated using the filtered random walk described above, except that now the temperature (T) will be decreased to zero as we iterate, according to some annealing schedule.

It can be noticed that the SA algorithm is characterized by its simple and elegant structure. However, a number of factors need to be considered to have an efficient implementation of the algorithm. These factors are, for example, the choice of cost function, the annealing schedule, and the algorithm termination condition.

33.4.1 Cost Function

The cost function is an application-dependent factor that measures the value of one solution relative to another [1,57]. In some applications the cost function can be of an analytical nature, which means that the structure of the function can be determined a priori [46]. In other cases, the cost function is nonanalytical and need to be determined in some indirect manner from observing the process that is being optimized [43].

Nevertheless, the value of the difference in the cost function (i.e., ΔE) is crucial for the success of the iterative process. The value of $e^{-\Delta E/T}$ suggests that for a state to be accepted with, say, probability of 0.85 at the initial T, the T must be at least six times higher than ΔE (i.e., $e^{-1/6} = 0.846$). As the ratio of T and ΔE decreases, so does the probability of accepting that state, which means that when we get to ratios of 1:7, 1:8 or greater, there is extremely low probability of accepting poor solutions [47].

33.4.2 Annealing Schedule

The annealing schedule determines the process by which T should be decreased, which influences the performance of the overall algorithm to a great extent. In this case, two issues need to be considered: the first is how T should be decremented over time, and the second is how many iterations should be computed at any given T.

In some complex problems, the annealing schedule needs to be designed specifically to suit the application [21,32,59]. A worst-case scheduling scheme that is quite slow was developed to guarantee the convergence of the SA algorithm [21]. Other research showed that the SA algorithm can converge faster than the worst-case situation [30]. Also, an adaptive cooling schedule was developed that is based on the characteristics of the cost distribution and the annealing curve [32].

Staying at the same T for a long period of time will guarantee finding the best solution since the SA algorithm is asymptotically optimal. This means that the longer the algorithm runs the better is the quality of the solution obtained [49]. However, this is not acceptable from a practical point of view.

The number of iterations that need to be computed at any given T can be determined in two different ways [47,55]. The number of iterations for any T can be given by

$$Y(T) = e^{(E_{\max} - E_{\min})/T} \tag{33.6}$$

where $E_{\max}$ and $E_{\min}$ are the highest and lowest values of the cost function obtained so far for the current T [32]. However, $Y(T)$ gets too large when T decreases, thus, requiring the introduction of an arbitrary upper bound. Another method, which is based on experimental data, allows a certain number of acceptances or rejections at any given T before allowing T to decrease. A ratio of 1:10 of accepts and rejects, respectively, was proposed in [17].

The next step is to decrease T. It is recommended that T is decreased by a factor which is less than one. In this case, the rate of decrease of T becomes exponential. The advantage of this kind of rate is that it becomes slow when the search process is near completion, which gives the system a chance to find the global minimum. Also, if one uses the formula $T_i = a^i T_{i-1}$, where $a < 1$ (ideally set between 0.5–0.99) this will have the same effect. The a in this case will control the rate of annealing.

Another important factor in designing the annealing schedule is the choice of the initial value of T. A high initial T is usually selected to ensure that most of the moves attempted are accepted because they lead to a lower cost. This allows a wider search of the solution space at the beginning. A method suggested in [36] to find the initial T can be outlined as follows:

1. Begin with a random T
2. Try a number of iterations at T, while keeping track of the percentage of the accepted moves (A) and rejected (R) ones
3. If $A/(A + R) < 0.8$, then update T to $2T$ and then goto (2) and repeat the process until the system is "warm" enough

However, one problem with this method is that the T might become too warm, which will consume much more time than what might be necessary.

33.4.3 Algorithm Termination

A number of schemes were proposed for estimating the termination (or freezing) temperature [47]. It was found that the value of T at which no further improvements can be made is [60]

$$T_f = \frac{E_{m'} - E_m}{\ln \nu} \tag{33.7}$$

where
E_m is the absolute minimum value of the cost function, which could be set to some predetermined value that depends on the application
$E_{m'}$ is the next largest value of the cost function (i.e., compared to E_m)
ν is the number of moves that takes to get from $E_{m'}$ to E_m

The freezing temperature determined by using Equation 33.7 represents the worst-case scenario, because by using T_f the annealing process will not stop too soon, that is it will not stop before the minimum has been found.

Another approach which is employed by a commercial SA package named TimberWolf^R [53] stops the annealing process when no new solution have been accepted after four consecutive decreases in T. However, it has been shown that this approach might stop the annealing process prematurely [47,55].

33.5   Convergence Conditions

The SA algorithm has a formal proof of convergence which depends on the annealing schedule. As seen from previous discussion, by manipulating the annealing schedule one can control the behavior of the algorithm. For any given T, a sufficient number of iterations always leads to equilibrium, at which point the temporal distribution of accepted states is stationary (the stationary distribution is Boltzmann). We also need to note, that at high T, almost any change is accepted. This means that the algorithm tries to span a very large neighborhood of the current state. At lower T, transitions to higher energy states becomes less frequent and the solution stabilizes.

The convergence of the SA algorithm to a global optimum can be proven by using Markov chain theory [12,33,38]. The sequence of perturbations (or moves) which are accepted by the algorithm form a Markov chain because the next state depends only on the current state and it is not influenced by past states—no record is kept for the past states. So, given that the current state is j, the probability that the next state is k is the product of two probabilities: the probability that state k is generated by one move from state j and the probability of accepting state k.

A Markov chain is irreducible if for every pair of states (j, k), there is a sequence of moves which allows k to be reached from j. If the Markov chain ensures irreducibility, then the sequence of states accepted at a given T forms a Markov chain with stationary distribution [1,49,57]. If T takes on a sequence of values approaching zero, where there are a sufficient number of iterations at each value, then the probability of being at a global optimum at the termination of the execution of the algorithm is one.

Over the last few years a number of studies dealt with the convergence problem since it a major issue in optimization problem [1,20,21,27,28,38,57]. Actually, if one is to compare the SA algorithm with other stochastic methods such as genetic algorithms and neural networks one finds in the SA literature more solid studies as far as issues of convergence are concerned [39,41,45]. Also, it has been shown that even if several iterations of the SA algorithm result in deteriorations of the cost function, that the algorithm will eventually recover and move towards the global maximum (or its neighborhoods) [6].

However, there are a number of underlying principles that govern the convergence of the SA algorithm [23,24,38]:

- The existence of a unique asymptotic probability distribution (stationary distribution) for the stationary Markov chain corresponding to each strictly positive value of an algorithm control parameter (i.e., T)
- The existence of stationary distribution limits as $T \to 0$
- The desired behavior of the stationary distribution limit (probability distribution with probability one) (see Equation 33.5)
- Sufficient conditions on the annealing schedule to ensure that the nonstationary algorithm asymptotically achieves the limiting distribution

The work in [21] proposes a version of the SA algorithm which is called the Gibbs sampler. It is shown that for temperature schedules of the form

$$T_n = \frac{c}{\log n} \quad (n\text{:large})$$

that if c is sufficiently large, then convergence will be guaranteed. Along the same lines, Gidas [23] considers the convergence of the SA algorithm and similar algorithms that are based on Markov chain sampling methods that are related to the Metropolis algorithm.

For more theoretical details on convergence of the SA algorithm, the reader is referred to [1,20,21,23,27,28,38,57] to name a few.

Before we conclude this section, it is important to note that the SA algorithm is closely related to another algorithm that has been used for global optimization and generated a lot of interest, which is called the Langevin algorithm [22]. This algorithm is based on the Langevin stochastic differential equation (proposed by Langevin in 1908 to describe the motion of a particle in a viscous fluid) given by

$$dx(t) = -\nabla g(x(t))dt + \gamma(t)dw(t) \tag{33.8}$$

where
∇g is the gradient of g
$w(t)$ is a standard r-dimensional Wiener process [44]

When $\gamma(t) \equiv \gamma_0(\gamma(t)$ is constant), then the probability density function of the solution process $x^{\gamma_0}(t)$ of Equation 33.8 approaches

$$e^{(2g(x)/\gamma_0^2)} \tag{33.9}$$

as $t \to \infty$. Equation 33.8 could also have a normalization constant. This distribution is exactly the Boltzmann distribution at temperature $\gamma_0^2/2$. This observation has generated many interesting studies [11,22].

However, only a few practical problems have been proposed along with some promising results. In [3], the authors use a modified Langevin algorithm which employs an interactive temperature schedule. They ran their tests on $g(\cdot)$ defined on $\Re^b$ where $b = 1, 2, \ldots, 14$. Other numerical results were reported by [23] that used a $g(\cdot)$ defined on $\Re$ with 400 local minima. Further, it was suggested that the Langevin algorithm be used with other multistart methods [31].

The comparison of different optimization algorithms is a very difficult task [20]. Some analytical results have been proposed to assist in such comparisons [52]. Also, a standard set of test functions have been developed in [16] to compare different optimization techniques. However, the applicability of these tests to the Langevin algorithm is rather questionable. This is due to the fact that such tests are more suited to compare algorithms in low dimensional spaces, but not the case of the Langevin algorithm which supposed to be used for functions in large dimensions. The structure and characteristics of such functions cannot be determined a priori.

33.6 Parallel Simulated Annealing Algorithms

One of the main drawbacks of the SA algorithm is the amount of time it takes to converge to a solution. As seen earlier, a number of approaches have been introduced to improve the computational efficiency of the SA algorithm [8,21,25,32,35,53]. However, most of these techniques used heuristics to simplify the search process by either limiting the cost function or educe the chances of generating future moves that are going to be rejected.

A more effective way to improve the speed of computations is to run SA algorithms by using parallel processor platforms. For a parallel SA algorithm to have the same desirable convergence properties as the sequential algorithm, one must either maintain the serial decision sequence, or employ a different decision sequence such that the generated Markov chains have the same probability distribution as the sequential algorithm [42,43].

Since, during the annealing process, a new state is generated by perturbing the previous state, it is natural to think that the SA algorithm is inherently sequential and cannot be parallelized. However, a number of ways have been used to remove this dependence between subsequent moves

[4,5,37,39,42,51,62]. A number of other issues need to be considered when parallelizing the SA algorithm, such as the division of the search space among processors, and most importantly the amount of speedup obtained with the parallel algorithm.

With the large multitude of parallel SA algorithms proposed in the literature over the last few years, one could observe that most of these algorithms don't have the same convergence characteristics as the sequential algorithm, which to some extent compromises the range of their applicability. Successful parallel algorithms tend to be application-dependent [9,14,63].

Two problem-independent algorithms were proposed in [37]. These two algorithms were called move-decomposition and parallel-moves. The former divides the move-evaluate-decide task into several subtasks, and maps these subtasks onto several processors, while the latter generates many moves in parallel but only chooses a serializable subset of these moves to be performed. This subset consists of a number of moves which, when applied to the current state in any order, always produces the same final state. In addition, both methods were used in a hybrid scheme that consists of applying move-decompose at higher temperatures and parallel-moves at lower temperatures. The hybrid algorithm was mapped onto parallel processor system. Overall, limited levels of parallelism were achieved by this work: the speedup approached saturation with four processors.

Another method was developed in [51]. In this work, the authors developed a parallel scheme in which the states have the same probability distribution as the sequential algorithm. The algorithm applies two modes of operations depending on the value of the moves' acceptance rate $\lambda(T)$ which is more formally defined as the ratios of accepted moves to the number of attempted moves for a given T. For m processors, the algorithm has the following two modes:

- High temperature mode: If $\lambda(T) \geq \frac{1}{m}$ each processor evaluates only one move and one of the accepted moves is chosen randomly. The processors' memories are updated with the new solution and the next step takes place. For the Markov chain to have the same acceptance rate as its sequential counterpart, the number of moves attempted is computed as follows: when m evaluations are performed in parallel and at least one move has been accepted, the algorithm assumes that $\frac{m+1}{m-l+1}$ moves have been attempted (l: number of rejected moves). On the other hand, if no moves have been accepted, the algorithm assumes that m moves have been attempted.

- Low temperature mode: In case of $\lambda(T) < \frac{1}{m}$, the different processors perform moves, in parallel, until a move is accepted. Then, the processors are synchronized and their memories updated with the new solution. Now, the next evaluation step takes place.

The high temperature mode is inefficient since many moves are not counted. In addition, since few moves are rejected at high temperatures, the value $\frac{m+1}{m-l+1}$, which represents speedup of the algorithm, approaches one. The results produced by the work showed that the computing time in this mode is higher than the sequential mode. It is important to note, that the time spent in this mode (i.e., high temperature) increases as the number of processor grows. In the low temperature mode, the algorithm is biased towards moves that can be generated rapidly.

The above problem was also encountered in the work by [37]. In their work, the authors evaluate the cost function for VLSI circuit placement in parallel, while simultaneously an additional processor is selecting the next state. However, the maximum speedup that was reported is bounded by $1+2\alpha+\eta$, where α is the average number of cells affected per move, and η is the average number of wires affected per move. Most cost function based computations exploit fine-grain parallelism, which lead to communication and synchronization dominating this type of algorithms [26], and so speedups due to parallelism are minimal.

Another approach to developing parallel SA is based on exploiting parallelism in making accept–reject decisions [62]. If a processor is assigned to each node in the tree shown below (Figure 33.3), then cost function evaluation for each suggested move can proceed in parallel.

The above algorithm maintains the serial decision sequence. The algorithm is based on the concurrency technique of speculative computation in which work is performed before it is known whether or not it is needed. As noted earlier, each move in the SA algorithm results in either an accept or a reject decision. Therefore, while one processor evaluates the current move, two other moves can be speculatively computed.

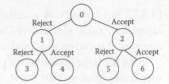

FIGURE 33.3　Decision tree.

For each temperature, the algorithm forms an unbalanced binary tree of processors which identifies the future work which is likely to be needed. Computation in the tree begins at the root and continues until a processor makes a decision but does not have a corresponding slave processor. The current solution is communicated to the root.

The main problem with the work proposed by [62] is the accompanied heavy communication cost. The root sends the solution to the reject node, generates a move, and then sends a new solution to the accept processor. Then, each node transfers the solutions to its slaves. After making a decision, the node at the end of the correct path sends its solution to the root processor. In the case of large networks of processors, the solution has to travel through several nodes to reach the root processor. In at least one experiment, the reported computation time of the sequential algorithm was lower than that of the parallel algorithm [61].

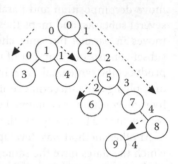

FIGURE 33.4　Moves transfer in a 10-processor unbalanced tree.

A more efficient variation of the above algorithm was developed in [42,43], which attempts to minimize the communication overhead. It is based on the observation that the difference between the solution at a node and any of its slaves and master is a maximum of one move. The gist of the work is to communicate only the new moves.

Initially, the root processor broadcasts a solution to all the processors in the network. Moreover, the algorithm ensures that all the nodes have the current solution at the beginning of each computation phase. The root generates the first move and then each other processor receives only the number of moves required for its operation.

Figure 33.4 shows an unbalanced tree which consists of 10 processors. Each node has a maximum of two slaves: the one on the left is the reject processor, while the one on the right is the accept processor. Node (0) is the root processor and the dotted arrows show the communication direction (i.e., they point to the destination processor). Note that the number attached to the arrow is the number of moves required to be transferred.

In Figure 33.4, node (1) is the root's reject processor—it does not require any moves. The same applies to node (3). The move generated by node (1) is sent to node (4). Node (2) requires the move generated by the root processor. It receives the move, generate its new move, and then sends both moves to node (5). The technique is applied to all the nodes in the network. A node evaluates its solution immediately after communicating with its slaves.

Once the correct path is identified, the node at the end of this path transfers the number of moves required to update its neighbors which, in turn, send the number of moves to update their neighbors. The maneuver is repeated until all the nodes are updated. The new iteration begins when the root processor completes communicating with its slaves.

Figure 33.5 illustrates the updating process in the 10-processor network described above. It is assumed that the correct path ends at node (8), hence the move generated by node (8) is accepted. Nodes (7) and (9) need to know only this move to update their solution. Node (5) requires the two moves generated by nodes (7) and (8). Simultaneously, node (5) updates the solution of its master and its reject processor. The same procedure is applied to all other nodes in the network.

The algorithm exhibits several salient advantages over the work of [61,62], for example,

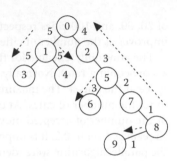

FIGURE 33.5 Updating the nodes of the 10-processor unbalanced tree.

- For problems of large size, the communication overhead is significantly reduced.

- The communication time depends only on the shape of the tree and the number of its nodes. Thus, in comparison to [61,62], as the size of the problem increases, the performance of the method improves as compared with a sequential solution.

- At low temperatures, less moves are required to be transferred because there are more rejects and, accordingly, the efficiency of the algorithm increases.

The two methods (i.e., [62] and [42]) were compared using the Traveling Salesman Problem as a benchmark [46]. Figures 33.6 through 33.8 below show the results of both algorithms in the case

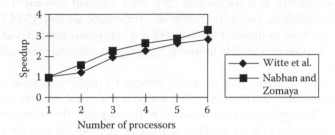

FIGURE 33.6 Speedup comparisons for 20 cities.

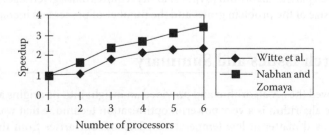

FIGURE 33.7 Speedup comparisons for 50 cities.

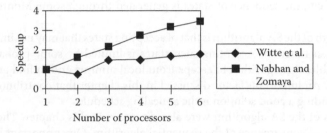

FIGURE 33.8 Speedup comparisons for 100 cities.

of 20, 50, and 100 cities respectively. The results show that the performance of the [42] algorithm improves as the number of cities increases.

The annealing schedule for the above experiments was $T_k = \eta^k T_{k-1}$, where $\eta = 0.99$ (to ensure a slow cooling rate). Normally, an initial high T is chosen to allow a good search of the solution space at the outset. The maximum number of moves (μ) attempted at each T was set equal to 100 times the number of cities. At each T a series of moves are generated until μ moves are computed, or the number of accepted moves equals to 0.1μ. The initial T (i.e., T_0) was set to 50, and the final (freezing) T set to 0.2. It is important to note, that in all the experiments, the solutions in the case of the parallel algorithm were identical to that of the sequential algorithm. For more details the reader is referred to [42].

The parallel SA algorithm developed in [43] was applied with success to wider range of problems such as solving the scheduling problem in parallel processor systems. In general, SA implementations based on speculative computation seem to produce high quality solutions in reasonable amounts of time [56].

Other parallel implementations of the SA algorithm have been proposed in the literature. These are based on the idea that the SA algorithm can tolerate, to some extent, increases or errors in the cost function without jeopardizing convergence to the correct global minimum. This type of parallel algorithms is known as asynchronous [26], since different processors can operate on old data, under certain conditions, and still converge to a reasonable solution [5,14,34,50]. In this case, the processors do not need to update their states in a synchronous fashion and can minimize the amount of information that they need to transmit. This helps in reducing the synchronization barrier quite considerably.

However, Greening [26] argues that one could construct a problem that can converge to a good solution in the sequential SA algorithm, but which converges to a local minimum in an asynchronous case. A number of other approaches can be found in the literature that employ both synchronous and asynchronous techniques [26]. Overall, synchronous methods seem to provide higher quality solutions.

In general, an efficient parallel SA algorithm should be scalable. Scalability is a general requirement for any well-designed parallel algorithm [64]. Scalability requires that a given algorithm be reasonably efficient when the size of the problem grows and the number of processors increases.

33.7 Research Issues and Summary

This chapter reviewed the SA algorithm and provided an insight into its origins and its more recent developments. The algorithm is a very powerful optimization technique that was motivated by the behavior of condensed matter at low temperatures. The analogy arises from the way that liquids freeze or metals cool and anneal.

The basic idea is to allow enough time for the states (atoms) of the system to rearrange themselves so that they achieve the lowest energy state (cost function value). Of course, in the case of an optimization problem, the sequence of states is generated through some Monte Carlo probability selection method.

The main strength of the SA algorithm is that of accepting states that may not improve immediately the value of the cost function. Accepting such states is limited by some probabilistic acceptance criterion. This enables the algorithm to escape from local minima and to find an optimal (globally minimal) solution. An important factor discussed in this chapter that contributes to the success of the algorithm in finding a good solution is the annealing schedule.

Parallel versions of the SA algorithm were also discussed in this chapter. These algorithms aim to speedup the rate of convergence of the sequential algorithm. One important issue that a parallel algorithm needs to maintain is the continuity of the Markov chain. If the Markov chain is broken, then there is no guarantee that the parallel algorithm will eventually converge, and most probably it will get trapped into local minima.

The SA algorithm (sequential or parallel) has been applied to a wide range of problems as can be seen from the list of references at the end of this chapter. However, there is still great potential in using the SA algorithm to solve more formidable problems that can be formulated as optimization problems. There is however, an even greater opportunity: using the SA algorithm in combination with other stochastic techniques such as neural networks and genetic algorithms to produce more powerful problem solving tools.

Another aspect that needs more research is the production of more efficient parallel SA algorithms. At this stage, the algorithms that provide high quality solutions within reasonable amounts of time are of the synchronous type. This means that their speedup is hindered by the need to synchronize and also to communicate massive amounts of data. These performance limitations don't occur in the case of asynchronous SA algorithms, however, asynchronous algorithms don't produce high quality solutions and might fail to converge. Therefore, there is a need to develop more efficient parallel SA algorithms that provide high quality solutions along with large speedup ratios.

33.8 Further Information

The list of references given in this chapter is quite extensive, and it shows that the use of the SA algorithm is quite ubiquitous. Actually, at the time of the writing of this chapter, one HotBot search on the Web returned 14,797 hits. This gives the reader some indication of the popularity of the SA algorithm.

There are no specific journals or conference proceedings that exclusively publish material related to the SA algorithm. However, it could be noticed from the list of references that the *IEEE Transactions on Computer-Aided Design* and the *IEEE Transactions on Parallel and Distributed Systems* often publish material related to the SA algorithm. Also, Proceedings of the IEEE/ACM Conference on Computer-Aided Design and the International Conference on Computer Design are good sources of material. The range of applications is quite diverse: parallel processing, graph drawing (Chapter 6 in *Algorithms and Theory of Computation Handbook, Second Edition: Special Topics and Techniques*), VLSI design (Chapter 8 in *Algorithms and Theory of Computation Handbook, Second Edition, Special Topics and Techniques.*), scheduling (Chapter 20 in *Algorithms and Theory of Computation Handbook, Second Edition, Special Topics and Techniques.*), to name a few.

A good starting point is the paper by Kirkpatrick et al. [36] and the books given in [1,57], after which one could proceed to more advanced topics, such as convergence issues and other theoretical themes [3,6,11,20,21,23,30,38] and parallelizing techniques [4,26]. One could also try to learn more bout the type of applications that the SA algorithm can be applied to solve by reading some of the references cited in this chapter. Furthermore, the reader might want to download some of the readily available software packages from Web. These could provide a valuable starting point to experiment with SA code.

Defining Terms

Asymptotic optimality: A given algorithm is called asymptotically optimal when the quality of the solution that can be obtained improves the more the algorithm is executed.

Detailed reversibility: The probability of a transition from state i to state j is equal to the probability in the reverse direction, from state j to state i.

Global optimization: Finding the lowest minimum of a nonlinear function $f(x)$ in some closed subregion U of $\Re^n$, in cases, where $f(x)$ may have multiple local minima in this subregion.

Irreducible Markov chain: A Markov chain is irreducible if for every two states i and j, there is a sequence of iterations or moves which enables j to be reached from i.

Local minimum: This is any point that has the lowest cost function value among all points in some open n-dimensional region around itself.

Markov chain: A sequence of trials where the outcome of any trial corresponds to the state of the system. A main characteristic of the Markov chain is that the new state depends only on the previous state and not on any earlier state.

Scalability: A parallel algorithm is scalable if it is capable of delivering an increase in performance proportional to the increase in the number of processors utilized.

Acknowledgment

Albert Y. Zomaya Acknowledges the support of the Australian Research Council grant no. 04/15/412/194.

References

1. Aarts, E. and Korst, J., *Simulated Annealing and Boltzmann Machines,* Wiley, Chichester, U.K., 1989.
2. Abramson, D., A very high speed architecture for simulated annealing, *IEEE Computer,* 25(5), 27–36, 1992.
3. Aluffi-Pentini, F., Parisi, V., and Zirilli, F., Global optimization and stochastic differential equations, *Journal of Optimization Theory and Applications,* 47, 1–16, 1985.
4. Azencott, R., Ed., *Simulated Annealing: Parallelization Techniques,* 1st edn., Wiley, New York, 1992.
5. Banerjee, P., Jones, M.H., and Sargent, J.S., Parallel simulated annealing algorithms for cell placement on hypercube multiprocessors, *IEEE Transactions on Parallel and Distributed Systems,* 1(1), 91–106, 1990.
6. Belisle, C.J.P., Convergence theorems for a class of simulated annealing algorithms on $\Re^d$, *Journal of Applied Probability,* 29(4), 885–895, 1992.
7. Boender, C.G.E. and Romeijn, H.E., Stochastic methods. In *Handbook of Global Optimization,* Horst, R. and Pardalos, P.M. (Eds.), Kluwer Academic Publishers, Dordrecht, the Netherlands, 1995, pp. 829–869.
8. Bohachevsky, I.O., Johnson, M.E., and Stein, M.L., Generalized simulated annealing for function optimization, *Technometrics,* 28, 209–217, 1986.
9. Casotto, A., Romeo, F., and Sangiovanni-Vincentelli, A., A parallel simulated annealing algorithm for the placement of macro-cells, *IEEE Transactions on Computer-Aided Design,* 6, 838–847, 1987.
10. Cerny, V., A Thermodynamic approach to the traveling salesman problem, *Journal of Optimization Theory and Applications,* 45, 41–51, 1985.
11. Chiang, T.S., Hwang, C.R., and Sheu, S.J., Diffusion for global optimization in $\Re^n$, *SIAM Journal on Control and Optimization,* 25, 737–753, 1987.
12. Cinlar, E., *Introduction to Stochastic Processes,* Prentice-Hall, Englewood Cliffs, NJ, 1975.
13. Corana, A., Marchesi, M., Martini, C., and Ridella, S., Minimizing multimodal functions for continuous variables with the "simulated annealing" algorithm, *ACM Transactions on Mathematical Software,* 13, 262–280, 1987.
14. Darema, F., Kirkpatrick, S., and Norton, A.V., Parallel algorithms for chip placement by simulated annealing, *IBM Journal of Research and Development,* 31, 259–260, 1987.
15. Dekkers, A. and Aarts, E., Global optimization and simulated annealing, *Mathematical Programming,* 50, 367–393, 1991.

16. Dixon, L.C.W. and Szego, G.P., *Towards Global Optimization*, North-Holland, Amsterdam, the Netherlands, 1978.

17. Donnett, J.G., Simulated annealing and code partitioning for distributed multimicroprocessors. Department of computer and information science, Queen's University, Kingston, Canada, 1987.

18. Dueck, G., New Optimization heuristics: The great deluge algorithm and the record-to-record travel, *Journal of Computational Physics*, 104, 86–92, 1993.

19. Dueck, G. and Scheuer, T., Threshold accepting: A general purpose optimization algorithm appearing superior to simulated annealing, *Journal of Computational Physics*, 90, 161–175, 1990.

20. Gelfand, S.B. and Mitter, S.K., Simulated annealing. In *Advanced School on Stochastics in Combinatorial Optimization*, Andreatta, G., Mason, F., and Serfami, P. (Eds.), World Scientific Publishing, Singapore, 1987, pp. 1–51.

21. Geman, S. and Geman, D., Stochastic relaxation, Gibbs distributions, and the Bayesian restoration of images, *IEEE Transactions on Pattern Analysis and Machine Intelligence*, 6(6), 721–736, 1984.

22. Geman, S. and Hwang, C.R., Diffusions for global optimization, *SIAM Journal of Control and Optimization*, 24, 1031–1043, 1986.

23. Gidas, B., Nonstationary Markov chains and convergence of the annealing algorithm, *Journal of Statistical Physics*, 39, 73–131, 1985.

24. Golden, B.L. and Skiscim, C.C., Using simulated annealing to solve routing and location problems, *Naval Research and Logistics Quarterly*, 33, 261–279, 1986.

25. Greene, J.W. and Supowit, K.J., Simulated annealing without rejected moves, *IEEE Transactions on Computer-Aided Design*, 5(1), 221–228, 1986.

26. Greening, D.R., Parallel simulated annealing techniques, *Physica D*, 42, 293–306, 1990.

27. Guus, C., Boender, E., and Romeijn, H.E., Stochastic methods. In *Handbook of Global Optimization*, Horst, R. and Pardalos, P.M. (Eds.), Kluwer Academic Publishers, Dordrecht, the Netherlands, 1995, pp. 829–869.

28. Hajek, B., Cooling schedules for optimal annealing, *Mathematics of Operations Research*, 13, 311–329, 1988.

29. Hansen, P. and Jaumard, B., Lipschitz optimization. In *Handbook of Global Optimization*, Horst, R. and Pardalos, P.M. (Eds.), Kluwer Academic Publishers, Dordrecht, the Netherlands, 1995, pp. 407–493.

30. Hastings, H.M., Convergence of simulated annealing, *ACM SIGACT*, 17(2), 52–63, 1985.

31. Horst, R. and Pardalos, P.M., Eds., *Handbook of Global Optimization*, 1st edn., Kluwer Academic Publishers, Dordrecht, the Netherlands, 1995.

32. Huang, M.D., Romeo, F., and Sangiovanni-Vincentelli, A., An efficient general cooling schedule for simulated annealing, University of California, Berkeley, CA, 1984.

33. Isaacson, D.L. and Madsen, R.W., *Markov Chains Theory and Applications*, Wiley, New York, 1976.

34. Jayaraman, R. and Dareme, F., Error tolerance in parallel simulated annealing techniques, *Proceedings of the International Conference on Computer Design*, pp. 545–548, 1988.

35. Jones, M. and Banerjee, P., An improved simulated annealing algorithm for standard cell placement, *Proceedings of the International Conference on Computer Design*, pp. 83–86, 1987.

36. Kirkpatrick, S., Gelatt, Jr., C.D., and Vecchi, M.P., Optimization by simulated annealing, *Science*, 220(4598), 671–680, 1983.

37. Kravitz, S.A. and Rutenbar, R., Placement by simulated annealing on a multiprocessor, *IEEE Transactions on Computer-Aided Design*, 6, 534–549, 1987.

38. Lundy, M. and Mees, A., Convergence of an annealing algorithm, *Mathematical Programming*, 34, 111–124, 1986.

39. Mahfoud, S.W. and Goldberg, D.E., Parallel recombinative simulated annealing: A genetic algorithm, *Parallel Computing*, 21(1), 1–28, 1995.

40. Metropolis, N., Rosenbluth, A.W., Rosenbluth, M.N., Teller, A.H., and Teller, E., Equations of state calculation by fast computing machines, *Journal of Chemical Physics*, 21(6), 1087–1092, 1953.

41. Mills, P.M., Zomaya, A.Y., and Tade, M., *Neuro-Adaptive Process Control: A Practical Approach,* Wiley, New York, 1996.
42. Nabhan, T.M. and Zomaya, A.Y., A parallel simulated annealing algorithm with low communication overhead, *IEEE Transactions on Parallel and Distributed Systems,* 6(12), 1226–1233, 1995.
43. Nabhan, T.M. and Zomaya, A.Y., A parallel computing engine for a class of time critical processes, *IEEE Transactions on Systems, Man and Cybernetics, Part B,* 27(4), 27(5), 774–786, 1997.
44. Narayan Bhat, U., *Elements of Applied Stochastic Processes,* 2nd edn., Wiley, New York, 1984.
45. Patterson, D.W., *Artificial Neural Networks,* 1st edn., Prentice-Hall, Singapore, 1996.
46. Press, W.H., Flanner, B.P., Teuklosky, S.A., and Vetterling, W.T., *Numerical Recipes in C: The Art of Scientific Computing,* Cambridge University Press, Cambridge, MA, 1988.
47. Quadrel, R.W., Woodbury, R.F., Fenves, S.J., and Talukdar, S.N., Controlling asynchronous team design environments by simulated annealing, *Research in Engineering Design,* 5, 88–104, 1993.
48. Ratschek, H. and Rokne, J., *New Computer Methods for Global Optimizations,* Ellis-Horwood, Chichester, U.K., 1988.
49. Romeo, F., Sechen, C., and Sangiovanni-Vincentelli, A., Simulated annealing research at Berkley, *Proceedings of the International Conference on Computer Design,* pp. 652–657, 1984.
50. Rose, J.S., Snelgrove, W.M., and Vranesic, Z.G., Parallel standard cell placement algorithms with quality equivalent to simulated annealing, *IEEE Transactions on Computer-Aided Design,* 7, 387–396, 1988.
51. Roussel-Ragot, P. and Dreyfus, G., A problem independent parallel implementation of simulated annealing: Models and experiments, *IEEE Transactions on Computer-Aided Design,* 9(8), 827–835, 1990.
52. Rubenstein, R., *Simulation and the Monte Carlo Method,* Wiley, New York, 1981.
53. Sechen, C. and Sangiovanni-Vincentelli, A., The timberwolf placement and routing package, *Proceedings of the Custom Integrated Circuits Conference,* pp. 522–527, 1984.
54. Siarry, P., Bergonzi, L., and Dreyfus, G., Thermodynamics optimization of block placement, *IEEE Transactions on Computer-Aided Design,* 6(2), 211–221, 1987.
55. Slagle, J., Bose, A., Busalacchi, P., Park, B., and Wee, C., *Enhanced Simulated Annealing for Automatic Reconfiguration of Multiprocessors in Space,* Department of Computer Science, University of Minnesota, Minneapolis, MN, 1989.
56. Sohn, A., Parallel N-ary speculative computation of simulated annealing, *IEEE Transactions on Parallel and Distributed Systems,* 6, 997–1005, 1995.
57. van Laarhoven, P.J.M. and Aarts, E.H.L., *Simulated Annealing: Theory and Applications,* Kluwer Academic Publishers, Boston, MA, 1987.
58. Vanderbilt, D. and Louie, S.G., A Monte Carlo simulated annealing approach to optimization over continuous variables, *Journal of Computational Physics,* 56, 259–271, 1984.
59. Vecchi, M.P. and Kirkpatrick, S., Global wiring by simulated annealing, *IEEE Transactions on Computer-Aided Design,* 2(4), 215–222, 1983.
60. White, S., Concepts of scale in simulated annealing, *Proceedings of the International Conference on Computer Design,* pp. 646–651, 1984.
61. Witte, E.E., Parallel simulated annealing using speculative computation. MS thesis, Washington University, St. Louis, MO, 1990.
62. Witte, E.E., Chamberlain, R.D., and Franklin, M.A., Parallel simulated annealing using speculative computation, *IEEE Transactions on Parallel and Distributed Systems,* 2(4), 483–494, 1991.
63. Wong, D.F., Leong, H.W., and Liu, C.L., *Simulated Annealing for VLSI Design,* 1st edn., Kluwer Academic Publishers, Dordrecht, the Netherlands, 1988.
64. Zomaya, A.Y., Edn., *Parallel and Distributed Computing Handbook,* 1st edn., McGraw-Hill, New York, 1996.

34

Approximation Algorithms for NP-Hard Optimization Problems

Philip N. Klein
Brown University

Neal E. Young
University of California

34.1 Introduction

In this chapter, we discuss approximation algorithms for optimization problems. An optimization problem consists in finding the best (cheapest, heaviest, etc.) element in a large set $\mathcal{P}$, called the feasible region and usually specified implicitly, where the quality of elements of the set are evaluated using a function $f(x)$, the objective function, usually something fairly simple. The element that minimizes (or maximizes) this function is said to be an optimal solution and the value of the objective function at this element is the optimal value.

$$\text{optimal value} = \min\{f(x) \mid x \in \mathcal{P}\} \qquad (34.1)$$

An example of an optimization problem familiar to computer scientists is that of finding a minimum-cost spanning tree of a graph with edge costs. For this problem, the feasible region $\mathcal{P}$, the set over which we optimize, consists of spanning trees; recall that a spanning tree is a set of edges that connect all the vertices but forms no cycles. The value $f(T)$ of the objective function applied to a spanning tree T is the sum of the costs of the edges in the spanning tree.

The minimum-cost spanning tree problem is familiar to computer scientists because there are several good algorithms for solving it—procedures that, for a given graph, quickly determine the minimum-cost spanning tree. No matter what graph is provided as input, the time required for each of these algorithms is guaranteed to be no more than a slowly-growing function of the number of vertices n and edges m (e.g., $O(m \log n)$).

For most optimization problems, in contrast to the minimum-cost spanning tree problem, there is no known algorithm that solves all instances quickly in this sense. Furthermore, there is not likely to be such an algorithm ever discovered, for many of these problems are NP-hard, and such an algorithm would imply that every problem in NP could be solved quickly (i.e., $P = NP$), which is considered unlikely.* One option in such a case is to seek an approximation algorithm—an algorithm that is guaranteed to run quickly (in time polynomial in the input size) and to produce a solution for which the value of the objective function is quantifiably close to the optimal value.

Considerable progress has been made toward understanding which combinatorial-optimization problems can be approximately solved, and to what accuracy. The theory of NP-completeness can provide evidence not only that a problem is hard to solve precisely but also that it is hard to approximate to within a particular accuracy. Furthermore, for many natural NP-hard optimization problems, approximation algorithms have been developed whose accuracy nearly matches the best achievable according to the theory of NP-completeness. Thus optimization problems can be categorized according to the best accuracy achievable by a polynomial-time approximation algorithm for each problem.

This chapter, which focuses on discrete (rather than continuous) NP-hard optimization problems, is organized according to these categories; for each category, we describe a representative problem, an algorithm for the problem, and the analysis of the algorithm. Along the way we demonstrate some of the ideas and methods common to many approximation algorithms. Also, to illustrate the diversity of the problems that have been studied, we briefly mention a few additional problems as we go. We provide a sampling, rather than a compendium, of the field—many important results, and even areas, are not presented. In Section 34.12 we mention some of the areas that we do not cover, and we direct the interested reader to more comprehensive and technically detailed sources, such as the books [7,15]. Because of limits on space for references, we do not cite the original sources for algorithms covered in [7].

34.2 Underlying Principles

Our focus is on combinatorial optimization problems, problems where the feasible region $\mathcal{P}$ is finite (though typically huge). Furthermore, we focus primarily on optimization problems that are NP-hard. As our main organizing principle, we restrict our attention to algorithms that are provably good in the following sense: for any input, the algorithm runs in time polynomial in the length of the input and returns a solution (i.e., a member of the feasible region) whose value (i.e., objective function value) is guaranteed to be near-optimal in some well-defined sense.[†] Such a guarantee is called the performance guarantee. Performance guarantees may be absolute, meaning that the additive difference between the optimal value and the value found by the algorithm is bounded. More commonly, performance guarantees are relative, meaning that the value found by the algorithm is within a multiplicative factor of the optimal value.

---

* For those unfamiliar with the theory of NP-completeness, see Chapters 22 and 23 or [5].

[†] An alternative to this worst-case analysis is average-case analysis. See Chapter 11.

When an algorithm with a performance guarantee returns a solution, it has implicitly discovered a bound on the exact optimal value for the problem. Obtaining such bounds is perhaps the most basic challenge in designing approximation algorithms. If one can't compute the optimal value, how can one expect to prove that the output of an algorithm is near it? Three common techniques are what which we shall call witnesses, relaxation, and coarsening.

Intuitively, a witness encodes a short, easily verified proof that the optimal value is at least, or at most, a certain value. Witnesses provide a dual role to feasible solutions to a problem. For example, for a maximization problem, where any feasible solution provides a lower bound to the optimal value, a witness would provide an upper bound on the optimal value. Typically, an approximation algorithm will produce not only a feasible solution, but also a witness. The performance guarantee is typically proven with respect to the two bounds—the upper bound provided by the witness and the lower bound provided by the feasible solution. Since the optimal value is between the two bounds, the performance guarantee also holds with respect to the optimal value.

Relaxation is another way to obtain a lower bound on the minimum value (or an upper bound in the case of a maximization problem). One formulates a new optimization problem, called a relaxation of the original problem, using the same objective function but a larger feasible region $\mathcal{P}'$ that includes $\mathcal{P}$ as a subset. Because $\mathcal{P}'$ contains $\mathcal{P}$, any $x \in \mathcal{P}$ (including the optimal element x) belongs to $\mathcal{P}'$ as well. Hence the optimal value of the relaxation, $\min\{f(x) \mid x \in \mathcal{P}'\}$, is less than or equal to the optimal value of the original optimization problem. The intent is that the optimal value of the relaxation should be easy to calculate and should be reasonably close to the optimal value of the original problem.

Linear programming can provide both witnesses and relaxations, and is therefore an important technique in the design and analysis of approximation algorithms. Randomized rounding is a general approach, based on the probabilistic method, for converting a solution to a relaxed problem into an approximate solution to the original problem.

To coarsen a problem instance is to alter it, typically restricting to a less complex feasible region or objective function, so that the resulting problem can be efficiently solved, typically by dynamic programming. For coarsening to be useful, the coarsened problem must approximate the original problem, in that there is a rough correspondence between feasible solutions of the two problems, a correspondence that approximately preserves cost. We use the term coarsening rather loosely to describe a wide variety of algorithms that work in this spirit.

34.3   Approximation Algorithms with Small Additive Error

34.3.1   Minimum-Degree Spanning Tree

For our first example, consider a slight variant on the minimum-cost spanning tree problem, the minimum-degree spanning tree problem. As before, the feasible region $\mathcal{P}$ consists of spanning trees of the input graph, but this time the objective is to find a spanning tree whose degree is minimum. The degree of a vertex of a spanning tree (or, indeed, of any graph) is the number of edges incident to that vertex, and the degree of the spanning tree is the maximum of the degrees of its vertices. Thus minimizing the degree of a spanning tree amounts to finding a smallest integer k for which there exists a spanning tree in which each vertex has at most k incident edges.

Any procedure for finding a minimum-degree spanning tree in a graph could be used to find a Hamiltonian path in any graph that has one, for a Hamiltonian path is a degree-two spanning tree. (A Hamiltonian path of a graph is a path through that graph that visits each vertex of the graph exactly once.) Since it is NP-hard even to determine whether a graph has a Hamiltonian path, even determining whether the minimum-degree spanning tree has degree two is presumed to be computationally difficult.

34.3.2 An Approximation Algorithm for Minimum-Degree Spanning Tree

Nonetheless, the minimum-degree spanning-tree problem has a remarkably good approximation algorithm [7, Chapter 7]. For an input graph with m edges and n vertices, the algorithm requires time slightly more than the product of m and n. The output is a spanning tree whose degree is guaranteed to be at most one more than the minimum degree. For example, if the graph has a Hamiltonian path, the output is either such a path or a spanning tree of degree three.

Given a graph G, the algorithm naturally finds the desired spanning tree T of G. The algorithm also finds a witness—in this case, a set S of vertices proving that T's degree is nearly optimal. Namely, let k denote the degree of T, and let $T_1, T_2, \ldots, T_r$ be the subtrees that would result from T if the vertices of S were deleted. The following two properties are enough to show that T's degree is nearly optimal (Figures 34.1 and 34.2):

1. There are no edges of the graph G between distinct trees T_i.
2. The number r of trees T_i is at least $|S|(k-1) - 2(|S|-1)$.

To show that T's degree is nearly optimal, let V_i denote the set of vertices comprising subtree T_i ($i = 1, \ldots, r$). Any spanning tree T^* at all must connect up the sets $V_1, V_2, \ldots, V_r$ and the vertices $y_1, y_2, \ldots, y_{|S|} \in S$, and must use at least $r + |S| - 1$ edges to do so. Furthermore, since no edges go between distinct sets V_i, all these edges must be incident to the vertices of S (Figure 34.3).

Hence we obtain

$$\sum \left\{ \deg_{T^*}(y) \mid y \in S \right\} \geq r + |S| - 1$$
$$\geq |S|(k-1) - 2(|S|-1) + |S| - 1$$
$$= |S|(k-1) - (|S|-1) \tag{34.2}$$

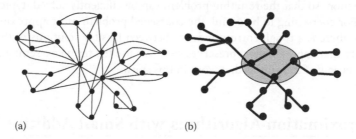

(a) (b)

FIGURE 34.1 (a) An example input graph G. (b) A spanning tree T that might be found by the approximation algorithm. The shaded circle indicates the nodes in the witness set S.

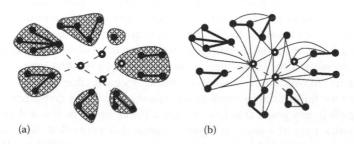

(a) (b)

FIGURE 34.2 (a) r trees $T_1, \ldots, T_r$ obtained from T by deleting the nodes of S. Each tree is indicated by a shaded region. (b) No edges of the input graph G connect different trees T_i.

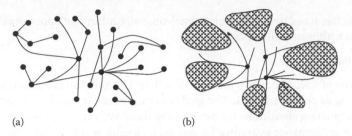

FIGURE 34.3 (a) An arbitrary spanning tree T^* for the same input graph G. (b) r shaded regions, one for each subset V_i of nodes corresponding to a tree T_i in Figure 34.2. The proof of the algorithm's performance guarantee is based on the observation that at least $r + |S| - 1$ edges are needed to connect up the V_i's and the nodes in S.

where $\deg_{T^*}(y)$ denotes the degree of y in the tree T^*. Thus the average of the degrees of vertices in S is at least $\frac{|S|(k-1)-(|S|-1)}{|S|}$, which is strictly greater than $k-2$. Since the average of the degrees of vertices in S is greater than $k-2$, it follows that at least one vertex has degree at least $k-1$.

We have shown that for every spanning tree T^*, there is at least one vertex with degree at least $k-1$. Hence the minimum degree is at least $k-1$.

We have not explained how the algorithm obtains both the spanning tree T and the set S of vertices, only how the set S shows that the spanning tree is nearly optimal. The basic idea is as follows. Start with any spanning tree T, and let d denote its degree. Let S be the set of vertices having degree d or $d-1$ in the current spanning tree. Let $T_1, \ldots, T_r$ be the subtrees comprising $T - S$. If there are no edges between these subtrees, the set S satisfies property 1 and one can show it also satisfies property 2; in this case the algorithm terminates. If on the other hand there is an edge between two distinct subtrees T_i and T_j, inserting this edge in T and removing another edge from T results in a spanning tree with fewer vertices having degree at least $d-1$. Repeat this process on the new spanning tree; in subsequent iterations the improvement steps are somewhat more complicated but follow the same lines. One can prove that the number of iterations is $O(n \log n)$.

We summarize our brief sketch of the algorithm as follows: either the current set S is a witness to the near-optimality of the current spanning tree T, or there is a slight modification to the set and the spanning tree that improve them. The algorithm terminates after a relatively small number of improvements.

This algorithm is remarkable not only for its simplicity and elegance but also for the quality of the approximation achieved. As we shall see, for most NP-hard optimization problems, we must settle for approximation algorithms that have much weaker guarantees.

34.3.3 Other Problems Having Small-Additive-Error Algorithms

There are a few other natural combinatorial-optimization problems for which approximation algorithms with similar performance guarantees are known. Here are two examples:

34.3.3.1 Edge Coloring

Given a graph, color its edges with a minimum number of colors so that, for each vertex, the edges incident to that vertex are all different colors. For this problem, it is easy to find a witness. For any graph G, let v be the vertex of highest degree in G. Clearly one needs to assign at least $\deg_G(v)$ colors to the edges of G, for otherwise there would be two edges with the same color incident to v. For any graph G, there is an edge coloring using a number of colors equal to one plus the degree of G.

The proof of this fact translates into a polynomial-time algorithm that approximates the minimum edge-coloring to within an additive error of 1.

34.3.3.2 Bin Packing

The input consists of a set of positive numbers less than 1. A solution is a partition of the numbers into sets summing to no more than 1. The goal is to minimize the number of sets in the partition. There are approximation algorithms for bin packing that have very good performance guarantees. For example, the performance guarantee for one such algorithm is as follows: for any input set I of item weights, it finds a packing that uses at most $\text{OPT}(I) + O(\log^2 \text{OPT}(I))$ bins, where $\text{OPT}(I)$ is the number of bins used by the best packing, i.e., the optimal value.

34.4 Randomized Rounding and Linear Programming

A linear programming problem is any optimization problem in which the feasible region corresponds to assignments of values to variables meeting a set of linear inequalities and in which the objective function is a linear function. An instance is determined by specifying the set of variables, the objective function, and the set of inequalities. Linear programs (LPs) are capable of representing a large variety of problems and have been studied for decades in combinatorial optimization and have a tremendous literature (see e.g., Chapters 30 and 31). Any LP can be solved—that is, a point in the feasible region maximizing or minimizing the objective function can be found—in time bounded by a polynomial in the size of the input.

A (mixed) integer linear programming problem is a linear programming problem augmented with additional constraints specifying that (some of) the variables must take on integer values. Such constraints make integer linear programming even more general than linear programming—in general, solving integer LPs is NP-hard.

For example, consider the following balanced matching problem: The input is a bipartite graph $G = (V, W, E)$. The goal is to choose an edge incident to each vertex in V ($|V|$ edges in total), while minimizing the maximum load of (number of chosen edges adjacent to) any vertex in W. The vertices in V might represent tasks, the vertices in W might represent people, while the presence of edge $\{v, w\}$ indicates that person w is competent to perform task v. The problem is then to assign each task to a person competent to perform it, while minimizing the maximum number of tasks assigned to any person.*

This balanced matching problem can be formulated as the following integer LP:

$$\text{minimize } \Delta$$
$$\text{subject to } \begin{cases} \sum_{u \in N(v)} x(u, v) = 1 & \forall v \in V \\ \sum_{v \in N(u)} x(u, v) \leq \Delta & \forall w \in W \\ x(u, v) \in \{0, 1\} & \forall (u, v) \in E. \end{cases}$$

Here $N(x)$ denotes the set of neighbors of vertex x in the graph. For each edge (u, v) the variable $x(u, v)$ determines whether the edge (u, v) is chosen. The variable Δ measures the maximum load.

Relaxing the integrality constraints (i.e., replacing them as well as we can by linear inequalities) yields the LP:

---

* Typically, randomized rounding is applied to NP-hard problems, whereas the balanced matching problem here is actually solvable in polynomial time. We use it as an example for simplicity—the analysis captures the essential spirit of a similar analysis for the well-studied integer multicommodity flow problem. (A simple version of that problem is, "Given a network and a set of commodities (each a pair of vertices), choose a path for each commodity minimizing the maximum congestion on any edge.")

$$\text{minimize } \Delta$$

$$\text{subject to } \begin{cases} \sum_{u \in N(v)} x(u, v) = 1 & \forall v \in V \\ \sum_{v \in N(u)} x(u, v) \leq \Delta & \forall w \in W \\ x(u, v) \geq 0 & \forall (u, v) \in E. \end{cases}$$

Rounding a fractional solution to a true solution. This relaxed problem can be solved in polynomial time simply because it is a LP. Suppose we have an optimal solution x^*, where each $x^*(e)$ is a fraction between 0 and 1. How can we convert such an optimal fractional solution into an approximately optimal integer solution? Randomized rounding is a general approach for doing just this [11, Chapter 5].

Consider the following polynomial-time randomized algorithm to find an integer solution $\hat{x}$ from the optimal solution x^* to the LP:

1. Solve the LP to obtain a fractional solution x^* of load Δ^*.

2. For each vertex $v \in V$:

   (a) Choose a single edge incident to v at random, so that the probability that a given edge (u, v) is chosen is $x^*(u, v)$. (Note that $\sum_{u \in N(v)} x^*(u, v) = 1$.)

   (b) Let $\hat{x}(u, v) \leftarrow 1$.

   (c) For all other edges (u', v) incident to v, let $\hat{x}(u', v) \leftarrow 0$.

The algorithm will always choose one edge adjacent to each vertex in V. Thus, $\hat{x}$ is a feasible solution to the original integer program. What can we say about the load? For any particular vertex $w \in W$, the load on w is $\sum_{u \in N(v)} \hat{x}(u, v)$. For any particular edge $(u, v) \in E$, the probability that $\hat{x}(u, v) = 1$ is $x^*(u, v)$. Thus the expected value of the load on a vertex $u \in U$ is $\sum_{v \in N(u)} x^*(u, v)$, which is at most Δ^*. This is a good start. Of course, the maximum load over all $u \in U$ is likely to be larger. How much larger?

To answer this, we need to know more about the distribution of the load on v than just the expected value. The key fact that we need to observe is that the load on any $v \in V$ is a sum of independent $\{0, 1\}$-random variables. This means it is not likely to deviate much from its expected value. Precise estimates come from standard bounds, called "Chernoff" or "Hoeffding" bounds, such as the following:

THEOREM 34.1   *Let X be the sum of independent $\{0, 1\}$ random variables. Let $\mu > 0$ be the expected value of X. Then for any $\epsilon > 0$,*

$$\Pr[X \geq (1 + \epsilon)\mu] < \exp\left(-\mu \min\left\{\epsilon, \epsilon^2\right\}/3\right).$$

(See, e.g.,[11, Chapter 4.1].) This is enough to analyze the performance guarantee of the algorithm. It is slightly complicated, but not too bad:

Claim: *With probability at least 1/2, the maximum load induced by $\hat{x}$ exceeds the optimal by at most an additive error of*

$$\max\left\{3\ln(2m), \sqrt{3\ln(2m)\Delta^*}\right\},$$

where $m = |W|$.

Proof sketch: As observed previously, for any particular v, the load on v is a sum (of independent random $\{0, 1\}$-variables) with expectation bounded by Δ^*. Let ϵ be just large enough so that $\exp(-\Delta^* \min\{\epsilon, \epsilon^2\}/3) = 1/(2m)$. By the Chernoff-type bound above, the probability that the load exceeds $(1 + \epsilon)\Delta^*$ is then less than $1/(2m)$. Thus, by the naive union bound,[*] the probability that the maximum load on any $v \in V$ is more than $\Delta^*(1 + \epsilon) = \Delta^* + \epsilon\Delta^*$ is less then $1/2$. We leave it to the reader to verify that the choice of ϵ makes $\epsilon\Delta^*$ equal the expression in the statement of the claim.

Summary. This is the general randomized-rounding recipe:

1. Formulate the original NP-hard problem as an integer linear programming problem (IP).
2. Relax the program IP to obtain a LP.
3. Solve the LP, obtaining a fractional solution.
4. Randomly round the fractional solution to obtain an approximately optimal integer solution.

34.5 Performance Ratios and ρ-Approximation

Relative (multiplicative) performance guarantees are more common than absolute (additive) performance guarantees. One reason is that many NP-hard optimization problems are rescalable: given an instance of the problem, one can construct a new, equivalent instance by scaling the objective function. For instance, the traveling salesman problem is rescalable—given an instance, multiplying the edge weights by any $\lambda > 0$ yields an equivalent problem with the objective function scaled by λ. For rescalable problems, the best one can hope for is a relative performance guarantee [14].

A ρ-approximation algorithm is an algorithm that returns a feasible solution whose objective function value is at most ρ times the minimum (or, in the case of a maximization problem, the objective function value is at least ρ times the maximum). We say that the performance ratio of the algorithm is ρ.[†]

34.6 Polynomial Approximation Schemes

The knapsack problem is an example of a rescalable NP-hard problem. An instance consists of a set of pairs of numbers (weight$_i$, profit$_i$), and the goal is to select a subset of pairs for which the sum of weights is at most 1 so as to maximize the sum of profits. (Which items should one put in a knapsack of capacity 1 so as to maximize profit?)

Since the knapsack problem is rescalable and NP-hard, we assume that there is no approximation algorithm achieving, say, a fixed absolute error. One is therefore led to ask, what is the best performance ratio achievable by a polynomial-time approximation algorithm? In fact (assuming $P \neq NP$), there is no such best performance ratio: for any given $\epsilon > 0$, there is a polynomial approximation algorithm whose performance ratio is $1 + \epsilon$. The smaller the value of ϵ, however, the greater the running time of the corresponding approximation algorithm. Such a collection of approximation algorithms, one for each $\epsilon > 0$, is called a (polynomial) approximation scheme.

Think of an approximation scheme as an algorithm that takes an additional parameter, the value of ϵ, in addition to the input specifying the instance of some optimization problem. The running time of this algorithm is bounded in terms of the size of the input and in terms of ϵ. For example, there is an approximation scheme for the knapsack problem that requires time $O(n \log(1/\epsilon) + 1/\epsilon^4)$

---

[*] The probability that any of several events happens is at most the sum of the probabilities of the individual events.

[†] This terminology is the most frequently used, but one also finds alternative terminology in the literature. Confusingly, some authors have used the term $1/\rho$-approximation algorithm or $(1 - \rho)$-approximation algorithm to refer to what we call a ρ-approximation algorithm.

for instances with n items. Below we sketch a much simplified version of this algorithm that requires time $O(n^3/\epsilon)$. The algorithm works by coarsening.

The algorithm is given the pairs $(\text{weight}_1, \text{profit}_1), \ldots, (\text{weight}_n, \text{profit}_n)$, and the parameter ϵ. We assume without loss of generality that each weight is less than or equal to 1. Let $\text{profit}_{max} = \max_i \text{profit}_i$. Let OPT denote the (unknown) optimal value. Since the item of greatest profit itself constitutes a solution, albeit not usually a very good one, we have $\text{profit}_{max} \leq$ OPT. In order to achieve a relative error of at most ϵ, therefore, it suffices to achieve an absolute error of at most $\epsilon \, \text{profit}_{max}$.

We transform the given instance into a coarsened instance by rounding each profit down to a multiple of $K = \epsilon \, \text{profit}_{max}/n$. In so doing, we reduce each profit by less than $\epsilon \, \text{profit}_{max}/n$. Consequently, since the optimal solution consists of no more than n items, the profit of this optimal solution is reduced by less than $\epsilon \, \text{profit}_{max}$ in total. Thus, the optimal value for the coarsened instance is at least OPT $- \epsilon \, \text{profit}_{max}$, which is in turn at least $(1 - \epsilon)$OPT. The corresponding solution, when measured according to the original profits, has value at least this much. Thus we need only solve the coarsened instance optimally in order to get a performance guarantee of $1 - \epsilon$.

Before addressing the solution of the coarsened instance, note that the optimal value is the sum of at most n profits, each at most profit_{max}. Thus OPT $\leq n^2 K/\epsilon$. The optimal value for the coarsened instance is therefore also at most $n^2 K/\epsilon$.

To solve the coarsened instance optimally, we use dynamic programming. Note that for the coarsened instance, each achievable total profit can be written as $i \cdot K$ for some integer $i \leq n^2/\epsilon$. The dynamic-programming algorithm constructs an $\lceil n^2/\epsilon \rceil \times (n+1)$ table $T[i,j]$ whose i,j entry is the minimum weight required to achieve profit $i \cdot K$ using a subset of the items 1 through j. The entry is infinity if there is no such way to achieve that profit.

To fill in the table, the algorithm initializes the entries $T[i, 0]$ to infinity, then executes the following step for $j = 1, 2, \ldots, n$:

$$\text{For each } i, \text{ set } T[i,j] := \min\left\{ T[i, j-1], \text{weight}_j + T\left[i - \left(\widehat{\text{profit}}_j/K\right), j-1\right]\right\}$$

where $\widehat{\text{profit}}_j$ is the profit of item j in the rounded-down instance. A simple induction on j shows that the calculated values are correct. The optimal value for the coarsened instance is

$$\widehat{\text{OPT}} = \max\{iK \mid T[i, n] \leq 1\}.$$

The above calculates the optimal value for the coarsened instance; as usual in dynamic programming, a corresponding feasible solution can easily be computed if desired.

34.6.1 Other Problems Having Polynomial Approximation Schemes

The running time of the knapsack approximation scheme depends polynomially on $1/\epsilon$. Such a scheme is called a fully polynomial approximation scheme. Most natural NP-complete optimization problems are strongly NP-hard, meaning essentially that the problems are NP-hard even when the numbers appearing in the input are restricted to be no larger in magnitude than the size of the input. For such a problem, we cannot expect a fully polynomial approximation scheme to exist [5, Section 4.2]. On the other hand, a variety of NP-hard problems in fixed-dimensional Euclidean space have approximation schemes. For instance, given a set of points in the plane:

Covering with disks: Find a minimum set of area-1 disks (or squares, etc.) covering all the points [7, Section 9.3.3].

Euclidean traveling salesman: Find a closed loop passing through each of the points and having minimum total arc length [16].

Euclidean Steiner tree: Find a minimum-length set of segments connecting up all the points [16].

Similarly, many problems in planar graphs or graphs of fixed genus can have polynomial approximation schemes [7, Section 9.3.3], For instance, given a planar graph with weights assigned to its vertices:

Maximum-weight independent set: Find a maximum-weight set of vertices, no two of which are adjacent.

Minimum-weight vertex cover: Find a minimum-weight set of vertices such that every edge is incident to at least one of the vertices in the set.

Euclidean traveling salesman: Find a closed walk passing through each vertex and having minimum total length [17].

Euclidean Steiner tree: Find a minimum-length subgraph spanning a given set of vertices [17].

The above algorithms use relatively more sophisticated and varied coarsening techniques.

34.7   Constant-Factor Performance Guarantees

We have seen that, assuming $P \neq NP$, rescalable NP-hard problems do not have polynomial-time approximation algorithms with small absolute errors but may have fully polynomial approximation schemes, while strongly NP-hard problems do not have fully polynomial approximation schemes but may have polynomial approximation schemes. Further, there is a class of problems that do not have approximation schemes: for each such problem there is a constant c such that any polynomial-time approximation algorithm for the problem has relative error at least c (assuming $P \neq NP$). For such a problem, the best one can hope for is an approximation algorithm with constant performance ratio.

Our example of such a problem is the vertex cover problem: given a graph G, find a minimum-size set C (a vertex cover) of vertices such that every edge in the graph is incident to some vertex in C. Here the feasible region $\mathcal{P}$ consists of the vertex covers in G, while the objective function is the size of the cover. Here is a simple approximation algorithm [7]:

1. Find a maximal independent set S of edges in G
2. Let C be the vertices incident to edges in S

(A set S of edges is independent if no two edges in S share an endpoint. The set S is maximal if no larger independent set contains S.) The reader may wish to verify that the set S can be found in linear time, and that because S is maximal, C is necessarily a cover.

What performance guarantee can we show? Since the edges in S are independent, any cover must have at least one vertex for each edge in S. Thus S is a witness proving that any cover has at least $|S|$ vertices. On the other hand, the cover C has $2|S|$ vertices. Thus the cover returned by the algorithm is at most twice the size of the optimal vertex cover.

The weighted vertex cover problem. The weighted vertex cover problem is a generalization of the vertex cover problem. An instance is specified by giving a graph $G = (V, E)$ and, for each vertex v in the graph, a number $wt(v)$ called its weight. The goal is to find a vertex cover minimizing the total weight of the vertices in the cover. Here is one way to represent the problem as an integer LP:

$$\text{minimize} \sum_{v \in V} wt(v)x(v)$$

$$\text{subject to} \begin{cases} x(u) + x(v) \geq 1 & \forall\{u, v\} \in E \\ x(v) \in \{0, 1\} & \forall v \in V. \end{cases}$$

There is one $\{0, 1\}$-variable $x(v)$ for each vertex v representing whether v is in the cover or not, and there are constraints for the edges that model the covering requirement. The feasible region of this program corresponds to the set of vertex covers. The objective function corresponds to the total weight of the vertices in the cover. Relaxing the integrality constraints yields

$$\text{minimize} \sum_{v \in V} \text{wt}(v)x(v)$$

$$\text{subject to} \begin{cases} x(u) + x(v) \geq 1 & \forall \{u, v\} \in E \\ x(v) \geq 0 & \forall v \in V. \end{cases}$$

This relaxed problem is called the fractional weighted vertex cover problem; feasible solutions to it are called fractional vertex covers.*

Rounding a fractional solution to a true solution. By solving this LP, an optimal fractional cover can be found in polynomial time. For this problem, it is possible to convert a fractional cover into an approximately optimal true cover by rounding the fractional cover in a simple way:

1. Solve the LP to obtain an optimal fractional cover x^*
2. Let $C = \{v \in V : x^*(v) \geq \frac{1}{2}\}$

The set C is a cover because for any edge, at least one of the endpoints must have fractional weight at least $1/2$. The reader can verify that the total weight of vertices in C is at most twice the total weight of the fractional cover x^*. Since the fractional solution was an optimal solution to a relaxation of the original problem, this is a two-approximation algorithm [7].

For most problems, this simple kind of rounding is not sufficient. The previously discussed technique called randomized rounding is more generally useful.

Primal-dual algorithms—witnesses via duality. For the purposes of approximation, solving a LP exactly is often unnecessary. One can often design a faster algorithm based on the witness technique, using the fact that every LP has a well-defined notion of "witness." The witnesses for a LP P are the feasible solutions to another related LP called the dual of P.

Suppose our original problem is a minimization problem. Then for each point y in the feasible region of the dual problem, the value of the objective function at y is a lower bound on the value of the optimal value of the original LP. That is, any feasible solution to the dual problem is a possible witness—both for the original integer LP and its relaxation. For the weighted vertex cover problem, the dual is the following:

$$\text{maximize} \sum_{e \in E} y(e)$$

$$\text{subject to} \begin{cases} \sum_{e \ni v} y(e) \leq \text{wt}(v) & \forall v \in V \\ y(e) \geq 0 & \forall e \in E. \end{cases}$$

A feasible solution to this LP is called an edge packing. The constraints for the vertices are called packing constraints.

Recall the original approximation algorithm for the unweighted vertex cover problem: find a maximal independent set of edges S; let C be the vertices incident to edges in S. In the analysis, the set S was the witness.

Edge packings generalize independent sets of edges. This observation allows us to generalize the algorithm for the unweighted problem. Say an edge packing is maximal if, for every edge, one of the edge's vertices has its packing constraint met. Here is the algorithm:

---

* The reader may wonder whether additional constraints of the form $x(v) \leq 1$ are necessary. In fact, assuming the vertex weights are nonnegative, there is no incentive to make any $x(v)$ larger than 1, so such constraints would be redundant.

1. Find a maximal edge packing y

2. Let C be the vertices whose packing constraints are tight for y

The reader may wish to verify that a maximal edge packing can easily be found in linear time and that the set C is a cover because y is maximal.

What about the performance guarantee? Since only vertices whose packing constraints are tight are in C, and each edge has only two vertices, we have

$$\sum_{v \in C} \text{wt}(v) = \sum_{v \in C} \sum_{e \ni v} y(e) \le 2 \sum_{e \in E} y(e).$$

Since y is a solution to the dual, $\sum_e y(e)$ is a lower bound on the weight of any vertex cover, fractional or otherwise. Thus, the algorithm is a two-approximation algorithm.

Summary. This is the general primal-dual recipe:

1. Formulate the original NP-hard problem as an integer linear programming problem (IP)

2. Relax the program IP to obtain a LP

3. Use the dual (DLP) of LP as a source of witnesses

Beyond these general guidelines, the algorithm designer is still left with the task of figuring out how to find a good solution and witness. See [7, Chapter 4] for an approach that works for a wide class of problems.

34.7.1 Other Optimization Problems with Constant-Factor Approximations

Constant-factor approximation algorithms are known for problems from many areas. In this section, we describe a sampling of these problems. For each of the problems described here, there is no polynomial approximation scheme (unless $P = \text{NP}$); thus constant-factor approximation algorithms are the best we can hope for. For a typical problem, there will be a simple algorithm achieving a small constant factor while there may be more involved algorithms achieving better factors. The factors known to be achievable typically come close to, but do not meet, the best lower bounds known (assuming $P \ne \text{NP}$).

For the problems below, we omit discussion of the techniques used; many of the problems are solved using a relaxation of some form, and (possibly implicitly) the primal-dual recipe. Many of these problems have polynomial approximation schemes if restricted to graphs induced by points in the plane or constant-dimensional Euclidean space (see Section 34.6.1).

MAX-SAT: Given a propositional formula in conjunctive normal form (an "and" of "or"'s of possibly negated Boolean variables), find a truth assignment to the variables that maximizes the number of clauses (groups of "or"'ed variables in the formula) that are true under the assignment. A variant called MAX-3SAT restricts to the formula to have three variables per clause. MAX-3SAT is a canonical example of a problem in the complexity class MAX-SNP [7, Section 10.3].

MAX-CUT: Given a graph, partition the vertices of the input graph into two sets so as to maximize the number of edges with endpoints in distinct sets. For MAX-CUT and MAX-SAT problems, the best approximation algorithms currently known rely on randomized rounding and a generalization of linear programming called semidefinite programming [7, Section 11.3].

Shortest superstring: Given a set of strings $\sigma_1, \ldots, \sigma_k$, find a minimum-length string containing all σ_i's. This problem has applications in computational biology [3,9].

K-Cluster: Given a graph with weighted edges and given a parameter k, partition the vertices into k clusters so as to minimize the maximum distance between any two vertices in the same cluster. For this and related problems see [7, Section 9.4].

Traveling salesman: Given a complete graph with edge weights satisfying the triangle inequality, find a minimum-length path that visits every vertex of the graph [7, Chapter 8].

Edge and vertex connectivity: Given a weighted graph $G = (V, E)$ and an integer k, find a minimum-weight edge set $E' \subseteq E$ such that between any pair of vertices, there are k edge-disjoint paths in the graph $G' = (V, E')$. Similar algorithms handle the goal of k vertex-disjoint paths and the goal of augmenting a given graph to achieve a given connectivity [7, Chapter 6]

Steiner tree: Given an undirected graph with positive edge-weights and a subset of the vertices called terminals, find a minimum-weight set of edges through which all the terminals (and possibly other vertices) are connected [7, Chapter 8]. The Euclidean version of the problem is "Given a set of points in R^n, find a minimum-total-length union of line segments (with arbitrary endpoints) that is connected and contains all the given points."

Steiner forest: Given a weighted graph and a collection of groups of terminals, find a minimum-weight set of edges through which every pair of terminals within each group are connected [7, Chapter 4]. The algorithm for this problem is based on a primal-dual framework that has been adapted to a wide variety of network design problems. See Section 34.8.1.

34.8 Logarithmic Performance Guarantees

When a constant-ratio performance guarantee is not possible, a slowly-growing ratio is the next best thing. The canonical example of this is the set cover problem: Given a family of sets $\mathcal{F}$ over a universe $\mathcal{U}$, find a minimum-cardinality set cover C—a collection of the sets that collectively contain all elements in $\mathcal{U}$. In the weighted version of the problem, each set also has a weight and the goal is to find a set cover of minimum total weight. This problem is important due to its generality. For instance, it generalizes the vertex cover problem.

Here is a simple greedy algorithm:

1. Let $C \leftarrow \emptyset$.
2. Repeat until all elements are covered: add a set S to C maximizing

$$\frac{\text{the number of elements in } S \text{ not in any set in } C}{\text{wt}(S)}.$$

3. Return C.

The algorithm has the following performance guarantee [7, Section 3.2]:

THEOREM 34.2 *The greedy algorithm for the weighted set cover problem is an H_s-approximation algorithm, where s is the maximum size of any set in $\mathcal{F}$.*

By definition $H_s = 1 + \frac{1}{2} + \frac{1}{3} + \cdots + \frac{1}{s}$; also, $H_s \leq 1 + \ln s$.

We will give a direct argument for the performance guarantee and then relate it to the general primal-dual recipe. Imagine that as the algorithm proceeds, it assigns charges to the elements as they are covered. Specifically, when a set S is added to the cover C, if there are k elements in S not previously covered, assign each such element a charge of $\text{wt}(S)/k$. Note that the total charge assigned over the course of the algorithm equals the weight of the final cover C.

Next we argue that the total charge assigned over the course of the algorithm is a lower bound on H_s times the weight of the optimal vertex cover. These two facts together prove the theorem.

Suppose we could prove that for any set T in the optimal cover C^*, the elements in T are assigned a total charge of at most $\text{wt}(T)H_s$. Then we would be done, because every element is in at least one set in the optimal cover:

$$\sum_{i \in U} \text{charge}(i) \leq \sum_{T \in C^*} \sum_{i \in T} \text{charge}(i) \leq \sum_{T \in C^*} \text{wt}(T)H_s.$$

So, consider, for example, a set $T = \{a, b, c, d, e, f\}$ with $\text{wt}(T) = 3$. For convenience, assume that the greedy algorithm covers elements in T in alphabetical order. What can we say about the charge assigned to a? Consider the iteration when a was first covered and assigned a charge. At the beginning of that iteration, T was not yet chosen and none of the 6 elements in T were yet covered. Since the greedy algorithm had the option of choosing T, whatever set it did choose resulted in a charge to a of at most $\text{wt}(T)/|T| = 3/6$.

What about the element b? When b was first covered, T was not yet chosen, and at least 5 elements in T remained uncovered. Consequently, the charge assigned to b was at most $3/5$. Reasoning similarly, the elements c, d, e, and f were assigned charges of at most $3/4$, $3/3$, $3/2$, and $3/1$, respectively. The total charge to elements in T is at most

$$3 \times (1/6 + 1/5 + 1/4 + 1/3 + 1/2 + 1/1) = \text{wt}(T)H_{|T|} \leq \text{wt}(T)H_s.$$

This line of reasoning easily generalizes to show that for any set T, the elements in T are assigned a total charge of at most $\text{wt}(T)H_s$.

Underlying duality. What role does duality and the primal-dual recipe play in the above analysis? A natural integer LP for the weighted set cover problem is

$$\text{minimize} \sum_{S \in \mathcal{F}} \text{wt}(S)x(S)$$

$$\text{subject to} \begin{cases} \sum_{S \ni i} x(S) \geq 1 & \forall i \in U \\ x(S) \in \{0, 1\} & \forall S \in \mathcal{F}. \end{cases}$$

Relaxing this integer LP yields the LP

$$\text{minimize} \sum_{S \in \mathcal{F}} \text{wt}(S)x(S)$$

$$\text{subject to} \begin{cases} \sum_{S \ni i} x(S) \geq 1 & \forall i \in U \\ x(S) \geq 0 & \forall S \in \mathcal{F}. \end{cases}$$

A solution to this LP is called a fractional set cover. The dual is

$$\text{minimize} \sum_{i \in U} y(i)$$

$$\text{subject to} \begin{cases} \sum_{i \in S} y(i) \leq \text{wt}(S) & \forall S \in \mathcal{F} \\ y(i) \geq 0 & \forall i \in U. \end{cases}$$

The inequalities for the sets are called packing constraints. A solution to this dual LP is called an element packing. In fact, the "charging" scheme in the analysis is just an element packing y, where $y(i)$ is the charge assigned to i divided by H_s. In this light, the previous analysis is simply constructing a dual solution and using it as a witness to show the performance guarantee.

34.8.1 Other Problems Having Poly-Logarithmic Performance Guarantees

Minimizing a linear function subject to a submodular constraint. This is a natural generalization of the weighted set cover problem. Rather than state the general problem, we give the following special case as an example: Given a family $\mathcal{F}$ of sets of n-vectors, with each set in $\mathcal{F}$ having a cost, find a subfamily of sets of minimum total cost whose union has rank n. A natural generalization of the greedy set cover algorithm gives a logarithmic performance guarantee [12].

Vertex-weighted network steiner tree. Like the network Steiner tree problem described in Section 34.7.1, an instance consists of a graph and a set of terminals; in this case, however, the graph can have vertex weights in addition to edge weights. An adaptation of the greedy algorithm achieves a logarithmic performance ratio.

Network design problems. This is a large class of problems generalizing the Steiner forest problem (see Section 34.7.1). An example of a problem in this class is survivable network design: given a weighted graph $G = (V, E)$ and a nonnegative integer r_{uv} for each pair of vertices, find a minimum-cost set of edges $E' \subseteq E$ such that for every pair of vertices u and v, there are at least r_{uv} edge-disjoint paths connecting u and v in the graph $G = (V, E')$. A primal-dual approach, generalized from an algorithm for the Steiner forest problem, yields good performance guarantees for problems in this class. The performance guarantee depends on the particular problem; in some cases it is known to be bounded only logarithmically [7, Chapter 4]. For a commercial application of this work see [10].

Graph bisection. Given a graph, partition the nodes into two sets of equal size so as to minimize the number of edges with endpoints in different sets. An algorithm to find an approximately minimum-weight bisector would be remarkably useful, since it would provide the basis for a divide-and-conquer approach to many other graph optimization problems. In fact, a solution to a related but easier problem suffices.

Define a $\frac{1}{3}$-balanced cut to be a partition of the vertices of a graph into two sets each containing at least one-third of the vertices; its weight is the total weight of edges connecting the two sets. There is an algorithm to find a $\frac{1}{3}$-balanced cut whose weight is $O(\log n)$ times the minimum weight of a bisector. Note that this algorithm is not, strictly speaking, an approximation algorithm for any one optimization problem: the output of the algorithm is a solution to one problem while the quality of the output is measured against the optimal value for another. (We call this kind of performance guarantee a "bait-and-switch" guarantee.) Nevertheless, the algorithm is nearly as useful as a true approximation algorithm would be because in many divide-and-conquer algorithms the precise balance is not critical. One can make use of the balanced-cut algorithm to obtain approximation algorithms for many problems, including the following.

Optimal linear arrangement. Assign vertices of a graph to distinct integral points on the real number line so as to minimize the total length of edges.

Minimizing time and space for sparse Gaussian elimination. Given a sparse, positive-semidefinite linear system, the order in which variables are eliminated affects the time and storage space required for solving the system; choose an ordering to simultaneously minimize both time and storage space required.

Crossing number. Embed a graph in the plane so as to minimize the number of edge-crossings.

The approximation algorithms for the above three problems have performance guarantees that depend on the performance guarantee of the balanced-separator algorithm. It is not known whether the latter performance guarantee can be improved: there might be an algorithm for balanced separators that has a constant performance ratio.

There are several other graph-separation problems for which approximation algorithms are known, e.g., problems involving directed graphs. All these approximation algorithms for cut problems make use of linear-programming relaxation. See [7, Chapter 5].

34.9 Multicriteria Problems

In many applications, there are two or more objective functions to be considered. There have been some approximation algorithms developed for such multicriteria optimization problems (though much work remains to be done). Several problems in previous sections, such as the k-cluster problem described in Section 34.7.1, can be viewed as a bicriteria problem: there is a budget imposed on one resource (the number of clusters), and the algorithm is required to approximately optimize use of another resource (cluster diameter) subject to that budget constraint. Another example is scheduling unrelated parallel machines with costs: for a given budget on cost, jobs are assigned to machines in such a way that the cost of the assignment is under budget and the makespan of the schedule is nearly minimum.

Other approximation algorithms for bicriteria problems use the bait-and-switch idea mentioned in Section 34.8.1. For example, there is a polynomial approximation scheme for variant of the minimum-spanning-tree problem in which there are two unrelated costs per edge, say weight and length: given a budget L on length, the algorithm finds a spanning tree whose length is at most $(1 + \epsilon)L$ and whose weight is no more than the minimum weight of a spanning tree having length at most L [13].

34.10 Hard-to-Approximate Problems

For some optimization problems, worst-case performance guarantees are unlikely to be possible: it is NP-hard to approximate these problems even if one is willing to accept very poor performance guarantees. Following are some examples [7, Sections 10.5 and 10.6].

Maximum clique. Given a graph, find a largest set of vertices that are pairwise adjacent (see also [6]).

Minimum vertex coloring. Given a graph, color the vertices with a minimum number of colors so that adjacent vertices receive distinct colors.

Longest path. Given a graph, find a longest simple path.

Max linear satisfy. Given a set of linear equations, find a largest possible subset that are simultaneously satisfiable.

Nearest codeword. Given a linear error-correcting code specified by a matrix, and given a vector, find the codeword closest in Hamming distance to the vector.

Nearest lattice vector. Given a set of vectors $v_1, \ldots, v_n$ and a vector v, find an integer linear combination of the v_i that is nearest in Euclidean distance to v.

34.11 Research Issues and Summary

We have given examples for the techniques most frequently used to obtain approximation algorithms with provable performance guarantees, the use of witnesses, relaxation, and coarsening. We have categorized NP-hard optimization problems according to the performance guarantees achievable in polynomial time:

1. A small additive error
2. A relative error of ϵ for any fixed positive ϵ
3. A constant-factor performance guarantee
4. A logarithmic- or polylogarithmic-factor performance guarantee
5. No significant performance guarantee

The ability to categorize problems in this way has been greatly aided by recent research developments in complexity theory. Novel techniques have been developed for proving the hardness of approximation of optimization problems. For many fundamental problems, we can state with considerable precision how good a performance guarantee can be achieved in polynomial time: known lower and upper bounds match or nearly match. Research toward proving matching bounds continues. In particular, for several problems for which there are logarithmic-factor performance guarantees (e.g., balanced cuts in graphs), researchers have so far not ruled out the existence of constant-factor performance guarantees.

Another challenge in research is methodological in nature. This chapter has presented methods of worst-case analysis: ways of universally bounding the error (relative or absolute) of an approximation algorithm. This theory has led to the development of many interesting and useful algorithms, and has proved useful in making distinctions between algorithms and between optimization problems. However, worst-case bounds are clearly not the whole story. Another approach is to develop algorithms tuned for a particular probability distribution of inputs, e.g., the uniform distribution. This approach is of limited usefulness because the distribution of inputs arising in a particular application rarely matches that for which the algorithm was tuned. Perhaps the most promising approach would address a hybrid of the worst-case and probabilistic models. The performance of an approximation algorithm would be defined as the probabilistic performance on a probability distribution selected by an adversary from among a large class of distributions. Blum [2] has presented an analysis of this kind in the context of graph coloring, and others (see [7, Section 13.7]) have addressed similar issues in the context of on-line algorithms.

34.12   Further Information

For an excellent survey of the field of approximation algorithms, focusing on recent results and research issues, see the survey by David Shmoys [14]. Further details on almost all of the topics in this chapter, including algorithms and hardness results, can be found in the book by Vijay V. Vazirani [15]. NP-completeness is the subject of the classic book by Michael Garey and David Johnson [5]. An article by Johnson anticipated many of the issues and methods subsequently developed [8]. Randomized rounding and other probabilistic techniques used in algorithms are the subject of an excellent text by [11]. As of this writing, a searchable compendium of approximation algorithms and hardness results, by Crescenzi and Kann, is available online [4].

Defining Terms

ρ-**Approximation algorithm:** An approximation algorithm that is guaranteed to find a solution whose value is at most (or at least, as appropriate) ρ times the optimum. The ratio ρ is the performance ratio of the algorithm.

Absolute performance guarantee: An approximation algorithm with an absolute performance guarantee is guaranteed to return a feasible solution whose value differs additively from the optimal value by a bounded amount.

Approximation algorithm: For solving an optimization problem. An algorithm that runs in time polynomial in the length of the input and outputs a feasible solution that is guaranteed to be nearly optimal in some well-defined sense called the performance guarantee.

Coarsening: To coarsen a problem instance is to alter it, typically restricting to a less complex feasible region or objective function, so that the resulting problem can be efficiently solved, typically by dynamic programming. This is not standard terminology.

Dual linear program: Every LP has a corresponding LP called the dual. For the LP under LP, the dual is $\max_y \{b \cdot y : A^T y \leq c \text{ and } y \geq \overline{0}\}$. For any solution x to the original LP and any solution y to the dual, we have $c \cdot x \geq (A^T y)^T x = y^T(Ax) \geq y \cdot b$. For optimal x and y, equality holds. For a problem formulated as an integer LP, feasible solutions to the dual of a relaxation of the program can serve as witnesses.

Feasible region: See optimization problem.

Feasible solution: Any element of the feasible region of an optimization problem.

Fractional solution: Typically, a solution to a relaxation of a problem.

Fully polynomial approximation scheme: An approximation scheme in which the running time of A_ϵ is bounded by a polynomial in the length of the input and $1/\epsilon$.

Integer linear program: A LP augmented with additional constraints specifying that the variables must take on integer values. Solving such problems is NP-hard.

Linear program: A problem expressible in the following form. Given an $n \times m$ real matrix A, m-vector b and n-vector c, determine $\min_x \{c \cdot x : Ax \geq b \text{ and } x \geq \overline{0}\}$ where x ranges over all n-vectors and the inequalities are interpreted component-wise (i.e., $x \geq \overline{0}$ means that the entries of x are nonnegative).

MAX-SNP: A complexity class consisting of problems that have constant-factor approximation algorithms, but no approximation schemes unless $P = NP$.

Mixed integer linear program: A LP augmented with additional constraints specifying that some of the variables must take on integer values. Solving such problems is NP-hard.

Objective function: See optimization problem.

Optimal solution: To an optimization problem. A feasible solution minimizing (or possibly maximizing) the value of the objective function.

Optimal value: The minimum (or possibly maximum) value taken on by the objective function over the feasible region of an optimization problem.

Optimization problem: An optimization problem consists of a set $\mathcal{P}$, called the feasible region and usually specified implicitly, and a function $f : \mathcal{P} \to \mathbb{R}$, the objective function.

Performance guarantee: See approximation algorithm.

Performance ratio: See ρ-approximation algorithm.

Polynomial approximation scheme: A collection of algorithms $\{A_\epsilon : \epsilon > 0\}$, where each A_ϵ is a $(1 + \epsilon)$-approximation algorithm running in time polynomial in the length of the input. There is no restriction on the dependence of the running time on ϵ.

Randomized rounding: A technique that uses the probabilistic method to convert a solution to a relaxed problem into an approximate solution to the original problem.

Relative performance guarantee: An approximation algorithm with a relative performance guarantee is guaranteed to return a feasible solution whose value is bounded by a multiplicative factor times the optimal value.

Relaxation: A relaxation of an optimization problem with feasible region $\mathcal{P}$ is another optimization problem with feasible region $\mathcal{P}' \supset \mathcal{P}$ and whose objective function is an extension of the original problem's objective function. The relaxed problem is typically easier to solve. Its value provides a bound on the value of the original problem.

Rescalable: An optimization problem is rescalable if, given any instance of the problem and integer $\lambda > 0$, there is an easily computed second instance that is the same except that the objective function for the second instance is (element-wise) λ times the objective function of the first instance. For such problems, the best one can hope for is a multiplicative performance guarantee, not an absolute one.

Semidefinite programming: A generalization of linear programming in which any subset of the variables may be constrained to form a semidefinite matrix. Used in recent results obtaining better approximation algorithms for cut, satisfiability, and coloring problems.

Strongly NP-hard: A problem is strongly NP-hard if it is NP-hard even when any numbers appearing in the input are bounded by some polynomial in the length of the input.

Triangle inequality: A complete weighted graph satisfies the triangle inequality if $\text{wt}(u, v) \leq \text{wt}(u, w) + \text{wt}(w, v)$ for all vertices u, v, and w. This will hold for any graph representing points in a metric space. Many problems involving edge-weighted graphs have better approximation algorithms if the problem is restricted to weights satisfying the triangle inequality.

Witness: A structure providing an easily verified bound on the optimal value of an optimization problem. Typically used in the analysis of an approximation algorithm to prove the performance guarantee.

References

1. Arora, S., Polynomial time approximation schemes for Euclidean TSP and other geometric problems. *Journal of the ACM*, 45(5), 753–782, 1998.
2. Blum, A., Algorithms for approximate graph coloring. PhD thesis, Massachusetts Institute of Technology. MIT Laboratory for Computer Science Technical Report MIT/LCS/TR-506, June 1991.
3. Blum, A., Jiang, T., Li, M., Tromp, J., and Yannakakis, M., Linear approximation of shortest superstrings. *Journal of the ACM*, 41(4), 630–647, 1994.
4. Crescenzi, P. and Kann, V. (Eds.), A compendium of NP optimization problems, http://www.nada.kth.se/~viggo/wwwcompendium/.
5. Garey, M.R. and Johnson, D.S., *Computers and Intractibility: A Guide to the Theory of NP-Completeness*. W.H. Freeman, New York, 1979.
6. Håstad, J., Clique is hard to approximate within $n^{1-\epsilon}$. In *IEEE, 37th Annual Symposium on Foundations of Computer Science*, Burlington, VT, pp. 627–636, 1996.
7. Hochbaum, D.S., Ed., *Approximation Algorithms for NP-Hard Problems*. PWS Publishing, Boston, MA, 1997.
8. Johnson, D.S., Approximation algorithms for combinatorial problems. *Journal of Computer and System Sciences*, 9, 256–278, 1974.
9. Li, M., Towards a DNA sequencing theory (learning a string) (preliminary version). In *31st Annual Symposium on Foundations of Computer Science*, volume I, St. Louis, MO, pp. 125–134, IEEE, 1990.
10. Mihail, M., Shallcross, D., Dean, N., and Mostrel, M., A commercial application of survivable network design: ITP/INPLANS CCS network topology analyzer. In *Proceedings of the Seventh Annual ACM-SIAM Symposium on Discrete Algorithms*, Atlanta, GA, pp. 279–287, 1996.

11. Motwani, R. and Raghavan, P., *Randomized Algorithms.* Cambridge University Press, New York, 1995.

12. Nemhauser, G.L. and Wolsey, L.A., *Integer and Combinatorial Optimization.* John Wiley & Sons, New York, 1988.

13. Ravi, R. and Goemans, M.X., The constrained minimum spanning tree problem. In *Proceedings Fifth Scandinavian Workshop Algorithm Theory,* number 1097 in *LNCS,* Springer-Verlag, Berlin, Heidelberg, pp. 66–75, 1996.

14. Shmoys, D.B., Computing near-optimal solutions to combinatorial optimization problems. In Cook, W., Lovasz, L., and Seymour, P. (Eds.), *Combinatorial Optimization,* volume 20 of *DIMACS Series in Discrete Mathematics and Computer Science,* AMS, Providence, RI, pp. 355–397, 1995.

15. Vazirani, V.V., *Approximation Algorithms,* Springer-Verlag, Berlin, Heidelberg, New York, 2004.

16. Mitchell, J.S.B., Guillotine subdivisions approximate polygonal subdivisions: a simple polygonomial-time approximation scheme for geometric TSP, K-MST, and related problems, *SIAM Journal on Computing,* 28(4), 1298–1309, 1999.

17. Borradaile. G., Kenyon-Mathieu, C., and Klein, P.N., A polynomial-time approximation scheme for Steiner tree in planar graphs, *ACM Transactions on Algorithms,* 5(31), 31-1–31-31, 2009.

Index

G

H

T - #0290 - 101024 - C0 - 241/171/55 [57] - CB - 9781584888222 - Gloss Lamination